Engineering Mathematics-II

E. Rukmangadachari
Professor of Mathematics,
Department of Humanities and Sciences,
Malla Reddy Engineering College,
Secunderabad

PEARSON
Chennai • Delhi • Chandigarh

Editor—Acquisitions: Sandhya Jayadev
Associate Editor—Production: M. Balakrishnan

ISBN 978-81-317-8495-2

First Impression

Published by Dorling Kindersley (India) Pvt. Ltd, licensees of Pearson Education in South Asia.

Head Office: 7th Floor, Knowledge Boulevard, A-8(A), Sector 62, Noida 201 309, UP, India.
Registered Office: 11 Community Centre, Panchsheel Park, New Delhi 110 017, India.

Compositor: BeSpoke Integrated Solutions, Puducherry.
Printed in India by Pushp Print Services.

To
My Beloved Parents
Enikapati Krishnamachari,
Rajamma, Ademma

About the Author

E. Rukmangadachari is former head of Computer Science and Engineering as well as Humanities and Sciences at Malla Reddy Engineering College, Secunderabad. Earlier, he was a reader in mathematics (PG course) at Government College, Rajahmundry. He is an M.A. from Osmania University, Hyderabad, and an M.Phil. and Ph.D. degree holder from Sri. Venkateswara University, Tirupathi.

A recent recipient of the Andhra Pradesh State Meritorius Teacher's award in 1981, Professor Rukmangadachari has published over 40 research papers in national and international journals. With a rich repertoire of over 45 years' experience in teaching mathematics to undergraduate, postgraduate and engineering students, he is currently the vice president of the Andhra Pradesh Society of Mathematical Sciences, Hyderabad. An ace planner with fine managerial skills, he was the organizing secretary for the conduct of the 17th Congress of the Andhra Pradesh Society for Mathematical Sciences, Hyderabad.

Contents

5 Curve Fitting

6 Numerical Differentiation and Integration

7 Numerical Solution of Ordinary Differential Equations

8. Fourier Series

9 Fourier Integral Transforms

10 Partial Differential Equations

11 Z-Transforms and Solution of Difference Equations

12 Special Functions

Preface

I am pleased to present this edition of *Engineering Mathematics-II* to the first year B.Tech. students.

The topics have been dealt with in a coherent manner, supported by illustrations for better comprehension. Each chapter is replete with examples and exercises, along with solutions and hints, wherever necessary.

The book also has a question bank, with numerous Multiple Choice Questions, Fill in the Blanks, and Match the Following and True or False statements, thus providing the student with an abundant repository of exam specific problems.

Suggestions for the improvement of the book are welcome and will be gratefully acknowledged.

Acknowledgements

I express my deep sense of gratitude to Sri. Ch. Malla Reddy, Chairman, and Sri. Ch. Mahender Reddy, Secretary, Malla Reddy Group of Institutions (MGRI), whose patronage has given me the opportunity to write a book.

I am thankful to Prof. Madan Mohan, Director (Academics) and Col. G. Ram Reddy, Director (Administration), MRGI; and Dr R. K. Murthy, Principal, Malla Reddy Engineering College, Secunderabad, for their kindness, guidance and encouragement.

E. Rukmangadachari

Eigenvalues and Eigenvectors

1

1.1 INTRODUCTION

In this chapter we will consider polynomials whose coefficients are not real numbers but n-square matrices, find eigenvalues and eigenvectors of a matrix, state and prove an important theorem known as the Cayley–Hamilton Theorem and see how it can be used to find powers of square matrices and inverses of nonsingular matrices.

1.1.1 Matrix Polynomial

If $P_0, P_1, \ldots, P_m$ $(\neq 0)$ are n-square matrices, then an expression of the form

$$\sum_{r=0}^{m} P_r x^r = P_0 + P_1 x + P_2 x^2 + \cdots + P_m x^m \tag{1.1}$$

is called a matrix polynomial of degree 'm'.

E.g. Let $P_0 = \begin{bmatrix} -1 & 1 \\ 0 & 3 \end{bmatrix}$, $P_1 = \begin{bmatrix} 0 & 7 \\ 3 & 2 \end{bmatrix}$, $P_2 = \begin{bmatrix} 5 & -4 \\ -3 & 6 \end{bmatrix}$

Then

$$P_0 + P_1 x + P_2 x^2 = \begin{bmatrix} -1 & 1 \\ 0 & 3 \end{bmatrix} + \begin{bmatrix} 0 & 7 \\ 3 & 2 \end{bmatrix} x + \begin{bmatrix} 5 & -4 \\ -3 & 6 \end{bmatrix} x^2 = \begin{bmatrix} -1+5x^2 & 1+7x-4x^2 \\ 3x-3x^2 & 3+2x+6x^2 \end{bmatrix} \tag{1.2}$$

is a matrix polynomial of degree 2.

1.2 LINEAR TRANSFORMATION

An $n \times n$ real matrix can be considered as a linear transformation from vector space V_n to V_n. The simplest type of linear transformation is the one that merely multiplies all vectors X in V_n by a fixed scalar, *i.e.*, $AX = \lambda X$ for some fixed λ. We will now deal with complex vector spaces and complex scalars; consequently the matrices may have complex entries also.

Consider an n-square matrix

$$A = [a_{ij}] = \begin{bmatrix} a_{11} & a_{12} & \cdots & a_{1n} \\ a_{21} & a_{22} & \cdots & a_{2n} \\ \cdots & \cdots & \cdots & \cdots \\ a_{n1} & a_{n2} & \cdots & a_{nn} \end{bmatrix} \tag{1.3}$$

Our aim is to find out whether for some nonzero vectors $X = [x_1, x_2, \ldots, x_n]^T$ it is possible to have

$$AX = \lambda X \tag{1.4}$$

for some suitable scalar λ, *i.e.*, the linear transformation $Y = AX$ transforms X into a scalar multiple of X, *i.e.*, X is invariant.

1.3 CHARACTERISTIC VALUE PROBLEM

Given a square matrix $A = [a_{ij}]_{n\times n}$ the problem of finding scalars λ and nonzero vectors $X = [x_1, x_2, \cdots x_n]^T_{n\times 1}$ which satisfy equation (1.4) is known as the characteristic value Problem. We will now take up the solution of the problem.

From equation (1.4) we have $AX - \lambda IX = 0$

$$\text{or}\quad \begin{bmatrix} a_{11}-\lambda & a_{12} & \cdots & a_{1n} \\ a_{21} & a_{22}-\lambda & \cdots & a_{2n} \\ \cdots & \cdots & \cdots & \cdots \\ a_{n1} & a_{n2} & \cdots & a_{nn}-\lambda \end{bmatrix}\begin{bmatrix} x_1 \\ x_2 \\ \vdots \\ x_n \end{bmatrix} = 0 \quad (1.5)$$

This is a system of n homogeneous linear equations which has a nontrivial solution if

$$P(\lambda) = |A - \lambda I| = 0$$

$$\text{or}\quad \begin{vmatrix} a_{11}-\lambda & a_{12} & \cdots & a_{1n} \\ a_{21} & a_{22}-\lambda & \cdots & a_{2n} \\ \cdots & \cdots & \cdots & \cdots \\ a_{n1} & a_{n2} & \cdots & a_{nn}-\lambda \end{vmatrix} = 0 \quad (1.6)$$

1.3.1 Characteristic Equation of Matrix *A*

Expanding the determinant in (1.6) we get a polynomial $P(\lambda)$ of degree n, called the *characteristic polynomial* of A and therefore (1.6) has n roots, $\lambda_1, \lambda_2, \ldots, \lambda_n$. For each of these λ's equation (1.4) has a nontrivial solution; λ_i are called the characteristic values or eigenvalues or proper values or latent values of matrix A; and each vector is called an eigenvector, characteristic vector or proper vector or latent vector. Equation (1.6) is called the *characteristic equation of A*.

$$Deg\ P(\lambda) = \text{order of } A$$

Note: Let a_{ij} be a diagonal element of A and M_{ij} its corresponding minor and let $a = \sum a_{ij} = \text{tr } A$, $b = \sum M_{ii}$ and $c = |A|$. Then the characteristic equation of A is

$$P(\lambda) = |A - \lambda I| = \lambda^2 - a\lambda + c = 0, \text{ if } A \text{ is of order 2;}$$

$$= -\lambda^3 + a\lambda^2 - b\lambda + c = 0, \text{ if } A \text{ is of order 3.}$$

This applies to matrices of orders 2 and 3 only.

1.3.2 Spectrum of *A*

The set of all eigenvalues of A is called the spectrum of A.

Note 1 If the eigenvalues of A are distinct then the corresponding eigenvectors are a set of n linearly independent vectors.

Note 2 In the case of repeated eigenvalues we may not have a linearly independent set of n eigenvectors.

Note 3 The algebraic multiplicity of an eigenvalue λ is the order of the eigenvalue as a root of the characteristic polynomial (*i.e.*, if λ is a double root then its algebraic multiplicity is 2).

Note 4 The geometric multiplicity of λ is the number of linearly independent eigenvectors corresponding to λ.

1.3.3 Procedure for Finding Eigenvalues and Eigenvectors

1. Solve the characteristic equation $P(\lambda) = \det(A - \lambda I) = 0$ for eigenvalues λ_i. If A is an n-square matrix then the eigenvalues will be n; repeated roots being counted according to their multiplicity.
2. Substituting a specific value λ_i solve the system of linear homogeneous equations $(A - \lambda_i I)X_i = 0$ to obtain the corresponding nonzero eigenvector X_i.

Orthogonal matrix: A real square matrix A is called an orthogonal matrix if $A^T A = I = AA^T$ or equivalently $A^T = A^{-1}$.

Since $1 = |I| = |A^T| = |A^T||A| = |A|^2$, $\therefore\ |A| = \pm 1$. A is nonsingular and A^{-1} exists. The linear transformation given by $Y = AX$ is called an orthogonal transformation if the matrix A is an orthogonal matrix.

Example 1.1

Find the eigenvalues and eigenvectors of $A = \begin{bmatrix} -5 & 2 \\ 2 & -2 \end{bmatrix}$.

Solution The characteristic equation of A is

$$|A - \lambda I| = 0 \quad \Rightarrow \quad \begin{vmatrix} -5-\lambda & 2 \\ 2 & -2-\lambda \end{vmatrix} = 0$$

$$\Rightarrow \quad (-5-\lambda)(-2-\lambda) - 4 = 0$$

$$\Rightarrow \quad \lambda^2 + 7\lambda + 6 = (\lambda+1)(\lambda+6) = 0$$

$$\Rightarrow \quad \lambda_1 = -1,\ \lambda_2 = -6 \qquad (1), (2)$$

are the eigen-values of A

Aliter: Here $a = \sum a_{ii} = \text{tr } A = -5 - 2 = -7$
$c = |A| = (-5)(-2) - 2 \cdot 2 = 6$

$\therefore$ The characteristic equation is

$$P(\lambda) = \lambda_2 - (-7)\lambda + 6 = 0$$

$$= \lambda_2 + 7\lambda + 6 = 0 \Rightarrow \lambda_1 = -1,\ \lambda_2 = -6$$

Let $X = [x_1, x_2]^T$ be the eigenvector corresponding to eigenvalue λ. Then we have to solve

$$(A - \lambda I)X = 0 \Rightarrow \begin{bmatrix} -5-\lambda & 2 \\ 2 & -2-\lambda \end{bmatrix}\begin{bmatrix} x_1 \\ x_2 \end{bmatrix} = \begin{bmatrix} 0 \\ 0 \end{bmatrix} \quad (3)$$

For $\lambda_1 = -1$

$$\begin{bmatrix} -5+1 & 2 \\ 2 & -2+1 \end{bmatrix}\begin{bmatrix} x_1 \\ x_2 \end{bmatrix} = \begin{bmatrix} 0 \\ 0 \end{bmatrix} \Rightarrow \left.\begin{matrix} -4x_1 + 2x_2 = 0 \\ 2x_1 - x_2 = 0 \end{matrix}\right\}$$

$$\Rightarrow \frac{x_1}{1} = \frac{x_2}{2} = c_1 \text{(say)} \Rightarrow X_1 = \begin{bmatrix} x_1 \\ x_2 \end{bmatrix} = c_1\begin{bmatrix} 1 \\ 2 \end{bmatrix} \quad (4)$$

is the corresponding eigenvector of A

For $\lambda_2 = -6$

$$\begin{bmatrix} -5+6 & 2 \\ 2 & -2+6 \end{bmatrix}\begin{bmatrix} x_1 \\ x_2 \end{bmatrix} = \begin{bmatrix} 0 \\ 0 \end{bmatrix} \Rightarrow \left.\begin{matrix} x_1 + 2x_2 = 0 \\ 2x_1 + 4x_2 = 0 \end{matrix}\right\}$$

$$\Rightarrow \frac{x_1}{2} = \frac{x_2}{-1} = c_2 \text{(say)} \Rightarrow X_2 = \begin{bmatrix} x_1 \\ x_2 \end{bmatrix} = c_2\begin{bmatrix} 2 \\ -1 \end{bmatrix} \quad (5)$$

is the corresponding eigenvector of A

Example 1.2

Find the eigenvalues and eigenvectors of $B = \begin{bmatrix} -15 & 6 \\ 6 & -6 \end{bmatrix}$.

Solution The characteristic equation of B is $|B - \lambda I| = 0$

$$\Rightarrow \begin{vmatrix} -15-\lambda & 6 \\ 6 & -6-\lambda \end{vmatrix} \begin{aligned} &= \lambda^2 + 21\lambda + 54 = 0 \\ &\Rightarrow (\lambda+3)(\lambda+18) = 0 \\ &\Rightarrow \lambda_1 = -3, \lambda_2 = -18 \end{aligned}$$

Aliter: Here $a = \Sigma a_{ii} = \text{tr } A = -15 - 6 = -21$; $c = |A| = (-15)(-6) - 6.6 = 54$

The characteristic equation is

$P(\lambda) = \lambda^2 - (-21)\lambda + 54 = 0 \Rightarrow \lambda^2 + 21\lambda + 54 = 0$
$\Rightarrow (\lambda + 3)(\lambda + 18) = 0 \quad \Rightarrow \lambda_1 = -3, \lambda_2 = -18$

Let $X = [x_1, x_2]^T$ be the eigenvector corresponding to eigenvalue λ. We have to solve

$$(B - \lambda I)X = 0 \Rightarrow \begin{bmatrix} -15-\lambda & 6 \\ 6 & -6-\lambda \end{bmatrix}\begin{bmatrix} x_1 \\ x_2 \end{bmatrix} = \begin{bmatrix} 0 \\ 0 \end{bmatrix}$$

For $\lambda = -3$

$$\begin{bmatrix} -15+3 & 6 \\ 6 & -6+3 \end{bmatrix}\begin{bmatrix} x_1 \\ x_2 \end{bmatrix} = \begin{bmatrix} 0 \\ 0 \end{bmatrix} \Rightarrow \left\{\begin{matrix} -12x_1 + 6x_2 = 0 \\ 6x_1 - 3x_2 = 0 \end{matrix}\right.$$

$$\Rightarrow \frac{x_1}{1} = \frac{x_2}{2} = c_1 \text{(say)} \Rightarrow X_1 = \begin{bmatrix} x_1 \\ x_2 \end{bmatrix} = c_1\begin{bmatrix} 1 \\ 2 \end{bmatrix}$$

For $\lambda = -18$

$$\begin{bmatrix} -15+18 & 6 \\ 6 & -6+18 \end{bmatrix}\begin{bmatrix} x_1 \\ x_2 \end{bmatrix} = \begin{bmatrix} 0 \\ 0 \end{bmatrix} \Rightarrow \left\{\begin{matrix} 3x_1 + 6x_2 = 0 \\ 6x_1 + 12x_2 = 0 \end{matrix}\right.$$

$$\Rightarrow \frac{x_1}{2} = \frac{x_2}{-1} = c_2 \text{(say)} \Rightarrow X_2 = \begin{bmatrix} x_1 \\ x_2 \end{bmatrix} = c_2\begin{bmatrix} 2 \\ -1 \end{bmatrix}$$

Note 1 From Examples 1.1 and 1.2 we observe that $B = 3A$. The eigenvalues of B are three times the eigenvalues of A. But the eigenvectors are same.

Also, tr. A = sum of eigenvalues and $|A|$ = Product of eigenvalues.

Example 1.3

Find the eigenvalues and eigenvectors of $A = \begin{bmatrix} 1 & 2 \\ 1 & 0 \end{bmatrix}$.

Also find those of $B = \begin{bmatrix} 2 & 4 \\ 2 & 0 \end{bmatrix}$ and $C = \begin{bmatrix} 3 & 2 \\ 1 & .2 \end{bmatrix}$. What do you observe?

Solution The characteristic equation of A is

$$|A - \lambda I| = 0 \Rightarrow \begin{vmatrix} 1-\lambda & 2 \\ 1 & 0-\lambda \end{vmatrix} = 0$$

$$\Rightarrow \lambda^2 - \lambda - 2 = (\lambda+1)(\lambda-2) = 0$$

$\Rightarrow \quad \lambda_1 = -1, \lambda_2 = 2$ are the eigenvalues of A.

Aliter: Here $a = \Sigma a_{ii} = \text{tr } A = 1 + 0 = 1$; $c = |A| = 1.0 - 1.2 = -2$

The characteristic equation is

$P(\lambda) = \lambda^2 - \lambda - 2 = 0 \quad \Rightarrow (\lambda + 1)(\lambda - 2) = 0$
$\Rightarrow (\lambda + 1)(\lambda - 2) = 0 \quad \Rightarrow \lambda_1 = -1, \lambda_2 = 2$

Let $X = [x_1, x_2]^T$ be the eigenvector corresponding to the eigenvalue λ.

We have to solve

$$(A - \lambda I)X = 0 \Rightarrow \begin{bmatrix} 1-\lambda & 2 \\ 1 & 0-\lambda \end{bmatrix}\begin{bmatrix} x_1 \\ x_2 \end{bmatrix} = \begin{bmatrix} 0 \\ 0 \end{bmatrix}$$

For $\lambda_1 = -1$

$$(A-\lambda I)X = 0 \;\Rightarrow\; \begin{bmatrix} 1+1 & 2 \\ 1 & 1 \end{bmatrix}\begin{bmatrix} x_1 \\ x_2 \end{bmatrix} = \begin{bmatrix} 0 \\ 0 \end{bmatrix}$$

$$\Rightarrow \begin{array}{l} 2x_1 + 2x_2 = 0 \\ x_1 + x_2 = 0 \end{array}$$

$$\Rightarrow \frac{x_1}{1} = \frac{x_2}{-1} = c_1 \;\Rightarrow\; X_1 = \begin{bmatrix} x_1 \\ x_2 \end{bmatrix} = c_1\begin{bmatrix} 1 \\ -1 \end{bmatrix}$$

is the corresponding eigenvector of A.

For $\lambda_2 = 2$

$$\begin{bmatrix} 1-2 & 2 \\ 1 & -2 \end{bmatrix}\begin{bmatrix} x_1 \\ x_2 \end{bmatrix} = \begin{bmatrix} 0 \\ 0 \end{bmatrix} \;\Rightarrow\; \begin{array}{l} -x_1 + 2x_2 = 0 \\ x_1 - 2x_2 = 0 \end{array}$$

$$\Rightarrow \frac{x_1}{2} = \frac{x_2}{1} = c_2 \;\Rightarrow\; X_2 = \begin{bmatrix} x_1 \\ x_2 \end{bmatrix} = c_2\begin{bmatrix} 2 \\ 1 \end{bmatrix}$$

is the corresponding eigenvector of A.

The characteristic equation of B is

$$|B-\lambda I| = 0 \;\Rightarrow\; \begin{vmatrix} 2-\lambda & 4 \\ 2 & 0-\lambda \end{vmatrix} = 0$$

$$\Rightarrow \lambda^2 - 2\lambda - 8 = (\lambda+2)(\lambda-4) = 0$$

$\Rightarrow \lambda_1 = -2, \lambda_2 = 4$ are the eigenvalues of B.

Let $X = [x_1, x_2]^T$ be the eigenvector corresponding to the eigenvalue λ of B.

We have to solve $(B - \lambda I)X = 0$

$$\Rightarrow \begin{bmatrix} 2-\lambda & 4 \\ 2 & -\lambda \end{bmatrix}\begin{bmatrix} x_1 \\ x_2 \end{bmatrix} = \begin{bmatrix} 0 \\ 0 \end{bmatrix}$$

For $\lambda_1 = -2$

$$\begin{bmatrix} 2+2 & 4 \\ 2 & 2 \end{bmatrix}\begin{bmatrix} x_1 \\ x_2 \end{bmatrix} = \begin{bmatrix} 0 \\ 0 \end{bmatrix} \;\Rightarrow\; \begin{array}{l} 4x_1 + 4x_2 = 0 \\ 2x_1 + 2x_2 = 0 \end{array}$$

$$\Rightarrow \frac{x_1}{1} = \frac{x_2}{-1} = c_1(\text{say}) \;\Rightarrow\; X_1 = [x_1, x_2]^T$$

$= c_1\,[1 - 1]^T$ is the corresponding eigenvector of B.

For $\lambda_2 = 4$

$$\begin{bmatrix} 2-4 & 4 \\ 2 & -4 \end{bmatrix}\begin{bmatrix} x_1 \\ x_2 \end{bmatrix} = \begin{bmatrix} 0 \\ 0 \end{bmatrix} \;\Rightarrow\; \begin{array}{l} -2x_1 + 4x_2 = 0 \\ 2x_1 - 4x_2 = 0 \end{array}$$

$$\Rightarrow \frac{x_1}{2} = \frac{x_2}{-1} = c_2(\text{say})$$

$$\Rightarrow X_2 = [x_1, x_2]^T = c_2\,[2\;\;1]^T$$

is the eigenvector corresponding to eigenvalue $\lambda = 4$ of A.

The characteristic equation of C is

$$|C-\lambda I| = 0 \;\Rightarrow\; \begin{vmatrix} 3-\lambda & 2 \\ 1 & 2-\lambda \end{vmatrix} = 0$$

$$\Rightarrow \lambda^2 - 5\lambda + 6 - 2 = 0 \;\Rightarrow\; (\lambda-1)\,(\lambda-4) = 0$$

$\lambda_1 = 1, \lambda_2 = 4$ are the eigenvalues of C.

Let $X = [x_1, x_2]^T$ be the eigenvector corresponding to the eigenvalue λ of C.

We have to solve

$$(C-\lambda I)X = 0 \;\Rightarrow\; \begin{bmatrix} 3-\lambda & 2 \\ 1 & 2-\lambda \end{bmatrix}\begin{bmatrix} x_1 \\ x_2 \end{bmatrix} = \begin{bmatrix} 0 \\ 0 \end{bmatrix}$$

For $\lambda_1 = 1$

$$\begin{bmatrix} 3-1 & 2 \\ 1 & 2-1 \end{bmatrix}\begin{bmatrix} x_1 \\ x_2 \end{bmatrix} = \begin{bmatrix} 0 \\ 0 \end{bmatrix} \;\Rightarrow\; \begin{array}{l} 2x_1 + 2x_2 = 0 \\ x_1 + x_2 = 0 \end{array}$$

$$\Rightarrow \frac{x_1}{1} = \frac{x_2}{-1} = c_1(\text{say})$$

$$\Rightarrow X_1 = [x_1, x_2]^T = c_1\,[1, -1]^T$$

is the corresponding eigenvector of C.

For $\lambda_2 = 4$

$$\begin{bmatrix} 3-4 & 2 \\ 1 & 2-4 \end{bmatrix}\begin{bmatrix} x_1 \\ x_2 \end{bmatrix} = \begin{bmatrix} 0 \\ 0 \end{bmatrix} \;\Rightarrow\; \begin{array}{l} -x_1 + 2x_2 = 0 \\ x_1 - 2x_2 = 0 \end{array}$$

$$\Rightarrow \frac{x_1}{2} = \frac{x_2}{1} = c_2(\text{say})$$

$$\Rightarrow X_2 = [x_1, x_2] = c_2\,[2, 1]^T$$

is the corresponding eigenvector of C.

Observation

We observe that $B = 2A$, $C = A^2$.

The eigenvalues of B are $(2\lambda_1, 2\lambda_2) = (2(-1), 2(2)) = (-2, 4)$.

The eigenvalues of C are $(\lambda_1^2, \lambda_2^2) = ((-1)^2, 2^2) = (1, 4)$, where λ_1, λ_2 are the eigenvalues of A, while the eigenvectors of A, B, C are the same.

Example 1.4

Find the eigenvalues and eigenvectors of

$$A = \begin{bmatrix} 2 & -2 & 2 \\ 1 & 1 & 1 \\ 1 & 3 & -1 \end{bmatrix}.$$

Solution The characteristic equation of A is

$$P(\lambda)=|A-\lambda I|=\begin{vmatrix}2-\lambda & -2 & 2\\ 1 & 1-\lambda & 1\\ 1 & 3 & -1-\lambda\end{vmatrix}$$

$$=(2-\lambda)(\lambda^2-4)+2(-2-\lambda)+2(2+\lambda)=0$$
$$=-(\lambda+2)(\lambda-2)^2=0$$
$$\Rightarrow \lambda_1=-2,\ \lambda_2=2,\ \lambda_3=2$$

are eigenvalues of A.

Aliter: Here $a=\Sigma a_{ii}=\text{tr } A=2+1-1=2;$

$$b=\Sigma M_{ii}=\begin{vmatrix}1 & 1\\ 3 & -1\end{vmatrix}+\begin{vmatrix}2 & 2\\ 1 & -1\end{vmatrix}+\begin{vmatrix}2 & -2\\ 1 & 1\end{vmatrix}$$
$$=-4-4+4=-4$$

$$c=|A|=2\begin{vmatrix}1 & 1\\ 3 & -1\end{vmatrix}-1\begin{vmatrix}-2 & 2\\ 3 & -1\end{vmatrix}+1\begin{vmatrix}-2 & 2\\ 1 & 1\end{vmatrix}$$
$$=2(-4)-(-4)-4=-8$$

The characteristic equation is

$$P(\lambda)=-\lambda^3-a\lambda^2-b\,\lambda+c=-\lambda^3-2\lambda^2-4\lambda-8=0$$
$$\Rightarrow-(\lambda+2)\,(\lambda-2)^2=0\quad\Rightarrow\lambda_1=-2,\ \lambda_2=2,\ \lambda_3=2$$

To find the eigenvectors we have to solve $(A-\lambda I)X=0$ where $X=[x_1, x_2, x_3]^T$.

For $\lambda_1=-2$

$$\begin{bmatrix}4 & -2 & 2\\ 1 & 3 & 1\\ 1 & 3 & 1\end{bmatrix}\begin{bmatrix}x_1\\ x_2\\ x_3\end{bmatrix}=\begin{bmatrix}0\\ 0\\ 0\end{bmatrix}\sim$$

$$e_1: 2x_1-x_2+x_3=0 \qquad (1)$$
$$e_2: x_1+3x_2+x_3=0 \qquad (2)$$
$$(e_1)-(e_2)\ :\ x_1-4x_2=0 \qquad (3)$$

$$\Rightarrow \quad \frac{x_1}{4}=\frac{x_2}{1}=\frac{x_3}{-7}=c_1(\text{say})$$
$$\Rightarrow \quad X_1=c_1\ [4\ \ 1\ \ -7]^T$$

For $\lambda_2=2$

$$\begin{bmatrix}0 & -2 & 2\\ 1 & -1 & 1\\ 1 & 3 & -3\end{bmatrix}\begin{bmatrix}x_1\\ x_2\\ x_3\end{bmatrix}=\begin{bmatrix}0\\ 0\\ 0\end{bmatrix}$$

$$-x_2+x_3=0 \qquad (1)$$
$$\Rightarrow \quad x_1-x_2+x_3=0 \qquad (2)$$
$$x_1+3x_2-3x_3=0 \qquad (3)$$

$$\Rightarrow \quad \frac{x_1}{0}=\frac{x_2}{1}=\frac{x_3}{1}=c_2(\text{say})$$
$$\Rightarrow \quad X_2=c_2\ [0\ \ 1\ \ 1]^T$$

Two eigenvalues are equal. ∴ algebraic multiplicity = 2; The geometric multiplicity = 1, since there is one eigenvector only corresponding to $\lambda_2=\lambda_3=2$.

Example 1.5

If $A=\begin{bmatrix}3 & -2 & -5\\ 4 & -1 & -5\\ -2 & -1 & -3\end{bmatrix}$, find the eigenvalues and eigenvectors of A.

Solution The characteristic equation of A is $P(\lambda)=|A-\lambda I|=0$

$$\Rightarrow \begin{vmatrix}3-\lambda & -2 & -5\\ 4 & -1-\lambda & -5\\ -2 & -1 & -3-\lambda\end{vmatrix}=0$$

$$=(3-\lambda)(\lambda^2+4\lambda-2)$$
$$+\ 2(-22-4\lambda)-5(-6-2\lambda)$$
$$=-\lambda^3-\lambda^2+16\lambda-20=0$$

Aliter: Here $a=\Sigma a_{ii}=\text{tr } A=3-1-3=-1;$

$$b=\Sigma M_{ii}=\begin{vmatrix}-1 & -5\\ -1 & -3\end{vmatrix}+\begin{vmatrix}3 & -5\\ -2 & -3\end{vmatrix}+\begin{vmatrix}3 & -2\\ 4 & -1\end{vmatrix}$$
$$=-2-19+5=-16$$

$$c=|A|=3\begin{vmatrix}-1 & -5\\ -1 & -3\end{vmatrix}-4\begin{vmatrix}-2 & -5\\ -1 & -3\end{vmatrix}-2\begin{vmatrix}-2 & -5\\ -1 & -5\end{vmatrix}$$
$$=3(-2)-4(1)-2(5)=-6-4-10=-20$$

The characteristic equation is

$$P(\lambda)=-\lambda^3+a\lambda^2-b\,\lambda+c=0\Rightarrow$$
$$-\lambda^3-\lambda^2+16\lambda-20=0$$
$$\Rightarrow \lambda_1=-5,\ \lambda_2=2,\ \lambda_3=2$$

$$\begin{array}{r|rrrr} 2 & -1 & -1 & 16 & -20 \\ & & -2 & -6 & 20 \\ \hline 2 & -1 & -3 & 10 & \underline{|0} \\ & & -2 & -10 & \\ \hline & -1 & -5 & \underline{|0} & \end{array}$$

To find the eigenvector $X \neq 0$ corresponding to the eigenvalue λ, we have to solve $(A - \lambda I)X = 0$.

For $\lambda_1 = -5$

$$(A - \lambda_1 I) = \begin{bmatrix} 8 & -2 & -5 \\ 4 & 4 & -5 \\ -2 & -1 & 2 \end{bmatrix} \overset{R_1 - R_2,\ R_2 + 2R_3}{\sim} \begin{bmatrix} 4 & -6 & 0 \\ 0 & 2 & -1 \\ -2 & -1 & 2 \end{bmatrix}$$

The system is equivalent to

$$\begin{aligned} 2x_1 - 3x_2 &= 0 \\ 2x_2 - x_3 &= 0 \quad \Rightarrow \quad \frac{x_1}{3} = \frac{x_2}{2} = \frac{x_3}{4} = c_1 \text{(say)} \\ -2x_1 - x_2 + 2x_3 &= 0 \end{aligned}$$

$\Rightarrow \quad X_1 = c_1\,[3 \quad 2 \quad 4]^T$

For $\lambda_2 = 2$

$$(A - \lambda_2 I) = \begin{bmatrix} 1 & -2 & -5 \\ 4 & -3 & -5 \\ -2 & -1 & -5 \end{bmatrix} \overset{R_3 - R_1,\ R_2 + 2R_3}{\sim} \begin{bmatrix} 1 & -2 & -5 \\ 0 & -5 & -15 \\ -3 & 1 & 0 \end{bmatrix}$$

$$\sim \left\{ \begin{aligned} x_1 - 2x_2 - 5x_3 &= 0 \\ x_2 + 3x_3 &= 0 \\ -3x_1 + x_2 &= 0 \end{aligned} \right.$$

$\Rightarrow \quad \dfrac{x_1}{1} = \dfrac{x_2}{3} = \dfrac{x_3}{-1} = c_2 \text{(say)}$

$\Rightarrow \quad X_2 = c_2\,[1 \quad 3 \quad -1]^T$

The algebraic multiplicity = 2 $\because$ two eigenvalues are equal and the geometric multiplicity = 1 since there is only one eigenvector corresponding to the eigenvalues $\lambda_2 = \lambda_3 = 2$.

Example 1.6

Find the eigenvalues and eigenvectors of $A = \begin{bmatrix} 1 & 0 & 0 \\ 0 & 1 & 0 \\ 0 & 0 & 1 \end{bmatrix}$. Find the algebraic and geometric multiplicity.

Solution The characteristic equation of A is

$$|A - \lambda I| = \begin{vmatrix} 1-\lambda & 0 & 0 \\ 0 & 1-\lambda & 0 \\ 0 & 0 & 1-\lambda \end{vmatrix} = (1-\lambda)^3 = 0$$

$\Rightarrow \quad \lambda_1, \lambda_2, \lambda_3 = 1$

Algebraic multiplicity = 3

$\lambda = 1$

$$\begin{bmatrix} 1-1 & 0 & 0 \\ 0 & 1-1 & 0 \\ 0 & 0 & 1-1 \end{bmatrix} \begin{bmatrix} x_1 \\ x_2 \\ x_3 \end{bmatrix} = 0 \Rightarrow \begin{bmatrix} 0 & 0 & 0 \\ 0 & 0 & 0 \\ 0 & 0 & 0 \end{bmatrix} \begin{bmatrix} x_1 \\ x_2 \\ x_3 \end{bmatrix} = 0$$

$\Rightarrow \quad x_1, x_2, x_3$ are arbitrary.

$X_1 = k_1 \begin{bmatrix} 1 \\ 0 \\ 0 \end{bmatrix}$; $\quad X_2 = k_2 \begin{bmatrix} 0 \\ 1 \\ 0 \end{bmatrix}$; $\quad X_3 = k_3 \begin{bmatrix} 0 \\ 0 \\ 1 \end{bmatrix}$ are three L. I nonzero eigenvectors corresponding to eigenvalue $\lambda = 1$.

$\therefore \quad$ The geometric multiplicity = 3.

EXERCISE 1.1

1. Find the eigenvalues and eigenvectors of $A = \begin{bmatrix} 1 & 3 & 4 \\ 0 & 2 & 5 \\ 0 & 0 & 3 \end{bmatrix}$ and those of its inverse.

   [JNTU 2000]

   Ans: Eigenvalues of A are 1, 2, 3; those of A^{-1} are $1, \frac{1}{2}, \frac{1}{3}$; eigenvectors of A, A^{-1} are same

   $X_1 = c_1 \begin{bmatrix} 1 \\ 0 \\ 0 \end{bmatrix}$, $X_2 = c_2 \begin{bmatrix} 3 \\ 1 \\ 0 \end{bmatrix}$, $X_3 = c_3 = \begin{bmatrix} 19 \\ 10 \\ 2 \end{bmatrix}$

2. Find the eigenvalues and eigenvectors of $A = \begin{bmatrix} 8 & -4 \\ 2 & 2 \end{bmatrix}$.

   [JNTU 2004 S (2)]

   Ans: $\lambda_1 = 4$, $\lambda_2 = 6$; $X_1 = \begin{bmatrix} 1 \\ 1 \end{bmatrix}$; $X_2 = \begin{bmatrix} 2 \\ 1 \end{bmatrix}$

3. Find the eigenvalues and eigenvectors of matrix

   [JNTU 1995, 1998]

   (a) $A = \begin{bmatrix} -2 & 2 & -3 \\ 2 & 1 & -6 \\ -1 & -2 & 0 \end{bmatrix}$ $\qquad$ (b) $B = \begin{bmatrix} 3 & -6 & 3 \\ 1 & 0 & -1 \\ 1 & 2 & -3 \end{bmatrix}$

(c) $C=\begin{bmatrix}3 & 1 & 4\\0 & 2 & 6\\0 & 0 & 5\end{bmatrix}$ (d) $D=\begin{bmatrix}0 & 1 & 0\\0 & 0 & 1\\1 & -3 & 3\end{bmatrix}$

(e) $E=\begin{bmatrix}2 & 0 & 0\\3 & 1 & 0\\-1 & 2 & 3\end{bmatrix}$

Ans: (a) $\lambda_1=-3, \lambda_2=-3, \lambda_3=5$;
Algebraic multiplicity = 2;
Geometric multiplicity = 2

$$X_1=\begin{bmatrix}-2\\1\\0\end{bmatrix}; X_2=\begin{bmatrix}3\\0\\1\end{bmatrix}; X_3=\begin{bmatrix}1\\2\\-1\end{bmatrix}$$

(b) $\lambda_1=0, \lambda_2=2, \lambda_3=-2$;

$$X_1=\begin{bmatrix}1\\1\\1\end{bmatrix}; X_2=\begin{bmatrix}3\\1\\1\end{bmatrix}; X_3=\begin{bmatrix}0\\1\\2\end{bmatrix}$$

(c) $\lambda_1=2, \lambda_2=3, \lambda_3=5$;

$$X_1=\begin{bmatrix}1\\-1\\0\end{bmatrix}; X_2=\begin{bmatrix}1\\0\\0\end{bmatrix}; X_3=\begin{bmatrix}3\\2\\1\end{bmatrix}$$

(d) $\lambda_1=1=\lambda_2=\lambda_3, X_1=X_2=X_3=\begin{bmatrix}1\\1\\1\end{bmatrix}$

(e) $\lambda_1=1, \lambda_2=2, \lambda_3=3$,

$$X_1=\begin{bmatrix}0\\1\\-1\end{bmatrix}; X_2=\begin{bmatrix}1\\3\\-5\end{bmatrix}; X_3=\begin{bmatrix}0\\0\\1\end{bmatrix}$$

4. Find the eigenvalues and eigenvectors of $A=\begin{bmatrix}-3 & -7 & -5\\2 & 4 & 3\\1 & 2 & 2\end{bmatrix}$

Ans: Eigenvalues $\lambda_1=\lambda_2=\lambda_3=1$; eigenvector

$X=\begin{bmatrix}-3\\1\\1\end{bmatrix}$; Algebraic multiplicity = 3;

Geometric multiplicity = 1

5. Find the eigenvalues and eigenvectors of $A=\begin{bmatrix}2 & 2 & 1\\1 & 3 & 1\\1 & 2 & 2\end{bmatrix}$

Ans: $\lambda_1=1, \lambda_2=1, \lambda_3=5$

$$X_1=c_1\begin{bmatrix}2\\-1\\0\end{bmatrix}; X_2=c_2\begin{bmatrix}1\\0\\-1\end{bmatrix}; X_3=c_3\begin{bmatrix}1\\1\\1\end{bmatrix}$$

6. Find the eigenvalues and eigenvectors of $A=\begin{bmatrix}3 & 10 & 5\\-2 & -3 & -4\\3 & 5 & 7\end{bmatrix}$

Ans: $\lambda_1=2, \lambda_2=2, \lambda_3=3; X_1=c_1\begin{bmatrix}5\\2\\-5\end{bmatrix}; X_3=c_3\begin{bmatrix}1\\1\\-2\end{bmatrix}$

1.4 Properties of Eigenvalues and Eigenvectors

1.4.1 Characteristic Polynomial $P_n(\lambda)$

Let A be an n-square matrix. Then the characteristic polynomial of A is

$$P_n(\lambda)=|A-\lambda I|=\begin{vmatrix}a_{11}-\lambda & a_{12} & \cdots & a_{1n}\\a_{21} & a_{22}-\lambda & \cdots & a_{2n}\\\cdots & \cdots & \cdots & \cdots\\a_{n1} & a_{n2} & \cdots & a_{nn}-\lambda\end{vmatrix}$$
$$=(-\lambda)^n+p_1(-\lambda)^{n-1}+\cdots+p_{n-1}(-\lambda)+p_n \tag{1.7}$$

This is an nth degree polynomial in λ, called the *characteristic polynomial of matrix A*. The coefficients p_i are expressible in terms of a_i. In particular, p_1 = coefficient of $(-\lambda)^{n-1}$ in the expansion of $|A-\lambda I| = a_{11}+a_{22}+\cdots+a_{nn}$ and $p_n=|A|$, obtained by putting $\lambda=0$.

The roots of the characteristic equation $P_n(\lambda)=|A-\lambda I|=0$ are called *characteristic roots, latent roots, eigenvalues or proper values* of the matrix A.

1. If A is real then the eigenvalues will be real or complex conjugate pairs.

   Let $\lambda_1, \lambda_2, \cdots, \lambda_n$ be the eigenvalues of A *i.e.*, the roots of the characteristic polynomial $P_n(\lambda)$. Then

$$P_n(\lambda) = (\lambda_1 - \lambda)(\lambda_2 - \lambda)\cdots(\lambda_n - \lambda)$$
$$= (-1)^n(\lambda - \lambda_1)(\lambda - \lambda_2)\cdots(\lambda - \lambda_n)$$
$$= (-1)^n[\lambda^n + (-1)(\lambda_1 + \lambda_2 + \cdots + \lambda_n)\lambda^{n-1} + \cdots + (-1)^n \lambda_1 \cdot \lambda_2 \cdots \lambda_n]$$

2. The sum of the eigenvalues of a matrix is the sum of the elements of the principal diagonal. That is, $\Sigma \lambda_i = \text{tr } A$.
3. The product of the eigenvalues of a matrix A is equal to its determinant

   $\det A = |A| = P_n(0) = \lambda_1\lambda_2 \cdots \lambda_n =$ Product of the eigenvalues of A
4. **Eigenvalues of A^T:** Since the diagonal elements of A and A^T are the same and $|A| = |A^T|$, $\det|A - \lambda I| = \det|A^T - \lambda I|$ and hence the eigenvalues of A^T and A are the same.
5. **Singular matrix A:** If at least one of the eigenvalues of a matrix A is zero then $\det A = 0$ so that A is a singular matrix. Conversely, if A is singular then $|A| = 0$ which implies that at least one eigenvalue of A is zero.
6. **Nonsingular matrix A:** If every eigenvalue of a matrix A is nonzero then the matrix A is nonsingular since $\det A \neq 0$. Conversely, if A is nonsingular then $|A| \neq 0$ so that none of the eigenvalues of A is zero.
7. **Inverse of matrix A:** If every eigenvalue of a matrix A is not zero then A is nonsingular so that A^{-1} exists.

   Eigenvalues of A^{-1} are $\frac{1}{\lambda_1}, \frac{1}{\lambda_2}, \ldots, \frac{1}{\lambda_n}$ if the eigenvalues of A are $\lambda_1, \lambda_2, \ldots, \lambda_n$, *i.e.*, the eigenvalues of A^{-1} are the reciprocals of those of A.

   $$AX = \lambda X \Rightarrow A^{-1}(AX) = A^{-1}(\lambda X)$$
   $$\Rightarrow (A^{-1}A)X = \lambda(A^{-1}X)$$
   $$\Rightarrow IX = \lambda(A^{-1}X) \Rightarrow A^{-1}X = \frac{1}{\lambda}X$$
8. **Spectral shift:** If the eigenvalues of A are $\lambda_1, \lambda_2, \ldots, \lambda_n$ then the eigenvalues of $(A \mp kI)$ are $\lambda_i \pm k$: and the eigenvectors of $(A \mp kI)$ are same as those of A.
9. **Scalar multiple of A:** If the eigenvalues of A are $\lambda_1, \lambda_2, \ldots, \lambda_n$ then the eigenvalues of a scalar multiple of matrix kA are $k\lambda_i$ $(i = 1, 2, \ldots, n)$

   $$k(A - \lambda I)X = 0 \quad \Rightarrow \quad |kA - k\lambda I| = 0$$

   $\therefore$ Matrix kA has eigenvalues $k\lambda_i$ $(i = 1, 2, \ldots \lambda_n)$
10. **Powers of A:** If the eigenvalues of A are $\lambda_1, \lambda_2, \ldots \lambda_n$ then the eigenvalues of A^m $(m \in N)$ are $\lambda_i^m (i = 1, 2, \ldots, n)$

    Assume that $A^k X_i = \lambda_i^k X_i$ is true
    $$\Rightarrow A(A^k X_i) = A(\lambda_i^k X_i)$$
    $$\Rightarrow A^{k+1} X_i = \lambda_i^k (AX_i) = \lambda_i^{k+1} X_i$$
    $$\because AX_i = \lambda_i X_i.$$

    For $m = 1$, $A^1 x_i = \lambda_i^1 X_i$ is true

    By induction $A^m X_i = \lambda_i^m X_i$ is true for all $m \in N$.
11. **Spectral mapping theorem:** If $\lambda_1, \lambda_2, \ldots, \lambda_n$ are the eigenvalues of a matrix A then $P(\lambda_i) = k_m\lambda_i^m + k_{m-1}\lambda_i^{m-1} + \cdots + k_1\lambda_i + k_0$ $(i = 1, 2, \ldots, n)$ are the eigenvalues of the polynomial matrix

    $$P(A) = k_m A^m + k_{m-1}A^{m-1} + \cdots + k_1 A + k_0$$

    By 9 and 10 above, kA^n has eigenvalues $k\lambda^n$, $n = 1, 2, \ldots, m$.

    For $m = 2$, $(k_2A^2 + k_1A + k_0I)X$
    $$= k_2A^2X + k_1AX + k_0IX$$
    $$= k_2\lambda^2 X + k_1\lambda X + k_0 IX$$
    $$= (k_2\lambda^2 + k_1\lambda + k_0)X$$
12. **An eigenvector cannot correspond to two distinct eigenvalues:**

    Let $X \neq 0$ be the eigenvector corresponding to two eigenvalues λ_1, λ_2 $(\lambda_1 \neq \lambda_2)$. Then
    $$(A - \lambda_1 I)X = 0 \quad \text{and} \quad (A - \lambda_2 I)X = 0$$
    $$\Rightarrow (A - \lambda_1 I)X - (A - \lambda_2 I)X = (\lambda_1 - \lambda_2)IX = 0$$
    $$\Rightarrow X = 0 \quad \because \quad \lambda_1 \neq \lambda_2$$

    But $X \neq 0$. Hence there is a contradiction. This proves that an eigenvector cannot correspond to two distinct eigenvalues of matrix A.

13. **Eigenvalues of a diagonal matrix and triangular matrix:**
Let $A = \text{diag}\,[a_{11}, a_{22}, \ldots, a_{nn}]$ $(A - \lambda I) = \text{diag}\,[a_{11} - \lambda, a_{22} - \lambda, \ldots, a_{nn} - \lambda]$
The characteristic polynomial
$$P_n(\lambda) = \det (A - \lambda I) = |A - \lambda I|$$
$$= |\text{diag}\,[a_{11} - \lambda, a_{22} - \lambda, \cdots, a_{nn} - \lambda]|$$
$\therefore$ characteristic eqn. $P_n(\lambda) = 0$
$\Rightarrow$ $(a_{11} - \lambda)(a_{22} - \lambda) \cdots (a_{nn} - \lambda) = 0$
Hence the eigenvalues are the diagonal elements $a_{11}, a_{22}, \ldots, a_{nn}$.

14. **Orthogonal matrix and its eigenvalues:**
If λ is an eigenvalue of an orthogonal matrix A then its reciprocal $1/\lambda$ is also an eigenvalue of A. By definition, an orthogonal matrix A is a square matrix such that $A^T A = I = AA^T$ or $A^T = A^{-1}$. But A and A^T have the same eigenvalues. Also, the eigenvalues of A^{-1} are the reciprocals of those of A. Hence if the eigenvalue of A is λ then $1/\lambda$ will be the reciprocal of λ and will be the eigenvalue of A^T.

15. **Orthogonal vectors:** Two vectors X and Y are said to be orthogonal if $X^T Y = Y^T X = 0$.

16. **Orthonormal set of vectors:** A set of vectors $X_i (i = 1, 2, \ldots, n)$ is said to be an orthonormal set
if $$X_i^T X_j = \delta_{ij} = \begin{Bmatrix} 1 & \text{if} & i = j \\ 0 & \text{if} & i \neq j \end{Bmatrix}$$

Note 1 Column or row vectors of an orthogonal matrix form an orthogonal system

1.5 CAYLEY–HAMILTON[1,2] THEOREM

We now state and prove an important theorem called the Cayley–Hamilton Theorem.

Theorem 1.1 Every square matrix satisfies its own characteristic equation.

Proof Let A be an n-square matrix and let $D(\lambda)$ be the characteristic polynomial of A so that
$$D(\lambda) = |\lambda I - A| = \lambda^n + p_{n-1}\lambda^{n-1} + p_{n-2}\lambda^{n-2} + \cdots + p_1\lambda + p_0 \qquad (1.8)$$

Let $B(\lambda)$ be the adjoint of $(\lambda I - A)$. The elements of $B(\lambda)$ are the cofactors of the elements of the matrix $(\lambda I - A)$ and are polynomials in λ of degree not exceeding $(n-1)$. Thus
$$B(\lambda) = B_{n-1}\lambda^{n-1} + B_{n-2}\lambda^{n-2} + \cdots + B_1\lambda + B_0 \qquad (1.9)$$
where B's are n-square matrices whose elements are functions of the elements of A and are independent of λ. We know that

Product of a matrix and its adjoint = determinant of the matrix × Unit matrix
$$(\lambda I - A).\,\text{adj}\,(\lambda I - A) = |\lambda I - A| I$$
$$\Rightarrow \quad (\lambda I - A).B(\lambda) = |\lambda I - A|.I$$

From (1.8) and (1.9) we have
$$(\lambda I - A)(B_{n-1}\lambda^{n-1} + B_{n-2}\lambda^{n-2} + \cdots + B_1\lambda + B_0) = I(\lambda^n + p_{n-1}\lambda^{n-1} + \cdots + p_1\lambda + p_0) \qquad (1.10)$$

Equating the like powers of λ on both sides of (1.10) we get
$$B_{n-1} = I$$
$$B_{n-2} - AB_{n-1} = p_{n-1} I$$
$$B_{n-3} - AB_{n-2} = p_{n-2} I$$
$$\ldots \ \ldots \ \ldots$$
$$B_0 - AB_1 = p_1 I$$
$$-AB_0 = p_0 I$$

Multiplying these equations by $A^n, A^{n-1}, \ldots, A; I$, respectively, and adding we get
$$0 = A^n + p_{n-1}A^{n-1} + p_{n-2}A^{n-2} + p_1 A + p_0 I \qquad (1.11)$$
This proves the theorem.

1.5.1 Inverse of a Matrix by Cayley–Hamilton Theorem

By the Cayley–Hamilton Theorem every square matrix A satisfies its characteristic equation
$$A^n + p_{n-1}A^{n-1} + p_{n-2}A^{n-2} + \cdots + p_1 A + p_0 I = 0 \qquad (1.12)$$

[1] CAYLEY, Arthur (1821–1895) was a professor at Cambridge and is known for his important contributions to algebra, matrices and differential equations. He contributed to the theory of algebraic invariants and higher dimension and geometry.

[2] HAMILTON, William Rowan (1805–1865), a neat Irish mathematician, is known for his work in dynamics, is an astronomer and also a physicist.

A^{-1} exists $\Leftrightarrow$ A is nonsingular
$\Leftrightarrow$ $\det A = |A| = D(0) = p_0 \neq 0$

Multiplying (1.12) by A^{-1}

$$A^{-1}(A^n + p_{n-1}A^{n-1} + p_{n-2}A^{n-2} + \cdots + p_1 A + p_0 I) = 0$$

$$\Rightarrow \quad A^{n-1} + p_{n-1}A^{n-2} + \cdots + p_1 I + p_0 A^{-1} = 0$$

$$\Rightarrow \quad A^{-1} = -\frac{1}{p_0}\left[A^{n-1} + p_{n-1}A^{n-2} + \cdots + p_1 I\right] \tag{1.13}$$

This result gives us the inverse of A in terms of $(n-1)$ powers of A. It is considered as a practical method for the computation of the inverses of large matrices. As a by-product of the computation, the characteristic equation and the determinant of the matrix are obtained.

Synthetic method For verification of the Cayley–Hamilton Theorem and finding A^{-1}, the characteristic equation (1.12) can be written in the form

$$((\cdots (A + p_{n-1} I) A + p_{n-2} I) A + \cdots + p_1 I)A + p_0 I = 0 \; (p_0 \neq 0) \tag{1.14}$$

$$\text{or} \quad BA = -p_0 I \tag{1.15}$$

where

$$B = (\cdots (A + p_{n-1} I)A + p_{n-2} I)A + \cdots + p_1 I$$

$$\Rightarrow \quad A^{-1} = \left(-\frac{1}{p_0}\right)B \tag{1.16}$$

Equation (1.14) is more convenient for the verification of the Cayley–Hamilton Theorem; and Equation (1.16) for finding A^{-1}.

Example 1.7

Verify the Cayley–Hamilton Theorem for $A = \begin{bmatrix} 1 & 4 \\ 2 & 3 \end{bmatrix}$.

Find A^{-1} and $B = A^5 - 4A^4 - 7A^3 + 11A^2 - A - 10I$.

Solution The characteristic equation of A is

$$P(\lambda) = \begin{vmatrix} 1-\lambda & 4 \\ 2 & 3-\lambda \end{vmatrix} = \lambda^2 - 4\lambda - 5 = 0$$

$$A^2 = \begin{bmatrix} 1 & 4 \\ 2 & 3 \end{bmatrix}\begin{bmatrix} 1 & 4 \\ 2 & 3 \end{bmatrix} = \begin{bmatrix} 9 & 16 \\ 8 & 17 \end{bmatrix};$$

$$A^2 - 4A - 5I = \begin{bmatrix} 9 & 16 \\ 8 & 17 \end{bmatrix} - \begin{bmatrix} 4 & 16 \\ 8 & 12 \end{bmatrix} - \begin{bmatrix} 5 & 0 \\ 0 & 5 \end{bmatrix} = \begin{bmatrix} 0 & 0 \\ 0 & 0 \end{bmatrix}$$

The Cayley–Hamilton Theorem is verified.

$$\det A = P(0) = -5 \neq 0 \quad \therefore \quad A^{-1} \text{exists.}$$

$$A^2 - 4A - 5I = 0$$

$$\Rightarrow \quad A^{-1}(A^2 - 4A - 5I) = A - 4I - 5A^{-1} = 0$$

$$\Rightarrow \quad A^{-1} = \frac{1}{5}(A - 4I) = \frac{1}{5}\left\{\begin{bmatrix} 1 & 4 \\ 2 & 3 \end{bmatrix} - \begin{bmatrix} 4 & 0 \\ 0 & 4 \end{bmatrix}\right\}$$

$$= \frac{1}{5}\begin{bmatrix} -3 & 4 \\ 2 & -1 \end{bmatrix}$$

$$A^3(A^2 - 4A - 5I) = A^5 - 4A^4 - 5A^3;$$
$$-2A(A^2 - 4A - 5I) = -2A^3 + 8A^2 + 10A$$
$$3(A^2 - 4A - 5I) = (3A^2 - 12A - 15I);$$
$$A + 5I = A + 5I$$

Adding we get $A^3.0 - 2A.0 + 3.0 + A + 5I = B$

$$\therefore B = A + 5I = \begin{bmatrix} 1 & 4 \\ 2 & 3 \end{bmatrix} + \begin{bmatrix} 5 & 0 \\ 0 & 5 \end{bmatrix} = \begin{bmatrix} 6 & 4 \\ 2 & 8 \end{bmatrix}$$

Synthetic method For verification of the Cayley–Hamilton Theorem we have to show that

$$5I = A^2 - 4A = (A - 4I)A = BA$$

$$\text{where} \quad B = A - 4I = \begin{bmatrix} 1 & 4 \\ 2 & 3 \end{bmatrix} - 4\begin{bmatrix} 1 & 0 \\ 0 & 1 \end{bmatrix} = \begin{bmatrix} -3 & 4 \\ 2 & -1 \end{bmatrix}$$

$$BA = \begin{bmatrix} -3 & 4 \\ 2 & -1 \end{bmatrix}\begin{bmatrix} 1 & 4 \\ 2 & 3 \end{bmatrix} = \begin{bmatrix} 5 & 0 \\ 0 & 5 \end{bmatrix} = 5I; \text{ and}$$

The Cayley–Hamilton Theorem is verified.

$$\text{Also,} \quad A^{-1} = \frac{1}{5}B = \frac{1}{5}\begin{bmatrix} -3 & 4 \\ 2 & -1 \end{bmatrix}$$

Example 1.8

If $A = \begin{bmatrix} 2 & 1 & 1 \\ 0 & 1 & 0 \\ 1 & 1 & 2 \end{bmatrix}$ find A^{-1} and $B = A^8 - 5A^7 + 7A^6 - 3A^5 + A^4 - 5A^3 + 8A^2 - 2A + I$.

Solution The characteristic equation of A is

$$P(\lambda) = \begin{vmatrix} 2-\lambda & 1 & 1 \\ 0 & 1-\lambda & 0 \\ 1 & 1 & 2-\lambda \end{vmatrix} = (1-\lambda)(\lambda^2 - 4\lambda + 3), \text{ expanding by } R_2 \tag{1}$$

$$= -\lambda^3 + 5\lambda^2 - 7\lambda + 3 = 0 \quad (2)$$

By the Cayley–Hamilton Theorem, A satisfies its characteristic equation so that we have

$$A^3 - 5A^2 + 7A - 3I = 0 \quad (3)$$

Since det $A = P(0) = 3 \neq 0$, A^{-1} exists.
Multiplying (3) by A^{-1}

$$\therefore \quad A^{-1}(A^3 - 5A^2 + 7A - 3I) = 0$$

$$\Rightarrow \quad A^2 - 5A + 7I - 3A^{-1} = 0$$

$$\Rightarrow \quad 3A^{-1} = A^2 - 5A + 7I$$

$$\text{Now,} \quad A^2 = \begin{bmatrix} 2 & 1 & 1 \\ 0 & 1 & 0 \\ 1 & 1 & 2 \end{bmatrix}\begin{bmatrix} 2 & 1 & 1 \\ 0 & 1 & 0 \\ 1 & 1 & 2 \end{bmatrix} = \begin{bmatrix} 5 & 4 & 4 \\ 0 & 1 & 0 \\ 4 & 4 & 5 \end{bmatrix}$$

$$\therefore 3A^{-1} = \begin{bmatrix} 5 & 4 & 4 \\ 0 & 1 & 0 \\ 4 & 4 & 5 \end{bmatrix} - \begin{bmatrix} 10 & 5 & 5 \\ 0 & 5 & 0 \\ 5 & 5 & 10 \end{bmatrix} + \begin{bmatrix} 7 & 0 & 0 \\ 0 & 7 & 0 \\ 0 & 0 & 7 \end{bmatrix}$$

$$= \begin{bmatrix} 2 & -1 & -1 \\ 0 & 3 & 0 \\ -1 & -1 & 2 \end{bmatrix}$$

$$A^{-1} = \frac{1}{3}\begin{bmatrix} 2 & -1 & -1 \\ 0 & 3 & 0 \\ -1 & -1 & 2 \end{bmatrix} \quad (4)$$

$$B = A^8 - 5A^7 + 7A^6 - 3A^5 + A^4 - 5A^3 + 8A^2 - 2A + I$$

$$= A^5(A^3 - 5A^2 + 7A - 3I) + A(A^3 - 5A^2 + 7A - 3I)$$

$$+ A^2 + A + I$$

$$= A^5(0) + A(0) + A^2 + A + I$$

$$= \begin{bmatrix} 5 & 4 & 4 \\ 0 & 1 & 0 \\ 4 & 4 & 5 \end{bmatrix} + \begin{bmatrix} 2 & 1 & 1 \\ 0 & 1 & 0 \\ 1 & 1 & 2 \end{bmatrix} + \begin{bmatrix} 1 & 0 & 0 \\ 0 & 1 & 0 \\ 0 & 0 & 1 \end{bmatrix} = \begin{bmatrix} 8 & 5 & 5 \\ 0 & 3 & 0 \\ 5 & 5 & 8 \end{bmatrix} \quad (5)$$

Synthetic method For verification of the Cayley–Hamilton Theorem we have to show that

$$3I = A^3 - 5A^2 + 7A = ((A - 5I)A + 7I)A = BA$$

where $B = ((A - 5I)A + 7I$

$$= \begin{bmatrix} -3 & 1 & 1 \\ 0 & -4 & 0 \\ 1 & 1 & -3 \end{bmatrix}\begin{bmatrix} 2 & 1 & 1 \\ 0 & 1 & 0 \\ 1 & 1 & 2 \end{bmatrix} + 7\begin{bmatrix} 1 & 0 & 0 \\ 0 & 1 & 0 \\ 0 & 0 & 1 \end{bmatrix}$$

$$= \begin{bmatrix} -5 & -1 & -1 \\ 0 & -4 & 0 \\ -1 & -1 & -5 \end{bmatrix} + 7\begin{bmatrix} 1 & 0 & 0 \\ 0 & 1 & 0 \\ 0 & 0 & 1 \end{bmatrix}$$

$$= \begin{bmatrix} 2 & -1 & -1 \\ 0 & 3 & 0 \\ -1 & -1 & 2 \end{bmatrix}$$

$$BA = \begin{bmatrix} 2 & -1 & -1 \\ 0 & 3 & 0 \\ -1 & -1 & 2 \end{bmatrix}\begin{bmatrix} 2 & 1 & 1 \\ 0 & 1 & 0 \\ 1 & 1 & 2 \end{bmatrix} = \begin{bmatrix} 3 & 0 & 0 \\ 0 & 3 & 0 \\ 0 & 0 & 3 \end{bmatrix} = 3I$$

$\therefore$ The Cayley–Hamilton Theorem is verified

$$\text{Also, } A^{-1} = \frac{1}{3}B = \frac{1}{3}\begin{bmatrix} 2 & -1 & -1 \\ 0 & 3 & 0 \\ -1 & -1 & 2 \end{bmatrix}$$

Example 1.9

Find A^{-1} if $A = \begin{bmatrix} 7 & 2 & -2 \\ -6 & -1 & 2 \\ 6 & 2 & -1 \end{bmatrix}$.

Solution The characteristic equation of A is

$$P(\lambda) = \begin{vmatrix} 7-\lambda & 2 & -2 \\ -6 & -1-\lambda & 2 \\ 6 & 2 & -1-\lambda \end{vmatrix}$$

$$= (7-\lambda)(\lambda^2 + 2\lambda - 3) + 6(-2 - 2\lambda + 4)$$

$$+ 6\{4 + (-2 - 2\lambda)\} \quad (1)$$

$$= -\lambda^3 + 5\lambda^2 - 7\lambda + 3 = 0 \quad (2)$$

By the Cayley–Hamilton Theorem, A satisfies its characteristic equation so that

$$A^3 - 5A^2 + 7A - 3I = 0 \quad (3)$$

Also, det $A = P(0) = 3 \neq 0$ and hence A^{-1} exists.
Multiplying (1) by A^{-1}

$$A^{-1}(A^3 - 5A^2 + 7A - 3I) = 0$$
$$\Rightarrow \quad A^2 - 5A + 7I - 3A^{-1} = 0$$
$$\Rightarrow \quad 3A^{-1} = A^2 - 5A + 7I \qquad (4)$$

Now, $A^2 = \begin{bmatrix} 7 & 2 & -2 \\ -6 & -1 & 2 \\ 6 & 2 & -1 \end{bmatrix}\begin{bmatrix} 7 & 2 & -2 \\ -6 & -1 & 2 \\ 6 & 2 & -1 \end{bmatrix}$

$$= \begin{bmatrix} 25 & 8 & -8 \\ -24 & -7 & 8 \\ 24 & 8 & -7 \end{bmatrix}$$

$$3A^{-1} = \begin{bmatrix} 25 & 8 & -8 \\ -24 & -7 & 8 \\ 24 & 8 & -7 \end{bmatrix} - \begin{bmatrix} 35 & 10 & -10 \\ -30 & -5 & 10 \\ 30 & 10 & -5 \end{bmatrix}$$
$$+ \begin{bmatrix} 7 & 0 & 0 \\ 0 & 7 & 0 \\ 0 & 0 & 7 \end{bmatrix} = \begin{bmatrix} -3 & -2 & 2 \\ 6 & 5 & -2 \\ -6 & -2 & 5 \end{bmatrix};$$

$$A^{-1} = \frac{1}{3}\begin{bmatrix} -3 & -2 & 2 \\ 6 & 5 & -2 \\ -6 & -2 & 5 \end{bmatrix} \qquad (5)$$

Synthetic method For verification of the Cayley–Hamilton Theorem we have to show that

$$3I = A^3 - 5A^2 + 7A = ((A - 5I)A + 7I)A = BA$$

where $B = (A - 5I)A + 7I$

$$= \begin{bmatrix} 2 & 2 & -2 \\ -6 & -6 & 2 \\ 6 & 2 & -6 \end{bmatrix}\begin{bmatrix} 7 & 2 & -2 \\ -6 & -1 & 2 \\ 6 & 2 & -1 \end{bmatrix} + 7I$$

$$= \begin{bmatrix} -10 & -2 & 2 \\ 6 & -2 & -2 \\ -6 & -2 & -2 \end{bmatrix} + 7\begin{bmatrix} 1 & 0 & 0 \\ 0 & 1 & 0 \\ 0 & 0 & 1 \end{bmatrix} = \begin{bmatrix} -3 & -2 & 2 \\ 6 & 5 & -2 \\ -6 & -2 & 5 \end{bmatrix}$$

$$BA = \begin{bmatrix} -3 & -2 & 2 \\ 6 & 5 & -2 \\ -6 & -2 & 5 \end{bmatrix}\begin{bmatrix} 7 & 2 & -2 \\ -6 & -1 & 2 \\ 6 & 2 & -1 \end{bmatrix} = 3I$$

Also, $A^{-1} = \frac{1}{3}B = \frac{1}{3}\begin{bmatrix} -3 & -2 & 2 \\ 6 & 5 & -2 \\ -6 & -2 & 5 \end{bmatrix}$

Remarks The matrix (Example 1.8) $\begin{bmatrix} 2 & 1 & 1 \\ 0 & 1 & 0 \\ 1 & 1 & 2 \end{bmatrix}$ and the matrix (Example 1.9) $\begin{bmatrix} 7 & 2 & -2 \\ -6 & -1 & 2 \\ -6 & 2 & -1 \end{bmatrix}$ have the same characteristic equation and the same eigenvalues. Such matrices are called 'similar'. We consider similarity of matrices below.

EXERCISE 1.2

1. Using the Cayley–Hamilton Theorem find A^8 if $A = \begin{bmatrix} 1 & 2 \\ 2 & -1 \end{bmatrix}$. [JNTU 2003]

   Ans: $A^8 = 625I$

2. Use the Cayley–Hamilton Theorem to find A^3 and A^{-3} if $A = \begin{bmatrix} 2 & 4 \\ 1 & 1 \end{bmatrix}$. [JNTU 2002 S]

   Ans: $A^3 = \begin{bmatrix} 28 & 44 \\ 11 & 17 \end{bmatrix}$, $A^{-3} = \frac{1}{8}\begin{bmatrix} -17 & 44 \\ 11 & -28 \end{bmatrix}$

3. Use the Cayley–Hamilton Theorem to express $2A^5 - 3A^4 + A^2 - 4I = aA + bI$ if $A = \begin{bmatrix} 3 & 1 \\ -1 & 2 \end{bmatrix}$.

   Ans: $a = 138$, $b = -403$

4. Show that $A = \begin{bmatrix} 1 & -2 & 2 \\ 1 & 2 & 3 \\ 0 & -1 & 2 \end{bmatrix}$ satisfies its characteristic equation. Find A^{-1}. [JNTU 2002]

   Ans: $A^{-1} = \frac{1}{9}\begin{bmatrix} 7 & 2 & -10 \\ -2 & 2 & -1 \\ -1 & 1 & 4 \end{bmatrix}$

   characteristic equation is $\lambda^3 - 5\lambda^2 + 13\lambda - 9 = 0$

5. Show that Adj $(A - \lambda I) = B_0 + \lambda B_1 + \lambda^2 I$ where $B_0 = \begin{bmatrix} -4 & 4 & -2 \\ -2 & -2 & -2 \\ 1 & -2 & 1 \end{bmatrix}$ and $B_1 = \begin{bmatrix} -1 & 0 & 2 \\ 0 & -1 & 2 \\ -1 & 2 & -2 \end{bmatrix}$

   if $A = \begin{bmatrix} 1 & 0 & 2 \\ 0 & 1 & 2 \\ 1 & 2 & 0 \end{bmatrix}$

6. State the Cayley–Hamilton Theorem and verify for $A=\begin{bmatrix}3&0&0\\5&4&0\\3&6&1\end{bmatrix}$. [JNTU 1999]

7. Verify the Cayley–Hamilton Theorem for $A=\begin{bmatrix}3&1&1\\-1&5&-1\\1&-1&3\end{bmatrix}$. Find A^{-1}. [JNTU 2002]

Ans: $A^{-1}=\frac{1}{20}\begin{bmatrix}7&2&-3\\10&4&1\\-2&2&8\end{bmatrix}$

8. Verify the Cayley–Hamilton Theorem for $A=\begin{bmatrix}1&1&3\\1&3&-3\\-2&-4&-4\end{bmatrix}$. [JNTU November 2003]

9. Verify the Cayley–Hamilton Theorem for $A=\begin{bmatrix}2&-1&1\\-1&2&-1\\1&-1&1\end{bmatrix}$. Find A^{-1}. [JNTU 2000 S, 2002]

Ans: $A^{-1}=\frac{1}{4}\begin{bmatrix}3&1&-1\\1&3&1\\-1&1&3\end{bmatrix}$

10. Verify the Cayley–Hamilton Theorem for $A=\begin{bmatrix}1&0&3\\2&1&-1\\1&-1&1\end{bmatrix}$. [JNTU 2002, 1999 S]

Ans: $A^{-1}=\frac{1}{9}\begin{bmatrix}0&3&3\\3&2&-7\\3&-1&-1\end{bmatrix}$

11. Find the characteristic equation of $A=\begin{bmatrix}0&c&-b\\-c&0&a\\b&-a&0\end{bmatrix}$.

Ans: Characteristic equation is $\lambda^3+\lambda(a^2+b^2+c^2)=0$;

$$A^3=(a^2+b^2+c^2)\begin{bmatrix}0&-c&b\\c&0&-a\\-b&a&0\end{bmatrix}$$

12. Find the characteristic equation of $A=\begin{bmatrix}0&3&-2\\-3&0&1\\2&-1&0\end{bmatrix}$.

Verify that A satisfies it.

Ans: Characteristic equation is $\lambda^3+14\lambda=0$;

$$A^3=14\begin{bmatrix}0&-3&2\\3&0&-1\\-2&1&0\end{bmatrix}$$

13. By applying the Cayley–Hamilton Theorem to $A=\begin{bmatrix}7&-1&3\\6&1&4\\2&4&8\end{bmatrix}$, find A^{-1}.

Ans: $A^{-1}=\begin{bmatrix}-1.60&0.40&-0.14\\-0.80&1.00&-0.20\\0.44&-0.60&0.26\end{bmatrix}$

14. Verify the Cayley–Hamilton Theorem and hence find A^{-1} if $A=\begin{bmatrix}0&1&0&0\\0&0&0&1\\0&0&1&0\\1&0&0&0\end{bmatrix}$

Ans: Characteristic equation is $\lambda^4-\lambda^3-\lambda+1=0$

$$A^{-1}=\begin{bmatrix}0&0&0&1\\1&0&0&0\\0&0&1&0\\0&1&0&0\end{bmatrix}$$

15. Using the Cayley–Hamilton Theorem find A^{-1} and A^4 if $A=\begin{bmatrix}7&2&-2\\-6&-1&2\\6&2&-1\end{bmatrix}$. [JNTU 2000]

Ans: $A^{-1}=\frac{1}{3}\begin{bmatrix}-3&-2&2\\6&5&-2\\-6&-2&5\end{bmatrix}$; $A^4=\begin{bmatrix}241&80&-80\\-240&-79&80\\240&80&-79\end{bmatrix}$

16. By the Cayley–Hamilton Theorem find A^{-1} if $A=\begin{bmatrix}0&0&1\\3&1&0\\-2&1&4\end{bmatrix}$. [JNTU 1993]

Ans: $A^{-1}=\frac{1}{5}\begin{bmatrix}4&1&-1\\-12&2&3\\5&0&0\end{bmatrix}$

17. Show that in addition to A, *Adj A* also satisfies the characteristic equation for $A=\begin{bmatrix}-4&-3&-3\\1&0&1\\4&4&3\end{bmatrix}$.

[**Hint**: *Adj* $A=A$.]

18. Show that $A = \begin{bmatrix} a & 0 & 0 \\ 0 & b & 0 \\ d & 0 & c \end{bmatrix}$ satisfies its characteristic equation
$p(\lambda) = s_0 s_1 - (s_0 + s_1 s_2)\lambda + (s_1 + s_2)\lambda^2 - \lambda^3 = 0$
where $s_0 = bc$, $s_1 = a + d$ and $s_2 = b + c$.

19. Verify the Cayley–Hamilton Theorem for $A = \begin{bmatrix} 2 & -1 & 1 \\ -1 & 2 & -1 \\ 1 & -1 & 2 \end{bmatrix}$. Hence compute A^{-1}.

Ans: $A^{-1} = \frac{1}{4}\begin{bmatrix} 3 & 1 & -1 \\ 1 & 3 & 1 \\ -1 & 1 & 3 \end{bmatrix}$

Characteristic equation is $\lambda^3 - 6\lambda^2 + 9\lambda - 4 = 0$

20. Find the characterstic equation of $A = \begin{bmatrix} 1 & 3 & 7 \\ 4 & 2 & 3 \\ 1 & 2 & 1 \end{bmatrix}$.

Show that it is satisfied by A. Find A^{-1}.

Ans: $\lambda^3 - 4\lambda^2 - 20\lambda - 35 = 0$;

$$A^{-1} = \frac{1}{35}\begin{bmatrix} -4 & 11 & -5 \\ -1 & -6 & 25 \\ 6 & 1 & -10 \end{bmatrix}$$

21. Verify the Cayley–Hamilton Theorem for A if $A = \begin{bmatrix} 1 & 2 & 3 \\ 2 & -1 & 4 \\ 3 & 1 & -1 \end{bmatrix}$ and find A^{-1}

Ans: Characteristic equation is $\lambda^3 + \lambda^2 - 18\lambda - 40 = 0$

$$A^{-1} = \frac{1}{40}\begin{bmatrix} -3 & 5 & 11 \\ 14 & -10 & 2 \\ 5 & 5 & -5 \end{bmatrix}$$

22. Using the Cayley–Hamilton Theorem find A^{-1} if $A = \begin{bmatrix} 1 & 2 & 1 \\ 2 & 3 & 5 \\ 0 & -1 & 4 \end{bmatrix}$.

Ans: $A^{-1} = \begin{bmatrix} -17 & 9 & -7 \\ 8 & -4 & 3 \\ 2 & -1 & 1 \end{bmatrix}$

23. Using the Cayley–Hamilton Theorem find A^{-1} if $A = \begin{bmatrix} 4 & 1 & 2 \\ 1 & 2 & 0 \\ 0 & -1 & 3 \end{bmatrix}$.

Ans: $A^{-1} = \frac{1}{19}\begin{bmatrix} 6 & -5 & -4 \\ -3 & 12 & 2 \\ -1 & 4 & 7 \end{bmatrix}$

24. If $A = \begin{bmatrix} 0 & \omega^2 & -\omega \\ -\omega^2 & 0 & 1 \\ \omega & -1 & 0 \end{bmatrix}$ where ω is a cube root of unity, by using the Cayley–Hamilton Theorem show that A is nilpotent of index 3.

1.6 Reduction of a Square Matrix to Diagonal Form

1.6.1 Diagonalisation—Powers of a Square Matrix A

Theorem 1.2 If a square matrix A of order n has n (≥ 2) linearly independent eigenvectors, then a nonsingular matrix P can be found such that $P^{-1}AP$ is a diagonal matrix.

Proof Let $X_1, X_2, \ldots, X_n$ be the $n \geq 2$ linearly independent eigenvectors of A corresponding to its n eigenvalues $\lambda_1, \lambda_2, \ldots, \lambda_n$.

Construct matrix P with $X_1, X_2, \ldots, X_n$ as n column vectors.

Let $P_{n \times n} = [X_1, X_2 \ldots X_n]$ where $X_j = [x_{ij}]$, $i = 1, 2, \ldots, n$. Since $X_1, X_2, \ldots, X_n$ are linearly independent P is nonsingular and therefore P^{-1} exists.

Consider $AP = A\,[X_1, X_2 \ldots X_n] = [AX_1, AX_2 \ldots AX_n] = [\lambda_1 X_1, \lambda_2 X_2 \ldots \lambda_n X_n]$

$$= \begin{bmatrix} \lambda_1 x_{11} & \lambda_2 x_{12} & \cdots & \lambda_j x_{1j} & \cdots & \lambda_n x_{1n} \\ \lambda_1 x_{21} & \lambda_2 x_{22} & \cdots & \lambda_j x_{2j} & \cdots & \lambda_n x_{2n} \\ \cdots & \cdots & \cdots & \cdots & \cdots & \cdots \\ \lambda_1 x_{n1} & \lambda_2 x_{n2} & \cdots & \lambda_j x_{nj} & \cdots & \lambda_n x_{nn} \end{bmatrix}$$

$$= \begin{bmatrix} x_{11} & x_{12} & \cdots & x_{1j} & \cdots & x_{1n} \\ x_{21} & x_{22} & \cdots & x_{2j} & \cdots & x_{2n} \\ \cdots & \cdots & \cdots & \cdots & \cdots & \cdots \\ x_{n1} & x_{n2} & \cdots & x_{nj} & \cdots & x_{nn} \end{bmatrix} \begin{bmatrix} \lambda_1 & 0 & \cdots & 0 \\ 0 & \lambda_2 & \cdots & 0 \\ \cdots & \cdots & \cdots & \cdots \\ 0 & 0 & \cdots & \lambda_n \end{bmatrix}$$

$= PD$ where $D = \text{diag}\,[\lambda_1, \lambda_2, \cdots \lambda_n]$

$\Rightarrow \quad P^{-1} AP = P^{-1} PD = ID = D$

1.6.2 Modal Matrix and Spectral Matrix of a Square Matrix *A*

The matrix P which diagonalises A is called a *modal matrix* of A and the resulting diagonal matrix D is known as the *spectral matrix* of A.

The matrix P is found by grouping the eigenvectors of A into a square matrix and the diagonal matrix has the eigenvalues of A as the principal diagonal elements. If $n = 1$ then P does not exist.

1.6.3 Similarity of Matrices

Let A be a square matrix of order n. A square matrix B of order n is said to be similar to A if there exists a nonsingular matrix P such that $AP = PB$ or $B = P^{-1} AP$. In such a case we say that A and B are *similar matrices*.

This relation defines a transformation of a matrix A into another matrix B and the transformation is called a *similarity transformation*.

Similarity of matrices is an equivalence relation on the set of n-square matrices.

Theorem 1.3 Similar matrices have the same eigenvalues.

Proof Let B be similar to matrix A. Then there exists a nonsingular matrix P such that $B = P^{-1} AP$.

Consider the characteristic polynomial of B

$$\begin{aligned}|B - \lambda I| &= |P^{-1}AP - \lambda I| = |P^{-1}AP - \lambda P^{-1}IP| \\ &= |P^{-1}(A - \lambda I)P| = |P^{-1}||A - \lambda I||P| \\ &\quad |A - \lambda I|,\end{aligned}$$

$$\text{since} \quad |P^{-1}||P| = |P^{-1}P| = |I| = 1$$

Thus A and B have the same characteristic polynomial and hence the same eigenvalues.

Theorem 1.4 If X is an eigenvector of a matrix A corresponding to an eigenvalue λ of A then the eigenvector of B, which is similar to A, is $Y = P^{-1}X$.

Proof Let X be an eigenvector corresponding to an eigenvalue λ of A so that $AX = \lambda X$.

Since B is similar to A we have $B = P^{-1}AP$ for some nonsingular matrix P. Now

$$BP^{-1} = (P^{-1}AP)\,P^{-1} = P^{-1}A\,(PP^{-1}) = P^{-1}A;$$

post-multiplying this by X,

$$\begin{aligned}(BP^{-1})X &= (P^{-1}A)\,X = P^{-1}(AX) = P^{-1}(\lambda X) \\ &= \lambda(P^{-1}X) \quad \Rightarrow \quad B(P^{-1}X) = \lambda(P^{-1}X)\end{aligned}$$

Thus $P^{-1}X$ is an eigenvector of B corresponding to the eigenvalue λ.

1.6.4 Diagonalisation—Conditions for Diagonalisability of a Matrix *A*

The existence of *distinct eigenvalues* is only a *sufficient condition* but not a necessary condition for a square matrix A to be diagonalisable.

The real condition for diagonalisability of a square matrix A of order n is the existence of n linearly independent eigenvectors, which may exist even if some of the eigenvalues of a matrix are repeated (algebraic multiplicity of an eigenvalue is greater then 1).

We know that, in general,

Algebraic multiplicity (A.M.) $\geq$ Geometic multiplicity (G.M.) of eigenvalue λ of a matrix A.

If G.M. < A.M. of an eigenvalue in respect of a square matrix A then we cannot find n linearly independent eigenvectors and hence A is not diagonalisable.

If G.M. and A.M. of an eigenvalue in respect of a square matrix A are equal then we can find n linearly independent eigenvectors and hence A is diagonalisable.

Example 1.10

Prove that $A = \begin{bmatrix} 1 & 1 \\ 0 & 1 \end{bmatrix}$ is *not* diagonalisable.

Solution The characteristic equation is $|A - \lambda I|$

$$= \begin{bmatrix} 1-\lambda & 1 \\ 0 & 1-\lambda \end{bmatrix} = (\lambda - 1)^2 = 0 \quad \Rightarrow \quad \lambda = 1, 1$$

To find the eigenvectors we have to solve

$$(A - \lambda I)\,X = 0 \quad \text{for} \quad \lambda = 1, 1$$

$\lambda = 1$

We have to solve $(A - I)X = 0$

$$\Rightarrow \begin{bmatrix} 0 & 1 \\ 0 & 0 \end{bmatrix} \begin{bmatrix} x_1 \\ x_2 \end{bmatrix} = \begin{bmatrix} 0 \\ 0 \end{bmatrix} \Rightarrow 0.x_1 + 1.x_2 = 0$$

Solving, we get $\frac{x}{1} = \frac{x_2}{0} \Rightarrow X_1 = c_1 \begin{bmatrix} 1 \\ 0 \end{bmatrix}$

We cannot find a second L.I. eigenvector.

Algebraic multiplicity of $(\lambda = 1) = 2$; Geometric Multiplicity of $(\lambda = 1) = 1$

$\therefore$ The matrix is not diagonalisable.

Example 1.11

Diagonalise the matrix $A = \begin{bmatrix} -2 & 2 & -3 \\ 2 & 1 & -6 \\ -1 & -2 & 0 \end{bmatrix}$

Solution The characteristic equation of A is $|A - \lambda I| = 0$

$$\Rightarrow \begin{vmatrix} -2-\lambda & 2 & -3 \\ 2 & 1-\lambda & -6 \\ -1 & -2 & 0-\lambda \end{vmatrix}$$

$$= (-2-\lambda)\begin{vmatrix} 1-\lambda & -6 \\ -2 & -\lambda \end{vmatrix} - 2\begin{vmatrix} 2 & -3 \\ -2 & -\lambda \end{vmatrix} - 1\begin{vmatrix} 2 & -3 \\ 1-\lambda & -6 \end{vmatrix}$$

$$= -\lambda^3 - \lambda^2 + 21\lambda + 45 = 0$$

Characteristic roots are $\lambda_1 = -3$, $\lambda_2 = -3$, $\lambda_3 = 5$.

To find eigenvectors we have to solve $(A - \lambda I)X = 0$.

$\lambda = -3$

$$(A + 3I) = \begin{bmatrix} -2+3 & 2 & -3 \\ 2 & 1+3 & -6 \\ -1 & -2 & 3 \end{bmatrix}$$

$$= \begin{bmatrix} 1 & 2 & -3 \\ 2 & 4 & -6 \\ -1 & -2 & 3 \end{bmatrix} \sim \begin{bmatrix} 1 & 2 & -3 \\ 0 & 0 & 0 \\ 0 & 0 & 0 \end{bmatrix} \begin{matrix} R_2 - 2R_1 \\ R_3 + R_1 \end{matrix}$$

Equivalent equation: $x_1 + 2x_2 - 3x_3 = 0$

Since the number of unknowns $n = 3$ and rank $r = 1$ $\Big\}$ $n - r = 3 - 1 = 2$ unknowns can be chosen arbitrarily.

We choose the eigenvectors so that they are *L.I.*

$$\Rightarrow \frac{x_1}{2} = \frac{x_2}{-1} = \frac{x_3}{0}; \quad \frac{x_1}{3} = \frac{x_2}{0} = \frac{x_3}{1}$$

$X_1 = [2 \ \ -1 \ \ 0]^T$; $X_2 = [3 \ \ 0 \ \ 1]^T$

taking $c_1 = c_2 = 1$

$\lambda = 5$

$$(A - 5I) = \begin{bmatrix} -2-5 & 2 & -3 \\ 2 & 1-5 & -6 \\ -1 & -2 & -5 \end{bmatrix}$$

$$= \begin{bmatrix} -7 & 2 & -3 \\ 2 & -4 & -6 \\ -1 & -2 & -5 \end{bmatrix} \overset{R_2+2R_3, R_1-7R_3}{\sim} \begin{bmatrix} 0 & 16 & 32 \\ 0 & -8 & -16 \\ -1 & -2 & -5 \end{bmatrix}$$

Equivalent system

$$\sim \begin{matrix} x_2 + 2x_3 = 0 \\ x_1 + x_3 = 0 \end{matrix} \Rightarrow \frac{x_1}{1} = \frac{x_2}{2} = \frac{x_3}{-1}$$

$$\Rightarrow X_3 = [1 \ \ 2 \ \ -1]^T$$

Modal matrix:

$$P = \begin{bmatrix} 2 & 3 & 1 \\ -1 & 0 & 2 \\ 0 & 1 & -1 \end{bmatrix} = \begin{bmatrix} a_1 & b_1 & c_1 \\ a_2 & b_2 & c_2 \\ a_3 & b_3 & c_3 \end{bmatrix} \text{ (say);}$$

$$Adj\ P = \begin{bmatrix} A_1 & A_2 & A_3 \\ B_1 & B_2 & B_3 \\ C_1 & C_2 & C_3 \end{bmatrix} = \begin{bmatrix} -2 & 4 & 6 \\ -1 & -2 & -5 \\ -1 & -2 & 3 \end{bmatrix}$$

$$P = a_1A_1 + a_2A_2 + a_3A_3 = 2(-2) + (-1)4 + 0 = -8$$

$$\therefore P^{-1} = \frac{1}{8}\begin{bmatrix} 2 & -4 & -6 \\ 1 & 2 & 5 \\ 1 & 2 & -3 \end{bmatrix}$$

$$\Rightarrow P^{-1}AP = D = \begin{bmatrix} -3 & 0 & 0 \\ 0 & -3 & 0 \\ 0 & 0 & 5 \end{bmatrix}$$

1.6.5 Orthogonalisation of a Symmetric Matrix

Let A be a real symmetric matrix with n distinct eigenvalues $\lambda_1, \lambda_2, \ldots, \lambda_n$. Then the corresponding eigenvectors are pairwise orthogonal. One can construct modal matrix P with the normalised eigenvectors, $\left(\frac{X_1}{\|X_1\|}, \frac{X_2}{\|X_2\|}, \ldots, \frac{X_n}{\|X_n\|}\right)$ as its columns. Such a matrix is called an orthogonal matrix. It satisfies $P^TP = I$ or $P^{-1} = P^T$. The diagonalisation of a square matrix A by an orthogonal matrix P is called *orthogonalisation of A*.

Example 1.12
Find a modal matrix (orthogonal matrix) P and diagonalise the symmetric matrix $A = \begin{bmatrix} 1 & 1 \\ 1 & 1 \end{bmatrix}$.

Solution The characteristic equation of A is

$$P(\lambda) = \begin{vmatrix} 1-\lambda & 1 \\ 1 & 1-\lambda \end{vmatrix} = (1-\lambda)^2 - 1$$
$$= \lambda(\lambda - 2) = 0 \quad \Rightarrow \quad \lambda_1 = 0, \ \lambda_2 = 2$$

$\lambda_1 = 0$

$$\begin{bmatrix} 1 & 1 \\ 1 & 1 \end{bmatrix}\begin{bmatrix} x_1 \\ x_2 \end{bmatrix} = \begin{bmatrix} 0 \\ 0 \end{bmatrix} \sim \quad x_1 + x_2 = 0$$
$$\Rightarrow \quad \frac{x_1}{1} = \frac{x_2}{-1} = c_1; \ X_1 = c_1\begin{bmatrix} 1 \\ -1 \end{bmatrix}$$

$\lambda_2 = 2$

$$\begin{bmatrix} -1 & 1 \\ 1 & -1 \end{bmatrix}\begin{bmatrix} x_1 \\ x_2 \end{bmatrix} = \begin{bmatrix} 0 \\ 0 \end{bmatrix} \sim \quad -x_1 + x_2 =$$
$$\Rightarrow \quad \frac{x_1}{1} = \frac{x_2}{1} = c_2; \ X_2 = c_2\begin{bmatrix} 1 \\ 1 \end{bmatrix}$$

The eigenvectors $X_1 = \begin{bmatrix} 1 \\ -1 \end{bmatrix}$ and $X_2 = \begin{bmatrix} 1 \\ 1 \end{bmatrix}$ $(c_1 = c_2 = 1)$ are *orthogonal*. $\because X_1^T . X_2 = [1, -1]\begin{bmatrix} 1 \\ 1 \end{bmatrix} = 0$

We construct modal matrix P with *orthonormal* vectors $\dfrac{X_1}{\|X_1\|}$ and $\dfrac{X_2}{\|X_2\|}$.

$$P = \begin{bmatrix} \frac{1}{\sqrt{2}} & \frac{1}{\sqrt{2}} \\ \frac{-1}{\sqrt{2}} & \frac{1}{\sqrt{2}} \end{bmatrix}; \ P^{-1} = P^T = \begin{bmatrix} \frac{1}{\sqrt{2}} & \frac{-1}{\sqrt{2}} \\ \frac{1}{\sqrt{2}} & \frac{1}{\sqrt{2}} \end{bmatrix}$$

(P is an orthogonal matrix satisfying $P^T P = I$ or $P^{-1} = P^T$.)

Now

$$P^{-1}AP = \begin{bmatrix} \frac{1}{\sqrt{2}} & \frac{-1}{\sqrt{2}} \\ \frac{1}{\sqrt{2}} & \frac{1}{\sqrt{2}} \end{bmatrix}\begin{bmatrix} 1 & 1 \\ 1 & 1 \end{bmatrix}\begin{bmatrix} \frac{1}{\sqrt{2}} & \frac{1}{\sqrt{2}} \\ \frac{-1}{\sqrt{2}} & \frac{1}{\sqrt{2}} \end{bmatrix}$$
$$= \frac{1}{2}\begin{bmatrix} 1 & -1 \\ 1 & 1 \end{bmatrix}\begin{bmatrix} 1 & 1 \\ 1 & 1 \end{bmatrix}\begin{bmatrix} 1 & 1 \\ -1 & 1 \end{bmatrix}$$
$$= \frac{1}{2}\begin{bmatrix} 0 & 0 \\ 2 & 2 \end{bmatrix}\begin{bmatrix} 1 & 1 \\ -1 & 1 \end{bmatrix} = \frac{1}{2}\begin{bmatrix} 0 & 0 \\ 0 & 4 \end{bmatrix} = \begin{bmatrix} 0 & 0 \\ 0 & 2 \end{bmatrix}$$
$$= D$$

Along the main diagonal of D we find the eigenvalues of A.

Example 1.13
Find a modal matrix (orthogonal matrix) P and diagonalise the symmetric matrix $A = \begin{bmatrix} 1 & 0 & 0 \\ 0 & 3 & -1 \\ 0 & -1 & 3 \end{bmatrix}$.

Solution The characteristic equation of A is

$$|A - \lambda I| = \begin{bmatrix} 1-\lambda & 0 & 0 \\ 0 & 3-\lambda & -1 \\ 0 & -1 & 3-\lambda \end{bmatrix} = (1-\lambda)(\lambda^2 - 6\lambda + 8)$$
$$= (1-\lambda)(\lambda - 2)(\lambda - 4) = 0$$
$$\Rightarrow \lambda = 1, 2, 4$$

To find the eigenvectors corresponding to the eigenvalue λ we have to solve $(A - I)X = 0$.

For $\lambda = 1$
We have to solve $(A - I)X = 0$

$$\Rightarrow \begin{bmatrix} 0 & 0 & 0 \\ 0 & 2 & -1 \\ 0 & -1 & 2 \end{bmatrix}\begin{bmatrix} x_1 \\ x_2 \\ x_3 \end{bmatrix} = \begin{bmatrix} 0 \\ 0 \\ 0 \end{bmatrix} \sim \begin{aligned} 0.x_1 + 0.x_2 + 0.x_3 &= 0 \\ 0.x_1 + 2x_2 - x_3 &= 0 \\ 0.x_1 + (-1)x_2 + 2x_3 &= 0 \end{aligned}$$
$$\Rightarrow \quad x_2 = 0, x_3 = 0 \text{ and } x_1 = c_1 \text{ (arbitrary)}$$
$$\Rightarrow \quad X_1 = c_1 [1 \quad 0 \quad 0]^T$$

For $\lambda = 2$
We have to solve $(A - 2I)X = 0$

$$\Rightarrow \begin{bmatrix} -1 & 0 & 0 \\ 0 & 1 & -1 \\ 0 & -1 & 1 \end{bmatrix}\begin{bmatrix} x_1 \\ x_2 \\ x_3 \end{bmatrix} = \begin{bmatrix} 0 \\ 0 \\ 0 \end{bmatrix}$$

$$\begin{aligned} 1x_1 + 0.x_2 + 0.x_3 &= 0 \\ \sim \quad 0.x_1 + 1x_2 - x_3 &= 0 \\ 0.x_1 - x_2 + x_3 &= 0 \\ \Rightarrow \quad x_1 = 0,\ x_2 = x_3 &= c_2 \\ \Rightarrow \quad X_2 &= c_2\,[0\ 1\ 1]^T \end{aligned}$$

For $\lambda = 4$

We have to solve $(A - 4I)X = 0$

$$\Rightarrow \begin{bmatrix} -3 & 0 & 0 \\ 0 & -1 & -1 \\ 0 & -1 & -1 \end{bmatrix}\begin{bmatrix} x_1 \\ x_2 \\ x_3 \end{bmatrix} = \begin{bmatrix} 0 \\ 0 \\ 0 \end{bmatrix}$$

$$\begin{aligned} -3x_1 + 0.x_2 + 0.x_3 &= 0 \\ \sim \quad 0.x_1 - x_2 + x_3 &= 0 \\ \Rightarrow \quad x_1 = 0,\ x_2 = -x_3 &= c_3 \\ \Rightarrow \quad X_3 &= c_3\,[0\ 1\ -1]^T \end{aligned}$$

Normalising these mutually orthogonal vectors

we write $P = \left[\dfrac{X_1}{\|X_1\|}\ \dfrac{X_2}{\|X_2\|}\ \dfrac{X_3}{\|X_3\|}\right]$

or

$$P = \begin{bmatrix} 1 & 0 & 0 \\ 0 & \frac{1}{\sqrt{2}} & \frac{1}{\sqrt{2}} \\ 0 & \frac{1}{\sqrt{2}} & -\frac{1}{\sqrt{2}} \end{bmatrix}$$

$$P^{-1} = P^T = P$$

($\because$ P is symmetric)

$P^{-1}AP =$

$$\begin{bmatrix} 1 & 0 & 0 \\ 0 & \frac{1}{\sqrt{2}} & \frac{1}{\sqrt{2}} \\ 0 & \frac{1}{\sqrt{2}} & -\frac{1}{\sqrt{2}} \end{bmatrix}\begin{bmatrix} 1 & 0 & 0 \\ 0 & 3 & -1 \\ 0 & -1 & 3 \end{bmatrix}\begin{bmatrix} 1 & 0 & 0 \\ 0 & \frac{1}{\sqrt{2}} & \frac{1}{\sqrt{2}} \\ 0 & \frac{1}{\sqrt{2}} & -\frac{1}{\sqrt{2}} \end{bmatrix}$$

$= \text{diag}\,[1,\ 2,\ 4] = D$

1.7 Powers of a Square Matrix A—Finding of Modal Matrix P and Inverse Matrix A^{-1}

Let A be an n-square matrix and D, the diagonal matrix similar to A. So, $D = P^{-1}AP$

Then $D^2 = (P^{-1}AP)(P^{-1}AP) = P^{-1}A\,(PP^{-1})\,AP$
$= P^{-1}AI\,AP = P^{-1}AAP = P^{-1}A^2P$
$\Rightarrow\quad D^n = P^{-1}A^nP$, by induction

$$\therefore \quad PD^nP^{-1} = P(P^{-1}A^nP)P^{-1} = (PP^{-1})\,A^n\,(PP^{-1}) = IA^nI = A^n$$

or $\quad A^n = PD^nP^{-1}$

Example 1.14

Find a matrix P which diagonalises the matrix $A = \begin{bmatrix} 4 & 1 \\ 2 & 3 \end{bmatrix}$. Verify that $P^{-1}AP = D$, where D is a diagonal matrix. Hence find A^6.

Solution The matrix A can be diagonalised by P whose columns are linearly independent eigenvectors of A, if A has at least two linearly independent eigenvectors.

To find the eigenvectors of A we write the characteristic equation of A.

$$P(\lambda) = |A - \lambda I| = \begin{vmatrix} 4-\lambda & 1 \\ 2 & 3-\lambda \end{vmatrix} = \lambda^2 - 7\lambda + 10 = (\lambda - 2)(\lambda - 5) = 0;$$

$\lambda = 2, 5$ are two distinct eigenvalues of A.

$$\boldsymbol{\lambda = 2:}\ \begin{bmatrix} 2 & 1 \\ 2 & 1 \end{bmatrix}\begin{bmatrix} x_1 \\ x_2 \end{bmatrix} = 0 \quad \Rightarrow \quad 2x_1 + x_2 = 0$$

$$\Rightarrow \quad \frac{x_1}{1} = \frac{x_2}{-2} \quad \Rightarrow \quad X_1 = c_1\begin{bmatrix} 1 \\ -2 \end{bmatrix}$$

$$\boldsymbol{\lambda = 5:}\ \begin{bmatrix} -1 & 1 \\ 2 & -2 \end{bmatrix}\begin{bmatrix} x_1 \\ x_2 \end{bmatrix} = 0 \quad \Rightarrow \quad -x_1 + x_2 = 0$$

$$\Rightarrow \quad \frac{x_1}{1} = \frac{x_2}{1} \quad \Rightarrow \quad X_2 = c_2\begin{bmatrix} 1 \\ 1 \end{bmatrix}$$

$X_1 = \begin{bmatrix} 1 \\ -2 \end{bmatrix}$ and $X_2 = \begin{bmatrix} 1 \\ 1 \end{bmatrix}$ are two linearly independent eigenvectors of A.

$\therefore$ The modal matrix which diagonalises A is $P = \begin{bmatrix} 1 & 1 \\ -2 & 1 \end{bmatrix}$.

Verification

$$P^{-1} = \frac{1}{3}\begin{bmatrix} 1 & 1 \\ -2 & 1 \end{bmatrix}$$

$$\Rightarrow \quad P^{-1}AP = \frac{1}{3}\begin{bmatrix} 1 & 1 \\ -2 & 1 \end{bmatrix}\begin{bmatrix} 4 & 1 \\ 2 & 3 \end{bmatrix}\begin{bmatrix} 1 & 1 \\ -2 & 1 \end{bmatrix}$$

$$= \frac{1}{3}\begin{bmatrix} 2 & -2 \\ 0 & 5 \end{bmatrix}\begin{bmatrix} 1 & 1 \\ -2 & 1 \end{bmatrix}$$

$$= \frac{1}{3}\begin{bmatrix} 6 & 0 \\ 0 & 15 \end{bmatrix} = \begin{bmatrix} 2 & 0 \\ 0 & 5 \end{bmatrix} = D,$$

the diagonal matrix containing the eigenvalues 2 and 5 of A as the principal diagonal elements.

To find A^6

$$A^6 = PD^6P^{-1} = \frac{1}{3}\begin{bmatrix} 1 & 1 \\ -2 & 1 \end{bmatrix}\begin{bmatrix} 2^6 & 0 \\ 0 & 5^6 \end{bmatrix}\begin{bmatrix} 1 & -1 \\ 2 & 1 \end{bmatrix}$$

$$= \frac{1}{3}\begin{bmatrix} 1 & 1 \\ -2 & 1 \end{bmatrix}\begin{bmatrix} 64 & 0 \\ 0 & 6250 \end{bmatrix}\begin{bmatrix} 1 & -1 \\ 2 & 1 \end{bmatrix}$$

$$= \frac{1}{3}\begin{bmatrix} 64 & 15625 \\ -128 & 115625 \end{bmatrix}\begin{bmatrix} 1 & -1 \\ 2 & 1 \end{bmatrix}$$

$$= \frac{1}{3}\begin{bmatrix} 31314 & 15561 \\ 31122 & 15753 \end{bmatrix} = \begin{bmatrix} 10438 & 5187 \\ 10374 & 5251 \end{bmatrix}$$

Example 1.15

Diagonalise $A = \begin{bmatrix} 1 & 6 & 1 \\ 1 & 2 & 0 \\ 0 & 0 & 3 \end{bmatrix}$ and hence find A^8. Find the modal matrix.

Solution The matrix A can be diagonalised by constructing the modal matrix P whose column vectors are three linearly independent eigenvectors of A.

To find the eigenvalues of A we write the characteristic equation of A.

$$P(\lambda) = \begin{bmatrix} 1-\lambda & 6 & 1 \\ 1 & 2-\lambda & 0 \\ 0 & 0 & 3-\lambda \end{bmatrix}$$

$$= (3-\lambda)(\lambda^2 - 3\lambda - 4)$$

$$= (3-\lambda)(\lambda+1)(\lambda-4) = 0;$$

$\lambda = -1, 3, 4$ are the eigenvalues of A.

$\lambda = -1$

$$\begin{bmatrix} 2 & 6 & 1 \\ 1 & 3 & 0 \\ 0 & 0 & 4 \end{bmatrix}\begin{bmatrix} x_1 \\ x_2 \\ x_3 \end{bmatrix} = 0 \qquad \begin{aligned} 2x_1 + 6x_2 + x_3 &= 0 \\ x_1 + 3x_2 &= 0 \\ 4x_3 &= 0 \end{aligned}$$

$$\frac{x_1}{-3} = \frac{x_2}{1} = \frac{x_3}{0} \qquad X_1 = c_1[-3 \ \ 1 \ \ 0]^T$$

$\lambda = 3$

$$\begin{bmatrix} -2 & 6 & 1 \\ 1 & -1 & 0 \\ 0 & 0 & 0 \end{bmatrix}\begin{bmatrix} x_1 \\ x_2 \\ x_3 \end{bmatrix} = 0 \qquad \begin{aligned} -2x_1 + 6x_2 + x_3 &= 0 \\ x_1 - x_2 &= 0 \end{aligned}$$

$$\frac{x_1}{1} = \frac{x_2}{1} = \frac{x_3}{-4} \qquad X_2 = c_2[1 \ \ 1 \ \ -4]^T$$

$\lambda = 4$

$$\begin{bmatrix} -3 & 6 & 1 \\ 1 & -2 & 0 \\ 0 & 0 & -1 \end{bmatrix}\begin{bmatrix} x_1 \\ x_2 \\ x_3 \end{bmatrix} = 0 \qquad \begin{aligned} -3x_1 + 6x_2 + x_3 &= 0 \\ x_1 - 2x_2 &= 0 \\ -x_3 &= 0 \end{aligned}$$

$$\frac{x_1}{2} = \frac{x_2}{1} = \frac{x_3}{0} \qquad X_3 = c_3[2 \ \ 1 \ \ 0]^T$$

$\therefore$ The modal matrix which diagonalises A is

$$P = \begin{bmatrix} -3 & 1 & 2 \\ 1 & 1 & 1 \\ 0 & -4 & 0 \end{bmatrix} = \begin{bmatrix} a_1 & b_1 & c_1 \\ a_2 & b_2 & c_2 \\ a_3 & b_3 & c_3 \end{bmatrix} \text{ (say)}$$

To find P^{-1}

To find the inverse of P we write $[P|I]$ and apply elementary transformations to reduce it to the form $[I|P^{-1}]$.

Diagonalisation of matrix A

$$D = P^{-1}AP$$

$$= \frac{1}{20}\begin{bmatrix} -4 & 8 & 1 \\ 0 & 0 & -5 \\ 4 & 12 & 4 \end{bmatrix}\begin{bmatrix} 1 & 6 & 1 \\ 1 & 2 & 0 \\ 0 & 0 & 3 \end{bmatrix}\begin{bmatrix} -3 & 1 & 2 \\ 1 & 1 & 1 \\ 0 & -4 & 0 \end{bmatrix}$$

$$= \frac{1}{20}\begin{bmatrix} 4 & -8 & -1 \\ 0 & 0 & -15 \\ 16 & 48 & 16 \end{bmatrix}\begin{bmatrix} -3 & 1 & 2 \\ 1 & 1 & 1 \\ 0 & -4 & 0 \end{bmatrix}$$

$$= \frac{1}{20}\begin{bmatrix} -20 & 0 & 0 \\ 0 & 60 & 0 \\ 0 & 0 & 80 \end{bmatrix} = \begin{bmatrix} -1 & 0 & 0 \\ 0 & 3 & 0 \\ 0 & 0 & 4 \end{bmatrix}$$

Note that the diagonal elements are the eigenvalues of A.

To find A^8

$$A^8 = PD^8P^{-1} = \begin{bmatrix} -3 & 1 & 2 \\ 1 & 1 & 1 \\ 0 & -4 & 0 \end{bmatrix}\begin{bmatrix} 1 & 0 & 0 \\ 0 & 6561 & 0 \\ 0 & 0 & 65536 \end{bmatrix} \frac{1}{20}\begin{bmatrix} -4 & 8 & 1 \\ 0 & 0 & -5 \\ 4 & 12 & 4 \end{bmatrix}$$

$$= \frac{1}{20}\begin{bmatrix} -3 & 6561 & 131072 \\ 1 & 6561 & 65536 \\ 0 & -26244 & 0 \end{bmatrix}\begin{bmatrix} -4 & 8 & 1 \\ 0 & 0 & -5 \\ 4 & 12 & 4 \end{bmatrix}$$

$$= \begin{bmatrix} 26215 & 78642 & 24574 \\ 13107 & 39322 & 11467 \\ 0 & 0 & 6561 \end{bmatrix}$$

Example 1.16

Find matrix P which diagonalises

$$A = \begin{bmatrix} 7.3 & 0.2 & -3.7 \\ -11.5 & 1.0 & 5.5 \\ 17.7 & 1.8 & -9.3 \end{bmatrix}.$$

Solution The characterstic equation is

$$P(\lambda) = \begin{vmatrix} 7.3-\lambda & 0.2 & -3.7 \\ -11.5 & 1.0-\lambda & 5.5 \\ 17.7 & 1.8 & -9.3-\lambda \end{vmatrix} = 0$$

It can be verified that for $\lambda = 0, 3, -4$ the above determinant vanishes and hence the characteristic equation is $\lambda^3 + \lambda^2 - 12\lambda = 0$.

$$P(0) = \begin{vmatrix} 7.3 & 0.2 & -3.7 \\ -11.5 & 1.0 & 5.5 \\ 17.7 & 1.8 & -9.3 \end{vmatrix}$$

$$\overset{R_2-5R_1,\ R_3-9R_1}{=} \begin{vmatrix} 7.3 & 0.2 & -3.7 \\ -48 & 0 & 24 \\ -48 & 0 & 24 \end{vmatrix} = 0; \quad \therefore \quad \lambda = 0 \text{ is a root}$$

$$P(3) = \begin{vmatrix} 4.3 & 0.2 & -3.7 \\ -11.5 & -2 & 5.5 \\ 17.7 & 1.8 & -12.3 \end{vmatrix}$$

$$\overset{R_2+10R_1,\ R_3-9R_1}{=} \begin{vmatrix} 4.3 & 0.2 & -3.7 \\ 31.5 & 0 & -31.5 \\ -21.0 & 0 & 21.0 \end{vmatrix} = 0; \therefore \quad \lambda = 3 \text{ is a root}$$

$$P(-4) = \begin{vmatrix} 11.3 & 0.2 & -3.7 \\ -11.5 & -2 & 5.5 \\ 17.7 & 1.8 & -5.3 \end{vmatrix}$$

$$\overset{R_2+10R_1,\ R_3-9R_1}{=} \begin{vmatrix} 11.3 & 0.2 & -3.7 \\ -294 & 0 & 98 \\ -84 & 0 & 28 \end{vmatrix} = 0; \therefore \ \lambda = -4 \text{ is a root}$$

$\lambda = 0$

$$\begin{bmatrix} 7.3 & 0.2 & -3.7 \\ -11.5 & 1.0 & 5.5 \\ 17.7 & 1.8 & -5.3 \end{bmatrix}\begin{bmatrix} x_1 \\ x_2 \\ x_3 \end{bmatrix} = 0$$

$$\sim \begin{cases} 7.3x_1 + 0.2x_2 - 3.7x_3 = 0 & \text{(i)} \\ -2x_1 + x_3 = 0 & \text{(ii)}, \quad \text{by } R_2 - 5R_1 \end{cases}$$

Since (ii) $\Rightarrow$ $x_3 = 2x$,

(i) $\Rightarrow$ $7.3x_1 - 2(3.7)x_1 + 0.2x_2 = 0$

$\Rightarrow$ $-x_1 + 2x_2 = 0$ (iii)

Solving (ii) and (iii) we get

$$\frac{x_1}{2} = \frac{x_2}{1} = \frac{x_3}{4} \quad \Rightarrow \quad X_1 = c_1[2 \quad 1 \quad 4]^T$$

$\lambda = 3$

$$\begin{bmatrix} 4.3 & 0.2 & -3.7 \\ -11.5 & -2 & 5.5 \\ 17.7 & 1.8 & -12.3 \end{bmatrix}\begin{bmatrix} x_1 \\ x_2 \\ x_3 \end{bmatrix} = 0$$

$$\sim \begin{cases} 4.3x_1 + 0.2x_2 - 3.7x_3 = 0 & \text{(i)} \\ -x_1 + x_3 = 0 & \text{(ii)} \end{cases}$$

Eliminating x_3 from (i) and (ii) we get

$$3x_1 + x_2 = 0 \quad \text{(iii)}$$

Solving (ii) and (iii) we get

$$\frac{x_1}{1} = \frac{x_2}{-3} = \frac{x_3}{1} \quad \Rightarrow \quad X_2 = c_2[1 \quad -3 \quad 1]^T$$

$\lambda = -4$

$$\begin{bmatrix} 11.3 & 0.2 & -3.7 \\ -11.5 & 5.0 & 5.5 \\ 17.7 & 1.8 & -5.3 \end{bmatrix}\begin{bmatrix} x_1 \\ x_2 \\ x_3 \end{bmatrix} = 0$$

$$\sim \begin{cases} 11.3x_1 + 0.2x_2 - 3.7x_3 = 0 & \text{(i)} \\ -3x_1 + x_3 = 0 & \text{(ii)} \end{cases}$$

Eliminating x_3 from (i) and (ii)

$$[11.3 + 3(-3.7)]x_1 + 0.2x_2 = 0\ x_1 + x_2 = 0 \quad \text{(iii)}$$

Solving (ii) and (iii) we get

$$\frac{x_1}{1} = \frac{x_2}{-1} = \frac{x_3}{3} \quad \Rightarrow \quad X_3 = c_3[1 \quad -1 \quad 3]^T$$

$$\therefore \quad P = \begin{bmatrix} 2 & 1 & 1 \\ 1 & -3 & -1 \\ 4 & 1 & 3 \end{bmatrix}$$

To find P^{-1} we write $[P\,|I]$ and transform it to $[I|\,P^{-1}]$ form by elementary transformations.

$$[P\,|\,I] = \left[\begin{array}{ccc|ccc} 2 & 1 & 1 & 1 & 0 & 0 \\ 1 & -3 & -1 & 0 & 1 & 0 \\ 4 & 1 & 3 & 0 & 0 & 1 \end{array}\right]$$

$$\overset{R_{12}}{\sim} \left[\begin{array}{ccc|ccc} 1 & -3 & -1 & 0 & 1 & 0 \\ 2 & 1 & 1 & 1 & 0 & 0 \\ 4 & 1 & 3 & 0 & 0 & 1 \end{array}\right]$$

$$\overset{R_2-2R_1,\ R_3-2R_2}{\sim} \left[\begin{array}{ccc|ccc} 1 & -3 & -1 & 0 & 1 & 0 \\ 0 & 7 & 3 & 1 & -2 & 0 \\ 0 & -1 & 1 & -2 & 0 & 1 \end{array}\right]$$

$$\overset{R_2-3R_3}{\sim} \left[\begin{array}{ccc|ccc} 1 & -3 & -1 & 0 & 1 & 0 \\ 0 & 10 & 0 & 7 & -2 & -3 \\ 0 & -1 & 1 & -2 & 0 & 1 \end{array}\right]$$

$$\overset{\frac{1}{10}R_2}{\sim} \left[\begin{array}{ccc|ccc} 1 & -3 & -1 & 0 & 1 & 0 \\ 0 & 1 & 0 & 7/10 & -2/10 & -3/10 \\ 0 & -1 & 1 & -2 & 0 & 1 \end{array}\right]$$

$$\overset{R_1+4R_2+R_3,\ R_3+R_2}{\sim} \left[\begin{array}{ccc|ccc} 1 & 0 & 0 & 4/5 & 1/5 & -1/5 \\ 0 & 1 & 0 & 7/10 & -2/10 & -3/10 \\ 0 & 0 & 1 & -13/10 & -2/10 & 7/10 \end{array}\right] = [I\,|\,P^{-1}]$$

$$P^{-1} = \frac{1}{10}\begin{bmatrix} 8 & 2 & -2 \\ 7 & -2 & -3 \\ -13 & -2 & 7 \end{bmatrix}$$

$$D = P^{-1}AP$$

$$= \frac{1}{10}\begin{bmatrix} 8 & 2 & -2 \\ 7 & -2 & -3 \\ -13 & -2 & 7 \end{bmatrix}\begin{bmatrix} 7.3 & 0.2 & -3.7 \\ -11.5 & 1.0 & 5.5 \\ 17.7 & 1.8 & -9.3 \end{bmatrix}\begin{bmatrix} 2 & 1 & 1 \\ 1 & -3 & -1 \\ 4 & 1 & 3 \end{bmatrix}$$

$$= \frac{1}{10}\begin{bmatrix} 8 & 2 & -2 \\ 7 & -2 & -3 \\ -13 & -2 & 7 \end{bmatrix}\begin{bmatrix} 0 & 3 & -4 \\ 0 & -9 & 4 \\ 0 & 3 & -12 \end{bmatrix}$$

$$= \frac{1}{10}\begin{bmatrix} 0 & 0 & 0 \\ 0 & 30 & 0 \\ 0 & 0 & -40 \end{bmatrix} = \begin{bmatrix} 0 & 0 & 0 \\ 0 & 3 & 0 \\ 0 & 0 & -4 \end{bmatrix}$$

Example 1.17

Find matrix P that diagonalizes $A = \begin{bmatrix} 1 & 1 & 3 \\ 1 & 5 & 1 \\ 3 & 1 & 1 \end{bmatrix}$ and hence find A^4.

Solution The characteristic equation is

$$P(\lambda)=\begin{bmatrix}1-\lambda & 1 & 3\\ 1 & 5-\lambda & 1\\ 3 & 1 & 1-\lambda\end{bmatrix}$$

$$=(1-\lambda)(\lambda^2-6\lambda+4)-1(1-\lambda-3)$$
$$+3(1-15+3\lambda)$$
$$=-\lambda^3+7\lambda^2-36=-(\lambda+2)(\lambda-3)(\lambda-6)=0$$

$$\begin{array}{r|rrrr}
-2 & -1 & 7 & 0 & -36\\
 & & 2 & -18 & 36\\ \hline
3 & -1 & 9 & -18 & \underline{|0}\\
 & & -3 & 18 & \\ \hline
 & -1 & 6 & \underline{|0} &
\end{array}$$

$\lambda = -2$

$$\begin{bmatrix}3 & 1 & 3\\ 1 & 7 & 1\\ 3 & 1 & 3\end{bmatrix}\begin{bmatrix}x_1\\ x_2\\ x_3\end{bmatrix}=0;$$

$$\sim\left.\begin{array}{r}3x_1+x_2+3x_3=0\\ x_1+7x_2+x_3=0\end{array}\right\}\ \frac{x_1}{-1}=\frac{x_2}{0}=\frac{x_3}{1}$$

$\lambda = 3$

$$\begin{bmatrix}-2 & 1 & 3\\ 1 & 2 & 1\\ 3 & 1 & -2\end{bmatrix}\begin{bmatrix}x_1\\ x_2\\ x_3\end{bmatrix}=0;$$

$$\sim\left\{\begin{array}{rl|l}-2x_1+x_2+3x_3=0 & \text{(i)} & \text{(iii)}-\text{(i)}\\ x_1+2x_2+x_3=0 & \text{(ii)} & 5x_1-5x_3=0\Longrightarrow x_3=x_1;\\ 3x_1+x_2-2x_3=0 & \text{(iii)} & \text{using in (ii), } x_1+x_2=0\end{array}\right.$$

$$\frac{x_1}{1}=\frac{x_2}{-1}=\frac{x_3}{1}$$

$\lambda = 6$

$$\begin{bmatrix}-5 & 1 & 3\\ 1 & -1 & 1\\ 3 & 1 & -5\end{bmatrix}\begin{bmatrix}x_1\\ x_2\\ x_3\end{bmatrix}=0;$$

$$\sim\left\{\begin{array}{l}8x_1-8x_3=0,\ \text{by } R_3-R_1\\ 2x_1-x_2=0,\ \text{by } R_3+5R_2\end{array}\right\}$$

Solving we get $\dfrac{x_1}{1}=\dfrac{x_2}{2}=\dfrac{x_3}{1}$

$$\therefore\quad P=\begin{bmatrix}-1 & 1 & 1\\ 0 & -1 & 2\\ 1 & 1 & 1\end{bmatrix}=\begin{bmatrix}a_1 & b_1 & c_1\\ a_2 & b_2 & c_2\\ a_3 & b_3 & c_3\end{bmatrix}$$

$$\Rightarrow\quad Adj\,P=\begin{bmatrix}A_1 & A_2 & A_3\\ B_1 & B_2 & B_3\\ C_1 & C_2 & C_3\end{bmatrix}=\begin{bmatrix}-3 & 0 & 3\\ 2 & -2 & 2\\ 1 & 2 & 1\end{bmatrix}$$

$$|P|=a_1A_1+a_2A_2+a_3A_3=(-1)(-3)+0+1\cdot 3=6$$

$$D=P^{-1}AP$$

$$=\frac{1}{6}\begin{bmatrix}-3 & 0 & 3\\ 2 & -2 & 2\\ 1 & 2 & 1\end{bmatrix}\begin{bmatrix}1 & 1 & 3\\ 1 & 5 & 1\\ 3 & 1 & 1\end{bmatrix}\begin{bmatrix}-1 & 1 & 1\\ 0 & -1 & 2\\ 1 & 1 & 1\end{bmatrix}$$

$$=\frac{1}{6}\begin{bmatrix}-3 & 0 & 3\\ 2 & -2 & 2\\ 1 & 2 & 1\end{bmatrix}\begin{bmatrix}2 & 3 & 6\\ 0 & -3 & 12\\ -2 & 3 & 6\end{bmatrix}$$

$$=\frac{1}{6}\begin{bmatrix}-12 & 0 & 0\\ 0 & 18 & 0\\ 0 & 0 & 36\end{bmatrix}=\begin{bmatrix}-2 & 0 & 0\\ 0 & 3 & 0\\ 0 & 0 & 6\end{bmatrix}$$

$$A^4=PD^4P^{-1}$$

$$=\begin{bmatrix}-1 & 1 & 1\\ 0 & -1 & 2\\ 1 & 1 & 1\end{bmatrix}\begin{bmatrix}(-2)^4 & 0 & 0\\ 0 & 3^4 & 0\\ 0 & 0 & 6^4\end{bmatrix}\frac{1}{6}\begin{bmatrix}-3 & 0 & 3\\ 2 & -2 & 2\\ 1 & 2 & 1\end{bmatrix}$$

$$=\begin{bmatrix}-1 & 1 & 1\\ 0 & -1 & 2\\ 1 & 1 & 1\end{bmatrix}\begin{bmatrix}16 & 0 & 0\\ 0 & 81 & 0\\ 0 & 0 & 1296\end{bmatrix}\frac{1}{6}\begin{bmatrix}-3 & 0 & 3\\ 2 & -2 & 2\\ 1 & 2 & 1\end{bmatrix}$$

$$=\frac{1}{6}\begin{bmatrix}-16 & 81 & 1296\\ 0 & -81 & 2592\\ 16 & 81 & 1269\end{bmatrix}\begin{bmatrix}-3 & 0 & 3\\ 2 & -2 & 2\\ 1 & 2 & 1\end{bmatrix}$$

$$=\begin{bmatrix}251 & 405 & 235\\ 405 & 837 & 405\\ 235 & 405 & 251\end{bmatrix}$$

Example 1.18

Find the matrix P that diagonalizes $A=\begin{bmatrix}8 & -6 & 2\\ -6 & 7 & -4\\ 2 & -4 & 3\end{bmatrix}$ and also find the diagonal matrix.

Solution Characteristic equation of A is

$$P(\lambda)=\begin{vmatrix}8-\lambda & -6 & 2\\ -6 & 7-\lambda & -4\\ 2 & -4 & 3-\lambda\end{vmatrix}=0$$

$$\Rightarrow (8-\lambda)(\lambda^2-10\lambda+5)+6(-10+6\lambda)+2(10+2\lambda)$$
$$=-\lambda^3+\lambda^2(8+10)+\lambda(-80-5+36+4)+40-60+20$$
$$=-\lambda^3+18\lambda^2-45\lambda=-\lambda(\lambda-3)(\lambda-15)=0$$

Aliter: $P(\lambda)=-\lambda^3+(\text{tr } A)\,\lambda^2-(\Sigma M_{ii})\lambda+|A|=0$
$\Rightarrow -\lambda^3+18\,\lambda^2+45\,\lambda=0$

$\because$ tr $A=8+7+3=18$;

$$\Sigma M_{ii}=\begin{vmatrix}7 & -4\\ -4 & 3\end{vmatrix}+\begin{vmatrix}8 & 2\\ 2 & 3\end{vmatrix}+\begin{vmatrix}8 & -6\\ -6 & 7\end{vmatrix}$$
$$=5+20+20=45$$

$$|A|=8\begin{vmatrix}7 & -4\\ -4 & 3\end{vmatrix}+6\begin{vmatrix}-6 & 2\\ -4 & 3\end{vmatrix}+2\begin{vmatrix}-6 & 2\\ 7 & -4\end{vmatrix}=0$$

To find the eigenvector $X=[x_1, x_2, x_3]^T$ corresponding to eigenvalue λ we have to solve $(A-\lambda I)X=0$. Denote the equations by e_1, e_2, e_3.

$\lambda = 0$

$$\begin{bmatrix}8 & -6 & 2\\ -6 & 7 & -4\\ 2 & -4 & 3\end{bmatrix}\begin{bmatrix}x_1\\ x_2\\ x_3\end{bmatrix}=0$$

$$\left\{\begin{array}{lll}e_1+e_2-e_3 & \Rightarrow & 5x_2-5x_3=0\\ & \Rightarrow & x_2=x_3\\ \text{Now } e_3 & \Rightarrow & 2x_1-x_2=0\end{array}\right\} \quad \text{Solving we get } \frac{x_1}{1}=\frac{x_2}{2}=\frac{x_3}{2}$$

$\lambda = 3$

$$\begin{bmatrix}5 & -6 & 2\\ -6 & 4 & -4\\ 2 & -4 & 0\end{bmatrix}\begin{bmatrix}x_1\\ x_2\\ x_3\end{bmatrix}=0$$

$$\left\{\begin{array}{lll}e_3 & \Rightarrow & x_1=2x_2\\ \text{Now } e_1 & \Rightarrow & 2x_2+x_3=0\end{array}\right\} \quad \text{Solving we get } \frac{x_1}{2}=\frac{x_2}{1}=\frac{x_3}{-2}$$

$\lambda = 15$

$$\begin{bmatrix}-7 & -6 & 2\\ -6 & -8 & -4\\ 2 & -4 & -12\end{bmatrix}\begin{bmatrix}x_1\\ x_2\\ x_3\end{bmatrix}=0$$

$$\left\{\begin{array}{lll}e_2+2e_1 & \Rightarrow & x_1=-x_2\\ e_2-2e_3 & \Rightarrow & x_1=2x_3\end{array}\right\} \quad \text{Solving we get } \frac{x_1}{2}=\frac{x_2}{-2}=\frac{x_3}{1}$$

We can easily see that the eigenvectors

$$[X_1, X_2, X_3]=\left[\begin{bmatrix}1\\ 2\\ 2\end{bmatrix}, \begin{bmatrix}2\\ 1\\ -2\end{bmatrix}, \begin{bmatrix}2\\ -2\\ 1\end{bmatrix}\right]$$ are pairwise orthogonal, each with magnitude = 3.

Normalised modal matrix

$$\therefore \quad P=\frac{1}{3}\begin{bmatrix}1 & 2 & 2\\ 2 & 1 & -2\\ 2 & -2 & 1\end{bmatrix} \text{ and its inverse } P^{-1}=P^T=P$$

since P is orthogonal and symmetric.

$$\text{Now,}\quad AP=\frac{1}{3}\begin{bmatrix}8 & -6 & 2\\ -6 & 7 & -4\\ 2 & -4 & 3\end{bmatrix}\begin{bmatrix}1 & 2 & 2\\ 2 & 1 & -2\\ 2 & -2 & 1\end{bmatrix}$$
$$=\frac{1}{3}\begin{bmatrix}0 & 6 & 30\\ 0 & 3 & -30\\ 0 & -6 & 15\end{bmatrix}$$

$$\Rightarrow \quad D=P^{-1}AP=\frac{1}{3}\begin{bmatrix}1 & 2 & 2\\ 2 & 1 & -2\\ 2 & -2 & 1\end{bmatrix}\begin{bmatrix}0 & 6 & 30\\ 0 & 3 & -30\\ 0 & -6 & 15\end{bmatrix}$$
$$=\begin{bmatrix}0 & 0 & 0\\ 0 & 3 & 0\\ 0 & 0 & 15\end{bmatrix}$$

1.7.1 Solution over the Complex Field—Eigenvectors of aReal Matrix Over Complex Field

Example 1.19

Find the eigenvalues and eigenvectors and diagonalise $A=\begin{bmatrix}1 & 1\\ -1 & 1\end{bmatrix}$

Solution The characterstic equation is

$$P(\lambda)=\begin{vmatrix}1-\lambda & 1\\ -1 & 1-\lambda\end{vmatrix}=\lambda^2-2\lambda+2=0$$
$$\Rightarrow \quad \lambda_{1,2}=1\pm i$$

There are no real eigenvalues. Hence matrix A is not diagonalisable over *reals*.

$\lambda_1 = 1 + i$

$$\begin{bmatrix} -i & 1 \\ -1 & -i \end{bmatrix}\begin{bmatrix} x_1 \\ x_2 \end{bmatrix} = 0; \quad -ix_1 + x_2 = 0$$

$$\Rightarrow \quad \frac{x_1}{1} = \frac{x_2}{i}; \quad X_1 = c_1\begin{bmatrix} 1 \\ i \end{bmatrix}$$

$\lambda_2 = 1 - i$

$$\begin{bmatrix} i & 1 \\ -1 & i \end{bmatrix}\begin{bmatrix} x_1 \\ x_2 \end{bmatrix} = 0; \quad ix_1 + x_2 = 0$$

$$\Rightarrow \quad \frac{x_1}{i} = \frac{x_2}{1}; \quad X_2 = c_2\begin{bmatrix} i \\ 1 \end{bmatrix}$$

$$\therefore \quad P = \begin{bmatrix} 1 & i \\ i & 1 \end{bmatrix} \quad \text{and} \quad P^{-1} = \frac{1}{2}\begin{bmatrix} 1 & -i \\ -i & 1 \end{bmatrix}$$

$$D = P^{-1}AP = \frac{1}{2}\begin{bmatrix} 1 & -i \\ -i & 1 \end{bmatrix}\begin{bmatrix} 1 & 1 \\ -1 & 1 \end{bmatrix}\begin{bmatrix} 1 & i \\ i & 1 \end{bmatrix}$$

$$= \frac{1}{2}\begin{bmatrix} 1+i & 1-i \\ -1-i & 1-i \end{bmatrix}\begin{bmatrix} 1 & i \\ i & 1 \end{bmatrix} = \begin{bmatrix} 1+i & 0 \\ 0 & 1-i \end{bmatrix}$$

EXERCISE 1.3

Find the eigenvalues and eigenvectors of matrix $A =$

(Qns. 1–4)

1. $\begin{bmatrix} 4 & -2 \\ 1 & 1 \end{bmatrix}$.

Ans: 2, 3; $\overset{X_1}{\begin{bmatrix} 1 \\ 1 \end{bmatrix}}, \overset{X_2}{\begin{bmatrix} 2 \\ 1 \end{bmatrix}}$

2. $\begin{bmatrix} 2 & 4 \\ 1 & 5 \end{bmatrix}$.

Ans: 1, 6; $\overset{X_1}{\begin{bmatrix} -4 \\ 1 \end{bmatrix}}, \overset{X_2}{\begin{bmatrix} 1 \\ 1 \end{bmatrix}}$

3. $\begin{bmatrix} 6 & -2 & 2 \\ -2 & 3 & -1 \\ 2 & -1 & 3 \end{bmatrix}$. [JNTU 2003 (1) 2005 S(3)]

Ans: 2, 2, 8; $\overset{X_1}{\begin{bmatrix} 1 \\ 1 \\ -1 \end{bmatrix}}, \overset{X_2}{\begin{bmatrix} 0 \\ 1 \\ 1 \end{bmatrix}} \overset{X_3}{\begin{bmatrix} 2 \\ -1 \\ 1 \end{bmatrix}}$

4. $\begin{bmatrix} 2 & 0 & 1 \\ 0 & 2 & 0 \\ 1 & 0 & 2 \end{bmatrix}$. [JNTU 2000 S]

Ans: 1, 2, 3; $\overset{X_1}{\begin{bmatrix} 1 \\ 0 \\ 1 \end{bmatrix}}, \overset{X_2}{\begin{bmatrix} 1 \\ 0 \\ -1 \end{bmatrix}} \overset{X_3}{\begin{bmatrix} 0 \\ 1 \\ 0 \end{bmatrix}}$

Diagonalise, if possible, the matrix $A =$

(Qns. 5–10)

5. $\begin{bmatrix} -2 & 2 & -3 \\ 2 & 1 & -6 \\ -1 & -2 & 0 \end{bmatrix}$. [JNTU 1995, 2005 S (4)]

Ans: –3, –3, 5; $\overset{X_1}{\begin{bmatrix} 2 \\ -1 \\ 0 \end{bmatrix}}, \overset{X_2}{\begin{bmatrix} 3 \\ 0 \\ 1 \end{bmatrix}} \overset{X_3}{\begin{bmatrix} 1 \\ 2 \\ -1 \end{bmatrix}}$

6. $\begin{bmatrix} 4 & 1 \\ 3 & 2 \end{bmatrix}$. Also, find A^5.

Ans: 1, 5; $P = \begin{bmatrix} 1 & 1 \\ -3 & 1 \end{bmatrix}$; $A^5 = \begin{bmatrix} 2344 & 781 \\ 2343 & 782 \end{bmatrix}$

7. $\begin{bmatrix} 1 & 2 \\ 4 & 3 \end{bmatrix}$. Also, find A^3.

Ans: 5, –1; $D = \begin{bmatrix} 5 & 0 \\ 0 & -1 \end{bmatrix}$; $A^3 = \begin{bmatrix} 41 & 42 \\ 84 & 83 \end{bmatrix}$

8. $\begin{bmatrix} -1 & 2 & 2 \\ 1 & 2 & 1 \\ -1 & -1 & 0 \end{bmatrix}$. [JNTU 2004]

Ans: –1, 1, 1; $X_1 = \begin{bmatrix} 2 \\ -1 \\ 1 \end{bmatrix}$; $X_2 = \begin{bmatrix} 0 \\ 1 \\ -1 \end{bmatrix}$

Geometric multiplicity of $\lambda = 1$ is 1; hence not diagonalizable.

9. $\begin{bmatrix} 1 & 6 & 1 \\ 1 & 2 & 0 \\ 0 & 0 & 3 \end{bmatrix}$.

Ans: $D = \begin{bmatrix} -1 & 0 & 0 \\ 0 & 3 & 0 \\ 0 & 0 & 4 \end{bmatrix}; P = \begin{bmatrix} -3 & 1 & 2 \\ 1 & 1 & 1 \\ 0 & -4 & 0 \end{bmatrix}$

10. $\begin{bmatrix} 1 & 0 & 0 \\ 0 & 3 & -1 \\ 0 & -1 & 3 \end{bmatrix}$. [JNTU 2003(4)]

Ans: $D = \begin{bmatrix} 1 & 0 & 0 \\ 0 & 2 & 0 \\ 0 & 0 & 4 \end{bmatrix}; P = \begin{bmatrix} 1 & 0 & 0 \\ 0 & 1 & 1 \\ 0 & 1 & -1 \end{bmatrix}$

Determine the eigenvalues of B if $A = \begin{bmatrix} 4 & -2 \\ 1 & 1 \end{bmatrix}$ and $B =$

(Qns. 11–14)

11. $2A$.

Ans: The eigenvalues of A, are (2,3) and those of $2A$ are $2(2, 3) = (4, 6)$

12. $8A^2 - A + 3I$.

[**Hint:** $8(2^2, 3^2) - (2, 3) + (3, 3) = (33, 72)$.]

Ans: (33, 72)

13. A^{-1}.

Ans: $\frac{1}{2}, \frac{1}{3}$

14. $(A + A^{-1})^2$.

Ans: Eigenvalues of $A^2 + A^{-2} + 2I$

$$(2^2, 3^2) + \left(\frac{1}{2^2}, \frac{1}{3^2}\right) + 2(1, 1) = \left(6\frac{1}{4}, 11\frac{1}{9}\right)$$

Verify the Cayley–Hamilton Theorem. Find A^{-1} if $A =$

(Qns. 15–19)

15. $\begin{bmatrix} 1 & 2 \\ -1 & 3 \end{bmatrix}$. Also find A^4.

Ans: $A^{-1} = \frac{1}{5}\begin{bmatrix} 3 & -2 \\ 1 & 1 \end{bmatrix}; A^4 = \begin{bmatrix} -31 & 48 \\ -24 & 17 \end{bmatrix}$

16. $\begin{bmatrix} 1 & 2 & 3 \\ 2 & 4 & 5 \\ 3 & 5 & 6 \end{bmatrix}$. [JNTU 2005 S (1)]

Ans: $A^{-1} = \begin{bmatrix} 1 & -3 & 2 \\ -3 & 3 & -1 \\ 2 & -1 & 0 \end{bmatrix}$

17. $\begin{bmatrix} 2 & -1 & 1 \\ -1 & 2 & -1 \\ 1 & -1 & 2 \end{bmatrix}$. [JNTU 2000 S (1), 2001]

Ans: $A^{-1} = \frac{1}{4}\begin{bmatrix} 3 & 1 & -1 \\ 1 & 3 & 1 \\ 1 & 1 & 3 \end{bmatrix}$

18. $\begin{bmatrix} 1 & 0 & -1 \\ 1 & 2 & 1 \\ 2 & 2 & 3 \end{bmatrix}$.

Ans: $D = \begin{bmatrix} 1 & 0 & 0 \\ 0 & 2 & 0 \\ 0 & 0 & 3 \end{bmatrix}$

19. $\begin{bmatrix} 2 & 2 & 1 \\ 1 & 3 & 1 \\ 1 & 2 & 2 \end{bmatrix}$.

Ans: $A^{-1} = \frac{1}{5}\begin{bmatrix} 4 & -2 & -1 \\ -1 & 3 & -1 \\ -1 & -2 & 4 \end{bmatrix}$

Prove that the matrices are pairwise similar.

20. $A = \begin{bmatrix} 2 & 0 & 1 \\ 0 & 2 & 0 \\ 1 & 0 & 2 \end{bmatrix}$ $B = \begin{bmatrix} 1 & 0 & -1 \\ 1 & 2 & 1 \\ 2 & 2 & 3 \end{bmatrix}$

$C = \begin{bmatrix} 1 & 3 & 4 \\ 0 & 2 & 5 \\ 0 & 0 & 3 \end{bmatrix}$ $D = \begin{bmatrix} 2 & 0 & 0 \\ 3 & 1 & 0 \\ -1 & 2 & 3 \end{bmatrix}$

[**Hint:** Each of these matrices has the same characterstic equation $\lambda^3 - 6\lambda^2 + 11\lambda - 6 = 0$ and the same eigenvalues $\lambda = 1, 2, 3$.]

Quadratic Forms

2

2.1 Introduction

Second-degree homogeneous expressions are called quadratic forms. They occur in physics and in geometry. In analytical geometry, for instance, a quadratic form has to be transformed into its principal-axes-form so as to determine the nature of the conic section such as parabola, ellipse or hyperbola, *etc*., if it involves two variables, and of the quadratic surface such as paraboloid, ellipsoid or hyperboloid, *etc*., if it involves three variables. A quadratic form can be represented by

$$X^T AX = \sum_{i,j=1}^{n} a_{ij}x_i x_j \qquad (2.1)$$

where X is a column n-vector and A is a symmetric matrix of the coefficients. We study here the method of transformation of a quadratic form (2.1) to sum-of-squares (Canonical) form to determine its nature.

2.2 Quadratic Forms

2.2.1 Quadratic Form: Definition

A quadratic form in n variables $x_1, x_2, \ldots, x_n$ is an expression of the form

$$\begin{aligned} Q = X^T AX &= \sum_{i=1}^{n}\sum_{j=1}^{n} a_{ij}x_i x_j \\ &= a_{11}x_1^2 + a_{12}x_1x_2 + \cdots + a_{1n}x_1x_n \\ &\quad + a_{21}x_2x_1 + a_{22}x_2^2 + \cdots + a_{2n}x_2x_n \\ &\quad + \cdots + a_{n1}x_nx_1 + a_{n2}x_nx_2 + \cdots + a_{nn}x_n^2 \end{aligned}$$

Here A is called the coefficient matrix. Rewriting

$$\begin{aligned} Q = X^T AX = a_{11}x_1^2 + (a_{12}+a_{21})x_1x_2 + \cdots + \\ (a_{1n}+a_{n1})x_1x_n + a_{22}x_2^2 + (a_{23}+a_{32})x_2x_3 \\ + \cdots + (a_{2n}+a_{n2})x_2x_n + \cdots + a_{nn}x_n^2 \end{aligned}$$

Putting $c_{ij} = \frac{1}{2}(a_{ij}+a_{ji}) = c_{ji}$ and $c_{ij}+c_{ji} = a_{ij}+a_{ji}$, a quadratic form can be written as

$$Q = \sum_{i=1}^{n}\sum_{j=1}^{n} c_{ij}x_i x_j$$

where $C = \left[c_{ij}\right] = \frac{1}{2}\left[A + A^T\right]$ is a real symmetric matrix.

Thus, the *coefficient matrix* in a quadratic form which is called the matrix of the quadratic form can always be assumed to be or can be constructed in the form of a real symmetric matrix.

Note 1 The rank of a quadratic form is the rank of the coefficient matrix. The quadratic form is singular if $n > r$ where n is the number of variables and r is the rank.

Example 2.1

Express $2x^2 - 5xy - 3y^2$, a quadratic form in two variables x and y in the form $X^T AX$.

Solution

$$\begin{aligned} 2x^2 - 5xy - 3y^2 &= 2x^2 - \frac{5}{2}xy - \frac{5}{2}yx - 3y^2 \\ &= [x \quad y]\begin{bmatrix} 2 & -5/2 \\ -5/2 & -3 \end{bmatrix}\begin{bmatrix} x \\ y \end{bmatrix} \end{aligned}$$

Example 2.2

Write $ax^2 + by^2 + cz^2 + 2fyz + 2gzx + 2hxy$, a general quadratic form in three variables x, y and z in the form $X^T AX$.

Solution

If $A=\begin{bmatrix} a & h & g \\ h & b & f \\ g & f & c \end{bmatrix}$ $X=\begin{bmatrix} x \\ y \\ z \end{bmatrix}, X^T=[x \quad y \quad z]$

then the quadratic form is $q = X^T AX = ax^2 + by^2 + cz^2 + 2fyz + 2gzx + 2hxy$, which is given above.

Example 2.3

Obtain the matrix of the quadratic form $x_1^2 - 2x_1x_2 + 2x_2^2$.

Solution

If $X^T AX$ is the quadratic form then the coefficient matrix is $A=\begin{bmatrix} 1 & -1 \\ -1 & 2 \end{bmatrix}$ since the quadratic form

$$= x_1^2 - x_1x_2 - x_2x_1 + 2x_2 = [x_1 \quad x_2]\begin{bmatrix} 1 & -1 \\ -1 & 2 \end{bmatrix}\begin{bmatrix} x_1 \\ x_2 \end{bmatrix}$$

Example 2.4

Obtain the matrix of the quadratic form $x^2 + 2y^2 + 3z^2 + 4xy + 5yz + 6zx$.

Solution

$$X^T AX = [x \quad y \quad z]\begin{bmatrix} 1 & 2 & 3 \\ 2 & 2 & 5/2 \\ 3 & 5/2 & 3 \end{bmatrix}\begin{bmatrix} x \\ y \\ z \end{bmatrix}$$

∴ matrix of the quadratic form is

$$A=\begin{bmatrix} 1 & 2 & 3 \\ 2 & 2 & 5/2 \\ 3 & 5/2 & 3 \end{bmatrix}$$

Example 2.5

Find a real symmetric matrix C of the quadratic form $Q = x_1^2 + 3x_2^2 + 2x_3^2 + 2x_1x_2 + 6x_2x_3$.

Solution

The coefficient matrix of Q is $A=\begin{bmatrix} 1 & 2 & 0 \\ 0 & 3 & 6 \\ 0 & 0 & 2 \end{bmatrix}$

since $Q = x_1^2 + 2x_1x_2 + 0.x_1x_3 + 0.x_2x_1 + 3x_2^2 + 6x_2x_3$

$= + 0.x_3x_1 + 0.x_3x_2 + 2x_3^2$

Symmetric matrix

$$C=\frac{1}{2}(A+A^T)$$

$$=\frac{1}{2}\left(\begin{bmatrix} 1 & 2 & 0 \\ 0 & 3 & 6 \\ 0 & 0 & 2 \end{bmatrix}+\begin{bmatrix} 1 & 0 & 0 \\ 2 & 3 & 0 \\ 0 & 6 & 2 \end{bmatrix}\right)=\begin{bmatrix} 1 & 1 & 0 \\ 1 & 3 & 3 \\ 0 & 3 & 2 \end{bmatrix}$$

The simplest way to write matrix C *is*

1. Put the coefficients of square terms as the diagonal elements.
2. Place $\frac{1}{2}$ of a_{ij}, the coefficient of x_ix_j at c_{ij} and the remaining half of a_{ji} at c_{ji}, *i.e.*, $c_{ij} = \frac{1}{2}a_{ij}$ so that $c_{ij} + c_{ji} = \frac{1}{2}(a_{ij} + a_{ji}) = a_{ij}$. For example, the coefficient of x_2x_3 in Example 2.5 is distributed equally as 3 and 3 to c_{23} and c_{32}.

Example 2.6

Find real symmetric matrix C such that $Q = X^TCX$ where $Q = 6x_1^2 - 4x_1x_2 + 2x_2^2$.

Solution

$$Q = 6x_1^2 - 4x_1x_2 + 0.x_2x_1 + 2x_2^2$$

The matrix of the quadratic form is

$$A=\begin{bmatrix} 6 & -4 \\ 0 & 2 \end{bmatrix} \Rightarrow A^T = \begin{bmatrix} 6 & 0 \\ -4 & 2 \end{bmatrix}$$

∴ Symmetric matrix

$$C=\frac{1}{2}(A+A^T)=\frac{1}{2}\begin{bmatrix} 12 & -4 \\ -4 & 4 \end{bmatrix}=\begin{bmatrix} 6 & -2 \\ -2 & 2 \end{bmatrix}$$

Example 2.7

Find the real symmetric matrix C such that $Q = X^TCX$ where $Q = 2(x_1 - x_2)^2 = 2x_1^2 - 4x_1x_2 + 2x_2^2$.

Solution

$$A=\begin{bmatrix} 2 & -4 \\ 0 & 2 \end{bmatrix} \Rightarrow A^T = \begin{bmatrix} 2 & 0 \\ -4 & 2 \end{bmatrix}$$

$$\text{Symmetric matrix } C=\frac{1}{2}(A+A^T)=\begin{bmatrix} 2 & -2 \\ -2 & 2 \end{bmatrix}$$

Example 2.8

Determine the real symmetric matrix C such that $Q = X^T CX$ where $Q = (x_1 + x_2 + x_3)^2$.

Solution We have

$$Q = x_1^2 + 2x_1x_2 + 2x_1x_3 + 0.x_2x_1 + x_2^2 + 2x_2x_3 + 0.x_3x_1 + 0.x_3x_2 + x_3^2$$

$$\text{Here } A = \begin{bmatrix} 1 & 2 & 2 \\ 0 & 1 & 2 \\ 0 & 0 & 1 \end{bmatrix} \text{ and } A^T = \begin{bmatrix} 1 & 0 & 0 \\ 2 & 1 & 0 \\ 2 & 2 & 1 \end{bmatrix}$$

$$\text{Symmetric matrix } C = \frac{1}{2}(A + A^T) = \begin{bmatrix} 1 & 1 & 1 \\ 1 & 1 & 1 \\ 1 & 1 & 1 \end{bmatrix}$$

Example 2.9

Obtain the real symmetric matrix C such that $Q = X^T CX$ where $Q = 4x_1x_3 + 2x_2x_3 + x_3^2$.

Solution We have

$$Q = 0.x_1^2 + 0.x_1x_2 + 4x_1x_3 + 0.x_2x_1 + 0.x_2^2 + 2x_2x_3 + 0.x_3x_1 + 0.x_3x_2 + 1.x_3^2$$

$$\text{Here } A = \begin{bmatrix} 0 & 0 & 4 \\ 0 & 0 & 2 \\ 0 & 0 & 1 \end{bmatrix} \text{ and } A^T = \begin{bmatrix} 0 & 0 & 0 \\ 0 & 0 & 0 \\ 4 & 2 & 1 \end{bmatrix}$$

$$\therefore C = \frac{1}{2}(A + A^T) = \begin{bmatrix} 0 & 0 & 2 \\ 0 & 0 & 1 \\ 2 & 1 & 1 \end{bmatrix}$$

2.3 Canonical Form (or) Sum of the Squares Form

Consider a real quadratic form

$$Q = X^T AX = \sum_{i=1}^{n}\sum_{j=1}^{n} a_{ij}x_ix_j \quad (2.1)$$

where A is the coefficient matrix, $X = [x_1, x_2, \ldots, x_n]^T$ and $X^T = [x_1, x_2, \ldots, x_n]$ are, respectively, the column and row vectors of n variables.

By the application of an orthogonal transformation defined by $X = PY$ the quadratic form (2.1) can be transformed to the following sum of the squares form or canonical form

$$Y^T DY = \lambda_1 y_1^2 + \lambda_2 y_2^2 + \cdots + \lambda_n y_x^2$$

where P is known as a *modal matrix* and D is called a *spectral matrix*. D is actually a diagonal matrix consisting of the eigenvalues of A as its diagonal elements.

The matrix P which is the matrix of transformation is an orthogonal matrix here, and therefore the above method of reduction is called the *orthogonal transformation*, which is explained in Sections 2.3 and 2.4. There are two more methods: Diagonalisation by simultaneous row/column transformations and Lagrange's method, explained in Section 2.4 through examples.

2.3.1 Index and Signature of a Real Quadratic Form

Let r be the rank of a matrix A in the quadratic form (2.1).

Index s: The index s of the real quadratic form (2.1) is the number of positive square terms in the canonical form of a real quadratic form, *e.g.*, if the canonical form of a real quadratic form is $2y_1^2 - y_2^2 - y_3^2$ then its index = 1.

Signature: The signature of a real quadratic form is the number of the positive terms minus the number of negative terms in the canonical form. *e.g.*, if the canonical form of a real quadratic form is $2y_1^2 - y_2^2 - y_3^2$ then its signature = 1 − 2 = −1.

2.4 Nature of Real Quadratic Forms

A real nonsingular quadratic form $Q = X^T AX$ (with $|A| \neq 0$) is said to be:

2.4.1 Positive Definite

If the rank and index of Q are equal to the number of variables, *i.e.*, $r = s = n$ or what is the same thing, if each of the eigenvalues of $A > 0$ then Q is positive definite.

2.4.2 Negative Definite

If the index of Q equals zero, *i.e.*, $s = 0$ and $r = n$ or what is the same thing if each of the eigenvalues of $A < 0$ then Q is negative definite.

2.4.3 Positive Semi-Definite

If rank and index of Q are equal but less than the number of variables, *i.e.*,

$$s = r < n \ (|A| = 0)$$

or, what is the same thing, if all of the eigenvalues of A are nonnegative, *i.e.*, $\lambda_i \ge 0$ and at least one eigenvalue is 0 then Q is positive semi-definite.

2.4.4 Negative Semi-Definite

If the index is zero and the rank $r < n$.

$$i.e., s = 0 \quad \text{and} \quad r < n \ (|A| = 0)$$

or what is the same thing, if all of the eigenvalues of A are nonpositive, *i.e.*, $\lambda_i \le 0$ and at least one eigenvalue is 0 then Q is positive semi-definite.

2.4.5 Indefinite

If Q is not of any type described above or what is the same thing, A has some positive and some negative eigenvalues, then Q is indefinite.

Note 1 If Q is negative definite (semi-definite) then $-Q$ is positive definite (semi-definite).

Example 2.10

Determine the nature, index and signature of the quadratic form

$$2x_1^2 + 2x_2^2 + 3x_3^2 + 2x_1x_2 - 4x_1x_3 - 4x_2x_3$$

Solution The real symmetric matrix A associated with the quadratic form is

$$A = \begin{bmatrix} 2 & 1 & -2 \\ 1 & 2 & -2 \\ -2 & -2 & 3 \end{bmatrix}$$

The characteristic equation of matrix A is

$$P(\lambda) = \begin{vmatrix} 2-\lambda & 1 & -2 \\ 1 & 2-\lambda & -2 \\ -2 & -2 & 3-\lambda \end{vmatrix}$$

$$= (2-\lambda)(\lambda^2 - 5\lambda + 2) - 1(-\lambda - 1) - 2(2 - 2\lambda) = 0$$

$$\Rightarrow -\lambda^3 + 7\lambda^2 - 7\lambda + 1 = 0$$

$$\Rightarrow \lambda = 1 \quad \text{or} \quad \lambda^2 - 6\lambda + 1 = 0$$

$$\therefore \quad \lambda = 1,\ 3 + 2\sqrt{2},\ 3 - 2\sqrt{2}, \text{ which are all positive.}$$

$$\begin{array}{r|rrrr} 1 & -1 & 7 & -7 & 1 \\ & & -1 & 6 & -1 \\ \hline & -1 & 6 & -1 & \underline{|0} \end{array}$$

So, the quadratic form is positive definite. Index = no. of positive eigenvalues = 3.

Signature = no. of positive eigenvalues – number of negative eigenvalues = 3 – 0 = 3.

Example 2.11

Find the nature, index and signature of the quadratic form $2x_1x_2 + 2x_1x_3 + 2x_2x_3$.

Solution The real symmetric matrix A associated with the quadratic form is

$$A = \begin{bmatrix} 0 & 1 & 1 \\ 1 & 0 & 1 \\ 1 & 1 & 0 \end{bmatrix}$$

The characteristic equation of A is

$$P(\lambda) = \begin{vmatrix} 0-\lambda & 1 & 1 \\ 1 & 0-\lambda & 1 \\ 1 & 1 & 0-\lambda \end{vmatrix}$$

$$= 0 - \lambda(\lambda^2 - 1) - 1(-\lambda - 1) + 1(1 + \lambda) = 0$$

$$\Rightarrow -\lambda^3 + 3\lambda + 2 = 0 \quad \Rightarrow \lambda = -1,\ -1,\ 2$$

$$\begin{array}{r|rrrr} -1 & -1 & 0 & 3 & 2 \\ & & 1 & -1 & -2 \\ \hline -1 & -1 & 1 & 2 & \underline{|0} \\ & & 1 & -2 & \\ \hline & -1 & 2 & \underline{|0} & \end{array}$$

Since one of the eigenvalues is positive, and the others are negative the quadratic form is indefinite. Index = 1, signature 1 – 2 = – 1

Example 2.12

Identify the nature, index and signature of the quadratic form

$$x_1^2 + 4x_2^2 + x_3^2 - 4x_1x_2 + 2x_3x_1 - 4x_2x_3$$

Solution The real symmetric matrix A associated with the quadratic form is

$$A = \begin{bmatrix} 1 & -2 & 1 \\ -2 & 4 & -2 \\ 1 & -2 & 1 \end{bmatrix}$$

The characteristic equation of A is

$$P(\lambda)=\begin{vmatrix}1-\lambda & -2 & 1\\ -2 & 4-\lambda & -2\\ 1 & -2 & 1-\lambda\end{vmatrix}$$

$$=(1-\lambda)(\lambda^2-5\lambda)+2(2\lambda)+(\lambda)=0$$

$$\Rightarrow -\lambda^3+6\lambda^2=0,\ \lambda=0,\ 0,\ 6$$

Here, rank = r = 1; index = s = 1; signature = 2s − r = 2 × 1 − 1 = 1. So, the quadratic form is positive semi-definite.

Example 2.13

Classify the quadratic form and find the index and signature of

$$-3x_1^2-3x_2^2-3x_3^2-2x_1x_2-2x_1x_3+2x_2x_3$$

Solution

$$A=\begin{bmatrix}-3 & -1 & -1\\ -1 & -3 & 1\\ -1 & 1 & -3\end{bmatrix}$$

The characteristic equation of A is

$$P(\lambda)=\begin{vmatrix}-3-\lambda & -1 & -1\\ -1 & -3-\lambda & 1\\ -1 & 1 & -3-\lambda\end{vmatrix}$$

$$=(-3-\lambda)(\lambda^2+6\lambda+8)+(4+\lambda)$$
$$+(4+\lambda)=0$$

$$\Rightarrow \lambda^3+9\lambda^2+24\lambda+16=0$$

$$\Rightarrow \lambda=-1,\ -4,\ -4$$

$$\begin{array}{r|rrrr} -1 & 1 & 9 & 24 & 16\\ & & -1 & -8 & -16\\ \hline -4 & 1 & 8 & 16 & \underline{|0}\\ & & -4 & -16 & \\ \hline & 1 & 4 & \underline{|0} & \end{array}$$

Since all the eigenvalues are negative the quadratic form is negative definite.

Index = 0, signature = 0 − 3= −3

Note 1 If we denote the above quadratic form by Q then

$$-Q=3x_1^2+3x_2^2+3x_3^2+2x_1x_2+2x_1x_3-2x_2x_3$$

is positive definite.

2.5 Reduction of a Quadratic Form to Canonical Form

We know that a homogeneous expression of second degree in any number of variables is called a quadratic form. For instance, let

$$A=\begin{bmatrix}a & h & g\\ h & b & f\\ g & f & c\end{bmatrix},\ X=\begin{bmatrix}x\\ y\\ z\end{bmatrix},\ X^T=[x\quad y\quad z] \quad (2.2)$$

Then the expression

$$X^TAX=ax^2+by^2+cz^2+2fyz+2gzx+2hxy \quad (2.3)$$

is a quadratic form.

Let λ_1, λ_2, λ_3 be the eigenvalues of the matrix A and

$$X_1=\begin{bmatrix}x_1\\ y_1\\ z_1\end{bmatrix},\ X_2=\begin{bmatrix}x_2\\ y_2\\ z_2\end{bmatrix},\ X_3=\begin{bmatrix}x_3\\ y_3\\ z_3\end{bmatrix} \quad (2.4)$$

be the corresponding eigenvectors in the normalized form, *i.e.*, in the form in which each component is divided by the length of the vector.

Denote the square matrix

$$[X_1\quad X_2\quad X_3]=\begin{bmatrix}x_1 & x_2 & x_3\\ y_1 & y_2 & y_3\\ z_1 & z_2 & z_3\end{bmatrix} \quad \text{by } P$$

We have

$$AP=A[X_1\quad X_2\quad X_3]=[AX_1\quad AX_2\quad AX_3]$$
$$=[\lambda_1X_1,\ \lambda_2X_2,\ \lambda_3X_3]$$
$$=[X_1\quad X_2\quad X_3]\begin{bmatrix}\lambda_1 & 0 & 0\\ 0 & \lambda_2 & 0\\ 0 & 0 & \lambda_3\end{bmatrix}=PD$$

(2.5)

Let Q be the quadratic form given by (2.1). The coefficient matrix A is a real symmetric matrix. So, it has a set of n L.I orthogonal eigenvectors as n columns corresponding to the n eigenvalues which may not be distinct.

Let $\hat{P}$ be a normalized modal matrix of A. Then $\hat{P}$ is an orthogonal matrix, *i.e.*, a matrix for which $\hat{P}^{-1}=\hat{P}^T$.

Thus, the transformation $X = \hat{P}Y$ is an orthogonal transformation. This transforms the quadratic form into a canonical form as follows.

We know that P diagonalises A. Thus

$$\hat{P}^{-1}A\hat{P} = D \text{ or } A = \hat{P}D\hat{P}^{-1} = \hat{P}D\hat{P}^{T} \quad (2.6)$$

since $\hat{P}$ is an orthogonal matrix satisfying $\hat{P}^{-1} = \hat{P}^{T}$

Substituting (2.6) and (2.1) we have

$$Q = X^T AX = X^T \hat{P}D\hat{P}^T X = (X^T\hat{P})(D)(\hat{P}^T X) \quad (2.7)$$

remultiplying $X = \hat{P}Y$ by $\hat{P}^{-1}$ we get $\hat{P}^{-1}X = \hat{P}^{-1}\hat{P}Y = Y \quad \Rightarrow \quad Y = \hat{P}^{-1}X = \hat{P}^{T}X$ (2.8)

Taking transposes of equation (2.8)

$$Y^T = (\hat{P}^T X)^T = X^T\hat{P} \quad (2.9)$$

Using (2.8) and (2.9) in (2.7) we have

$$Q = X^T AX = Y^T DY =$$

$$[y_1 \quad y_2 \quad \cdots \quad y_n]\begin{bmatrix} \lambda_1 & 0 & \cdots & 0 \\ 0 & \lambda_2 & \cdots & 0 \\ \cdots & \cdots & \cdots & \cdots \\ 0 & 0 & \cdots & \lambda_n \end{bmatrix}$$

or $Q = \lambda_1 y_1^2 + \lambda_2 y_2^2 + \cdots + \lambda_n y_n^2,$ (2.10)

which is known as the canonical form or sum-of-the squares form or the principal axes form of the quadratic form.

2.6 Sylvestor's[1] Law of Inertia

Theorem 2.1 The signature of a real quadratic form is invariant for all normal reductions (Canonical forms).

2.7 Methods of Reduction of a Quadratic Form to a Canonical Form

A given quadratic form (Q.F) can be reduced to a canonical form (C.F) or sum-of-squares form by one of the following methods.

1. Diagonalisation (by simultaneous application of row and column transformations).
2. Orthogonalisation.
3. Lagrange's method of reduction (completion of squares).

We illustrate these methods one by one through solved examples.

2.7.1. Diagonalisation (by Simultaneous Application of Row and Column Transformations)

Example 2.14

Reduce the quadratic form $-21x_1^2 + 30x_1x_2 - 12x_1x_3 - 11x_2^2 + 8x_2x_3 - 2x_3^2$ to canonical form and find its nature.

Solution The given quadratic form is

$$-21x_1^2 - 11x_2^2 - 2x_3^2 + 30x_1x_2 - 12x_1x_3 + 8x_2x_3$$

$$= X^T AX = [x_1 \quad x_2 \quad x_3]\begin{bmatrix} -21 & 15 & -6 \\ 15 & -11 & 4 \\ -6 & 4 & -2 \end{bmatrix}\begin{bmatrix} x_1 \\ x_2 \\ x_3 \end{bmatrix}$$

Let us write A as

$$\begin{bmatrix} -21 & 15 & -6 \\ 15 & -11 & 4 \\ -6 & 4 & -2 \end{bmatrix} = \begin{bmatrix} 1 & 0 & 0 \\ 0 & 1 & 0 \\ 0 & 0 & 1 \end{bmatrix} A \begin{bmatrix} 1 & 0 & 0 \\ 0 & 1 & 0 \\ 0 & 0 & 1 \end{bmatrix}$$

$$R_2 + \frac{5}{7}R_1,\ R_3 - \frac{2}{7}R_1,\ C_2 + \frac{5}{7}C_1,\ C_3 - \frac{2}{7}C_1$$

$$\begin{bmatrix} -21 & 0 & 0 \\ 0 & -\frac{2}{7} & -\frac{2}{7} \\ 0 & -\frac{2}{7} & -\frac{2}{7} \end{bmatrix} = \begin{bmatrix} 1 & 0 & 0 \\ \frac{5}{7} & 1 & 0 \\ -\frac{2}{7} & 0 & 1 \end{bmatrix} A \begin{bmatrix} 1 & \frac{5}{7} & -\frac{2}{7} \\ 0 & 1 & 0 \\ 0 & 0 & 1 \end{bmatrix}$$

$$R_3 - R_2,\ C_3 - C_2$$

$$D = \begin{bmatrix} -21 & 0 & 0 \\ 0 & -\frac{2}{7} & 0 \\ 0 & 0 & 0 \end{bmatrix}$$

[1] SYLVESTOR, James Joseph (1814–1897), English algebraist, combinatorist, geometer, number theorist and poet; cofounder with Cayley of theory of invariants (anticipated to some extent by Boole and Lagrange); spent two periods in the U.S. where he was a stimulant to mathematical research. In 1850, he introduced for the first time the word 'matrix', in the sense of 'the mother of determinants'.

$$\begin{bmatrix} 1 & 0 & 0 \\ \frac{5}{7} & 1 & 0 \\ -1 & -1 & 1 \end{bmatrix} A \begin{bmatrix} 1 & \frac{5}{7} & -1 \\ 0 & 1 & -1 \\ 0 & 0 & 1 \end{bmatrix} = P^T AP$$

The Linear Transformation is given by

$$X = PY \quad \Rightarrow \quad \begin{bmatrix} x_1 \\ x_2 \\ x_3 \end{bmatrix} = \begin{bmatrix} 1 & \frac{5}{7} & -1 \\ 0 & 1 & -1 \\ 0 & 0 & 1 \end{bmatrix} \begin{bmatrix} y_1 \\ y_2 \\ y_3 \end{bmatrix}$$

The canonical form is $-21y_1^2 - \frac{2}{7}y_2^2 + 0.y_3^2$. The number of variables $= n = 3$. The rank of the quadratic form $q = r = 2$ or $r(A) = 2$. The index $= s =$ number of positive terms $= 0$. The signature $= 2s - r = 2 \times 0 - 2 = -2$. Since $s = 0$ and $r = 2 < n = 3$ the nature of the quadratic form is negative semi-definite.

Example 2.15

Reduce to canonical form and find the rank, signature and nature of the quadratic form $x_1^2 - 2x_2^2 + 3x_3^2 - 4x_2x_3 + 6x_3x_1$.

Solution By comparing the given quadratic form with

$$X^T AX = \sum a_{ij} x_i x_j$$

$$= [x_1 \quad x_2 \quad x_3] \begin{bmatrix} 1 & 0 & 3 \\ 0 & -2 & -2 \\ 3 & -2 & 3 \end{bmatrix} \begin{bmatrix} x_1 \\ x_2 \\ x_3 \end{bmatrix}$$

and writing the matrix A as

$$\begin{bmatrix} 1 & 0 & 3 \\ 0 & -2 & -2 \\ 3 & -2 & 3 \end{bmatrix} = \begin{bmatrix} 1 & 0 & 0 \\ 0 & 1 & 0 \\ 0 & 0 & 1 \end{bmatrix} A \begin{bmatrix} 1 & 0 & 0 \\ 0 & 1 & 0 \\ 0 & 0 & 1 \end{bmatrix}$$

and applying $R_3 - 3R_1$, $C_3 - 3C_1$, we have

$$\begin{bmatrix} 1 & 0 & 0 \\ 0 & -2 & -2 \\ 0 & -2 & -6 \end{bmatrix} = \begin{bmatrix} 1 & 0 & 0 \\ 0 & 1 & 0 \\ -3 & 0 & 1 \end{bmatrix} A \begin{bmatrix} 1 & 0 & -3 \\ 0 & 1 & 0 \\ 0 & 0 & 1 \end{bmatrix}$$

Applying $R_3 - R_2$, $C_3 - C_2$

$$\begin{bmatrix} 1 & 0 & 0 \\ 0 & -2 & 0 \\ 0 & 0 & -4 \end{bmatrix} = \begin{bmatrix} 1 & 0 & 0 \\ 0 & 1 & 0 \\ -3 & -1 & 1 \end{bmatrix} A \begin{bmatrix} 1 & 0 & -3 \\ 0 & 1 & -1 \\ 0 & 0 & 1 \end{bmatrix}$$

So, the Linear Transformation is given by $X = PY$

$$\Rightarrow \begin{bmatrix} x_1 \\ x_2 \\ x_3 \end{bmatrix} = \begin{bmatrix} 1 & 0 & -3 \\ 0 & 1 & -1 \\ 0 & 0 & 1 \end{bmatrix} \begin{bmatrix} y_1 \\ y_2 \\ y_3 \end{bmatrix}$$ and the canonical

form is $q = y_1^2 - 2y_2^2 - 4y_3^2$

The number of variables $n = 3$. The rank of the quadratic form $= r = \rho(A) = 3$. The index of the quadratic form = number of positive terms $= s = 1$. The signature $= 2s - r = 2 \times 1 - 3 = -1$. Since $r = n$ it is either positive or negative definite. But s is neither equal to n nor o, so it is not definite. Hence the quadratic form is indefinite.

2.7.2 Orthogonalisation

Example 2.16

Reduce the quadratic form $x_1^2 + 3x_2^2 + 3x_3^2 - 2x_2x_3$ by orthogonal transformation. Find its rank, signature and nature.

Solution The quadratic form is

$$X^T AX = [x_1 \quad x_2 \quad x_3] \begin{bmatrix} 1 & 0 & 0 \\ 0 & 3 & -1 \\ 0 & -1 & 3 \end{bmatrix} \begin{bmatrix} x_1 \\ x_2 \\ x_3 \end{bmatrix}$$

The eigenvalues of the matrix A are $\lambda_1 = 1$, $\lambda_2 = 2$, $\lambda_3 = 4$ (please refer to Example 2.13, p. 17)

The orthogonal modal matrix is

$$P = \begin{bmatrix} 1 & 0 & 0 \\ 0 & \frac{1}{\sqrt{2}} & \frac{1}{\sqrt{2}} \\ 0 & \frac{1}{\sqrt{2}} & -\frac{1}{\sqrt{2}} \end{bmatrix}$$

The canonical form of q is $y_1^2 + 2y_2^2 + 4y_3^2$. The number of variables $= n = 3$. The rank of the quadratic form $= r = \rho(A) = 3$. The index $= s =$ the number of positive terms $= 3$. The signature $= 2s - r = 2 \times 3 - 3 = 3$. $\therefore$ The quadratic form is positive definite.

Example 2.17

Reduce the quadratic form $q = x^2 + y^2 + z^2 + 4yz - 4zx + 4xy$ to canonical form by orthogonalisation and find the rank, index, signature and nature of it.

Solution The given quadratic form is

$$X^T AX = [x \quad y \quad z]\begin{bmatrix} 1 & 2 & -2 \\ 2 & 1 & 2 \\ -2 & 2 & 1 \end{bmatrix}\begin{bmatrix} x \\ y \\ z \end{bmatrix}$$

The characteristic roots of A are obtained from

$$|A-\lambda I| = \begin{vmatrix} 1-\lambda & 2 & -2 \\ 2 & 1-\lambda & 2 \\ -2 & 2 & 1-\lambda \end{vmatrix} = \begin{vmatrix} 1-\lambda & 2 & -2 \\ 3-\lambda & 3-\lambda & 0 \\ 0 & 3-\lambda & 3-\lambda \end{vmatrix}$$

$$= (3-\lambda)^2 \begin{vmatrix} 1-\lambda & 2 & -2 \\ 1 & 1 & 0 \\ 0 & 1 & 1 \end{vmatrix} \text{ by } R_3+R_2 \quad R_2+R_1$$

$$(3-\lambda)^2(3+\lambda) = 0 \qquad \Rightarrow \lambda = 3, 3, -3$$

The eigenvalues are not distinct. So, we have to see if we can construct two Linearly Independent eigenvectors corresponding to $\lambda = 3$.

To find the eigenvectors we have to solve $(A - \lambda I)\, X = 0$.

For $\lambda = 3$ We have to solve

$$\begin{bmatrix} -2 & 2 & -2 \\ 2 & -2 & 2 \\ -2 & 2 & -2 \end{bmatrix}\begin{bmatrix} x_1 \\ x_2 \\ x_3 \end{bmatrix} = \begin{bmatrix} 0 \\ 0 \\ 0 \end{bmatrix}$$

$$\sim x_1 - x_2 + x_3 = 0 \quad \Rightarrow \quad x_1 = x_2 - x_3$$

We may choose two linearly independent vectors (orthogonal)

as (0, 1, 1), (2, 1, –1); normalised vectors are

$$\left(0, \frac{1}{\sqrt{2}}, \frac{1}{\sqrt{2}}\right) \text{ and } \left(\frac{2}{\sqrt{6}}, \frac{1}{\sqrt{6}}, \frac{-1}{\sqrt{6}}\right)$$

For $\lambda = -3$ We have to solve

$$\begin{bmatrix} 4 & 2 & -2 \\ 2 & 4 & 2 \\ -2 & 2 & 4 \end{bmatrix}\begin{bmatrix} x_1 \\ x_2 \\ x_3 \end{bmatrix} = \begin{bmatrix} 0 \\ 0 \\ 0 \end{bmatrix}$$

$$\sim \begin{matrix} x_1 + x_2 = 0 \\ x_2 + x_3 = 0 \end{matrix} \quad \frac{x_1}{1} = \frac{x_2}{-1} = \frac{x_3}{1}$$

Normalized vector is $\left(\frac{1}{\sqrt{3}}, \frac{-1}{\sqrt{3}}, \frac{1}{\sqrt{3}}\right)$

The orthogonal modal matrix is

$$P = \begin{bmatrix} 0 & \frac{\sqrt{2}}{\sqrt{3}} & \frac{1}{\sqrt{3}} \\ \frac{1}{\sqrt{2}} & \frac{1}{\sqrt{6}} & -\frac{1}{\sqrt{3}} \\ \frac{1}{\sqrt{2}} & -\frac{1}{\sqrt{6}} & \frac{1}{\sqrt{3}} \end{bmatrix}$$

$$P^T AP = \begin{bmatrix} 3 & 0 & 0 \\ 0 & 3 & 0 \\ 0 & 0 & -3 \end{bmatrix} = D$$

The canonical form of the quadratic form is $3y_1^2 + 3y_2^2 - 3y_3^2$.

The number of variables $= n = 3$; the rank of the matrix $= r = \rho(A) = 3$.

The index = number of positive terms $= s = 2$; The signature $= 2s - r = 2 \times 2 - 3 = 1$.

Since $r = n = 3$ the quadratic form must be either positive definite or negative definite.

But since $s \neq n$ or 0, it is neither. Hence q is indefinite.

2.7.3 Lagrange's Method of Reduction (Completing Squares)

Example 2.18

Reduce the quadratic form $x_1^2 + 2x_2^2 - 7x_3^2 - 4x_1x_2 + 8x_1x_3$ to canonical form by Lagrange's method of reduction.

Solution The given quadratic form is $x_1^2 - 2x_1(2x_2 - 4x_3) + 2x_2^2 - 7x_3^2$

$$= [x_1 - 2(x_2 - 2x_3)]^2 - 2x_2^2 - 23x_3^2 + 16x_2x_3$$

$$= [x_1 - 2(x_2 - 2x_3)]^2 - 2\left[x_2^2 - 2x_2 \cdot 4x_3 + 4^2 x_3^2\right] + 9x_3^2$$

$$= (x_1 - 2x_2 + 4x_3)^2 - 2(x_2 - 4x_3)^2 + 9x_3^2$$

$$= y_1^2 - 2y_2^2 + 9y_3^2 \quad \text{(say)}$$

Linear Transformation is $X = PY$

$$\Rightarrow \begin{bmatrix} x_1 \\ x_2 \\ x_3 \end{bmatrix} = \begin{bmatrix} 1 & 2 & 4 \\ 0 & 1 & 4 \\ 0 & 0 & 1 \end{bmatrix} \begin{bmatrix} y_1 \\ y_2 \\ y_3 \end{bmatrix}$$

Here, n = number of variables = 3. Rank of the quadratic form $r = \rho(A) = 3$. Number of positive terms = $s = 2$. Signature = $2s - r = 2 \times 2 - 3 = 1$. Since $r = n = 3$ the quadratic form must be positive or negative definite. But since $s \neq n$, $s \neq 0$ it is neither positive definite nor negative definite. Hence it is indefinite.

Example 2.19

Reduce the quadratic form $q = x^2 - 4y^2 + 6z^2 + 2xy - 4xz$ to canonical form by Lagrange's method. Find its nature.

Solution The given quadratic form can be written as

$$\begin{aligned} q &= x^2 + 2x(y - 2z) - 4y^2 + 6z^2 \\ &= [x^2 + 2x(y - 2z) + (y - 2z)^2] \\ &\quad - (y^2 + 4z^2 - 4yz) - 4y^2 + 6z^2 \\ &= (x + y - 2z)^2 - 5y^2 + 2z^2 + 4yz \\ &= (x + y - 2z)^2 + 2(z^2 + 2yz + y^2) - 7y^2 \\ &= (x + y - 2z)^2 + 2(y + z)^2 - 7y^2 \\ &= u^2 - 7v^2 + 2w^2 \end{aligned}$$

where $u = x + y - 2z$, $v = y$, $w = y + z$; L.T. is $X = PY$

$$\text{where } P = \begin{bmatrix} 1 & -3 & 2 \\ 0 & 1 & 0 \\ 0 & -1 & 1 \end{bmatrix}, \quad X = [x\ y\ z]^T, \quad Y = [u\ v\ w]^T$$

Here number of variables $n = 3$, Rank of $q = r = \rho(A) = 3$; Index $= s = 2$. Signature $= 2s - r = 2 \times 2 - 3 = 1$. Since $r = n$, q must either be positive definite or negative definite. But $s \neq n$, $s \neq 0$ and so, it is neither. Hence q is indefinite.

2.8 Singular Value Decomposition of a Matrix

Introduction

So far, we have considered square matrices only. In sec. 1·15·5 we have considered LU decomposition of a square matrix. In this, we have seen how a square matrix can be written as the product of two triangular matrices one lower and the other upper. A somewhat similar decomposition is possible for a rectangular matrix. It is called a *singular value decomposition* (SVD). The SVD of a rectangular matrix is of great importance in matrix theory. It is useful in finding the generalized inverse of a singular matrix. It has applications in least square fitting, matrix approximation theory, in determining the rank of a matrix and in image processing etc.

SVD of a Rectangular Matrix

Let A be a real matrix with m rows and n columns ($m \geq n$). Then $B = A^TA$ and $C = AA^T$ are square matrices of orders m and n respectively.
Further,

$$B^T = (A^TA)^T = A^T(A^T)^T = A^TA = B, \qquad (1)$$

$$\text{and } C^T = (AA^T)^T = (A^T)^T A^T = AA^T = C, \qquad (2)$$

so that B and C are symmetric matrices.

It can be seen from the following examples and it can be established generally also that A^TA and AA^T have the same characteristic equations and hence the same eigenvalues λ_n, say, with the exception of $(m - n)$ zero eigenvalues. Let these eigenvalues be arranged in decreasing order of their magnitudes. Then $s_r = \sqrt{\lambda_r}$ $(r = 1, 2, 3, \ldots n)$ are called the *singular values of matrix A*. We construct a diagonal matrix D by placing these singular values, with decreasing. order of their magnitude, along its main diagonal.
Thus

$$D = \begin{bmatrix} s_1 & 0 & 0 & \cdots & 0 \\ 0 & s_2 & 0 & \cdots & 0 \\ \ldots & \ldots & \ldots & \ldots & \ldots \\ 0 & 0 & 0 & 0 & s_n \end{bmatrix} \qquad (3)$$

where

$$s_1 \geq s_2 \geq s_3 \geq \cdots \geq s_n \geq 0 \qquad (4)$$

If the rank of A is $r < n$ then

$$s_{r+1} = s_{r+2} = \cdots s_n = 0 \qquad (5)$$

After obtaining the eigenvalues λ_n $(n = 1, 2, \ldots n)$ of $B = A^TA$ we can obtain eigenvectors X_i such that $X_i \cdot X_j = 0$ if $i \neq j$ and $= 1$ if $i = j$; i.e. they are pairwise orthogonal and each is of norm unity. In short, the eigenvectors X_i are orthonormalized eigenvectors of $B = A^T A$. Then we have

$$BX_i = A^T AX_i = \lambda_i X_i \quad (i = 1, 2, 3, \dots n) \tag{6}$$

If Y_i are the n orthonormalized eigenvectors of $C = AA^T$ then

$$CY_i = AA^T Y_i = \lambda_i Y_i \quad (i = 1, 2, 3, \dots n) \tag{7}$$

The given matrix A can now be decomposed into the following form

$$A = UDV^T \tag{8}$$

where V is the matrix of order n constructed with X_i, the n orthonormalized eigenvectors of $B = A^T A$, as its columns.

U and V are such that

$$U^T U = V^T V = VV^T = I_n \tag{9}$$

and

$$D = \text{diag}\,(s_1, s_2, \dots s_n) \tag{10}$$

It follows then that Y_i, then eigenvectors of $C = AA^T$ are given by

$$y_i = \frac{1}{\sqrt{\lambda_i}} AX_i \quad (i = 1, 2, 3, \dots n) \tag{11}$$

The decomposition defined by (8) is called the singular value decomposition (SVD) of the matrix A.

It can be shown that the SVD of A is *unique* if the λ_i are distinct and are such that eq. (4) holds.

In case, A is a square matrix of order n, then the matrices U, D and V are also square matrices of the same size. Now, the inverse of A can be formed by the relation.

$$A^{-1} = VD^{-1}.U^T \tag{12}$$

where

$$D^{-1} = \text{diag}\left(\frac{1}{s_1}, \frac{1}{s_2}, \frac{1}{s_3}, \dots \frac{1}{s_n}\right) \tag{13}$$

If any of the λ_i's are zero then the matrix A is singular. Also, if λ_i are very small then A is very nearly singular. Thus, the SVD of a matrix gives a clear indication whether the matrix is singular or very nearly singular.

The following examples illustrate the method.

Example 2.20

Find the eigenvalues and the corresponding orthonormalized eigenvectors of the matrix $A^T A$ if $A = \begin{bmatrix} 2\cdot0 & 0\cdot0 \\ 1\cdot5 & 2\cdot5 \end{bmatrix}$. Hence find the SVD of A.

Solution Let us denote $A^T A$ by B and AA^T by C. We have

$$B = A^T A = \begin{bmatrix} 2\cdot0 & 1\cdot5 \\ 0\cdot0 & 2\cdot5 \end{bmatrix}\begin{bmatrix} 2\cdot0 & 0\cdot0 \\ 1\cdot5 & 2\cdot5 \end{bmatrix} = \begin{bmatrix} 6\cdot25 & 3\cdot75 \\ 3\cdot75 & 6\cdot25 \end{bmatrix}$$

The characteristic equation of B is $|B - \lambda I| = 0$

$$\Rightarrow \lambda^2 - (6\cdot25 + 6\cdot25)\lambda + (6\cdot25)^2 - (3\cdot75)^2 = 0$$

$$\Rightarrow \lambda^2 - 12\cdot5\lambda + 25 = 0$$

$$\Rightarrow (\lambda - 10)(\lambda - 2.5) = 0$$

$\Rightarrow$ The eigenvalues are $\lambda_1 = 10$, $\lambda_2 = 2\cdot5$

Also, the singular values s_1, s_2 of A in decreasing order of their magnitudes are

$$s_1 = \sqrt{\lambda_1} = \sqrt{10} = 3.1623 \quad \therefore \frac{1}{s_1} = \frac{1}{\sqrt{10}} = 0.3162$$

$$s_2 = \sqrt{\lambda_2} = \sqrt{2\cdot5} = 1.5811 \quad \frac{1}{s_2} = \frac{1}{\sqrt{2\cdot5}} = 0.6324$$

Now, the diagonal matrix D and its inverse D^{-1} are

$$D = \begin{bmatrix} \sqrt{10} & 0 \\ 0 & \sqrt{2.5} \end{bmatrix} = \begin{bmatrix} 3.1623 & 0 \\ 0 & 1.5811 \end{bmatrix};$$

$$D^{-1} = \begin{bmatrix} \frac{1}{\sqrt{10}} & 0 \\ 0 & \frac{1}{\sqrt{2.5}} \end{bmatrix} = \begin{bmatrix} 0.3162 & 0 \\ 0 & 0.6324 \end{bmatrix}$$

The eigenvector $X \neq 0$ corresponding to an eigenvalue λ of B is obtained by solving the system $(B - \lambda I)X = 0$.

For $\lambda_1 = 10$

We have to solve

$$\begin{bmatrix} -3.75 & 3.75 \\ 3.75 & -3.75 \end{bmatrix}\begin{bmatrix} x_1 \\ x_2 \end{bmatrix} = 0 \Rightarrow -x_1 + x_2 = 0 \Rightarrow \frac{x_1}{1} = \frac{x_2}{1}$$

$$\therefore \frac{X_1}{\|X_1\|} = \begin{bmatrix} \frac{1}{\sqrt{2}} \\ \frac{1}{\sqrt{2}} \end{bmatrix}$$

For $\lambda_2 = 2.5$

We have to solve

$$\begin{bmatrix} 3.75 & 3.75 \\ 3.75 & 3.75 \end{bmatrix}\begin{bmatrix} x_1 \\ x_2 \end{bmatrix} = 0 \Rightarrow x_1 + x_2 = 0 \Rightarrow \frac{x_1}{1} = \frac{x_2}{-1}$$

$$\therefore \frac{X_2}{\|X_2\|} = \begin{bmatrix} \frac{1}{\sqrt{2}} \\ -\frac{1}{\sqrt{2}} \end{bmatrix}$$

With these orthonormalized eigenvectors as columns we construct V. Thus we have

$$V = \begin{bmatrix} \frac{1}{\sqrt{2}} & \frac{1}{\sqrt{2}} \\ \frac{1}{\sqrt{2}} & -\frac{1}{\sqrt{2}} \end{bmatrix} = V^T.$$

Now,

$$C = AA^T = \begin{bmatrix} 2.0 & 0.0 \\ 1.5 & 2.5 \end{bmatrix}\begin{bmatrix} 2.0 & 1.5 \\ 0.0 & 2.5 \end{bmatrix} = \begin{bmatrix} 4.0 & 3.0 \\ 3.0 & 8.5 \end{bmatrix}$$

The characteristic equation of C is $|C - \lambda I| = 0$

$$\Rightarrow \lambda^2 - (4.0 + 8.5)\lambda + 4.0 \times 8.5 - 3.0 \times 3.0 = 0$$

$$\Rightarrow \lambda^2 - 12 \cdot 5\lambda + 25 = 0$$

The eigenvalues of C are $\lambda_1 = 10$, $\lambda_2 = 2.5$

Note that the characteristic equations and the eigenvalues of both $B = A^TA$ and $C = AA^T$ are the same.

To find the eigenvector, we have to solve $(C - \lambda I)X = 0$ for each of the eigenvalues.

For $\lambda_1 = 10$

We have to solve

$$\begin{bmatrix} -6 & 3 \\ 3 & -1.5 \end{bmatrix}\begin{bmatrix} x_1 \\ x_2 \end{bmatrix} = 0 \Rightarrow -2x_1 + x_2 = 0 \Rightarrow \frac{x_1}{1} = \frac{x_2}{2}$$

$$\therefore \frac{X_1}{\|X_1\|} = \begin{bmatrix} \frac{1}{\sqrt{5}} \\ \frac{2}{\sqrt{5}} \end{bmatrix}$$

For $\lambda_2 = 2.5$

We have to solve

$$\begin{bmatrix} 1.5 & 3 \\ 3 & 6 \end{bmatrix}\begin{bmatrix} x_1 \\ x_2 \end{bmatrix} = 0 \Rightarrow x_1 + 2x_2 = 0 \Rightarrow \frac{x_1}{2} = \frac{x_1}{-1}$$

$$\therefore \frac{X_2}{\|X_2\|} = \begin{bmatrix} \frac{2}{\sqrt{5}} \\ -\frac{1}{\sqrt{5}} \end{bmatrix}$$

Now,

$$U = \begin{bmatrix} \frac{1}{\sqrt{5}} & \frac{2}{\sqrt{5}} \\ \frac{2}{\sqrt{5}} & -\frac{1}{\sqrt{5}} \end{bmatrix}$$ and the S. V. D of A is UDVT

$$= \frac{1}{\sqrt{5}}\begin{bmatrix} 1 & 2 \\ 2 & -1 \end{bmatrix} \cdot \frac{\sqrt{10}}{2}\begin{bmatrix} 2 & 0 \\ 0 & 1 \end{bmatrix} \cdot \frac{1}{\sqrt{2}}\begin{bmatrix} 1 & 1 \\ 1 & -1 \end{bmatrix}$$

$$= \frac{1}{\sqrt{5}} \cdot \frac{\sqrt{10}}{2} \cdot \frac{1}{\sqrt{2}}\begin{bmatrix} 1 & 2 \\ 2 & -1 \end{bmatrix}\begin{bmatrix} 2 & 2 \\ 1 & -1 \end{bmatrix}$$

$$= \frac{1}{2}\begin{bmatrix} 4 & 0 \\ 3 & 5 \end{bmatrix} = \begin{bmatrix} 2.0 & 0.0 \\ 1.5 & 2.5 \end{bmatrix} = A$$

Not that U can also be obtained by $U = AVD^{-1}$

$$= \begin{bmatrix} 2.0 & 0.0 \\ 1.5 & 2.5 \end{bmatrix}\begin{bmatrix} \frac{1}{\sqrt{2}} & \frac{1}{\sqrt{2}} \\ \frac{1}{\sqrt{2}} & -\frac{1}{\sqrt{2}} \end{bmatrix}\begin{bmatrix} \frac{1}{\sqrt{10}} & 0 \\ 0 & \frac{2}{\sqrt{10}} \end{bmatrix}$$

$$= \frac{1}{4\sqrt{5}}\begin{bmatrix} 4 & 0 \\ 3 & 5 \end{bmatrix}\begin{bmatrix} 1 & 2 \\ 1 & -2 \end{bmatrix} = \begin{bmatrix} \frac{1}{\sqrt{5}} & \frac{2}{\sqrt{5}} \\ \frac{2}{\sqrt{5}} & \frac{-1}{\sqrt{5}} \end{bmatrix}$$

Numerical evaluation of SVD of A is UDVT

$$= \begin{bmatrix} 0.4471 & 0.8942 \\ 0.8942 & -0.4471 \end{bmatrix}\begin{bmatrix} 3.1623 & 0 \\ 0 & 1.5811 \end{bmatrix}$$

$$\begin{bmatrix} 0.7071 & 0.7071 \\ 0.7071 & -0.7071 \end{bmatrix}$$

$$= \begin{bmatrix} 1.4139 & 1.4138 \\ 2.8277 & -0.7069 \end{bmatrix}\begin{bmatrix} 0.7071 & 0.7071 \\ 0.7071 & -0.7071 \end{bmatrix}$$

$$= \begin{bmatrix} 1.9995 & 0 \\ 1.4996 & 2.4993 \end{bmatrix} \simeq \begin{bmatrix} 2 & 0 \\ 1.5 & 2.5 \end{bmatrix}$$

$= A$

Example 2.21

Find the eigenvalues and the corresponding orthonormalized eigenvectors of A^TA where

$A=\begin{bmatrix}5 & 0\\4 & 3\end{bmatrix}$. Hence determine the SVD of A.

Solution Let B and C denote the matrices A^TA and AA^T respectively

We have

$$B=A^TA=\begin{bmatrix}5 & 4\\0 & 3\end{bmatrix}\begin{bmatrix}5 & 0\\4 & 3\end{bmatrix}=\begin{bmatrix}41 & 12\\12 & 9\end{bmatrix}$$

The characteristic equation of B is $|B-\lambda I|=0$

$$\Rightarrow \lambda^2+(41+9)\lambda+41\times 9-12^2=0$$

$$\Rightarrow \lambda^2-50\lambda+225=0$$

$$\Rightarrow (\lambda-45)(\lambda-5)=0$$

The eigenvalues of B are

$$\lambda_1=45 \text{ and } \lambda_2=5$$

The singular values s_1, s_2 of A in the order of their magnitudes and their reciprocals are

$$s_1=\sqrt{\lambda_1}=\sqrt{45}=6.7082;\ \frac{1}{s_1}=0.1491$$

$$s_2=\sqrt{\lambda_2}=\sqrt{5}=2.2361;\ \frac{1}{s_2}=0.4472$$

The diagonal matrix D and its inverse D^{-1} are

$$D=\begin{bmatrix}3\sqrt{5} & 0\\0 & \sqrt{5}\end{bmatrix}=\begin{bmatrix}6.7082 & 0\\0 & 2.2361\end{bmatrix};$$

$$D^{-1}=\begin{bmatrix}0.1491 & 0\\0 & 0.4472\end{bmatrix}$$

To find the eigenvector X corresponding to an eigenvalue λ we have to solve the homogeneous system $(B-\lambda I)X=0$

For $\lambda_1=45$

We have to solve

$$\begin{bmatrix}-4 & 12\\12 & -36\end{bmatrix}\begin{bmatrix}x_1\\x_2\end{bmatrix}=0\Rightarrow\frac{x_1}{3}=\frac{x_2}{1};\ \therefore\frac{X_1}{\|X_1\|}=\begin{bmatrix}\frac{3}{\sqrt{10}}\\\frac{1}{\sqrt{10}}\end{bmatrix}$$

For $\lambda_2=5$

We have to solve

$$\begin{bmatrix}36 & 12\\12 & 4\end{bmatrix}\begin{bmatrix}x_1\\x_2\end{bmatrix}=0\Rightarrow\frac{x_1}{1}=\frac{x_2}{-3};\therefore\frac{X_2}{\|X_2\|}=\begin{bmatrix}\frac{1}{\sqrt{10}}\\-\frac{3}{\sqrt{10}}\end{bmatrix}$$

Taking the normalized eigenvectors as column vectors V is constructed as

$$V=\begin{bmatrix}\frac{3}{\sqrt{10}} & \frac{1}{\sqrt{10}}\\\frac{1}{\sqrt{10}} & \frac{-3}{\sqrt{10}}\end{bmatrix}=V^T$$

$$C=AA^T=\begin{bmatrix}5 & 0\\4 & 3\end{bmatrix}\begin{bmatrix}5 & 4\\0 & 3\end{bmatrix}=\begin{bmatrix}25 & 20\\20 & 25\end{bmatrix}$$

Clearly $A^TA\neq AA^T$. Now the characteristic equation of C is $|C-\lambda I|=0\Rightarrow\lambda^2-(25+25)\lambda+25^2-20^2=0$

$$\Rightarrow\lambda^2-50\lambda+225=0\Rightarrow(\lambda-45)(\lambda-5)=0$$

The eigenvalues of C are $\lambda_1=45$, $\lambda_2=5$. The characteristic equation and hence the eigenvalues of both A^TA and AA^T are the same.

The eigenvector corresponding to the eigenvalue $\lambda_1=45$ is obtained by solving the system $(C-\lambda_1 I)X_1=0$

$$\Rightarrow\begin{bmatrix}-20 & 20\\20 & -20\end{bmatrix}\begin{bmatrix}x_1\\x_2\end{bmatrix}=0\Rightarrow -x_1+x_2=0\Rightarrow\frac{x_1}{1}=\frac{x_2}{1}$$

$$\therefore\frac{X_1}{\|X_1\|}=\begin{bmatrix}\frac{1}{\sqrt{2}}\\\frac{1}{\sqrt{2}}\end{bmatrix}$$

The eigenvector corresponding to the eigenvalue $\lambda_2=5$ is obtained by solving the system $(C-\lambda_2 I)X_2=0$

$$\Rightarrow\begin{bmatrix}20 & 20\\20 & 20\end{bmatrix}\begin{bmatrix}x_1\\x_2\end{bmatrix}=0\Rightarrow x_1+x_2=0\Rightarrow\frac{x_1}{1}=\frac{x_2}{-1}$$

$$\therefore\frac{X_2}{\|X_2\|}=\begin{bmatrix}\frac{1}{\sqrt{2}}\\-\frac{1}{\sqrt{2}}\end{bmatrix}$$

$$\therefore U=\begin{bmatrix}\frac{1}{\sqrt{2}} & \frac{1}{\sqrt{2}}\\\frac{1}{\sqrt{2}} & -\frac{1}{\sqrt{2}}\end{bmatrix}=\begin{bmatrix}0.7071 & 0.7071\\0.7071 & -0.7071\end{bmatrix}$$

We may write the singular value decomposition of A as

$$UDV^T = \begin{bmatrix} \frac{1}{\sqrt{2}} & \frac{1}{\sqrt{2}} \\ \frac{1}{\sqrt{2}} & -\frac{1}{\sqrt{2}} \end{bmatrix} \begin{bmatrix} 3\sqrt{5} & 0 \\ 0 & \sqrt{5} \end{bmatrix} \begin{bmatrix} \frac{3}{\sqrt{10}} & \frac{1}{\sqrt{10}} \\ \frac{1}{\sqrt{10}} & -\frac{3}{\sqrt{10}} \end{bmatrix}$$

$$= \frac{1}{\sqrt{2}} \cdot \sqrt{5} \cdot \frac{1}{\sqrt{10}} \begin{bmatrix} 1 & 1 \\ 1 & -1 \end{bmatrix} \begin{bmatrix} 3 & 0 \\ 0 & 1 \end{bmatrix} \begin{bmatrix} 3 & 1 \\ 1 & -3 \end{bmatrix}$$

$$= \frac{1}{2} \begin{bmatrix} 3 & 1 \\ 3 & -1 \end{bmatrix} \begin{bmatrix} 3 & 1 \\ 1 & -3 \end{bmatrix}$$

$$= \frac{1}{2} \begin{bmatrix} 10 & 0 \\ 8 & 6 \end{bmatrix} = \begin{bmatrix} 5 & 0 \\ 4 & 3 \end{bmatrix} = A.$$

Example 2.22

Find the eigenvalues and the corresponding orthonormalized eigenvectors of the matrix A^TA if $A = \begin{bmatrix} 2 & 0 \\ 1 & 1 \end{bmatrix}$. Hence find the SVD of A.

Solution Let us denote A^TA by B and AA^T by C.

$$\therefore \text{We have } B = A^T A = \begin{bmatrix} 2 & 1 \\ 0 & 1 \end{bmatrix} \begin{bmatrix} 2 & 0 \\ 1 & 1 \end{bmatrix} = \begin{bmatrix} 5 & 1 \\ 1 & 1 \end{bmatrix} \quad (1)$$

Now, the characteristic equation of B is $|B - \lambda I| = 0$

$$\Rightarrow \lambda^2 - (5+1)\lambda + 5\times 1 - 1\times 1 = \lambda^2 - 6\lambda + 4 = 0 \quad (2)$$

Solving by formula $\lambda = \dfrac{-b \pm \sqrt{b^2 - 4ac}}{2a} = 3 \pm \sqrt{5}$

We have the eigenvalues of B as

$$\left.\begin{aligned} \lambda_1 &= 3 + \sqrt{5} = 5.2361 \\ \lambda_2 &= 3 - \sqrt{5} = 0.7639 \end{aligned}\right\} \quad (3)$$

Also, the singular values s_1 and s_2 of A and their reciprocals are

$$s_1 = \sqrt{\lambda_1} = \sqrt{3+\sqrt{5}} \qquad \frac{1}{s_1} = \frac{1}{\sqrt{\lambda_1}} = \frac{\sqrt{5}-1}{2\sqrt{2}} = 0.4370$$

$$= \frac{\sqrt{5}+1}{\sqrt{2}} = 2.2882$$

$$s_2 = \sqrt{\lambda_2} = \sqrt{3-\sqrt{5}} \qquad \frac{1}{s_2} = \frac{1}{\lambda_2} = \frac{\sqrt{5}+1}{2\sqrt{2}} = 1.1441$$

$$= \frac{\sqrt{5}-1}{\sqrt{2}} = 0.8740$$

Now, we write down the diagonal matrix D by placing along its main diagonal the singular values s_1 and s_2 in decreasing order of their magnitudes and its inverse.

$$D = \frac{1}{\sqrt{2}} \begin{bmatrix} \sqrt{5}+1 & 0 \\ 0 & \sqrt{5}-1 \end{bmatrix} = \begin{bmatrix} 2.2882 & 0 \\ 0 & 0.8740 \end{bmatrix};$$

$$D^{-1} = \frac{1}{2\sqrt{2}} \begin{bmatrix} \sqrt{5}-1 & 0 \\ 0 & \sqrt{5}+1 \end{bmatrix} = \begin{bmatrix} 0.4370 & 0 \\ 0 & 1.1441 \end{bmatrix}$$

To find the eigenvector $X \neq 0$ corresponding to an eigenvalue λ of we have to solve the system $(B - \lambda Z)X = 0$

For $\lambda_1 = 3 + \sqrt{5}$

We have to solve

$$\begin{bmatrix} 5-(3+\sqrt{5}) & 1 \\ 1 & 1-(3+\sqrt{5}) \end{bmatrix} \begin{bmatrix} x_1 \\ x_2 \end{bmatrix} = 0$$

$$\Rightarrow \begin{bmatrix} 2-\sqrt{5} & 1 \\ 1 & -2-\sqrt{5} \end{bmatrix} \begin{bmatrix} x_1 \\ x_2 \end{bmatrix} = 0$$

$$\Rightarrow \left.\begin{aligned} (2-\sqrt{5})x_1 + x_2 &= 0 \\ x_1 - (2+\sqrt{5})x_2 &= 0 \end{aligned}\right]$$

$$\Rightarrow \frac{x_1}{2+\sqrt{5}} = \frac{x_2}{1} \Rightarrow X_1 = \begin{bmatrix} 2+\sqrt{5} \\ 1 \end{bmatrix} = \begin{bmatrix} 4.2361 \\ 1 \end{bmatrix}$$

For $\lambda_2 = 3 - \sqrt{5}$

We have to solve

$$\begin{bmatrix} 5-(3-\sqrt{5}) & 1 \\ 1 & 1-(3-\sqrt{5}) \end{bmatrix} \begin{bmatrix} x_1 \\ x_2 \end{bmatrix} = 0$$

$$\Rightarrow \begin{bmatrix} 2+\sqrt{5} & 1 \\ 1 & -2+\sqrt{5} \end{bmatrix} \begin{bmatrix} x_1 \\ x_2 \end{bmatrix} = 0$$

$$\Rightarrow \left.\begin{aligned} (2+\sqrt{5})x_1 + x_2 &= 0 \\ x_1 + (-2+\sqrt{5})x_2 &= 0 \end{aligned}\right] \Rightarrow \frac{x_1}{-1} = \frac{x_2}{2+\sqrt{5}}$$

$$\Rightarrow X_2 = \begin{bmatrix} -1 \\ 2+\sqrt{5} \end{bmatrix} = \begin{bmatrix} 1 \\ 4.2361 \end{bmatrix}$$

$$\|X_1\| = \|X_2\| = \sqrt{(2+\sqrt{5})^2 + 1^2} = \sqrt{10+4\sqrt{5}} = 4.3525$$

With these orthonormalized eigenvectors $\dfrac{X_1}{\|X_1\|}$ and $\dfrac{X_2}{\|X_2\|}$

We construct V thus

$$V = \frac{1}{\sqrt{10+4\sqrt{5}}}\begin{bmatrix} 2+\sqrt{5} & -1 \\ 1 & 2+\sqrt{5} \end{bmatrix}$$

$$= \frac{1}{\sqrt{2}\sqrt{10+2\sqrt{5}}}\begin{bmatrix} 3+\sqrt{5} & -\left(\sqrt{5}-1\right) \\ \sqrt{5}-1 & 3+\sqrt{5} \end{bmatrix}$$

$$V^T = \frac{1}{\sqrt{2}\sqrt{10+2\sqrt{5}}}\begin{bmatrix} 3+\sqrt{5} & \sqrt{5}-1 \\ -\sqrt{5}+1 & 3+\sqrt{5} \end{bmatrix}$$

Consider $(10+2\sqrt{5})(\sqrt{5}+1)^2$

$$= (10+2\sqrt{5})(6+2\sqrt{5}) = 80+32\sqrt{5}$$

$$= 8(10+4\sqrt{5})$$

$$\Rightarrow \frac{1}{\sqrt{10+4\sqrt{5}}} = \frac{1}{\sqrt{10+2\sqrt{5}}}\cdot\frac{2\sqrt{2}}{\sqrt{5}+1}$$

$$= \frac{1}{\sqrt{10+2\sqrt{5}}}\left(\frac{\sqrt{5}-1}{\sqrt{2}}\right)$$

Now $C = AA^T = \begin{bmatrix} 2 & 0 \\ 1 & 1 \end{bmatrix}\begin{bmatrix} 2 & 1 \\ 0 & 1 \end{bmatrix} = \begin{bmatrix} 4 & 2 \\ 2 & 2 \end{bmatrix}$

The characteristic equation of C is $|C-\lambda I| = 0$

$\Rightarrow \lambda^2 - 6\lambda + 4 = 0$; and the eigenvalues of C are $\lambda = 3+\sqrt{5}$ and $\lambda_2 = 3-\sqrt{5}$, which are the same as for $B = A^TA$.

We find the eigenvectors corresponding to these eigenvalues of C by solving $(C-\lambda I)X = 0$

For $\lambda_2 = 3+\sqrt{5}$

We have to solve

$$\begin{bmatrix} 1-\sqrt{5} & 2 \\ 2 & -1-\sqrt{5} \end{bmatrix}\begin{bmatrix} x_1 \\ x_2 \end{bmatrix} = 0 \Rightarrow \begin{matrix} (1-\sqrt{5})x_1 + 2x_2 = 0 \\ 2x_1 - (1+\sqrt{5})x_2 = 0 \end{matrix}$$

$$\Rightarrow \frac{x_1}{\sqrt{5}+1} = \frac{x_2}{2} \qquad X_1 = \begin{bmatrix} \sqrt{5}+1 \\ 2 \end{bmatrix}$$

For $\lambda_2 = 3-\sqrt{5}$

We have to solve

$$\begin{bmatrix} 1+\sqrt{5} & 2 \\ 2 & \sqrt{5}-1 \end{bmatrix}\begin{bmatrix} x_1 \\ x_2 \end{bmatrix} = 0 \Rightarrow \begin{matrix} (1-\sqrt{5})x_1 + 2x_2 = 0 \\ 2x_1 + (\sqrt{5}-1)x_2 = 0 \end{matrix}$$

$$\Rightarrow \frac{x_1}{-2} = \frac{x_2}{\sqrt{5}+1} \qquad X_2 = \begin{bmatrix} 2 \\ \sqrt{5}+1 \end{bmatrix}$$

$$\|X_1\| = \|X_2\| = \sqrt{(\sqrt{5}+1)^2 + 2^2} = \sqrt{10+2\sqrt{5}}$$

With these eigenvectors we constant U

$$U = \frac{1}{\sqrt{10+2\sqrt{5}}}\begin{bmatrix} \sqrt{5}+1 & -2 \\ 2 & \sqrt{5}+1 \end{bmatrix}; \text{ Also,}$$

V can also be constructed by $V = AUD^{-1}$

$$= \begin{bmatrix} 2 & 0 \\ 1 & 1 \end{bmatrix}\frac{1}{\sqrt{2}\sqrt{10+2\sqrt{5}}}$$

$$\begin{bmatrix} 3+\sqrt{5} & -(\sqrt{5}-1) \\ \sqrt{5}-1 & 3+\sqrt{5} \end{bmatrix}\cdot\frac{1}{2\sqrt{2}}\begin{bmatrix} \sqrt{5}-1 & 0 \\ 0 & \sqrt{5}+1 \end{bmatrix}$$

$$= \frac{1}{4\sqrt{10+2\sqrt{5}}}\begin{bmatrix} 2 & 0 \\ 1 & 1 \end{bmatrix}\begin{bmatrix} 2+2\sqrt{5} & -4 \\ 6-2\sqrt{5} & 8+4\sqrt{5} \end{bmatrix}$$

$$= \frac{1}{2\sqrt{10+2\sqrt{5}}}\begin{bmatrix} 2 & 0 \\ 1 & 1 \end{bmatrix}\begin{bmatrix} \sqrt{5}+1 & -2 \\ 3-\sqrt{5} & 4+2\sqrt{5} \end{bmatrix}$$

$$= \frac{1}{2\sqrt{10+2\sqrt{5}}}\begin{bmatrix} 2(\sqrt{5}+1) & -4 \\ 4 & 2+2\sqrt{5} \end{bmatrix}$$

$$= \frac{1}{\sqrt{10+2\sqrt{5}}}\begin{bmatrix} \sqrt{5}+1 & -2 \\ 2 & \sqrt{5}+1 \end{bmatrix}$$

SVD of A is $A = UDV^T$

$$UDV^T = \frac{1}{\sqrt{10+2\sqrt{5}}}\begin{bmatrix} \sqrt{5}+1 & -2 \\ 2 & \sqrt{5}+1 \end{bmatrix}$$

$$\cdot\frac{1}{\sqrt{2}}\begin{bmatrix} \sqrt{5}+1 & 0 \\ 0 & \sqrt{5}+1 \end{bmatrix}$$

$$\cdot\frac{1}{\sqrt{2}\sqrt{10+2\sqrt{5}}}\begin{bmatrix} 3+\sqrt{5} & \sqrt{5}-1 \\ -\sqrt{5}+1 & 3+\sqrt{5} \end{bmatrix}$$

$$= \frac{1}{20+4\sqrt{5}}\begin{bmatrix} 6+2\sqrt{5} & -2\sqrt{5}+2 \\ 2\sqrt{5}+2 & 4 \end{bmatrix}$$

$$\begin{bmatrix} 3+2\sqrt{5} & \sqrt{5}-1 \\ -\sqrt{5}+1 & 3+\sqrt{5} \end{bmatrix}$$

$$= \frac{1}{20+4\sqrt{5}}\begin{bmatrix} 40+8\sqrt{5} & 0 \\ 20+4\sqrt{5} & 20+4\sqrt{5} \end{bmatrix}$$

$$= \begin{bmatrix} 2 & 0 \\ 1 & 1 \end{bmatrix} = A.$$

Example 2.23

Find the eigenvalues and the corresponding orthonormalized eigenvectors of the matrix A^TA if $A=\begin{bmatrix}2 & 1 & 0\\1 & 0 & 1\end{bmatrix}$. Hence find the S. V. D of A.

Solution Let us denote the matrix A^TA by B and AA^T by C. We have

$$C = AA^T = \begin{bmatrix}2 & 1 & 0\\1 & 0 & 1\end{bmatrix}\begin{bmatrix}2 & 1\\1 & 0\\0 & 2\end{bmatrix} = \begin{bmatrix}5 & 2\\2 & 2\end{bmatrix}$$

The characteristic equation of C is $|C-\lambda I|=0$

$$\Rightarrow \lambda^2-(5+2)\lambda+5\times 2-2\times 2=0 \Rightarrow \lambda^2-7\lambda+6=0$$

The eigenvalues of C are $\lambda_1 = 6$ and $\lambda_2 = 1$.

The singular values s_1 and s_2 of A are

$s_1=\sqrt{\lambda_1}=\sqrt{6}=$	Also, $\frac{1}{s_1}=\frac{1}{\sqrt{6}}=$
$s_2=\sqrt{\lambda_2}=1=1.0000$	$\frac{1}{s_2}=1=1.0000$

The diagonal matrix D is obtained by placing s_1 and s_2 $(s_1 > s_2)$ along its main diagonal. Thus

$$D=\begin{bmatrix}\sqrt{6} & 0\\0 & 1\end{bmatrix} \quad D^{-1}=\begin{bmatrix}\frac{1}{\sqrt{6}} & 0\\0 & 1\end{bmatrix}$$

The eigenvector $X \neq 0$ corresponding to an eigenvalue λ is found by solving the system $(C-\lambda I)X=0$

For $\lambda_1 = 6$

We have to solve

$$\begin{bmatrix}-1 & 2\\2 & -4\end{bmatrix}\begin{bmatrix}x_1\\x_2\end{bmatrix}=0 \Rightarrow -x_1+2x_2=0 \Rightarrow \frac{x_1}{2}=\frac{x_2}{1}$$

$$\therefore \frac{X_1}{\|X_1\|}=\frac{1}{\sqrt{5}}\begin{bmatrix}2\\1\end{bmatrix}$$

For $\lambda_2 = 1$

We have to solve

$$\begin{bmatrix}4 & 2\\2 & 1\end{bmatrix}\begin{bmatrix}x_1\\x_2\end{bmatrix}=0 \Rightarrow 2x_1+x_2=0 \Rightarrow \frac{x_1}{1}=\frac{x_2}{-2}$$

$$\therefore \frac{X_2}{\|X_2\|}=\frac{1}{\sqrt{5}}\begin{bmatrix}1\\-2\end{bmatrix}$$

With these orthonormalized eigenvectors we construct the matrix U; thus,

$$U=\frac{1}{\sqrt{5}}\begin{bmatrix}2 & 1\\1 & -2\end{bmatrix}$$

$$\text{Now, } B=A^TA=\begin{bmatrix}2 & 1\\1 & 0\\0 & 1\end{bmatrix}\begin{bmatrix}2 & 1 & 0\\1 & 0 & 1\end{bmatrix}=\begin{bmatrix}5 & 2 & 1\\2 & 1 & 0\\1 & 0 & 1\end{bmatrix}$$

The characteristic equation of B is $|B-\lambda I|=0$

$$\Rightarrow -\lambda^3+(5+1+1)\lambda^2-\lambda(1+4+1)+[(-1)+5.1-2.2]=0$$

$\Rightarrow -\lambda(\lambda^2-7\lambda+6)=0$ The eigenvalues of B are $\lambda_1=6, \lambda_2=1, \lambda_3=0.$

To find the eigenvectors we have to solve $(B-\lambda I)X=0$

For $\lambda_1 = 6$

We have to solve

$$\begin{bmatrix}-1 & 2 & 1\\2 & -5 & 0\\1 & 0 & -5\end{bmatrix}\begin{bmatrix}x_1\\x_2\\x_3\end{bmatrix}=0 \Rightarrow \begin{aligned}-x_1+2x_2+x_3&=0\\2x_1-5x_2&=0\\x_1-5x_3&=0\end{aligned}$$

$$\therefore \frac{X_1}{\|X_1\|}=\frac{1}{\sqrt{30}}\begin{bmatrix}5\\2\\1\end{bmatrix}$$

For $\lambda_2 = 1$

We have to solve

$$\begin{bmatrix}4 & 2 & 1\\2 & 0 & 0\\1 & 0 & 0\end{bmatrix}\begin{bmatrix}x_1\\x_2\\x_3\end{bmatrix}=0 \Rightarrow \begin{aligned}4x_1+2x_2+x_3&=0\\x_1&=0\end{aligned}$$

$$\therefore \frac{X_2}{\|X_2\|}=\frac{1}{\sqrt{5}}\begin{bmatrix}0\\1\\-2\end{bmatrix}$$

For $\lambda_3 = 0$

We have to solve

$$\begin{bmatrix}5 & 2 & 1\\2 & 1 & 0\\1 & 0 & 1\end{bmatrix}\begin{bmatrix}x_1\\x_2\\x_3\end{bmatrix}=0 \Rightarrow \begin{aligned}5x_1+2x_2+x_3&=0\\2x_1+x_2&=0\\x_1+x_3&=0\end{aligned}$$

$$\therefore \frac{X_3}{\|X_3\|}=\frac{1}{\sqrt{6}}\begin{bmatrix}-1\\2\\1\end{bmatrix}$$

Now the matrix V and its transpose V^T are

$$V = \begin{bmatrix} \frac{5}{\sqrt{30}} & 0 & -\frac{1}{\sqrt{6}} \\ \frac{2}{\sqrt{30}} & \frac{1}{\sqrt{5}} & \frac{2}{\sqrt{6}} \\ \frac{1}{\sqrt{30}} & \frac{-2}{\sqrt{5}} & \frac{1}{\sqrt{6}} \end{bmatrix}; V^T = \begin{bmatrix} \frac{5}{\sqrt{30}} & \frac{2}{\sqrt{30}} & \frac{1}{\sqrt{30}} \\ 0 & \frac{1}{\sqrt{5}} & -\frac{2}{\sqrt{5}} \\ -\frac{1}{\sqrt{6}} & \frac{2}{\sqrt{6}} & \frac{1}{\sqrt{6}} \end{bmatrix}$$

Now the S. V. D of A is UDV^T

$$= \frac{1}{\sqrt{5}}\begin{bmatrix} 2 & 1 \\ 1 & -2 \end{bmatrix}\begin{bmatrix} \sqrt{6} & 0 & 0 \\ 0 & 1 & 0 \end{bmatrix}\begin{bmatrix} \frac{5}{\sqrt{30}} & \frac{2}{\sqrt{30}} & \frac{1}{\sqrt{30}} \\ 0 & \frac{1}{\sqrt{5}} & -\frac{2}{\sqrt{5}} \\ -\frac{1}{\sqrt{6}} & \frac{2}{\sqrt{6}} & \frac{1}{\sqrt{6}} \end{bmatrix}$$

$$= \begin{bmatrix} \frac{2\sqrt{6}}{\sqrt{5}} & \frac{1}{\sqrt{5}} & 0 \\ \frac{\sqrt{6}}{\sqrt{5}} & -\frac{2}{\sqrt{5}} & 0 \end{bmatrix}\begin{bmatrix} \frac{5}{\sqrt{30}} & \frac{2}{\sqrt{30}} & \frac{1}{\sqrt{30}} \\ 0 & \frac{1}{\sqrt{5}} & -\frac{2}{\sqrt{5}} \\ -\frac{1}{\sqrt{6}} & \frac{2}{\sqrt{6}} & \frac{1}{\sqrt{6}} \end{bmatrix}$$

$$= \begin{bmatrix} 2 & 1 & 0 \\ 1 & 0 & 1 \end{bmatrix} = A.$$

EXERCISE 2.1

Write down the matrix of the quadratic form in each case: (Qns. 1–4)

1. $xy + yz + zx$

Ans: $\begin{bmatrix} 0 & \frac{1}{2} & \frac{1}{2} \\ \frac{1}{2} & 0 & \frac{1}{2} \\ \frac{1}{2} & \frac{1}{2} & 0 \end{bmatrix}$

2. $x^2 + y^2 + z^2 - 2xy + 2xz - 2yz$

Ans: $\begin{bmatrix} 1 & -1 & 1 \\ -1 & 1 & -1 \\ 1 & -1 & 1 \end{bmatrix}$

3. $x^2 + y^2 + 2xy - 2xz + 2yz$

Ans: $\begin{bmatrix} 1 & 1 & -1 \\ 1 & 1 & 1 \\ -1 & 1 & 0 \end{bmatrix}$

4. $-x^2 + 4xy + 4y^2 + 6xz + 5z^2$

Ans: $\begin{bmatrix} -1 & 2 & 3 \\ 2 & 4 & 0 \\ 3 & 0 & 5 \end{bmatrix}$

Determine the quadratic form corresponding to the symmetric matrix $A =$ (Qns. 5–8)

5. $\begin{bmatrix} 0 & 0 & 1 \\ 0 & 1 & -1 \\ 0 & -1 & 0 \end{bmatrix}$

Ans: $y^2 + 2xz - 2yz$

6. $\begin{bmatrix} 2 & -1 & 0 \\ -1 & -2 & 5 \\ 0 & 5 & 1 \end{bmatrix}$

Ans: $2x^2 - 2y^2 + z^2 - 2xy + 10yz$

7. $\begin{bmatrix} 1 & 1 & 1 \\ 1 & 1 & 1 \\ 1 & 1 & 1 \end{bmatrix}$

Ans: $(x + y + z)^2$

8. $\begin{bmatrix} 4 & -1 & 0 \\ -1 & 1 & 3 \\ 0 & 3 & -3 \end{bmatrix}$

Ans: $4x^2 - 2xy + 6yz + y^2 - 3z^2$

Reduce the following quadratic form to canonical form by diagonalisation. Find its rank, index, signature (Qns. 9–12)

9. $6x_1^2 + 3x_2^2 + 3x_3^2 - 4x_1x_2 + 4x_1x_3 - 2x_2x_3$ [JNTU 2000]

Ans: $2y_1^2 + 2y_2^2 + 3y_3^2$ rank $= r = 3$, index $= s = 3$; signature $= 3$; Positive definite

10. $x_1^2 + 2x_2^2 + 3x_3^2 + 4x_1x_2 - 2x_2x_3 + 6x_1x_3$

Ans: $y_1^2 - 2y_2^2 + \frac{37}{2}y_3^2$

L.T. is $X = PY$, $P = \begin{bmatrix} 1 & -2 & 4 \\ 0 & 1 & -\frac{7}{2} \\ 0 & 0 & 1 \end{bmatrix}$ rank $= r = 3$, index $= s = 2$, Signature = 1; Indefinite

11. $q = X^T AX$, $X = [x \quad y \quad z]^T$ $A = \begin{bmatrix} 1 & -2 & 1 \\ -2 & 3 & -1 \\ 1 & 2 & 5 \end{bmatrix}$

Ans:

$y_1^2 - y_2^2 + 20y_3^2$; $P = \begin{bmatrix} 1 & 2 & 7 \\ 0 & 1 & 4 \\ 0 & 0 & 1 \end{bmatrix}$ rank $= r = 3$, index $= s = 2$, Signature = 1; Indefinite

12. $q = X^T AX, X = [x \quad y \quad z]^T \quad A = \begin{bmatrix} 6 & -2 & 2 \\ -2 & 3 & -1 \\ 2 & -1 & 3 \end{bmatrix}$

Ans: $6y_1^2 + \frac{7}{3}y_2^2 + \frac{16}{7}y_3^2 \quad P = \begin{bmatrix} 1 & 1/3 & -2/7 \\ 0 & 1 & 1/7 \\ 0 & 0 & 1 \end{bmatrix}$

rank = $r = 3$
index = $s = 3$
signature = 3; Positive definite

Reduce the following quadratic form to canonical form. Find its rank index, signature and nature. (Qns. 13–16)

13. $3x^2 + 5y^2 + 3z^2 - 2yz + 2zx - 2xy$ [JNTU 1999S Dec 2002]

Ans: $\lambda_1 = 2, \lambda_2 = 3, \lambda_3 = 6$
$q = 2y_1^2 + 3y_2^2 + 6y_3^2 \quad P = \begin{bmatrix} -\frac{1}{\sqrt{2}} & \frac{1}{\sqrt{3}} & \frac{1}{\sqrt{6}} \\ 0 & \frac{1}{\sqrt{3}} & -\frac{2}{\sqrt{6}} \\ \frac{1}{\sqrt{2}} & \frac{1}{\sqrt{3}} & \frac{1}{\sqrt{6}} \end{bmatrix}$

rank = $r = 3$
index = $s = 3$
signature = $2s - r = 3$; Positive definite

14. $x^2 + 3y^2 + 4z^2 - 2yz$ [JNTU May/June 2004]

Ans: $\lambda_1 = 2, \lambda = 2, \lambda_3 = 4$
$q = y_1^2 + 2y_2^2 + 4y_3^2 \quad P = \begin{bmatrix} 1 & 0 & 0 \\ 0 & \frac{1}{\sqrt{2}} & \frac{1}{\sqrt{2}} \\ 0 & \frac{1}{\sqrt{2}} & -\frac{1}{\sqrt{2}} \end{bmatrix}$

rank = $r = 3$
index = $s = 3$
signature = $(2s - r) = 3$; Positive definite

15. $3x_1^2 + 3x_2^2 + 3x_3^2 + 2x_1x_2 - 2x_2x_3 + 2x_1x_3$ [JNTU 2003]

Ans: $\lambda_1 = 1, \lambda_2 = 4, \lambda_3 = 4$
$q = y_1^2 + 4y_2^2 + 4y_3^2 \quad P = \begin{bmatrix} -\frac{1}{\sqrt{3}} & 0 & \frac{2}{\sqrt{6}} \\ \frac{1}{\sqrt{3}} & \frac{1}{\sqrt{2}} & \frac{1}{\sqrt{6}} \\ \frac{1}{\sqrt{3}} & -\frac{1}{\sqrt{2}} & \frac{1}{\sqrt{6}} \end{bmatrix}$

rank = $r = 3$
index = $s = 3$
signature = $2s - r = 3$; Positive definite

16. $6x_1^2 + 3x_2^2 + 3x_3^2 - 4x_1x_2 - 2x_2x_3 + 4x_3x_1$

Ans: $\lambda_1 = 2, \lambda_2 = 2, \lambda_3 = 8$
$q = 2y_1^2 + 2y_2^2 + 8y_3^2 \quad P = \begin{bmatrix} \frac{1}{\sqrt{5}} & \frac{2}{\sqrt{30}} & \frac{2}{\sqrt{6}} \\ 0 & \frac{5}{\sqrt{30}} & -\frac{1}{\sqrt{6}} \\ -\frac{2}{\sqrt{5}} & \frac{1}{\sqrt{30}} & \frac{1}{\sqrt{6}} \end{bmatrix}$

rank = $r = 3$
index = $s = 3$
signature = $2s - r = 3$; Positive definite

Reduce each of the following quadratic forms to canonical form by Lagrange's method of reduction. (Qns. 17–20)

17. $2x_1^2 + 7x_2^2 + 5x_3^2 - 8x_1x_2 - 10x_2x_3 + 4x_1x_3$

Ans: $2y_1^2 - y_2^2 + 4y_3^3 \quad P = \begin{bmatrix} 1 & 2 & -1 \\ 0 & 1 & -1 \\ 0 & 0 & 1 \end{bmatrix}$

rank = $r = 3$
index = $s = 2$
signature = $2s - r = 1$; Indefinite

18. $x_1^2 + 2x_2^2 - 7x_3^2 - 4x_1x_2 + 8x_1x_3$ [JNTU 2003S]

Ans: $y_1^2 - 2y_2^2 + 9y_3^2; \quad P = \begin{bmatrix} 1 & 2 & -12 \\ 0 & 1 & -4 \\ 0 & 0 & 1 \end{bmatrix}$

19. $q = X^T AX$, where $X = [x \quad y \quad z]^T$,

$A = \begin{bmatrix} 1 & 2 & 4 \\ 2 & 6 & -2 \\ 4 & -2 & 18 \end{bmatrix}$ [JNTU 2003S]

Ans: $y_1^2 + 2y_2^2 + 48y_3^2; \quad P = \begin{bmatrix} 1 & 2 & -14 \\ 0 & 1 & 5 \\ 0 & 0 & 1 \end{bmatrix}$

20. $x_1x_2 + x_2^2 + 4x_1x_3 + x_3^2$

Ans: $y_1^2 + y_2^2 - \frac{17}{4}y_3^2$

21. Find the singular value decomposition (SVD) for the matrix $A = \begin{bmatrix} 4 & 3 \\ 0 & -5 \end{bmatrix}$

Ans: $U = \begin{bmatrix} 0.447 & 0.894 \\ 0.894 & -0.447 \end{bmatrix}$

$V = \begin{bmatrix} 0.707 & 0.707 \\ -0.707 & 0.707 \end{bmatrix}$

22. Find the eigenvalues and the corresponding ortho-normalized eigenvectors of the matrix $A^T A$ if $A = \begin{bmatrix} 20 & 4 \\ 10 & 14 \\ 5 & 5 \end{bmatrix}$. Hence find the SVD of A.

Ans: $\lambda_1 = 665,\ \lambda_2 = 97$

$$X_1 = \begin{bmatrix} 0.87 \\ 0.49 \end{bmatrix} \qquad X_2 = \begin{bmatrix} -0.49 \\ 0.87 \end{bmatrix}$$

$$A = UDV^T = \begin{bmatrix} 0.76 & -0.65 \\ 0.60 & 0.74 \\ 0.26 & 0.19 \end{bmatrix} \begin{bmatrix} 25.78 & 0 \\ 0 & 9.85 \end{bmatrix} \begin{bmatrix} 0.87 & 0.49 \\ -0.49 & 0.87 \end{bmatrix}$$

23. Obtain the singular value decomposition (SVD) of $A = \begin{bmatrix} 1 & 2 \\ 1 & 1 \\ 1 & 3 \end{bmatrix}$

Ans: $\lambda_1 = 16.64,\ \lambda_2 = 0.36$

$$X_1 = \begin{bmatrix} 0.4033 \\ 0.9166 \end{bmatrix}, \qquad X_2 = \begin{bmatrix} 0.9166 \\ -0.4033 \end{bmatrix}$$

$$A = UDV^T = \begin{bmatrix} 0.5480 & 0.1833 \\ 0.3235 & 0.8555 \\ 0.7727 & -0.4889 \end{bmatrix} \begin{bmatrix} 4.080 & 0 \\ 0 & 0.60 \end{bmatrix} \begin{bmatrix} 0.4033 & 0.9166 \\ 0.9166 & -0.4033 \end{bmatrix}$$

Solution of Algebraic and Transcendental Equations 3

3.1 Introduction to Numerical Methods

Analytical methods fail to produce desired results in many practical problems. So, the need for numerical methods—methods of solution of problems in terms of numbers on a computer or a calculator—has been greatly felt. With the advent of high-speed computers the importance of numerical techniques has enormously increased. Engineering students have some basic knowledge of concepts like algorithms, errors, computation, convergence and stability.

Numerical methods are used to solve problems on computer or calculators by numerical calculations giving a table of numbers or figures. In solving a problem numerically, the following steps are involved: Mathematical modelling of the problem in the form of an integral or differential or integral equation, selecting a mathematical method together with error estimation, writing an algorithm, programming in C, C++, *etc.*, and running the program successfully on a computer and finally interpreting the results obtained.

An *algorithm* is a finite sequence of rules for doing computations on a computer. It has to be stable by which we mean that small changes in the initial data do not produce large changes in the final results.

Computers have two ways of representing numbers:

1. *Fixed point system* in which numbers are represented with a fixed number of decimal places, *e.g.* 32.105, 1.012, *etc.*
2. *Floating point system* in which the number of significant digits is kept fixed while the decimal point is floating, *e.g.* 0.5371×10^2, -0.1314×10^{-3}, *etc.*

Theoretically any nonzero number a can be represented as

1. $a = \pm m \cdot 10^e$, where $0.1 \le m < 1$; e: integer.
2. $\bar{a} = \pm \overline{m} \cdot 10^e$ on a computer, m is limited to t digits ($t = 8$) and e is limited.

$$\overline{m} = 0 \cdot d_1 d_2 \cdots d_t \quad d_1 > 0; |e| < m$$

3.2 Errors and their Computation

3.2.1 Exact and Approximate Numbers

Exact numbers are those which have definite value:

E.g. $2, 3, 7, \frac{1}{2}, \frac{1}{3}, \pi, e,$ *etc.* Approximate numbers are those that represent the exact numbers to a certain degree of accuracy:

E.g. $\sqrt{2} = 1.4, \pi = 22/7$ or if better accuracy is required $\sqrt{2} = 1.414, \pi = 3.1416$.

3.2.2 Significant Digits

The digits that are used to express a number are called significant digits or figures. The numbers 3.1416, 0.66667, 40.687 contain 5 significant digits. The number 0.00023 has only two significant digits 2 and 3, since the zeros serve only to fix the position of the decimal point.

3.2.3 Loss of Significant Digits

Loss of significant digits means that a result of a calculation has fewer correct digits than the number from which it was obtained.

Example 3.1
Find the roots of the following equation using four significant digits in computation:

$$x^2 - 40x + 2 = 0 \quad (1)$$

Solution We have

$$x_1 = \frac{1}{2a}(-b + \sqrt{b^2 - 4ac}), \quad (2)$$

$$x_2 = \frac{1}{2a}(-b - \sqrt{b^2 - 4ac}) \quad (3)$$

Also $$x_1 x_2 = \frac{c}{a} = 2 \quad (4)$$

From (2) we get $x = 20 \pm \sqrt{398} = 20 \pm 19.95 \Rightarrow x_1 = 39.95$, $x_2 = 0.05$. This result is poor as there is loss of significant digits.

Instead, if we use (4) $x_2 = \frac{2}{x_1} = \frac{2}{39.95} = 0.05006$, which is in error by less than one unit of the last digit.

3.2.4 Rounding off

Cutting off of some of the end digits of a number and retaining a fixed number of digits in numerical computations is called rounding off.

3.2.5 Rules for Rounding off

To round off a number to n significant digits.

1. Discard the digits after the nth digit if the discarded number
   (a) < 5; (b) = 5 and the nth digit is even;
2. increase the nth digit by 1 if the discarded number
   (a) > 5; (b) = 5 and the nth digit is odd.

Thus, the number of terms obtained is said to be correct to n significant figures.

The value of π correct to 1, 2, 3, 4, 5,... significant figures is 3, 3.1, 3.14, 3.142, 3.1416 ..., respectively.

Numbers 1.6583, 1.6586, 0.859378, 1.7255, 1.7245, if rounded to four significant figures, are 1.658, 1.659, 0.8594, 1.726, 1.724, respectively.

A useful rule to reduce round-off errors in computations is to retain at least one more significant digit than that given in the data, perform the last operation and then round-off.

3.2.6 Absolute, Relative and Percentage Errors

Absolute error: e_a = |True Value – Approximate Value | $= |x - \bar{x}| = \delta x$.

Relative error: $e_r = \frac{e_a}{x} = \frac{\delta x}{x}$.

Percentage error: $e_p = 100e_r$.

Bound

A number Δx such that $|x - \bar{x}| = |\Delta x| \le \Delta x$ is an upper bound for the absolute error and it measures the *absolute accuracy* and $\frac{\Delta x}{|x|} \simeq \frac{\Delta x}{|\bar{x}|}$ measures the *relative accuracy*.

3.3 Formulas for Errors

Error of $\tilde{a}$ is an approximate value of a quantity whose exact value is 'a', we call the difference $\in = a - \tilde{a}$, the error of $\tilde{a}$. Hence $a = \tilde{a} + \in$ (True Value = Approx + error). For instance, if $\tilde{a} = 10.8$ is an approximation of $a = 10.2$ its error is $\in = -0.6$. The error of an approximation $\tilde{a} = 1.84$ is $\in = 0.34$.

3.3.1 Relative Error

$\in_r$ of $\tilde{a}$ is defined by $\in_r = \frac{\in}{a} = \frac{a - \tilde{a}}{a}$

$$= \frac{\text{Error}}{\text{True Value}} (a \neq 0) \simeq \frac{\in}{\tilde{a}} \quad \text{if } |\in| << |\tilde{a}|$$

3.3.2 Error Bound for $\tilde{a}$

In practice we can obtain a number β such that $|\in| \le \beta$ hence $|a - \tilde{a}| \le \beta$ for relative error $\in_r$, $|\in_r| \le \beta_r$ hence $\left|\frac{a - \tilde{a}}{a}\right| \le \beta$.

3.3.3 Error Propagation

Error propagation refers to how errors at the beginning and in later steps propagate into the computation and affect accuracy.

Bound

In addition and subtraction an error bound for the results is given by the sum of the error bounds: $|\epsilon| \le \beta_1 + \beta_2$.

In multiplication and division a bound for the relative error of the results is given by the sum of the bounds for the relative errors of the given numbers: $|\epsilon_r| \le \beta_{r1} + \beta_{r2}$.

3·3·4 Error in Rounding

Let $\tilde{a} = fl(a)$ be the floating point computer approximation of a obtained by round-off. The round-off rule gives

$$|m - \bar{m}| \le \frac{1}{2}10^{-t} \Rightarrow \left|\frac{a-\bar{a}}{a}\right| \simeq \left|\frac{m-\bar{m}}{m}\right| \le \frac{1}{2}\cdot 10^{1-t} \quad (3.1)$$

The RHS $u = \frac{1}{2}\cdot 10^{1-t}$ is called the *rounding unit*. If we write $\bar{a} = a(1+\delta)$ we have $\frac{\bar{a}-a}{a} = \delta$. Hence $|\delta| \le u$ by (3.1).

This shows that the rounding unit u is an error bound in rounding. These errors become more pronounced with more arithmetical operations. So we have to analyze computational programs for expected rounding errors and see that their effect is as small as possible.

3·3·5 Programming Errors

A common name for all kinds of errors in a computer program is bugs. Debugging means locating and removing bugs.

3·3·6 Errors of Numerical Results

Final results of computations of unknown quantities are generally approximations and involve errors. Round-off errors result from rounding. Experimental errors are errors of given data. Truncating errors result from truncating (premaurely breaking off), *e.g.* if we replace Taylor's series by the sum of its first few terms. These errors depend on the computational method used.

3·4 Mathematical Pre-requisites

The following theorems, which are required while dealing with numerical methods are stated without proofs.

Theorem 3.1 If $f(x)$ is continuous in the closed interval $[a, b]$ and if $f(a)$ and $f(b)$ are of opposite signs then there exists at least one point ξ (or an odd number of points) in (a, b) such that $f(\xi) = 0$.

Theorem 3.2 (Rolle's[1] Theorem). If $f(x)$ is continuous in the closed interval $[a, b]$, f is derivable in the open interval (a, b) and $f(a) = f(b)$ then there exists at least one point ξ in (a, b) such that $f'(\xi) = 0$.

Theorem 3.3 (Generalised Rolle's Theorem). Let $f(x)$ be a function which is n times derivable on $[a, b]$. If $f(x)$ vanishes at the $(n + 1)$ distinct points $x_0, x_1, \ldots, x_n$ in (a, b) then there exists a number ξ in (a, b) such that $f^{(n)}(\xi) = 0$.

Theorem 3.4 (Intermediate Value Theorem). Let $f(x)$ be continuous in $[a, b]$ and let k be any number between $f(a)$ and $f(b)$. Then there exists a number ξ in (a, b) such that $f(\xi) = k$ (Figure 3.1).

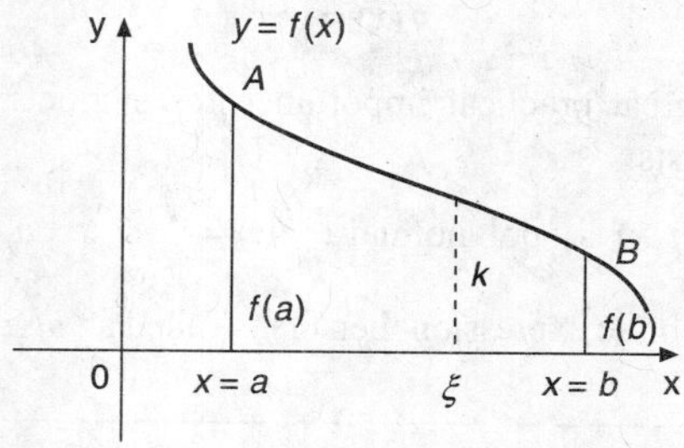

Figure 3.1 A function continous on $[a, b]$ attains an intermediate value at some point in $[a, b]$

Theorem 3.5 (Lagrange's[2] Mean Value Theorem). If $f(x)$ is continuous in $[a, b]$ and derivable in (a, b) then there exists at least one value of x say ξ between a and b such that $f'(\xi) = [f(b)-f(a)]/(b, a)$, $(a < \xi < b)$. If we set $b = a + h$ then $f(a + h) = f(a) + hf'(a + \theta h)$ $(0 < \theta < 1)$.

---

[1] ROLLE, Michel (1652–1719), French analyst, algebraist and geometer.

[2] LAGRANGE, Joseph Louis (1736–1813), great French algebraist, analyst, number theorist, probabilist, physicist and astronomer. His contributions are especially to calculus of variations, analytical mechanics and astronomy. He became professor at Military Academy, Twain at 19 and later Director of Berlin Academy in 1766 at 30.

Theorem 3.6 (Taylor's[3] Theorem). If $f(x)$ is continuous and has continuous derivatives of order n in an interval I containing a then in I

$$f(x)=f(a)+(x-a)f'(a)+\frac{(x-a)^2}{2!}f''(a)+\cdots+\frac{(x-a)^{n-1}}{(n-1)!}f^{(n-1)}(a)+R_n(x)$$

where $R_n(x)=\dfrac{(x-a)^n}{n!}f^{(n)}(\xi)\quad (a<\xi<x)$ is the remainder after n terms.

3.5 Solution of Algebraic and Transcendental Equations

3.3.1 Introduction

Determination of values of x satisfying equations of the form

$$f(x)=0 \tag{3.2}$$

is of great practical importance for engineers and scientists.

If $f(x)$ is a polynomial $\left(f(x)=\sum_{r=0}^{n}a_rx^r,\ a_r\in \mathrm{R}\right)$ or algebraic expression then (1) is called an *algebraic equation* and if $f(x)$ involves trigonometric, logarithmic, exponential or hyperbolic functions then it is called a *transcendental equation.*

$3x^2-5x+2=0,\ x^3-2\sqrt{x}+1=0$ are algebraic equations;

$xe^x=1$, $2\sin x=1$, $\sin x=2$, $\sin x=x/2$ are transcendental equations.

3.3.2 Zero or Root of a Function

A number α is called a zero of a function if $f(\alpha)=0$. A zero is called a root or a solution of $f(x)$ or of the equation $f(x)=0$.

The process of finding roots of $f(x)=0$ is called solving the equation. An algebraic equation

$$f(x)=\sum_{r=0}^{n}a_rx^r\quad (a_r\in \mathrm{R})$$

of degree n has exactly n roots, distinct or repeated. But the transcendental equations are different in this respect.

$y=\sin x=2$ has no real roots (if x is complex it has infinitely many roots). $y=\sin x=1/2$ has infinitely many real roots. $y=\sin x=x/2$ has exactly three roots.

The graph in Figure 3.2 illustrates these points.

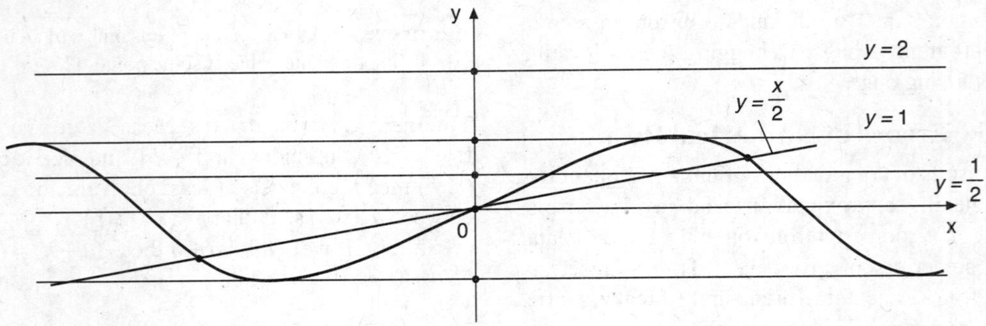

Figure 3.2 Infinite number of roots of $\sin x=1/2$ and finite number (three) of roots of $\sin x=x/2$

[3] TAYLOR, Brook (1685–1731), English analyst, geometer, and philosopher. Published books on perspective and on calculus of finite differences.

3.6 Direct Methods of Solution

Algebraic equations of degree $n\le 4$ with arbitrary coefficients can be solved completely. Higher degree algebraic equations and the transcendental equations require numerical methods for finding the desired roots.

Geometrically a root of equation (3.1) where $f(x)$ is a continuous function is that value of x where the graph of $y = f(x)$ crosses the x-axis.

Every equation with real coefficients and of odd degree has at least one real root.

3.6.1 Descartes'[4] Rule of Signs

A polynomial equation of the form $f(x) = 0$ cannot have more positive roots than the changes of signs in $f(x)$ and more negative roots than the changes of signs in $f(-x)$.

$$E.g.\quad f(x) = x^5 - 6x^2 - 4x + 5 = 0$$

The sequences of signs in $f(x)$ and $f(-x)$ are $+--+$ and $--++$, respectively.

There are two changes of signs in $f(x)$ and so, there cannot be more than two positive roots and there is only one change of sign in $f(-x)$ and so, there cannot be more than one negative root of $f(x)$.

Observe the following table:

x	$-\infty$	-2	-1	0	1	2	$+\infty$
$f(x)$	$-$	$-$	$+$	$+$	$-$	$+$	$+$

Since $f(-2)\,f(-1) < 0$ there lies a negative root in $(-2, -1)$. Also, since $f(0)\,f(1) < 0$ and $f(1)\,f(2) < 0$ there lie two positive roots one in each of the intervals $(0, 1)$ and $(1, 2)$.

3.7 Numerical Methods of Solution of Equations of the Form $f(x) = 0$

There are several numerical methods of solution of equations of the form $f(x) = 0$ where $f(x)$ is an algebraic or a transcendental function.

1. Fixed point iteration (Successive approximations) method.
2. Bisection or Bolzano (Interval-halving) method.
3. Newton–Raphson (Tangents) method.
4. Secants (Chords) method.
5. False position (Regula Falsi) method.

3.7.1 Fixed Point Iteration (Successive Approximation) Method

The word iterate means to do again or to repeat. It is derived from the Latin word 'iterum' which means 'again'. The iterative method is a process in which we start from an initial guess x_0 and compute step-by-step approximations $x_1, x_2, \ldots$ of an unknown solution of an equation of the form

$$f(x) = 0 \tag{3.3}$$

In general, iteration methods are easy to program since the computational operations are the same in each step, only the data change from step to step. More important, if in a particular case a method converges then it is, in general, stable.

Suppose equation (3.3) is transformed to the form

$$x = g(x) \tag{3.4}$$

We choose an x_0 and compute

$$x_1 = g(x_0),\ x_2 = g(x_1),\ \ldots \text{ and generally,}$$

$$x_{n+1} = g(x_n),\ n = 0, 1, 2, \ldots \tag{3.5}$$

A solution of equation (3.4) is called a fixed point of g, and hence the method is called the fixed point iteration. This is a solution of (3.3) since from (3.4) we can get back to (3.3). From equation (3.3) we may get several different forms of (3.4). The behaviour of corresponding iterative sequences $x_0, x_1, x_2, \ldots$ may differ in particular, with respect to the speed of convergence.

Convergence of Iterative Process

An iterative process defined by equation (3.4) is called convergent if, for an x_0, the corresponding sequence $x_0, x_1, x_2, \ldots$ is convergent.

A sufficient condition for convergence is given in the following theorem which has many practical applications.

[4] DESCARTES, Rene (1596–1650), philosopher and mathematician. Born in France, lived in several countries of Western Europe, settled in Holland. Founded analytic geometry (an honour shared with Fermat) which is commonly called Cartesian Geometry.

A Sufficient Condition for Convergence

Theorem 3.7 (Convergence of Fixed Point Iteration) Let $x = \alpha$ be a solution of $x = g(x)$ and suppose that g has a continous derivative in some interval I containing α. Then if

$$|g'(x)| \le M < 1 \quad \text{in I} \tag{3.6}$$

the iteration process defined in

$$x_{n+1} = g(x_n) \qquad (n = 0, 1, 2, \ldots) \tag{3.7}$$

converges for any x_0 in I.

Proof By Lagrange's mean value theorem of differential calculus there is a a $\xi \in (x, \alpha)$ such that

$$g(x) - g(\alpha) = g'(\xi)\,(x - \alpha), \; (x \in \text{I}) \tag{3.8}$$

$$\left.\begin{array}{l}\text{Since } g(\alpha) = \alpha \quad \text{and} \quad x_1 = g(x_0), x_2 = g(x_1) \\ \text{and } x_n = g(x_{n-1})\end{array}\right\} \tag{3.9}$$

we obtain from (3.6) and (3.8)

$$\begin{aligned} |x_n - \alpha| &= |g(x_{n-1}) - g(\alpha)| \\ &= |g'(\xi)||x_{n-1} - \alpha| \qquad \text{by (5.8)} \\ &\le M|x_{n-1} - \alpha| = M|g(x_{n-2}) - g(\alpha)| \\ &= M|g'(\xi)||x_{n-2} - \alpha| \le M^2|x_{n-2} - \alpha| \\ &\cdots \qquad \text{proceeding succesively} \\ &\le M^n|x_0 - \alpha| \end{aligned}$$

Since $M < 1$we have $M^n \to 0$ and hence $|x_n - \alpha| \to 0$.

Note 1 An equation may be written in the form $x = g(x)$ in more than one way. So, we have to choose that form which ensures convergence for which the condition is $|g'(x)| < 1$; the smaller the value of $|g'(x)|$ the faster the convergence.

Rate of Convergence

As pointed out above there are different approximate methods for the solution of equations of the form $f(x) = 0$. We have to choose an efficient method depending upon the available means and other constraints. One consideration for selecting a particular method is the rate of convergence of the sequence $\langle x_n \rangle$ of iterates by that method. The deviation of an approximate value from the actual value is called the error. That is, if 'e' denotes the error and α denotes the actual root of $f(x) = 0$ then

$$e_n = \alpha - x_n = x_{n+1} - x_n$$

If e_{n+1}/e_n^p is nearly constant then the convergence of the method is said to be of order p.

If $p = 1$ then the convergence is said to be linear, which is used for slow convergence.

Example 3.2

Find the larger root of $x^2 - 87x + 1 = 0$ by iteration.

Solution The given equation is

$$f(x) = x^2 - 87x + 1 = 0 \tag{1}$$

which is a quadratic equation of the form

$$ax^2 + bx + c = 0 \tag{2}$$

whose analytical solutions are given by

$$\alpha = \left(-b + \sqrt{b^2 - 4ac}\right)\Big/2a, \tag{3}$$

$$\beta = \left(-b - \sqrt{b^2 - 4ac}\right)\Big/2a \tag{4}$$

Here $a = 1, b = -87, c = 1$

$$\alpha = \left(87 + \sqrt{87^2 - 4.1.1}\right)\Big/2 = 86.988504 \tag{5}$$

If we calculate β from (4)

$$\beta = \left(87 - \sqrt{87^2 - 4.1.1}\right)\Big/2 = 0.011495771 \tag{6}$$

Writing (1) in the form

$$x = g(x) = 87 - \frac{1}{x}$$

$$\Rightarrow \quad x_{n+1} = 87 - \frac{1}{x_n} \quad \text{(Iteration formula)} \tag{7}$$

and taking $x_0 = 87$ as an initial approximation we obtain

$$x_1 = 87 - \frac{1}{x_0} = 86.988506 \tag{8}$$

$$x_2 = 87 - \frac{1}{x_1} = 86.988504 \qquad (9)$$

The smaller root can be obtained from the fact that the product of the roots = 1. Thus

$$\beta = \frac{1}{\alpha} = \frac{1}{86.988504} = 0.011495771 \qquad (10)$$

It is advisable to adopt this method since the method of iteration does not give the smaller root. Convergence of the iterates may not be there. Further, in calculating β from the formula (6) there may be loss of significant digits β being the difference of two almost equal numbers.

Example 3.3
Solve the equation $x^2 - 3x + 1 = 0$ by iteration.

Solution The given equation is

$$f(x) = x^2 - 3x + 1 = 0 \qquad (1)$$

which is a quadratic equation of the form

$$ax^2 + bx + c = 0 \qquad (2)$$

whose analytical solutions are given by

$$\alpha = \left(-b + \sqrt{b^2 - 4ac}\right)/2a, \qquad (3)$$

$$\beta = \left(-b - \sqrt{b^2 - 4ac}\right)/2a \qquad (4)$$

In the present case we have

$$a = 1, b = -3, c = 1 \qquad (5)$$

so that the larger and smaller roots are

$$\alpha = \left(3 + \sqrt{3^2 - 4}\right)/2 = 2.618034, \qquad (6)$$

$$\beta = \left(3 - \sqrt{3^2 - 4}\right)/2 = 0.381966 \qquad (7)$$

For applying the method of iteration we have to write $f(x) = 0$ in the form $x = g(x)$

This can be done in more than one way, we consider the following two forms:

(1) $x = g_1(x) = \frac{1}{3}(x^2 + 1) \quad \Rightarrow \quad x_{n+1} = \frac{1}{3}(x_n^2 + 1)$

(Iteration formula) (8)

(2) $x = g_2(x) = 3 - \frac{1}{3x} \quad \Rightarrow \quad x_{n+1} = 3 - \frac{1}{x_n}$ (9)

Case (1) If we choose an initial approximation $x_0 = 1.000$, we obtain the following sequence of iterates:

$x_1 = 0.666667$; $x_2 = 0.481481$; $x_3 = 0.410608$;

$x_4 = 0.389533$; $x_5 = 0.383912$; $x_6 = 0.382463$;

$x_7 = 0.382092$; $x_8 = 0.381998$; $x_9 = 0.381974$;

$x_{10} = 0.381968$; $x_{11} = 0.381966$;

$x_{12} = 0.381966 \rightarrow$ smaller root β

If we choose an initial approximation $x_0 = 2.000$ we obtain the sequence of iterates:

$x_1 = 1.666667$; $x_2 = 1.259259$; $x_3 = 0.861911$;

$x_4 = 0.580964$; $x_5 = 0.445840$ $x_6 = 0.399591$;

$x_7 = 0.386558$; $x_8 = 0.383142$; $x_9 = 0.382266$;

$x_{10} = 0.382042$ $x_{11} = 0.381985$; $x_{12} = 0.381971$;

$x_{13} = 0.381967$; $x_{14} = 0.381967$; $x_{15} = 0.381966$

$x_{16} = 0.381966 \rightarrow$ smaller root β

If we choose an initial approximation $x_0 = 3.000$ we obtain the sequence of iterates:

$x_1 = 3.333333$; $x_2 = 4.037037$; $x_3 = 5.765889$;

$x_4 = 11.415160$; $x_5 = 43.768626$;

$x_6 = 638.897535, \ldots$

which diverges; and does not converge to a root (upper part of Figure 3.3(a)).

Case (2) If we choose an initial approximation as $x_0 = 1$ or 2 we obtain the same sequence of iterates barring an extra term at the start when we take $x_0 = 1$ (Figure 3.3(b)):

$x_0 = 2.000000$; $x_1 = 2.500000$; $x_2 = 2.600000$;

$x_3 = 2.615385$; $x_4 = 2.617647$; $x_5 = 2.617978$;

$x_6 = 2.618026$; $x_7 = 2.618033$; $x_8 = 2.618034$;

$x_9 = 2.618034 \rightarrow$ larger root α

If we choose an initial approximation as $x_0 = 3.000$ we obtain the sequence of iterates:

$x_0 = 3.000000$; $x_1 = 2.666667$; $x_2 = 2.625000$;

$x_3 = 2.619048$; $x_4 = 2.618182$; $x_5 = 2.618056$;

$x_6 = 2.618037$; $x_7 = 2.618034$;

$x_8 = 2.618034 \rightarrow$ larger root α

The smaller root cannot be obtained in this case by the above procedure. But we can get it from the fact that $\alpha\beta = 1 \quad \Rightarrow \quad \beta = \dfrac{1}{\alpha} = \dfrac{1}{2.618034} = 0.381966$

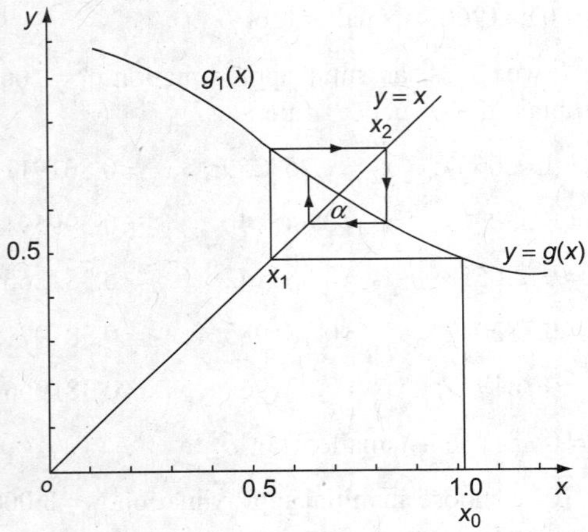

Figure 3.3 Iteration method

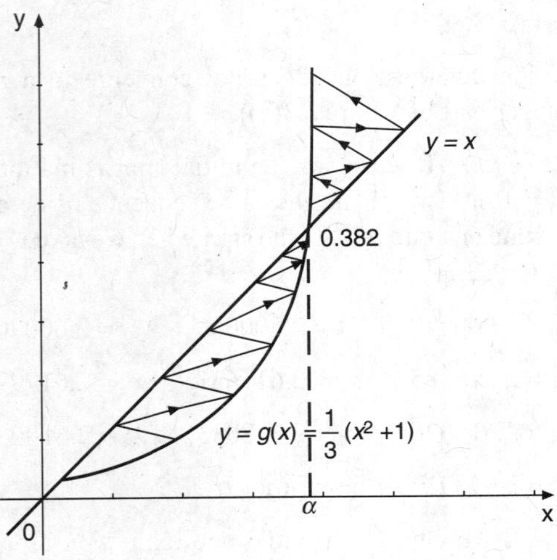

Figure 3.3(a) Iteration method for $f(x) = x^2 - 3x + 1 = 0$

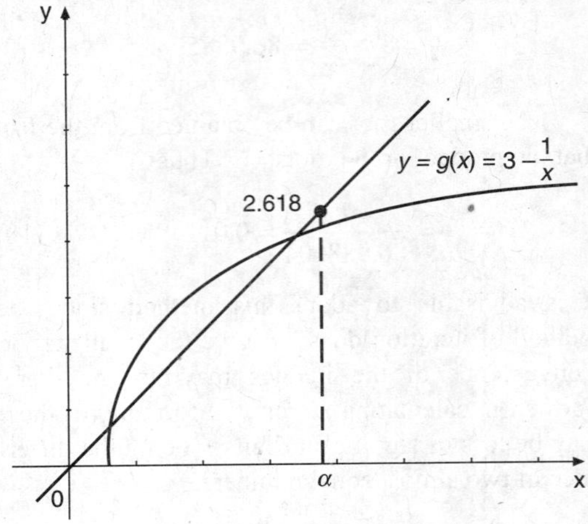

Figure 3.3(b) Iteration method for $f(x) = x^2 - 3x + 1 = 0$

In Figure 3.3(a) the slope of $g_1(x) <$ the slope of the straight line $y = x$, thus $|g'_1(x)| < 1$ and we have convergence.

In the upper part $g_1(x)$ is steeper and $g'_1(x) > 1$ so that we have divergence.

In Figure 3.3(b) the slope of $g_2(x)$ is less near the intersection point $x = 2.618$ which is the fixed point of $g_2(x)$ and is the solution of $f(x) = 0$. Both the sequences are converging. We conclude that convergence depends on the fact that in a neighbourhood of a solution the curve of $g(x)$ is less steep than $y = x$. We observe that $|g'(x)| < 1$ is a sufficient condition for the convergence of the sequence of iterates to a root of the equation.

The above problem shows that the iterative method has to be applied only after ensuring convergence of the sequence of iterates, which depends upon (1) the form $x = g(x)$ and (2) the initial approximation x_0 with which we start the iteration.

Example 3.4

Find the real root of the equation $x^3 + x - 1 = 0$ using the method of iteration.

Solution

Let $\qquad f(x) = x^3 + x - 1 = 0 \qquad (1)$

$$f\left(\frac{1}{2}\right) = \frac{1}{2^3} + \frac{1}{2} - 1 = -\frac{3}{8} = -0.375 < 0; \quad f(1) = 1 > 0$$

By the Intermediate Value Theorem, a root of (1) lies in $\left(\frac{1}{2}, 1\right)$. From equation (1) we have

$$x(x^2+1)=1 \quad \Rightarrow \quad x=\frac{1}{1+x^2} \tag{2}$$

We set $x=g(x)=\frac{1}{1+x^2}$ (3)

so that the iterative formula is

$$x_{n+1}=g(x_n)=\frac{1}{1+x_n^2} \tag{4}$$

Further

$$g'(x)=-\frac{2x}{(1+x^2)^2} \quad \Rightarrow \quad |g'(x)|=\frac{2|x|}{(1+x^2)^2}<1 \tag{5}$$

Indeed, for any x,

$$\frac{4x^2}{(1+x^2)^4}=\frac{4x^2}{1+4x^2+\cdots}<1 \tag{6}$$

Hence for any initial guess x_0 the sequence of iterates $\langle x_n \rangle$ converges. (Fig. 3.3)

Table 3.1 Choose $x_0 = 1$

$x_1=g(x_0)=\frac{1}{1+x_0^2}=\frac{1}{1+1^2}=0.500$
$x_2=g(x_1)=\frac{1}{1+x_1^2}=\frac{1}{1+(0.5)^2}=0.800$
$x_3=g(x_2)=\frac{1}{1+x_2^2}=\frac{1}{1+(0.800)^2}=0.6098 \simeq 0.61$
$x_4=g(x_3)=\frac{1}{1+x_3^2}=\frac{1}{1+(0.61)^2}=0.7288 \simeq 0.729$
$x_5=g(x_4)=\frac{1}{1+x_4^2}=\frac{1}{1+(0.729)^2}=0.65297 \simeq 0.653$
$x_6=g(x_5)=\frac{1}{1+x_5^2}=\frac{1}{1+(0.653)^2}=0.70106 \simeq 0.701$

The solution correct to six decimal places is 0.682328.

The reader may compare the solution obtained by Newton's method.

Example 3.5

Using the method of iteration find a positive root of the equation $xe^x = 1$ that lies between 0 and 1.

Solution We write the equation

$$f(x)=xe^x-1=0 \tag{1}$$

in the form

$$x=g(x)=e^{-x} \tag{2}$$

Iteratative formula is $x_{n+1}=g(x_n)=e^{-x_n}$ (3)

We find $g(x)=e^{-x}$ and $g'(x)=-e^{-x}$

so that $|g'(x)| < 1$ for $x < 1$. Thus, the convergence of iterative formula (3) is assured.

Let us take the starting value $x_0 = 1$. We find the successive iterates in Table 3.2, which are accurate up to four decimal places.

Table 3.2 Root of $xe^x = 1$ in (0, 1)

$x_1=e^{-x0}=e^{-1}=1/e=0.3678794$	$x_{11}=0.566414733$
$x_2=e^{-x1}=0.692200656$	$x_{12}=0.567556639$
$x_3=e^{-x2}=0.500473486$	$x_{13}=0.56690891$
$x_4=e^{-x3}=0.606243543$	$x_{14}=0.567276232$
$x_5=e^{-x4}=0.545395781$	$x_{15}=0.567267898$
$x_6=e^{-x5}=0.579612338$	$x_{16}=0.56718605$
$x_7=e^{-x6}=0.560115459$	$x_{17}=0.56711904$
$x_8=e^{-x7}=0.571143116$	$x_{18}=0.567157044$
$x_9=e^{-x8}=0.564879346$	$x_{19}=0.56713549$
$x_{10}=e^{-x9}=0.568428725$	$x_{20}=0.567147714$

3.7.2 Bolzano's[5] (Bisection or Interval-Halving) Method

This method is very simple and easy to implement. It is based on repeated application of the intermediate value theorem. If we know that a root of $f(x) = 0$ lies in the interval $I_0 = (a_0, b_0)$, we bisect I_0 at the point $m_1 = (a_0 + b_0)/2$.

Let

$$I_1 = \begin{cases} (a_0, m_1), \text{ if } f(a_0)f(m_1) < 0 \\ (m_1, b_0), \text{ if } f(m_1)f(b_0) < 0 \end{cases}$$

In either case I_1 also contains the root. We bisect the interval I_1 and get a subinterval I_2 at

[5] BOLZANO, Bernhard (1781–1848), is a Czechoslovakian analyst.

whose end points, $f(x)$ takes values of opposite signs and hence contains the root. Continuing this process we obtain a sequence of nested intervals $\langle I_n \rangle$ with

$$I_0 \supset I_1 \supset I_2 \supset \cdots$$

such that each of these intervals contains the root of $f(x) = 0$.

After repeating the bisection process p times we either find the root or find the interval I_p of length $\frac{(b_0 - a_0)}{2^p}$ which contains the root. We take the mid-point of the last sub-interval as the desired approximation to the root.

This root has an error not greater than one half of the length of the interval of which it is the mid-point. Thus, we have

$$m_{k+1} = a_k + \frac{1}{2}(b_k - a_k) \quad k = 0, 1, 2, \ldots$$

where,

$$(a_{k+1}, b_{k+1}) = \begin{cases} (a_k, m_{k+1}) & \text{if } f(a_k)f(m_{k+1}) < 0 \\ (m_{k+1}, b_k) & \text{if } f(m_{k+1})f(b_k) < 0 \end{cases}$$

We observe that this method uses only the end points of the interval $[a_k, b_k]$ for which $f(a_k)\,f(b_k) < 0$ and not the values of $f(x)$ at these end points to obtain the next approximation to the root.

Further, the sequence of approximations always converges to the root for any function $f(x)$ which is continuous in the interval that contains the root.

Suppose the permissible error is $\in$ then the approximate number of iterations required may be determined from the relation

$$\frac{b_0 - a_0}{2^n} \le \in \quad \text{or} \quad n \ge \frac{\log(b_0 - a_0) - \log \in}{\log 2}$$

The minimum number of iterations required for the sequence of iterates to converge to a root in the interval (0,1) for a given $\in$ is given in Table 3.3.

Table 3.3 Tally of Number of Iterations n and Permissible Error $\in$

$\in$	10^{-2}	10^{-3}	10^{-4}	10^{-5}	10^{-6}	10^{-7}
n	7	10	14	17	20	24

Thus, the bisection method requires a large number of iterations to achieve a reasonable degree of accuracy for the root.

Algorithm BOLZANO (f_1, f_2, a, b, $\in$, N)

This algorithm computes a solution of $f(x) = 0$ given two initial approximations $a = x_1$ and $b = x_2$ such that $f_1 \times f_2 < 0$ where $f_1 = f(x_1)$ and $f_2 = f(x_2)$.

INPUT

$$a = x_1, \quad b = x_2$$

are initial approximations, tolerance $\in$ and maximum number of iterations N.

OUTPUT Approximate solution x_n $(n \le N)$ or message that root does not exist in the stated range

For $n = 1, 2, 3, (N-1)$ *do*

1. Compute $f_1 = f(x_1)$, $f_2 = f(x_2)$
2. If $f_1 \times f_2 > 0$ then OUTPUT no root lies between x_1 and x_2. Stop
3. Else compute
$$x_{n+2} = x_n + \frac{1}{2}(x_{n+1} - x_n), \quad n = 1, 2, \ldots, N-1$$
4. If $|x_{n+1} - x_n| < \in$, then OUTPUT $\alpha = (x_n - x_{n+1})/2$.

Stop

Else

OUTPUT No root lies between x_n and x_{n-1}. Stop

Example 3.6

Perform five iterations of the bisection method and obtain the smallest positive root of the equation $f(x) = x^3 - 5x + 1 = 0$.

Solution

Let $\quad f(x) = x^3 - 5x + 1 = 0 \quad (1)$

Since $f(0) = 1 > 0$ and $f(1) = -3 < 0$, the smallest positive root lies in the interval (0, 1).

Taking $a_0 = 0$ and $b_0 = 1$ we have

$$m_1 = \frac{1}{2}(a_0 + b_0) = \frac{1}{2}(0+1) = 0.5$$

$$f(m_1) = f(0.5) = -1.375 \text{ and } f(a_0)\,f(m_1) = f(0) \cdot f(0.5) = -1.375 < 0$$

Thus the root lies in the interval (0, 0.5).
Taking $a_1 = 0$ and $b_1 = 0.5$ we have

$$m_2 = \frac{1}{2}(a_1 + b_1) = \frac{1}{2}(0 + 0.5) = 0.25$$

$$f(m_2) = f(0.25) = -0.234375 \text{ and}$$

$$f(a_1)\,f(m_2) = f(0)\,f(0.25) = -0.234375 < 0$$

Thus the root lies in the interval (0, 0.25). The sequence of intervals is given in Table 3.4.

Table 3.4 Table of Sequence of Intervals for the Bisection Method

k	a_{k-1}	b_{k-1}	m_k	$f(m_k)\,f(a_{k-1})$
1	0	1	0.5	< 0
2	0	0.5	0.25	< 0
3	0	0.25	0.125	> 0
4	0.125	0.25	0.1875	> 0
5	0.1875	0.25	0.21875	< 0

Hence the root lies in (0.1875, 0.21875). The approximate root is taken as the midpoint of this interval, that is, 0.203125.

Example 3.7

Using Bolzano's Bisection method, find the real root of the equation $f(x) = x^3 - 4x - 9 = 0$ lying between 2 and 3.

Solution

Let $$f(x) = x^3 - 4x - 9 = 0 \qquad (1)$$

Since $f(a) = f(2) = 2^3 - 4.2 - 9 = -9 < 0$,

$$f(b) = f(3) = 3^3 - 4.3 - 9 = 6 > 0$$

therefore $f(2)$ and $f(3)$ are of opposite signs.

Hence there lies a root of equation (1) in the interval $(a, b) = (2, 3)$.

Starting with $a = x_0 = 2$ and $b = x_1 = 3$ as the initial approximations we compute successive iterates using the following iteration formula of the bisection method:

$$x_{n+1} = (x_{n-1} + x_n)/2 \quad (n = 1, 2, 3, \ldots) \qquad (2)$$

∴ First approximation to the root is $x_2 = \frac{1}{2}(2+3) = 2.5$.

$$\text{Now, } f(x_2) = x_2\,(x_2^2 - 4) - 9 = 2.5\,(2.5^2 - 4) - 9 = -3.375 < 0$$

∴ The root lies in the interval $(x_2, x_1) = (2.5, 3)$

So, the second approximation to the root is $x_3 = \frac{1}{2}(2.5+3) = 2.75$.

$$\text{Now, } f(x_3) = x_3\,(x_3^2 - 4) - 9 = 2.75\,(2.75^2 - 4) - 9 = 0.797 > 0$$

∴ The root lies in the interval $(x_2, x_3) = (2.5, 2.75)$. Proceeding like this we compute successive iterates checking if $f(x)$ is positive or negative and choosing the subinterval at whose end points $f(x)$ has opposite signs.

The results of the computations are shown in Table 3.5.

Table 3.5 Bisection Method: Successive Iterates for the Root of $f(x) = x^3 - 4x - 9 = 0$

n	x_{n+1}	$f(x_{n+1})$	Succeeding interval with the signs of functional values
0	$a = x_0 = 2, b = x_1 = 3$	$f(x) = x(x^2 - 4) - 9$ $f(x_0) = -9, f(x_1) = 6$	$(x_0, x_1) = \left(\overset{-}{2}, \overset{+}{3}\right)$
1	$x_2 = \frac{2+3}{2} = 2.5$	$f(x_2) = 2.5\,(2.5^2 - 4) - 9 = -3.375$	$(x_2, x_1) = \left(\overset{-}{2.5}, \overset{+}{3}\right)$
2	$x_3 = \frac{2.5+3}{2} = 2.75$	$f(x_3) = 2.75\,(2.75^2 - 4) - 9 = 0.797$	$(x_2, x_3) = \left(\overset{-}{2.5}, \overset{+}{2.75}\right)$

(Continued)

Table 3.5 *Continued*

n	x_{n+1}	$f(x_{n+1})$	Succeeding interval with the signs of functional values
3	$x_4 = \frac{2.5+2.75}{2} = 2.63$	$f(x_4) = 2.63\,(2.63^2 - 4) - 9 = -1.329$	$(x_4, x_3) = \left(\overset{-}{2.63}, \overset{+}{2.75}\right)$
4	$x_5 = \frac{2.63+2.75}{2} = 2.69$	$f(x_5) = 2.69\,(2.69^2 - 4) - 9 = -0.295$	$(x_5, x_3) = \left(\overset{-}{2.69}, \overset{+}{2.75}\right)$
5	$x_6 = \frac{2.69+2.75}{2} = 2.72$	$f(x_6) = 2.72\,(2.72^2 - 4) - 9 = 0.244$	$(x_5, x_6) = \left(\overset{-}{2.69}, \overset{+}{2.72}\right)$
6	$x_7 = \frac{2.69+2.72}{2} = 2.705$	$f(x_7) = 2.705\,(2.705^2 - 4) - 9 = -0.02745$	$(x_7, x_6) = \left(\overset{-}{2.705}, \overset{+}{2.72}\right)$
7	$x_8 = \frac{2.705+2.720}{2} = 2.7125$	$f(x_8) = 2.7125\,(2.7125^2 - 4) - 9 = 0.10764$	$(x_7, x_8) = \left(\overset{-}{2.705}, \overset{+}{2.7125},\right)$
8	$x_9 = \frac{2.705+2.7125}{2} = 2.70875$	$f(x_9) = 2.70875\,(2.70875^2 - 4) - 9 = 0.03998$	$(x_7, x_9) = \left(\overset{-}{2.705}, \overset{+}{2.70875},\right)$
9	$x_{10} = \frac{2.705+2.70875}{2} = 2.706875$	$f(x_{10}) = 2.706875\,(2.706875^2 - 4) - 9 = 0.00624$	$(x_7, x_{10}) = \left(\overset{-}{2.705}, \overset{+}{2.706875},\right)$
10	$x_{11} = \frac{2.705+2.706875}{2} = 2.7059375$	$f(x_{11}) = -0.00750$	$(x_{11}, x_{10}) = \left(\overset{-}{2.7059375}, \overset{+}{2.706875},\right)$
11	$x_{12} = (2.7059375 + 2.706875)/2 = 2.70640625$	$f(x_{12}) = -003746$	$\left(\overset{-}{x_{12}}, \overset{+}{x_{10}}\right)$
12	$x_{13} = (2.70640625 + 2.706875)/2 = 2.706640625$	$f(x_{13}) = 0.002025$ $f(\alpha) = f\left(\frac{x_{12}+x_{13}}{2}\right) = -0.000081187$	$\left(\overset{-}{x_{12}}, \overset{+}{x_{13}}\right)$

The approximate value of the root $\alpha = \frac{x_{12}+x_{13}}{2} = 2.706523438$

Example 3.8

Find the positive root of the equation $xe^x = 1$ which lies in the interval (0, 1) by the Bisection method.

Solution $\qquad f(x) = xe^x - 1 = 0 \qquad (1)$

Since $f(a) = f(0) = -1 < 0$ and $f(b) = f(1) = 1.718 > 0$, a root of equation (1) lies in the interval $(a, b) = (0, 1)$.

Starting with $a = x_0 = 0$ and $b = x_1 = 1$ as initial approximations we compute successive iterates using the Bisection iteration formula

$$x_{n+1} = \frac{x_{n-1} + x_n}{2} \qquad n = 1, 2, 3, \ldots \qquad (2)$$

and they are entered in Table 3.6.

After 12 iterations the error $\in$ is within the tolerance of 0.05%. Hence the required root is 0.567, correct to three decimal places.

Table 3.6 Bisection Method: Succesive Iterates for the Root of $f(x) = xe^x - 1 = 0$

n	a	b	x	$f(x)$	n	a	b	x	$f(x)$
1	0	1	0.5	–	7	0.5625	0.5781	0.5703	+
2	0.5	1	0.75	+	8	0.5625	0.5703	0.5664	–
3	0.5	0.75	0.625	+	9	0.5664	0.5703	0.5684	+
4	0.5	0.625	0.5625	–	10	0.5664	0.5684	0.5674	+
5	0.5625	0.625	0.59375	+	11	0.5664	0.5674	0.5669	–
6	0.5625	0.59375	0.5781	+	12	0.5669	0.5674	0.5671	–

3.7.3 Newton–Raphson[6] Method

This method is simple and is therefore most commonly used to solve an equation of the form $f(x) = 0$. Here f is assumed to be continuous and to have a continuous derivative f' in an interval (a, b).

The underlying idea is that we approximate the graph of f by suitable tangents. Using an approximate value x_0 obtained from the graph of f we let x_1

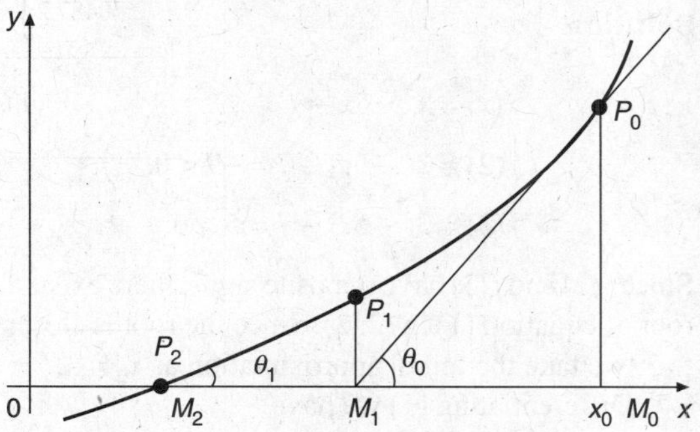

Figure 3.4 Newton–Raphson method

be the point of intersection of the x-axis and the tangent to the curve of f at x_0 (Figure 3.4).

From Figure 3.4, we have

$$\tan\theta_0 = f'(x_0) = \frac{f(x_0)}{x_0 - x_1} \tag{3.10}$$

$$\Rightarrow \quad x_1 = x_0 - \frac{f(x_0)}{f'(x_0)} \tag{3.11}$$

From this we write the Newton–Raphson iteration formula

$$x_{n+1} = x_n - \frac{f(x_n)}{f'(x_n)} \tag{3.12}$$

This formula can also be obtained using Taylor series expansion of f at $x = x_0$:

$$f(x) = f(x_0) + (x - x_0) f'(x_0) + \frac{(x - x_0)^2}{2!} f''(x_0) + \cdots \tag{3.13}$$

For a first approximation x_1 of x we truncate the series and obtain a linear equation

$$f(x_1) = f(x_0) + (x_1 - x_0) f'(x_0) = 0$$

$$\Rightarrow \quad x_1 = x_0 - \frac{f(x_0)}{f'(x_0)} \tag{3.14}$$

From this we obtain iteration formula (3.12).

Newton–Raphson method for solving equation $f(x) = 0$

Algorithm NEWTON (f, f', x_0, $\in$, N)

This algorithm computes a solution of $f(x) = 0$ given an initial approximation x_0. x_0 is the starting value of iteration.

INPUT f, f'

Initial approximation x_0
tolerance $\in > 0$

maximum number of iterations N

---

[6] NEWTON, Sir Isaac (1642–1727), is a great English mathematician, physicist and astronomer. With Archimedes and Gauss, he is one of the three greatest mathematicians of all time. He and Leibnitz invented calculus independently.

OUTPUT approximate solution $x_n (n \le N)$
or message of 'root does not lie in the range'
For $n = 0, 1, 2, \ldots N-1$ do

1. Compute $f'(x_n)$
2. If $f'(x_n) = 0$ then
   OUTPUT 'Root cannot be found' stop
3. Else Compute
   $$x_{n+1} = x_n - \frac{f(x_n)}{f'(x_n)}$$
4. If $|x_{n+1} - x_n| \le \in$ then
   OUTPUT x_{n+1}. Stop
   END
5. OUTPUT 'Root does not lie in the range'.
   Stop
   END NEWTON

The inequality $|x_{n+1} - x_n| \le \in$ in the above algorithm is a *termination criterion.*

If the sequence $\langle x_n \rangle$ converges and the criterion holds we have reached the desired degree of accuracy and we stop. Note that the criterion by itself does not imply convergence.

For example, the harmonic series $\sum \frac{1}{n}$ as we know diverges although its sequence of partial sums $\langle x_n \rangle$ given by $x_n = \sum_{k=1}^{n} \frac{1}{k} = 1 + \frac{1}{2} + \cdots + \frac{1}{n}$ satisfies the criterion because $\lim (x_{n+1} - x_n) = \lim \frac{1}{n+1} = 0.$

Line 5 in the algorithm gives another termination criterion. It is needed because Newton's method may diverge or due to improper choice of x_0 may not reach the degree of accuracy by a reasonable number of iterations. Then we may try another x_0.

If $f(x) = 0$ has more than one solution, different choices of x_0 may produce different solutions. Also, an iterative sequence may sometimes converge to a solution different from the expected one.

Newton's Method of Quadratic Convergence

Let $x_n = \alpha + \in_n$ and $x_{n+1} = \alpha + \in_{n+1}$ so that

$$\in_{n+1} = \in_n - \frac{f(\alpha + \in_n)}{f'(\alpha + \in_n)}$$

$$= \in_n - \frac{f(\alpha) + \in_n f'(\alpha) + \frac{1}{2!} \in_n^2 f''(\alpha) + \cdots}{f'(\alpha) + \in_n f''(\alpha) + \cdots},$$

by Taylor's Theorem

$$= \frac{\in_n^2}{2} \cdot \frac{f''(\alpha)}{f'(\alpha)} \quad \because \quad f(\alpha) = 0;$$

neglecting $\in_n^3, \in_n^4$ being small

(3.15)

This shows that the error at each step is proportional to the square of the error at the previous step and as such the convergence is quadratic, *i.e.* of second order.

Example 3.9

Find the positive real root of the equation $x^3 - 5x - 7 = 0$ by the Newton–Raphson method.

Solution

Let $\quad f(x) = x^3 - 5x - 7 \quad (1)$

$f(2) = 2^3 - 5.2 - 7 = -9 < 0,$

$f(3) = 3^3 - 5.3 - 7 = 5 > 0$

Since $f(2)$ and $f(3)$ have opposite signs, there exists a root of equation (1) in (2, 3). Since the root is closer to 3, we take the initial approximation as $x_0 = 3$.

Differentiating (1) we have

$$f'(x) = 3x^2 - 5 \quad (2)$$

The Newton–Raphson iterative formula is

$$x_{n+1} = x_n - \frac{f(x_n)}{f'(x_n)} \quad n = 0, 1, 2, \ldots \quad (3)$$

$$= x_n - \frac{x_n^3 - 5x_n - 7}{3x_n^2 - 5} = \frac{2x_n^3 + 7}{3x_n^2 - 5} \quad (4)$$

$$x_0 = 3, \quad x_1 = \frac{2x_0^3 + 7}{3x_0^2 - 5} = \frac{61}{22} = 2.7727;$$

$$x_2 = \frac{2x_1^3 + 7}{3x_1^2 - 5} = \frac{2 \cdot (2.7727)^3 + 7}{3 \cdot (2.7727)^2 - 5} = 2.7477;$$

$$x_3 = \frac{2x_2^3+7}{3x_2^2-5} = \frac{2(2.7477)^3+7}{2(2.7477)^2-5} = 2.7474;$$

$$x_4 = \frac{2x_3^3+7}{3x_3^2-5} = \frac{2(2.7474)^3+7}{3(2.7474)^2-5} = 2.7473;$$

$$x_5 = \frac{2x_4^3+7}{3x_4^2-5} = \frac{2(2.7473)^3+7}{3(2.7473)^2-5} = 2.7473$$

We may stop at this stage since iterates x_4 and x_5 agree up to four decimal places.

∴ The required positive root of the given equation is

$$\alpha = 2.7473$$

Example 3.10

Solve the equation $x^3 - 9x + 1 = 0$ for the root between 2 and 3 by the Newton–Raphson method.

Solution

Let $\quad f(x) = x^3 - 9x + 1 = 0 \quad (1)$

Then differentiating $f(x)$ with respect to x we get $\quad f'(x) = 3x^2 - 9 \quad (2)$

Newton's iterative formula is

$$x_{n+1} = x_n = \frac{f(x_n)}{f'(x_n)} \quad (3)$$

Since $\quad f(a) = f(2) = 2(2^2 - 9) + 1 = -9 < 0;$

$$f(b) = f(3) = 3(3^2 - 9) + 1 = 1 > 0 \quad (4)$$

there exists a root of $f(x) = 0$ in the interval $(a, b) = (2, 3)$. Clearly the root lies closer to $x = 3$. Taking $x_0 = 3$ as an initial approximation to the root, we compute successive iterates from the formula

$$x_{n+1} = x_n - \frac{x_n^3 - 9x_n + 1}{3x_n^2 - 9} = \frac{2x_n^3 - 1}{3x_n^2 - 9} \quad (5)$$

First, second, third approximations to the root are

$$\therefore \quad x_1 = \frac{2x_0^3 - 1}{3x_0^2 - 9} = \frac{2.3^3 - 1}{3.3^2 - 9} = \frac{53}{18} = 2.94;$$

$$x_2 = \frac{2x_1^3 - 1}{3x_1^2 - 9} = \frac{50.05521262}{17.00925925} = 2.94282142;$$

$$x_3 = \frac{2x_2^3 - 1}{3x_2^2 - 9} = \frac{49.97083229}{16.98059389} = 2.942820058$$

Since $x_2 = x_3$ up to five decimal places, the approximate value of the root is $\alpha = 2.94282$.

Example 3.11

Find a real root of the equation $x = e^{-x}$ using the Newton–Raphson method.

Solution We write the equation as

$$f(x) = xe^x - 1 = 0 \quad (1)$$

so that $\quad f'(x) = (x+1)e^x \quad (2)$

The Newton–Raphson iteration formula is

$$x_{n+1} = x_n - \frac{f(x_n)}{f'(x_n)}, \quad n = 0, 1, 2, \ldots \quad (3)$$

$$\Rightarrow \quad x_{n+1} = x_n - \frac{x_n e^{x_n} - 1}{(x_n+1)e^{x_n}} = \frac{x_n^2 e^{x_n} + 1}{(x_n+1)e^{x_n}} \quad (4)$$

Since $\quad f(a) = f(0) = 0.e^0 - 1 = -1 < 0;$

$$f(b) = f(1) = 1.e^1 - 1 = 1.71828 > 0$$

there lies a root of equation (1) in the interval $(a, b) = (0, 1)$.

We take $a = x_0 = 0$ as an initial guess for the root. The successive approximations are

$$x_1 = \frac{x_0^2 e^{x_0} + 1}{(x_0+1)e^{x_0}} = 1; \qquad x_2 = \frac{x_1^2 e^{x_1} + 1}{(x_1+1)e^{x_1}} = 0.68394;$$

$$x_3 = \frac{x_2^2 e^{x_2} + 1}{(x_2+1)e^{x_2}} = 0.57745; \quad x_4 = \frac{x_3^2 e^{x_3} + 1}{(x_3+1)e^{x_3}} = 0.56723;$$

$$x_5 = \frac{x_4^2 e^{x_4} + 1}{(x_4+1)e^{x_4}} = 0.56714; \quad x_6 = \frac{x_5^2 e^{x_5} + 1}{(x_5+1)e^{x_5}} = 0.56714$$

Since $x_5 = x_6$ up to five decimal places the approximate root is $\alpha = 0.56714$.

Example 3.12

Using the Newton–Raphson method, evaluate to two decimal places the root of the transcendental equation $f(x) = e^x - 3x = 0$ lying between 0 and 1.

Solution

Let $\quad f(x) = e^x - 3x = 0 \quad \Rightarrow \quad f'(x) = e^x - 3 \quad (1), (2)$

The Newton–Raphson Iteration formula is

$$x_{n+1} = x_n - \frac{f(x_n)}{f'(x_n)} \quad (3)$$

$$\Rightarrow \quad x_{n+1} = x_n - \frac{e^{x_n} - 3x_n}{e^{x_n} - 3} \quad (4)$$

Since $f(a) = f(0) = e^0 - 0 = 1 > 0;$

$$f(b) = f(1) = e^1 - 3 = -0.28 < 0 \quad (5), (6)$$

there lies a root of equation (1) in the interval $(a, b) = (0, 1)$.

Taking $a = x_0 = 0$ as an initial guess, we get successive approximations:

$$x_1 = x_0 - \frac{f_0}{f_0'} = 0 - \frac{1}{-2} = 0.5;$$

$$x_2 = x_1 - \frac{f_1}{f_1'} = 0.5 - \frac{.15}{-1.35} = 0.61$$

Proceeding like this we get successive iterates and these are entered in Table 3.7.

Table 3.7 Successive Iterates by Newton's Method for the Root of $f(x) = e^x - 3x = 0$

n	x_n	e^{x_n}	f_n	f'_n
0	0	1	1	–2
1	0.5	1.65	0.15	–1.35
2	0.61	1.84	0.01	–1.16
3	0.62	1.86	0.00	

The required root $\alpha = 0.62$ to two decimal places.

Example 3.13

Derive a formula for finding the pth root of a positive number N. Hence evaluate $\sqrt{28}$ to two decimal places.

Solution We have to solve the equation

$$f(x) = x^p - N = 0, \quad \Rightarrow \quad f'(x) = px^{p-1} \quad (1), (2)$$

By Newton's formula we have

$$x_{n+1} = x_n - \frac{f(x_n)}{f'(x_n)} = x_n - \frac{x_n^p - N}{p{x_n}^{p-1}}$$

$$= \frac{1}{p}\left[(p-1)x_n + \frac{N}{x_n^{p-1}}\right] \quad (3)$$

For a square root $p = 2$, cube root $p = 3$ and the formulas are

$$x_{n+1} = \frac{1}{2}\left(x_n + \frac{N}{x_n}\right), \quad x_{n+1} = \frac{1}{3}\left(2x_n + \frac{N}{x_n^2}\right) \quad (4), (5)$$

Now to evaluate $\sqrt{28}$ we set $N = 28$ in (4)

$$x_{n+1} = \frac{1}{2}\left(x_n + \frac{28}{x_n}\right) = \frac{1}{2} \cdot x_n + \frac{14}{x_n}$$

Take $x_0 = 5$ as an initial guess. Then first, second and third approximations are

$$x_1 = \frac{1}{2} \cdot x_0 + \frac{14}{x_0} = \frac{5}{2} + \frac{14}{5} = 2.5 + 2.8 = 5.3;$$

$$x_2 = \frac{1}{2} \cdot x_1 + \frac{14}{x_1} = \frac{5.3}{2} + \frac{14}{5.3} = 5.2915$$

$$x_3 = \frac{1}{2} \cdot x_2 + \frac{14}{x_2} = 5.2915$$

$$\therefore \quad \sqrt{28} = 5.2915$$

To find the reciprocal of a number: Newton–Raphson method

Example 3.14

Derive a formula for finding the reciprocal of a number. Hence find the reciprocal of (a) 22, (b) 18.

Solution Suppose we have to find the reciprocal of a number N. We have to solve the equation

$$f(x) = \frac{1}{x} - N = 0 \quad \Rightarrow \quad f'(x) = -\frac{1}{x^2} \quad (1), (2)$$

Writing the Newton–Raphson Iteration formula

$$x_{n+1} = x_n - \frac{f(x_n)}{f'(x_n)} \quad \Rightarrow \quad x_{n+1} = x_n - \frac{\frac{1}{x_n} - N}{-\frac{1}{x_n^2}}$$

$$\Rightarrow \quad x_{n+1} = x_n(2 - Nx_n), \quad n = 0, 1, 2, 3, \ldots \quad (3)$$

(a) Here $N = 22$. Taking an initial guess as $x_0 = 0.045$, we compute the iterates and construct Table 3.8.

Table 3.8 Table of Values for Finding Reciprocal of $N=22$

n	x_n	Nx_n	$2-Nx_n$	$x_{n+1}=x_n(2-Nx_n)$
0	0.045	0.99	1.01	0.04545
1	0.04545	0.9999	1.0001	0.0454545
2	0.0454545	0.999999	1.00001	0.045454545
…	…	…	…	…

Hence $1/22 = 0.045454545$ approximately.

(b) Here $N = 18$. Taking an initial guess as $x_0 = 0.055$, we have iterates as in Table 3.9.

Table 3.9 Table of Values for Finding Reciprocal of $N=18$

n	x_n	Nx_n	$2-Nx_n$	$x_{n+1}=x_n(2-Nx_n)$
0	0.055	0.99	1.01	0.055555
1	0.05555	0.9999	1.0001	0.05555555
2	0.0555555	0.99999999	1.00000001	0.05555555

Hence $1/18 = 0.05555555$ approximately.

3.7.4 Secants Method (or Chords Method)

Newton's method for the solution of an equation of the form $f(x) = 0$ is very powerful. But the presence of the derivative $f'(x)$ in its formula may pose problems sometimes. To overcome this situation secant method (chord method) is used which replaces the slope $f'(x)$ of the tangent to the curve by the slope $\dfrac{f(b)-f(a)}{b-a}$ of the secant AB joining points $A(a, f(a))$ and $B(b, f(b))$ of the curve (Figure 3.5 (a)).

Geometrically, this means that an approximation to the root is obtained by finding the x-coordinate of the point where the secant AB cuts the x-axis. Now the equation of the secant AB through the points $A = (a, f(a))$ and $B = (b, f(b))$ on the curve $y = f(x)$ is

$$y - f(a) = \frac{f(b)-f(a)}{b-a}(x-a) \quad \text{or}$$

$$y - f_a = \frac{f_b - f_a}{b-a}(x-a) \tag{3.16}$$

Putting $y = 0$ in (3.16) and solving for x we get a first approximation to the root of $f(x) = 0$ as

$$x_1 = a - \frac{b-a}{f_b - f_a} f_a \tag{3.17}$$

A second approximation to the root is obtained from

$$x_2 = x_0 - \frac{x_1 - x_0}{f(x_1) - f(x_0)} f(x_0) \tag{3.18}$$

where $x_0 = b$.

The general iterative formula of the secant method (chord method) is

$$x_{n+1} = x_{n-1} - \frac{x_n - x_{n-1}}{f(x_n) - f(x_{n-1})} f(x_{n-1}) \, (n = 1, 2, 3, \ldots) \tag{3.19}$$

which is sometimes written as

$$x_{n+1} = \frac{x_{n-1} f(x_n) - x_n f(x_{n-1})}{f(x_n) - f(x_{n-1})} \quad (n = 1, 2, 3, \ldots) \tag{3.20}$$

It is not advisable to use (3.20) in place of (3.19) as it may lead to loss of significant digits if x_n and x_{n-1} are nearly equal.

Also, if at any iteration $f(x_n) = f(x_{n-1})$ the method fails and shows that it does not converge. This is the drawback of the secant method over the method of Regula Falsi discussed below, which always converges. But if this method converges, the rate of convergence is 1.628 which is faster than that of Regula Falsi.

Further, at each iteration, two most recent approximations to the root are used to find the next approximation. Also, it is not neccessary that the interval at each step must contain the root. That is to say, we do not check at each step $f_n f_{n-1} < 0$ as in other methods.

If the iterative formula (3.19) is applied together with the condition that $f_n f_{n-1} < 0$ at each step then the method is called the method of false position (or Regula Falsi method).

3.7.5 Method of False Position (Regula Falsi)

This is one of the oldest methods used for solving an equation of the form $f(x) = 0$. In this method we choose two numbers a and b such that $f(a)$ and $f(b)$ are of opposite signs so that the graph $y = f(x)$ crosses the x-axis between $x = a$ and $x = b$. This shows that a root lies between a and b and we have $f(a)\,f(b) < 0$ (Figure 3.5(b)).

The equation of the chord joining the points $A = (a, f(a))$ and $B = (b, f(b))$ is

$$y - f(a) = \frac{f(b) - f(a)}{b - a}(x - a) \tag{3.21}$$

The method consists of replacing the curve $y = f(x)$ between A and B by the chord AB and taking the x-coordinate of the point of intersection of the chord with the x-axis as an approximation to the root. Thus a first approximation to the root is given by

$$x_1 = a - \frac{b - a}{f(b) - f(a)} \cdot f(a) \tag{3.22}$$

Now if $f(a)\,f(x_1) < 0$ then a second approximation to the root is given by

$$x_2 = a - \frac{x_1 - a}{f(x_1) - f(a)} \cdot f(a) \tag{3.23}$$

replacing $B\,(b, f(b))$ by $C_1\,(x_1, f(x_1))$.

If $f(a)\,f(x_1) > 0$ then $f(b)\,f(x_1) < 0$, we proceed with b in place of a.

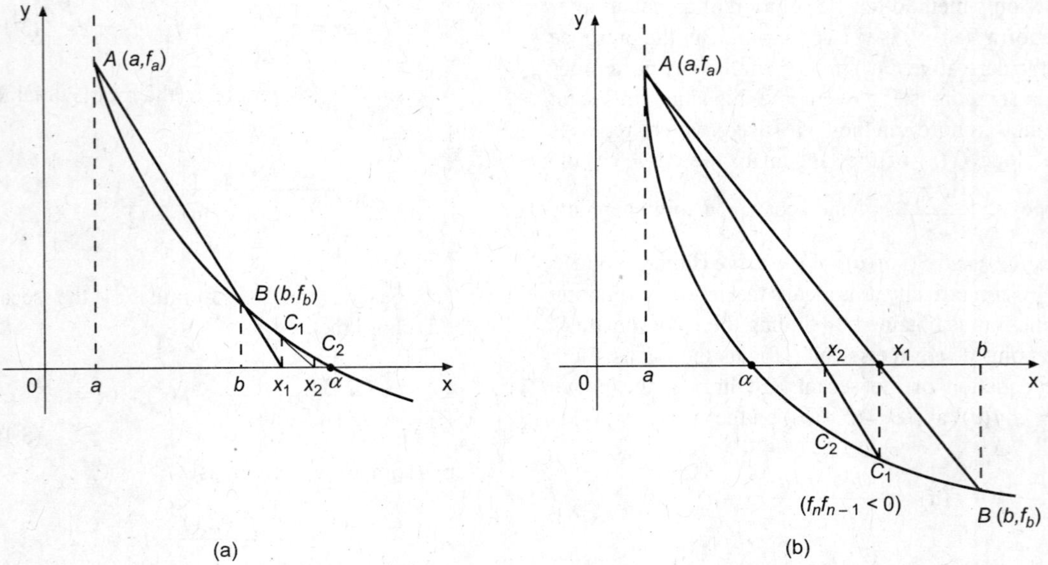

Figure 3.5 (a) Secant method and (b) Regula Falsi method

The general iteration formula of the Regula Falsi method is

$$x_{n+1} = x_n - \frac{x_n - x_{n-1}}{f(x_n) - f(x_{n-1})} \cdot f(x_n) \quad (n = 1, 2, 3, \ldots) \tag{3.24}$$

together with the condition $f_n \cdot f_{n-1} < 0$ to be satisfied at each step.

Since (x_{n-1}, f_{n-1}) and (x_n, f_n) are known before the starting of the iteration, the secant method or Regula Falsi method requires one function evaluation for each iteration.

Example 3.15

Perform four iterations of (a) the secant method and (b) the Regula Falsi method and obtain the root of $x^3 - 5x + 1 = 0$ in (0,1).

Solution The given equation is

$$f(x) = x^3 - 5x + 1 = 0 \tag{1}$$

We have $x_0 = 0$ and $x_1 = 1$

$$\therefore \quad f_0 = f(x_0) = 1 > 0 \quad \text{and} \quad f_1 = f(x_1) = -3 < 0$$

Since $f(x_0)\, f(x_1) = f(0)\, f(1) < 0$ a root of (1) lies in (0, 1). (2)

(a) Secant Method

$$x_2 = x_1 - \frac{x_1 - x_0}{f_1 - f_0} \cdot f_1 = 0.25;$$

$$f_2 = f(x_2) = -0.234375 \tag{3}$$

$$x_1 = 1, \quad f_1 = -3 < 0, \quad x_2 = 0.25,$$
$$f_2 = -0.234375 < 0 \tag{4}$$

(In the secant method f_1 and f_2 may have the same signs as above):

$$x_3 = x_2 - \frac{x_2 - x_1}{f_2 - f_1} \cdot f_2$$

$$= 0.25 - \frac{-0.75}{2.765625} \cdot (-0.234375) = 0.186441 \tag{5}$$

$$\Rightarrow \quad f_3 = 0.074276$$

$$x_2 = 0.25, \quad f_2 = -0.234375 < 0$$

$$x_3 = 0.186441, \quad f_3 = 0.074276 > 0$$

$$x_4 = x_3 - \frac{x_3 - x_2}{f_3 - f_2} \cdot f_3$$

$$= 0.186441 - \frac{-0.063559}{0.308651} \cdot (0.074276) = 0.201736 \tag{6}$$

$$\Rightarrow \quad f_4 = -0.000470$$

$$x_3 = 0.186441, \quad f_3 = 0.074276 > 0$$

$$x_4 = 0.201736, \quad f_4 = -0.000470 < 0$$

$$x_5 = x_4 - \frac{x_4 - x_3}{f_4 - f_3} \cdot f_4$$

$$= 0.201736 - \frac{0.015295}{-0.074746} \cdot (-0.000470) = 0.201640 \tag{7}$$

(b) Regula Falsi

$$x_2 = x_1 - \frac{x_1 - x_0}{f_1 - f_0} \cdot f_1 = 0.25; \quad f_2 = -0.234375$$

$$x_0 = 0; \quad f_0 = 1 > 0; \quad x_2 = 0.25; \quad f_2 = -0.234375 < 0 \tag{8}$$

In Regula Falsi we have to check the condition $f(a)\, f(b) < 0$ and select the points. Here $f(x_0)\, f(x_2) < 0$ so that the required root $\alpha \in (x_0, x_2)$:

$$x_3 = x_2 - \frac{x_2 - x_0}{f_2 - f_0} \cdot f_2 = 0.202532; \quad f_3 = -0.0043523 \tag{9}$$

Here $f(x_0)\, f(x_3) < 0$ so that the required root $\alpha \in (x_0, x_3)$:

$$x_4 = x_3 - \frac{x_3 - x_0}{f_3 - f_0} \cdot f_3 = 0.201654; \quad f_4 = -0.0000699 \tag{10}$$

Here $f(x_0)\, f(x_4) < 0$ so that the required root $\alpha \in (x_0, x_4)$:

$$x_5 = x_4 - \frac{x_4 - x_0}{f_4 - f_0} \cdot f_4 = 0.201640 \tag{11}$$

Example 3.16

By the secant and the Regula Falsi methods, find the root of the equation:

$$\cos x - xe^x = 0$$

Solution The given equation is

$$f(x) = \cos x - xe^x = 0 \quad (1)$$

Taking the initial approximations as $x_0 = 0$ and $x_1 = 1$ we have

$$f_0 = f(x_0) = f(0) = 1 \quad \text{and} \quad f_1 = f(x_1)$$

$$= f(1) = \cos 1 - e = -2.17798 \quad (2)$$

Secant Method

$$x_0 = 0; \quad f_0 = 1; \quad x_1 = 1; \quad f_1 = -2.17798 \quad (3)$$

$$x_2 = x_1 - \frac{x_1 - x_0}{f_1 - f_0} \cdot f_1$$

$$= 1 - \frac{1}{-3.17798} \cdot (-2.17798) = 0.31467$$

$$x_1 = 1; \quad f_1 = 1; \quad x_2 = 0.31467;$$

$$f_2 = 0.95090 - 0.43104 = 0.51986 \quad (4)$$

$$x_3 = x_2 - \frac{x_2 - x_1}{f_2 - f_1} \cdot f_2$$

$$= 0.31467 - \frac{-0.68533}{2.69784} \cdot (0.51986) = 0.44673 \quad (5)$$

$$x_2 = 0.314671, \quad f_2 = 0.51986; \quad x_3 = 0.44673,$$

$$f_3 = 0.90186 - 0.69832 = 0.20354$$

$$x_4 = x_3 - \frac{x_3 - x_2}{f_3 - f_2} \cdot f_3$$

$$= 0.44673 - \frac{-0.13206}{-0.31632} \cdot (0.20354) = 0.53171 \quad (6)$$

Regula Falsi

$$x_2 = x_1 - \frac{x_1 - x_0}{f_1 - f_0} \cdot f_1 = 0.31467, \; f_2 = 0.51987 \quad (7)$$

Here $f(x_1) \cdot f(x_2) < 0$ so that the root $\alpha \in (x_1, x_2)$

$$x_3 = x_2 - \frac{x_2 - x_1}{f_2 - f_1} \cdot f_2 = 0.44673; \; f_3 = 0.20354 \quad (8)$$

Here $f(x_1) \cdot f(x_3) < 0$ so that the root $\alpha \in (x_1, x_3)$

$$x_4 = x_3 - \frac{x_3 - x_1}{f_3 - f_1} \cdot f_3 = 0.49402 \quad (9)$$

Proceeding like this we get the successive iterates:

$$x_5 = 0.50995; \quad x_6 = 0.51520; \quad x_7 = 0.51692;$$
$$x_8 = 0.51748; \quad x_9 = 0.51767; \quad x_{10} = 0.51775; \quad (10)$$

Hence $\alpha = 0.5177$; correct to four decimal places.

EXERCISE 3.1

Fixed point Iteration

1. Solve the following equation by the method of fixed point iteration and find a positive root in the interval noted.
   (a) $x^3 - 3x - 5 = 0$ (2, 3) (b) $x \sin x = 1$ (1, 2)
   (c) $\tan x = x$ (4, 5) (d) $\cos x = 3x - 1$ (0, 1)
   Ans: (a) 2.28 (b) 1.068 (c) 4.4346 (d) 0.6071
2. Evaluate $\sqrt{30}$ by the iteration method with $x_0 = 5$ as an initial guess.
   Ans: 5.4772
3. Find the positive root of each of the following equations by the *Bisection method.*
   (a) $x^3 - 4x - 9 = 0$ (0, 1) (b) $x^3 - x - 11 = 0$ (2, 3)
   (c) $xe^x = \cos x$ (0, 1) (d) $x \log_{10} x = 1.2$ (2, 3)
   Ans: (a) 0.765 (b) 2.375 (c) 0.51776 (d) 2.74
4. Find the root of each equation by applying *Newton's method.*
   (a) $x^3 - 3x + 1 = 0$ (1, 2) (b) $x + \log x = 2$ (1, 2)
   (c) $\log x = x - 3$ (0, 1) (d) $xe^x = 2$ (0, 1)
   Ans: (a) 1.532 (b) 1.8432 (c) 0.3241 (d) 0.853
5. Evaluate (a) $\sqrt{41}$ (b) $3\sqrt{2}$ by Newton's method.
   Ans: 6.4032, 1.259921
6. Using the *Regula Falsi* method find the root of
   (a) $x^2 - \log_e x = 12$ (3, 4) (b) $x = \cos x$ (0, 1)
   (c) $x^3 - 4x - 9$ (2, 3)
   Ans: (a) 3.6461 (b) 0.739079 ≃ 0.7391 (c) 2.7065

4 Interpolation

4.1 Introduction

Let the values of a function $y = f(x)$ be given at different x values; thus,

$$\begin{matrix} x_0 & x_1 & x_2 & \cdots & x_n \\ f(x_0) & f(x_1) & f(x_2) & \cdots & f(x_n) \end{matrix}$$

Then *interpolation* means to find an approximate value of $f(x)$ for an x between two x-values in (x_0, x_n). This is done by finding a polynomial called an *interpolating polynomial* which agrees with the function at the nodal points x_i $(i = 0, 1, 2, \ldots, n)$. In finding a suitable interpolating polynomial we need to have a formula for errors in polynomial approximation.

4.1.1 Formula for Errors in Polynomial Interpolation

Let $y = f(x)$ be a function defined at the $(n + 1)$ points

$$(x_i, y_i) \quad (i = 0, 1, 2, \ldots, n) \tag{4.1}$$

Suppose $f(x)$ is continuous and differentiable $(n + 1)$ times in (x_0, x_n). If $p_n(x)$ is an interpolating polynomial of degree utmost n satisfying the conditions

$$p_n(x_i) = f_i \quad (i = 0, 1, 2, \ldots, n) \tag{4.2}$$

then $f(x) - p_n(x) = 0$ for $x = x_i$ $(i = 0, 1, 2 \ldots n)$ (4.3)

so that we can write

$$f(x) - p_n(x) = A \prod_{i=0}^{n}(x - x_i) \tag{4.4}$$

where A is a constant to be determined such that

$$f(\bar{x}) - p_n(\bar{x}) = A \prod_{i=0}^{n}(\bar{x} - x_i) \quad \text{for} \quad \bar{x} \in (x_0, x_n) \tag{4.5}$$

$$\Rightarrow \quad A = \frac{f(\bar{x}) - p_n(\bar{x})}{\prod_{i=0}^{n}(\bar{x} - x_i)} \tag{4.6}$$

Let

$$F(x) = f(x) - p_n(x) - A \prod_{i=0}^{n}(x - x_i) \tag{4.7}$$

where A is given by (4.6). Now

$$\text{Also,} \quad \left.\begin{aligned} F(x_0) = F(x_1) = \cdots = F(x_n) = 0 \\ F(\bar{x}) = 0 \qquad x \in (x_0, x_n) \end{aligned}\right\} \tag{4.8}$$

Thus F vanishes at $(n + 2)$ points $\bar{x}, x_0, x_1, \ldots x_n$ in $[x_0, x_n]$.

By repeated application of Rolle's theorem we conclude that in $[x_0, x_n]$

$$\begin{aligned} F'(x) &= 0 \quad (n+1) \quad \text{times} \\ F''(x) &= 0 \quad n \quad \text{times} \\ &\cdots \quad \cdots \quad \cdots \quad \cdots \\ F^{(n+1)}(x) &= 0 \quad \cdots \text{once} \end{aligned}$$

Let this point be $x = \xi$ in (x_0, x_n).

Now, differentiating (4.7) $(n + 1)$ times with respect to x and putting $x = \xi$

$$f^{(n+1)}(\xi) = A(n+1)! \quad \Rightarrow \quad A = \frac{f^{(n+1)}(\xi)}{(n+1)!} \tag{4.9}$$

Comparing (4.6) and (4.9)

$$f(\bar{x}) - p_n(\bar{x}) = \frac{f^{(n+1)}(\xi)}{(n+1)!} \prod_{i=0}^{n}(\bar{x} - x_i)$$

$$\Rightarrow \quad f(x) - p_n(x) = \frac{1}{(n+1)!} \prod_{i=0}^{n}(x - x_i) f^{(n+1)}(\xi)$$

$$(x_0 < \xi < x_n) \tag{4.10}$$

This gives the required expression for error in polynomial interpolation.

4.2 Interpolation with Equal Intervals

4.2.1 Finite Differences

Introduction: Let a function $y = f(x)$ be given. In numerical analysis the independent variable x is also called the argument and the dependent variable y is called the entry. Finite differences are calculated and exhibited in the form of a table after tabulating the values, $x_0, x_1, x_2, \ldots$ of the independent variable x and the corresponding functional values $y_0, y_1, y_2, \ldots$ of the dependent variable y. The difference between two consecutive values of the independent variable is called the interval of differencing. If the interval is uniform, it will be denoted by h. For determining the value of $f(x)$ at some intermediate value of x, three types of differences are used: (1) forward differences, (2) backward differences and (3) central differences.

Divided differences are used when the interval of differencing is not uniform.

4.2.2 Forward (Advancing) Difference Operator Δ

The *forward difference operator* Δ (delta) is defined by the equation

$$\Delta f(x) = f(x+h) - f(x) \tag{4.11}$$

The quantity $\Delta f(x)$ is called the first forward difference of $f(x)$. Since $\Delta f(x)$ is a function of x we can repeat the operation of differencing to obtain the second difference of $f(x)$

$$\Delta^2 f(x) = \Delta(\Delta f(x)) =. \Delta f(x+h) - \Delta f(x) \tag{4.12}$$

nth Forward difference of $f(x)$: Recursion relation

In general, the nth difference of $f(x)$ will be defined by the recursion relation

$$\begin{aligned}\Delta^n f(x) &= \Delta[\Delta^{n-1} f(x)]\\ &= \Delta^{n-1}[\Delta f(x)] = \Delta^{n-1} f(x+h) - \Delta^{n-1} f(x)\end{aligned} \tag{4.13}$$

Table 4.1 shows how forward differences of a function $y = f(x)$ are exhibited in a tabular form

Table 4.1 Forward (Advancing) Differences Table

Argument x	Entry y	First forward difference Δy	Second forward difference $\Delta^2 y$	Third forward difference $\Delta^3 y$	Fourth forward difference $\Delta^4 y$
x_0	y_0				
		Δy_0			
x_1	y_1		$\Delta^2 y_0$		
		Δy_1		$\Delta^3 y_0$	
x_2	y_2		$\Delta^2 y_1$		$\Delta^4 y_0$
		Δy_2		$\Delta^3 y_1$	
x_3	y_3		$\Delta^2 y_2$		
		Δy_3			
x_4	y_4				

y_0, the first entry, is known as the leading term and $\Delta y_0, \Delta^2 y_0, \Delta^3 y_0, \Delta^4 y_0, \ldots$ are known as the leading differences.

Example 4.1

Build a difference table for

$$f(x) = x^3 + 7x + 2 \tag{1}$$

starting with $x = 0$ as the initial value and with the functional values for 0, 2, 4, 6, 8.

Find $f(10)$ by extending the table.

Solution In this example, the analytical formula for $\Delta f(x)$ is

$$\begin{aligned}\Delta f(x) &= (x+2)^3 + 7(x+2) + 2\\ &\quad - (x^3 + 7x + 2)\\ &= 6x^2 + 12x + 22\end{aligned} \tag{2}$$

Similarly

$$\begin{aligned}\Delta^2 f(x) &= 6(x+2)^2 + 12(x+2)\\ &\quad + 22 - (6x^2 + 12x + 22)\\ &= 24x + 48\end{aligned} \tag{3}$$

and $\quad \Delta^3 f(x) = 24(x+2) + 48 - (24x + 48) = 48 \quad (4)$

Clearly $\quad \Delta^n f(x) = 0$ for $n \geq 4 \quad (5)$

Thus, the property that $\Delta^3 f(x)$ is equal to a constant for all x may be used to extend the table of functional values as far as we please. This is done in Table 4.2 to give $f(10)$.

Table 4.2 Difference Table for $f(x) = x^3 + 7x + 2$

x	$f(x)$	$\Delta f(x)$	$\Delta^2 f(x)$	$\Delta^3 f(x)$
0	2			
		22		
2	24		48	
		70		48
4	94		96	48
		166		
6	260		144	48
		310		
8	570		192	
		502		
10	1072			

4.2.3 Properties Satisfied by Δ

(a) Linearity

From the definition, it follows that the operator Δ obeys the linearity property

(i) $$\Delta[f(x) + g(x)] = \Delta f(x) + \Delta g(x) \tag{4.14}$$

(ii) $$\Delta[cf(x)] = c[\Delta f(x)] \tag{4.15}$$

(b) Index Law

$$\Delta^m[\Delta^n f(x)] = \Delta^{m+n} f(x) = \Delta^n[\Delta^m f(x)]$$

4.2.4 Backward Difference Operator ∇

We now introduce a second *operator the backward difference operator* ∇ (del) defined by the equation

$$\nabla f(x) = f(x) - f(x-h)$$

The quantity $\nabla f(x)$ is called the first backward difference of $f(x)$. Since $\nabla f(x)$ is a function of x we can repeat the operation of differencing to obtain the second difference of $f(x)$.

$$\nabla^2 f(x) = \nabla(\nabla f(x)) = \nabla[f(x) - f(x-h)]$$

***n*th Backward difference of *f*(*x*): Recursion relation**

In general, the nth difference of $f(x)$ will be defined by the recursion relation

$$\nabla^n f(x) = \nabla[\nabla^{n-1} f(x) = \nabla^{n-1}[\nabla f(x)]$$
$$= \nabla^{n-1} f(x) - \nabla^{n-1} f(x-h)$$

Note 1 ∇ also satisfies the properties satisfied by Δ.

We present Table 4.3 which is the table of backward differences.

Theorem 4.1 If $f(x)$ is a polynomial of degree n

$$f(x) = \sum_{r=0}^{n} a_r x^r \tag{4.16}$$

then $\quad \Delta^n f(x) = a_n n!\, h^n \quad$ (constant) $\quad (4.17)$

Proof We prove the theorem by induction. For $n = 1$

$$f(x) = a_1 x + a_0 \tag{4.18}$$

Table 4.3 Backward Differences Table

Argument	Entry	First backward difference	Second backward difference	Third backward difference	Fourth backward difference
x	y	∇y	$\nabla^2 y$	$\nabla^3 y$	$\nabla^4 y$
(1)	(2)	(3)	(4)	(5)	(6)
x_0	y_0				
		∇y_1			
x_1	y_1		$\nabla^2 y_2$		
		∇y_2		$\nabla^3 y_3$	
x_2	y_2		$\nabla^2 y_3$		$\nabla^4 y_4$
		∇y_3		$\nabla^3 y_4$	
x_3	y_3		$\nabla^2 y_4$		
		∇y_4			
x_4	y_4				

The last entry y_4 is called the leading term and ∇y_4, $\nabla^2 y_4$, $\nabla^3 y_4$, $\nabla^4 y_4$, ... are known as *the leading backward differences*.

and $\quad \Delta f(x) = \Delta(a_1 x + a_0) = a_1(\Delta x) + 0$
$$= a_1[x + h - x] = a_1 h \qquad (4.19)$$
so that (4.17) is true for $n = 1$.

Assume that the theorem is true for all degrees 1, 2, 3, …, $(n - 1)$ (4.20)

so that $\Delta^n\left(\sum_{r=0}^{n-1} a_r x^r\right) = \Delta\left(\Delta^{n-1}\sum_{r=0}^{n-1} a_r x^r\right) =$
$$\Delta \text{ (constant)} = 0, \quad \text{by (4.20)} \qquad (4.21)$$

Thus
$$\Delta^n f(x) = \Delta^n\left(\sum_{r=0}^{n} a_r x^r\right) = \Delta^n\left(\sum_{r=0}^{n-1} a_r x^r\right) + \Delta^n(a_n x^n)$$
$$= 0 + a_n \Delta^{n-1}(\Delta x^n) \quad \text{by (4.21)}$$
$$= a_n \Delta^{n-1}[(x+h)^n - x^n]$$
$$= a_n \Delta^{n-1} nhx^{n-1} + g(x)$$
where $g(x)$ is a polynomial of degree $< (n - 1)$.

Hence, by applying the induction hypothesis (4.20) again, we find
$$\Delta^n f(x) = a_n \Delta^{n-1}(rhx^{n-1}) \qquad (4.22)$$
$$= a_n.nh.(n-1)!h^{n-1} = a_n n! h^n \qquad (4.23)$$
This proves the theorem.

Note 1 The converse of the theorem is also true. If the *n*th differences of a function tabulated at equally spaced intervals are constant, the function is a polynomial of degree *n*.

Note 2 For convenience, the argument *x* is denoted as a subscript. Thus instead of $f(x)$ we could write f_x.

4.3 Symbolic Relations and Separation of Symbols

4.3.1 Factorial Function

The product of *r* factors starting with *x* and each successive factor decreasing by a constant '*h*' is called 'the factorial function of degree *r*' and is denoted by $x^{(r)}$.

Thus $x^{(r)}$ is defined by
$$x^{(r)} = x(x-h)(x-2h)\cdots(x-\overline{r-1}h) \qquad (4.24)$$
$$x^{(0)} = 1 \quad \text{and} \quad x^{(1)} = x \qquad (4.25)$$

Now
$$\Delta x^{(r)} = (x+h)^{(r)} - x^{(r)}, \quad \text{by definition}$$
$$= (x+h)x(x-h)\cdots(x-\overline{r-2}h) - x(x-h)\cdots(x-\overline{r-2}h)(x-\overline{r-1}h)$$
$$= x(x-h)\cdots(x-\overline{r-2}h)[x+h-(x-\overline{r-1}h)]$$
$$= rhx^{(r-1)} \qquad (4.26)$$

Similarly
$$\Delta^2 x^{(r)} = \Delta(\Delta x^{(r)}) = \Delta(rhx^{(r-1)}) \quad \text{by (4.26)}$$
$$= rh\cdot[\Delta x^{(r-1)}] = r(r-1)h^2 x^{(r-2)}$$

In general
$$\Delta^m x^{(r)} = r(r-1)(r-2)\cdots(r-\overline{m-1})h^m x^{(r-m)} \quad (m \le r) \qquad (4.27)$$

Compare Δ^m with the differential operator $D^m \equiv \dfrac{d^m}{dx^m}$
$$D^m(x^n) = n.(n-1)(n-2)\cdots(n-m+1)x^{n-m}$$

Example 4.2

Represent $f(x) = x^3 + 3x + 7$ in factorial notation.

Solution Let $\quad f(x) = x^3 + 3x + 7$
$$\equiv x(x-1)(x-2) + Bx(x-1) + Cx + D$$
where *A*, *B*, *C* and *D* are constants to be determined. Equating the coefficients of like powers of *x* on both sides

Coefficient of x^2, $0 = -3 + B \quad \Rightarrow \quad B = 3$

Coefficient of x, $3 = 2 - B + C \quad \Rightarrow \quad C = 4$

x^0, $7 = D \qquad$ Also $A = 1$

$\therefore \quad x^3 + 3x + 7 = x^{(3)} + 3x^{(2)} + 4x^{(1)} + 7$

Note 1 Every polynomial of degree *n* can be expressed as a factorial polynomial of the same degree and vice versa.

A simpler procedure to express a polynomial in the factorial notation is as follows.

Arrange the coeffficients of the powers of *x* in decreasing order, putting a zero for the coefficient of a missing power of *x* and divide by $x - 1$, $x - 2$, *etc.*, successively.

Example 4.3

Write $f(x) = x^3 + 3x + 7$ in factorial notation, and find the differences.

Solution Let $f(x) = x^3 + 3x + 7 = Ax^{(3)} + Bx^{(2)} + Cx^{(1)} + D$; clearly $D = 7$.

Writing the coefficients in order and dividing by $x - 1$, $x - 2$, $x - 3$ in turn

1	1	0	3	$\lfloor 7 = D$
		1	1	
2	1	1	$\lfloor 4 = C$	
		2		
3	1	$\lfloor 3 = B$		
	$1 = A$			

Hence $f(x) = x^3 + 3x + 7 = x^{(3)} + 3x^{(2)} + 4x^{(1)} + 7$

Now
$$\Delta f(x) = \Delta\,[x^{(3)} + 3x^{(2)} + 4x^{(1)} + 7]$$
$$= \Delta x^{(3)} + 3\Delta x^{(2)} + 4x^{(1)} + \Delta(7) = 3x^{(2)} + 3.2x^{(1)} + 4 + 0 = 3x^{(2)} + 6x^{(1)} + 4$$
$$\Delta^2 f(x) = \Delta\,[3x^{(2)} + 6x^{(1)} + 4]$$
$$= 3\Delta x^{(2)} + 6\Delta x^{(1)} + \Delta(4) = 3.2.x^{(1)} + 6.1 + 0 = 6x^{(1)} + 6$$
$$\Delta^3 f(x) = \Delta\,[6x^{(1)} + 6] = 6$$

Example 4.4

Determine the function whose first difference is $9x^2 + 11x + 5$.

Solution Let $f(x)$ be the required function. We have

$$\Delta f(x) = 9x^2 + 11x + 5 = 9x\,(x-1) + Ax + B$$

Put $x = 0$, $5 = B$ and $x = 1$, $9 + 11 + 5 = A + B$

$$\Rightarrow \quad A = 20$$
$$\Delta f(x) = 9x^{(2)} + 20x^{(1)} + 5$$

Integrating we get

$$f(x) = 9\frac{x^{(3)}}{3} + 20\frac{x^{(2)}}{2} + 5x^{(1)} + c$$
$$= 3x(x-1)(x-2) + x(x-1) + 5x + c$$
$$= 3(x^3 - 3x^2 + 2x) + 10(x^2 - x) + 5x + c$$
$$= 3x^3 + x^2 + x + c$$

where c is an arbitrary constant.

4.3.2 The Enlargement or Displacement or Shift Operator E

We introduce a third operator E called the *enlargement or displacement or shift operator* defined by

$$Ef_x = f_{x+h} \tag{4.28}$$

E moves forward a value f_x to the next higher value f_{x+h}. Applying E again

$$E^2 f_x = E(Ef_x) = Ef_{x+h} = f_{x+2h} \tag{4.29}$$

In general

$$E^n f_x = E(E^{n-1} f_x) = Ef_{x+\overline{n-1}h} \tag{4.30}$$

We also define the inverse operator E^{-1} by $E^{-1} f_x = f_{x-h}$, so that $E^{-2} f_x = f_{x-2h}$, *etc.*

The Identity Operator 1

We introduce a fourth operator 1 called *the identity operator*, which is defined by

$$1f_x = f_x \tag{4.31}$$

4.3.3 The Relations Between Δ, ∇, E and 1

Since

$$\Delta f_x = f_{x+h} - f_x = E.f_x - 1.f_x \quad \text{or}$$
$$Ef_x = 1f_x + \Delta f_x = (1 + \Delta)f_x \tag{4.32}$$

Thus, E and $1 + \Delta$ when applied on a function f_x produce the same result. Hence, they may be treated as equivalent operators:

$$E = 1 + \Delta \tag{4.33}$$

Note that they have no existence or meaning as separate entities. Now we find higher differences in terms of E treating 1, Δ and E as mathematical objects obeying all algebraic laws. Formally we have the following results:

$$\Delta f_x = f_{x+h} - f_x \tag{4.34}$$
$$\Delta^2 f_x = \Delta f_{x+h} - \Delta f_x = (f_{x+2h} - f_{x+h}) - (f_{x+h} - f_x)$$
$$= f_{x+2h} - 2f_{x+h} + f_x = E^2 f_x - 2Ef_x + 1f_x$$
$$= (E-1)^2 f_x \tag{4.35}$$

Similarly $\quad \Delta^3 f_x = (E-1)^3 f_x,\ \Delta^4 f_x = (E-1)^4 f_x$, *etc.*

Conversely $f_{x+3h} = E^3 f_x = (1 + \Delta)^3 f_x$
$$= (1 + 3\Delta + 3\Delta^2 + \Delta^3)f_x$$
$$= 1f_x + 3\Delta f_x + 3\Delta^2 f_x + \Delta^3 f_x \tag{4.36}$$

Expanding formally by Binomial Theorem we obtain

$$f_{x+nh} = E^n f_x = (1+\Delta)^n f_x = \sum_{r=0}^{\infty}\binom{n}{r}\Delta^r f_x \tag{4.37}$$

Note 1 Formula (4.37) holds not only for positive integral values of n but for all values of n provided the RHS series converges.

Example 4.5

Prove that $\nabla = 1 - E^{-1} = \dfrac{E-1}{E}$.

Solution

$$\nabla f_x = f_x - f_{x-h} = 1f_x - E^{-1}f_x$$
$$= (1 - E^{-1})f_x, \quad \Rightarrow \quad \nabla = 1 - E^{-1}$$

Also
$$\nabla f_x = Ef_{x-h} - 1f_{x-h} = (E-1)f_{x-h}$$
$$= (E-1)\frac{1}{E}f_x \quad \Rightarrow \quad \nabla = \frac{E-1}{E}$$

Example 4.6

Prove that $\nabla E = E\nabla = \Delta$.

Solution

$$\nabla Ef_x = \nabla f_{x+h} = f_{x+h} - f_x = \Delta f_x$$
$$E\nabla f_x = E(f_x - f_{x-h}) = Ef_x - Ef_{x-h} = f_{x-h} - f_x$$
$$= \Delta f_x \quad \Rightarrow \quad \nabla E = E\nabla = \Delta$$

Example 4.7

Prove that $EE^{-1} = E^{-1}E = 1$.

Solution

$$1.f_x = Ef_{x-h} = E(E^{-1}f_x) = (EE^{-1})f_x$$
$$= E^{-1}f_{x+h} = E^{-1}(Ef_x) = (E^{-1}E)f_x$$
$$\Rightarrow \quad EE^{-1} = E^{-1}E = 1$$

Example 4.8

Prove that $(1+\Delta)(1-\nabla) = 1$.

Solution

$$Ef_x = f_{x+h} = f_x + (f_{x+h} - f_x) = 1f_x + \Delta f_x = (1+\Delta)f_x$$
$$E^{-1}f_x = f_{x-h} = f_x - (f_x - f_{x-h}) = 1f_x - \nabla f_x = (1-\nabla)f_x$$
$$\therefore \quad 1f_x = EE^{-1}f_x = (1+\Delta)(1-\nabla)f_x$$
$$\Rightarrow 1 = (1+\Delta)(1-\nabla)$$

Example 4.9

Prove that $\Delta\nabla = \Delta - \nabla$.

Solution

Applying Δ both sides of $\nabla f_x = f_x - f_{x-h}$ we have

$$\Delta\nabla f_x = \Delta f_x - \Delta f_{x-h} = f_{x+h} - f_x - f_x + f_{x-h}$$
$$= \Delta f_x - \nabla f_x = (\Delta - \nabla)f_x \quad \Rightarrow \quad \Delta\nabla = \Delta - \nabla$$

Example 4.10

Find the next two terms of the sequence $u_0 = 5$, $u_1 = 11$, $u_2 = 22$, $u_3 = 40$, $u_4 = 40$, $u_5 = 140$, $u_6 = 261$, $u_7 = 467$, given that the general term u_x is represented by the fourth degree polynomial in n.

Solution

Since u_x is a polynomial of fourth degree, $\Delta^4 u_x = 7$ (constant) as seen from column 6 of Table 4.4. Adding 7 to 30 in column 5 we obtain 37; 85 + 37 = 142; 206 + 142 = 348; 467 + 348 = 815. We extend the table by working out figures like this. Similarly for the next row.

Table 4.4 Forward Difference Table

x	u_x	Δu_x	$\Delta^2 u_x$	$\Delta^3 u_x$	$\Delta^4 u_x$
1	2	3	4	5	6
0	5				
		6	5		
1	11				
		11		2	
			7		7
2	22			9	
		18	16		7
3	40			16	
		34	32		7
4	74			23	
		66	55		7
5	140			30	
		121	85		7
6	261			37	
		206	142		7
7	467			44	
		348	186		
8	815				
		534			
9	1349				

Example 4.11

Find $\Delta^3(1-ax)(1-bx^2)$.

Solution

$$\Delta^3(1-ax)(1-bx^2) = \Delta^3(1 - ax - bx^2 + abx^3)$$
$$= \Delta^3(1) - a\Delta^3(x) - b\Delta^3(x^2) + ab\Delta^3(x^3)$$
$$= 0 - a.0 - b.0 + ab.3! = 6ab$$

Example 4.12

Show that $\Delta^n(e^{ax+b}) = (e^{ah} - 1)^n e^{ax+b}$.

Solution We prove the result by induction

For $n = 1$, $\Delta^1(e^{ax+b}) = e^{a(x+h)+b} - e^{ax+b}$ (1)

$= e^{ax+b}(e^{ah} - 1)$ (2)

It is true.

Assume that the result holds true for $n = m$ so that $\Delta^m (e^{ax+b}) = (e^{ah} - 1)^m e^{ax+b}$ (3)

$$\Delta^{m+1} (e^{ax+b}) = \Delta[\Delta^m e^{ax+b}] = \Delta[(e^{ah} - 1)^m e^{ax+b}] \text{ by (3)}$$
$$= (e^{ah} - 1)^m [\Delta e^{ax+b}] = (e^{ah} - 1)^m$$
$$(e^{ah} - 1) e^{ax+b} = (e^{ah} - 1)^{m+1} e^{ax+b}$$

By the principle of mathematical induction, the result is true for all $n \in N$.

Example 4.13

Show that $\left(\frac{\Delta^2}{E}\right)x^3 = 6x.$

Solution

$$\left(\frac{\Delta^2}{E}\right)x^3 = \frac{(E-1)^2}{E}x^3 = \left(\frac{E^2-2E+1}{E}\right)x^3$$
$$= (E-2+E^{-1})x^3 = Ex^3 + E^{-1}x^3 - 2x^3$$
$$= (x+1)^3 + (x-1)^3 - 2x^3$$
$$= 2(x^3+3x) - 2x^3 = 6x$$

Example 4.14

Prove that $u_1x + u_2x^2 + u_3x^3 + \cdots = \frac{x}{1-x}u_1 + \left(\frac{x}{1-x}\right)^2 \Delta u_1 + \left(\frac{x}{1-x}\right)^3 \Delta^2 u_1 + \cdots$

Solution

$$\text{LHS} = xu_1 + x^2Eu_1 + x^3E^2u_1 + \cdots$$
$$= x\left[1+(xE)+(xE)^2+\cdots\right]u_1$$
$$= x(1-xE)^{-1}u_1 = \frac{x}{1-xE}u_1 = \frac{x}{1-x-x\Delta}u_1$$
$(\because E = 1+\Delta)$
$$= \left(\frac{\frac{x}{1-x}}{1-\frac{x}{1-x}\Delta}\right)u_1 = \frac{x}{1-x}\left(1-\frac{x}{1-x}\Delta\right)^{-1}$$
$$= \frac{x}{1-x}\left[\left(1+\frac{x}{1-x}\Delta+\frac{x^2\Delta^2}{(1-x)^2}+\cdots\right)\right]u_1$$
$$= \frac{x}{1-x}u_1 + \left(\frac{x}{1-x}\right)^2\Delta u_1 + \left(\frac{x}{1-x}\right)^3\Delta^2 u_1 + \cdots$$

Example 4.15

If $f(x) = \frac{2x+1}{x(x+1)}$ find $\Delta f(x)$ and $\Delta^2 f(x)$.

Solution Putting the rational function $f(x) = \frac{2x+1}{x(x+1)}$ into partial fractions we have

$$f(x) = \frac{2x+1}{x(x+1)} = \frac{1}{x} + \frac{1}{x+1} \tag{1}$$
$$\Delta f(x) = \Delta\left(\frac{1}{x}+\frac{1}{x+1}\right) = \Delta\left(\frac{1}{x}\right) + \Delta\left(\frac{1}{x+1}\right)$$
$$= \left(\frac{1}{x+1}-\frac{1}{x}\right) + \left(\frac{1}{x+2}-\frac{1}{x+1}\right) = \frac{1}{x+2}-\frac{1}{x}$$
$$= \frac{-2}{x(x+2)} \tag{2}$$
$$\Delta^2 f(x) = \Delta\left[\frac{-2}{x(x+2)}\right] = -2.\Delta\left(\frac{1}{x(x+2)}\right)$$
$$= -2.\left[\frac{1}{(x+1)(x+3)} - \frac{1}{x(x+2)}\right]$$
$$= -2\frac{x^2+2x-(x^2+4x+3)}{x(x+1)(x+2)(x+3)}$$
$$= \frac{4x+6}{x(x+1)(x+2)(x+3)} \tag{3}$$

Example 4.16

If $f(x) = \frac{1}{x^2+5x+6}$ find $\Delta f(x)$ and $\Delta^2 f(x)$.

Solution

$$\Delta f(x) = \Delta\left(\frac{1}{x^2+5x+6}\right) = \Delta\left(\frac{1}{x+2}\right) - \Delta\left(\frac{1}{x+3}\right)$$
$$= \left(\frac{1}{x+3}-\frac{1}{x+2}\right) - \left(\frac{1}{x+4}-\frac{1}{x+3}\right)$$
$$= \frac{-1}{(x+2)(x+3)} + \frac{1}{(x+3)(x+4)}$$
$$\Delta^2 f(x) = (-1)\Delta\left[\frac{1}{(x+2)(x+3)}\right] + \Delta\left[\frac{1}{(x+3)(x+4)}\right]$$
$$= (-1)\left[\frac{1}{(x+3)(x+4)} - \frac{1}{(x+2)(x+3)}\right]$$
$$+ \left[\frac{1}{(x+4)(x+5)} - \frac{1}{(x+3)(x+4)}\right]$$

$$= \frac{(-1)(-2)}{(x+2)(x+3)(x+4)} + \frac{-2}{(x+3)(x+4)(x+5)}$$

$$= 2\left[\frac{x+5-x-2}{(x+2)(x+3)(x+4)(x+5)}\right]$$

$$= \frac{6}{(x+2)(x+3)(x+4)(x+5)}$$

EXERCISE 4.1

1. Find $\Delta^3(1-ax)(1-bx)(1-cx)$.

   Ans: $-6abc$

2. Prove that $\Delta(u_x v_x) = u_x \Delta v_x + v_{x+1}\Delta u_x = u_{x+1}\Delta v_x + v_x \Delta u_x$.

3. Show that $\Delta\left(\frac{1}{f(x)}\right) = \frac{-\Delta f(x)}{f(x)f(x+1)}$.

6. Show that $\Delta \log f(x) = \log\left[1 + \frac{\Delta f(x)}{f(x)}\right]$.

5. Show that $\Delta^2\left(\frac{1}{x^2+5x+6}\right) = \frac{6}{(x+2)(x+3)(x+4)(x+5)}$.

4. Prove that $\Delta^n\left(\frac{1}{x}\right) = \frac{(-1)^n n!}{x^{(n+1)}}$.

7. Prove that $u_0 + \frac{u_1 x}{1!} + \frac{u_2 x^2}{2!} + \frac{u_3 x^3}{3!} + \cdots = e^x\left(u_0 + x\Delta u_0 + \frac{x^2}{2!}\Delta^2 u_0 + \cdots\right)$.

8. Prove that $u_{x+n} = u_n + \binom{x}{1}\Delta u_{n-1} + \binom{x+1}{2}\Delta^2 u_{n-2} + \binom{x+2}{3}\Delta^3 u_{n-3} + \cdots$

   Hint: $u_{x+n} = E^x u_n = (E^{-1})^{-x} u_n = (1-\Delta E^{-1})^{-x} u_n$

9. Prove that $\binom{n}{0}x - \binom{n}{1}(x-1) + \binom{n}{2}(x-2) - \binom{n}{3}(x-3) + \cdots = 0$.

10. Prove that $u_1 + u_2 + \cdots u_n = \binom{n}{1}u_1 + \binom{n}{2}\Delta u_1 + \cdots + \binom{n}{n}\Delta^{n-1}u_1$.

4.4 INTERPOLATION

Suppose that a function is defined by a table of values (x_r, y_r), $(r = 0, 1, 2, \ldots, n)$ or is tabulated for a number of equidistant values of the argument; for example, the tabulated values of logarithmic function or trigonometric function. We may have to estimate the value of the function at a point not coinciding with the given points. The process of estimating the value of y for an x belonging to the range $[x_0, x_n]$ is known as *interpolation.*

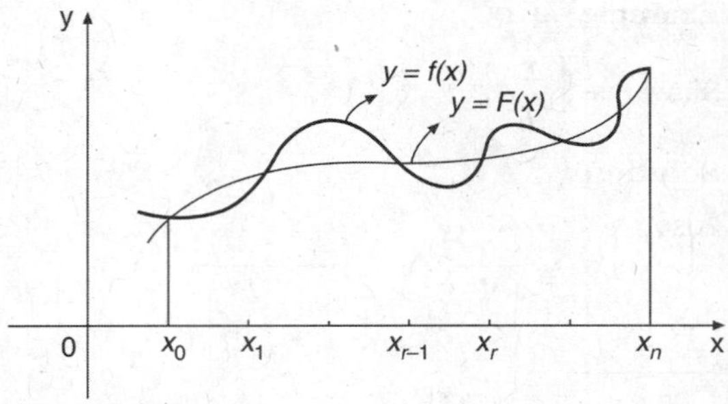

Figure 4.1 Graph of an interpolating curve

Here the points $x_0, x_1, \ldots, x_n$ are called the interpolation points (Figure 4.1).

The process of estimating the value of y for an x outside the range $[x_0, x_n]$ is known as *extrapolation.* Interpolation is to construct a new function $F(x)$ which coincides with the known function $f(x)$ at the tabulated $(n+1)$ interpolation points.

Geometrically this means that the graphs of $y = f(x)$ and $y = F(x)$ coincide at the $(n+1)$ points. We know that a unique straight line can be drawn through two given points, a unique parabola through three given points and a unique nth degree polynomial curve through $(n+1)$ points, the process of which is called interpolation. The interpolating polynomial will be expressed in terms of finite differences which lead to Newton–Gregory forward and backward difference formulas; Central difference formulas of Gauss, Stirling, Bessel, Everette and also Lagrange's and Newton's divided difference formulas.

4.4.1 The Differential Operator *D*: Relation Between Δ, ∇, *E* and *D*

If $D^n y_x = \frac{d^n y_x}{dx^n}$ is the nth derivative of y_x, we have by Taylor's theorem

$$Ey_x = y_{x+h} = y_x + hDy_x + \frac{h^2}{2!}D^2 y_x + \cdots$$
$$+\frac{h^n}{n!}D^n y_n + \cdots = \left(1 + \frac{hD}{1!} + \frac{h^2D^2}{2!} + \cdots + \frac{h^n D^n}{n!} + \cdots\right) y_x = e^{hD} y_x \quad (4.38)$$

where h is the interval of differencing.

Hence we have the symbolic equivalence

$$E = 1 + \Delta = e^{hD}$$
$$\Rightarrow \quad hD = \log E = \log(1+\Delta) = \Delta - \frac{1}{2}\Delta^2 + \frac{1}{3}\Delta^3 - \cdots \quad (4.39)$$

Also $e^{-hD} = E^{-1} = 1 - \nabla \Rightarrow hD = -\log(1-\nabla)$

$$= \nabla + \frac{1}{2}\nabla^2 + \frac{1}{3}\nabla^3 + \cdots \quad (4.40)$$

4.5 Interpolation Formulas for Equal Intervals

4.5.1 Newton[1]–Gregory[2] Forward Interpolation Formula

Let the values of $y_r = f(x_r)$ be given for equally spaced values of the independent variable

$$x_r = x_0 + rh, \qquad r = 0, 1, 2, \ldots, n \quad (4.41)$$

Suppose that the nth degree interpolating polynomial is

$$\phi(x) = a_0 + a_1(x - x_0) + a_2(x - x_0)(x - x_1) + \cdots + a_n(x - x_0)(x - x_1)\cdots(x - x_{n-1}) \quad (4.42)$$

Since the curve $y = \phi(x)$ passes through the points $[x_r, y_r]$, $r = 0 \cdots n$, we get the following equations for determining the constants $a_0, a_1, \ldots, a_n$.

$$\left.\begin{aligned}
&x = x_0 \Rightarrow y_0 = \phi(x_0) = a_0 \Rightarrow a_0 = y_0\\
&x = x_1 \Rightarrow y_1 = \phi(x_1) = a_0 + a_1(x_1 - x_0) \Rightarrow a_1 = \frac{y_1 - y_0}{x_1 - x_0} = \frac{1}{h}\Delta y_0 \because x_1 - x_0 = h\\
&x = x_2 \Rightarrow y_2 = \phi(x_2) = a_0 + a_1(x_1 - x_0) + a_2(x_2 - x_0)(x_2 - x_1)\\
&= a_0 + a_1.2h + a_2.2h.h \Rightarrow a_2 = \frac{y_2 - 2y_1 + y_0}{2h^2} = \frac{1}{2h^2}\Delta^2 y_0\\
&x = x_3 \Rightarrow y_3 = \phi(x_3) = a_0 + a_1(x_3 - x_0) + a_2(x_3 - x_0)(x_3 - x_1)\\
&\qquad + a_3(x_3 - x_0)(x_3 - x_1)(x_3 - x_2) = a_0 + a_1.3h + a_2.3h.2h + a_3.3h.2h.h\\
&\Rightarrow a_3 = \frac{y_3 - 3y_2 + 3y_1 - y_0}{3!h^3} = \frac{1}{3!h^3}\Delta^3 y_0
\end{aligned}\right] \quad (4.43)$$

Generally $a_r = \dfrac{1}{r!h^r}\Delta^r y_0 \qquad r = 0, 1, 2, \ldots, n \quad (4.44)$

Substituting these values of a_r in (4.42) we get the Newton–Gregory forward interpolation formula

$$y_x = \phi(x) = y_0 + \frac{(x - x_0)}{h}\Delta y_0 + \frac{(x - x_0)(x - x_1)}{2!h^2}\Delta^2 y_0 + \frac{(x - x_0)(x - x_1)(x - x_2)}{3!h^3}\Delta^3 y_0 + \cdots + \frac{(x - x_0)(x - x_1)(x - x_2)\cdots(x - x_{n-1})}{n!h^n}\Delta^n y_0 \quad (4.45)$$

[1] NEWTON, Sir Isaac (1642–1727), the great English mathematician, physicist and astronomer. He and Leibnitz invented Calculus independently. He became professor at Cambridge in 1699. His mathematical principles of natural philosophy was completed in (1687).

[2] GREGORY, James (1638–1675), was a Scottish astronomer, algebraist and analyst.

Putting $x = x_0 + ph, y(x_0) = y_0$

formula (4.45) can be put in a simpler form as

$$E^p y_0 = y_p = \phi(x_0 + ph) = y_0 + \frac{p}{1!}\Delta y_0 + \frac{p(p-1)}{2!}\Delta^2 y_0 + \frac{p(p-1)(p-2)}{3!}\Delta^3 y_0 + \cdots + \frac{p(p-1)(p-2)\cdots(p-n+1)}{n!h^n}\Delta^n y_0$$

$$= \sum_{r=0}^{\infty}\binom{p}{r}\Delta^r y_0 \quad (4.46)$$

Newton's formula (4.46) is most suited for interpolation at the beginning of the table and also for extrapolation to the left of y_0.

For $n = 1$ and 2 we get linear and quadratic (parabolic) interpolation formulas as

$$p_1(x) = y_0 + p\Delta y_0, \quad (4.47)$$

$$p_2(x) = y_0 + p\Delta y_0 + \frac{1}{2}p(p-1)\Delta^2 y_0 \quad (4.48)$$

respectively.

4.5.2 Newton–Gregory Backward Interpolation Formula

Assuming the polynomial as

$$y = \phi(x) = a_0 + a_1(x - x_n) + a_2(x - x_n)(x - x_{n-1}) + a_3(x - x_n)(x - x_{n-1})(x - x_{n-2}) + \cdots + a_n(x - x_n)(x - x_{n-1})(x - x_{n-2})\cdots(x - x_1) \quad (4.49)$$

and determining the constants as above we obtain the Newton–Gregory Backward Interpolation formula as

$$y_p = \phi(x_n + ph) = y_n + p\nabla y_n + \frac{p(p+1)}{2!}\nabla^2 y_n + \cdots + \frac{p(p+1)(p+2)\cdots(p+n-1)}{n!}\nabla^n y_n \quad \text{or}$$

$$E^p y_n = (E^{-1})^{-p} y_n = (1-\nabla)^{-p} y_n = \sum_{n=0}^{n}\frac{p(p+1)\cdots(p+r-1)}{r!}\nabla^r y_n \quad (4.50)$$

This formula is most suited for interpolation at the end of the table and also for extrapolation to the right of y_n.

Example 4.17

Find the cubic polynomial y_x which takes on the values

$y_0 = -5, \quad y_1 = 1, \quad y_2 = 9, \quad y_3 = 25, \quad y_4 = 55, \quad y_5 = 105$

Use the difference table to calculate $y_{3.2}$.

Solution Constructing the difference table for the function we have

x	y_x	Δy_x	$\Delta^2 y_x$	$\Delta^3 y_x$
0	–5			
		6		
1	1		2	
		8		6
2	9		8	
		16		6
3	25		14	
		30		6
4	55		20	
		50		
5	105			

$$y_x = E^x y_0 = (1+\Delta)^x y_0 = \left[1 + x\Delta + \frac{x(x-1)}{2!}\Delta^2 + \frac{x(x-1)(x-2)}{3!}\Delta^3\right]y_0$$

$$= y_0 + x\Delta y_0 + \frac{x_2 - x}{2}\Delta^2 y_0 + \frac{x_3 - 3x_2 + 2}{6}\Delta^3 y_0$$

Substituting $y_0 = -5, \Delta y_0 = 6, \Delta^2 y_0 = 2, \Delta^3 y_0 = 6$, we obtain

$$y_x = x^3 - 2x^2 + 7x - 5$$

Interpolating to obtain $y_{3.2}$ we find

$$y_{3.2} = E^{0.2}y_3 = (1 + \Delta)^{0.2}y_3 = (1 + 0.2\Delta - 0.08\Delta^2 + 0.048\Delta^3)y_3$$

$$= y_3 + (0.2)\Delta y_3 - (0.08)\Delta^2 y_3 + (0.048)$$

$$\Delta^3 y_3 = 25 + (0.2)30 - (0.08)20 + (0.048)6 = 29.688$$

Example 4.18

The population of a country in the decennial census is as shown below. Estimate the population for the year 1925.

Year	x	1891	1901	1911	1921	1931
Population (in 1000s)	y	46	66	81	93	101

Solution Constructing the difference table, we have

x	y	∇y	$\nabla^2 y$	$\nabla^3 y$	$\nabla^4 y$
1891	46				
		20			
1901	66		−5		
		15		2	
1911	81		−3		−3
		12		−1	
1921	93		−4		
		8			
1931	101				

Since five values are given we must assume that the fourth differences are constant. We require the entry for $x = 1925$.

We use Newton–Gregory Backward Interpolation formula (4.50).

Here $p = \dfrac{x - x_n}{h} = \dfrac{1925 - 1931}{10} = -0.6$

$$\therefore\ y_p = 101 + (-0.6)(8) + \frac{(-0.6)(0.4)(-4)}{2} + \frac{(-0.6)(0.4)(1.4)(-1)}{6} + \frac{(-0.6)(0.4)(1.4)(2.4)(-3)}{24} \cdots$$

$$= 101 - 4.8 + 0.48 + 0.056 - 0.1008 = 96.8368$$

Hence, the population for 1925 is estimated at 96.84 thousands.

Example 4.19

Given log 100 = 2, log 101 = 2.0043, log 103 = 2.0128, log 104 = 2.0170, find log 102.

Solution Let the missing value be y_2. Then the difference table is

x	$y = \log x$	Δy	$\Delta^2 y$	$\Delta^3 y$	$\Delta^4 y$
100	2.0000				
		0.0043			
101	2.0043		$y_2 - 2.0086$		
		$y_2 - 2.0043$		$6.0257 - 3y_2$	
102	y_2		$4.0171 - 2y_2$		$6y_2 - 12.0514$
		$2.0128 - y_2$		$3y_2 - 6.0257$	
103	2.0128		$y_2 - 2.0086$		
		0.0042			
104	2.0170				

Since, by assumption, fourth differences must be zero we have $6y_2 = 12.0514$

$$\Rightarrow \quad \log 102 = \frac{12.0514}{6} = 2.0086$$

Example 4.20

Find the missing values in the table

x	45	50	55	60	65
y	3.0	–	2.0	–	−2.4

Solution Since only three entries $y_0 = 3$, $y_2 = 2$ and $y_4 = -2.4$ are given, the function y can be represented by a second degree polynomial

$$\Rightarrow \Delta^3 y_0 = 0;\ \Delta^3 y_1 = 0 \Rightarrow (E-1)^3 y_0;\ (E-1)^3 y_1 = 0$$

$$\Rightarrow (E^3 - 3E^2 + 3E - 1)\, y_0 = 0,\ (E^3 - 3E^2 + 3E - 1)\, y_1 = 0$$

$$\Rightarrow y_3 - 3y_2 + 3y_1 - y_0 = 0;\ y_4 - 3y_3 + 3y_2 - y_1 = 0$$

$$\Rightarrow y_3 + 3y_1 = 3y_2 + y_0 = 9;\ 3y_3 + y_1 = y_4 + 3y_2 = 3.6,$$

substituting for y_0, y_2 and y_4. Solving the pair of equations, we obtain the missing values $y_1 = 2.925$ and $y_3 = 0.225$.

Example 4.21

From the table, estimate the number of students who obtained marks between 40 and 45.

Marks	30–40	40–50	50–60	60–70	70–80
No. of students	31	42	51	35	31

Solution First we have to prepare a cumulative frequency table for the data.

Marks less than	x	40	50	60	70	80
No. of students	y_x	31	73	124	159	190

We now construct a difference table for the data.

x	y	Δy_x	$\Delta^2 y_x$	$\Delta^3 y_x$	$\Delta^4 y_x$
40	31				
		42			
50	73		9		
		51		−25	
60	124		−16		37
		35		12	
70	159		−4		
		31			
80	190				

Now we find y_{45}, the number of students with marks less than 45 by taking $x_0 = 40$, $x = 45$, $h = 10$ so that $p = \dfrac{x - x_0}{h} = \dfrac{45-40}{10} = 0.5.$

Using Newton–Gregory forward interpolation formula (4.46)

$$y_{0.5} = 31 + 0.5 \times 42 + \frac{0.5(-0.5)}{2} \times 9 + \frac{0.5(-0.5)(-1.5)}{6}(-25) + \frac{0.5(-0.5)(-1.5)(-2.5)}{24} \times 37 = 47.87$$

∴ The number of students with marks less than 45 is 48 (rounded figure of 47.87).

But the number of students with marks less than 40 is 31. Hence, the number of students who get marks between 40 and 45 = 48 – 31 = 17.

Example 4.22

Given sin 45° = 0.7071, sin 50° = 0.7660, sin 55° = 0.8192, sin 60° = 0.8660 find sin 52° using Newton's forward interpolation formula.

Solution Let x denote the angle in degrees and let $y = f(x) = \sin x$.

Here $x_0 = 45$, $h = 5$, $x = x_0 + ph = 52$ is nearer the starting of the table.

$$p = \frac{x - x_0}{h} = \frac{52-45}{5} = 1.4$$

Constructing the forward difference table we have

x	y	Δy	$\Delta^2 y$	$\Delta^3 y$
45	0.7071			
		0.0509		
50	0.7660		−0.0057	
		0.0532		−0.0007
55	0.8192		−0.0064	
		0.0468		
60	0.8660			

Using Newton–Gregory forward interpolation formula (4.46)

$$y_{1.4} = 0.7071 + (1.4)(0.0589) + \frac{(1.4)(0.4)}{2}(-0.0057) + \frac{(1.4)(0.4)(-0.6)}{6}(-0.0007)$$

$$= 0.7071 + 0.08246 - 0.00160 + 0.0004$$

$$= 0.7880 \text{ approx.}$$

∴ sin 52° = 0.7880, correct to four decimal places.

Example 4.23

The following data give the melting point of an alloy of lead and zinc where t is the temperature in °C, and x is the percentage of lead in the alloy.

(x%)	60	70	80	90
t	226	250	276	304

Find the melting point of the alloy containing 84% of lead using Newton's interpolation formula.

Solution Here x = 84 is at the end of the table. We write the backward difference table with $x_n = 90$, $h = 10$, $p = \dfrac{x - x_n}{h} = \dfrac{84-90}{10} = -0.6$

x	$t_{(y)}$	Δt	$\Delta^2 t$	$\Delta^3 t$
60	226			
		24		
70	250		2	
		26		0
80	276		2	
		28		
90	304			

$$y_{-0.6} = t_0 + p\nabla t_0 + \frac{p(p+1)}{2}\nabla^2 t_0$$

$$\Rightarrow \quad f(84) = 304 - 0.6(28) + \frac{-0.6(0.4)}{2}(2)$$

$$= 304 - 16.8 - 0.24 = 286.96$$

EXERCISE 4.2

1. Find the missing term in the table

x	16	18	20	22	24	26
y	39	85	–	151	264	388

Ans: 96.4

2. The values of $y_x = \frac{1}{\sqrt{2\pi}}\int_{-\infty}^{x} e^{-t^2/2}dt$ are tabulated below for $x = 0.30\ (0.02)0.40$. Evaluate $y_{0.316}$.

x	0.30	0.32	0.34	0.36	0.38	0.40
y_x	0.61791144	0.6255158	0.6330717	0.6405764	0.6480273	0.6554217

Ans: 0.6230087

Note 0.30 (0.02) 0.40 means that x ranges from 0.30 to 0.40 with an increment of 0.02 each time.

3. Find $f(1.28)$ if $f(1.15) = 1.0723, f(1.20) = 1.0954, f(1.25) = 1.1180$ and $f(1.30) = 1.1401$. [Bangalore 1993]

Ans: 1.1312

4. The tabulated values of the elliptic integral $y(x) = \int_0^x \left(1 - \frac{1}{2}\sin^2 t\right)^{-1/2} dt$ for $x = 21(1)\ 26$ are aș follows. Find $y\ (23.5)$ by Newton–Gregory forward and backward formulas.

x	21	22	23	24	25	26
y_x	0.3706	0.3887	0.4068	0.4250	0.4433	0.4616

Ans: 0.4159, 0.41588

5. Find $\log_{10} 1044$ using Newton's backward difference formula given the data

x	1000	1010	1020	1030	1040	1050
$\log_{10} x = y_x$	3	3.0043214	3.0086002	3.0128372	3.017033	3.601193

Ans: 3.0187005

6. Fit a cubic polynomial which takes the values

x	3	4	5	6
y	6	24	60	120

Ans: $y(x) = x^3 - 3x^2 + 2x$

7. Find the fourth degree polynomial which fits the table of values

x	0	1	2	3	4	5
y	1	–1	5	55	209	551

Ans: $y(x) = x^4 - 3x^2 + 1$

8. Using the quadratic interpolation find y (2.72) for $f(x) = \frac{1}{x}$ with the data

x	2.7	2.8	2.9
y	0.3704	0.3571	0.3448

Ans: 0.3677

4.6 Interpolation with Unequal Intervals

So far we have considered interpolation formulas for equally spaced values of x. We now develop two interpolation formulas for unequally spaced values of x.

4.4.1 Divided Differences

Suppose that the function y_x is given for values $x = a, b, c, d, \ldots$ where the intervals $b - a$, $c - b$, $d - c$ are not necessarily equal. We define the divided difference of y_a at b by

$$\underset{b}{\Delta} y_a = \frac{y_b - y_a}{b - a} \tag{4.51}$$

Then the difference table is written as follows

x	y_x	$\Delta' y_x$	$\Delta'^2 y_x$	$\Delta'^3 y_x$
a	y_a			
		$\underset{b}{\Delta} y_a$		
b	y_b		$\underset{bc}{\Delta^2} y_a$	
		$\underset{c}{\Delta} y_b$		$\underset{bcd}{\Delta^3} y_a$
c	y_c		$\underset{cd}{\Delta^2} y_b$	
		$\underset{d}{\Delta} y_c$		
d	y_d			

(4.52)

Also, second, third and higher divided differences in terms of functional values

$$\underset{bc}{\Delta^2} y_a = \frac{\underset{c}{\Delta} y_b - \underset{b}{\Delta} y_a}{c - a} = \frac{\frac{y_c - y_b}{c - b} - \frac{y_b - y_a}{b - a}}{c - a}$$

$$= \frac{y_a}{(a-b)(a-c)} + \frac{y_b}{(b-c)(b-a)} + \frac{y_c}{(c-a)(c-b)} \tag{4.53}$$

and generally we can prove Theorem 4.2 by induction.

Theorem 4.2 If $a, b, c, \ldots$ are values of the argument x then

$$\underset{bc\cdots jk}{\Delta^r} y_a = \frac{y_a}{(a-b)(a-c)\cdots(a-k)} + \cdots + \frac{y_k}{(k-a)(k-b)\cdots(k-j)} \tag{4.54}$$

Example 4.24

Construct a divided difference table for the data $y_{-2} = 5, y_0 = 3, y_3 = 15, y_4 = 47, y_9 = 687$.

x	y_x	$\Delta' y_x$	$\Delta'^2 y_x$	$\Delta'^3 y_x$
−2	5			
		−1		
0	3		+1	
		+4		+1
3	15		+7	
		+32		+1
4	47		+6	
		+128		
9	687			

It may be noted that the triangular arrangement of a divided difference table is most convenient. We have

$$\underset{3,4}{\Delta^2} y_0 = \frac{32 - 4}{4 - 0}$$

and the triangle drawn points to the proper divider $4 - 0$.

4.4.2 Divided Differences: Notation

There are several notations used for dealing with divided differences. Each has certain advantages and certain disadvantages. The above notation is one. Another notation will be given below, and will be used where it is convenient.

Let the function $y = f(x)$ consist of tabulated values $[x_r, y_r], \ldots, [x_n, y_n] \ldots$ with unequal intervals $x_{r+1} - x_r \neq 0$ for $r = 0, 1, 2, \ldots$. Then the first order divided difference denoted by $[x_r, x_{r+1}]$ is the ratio given by

$$[x_r, x_{r+1}] = \frac{y_{r+1} - y_r}{x_{r+1} - x_r} \qquad r = 0, 1, 2, \ldots \tag{4.55}$$

In particular, first order divided difference

$$[x_0, x_1] = \frac{y_1 - y_0}{x_1 - x_0}, \; [x_1, x_2] = \frac{y_2 - y_1}{x_2 - x_1}, \; etc. \tag{4.56}$$

second order divided difference

$$[x_0, x_1, x_2] = \frac{[x_1, x_2] - [x_0, x_1]}{x_2 - x_0} \tag{4.57}$$

nth order divided difference

$$[x_r, x_{r+1}, \ldots, x_{r+n}] = \frac{[x_{r+1}, x_{r+2} \ldots x_{r+n}] - [x_r, x_{r+1}, \ldots x_{r+n-1}]}{x_{n+r-x_r}}$$

$$n = 1, 2, 3, \ldots, \quad r = 0, 1, 2, \ldots \tag{4.58}$$

4.7 Properties Satisfied by Δ'

4.7.1 Linearity Property

The divided difference operator Δ' is linear (a) $\Delta'[f(x)+g(x)] = \Delta' f(x) + \Delta' g(x)$; (b) $\Delta'[cf(x)] = c\Delta'[f(x)]$.

4.7.2 Symmetrical Property

The divided differences are symmetrical in the symbols.

$$[x_0, x_1] = \frac{y_1 - y_0}{x_1 - x_0} = \frac{y_0}{x_0 - x_1} + \frac{y_1}{x_1 - x_0}$$

$$= \frac{y_0 - y_1}{x_0 - x_1} = [x_1, x_0] \tag{4.59}$$

$$[x_0, x_1, x_2] = \frac{[x_1, x_2] - [x_0, x_1]}{x_2 - x_0}$$

$$= \frac{1}{x_2 - x_0}\left[\frac{y_1}{x_1 - x_2} + \frac{y_2}{x_2 - x_1} - \frac{y_0}{x_0 - x_1} - \frac{y_1}{x_1 - x_0}\right]$$

$$= \frac{y_0}{(x_0 - x_1)(x_0 - x_2)} + \frac{y_1}{(x_1 - x_0)(x_1 - x_2)}$$

$$+ \frac{y_2}{(x_2 - x_0)(x_2 - x_1)} = [x_1, x_2, x_0] = [x_2, x_0, x_1] \tag{4.60}$$

4.7.3 Vanishing of (n + 1) Divided Differences

If $y = y(x)$ is an nth degree polynomial then its nth order divided difference $\Delta'^n y_x$ is constant and higher order divided differences vanish.

Theorem 4.3 If y_x is a polynomial of degree n then $\Delta'^n y_x$ is constant.

Proof

$$\underset{x-h}{\Delta}\, x^n = \frac{(x+h)^n - x^n}{x + h - x} = \frac{nhx^{n-1} + \cdots}{n} = \text{a polynomial}$$

of degree $(n-1)$.

Now Δ' is a linear operator. Hence it follows that the first divided difference of $y_x = a_0 + a_1 x + \cdots + a_n x^n$ is a polynomial of degree $(n-1)$; the second divided difference is a polynomial of degree $(n-2)$; the nth divided difference is constant, and all higher divided differences are zero.

4.7.4 Special Case: Equally spaced Divided Differences

If $x_{r+1} - x_r = b$ (constant) then the variable x is equally spaced and the divided differences become ordinary forward (backward) differences,

$$[x_0, x_1] = \frac{y_1 - y_0}{x_1 - x_0} = \frac{1}{h}\Delta y_0$$

$$[x_0, x_1, x_2] = \frac{[x_1, x_2] - [x_0, x_1]}{x_2 - x_0} = \frac{1}{2h}\left(\frac{\Delta y_1}{h} - \frac{\Delta y_0}{h}\right)$$

$$= \frac{1}{h^2 2!}\Delta^2 y_0 \tag{4.61}$$

In general

$$[x_0, x_1, x_2 \ldots x_n] = \frac{1}{h^n n!}\Delta^n y_0 \tag{4.62}$$

Theorem 4.4 Let the divided differences be taken at arguments $x_0, x_1, \ldots, x_n, x_{n+1}$ with $x_{r+1} - x_r = h$ for $r = 0, 1, 2, \ldots, n$. Prove that

$$[x_0, x_1, \ldots, x_n] = \frac{1}{h^n n!}\Delta^n y_0 \tag{4.63}$$

Proof For $n = 1$ we have

$$[x_0, x_1] = \frac{y_1 - y_0}{x_1 - x_0} = \frac{1}{h}\Delta y_0 \tag{4.64}$$

$\therefore$ (4.63) is true for $n = 1$.

Assume that (4.63) is true for $n = m$. Then

$$[x_0, x_1, \ldots, x_m] = \frac{1}{h^m m!}\Delta^m y_0 \tag{4.65}$$

Now $[x_0, x_1, \ldots, x_m, x_{m+1}]$

$$= \frac{[x_1, x_2, \ldots, x_{m+1}] - [x_0, x_1, \ldots, x_m]}{x_m - x_0}$$

$$= \frac{1}{(m+1)h}\left[\frac{\Delta^m y_1}{h^m m!} - \frac{\Delta^m y_0}{h^m m!}\right], \quad \text{by (4.65)}$$

$$= \frac{1}{(m+1)!\,h^{m+1}}\Delta^{m+1} y_0 \tag{4.66}$$

Hence, the theorem holds for all $n \in N$ by the principle of mathematical induction.

4.8 Divided Difference Interpolation Formula

4.8.1 Newton's Divided Difference Formula

Consider the function y_x for the arguments $x, a, b, c, d, \ldots, j, k$. Then

$$\underset{a}{\Delta'} y_x = \underset{x}{\Delta'} y_a = \frac{y_x - y_a}{x-a} \quad \Rightarrow \quad y_x = y_a + (x-a)\underset{x}{\Delta'} y_a \tag{4.67}$$

By the symmetry property

$$\underset{bx}{\Delta'^2} y_a = \underset{ab}{\Delta'^2} y_x = \frac{\underset{a}{\Delta'} y_x - \underset{b}{\Delta'} y_a}{x-b} \tag{4.68}$$

from which on using the symmetry property, we obtain

$$\underset{x}{\Delta'} y_a = \underset{b}{\Delta'} y_a + (x-b)\underset{bx}{\Delta'^2} y_a \tag{4.69}$$

Substituting into (4.67) we get

$$y_x = y_a + (x-a)\underset{b}{\Delta'} y_a + (x-a)(x-b)\underset{bx}{\Delta'^2} y_a \tag{4.70}$$

Finally

$$\underset{bcx}{\Delta'^3} y_a = \underset{abc}{\Delta'^3} y_x = \frac{\underset{ab}{\Delta'^2} y_x - \underset{bc}{\Delta'^3} y_a}{x-c}, \text{ and}$$

$$\underset{bx}{\Delta'^2} y_a = \underset{bc}{\Delta'^2} y_x + (x-c)\underset{bcx}{\Delta'^3} y_a \tag{4.71), (4.72}$$

Substituting these in (4.70) we obtain

$$y_x = y_a + (x-a)\underset{b}{\Delta'} y_a + (x-a)(x-b)\underset{bc}{\Delta'^2} y_a$$
$$+(x-a)(x-b)(x-c)\underset{bcx}{\Delta'^3} y_a \tag{4.73}$$

We continue this process until nth differences are reached. Assuming that y_x is represented by an nth degree polynomial, all higher differences vanish and we have Newton's divided difference formula:

$$y_x = y_a + A\underset{b}{\Delta'} y_a + AB\underset{bc}{\Delta'^2} y_a + ABC\underset{bcd}{\Delta'^3} y_a$$
$$+ \cdots + ABC\cdots J \underset{bc\cdots k}{\Delta'^{\,n}} y_a \tag{4.74}$$

where there are $(n + 1)$ arguments $a, b, c, \ldots, k$ and $A = (x - a), B = (x - b), C = (x - c), \ldots, K = (x - k)$.

Corollary If the arguments $a, b, c, \ldots$ are taken as $0, 1, 2, \ldots$ then $\Delta'^r y_a = \dfrac{\Delta^r y_0}{r!}$ and we obtain from equation (4.74)

$$y_x = y_0 + \binom{x}{1}\Delta y_0 + \binom{x}{2}\Delta^2 y_0 + \cdots + \binom{x}{n}\Delta^n y_0 \tag{4.75}$$

which is Newton's advancing difference formula.

4.8.2 Sheppard's[3] Zig-Zag Rule

Newton's divided difference formula (4.74) can be easily remembered by what is known as Sheppard's zig-zag rule. This will be explained now. Suppose we have a table of divided differences. Table 4.5 illustrates the case of six arguments a, b, c, d, e, f. Let us draw any zig-zag line connecting any argument with the highest difference in the table. We may only zig-zag from difference to an adjacent difference of the next higher order choosing one of the two paths at each stage. The letters a, b, c, d, e, f will appear in the differences along the solid zig-zag line in some order. The path illustrated in Table 4.5 connects points c, d, b, c, f, a.

Table 4.5 Sheppard's Zig-Zag Rule

y_a					
	$\underset{b}{\Delta'} y_a$				
y_b		$\underset{bc}{\Delta'^2} y_a$			
	$\underset{c}{\Delta'} y_b$		$\underset{bcd}{\Delta'^3} y_a$		
y_c		$\underset{cd}{\Delta'^2} y_b$		$\underset{bcde}{\Delta'^4} y_a$	
	$\underset{d}{\Delta'} y_c$		$\underset{cde}{\Delta'^3} y_b$		$\underset{bcdef}{\Delta'^5} y_a$
y_d		$\underset{de}{\Delta'^2} y_c$		$\underset{cdef}{\Delta'^4} y_b$	
	$\underset{e}{\Delta'} y_d$		$\underset{def}{\Delta'^3} y_c$		
y_e		$\underset{ef}{\Delta'^2} y_d$			
	$\underset{f}{\Delta'} y_e$				
y_f					

Sheppard's Rule is as follows. Expand y_x in terms of differences on the zig-zag line using coefficients $1, C, D, B, E, F$. That is, the capital letters appear in the same order as the small letters. Thus, the above diagram represents the expansion

$$y_x = y_c + (x-c)\underset{d}{\Delta'} y_c + (x-c)(x-d)\underset{cd}{\Delta'^2} y_b$$
$$+(x-c)(x-d)(x-b)\underset{cde}{\Delta'^3} y_b$$

[3] SHEPPARD, William Fleetwood (1863–1936), is an English probabilist and statistician.

$$+(x-c)(x-d)(x-b)(x-e)\underset{cdef}{\Delta'^4}\, y_b$$

$$+(x-c)(x-d)(x-b)(x-e)(x-f)\underset{bcdef}{\Delta'^5}\, y_a \tag{4.76}$$

The Newton divided difference formula at (4.74) corresponds to the dotted line in the table.

4.8.3 Lagrange's[4] Formula for Unequal Intervals

Theorem 4.5 If the functional values $f(x_r)$ are known at the arguments x_r $(r = 0, 1, 2, \ldots, n)$ then

$$f(x)=\frac{(x-x_1)(x-x_2)\cdots(x-x_n)}{(x_0-x_1)(x_0-x_2)\cdots(x_0-x_n)}f(x_0)+\frac{(x-x_0)(x-x_2)\cdots(x-x_n)}{(x_1-x_0)(x_1-x_2)\cdots(x_1-x_n)}f(x_1)+\cdots+\frac{(x-x_0)(x-x_1)\cdots(x-x_{n-1})}{(x_n-x_0)(x_n-x_1)\cdots(x_n-x_{n-1})}f(x_n) \tag{4.77}$$

Proof Let $y = f(x)$ be a function which is satisfied by the $(n + 1)$ points $(x_r, f(x_r))$, $(r = 0, 1, 2 \ldots n)$ then $f(x)$ is a polynomial of degree n, which can be assumed in the form

$$y=f(x)=\phi(x)=a_0(x-x_1)(x-x_2)\cdots(x-x_n)+a_1(x-x_0)(x-x_2)\cdots(x-x_n)+\cdots+a_n(x-x_0)(x-x_1)\cdots(x-x_{n-1}) \tag{4.78}$$

$$x=x_0 \Rightarrow y=y_0=f(x_0)=a_0(x_0-x_1)(x_0-x_1)\cdots(x_0-x_n)$$

$$\Rightarrow a_0=\frac{1}{(x_0-x_1)(x_0-x_2)\cdots(x_0-x_n)}f(x_0) \tag{4.79}$$

and, generally

$$a_r=\frac{1}{(x_r-x_0)(x_r-x_1)\cdots(x_r-x_{r-1})(x_r-x_{r+1})\cdots(x_r-x_n)}f(x_r)$$
$$r=0,1,2,\ldots,n \tag{4.80}$$

Substituting the values of the constants in (4.79) we get (4.76).

Note 1 If the points are equally spaced then Lagrange's interpolation polynomial coincides with that of Newton's interpolation.

Note 2 The function $y = f(x)$ can be split into partial fractions by dividing (4.77) by $(x - x_0)(x - x_1)\ldots(x - x_n)$ with $(n + 1)$ factors; thus

$$\frac{f(x)}{(x-x_0)(x-x_1)\cdots(x-x_n)}=\frac{f(x_0)}{(x_0-x_1)(x_0-x_2)\cdots(x_0-x_n)}\cdot\frac{1}{x-x_0}+\frac{f(x_1)}{(x_1-x_0)(x_1-x_2)\cdots(x_1-x_n)}\cdot\frac{1}{x-x_1}+\cdots+\frac{f(x_n)}{(x_n-x_0)(x_n-x_1)\cdots(x_n-x_{n-1})}\cdot\frac{1}{x-x_n} \tag{4.81}$$

Note 3 Lagrange's interpolation formula can be directly obtained by using the property that $(n + 1)$th divided difference of y vanishes, y being a polynomial of degree n.

4.9 Inverse Interpolation Using Lagrange's Interpolation Formula

In the inverse interpolation for a given value of y, the corresponding value of x is to be found. Interchanging the roles of x and $y = f(x)$ in Lagrange's interpolation formula we get Lagrange's inverse interpolation formula as

$$x=\frac{(y-y_1)(y-y_2)\cdots(y-y_n)}{(y_0-y_1)(y_0-y_2)\cdots(y_0-y_n)}x_0+\frac{(y-y_0)(y-y_2)\cdots(y-y_n)}{(y_1-y_0)(y_1-y_2)\cdots(y_1-y_n)}x_1+\cdots+\frac{(y-y_0)(y-y_1)\cdots(y-y_{n-1})}{(y_n-y_0)(y_b-y_1)\cdots(y_n-y_{n-1})}x_n \tag{4.82}$$

Example 4.25

Using Newton's divided difference formula find $f(2), f(8)$ and $f(15)$ from the table:

x	4	5	7	10	11	13
$f(x)$	48	100	204	900	1210	2028

Solution The divided differences table is

---

[4] LAGRANGE, Joseph Louis (1736–1813), the great French analyst, algebraist, number theorist, probablist, physicist and astronomer. His contributions are especially to calculus of variations, analytical mechanics and astronomy. He became professor at the Military Academy, Turin, at 19 and later in 1766 became the Director of Berlin Academy at 30.

x	$f(x)$	$\Delta f(x)$	$\Delta^2 f(x)$	$\Delta^3 f(x)$	$\Delta^4 f(x)$
4	48				
		$\frac{100-48}{5-4}=52$			
5	100		$\frac{97-52}{7-4}=15$		
		$\frac{294-100}{7-5}=97$		$\frac{21-15}{10-4}=1$	
7	294		$\frac{202-97}{10-5}=21$		0
		$\frac{900-294}{10-7}=202$		$\frac{27-21}{11-2}=1$	
10	900		$\frac{310-202}{11-7}=27$		0
		$\frac{1210-900}{11-10}=310$		$\frac{33-27}{13-7}=1$	
11	1210		$\frac{409-310}{13-10}=33$		
		$\frac{2028-1210}{13-11}=409$			
13	2028				

Newton's divided differences formula is

$$y=f(x)=f(x_0)+(x-x_0)\,[x_0,x_1]+(x-x_0)(x-x_1)\cdot[x_0,x_1,x_2]+(x-x_0)(x-x_1)(x-x_2)\cdot[x_0,x_1,x_2]+\cdots=48+52(x-4)+15(x-4)(x-5)+1(x-4)(x-5)(x-7)=x^2(x-1)$$

$$\therefore\quad f(2)=2^2(2-1)=4$$

$$f(8)=8^2(8-1)=448$$

$$f(15)=15^2(15-1)=3150$$

Example 4.26

From the table of values

x	5	7	11	13	17
$f(x)$	150	392	1452	2366	5202

evaluate $f(x)$ using (a) Lagrange's formula and (b) Newton's divided difference formula.

Solution (a) Here $x_0 = 5, x_1 = 7, x_2 = 11, x_3 = 13, x_4 = 17$ and $y_0 = 150, y_1 = 392, y_2 = 1452, y_3 = 2366, y_4 = 5202$. Putting $x = 9$ and substituting the above values in Lagrange's formula we obtain

$$f(9)=\frac{(9-7)(9-11)(9-13)(9-17)}{(5-7)(5-11)(5-13)(5-17)}\times 150+\frac{(9-5)(9-11)(9-13)(9-17)}{(7-5)(7-11)(7-13)(7-17)}\times 392+\frac{(9-5)(9-7)(9-13)(9-17)}{(11-5)(11-7)(11-13)(11-17)}\times 1452+\frac{(9-5)(9-7)(9-11)(9-17)}{(13-5)(13-7)(13-11)(13-17)}\times 2366+\frac{(9-5)(9-7)(9-11)(9-13)}{(17-5)(17-7)(17-11)(17-13)}\times 5202$$

$$=-\frac{50}{3}+\frac{3136}{15}+\frac{3872}{3}-\frac{2366}{3}+\frac{578}{5}=810$$

(b) Building the divided difference table

x	$f(x)$	$\Delta f(x)$	$\Delta^2 f(x)$	$\Delta^3 f(x)$
5	150			
		$\frac{392-150}{7-5}=121$		
7	392		$\frac{265-121}{11-5}=24$	
		$\frac{1452-392}{11-7}=265$		$\frac{32-24}{13-5}=1$
11	1452		$\frac{457-265}{13-7}=32$	
		$\frac{2366-1452}{13-11}=457$		$\frac{42-32}{17-7}=1$
13	2366		$\frac{709-457}{17-11}=42$	
		$\frac{5202-2366}{17-13}=709$		
17	5202			

Substituting the values from the table in Newton's divided difference formula, taking $x = 9$,

$$f(9) = 150 + (9-5)\times 121 + (9-5)(9-7)\times 24 + (9-5)(9-7)(9-11)\times 1$$

$$= 150 + 484 + 192 - 16 = 810$$

Example 4.27

Compute the value of x when $y = 8$ by Lagrange's inverse interpolation formula

x	−2	−1	1	2
y	−7	2	0	11

Solution From the given data we have

$x_0 = -2,\quad x_1 = -1,\quad x_2 = 1,\quad x_3 = 2$

$y_0 = -7,\quad y_1 = 2,\quad y_2 = 0,\quad y_3 = 11$

By Lagrange's inverse interpolation formula (4.82)

$$x = \frac{(y-2)(y-0)(y-11)}{(-7-2)(-7-0)(-7-11)}\times(-2) + \frac{(y+7)(y-0)(y-11)}{(2+7)(2-0)(2-11)}\times(-1) + \frac{(y+7)(y-2)(y-11)}{(0+7)(0-2)(0-11)}\times 1 + \frac{(y+7)(y-2)(y-0)}{(11+7)(11-2)(11-0)}\times 2$$

Putting $y = 8$ we obtain

$$x = \frac{6.8(-3)}{(-9)(-7)(-18)}(-2) + \frac{15.8.(-3)}{9.2.(-9)}(-1) + \frac{15.6.(-3)}{7.(-2).(-11)}(1) + \frac{15.6.8}{18.9.11}(2)$$

$$= -\frac{8}{21} - \frac{20}{9} - \frac{135}{77} + \frac{80}{99} = -3.5483$$

Example 4.28

Find $\log_{10} 656$ given that $\log_{10} 654 = 2.8156$, $\log_{10} 658 = 2.8182$, $\log_{10} 659 = 2.8189$ and $\log_{10} 661 = 2.8202$, using Lagrange's interpolation formula.

Solution Let $y = f(x) = \log_{10} x$ and

$$x_0 = 654, x_1 = 658, x_2 = 659, x_3 = 661, x = 656$$
$$y_0 = f(x_0) = 2.8156, y_1 = f(x_1) = 2.8182, y_2 = f(x_2) = 2.8189 \text{ and } y_3 = f(x_3) = 2.8202$$

Substituting these values in Lagrange's interpolation formula

$$y = f(x) = \log_{10} 656$$

$$= \frac{(656-658)(656-659)(656-661)}{(654-658)(654-659)(654-661)}\times 2.8156 + \frac{(656-654)(656-659)(656-661)}{(658-654)(658-659)(658-661)}\times 2.8182 + \frac{(656-654)(656-658)(656-661)}{(659-654)(659-658)(659-661)}\times 2.8189 + \frac{(656-654)(656-658)(656-659)}{(661-654)(661-658)(661-659)}\times 2.8202$$

$$= \frac{(-2)(-3)(-5)}{(-4)(-5)(-7)}\times 2.8156 + \frac{2\times(-3)(-5)}{4\times(-1)(-3)}\times 2.8182 + \frac{2\times(-2)(-5)}{5\times 1\times(-2)}\times 2.8189 + \frac{2\times(-2)(-3)}{7\times 3\times 2}\times 2.8202$$

$$= 0.6033 + 7.0455 - 5.6378 + 0.8059 = 2.8169$$

Example 4.29

The mode of a frequency curve $y = f(x)$ is very near $x = 9$ and the value of the frequency density $f(x)$ for $x = 8.9$, 9.0 and 9.3 are, respectively, 0.30, 0.35 and 0.25. Calculate the approximate value of the mode.

Solution We have

x	8.9	9.0	9.3
$f(x)$	0.30	0.35	0.25

First we will find the form of that frequency density $f(x)$. By Lagrange's formula

$$f(x) = \frac{(x-9)(x-9.3)}{(8.9-9)(8.9-9.3)}\times 0.30 + \frac{(x-8.9)(x-9.3)}{(9-8.9)(9-9.3)}\times 0.35 + \frac{(x-8.9)(x-9)}{(9.3-8.9)(9.3-9)}\times 0.25$$

$$f(x) = -\frac{25}{12}x^2 + \frac{453.5}{12}x - \frac{2052.3}{12}$$

which gives the frequency density. It will be maximum for the value of x lying in the range (8.9, 9.3) for which $f'(x) = 0$ and $f''(x)$ is negative.

$$f'(x) = 0 \Rightarrow -\frac{5}{12}(10x - 90.7) = 0 \text{ or } x = 9.07$$

and $f''(x) = -\frac{25}{6}$ which is negative. Hence the mode of the frequency curve $y = f(x)$ is 9.07.

Partial Fractions

Example 4.30
Express the function $\dfrac{3x^2-12x+11}{(x-1)(x-2)(x-3)}$ as a sum of partial fractions, using Lagrange's formula.

Solution Let $f(x) = 3x^2 - 12x + 11$. The zeros of the denominator are at $x_0 = 1$, $x_1 = 2$ and $x_2 = 3$. Evaluate $f(x)$ at these points.

$$f(x_0) = f(1) = 3.1^2 - 12.1 + 11 = 2,$$
$$f(x_1) = f(2) = 3.2^2 - 12.2 + 11 = -1$$
$$f(x_2) = f(3) = 3.3^2 - 12.3 + 11 = 2$$

We fit a polynomial using Lagrange's formula

$$f(x) = \frac{(x-2)(x-3)}{(1-2)(1-3)}\times 2 + \frac{(x-1)(x-3)}{(2-1)(2-3)}\times(-1) + \frac{(x-1)(x-2)}{(3-1)(3-2)}\times 2 = (x-2)(x-3)\cdot 1 + (x-1)(x-3)\cdot 1 + (x-1)(x-2)\cdot 1$$

Dividing both sides by $(x-1)(x-2)(x-3)$ we obtain

$$\frac{f(x)}{(x-1)(x-2)(x-3)} = \frac{3x^2-12x+11}{(x-1)(x-2)(x-3)} = \frac{1}{x-1} + \frac{1}{x-2} + \frac{1}{x-3}$$

Inverse Interpolation

Example 4.31
Determine $x(7)$ from the following data

x	1	3	4
y	4	12	19

Solution Let $y = f(x)$ and $x_0 = 1$, $x_1 = 3$, $x_2 = 4$; $y_0 = 4$, $y_1 = 12$, $y_2 = 19$.

By Lagrange's inverse interpolation formula (4.82),

$$x = \frac{(7-12)(7-19)}{(4-12)(4-19)}\times 1 + \frac{(7-4)(7-19)}{(12-4)(12-19)}\times 3 + \frac{(7-4)(7-12)}{(19-4)(19-12)}\times 4 = \frac{(-5)(-12)}{(-8)(-15)} + \frac{3(-12)}{8(-7)}\times 3 + \frac{3(-5)}{15.7}\times 4 = \frac{1}{2} + \frac{27}{14} - \frac{4}{7} = 1.8571$$

Newton's Formula

Example 4.32
Build up a divided difference table for the data $y_3 = 15$, $y_{-2} = 5$, $y_9 = 687$, $y_0 = 3$, $y_4 = 47$. Using Newton's divided difference formula, find y_2.

Solution Divided differences are symmetrical, and we can permute the letters without affecting the result. Building up a difference table we have

Substituting from the table into Newton's divided difference formula (4.74) we get

$$y = 5 + (x-(-2)).(-1) + (x-(-2))(x-0).1 + (x-(-2))(x-0)(x-3).1$$

x	y_x	$\Delta' y_x$	$\Delta'^2 y_x$	$\Delta'^3 y_x$
−2	5			
		$\frac{3-5}{0-(-2)} = -1$		
0	3		$\frac{4-(-1)}{3-(-2)} = +1$	
		$\frac{15-3}{3-0} = +4$		$\frac{7-1}{4-(-2)} = +1$
3	15		$\frac{32-4}{4-0} = +7$	
		$\frac{47-15}{4-3} = +32$		$\frac{16-7}{9-0} = +1$
4	47		$\frac{128-32}{9-3} = +16$	
		$\frac{687-47}{9-4} = +128$		
9	687			

$y = f(x) = x^3 - 5x + 3$ is the interpolating polynomial. Now, we have to find y_2

$$y_2 = 2^3 - 5.2 + 3 = 1$$

4.10 Central Difference Formulas

Newton-Gregory forward and backward difference formulas are applicable for interpolation near the beginning and end of the tabulated values, respectively. Central difference formulas are well suited for interpolation near the middle of the table.

Consider a function y_x tabulated for equally spaced arguments x_0, $x_0 + h$, $x_0 + 2h$, … Making the transformation $p = \dfrac{x-(x_0+rh)}{h}$ the value y_{x_0+rh} now becomes y_0 and the interval of differencing becomes 1. The difference table appears as follows

x	p	y_p	Δy_p	$\Delta^2 y_p$	$\Delta^3 y_p$	$\Delta^4 y_p$
$x_0+(r-2)h$	−2	y_{-2}				
			Δy_{-2}			
$x_0+(r-1)h$	−1	y_{-1}		$\Delta^2 y_{-2}$		
			Δy_{-1}		$\Delta^3 y_{-2}$	
x_0+rh	0	y_0		$\Delta^2 y_{-1}$		$\Delta^4 y_{-2}$
			Δy_0		$\Delta^3 y_{-1}$	
$x_0+(r+1)h$	1	y_1		$\Delta^2 y_0$		
			Δy_1			
$x_0+(r+2)h$	2	y_2				

We refer to the differences in the above table *central differences*; and also introduce a special *central difference operator* δ by the definition

$$\delta = E^{1/2} - E^{-1/2} \tag{4.83}$$

It follows that

$$\delta E^{1/2} = E - 1 = \Delta \tag{4.84}$$

Hence in δ notation we can write

$$\Delta y_{-2} = \delta E^{1/2} y_{-2} = \delta\, y_{-3/2};\ \Delta y_{-1} = \delta\, y_{-1/2};\ \Delta y_0 = \delta y_{1/2} \tag{4.85}$$

Second differences follow from

$$\delta^2 E = \Delta^2 \tag{4.86}$$

Thus

$$\Delta^2 y_{-2} = \delta^2 E y_{-2} = \delta^2 y_{-1};\quad \Delta^2 y_{-1} = \delta^2 y_0, \text{ etc.} \tag{4.87}$$

Similarly

$$\delta^3 E^{3/2} = \Delta^3, \text{ and } \Delta^3 y_{-2} = \delta^3 y_{-1/2};\ \ \Delta^3 y_{-1} = \delta^3 y_{1/2} \tag{4.88}$$

Table 4.6 Central Difference Notation

p	y_p	$\delta\, y_p$	$\delta^2 y_p$	$\delta^3 y_p$	$\delta^4 y_p$
−2	y_{-2}				
		$\delta y_{-3/2}$			
−1	y_{-1}		$\delta^2 y_{-1}$		
		$\delta y_{-1/2}$		$\delta^3 y_{-1/2}$	
0	y_0		$\delta^2 y_0$		$\delta^4 y_0$
		$\delta y_{1/2}$		$\delta^3 y_{1/2}$	
1	y_1		$\delta^2 y_1$		
		$\delta y_{3/2}$			
2	y_2				

It follows that in δ notation the difference table may be rewritten as in Table 4.6.

The great advantage of the δ notation is that, as seen from the above table, $\delta^r y_0$ appears on the line exactly opposite the entry y_0.

We introduce another operator the *averaging operator* μ defined by

$$\mu = \frac{1}{2}(E^{1/2} + E^{-1/2}) \tag{4.89}$$

Thus,

$$\mu y_0 = \frac{1}{2}(y_{1/2} + y_{-1/2}) \tag{4.90}$$

Various relations exist among the operators Δ, E, δ and μ. Two of the most important are

$$\mu^2 = \frac{1}{4}(E + E^{-1} + 2) = 1 + \frac{1}{4}\delta^2 \quad \text{and}$$

$$\mu\delta = \frac{1}{2}(E + E^{-1}) = \frac{1}{2}\Delta E^{-1} + \frac{1}{2}\Delta \tag{4.91}$$

Thus

$$\mu\delta y_0 = \frac{1}{2}(\delta y_{1/2} + \delta y_{-1/2}) = \frac{1}{2}(\Delta y_{-1} + \Delta y_0)$$

4.10.1 Gauss's[5] Interpolation Formulae

Making use of the divided differences, we will develop two formulae due to Gauss. Though they are not of great practical use they serve as intermediate

[5] Named after the great German mathematician Carl Friedrick Gauss (1777–1855) who made his first great discovery as a student at Gottingen. His important contributions are to algebra, number theory, mechanics, complex analysis, differential equations, differential geometry, noneuclidean geometry, numerical analysis, astronomy and electromagnetism. He became director of the observatory of Gottingen in 1807.

steps in obtaining important formulae due to Stirling and Bessel.

First we construct a central difference table in Δ notation

0	y_p	Δy_p	$\Delta^2 y_p$	$\Delta^3 y_p$	$\Delta^4 y_p$	$\Delta^5 y_p$
−3	y_{-3}					
		Δy_{-3}				
−2	y_{-2}		$\Delta^2 y_{-3}$			
		Δy_{-2}		$\Delta^3 y_{-3}$		
−1	y_{-1}		$\Delta^2 y_{-2}$		$\Delta^4 y_{-3}$	
		Δy_{-1}		$\Delta^3 y_{-2}$		$\Delta^5 y_{-3}$
0	y_0		$\Delta^2 y_{-1}$		$\Delta^4 y_{-2}$	
		Δy_0		$\Delta^3 y_{-1}$		
1	y_1		$\Delta^2 y_0$			
		Δy_1				
2	y_2					

The table can be extended in either direction. Since the interval of differencing is constant and equal to unity; the divided differences of y_x are given by the relation

$$\Delta'^r y_x = \frac{\Delta^r y_x}{r!} \tag{4.92}$$

So, applying Sheppard's zig-zag rule along the solid line we obtain

$$y_p = y_0 + p\frac{\Delta y_0}{1!} + p(p-1)\frac{\Delta^2 y_{-1}}{2!} + p(p-1)(p+1) \times \frac{\Delta^3 y_{-1}}{3!} + p(p-1)(p+1)(p-2)\frac{\Delta^4 y_{-2}}{4!} + p(p-1)(p+1)(p-2)(p+2)\frac{\Delta^5 y_{-2}}{5!} + \cdots \tag{4.93}$$

which gives the Gauss's Forward Formula

$$y_p = y_0 + \binom{p}{1}\Delta y_0 + \binom{p}{2}\Delta^2 y_{-1} + \binom{p+1}{3}\Delta^3 y_{-1} + \binom{p+1}{4}\Delta^4 y_{-2} + \binom{p+2}{5}\Delta^5 y_{-2} + \cdots \tag{4.94}$$

In δ notation, Gauss's Forward Formula is

$$y_p = y_0 + \binom{p}{1}\delta y_{1/2} + \binom{p}{2}\delta^2 y_0 + \binom{p+1}{3}\delta^3 y_{1/2} + \binom{p+1}{4}\delta^4 y_0 + \binom{p+2}{5}\delta^5 y_{1/2} + \cdots \tag{4.95}$$

On the other hand, using Sheppard's zig-zag rule along the dotted line and rearranging terms we have Gauss's Backward Formula

$$y_p = y_0 + \binom{p}{1}\Delta y_{-1} + \binom{p+1}{2}\Delta^2 y_{-1} + \binom{p+1}{3}\Delta^2 y_{-2} + \binom{p+2}{4}\Delta^4 y_{-2} + \binom{p+2}{5}\Delta^5 y_{-3} + \cdots \tag{4.96}$$

In δ notation, Gauss's Backward Formula is

$$y_p = y_0 + \binom{p}{1}\delta y_{-1/2} + \binom{p+1}{2}\delta^2 y_0 + \binom{p+1}{3}\delta^3 y_{-1/2} + \binom{p+2}{4}\delta^4 y_0 + \binom{p+2}{5}\delta^5 y_{-1/2} + \cdots \tag{4.97}$$

Thus, we have expansions in terms of the central differences of y_0 and of either $y_{1/2}$ or $y_{-1/2}$ as the case may be.

4.10.2 Stirling's[6] Formula

Now, if we take the average of Gauss's Forward and Backward Formulas (4.92) and (4.94), we get *Stirling's Formula*

$$y_p = y_0 + \binom{p}{1}\mu\delta y_0 + \frac{p}{2}\binom{p}{1}\delta^2 y_0 + \binom{p+1}{3}\mu\delta^3 y_0 + \frac{p}{4}\binom{p+1}{3}\delta^4 y_0 + \binom{p+2}{5}\mu\delta^5 y_0 + \cdots \tag{4.98}$$

[6] STIRLING, James (1692–1770), is a Scottish mathematician, who began his career at Oxford, Venice and London with brilliant academic work in mathematics, had friendship with Newton and Maclaurin and devoted most of his life after 1735 to industrial management.

or, alternatively,

$$y_p = y_0 + \frac{p}{1!}\mu\delta y_0 + \frac{p}{2!}\delta^2 y_0 + \frac{p(p^2-1^2)}{3!}\mu\delta^3 y_0 + \frac{p^2(p^2-1^2)}{4!}\delta^4 y_0 + \frac{p(p^2-1^2)(p^2-2^2)}{5!}\mu\delta^5 y_0 + \cdots \quad (4.99)$$

It is convenient to note that the differences that appear in Stirling's Formula lie on a horizontal line through y_0, as shown below:

$$y_0 \text{ —— } \mu\delta y_0 \text{ —— } \delta^2 y_0 \text{ —— } \mu\delta^3 \; y_0 \text{ —— } \mu\delta^3 \; y_0 \text{ —— } \cdots \quad (4.100)$$

Then even-ordered differences $\delta^{2n} y_0$ are actual tabular values while the odd-ordered mean differences

$$\mu\delta^{2n+1} y_0 = \frac{1}{2}(\delta^{2n+1} y_{-1/2} + \delta^{2n+1} y_{1/2} \quad (4.101)$$

are averages of adjacent tabular values.

4.10.3 Bessel[7] Formula

We rewrite Gauss's Backward Formula based on a zig-zag line one unit below the earlier dotted line. This means shifting the initial point from (x_0, y_0) to (x_1, y_1) and expressing y_x in terms of the differences of $y_{1/2}$ and y_1 instead of in terms of the differences of $y_{-1/2}$ and y_0. Gauss's Backward Formula then becomes

$$y_p = y_1 + \binom{p-1}{1}\delta y_{1/2} + \binom{p}{2}\delta^2 y_1 + \binom{p}{3}\delta^3 y_{1/2} + \binom{p+1}{4}\delta^4 y_1 + \binom{p+1}{5}\delta^5 y_{1/2} + \cdots \quad (4.102)$$

Taking the average of (4.92) and (4.102), we obtain Bessel's Formula

$$y_p = \mu y_{1/2} + \left(p - \frac{1}{2}\right)\delta y_{1/2} + \binom{p}{2}\mu\delta^2 y_{1/2} + \frac{1}{3}\left(p - \frac{1}{2}\right)\binom{p}{2}\delta^3 y_{1/2} + \binom{p+1}{4}\mu\delta^4 y_{1/2} + \frac{1}{5}\binom{p+1}{4}\delta^5 y_{1/2} + \cdots \quad (4.103)$$

Note that the differences and mean differences appearing in (4.100) all lie on the horizontal line through $y_{1/2}$.

Bessel's Formula, which is of very great importance, can be put in a more elegant form by setting $p = u + \frac{1}{2}$. We then obtain

$$y_{u+\frac{1}{2}} = \mu y_{1/2} + u\delta y_{1/2} + \frac{u^2 - \frac{1}{4}}{2!}\mu\delta^2 y_{1/2} + \frac{u(u^2 - \frac{1}{4})}{3!}\delta^2 y_{1/2} + \frac{(u^2 - \frac{1}{4})(u^2 - \frac{9}{4})}{4!}\mu\delta^4 y_{1/2} + \frac{u(u^2 - \frac{1}{4})(u^2 - \frac{9}{4})}{5!}\delta^5 y_{1/2} + \cdots \quad (4.104)$$

Putting $u = 0$ gives the important formula, useful for halving the interval of tabulation

$$y_{1/2} = \mu y_{1/2} - \frac{1}{8}\mu\delta^2 y_{1/2} + \frac{3}{128}\mu\delta^4 y_{1/2} - \cdots \quad (4.105)$$

We should note that, by proper choice of origin, p in (4.103) or any other central difference formula may be taken in the range $0 \le p \le 1$ or in the range $-1/2 \le p \le 1/2$.

Gauss's Forward Formula

Example 4.33

Estimate $f\left(\frac{1}{2}\right)$ by Gauss's formula given that $f(2) = 10, f(1) = 8, f(0) = 5, f(-1) = 10$.

Solution Writing the difference table we have

x	y	Δy	$\Delta^2 y$	$\Delta^3 y$
−1	10			
		−5		
0	5		8	
		3		−9
1	8		−1	
		2		
2	10			

[7] BESSEL, Friedrick Wilhelm (1784–1846), is a German mathematician and astronomer. His paper on Bessel functions appeared in 1824. He studied astronomy of his own and became Director of Konigsberg observatory.

In Gauss's forward formula (4.94) putting $p=\frac{1}{2}$ we get

$$y=5+\left(\frac{1}{2}\right)3+\frac{\frac{1}{2}(\frac{1}{2}-1)}{2}8+\frac{(\frac{1}{2}+1)\frac{1}{2}(\frac{1}{2}-1)}{6}(-9)$$

$$=5+\frac{3}{2}+\frac{1}{2}\left(-\frac{1}{2}\right)\frac{1}{2}\times 8+\frac{1}{2}\cdot\frac{1}{2}\left(-\frac{1}{2}\right)\cdot\left(\frac{-9}{2}\right)$$

$$=5+\frac{3}{2}-1+\frac{9}{16}=\frac{80+24-16+9}{16}=\frac{97}{16}=6.0625$$

Example 4.34

Estimate the value of y (2.5) using Gauss's forward formula given that

x	1.0	2.0	3.0	4.0
y_x	1.0	4.0	9.0	16.0

Solution Take the origin at 2.0, $h=1$ and $x=2.5$. Then $p=\frac{2.5-2.0}{1}=0.5$.

Writing the central difference table we have

x	p	y	Δy	$\Delta^2 y$	$\Delta^3 y$
1.0	21	1	3	2	
2.0	0	4	5	2	
			7		0
3.0	1	9			
4.0	2	16			

By Gauss's forward interpolation formula (4.97), we get

$$y_{2.5}=4+\frac{1}{2}\times 5+\frac{\frac{1}{2}(\frac{1}{2}-1)}{2}2=6.25$$

Gauss's Backward Formula

Example 4.35

Interpolate by means of Gauss's backward formula the population for the year 1936 given the following table:

Year	1901	1911	1921	1931	1941	1951
Pop. (1000s)	12	15	20	27	39	52

Solution Take 1931 as the origin and 10 years as the length of the interval, then the population is to be estimated for $p=\frac{1936-1931}{10}=\frac{5}{20}=\frac{1}{2}$. Writing the difference table we have

p	y	Δy	$\Delta^2 y$	$\Delta^3 y$	$\Delta^4 y$	$\Delta^5 y$
−3	12	3	2	0	+3	
−2	15	5	2	+3	−7	
−1	20	7	5	−4		−10
		12	1			
0	27	13				
1	39					
2	52					

Substituting from the table in the Gauss's Backward Formula (4.99), we obtain

$$y_{1/2}=27+\frac{1}{2}\times 7+\frac{\frac{3}{2}\times\frac{1}{2}}{2}\times 5+\frac{\frac{3}{2}\times\frac{1}{2}\times-\frac{1}{2}}{6}\times 3$$

$$+\frac{\frac{5}{2}\cdot\frac{3}{2}\cdot\frac{1}{2}\cdot(-\frac{1}{2})}{24}\times(-7)+\frac{\frac{5}{2}\cdot\frac{3}{2}\cdot\frac{1}{2}\cdot(-\frac{1}{2})(-\frac{3}{2})}{120}\times(-10)$$

$$=27+3.5+1.875-0.1875+0.2734-0.1172$$

$$=32.6284-0.3047$$

$$=32.3437 \text{ thousands.}$$

Stirling's Formula

Example 4.36

Use Stirling's Formula to find $f(35)$ given $f(20)=512$, $f(30)=439$, $f(40)=346$, $f(50)=243$.

Solution Take 30 as the origin and 10 as the unit. Then we have $x=35$, $x_0=30$, $h=10$

$$\therefore \quad p=\frac{35-30}{10}=0.5.$$

Writing the difference table

x	p	y	Δy	$\Delta^2 y$	$\Delta^3 y$
20	−1	512			
			−73		
30	0	439		−20	
			−93		10
40	1	346		−10	
			−103		
50	2	243			

Stirling's Formula

$$y_p = y_0 + p\frac{\Delta y_0 + \Delta y_{-1}}{2} + \frac{p^2}{2}\Delta^2 y_{-1} + \frac{p(p^2-1)}{6}\frac{\Delta^3 y_{-2} + \Delta^3 y_{-1}}{2} + \frac{p^2(p^2-1)}{24}\Delta^4 y_{-2} + \cdots$$

$$y_{\frac{1}{2}} = 439 + \frac{1}{2}\frac{-93-73}{2} + \frac{\frac{1}{4}}{\frac{1}{4}}(-20) = 439 - 41.5 - 2.5 = 395$$

Example 4.37

Apply Bessel's formula to obtain $f(25)$ given $f(20) = 2854, f(24) = 3162, f(28) = 3544, f(12) = 3992$.

Solution We take $x_0 = 24$ as the origin and $h = 4$ as the unit. Thus $p = \frac{x-24}{4}$. We have to obtain $f(1/4)$. The difference table for the data is (we put $y = f(x)$)

x	p	y	Δy	$\Delta^2 y$	$\Delta^3 y$
20	−1	2854			
			308		
24	0	3162		74	
			382		−8
28	1	3544		66	
			448		
12	2	3992			

Bessel's Formula

$$y_p = y_0 + p\Delta y_0 + \frac{p(p-1)}{2!}\left(\frac{\Delta^2 y_{-1} + \Delta^2 y_0}{2}\right) + \frac{(p-1/2)p(p-1)}{3!}\Delta^3 y_{-1} + \cdots$$

$$y_{1/4} = f\left(\frac{1}{4}\right) = (3162) + \frac{1}{4}\times 382 + \left(\frac{1}{4} - \frac{1}{2}\right)\times 382 + \frac{\frac{1}{4}\left(\frac{1}{4}-1\right)}{2!}\frac{74+66}{2} + \frac{\left(\frac{1}{4}-\frac{1}{2}\right)\frac{1}{4}\left(\frac{1}{4}-1\right)}{6}\times(-8)$$

$$= 3250.875$$

Example 4.38

Using the following table, find an estimate of $f(53)$ by Bessel's formula:

x	48	50	52	54	56
$f(x)$	0.48936170	0.47500468	0.46106887	0.44754192	0.43441182

Solution In this problem we have interval of differencing $h = 2$ and we have to interpolate for $x = 53$ which is the midpoint between $x = 52$ and $x = 54$. Hence, we have to use Bessel's formula

$$f\left(\frac{1}{2}\right) = \mu f\left(\frac{1}{2}\right) - \frac{1}{8}\mu\delta^2 f\left(\frac{1}{2}\right) + \frac{3}{120}\mu\delta^4 f\left(\frac{1}{2}\right)$$

If we set $p = \frac{x-52}{2}$ the origin is shifted to 52 and the length of the interval becomes unity. Writing the difference table with $y = f(x)$

p	y	Δy	$\Delta^2 y$	$\Delta^3 y$	$\Delta^4 y$
−2	0.48936170				
		−0.01435702			
−1	0.47500468		0.00042121		
		−0.01393581		−0.00001235	
0	0.46106887		0.00040886		0.00000034
		−0.1352645		−0.00001201	
1	0.44754192		0.00039685		
		−0.01313010			
2	0.43441182				

$$y_{\frac{1}{2}} = f\left(\frac{1}{2}\right) = \frac{1}{2}(0.46106887 + 0.44754192)$$

$$-\frac{1}{8} \times \frac{1}{2}(0.00040886 + 0.00039685)$$

$$+\frac{3}{128} \times \frac{1}{2}(0.00000034 + 0.00000034)$$

$$= 0.45425505$$

EXERCISE 4.3

1. Find the polynomial having values 3, 12, 15 and – 21 at $x = 3, 2, 1$ and -1, respectively. Find its value at $x = 1.5$.

   Ans: $P(x) = x^3 - 9x^2 + 17x + 6$; 14.625

2. Find Lagrange's polynomial using the data

x	0	1	2	5
$f(x)$	2	3	12	147

   Interpolate to find the value $f(2.25)$.

   Ans: $x^3 + x^2 - x + 2$, 16.203125

3. Using Newton's divided difference formula find $\log_{10} 310$ from the data given below:

x	300	304	305	307
$y_x = \log_{10} x$	2.4771	2.4829	2.4843	2.4871

   Ans: 2.4786

4. Use Lagrange's formula to express the following functions as the sums of partial fractions:

   (a) $\dfrac{x^2 + 6x + 1}{(x^2 - 1)(x - 4)(x - 6)}$ (b) $\dfrac{3x^2 + x + 1}{x^3 - 6x^3 + 11x - 6}$

   (c) $\dfrac{x^2 + 2x + 4}{x^2 - x + 12}$

   Ans: (a) $\dfrac{2}{35} \cdot \dfrac{1}{x+1} + \dfrac{4}{15} \cdot \dfrac{1}{x-1} - \dfrac{41}{30} \dfrac{1}{x-4}$

   (b) $\dfrac{5}{2} \cdot \dfrac{1}{x-1} - \dfrac{15}{x-2} + \dfrac{31}{2} \cdot \dfrac{1}{x-3}$

   Dividing once one gets

   (c) $1 + \dfrac{3x - 8}{(x-4)(x+3)} = 1 + \dfrac{4}{7} \cdot \dfrac{1}{x-4} + \dfrac{17}{7} \dfrac{1}{x+3}$

5. Estimate the value of $f(4)$ applying Lagrange's formula to the following data:

x	0	2	3	6
$f(x)$	707	819	866	966

   Ans: 906.44

6. From the following table, find x for $y = -0.510$, using Lagrange's inverse formula:

x	0.4	0.5	0.7	0.8
$f(x)$	–0.916	–0.693	–0.357	–0.223

   Ans: $x = 0.6$

7. Using Gauss's forward formula obtain the value of $e^{1.17}$ from the following data:

x	1.00	1.05	1.10	1.15	1.20	1.25	1.30
e^x	2.7183	2.8577	3.0042	3.1582	3.3201	3.4903	3.6613

   Ans: 3.2221 32.345 (thousands)

8. The census of a town is as given in the table below. Using Gauss's backward formula, find the population for the year 1974.

Year x	1939	1949	1959	1969	1979	1989
Population in 1000s	12	15	20	27	39	52

   Ans: 32.345 (thousands)

9. The following table gives the values of $y_x = 1 + \log_{10} \sin x$ for $x° = 10° (1°) 14°$. Using Stirling's formula, find $y_{12.2}$.

   Table of values of $y_x = 1 + \log_{10} \sin x$

x	10	11	12	13	14
y_x	0.2397	0.2806	0.3179	0.3521	0.3837

   Ans: 0.32497

10. Using Bessel's formula, find y_{25} given the following data:

x	20	24	28	32
y_x	2854	3162	3544	3992

   Ans: 3250.875

Curve Fitting

5

5.1 Introduction

The data obtained through measurement or observation may be plotted graphically, and a smooth curve is drawn joining the data points. Such graph is called an approximating curve for the data. The process of finding a curve that is "a best fit" for given data is called curve fitting. Several methods exist for curve fitting. We consider, in this chapter, curve fitting by a sum of exponential, linear/nonlinear weighted least squares approximation methods.

5.1.1 Curve Fitting: Method of Least Squares

So far we have employed polynomial interpolation where we have replaced the unknown function $y=f(x)$ by a polynomial of degree at most n such that

$$f(x_i) = p_n(x_i) \quad (i = 0, 1, 2, \ldots, n) \tag{5.1}$$

This gives satisfactory results for tabulated functions such as sin *x*, ln *x*, *etc*.

In the case of experimental data with observational errors, we have to look for a curve which is statistically best fit. The process of finding a mathematical relationship for a set of scattered experimental data is known as *Regression analysis or curve fitting.*

Let (x_i, y_i) $(i = 1, 2, \ldots, n)$ be the data and $e_i = y_i - f(x_i)$ be the error in the ith observation, which may be positive or negative. In order to obtain the best fit, we have to minimize the sum S of the squares of the errors given by

$$S = \sum e_i^2 = \sum (y_i - \bar{y}_i)^2$$

and the method of minimizing S is called the least squares regression or the least squares fit or the principle of least squares and was suggested by Legendre[1] in 1806.

Least Squares Fit

We now consider some standard approximating curves and discuss the method of least squares fit in respect of them.

5.1.2 Some Standard Approximating Curves

Straight Line (linear) Fit or Linear Regression

1. $y = a + bx$.

Parabolic (or Quadratic or Second Degree) Curve

2. $y = a + bx + cx^2$.

*n*th Degree (or Polynomial) Curve

3. $y = \sum_{i=0}^{n} a_i x^i = a_0 + a_1 x + a_2 x^2 + \cdots + a_n x^n$.

Exponential Curve

4. (a) $y = ab^x$ growth curve; (b) $y = ab^{-x}$ (decay curve).

Power Curve or Geometric Curve

5. $y = ax^b$.

---

[1] LEGENDRE, Adrien Marie (1752–1833), a French mathematician, famous for the differential equation $\frac{d}{dx}\left[(1-x^2)\frac{dy}{dx}\right] + n(n+1)y = 0$ and its solutions $P_n(x)$ and $Q_n(x)$ which bear his name. His other important contributions are in number theory, calculus of variations and elliptic integrals.

Hyperbolic Curve

6. $y = (a + bx)^{-1}$.

In the following the summation $\sum$ extends from $i = 1$ to n.

5.2 Curve Fitting by the Method of Least Squares

5.2.1 Least Squares Straight Line Fit or Linear Regression

In a least squares straight line fit (or linear regression) we take a straight line equation and set

$$y = a + bx = f(x) \tag{5.2}$$

to fit the given data

$$(x_i, y_i), (i = 1, 2, \dots, n) \tag{5.3}$$

where $y = f(x)$.

Then $$S = \sum e_i^2 = \sum (y_i - a - bx_i)^2 \tag{5.4}$$

For S to be minimum we must have

$$\frac{\partial S}{\partial a} = 0 \quad \text{and} \quad \frac{\partial S}{\partial b} = 0 \qquad (5.5), (5.6)$$

and these conditions imply that

$$-2\sum (y_i - a - bx_i) = 0$$

and $$-2\sum (y_i - a - bx_i)x_i = 0$$

$$\Rightarrow \quad \sum y_i = \sum a + b\sum x_i$$

and $$\sum y_i x_i = a\sum x_i + b\sum x_i^2$$

or $$\sum y_i = na + b\sum x_i,$$

$$\sum x_i y_i = a\sum x_i + b\sum x_i^2 \qquad (5.7), (5.8)$$

which are called the *normal equations* for a straight line. Dropping the suffixes for convenience, we may write the equations (5.7) and (5.8) as

$$\sum y = na + b\sum x, \quad \sum xy = a\sum x + b\sum x^2 \qquad (5.9), (5.10)$$

Solving these equations, we obtain the values of a and b.

Let $$\Delta = \begin{vmatrix} n & \sum x_i \\ \sum x_i & \sum x_i^2 \end{vmatrix}; \quad \Delta_a = \begin{vmatrix} \sum y_i & \sum x_i \\ \sum x_i y_i & \sum x_i^2 \end{vmatrix};$$

$$\Delta_b = \begin{vmatrix} n & \sum y_i \\ \sum x_i & \sum x_i y_i \end{vmatrix}$$

Then $$a = \frac{\Delta_a}{\Delta} = \frac{\sum x_i^2 \sum y_i - \sum x_i \sum x_i y_i}{\left(n\sum x_i^2 - \left(\sum x_i\right)^2\right)};$$

$$b = \frac{\Delta_b}{\Delta} = \frac{n\sum x_i y_i - \sum x_i \sum y_i}{n\sum x_i^2 - \left(\sum x_i\right)^2}$$

The curve is $$y = a + bx = a + \left(\frac{S_{xy}}{S_{xx}}\right) \cdot x \tag{5.11}$$

where $$S_{xy} = n\sum xy - \sum x \sum y \tag{5.12}$$

5.2.2 Least Squares Parabolic (Quadratic) Curve

Let us take the parabola

$$y = a + bx + cx^2 \tag{5.13}$$

as the approximating curve to fit a set of given data. Then the unknowns a, b and c are determined from the following normal equations obtained similarly:

$$\left.\begin{aligned} \sum y_i &= na + b\sum x_i + c\sum x_i^2 \\ \sum x_i y_i &= a\sum x_i + b\sum x_i^2 + c\sum x_i^3 \\ \sum x_i^2 y_i &= a\sum x_i^2 + b\sum x_i^3 + c\sum x_i^4 \end{aligned}\right|$$

or, dropping the suffix for convenience

$$\left.\begin{aligned} \sum y &= na + b\sum x + c\sum x^2 \\ \sum xy &= a\sum x + b\sum x^2 + c\sum x^3 \\ \sum x^2 y &= a\sum x^2 + b\sum x^3 + c\sum x^4 \end{aligned}\right\} \tag{5.14}$$

Solving, we get

$$a = \frac{\Delta_a}{\Delta}, \quad b = \frac{\Delta_b}{\Delta}, \quad c = \frac{\Delta_c}{\Delta} \tag{5.15}$$

where

$$\Delta = \begin{vmatrix} n & \sum x & \sum x^2 \\ \sum x & \sum x^2 & \sum x^3 \\ \sum x^2 & \sum x^3 & \sum x^4 \end{vmatrix} \tag{5.16}$$

Δ_a, Δ_b, Δ_c are obtained from Δ by replacing C_1, C_2, C_3, respectively, by the column vector $\left(\sum y, \sum xy, \sum x^2 y\right)^T$.

5.2.3 Nonlinear Curves

In the case of nonlinear curves given by equations (5.17), (5.18) and (5.19), we can transform the equations into linear form by an appropriate transformation.

$$y = ab^x \quad \text{(Exponential curve)} \qquad (5.17)$$

$$y = ax^b \quad \text{(Power curve)} \qquad (5.18)$$

and the curve $$y = \frac{1}{a+bx} \qquad (5.19)$$

For example, consider the exponential curves, (5.17) and (5.18).

Taking logarithms on both sides of the equations: (5.17) and (5.18)

$\ln y = \ln a + x \ln b$	$\ln y = \ln a + b \ln x$
or $Y = A + Bx$	or $Y = A + bX$

(5.20), (5.21)

where $A = \ln a$; $B = \ln b$; $X = \ln x$, $X = \text{In}\, x$; $Y = \ln y$.

Equation (5.19) can be transformed to linear form by putting

$$Y = \frac{1}{y} \text{ (Reciprocal curve)}$$

Note 1 The normal equations for the nth degree curve $y(x) = a_0 + a_1 x + \cdots + a_n x^n$ are obtained by multiplying both sides of this eqn. formally by 1, $x, x^2, \ldots, x^n$ and summing the terms. We get $(n + 1)$ normal equations from which the $(n + 1)$ unknowns $a_i (i = 0, 1, 2, \ldots, n)$ can be determined.

Linear interpolation is to find y corresponding to a value of x is the range $x_1 < x < x_n$ (extrapolation is to find y corresponding to an x outside the range $x_1 < x < x_n$).

The above normal equations are derived under the assumption that errors occur in the independent variable x. If the errors occur in the dependent variable y we have to change the roles of x and y.

Now we prove some standard results satisfied by the normal equations.

Example 5.1

Show that the least squares straight line always passes through the centroid of the data points (x_i, y_i) $(i = 1, 2, \ldots, n)$

Solution Let $y = a + bx$ be the least squares line (LSL) of y on x. Its normal equation is

$$\sum y = na + b \sum x \quad \Rightarrow \quad \frac{\sum y}{n} = a + b\frac{\sum x}{n},$$

(dividing by n)

$$\Rightarrow \quad \bar{y} = a + b\bar{x}$$

Note 1 Similarly we can prove for $x = a + by$, LSL of x on y.

Example 5.2

Prove that LSL y on x can be expressed as

$$Y = \left(\frac{\sum XY}{\sum X^2}\right) X$$

where $$X = x - \bar{x}, \; Y = y - \bar{y}$$

Solution Let the LSL be

$$y = a + bx \qquad (1)$$

Since it passes through $(\bar{x}, \bar{y})$ we have

$$\bar{y} = a + b\bar{x} \qquad (2)$$

From (1) and (2) we have $y - \bar{y} = b(x - \bar{x})$ or $Y = bX$ where

$$X = x - \bar{x}, \; Y = y - \bar{y} \qquad (3)$$

$$\text{or} \quad x = X + \bar{x}, y = Y + \bar{y}$$

Now

$$b = \frac{n\sum xy - \sum x \sum y}{n\sum x^2 - \left(\sum x\right)^2}$$

$$= \frac{n\sum (XY + \bar{x}Y + \bar{y}X + \bar{x}\,\bar{y}) - \left(\sum X + n\bar{x}\right)\left(\sum Y + x\bar{y}\right)}{n\sum X^2 + 2n\bar{x}\sum X + n^2\bar{x}^2 - \left(\sum X + \sum \bar{x}\right)^2}$$

But

$$\sum X = \sum (x - \bar{x}) = 0 \qquad \sum Y = \sum (y - \bar{y}) = 0$$

$$\therefore \quad b = \frac{n\sum XY + n^2\bar{x}\bar{y} - n^2\overline{xy}}{n\sum X^2 = n^2\bar{x}^2 - n^2\bar{x}^2} = \frac{\sum XY}{\sum X^2}$$

Corollary The LSL of x on y can be expressed as

$$X = \left(\frac{\sum XY}{\sum Y^2}\right) Y$$

Example 5.3

Fit a straight line to the following data:

x	1	2	3	4	5
y	12	25	40	50	65

Solution Let $y = a + bx$ be the equation of the straight line.

We construct the table of values $\sum x$, $\sum y$, $\sum x^2$, $\sum xy$.

	x	y	x^2	xy
	1	12	1	12
	2	25	4	50
	3	40	9	120
	4	50	16	200
	5	65	25	325
$n = 5$	15	192	55	707
Total	$\sum x$	$\sum y$	$\sum x^2$	$\sum xy$

The normal equations are

$$na + b\sum x = \sum y$$
$$a\sum x + b\sum x^2 = \sum xy$$

Substituting the values we have

$$5a + 15b = 192$$
$$15a + 55b = 707$$

These equations reduce to

$$\begin{array}{l|lll} a + 3b = 38.4 & \times 3 & \Rightarrow & 3a + 9b = 115.2 \\ 3a + 11b = 141.4 & \times 1 & \Rightarrow & 3a + 11b = 141.4 \end{array}$$

$$2b = 26.2$$

$$\Rightarrow \quad b = 13.1, a = 38.4 - 39.3 = -0.9$$

Alternative Method

(Method of Shifting the Origin)

The x-data are equally spaced. Put $X = x - 3$ and calculate the values $\sum X$, *etc.* (Table 5.1).

Table 5.1 $y = a + bx$

	x	X	y	X^2	Xy
	1	−2	12	4	−24
	2	−1	25	1	−25
	3	0	40	0	0
	4	1	50	1	50
$n = 5$	5	2	65	4	130
Total	15	0	192	10	131

The normal equations are

$$\sum y = na + b\sum X$$
$$\sum Xy = a\sum X + b\sum X^2$$

which become

$$5a + b \cdot 0 = 192$$
$$0 \cdot a + 10b = 131$$

which give

$$a = 38 \cdot 4, \quad b = 13 \cdot 1$$

The required straight line is

$$y = 38 \cdot 4 + 13 \cdot 1\,(x - 3)$$
$$\text{or} \quad y = 38 \cdot 4 - 39 \cdot 3 + 13 \cdot 1x$$
$$\text{or} \quad y = -0 \cdot 9 + 13 \cdot 1x$$

Example 5.4

Find a least squares straight line for the following data:

x	0	1	3	6	8
y	1	3	2	5	4

and estimate (predict) y at $x = 4$ and x at $y = 2.5$.

Solution Constructing a regression table for the data we have

x	y	x^2	y^2	xy
0	1	0	1	0
1	3	1	9	3
3	2	9	4	6
6	5	36	25	30
8	4	64	16	32
Total $\sum x = 18$	$\sum y = 15$	$\sum x^2 = 110$	$\sum y^2 = 55$	$\sum xy = 71$

Let us assume the regression line of y on x in the form

$$y = a + bx \quad (1)$$

Its normal equations are

$$\sum y = na + b\sum x \quad (2)$$

$$\sum xy = a\sum x + b\sum x^2 \quad (3)$$

Substituting the values we obtain

$$5a + 18b = 15 \quad (4)$$

$$18a + 110b = 71 \quad (5)$$

$$\left[\begin{array}{cc|c} 5 & 18 & 15 \\ 18 & 110 & 71 \end{array}\right] \xrightarrow{R_2 - \frac{18}{5}R_1} \left[\begin{array}{cc|c} 5 & 18 & 15 \\ 0 & 45.2 & 71 \end{array}\right]$$

$$b = 0.376, \quad a = 3 - \frac{18}{5}b = 1.646$$

$\therefore$ The equation of the least squares straight line of y on x is

$$y = 0.376 + 1.646x \quad (6)$$

Example 5.5

Fit a straight line for the following set of measured data:

x	1	2	4	5	6	8	9
y	2	5	7	10	12	15	19

Solution Let the straight line be

$$y = a + bx \quad (1)$$

The normal equations are

$$\sum y = na + b\sum x \quad (2)$$

$$\sum xy = a\sum x + b\sum x^2 \quad (3)$$

where

$\sum x = \sum x_i$, $\sum y_i = \sum y$, $\sum x^2 = \sum x_i^2$ and $\sum xy = \sum x_i y_i$ (4)

The values of x and y are tabulated below:

	x	y	xy	x^2
	1	2	2	1
	2	5	10	4
	4	7	28	16
	5	10	50	25
	6	12	72	36
	8	15	120	64
	9	19	171	81
$n = 9$	35	70	453	227
Total	$\sum x$	$\sum y$	$\sum xy$	$\sum x^2$

The normal equations become

$$7a + 35b = 70$$
$$35a + 227b = 453 \quad (5)$$

Solving for a and b we have

$$a = \frac{\begin{vmatrix} 70 & 35 \\ 453 & 227 \end{vmatrix}}{\begin{vmatrix} 7 & 35 \\ 35 & 227 \end{vmatrix}} = \frac{15,890 - 15,855}{1589 - 1225} = \frac{35}{364} = 0.096 \quad (6)$$

$$b = \frac{\begin{vmatrix} 7 & 70 \\ 35 & 453 \end{vmatrix}}{\begin{vmatrix} 7 & 35 \\ 35 & 227 \end{vmatrix}} = \frac{3571 - 2450}{1589 - 1225} = \frac{721}{364} = 1.98 \quad (7)$$

Hence the equation of the LSL is $y = 0.096 + 1.98x$. (8)

Example 5.6

Find the equation of a straight line which fits the data:

x	0	2	5	7
y	−1	5	12	20

Solution Let the equation of the LSL be

$$y = a + bx \quad (1)$$

The normal equations are

$$\sum y = na + b\sum x \quad (2)$$

$$\sum xy = a\sum x + b\sum x^2 \quad (3)$$

We tabulate the given data and calculate $\sum x, \sum y,$ *etc.*

	x	y	x^2	xy
	0	−1	0	0
	2	5	4	10
	5	12	25	60
	7	20	49	140
$n = 4$	14	36	78	210
Total	$\sum x$	$\sum y$	$\sum x^2$	$\sum xy$

Now the normal equations become

$$4a + 14b = 36 \quad (4)$$

$$14a + 78b = 210 \quad (5)$$

Solving these equations, we get

$$b = \frac{\begin{vmatrix} 4 & 36 \\ 14 & 210 \end{vmatrix}}{\begin{vmatrix} 4 & 14 \\ 14 & 78 \end{vmatrix}} = \frac{840 - 504}{312 - 196} = \frac{336}{116} = 2.8966 \quad (6)$$

Substituting this value in (4), we have

$$a = \frac{18 - 7b}{2} = \frac{-2.2762}{2} = -1.1381 \quad (7)$$

$$y_{est} = y \text{ (at } x = 4) = 3.15 \quad (8)$$

Similarly, assume the least squares straight line of x on y as

$$x = a + by \quad (9)$$

The normal equations are

$$\sum x = na + b\sum y$$

$$\sum xy = a\sum y + b\sum y^2$$

Subtituting the values, we obtain

$$5a + 15b = 18 \quad (10)$$

$$15a + 55b = 71 \quad (11)$$

$$\left[\begin{array}{cc|c} 5 & 15 & 18 \\ 15 & 55 & 71 \end{array}\right] \underset{\sim}{R_2 - 3R_1} \left[\begin{array}{cc|c} 5 & 15 & 18 \\ 15 & 55 & 71 \end{array}\right]$$

$$10b = 17 \quad \Rightarrow \quad b = 1.7 \quad (12)$$

$$a = \frac{18 - 15b}{5} = -1.5$$

Hence the least squares straight line of x on y is $x = 1.5 + 1.7y$

Now, $\quad x_{est} = x \quad (\text{at } y = 2.5)$

$$= 2.75 \quad (13)$$

Example 5.7

Fit a parabola to the data points given in the table below:

x	0.0	1.0	2.0
y	1.0	6.0	17.0

Solution Constructing a table of expressions required for quadratic regression, we have

	x	y	x^2	x^3	x^4	xy	x^2y
	0	1.0	0	0	0	0	0
	1.0	6.0	1	1	1	6	6
	2.0	17.0	4	8	16	34	68
	3.0	24.0	5	9	17	40	74
$n = 3$	$\sum x$	$\sum y$	$\sum x^2$	$\sum x^3$	$\sum x^4$	$\sum xy$	$\sum x^2y$

The normal equations are

$$\left.\begin{aligned} \sum y &= +na + b\sum x + c\sum x^2 \\ \sum xy &= a\sum x + b\sum x^2 + c\sum x^3 \\ \sum x^2y &= a\sum x^2 + b\sum x^3 + c\sum x^4 \end{aligned}\right\} \quad (1)$$

Substituting the values, we have

$$\left.\begin{aligned} 3a + 3b + 5c &= 24 \\ 3a + 5b + 9c &= 40 \\ 5a + 9b + 17c &= 74 \end{aligned}\right\} \quad (2)$$

$$\begin{bmatrix} 3 & 3 & 5 & \vdots & 24 \\ 3 & 5 & 9 & \vdots & 40 \\ 5 & 9 & 17 & \vdots & 74 \end{bmatrix} \underset{\sim}{\begin{matrix} R_2 - R_1 \\ R_3 - (5/3)R_2 \end{matrix}} \begin{bmatrix} 3 & 3 & 5 & \vdots & 24 \\ 0 & 2 & 4 & \vdots & 16 \\ 5 & 2/3 & 2 & \vdots & 22/3 \end{bmatrix}$$

$$\underset{\sim}{\frac{1}{2}R_2} \begin{bmatrix} 3 & 3 & 5 & \vdots & 24 \\ 0 & 1 & 2 & \vdots & 8 \\ 0 & 2/3 & 2 & \vdots & 22/3 \end{bmatrix}$$

$$\underset{\sim}{R_3 - \frac{2}{3}R_2} \begin{bmatrix} 3 & 3 & 5 & \vdots & 24 \\ 0 & 1 & 2 & \vdots & 8 \\ 0 & 0 & 1 & \vdots & 3 \end{bmatrix}$$

$$c = 3 \quad b = 2 \quad a = 1 \qquad (3)$$

The equation of the parabola is $y = 1 + 2x + 3x^2$. (4)

Example 5.8

Fit a parabola to the data points given in the table below:

x	0	1	2	3	4
y	1	1.8	1.3	2.5	6.3

Solution Let us construct a table of values of expressions $\sum x, \sum y$, *etc.* from the given data:

	x	y	x^2	x^3	x^4	xy	x^2y
	0	1.0	0	0	0	0	0
	1	1.8	1	1	1	1.8	1.8
	2	1.3	4	8	16	2.6	5.2
	3	2.5	9	27	81	7.5	22.5
	4	6.3	16	64	256	25.2	100.8
$n = 5$	10	12.9	30	100	354	37.1	130.3
	$\sum x$	$\sum y$	$\sum x^2$	$\sum x^3$	$\sum x^4$	$\sum xy$	$\sum x^2y$

Let the equation of the second-degree curve (parabola) be

$$y = a + bx + cx^2 \qquad (1)$$

Its normal equations are

$$\sum y = na + b\sum x + c\sum x^2$$
$$\sum xy = a\sum x + b\sum x^2 + c\sum x^3$$
$$\sum x^2y = a\sum x^2 + b\sum x^3 + c\sum x^4$$

Substituting the values, we have

$$5a + 10b + 30c = 12.9 \qquad \text{(i)}$$
$$10a + 30b + 100c = 37.1 \qquad \text{(ii)}$$
$$30a + 100b + 354c = 130.3 \qquad \text{(iii)} \qquad (2)$$

Dividing 2 (i), (ii) and (iii) by 5,10,10, respectively, and writing the augmented matrix for the system,

$$\begin{bmatrix} 1 & 2 & 6 & \vdots & 2.58 \\ 1 & 3 & 10 & \vdots & 3.71 \\ 3 & 10 & 35.4 & \vdots & 13.03 \end{bmatrix} \underset{\sim}{\begin{matrix} R_2 - R_1 \\ R_3 - 3R_2 \end{matrix}} \begin{bmatrix} 1 & 2 & 6 & \vdots & 2.58 \\ 0 & 1 & 4 & \vdots & 1.13 \\ 0 & 1 & 5.4 & \vdots & 1.90 \end{bmatrix}$$

$$\underset{\sim}{R_3 - R_2} \begin{bmatrix} 1 & 2 & 6 & 2.58 \\ 0 & 1 & 4 & 1.13 \\ 0 & 0 & 1.4 & 0.77 \end{bmatrix}$$

By back substitution $c = 0.55$

$$b = 1.13 - 4 \times 0.55 = -1.07 \qquad (3)$$
$$a = 2.58 - 6 \times 0.55 - 2 \times (-1.07) = 1.42$$

The required second-degree curve is

$$y = 1.42 - 1.07x + 0.55x^2 \qquad (4)$$

Example 5.9

Find the parabola of the form $y = a + bx + cx^2$ which fits most closely with the observations

x	−3	−2	−1	0	1	2	3
y	4.63	2.11	0.67	0.09	0.63	2.15	4.58

[JNTU 2003]

Solution We compute the values of $\sum x, \sum y$, *etc.* and tabulate below:

	x	y	x^2	x^3	x^4	xy	x^2y
	−3	4.63	9	−27	81	−13.87	41.67
	−2	2.11	4	−8	16	−4.22	8.44
	−1	0.67	1	−1	1	−0.67	0.67
	0	0.09	0	0	0	0	0
	1	0.63	1	1	1	0.63	0.63
	2	2.15	4	8	16	4.30	8.60
	3	4.58	9	27	81	13.74	41.22
$n=7$	0	14.86	28	0	196	−0.11	101.23
	$\sum x$	$\sum y$	$\sum x^2$	$\sum x^3$	$\sum x^4$	$\sum xy$	$\sum x^2y$

(1)

Equation of parabola is

$$y = a + bx + cx^2 \quad (2)$$

Its normal equations are

$$\sum y = na + b\sum x + c\sum x^2$$
$$\sum xy = a\sum x + b\sum x^2 + c\sum x^3$$
$$\sum x^2 y = a\sum x^2 + b\sum x^3 + c\sum x^4$$

Substituting the values, we have

$$7a + 0.b + 28c = 14.86 \quad \text{(i)}$$
$$0.a + 28b + 0.c = -0.11 \quad \text{(ii)}$$
$$28a + 0.b + 196c = 101.23 \quad \text{(iii)} \quad (3)$$

(3) (ii) $\Rightarrow \quad b = \dfrac{-0.11}{28} = -0.00393$ from (i) and (iii)

$$(196 - 4 \times 28)c = 101.23 - 4 \times 14.86 = 41.79$$

$$\Rightarrow \quad c = \frac{41.79}{84} = 0.49750 \quad (4)$$

$$a = \frac{14.86 - 28 \times 0.4975}{7} = 0.13286$$

The required equation of the parabola is

$$y = 0.13286 - 0.00393x + 0.49750x^2 \quad (5)$$

5.3 Curvilinear (or Nonlinear) Regression

The relationship between the dependent and independent variables may sometimes be nonlinear. In such cases we may have to consider higher degree polynomials or transcendental functions for curve fitting.

5.3.1 Polynomial Regression

Let

$$y = \sum_{i=0}^{n} a_i x^i \quad (5.22)$$

be a polynomial in x of degree n for a given set of n pairs of observations (x_i, y_i); the unknowns a_i (i = 0, 1, ..., n) are estimated by the least squares method by minimizing

$$S = \sum e_i^2 = \sum_{i=0}^{n}\left[y_i - \left(\sum_{i=0}^{n} a_i x^i\right)\right]^2 \quad (5.23)$$

Proceeding on the same lines as for linear regression, we obtain

Normal equations

$$\sum y_i = na_0 + a_1\sum x_i + a_2\sum x_i^2 + \cdots a_n\sum x_i^n$$
$$\sum x_i y_i = a_0\sum x_i + a_1\sum x_i^2 + \cdots + a_n\sum x_i^{n+1}$$
$$\sum x_i^n y_i = a_0\sum x_i^n + a_1\sum x_i^{n+1} + \cdots a_n\sum x_i^{2n}$$
$$(5.24)$$

Special Cases

1. $n = 2$ parabola (or quadratic curve) $y = a_0 + a_1x + a_2x^2$ (which we have considered above).
2. $n = 3$ cubic curve: $y = a_0 + a_1x + a_2x^2 + a_3x^3$.
3. $n = 4$ quadratic (or biquadratic) curve: $y = a_0 + a_1x + a_2x^2 + a_3x^3 + a_4x^4$.

5.3.2 Transcendental Curves

1. Exponential growth curve: $y = ab^x$.
2. Exponential decay curve: $y = ab^{-x}$.
3. Power (geometric) curve: $y = ax^b$.
4. Reciprocal curve: $y = \dfrac{1}{a + bx}$.

We can transform (1)–(3) to linear form by taking logarithms, and (4) can be transformed by putting $Y = \dfrac{1}{y}$ as already mentioned above.

Example 5.10

Fit an exponential curve of the form $y = ab^x$ for the data:

x	1	2	3	4
y	7	11	17	27

Solution Applying logarithms on both sides of the nonlinear exponential curve $y = ab^x$, we obtain

$$\ln y = \ln a + x \ln b \quad \text{or} \quad Y = A + Bx \quad (1)$$

where $Y = \ln y$, $A = \ln a$ and $B = \ln b$

We construct the table of values of $\sum X, \sum Y$, *etc.*

	x	y	$Y = \ln y$	x^2	xY
	1	7	1.946	1	1.946
	2	11	2.398	4	4.796
	3	17	2.833	9	8.499
	4	27	3.296	16	13.184
$n = 4$	10	–	10.473	30	28.425
	$\sum x$		$\sum Y$	$\sum x^2$	$\sum xY$

The normal equations are

$$\sum Y = nA + B\sum x$$
$$\sum xY = A\sum x + B\sum x^2 \quad (2)$$

Substituting the values, we obtain

$$4A + 10B = 10.473$$
$$10A + 30B = 28.425 \quad (3)$$

$$\Rightarrow \quad A + 3B = 2.84250$$
$$A + 2.5B = 2.61825$$

Subtracting $0.5B = 0.22425 \quad \therefore B = 0.4485$

Hence $b = e^B = 1.566$

$$A = 2.84250 - 1.3455 = 1.497 \quad (4)$$

Hence

$$a = e^A = 4.4683 \quad (5)$$

Therefore $y = 4.47(1.566)^x$.

Example 5.11
Fit a geometric curve to the following data:

x	1	2	4	6
y	6	4	2	2

Estimate y (2.5).

Solution Let the equation of the curve be $y = ax^b$. Taking logarithms on both sides, we have

$$\ln y = \ln a + b \ln x \quad \text{or} \quad Y = A + bX \quad (1)$$

where $X = \ln x$; $Y = \ln y$ and $A = \ln a$.

We construct a table of values $\sum x, \sum y$, *etc.*

	x	y	$X = \ln x$	$Y = \ln y$	X^2	XY
	1	6	0	1.792	0	0
	2	4	0.693	1.386	0.480	0.960
	4	2	1.386	0.693	1.921	0.960
	6	2	1.792	0.693	3.211	1.242
$n = 4$			3.871	4.564	5.612	3.162
			$\sum X$	$\sum Y$	$\sum X^2$	$\sum XY$

The normal equations are

$$\sum Y = nA + b\sum X$$
$$\sum XY = A\sum X + b\sum X^2 \quad (2)$$

Substituting the values, we obtain

$$4A + 3.871b = 4.564 \quad \text{or} \quad 4A + 3.87b = 4.56$$
$$3.871A + 5.612b = 3.162 \quad 3.87A + 5.61b = 3.16$$

(rounding off to two decimals) (3)

Solving these equations,

$$A = \frac{\begin{vmatrix} 4.56 & 3.87 \\ 3.16 & 5.61 \end{vmatrix}}{\begin{vmatrix} 4 & 3.87 \\ 3.87 & 5.61 \end{vmatrix}} = \frac{13.352}{7.463} = 1.789 \quad (4)$$

$$a = e^A = 5.983 \text{ and} \quad (5)$$

$$b = \frac{\begin{vmatrix} 4 & 4.56 \\ 3.87 & 3.16 \end{vmatrix}}{\begin{vmatrix} 4 & 3.87 \\ 3.87 & 5.61 \end{vmatrix}} = \frac{-5.007}{7.463} = -0.671 \quad (6)$$

Equation of the curve is $y = 5.983x^{-0.671}$ (7)

Example 5.12
Fit the ideal gas law $pv^\gamma = c$ to the following data:

Pressure p (lb/in^2)	(y)	16.6	39.7	78.5	115.5	195.3	546.1
Volume v (in^3)	(x)	50	30	20	15	10	5

Solution Applying logarithms on both sides of the equation

$$pv^{\gamma}=c \quad \text{or} \quad y=v^{-\gamma}(y=p, x=v)$$

we obtain

$$\ln y=\ln c-\gamma \ln x \quad \text{or} \quad Y=A+BX \qquad (1)$$

where $Y = \ln y = \ln p$; $A = \ln c$, $B = -\gamma$ and $X=\ln x=\ln v$

We construct a table of values of $\sum x, \sum y$, *etc.*

	x	y	$X=\ln x$	$Y=\ln y$	X^2	XY
	50	16.6	3.912	2.809	15.304	10.989
	30	39.7	3.401	3.681	11.567	12.519
	20	78.5	2.996	4.363	8.976	13.072
	15	115.5	2.708	4.749	5.333	12.860
	10	195.3	2.303	5.275	5.304	12.148
	5	546.1	1.609	6.303	2.589	10.142
$n=6$			16.929	27.180	51.073	71.730
			$\sum X$	$\sum Y$	$\sum X^2$	$\sum XY$

The normal equations are

$$nA+B\sum X=\sum Y \qquad (2)$$

$$A\sum X+B\sum X^2=\sum XY \qquad (3)$$

Substituting values, we have

$$6A+16.929B=27.180 \qquad (4)$$

$$16.929A+51.073B=71.730 \qquad (5)$$

$$(4) \Rightarrow A+2.822B=4.530$$

$$(5) \Rightarrow A+3.017B=4.237$$

$$\Rightarrow \quad B=\frac{-0.293}{0.195}=-1.5026 \qquad (6)$$

$$\left.\begin{aligned} A&=4.530-2.822(-1.5026) \\ c&=e^{A}=6440.344 \end{aligned}\right\} \qquad (7)$$

$$\gamma=1.5026 \qquad (8)$$

The best fit is $pv^{\gamma}=c$ with the values of c and γ at equations (7) and (8).

Example 5.13

Reciprocal function

Estimate y at $x=2.25$ by fitting an indifference curve of the form $xy=ax+b$ to the following data:

x	1	2	3	4
y	3	1.5	6	7.5

Solution The equation $xy=ax+b$ can be written as $y=a+bX$ where $X=1/x$. Writing the table of values of $\sum X, \sum y$, *etc*, we have

	x	X	y	X^2	Xy
	1	1.00	3	1.00	3.0000
	2	0.50	1.5	0.25	0.7500
	3	0.33	6	0.1111	1.9998
	4	0.25	7.5	0.0625	1.8750
$n=4$	Total	2.08	18.0	1.4236	7.6248
		$\sum X$	$\sum y$	$\sum X^2$	$\sum Xy$

The normal equations are

$$na+b\sum X=\sum y \qquad (1)$$

$$a\sum X+b\sum X^2=\sum Xy \qquad (2)$$

Substituting the values, we have

$$4a+2.08b=18.0 \qquad \text{(i)}$$

$$2.08a+1.4236b=7.6248 \quad \text{(ii)}$$

$$\times 2.08 \Rightarrow 8.32a+4.3264b=37.4400$$

$$\times 4 \quad \Rightarrow 8.32a+5.6944b=30.4992 \qquad (3)$$

Substracting $1.3680b=-6.9408$

$$\Rightarrow \quad b:-5.0737$$

Substituting for b in (i) we obtain $a=7.1383$.

The required equation of the curve is

$$y=7.1383-\frac{5.0737}{x} \quad \text{or} \quad xy=(7.1383)x-5.0737$$

Now estimate y at $x=2.25$ is

$$y_{2.25}=7.1383-\frac{5.0737}{2.25}=4.8833 \qquad (4)$$

5.4 Curve Fitting by a Sum of Exponentials

A problem that occurs frequently in Engineering and Physics is that of fitting a sum of exponentials of the form

$$y = f(x) = A_1e^{\lambda_1 x} + A_2e^{\lambda_2 x} + A_3e^{\lambda_3 x} + \cdots + A_n e^{\lambda_n x} \tag{5.25}$$

to a set of data points (x_i, y_i); $i = 1, 2, \ldots, n$.

We assume that n is known and A_i, λ_i are to be found. From the form of (5.22) we observe that $f(x)$ satisfies a differential equation of the form

$$\frac{d^n y}{dx^n} + a_1 \frac{d^{n-1} y}{dx^{n-1}} + a_2 \frac{d^{n-2} y}{dx^{n-2}} + \cdots + a_n y = 0 \tag{5.26}$$

where the coefficients a_i are unknown.

A method suggested by Froberg is to evaluate the derivatives $\frac{d^{ny}}{dx^n}, \frac{d^{n-1}}{dx^{n-1}} \ldots$ at the n data points and to substitute them in (5.22). Thus, we obtain a system of n linear equations; solving these we find a_i. Also, λ_i are the roots of the equation

$$\lambda^n + a_1 \lambda^{n-1} + \ldots + a_n = 0 \tag{5.27}$$

which, when solved, enables us to compute $A_1, A_2 \ldots A_n$ from (5.22) by the method of least squares. This method has a disadvantage. The numerical evaluation of the derivatives whose accuracy decreases with their increasing order leads to unreliable results.

We discuss now a method due to Moore which leads to more reliable results. We demonstrate it for the case $n = 2$.

Let the function to be fitted to the data be of the form

$$y = A_1e^{\lambda_1 x} + A_2e^{\lambda_2 x} \tag{5.28}$$

This satisfies a differential equation

$$\frac{d^2 y}{dx^2} = a_1 \frac{dy}{dx} + a_2 y \tag{5.29}$$

where a_1 and a_2 are constants to be determined. Assume that 'a' is the initial value of x. Integrating (5.29) from a to x, we obtain

$$y'(\mathrm{x}) - y'(\mathrm{a}) = \mathrm{a}_1 y(x) - a_1 y(a) + a_2 \int_a^x y(x)\, dx, \quad (y' = dy/dx) \tag{5.30}$$

Integrating (5.30) again we obtain

$$y(\mathrm{x}) - y'(\mathrm{a}) - y'(a)(x-a) = a_1 \int_a^x y(x)\, dx - a_1(x-a)y(\mathrm{a}) + a_2 \int_a^x \int_a^x y(x)\, dx\, dx \tag{5.31}$$

By the formula

$$\int_a^x \cdots \int_a^x f(x)\, dx \cdots dx = \frac{1}{(n-1)!} \int_a^x (x-t)^{n-1} f(t)\, dt \tag{5.32}$$

equation (5.31) becomes

$$y(x) - y(a) - (x-a)y'(a) = a_1 \int_a^x y(x)\, dx - a_1(x-a)y(a) + a_2 \int_a^x (x-t)y(t)\, dt \tag{5.33}$$

Now, to eliminate $y'(a)$ from equation (5.33), choose two data points x_1 and x_2 such that $a - x_1 = x_2 - a$ so that we have from (5.33)

$$y(x_1) - y(a) - (x_1 - a)y'(a) = a_1 \int_a^{x_1} y(x)\, dx - a_1(x_1 - a)y(a) + a_2 \int_a^{x_1} (x_1 - t)y(t)\, dt$$

$$y(x_2) - y(a) - (x_2 - a)y'(a) = a_1 \int_a^{x_2} y(x)\, dx - a_1(x_2 - a)y(a) + a_2 \int_a^{x_2} (x_2 - t)y(t)\, dt$$

Adding these equations and simplifying, we obtain

$$y(x_1) + y(x_2) - 2y(a) = a_1 \left[\int_a^{x_1} y(x)\, dx + \int_a^{x_2} y(x)\, dx \right] + a_2 \left[\int_a^{x_1} (x_1 - t)y(t)\, dt + \int_a^{x_2} (x_2 - t)y(t)\, dt \right] \tag{5.34}$$

Equation (5.34) can now be used to set up a linear system of equations for a_1 and a_2. λ_1 and λ_2 can be determined from the characteristic equation

$$\lambda^2 = a_1 \lambda + a_2 \tag{5.35}$$

A_1 and A_2 can then be obtained by the method of least squares.

Example 5.14

Fit a function $y = f(x) = A_1e^{\lambda_1 x} + A_2e^{\lambda_2 x}$, a sum of exponentials to the following data using Moore's technique:

x	2.0	2.2	2.4	2.6	2.8	3.0	3.2	3.4	3.6
y	3.63	4.46	5.47	6.70	8.19	10.02	12.25	14.97	18.29

Solution Choose $x_1 = 2.0, x_2 = 2.8$ and a such that $a - x_1 = x_2 - a \Rightarrow \quad a - 2.0 = 2.8 - a$
$\Rightarrow \quad a = 2.4$

Equation (5.34) gives

$$y(2) + y(2.8) - 2y(2.4) = a_1\left[\int_{2.4}^{2} y(x)\,dx + \int_{2.4}^{2.8} y(x)\,dx\right] + a_2\left[\int_{2.4}^{2} (2-t)y(t)\,dt + \int_{2.4}^{2.8} (2.8-t)y(t)\,dt\right] \quad (1)$$

The integrals on the RHS are evaluated using Simpson's $\frac{1}{3}$ rule on two subintervals with three points. For example

$$I_1 = -\int_{2}^{2.4} y(x)\,dx = -\frac{0.2}{3}[3.63 + 4(4.46) + 5.47] \quad (2)$$

$$I_2 = \int_{2.4}^{2.8} y(x)\,dx = \frac{0.2}{3}[5.47 + 4(4.46) + 8.19] \quad (3)$$

$$I_3 = -\int_{2}^{2.4} (2-t)y(t)\,dt = -\frac{0.2}{3}[(2-2)3.63 + 4(2-2.2)4.46 + (2-2.4)5.47] \quad (4)$$

$$I_4 = \frac{0.2}{3}[(2.8-2.4)5.47 + 4(2.8-2.6)6.70 + (2.8-2.8)8.19], \textit{ etc.} \quad (5)$$

After simplification, this yields one equation for a_1 and a_2 as

$$0.88 = 0.90a_1 + 0.887a_2 \quad (6)$$

Similarly choosing $x_1 = 2.8$, $x_2 = 3.6$ we get a from $a - x_1 = x_2 - a$, $a - 2.8 = 3.6 - a$ as $a = 3.2$.
Substituting the data in (5.34) we have

$$y(2.8) + y(3.6) - 2y(3.2) = a_1\left[\int_{3.2}^{2.8} y(x)\,dx + \int_{3.2}^{3.6} y(x)\,dx\right] + a_2\left[\int_{3.2}^{2.8} (2.8-t)y(t)\,dt + \int_{3.2}^{3.6} (3.6-t)y(t)\,dt\right] \quad (7)$$

After evaluation of the intergrals on the RHS using Simpson's $\frac{1}{3}$ rule, we get a second equation for a_1 and a_2:

$$1.98 = 0.6733\,a_1 + 1.986133\,a_2 \quad (8)$$

Solving (6) and (8) we get

$$a_1 = 0.0071115, \quad a_2 = 0.999324 \quad (9)$$

Substituting values of a_1, a_2 in

$$\lambda^2 - a_1\lambda - a_2 = 0 \text{ we have } \lambda^2 + 0.007\lambda - 0.999 = 0 \quad (10)$$

whose roots are

$$\lambda_1 = -1.0031, \quad \lambda_2 = 0.99615 \quad (11)$$

The equation takes the form

$$y = A_1e^{-1.0031x} + A_2e^{0.99615x}$$
$$\Rightarrow \quad y = A_1e^{-x} + A_2e^{0.99x} \quad (12)$$

The values of A_1 and A_2 are determined by the method of least squares by solving the two normal equations

$$A_1\sum_{k=1}^{9} e^{-2x_k} + A_2\sum_{k=1}^{9} e^{-0.01x_k} = \sum_{k=1}^{9} y_ke^{-x_k} \quad (13)$$

$$\text{and} \quad A_1\sum_{k=1}^{9} e^{-0.01x_k} + A_2\sum_{k=1}^{9} e^{1.98x_k} = \sum_{k=1}^{9} y_ke^{0.99x_k} \quad (14)$$

The values of A_1 and A_2 are

$$A_1 = -0.51, A_2 = 0.499 \quad (15)$$

$$y = f(x) = -0.51e^{-x} + 0.499e^{x} \quad (16)$$

The above data have been computed from the exact function $y = \sinh x$ whose values have been rounded off to two decimal places.

Example 5.15

Fit a function of the form

$$y = A_1e^{\lambda_1 x} + A_2e^{\lambda_2 x}$$

to the data

x	1.0	1.1	1.2	1.3	1.4	1.5	1.6	1.7	1.8
y	1.54	1.67	1.81	1.97	2.15	2.35	2.58	2.87	3.11

Solution Let us take

$$x_1 = 1.0, \quad a = 1.2, \quad x_2 = 1.4 \tag{1}$$

Equation (5.34) gives

$$0.07 = a_1\left[-\int_{1.0}^{1.2} y(x)\,dx + \int_{1.2}^{1.4} y(x)\,dx\right] + a_1\left[-\int_{1.0}^{1.2}(1.0-t)y(t)\,dx + \int_{1.2}^{1.4}(1.4-t)y(t)\,dt\right]$$

Evaluating the integrals by Simpson's $\frac{1}{3}$ rule and simplifying, we obtain

$$1.81a_1 + 2.180a_2 = 2.10 \tag{2}$$

Let us take

$$x_1 = 1.4, \quad a = 1.6, \quad x_2 = 1.8 \tag{3}$$

So, we get

$$2.88\,a_1 + 3.104\,a_2 = 3.00 \tag{4}$$

Solving (2) and (4) we obtain

$$a_1 = 0.03204 \quad \text{and} \quad a_2 = 0.9364$$

Equation $\lambda^2 - a_1\,\lambda - a_2 = 0$ now gives

$$\lambda^2 - 0.03204\lambda - 0.9364 = 0 \tag{5}$$

$$\Rightarrow \lambda_1 = 0.988 = 0.99 \quad \text{and} \quad \lambda_2 = -0.96 \tag{6}$$

By the method of least squares, we obtain

$$A_1 = 0.499, \quad A_2 = 0.491 \tag{7}$$

The above data have been constructed from $y = \cosh x$ so that

$$A_1 = A_2 = \frac{1}{2}, \quad \lambda_1 = 1.0, \quad \lambda_2 = -1.0 \tag{8}$$

5.5 Weighted Least Squares Approximation

In least squares approximation, we have minimized the sum of the errors. A more general procedure is to minimize the weighted sum of the squares of the errors taken over all data points. By a weight function on $[a, b]$ we mean a function $w(x) \geq 0$ for $x \in [a, b]$ and integrable there. The purpose of a weight function is to give varying degrees of importance to approximations on certain portions of the interval. If this sum is denoted by S then $S = \sum w_i\,[y_i - f(x_i)]^2 = \sum w_i e_i^2$.

We consider linear and nonlinear cases below.

5.5.1 Linear Weighted Least Squares Approximation

Let

$$y = a + bx \tag{5.36}$$

be the straight line to be fitted to the given data. Then

$$S(a,b) = \sum_{i=1}^{m} w_i\,[y_i - (a + bx_i)]^2 \tag{5.37}$$

The necessary conditions for (5.34) to be a minimum

$$\frac{\partial S}{\partial a} = 0 \quad \text{and} \quad \frac{\partial S}{\partial b} = 0$$

$$\Rightarrow \begin{cases} -2\sum w_i\,[y_i - (a + bx_i)]^2 = 0 & (5.38) \\ -2\sum w_i\,[y_i - (a + bx_i)]x_i^2 = 0 & (5.39) \end{cases}$$

from which we obtain the normal equations

$$a\sum w_i + b\sum w_i x_i = \sum w_i y_i \tag{5.40}$$

$$a\sum w_i x_i + b\sum w_i x_i^2 = \sum w_i x_i y_i \tag{5.41}$$

5.5.2 Nonlinear Weighted Least Squares Approximation

Now, consider the least squares approximation of a set of m data points (x_i, y_i), $i = 1, 2, \ldots, m$ by a polynomial

$$y = \sum_{r=0}^{n} a_r x^r \tag{5.42}$$

of degree $n < m$. We then have

$$S(a_0, a_1 \ldots a_n) = \sum_{i=0}^{m} w_i\left[y_i - \sum_{r=0}^{m} a_r x^r\right]^2 \tag{5.43}$$

For S to be minimum at $(a_0, a_1 \ldots a_n)$ we must have

$$\frac{\partial S}{\partial a_i} = 0 \tag{5.44}$$

These conditions yield the normal equations

$$\left.\begin{aligned} a_0\sum w_i + a_1\sum w_i x_i + \cdots + a_n\sum w_i x_i^n &= \sum w_i y_i \\ a_0\sum w_i x_i + a_1\sum w_i x_i^2 + \cdots + a_n\sum w_i x_i^{n+1} &= \sum w_i x_i y_i \\ \cdots \quad \cdots \quad \cdots \quad \cdots & \\ a_0\sum w_i x_i^n + a_1\sum w_i x_i^{n+1} + \cdots + a_n\sum w_i x_i^{2n} &= \sum w_i x_i^n y_i \end{aligned}\right\} \tag{5.45}$$

These are $(n + 1)$ equations in $(n + 1)$ unknowns a_i $(i = 0, 1 \ldots n)$. If x_i are distinct and $n < m$ then the equations have a unique solution.

Example 5.16

Fit a linear weighted least squares straight line to the data and estimate y (1).

x	−2	0	2	4	6
y	1	3	6	8	13
w	1	1	10	1	1

Solution Let $y = a + bx$ be the straight line. Normal equations are

$$a\sum_{i=1}^{5} w_i + b\sum_{i=0}^{5} w_i x_i = \sum_{i=0}^{5} w_i y_i \tag{1}$$

$$a\sum_{i=1}^{5} w_i x_i + b\sum_{i=1}^{5} w_i x_i^2 = \sum_{i=1}^{5} w_i x_i y_i \tag{2}$$

	x	y	w	wx	wx^2	wy	wxy
	−2	1	1	−2	4	1	−2
	0	3	1	0	0	3	0
	2	6	10	20	40	60	120
	4	8	1	4	16	3	32
	6	13	1	6	36	13	78
$n = 5$			14	28	96	85	230
			$\sum w$	$\sum wx$	$\sum wx^2$	$\sum wy$	$\sum wxy$

(3)

Substituting the values, we have

$$\begin{aligned} 14a + 28b &= 85 \quad \times 2 \quad \Rightarrow \quad 28a + 56b = 170 \\ 28a + 96b &= 230 \quad \times 1 \quad \Rightarrow \quad 28a + 96 = 230 \end{aligned} \tag{4}$$

Subtracting $-40b = -60 \quad \Rightarrow \quad b = 1.5$ (5)

$$\therefore \quad a = \frac{85 - 28 \times 1.5}{14} = 3.0714 \tag{6}$$

The equation of the straight line: $y = 3.07 + 1.5x$ (7)

$$\therefore \quad y\,(x = 1) = 3.07 + 1.5 \times 1 = 4.57 \tag{8}$$

EXERCISE 5.1

1. Find a least squares straight line for the following data:

x	1	2	3	4
y	5	3	6	7

Ans: $y = 3 + 0.9x$

2. Fit a straight line to the following data:

x	1	2	3	4	5
y	3	4	5	6	8

Ans: $y = 1.6 + 1.2x$

3. Fit a quadratic polynomial to the following data:

x	1	3	5	7	9
y	2	7	10	11	9

Ans: $y = -1.556 + 3.757x - 0.2857x^2$

4. Fit the following data to the law $pv^\gamma = c$:

p	200	100	50	30	20	10
v	1.0	1.7	2.9	4.8	5.9	10

Ans: $pv^{1.3} = 200$

5. Fit the curve $y = ae^{bx}$ to the following data:

x	0	2	4
y	5.1	10	31.1

[Andhra 1999]

Ans: $y = 4.1e^{0.43}$

6. Fit a least square geometric curve $y = ax^b$ to the data:

x	1	2	3	4	5
y	0.5	2	4.5	8	12.5

Ans: $y = 0.5012\,x^{1.9977}$

7. Fit a second-degree parabola to the following data:

x	0	1	2	3	4
y	1.0	1.8	1.3	2.5	6.3

[JNTU 1999]

Ans: $y = 1.48 - 1.07x + 0.55x^2$

8. Fit a curve $y = ab^x$ to the data:

x	2	4	6	8	10
y	3.07	12.85	31.47	57.38	91.20

Ans: $y = 1.922(1.513)x^2$

9. For the data given below find the best fitting curve of the form $y = ax^b$:

x	1	2	3	4	5	6
y	2.98	4.26	5.21	6.10	6.8	7.5

Ans: $y = 2.978x^{-0.544}$

10. Find the curve of best fit of the type $y = ae^{bx}$ to the data:

x	1	5	7	9	12
y	10	15	12	15	21

Ans: $y = 9.4754e^{0.059x}$

11. Growth of bacteria (N) in a culture after t hours is given in the following table:

t	0	1	2	3	4	5	6
N	32	47	65	92	132	190	275

Fit a curve of the form $N = ab^t$ and estimate N for $t = 7$. $a = 32.15$, $b = 1.43$, $N = 387$.

Ans: $a = 32.15$, $b = 1.43$, $N = 387$

12. The pressure and volume of a gas are related by $pv^\gamma = c$. Fit it to the data:

p(kg/cm^2)	0.5	1.0	1.5	2.0	2.5	3.0
v(litres)	1.62	1.00	0.75	0.62	0.52	0.46

Ans: $pv^{1.276} = 1.039$

13. Fit in a power function of the form $y = ax^b$ to the data:

x	1	2	3	4	5
y	0.5	2.0	4.5	8.0	12.5

(the actual function is $y = \frac{1}{2}x^2$)

Ans: $y = 0.50001x^{1.9998}$

14. Fit a linear weighted least squares straight line to the data:

x	−2	0	2	4	6
y	1	3	6	8	13
w	2	5	10	1	4

Ans: $y = 3.127 + 1.55x$

Numerical Differentiation and Integration

6

6.1 Introduction

Situations arise frequently when the conventional calculus methods of differentiation and integration cannot be applied to a function, for instance,

1. when we do not know the exact function but know only values of it at a set of discrete points; and
2. when the function is known but is too complicated to apply the conventional methods of calculus.

In such cases we apply numerical methods. The method of finding the derivative dy/dx / integral $\int_a^b f(x)\,dx$ of a function $y=f(x)$ using numerical techniques is called numerical differentiation/numerical integration.

6.1.1 Numerical Differentiation

Numerical differentiation is a process of computing the derivative of a function at some assigned value of x from a given set of data

$$(x_i, f_i) \quad f_i = f(x_i) \quad (i = 0, 1, 2, \ldots, n)$$

which corresponds to the values of an unknown function $y = f(x)$.

To find dy/dx we first replace $y = f(x)$ by an interpolating polynomial $y = \phi(x)$ and then differentiate it as many times as desired.

The choice of the interpolation formula depends on the value of x at which dy/dx is to be computed.

If the points are equally spaced, then we use Newton's formula. We use forward formula or backward formula depending on whether we require dy/dx at the beginning or at the end of the table.

If we require dy/dx at the middle, we use a central difference formula.

If the points are not equally spaced, we use Newton's divided difference formula or Lagrange's formula.

6.1.2 Numerical Differentiation by Newton's Forward Interpolation Formula

Let $y = f(x)$ be specified in an interval $[a, b]$ at equally spaced points $x_i = x_0 + ih\ (i = 0, 1, 2, \ldots, n)$ by means of the values $y_i = f(x_i)$. Suppose the tabulated function $y = f(x)$ is replaced by Newton's forward interpolation formula passing through the $(n + 1)$ points $x_i\ (i = 0, 1, 2, \ldots, n)$. Then

$$y(x) = y_0 + p\Delta y_0 + \frac{p(p-1)}{2!}\Delta^2 y_0 + \frac{p(p-1)(p-2)}{3!}\Delta^3 y_0 + \frac{p(p-1)(p-2)(p-3)}{4!}\Delta^4 y_0 + \cdots \tag{6.1}$$

where $p = \dfrac{x - x_0}{h}$ and $h = x_{i+1} - x_i$ for $i = 0, 1, \ldots$.
Here p is a function of x and $dp/dx = 1/h$. We can rewrite (6.1) as

$$y(x) = y_0 + p\Delta y_0 + \frac{p^2 - p}{2}\Delta^2 y_0 + \frac{p^3 - 3p^2 + 2p}{6}\Delta^3 y_0 + \frac{p^4 - 6p^3 + 11p^2 - 6p}{24}\Delta^4 y_0 + \cdots \tag{6.2}$$

Differentiating (6.2) with respect to x successively, we have

$$\frac{dy}{dx}=\frac{dy}{dp}\cdot\frac{dp}{dx}=\frac{1}{h}\frac{dy}{dp}$$

$$=\frac{1}{h}\left[\Delta y_0+\frac{2p-1}{2}\Delta^2 y_0+\frac{3p^2-6p+2}{6}\Delta^3 y_0+\frac{4p^3-18p^2+22p-6}{24}\Delta^4 y_0+\cdots\right] \quad (6.3)$$

$$\frac{d^2y}{dx^2}=\frac{d}{dx}\left(\frac{dy}{dx}\right)=\frac{1}{h^2}\frac{d^2y}{dp^2}=\frac{1}{h^2}\left[\Delta^2 y_0+(p-1)\Delta^3 y_0+\frac{6p^2-18p+11}{12}\Delta^4 y_0+\cdots\right] \quad (6.4)$$

and so on.

Special Case

When we require dy/dx at a basic tabulated point x_i we choose $x = x_0$ so that $p = 0$. Then

$$\left.\frac{dy}{dx}\right|_{x=x_0}=y'(x_0)=\frac{1}{h}\left[\Delta y_0-\frac{1}{2}\Delta^2 y_0+\frac{1}{3}\Delta^3 y_0-\frac{1}{4}\Delta^4 y_0+\frac{1}{5}\Delta^5 y_0-\frac{1}{6}\Delta^6 y_0+\cdots\right] \quad (6.5)$$

$$\left.\frac{d^2y}{dx^2}\right|_{x=x_0}=y''(x_0)=\frac{1}{h^2}\left[\Delta^2 y_0-\Delta^3 y_0+\frac{11}{12}\Delta^4 y_0-\frac{5}{6}\Delta^5 y_0+\cdots\right] \quad (6.6)$$

$$\left.\frac{d^3y}{dx^3}\right|_{x=x_0}=y'''(x_0)=\frac{1}{h^3}\left[\Delta^3 y_0-\frac{3}{2}\Delta^4 y_0+\frac{7}{4}\Delta^5 y_0-\cdots\right] \quad (6.7)$$

$$\left.\frac{d^4y}{dx^4}\right|_{x=x_0}=y''''(x_0)=\frac{1}{h^4}\left[\Delta^4 y_0-2\Delta^5 y_0+\cdots\right] \quad (6.8)$$

and so on.

6.1.3 Numerical Differentiation by Newton's Backward Interpolation Formula

Differentiating with respect to x Newton's Backward Interpolation Formula

$$y(x)=y_0+p\nabla y_n+\frac{p(p+1)}{2!}\nabla^2 y_n+\frac{p(p+1)(p+2)}{3!}\nabla^3 y_n+\cdots$$

we obtain

$$y'(x)=\frac{1}{h}\left[\nabla y_n+\frac{1}{2}(2p+1)\nabla^2 y_n+\frac{3p^2+6p+2}{6}\nabla^3 y_n+\cdots\right] \quad (6.9)$$

$$y''(x)=\frac{1}{h^2}\left[\nabla^2 y_n+(p+1)\nabla^3 y_n+\frac{6p^2+18p+11}{12}\nabla^4 y_n\cdots\right] \quad (6.10)$$

Special Case

When $x = x_n$, $p = 0$ and we have

$$y'(x_n)=\frac{1}{h}\left[\nabla y_n+\frac{1}{2}\nabla^2 y_n+\frac{1}{3}\nabla^3 y_n+\frac{1}{4}\nabla^4 y_n+\cdots\right] \quad (6.11)$$

$$y''(x_n)=\frac{1}{h^2}\left[\nabla^2 y_n+\nabla^3 y_n+\frac{11}{12}\nabla^4 y_n\cdots\right] \quad (6.12)$$

6.1.4 Numerical Differentiation by Stirling's Formula

Differentiating Stirling's formula

$$y(x)=y_0+\frac{p}{1!}\left(\frac{\Delta y_0+\Delta y_{-1}}{2}\right)+\frac{p^2}{2!}\Delta^2 y_{-1}+\frac{p(p^2-1)}{3!}\left(\frac{\Delta^3 y_{-1}+\Delta^3 y_{-2}}{2}\right)+\frac{p^2(p^2-1)}{4}\Delta^4 y_{-2}+\cdots \quad (6.13)$$

we obtain

$$\frac{dy}{dx} = y'(x) = \frac{1}{h}\left[\frac{\Delta y_0 + \Delta y_{-1}}{2} + p\Delta^2 y_{-1} + \frac{3p^2-1}{6}\left(\frac{\Delta^3 y_{-1} + \Delta^3 y_{-2}}{2}\right) + \frac{2p^3 - p}{12}\Delta^4 y_{-2} + \cdots\right] \quad (6.14)$$

$$\frac{d^2y}{dx^2} = y''(x) = \frac{1}{h^2}\left[\Delta^2 y_{-1} - p\left(\frac{\Delta^3 y_{-1} + \Delta^3 y_{-2}}{2}\right) + \frac{6p^2-1}{12}\Delta^4 y_{-2} + \cdots\right] \quad (6.15)$$

and so on.

Special Case

When $x = x_0, p = 0$

$$y'(x_0) = \frac{1}{h}\left[\frac{\Delta y_0 + \Delta y_{-1}}{2} - \frac{1}{6}\cdot\frac{\Delta^3 y_{-1} + \Delta^3 y_{-2}}{2} + \frac{1}{30}\frac{\Delta^5 . y_{-1} + \Delta^5 y_{-2}}{2} + \cdots\right] \quad (6.16)$$

$$y''(x_0) = \frac{1}{h^2}\left[\Delta^2 y_{-1} - \frac{1}{12}\Delta^4 y_{-2} + \frac{1}{90}\Delta^6 y_{-3} + \cdots\right] \quad (6.17)$$

The above formulae can be derived as follows. We know that

$$E = 1 + \Delta = e^{hD} \quad \Rightarrow \quad D = \frac{1}{h}\log(1+\Delta)$$

$$= \frac{1}{h}\left[\Delta - \frac{1}{2}\Delta^2 + \frac{1}{3}\Delta^3 - \frac{1}{4}\Delta^4 + \cdots\right]$$

$$D^2 = \frac{1}{h^2}\left(\Delta - \frac{1}{2}\Delta^2 + \frac{1}{3}\Delta^3 - \cdots\right)^2$$

$$= \frac{1}{h^2}\left(\Delta^2 - \Delta^3 + \frac{11}{12}\Delta^4 - \cdots\right)$$

$$\left(\frac{dy}{dx}\right)_{x=x_0} = \frac{1}{h}\left(\Delta y_0 - \frac{1}{2}\Delta^2 y_0 + \frac{1}{3}\Delta^3 y_0 + \cdots\right)$$

$$\left(\frac{d^2y}{dx^2}\right)_{x=x_0} = \frac{1}{h^2}\left(\Delta^2 y_0 - \Delta^3 y_0 + \frac{11}{12}\Delta^4 y_0 + \cdots\right)$$

Also, $E^{-1} = 1 - \nabla = e^{-hD} \Rightarrow D = -\frac{1}{h}\log(1-\nabla)$

$$= \frac{1}{h}\left[\nabla + \frac{1}{2}\nabla^2 + \frac{1}{3}\nabla^3 + \frac{1}{4}\nabla^4 + \cdots\right]$$

$$D^2 = \frac{1}{h^2}\left(\nabla + \frac{1}{2}\nabla^2 + \frac{1}{3}\nabla^3 - \cdots\right)^2$$

$$= \frac{1}{h^2}\left(\nabla^2 + \nabla^3 + \frac{11}{12}\nabla^4 - \cdots\right)$$

$$\left(\frac{dy}{dx}\right)_{x=x_0} = \frac{1}{h}\left(\nabla y_0 + \frac{1}{2}\nabla^2 y_0 + \frac{1}{3}\nabla^3 y_0 + \frac{1}{4}\nabla^4 y_0 \cdots\right)$$

$$\left(\frac{d^2y}{dx^2}\right)_{x=x_0} = \frac{1}{h^2}\left(\nabla^2 y_0 + \nabla^3 y_0 + \frac{11}{12}\nabla^4 y_0 + \cdots\right)$$

Example 6.1

Find the first and second derivatives of the function tabulated below at $x = 1.1$ and $x = 1.2$.

Solution We construct a forward difference table for the data

x	y	Δy	$\Delta^2 y$	$\Delta^3 y$	$\Delta^4 y$
1.0	0				
		0.128			
1.2	0.128		0.288		
		0.416		0.048	
1.4	0.544		0.336		0
		0.752		0.048	
1.6	1.296		0.384		0
		1.136		0.048	
1.8	2.432		0.432		
		1.568			
2.0	4.000				

$x = 1.1$

By Newton's forward interpolation formulas (6.3) and (6.4)

Take $x_0 = 1.0$; $p = \dfrac{x - x_0}{h} = \dfrac{1.1 - 1.0}{0.2} = 0.5$

$$y'(1.1) = \frac{1}{h}\left[\Delta y_0 + \frac{2p-1}{2}\Delta^2 y_0 + \frac{3p^2 - 6p + 2}{6}\Delta^3 y_0\right]$$

$$= \frac{1}{0.2}\left[0.128 + 0 + \frac{3(0.5)^2 - 6(0.5) + 2}{6}(0.048)\right]$$

$$= 0.63$$

$$y''(1.1) = \frac{1}{h^2}\left[\Delta^2 y_0 + (p-1)\Delta^3 y_0\right]$$

$$= \frac{1}{(0.2)^2}\left[0.288 + (0.5-1)0.048\right] = 6.6$$

$x = 1.2$

Take $x_0 = 1.2, p = 0$; by formulas (6.5) and (6.6)

$$y'(x_0)\Big|_{x_0=1.2} = \left\{\frac{1}{h}\left[\Delta y_0 - \frac{1}{2}\Delta^2 y_0 + \frac{1}{3}\Delta^3 y_0\right]_{x_0=1.2}\right\}$$

$$= \frac{1}{0.2}\left[0.416 - \frac{1}{2}(0.336) + \frac{1}{3}(0.048)\right] = 1.32$$

$$y''(x_0)\Big|_{x_0=1.2} = \left\{\frac{1}{h^2}\left[\Delta^2 y_0 - \Delta^3 y_0\right]_{x_0=1.2}\right\}$$

$$= \frac{1}{(0.2)^2}\left[0.336 - 0.048\right] = 7.2$$

Example 6.2

From the following data obtain $\frac{dy}{dx}$ and $\frac{d^2y}{dx^2}$ for $x = 0.2$

x	0.0	0.1	0.2	0.3	0.4	0.5
y	0.00000	0.10017	0.20134	0.30452	0.41075	0.52110

Solution We construct a forward difference table for the data

x	y	Δy	$\Delta^2 y$	$\Delta^3 y$	$\Delta^4 y$
0.0	0.00000				
		0.10017			
0.1	0.10017		0.00100		
		0.10117		0.00101	
0.2	0.20134		0.00201		0.00003
		0.10318		0.00104	
0.3	0.30452		0.00305		0.00003
		0.10623		0.00107	
0.4	0.41075		0.00412		
		0.11035			
0.5	0.52110				

We use Stirling formula for tabular values at $x = 0.2$

$$\frac{dy}{dx} = \frac{1}{h}\left[\frac{\Delta y_{-1} + \Delta y_0}{2} - \frac{1}{6}\frac{\Delta^3 y_{-2} + \Delta^3 y_{-1}}{2}\right]$$

$$= \frac{1}{0.1}\left[\frac{0.10117 + 0.10318}{2} - \frac{1}{12}(0.00101 + 0.00104)\right] = 1.020$$

$$\frac{d^2y}{dx^2} = \frac{1}{h^2}\left[\Delta^2 y_{-1} - \frac{1}{12}\Delta^4 y_{-2}\right]$$

$$= \frac{1}{(0.1)^2}\left[0.00201 - \frac{1}{12}(0.00003)\right] = 0.201$$

Example 6.3

From the following table of values of x and y find dy/dx and d^2y/dx^2 at $x = 1.6$:

x	1.0	1.2	1.4	1.6	1.8	2.0	2.2
y	2.7183	3.3201	4.0552	4.9530	6.0496	7.3891	9.0250

Solution Constructing the difference table (Table 6.1) for the data we have

Table 6.1 Difference Table for the Data of Example 6.3

x	y	Δy	$\Delta^2 y$	$\Delta^3 y$	$\Delta^4 y$	$\Delta^5 y$	$\Delta^6 y$
1.0	2.7183						
		0.6018					
1.2	3.3201		0.1333				
		0.7351		0.0294			
1.4	4.0552		0.1627		0.0067		
		0.8978		0.0361		0.0013	
1.6	4.9530		0.1988		0.0080		0.0001
		1.0966		0.0441		0.0014	
1.8	6.0496		0.2429		0.0094		
		1.3395		0.0535			
2.0	7.3891		0.2964				
		1.6359					
2.2	9.0250						

Stirling's formulas

$$\left(\frac{dy}{dx}\right)_{x=x_0} = \frac{1}{h}\left(\frac{\Delta y_{-1}+\Delta y_0}{2} - \frac{1}{6}\frac{\Delta^3 y_{-2}+\Delta^3 y_{-1}}{2} + \frac{1}{30}\frac{\Delta^5 y_{-3}+\Delta^5 y_{-2}}{2} + \cdots\right) \quad (1)$$

$$\left(\frac{d^2y}{dx^2}\right)_{x=x_0} = \frac{1}{h^2}\left(\Delta^2 y_{-1} - \frac{1}{12}\Delta^4 y_{-2} + \frac{1}{90}\Delta^6 y_{-3}\cdots\right) \quad (2)$$

Here $x_0 = 1.6$, $h = 0.2$

$$\left(\frac{dy}{dx}\right)_{x=1.6} = \frac{1}{0.2}\left(\frac{0.8978+1.0966}{2} - \frac{1}{2}\frac{0.361+0.0441}{2} + \frac{1}{30}\frac{0.0013+0.0014}{2}\right) = 4.9530$$

$$\left(\frac{d^2y}{dx^2}\right)_{x=1.6} = \frac{1}{0.04}\left(0.1988 - \frac{1}{12}(0.0080) + \frac{1}{90}(0.0001)\right) = 4.9525$$

Note 1 The tabulated function here is $y = e^x$. The error is more in the case of second derivatives, because the approximating polynomial agrees with the tabulated values but its slopes at the points differ.

Example 6.4

For the following data find $f'(5)$:

x	0	2	3	4	7	9
y	4	26	58	112	466	922

Solution Since the values of x are not equally spaced, we use Newton's divided difference formula

$$f(x) = f(x_0) + (x - x_0)f(x_0, x_1) + (x - x_0)(x - x_1) f(x_0, x_1, x_2) + (x - x_0)(x - x_1)(x - x_2) f(x_0, x_1, x_2, x_3) + (x - x_0)(x - x_1)(x - x_2) (x - x_3) f(x_0, x_1, x_2, x_3, x_4) \quad (1)$$

x	$f(x)$	$\Delta' f(x)$	$\Delta'^2 f(x)$	$\Delta'^3 f(x)$	$\Delta'^4 f(x)$
0	4				
		$\frac{26-4}{2-0}=11$			
2	26		$\frac{32-11}{3-0}=7$		
		$\frac{58-26}{3-2}=32$		$\frac{11-7}{4-0}=1$	
3	58		$\frac{54-32}{4-2}=11$		0
		$\frac{112-58}{4-3}=54$		$\frac{16-11}{7-2}=1$	
4	112		$\frac{118-54}{7-3}=16$		0
		$\frac{466-112}{7-4}=118$		$\frac{22-16}{9-3}=1$	
7	466		$\frac{228-118}{9-4}=22$		
		$\frac{922-466}{9-7}=228$			
9	922				

Differentiating (1) with respect to x we have

$$f'(x) = f(x_0, x_1) + (x - x_0 + x - x_1) f(x_0, x_1, x_2) + [(x - x_1)(x - x_2) + (x - x_0)(x - x_2) + (x - x_0)(x - x_1)] f(x_0, x_1, x_2, x_3)$$

Substituting the values $x = 5$, $x_0 = 0$, $x_1 = 2$, $x_2 = 3$, $x_3 = 4$, we obtain

$$f'(5) = 11 + 8 \times 7 + (6 + 15 + 10) \times 1 = 11 + 56 + 31 = 98$$

Note 1 This problem can be solved using Lagrange's formula also.

6.2 Errors in Numerical Differentiation

The numerical differentiation involves two types of errors

(1) truncation error (TE) and (2) rounding error (RE)

6.2.1 Truncation Error

Consider Stirling's formula

$$\left(\frac{dy}{dx}\right)_{x=x_0} = \frac{1}{h}\left(\frac{\Delta y_{-1} + \Delta y_0}{2} - \frac{1}{6}\frac{\Delta^3 y_{-2} + \Delta^3 y_{-1}}{2} + \frac{1}{30}\frac{\Delta^5 y_{-3} + \Delta^5 y_{-2}}{2} + \cdots\right) \quad (6.18)$$

It can be written as

$$\left(\frac{dy}{dx}\right)_{x=x_0} = \frac{\Delta y_{-1} + \Delta y_0}{2h} + T_1 = \frac{y_1 + y_{-1}}{2h} + T_1 \quad (6.19)$$

where T_1 is the TE given by

$$T_1 = \frac{1}{6h}\left|\frac{\Delta^3 y_{-2} + \Delta^3 y_{-1}}{2}\right| \quad (6.20)$$

Also, $$\left(\frac{d^2 y}{dx^2}\right)_{x=x_0} = \frac{1}{h^2}\Delta^2 y_{-1} + T_2 \quad (6.21)$$

where $$T_2 = \frac{1}{12h^2}\left|\Delta^4 y_{-2}\right| \quad (6.22)$$

6.2.2 Rounding Error

In Stirling's formula above, the RE is $2\varepsilon/2h = \varepsilon/h$, where ε = maximum error in y_i.

The formula

$$\left(\frac{dy}{dx}\right)_{x=x_0} = \frac{\Delta y_{-1} + \Delta y_0}{2h} - \frac{\Delta^3 y_{-2} + \Delta^3 y_{-1}}{12h} = \frac{y_{-2} - 8y_{-1} + 8y_1 - y_2}{12h} \quad (6.23)$$

has maximum RE $= \dfrac{18\varepsilon}{12h} = \dfrac{3\varepsilon}{2h}$ (6.24)

and $$\left(\frac{d^2 y}{dx^2}\right)_{x=x_0} = \frac{\Delta^2 y_{-1}}{h^2} + \cdots = \frac{y_{-1} - 2y_0 + y_1}{h^2} \quad (6.25)$$

has maximum RE $4\varepsilon/h^2$. It is clear that in the case of higher-order derivatives the RE increases rapidly.

Example 6.5

Assuming that the function values given in Table 6.1 above are correct to the accuracy given, estimate the errors in the values of dy/dx and d^2y/dx^2 at $x = 1.6$:

Solution Since the values are correct to 4D, it follows that

$$\varepsilon < 0.00005 = 0.5 \times 10^{-4}$$

Value of $\left(\dfrac{dy}{dx}\right)_{x=1.6}$

$$\text{Truncation error} = \frac{1}{6h}\left|\frac{\Delta^3 y_{-1} + \Delta^3 y_0}{2}\right| = \frac{1}{6(0.2)}\frac{0.0361 + 0.0441}{2} = 0.03342$$

$$\text{Rounding error} = \frac{3\varepsilon}{2h} = \frac{3(0.5)10^{-4}}{0.4} = 0.00038$$

$$\text{Total error} = 0.0334 + 0.0038 = 0.0338$$

From Stirling's formula (1) in Example 6.3 above

$$\left(\frac{dy}{dx}\right)_{x=1.6} = \frac{\Delta y_{-1} + \Delta y_0}{2h} = \frac{0.8978 + 1.0966}{0.4}$$

$$= \frac{1.9944}{0.4} = 4.9860$$

The exact value is 4.9530 so that error = 4.9860 – 4.9530 = 0.033, which agrees with the total error obtained.

From formula (2) of Example 6.3,

$$\left(\frac{d^2y}{dx^2}\right)_{x=1.6} = \frac{\Delta^2 y_{-1}}{h^2} = \frac{0.1888}{0.4} = 4.97$$

Error = 4.97 – 4.953 = 0.017

$$\text{Also truncation error} \;= \frac{1}{12h^2}\left|\Delta^4 y_{-2}\right|$$

$$= \frac{1}{n(0.04)} \times 0.0080 = 0.01667$$

$$\text{Rounding error} = \frac{4\varepsilon}{h^2} = \frac{4 \times 0.5 \times 10^{-4}}{0.04} = 0.005$$

$$\text{Total error is } \left(\frac{d^2y}{dx^2}\right)_{x=1.6} = 0.0167 + 0.005 = 0.0217$$

6.3 Maximum and Minimum Values of a Tabulated Function

We know that maximum and minimum values of a function are found by equating the first derivative to zero and solving for the variable. We can apply this procedure to a tabulated function as well.

Differentiating Newton's Forward interpolation formula

$$y = y_0 + p\Delta y_0 + \frac{p(p-1)}{2}\Delta^2 y_0 + \frac{p(p-1)(p-2)}{6}\Delta^3 y_0 + \cdots \tag{6.26}$$

we have

$$\frac{dy}{dp} = \Delta y_0 + \frac{2p-1}{2}\Delta^2 y_0 + \frac{3p^2-6p+2}{6}\Delta^3 y_0; \tag{6.27}$$

neglecting higher-order terms.

For an extreme value of y, we must have $dy/dp = 0$. This yields the quadratic in p:

$$a_0 + a_1 p + a_2 p^2 = 0 \tag{6.28}$$

with $\quad a_0 = \Delta y_0 - \frac{1}{2}\Delta^2 y_0 + \frac{1}{3}\Delta^3 y_0;$

$$a_1 = \Delta^2 y_0 - \Delta^3 y_0; \quad a_2 = \frac{1}{2}\Delta^3 y_0 \tag{6.29}$$

The values of x are found from $x = x_0 + ph$.

Example 6.6

Find x correct to two decimal places for which y is maximum from the table and find this value of y

x	1.2	1.3	1.4	1.5	1.6
y	0.9320	0.9636	0.9855	0.9975	0.9996

Solution Building a difference table we have

x	y	Δy	$\Delta^2 y$	$\Delta^3 y$
1.2	0.9320			
		0.0316		
1.3	0.9636		–0.0097	
		0.0219		–0.0002
1.4	0.9855		–0.0099	
		0.0120		–0.0000
1.5	0.9975		–0.0099	
		0.0021		
1.6	0.9996			

Let $x_0 = 1.2$. From (6.27), neglecting third and higher differences

$$0 = 0.0316 + \frac{2p-1}{2}(-0.0097) \quad \Rightarrow \quad p = 3.8$$

$$\therefore \quad x = x_0 + ph = 1.2 + 3.8\,(0.1) = 1.58$$

For this value of x, Newton's Backward difference formula at $x_n = 1.6$ gives

$$y(1.58) = 0.9996 - 0.2(0.0021) + \frac{-0.2(-0.2+1)}{2}(-0.0099)$$

$$= 0.9996 - 0.0004 + 0.0008 = 1.0$$

EXERCISE 6.1

1. Using the data given in the table find (a) $y'(1.1)$, $y''(1.1)$, (b) $y'(1.6)$ and $y''(1.6)$:

x	1.0	1.1	1.2	1.3	1.4	1.5	1.6
y	7.989	8.403	8.781	9.129	9.451	9.750	10.031

Ans: (a) $y'(1.1) = 3.946$, $y''(1.1) = -3.545$
(b) $y'(1.6) = 2.727$, $y''(1.6) = -1.703$

2. For the table of values of x and y find $\frac{dy}{dx}$ at $x = 4$

x	1	2	4	8	10
y	0	1	5	21	27

Ans: $y'(4) = 2.883$

3. Find the value of cos 1.74 using the values given in the table

x	1.70	1.74	1.78	1.82	1.86
$\sin x$	0.9916	0.9857	0.9781	0.9691	0.9584

[Madras 2000]

Ans: $\frac{d}{dx}(\sin x)\Big|\ x = 1.74 = -0.175$

4. From the following table find the value of x for which y is maximum and find this value of y:

x	1.2	1.3	1.4	1.5	1.6
y	0.9320	0.9636	0.9855	0.9975	0.9996

[Madras 1998]

Ans: $x = 1.58, y = 1.00$

5. For the following data find $\frac{dy}{dx}$ and $\frac{d^2y}{dx^2}$ at (a) $x = 1.2$ (b) $x = 1.6$ and (c) $x = 2.2$:

x	1.0	1.2	1.4	1.6	1.8	2.0	2.2
y	2.7183	3.3201	4.0552	4.9530	6.0496	7.3891	9.0250

Ans: (a) $y'(1.2) = 3.3205, y''(1.2) = 3.318$
(b) $y'(1.6) = 4.9530, y''(1.6) = 4.9525$
(c) $y'(2.2) = 9.0228, y''(2.2) = 8.992$

6. From the following table find the value of x for which y is maximum and find this value of y

x	1.2	1.3	1.4	1.5	1.6
y	0.9320	0.9636	0.9855	0.9975	0.9996

Ans: $y_{max} = 1.00$ at $x = 1.58$

7. Determine $y'(0)$ and $y''(0)$ from the following data:

x	0	1	2	3	4	5
y	4	8	15	7	6	2

Ans: $y'(0) = -27.9$, $y''(0) = 117.67$

6.4 NUMERICAL INTEGRATION: INTRODUCTION

We can find the integral of a function $y = f(x)$ defined and continuous on an interval $[a, b]$ if there exists a function $F(x)$ such that $F'(x) = f(x)$. According to the Fundamental Theorem of Integral Calculus we have

$$\int_a^b f(x)\,dx = F(b) - F(a)$$

Geometrically, it represents the area under the curve $y = f(x)$ and between the ordinates $x = a$ and $x = b$.

In applications, evaluation of the integral may prove to be very complicated and not practical:

1. when we cannot find the antiderivative $F(x)$ of $f(x)$ and
2. when the integrand $f(x)$ is a tabulated function.

In such cases we have to resort to numerical evaluation which is also called mechanical quadrature. The basic idea is to replace $f(x)$ by an interpolating polynomial $\phi(x)$ using a suitable interpolation formula. We derive a general formula for numerical integration using Newton's Forward difference formula.

6.4.1 Newton–Cotes Quadrature Formula

Let

$$I = \int_a^b f(x)\,dx \tag{6.30}$$

Divide $[a, b]$ into n subintervals of equal length $h = \frac{b-a}{n}$ by means of points

$$a = x_0 < x_1 < \cdots < x_{n-1} < x_n = b \tag{6.31}$$

Hence we have

$$I = \int_{x_0}^{x_n} y\,dx = \int_{x_0}^{x_n}\left[y_0 + p\Delta y_0 + \frac{p(p-1)}{2}\Delta^2 y_0 + \frac{p(p-1)(p-2)}{6}\Delta^3 y_0 + \cdots\right]dx \tag{6.32}$$

where y is approximated by Newton's forward difference formula.

Since $x = x_0 + ph$, $dx = h\,dp$ the above integral becomes

$$I = h\int_0^n\left[y_0 + p\Delta y_0 + \frac{p(p-1)}{2}\Delta^2 y_0 + \frac{p(p-1)(p-2)}{6}\Delta^3 y_0 + \cdots\right]dp \tag{6.33}$$

After evaluation of the integral and simplification, we obtain

$$I = \int_{x_0}^{x_n} y\,dx = nh\left[y_0 + \frac{n}{2}\Delta y_0 + \frac{n(2n-3)}{12}\Delta^2 y_0 + \frac{n(n-2)^2}{24}\Delta^3 y_0 + \cdots\right] \tag{6.34}$$

which is called Newton-Cotes quadrature formula.

We obtain various integration formulae from the general formula (6.34) by taking $n = 1, 2, 3, \ldots$. But for practical problems, the trapezoidal rule for $n = 1$ and Simpson's rule for $n = 2$ are simple and most useful; and better accuracy can be ensured by using composite trapezoidal rule and composite Simpson's rule with smaller step size h.

6.4.2 Trapezoidal Rule

Taking $n = 1$ in (6.34) and neglecting all differences higher than the first, we obtain

$$I_1 = \int_{x_0}^{x_1} y\,dx = h\left(y_0 + \frac{1}{2}\Delta y_0\right) = \frac{h}{2}\left[y_0 + \frac{(y_1 - y_0)}{2}\right] = \frac{h}{2}(y_0 + y_1) \tag{6.35}$$

(Here the curve $y = f(x)$ is replaced by a straight line segment.)

Similarly for

$$[x_1, x_2],\ I_2 = \int_{x_1}^{x_2} y\,dx = \frac{h}{2}(y_1 + y_2), \cdots;$$

$$\text{for}\quad [x_{n-1}, x_n],\ I_n = \int_{x_{n-1}}^{x_n} y\,dx = \frac{h}{2}(y_{n-1} + y_n)$$

Adding $I_1, I_2, \ldots, I_n$ we obtain

General (or Composite) Formula

$$\begin{aligned} I &= \int_{x_0}^{x_n} y\,dx = \frac{h}{2}[y_0 + 2(y_1 + y_2 + \cdots + y_{n-1}) + y_n] \\ &= \frac{h}{2}\left[(y_0 + y_n) + 2(y_1 + y_2 + \cdots + y_{n-1})\right] \\ &= \frac{h}{2}(X + 2R) \end{aligned} \tag{6.36}$$

where X = sum of the extreme ordinates and R = sum of the remaining ordinates

(Here, the curve $y = f(x)$ is replaced by n jointed line segments.)

$$\text{Total error} = -\frac{(b-a)}{12}h^2 y^{(2)}(\xi),$$

$$\text{where } y^{(2)}(\xi) = \max\left(y_0^{(2)}, y_1^{(2)}, \ldots y_{n-1}^{(2)}\right) \tag{6.37}$$

The Trapezoidal Rule can be applied to any number of subintervals, that is, when n is even or odd.

6.4.3 Simpson's[1] $\frac{1}{3}$ Rule

Taking $n = 2$ in (6.34) and neglecting all differences higher than the second, we obtain

---

[1] SIMPSON, Thomas (1710–1767), proposed this rule in his mathematical dissertations on a variety of physical and analytical subjects (1743). But the rule appeared earlier in James Gregory's Exterminations Geometrical (1668) in a somewhat different form. He learnt mathematics by himself. In 1743, he became professor of mathematics at the Royal Military Academy at Woolwich and was elected to the Royal Society in 1745.

$$I_1 = \int_{x_0}^{x_2} y\,dx = 2h\left[y_0 + \Delta y_0 + \frac{1}{6}\Delta^2 y_0\right]$$

$$= 2h\left[y_0 + (y_1 - y_0) + \frac{1}{6}(y_2 - 2y_1 + y_0)\right]$$

$$= \frac{h}{3}[y_0 + 4y_1 + y_2] \qquad (6.38)$$

(Here the curve $y = f(x)$ is approximated by a parabola or a second-degree polynomial.)

Similarly

for $[x_2, x_4]$, $\quad I_2 = \int_{x_2}^{x_4} y\,dx = \frac{h}{3}[y_2 + 4y_3 + y_4], \cdots$

and for $\quad [x_{n-2}, x_n]$, $\quad I_m = \int_{x_{n-2}}^{x_n} y\,dx$

$$= \frac{h}{3}[y_{n-2} + 4y_{n-1} + y_n]$$

(n even, $n = 2m$)

Adding $I_1, I_2, \ldots, I_m$ we obtain

General (or Composite) Formula

$$I = \int_{x_0}^{x_n} y\,dx = \frac{h}{3}[y_0 + 4y_1 + 2y_2 + 4y_3 + 2y_4 + \cdots$$

$$+ 2y_{2m-2} + 4y_{2m-1} + y_{2m}]$$

$$= \frac{h}{3}[(y_0 + y_n) + 4(y_1 + y_3 + y_5 + \cdots)$$

$$+ 2(y_2 + y_4 + \cdots)]$$

$$= \frac{h}{3}[X + 4O + 2E] \qquad (6.39)$$

where X = Sum of the extreme ordinates; O = Sum of the odd ordinates; E = Sum of the even ordinates.

(Here, the curve is approximated by piecewise jointed curve consisting of $n/2$ parabolic arcs.)

$$\text{Total error} = -\frac{(b-a)}{180}h^4 y^{(4)}(\xi), y^{(4)}(\xi)$$

$$= \max\,(y_0^{(4)}, y_1^{(4)}, \ldots, y_{n-2}^{(4)}) \qquad (6.40)$$

Simpson's $\frac{1}{3}$ Rule can be applied when the number n of subintervals is even ($n = 2m$).

6.4.4 Simpson's $\frac{3}{8}$ Rule

Taking $n = 3$ in (6.34) and neglecting all differences higher than the third, we obtain

$$I_1 = \int_{x_0}^{x_3} y\,dx = 3h\left[y_0 + \frac{3}{2}\Delta y_0 + \frac{3}{4}\Delta^2 y_0 + \frac{1}{8}\Delta^3 y_0\right]$$

$$= 3h\left[y_0 + \frac{3}{2}(y_1 - y_0) + \frac{3}{4}(y_2 - 2y_1 + y_0)\right.$$

$$\left. + \frac{1}{8}(y_3 - 3y_2 + 3y_1 - y_0)\right]$$

$$= \frac{3h}{8}[y_0 + 3y_1 + 3y_2 + y_3] \qquad (6.41)$$

(Here the curve is approximated by a third-degree polynomial.)

Similarly

for $[x_3, x_6]$, $I_2 = \int_{x_3}^{x_6} y\,dx = \frac{3h}{8}[y_3 + 3y_4 + 3y_5 + y_6], \cdots$

and for $[x_{3m-3}, x_{3m}]$,

$$I_m = \int_{x_{3m-3}}^{x_{3m}} y\,dx = \frac{3h}{8}[y_{3m-3}$$

$$+ 3y_{3m-2} + 3y_{3m-1} + y_{3m}]$$

(n must be a multiple of 3, $n = 3m$)

Adding $I_1, I_2, \ldots, I_m$ we obtain

General (or Composite) Formula

$$I = \int_{x_0}^{x_n} y\,dx = \frac{3h}{8}[(y_0 + 3y_1 + 3y_2 + y_3)$$

$$+ (y_3 + 3y_4 + 3y_5 + y_6) + \cdots$$

$$+ (y_{3m-3} + 3y_{3m-2} + 3y_{3m-2} + 3y_{3m-1} + y_{3m})]$$

$$= \frac{3h}{8}[(y_0 + y_n) + 2(y_3 + y_6 + \cdots y_{n-3})$$

$$+ 3(y_1 + y_2 + y_4 + y_5 + \cdots + y_{n-2} + y_{n-1})]$$

$$= \frac{3h}{8}[X + 2Y + R] \qquad (6.42)$$

where X = Sum of the extreme ordinates; Y = Sum of the ordinates y_{3m}, with suffixes divisible by 3; R = Sum of the remaining ordinates.

(Here the curve is approximated by a piecewise jointed curve consisting of $n/3$ third-degree polynomial arcs.)

$$\text{Total error} = -\frac{3}{80}(b-a)h^4 y^{(4)}(\xi);$$

$$y^{(4)}(\xi) = \max\,(y_0^{(4)}, y_1^{(4)}, \ldots y_{n-3}^{(4)}) \qquad (6.43)$$

Simpson's $\frac{3}{8}$ Rule can be applied when the number n of subintervals is a multiple of 3 ($n = 3m$).

6.4.5 Boole's[2] Rule

Taking $n = 4$ in (6.34) and neglecting all the differences higher than the fourth, we obtain

$$I_1 = \int_{x_0}^{x_4} y\,dx = 4h\left[y_0 + 2\Delta y_0 + \frac{5}{3}\Delta^2 y_0 + \frac{2}{3}\Delta^3 y_0 + \frac{7}{90}\Delta^4 y_0\right] \qquad (6.44)$$

$$= \frac{2h}{45}(7y_0 + 32y_1 + 12y_2 + 32y_3 + 7y_4)$$

(Here the curve is approximated by a fourth-degree polynomial; n must be a multiple of 4, *i.e.*, $n = 4m$.)

$$\textbf{Total error} = -\frac{8}{945}(b-a)h^6 y^{(6)}(\xi);$$

$$\text{where } y^{(6)}(\xi) = \max\,(y_0^{(6)}, y_1^{(6)}, \ldots y_{4(m-1)}^{(6)}) \qquad (6.45)$$

6.4.6 Weddle's[3] Rule

Taking $n = 6$ in (6.34) and neglecting all the differences higher than the sixth, we obtain

$$I = \int_{x_0}^{x_6} y\,dx = 6h\left[y_0 + 3\Delta y_0 + \frac{9}{2}\Delta^2 y_0 + 4\Delta^3 y_0 + \frac{123}{60}\Delta^4 y_0 + \frac{11}{20}\Delta^5 y_0 + \frac{1}{6}\cdot\frac{41}{140}\Delta^6 y_0\right] \qquad (6.46)$$

If we replace $\frac{41}{140}$ in the last term by $\frac{3}{10}$, the error made will be $\frac{3}{10} - \frac{41}{140} = \frac{1}{140} \approx 0.007$ which is negligible when h and $\Delta^6 y_0$ are small.

Substituting for Δy_0, $\Delta^2 y_0$, … and simplifying, we obtain

$$I = \int_{x_0}^{x_6} y\,dx = \frac{3h}{10}[y_0 + 5y_1 + y_2 + 6y_3 + y_4 + 5y_5 + y_6] \qquad (6.47)$$

(Here the curve is approximated by a sixth-degree polynomial curve and n must be a multiple of 6, $n = 6m$.)

$$\text{Total error} = -\frac{(b-a)}{140}h^6 y^{(6)}(\xi);$$

$$\text{where } y^{(6)}(\xi) = \max\,(y_0^{(6)}, y_1^{(6)}, \ldots y_{n-1}^{(6)}) \qquad (6.48)$$

Example 6.7

Evaluate (a) $\int_0^4 x^2\,dx$ (b) $\int_0^4 x^3\,dx$ (c) $\int_0^4 x^4\,dx$ by (i) Trapezoidal rule and (ii) Simpson's rule.

Solution Writing the table of values

x	x^2	x^3	x^4
$x_0 = 0$	$y_0 = 0$	0	0
$x_1 = 1$	$y_1 = 1$	1	1
$x_2 = 2$	$y_2 = 4$	8	16
$x_3 = 3$	$y_3 = 9$	27	81
$x_4 = 4$	$y_4 = 16$	64	256

(i) Trapezoidal Rule

(a) $$\int_0^4 x^2\,dx = \frac{h}{2}[(y_0 + y_4) + 2(y_1 + y_2 + y_3)]$$

$$= \frac{1}{2}[(0+16) + 2(1+4+9)] = 22$$

[2] BOOLE, George (1815–1864), was a pioneering British Logician. He also worked in Algebra, developed an algebra called Boolean algebra. He also worked in analysis, calculus of variations and probability theory.

[3] WEDDLE, Thomas (1817–1853), was an English analyst and geometer.

(b) $\int_0^4 x^3\, dx = \frac{1}{2}[(0+64)+2(1+8+27)] = 68$

(c) $\int_0^4 x^4\, dx = \frac{1}{2}[(0+256)+2(1+16+81)] = 226$

(ii) Simpson's Rule

(a) $\int_0^4 x^2\, dx = \frac{h}{3}[(y_0+y_4)+4(y_1+y_3)+2y_2]$

$= \frac{1}{3}[(0+16)+4(1+9)+2\times 4]$

$= \frac{64}{3} = 21.3333$

(b) $\int_0^4 x^3\, dx = \frac{1}{3}[(0+64)+4(1+27)+2\times 8]$

$= \frac{192}{3} = 64$

(c) $\int_0^4 x^4\, dx = \frac{1}{3}[(0+256)+4(1+81)+2\times 16]$

$= \frac{616}{3} = 205.3333$

Exact Values

(a) $\int_0^4 x^2\, dx = \left.\frac{x^3}{3}\right|_0^4 = \frac{64}{3} = 21.3333$

(b) $\int_0^4 x^3\, dx = \left.\frac{x^4}{4}\right|_0^4 = 64$

(c) $\int_0^4 x^4\, dx = \left.\frac{x^5}{5}\right|_0^4 = \frac{4^5}{5} = \frac{1024}{5} = 204.8$

Thus, Simpson's rule gives exact values of $\int_0^n x^m\, dx$ for $m = 0, 1, 2, 3$. There is an error of 0.533 for $m = 4$.

Example 6.8

Find, from the following table, the area bounded by the curve $y = f(x)$ and the x-axis from $x = 7.47$ to $x = 7.52$:

i	0	1	2	3	4	5
x	7.47	7.48	7.49	7.50	7.51	7.52
y	1.93	1.95	1.98	2.01	2.03	2.06

Solution Area under the curve $y = f(x)$ is

$$I = \int_{7.47}^{7.52} f(x)\, dx$$

The trapezoidal rule is

$$\int_{x_0}^{x_5} y\, dx = \frac{h}{2}[(y_0+y_5)+2(y_1+y_2+y_3+y_4)]$$

We have $h = 0.01$

$$I = \int_{7.47}^{7.52} f(x)\, dx = \frac{0.01}{2}[(1.93+2.06) + 2(1.95+1.98+2.01+2.03)]$$

$$= \frac{0.01}{2}(3.99+2\times 7.97) = \frac{0.01}{2}(3.99+15.94)$$

$$= \frac{0.01}{2}\times 19.93 = 0.09965$$

Example 6.9

Evaluate $\int_1^3 \frac{1}{x}\, dx$ over four subintervals using

(i) Trapezoidal rule and (ii) Simpson's rule.

Solution Table of values

	0	1	2	3	4
x_i	1.0	1.5	2.0	2.5	3.0
y_i	1.0000	0.6667	0.5000	0.4000	0.3333
		(2/3)	(1/2)	(2/5)	1/3

(i) Trapezoidal Rule

$$h = 0.5 = \frac{1}{2} \int_{x_0}^{x_4} y\, dx = \frac{h}{2}[y_0+2y_1+2y_2+2y_3+y_4]$$

$$= \frac{1/2}{2}\left[1+\frac{1}{3}+2\left(\frac{2}{3}+\frac{1}{2}+\frac{2}{5}\right)\right] = \frac{67}{60} = 1.0117$$

(ii) Simpson's Rule

$$\int_{x_0}^{x4} y\,dx = \frac{h}{3}[y_0 + 4y_1 + 2y_2 + 4y_3 + y_4]$$

$$= \frac{1/2}{3}\left[1 + \frac{8}{3} + 2\cdot\frac{1}{2} + 4\cdot\frac{2}{5} + \frac{1}{3}\right] = \frac{1}{6} \times \frac{33}{5} = \frac{11}{10} = 1.1$$

Exact value $\int_1^3 \frac{1}{x}\,dx = \log x\Big|_1^3 = \log 3 = 1.0986$

Error in Simpson's result = 0.0014.

Example 6.10

Evaluate $\int_0^4 e^x\,dx$ by Simpson's rule. Compare it with the true value.

Solution Here $b = 4$, $a = 0$, $n = 4$ and

$$\therefore\ h = \frac{b-a}{n} = \frac{4-0}{4} = 1$$

We tabulate the values of the function.

x	$x_0 = 0$	$x_1 = 1$	$x_2 = 2$	$x_3 = 3$	$x_4 = 4$
$y = e^x$	$y_0 = 1.0000$	$y_1 = e^1 = 2.7183$	$y_2 = e^2 = 7.3891$	$y_3 = e^3 = 20.0855$	$y_4 = e^4 = 54.5982$

Simpson's rule is

$$\int_{x_0}^{x_4} y\,dx = \frac{h}{3}[(y_0 + y_4) + 4(y_1 + y_3) + 2y_2]$$

$$= \frac{1}{3}[(1 + 54.5982) + 4(2.7183 + 20.0855) + 2 \times 7.3891]$$

$$= \frac{1}{3}[55.5982 + 4(22.8038) + 2 \times 7.3891]$$

$$= \frac{1}{3}(161.5916) = 53.86387$$

True value

$$\int_0^4 e^x\,dx = e^x\Big|_0^4 = e^4 - 1 = 54.59815 - 1 = 53.59815$$

The two values agree up to integers.

Example 6.11

A dam with 2 m thickness has to be built across a river 100 m wide. The depth y m at a distance x m from one bank is as given below:

x (m)	0	10	20	30	40	50	60	70	80	90	100
y (m)	0	4	7	9	12	16	14	8	7	4	2

Estimate the cost of construction if the cost per cubic meter is Rs. 2000.

Solution Simpson's rule $h = 10$

$$\text{Area of cr. sec.} = \int_a^b y\,dx = \frac{h}{3}[(y_0 + y_{10}) + 4(y_1 + y_3 + y_5 + y_7 + y_9) + 2(y_2 + y_4 + y_6 + y_8)]$$

$$= \frac{10}{3}[(0+2) + 4(4+9+16+8+4) + 2(7+12+14+7)] = \frac{10}{3}[2 + 164 + 80]$$

$$= \frac{10}{3} \times 246 = 820\ \text{m}^2$$

Volume of the dam = 820 × 2 = 1640 m^3.

Cost of construction = 1640 × 2000 = Rs. 32,80,000.

Example 6.12

Find, from the following table, the area bounded by the curve $y = f(x)$ and the x-axis from $x = 5.1$ to $x = 5.7$ using (i) Trapezoidal rule, (ii) Simpson's rule and (iii) Weddle's rule. Compare the results with true value if $y = \log x$

x	5.1	5.2	5.3	5.4	5.5	5.6	5.7
y	1.629	1.649	1.668	1.686	1.705	1.723	1.740
	y_0	y_1	y_2	y_3	y_4	y_5	y_6

Solution Here $h = \dfrac{x_6 - x_0}{6} = \dfrac{5.7 - 5.1}{6} = 0.1$

(i) Trapezoidal Rule

$$\int_{x_0}^{x_6} y\,dx = \frac{h}{2}[(y_0 + y_6) + 2(y_1 + y_2 + y_3 + y_4 + y_5)]$$
$$= \frac{0.1}{2}[(1.629 + 1.740) + 2(1.649 + 1.668 + 1.686 + 1.705 + 1.723)]$$
$$= \frac{0.1}{2}[3.369 + 16.862] = \frac{0.1}{2} \times 20.231$$
$$= \frac{2.0231}{2} = 1.01155$$

(ii) Simpson's Rule

$$\int_{x_0}^{x_6} y\,dx = \frac{h}{2}[(y_0 + y_6) + 4(y_1 + y_3 + y_5) + 2(y_2 + y_4)]$$
$$= \frac{0.1}{3}[(1.629 + 1.740) + 4(1.649 + 1.686 + 1.723) + 2(1.668 + 1.705)]$$
$$= \frac{0.1}{3}[3.369 + 20.232 + 6.746]$$
$$= \frac{0.1}{3} \times 30.347 = 1.0115666$$

(iii) Weddle's Rule

$$\int_{x_0}^{x_6} y\,dx = \frac{3h}{10}[(y_0 + y_6) + 5(y_1 + y_5) + (y_2 + y_4) + 6y_3]$$
$$= \frac{3 \times 0.1}{10}[3.369 + 5 \times (3.372) + 3.373 + 6 \times 1.686]$$
$$= 0.03[33.718] = 1.01154$$

Exact Value

$$\int_{5.1}^{5.7} \log_e x\,dx = [x(\log x - 1)]_{5.1}^{5.7} = 5.7(\log_e 5.7 - 1) - 5.1(\log_e 5.1 - 1)$$
$$= 5.7 \times 0.740466175 - 5.1 \times 0.62924054$$
$$= 4.220657198 - 3.209126754$$
$$= 1.011530444 \simeq 1.01153$$

Thus, we see that the result obtained by Weddle's rule is closer to the exact value.

Example 6.13

Evaluate $\int_0^{\pi/2} \sin x\,dx$ by (i) Trapezoidal rule and (ii) Simpson's rule by taking $h = \pi/20$.

Solution We compute the following table of values:

	0	1	2	3	4	5	6	7	8	9	10
i	0	$\pi/20$	$2\pi/20$	$3\pi/20$	$4\pi/20$	$5\pi/20$	$6\pi/20$	$7\pi/20$	$8\pi/20$	$9\pi/20$	$10\pi/20$
x	0	0.1571	0.3142	0.4712	0.6283	0.7854	0.9425	1.0996	1.2566	1.4137	1.5708
y	0	0.1565	0.3091	0.4540	0.5878	0.7071	0.8090	0.8910	0.9510	0.9877	1.0000

(i) Trapezoidal Rule

$$\int_{x_0}^{x_{10}} y\,dx = \frac{h}{2}[(y_0 + y_{10}) + 2(y_1 + y_2 + y_3 + y_4 + y_5 + y_6 + y_7 + y_8 + y_9)]$$
$$= \frac{0.1571}{2}[(0+1) + 2(0.1565 + 0.3091 + 0.4540 + 0.5878 + 0.7071 + 08090 + 0.8910 + 0.9510 + 0.9877)]$$
$$= \frac{0.1571}{2} \times 12.7064 = 0.99808772 \simeq 0.9981$$

(ii) Simpson's Rule

$$\int_{x_0}^{x_{10}} y\,dx = \frac{h}{3}[(y_0 + y_{10}) + 4(y_1 + y_3 + y_5 + y_7 + y_9) + 2(y_2 + y_4 + y_6 + y_8)]$$

$$= \frac{0.1571}{3}[(0+1) + 4(0.1565 + 0.4540 + 0.7071 + 0.8910 + 0.9877) + 2(0.3091 + 0.5878 + 0.8090 + 0.9510)]$$

$$= \frac{0.1571}{3} \times 19.0990 = 1.000150967 \simeq 1.00015$$

Exact Value

$$\int_0^{\pi/2} \sin x\,dx = (-\cos x)_0^{\pi/2} = 0 - (-1) = 1$$

Thus, Simpson's rule gives a better approximation.

Example 6.14

The velocity v (km/hr) of a car which starts from rest is measured on an hourly basis and is recorded. Estimate the distance travelled in 6 hours.

t (hr)	0	1	2	3	4	5	6
v (km/hr)	0	45	50	40	30	35	20

Solution Let S denote the distance in kilometer covered in t hours. Then $ds/dt = v$. Integrating both sides between $t = 0$ and $t = 6$, we have by Simpson's rule

$$s\big|_{t=0}^{6} = \int_0^6 v\,dt = \frac{h}{3}[(v_0 + v_6) + 4(v_1 + v_3 + v_5) + 2(v_2 + v_4)]$$

$$= \frac{1}{3}[(0+20) + 4(45+40+35) + 2(50+30)] = 220 \text{ km}$$

Example 6.15

Find the approximate area bounded by the curve $y = \sqrt{1-x^2}$ and the x-axis by (i) Trapezoidal rule and (ii) Simpson's rule.

Solution We have to evaluate the integral $\int_0^1 \sqrt{1-x^2}\,dx$. Let us divide the interval into four subintervals with width $h = \frac{1}{4}$. We compute the values of the integrand $y = \sqrt{1-x^2}$ at 0, 0.25, 0.50, 0.75 and 1.0.

x	$x_0 = 0$	$x_1 = 0.25$	$x_2 = 0.50$	$x_3 = 0.75$	$x_4 = 1.00$
$y = \sqrt{1-x^2}$	$y_0 = 1.0000$	$y_1 = 0.9682$	$y_2 = 0.8660$	$y_3 = 0.6614$	$y_4 = 0$

(i) Trapezoidal Rule

$$\int_{x_0}^{x_4} y\,dx = \frac{h}{2}\left[(y_0 + y_4) + 2(y_1 + y_2 + y_3)\right]$$

$$\int_0^1 \sqrt{1-x^2}\,dx = \frac{1/4}{2}[(1+0) + 2(0.9682 + 0.8660 + 0.6614)] = 0.7488$$

(ii) Simpson's Rule

$$\int_{x_0}^{x_4} y\,dx = \frac{h}{3}[(y_0 + y_4) + 4(y_1 + y_3) + 2y_2]$$

$$\int_0^1 \sqrt{1-x^2}\,dx = \frac{1/4}{3}[(1+0) + 4(0.9682 + 0.6614) + 2(0.8660)]$$

$$= \frac{1}{12}[1 + 6.5284 + 1.7520]$$

$$= \frac{1}{12} \times 9.2504 = 0.770866 = 0.7709$$

True Value

$$\int_0^1 \sqrt{1-x^2}\,dx = \left[\frac{x}{2}\sqrt{1-x^2} + \frac{1}{2}\sin^{-1} x\right]_0^1$$

$$= \frac{1}{2}\sin^{-1} 1 = \frac{\pi}{2} = 0.785398163 \simeq 0.7854$$

Example 6.16

Evaluate $I = \int_0^1 \frac{dx}{1+x}$ using (i) Trapezoidal rule (ii) Simpson's rule with (a) $h = 0.5$ (b) $h = 0.25$.

Solution (a) 0(0.5)1

i	0	1	2
x	0.0	0.5	1.0
y	1.0000	0.6667	0.5000

(i) Trapezoidal Rule $h = 0.5$,

$$I = \frac{1}{4}[1.0000 + 2(0.6667) + 0.5000] = 0.70835$$

(ii) Simpson's Rule

$$I = \frac{1}{6}[1.0000 + 4(0.6667) + 0.5000] = 0.6945$$

(b) $h = 0.25$

i	0	1	2	3	4
x	0.00	0.25	0.50	0.75	1.00
y	1.0000	0.5000	0.6667	0.5714	0.5000

(i) Trapezoidal Rule

$$I = \frac{1}{8}[(1.0) + 2(0.8000 + 0.6667 + 0.5714) + 0.5] = 0.6970$$

(ii) Simpson's Rule

$$I = \frac{1}{12}[1.0 + 4(0.8000 + 0.5714) + 2(0.667) + 0.5] = 0.6932$$

Example 6.17

Evaluate $\int_0^1 \frac{dx}{1+x}$ by (i) Trapezoidal rule and (ii) Simpson's rule with $h = 0.125$

i	0	1	2	3	4	5	6	7	8
x	0.000	0.125	0.250	0.375	0.500	0.625	0.750	0.875	1.000
y	1.000	0.8889	0.8000	0.7273	0.6667	0.6154	0.5714	0.5333	0.500

(i) Trapezoidal Rule

$$I = \frac{h}{2}[y_0 + 2(y_1 + y_2 + y_3 + y_4 + y_5 + y_6 + y_7) + y_8]$$

$$= \frac{1}{16}[1.0 + 2(2.7649 + 2.0381) + 0.5] = 0.6941$$

(ii) Simpson's Rule

$$I = \frac{h}{3}[y_0 + y_8 + 4(y_1 + y_3 + y_5 + y_7) + 2(y_2 + y_4 + y_6 + y_8)]$$

$$= \frac{1}{24}[1.0 + 4(2.7649) + 2(2.0381)]$$

$$= 0.693158333 \simeq 0.6932$$

The exact value of $I = \log_e 2 = 0.693147\ldots$. The value obtained by Simpson's rule is close to this. Hence Simpson's rule gives a better approximation than the Trapezoidal rule.

Example 6.18

Evaluate $\int_0^1 (1 + e^{-x} \sin 4x)\, dx$ with equally spaced quadrature nodes x_0, x_1, x_2, x_3 and x_4 separated by distance $h = \frac{1}{4}$ using (i) Trapezoidal rule, (ii) Simpson's and (iii) Boole's rules.

Solution Constructing the table of values of the integrand at the given points

i	0	1	2	3	4
x	0.00	0.25	0.50	0.75	1.00
y	1.00000	1.65534	1.55152	1.06667	0.72159

Here $a = 0$, $b = 1$, $h = \frac{1-0}{4} = \frac{1}{4}$

(i) Trapezoidal Rule

$$\int_a^b y\,dx = \frac{h}{2}[y_0 + 2(y_1 + y_2 + y_3) + y_4]$$

$$\int_a^b y\,dx = \frac{1/4}{2}[1.00000 + 2(1.65534 + 1.55152 + 1.06666) + 0.72159]$$

$$= 1.28357875$$

(ii) Simpson's Rule

$$\int_a^b y\,dx = \frac{h}{3}[y_0 + y_4 + 4(y_1 + y_3) + 2y_2]$$

$$\int_a^b y\,dx = \frac{1/4}{3}[1.72159 + 4(1.65534 + 1.06666) + 2(1.55152)] = \frac{1}{12}[1.72159 + 10.888 + 3.10304] = 1.309385833$$

(iii) Boole's Rule

$$\int_{x_0}^{x_4} y\,dx = \int_a^b y\,dx = \frac{2h}{45}[7y_0 + 32y_1 + 12y_2 + 32y_3 + 7y_4]$$

$$\int_0^1 y\,dx = \frac{2\times\frac{1}{4}}{45}[7\times 1 + 32\times 1.65534 + 12\times 1.55152 + 32\times 1.06666 + 7\times 0.72159]$$

$$= \frac{1}{90}[7\times 1.72159 + 32\times(2.72200) + 12\times(1.55152)] = 1.308593$$

True Value

$$\int_0^1 (1 + e^{-x}\sin 4x)\,dx$$

$$= 1 + \left[\frac{e^{-x}}{17}(-\sin 4x - 4\cos 4x)\right]_0^1$$

$$= 1 + \frac{e^{-1}}{17}(-\sin 4 - 4\cos 4) - \frac{1}{17}(-4)$$

$$= \frac{21}{17} - \frac{e^{-1}}{17}(\sin 4 + 4\cos 4)$$

$$= \frac{21e - (\sin 4 + 4\cos 4)}{17e} = 1.235294118$$

Example 6.19

Evaluate $\int_0^6 \frac{dx}{1+x^2}$ by using (i) Trapezoidal rule, (ii) Simpson's $\frac{1}{3}$ rule (iii) Simpson's $\frac{3}{8}$ rule, (iv) Boole's rule and (v) Weddle's rule. Compare the results with its true value.

Solution We divide the interval [0, 6] into six subintervals each of width $h = 1$. We tabulate below the values of $y = f(x) = \frac{1}{1+x^2}$:

x	0	1	2	3	4	5	6
$y = f(x)$	1	0.5	0.2	0.1	0.0588	0.0385	0.027
	y_0	y_1	y_2	y_3	y_4	y_5	y_6

(i) Trapezoidal Rule

$$\int_{x_0}^{x_6} y\,dx = \frac{h}{2}[(y_0 + y_6) + 2(y_1 + y_2 + y_3 + y_4 + y_5 + y_6)]$$

$$\int_0^6 \frac{dx}{1+x^2} = \frac{1}{2}[(1 + 0.027) + 2(0.5 + 0.2 + 0.1 + 0.0588 + 0.0385)]$$

$$= 1.4108$$

(ii) Simpson's $\frac{1}{3}$ Rule

$$\int_{x_0}^{x_6} y\,dx = \frac{h}{3}[(y_0 + y_6) + 4(y_1 + y_3 + y_5) + 2(y_2 + y_4 + y_6)]$$

$$\int_0^6 \frac{dx}{1+x^2} = \frac{1}{3}[(1 + 0.027) + 4(0.5 + 0.1 + 0.0385) + 2(0.2 + 0.0588)] = 1.3662$$

(iii) Simpson's $\frac{3}{8}$ Rule

$$\int_{x_0}^{x_6} y\,dx = \frac{3h}{8}[(y_0 + y_6) + 3(y_1 + y_2 + y_4 + y_5) + 2y_3]$$

$$\int_0^6 \frac{dx}{1+x^2} = \frac{3}{8}[(1 + 0.027) + 3(0.5 + 0.2 + 0.0588 + 0.0385) + 2(0.1) = 1.3571$$

(iv) Boole's Rule

For the application of Boole's rule, the interval has to be subdivided into a number subintervals, which is a multiple of 4.

If we divide the interval 0 to 6 into four subintervals then the nodes are $x_0 = 0$, $x_1 = 1.5$, $x_2 = 3.0$, $x_3 = 4.5$ and $x_4 = 6.0$.

x	0	1.5	3.0	4.5	6.0
y	1	0.3077	0.1	0.0471	0.027
	y_0	y_1	y_2	y_3	y_4

Boole's Rule

$$\int_{x_0}^{x_6} y\,dx = \frac{2h}{45}[7y_0 + 32y_1 + 12y_2 + 32y_3 + 7y_4]$$

$$\int_0^6 \frac{dx}{1+x^2} = \frac{2\times 1.5}{45}[7\times 1 + 32\times 0.3077 + 12\times 0.1 + 32\times 0.0471 + 7\times 0.027]$$

$$= \frac{1}{15}[7 + 9.8464 + 1.2 + 1.5072 + 0.189]$$

$$= 1.3161733 \simeq 1.3162$$

(v) Weddle's Rule

$$\int_{x_0}^{x_6} y\,dx = \frac{3h}{10}[y_0 + 5y_1 + y_2 + 6y_3 + y_4 + 5y_5 + y_6]$$

$$\int_0^6 \frac{dx}{1+x^2} = \frac{3\times 1}{10}[1 + 5\times 0.5 + 0.2 + 6\times 0.1 + 0.0588 + 5\times 0.0385 + 0.027]$$

$$= 1.37349 \simeq 1.3735$$

True value is obtained by evaluating

$$\int_0^6 \frac{dx}{1+x^2} = \tan^{-1} x\Big|_0^6$$

$$= \tan^{-1} 6 = 1.405647649 \simeq 1.4056$$

The value obtained by Weddle's rule is closer to the true value.

Example 6.20

A solid of revolution is formed by rotating about the x-axis, the area between the x-axis, the lines $x = 0$ and $x = 1$ and a curve through the points with the coordinates

x	0.00	0.25	0.50	0.75	1.00
y	1.0000	0.9896	0.9589	0.9089	0.8415

Find its volume.

Solution Here $h = 0.25$, $y_0 = 1.0000$, $y_1 = 0.9896$, $y_2 = 0.9589$, $y_3 = 0.9089$, $y_4 = 0.08415$.

Volume of the solid of revolution $= V$

$$= \pi\int_0^1 y^2 dx = \pi\frac{h}{3}\left[(y_0^2 + y_4^2) + 4(y_1^2 + y_3^2) + 2y_2^2\right]$$

$$= \frac{0.25\pi}{3}[1 + (0.8415)^2 + 4(0.9896)^2 + 4(0.9089)^2 + 2(0.9589)^2]$$

$$= \frac{0.25}{3}\times\pi[1.7081 + 3.9172 + 3.3044 + 1.839]$$

$$= \frac{0.25\times\pi}{3}\times 10.7687 = 0.89739\pi = 2.81924$$

Example 6.21

The heights of the upper and the lower boundary curves from a central line across the map of an island drawn to the scale 1 cm = 10 km are given below

x	0	1	2	3	4
y	1	1.2	1.4	0.8	0.2
	0.3	0.8	0.6	0.4	0

Estimate the area of the island.

Solution Simpson's rule is

$$\int_0^4 y\,dx = \frac{h}{3}[(y_0 + y_4) + 4(y_1 + y_3) + 2y_2]$$

$$\underset{\text{(upper area)}}{A_1} = \int_0^4 y\,dx = \frac{1}{3}[(1 + 0.2) + 4(1.2 + 0.8) + 2(1.4)]$$

$$= \frac{1}{3}(1.2 + 8.0 + 2.8) = \frac{12}{3} = 4\text{ cm}^2$$

$$\underset{\text{(lower area)}}{A_2} = \int_0^4 y\,dx = \frac{1}{3}[(0.3 + 0) + 4(0.8 + 0.4) + 2\times 0.6]$$

$$= \frac{1}{3}[0.3 + 4.8 + 1.2] = 2.10\text{ cm}^2$$

Total area in the map $= A_1 + A_2 = 4 + 21$ $= 6.10$ km^2.

Scale 1 cm = 10 km $\Rightarrow$1 cm$^2 = 10 \times 10$ $= 100$ km^2.

Total area of the island $= 6.10 \times 100 = 610$ km^2.

Example 6.22

The velocity v (km/hr) of a car which starts from rest is measured at fixed intervals of time t (hours) as shown below:

t (hr)	0	1	2	3	4	5	6
v (km/hr)	0	45	50	40	30	35	20

Estimate the distance covered in 6 hours.

Solution Let s denote the distance covered in km in time t hr. Then $\frac{ds}{dt} = v$

$$s\Big|_{t=0}^{6} = \int_0^6 v\,dt = \frac{h}{3}[X + 4\times 0 + 2\times E], \text{ by Simpson's rule}$$

$$X = 0 + 20 = 20; 0 = 45 + 40 + 35$$

$$= 120; E = 50 + 30 = 80$$

$$= \frac{1}{3}[(0+10) + 4(45+40+35) + 2(50+30)]$$

$$= \frac{1}{3}(10+480+160) = 220 \text{ km}$$

Example 6.23

Estimate the length of the arc of the curve $3y = x^3$ from (0, 0) to (1, 1/3) using Simpson's rule taking eight subintervals.

Solution The arc length 's' is given by

$$s = \int_0^1 \sqrt{1+\left(\frac{dy}{dx}\right)^2}\,dx$$

The equation of the curve is $y = \frac{1}{3}x^3$ so that

$$\frac{dy}{dx} = x^2; \; s = \int_0^1 \sqrt{1+x^4}\,dx, \; h = \frac{1-0}{8} = 0.125$$

Table of values are in Table 6.2.

Table 6.2 Table of Values

i	0	1	2	3	4	5	6	7	8
x	0	0.125	0.25	0.375	0.5	0.625	0.750	0.875	1.000
y	1.00000	1.00012	1.00195	1.01958	1.03078	1.07359	1.14735	1.25944	1.41421

Simpson's Rule

$$\int_{x_0}^{x_8} y\,dx = \frac{h}{3}[(y_0 + y_8)$$

$$+4(y_1 + y_3 + y_5 + y_7) + 2(y_2 + y_4 + y_6)]$$

$$= \frac{1/8}{3}[(1+1.41421) + 4(1.00012+1.01958$$

$$+1.07359+1.25944) + 2(1.00195+1.03078$$

$$+1.14735)] = \frac{1}{24}[2.41421+17.41048$$

$$+6.36016] = 1.091035417 \approx 1.09104$$

Example 6.24

Evaluate $\int_1^2 \frac{1}{x}dx$ by (i) Trapezoidal rule and (ii) Simpson's rule for (a) $n = 2$ and (b) $n = 10$ and compare with the exact value.

Solution (a) Table of values for $n = 2$, $h = \frac{2-1}{2} = 0.5$ are in Table 6.3.

Table 6.3 Table of Values

x	$y = \frac{1}{x}$
$x_0 = 1.0$	$y_0 = 1$
$x_1 = 1.5 = \frac{3}{2}$	$y_1 = 0.667 = \frac{2}{3}$
$x_2 = 2.0$	$y_2 = 0.5000 = \frac{1}{2}$

(i) Trapezoidal Rule

$$\int_1^2 \frac{1}{x}dx = \frac{h}{2}[y_0 + 2y_1 + y_2]$$

$$= \frac{1/2}{2}\left[1 + 2\left(\frac{2}{3}\right) + \frac{1}{2}\right] = 0.7083$$

(ii) Simpson's Rule

$$\int_1^2 \frac{1}{x}\,dx = \frac{h}{3}\left[y_0 + 4y_1 + y_2\right]$$

$$= \frac{1/2}{3}\left[1 + 4\left(\frac{2}{3}\right) + \frac{1}{2}\right] = 0.6944$$

True Value

$$\int_1^2 \frac{1}{x}\,dx = \log_e x\Big|_1^2 = \log_e 2 = 0.69314718 \simeq 0.6931$$

The results by Simpson's rule agree with the true value up to four decimal places.

(b) Table of values for $n = 10,\ h = -1/10 = 0.1$

x	$y = \frac{1}{x}$		
$x_0 = 1.0$	$y_0 = 1.0$		
$x_1 = 1.1$		$y_1 = 0.9090909$	
$x_2 = 1.2$			$y_2 = 0.8333333$
$x_3 = 1.3$		$y_3 = 0.7692308$	
$x_4 = 1.4$			$y_4 = 0.7142857$
$x_5 = 1.5$		$y_5 = 0.6666667$	
$x_6 = 1.6$			$y_6 = 0.6250000$
$x_7 = 1.7$		$y_7 = 0.5832353$	
$x_8 = 1.8$			$y_8 = 0.5555556$
$x_9 = 1.9$		$y_9 = 0.5263158$	
$x_{10} = 2.0$	$y_{10} = 0.5$		
	1.5	3.4595395	2.728174633

(i) Trapezoidal Rule

$$\int_{x_0}^{x_{10}} y\,dx = \frac{h}{2}[(y_0 + y_{10}) + 2(y_1 + y_2 + \cdots + y_9)]$$

$$= \frac{1}{20}[1.5 + 2(3.4595395 + 2.728174633)]$$

$$= 6.9377141$$

(ii) Simpson's Rule

$$\int_{x_0}^{x_{10}} y\,dx = \frac{h}{3}[(y_0 + y_{10}) + 4(y_1 + y_3 + y_5 + y_7 + y_9)$$
$$+ 2(y_2 + y_4 + y_6 + y_8)]$$

$$= \frac{1}{30}[1.5 + 4(3.4595395) + 2(2.728174633)]$$

$$= 0.69315024$$

Example 6.25

Evaluate $\int_0^1 e^{-x^2}\,dx$ using (i) Trapezoidal rule and (ii) Simpson's rule taking $h = 0.2$

Solution We will compute the functional values and tabulate in Table 6.4.

(i) Trapezoidal Rule

$$\int_{x_0}^{x_5} y\,dx = \frac{h}{2}[(y_0 + y_5) + 2(y_1 + y_2 + y_3 + y_4)]$$

$$= \frac{0.2}{2}[1.367879 + 2(1.658465 + 1.379436)]$$

$$= 0.7443681 \simeq 0.7444$$

Table 6.4 Table of Values

i	$x_0 = 0$	$x_1 = 1$	$x_2 = 2$	$x_3 = 3$	$x_4 = 4$	$x_5 = 5$	
x_i	0.0	0.2	0.4	0.6	0.8	1.0	
x_i^2	0.0	0.04	0.16	0.36	0.64	1.00	
	1.000000					0.367879	1.367879
$y = e^{-x_i^2}$		0.960789		0.697676			1.658465
			0.852144		0.527292		1.379436
	y_0	y_1	y_2	y_3	y_4	y_5	

(ii) Simpson's Rule

$$\int_{x_0}^{x_{10}} y\,dx = \frac{h}{3}[(y_0+y_5)+4(y_1+y_3)+2(y_2+y_4)]$$
$$= \frac{0.2}{3}[1.367879+4(1.658465)+2(1.379436)]$$
$$= \frac{1}{15}[1.367879+6.633860+2.758872]$$
$$= 0.717374066 \simeq 0.7174$$

EXERCISE 6.2

1. Using the data given below, evaluate $\int_{0.5}^{1.1} y\,dx$ by (i) Trapezoidal rule, (ii) Simpson's $\frac{1}{3}$ rule, (iii) Simpson's $\frac{3}{8}$ rule and (iv) Weddle's rule

x	0.5	0.6	0.7	0.8	0.9	1.0	1.1
y	0.4804	0.5669	0.6490	0.7262	0.7985	0.8658	0.9281

Ans: (i) 0.431065 (ii) 0.43131 (iii) 0.431306 (iv) 0.431301

2. Evaluate $\int_1^7 \sin x^2\,dx$ taking seven ordinates using (i) Trapezoidal rule and (ii) Simpson's rule.

Ans: (i) −1.81285 (ii) −2.56622

3. Evaluate $\int_0^1 e^{-x^2}\,dx$ taking $h = 1/6$ by (i) Trapezoidal rule and (ii) Simpson's rule.

Ans: (i) 0.7451 (ii) 0.7468

4. Evaluate $\int_4^{5.2} \log x\,dx$ taking seven grid points by (i) Simpson's $\frac{1}{3}$ rule, (ii) Simpson's $\frac{3}{8}$ rule and (iii) Weddle's rule.

Ans: (i) 1.8278472 (ii) 1.8278475 (iii) 1.8278474

5. Evaluate $\int_1^5 \frac{dx}{x}$ by Boole's rule. Compare with the exact value.

Ans: 1.61777; Exact value = 1.60943

6. Evaluate $\int_0^1 \frac{dx}{1+x^2}$ $(n = 10)$ by Trapezoidal rule.

Ans: 0.6547

7. Evaluate $\log_e 7$ by Simpson's rule.

$\left[\text{Hint: } I = \int_0^6 \frac{dx}{1+x}.\right]$

Ans: 1.9588

8. Find the value of $\int_0^{\frac{\pi}{2}} \sqrt{\sin x}.\,dx$

$\left[\text{Hint: } n = 6, h = \frac{\pi}{2}.\right]$

Ans: 1.18728

9. Evaluate (a) $\int_0^1 \sin(\pi x)\,dx$ (b) $\int_0^1 \sin\sqrt{x}\,dx$ with $h = 1/4$ by (i) Trapezoidal rule (ii) Simpson's $\frac{1}{3}$ rule, (iii) Boole's rule.

Ans: (a) (i) 0.603553; (ii) 0.638071; (iii) 0.636165 (b) (i) 0.577889; (ii) 0.592124; (iii) 0.593376

10. Evaluate $\int_1^6 (2+\sin 2\sqrt{x})\,dx$ by Trapezoidal rule with 11 sample points.

Ans: 8.19385457

6.5 Cubic Splines

6.5.1 Interpolation by Spline Functions

It has been observed that, contrary to our expectations, interpolating polynomials may tend to oscillate more and more between nodes (end points of subintervals) as n increases. Such oscillations are avoided by the method of Splines, which was initiated by I.J. Schoenberg (1946) and is now widely used. The name "Spline" is borrowed from a drafts-man's spline which is an elastic band bent to conform to the points.

The mathematical idea underlying the method is to subdivide the interval $I = [a, b]$ and use several lower-degree polynomials, one over each subinterval and get an interpolating function called a spline by fitting them together into a single curve.

Thus, spline interpolation is piecewise polynomial interpolation.

We subdivide I into subintervals with common end points called *nodes*:

$$a = x_0 < x_1 < x_2 < \cdots < x_{n-1} < x_n = b \qquad (6.49)$$

The method consists of piecing together the graphs of lower-degree polynomials $S_i(x)$ and interpolating between the successive nodes (x_{i-1}, y_{i-1}) and (x_i, y_i). The two adjacent portions of the curve $y = S_{i-1}(x)$ and $y = S_i(x)$ lying above (x_{i-1}, x_i) and (x_i, x_{i+1}), respectively, pass through a common knot (x_i, y_i). The two curves are tied together at the knot (x_i, y_i) and the set of functions $\{S_i(x)\}$ forms a piecewise polynomial curve denoted by $S(x)$.

Let $S(x)$ be the interpolating polynomial for $y = f(x)$ on I. We require that

$$S(x_i) = f(x_i) = f_i; \qquad i = 0,1,2,\ldots,n \qquad (6.50)$$

and $S(x)$ to be given by a polynomial on each subinterval, and $S(x)$ is several times differentiable.

It has been observed that linear and quadratic polynomials have their own short comings. Cubic splines are found to give satisfactory results. Hence, we describe a cubic spline interpolation method below. A cubic spline $S(x)$ on I corresponding to a partition (6.49) is a continuous function $S(x)$ that has continuous first and second derivatives everywhere in I. In each subinterval it is represented by a polynomial of degree ≤ 3. Also $S(x)$ consists of n such polynomials and is such that

$$S'(x_0) = k_0, \quad S'(x_n) = k_n \qquad (6.51)$$

where k_0 and k_n are given numbers prescribing the tangent directions at $x = a$ and $x = b$, respectively.

Let the given data points be

$$(x_i, y_i) \quad i = 0,1,2,\ldots,n \qquad (6.52)$$

where x_i satisfy (6.49)

and let $\quad h_i = x_i - x_{i-1}, \quad i = 1,2,\ldots,n \qquad (6.53)$

be the length of the ith subinterval.

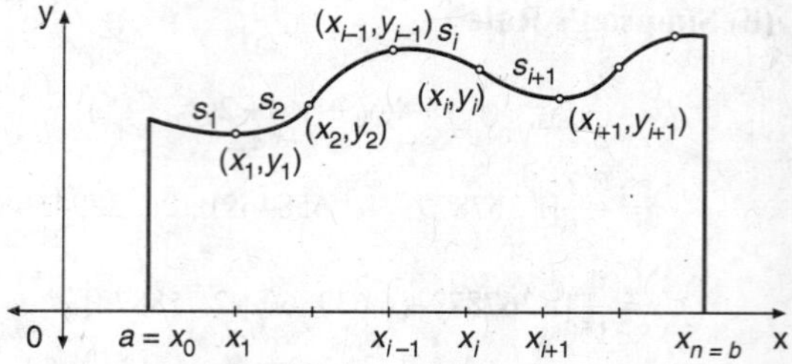

Figure 6.1 Graph of a cubic spline polynomial curve

Suppose $S_i(x)$ is the cubic spline defined in the interval $[x_{i-1}, x_i]$. The conditions for a natural spline are

1. $S_i(x)$ is a polynomial of degree ≤ 3 in each subinterval $[x_{i-1}, x_i]$, $\quad i = 1, 2, \ldots, n$.
2. $S_i(x_i) = y_i$, $i = 0, 1, 2, \ldots, n$.
3. $S_i(x)$, $S_i'(x)$ and $S_i''(x)$ are continuous in $[x_0, x_n]$.
4. $S_i''(x_0) = S_i''(x_n) = 0. \qquad (6.54)$

In order to derive the governing equations of the cubic spline, we observe that the second derivatives of the spline must be linear. Hence, we have in $[x_{i-1}, x_i]$

$$S_i''(x) = \frac{1}{h_i}[(x_i - x)M_{i-1} + (x - x_{i-1})M_i] \qquad (6.55)$$

where $\quad h_i = x_i - x_{i-1}$ and $S''_i(x_i) = M_i$ for all i.

Clearly the second derivatives of the spline are continuous. Integrating (6.55) twice with respect to x we obtain

$$S_i(x) = \frac{1}{h_i}\left[\frac{(x_i - x)^3}{6}M_{i-1} + \frac{(x - x_{i-1})^3}{6}M_i\right] + c_i(x_i - x) + d_i(x - x_{i-1}) \qquad (6.56)$$

where c_i and d_i are constants to be determined.

Using the conditions $S_i(x_{i-1}) = y_{i-1}$ and $S_i(x_i) = y_i$, we readily obtain

$$c_i = \frac{1}{h_i}\left(y_{i-1} - \frac{h_i^2}{6}M_{i-1}\right) \text{ and } d_i = \frac{1}{h_i}\left(y_i - \frac{h_i^2}{6}M_i\right) \qquad (6.57)$$

Substituting for c_i and d_i in (6.56) we obtain

$$S_i(x)=\frac{1}{h_i}\left[\frac{(x_i-x)^3}{6}M_{i-1}+\frac{(x-x_{i-1})^3}{6}M_i\right.$$
$$+\left(y_{i-1}-\frac{h_i^2}{6}M_{i-1}\right)(x_i-x)$$
$$\left.+\left(y_i-\frac{h_i^2}{6}M_i\right)(x-x_{i-1})\right] \quad (6.58)$$

In (6.58) the second derivatives M_i of the spline are still not known. To determine them, we use the condition of continuity of $S_i'(x)$.

Differentiating (6.58) with respect to x we obtain

$$S_i'(x)=\frac{1}{h_i}\left[\frac{-3(x_i-x)^2}{6}M_{i-1}+\frac{3(x-x_i)^2}{6}M_i\right.$$
$$\left.-\left(y_{i-1}-\frac{h_i^2}{6}M_{i-1}\right)+\left(y_i-\frac{h_i^2}{6}M_i\right)\right] \quad (6.59)$$

Putting $x = x_i$ in (6.59) we obtain the LH derivative

$$S_i'(x_i-)=\frac{h_i}{2}M_i-\frac{1}{h_i}\left(y_{i-1}-\frac{h_i^2}{6}M_{i-1}\right)+\frac{1}{h_i}\left(y_i-\frac{h_i^2}{6}M_i\right)$$
$$=\frac{1}{h_i}(y_i-y_{i-1})+\frac{h_i}{6}M_{i-1}+\frac{h_i}{3}M_i \quad (i=1,2,3,\ldots) \quad (6.60)$$

In order to obtain the RH derivative we have to write the equation of the cubic spline in the sub-interval $[x_i, x_{i+1}]$. This is obtained by replacing i by $i+1$ in (6.58)

$$S_{i+1}(x)=\frac{1}{h_{i+1}}\left[\frac{(x_{i+1}-x)^2}{6}M_i+\frac{(x-x_i)^2}{6}M_{i+1}\right.$$
$$+\left(y_i+\frac{h_{i+1}}{6}M_i\right)(x_{i+1}-x)$$
$$\left.+\left(y_{i+1}-\frac{h_{i+1}}{6}M_{i+1}\right)(x-x_i)\right] \quad (6.61)$$

where $h_{i+1} = x_{i+1} - x_i$. Differentiating (6.61) and putting $x = x_i$ we obtain the RH derivative at $x = x_i$ as

$$S_{i+1}'(x_i+)=\frac{1}{h_{i+1}}(y_{i+1}-y_i)-\frac{h_{i+1}}{3}M_i$$
$$-\frac{h_{i+1}}{6}M_{i+1} \quad (i=0,1,2,\ldots) \quad (6.62)$$

Equating (6.60) and (6.62), we get the recurrence relation

$$\frac{h_i}{6}M_{i-1}+\frac{1}{3}(h_i+h_{i+1})M_i+\frac{h_{i+1}}{6}M_{i+1}$$
$$=\frac{y_{i+1}-y_i}{h_{i+1}}-\frac{y_i-y_{i-1}}{h_i}, \quad (i=1,2,\ldots n) \quad (6.63)$$

For equal intervals we have $h_i = h_{i+1} = h$ and equation (6.63) simplifies to

$$M_{i-1}+4M_i+M_{i+1}=\frac{6}{h^2}(y_{i+1}-2y_i+y_{i-1}) \quad (6.64)$$
$$(i=1,2,3,\ldots)$$

The system of equations (6.63) has some special significance. If M_0 and M_n are known, then the system can be written as

$$\left.\begin{aligned}
2(h_1+h_2)M_1+h_2M_2&=6\left(\frac{y_2-y_1}{h_2}-\frac{y_1-y}{h_1}\right)-h_1M_0\\
h_2M_1+2(h_2+h_3)M_2+h_3M_3&=6\left(\frac{y_3-y_2}{h_3}-\frac{y_2-y_1}{h_2}\right)\\
h_1M_2+2(h_3+h_4)M_3+h_4M_4&=6\left(\frac{y_4-y_3}{h_4}-\frac{y_3-y_2}{h_3}\right)\\
h_{n-1}M_{n-2}+2(h_{n-1}+h_n)M_{n-1}&=6\left[\frac{y_n-y_{n-1}}{h_n}-\frac{y_{n-1}-y_{n-2}}{y_{n-1}}\right]+h_nM_n
\end{aligned}\right] \quad (6.65)$$

Equations (6.63) or (6.64) constitute a system of $(n-1)$ equations and with the two conditions at (iv) in (6.54) for the natural spline, we have a complete system which can be solved for M_i. Systems of the type are called *tridiagonal* systems which can be solved and a unique solution obtained. When the M_i are known, equation (6.58) then gives the required cubic spline in the subinterval $[x_{i-1}, x_i]$. Also, the y_i are obtained from equations (6.60) and (6.62).

Example 6.26

Determine the cubic spline for the data points (1, –8) (2, –1) and (3, 18) satisfying the function $y = f(x)$. Find the approximate values of $y(2.5)$ and $y'(2.0)$.

Solution Here we have

$$n = 2 \quad \text{and} \quad M_0 = M_2 = 0. \qquad (1)$$

The data points are $(x_0, y_0) = (1,-8)$, $(x_1, y_1) = (2,-1)$ and $(x_2, y_2) = (3,18)$

$$\therefore \qquad h = x_1 - x_0 = 2 - 1 = 1 \qquad (2)$$

Recurrence relation (6.64) for $i = 1$ is

$$M_0 + 4M_1 + M_2 = 6(18 - 2(-1) - 8) = 6.12 \qquad (3)$$

$$\Rightarrow \qquad M_1 = 18 \qquad (4)$$

$$M_0 = M_2 = 0 \text{ and } y_2 = 18, y_1 = -1, y_0 = -1 \qquad (5)$$

Let the cubic spline in the interval $1 \le x \le 2$ be $S_1(x)$ and the cubic spline in the interval $2 \le x \le 3$ be $S_2(x)$.

Putting $i = 1$ in equation (6.58)

$$S_1(x) = \frac{(x_1 - x)^3}{6}M_0 + \frac{(x - x_0)^3}{6}M_1 + \left(y_0 - \frac{1}{6}M_0\right)(x_1 - x) + \left(y_1 - \frac{1}{6}M_1\right)(x - x_0) \quad \because h_1 = 1 \qquad (6)$$

Substituting $M_0 = 0, x_0 = 1, x_1 = 2, y_0 = -8$, and $y_1 = -1$ we get $M_1 = 18$

$$S_1(x) = 3(x-1)^3 - 8(2 - x) + (-1 - 3)(x - 1) \qquad (7)$$

$$S_1(x) = 3(x-1)^3 + 4x - 12 \qquad (8)$$

Putting $i = 2$ in equation (6.58)

$$S_2(x) = \frac{(x_2 - x)^3}{6}M_1 + \frac{(x - x_1)^3}{6}M_2 + \left(y_1 - \frac{1}{6}M_1\right)(x_2 - x) + \left(y_2 - \frac{1}{6}M_2\right)(x - x_1) \quad \because h_2 = 1 \qquad (9)$$

Substituting $M_1 = 18$, $M_2 = 0$, $x_1 = 2$, $x_2 = 3$, $y_1 = -1, y_2 = 1$

$$S_2(x) = \frac{(3-x)}{6}\times 18 + 0 + \left(-1 - \frac{1}{6}.18\right)(3-x) + \left(18 - \frac{1}{6}.0\right)(x-2)$$

$$S_2(x) = 3(3-x)^3 - 4(3-x) + 18(x-2)$$

$$S_2(x) = 3(3-x)^3 + 22x - 48;$$

$$S_2'(x) = -9(3-x)^2 + 22 \qquad (10)$$

Therefore, since 2.5 lies in the interval $2 \le x \le 3$, using $S_2(x)$

$$y(2.5) \simeq S_2(2.5) = \frac{3}{8} + 7 = 7.375 \qquad (11)$$

$$\text{and} \quad y'(2.0) \simeq S_2'(2.0) = -9(3-2)^2 + 22 = 13 \qquad (12)$$

Here the tabulated function is $y = x^3 - 9$. Hence the exact values of $y(2.5)$ and $y'(2.0)$ are 6.625 and 12.0, respectively.

Example 6.27

Fit a natural cubic spline in the interval [0, 3] to the data

x	0	1	2	3
y	1	2	33	244

Estimate $y(x)$, $y'(x)$ at $x = 0.5$, 1.5 and 2.5.

Solution Here we have uniform interval length $h = 1$

Also, since the cubic spline is natural, $M_0 = M_3 = 0$.

Putting $i = 1$ in the Recurrence Relation (6.63) we have

$$M_0+4M_1+M_2=\frac{6}{1}(y_2-2y_1+y_0)$$
$$=6(33-2\times 2+1) \quad (1)$$
$$\because \quad y_2=33, \quad y_1=2, \quad y_0=1$$
$$4M_1+M_2=180 \quad \because M_0=0$$

Putting $i = 2$ in the Recurrence Relation (6.63) we have

$$M_1+4M_2+M_3=\frac{6}{1}(y_3-2y_2+y_1)$$
$$=6(244-2\times 33+2)$$
$$\because \quad y_3=244, \quad y_2=33, \quad y_1=2$$
$$M_1+4M_2=1080 \quad \because M_3=0 \quad (2)$$

Add (1) and (2)

$$5M_1+5M_2=1080+180=1260$$
$$\Rightarrow \quad M_1+M_2=252 \quad (3)$$

Subtract (1) from (2)

$$3M_2-3M_1=1080-180=900$$
$$\Rightarrow \quad M_2-M_1=300 \quad (4)$$

Add (3) and (4) $2M_2 = 252 + 300 = 552$ $\Rightarrow M_2 = 276$

Subtract (4) from (3) $2M_1 = 252 - 300 = -48$ $\Rightarrow M_1 = -24$

Equation (6.58) with $h_i = 1$

$i = 1$

$$S_i(x)=\frac{(x_i-x)^3}{6}M_{i-1}+\frac{(x-x_{i-1})^3}{6}M_i+\left(y_{i-1}-\frac{1}{6}M_{i-1}\right)(x_i-x)+\left(y_i-\frac{1}{6}M_i\right)(x-x_{i-1}) \quad (5)$$

$$S_1(x)=\frac{(x_1-x)^3}{6}M_0+\frac{(x-x_0)^3}{6}M_1+\left(y_0-\frac{1}{6}M_0\right)(x_i-x)+\left(y_1-\frac{1}{6}M_1\right)(x-x_0)$$
$$=0+\frac{(x-0)^3}{6}(-24)+\left(1-\frac{1}{6}.0\right)(1-x)-(2-5(-24))(x-0)$$
$$=-4x^3+(1-x)+6x=1+5x-4x^3 \quad (6)$$

$i = 2$

$$S_2(x)=\frac{(x_2-x)^3}{6}M_1+\frac{(x-x_1)^3}{6}M_2+\left(y_1-\frac{1}{6}M_1\right)(x_2-x)+\left(y_2-\frac{1}{6}M_2\right)(x-x_1)$$
$$=-\frac{24}{6}(2-x)^3+\frac{276}{6}(x-1)^3+\left(2-\frac{1}{6}.(-24)\right)(2-x)-\left(33-\frac{1}{6}.276\right)(x-1)$$
$$=-4(2-x)^3+46(x-1)^3+6(2-x)-13(x-1) \quad (7)$$

$$S_3(x)=\frac{(x_3-x)^3}{6}M_2+\frac{(x-x_2)^3}{6}M_3+\left(y_2-\frac{1}{6}M_3\right)(x_3-x)+\left(y_3-\frac{1}{6}M_3\right)(x-x_2)$$
$$=46(3-x)^3+0+\left(33-\frac{276}{6}\right)(3-x)(244-0)(x-2)$$
$$=46(3-x)^3-13(3-x)+244(x-2) \quad (8)$$

$$S_1(0.5)=1+\frac{5}{2}-4\cdot\frac{1}{8}=3;$$
$$S_2(1.5)=-4\cdot\frac{1}{8}+46\cdot\frac{1}{8}+25-19\cdot\frac{3}{2}=\frac{7}{4} \quad (9)$$
$$S_3(2.5)=46\cdot\frac{1}{8}+257\cdot\frac{5}{2}-527=\frac{485}{4} \quad (10)$$
$$S_1'(0.5)=2; \quad S_2'(1.5)=\frac{31}{2}; \quad S_3'(2.5)=\frac{445}{2} \quad (11)$$

6.5.2 Numerical Differentiation: Cubic Spline Method

The cubic spline can be used to compute the first and second derivatives of a function. For a natural cubic spline, the recurrence formula (6.63) or (6.64) may be used for computing the second derivatives depending on whether the subintervals are of unequal length or equal length. Then (6.61) gives the spline function in the interval $[x_i, x_{i+1}]$ from which the first derivatives are obtained. For the first derivatives at tabular points, it is easier to use formulae (6.60) and (6.61) directly.

Example 6.28

Estimate $y'\left(\frac{\pi}{4}\right)$ and $y''\left(\frac{\pi}{4}\right)$ using spline formulae in numerical differentiation for the function $y(x) = \sin x$ in $[0, \pi]$.

Solution Here the boundary conditions are $M_0 = M_n = 0$

Let $n = 2$ so that $h = \frac{\pi - 0}{2} = \frac{\pi}{2}$. Also, the table of values is

i	0	1	2
x	0	$\pi/2$	π
$y = \sin x$	0	1	0

(1)

Using formula (6.64) we obtain

$$M_0 + 4M_1 + M_2 = \frac{6}{h^2}(y_2 - 2y_1 + y_0)$$

$$\Rightarrow 0 + 4M_1 + 0 = \frac{6}{\left(\frac{\pi}{2}\right)^2}(0 - 2\times 1 + 0)$$

$$\Rightarrow \quad M_1 = -\frac{12}{\pi^2} \tag{2}$$

Now formula (6.61) for $i = 0$ gives in $0 \le x \le \frac{\pi}{2}$

$$S(x) = \frac{1}{h}\left[\frac{(x_1 - x)^3}{6}M_0 + \frac{(x - x_0)^3}{6}M_1 + \left(y_0 - \frac{h^2}{6}M_0\right)(x_2 - x) + \left(y_1 - \frac{h^2}{6}M_1\right)(x - x_0)\right] \tag{3}$$

$$S(x) = \frac{2}{\pi}\left[\frac{x^3}{6}\left(-\frac{12}{\pi^2}\right) + 0 + \left(1 - \frac{1}{6}\left(-\frac{12}{\pi^2}\right)\frac{\pi^2}{4}\right)x\right]$$

$$= \frac{2}{\pi}\left[-\frac{2x^3}{\pi^2} + \frac{3x}{2}\right] \tag{4}$$

Differentiating with respect to x, we obtain

$$S'(x) = \frac{2}{\pi}\left(-\frac{6x^2}{\pi^2} + \frac{3}{2}\right); \quad s''(x) = -\frac{24x}{\pi^3} \tag{5}$$

$$S'\left(\frac{\pi}{4}\right) = \frac{2}{\pi}\left(-\frac{6}{\pi^2}\cdot\frac{\pi^2}{16} + \frac{3}{2}\right) = \frac{9}{4\pi} = 0.71619725 \tag{6}$$

$$S''\left(\frac{\pi}{4}\right) = -\frac{24}{\pi^3}\cdot\frac{\pi}{4} = -\frac{6}{\pi^2} = -0.60792710 \tag{7}$$

Exact values are

$$S'\left(\frac{\pi}{4}\right) = \cos\frac{\pi}{4} = \frac{1}{\sqrt{2}} = 0.7 = 710 \tag{8}$$

$$S''\left(\frac{\pi}{4}\right) = -\sin\frac{\pi}{4} = -\frac{1}{\sqrt{2}} = -0.707 \tag{9}$$

% error in $S'(x)$: 1.28% and % error in $S''(x)$: 14.03% (10)

6.5.3 Numerical Integration: Cubic Spline Method

If $S(x)$ is the cubic spline in the interval $[x_{i-1}, x_i]$ then we have, on integrating equation (6.58) with $h_i = h$ (uniform interval length),

$$I = \int_{x_0}^{x_n} y\,dx \simeq \sum_{i=1}^{n}\int_{x_{i-1}}^{x_i} s(x)\,dx$$

$$= \sum_{i=1}^{n}\int_{x_{i-1}}^{x_i}\left\{\frac{1}{6h}\left[(x_i - x)^3 M_{i-1} + (x - x_{i-1})^3 M_i\right] + \frac{1}{h}(x_i - x)\left(y_{i-1} - \frac{h^2}{6}M_{i-1}\right) + \frac{1}{h}(x - x_{i-1})\left(y_i - \frac{h^2}{6}M_i\right)\right\}dx \tag{6.66}$$

Carrying out the integration and simplifying, we obtain

$$I = \sum_{i=1}^{n}\left[\frac{h}{2}(y_{i-1} + y_i) - \frac{h^3}{24}(M_{i-1} + M_i)\right] \tag{6.67}$$

where M_i the second derivatives of the spline function are calculated from the recurrence relations

$$M_{i-1}+4M_i+M_{i+1}=\frac{6}{h^2}(y_{i-1}-2y_i+y_{i+1})$$
$$i=1,2,3,\ldots,(n-1) \qquad (6.68)$$

Example 6.29

Evaluate $I=\int_0^1 \sin \pi x\, dx$ using the cubic spline method.

Solution Let us take $n=2$ so that $h=\frac{1}{2}$. In this case the table of values of x and y is

i	0	1	2
x	0	0.5	1.0
$y=\sin \pi x$	0	1.0	0

(1)

Using equation (6.64) with $M_0=M_2=0$ we have

$$M_0+4M_1+M_2=\frac{6}{h}(y_0-2y_1+y_2)\Rightarrow M_1=-12 \qquad (2)$$

By cubic spline formula for integration (6.67) we have

$$I=\frac{h}{2}\cdot(y_0+y_1)-\frac{h^3}{24}\cdot(M_0+M_1)$$
$$+\frac{h}{2}\cdot(y_1+y_2)-\frac{h^3}{24}\cdot(M_1+M_2)$$
$$=\frac{1}{4}(0+1)-\frac{1}{8\times 24}(0-12)$$
$$+\frac{1}{4}(1+0)-\frac{1}{8\times 24}(-12+0)$$
$$=\frac{1}{4}+\frac{1}{16}+\frac{1}{4}+\frac{1}{16}=\frac{5}{8}=0.625 \qquad (3)$$

The exact value is

$$-\frac{1}{\pi}\cos \pi x\Big|_0^1=\frac{2}{\pi}=0.63661978 \qquad (4)$$

The absolute error in natural spline solution = 0.01161978 (5)

6.6 Gaussian Integration

Consider numerical evaluation of the integral

$$I=\int_a^h f(x)dX \qquad (6.69)$$

In the previous sections we have derived Trapezoidal and Simpsons rules, which are of the form:

Trapezoidal rule

$$\int_{x_0}^{x_n} f(x)dx=h\left[\left(\frac{1}{2}\right)f_0+f_1+f_2+\cdots+f_{n-2}+f_{n-1}+\left(\frac{1}{2}\right)f_n\right] \qquad (6.70)$$

Simpsons $\frac{1}{3}$ rule

$$\int_{x_0}^{x_n} f(x)dx=h\left[\left(\frac{1}{3}\right)f_0+\left(\frac{4}{3}\right)f_1+\left(\frac{2}{3}\right)f_2+\cdots+\left(\frac{2}{3}\right)f_{n-2}+\left(\frac{4}{3}\right)f_{n-1}+\left(\frac{1}{3}\right)f_n\right] \qquad (6.71)$$

These formulas require values of the function at predetermined and equally spaced points *xi*, which are called the nodes. They give exact results for polynomials not exceeding a certain degree, which is called '*the degree of precision*.'

Observe that in the above formulas the function values are multiplied by certain constants which are such that equidistant constants from both ends are same.

Gauss derived a formula which uses the same number of function values but with different spacing, which gives a better accuracy.

First we notice that the general interval $[a, h]$ of integration can be reduced to $[-1, 1]$ symmetrical w.r.t. the origin, by a change of the variable as follows

$$x=\frac{a}{2}(1-u)+\frac{b}{2}(1+u)=\frac{b-a}{2}u+\frac{b+a}{2}$$
$$x=a\Leftrightarrow ut=-1,\ x=b\Leftrightarrow ut=1.$$

Gauss's formula is written in the form

$$\int_{-1}^{1}\phi(u)du=w_1\phi(u_1)+w_2\phi(u_2)+\cdots+w_n\phi_n(u_n)$$
$$=\sum_{i=1}^{n}w_i\phi_i(u) \qquad (6.72)$$

where the constants w_i are called the weight and u_i are called the abscissas (or nodes).

Here n is fixed. We determine the n coefficients $w_1, w_2, \ldots, w_n$ (called the weights) and n nodes $u_1, u_2, \ldots, u_n$ so that (6...) gives exact results for polynomial of degree k as high as possible. Since polynomial of degree $(2n-1)$ has $2n$ coefficients. So, if follows that $k \le (2n-1)$, Gauss has shown that exactness for polynomials of degree not exceeding $(2n-1)$ can be attained. Further, he has shown that

Let us consider the case $n = 2$. In this case the formula is

$$\int_{-1}^{1} \phi(u)du = w_1\phi(u_1) + w_2\phi(u_2) \tag{6.73}$$

This formula is exact when $F(u)$ is a polynomial of degree ≤ 3 we put successively $\phi(u) = 1, u, u^2$ and u^3. Then (6...) gives the four equations

$$w_1 + w_2 = \int_{-1}^{1} 1\,du = u|_{-1}^{1} = 2 \tag{i}$$

$$w_1u_1 + w_2u_2 = \int_{-1}^{1} u\,du = \left.\frac{u^2}{2}\right|_{-1}^{1} = 0 \tag{ii}$$

$$w_1u_1^2 + w_2u_2^2 = \int_{-1}^{1} u^2du = \left.\frac{1}{3}u^3\right|_{-1}^{1} = \frac{2}{3} \tag{iii}$$

$$w_1u_1^3 + w_2u_2^3 = \int_{-1}^{1} u^3du = \left.\frac{1}{4}u^4\right|_{-1}^{1} = 0 \tag{iv}$$

Solving these equations we obtain
From (ii) and (iv) we obtain

$$\frac{w_1}{w_2} = -\frac{u_2}{u_1} = -\frac{u_2^3}{u_1^3} \Rightarrow \left[\left(\frac{u_2}{u_1}\right)^2 - 1\right]\left(\frac{u_2}{u_1}\right) = C$$

$$\left(\frac{u_2}{u_1}\right)^2 = 1 \because u_2, u_1 \text{ are non zero } \frac{w_1}{w_2} = -\frac{u_2}{u_1} = -\frac{u_2^3}{u_1^3}$$

$$a^3 = a \Rightarrow a(a^2-1) = 0$$

$$a = 0 \text{ or } a = \pm 1$$

if $u_2 = u_1$ then from $w_1 = -w_2$
Hence $u_2 = -u_1$ and $w_1 = w_2 = 1$

From eq. (iii) $2\cdot u_2^2 = \frac{2}{3} \Rightarrow u_2 = \frac{1}{\sqrt{3}}, u_1 = -\frac{1}{\sqrt{3}}$

In the general case, eventhough we can formulate the system of equations, as in the particular case of $n = 2$, the solution of the nonlinear equations proves extremely difficult. Gauss has instead, shown that the u_i are the zeros of the $(n+1)$ m Legendre polynomial $P_{n+1}(u)$ which satisfies the Recurrence Relation

$$(n+1)P_{n+1}(u) = (2n+1)uP_n(u) - nP_{n-1}(u)$$

where $P_1(u) = 1$ and $P_1(u) = u$. The first five legendre polynomials are

$$P_0(u) = 1, P_1(u) = u, P_2(u) = \frac{1}{2}(3u^2-1)$$

$$P_3(u) = \frac{1}{2}(5u^3-3u), P_4(u) = \frac{1}{8}(35u^4-30u^2+3)$$

It can also be shown that the corresponding weights w_i are given by

$$w_i = \int_{-1}^{1} \prod_{j=0,\, j\ne i}^{n} \frac{u-u_j}{u_i-u_j}du$$

where u_i are the nodes (abscissæ)

As an example when $n = 1$ we solve $P_2(u) = 0$

$$\Rightarrow \frac{1}{2}(3u^2-1) = 0 \Rightarrow u = \pm\frac{1}{\sqrt{3}}$$

Thus

$$u_0 = -\frac{1}{\sqrt{3}} = -\frac{\sqrt{3}}{3} \text{ and } u_1 = \frac{1}{\sqrt{3}} = \frac{\sqrt{3}}{3}$$

The corresponding weights are given by

$$w_0 = \int_{-1}^{1} \frac{u-u_1}{u_0-u_1}du = \frac{1}{u_0-u_1}\left(\frac{1}{2}u^2 - u_1u\right)_{-1}^{1}$$

$$= \frac{1}{u_0-u_1}(0-2u_1)$$

$$= \frac{3}{-\sqrt{3}-\sqrt{3}}\left(-2\cdot\frac{\sqrt{3}}{3}\right) = 1$$

$$w_1 = \int_{-1}^{1} \frac{u-u_0}{u_1-u_0}du = \frac{1}{u_1-u_0}\left(\frac{1}{2}u^2 - u_0u\right)_{-1}^{1}$$

$$= \frac{1}{u_1-u_0}(0-2u_0)$$

$$= \frac{1}{2\frac{\sqrt{3}}{3}}\left(2\frac{\sqrt{3}}{3}\right) = 1$$

For $n = 2$ we have already obtained w_i and w_0. For $n = 3$ we can proceed as explained above and find w_i and w_0. But u_i and w_i are tabulated for large number of values n. Here, we given below, a short table of these values.

Table 6.5 Nodes and Weights for Gaussian Integration

n	u_i	w_i	*Degree of precision*
2	$\pm\frac{1}{\sqrt{3}} = \pm 0\cdot5773502692$	1·0000	3
3	0·0 $\pm\sqrt{\frac{3}{5}} = \pm 0\cdot7745966692$	0·8888888889 0·5555555556	5
4	0·3399810436 0·8611363116	0·6521451549 0·3478548451	7

Example 6.30

Evaluate $\int_0^1 x\,dX$ by Gauss's formula

Solution First we have to change the limits of integration from $[a, b] = [0, 1]$ to $[-1, 1]$

Put $x = \frac{u}{2}(b-a) + \frac{1}{2}(b+a) = \frac{u}{2}(1-0) + \frac{1}{2}(1+0)$

$= \frac{1}{2}(u+1)$

∴ We have

$$I = \int_0^1 x\,dx = \frac{1}{4}\int_{-1}^{1}(u+1)du = \frac{1}{4}\sum_{i=1}^{n} w_i\phi(u_i)$$

where $\phi(u_i) = u_i + 1$. For the sake of simplicity, we take $n = 4$, use the values give in the table and obtain

$$I = \frac{1}{4}\begin{bmatrix}(-0\cdot86114+1)(0\cdot34785) \\ +(-0\cdot33998+1)(0\cdot65214) \\ +(0\cdot33998+1)(0\cdot65214) \\ +(0\cdot86114+1)(0\cdot34785)\end{bmatrix}$$

$= 0\cdot49999.$

Here we have taken the values of nodes and weights rounded to five decimal places. The true value is $\frac{1}{2} = 0\cdot5$. So, the error is 0·00001.

Example 6.31 Evaluate $\int_{-1}^{1}\frac{dy}{1+x^2}$ by Gauss's integration formula with $n = 2$

Solution $f(x) = \frac{1}{1+x^2}$ The integral is $[-1, 1]$ Gauss's 2 point formula is

$$\int_{-1}^{1} f(x)dy = f\left(-\frac{1}{\sqrt{3}}\right) + f\left(\frac{1}{\sqrt{3}}\right) = \frac{1}{1+\left(\frac{1}{\sqrt{3}}\right)^2}$$

$$+\frac{1}{1+\left(\frac{1}{\sqrt{3}}\right)^2} = \frac{3}{4} + \frac{3}{4} = \frac{3}{2} = 1.5$$

Exact Value:

$$\int_{-1}^{1}\frac{dy}{1+x^2} = \tan^{-1}x\Big|_{-1}^{1} = 2\tan^{-1}1 = \frac{\pi}{2} = 1\cdot570796$$

Error = (Absolute Value – Approx Value) = $1\cdot570796 - 1\cdot5 = 0\cdot070796$

net error = 0·045

Example 6.32 Evaluate $\int_0^1 (6x^2 + 7x^4)dx$ by Gauss's 3point formula.

Solution Gauss's 3 point formula is

$$\int_{-1}^{1} f(x)dx = \frac{1}{9}\left[8f(0) + 5\left\{f\left(-\sqrt{\frac{3}{5}}\right) + f\left(\sqrt{\frac{3}{5}}\right)\right\}\right]$$

Here the integrand $f(x) = 6x^2 + 7x^4$ is an even function

$$\therefore \int_0^1 f(x)dx = \frac{1}{2}\int_{-1}^{1}(x)dx$$

$$= \frac{1}{18}\left[8f(0) + 5\left\{f\left(-\sqrt{\frac{3}{5}}\right) + f\left(\sqrt{\frac{3}{5}}\right)\right\}\right]$$

$$= \frac{1}{18}\left[8(0) + 5\times 2\left\{6\left(\sqrt{\frac{3}{5}}\right)^2 + 7\left(\sqrt{\frac{3}{5}}\right)^2\right\}\right]$$

$$= \frac{1}{18}(61.2) = 3.4$$

Numerical Evaluation of Singular Integrals

The numerical integration formulae discussed in the previous sections are applicable if the integrand $f(x)$ can be represented by a polynomial or can be expanded in a Taylor's series. A function $f(x)$ is said to have a *singularity* at a point if $f(x)$ or any of its derivatives is infinite at that point. If the integrand has a singularity at a point in the interval of integration then the integral is called a *singular integral*. In such situations, the methods of integration we have so far studied are not useful and special methods need to be developed. We therefore consider some methods which can be applied in such situations.

Before we take up methods of evaluation of such integrals, let us consider an example of a singular integral.

Let $I = \int_{-1}^{1} \frac{dx}{x^3}$

We observe that $x = 0$ is a point where the integrand $f(x) = \frac{1}{x^3}$ is not defined. This is a singular integral. By definition, the value of the integral, if it exists, is

$$\lim_{\epsilon \to 0}\left[\int_{-1}^{-\epsilon} \frac{dx}{x^3} + \int_{\epsilon}^{1} \frac{dx}{x^3}\right] = \left(-\frac{1}{2}\right)\lim_{\epsilon \to 0}\left[\frac{1}{x^2}\bigg|_{-1}^{-\epsilon} + \frac{1}{x^2}\bigg|_{\epsilon}^{1}\right] = 0$$

where ϵ is a small positive number. The limit exists and is equal to 0. It is called the principal value of the integral. It is denoted by

$$P(I) = P\int_{-1}^{1} \frac{dx}{x^3} = 0.$$

Evaluation of Principal Value Integrals

Consider numerical evaluation of

$$I(f) = \int_a^b \frac{f(x)dx}{x-t} \tag{1}$$

The integrand has a singularity at $t = x$.

The principal value $P(I)$ of the integral is, by definition,

$$P(I) = \lim_{\epsilon \to 0}\left[\int_a^{t-\epsilon} \frac{f(x)dx}{x-t} + \int_{t+\epsilon}^{b} \frac{f(x)dx}{x-t}\right], \quad (a < t < b) \tag{2(a)}$$

$$= I(f) \text{ for } t < a \text{ or } t > b \tag{2(b)}$$

Put $x = a + uh$ and $t = a + kh$ in (1). We have $x - t = (u - k)h$, $dx = h\,du$. The limits are $x = a \Rightarrow u = 0$

$$x = t \Rightarrow u = \frac{t-a}{h} = p, \text{ say.}$$

We obtain

$$P(I) = P\int_0^p \frac{f(a+uh)}{(u-k)h} h\,du = P\int_0^p \frac{f(a+uh)}{u-k} du \tag{3}$$

We now replace $f(a + uh)$ by Newton's forward interpolation polynomial.

$$f(a+uh) = f(a) + u^{(1)}\frac{\Delta}{1!}f(a) + u^{(2)}\frac{\Delta^2}{2!}f(a) + u^{(3)}\frac{\Delta^3 f(a)}{3!} + \cdots\cdots + \frac{u^{(n)}\Delta^n f(a)}{n!}$$

$$= \sum_{r=0}^{\infty} u^{(r)}\frac{\Delta^r f(a)}{r!} \tag{4}$$

where $u^{(r)} = u(u-1)(u-2)\cdots(u-r+1)$.

Substituting (4) in (3) we get

$$P(I) = \sum_{r=0}^{\infty} \frac{\Delta^r f(a)}{r!}\int_0^p \frac{u^{(r)}}{u-k} du \tag{5}$$

We can obtain various approximation formulæ by truncating the RHS series in (5). Thus, by truncating at n we get

$$I_n(f) = \sum_{r=0}^{n} c_r \frac{\Delta^r f(a)}{r!} \tag{6}$$

where c_r are the constants given by

$$c_r = \int_0^p \frac{u^{(r)}}{u-k} du \tag{7}$$

We obtain rules of orders 1, 2, 3, ... by taking $n = 1, 2, 3, \ldots$ respectively

(a) $n = 1$, *Two-point Rule*

$$I_1(f) = \sum_{r=0}^{1} c_r \frac{\Delta^r f(a)}{r!} = c_0 f(a) + c_1 \Delta f(a)$$

$$= (c_0 - c_1) f(a) + c_1 f(a+h)$$

$$[\because \Delta f(a) = f(a+h) - f(a)]$$

(b) $n = 2$, *Three point Rule*

$$I_2(f) = \sum_{r=0}^{2} c_r \frac{\Delta^r f(a)}{r!} = c_0 f(a) + c_1 \Delta f(a) + c_2 \Delta^2 f(a)$$

$$= \left(c_0 - c_1 + \frac{1}{2}c_2\right) f(a) + (c_1 - c_2) f(a+h) + \frac{1}{2} c_2 f(a+2h)$$

$$[\because \Delta^2 f(a) = f(a+2h) - 2f(a+h) + f(a)]$$

In the above relations, the value of c_r are given by

$$c_0 = \int_0^p \frac{du}{u-k} = \log(u-k)|_0^p = \log\left|\frac{p-k}{k}\right|$$

$$c_1 = \int_0^p \frac{u\,du}{u-k} = [u + k\log|u-k|]_0^p$$

$$= p + k\log\left|\frac{p-k}{k}\right| = p + c_0 k$$

$$c_2 = \int_0^p \frac{(u^2 - u)du}{u-k}$$

$$= \left[\frac{1}{2}u^2 + (k-1)u + k(k-1)\log|u-k|\right]_0^p$$

$$= \frac{1}{2}p^2 + p(k-1) + c_0 k(k-1)$$

Generalized Quadrature

Quite often singular integrals occur in applications. Here is therefore need to develop special integration methods to deal with such integrals.
We therefore consider here numerical integration of integrals of the form

$$I(s) = \int_a^b f(t)\phi(t-s)dt \tag{6.74}$$

where $f(t)$ is continuous but $\phi(t - s)$ may have an integrable singularity of the type $\log|s-t|$ or $|s-t|^\alpha$ where $\alpha > -1$. For numerical integration, we divide the internal of integration $[a, b]$ by the partition $t_j = a + jh$ $(j = 0, 1, 2, \dots, n)$ with $b = a + nh$. The integral $I(s)$ can now be written as

$$I(s) = \sum_{j=0}^{n-1} \int_{t_j}^{t_{j+1}} f(t)\phi(t-s)dt \tag{6.75}$$

We approximate $f(t)$ by the linear interpolating polynomial function $f_n(t)$ given by

$$f_n(t) = \frac{1}{h}\left[(t_{j+1} - t)f(t_j) + (t - t_j)f(t_{j+1})\right] \tag{6.76}$$

Replacing $f(t)$ by $f_n(t)$ given by (6...), we obtain

$$I(s) = \frac{1}{h}\sum_{j=0}^{n-1}\int_{t_j}^{t_{j+1}}\left[(t_{j+1} - t)f(t_j) + (t - t_j)f(t_{j+1})\right] \phi(t-s)dt \tag{6.77}$$

If we now set $t = t_j + ph \Rightarrow dt = h\,dp$ and the limits of integration for p are $p = 0$ and 1

$$I(s) = h\sum_{j=0}^{n-1}\int_0^1\left[(1-p)f(t_j) + pf(t_{j+1})\phi(t_j + ph - s)\right]dp$$

$$\text{or } I(s) = h\sum_{j=0}^{n-1}\left[\alpha_j f(t_j) + \beta_j f(t_{j+1})\right] \tag{6.78}$$

$$\text{where } \alpha_j = h\int_0^1 (1-p)\phi(t_j + ph - s)dp \tag{6.79}$$

$$\beta_j = h\int_0^1 p\phi(t_j + ph - s)dp \tag{6.80}$$

Corollary: If $\phi(u) \equiv 1$ then $\alpha_j = \beta_j = h_2$ and the integral becomes

$$I(s) = \frac{h}{\Sigma}[f(t_0) + 2f(t_1) + 2f(t_1) + \cdots + 2f(t_{n-1}) + f(t_n)]$$

which is the trapezoidal rule. The rule defined by (5) together with eqns. (6) and (7) is called the generalized *trapezoidal rule.*

Similarly, we can find the generalized *Simpson's rule* if we approximate $f(b)$ by a quadratic in the interval $[t_j, t_{j+1}]$.

EXERCISE 6.3

1. Estimate the values of $y(3.5)$ from the following data using cubic spline method:

x	1	2	3	4	5
y	30	15	32	18	25

Ans: $y(3.5) = 29.82$

2. Use cubic splines to estimate y (7) from the following data:

x	4	9	16
y	2	3	4

Ans: $S_1(7) = \frac{1844}{700} = 2.7343$

$$S_1(x) = \frac{1}{5}\left[\frac{-(x-4)^3}{420} + 2(9-x) + \frac{257}{84}(x-4)\right]$$

$$S_2(x) = \left[\frac{-(6-x)^3}{420} + \frac{187}{60}(16-x) + 16(x-4)\right]$$

3. Fit a natural cubic spline to the following data. Hence compute y (1.5) and $y'(2.0)$:

x	1	2	3	4
y	1	5	11	8

Ans: $y(1.5) = \frac{103}{40} = 2.575$

$$y'(2.0) = \frac{94}{15} = 6.267$$

4. Fit a natural cubic spline to the following data. Estimate y (0.5) and y (1.5):

x	0	1	2
y	4	1	2

Ans: $S_1(x) = 4 - 4x + x^3$ in [0, 1]

$S_2(x) = 1 - (x-1) + 3(x-1)^2 - (x-1)^3$ in [1, 2]

$$y(0.5) = \frac{17}{8} = 2.125$$

$$y(1.5) = \frac{9}{8} = 1.125$$

5. Evaluate $\int_0^1 \frac{dx}{1+x}$ by Gauss's formula with $n = 3$

Ans: 0·6931

6. Find the value of $\int_0^{\frac{\pi}{2}} \sin x\, dx$ using Gauss-Legendre formula with $n = 2$.

Ans: 1·00002

7. Using Gauss's 3point formula, find $\int_1^2 \frac{x\, dx}{1+x^4}$

Ans: 0·2703

8. By Gauss's 3point quadrative formula evaluate $\int_2^3 \frac{\cos 2x}{1+\sin x} dY$

Ans: 0·2027

Numerical Solution of Ordinary Differential Equations

7

7.1 Introduction

Many engineering and scientific problems are modelled as differential equations. Forming differential equation, finding its solution and interpreting the results obtained are of utmost importance for engineers and scientists. Exact solutions of differential equations can be found only in a limited number of cases. Hence, there is need for numerical solution. Even in cases where closed-form solutions are possible to assess and analyse them we have to compute. Instead, we could solve the differential equation itself numerically and have a better understanding of the solution, especially now with the availability of high-speed digital computers.

7.1.1 Ordinary Differential Equation

An equation of the form

$$F(x, y, y', y'', \dots, y^{(n)}) = 0 \tag{7.1}$$

where $y' = \frac{dy}{dx}, y'' = \frac{d^2 y}{dx^2}, \dots, y^{(n)} = \frac{d^{(n)}y}{dx^{(n)}}$ is called an nth order ordinary differential equation.

When we rewrite (7.1) in the form

$$y^{(n)}(x) = f(x, y, y', y'', \dots, y^{(n-1)}) \tag{7.2}$$

it is called a canonical representation of equation (7.1). A relation of the form

$$\phi(x, y, c_1, c_2, \dots, c_n) \tag{7.3}$$

involving n arbitary constants and satisfying (7.1) or (7.2) identically is called the general solution of (7.1).

7.1.2 Initial or Boundary Value Problem

Together with the differential equation (7.1), if conditions for determining a particular solution are specified at a fixed point, called the initial point, then the problem is called an initial value problem (IVP), and if they are specified at more than one point then it is called a boundary value problem (BVP).

Since higher-order differential equations can be reduced to a system of first order and first-degree equations, we now consider the numerical solution of the IVP:

$$\frac{dy}{dx} = f(x, y), \; y(x_0) = y_0 \tag{7.4}$$

We will discuss methods for computing numerical values of the solutions. These methods are step-by-step methods. That is, we start from the given $y_0 = y(x_0)$ and proceed step-wise computing approximate values of the solution $y(x)$ at the "mesh points": $x_1 = x_0 + h$, $x_2 = x_0 + 2h$, $x_3 = x_0 + 3h$, ... $x_n = x_0 + nh$, ... where the *step size* h is a fixed number, for instance 0.2 or 0.1, or 0.01 as the case may be, depending upon the error of the solution and the tolerance level given in advance.

7.2 Methods of Solution

In this chapter, we discuss the following numerical methods of solution of equation (7.4) with the condition $y(x_0) = y_0$:

1. Taylor's series method.
2. Picard's method of successive approximations.
3. Euler's method.

4. Euler's modified (or Heun's) method.
5. Runge–Kutta (RK) method; fourth-order RK method; first, second, Third-order RK methods.

Predictor–Corrector Methods

6. (a) Milne–Simpson's method.
   (b) Adams–Bashforth–Moulton (ABM) method.

Numerical methods, yield the solution in one of the two forms:

(a) A series for y in terms of powers of x from which the value of y can be obtained by direct substitution; *e.g.*, Taylor's method or Picard's method of successive approximations.

(b) A set of tabulated values of x and y, *e.g.*, Euler's method, RK method, or the ABM method.

7.2.1 Method 1: Taylor's Series Method

Consider the IVP

$$y' = f(x, y), \quad y(x_0) = y_0 \qquad (7.5(a)), (7.5(b))$$

Let $y(x)$ be the exact solution of (7.5). Then the Taylor series expansion for $y(x)$ about $x = x_0$ is

$$y(x) = y_0 + \frac{(x-x_0)}{1!}y_0' + \frac{(x-x_0)^2}{2!}y_0'' + \cdots \qquad (7.6)$$

If the values of $y_0', y_0'', \ldots$ are known then (7.6) gives a power series for y. Using the formula for total derivatives, we have, on differentiating (7.5),

$$y'' = f' = f_x + y'f_y = f_x + ff_y$$

where the suffixes denote partial derivatives with respect to the indicated variables. Similarly we obtain

$$\begin{aligned} y''' = f'' &= f_{xx} + f_{xy}f + f(f_{yx} + f_{yy}f) + f_y(f_x + f_y f) \\ &= f_{xx} + 2ff_{xy} + ff_{yy} + f_x f_y + ff_y^2 \end{aligned}$$

and so on.

Example 7.1

Solve $y' = x - y^2$, $y(0) = 1$, by Taylor's series method and find $y(0.1)$ correct to four decimal places.

Solution Taylor's series expansion for $y(x)$ about the point $x_0 = 0$ is given by

$$y(x) = 1 + xy_0' + \frac{1}{2}x^2 y_0'' + \frac{x^3}{6}y_0''' + \frac{x^4}{24}y^{iv} + \frac{x^5}{120}y^{v} + \cdots$$

where we have used $y_0 = 1$

We compute $y_0', y_0'', \ldots$ below:

$$y'(x) = x - y^2, \quad y''(x) = 1 - 2yy', \quad y'''(x) = -2yy'' - 2y'^2,$$

$$y_0' = -1, \qquad y_0'' = 3, \qquad y_0''' = -8,$$

$$y^{iv}(x) = -2yy''' - 6y'y'', \quad y^{v}(x) = -2yy^{iv} - 8y'y''' - 6y''^2,$$

$$y_0^{iv} = 34, \qquad y_0^{v} = -186$$

Substituting these values, we obtain

$$y(x) = 1 - x + \frac{3}{2}x^2 - \frac{4}{3}x^3 + \frac{17}{12}x^4 - \frac{31}{20}x^5 + \cdots$$

To obtain the value of $y(0.1)$ correct to four decimal places we have to consider terms up to x^4:

$$\therefore \quad y(0.1) = 1 - 0.1 + 1.5 \times (0.01) - (1.3333) \times (0.001) + 1.4167 \times 0.0001 = 0.9138$$

Example 7.2

Given $\frac{dy}{dx} - 1 = xy$, $y(0) = 1$, obtain the Taylor's series for $y(x)$ and compute $y(0.1)$ correct to four decimal places.

Solution Taylor's series for $y(x)$ about $x_0 = 0$ is

$$y(x) = 1 + xy_0' + \frac{x^2}{2}y_0'' + \frac{x^3}{6}y_0''' + \frac{x^4}{24}y_0^{iv} + \frac{x^5}{120}y^{v} + \cdots$$

We compute the values of $y_0', y_0'', \ldots$ below:

$$y'(x) = 1 + xy, \quad y''(x) = xy' + y, \quad y'''(x) = xy'' + 2y',$$

$$y^{iv}(x) - xy''' + 3y'', \quad y^{v}(x) = xy^{iv} + 4y''',$$

$$y'_0 = 1, \quad y''_0 = 1, \quad y'''_0 = 2, \quad y_0^{iv} = 3, y^{v} = 8$$

Substituting the above values, we obtain

$$y(x) = 1 + x + \frac{x^2}{2} + \frac{x^3}{3} + \frac{x^4}{8} + \frac{x^5}{15} + \cdots$$

$$y(0.1) = 1 + 0.1 + \frac{1}{2}(0.01) + \frac{1}{3}(0.001) + \frac{1}{8}(0.0001),$$

retaining up to x^4

$$= 1.1 + 0.005 + 0.00033 + 0.0000125$$

$$= 1.1053 \text{ correct to four decimal places.}$$

7.2.2 Method 2: Picard's Method of Successive Approximations

The differential equation in the IVP:

$$\frac{dy}{dx} = f(x, y),\ y(x_0) = y_0 \tag{7.7}$$

can be converted into an integral equation; by integrating (7.7) from x_0 to x we obtain

$$y = y_0 + \int_{x_0}^{x} f(x, y)\,dx \tag{7.8}$$

which is equivalent to (7.7) and can be solved by the method of successive approximations. The first approximation to y is obtained by putting y_0 for y on the RHS of (7.8). Thus, we have

$$y^{(1)} = y_0 + \int_{x_0}^{x} f(x, y_0)\,dx \tag{7.9}$$

The integral on the RHS can now be evaluated and $y^{(1)}$ obtained. The value of $y^{(1)}$ thus obtained is substituted for y on the RHS of (7.8) to obtain the second approximation $y^{(2)}$

$$y^{(2)} = y_0 + \int_{x_0}^{x} f(x, y^{(1)})\,dx \tag{7.10}$$

Proceeding in this way we obtain $y^{(3)}, y^{(4)}, \ldots y^{(n-1)}$ and $y^{(n)}$ where

$$y^{(n)} = y_0 + \int_{x_0}^{x} f(x, y^{(n-1)})\,dx \text{ with } y^{(0)} = y_0 \tag{7.11}$$

The process of iteration is stopped whenever two consecutive iterates agree up to the desired degree of accuracy.

By this method we obtain a sequence of approximations $y^{(1)}, y^{(2)}, y^{(3)}, \ldots, y^{(n)}$ which converges to the unique solution by the *Existence and Uniqueness*.

Theorem 7.1 If $f(x, y)$ is continuous at all points (x, y) in some rectangle R: $|x - x_0| < a$, $|y - y_0| < b$ and is bounded in R (*i.e.*,) $|f(x, y)| \le K$ for all (x, y) in R then the IVP has a solution which is unique if the following Lipschitz condition is satisfied:

$$|f(x, y_2) - f(x, y_1)| \le M|y_2 - y_1| \quad (x, y_1), (x, y_2) \in R$$

where M is a positive constant.

Example 7.3

Solve the IVP

$$y' = x + y^2, \quad y(0) = 1$$

by the Picard's method of successive approximations.

Solution Let us start with $y^{(0)} = 1$. Substituting in $y = y_0 + \int_{x_0}^{x} f(x, y)\,dx$ we obtain

$$y^{(1)} = 1 + \int_0^x (x+1)\,dx = 1 + x + \frac{1}{2}x^2$$

Then the second approximation is

$$y^{(2)} = 1 + \int_0^x \left(x + \left(1 + x + \frac{1}{2}x^2\right)^2 \right) dx$$

$$= 1 + x + \frac{3}{2}x^2 + \frac{2}{3}x^3 + \frac{1}{4}x^4 + \frac{1}{20}x^5$$

We observe that the integrations are becoming more and more complicated as we proceed further.

Example 7.4

Solve, $y' = \dfrac{x^2}{(y^2+1)}$, $y(0) = 0$ by Picard's method of successive approximations. Find $y(0.25)$, $y(0.5)$ and $y(1.0)$ correct to three decimal places.

Solution We have

$$y = \int_0^x \frac{x^2}{y^2+1}\,dx$$

Setting $y^{(0)} = 0$ we obtain

$$y^{(1)} = \int_0^x x^2\,dx = \frac{1}{3}x^3$$

and

$$y^{(2)} = \int_0^x \frac{x^2}{\left(\frac{x^3}{3}\right)^2 + 1}\,dx = \tan^{-1}\left(\frac{x^3}{3}\right) = \frac{1}{3}x^3 - \frac{1}{81}x^9 + \cdots$$

$y^{(1)}$ and $y^{(2)}$ agree to the first term $\frac{1}{3}x^3$

To find the range of values of x so that the series with the term $\frac{1}{3}x^3$ alone gives the result correct to three decimal places, we put $\frac{1}{81}x^9 \le 0.0005 \Rightarrow x \le 0.77$.

Hence

$$y(0.25) = \tfrac{1}{3}(0.25)^3 = 0.005 \quad y(0.5) = \tfrac{1}{3}(0.5)^3 = 0.042$$

Since $x = 1 > 0.7$ we have to consider the next term also for $y(1.0)$

$$y(1.0) = \frac{1}{3} - \frac{1}{81} = 0.321$$

We have so far discussed the methods which give the solution of a differential equation in the form of a power series. We will now describe the methods which give the solution in the form of a set of tabulated values.

7.2.3 Method 3: Euler's Method

Suppose that we wish to solve IVP

$$y' = f(x, y), \quad y(x_0) = y_0 \tag{7.12}$$

for values of y at $x = x_r = x_0 + rh$ $(r = 1, 2, 3, ...)$

Integrating (7.12) with respect to x from x_0 to x_1, we obtain

$$y_1 = y_0 + \int_{x_0}^{x_1} f(x, y)\, dx \tag{7.13}$$

Assuming that $f(x{\cdot}y) = f(x_0, y_0)$ in $[x_0, x_1]$ this gives Euler's formula

$$y_1 \simeq y_0 + hf(x_0, y_0) \tag{7.14}$$

Similary for the subinterval $[x_1, x_2]$ we have

$$y_2 = y_1 + \int_{x_1}^{x_2} f(x, y)\, dx$$

Substituting $f(x_1, y_1)$ for $f(x, y)$ in $[x_1, x_2]$, we obtain

$$y_2 = y_1 + hf(x_1, y_1) \tag{7.15}$$

Proceeding in this way we obtain the general formula

$$y_{n+1} = y_n + hf(x_n, y_n), \quad n = 0, 1, 2, \ldots \tag{7.16}$$

This method is very slow and to obtain a reasonable accuracy by Euler's method, we have to take h much small. In view of this restriction on h, it is not suitable for practical use.

A modified version of Euler's method which gives more accurate results will be described below, in Section 7.2.4.

Geometric Interpretation

In Euler's method of solution of IVP (7.12) instead of finding a differentiable function that satisfies (7.12) we generate a set of points (x_r, y_r) which are used for an approximation, *i.e.*, $y(x_r) \simeq y_r$. We start at the point (x_0, y_0). Compute the values of the slope $m_0 = f(x_0, y_0)$ and move horizontally through a distance h and vertically through $hf(x_0, y_0)$. Thus, we are moving along the tangent line to the curve $y = y(x)$ and will reach the point (x_1, y_1) and so on (Figure 7.1).

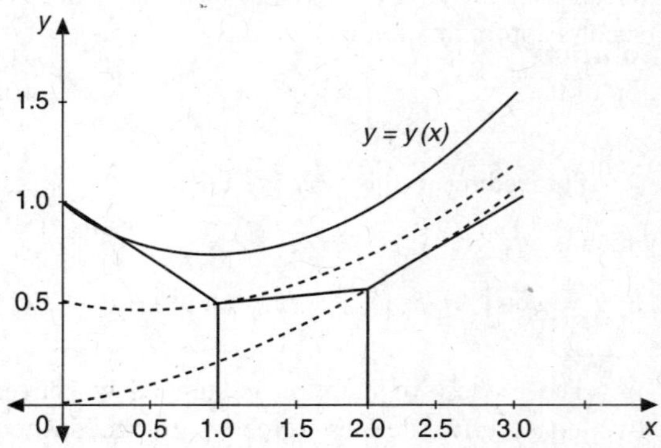

Figure 7.1 Euler's approximation $y_{n+1} = y_n + hf(x_n, y_n)$

Example 7.5

Solve $y' = -y$, $y(0) = 1$ by Euler's method.

Solution The iterative formula is

$$y_{n+1} = y_n + hf(x_n, y_n), \quad n = 0, 1, 2 \ldots$$

Here

$$f(x, y) = -y,\ h = 0.01 \quad \text{and} \quad y_0 = 1, f(x_r, y_r) = -y_r$$

$$n = 0, \quad f(x_0, y_0) = -y_0 = -1$$

$$y_1 = y(0.01) = y_0 + hf(x_0, y_0) = 1 + 0.01(-1) = 0.99$$

$n = 1, \quad y_2 = y(0.02) = y_1 + hf(x_1, y_1)$

$= 0.99 + (0.01)(-0.99) = 0.9801$

$n = 2, \quad y_3 = y(0.03) = y_2 + hf(x_2, y_2)$

$= 0.9801 + (0.01)(-0.9801) = 0.9703$

$n = 3, \quad y_4 = y(0.04) = y_3 + hf(x_3, y_3)$

$= 0.9703 + (0.01)(-0.9703) = 0.9606$

The equation $y' + y = 0$ has solution $y = e^{-x}$ which satisfies the condition $y = 1$ when $x = 0$.

The value of $y = e^{-x}$ at $x = 0.04$ is 0.96080. The approximate value obtained above agrees with this up to three decimal places.

Example 7.6
Solve

$$\frac{dy}{dx} = \frac{y-x}{y+x}, \quad y(0) = 1$$

by Euler's method. Find $y(0.1)$ approximately, in five steps.

Solution Euler's iterative formula for the solution of the integral equation

$$y = y_0 + \int_{x_0}^{x} f(x, y)\, dx \qquad (1)$$

which is equivalent to the IVP

$$y' = f(x, y), \quad y(x_0) = y_0 \qquad (2)$$

is

$$y_{n+1} = y_n + hf(x_n, y_n), \quad n = 0, 1, 2, \ldots \qquad (3)$$

In the present problem

$$f(x, y) = \frac{y-x}{y+x}, \qquad y(x_0) = y(0) = 1, \quad n = 5$$

$$\Delta x = h = \frac{0.1-0}{5} = 0.02$$

In the first step

$$\Delta y = \frac{1-0}{1+0}\Delta x = 0.02$$

$\therefore \quad y = y_0 + \Delta y = 1 + 0.02 = 1.02 \quad \text{at } x = 0.02$

In the second step

$$\Delta y = \frac{1.02-0.02}{1.02+0.02}\Delta x = 0.0192$$

$\therefore \quad y = 1.02 + 0.0192 = 1.0392 \quad \text{at } x = 0.04$

In the third step

$$\Delta y = \frac{1.0392-0.04}{1.0392+0.04}\Delta x = 0.0185$$

$\therefore \quad y = 1.0392 + 0.0185 = 1.0577 \quad \text{at } x = 0.06$

In the fourth step

$$\Delta y = \frac{1.0577-0.06}{1.0577+0.06}\Delta x = 0.0179$$

$\therefore \quad y = 1.0577 + 0.0179 = 1.0756 \quad \text{at } x = 0.08$

In the fifth step

$$\Delta y = \frac{1.0756-0.08}{1.0756+0.08}\Delta x = 0.0172$$

$\therefore \quad y = 1.0756 + 0.0172 = 1.0928 \quad \text{at } x = 0.10$

7.2.4 Method 4: Euler's Modified Method or Heun's Method

In this method first we convert the differential equation of the IVP

$$y'(x) = f(x, y(x)), \quad y(x_0) = y_0 \qquad (7.17)$$

into an integral equation by integrating both sides of (7.17)

$$\int_{x_0}^{x_1} f(x, y(x))\, dx = \int_{x_0}^{x_1} y'(x)\, dx = y(x_1) - y(x_0) \qquad (7.18)$$

where the antiderivative of $y'(x)$ is the desired function $y(x)$. When equation (7.18) is solved for $y(x_1)$ the result is

$$y(x_1) = y(x_0) + \int_{x_0}^{x_1} f(x, y(x))\, dx \qquad (7.19)$$

Now a numerical integration method can be used to approximate the definite integral in (7.19). If the trapezoidal rule is used with step size $h = x_1 - x_0$, then we obtain

$$y_1^{(1)} = y(x_1) = y_0 + \frac{h}{2}(f(x_0, y_0) + f(x_1, y_1)) \qquad (7.20)$$

Observe that the formula on the RHS of (7.20) contains the yet-to-be determined value $y(x_1)$. We have to use an estimate for $y(x_1)$. Euler's solution provides it. After it is substituted into (7.20) the following formula for finding (x_1, y_1) is obtained which is called the Euler's modified method or Heun's method:

$$y_1^{(1)} = y_0 + \frac{h}{2}[f(x_0, y_0) + f(x_1, y_0) + hf(x_0, y_0)]$$

$$= y_0 + \frac{h}{2}\left[f(x_0, y_0) + f(x_1, y_1^{(0)})\right]$$

The iteration formula is

$$y_1^{(n+1)} = y_0 + \frac{h}{2}\left[f(x_0, y_0) + f(x_1, y_1^{(n)})\right] \quad (7.21)$$

The process is repeated and it generates a sequence of points that approximates the solution $y = y(x)$. At each step, Euler's method is used as a prediction, and then the trapezoidal rule is used to make a correction to obtain the final value.

The general step for the method is

$$\left.\begin{aligned} y_{r+1}^{(n+1)} &= y_r + \frac{h}{2}\left[f(x_r, y_r) + f(x_{r+1}, y_{r+1}^{(n)})\right]x \\ x_{r+1} &= x_r + h \end{aligned}\right\} (7.22)$$

$$n = 0, 1, 2, \ldots, \quad \therefore r = 0, 1, 2, \ldots$$

Working Rule

1. Euler's Modified Formula

$$y_{r+1}^{(n+1)} = y_r + \frac{h}{2}\left[f(x_r, y_r) + f(x_{r+1}, y_{r+1}^{(n)})\right]$$

2. When $n = 0$, $y_{n+1}^{(0)}$ is calculated from Euler's method.
3. The values $r = 0, 1, 2, 3 \ldots$ give the number of iterations, and the values $n = 0, 1, 2, 3, \ldots$ give the number of times a particular iteration r is repeated.

Example 7.7

By Euler's method, find $y(1)$ given that

$$\frac{dy}{dx} = x + y, \; y(0) = 1$$

Solution The given IVP

$$y' = x + y, \quad y(0) = 1$$

Here $x_0 = 0, y_0 = 1$ and $f(x, y) = x + y$.
We take $h = 0.1$.

By Euler's method

$$y_1 = y_0 + hf(x_0, y_0) = y_0 + h(x_0, y_0)$$
$$= 1 + 0.1(0 + 1) = 1.10$$

$$y_2 = y_1 + hf(x_1, y_1) = y_1 + h(x_1 + y_1)$$
$$= 1.10 + 0.1(0.1 + 1.10) = 1.22$$
$$\because \; x_1 = x_0 + h = 0 + 0.1$$

$$y_3 = y_2 + hf(x_2, y_2) = y_2 + h(x_2 + y_2)$$
$$= 1.22 + 0.1(0.2 + 1.22) = 1.362$$
$$\because \; x_2 = x1 + h = 0.1 + 0.1$$

We compute the values $y_4, y_5, \ldots$ similarly and enter them in a tabular form

x	y	$x + y = \frac{dy}{dx}$	$y_{old}(0.1)\left(\frac{dy}{dx}\right) = y_{new}$
0.0	1.00	1.00	1.00 + 0.1(1.00) = 1.10
0.1	1.10	1.20	1.10 + 0.1(1.20) = 1.22
0.2	1.22	1.42	1.22 + 0.1(1.42) = 1.36
0.3	1.36	1.66	1.36 + 0.1(1.66) = 1.53
0.4	1.53	1.93	1.53 + 0.1(1.93) = 1.72
0.5	1.72	2.22	1.72 + 0.1(2.22) = 1.94
0.6	1.94	2.54	1.94 + 0.1(2.54) = 2.19
0.7	2.19	2.89	2.19 + 0.1(2.89) = 2.48
0.8	2.48	3.29	2.48 + 0.1(3.29) = 2.81
0.9	2.81	3.71	2.81 + 0.1(3.71) = 3.18
1.0	3.18	–	–

Example 7.8

By Euler's modified method, find $y(0.3)$ given that

$$\frac{dy}{dx} = x + y, \; y(0) = 1$$

Solution The given IVP is

$$y' = x + y, \quad y(0) = 1$$

Here $x_0 = 0$, $y_0 = 1$ and $f(x, y) = x + y$. We take $h = 0.1$ Euler's modified formula as

$$y_{r+1}^{(n+1)} = y_r + \frac{h}{2}\left[f(x_r, y_r) + f(x_{r+1}, y_{r+1}^{(n)})\right]$$

$n = 0, 1, 2, 3, \ldots$ for fixed r and $r = 0, 1, 2, 3, \ldots$

For first iteration taking $r = 0$ when $n = 0$ we have

$$y_1^{(1)} = y_0 + \frac{h}{2}\left[f(x_0, y_0) + f(x_1, y_1^{(0)})\right]$$

First we apply Euler's method to compute $y_1^{(0)}$

$$y_1^{(0)} = y_0 + hf(x_0, y_0) = y_0 + h(x_0 + y_0)$$
$$= 1 + 0.1(0+1) = 1.10$$

Next

$$y_1^{(1)} = y_0 + \frac{h}{2}\left[f(x_0, y_0) + f(x_1, y_1^{(0)})\right]$$
$$= 1 + \frac{0.1}{2}[(0+1) + (0.1 + 1.10)] = 1 + 0.05(2.2)$$
$$= 1.11$$

$$\because x_1 = x_0 + h = 0 + 0.1$$

$$y_1^{(2)} = y_0 + \frac{h}{2}\left[f(x_0, y_0) + f(x_1, y_1^{(1)})\right]$$
$$= 1 + \frac{0.1}{2}[(0+1) + (0.1 + 1.11)] = 1.1105$$

$$y_1^{(3)} = y_0 + \frac{h}{2}\left[f(x_0, y_0) + f(x_1, y_1^{(2)})\right]$$
$$= 1 + \frac{0.1}{2}[(0+1) + (0.1 + 1.1105)] = 1.1105$$

These computations are carried out until we find $y(0.3)$ and tabulate the values

x	$x + y = y'$	Mean slope	$y_{old} + 0.1(\text{Mean slope}) = y_{new}$
0.0	0 + 1 =	...	1.00 + 0.1(1.00) = 1.10
0.1	0.1 + 1.1	$\frac{1}{2}(1 + 1.20)$	1.00 + 0.1(1.10) = 1.11
0.1	0.1 + 1.11	$\frac{1}{2}(1 + 1.21)$	1.00 + 0.1(1.105) = 1.1105
0.1	0.1 + 1.1105	$\frac{1}{2}(1 + 1.2105)$	1.00 + 0.1(1.1052) = 1.1105
0.1	1.2105	...	1.1105 + 0.1(1.2105) = 1.2316
0.2	0.2 + 1.2316	$\frac{1}{2}(1.2105 + 1.4316)$	1.1105 + 0.1(1.3211) = 1.2426
0.2	0.2 + 1.2426	$\frac{1}{2}(1.2105 + 1.4426)$	1.1105 + 0.1(1.3211) = 1.2432
0.2	0.2 + 1.2432	$\frac{1}{2}(1.2105 + 1.4432)$	1.1105 + 0.1(1.3266) = 1.2432
0.2	1.4432	...	1.2432 + 0.1(1.4432) = 1.3875
0.3	0.3 + 1.3875	$\frac{1}{2}(1.4432 + 1.6875)$	1.2432 + 0.1(1.5654) = 1.3997
0.3	0.3 + 1.3997	$\frac{1}{2}(1.4432 + 1.6997)$	1.2432 + 0.1(1.575) = 1.4003
0.3	0.3 + 1.4003	$\frac{1}{2}(1.4432 + 1.7003)$	1.2432 + 0.1(1.5718) = 1.4004
0.3	0.3 + 1.4004	$\frac{1}{2}(1.4432 + 1.7004)$	1.2432 + 0.1(1.5718) = 1.4004

$\therefore \quad y(0.3) = 1.4004$, approximately

7.2.5 Method 5: Runge[1]–Kutta[2] Methods

Introduction

The Taylor series method has the disadvantage of computing higher derivatives, and Euler's method requires the step size h to be very small for a reasonable accuracy.

The methods developed by Runge and Kutta give better accuracy and are therefore preferable.

We will describe below the fourth-order RK method which is most commonly used in practical applications.

Fourth-order Runge–Kutta Method

Consider the differential equation

$$\frac{dy}{dx} = f(x, y) \tag{7.23}$$

By Taylor's Theorem we have

$$\begin{aligned}\Delta y &= (e^{hD} - 1)y \\ &= \left(hD + \frac{h^2D^2}{2!} + \frac{h^3D^3}{3!} + \frac{h^4D^4}{4!} + \cdots \right) y \\ &= \left(1 + \frac{hD}{2} + \frac{h^2D^2}{6} + \frac{h^3D^3}{24} + \cdots \right) h(Dy) \qquad (7.24) \\ &= \left(1 + \frac{hD}{2} + \frac{h^2D^2}{6} + \frac{h^3D^3}{24} + \cdots \right) hf \\ &\because Dy = f = f(x, y), \text{ by (7.23)} \\ &= \left(hf + \frac{h^2}{2}(Df) + \frac{h^2}{6}(D^2 f) + \frac{h^4}{24}(D^3 f) + \cdots \right)\end{aligned} \tag{7.25}$$

Computing the total derivatives

$$\left.\begin{aligned} Df &= f_x + f f_y \\ D^2 f &= f_{xx} + 2f f_{xy} + f^2 f_{yy} + f_y(f_x + f f_y) \\ D^3 f &= f_{xxx} + 3f f_{xxy} + 3f^2 f_{xyy} + f^3 f_{yyy} \\ &\quad + f_y(f_{xy} + 2f f_{xy} + f^2 f_{yyy}) \\ &\quad + 3(f_x + f f_y)(f_{xy} + f f_{yy}) + f_y^2(f_x + f f_y) \end{aligned}\right\} \tag{7.26}$$

and, using the following abbreviations:

$$\left.\begin{aligned} F_1 &= f_x + f f_y \\ F_2 &= f_{xx} + 2f f_{xy} + f^2 f_{yy} \\ F_3 &= f_{xxx} + 3f f_{xxy} + 3f^2 f_{xyy} + f^3 f_{yyy} \end{aligned}\right\} \tag{7.27}$$

we can write (7.25) as

$$\begin{aligned}\Delta y &= hf + f\frac{h^2}{2}F_1 + \frac{h^3}{6}(F_2 + f_2^2 F_1) \\ &= \frac{h^4}{24}(F_3 + f_2 F_2 + 3f_2' F_1 + f_2^2 F_1) + \cdots \end{aligned} \tag{7.28}$$

An RK formula of order n is a formula which expresses Δy in terms of n values of the function $f(x, y)$ in such a manner that the value obtained coincides with equation (7.28) as far as the terms involving h^n are concerned.

Now we express Δy in the form

$$\Delta y = ak_1 + bk_2 + ck_3 + dk_4 \tag{7.29}$$

where

$$\left.\begin{aligned} k_1 &= hf(x, y) = hf \\ k_2 &= hf(x + mh, y + mk_1) \\ &= h\left[f + mhF_1 + \frac{m^2h^2}{6}F_2 + \frac{m^3h^3}{6}F_3 + \cdots \right] \\ k_3 &= hf(x + nh, y + nk_2) \\ &= h\left[f + nhF_1 + \frac{h^2}{2}(n^2F_2 + 2mnf_2F_1) \right. \\ &\quad \left. + \frac{h^3}{6}(\, n^3 + F_3 + 3m^2nf_2F_2 + 6mn^2 f'_2F_1) + \cdots \right] \\ k_4 &= hf\left[x + ph, y + pk_3\right] \\ &= h\left[f + phF_1 + \frac{h^2}{2}(p^2F_2 + 2npf_2F_1) \right. \\ &\quad + \frac{h^3}{6}(p^3F_3 + 3n^2pf_2F_2) + 6np^2 \\ &\quad \left. + f_2' F_1 + 6mnpf_3^2 F_1) + \cdots \right] \end{aligned}\right\} \tag{7.30}$$

In (7.30) the functions involved have been expanded using Taylor's series for two variables.

---

[1] RUNGE, Carl David Tolme (1856–1927), was a German analyst.

[2] KUTTA, Wilhelm Martin (1867–1944), was a German applied mathematician.

Now, equating the expressions for Δy in (7.28) and (7.29) with (7.30) the following equations are obtained:

$$\left.\begin{aligned} a+b+c+d &= 1 & cmn+dnp &= \frac{1}{6} \\ bm+cn+dp &= \frac{1}{2} & cmn^2+dnp^2 &= \frac{1}{8} \\ bm^2+cn^2+dp^2 &= \frac{1}{3} & cm^2n+dn^2p &= \frac{1}{12} \\ bm^3+cn^3+dp^3 &= \frac{1}{4} & dmnp &= \frac{1}{24} \end{aligned}\right\} \quad (7.31)$$

Any solution of equation (7.31) will work. If we take

$$m=n=\frac{1}{2}, \quad p=1, \quad a=d=\frac{1}{6}, \quad b=c=\frac{1}{3} \quad (7.32)$$

then equations (7.31) are satisfied. We obtain

$$\Delta y=\frac{1}{6}(k_1+2k_2+2k_3+k_4) \quad (7.33)$$

$$\text{where} \quad \left.\begin{aligned} k_1 &= hf(x,y) \\ k_2 &= hf\left(x+\frac{h}{2}, y+\frac{k_1}{2}\right) \\ k_3 &= hf\left(x+\frac{h}{2}, y+\frac{k_2}{2}\right) \\ k_4 &= hf(x+h, y+k_3) \end{aligned}\right\} \quad (7.34)$$

That is, (7.33) expresses Δy in terms of the four ordinates $k_i (i = 1, 2, 3, 4)$ and the expression agrees with the exact expression (7.28) up to and including the terms containing h^4.

By taking the terms up to h, h^2 and h^3 in Taylor's expansions, we obtain the first, the second and the third-order RK formulas, respectively.

(i) First-order RK Formula

$$y_1 = y_0 + k_1 \quad \text{where } k_1 = hf(x_0, y_0) \quad (7.35)$$

(ii) Second-order RK Formula

$$y_1 = y_0 + \frac{1}{2}(k_1+k_2) \quad \text{where } k_1 = hf(x_0, y_0)$$
$$\text{and } k_2 = hf(x_0+h, y_0+k_1) \quad (7.36)$$

(iii) Third-order RK Formula

$$y_1 = y_0 + \frac{1}{6}(k_1+4k_2+k_3) \quad (7.37)$$

where

$$k_1 = hf(x_0, y_0),\ k_2 = hf\left(x_0+\frac{h}{2}, y_0+\frac{k_1}{2}\right),$$
$$k_3 = hf(x_0+h, y_0+2k_2-k_1) \quad (7.38)$$

If may be observed that the first and the second-order RK formulas agree, respectively, with Euler's and Euler's modified formulas.

Note that the formulas for k_2 and k_3 in second order, third and fourth-order RK formulas are different.

Example 7.9

Given $\dfrac{dy}{dx} = y - x$ where $y(0) = 2$ find $y(0.1)$ and $y(0.2)$ correct to four decimal places.

Solution We solve the problem using RK second and fourth-order formulas.

1. RK Second-order Formula

Second-order RK formula is

$$y_1 = y_0 + \frac{1}{2}(k_1+k_2)$$

where $k_1 = hf(x_0, y_0)$, $k_2 = hf(x_0 + h, y_0 + k_1)$

Here $f(x, y) = y - x$; $x_0 = 0$ and $y_0 = 2$ and we take $h = 0.1$

$$\therefore \quad k_1 = h(y_0 - x_0) = 0.1(2 - 0) = 0.2,$$

$$k_2 = hf(x_0+h, y_0+k_1) = h(y_0+k_1-x_0-h) = (0.21)$$

Hence

$$y_1 = y_0 + \frac{1}{2}(k_1+k_2) \quad \Rightarrow \quad y_1 = y(0.1)$$
$$= 2 + \frac{1}{2}(0.20+0.21) = 2+0.2050 = 2.2050$$

To determine $y_2 = y(0.2)$ we note that

$$x_0 = 0.1 \quad \text{and} \quad y_0 = 2.2050$$
$$k_1 = hf(x_0, y_0) = h(y_0 - x_0) = 0.1(2.2050-0.1)$$
$$= 0.1(2.105) = 0.2105$$
$$k_2 = hf(x_0+h, y_0+k_1) = h(y_0+k_1-x_0-h)$$
$$= 0.1(2.205+0.2105-0.1-0.1)$$

$$= 0.1\times 2.2155 = 0.22155$$

$$y_2 = y(0.2) = y_1 + \frac{1}{2}(k_1 + k_2)$$

$$= 2.2050 + \frac{1}{2}(0.2105 + 0.22155) = 2.4210$$

Proceeding like this, we obtain

$y_3 = y(0.3) = 2.6492$ and $y_4 = y(0.4) = 2.8909$

We next choose $h = 0.2$ and compute $y(0.2)$ and $y(0.4)$ directly.

With $h = 0.2$, $x_0 = 0$ and $y_0 = 2$

$$k_1 = hf(x_0, y_0) = h(y_0 - x_0) = 0.2(2 - 0) = 0.4$$

$$k_2 = hf(x_0 + h, y_0 + k_1) = h(y_0 + k_1 - x_0 - h)$$

$$= 0.2(2 + 0.4 - 0 - 0.2) = 0.2 \times 2.2 = 0.44$$

Hence

$$y(0.2) = y_0 + \frac{1}{2}(k_1 + k_2)$$

$$= 2 + \frac{1}{2}(0.4 + 0.44) = 2 + 0.42 = 2.42$$

Likewise we also obtain

$$y(0.4) = 2.8880$$

Analytical Solution

The given equation is a first-order linear differential equation $dy/dx - y = -x$; $P = -1$, $R = x$; Integrating Factor $= e^{-x}$.

Its general solution is

$$ye^{-x} = c - \int xe^{-x}dx = c + (x+1)e^{-x} = 1 + (x+1)e^{-x}$$

Since $y(0) = 2$ we have $c = 1$ so that the analytical solution is $y = x + 1 + e^x$. The exact values of $y(0.2)$ and $y(0.4)$ are, respectively, 2.4214 and 2.8918.

2. RK Fourth-order Formula

We have $x_0 = 0$, $y_0 = 2$ and $h = 0.1$; we determine $y(0.1)$

$$k_1 = hf(x_0, y_0) = h(y_0 - x_0) = 0.1(2 - 0) = 0.2$$

$$k_2 = hf\left(x_0 + \frac{h}{2}, y_0 + \frac{k_1}{2}\right) = h\left(y_0 - x_0 + \frac{k_1 - h}{2}\right)$$

$$= 0.1\left(2 - 0 + \frac{0.2 - 0.1}{2}\right) = 0.205$$

$$k_3 = hf\left(x_0 + \frac{h}{2}, y_0 + \frac{k_2}{2}\right) = h\left(y_0 - x_0 + \frac{k_2 - h}{2}\right)$$

$$= 0.1\left(2 - 0 + \frac{0.205 - 0.1}{2}\right) = 0.20525$$

$$k_4 = hf(x_0 + h, y_0 + k_3) = h(y_0 - x_0 + k_3 - h)$$

$$= 0.1(2 - 0 + 0.20525 - 0.1) = 0.210525$$

Hence

$$y_1 = y_0 + \frac{1}{6}(k_1 + 2k_2 + 2k_3 + k_4)$$

$$\Rightarrow y(0.1) = 2 + \frac{1}{6}[0.2 + 2(0.205 + 0.20525) + 0.210525]$$

$$= 2.2052$$

Proceeding in a like manner we obtain $y(0.2) = 2.4214$

Example 7.10

Use the RK method to solve $y' = xy$ for $x = 1.2$. Initially $y = 2$ when $x = 1$. Take $h = 0.2$.

Solution Here $f(x, y) = xy$, $x_0 = 1$, $y_0 = 2$ and $h = 0.2$

$$k_1 = hf(x_0, y_0) = hx_0y_0 = 0.2(1 \times 2) = 0.4$$

$$k_2 = hf\left(x_0 + \frac{h}{2}, y_0 + \frac{k_1}{2}\right) = h\left(x_0 + \frac{h}{2}\right)\left(y_0 + \frac{k_1}{2}\right)$$

$$= 0.2 \times 1.1 \times 2.2 = 0.4840$$

$$k_3 = hf\left(x_0 + \frac{h}{2}, y_0 + \frac{k_2}{2}\right) = h\left(x_0 + \frac{h}{2}\right)\left(y_0 + \frac{k_2}{2}\right)$$

$$= 0.2(1.1)(2.242) = 0.49324$$

$$k_4 = hf(x_0 + h, y_0 + k_3) = h(x_0 + h)(y_0 + k_3)$$

$$= 0.2(1.2)(2.49324) = 0.5983776$$

$$\therefore\ y_1 = y(1.2) = y_0 + \frac{1}{6}(k_1 + 2k_2 + 2k_3 + k_4)$$

$$= 2 + \frac{1}{6}[0.4 + 2(0.484 + 0.49324) + 0.5983776]$$

$$= 2.4921429$$

Example 7.11

Given the IVP

$$y' = -2xy^2, \quad y(0) = 1$$

estimate $y(0.4)$ using Heun's method (Euler's modified method) with $h = 0.2$. Compare it with the exact solution.

Solution The iteration formula of Heun's method (or Euler's modified method) is

$$y_{n+1} = y_n + \frac{1}{2}(k_1 + k_2), \quad n = 0, 1, 2, \ldots \quad (1)$$

$$\text{where} \quad \left.\begin{aligned} k_1 &= hf(x_n, y_n) \\ k_2 &= hf(x_n + h, y_n + k_1) \end{aligned}\right\} \quad (2)$$

In the present case, we have

For $n = 0$

$$f(x, y) = -2xy^2, \ x_0 = 0, \ y_0 = 1, \ h = 0.2$$

$$\begin{aligned} k_1 &= hf(x_0, y_0) = h(-2x_0 y_0^2) \\ &= -2 \times 0.2 \times 0 \times 1^2 = 0 \\ k_2 &= hf(x_0 + h, y_0 + k_1) \\ &= h(-2)(x_0 + h)(y_0 + k_1)^2 \\ &= -0.4(0 + 0.2)(1 + 0)^2 = -0.08 \end{aligned}$$

$$\therefore \quad y(0.2) \simeq y_1 = y_0 + \frac{1}{2}(k_1 + k_2)$$

$$= 1 + \frac{1}{2}(0 - 0.08) = 0.96$$

For $n = 1$

$$f(x, y) = -2xy^2, \ x_1 = x_0 + h = 0 + 0.2 = 2$$

$$y_1 = 0.96, \ h = 0.2$$

$$\begin{aligned} k_1 &= hf(x_1, y_1) = h(-2x_1 y_1^2) = -0.4(0.2)(0.96)^2 \\ &= -0.08(0.96)^2 = -0.073728 \\ k_2 &= hf(x_1 + h, y_1 + k_1) = h(-2)(x_1 + h)(y_1 + k_1)^2 \\ &= -0.4(0.2 + 0.2)(0.96 - 0.073728)^2 \\ &= -0.16 \times 0.785478 = -0.125676 \end{aligned}$$

$$\therefore \quad y(0.4) \simeq y_2 = y_1 + \frac{1}{2}(k_1 + k_2)$$

$$= 0.96 + \frac{1}{2}(-0.073728 - 0.125676 = 0.860298)$$

Analytical Solution

Separating the variables we have $y'/y^2 = -2x$

Integrating we obtain $-1/y = -x^2 + c$; since $x = 0$

$$\Rightarrow \quad y = 1 \quad \therefore \quad c = -1$$

The required solution is $y = \dfrac{1}{1 + x^2}$.

$$\therefore \quad y(0.2) = 0.961538 \text{ and } y(0.4) = 0.862069$$

Absolute errors are

$$\in(0.2) = |\text{True value} - \text{Approx. value}| = 0.001538$$

$$\in(0.4) = 0.001771.$$

Example 7.12

Solve the IVP $y' = -2xy^2$, $y(0) = 1$ with $h = 0.2$ on the interval [0, 0.4] using the fourth-order RK method. Compare the result with the exact solution.

Solution The fourth-order RK iteration formula is

$$y_{n+1} = y_n + \frac{1}{2}(k_1 + 2k_2 + 2k_3 + k_4) \quad (1)$$

$$\text{where} \quad \left.\begin{aligned} k_1 &= hf(x_n, y_n) \\ k_2 &= hf\left(x_n + \frac{h}{2}, y_n + \frac{k_1}{2}\right) \\ k_3 &= hf\left(x_n + \frac{h}{2}, y_n + \frac{k_2}{2}\right) \\ k_4 &= hf(x_n + h, y_n + k_3) \end{aligned}\right\} \quad (2)$$

In the present problem

$$f(x, y) = -2xy^2$$

For $n = 0$

We have $x_0 = 0, \ y_0 = 1, \ h = 0.2$

$$\begin{aligned} k_1 &= hf(x_0, y_0) = h(-2x_0 y_0^2) \\ &= -2hx_0 y_0^2 = 0 \quad \because x_0 = 0 \\ k_2 &= hf\left(x_0 + \frac{h}{2}, y_0 + \frac{k_1}{2}\right) = -2h\left(x_0 + \frac{h}{2}\right)\left(y_0 + \frac{k_1}{2}\right)^2 \\ &= -2(0.2)\left(0 + \frac{0.2}{2}\right)(1 + 0)^2 = -0.04 \\ k_3 &= hf\left(x_0 + \frac{h}{2}, y_0 + \frac{k_2}{2}\right) = -2h\left(x_0 + \frac{h}{2}\right)\left(y_0 + \frac{k_2}{2}\right)^2 \\ &= -2(0.2)\left(0 + \frac{0.2}{2}\right)\left(1 - \frac{0.4}{2}\right)^2 = -0.4(0.1)(0.98)^2 \\ &= -0.038416 \end{aligned}$$

$$k_4 = hf(x_0+h, y_0+k_3) = -2h(x_0+h)(y_0+k_3)^2$$
$$= -2(0.2)(0+0.2)(1-0.038416)^2$$
$$= -0.08(0.961584)^2 = -0.073972$$

$$y(0.2) \simeq y_1 = 1+\frac{1}{6}(k_1+2k_2+2k_3+k_4)$$
$$= 1+\frac{1}{6}[0-0.08-0.076832-0.073972]$$
$$= 1-\frac{1}{6}(0.230804) = 0.961533$$

For $n = 1$

We have $x_1 = x_0 + h = 0 + 0.2 = 0.2$, $y_1 = 0.961533$, $h = 0.2$

$$k_1 = hf(x_1, y_1) = h(-2x_1y_1^2) = -2(0.2)(0.2)(0.961533)^2$$
$$= -0.08 \times 0.924545 = -0.073964$$

$$k_2 = hf\left(x_1+\frac{h}{2}, y_1+\frac{k_1}{2}\right) = h(-2)\left(x_1+\frac{h}{2}\right)\left(y_1+\frac{k_1}{2}\right)^2$$
$$= -2(0.2)\left(0.2+\frac{0.2}{2}\right)\left(0.961533-\frac{0.073964}{2}\right)^2$$
$$= -0.12(0.961533-0.0366982)^2$$
$$= -0.12\times(0.924551)^2 = -0.102575$$

$$k_3 = hf\left(x_1+\frac{h}{2}, y_1+\frac{k_2}{2}\right) = (-2)\left(x_1+\frac{h}{2}\right)\left(y_1+\frac{k_2}{2}\right)^2$$
$$= -2(0.2)\left(0.2+\frac{0.2}{2}\right)\left(0.961533-\frac{0.102575}{2}\right)^2$$
$$= -0.12(0.961533-0.051288)^2$$
$$= -0.12\times(0.910245)^2$$
$$= -0.12(0.828546) = -0.099426$$

$$k_4 = hf(x_1+h, y_1+k_3) = h(-2)(x_1+h)(y_1+k_3)^2$$
$$= -2(0.2)\,(0.2+0.2)(0.961533-0.099426)$$
$$= -0.16(0.862107)^2 = -0.16(0.743229)$$
$$= -0.118917$$

$$\therefore \quad y(0.4) \simeq y_2 = y_1+\frac{1}{6}(k_1+2k_2+2k_3+k_4)$$
$$= 0.961533+\frac{1}{6}[-0.073963-2\times 0.102575-2\times 0.099426-0.118917]$$
$$= 0.961533-\frac{1}{6}(0.596882) = 0.862052$$

The analytical solution of the problem is

$$y = \frac{1}{1+x^2}$$

$$\therefore \quad y(0.2) = 0.961538 \text{ and } y(0.4) = 0.862069$$

Hence, the absolute errors in the numerical calculations are

$$\in(0.2) = |\text{True value} - \text{Approx. value}|$$
$$= |0.961538 - 0.961533| = 0.000005$$

$$\in(0.4) = |0.862069 - 0.862052| = 0.000017$$

Additional Examples

Example 7.13

Using Taylor's Series method solve the IVP

$$\frac{dy}{dx} = x - y, \quad y(0) = 1$$

at $x = 0.1$, 0.2, 0.3 and 0.4 and compare the value with the exact solution.

Solution The given IVP is

$$y' = x^2 - y, y(0) = 1 \tag{1}$$

The initial condition is $y_0 = 1$ when $x_0 = 0$. The size of the interval $h = 0.1$ and the points are $x_1 = 0.1$, $x_2 = 0.2$, $x_3 = 0.3$ and $x_4 = 0.4$. (2)

Now we have, on successive differentiation,

$$\begin{aligned} y' &= x^2 - y &&\Rightarrow & y_0' &= x_0^2 - y_0' = -1 \\ y'' &= 2x - y' &&\Rightarrow & y_0'' &= 2x_0 - y_0' = 1 \\ y''' &= 2 - y'' &&\Rightarrow & y_0''' &= 2 - y''_0 = 1 \\ y^{iv} &= -y''' &&\Rightarrow & y_0^{iv} &= -y_0''' = -1 \end{aligned} \tag{3}$$

By Taylor's series

$$y_1 = y_0 + hy_0' + \frac{h^2}{2!}y_0'' + \frac{h^3}{3!}y_0''' + \frac{h^4}{4!}y^{iv} + \cdots$$

$$\Rightarrow \quad y(0.1) \simeq y_1 = 1 + (0.1)(-1) + \frac{(0.1)^2}{2}\times 1 + \frac{(0.1)^3}{6}\times 1 + \frac{(0.1)^4}{24} + \cdots$$
$$= 1 - 0.1 + 0.005 + 0.0001667 - 0.0000417 + \cdots = 0.9051 \tag{4}$$

Substituting in the above equations, we obtain

$$y_1' = x_1^2 - y_1 = (0.1)^2 - 0.9051 = -0.8951$$
$$y_1'' = 2x_1 - y_1' = 2(0.1) - (-0.8951) = 1.0951$$
$$y_1''' = 2 - y_1'' = 2 - 1.0951 = 0.9049$$
$$y_1^{iv} = y_1^{iv} = -0.9049 \quad (5)$$

Now

$$y(0.2) \simeq y_2 = y_1 + hy_1' + \frac{h^2}{2!}y_1'' + \frac{h^3}{3!}y_1''' + \frac{h^4}{4!}y_1^{iv} + \cdots$$
$$= 0.9051 + (0.1)(-0.8951)$$
$$+ \frac{(0.1)^2}{2!}(1.0951) + \frac{(0.1)^3}{6}(0.9049)$$
$$+ \frac{(0.1)^4}{24}(-0.9049) = 0.8212 \quad (6)$$

Substituting in the above equations

$$y_2' = x_2^2 - y_2 = (0.2)^2 - 0.8212 = -0.7812$$
$$y_2'' = 2x_2 - y_2' = 2(0.2) - (-0.7812) = 1.1812$$
$$y_2''' = 2 - y_2'' = 2 - 1.1812 = 0.8188$$
$$y_2^{iv} = -y_2''' = -0.3138 \quad (7)$$

Now

$$y(0.3) \simeq y_3 = y_2 + hy_2' + \frac{h^2}{2!}y_2'' + \frac{h^3}{3!}y_2''' + \frac{h^4}{4!}y_2^{iv} + \cdots$$
$$= 0.8212 + (0.1)(-0.7812)$$
$$+ \frac{(0.1)^2}{2!}(0.1812) + \frac{(0.1)^3}{3!}(0.8188)$$
$$+ \frac{(0.1)^4}{4!}(-0.8188) + \cdots = 0.7492 \quad (8)$$

Again,
$$y_3' = x_3^2 - y_3 = (0.3)^2 - (0.7492) = -0.6592$$
$$y_3'' = 2x_3 - y_3' = 0.6 - (+0.6592) = 1.2592$$
$$y_3''' = 2 - y_3'' = 0.7408$$
$$y_3^{iv} = -y_3''' = -0.7408 \quad (9)$$

$$y(0.4) \simeq y_4 = y_3 + \frac{h}{1!}y_3' + \frac{h^2}{2!}y_3'' + \frac{h^3}{3!}y_3'''$$
$$+ \frac{h^4}{4!}y_3^{iv} + \cdots$$
$$= 0.7492 + (0.1)(-0.6592) + \frac{(0.1)^2}{2!}(1.2592)$$
$$+ \frac{(0.1)^3}{6}(0.7408) + \frac{(0.1)^4}{4!}(-0.7408) + \cdots$$
$$= 0.6897 \quad (10)$$

Analytical Solution

The given differential equation $dy/dx = x^2 - y$, equation (1) can be written as $dy/dx + y = x^2$ which is a first-order linear equation.

Integrating factor $= e^{\int p\,dx} = e^{\int 1\,dx} e^x$. Hence, the general solution is

$$ye^x = c + \int x^2 e^x\,dx \Rightarrow ye^x = c + (x^2 - 2x + 2)e^x$$

Initial condition is $y = 1$ when $x = 0 \Rightarrow c = -1$

∴ The required particular solution is

$$y = -e^{-x} + x^2 - 2x + 2$$

We have $y(0.1) \simeq y_1 = 0.9052$; $y(0.2) \simeq y_2 = 0.8213$; $y(0.3) \simeq y_3 = 0.7492$; $y(0.4) \simeq y_4 = 0.6897$.

The approximate values computed by the Taylor series method agree up to three decimal places with the exact values.

Example 7.14

Use Picard's method to obtain $y(0.2)$ correct to five decimal places by solving the IVP

$$\frac{dy}{dx} = x - y, \quad y(0) = 1$$

Compare the result with the exact solution.

Solution The given differential equation is equivalent to

$$y = y_0 + \int_{x_0}^{x} f(x, y)\,dx \quad (1)$$

Here $x_0 = 0$, $y_0 = 1$ and $f(x, y) = x - y$.

Putting $y = y_0 = 1$ in the integrand, a first approximation to the solution is obtained as

$$y^{(1)} = 1 + \int_0^x (x-1)\,dx = 1 - x + \frac{x^2}{2} \quad (2)$$

Putting $y = y^{(1)} = 1 - x + x^2/2$ in the integrand, a second approximation is obtained as

$$y^{(2)} = 1 + \int_0^x \left[x - \left(1 - x + \frac{x^2}{2}\right)\right] dx = 1 - x + x^2 - \frac{x^3}{6} \quad (3)$$

Putting $y = y^{(2)} = 1 - x + x^2 - x^3/6$ in the integrand, a third approximation is

$$y^{(3)} = 1 + \int_0^x \left[x - \left(1 - x + x^2 - \frac{x^2}{6}\right)\right] dx$$

$$= 1 - x + x^2 - \frac{x^3}{3} + \frac{x^4}{24} \quad (4)$$

Substituting this value $y^{(3)}$ for y in the integrand, a fourth approximation is

$$y^{(4)} = 1 + \int_0^x \left[x - \left(1 - x + x^2 - \frac{x^3}{3} + \frac{x^4}{24}\right)\right] dx$$

$$= 1 - x + x^2 - \frac{x^3}{3} + \frac{x^4}{12} - \frac{x^5}{120} \quad (5)$$

Substituting this value $y^{(4)}$ for y in the integrand, a fifth approximation is

$$y^{(5)} = 1 + \int_0^x \left[x - \left(1 - x + x^2 - \frac{x^3}{3} + \frac{x^4}{24} - \frac{x^5}{120}\right)\right] dx$$

$$= 1 - x + x^2 - \frac{x^3}{3} + \frac{x^4}{12} - \frac{x^5}{60} + \frac{x^6}{720} \quad (6)$$

Computing the approximate values of y for $x = 0.2$ we have $y^{(1)} = 0.82$, $y^{(2)} = 0.83867$, $y^{(3)} = 0.83740$, $y^{(4)} = 0.83746$ and $y^{(5)} = 0.83746$.

$\therefore$ The approximate value of y at $x = 0.2$ is

$$y = 0.83746$$

upto five decimal places.

Analytical Solution

The given differential equation is

$$\frac{dy}{dx} + y = x \quad (7)$$

which is a first-order linear equation. The intergrating factor is e^x $(IF = e^{\int pdx} = e^{\int idx} = e^x)$ The general solution is

$$y.e^x = c + \int xe^x dx = c + (x-1)e^x \quad (8)$$

It is given that $y = 1$ when $x = 0$ (the initial condition)

$$\therefore \quad I.e^0 = c + (0-1)e^0 \quad \Rightarrow \quad c = 2$$

The required particular solution of the problem is

$$y = 2e^{-x} + x - 1 \quad (9)$$

The exact solution is

$$y(0.2) = 2e^{-0.2} + 0.2 - 1$$

or $$y(0.2) = 0.83746 \quad (10)$$

up to five decimal places. The approximate solution obtained by Picard's method agrees with this result.

Example 7.15

Use Euler's method to solve the IVP $y' = -2xy^2$ $y(0) = 1$ with $h = 0.2$, 0.1 and 0.5 on the interval $[0, 1]$.

Solution Euler's Iteration formula is

$$y_{n+1} = y_n + hf(x_n, y_n), \quad n = 0, 1, 2, \ldots \quad (1)$$

Case (1) $h = 0.2$ In the present problem, we have $f(x, y) = -2xy^2$ (2)

$$y_{n+1} = y_n + hf(x_n, y_n) = y_n - 2hx_n y_n^2$$
$$= y_n(1 - 0.4x_n, y_n) \quad n = 0, 1, 2, \ldots \quad (3)$$

For $n = 0$

We have $x_0 = 0$, $y_0 = 1$ and $h = 0.2$

$$y(0.2) \simeq y_1 = y_0(1 - 0.4x_0y_0)$$
$$= 1(1 - 0.4 \times 0) = 1 \simeq x_0 = 0 \quad (4)$$

For $n = 1$ $\quad x_1 = 0.2, y_1 = 1$

$$y(0.4) \simeq y_2 = y_1(1 - 0.4x_1y_1)$$
$$= 1(1 - 0.4 \times 0.2 \times 1) = 0.92 \quad (5)$$

For $n = 2$ $\quad x_2 = 0.4, y_2 = 0.92$

$$y(0.6) \simeq y_3 = y_2(1 - 0.4x_2y_2)$$
$$= 0.92(1 - 0.4 \times 0.4 \times 0.92)$$
$$= 0.92 \times 0.8528 = 0.784576 \quad (6)$$

For $n = 3$ $x_3 = 0.6, y_3 = 0.784576$

$$y(0.8) \simeq y_4 = y_3(1 - 0.4x_3y_3)$$
$$= 0.784576(1 - 0.4 \times 0.6 \times 0.784576)$$
$$= 0.784576 \times 0.811702 = 0.636842 \quad (7)$$

For $n = 4$ $x_4 = 0.8, y_4 = 0.636842$

$$y(1.0) \simeq y_5 = y_4(1 - 0.4x_4y_4)$$
$$= 0.636842(1 - 0.4 \times 0.8 \times 0.636842)$$
$$= 0.636842 \times 0.796211 = 0.507060 \quad (8)$$

Case (2) $h = 0.1$

$$y_{n+1} = y_n + hf(x_n, y_n) = y_n + h(-2x_n y_n^2)$$
$$= y_n(1 - 0.2x_n, y_n) \quad n = 0, 1, 2, \cdots \quad (9)$$

For $n = 0$ $x_0 = 0, y_0 = 1, h = 0.1$

$$y(0.1) \simeq y_1 = y_0(1 - 0.2x_0y_0) = 1$$
$$\because \quad x_0 = 0 \quad (10)$$

For $n = 1$ $x_1 = x_0 + h = 0 + 0.1, y_1 = 1$

$$y(0.2) \simeq y_2 = y_1(1 - 0.2x_1y_1)$$
$$= 1(1 - 0.2 \times 0.1 \times 1) = 0.98 \quad (11)$$

For $n = 2$ $x_2 = x_0 + 2h = 0 + 2 \times 0.1 = 0.2$, $y_2 = 0.98$

$$y(0.3) \simeq y_3 = y_2(1 - 0.2x_2y_2)$$
$$= 0.98(1 - 0.2 \times 0.2 \times 0.98) \quad (12)$$
$$= 0.98 \times 0.9608 = 0.941584$$

For $n = 3$ $x_3 = x_0 + 3h = 0 + 3 \times 0.1 = 0.3$, $y_3 = 0.941584$

$$y(0.4) \simeq y_4 = y_3(1 - 0.2x_3y_3)$$
$$= 0.941584(1 - 0.2 \times 0.3 \times 0.941584) \quad (13)$$
$$= 0.941584 \times 0.943505 = 0.888389$$

For $n = 4$ $x_4 = x_0 + 4h = 0 + 4 \times 0.1 = 0.4$, $y_4 = 0.888389$

$$y(0.5) \simeq y_5 = y_4(1 - 0.2x_4y_4)$$
$$= 0.887010(1 - 0.2 \times 0.4 \times 0.888389) \quad (14)$$
$$= 0.888389 \times 0.928929 = 0.825250$$

For $n = 5$ $x_5 = x_0 + 5h = 0 + 5 \times 0.1 = 0.5$, $y_5 = 0.825250$

$$y(0.6) \simeq y_6 = y_5(1 - 0.2x_5y_5)$$
$$= 0.825250(1 - 0.2 \times 0.4 \times 0.825250) \quad (15)$$
$$= 0.825250 \times 0.917475 = 0.757150$$

For $n = 6$ $x_6 = x_0 + 6h = 0 + 4 \times 0.1 = 0.6$, $y_6 = 0.757150$

$$y(0.7) \simeq y_7 = y_6(1 - 0.2x_6y_6)$$
$$= 0.757150(1 - 0.12 \times 0.757150) \quad (16)$$
$$= 0.757150 \times 0.909142 = 0.688357$$

For $n = 7$ $x_7 = x_0 + 7h = 0 + 7 \times 0.1 = 0.7$, $y_7 = 0.688357$

$$y(0.8) \simeq y_8 = y_7(1 - 0.2x_7y_7)$$
$$= 0.688357(1 - 0.14 \times 0.688357) \quad (17)$$
$$= 0.688357 \times 0.903630 = 0.622020$$

For $n = 8$ $x_8 = x_0 + 8h = 0 + 8 \times 0.1 = 0.8$, $y_8 = 0.622020$

$$y(0.9) \simeq y_9 = y_8(1 - 0.2x_8y_8)$$
$$= 0.622020(1 - 0.16 \times 0.622020) \quad (18)$$
$$= 0.622020 \times 0.900477 = 0.560115$$

For $n = 9$ $x_9 = 0.9, y_9 = 0.560115$

$$y(1.0) \simeq y_{10} = y_9(1 - 0.2x_9y_9)$$
$$= 0.560115(1 - 0.18 \times 0.560115) \quad (19)$$
$$= 0.560115(0.899179) = 0.503644$$

Case (3) On the same lines, we compute the value for $h = 0.5$ and tabulate below:

x	y	x	y
0.05	0.1	0.10	0.995
0.15	0.9851	0.20	0.97054
0.25	0.9517	0.30	0.92906
0.35	0.90316	0.40	0.87461
0.45	0.84401	0.50	0.81195
0.55	0.77899	0.60	0.74561
0.65	0.71225	0.70	0.67928
0.75	0.64698	0.80	0.61559
0.85	0.58527	0.90	0.55615
0.95	0.52831	1.00	0.50179

(20)

Analytical Solution

Seperating the variables in $y' = -2xy^2$ and integrating, we obtain $-1/y = x^2 + c$ where $c = -1 \quad \because \quad y(0) = 1$

$$\therefore \quad y = \frac{1}{1+x^2} \text{ and hence } y(1.0) = 0.5$$

The absolute errors are

$$h = 0.2 \in (1.0) = |0.507060 - 0.5| = .00706$$

$$h = 0.1 \in (1.0) = |0.503644 - 0.5| = .003644$$

$$h = 0.05 \in (1.0) = |0.50179 - 0.5| = .00179$$

Example 7.16

Solve the IVP; $dy/dx = x + xy$ with $y(1) = 1$ for finding $y(1.1)$ by Euler's modified method with $h.0.05$. Compare it with the exact value.

Solution Here $f(x, y) = x + xy$

Euler's modified iteration formula is

$$y_{n+1}^{(k+1)} = y_n + \frac{h}{2}\left[f(x_n, y_n) + f(x_{n+1}, y_{n+1}^{(k)})\right]$$

$$n = 0, 1, 2, \ldots, \quad k = 0, 1, 2, \ldots$$

$$x_0 = 1, \quad y_0 = 1, \quad h = 0.05$$

By Euler's formula

$$f(x_0, y_0) = x_0 + x_0 y_0 = 1 + 1 \times 1 = 2$$

$$y(0.05) \simeq y_1^{(0)} = y_0 + hf(x_0, y_0) = 1 + 0.05 \times 2 = 1.1$$

$$x_1 = x_0 + h = 1.05$$

$$f(x_1, y_1^{(0)}) = x_1 + x_1 y_1^{(0)} = 1.05 + 1.05 \times 1.1$$

$$= 1.05 \times 2.1 = 2.205$$

***n* = 0**

$$y_1^{(1)} = y_0 + \frac{h}{2}\left[f(x_0, y_0) + f(x_1, y_1^{(0)})\right]$$

$$= 1 + \frac{0.05}{2}[2 + 2.205]$$

$$= 1 + 0.025 \times 4.205 = 1.11051$$

$$f(x_1, y_1^{(1)}) = x_1 + x_1 y_1^{(1)} 1.05(1 + 1051)$$

$$= 1.05 \times 2.1051 - 2.2104$$

***n* = 1**

$$y_1^{(2)} = y_0 + \frac{h}{2}\left[f(x_0, y_0) + f(x_1, y_1^{(1)})\right]$$

$$= 1 + \frac{0.05}{2}[2 + 2.2104] = 1.1053$$

Example 7.17

Apply the RK method to find $y(0.1)$ where $dy/dx = x + y$, $y(0) = 1$.

Solution Here $x_0 = 0$, $y_0 = 1$ and $f(x, y) = x + y$. Choosing $h = 0.2$, we compute k_1, k_2, k_3 and k_4.

$$k_1 = hf(x_0, y_0) = h(x_0 + y_0) = 0.2(0 + 1) = 0.2$$

$$k_2 = hf\left(x_0 + \frac{h}{2}, y_0 + \frac{k_1}{2}\right) = h\left(x_0 + \frac{h}{2} + y_0 + \frac{k_1}{2}\right)$$

$$= 0.2\left(0 + \frac{0.2}{2} + 1\frac{0.2}{2}\right) = 0.24$$

$$k_3 = hf\left(x_0 + \frac{h}{2}, y_0 + \frac{k_2}{2}\right) = h\left(x_0 + \frac{h}{2} + y_0 + \frac{k_2}{2}\right)$$

$$= 0.2\left(0 + \frac{0.2}{2} + 1\frac{0.24}{2}\right) = 0.244$$

$$k_4 = hf(x_0 + h, y_0 + k_3) = h(x_0 + y_0 + h + k_3)$$

$$= 0.2(0 + 1 + 0.2 + 0.244) = 0.2888$$

$\therefore$ The solution at $x = 0.2$ is obtained as

$$y(0.2) \simeq y_1 = y_0 + \frac{1}{6}(k_1 + 2k_2 + 2k_3 + k_4)$$

$$= 1 + \frac{1}{6}[0.2 + 2(0.24) + 2(0.244) + 0.2888]$$

$$= 1 + \frac{1}{6}(1.4568) = 1.2428$$

Analytical Solution

$dy/dx - y = x$. This is a first-order linear equation

Integrating factor $= e^{-x}$ of multiplying the equation by e^{-x} and integrating

$$ye^{-x} = c + \int xe^{-x}dx = c - xe^{-x} - e^{-x}$$

When $x = 0$, $y = 1$

$$\therefore \quad 1.1 = c - 0 - 1 \quad \Rightarrow \quad c = 2$$

The analytical solution is $y = 2e^x - x - 1$

$$y(0.2) = 2e^{0.2} - 0.2 - 1$$

$$= 2 \times 1.221403758 - 0.2 - 1 = 1.242805516$$

We see that the approximate solution obtained by RK fourth-order method agrees with the exact solution up to four decimal places.

7.3 Predictor–Corrector Methods

7.3.1 Introduction

In the methods due to Taylor, Euler and Heun, we use values at one point y_n to compute the next value y_{n+1}. Hence, they are called *single-step methods*.

To improve the accuracy, multistep methods where the values at more than one station are used to compute the next value have been developed. We now describe two such methods called the predictor–corrector methods which make use of values of y at four previous points, computed by the other methods such as Picard's or Taylor's to compute the next value.

7.3.2 Milne–Simpson's Method

Consider the IVP

$$\frac{dy}{dx} = f(x,y), y(x_0) = y_0 \tag{7.39}$$

In order to find $y(x_n)$ where $x_n = x_0 + nh$ by Milne's method we proceed in the following way.

Since the value $y_0 = y(x_0)$ is given to us, we compute $y_1 = y(x_1) = y(x_0 + h)$, $y_2 = y(x_2) = y(x_0 + 2h)$, $y_3 = y(x_3) = y(x_0 + 3h)$ by Picard's or Taylor's method. Next we calculate

$$f_0 = f(x_0, y_0) \quad f_1 = f(x_1, y_1) = f(x_0 + h, y_1)$$
$$f_2 = f(x_2, y_2) = f(x_0 + 2h, y_2)$$
$$f_3 = f(x_3, y_3) = f(x_0 + 3h, y_3)$$

For computing $y_4 = y(x_0 + 4h)$, we make use of the integration formula

$$y_4 = y_0 + \int_{x_0}^{x_0+4h} f(x,y)\,dx$$
$$= y_0 + \int_{x_0}^{x_0+4h} \left(f_0 + n\Delta f_0 + \frac{n(n-1)}{2!}\Delta^2 f_0 + \frac{n(n-1)(n-2)}{3!}\Delta^3 f_0 + \cdots \right) dx \tag{7.40}$$

by Newton's forward interpolation formula.

Putting $x = x_0 + nh$, $dx = h\,dn$ and neglecting the fourth and higher-order differences, we obtain

$$y_4 = y_0 + h\int_0^4 \left(f_0 + n\Delta f_0 + \frac{(n^2 - n)}{2}\Delta^2 f_0 + \frac{x^3 - 3n^2 + 2n}{6}\Delta^3 f_0 \right) dn \tag{7.41}$$

$$= y_0 + h\left(4f_0 + 8\Delta f_0 + \frac{20}{3}\Delta^2 f_0 + \frac{8}{3}\Delta^3 f_0 \right) \tag{7.42}$$

After replacing the differences by their functional values and simplifying, we obtain

$$y_4 = y_0 + \frac{4h}{3}(2f_1 - f_2 + 2f_3) \tag{7.43}$$

which is called a *predictor*. With this y_4 a first approximation to

$$f_4 = f(x_0 + 4h, y_4)$$

is computed and then a better value of y_4 is found by Simpson's rule

$$y_4 = y_2 + \frac{h}{3}(f_2 + 4f_3 + f_4) \tag{7.44}$$

which is called a *corrector*.

Next we calculate an improved value of f_4, and the corrector is used to find still better value of y_4. We repeat this step until y_4 remains unchanged.

When y_4 and f_4 are obtained to a desired degree of accuracy, then $y_5 = y(x_0 + 5h)$ is found from the predictor as

$$y_5 = y_0 + \frac{4h}{3}(2f_2 - 4f_3 + 2f_4) \tag{7.45}$$

and $f_5 = f(x_0 + 5h, y_5)$ is calculated. Then a better approximation to the value of y_5 is obtained from the *corrector* as

$$y_5 = y_3 + \frac{h}{3}(f_3 + 4f_4 + f_5) \tag{7.46}$$

We repeat this step until y_5 remains unchanged. We then proceed to calculate the next value y_6, and the process continues.

This is the Milne–Simpson's *predictor–corrector* method. To ensure greater accuracy, we must first improve the accuracy of the starting values and then subdivide the interval.

7.3.3 Adams–Bashforth–Moulton (ABM) Method

Consider the IVP

$$\frac{dy}{dx} = f(x, y), y(x_0) = y_0 \tag{7.47}$$

We first compute

$$y_{-1} = y(x_0 - h),\ y_{-2} = y(x_0 - 2h),$$
$$y_{-3} = y(x_0 - 3h) \tag{7.48}$$

by any one of the single-step methods (Taylor, Euler or RK) and then calculate

$$f_{-1} = f(x_0 - h, y_{-1}),\quad f_{-2} = f(x_0 - 2h, y_{-2})$$
$$f_{-3} = f(x_0 - 3h, y_{-3})$$

For computing $y_1 = y(x_0 + h)$, we make use of the integration for formula

$$y_1 = y_0 + \int_{x_0}^{x_0+h} f(x, y)\, dx$$

$$= y_0 + \int_{x_0}^{x_0+h}\left(f_0 + h\nabla f_0 + \frac{n(n+1)}{2}\nabla^2 f_0 + \frac{n(n+1)\,(n+2)}{6}\nabla^3 f_0 + \cdots \right) dx \tag{7.49}$$

by Newton's backward interpolation formula.

Putting $x = x_0 + nh$, $dx = h\, dn$ and neglecting fourth and higher-order differences, we obtain

$$y_1 = y_0 + h\int_0^1\left(f_0 + n\nabla f_0 + \frac{n(n+1)}{2}\nabla^2 f_0 + \frac{n(n+1)\,(n+2)}{6}\nabla^3 f_0 + \cdots \right) dn$$

$$= y_0 + h\left(f_0 + \frac{1}{2}\nabla f_0 + \frac{5}{12}\nabla^2 f_0 + \frac{3}{8}\Delta^3 f_0 \right) \tag{7.50}$$

After substituting the function values and simplifying, we obtain

$$y_1 = y_0 + \frac{h}{24}(55 f_0 - 59 f_{-1} + 37 f_{-2} - 9 f_{-3}) \tag{7.51}$$

which is the Adams–Beshforth *predictor* formula.

With this y_1 a first approximation to $f_1 = f(x_0 + h, y_1)$ is computed and then a better value of y_1 is found by a corrector formula, which we derive by Newton's backward interpolation formula at f_1 as

$$f(x, y) = f_1 + n\nabla f_1 + \frac{n(n+1)}{2}\nabla^2 f_1 + \frac{n(n+1)\,(n+2)}{6}\nabla^2 f_1 + \cdots \tag{7.52}$$

Hence

$$y_1 = y_0 + \int_{x_0}^{x_1}\left(f_1 + n\nabla^2 f_1 + \frac{n(n+1)}{2}\nabla^2 f_1 + \frac{n(n+1)(n+2)}{6}\nabla^3 f_1 + \cdots \right) dx \tag{7.53}$$

Putting $x = x_1 + nh$, $dx = h\, dn$ and neglecting fourth and higher-order differences and replacing the differences by the functional values, we obtain

$$y_1 = y_0 + \frac{h}{24}(9 f_1 + 19 f_0 - 5 f_{-1} f_{-2}) \tag{7.54}$$

which is called the *Adams–Moulton corrector* formula. Then an improved value of f_1 is calculated and again the corrector (7.54) is applied to find a still better value y_1. This step is repeated until y_1 remains unchanged. Then we proceed to calculate the next value y_2 and the process continues.

Example 7.18

Solve the IVP $dy/dx = x^2 - y^2, y(0) = 0$ in the interval [0.1] by Milne–Simpson's method.

Solution We first compute the values of y and f at four points $x_0 = 0, x_1 = 0.2, x_2 = 0.4$ and $x_3 = 0.6$ with $h = 0.2$ by Picard's method. Here $f(x, y) = x - y^2$. The integral formula to be used is

$$y = y(0) + \int_0^x f(x, y)\, dx \tag{1}$$

To obtain a first approximation, we use the initial condition $y(x_0) = 0$ at $x_0 = 0$ in $f(x, y) = x - y^2 = x$.

This gives

$$y_1 = 0 + \int_0^x x\, dx = \frac{x^2}{2} \tag{2}$$

To obtain a second approximation, we put $y = x^2/2$ in $f(x, y) = x - y^2$. This gives

$$y_2 = \int_0^x \left(x - \frac{x^4}{4}\right) dx = \frac{x^2}{2} - \frac{x^5}{20} \quad (3)$$

To obtain a third approximation, we put this value of y in $f(x, y) = x - y^2$ so that $f(x, y) = x - \left(\frac{x^2}{2} - \frac{x^5}{20}\right)^2$. This gives

$$y_3 = \int_0^x \left(x - \frac{x^4}{4} + \frac{x^7}{20} + \frac{x^{10}}{400}\right) dx$$

$$= \frac{x^2}{2} - \frac{x^5}{20} + \frac{x^8}{160} - \frac{x^{11}}{4400} \quad (4)$$

Now we compute the starting values for Milne's method with $h = 0.2$:

n	x	y	f
0	$x_0 = 0.0$	$y_0 = 0.0000$	$f_0 = f(x_0, y_0) = x_0 - y_0^2 = 0$
1	$x_1 = 0.2$	$y_1 = 0.0200$	$f_1 = f(x_1, y_1) = x_1 - y_1^2 = 0.2 - (0.020)^2 = 0.1996$
2	$x_2 = 0.4$	$y_2 = 0.079488$	$f_2 = 0.4 - (0.079488)^2 = 0.393682$
3	$x_3 = 0.6$	$y_3 = 0.176216$	$f_3 = 0.6 - (0.176216)^2 = 0.568948$

(5)

Now we use the predictor formula and compute y_4 at

$$y_4 = y_0 + \frac{4h}{3}(2f_1 - f_2 + 2f_3)$$

$$x = 0.8 \quad y_4 = 0 + \frac{0.8}{3}(0.3992 - 0.393682 + 1.137896) = 0.3049104 \quad (6)$$

$$f_4 = 0.8 - (0.304910)^2 = 0.707030 \quad (7)$$

We use the corrector formula

$$y_4 = y_2 + \frac{h}{3}(f_2 + 4f_3 + f_{4y})$$

and compute

$$y_4 = 0.79488 + \frac{0.2}{3}(0.393682 + 2.275579 + 0.707030) = 0.304574 \quad (8)$$

$$f_4 = 0.8 - (0.304574)^2 = 0.707235 \quad (9)$$

Using the corrector again

$y_4 = 0.304588$

$f_4 = 0.8 - (0.304588)^2 = 0.707226$

$y_4 = 0.304587$

$f_4 = 0.8 - (0.304587)^2 = 0.707226 \quad (10)$

$y_4 = 0.304587$, which agrees with the last computed value

Now, using the predictor

$$y_5 = y_1 + \frac{4h}{3}(2f_2 - f_3 + 2f_4)$$

$$x = 1.0 \quad y_5 = 0.02 + \frac{0.8}{3}(0.787364 - 0.568948 + 1.414452) = 0.455431 \quad (11)$$

$$f_5 = 1.0 - (0.455431)^2 = 0.792582$$

corrector … $y_5 = y_3 + \frac{h}{3}(f_3 + 4f_4 + f_5)$

$$= 0.176216 + \frac{0.2}{3}(0.568948 - 2.828904 + 0.782582) = 0.455578 \quad (12)$$

$$f_5 = 1.0 - (0.455578)^2 = 0.792582$$

corrector … $y_5 = 0.455569$

$f_5 = 1.0 - (0.455569)^2 = 0.792457$

corrector … $y_5 = 0.455570$

$f_5 = 1.0 - (0.455570)^2 = 0.792456$

corrector … $y_5 = 0.45556987$

$f_5 = 0.792456 \quad (13)$

corrector … $y_5 = 0.45556987$. This value agrees with the earlier value

$$\therefore \quad y(1.0) = 0.45556987 \quad (14)$$

Example 7.19

Solve the IVP $dy/dx = 1 + y^2$, $y(0) = 0$ and compute $y(0.8)$ by the ABM method.

Solution We first compute $y(0.2)$, $y(0.4)$ and $y(0.6)$ by the RK method. Formulas for fourth-order RK method are

$$y_1 = y_0 + \frac{1}{6}(k_1 + 2k_2 + 2k_3 + k_4) \quad (1)$$

where

$$\left.\begin{aligned} &k_1 = hf(x_0, y_0), \quad k_2 = hf\left(x_0 + \frac{h}{2}, y_0 + \frac{k_1}{2}\right) \\ &k_3 = hf\left(x_0 + \frac{h}{2}, y_0 + \frac{k_2}{2}\right), \\ &k_4 = hf(x_0 + h, y_0 + k_3) \quad \text{and} \quad f(x, y) = 1 + y^2 \end{aligned}\right] \quad (2)$$

To find y(0.2)

We take $h = 0.2$ and $x_0 = y_0 = 0$. We obtain

$$k_1 = 0.2 \times 1 = 0.2 \quad \because f(x_0, y_0) = 1 + 0^2 = 1$$

$$k_2 = 0.2(1.01) = 0.202$$

$$\because f\left(x_0 + \frac{h}{2}, y_0 + \frac{k_1}{2}\right) = 1 + (0.1)^2 = 1.01$$

$$k_3 = 0.2(1.01020) = 0.20204$$

$$\because f\left(x_0 + \frac{h}{2}, y_0 + \frac{k_2}{2}\right) = 1 + (0.101)^2 = 1.010201$$

$$k_4 = 0.2(1.040820) = 0.208164$$

$$\because f(x_0 + h, y_0 + k_3) = 1 + (0.20204)^2 = 1.040820 \quad (3)$$

$$y(0.2) \simeq y_2 = 0 + \frac{1}{6}(0.2 + 2 \times 0.202 + 2 \times 0.20204 + 0.208164) = 0.202707 \quad (4)$$

To find y(0.4)

We next compute $y(0.4)$, taking $x_0 = 0.2$, $y_0 = 0.202707$ and $h = 0.2$:

$$k_1 = hf(x_0, y_0) = h(1 + y_0^2) \simeq 0.2\ [1 + (0.202707)^2] = 0.208218$$

$$k_2 = h\left[1 + \left(y_0 + \frac{k_1}{2}\right)^2\right] = 0.2\left[1 + (0.306816)^2\right] = 0.218827$$

$$k_3 = h\left[1 + \left(y_0 + \frac{k_2}{2}\right)^2\right] = 0.2\left[1 + (0.31212)^2\right] = 0.2194838$$

$$k_4 = h[1 + (y_0 + k_3)^2] = 0.2\ [1 + (0.422191)^2] = 0.235649$$

$$\begin{aligned} y(0.4) &= y_0 + \frac{1}{6}(k_1 + 2k_2 + 2k_3 + k_4) \\ &= 0.202707 + \frac{1}{6}[0.208218 + 2(0.218827) + 2(0.219484) + 0.235649] \\ &= 0.202707 + 0.220082 = 0.422788 \end{aligned} \quad (5)$$

To find y(0.6)

We next compute $y(0.6)$ taking $x_0 = 0.4$, $y_0 = 0.422788$ and $h = 0.2$ and $f(x, y) = 1 + y^2$

$$k_1 = h(1 + y_0^2) = 0.2\left[1 + (0.422788)^2\right] = 0.235750$$

$$k_2 = h\left[1 + \left(y_0 + \frac{k_1}{2}\right)^2\right] = 0.2\left[1 + (0.540663)^2\right] = 0.258463$$

$$k_3 = h\left[1 + \left(y_0 + \frac{k_2}{2}\right)^2\right] = 0.2\left[1 + (0.552019)^2\right] = 0.260945$$

$$k_4 = h\left[1 + (y_0 + k_3)^2\right] = 0.2\left[1 + (0.683733)^2\right] = 0.293498 \quad (6)$$

$$\begin{aligned} y(0.6) &= y_0 + \frac{1}{6}[k_1 + 2k_2 + 2k_3 + k_4] \\ &= 0.422788 + \frac{1}{6}[0.235750 + 2(0.258464) + 2(0.260945) + 0.293498] \\ &= 0.422788 + \frac{1}{6}(0.235750 + 0.516928 + 0.521890 + 0.293498) \\ &= 0.422788 + 0.2613448 = 0.684132 \end{aligned} \quad (7)$$

The predictor for an ABM method is

$$y_1^P = y_0 + \frac{h}{24}[55f_0 - 59f_{-1} + 37f_{-2} - 9f_{-3}]$$

We now tabulate the computed values y_i and compare f_i to be used in ABM predictor y_1^P. $f(x, y) = 1 + y^2$; $h = 0.2$

$y_0 = y(0.6) = 0.684132;$
$f_0 = 1 + y_0^2 = 1 + (0.684132)^2 = 1.468036$
$y_{-1} = y(0.4) = 0.422788;$
$f_{-1} = 1 + y_{-1}^2 = 1 + (0.422788)^2 = 1.17875D$
$y_{-2} = y(0.2) = 0.202707;$
$f_{-2} = 1 + y_{-2}^2 = 1 + (0.202707)^2 = 1.041090$
$y_{-3} = y(0.0) = 0.000000;$
$f_{-3} = 1 + y_{-3}^2 = 1 + (0)^2 = 1.00000$ (8)

Substituting these values in y_1^p above and simplifying

$$y_1^p = 0.684132 + \frac{0.2}{24}[55 \times 1.468036 - 59 \times 1.178750 + 37 \times 1.041090 - 9 \times 1]$$
$$= 0.684132 + 0.339300 = 1.023432 \tag{9}$$

Now we use the Adams–Moulton corrector

$$y_1^c = y_0 + \frac{h}{24}\left[9f_1^p + 19f_0 - 5f_{-1} + f_{-2}\right]$$

Computing f_1^p and using y_0, f_0, f_{-1} and f_{-2} already computed and substituting in the above corrector y_1^c we obtain $y(0.8)$.

Now, the Adams–Bashforth predictor

$$f_1^p = 1 + (y_1^p)^2 = 1 + (1.023432)^2 = 2.047413$$

and $f_0 = 1.468036$, $f_{-1} = 1.178750$ and $f_{-2} = 1.041090$

Substituting these figures in y_1^c above

$$y_1^c = 0.684132 + \frac{0.2}{24}[9(2.047413) + 19(1.468036) - 5(1.178750) + 1.041090]$$
$$= 0.684132 + 0.345556 = 1.029688 \tag{10}$$
$$\therefore \quad y(0.8) = 1.029688 \tag{11}$$

Analytical Solution

The given differential equation is

$$\frac{dy}{dx} = 1 + y^2 \quad \text{with } y(0) = 0 \tag{12}$$

Seperating the variables and integrating

$$\int dx = \int \frac{dy}{1+y^2} \Rightarrow x + c = \tan^{-1} y$$

But $y = 0$ when $x = 0$ which gives $c = 0$

∴ The required solution is

$$y = \tan x \tag{13}$$

Hence, the exact value

$$y(0.8) = 1.029639 \tag{14}$$

Absolute error = |True value − Approx. value| = 0.000048

Relative error = 0.0000466

Percentage error = 0.00466 %

EXERCISE 7.1

1. Use the Taylor series method to solve the IVP $y' = x^2 + y^2$; $y(0) = 0$ for the interval [0, 0.4] using two subintervals of size $h = 0.2$.

   Ans: 0.1021352, 0.021333

2. Using Picard's method find $y(0.1)$ if $y' = \frac{y-x}{y+x}$, $y(0) = 1$.

   Ans: $y(0.1) = 0.9828$

3. Given $dy/dx = x^2 + y$, $y(0) = 1$ find $y(0.02)$, $y(0.04)$ and $y(0.6)$ by Euler's modified method.

   Ans: 1.0202; 1.0408; 1.0619

4. Solve by Euler's method $dy/dx = x + y$, $y(0) = 0$ with $h = 0.2$ and compute $y(0.4)$ and $y(0.6)$.

   Ans: 0.0938, 0.2258

5. Use the fourth-order RK method to find the value of $y(1.0)$ given that $y(0) = 1$ and $dy/dx = y - x/y + x$.

   Ans: 1.4983

6. Using Milne's method, find $y(0.8)$ if $dy/dx = x - y^2$, $y(0) = 0$, $y(0.2) = 0.02$, $y(0.4) = 0.0795$ and $y(0.6) = 0.1762$.

7. By Milne–Simpson's method, find $y(0.8)$ if $y' = y + x^2$, $y(0) = 1$ with $h = 0.2$.

   Ans: 2.4366

8. Compute $y(0.8)$ and $y(1.0)$ by Milne–Simpson's method if $dy/dx = 1 + y^2$; $y(0) = 0$, $y(0.2) = 0.2027$, $y(0.4) = 0.4228$ and $y(0.6) = 0.6846$. Solve by ABM method using the given initial values. Compare with the exact solution.

   Ans: $f_0 = 1$, $f_1 = 1.0411$, $f_2 = 1.1787$, $f_3 = 0.6841$
   $y^P(0.8) = 1.0239$, $f(0.8) = 2.048$

9. $y' = 2xy^2$, $y(0) = 1$, $h = 0.05$ find y on [0, 0.5]. Initial values are $y(0.05) = 1.0025$, $y(0.10) = 1.0101$, $y(0.15) = 1.0230$.

   Ans: $y(0.20) = 1.0417$, $y(0.25) = 1.0667$,
   $y(0.30) = 1.0989$, $y(0.40) = 1.1906$,
   $y(0.50) = 1.3334$

10. $y' = x^2 + y^2$, $y(0) = 0$, $h = 0.1$ find y on [0,12]. When the initial values are $y(0.2) = 0.0004$, $y(0.4) = 0.0064$ and $y(0.6) = 0.0325$, $y(1.0) = 0.2574$, $y(1.1) = 0.3856$, $y(12) = 0.5581$.

    Ans: $y(0.8) = 1.035$, $y(0.9) = 0.1669$,
    $y(1.0) = 0.2574$, $y(1.1) = 0.3836$,
    $y(1.2) = 0.5581$

11. $y' = y^2 \sin x$ $y(0) = 1$ on [0, 0.5] = 0.05. Find y on [0, 0.05] when the initial values are $y(0.05) = 1.00125$, $y(0.10) = 1.00502$, $y(0.15) = 1.01136$.

    [**Hint:** Exact solution $y(x) = \sec x$.]

    Ans: $y(0.2) = 1.02033$, $y(0.25) = 1.03209$,
    $y(0.3) = 1.04675$, $y(0.4) = 1.08570$,
    $y(0.5) = 1.13950$

Fourier Series

8

8.1 Introduction

Fourier series[1] is an important tool in solving problems in conduction of heat, electrical engineering, current and voltage in alternating circuits, electrodynamics, acoustics, *etc*. The basic idea is to represent periodic functions by a series involving sines and cosines. In this chapter, we find expansions of even/odd functions first in an interval of 2π and later of $2l$. Finally, we develop half-range Fourier series also.

First we begin with the notion of periodic functions; and even and odd functions.

8.2 Periodic Functions, Properties

The graph of a trigonometric function repeats itself in regular intervals. Such functions are called periodic functions.

Periodic Function: Definition A function $f(x)$ is called a periodic function if there exists a positive real number p such that $f(x+p) = f(x)\ \forall x \in \text{dom}\, f$. Generally dom $f = R$ (the set of all real numbers).

Examples of periodic functions in engineering: (i) emf in an AC circuit, (ii) output of half-wave rectifier, (iii) thrust on the piston, (iv) digital signals, *etc*.

The smallest value of p for which this holds is called the fundamental period or the primitive period or simply the period of f and is denoted by $P(f) = p$.

[1] Named after the French analyst and mathematical physicist, Baron de Jean Baptiste Joseph Fourier (1768–1830). While investigating heat conduction problems, he developed the theory of Fourier series and used it first in his memorable work Theorie Analytique de la Chaleur. This initiated a great mathematical activity and led to the development of a new branch of mathematics called Fourier Analysis.

Sin x, cos x, cosec x, sec x are of period 2π and tan x, cot x are of period π.

8.2.1 Properties of Periodic Functions

1. A constant function $f(x) = c$ is periodic. Any positive real number p is a period of f: it has no primitive period.
2. The period of $-f(x)$ is the same as that of $f(x)$.
3. If p is a period of a function f then np (a multiple of p) is also a period of f.
4. $P(f(x)) = p \Rightarrow P(f(ax+b)) = p/|a|$

   E.g. $P(\tan ax) = \pi/|a|$

   $P(\sin 3x) = 2\pi/3$
5. $P(f_1(x)) = P(f_2(x)) = p$

   $\Rightarrow P(c_1 f_1(x) + c_2 f_2(x)) = p,$

   where c_1, c_2 are constants.

   E.g. $P\left(3\tan\frac{x}{2} + 4\cot\frac{x}{2}\right) = 2\pi;$

   $P(6\tan 2x + 5\sin 4x) = \frac{\pi}{2}$
6. A linear combination of two periodic functions is a periodic function.

Let $P(f_1(x)) = p_1$ and $P(f_2(x)) = p_2$. Then $(c_1 f_1(x) + c_2 f_2(x) = k(m, n))$ where k is the least positive real number such that $p_1 = km$, $p_2 = kn$ and $(m, n) = \text{LCM}$ of m and $n \in N$.

E.g. $P\left(\cos\frac{x}{3}\right) = 6\pi = 18\left(\frac{\pi}{3}\right) = mk,$

$P\left(\tan\frac{3\pi}{4}\right) = \frac{4\pi}{3} = 4\left(\frac{\pi}{3}\right) = nk$

where $k = \pi/3$, $m = 18$ and $n = 4$.

$$P\left(4\cos\frac{x}{3} + 3\tan\frac{3\pi}{4}\right) = \frac{\pi}{3}(18,4)$$
$$= \frac{\pi}{3} \times 36 = 12\pi$$

8.3 Classifiable Functions—Even and Odd Functions

A function f defined on a set S is called a classifiable function if $-x \in S$ whenever $x \in S$.

E.g. x, $x + x^2$, $\sin x$, $\cos x$, $\tan x$, e^x, e^{-x}, $-x+$ $-x + \cos x$, $|x|$, $\log|x|$, $\log[(1-x)/(1+x)]$, *etc.*, are all classifiable functions, as they are defined for positive as well as negative real values of x.

Nonclassifiable Functions Functions which are not classifiable are called nonclassifiable functions. *E.g.* $\log x$, $\sqrt{x}$, which are defined for $x > 0$ only $\log(-x)$, $\sqrt{-x}$ which are defined for $x < 0$ only are nonclassifiable functions.

The set of classifiable functions can be divided into three mutually disjoint[2] subsets consisting of (1). Even functions, (2). Odd functions and (3). Neither even nor odd functions.

Even Function: Definition A function defined on a set S is an even function if (i) $-x \in S$ whenever $x \in S$ and (ii) $f(-x) = f(x) \forall \in S$:

$$E.g.\ x^2,\ \cos x,\ x\sin x,\ x^4\cos x,\ x\log\frac{1-x}{1+x},$$

Odd Function: Definition A function f defined on S is an odd function if (i) $-x \in S$ whenever $x \in S$ and (ii) $f(-x) = -f(x)$ $\forall x \in S$.

E.g. x, $\sin x$, $x^3\cos x$, $x^2 \sin x + 2x$

Note 1 The identity function $f(x) = 0$ for all $x \in R$ is the only function which is both even and odd.

[2] With the exception of the identity function $f(x) = 0$ for all $x \in R$, which is even as well as odd.

Note 2 There are classifable functions which are neither even nor odd. *E.g.* $x^2 + x$, $\sin x + \cos x$, e^x

Note 3 Every classifable function defined on a set S (which is symmetric with respect to $x = 0$) can be expressed as the sum of two functions, one even and the other odd. Let $f(x)$ be defined on S. Then

$$f(x) = \frac{1}{2}[f(x) + f(x)]$$
$$= \underbrace{\frac{1}{2}[f(x) + f(-x)]}_{\text{even}} + \underbrace{\frac{1}{2}[f(x) - f(-x)]}_{\text{odd}}$$

Note 4 The graph of an even function $f(x)$ is symmetric about the y-axis.

Note 5 An even function $f(x)$ contains even powers of x and possibly $\cos x$, $\sec x$, *etc.*, and odd powers of x multiplied by $\sin x$, $\operatorname{cosec} x$, *etc.*

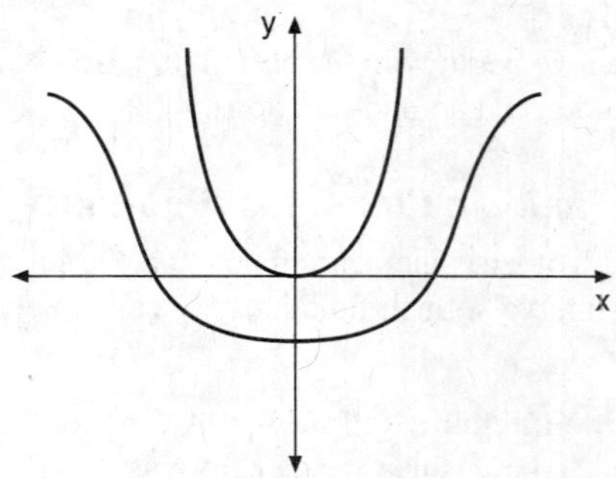

Figure 8.1 Symmetry of an even function w.r.t. the y-axis

Note 6 $\int_{-a}^{a} f(x)\,dx = 2\int_0^a f(x)\,dx$, if $f(x)$ is even.

Note 7 The sum or the product of two or more even functions is an even function.

Note 8 The product of an *even* number of odd functions is even.

Note 9 The graph of an odd function $f(x)$ is symmetric about the origin. (It lies in I and III or II and IV quadrants.)

Note 10 An odd function $f(x)$ contains odd powers of x and possibly $\sin x$, $\operatorname{cosec} x$, *etc.*

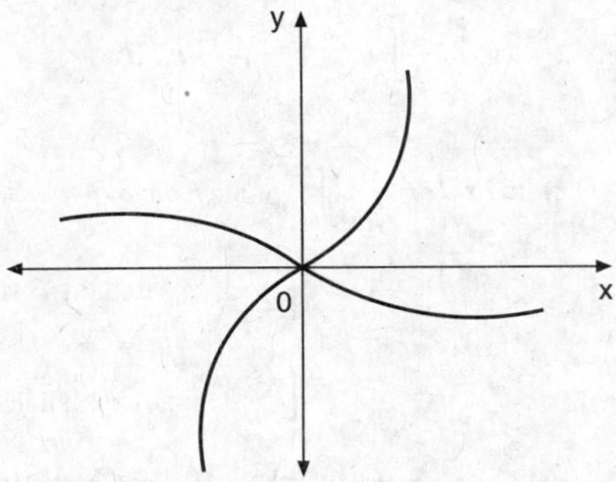

Figure 8.2 Symmetry of an odd function w.r.t. the origin

Note 11 $f(x)$ is odd $\Rightarrow \int_{-a}^{a} f(x)\,dx = 0$

Note 12 The sum of any number of odd functions is odd.

Note 13 The product of an odd number of odd functions is an odd function.

Note 14 The product of an odd and an even function is an odd function.

8.4 Fourier Series, Fourier Coefficients and Euler's Formulae in $(\alpha, \alpha + 2\pi)$

The Fourier series corresponding to a function $f(x)$ in the interval $\alpha < x < \alpha + 2\pi$ is the trigonometric series on the RHS of (8.1) below and we write

$$f(x) = \frac{1}{2}a_0 + \sum_{n=1}^{\infty}(a_n \cos nx + b_n \sin nx) \qquad (8.1)$$

where the constants a_0, a_n, b_n $(n = 1, 2, 3, \ldots)$ are called Fourier coefficients and are defined by Euler's formulae[3]

$$(a_0, a_n, b_n) = \frac{1}{\pi}\int_{\alpha}^{\alpha+2\pi}(1, \cos nx, \sin nx)\, f(x)\,dx \qquad (8.2)$$

[3] EULER, Leonhard (1707–1783), was a gifted Swiss mathematician, and the most prolific mathematician, who contributed to almost all branches of mathematics even after becoming totally blind in 1771. He studied under John Bernoulli and became a professor of mathematics at St. Petersburg, Russia. He is the first Modern Mathematical Universalist.

8.4.1 Determination of Fourier Coefficients

Derivation of Euler's Formulae

In order to establish (8.2) we require the following results:

1. $$\int_{\alpha}^{\alpha+2\pi} \cos nx\,dx = \left[\frac{\sin nx}{n}\right]_{\alpha}^{\alpha+2\pi} = \frac{\sin(n\alpha + 2n\pi) - \sin n\alpha}{n} = 0 \qquad (n \neq 0)$$

2. $$\int_{\alpha}^{\alpha+2\pi} \sin nx\,dx = \left[\frac{-\cos nx}{n}\right]_{\alpha}^{\alpha+2\pi} = \frac{-\cos(n\alpha + 2n\pi) + \cos n\alpha}{n} = 0 \qquad (n \neq 0)$$

3. $$\int_{\alpha}^{\alpha+2\pi} \cos mx \cos nx\,dx = \frac{1}{2}\int_{\alpha}^{\alpha+2\pi}[\cos(m+n)x + \cos(m-n)x]\,dx = \frac{1}{2}\left[\frac{\sin(m+n)x}{m+n} + \frac{\sin(m-n)x}{m-n}\right]_{\alpha}^{\alpha+2\pi} = 0 \qquad (m^2 \neq n^2)$$

4. $$\int_{\alpha}^{\alpha+2\pi} \cos^2 nx\,dx = \frac{1}{2}\int_{\alpha}^{\alpha+2\pi}(1 + \cos 2nx)\,dx = \left(\frac{x}{2} + \frac{\sin 2nx}{4n}\right)_{\alpha}^{\alpha+2\pi} = \pi \qquad (n \neq 0)$$

5. $$\int_{\alpha}^{\alpha+2\pi} \sin mx \cos nx\,dx = \frac{1}{2}\int_{\alpha}^{\alpha+2\pi}[\sin(m+n)x + \sin(m-n)x]\,dx = -\frac{1}{2}\left[\frac{\cos(m+n)x}{m+n} + \frac{\cos(m-n)x}{(m-n)}\right]_{\alpha}^{\alpha+2\pi} = 0 \qquad (m^2 \neq n^2)$$

6. $$\int_{\alpha}^{\alpha+2\pi} \sin nx \cos nx\,dx = \left(\frac{\sin^2 nx}{2n}\right)_{\alpha}^{\alpha+2\pi} = 0$$

7. $\int_{\alpha}^{\alpha+2\pi} \sin mx \sin nx\, dx$

$$= \frac{1}{2}\int_{\alpha}^{\alpha+2\pi} [\cos(m-n)x - \cos(m+n)x]\, dx$$

$$= \frac{1}{2}\left[\frac{\sin(m-n)x}{m-n} - \frac{\sin(m+n)x}{(m+n)}\right]_{\alpha}^{\alpha+2\pi} = 0$$

$$(m^2 \neq n^2)$$

8. $\int_{\alpha}^{\alpha+2\pi} \sin^2 nx\, dx = \left(\frac{x}{2} - \frac{\sin 2nx}{4n}\right)_{\alpha}^{\alpha+2\pi} = \pi$

$$(n \neq 0)$$

(8.3)

Let $f(x)$ be represented in the interval $(\alpha, \alpha+2\pi)$ by the Fourier Series (8.1)

We now find the Fourier coefficients a_0, a_n, b_n $(n = 1, 2, \ldots)$ assuming that the series (8.1) can be integrated term by term from $x = \alpha$ to $x = \alpha + 2\pi$. For this, it is suffcient that the series is uniformly convergent.

Integrate both sides of (8.1) from $x = \alpha$ to $x = \alpha + 2\pi$ after multiplying by 1, cos nx and sin nx to obtain a_0, a_n and b_n as follows:

$$\int_{\alpha}^{\alpha+2\pi} f(x)\, dx = \frac{1}{2}a_0\int_{\alpha}^{\alpha+2\pi} dx + \int_{\alpha}^{\alpha+2\pi}\left(\sum_{n=1}^{\infty} a_n \cos nx\right) dx + \int_{\alpha}^{\alpha+2\pi}\left(\sum_{n=1}^{\infty} b_n \sin nx\right) dx$$

$$= \frac{1}{2}a_0(\alpha + 2\pi - \alpha) + 0 + 0 = a_0\pi$$

by (1) and (2)

$$\therefore \quad a_0 = \frac{1}{\pi}\int_{\alpha}^{\alpha+2\pi} f(x)\, dx \qquad (8.4)$$

$$\int_{\alpha}^{\alpha+2\pi} f(x)\cos nx\, dx = \frac{1}{2}a_0\int_{\alpha}^{\alpha+2\pi} \cos nx\, dx + \int_{\alpha}^{\alpha+2\pi}\left(\sum_{n=1}^{\infty} a_n \cos nx\right)\cos nx\, dx + \int_{\alpha}^{\alpha+2\pi}\left(\sum_{n=1}^{\infty} b_n \sin nx\right)\cos nx\, dx$$

$$= 0 + \pi a_n + 0$$

by (1), (3), (4), (5), (6);

$$\therefore \quad a_n = \frac{1}{\pi}\int_{\alpha}^{\alpha+2\pi} f(x)\cos nx\, dx \qquad (8.5)$$

$$\int_{\alpha}^{\alpha+2\pi} f(x)\sin nx\, dx = \frac{1}{2}a_0\int_{\alpha}^{\alpha+2\pi} \sin nx\, dx + \int_{\alpha}^{\alpha+2\pi}\left(\sum_{n=1}^{\infty} a_n \cos nx\right)\sin nx\, dx + \int_{\alpha}^{\alpha+2\pi}\left(\sum_{n=1}^{\infty} b_n \sin nx\, dx\right)\sin nx\, dx$$

$$= 0 + 0 + \pi b_n \qquad (8.6)$$

by (2), (5), (6), (7), (8);

By choosing α as 0 and $-\pi$, we obtain expansions in the intervals $(0, 2\pi)$ and $(-\pi, \pi)$, respectively.

8.5 Dirichlet's Conditions for Fourier Series Expansion of a Function

We have seen above that if a function $f(x)$ can be formally expanded in a Fourier series then the coefficients are given by Euler's formulae (8.2). The theory concerning the conditions under which a function can be represented by a Fourier series and its convergence is complicated. However, we state here the well-known[4] Dirichlet's conditions, which cover almost all engineering applications.

Any function $f(x)$ can be developed as a Fourier series (8.1)

$$\frac{1}{2}a_0 + \sum_{n=1}^{\infty} a_n \cos nx + \sum_{n=1}^{\infty} b_n \sin nx$$

where a_0, a_n, b_n are constants given by Euler's formulae (8.2) mentioned above provided they satisfy the following sufficient conditions of Dirichlet.

(i) $f(x)$ is periodic; $f(x)$ and its integrals are single valued and finite (bounded).

(ii) $f(x)$ has at the most a finite number of discontinuities in any one period.

[4] DIRICHLET, Peter Gustav Lejune (1805–1859), is a German number theorist, analyst and applied mathematician. He showed that every arithmetical sequence $\langle a, a + b, a + 2b, \ldots\rangle$ where a and b are relatively prime integers, contains an infinite number of primes.

(iii) $f(x)$ has at the most a finite number of maxima/minima.

Functions Having Points of Discontinuity

In deriving Euler's formulae for a_0, a_n and b_n above, it was assumed that $f(x)$ was continuous. But a function may have a finite number of points of finite discontinuity, *i.e.*, its graph may consist of a finite number of different curves given by different equations. Such a function is also expressible as a Fourier series.

For instance, if $f(x)$ is defined in the interval $(\alpha, \alpha + 2\pi)$ by

$$\begin{aligned} f(x) &= \phi(x) \quad \alpha < x < c \\ &= \psi(x) \quad c < x\ \alpha + 2\pi \end{aligned} \tag{8.7}$$

i.e., $x = c$ is a point at finite discontinuity for $f(x)$.

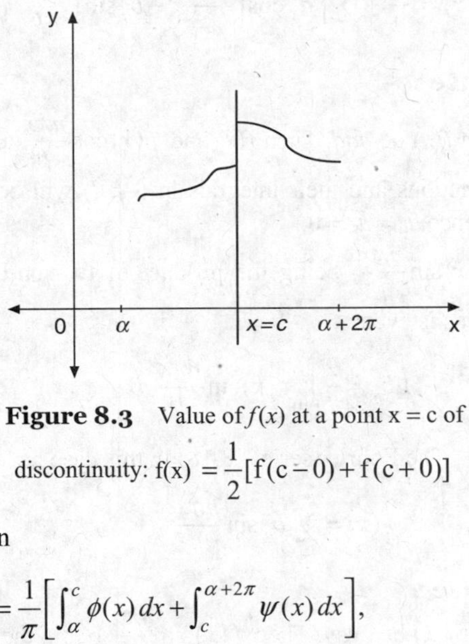

Figure 8.3 Value of $f(x)$ at a point x = c of discontinuity: $f(x) = \frac{1}{2}[f(c-0)+f(c+0)]$

Then

$$a_0 = \frac{1}{\pi}\left[\int_\alpha^c \phi(x)\,dx + \int_c^{\alpha+2\pi} \psi(x)\,dx\right],$$

$$a_n = \frac{1}{\pi}\left[\int_\alpha^c \phi(x)\cos nx\,dx + \int_c^{\alpha+2\pi} \psi(x)\cos nx\,dx\right],$$

$$b_n = \frac{1}{\pi}\left[\int_\alpha^c \phi(x)\sin nx\,dx + \int_c^{\alpha+2\pi} \psi(x)\sin nx\,dx\right] \tag{8.8}$$

At $x = c$ there is a finite jump discontinuity. Both the left and right limits $f(c - 0)$ and $f(c + 0)$ exist and are different.

At such a point Fourier series gives the value of $f(x)$ as the arithmetic mean of the two limits:

$$\text{At } x = c,\ f(x) = \frac{1}{2}[f(c-0)+(c+0)] \tag{8.9}$$

Functions Which have no Fourier Series Expansions

$\text{Sin}^{-1}x$ in $(-1, 1)$ is not single valued, $\tan x$ in $(0, 2\pi)$ has infinite discontinuity at $x = \pi/2, 3\pi/2$; and $\sin 1/x$ in $(-\pi, \pi)$ has infinite number of maxima and minima in the nbd of $x = 0$. So, the set of other such functions has no Fourier series expansions. Dirichlet's conditions for the existence of Fourier series expansions for functions are only suffcient but not necessary. This means that though a function does not satisfy the conditions it may have Fourier series expansion.

8.6 Fourier Series Expansions: Even/Odd Functions

The following properties are useful in deriving the Fourier series expansions of combinations of functions when we have derived the Fourier series expansions for the typical functions.

If the Fourier series expansions for $f_1(x)$ and $f_2(x)$ in an interval (a, b) are given by S_1 and S_2, respectively, then the Fourier series expansion for $c_1 f_1(x) + c_2 f_2(x)$ in the interval (a, b) is $c_1 S_1 + c_2 S_2$ where c_1 and c_2 are any constants.

From this we infer the following:

1. The Fourier series of constant times a function in (a, b) is constant times the Fourier series of that function in (a, b).
2. The Fourier series of the sum (or difference) of two functions in (a, b) is the sum (or difference) of their Fourier series in (a, b).

Thus, for example, if we want to find the Fourier series of the function $(\pi/2 - x/2)$ in $(0, 2\pi)$, we may proceed as follows. Suppose we have found Fourier series for $f(x) = x$ in $(0, 2\pi)$ as

$$x = \pi - 2\sum_{n=1}^{\infty}\frac{\sin nx}{n} \tag{8.10}$$

Multiplying both sides of (8.10) by $\frac{1}{2}$ and transposing, we obtain the Fourier series expansion of $(\pi/2 - x/2)$ in $(0, 2\pi)$ as

$$\frac{\pi}{2} - \frac{x}{2} = \sum_{n=1}^{\infty} \frac{\sin nx}{n} \quad (8.11)$$

Again, suppose we have found the Fourier series expansion for x^2 in $(0, 2\pi)$ as

$$x^2 = \frac{4\pi^2}{3} + 4\sum_{n=1}^{\infty} \frac{\cos nx}{n^2} - 4\pi \sum_{n=1}^{\infty} \frac{\sin nx}{n} \quad (8.12)$$

Using these, the Fourier series of $(x + x^2)$, $(x - x^2)$ and $\left(\frac{\pi - x}{2}\right)^2$ may be obtained as follows:

$$x + x^2 = \left(\pi + \frac{4\pi^2}{3}\right) + 4\sum_{n=1}^{\infty} \frac{\cos nx}{n^2} - (2 + 4\pi)\sum_{n=1}^{\infty} \frac{\sin nx}{n^2} \quad (8.13)$$

$$x - x^2 = \left(\pi - \frac{4\pi^2}{3}\right) - 4\sum_{n=1}^{\infty} \frac{\cos nx}{n^2} + (4\pi - 2)\sum_{n=1}^{\infty} \frac{\sin nx}{n^2} \quad (8.14)$$

$$\left(\frac{\pi - x}{2}\right)^2 = \frac{\pi^2}{4} - \frac{\pi x}{2} + \frac{x^2}{4} = \frac{\pi^2}{12} + \sum_{n=1}^{\infty} \frac{\cos nx}{n^2},$$

after simplication (8.15)

The Fourier series of two *different functions* in the *same* interval are evidently *different.* The Fourier series of the *same* function in *different* intervals may, in certain cases, be the same.

Fourier series expansion of $f(x) = x$ in $(0, 2\pi)$ is (8.10) but for the same function in $(-\pi, \pi)$ the Fourier series expansion is

$$x = 2\sum_{n=1}^{\infty} \frac{(-1)^{n-1} \sin nx}{n} \quad (8.16)$$

But the Fourier series expansion for $f(x) = \sqrt{1 - \cos x}$ in $(0, 2\pi)$ is

$$\sqrt{1 - \cos x} = \frac{2\sqrt{2}}{\pi} - \frac{4\sqrt{2}}{\pi} \sum_{n=1}^{\infty} \frac{(\cos nx)}{(2n-1)(2n+1)} \quad (8.17)$$

which is the same for $f(x) = \sqrt{1 - \cos x}$ in $(-\pi, \pi)$ where it is defined by

$$f(x) = \sqrt{1 - \cos x} = \begin{cases} -\sqrt{2} \sin \frac{x}{2} & (-\pi, 0) \\ \sqrt{2} \sin \frac{x}{2} & (0, \pi) \end{cases} \quad (8.17)$$

8.6.1 Fourier Series of Odd and Even Functions in the Interval (–l, l)

In Section 8.8, it is shown that Fourier series expansion for a function $f(x)$ with period $2l$ is given by

$$f(x) = \frac{a_0}{2} + \sum \left[a_n \cos\left(\frac{n\pi x}{l}\right) + b_n \sin\left(\frac{n\pi x}{l}\right) \right] \quad (8.18)$$

Case 1

Let $f(x)$ be *odd*. Then $f(x)$ and $f(x)\cos\frac{n\pi x}{l}$ are odd functions and their integrals in $(-l, l)$ will be zero. Hence $a_0 = a_n = 0$.

$f(x)\sin\frac{n\pi x}{l}$, being the product of two odd functions is even.

$$\therefore \quad b_n = \frac{2}{l}\int_0^l f(x)\sin\frac{n\pi x}{l}\,dx \quad (8.19)$$

The Fourier series (8.18) in this case becomes

$$f(x) = \sum_{n=1}^{\infty} b_n \sin\frac{n\pi x}{l} \quad (8.20)$$

Case 2

Let $f(x)$ be *even*. Then $f(x)\sin\frac{n\pi x}{l}$ being the product of an even and an odd function, is odd and its integral in $(-l, l)$ will be zero. Hence $b_n = 0$.

$f(x)$ and $f(x)\cos\frac{n\pi x}{l}$ are even. Consequently a_0 and a_n are given by

$$(a_0, a_n) = \frac{2}{l}\int_0^l f(x)\left(1, \cos\frac{n\pi x}{l}\right)dx \quad (8.21)$$

The Fourier series (8.18) in this case becomes

$$f(x)=\frac{a_0}{2}+\sum_{n=1}^{\infty}a_n\cos\frac{n\pi x}{l} \tag{8.22}$$

8.6.2 Fourier Series of Odd and Even Functions in the Interval $(-\pi, \pi)$

Case 1

Let $f(x)$ be odd. Taking $l=\pi$ in (8.19) we have

$$b_n=\frac{2}{\pi}\int_0^{\pi}f(x)\sin nx\,dx \tag{8.23}$$

The Fourier series is given by

$$f(x)=\sum_{n=1}^{\infty}b_n\sin nx \tag{8.24}$$

while $a_0=a_n=0$

Case 2

Let $f(x)$ be even. Taking $l=\pi$ in (8.21) we have

$$(a_0,\, a_n)=\frac{2}{\pi}\int_0^{\pi}f(x)(1,\,\cos nx)\,dx \tag{8.25}$$

The Fourier series (8.22) in this case becomes

$$f(x)=\frac{a_0}{2}+\sum_{n=1}^{\infty}a_n\cos nx \tag{8.26}$$

while $b_n=0$

8.7 Simply-Defined and Multiply-(Piecewise) Defined Functions

By a function $f: S\rightarrow R$ we mean a rule or a set of rules assigning to each element in S (domain) a unique element in R (codomain). If this assignment is by a simple (single) rule or law we call the function a *simply-defined* function and if this assignment is through a multiple rule or law, we call the function *multiply-defined* or *piecewise*-defined function.

1. Simply-Defined Functions

The functions $f, g, h: R\rightarrow R$ defined by

$$f(x)=x^2,\ g(x)=1+e^x\sin x,\ h(x)=\sin x+\cos x,$$

respectively, are simply-defined functions.

2. Multiply-Defined or Piecewise-Defined Functions

(a) $f: [-\pi, \pi]\rightarrow R$ defined by

$$f(x)=\begin{cases}0, & \text{in } (-\pi, 0)\\ \cos x, & \text{in } (0, \pi)\end{cases}$$

(b) $g: (-2, 2)\rightarrow R$ defined by

$$g(x)=\begin{cases}1, & \text{in } (-2,-1)\\ 0, & \text{in } (-1,0)\\ 1-x, & \text{in } (0,1)\\ 1+x, & \text{in } (1,2)\end{cases}$$

are multiply-defined or piecewise-defined functions.

While evaluating integrals of multiply- or piecewise-defined functions we have to split the interval of integration into sub-intervals and write the integral as the sum of integrals over the respective subintervals with the corresponding values of the function taken as the integrands. Thus,

$$\int_{-\pi}^{\pi}f(x)\,dx=\int_{-\pi}^{0}0.dx+\int_0^{\pi}\cos x\,dx;\ \int_{-2}^{2}g(x)\,dx$$

$$=\int_{-2}^{-1}1.dx+\int_{-1}^{0}0.dx+\int_0^1(1-x)\,dx+\int_1^2(1+x)\,dx$$

and so on.

Example 8.1

Find the Fourier series of the function $f(x)=x$ in $(0, 2\pi)$. [JNTU Nov. 2003]

Solution The Fourier series of the function

$$f(x)=x \text{ in } (0,\, 2\pi) \tag{1}$$

is given by

$$f(x)=x=\frac{a_0}{2}+\sum_{n=1}^{\infty}a_n\cos nx+\sum_{n=1}^{\infty}b_n\sin nx \tag{2}$$

where

$$a_0=\frac{1}{\pi}\int_0^{2\pi}f(x)\,dx=\frac{1}{\pi}\int_0^{2\pi}x\,dx=\frac{1}{\pi}\left(\frac{x^2}{2}\right)_0^{2\pi}$$

$$=\frac{1}{\pi}\cdot\frac{4\pi^2}{2}=2\pi \tag{3}$$

$$a_n = \frac{1}{\pi}\int_0^{2\pi} f(x)\cos nx\, dx = \frac{1}{\pi}\int_0^{2\pi} x\cos nx\, dx$$

$$= \frac{1}{\pi}\left[\frac{x\sin nx}{n} - 1\cdot\left(-\frac{\cos nx}{n^2}\right)\right]_0^{2\pi}$$

$$= \frac{1}{\pi}\left[\frac{1}{n}(2\pi\sin 2n\pi - 0.\sin 0)\right.$$

$$\left. + \frac{1}{n^2}(\cos 2n\pi - \cos 0)\right]$$

$$= \frac{1}{\pi}\left[\frac{1}{n}\cdot 0 + \frac{1}{n^2}\cdot 0\right] = 0 \tag{4}$$

$$b_n = \frac{1}{\pi}\int_0^{2\pi} f(x)\sin nx\, dx = \frac{1}{\pi}\int_0^{2\pi} x\sin nx\, dx$$

$$= \frac{1}{\pi}\left[x\left(-\frac{\cos nx}{n}\right) - 1\cdot\left(-\frac{\sin nx}{n^2}\right)\right]_0^{2\pi}$$

$$= \frac{1}{\pi}\left[\frac{1}{n}(-2\pi\cdot\cos 2n\pi + 0\cdot\cos 0)\right.$$

$$\left. + \frac{1}{n^2}(\sin 2n\pi - \sin 0)\right] = -\frac{2}{n} \tag{5}$$

Substituting the values of the Fourier coefficients a_0, a_n and b_n in (2), we have

$$x = \pi - 2\sum_{n=1}^{\infty}\frac{\sin nx}{n} \text{ in } (0, 2\pi) \tag{6}$$

Deductions

Multiplying (6) by b and adding a we obtain the Fourier series in $(0, 2\pi)$ for $a + bx$.

$$a + bx = a + b\pi - 2b\sum_{n=1}^{\infty}\frac{\sin nx}{n} \tag{7}$$

Putting $a = 1$, $b = 1$ in (7) we have the Fourier series for $(1 + x)$:

$$1 + x = 1 + \pi - 2\sum_{n=1}^{\infty}\frac{\sin nx}{n} \tag{8}$$

Putting $a = 1$, $b = -1$ in (7) we have the Fourier series for $(1 - x)$:

$$1 - x = 1 - \pi + 2\sum_{n=1}^{\infty}\frac{\sin nx}{n} \tag{9}$$

Putting $a = \frac{\pi}{2}$, $b = \frac{-1}{2}$ in (7) we have the Fourier series in $(0, 2\pi)$ for $\left(\frac{\pi - x}{2}\right)$:

$$\frac{\pi - x}{2} = \sum_{n=1}^{\infty}\frac{\sin nx}{n} = \sin x + \frac{\sin 2x}{2} + \frac{\sin 3x}{3} + \cdots \tag{10}$$

At $x = \pi/2$, a point of continuity we obtain

$$\frac{\pi}{2} - \frac{\pi}{4} = \frac{\pi}{4} = \sum_{n=1}^{\infty}\frac{1}{n}\sin\frac{n\pi}{2}$$

$$= \sum_{m=0}^{\infty}\frac{1}{2m+1}\sin\,(2m+1)\frac{\pi}{2} = \sum_{m=0}^{\infty}\frac{(-1)^m}{2m+1}$$

$$\left(\because \quad \sin\frac{n\pi}{2} = 0 \quad \text{when } n = 2m \text{ (even)}\right)$$

$$\Rightarrow \quad 1 - \frac{1}{3} + \frac{1}{5} - \frac{1}{7} + \cdots = \frac{\pi}{4}$$

Example 8.2

Find the Fourier series of the function $f(x) = x^2$ in $(0, 2\pi)$.

Solution The Fourier series of the function

$$f(x) = x^2 \text{ in } (0, 2\pi) \tag{1}$$

is given by

$$f(x) = x^2 = \frac{a_0}{2} + \sum_{n=1}^{\infty} a_n\cos nx + \sum_{n=1}^{\infty} b_n\sin nx \tag{2}$$

where

$$a_0 = \frac{1}{\pi}\int_0^{2\pi} f(x)\,dx = \frac{1}{\pi}\int_0^{2\pi} x^2\,dx = \frac{1}{\pi}\left(\frac{x^3}{3}\right)_0^{2\pi} \tag{3}$$

$$= \frac{1}{\pi}\cdot\frac{8\pi^3}{3} = \frac{8\pi^2}{3}$$

$$a_n = \frac{1}{\pi}\int_0^{2\pi} f(x)\cos nx\,dx = \frac{1}{\pi}\int_0^{2\pi} x^2\cos nx\,dx \tag{4}$$

$$= \frac{1}{\pi}\left[x^2\cdot\left(\frac{\sin nx}{n}\right) - 2x\cdot\left(-\frac{\cos nx}{n^2}\right)\right.$$

$$\left. + 2\cdot\left(-\frac{\sin nx}{n^3}\right)\right]_0^{2\pi}$$

$$= \frac{1}{\pi}\frac{2 \cdot 2\pi \cdot \cos 2n\pi}{n^2} = \frac{4}{n^2}; \qquad t_1 = t_3 = 0.$$

$\because \quad \sin nx = 0$ at $x = 2\pi,\ 0;$

$t_2 = 0$ at $x = 0$ and $\cos 2n\pi = 1$

$$b_n = \frac{1}{\pi}\int_0^{2\pi} f(x)\sin nx\, dx = \frac{1}{\pi}\int_0^{2\pi} x^2 \sin nx\, dx$$

$$= \frac{-1}{\pi}\left[x^2 \cdot \left(-\frac{\cos nx}{n}\right) - 2x \cdot \left(-\frac{\sin nx}{n^2}\right) + 2 \cdot \left(\frac{\cos nx}{n^3}\right)\right]_0^{2\pi}$$

$$= \frac{1}{\pi}\left[\frac{-1}{n}(4\pi^2 \cos 2n\pi - 0 \cdot \cos 0) + 0 + \frac{2}{n^3}(\cos 2n\pi - \cos 0)\right] = -\frac{4\pi}{n} \qquad (5)$$

$\because \quad \sin nx = 0$ at $x = 2\pi, 0;$

$t_2 = 0$ at $x = 0,$ and

$\cos 2n\pi = \cos 0 = 1$

Substituting the values of the Fourier coefficients a_0, a_n and b_n in (2), we have

$$x^2 = \frac{4\pi^2}{3} + 4\sum_{n=1}^{\infty}\frac{\cos nx}{n^2} - 4\pi\sum_{n=1}^{\infty}\frac{\sin nx}{n} \qquad (6)$$

$$\Rightarrow \quad \frac{x^2}{4} = \frac{\pi^2}{3} + \sum_{n=1}^{\infty}\frac{\cos nx}{n^2} - \pi\sum_{n=1}^{\infty}\frac{\sin nx}{n} \qquad (7)$$

The Fourier series expansion for x in $(0, 2\pi)$ is

$$x = \pi - 2\sum_{n=1}^{\infty}\frac{\sin nx}{n} \qquad (8)$$

We now have the Fourier series for $\left(\frac{\pi - x}{2}\right)^2$ in $(0, 2\pi)$

$$\left(\frac{\pi - x}{2}\right)^2 = \frac{\pi^2}{4} - \frac{\pi x}{2} + \frac{x^2}{4}$$

$$= \frac{\pi^2}{4} - \frac{\pi^2}{2} + \pi\sum_{n=1}^{\infty}\frac{\sin nx}{n} + \frac{\pi^2}{3} + \sum_{n=1}^{\infty}\frac{\cos nx}{n^2} - \pi\sum_{n=1}^{\infty}\frac{\sin nx}{n}, \quad \text{by (5) and (6)}$$

$$\Rightarrow \quad \left(\frac{\pi - x}{2}\right)^2 = \frac{\pi^2}{12} + \sum_{n=1}^{\infty}\frac{\cos nx}{n^2} \qquad (9)$$

Note 1 To obtain (9) directly take $f(x) = ax + bx^2$ and put $a = -\pi/2,\ b = 1/4$ and add $\pi^2/4$ on both the sides.

Example 8.3

Find the Fourier series of the function $f(x) = e^{ax}$ in $(0, 2\pi)$.

Solution The Fourier series of the function

$$f(x) = e^{ax} \text{ in } (0, 2\pi) \qquad (1)$$

is given by

$$f(x) = e^{ax} = \frac{a_0}{2} + \sum_{n=1}^{\infty} a_n \cos nx + \sum_{n=1}^{\infty} b_n \sin nx \qquad (2)$$

$$\text{where} \quad a_0 = \frac{1}{\pi}\int_0^{2\pi} f(x)\, dx = \frac{1}{\pi}\int_0^{2\pi} e^{ax} dx$$

$$= \frac{1}{\pi}\left(\frac{e^{ax}}{a}\right)_0^{2\pi} = \frac{1}{a\pi}(e^{2a\pi} - 1) \qquad (3)$$

$$a_n = \frac{1}{\pi}\int_0^{2\pi} f(x)\cos nx\, dx = \frac{1}{\pi}\int_0^{2\pi} e^{ax}\cos nx\, dx$$

$$= \frac{1}{\pi}\left[\frac{e^{ax}}{a^2 + n^2}(a\cos nx + n\sin nx)\right]_0^{2\pi}$$

$$= \frac{1}{\pi} \cdot \frac{1}{a^2 + n^2}\left[e^{2a\pi}(a\cos 2n\pi + n\sin 2n\pi) - e^0(a \cdot 1 + 0)\right] = \frac{1}{\pi} \cdot \frac{1}{a^2 + n^2} \cdot a \cdot (e^{2a\pi} - 1) \qquad (4)$$

$\because \quad \cos 2n\pi = 1,\ \sin 2n\pi = 0,\ e^0 = 1$

$$b_n = \frac{1}{\pi}\int_0^{2\pi} f(x)\sin nx\, dx = \frac{1}{\pi}\int_0^{2\pi} e^{ax}\sin nx\, dx$$

$$= \frac{1}{\pi}\left[\frac{e^{ax}}{a^2 + n^2}(a\sin nx - n\cos nx)\right]_0^{2\pi}$$

$$= \frac{1}{\pi} \cdot \frac{1}{a^2 + n^2}\left[e^{2a\pi}(a \cdot 0 - n \cdot 1) - e^0(a \cdot 0 - n \cdot 1)\right]$$

$$= \frac{1}{\pi} \cdot \frac{(-n)}{a^2 + n^2}(e^{2a\pi} - 1) \qquad (5)$$

Substituting the values of the Fourier coefficients a_0, a_n, and b_n in (2) we have

$$e^{ax} = \frac{1}{2a\pi}(e^{2a\pi}-1) + \frac{1}{\pi}(e^{2a\pi}-1)\sum_{n=1}^{\infty}\frac{1}{a^2+n^2}(a\cos nx - n\sin nx)$$

$$\text{or } e^{ax} = \frac{e^{2a\pi}-1}{\pi}\left[\frac{1}{2a}+\sum_{n=1}^{\infty}\left(\frac{a\cos nx - n\sin nx}{a^2+n^2}\right)\right] \quad (6)$$

Deductions

Replacing a by $-a$ in (6) we have the Fourier series for e^{-ax}:

$$1.\ e^{-ax} = \frac{1-e^{-2a\pi}}{\pi}\left[\frac{1}{2a}+\sum_{n=1}^{\infty}\left(\frac{a\cos nx + n\sin nx}{a^2+n^2}\right)\right] \quad (7)$$

From (6) and (7) the Fourier series for $\cosh ax = \frac{e^{ax}+e^{-ax}}{2}$ and $\sinh ax = \frac{e^{ax}-e^{-ax}}{2}$ are

$$2.\ \cosh ax = \frac{\sinh 2a\pi}{\pi}\left(\frac{1}{2a}+\sum_{n=1}^{\infty}\frac{a\cos nx}{a^2+n^2}\right) - \frac{\cosh 2a\pi - 1}{\pi}\sum_{n=1}^{\infty}\frac{n\sin nx}{a^2+n^2} \quad (8)$$

$$3.\ \sinh ax = \frac{\cosh 2a\pi - 1}{\pi}\left(\frac{1}{2a}+\sum_{n=1}^{\infty}\frac{a\cos nx}{a^2+n^2}\right) - \frac{\sinh 2a\pi}{\pi}\sum_{n=1}^{\infty}\frac{n\sin nx}{a^2+n^2} \quad (9)$$

Also, for $a = 1$ and $a = -1$ we have

$$4.\ e^{x} = \frac{(e^{2\pi}-1)}{\pi}\left[\frac{1}{2}+\sum_{n=1}^{\infty}\left(\frac{\cos nx - n\sin nx}{1+n^2}\right)\right] \quad (10)$$

$$5.\ e^{-x} = \frac{(1-e^{-2\pi})}{\pi}\left[\frac{1}{2}+\sum_{n=1}^{\infty}\left(\frac{\cos nx + n\sin nx}{1+n^2}\right)\right] \quad (11)$$

Example 8.4

Expand $f(x) = x\sin x$ as a Fourier series in $(0, 2\pi)$.

[JNTU, May/June 2004, 1993]

Solution The Fourier series of the function

$$f(x) = x\sin x \text{ in } (0, 2\pi) \quad (1)$$

is given by

$$f(x) = x\sin x = \frac{a_0}{2}+\sum_{n=1}^{\infty}a_n\cos nx + \sum_{n=1}^{\infty}b_n\sin nx \quad (2)$$

where

$$a_0 = \frac{1}{\pi}\int_0^{2\pi} f(x)\,dx = \frac{1}{\pi}\int_0^{2\pi} x\sin x\,dx$$

$$= \frac{1}{\pi}[x(-\cos x) - 1(-\sin x)]_0^{2\pi} = \frac{1}{\pi}[(-2\pi\cos 2\pi + 0\cdot\cos 0) + (\sin 2\pi - \sin 0)] = -2 \quad (3)$$

$\because\ \cos 2\pi = \cos 0 = 1,\ \sin 2\pi = \sin 0 = 0$

$$a_n = \frac{1}{\pi}\int_0^{2\pi} f(x)\cos nx\,dx = \frac{1}{\pi}\int_0^{2\pi} x\sin x\cos nx\,dx$$

$$= \frac{1}{2\pi}\int_0^{2\pi} x[\sin(n+1)x - \sin(n-1)x]\,dx$$

$$= \frac{1}{2\pi}\left[x\left(-\frac{\cos(n+1)x}{n+1}\right) - 1\cdot\left(-\frac{\sin(n+1)x}{(n+1)^2}\right)\right]_0^{2\pi} - \frac{1}{2\pi}\left[x\left(-\frac{\cos(n-1)x}{n-1}\right) - 1\cdot\left(-\frac{\sin(n-1)x}{(n-1)^2}\right)\right]_0^{2\pi};$$

$$(n \neq 1)$$

(All terms vanish except cosines at $x = 2\pi$

$\cos(n+1)\,2\pi = \cos(n-1)\,2\pi = 1$

$$= \frac{-1}{2\pi}\cdot 2\pi\left(\frac{1}{n+1}-\frac{1}{n-1}\right) = \frac{2}{n^2-1} \quad (4)$$

For $n = 1$

$$a_1 = \frac{1}{\pi}\int_0^{2\pi} x\sin x\cos x\,dx = \frac{1}{2\pi}\int_0^{2\pi} x\sin 2x\,dx$$

$$= \frac{1}{2\pi}\int_0^{2\pi}\left[x\left(-\frac{\cos 2x}{2}\right) - 1\left(-\frac{\sin 2x}{2^2}\right)\right]_0^{2\pi}$$

$$= \frac{1}{2\pi}\cdot 2\pi\left(-\frac{1}{2}\right) = -\frac{1}{2} \quad (5)$$

(All terms vanish except $\cos 2x = 1$ at $x = 2\pi$)

$$b_n = \frac{1}{\pi}\int_0^{2\pi} x\sin x \sin nx\, dx$$

$$= \frac{1}{2\pi}\int_0^{2\pi} x[\cos(n-1)x - \cos(n+1)x]\, dx$$

$$= \frac{1}{2\pi}\left[x\cdot\frac{\sin(n-1)x}{n-1} - 1\cdot\frac{-\cos(n-1)x}{(n-1)^2}\right.$$

$$\left. - x\cdot\frac{\sin(n+1)x}{n+1} - 1\cdot\frac{-\cos(n+1)x}{n+1}\right]_0^{2\pi} = 0,$$

$$(n \neq 1) \quad (6)$$

For $n = 1$

$$b_1 = \frac{1}{\pi}\int_0^{2\pi} x\sin x\cdot \sin x\, dx = \frac{1}{2\pi}\int_0^{2\pi}(x - x\cos 2x)\, dx$$

$$= \frac{1}{2\pi}\left[\frac{x^2}{2} - x\left(\frac{\sin 2x}{2}\right) + 1\cdot\left(-\frac{\cos 2x}{2^2}\right)\right]_0^{2\pi}$$

$$= \frac{1}{2\pi}\left[\frac{4\pi^2 - 0}{2} - \frac{2\pi\sin 4\pi - 0}{2} - \frac{1}{2^2}(\cos 4\pi - \cos 0)\right]$$

$$= \pi, \quad \because \quad \cos 4\pi = \cos 0 = 1; \quad \sin 4\pi = 0 \quad (7)$$

Substituting these values in (2) we obtain

$$x\sin x = -1 - \frac{1}{2}\cos x + \pi\cdot\sin x + 2\sum_{n=2}^{\infty}\frac{\cos nx}{n^2 - 1}\cdots \quad (8)$$

The function is continuous at $x = \pi/2$ and hence from (8) we obtain

$$\frac{\pi}{2}\sin\frac{\pi}{2} = -1 - \frac{1}{2}\cos\frac{\pi}{2} + \pi\cdot\sin\frac{\pi}{2} + 2\sum_{n=2}^{\infty}\frac{\cos(n\pi/2)}{(n-1)(n+1)}$$

$$\cos\frac{n\pi}{2} = \begin{cases} 0, & \text{if } n \text{ is odd}; \quad n = \frac{2m+1}{2}\pi \\ (-1)^m, & \text{if } n \text{ is even}; \quad n = 2m \end{cases}$$

$$\Rightarrow \quad \frac{\pi - 2}{4} = -\sum_{m=1}^{\infty}\frac{\cos m\pi}{(2m-1)(2m+1)} = \frac{1}{1.3} - \frac{1}{3.5} + \frac{1}{5.7}\cdots$$

$$(9)$$

Example 8.5

Obtain the Fourier series of $f(x) = \cos\alpha x$, in $(0, 2\pi)$ $(\alpha \notin Z)$.

Solution The Fourier series of the function

$$f(x) = \cos\alpha x, \quad \text{in } (0, 2\pi) \quad (\alpha \notin Z) \quad (1)$$

is given by

$$f(x) = \cos\alpha x = \frac{a_0}{2} + \sum_{n=1}^{\infty} a_n\cos nx + \sum_{n=1}^{\infty} b_n \sin nx \ (2)$$

where

$$a_0 = \frac{1}{\pi}\int_0^{2\pi} f(x)\, dx = \frac{1}{\pi}\int_0^{2\pi}\cos\alpha x = \frac{1}{\pi}\left(\frac{\sin\alpha x}{\alpha}\right)_0^{2\pi}$$

$$= \frac{1}{\pi\alpha}(\sin 2\alpha\pi - \sin 0) = \frac{\sin 2\alpha\pi}{\pi\alpha} \quad (3)$$

$$a_n = \frac{1}{\pi}\int_0^{2\pi} f(x)\cos nx\, dx = \frac{1}{\pi}\int_0^{2\pi}\cos\alpha x\cos nx\, dx$$

$$= \frac{1}{2\pi}\int_0^{2\pi}[\cos(n+\alpha)x + \cos(n-\alpha)x]\, dx$$

$$= \frac{1}{2\pi}\left[\frac{\sin(n+\alpha)x}{(n+\alpha)} + \frac{\sin(n-\alpha)x}{(n-\alpha)}\right]_0^{2\pi}$$

$$= \frac{1}{2\pi}\left[\frac{\sin(2n\pi + 2\pi\alpha)}{(n+\alpha)} + \frac{\sin(2n\pi - 2\pi\alpha)}{(n-\alpha)}\right] \quad (4)$$

$$= \frac{\sin 2\pi\alpha}{2\pi}\left[\frac{1}{n+\alpha} - \frac{1}{n-\alpha}\right] = -\frac{\alpha\sin 2\pi\alpha}{\pi(n^2 - \alpha^2)}$$

$$\left[\begin{array}{l} \because \ \sin(\pi \pm \alpha)2\pi \\ = \sin(2n\pi \pm 2\pi\alpha) \\ = \ \pm\sin 2\pi\alpha \end{array}\right]$$

$$b_n = \frac{1}{\pi}\int_0^{2\pi} f(x)\sin nx\, dx = \frac{1}{\pi}\int_0^{2\pi}\cos\alpha x\sin nx\, dx$$

$$= \frac{1}{2\pi}\int_0^{2\pi}[\sin(n+\alpha)x + \sin(n-\alpha)x]\, dx$$

$$= \frac{-1}{2\pi}\left[\frac{\cos(n+\alpha)x}{(n+\alpha)} + \frac{\cos(n-\alpha)x}{(n-\alpha)}\right]_0^{2\pi}$$

$$= \frac{-1}{2\pi}\left[\frac{\cos 2\pi\alpha - 1}{(n+\alpha)} + \frac{\cos 2\pi\alpha - 1}{(n-\alpha)}\right] \quad (5)$$

$$= \frac{(1 - \cos 2\pi\alpha)}{\pi}\left(\frac{n}{n^2 - \alpha^2}\right)$$

$$\left[\begin{array}{l} \because \ \cos(2n\pi \pm 2\pi\alpha) \\ = \cos 2\pi\alpha \end{array}\right]$$

Substituting these values the Fourier series is

$$\cos\alpha x = \frac{\sin 2\pi\alpha}{2\pi\alpha} - \frac{\alpha\sin 2\pi\alpha}{\pi}\sum_{n=1}^{\infty}\frac{\cos nx}{(n^2-\alpha^2)} + \frac{(1-\cos 2\pi\alpha)}{\pi}\sum_{n=1}^{\infty}\frac{n\sin nx}{(n^2-\alpha^2)} \qquad (6)$$

Example 8.6

Find a Fourier series for $f(x)=\sqrt{1-\cos x}$ in $(0, 2\pi)$ and deduce that $\frac{1}{1.3}+\frac{1}{3.5}+\frac{1}{5.7}+\cdots=\frac{1}{2}$.

Solution The Fourier series for

$$f(x)=\sqrt{1-\cos x} \text{ in } (0,2\pi) \qquad (1)$$

is given by

$$f(x)=\sqrt{1-\cos x}=\sqrt{2}\sin\frac{x}{2} = \frac{a_0}{2}+\sum_{n=1}^{\infty}a_n\cos nx+\sum_{n=1}^{\infty}b_n\sin nx \qquad (2)$$

where

$$a_0=\frac{1}{\pi}\int_0^{2\pi}\sqrt{2}\sin\frac{x}{2}dx=\frac{\sqrt{2}}{\pi}\left(-2\cos\frac{x}{2}\right)_0^{2\pi} = \frac{-2\sqrt{2}}{\pi}(\cos\pi-\cos 0)=\frac{4\sqrt{2}}{\pi} \qquad (3)$$

$$a_n=\frac{1}{\pi}\int_0^{2\pi}\sqrt{2}\sin\frac{x}{2}\cos nx\,dx$$

$$=\frac{1}{\sqrt{2}\pi}\int_0^{2\pi}\left[\sin\left(n+\frac{1}{2}\right)x-\sin\left(n-\frac{1}{2}\right)x\right]dx$$

$$=\frac{1}{\sqrt{2}\pi}\left[-\frac{\cos(n+\frac{1}{2})x}{(n+\frac{1}{2})}+\frac{\cos(n-\frac{1}{2})x}{(n-\frac{1}{2})}\right]_0^{2\pi}$$

$$=\frac{1}{\sqrt{2}\pi}\left[-\frac{\cos(2n+1)\pi-\cos 0}{(n+\frac{1}{2})}+\frac{\cos(2n-1)\pi-\cos 0}{(n-\frac{1}{2})}\right]$$

$$=\frac{1}{\sqrt{2}\pi}\left(\frac{4}{2n+1}-\frac{4}{2n-1}\right)=\frac{-4\sqrt{2}}{(4n^2-1)\pi}$$

$$\left[\because \cos(2n+1)\pi=\cos(2n-1)\pi=-1,\ \cos 0=1\right] \qquad (4)$$

$$b_n=\frac{1}{\pi}\int_0^{2\pi}\sqrt{2}\sin\frac{x}{2}\sin nx\,dx$$

$$=\frac{1}{\sqrt{2}\pi}\int_0^{2\pi}\left[\cos\left(n-\frac{1}{2}\right)x-\cos\left(n+\frac{1}{2}\right)x\right]dx$$

$$=\frac{1}{\sqrt{2}\pi}\left[\frac{\sin(n-\frac{1}{2})x}{(n-\frac{1}{2})}-\frac{\sin(n+\frac{1}{2})x}{(n+\frac{1}{2})}\right]_0^{2\pi}$$

$$=\frac{1}{\sqrt{2}\pi}\left[\frac{\sin(2n-1)\pi-\sin 0}{(n-\frac{1}{2})} - \frac{\sin(2n+1)\pi-\sin 0}{(n+\frac{1}{2})}\right]=0 \qquad (5)$$

$$[\because \sin(2n\pm 1)\pi=\sin 0=0]$$

Substituting these values in (2), we obtain

$$\sqrt{1-\cos x}=\frac{2\sqrt{2}}{\pi}-\frac{4\sqrt{2}}{\pi}\sum_{n=1}^{\infty}\frac{\cos nx}{4n^2-1} \qquad (6)$$

$f(x)=\sqrt{1-\cos x}$ is continuous at $x = 0$ and the series converges to $f(0)=0$

$$\therefore \quad 0=\frac{2\sqrt{2}}{\pi}-\frac{4\sqrt{2}}{\pi}\sum_{n=1}^{\infty}\frac{1}{(2n-1)(2n+1)} \qquad (7)$$

$$\Rightarrow \quad \frac{1}{2}=\frac{1}{1.3}+\frac{1}{3.5}+\frac{1}{5.7}+\cdots$$

Also, at $x=\frac{\pi}{2}$

$$1=\frac{2\sqrt{2}}{\pi}-\frac{4\sqrt{2}}{\pi}\sum_{n=1}^{\infty}\frac{\cos\frac{n\pi}{2}}{(2n-1)(2n+1)}; \qquad (8)$$

$$\cos\frac{n\pi}{2}=\begin{bmatrix}0, & \text{if } n=2m+1, \text{ (odd)}\\ (-1)^m, & \text{if } n=2m \text{ (even)}\end{bmatrix}$$

Multiplying by $\pi/4\sqrt{2}$ and transposing,

$$\frac{\pi}{4\sqrt{2}}-\frac{1}{2}=\sum_{m=1}^{\infty}\frac{-(-1)^m}{(4m-1)(4m+1)}$$

$$\Rightarrow \quad \frac{\pi}{4\sqrt{2}}-\frac{1}{2}=\frac{1}{3.5}-\frac{1}{7.9}+\frac{1}{11.13}-\frac{1}{15.17}+\cdots \qquad (9)$$

Example 8.7
Obtain a Fourier series for the function $f(x) = ax + bx^2$ in $(-\pi, \pi)$.

Solution The Fourier series of

$$f(x) = ax + bx^2 \text{ in } (-\pi, \pi) \quad (1)$$

is given by

$$f(x) = ax + bx^2 = \frac{a_0}{2} + \sum_{n=1}^{\infty} a_n \cos nx + \sum_{n=1}^{\infty} b_n \sin nx \quad (2)$$

where

$$a_0 = \frac{1}{\pi}\int_{-\pi}^{\pi}(ax+bx^2)\,dx = \frac{1}{\pi}\left(\frac{a}{2}x^2 + \frac{b}{3}x^3\right)_{-\pi}^{\pi} \quad (3)$$

$$= \frac{1}{\pi}\left[\frac{a}{2}(\pi^2 - (-\pi)^2) + \frac{b}{3}(\pi^3 - (-\pi)^3)\right] = \frac{2b\pi^2}{3}$$

$$a_n = \frac{1}{\pi}\int_{-\pi}^{\pi}(ax+bx^2)\cos nx\,dx = 0 + \frac{2b}{\pi}\int_0^{\pi} x^2 \cos nx\,dx$$

$$= \frac{2b}{\pi}\left[x^2\left(\frac{\sin nx}{n}\right) - 2x\cdot\left(-\frac{\cos nx}{n^2}\right) + 2\left(-\frac{\sin nx}{n^3}\right)\right]_0^{\pi}$$

[$x\cos nx$ is an odd function and $x^2 \cos nx$ is an even funtion]

$$= \frac{4b}{\pi n^2}(\pi\cos n\pi - 0\cdot\cos 0) = \frac{4b}{n^2}(-1)^n$$

$$[\because \sin nx = 0 \text{ at } \pi, 0;\ \cos n\pi = (-1)^n] \quad (4)$$

$$b_n = \frac{1}{\pi}\int_{-\pi}^{\pi}(ax+bx^2)\sin nx\,dx = \frac{1}{\pi}\cdot 2a\int_0^{\pi} x\sin nx\,dx + 0$$

$$= \frac{2a}{\pi}\left[x\left(-\frac{\cos nx}{n}\right) - 1\cdot\left(-\frac{\sin nx}{n^2}\right)\right]_0^{\pi}$$

[$x\sin nx$ is an even function and $x^2 \sin nx$ is an odd funtion]

$$= \frac{-2a}{\pi n}(\pi\cos n\pi - 0\cdot\cos 0) + 0 \quad (5)$$

$$= \frac{2a}{n}(-1)^{n+1};\ [\sin nx = 0 \text{ at } x = \pi, 0;\ \cos n\pi = (-1)^n]$$

Substituting these values, we have

$$ax + bx^2 = \frac{b\pi^2}{3} + 4b\sum_{n=1}^{\infty}\frac{(-1)^n \cos nx}{n^2} - 2a\sum_{n=1}^{\infty}\frac{(-1)^n \sin nx}{n} \quad (6)$$

Deductions

(i) $a = 1, b = 1$

$$x + x^2 = \frac{\pi^2}{3} + 4\sum_{n=1}^{\infty}\frac{(-1)^n \cos nx}{n^2} - 2\sum_{n=1}^{\infty}\frac{(-1)^n \sin nx}{n} \quad (7)$$

(ii) $a = 1, b = 21$

$$x - x^2 = -\frac{\pi^2}{3} + 4\sum_{n=1}^{\infty}\frac{(-1)^{n+1} \cos nx}{n^2} + 2\sum_{n=1}^{\infty}\frac{(-1)^{n+1} \sin nx}{n} \quad (8)$$

(iii) $f(x)$ is continuous at $x = 0$ and the series converges to $f(0)$. So from (8) we obtain

$$0 = \frac{-\pi^2}{3} + 4\sum_{n=1}^{\infty}\frac{(-1)^{n+1}}{n^2}$$

$$\Rightarrow \frac{\pi^2}{12} = \frac{1}{1^2} - \frac{1}{2^2} + \frac{1}{3^2} - \frac{1}{4^2} + \frac{1}{5^2} - \frac{1}{6^2} + \cdots \quad (9)$$

(iv) Putting $a = 0$, $b = -1/4$ and adding $\pi^2/12$ to both sides of (6) we obtain

$$\frac{\pi^2}{12} - \frac{x^2}{4} = \sum_{n=1}^{\infty}\frac{(-1)^{n+1}}{n^2}\cos nx \quad (10)$$

(v) Putting $a = -\pi/2$, $b = 1/4$ and adding $\pi^2/12$ to both sides of (6) we obtain

$$\left(\frac{\pi - x}{2}\right)^2 = \frac{\pi^2}{12} + \sum_{n=1}^{\infty} \frac{\cos nx}{n^2} \qquad (11)$$

Example 8.8

Find the Fourier series of $f(x) = e^{ax}$ in $(-\pi, \pi)$.

Solution The Fourier series of $f(x) = e^{ax}$ in $(-\pi, \pi)$ is (1)

$$f(x) = e^{ax} = \frac{a_0}{2} + \sum_{n=1}^{\infty} a_n \cos nx + \sum_{n=1}^{\infty} b_n \sin nx \qquad (2)$$

where $$a_0 = \frac{1}{\pi}\int_{-\pi}^{\pi} e^{ax}\, dx = \frac{1}{\pi}\left(\frac{e^{ax}}{a}\right)_{-\pi}^{\pi} \qquad (3)$$

$$= \frac{1}{\pi a}(e^{a\pi} - e^{-a\pi}) = \frac{2\sinh a\pi}{\pi a}$$

$$a_n = \frac{1}{\pi}\int_{-\pi}^{\pi} e^{ax} \cos nx\, dx$$

$$= \frac{1}{\pi}\left[\frac{e^{ax}}{a^2 + n^2}(a\cos nx + n \sin nx)\right]_{-\pi}^{\pi}$$

$$= \frac{1}{\pi}\cdot\frac{a}{a^2+n^2}\cdot\left[e^{a\pi}\cos n\pi - e^{-a\pi}\cos(-n\pi)\right]$$

$$= \frac{a}{\pi(a^2+n^2)}\cos n\pi\,(e^{a\pi} - e^{-a\pi})$$

$$= \frac{2a(-1)^n}{\pi(a^2+n^2)}\cdot \sinh a\pi \qquad (4)$$

$[\because \sin n\pi = 0, \quad \because \cos n\pi = (-1)^n]$

$$b_n = \frac{1}{\pi}\int_{-\pi}^{\pi} e^{ax} \sin nx\, dx$$

$$= \frac{1}{\pi}\left[\frac{e^{ax}}{a^2+n^2}(a \sin nx - n \cos nx)\right]_{-\pi}^{\pi}$$

$$= \frac{1}{\pi}\cdot\frac{(-n)(-1)^n}{a^2+n^2}\cdot(e^{a\pi} - e^{-a\pi}), \qquad (5)$$

$$= \frac{n(-1)^{n+1}}{\pi(a^2+n^2)} 2\sinh a\pi$$

$[\because \cos n\pi = (-1)^n, \ \sin n\pi = 0]$

Substituting these values in (2) we obtain

$$e^{ax} = \frac{2\sinh a\pi}{\pi}\left[\frac{1}{2a} + a\sum_{n=1}^{\infty}\frac{(-1)^n}{a^2+n^2}\cos nx - \sum_{n=1}^{\infty}\frac{n(-1)^n}{a^2+n^2}\sin nx\right] \qquad (6)$$

or $$\frac{\pi}{2\sinh a\pi}e^{ax} = \frac{1}{2a} + \sum_{n=1}^{\infty}\frac{(-1)^n}{a^2+n^2}(a\cos nx - n \sin nx) \qquad (7)$$

Taking $a = 1$ we have

$$\frac{\pi}{2\sinh\pi}e^{x} = \frac{1}{2} + \sum_{n=1}^{\infty}\frac{(-1)^n}{1+n^2}(\cos nx - n\sin nx)$$

[JNTU 1995] (8)

Half of the sum of the series

$$f[at\ x = \pm\,\pi] = \frac{1}{2}[f(-\pi+0) + f(\pi - 0)]$$

$$= \frac{1}{2}\frac{\pi}{2\sinh\pi}(e^{-\pi} + e^{\pi}) = \frac{\pi}{2}\coth\pi \qquad (9)$$

Example 8.9

Find the Fourier series for $f(x) = \sqrt{1-\cos x}$ in $(-\pi, \pi)$.

Solution The given function is

$$f(x) = \sqrt{1-\cos x} = \begin{cases} -\sqrt{2}\sin\frac{x}{2}, & (-\pi, 0) \\ \sqrt{2}\sin\frac{x}{2}, & (\pi, 0) \end{cases} \qquad (1)$$

The Fourier series is

$$f(x) = \frac{a_0}{2} + \sum_{n=1}^{\infty} a_n \cos nx + \sum_{n=1}^{\infty} b_n \sin nx \qquad (2)$$

where

$$a_0 = \frac{1}{\pi}\int_{-\pi}^{\pi} f(x)\,dx = \frac{1}{\pi}\int_{-\pi}^{0}\left(-\sqrt{2}\sin\frac{x}{2}\right)dx + \frac{1}{\pi}\int_{0}^{\pi}\left(\sqrt{2}\sin\frac{x}{2}\right)dx$$

$$= \frac{1}{\pi}\int_{0}^{\pi}\sqrt{2}\sin\frac{x}{2}\,dx + \frac{1}{\pi}\int_{0}^{\pi}\sqrt{2}\sin\frac{x}{2}\,dx \qquad (3)$$

$$= \frac{2\sqrt{2}}{\pi}\int_0^{\pi} \sin\frac{x}{2}\,dx = \frac{2\sqrt{2}}{\pi}\left(-2\cos\frac{x}{2}\right)_0^{\pi} = \frac{4\sqrt{2}}{\pi}$$

$$a_n = \frac{1}{\pi}\int_{-\pi}^{\pi} f(x)\cos nx\,dx$$

$$= \frac{1}{\pi}\int_{-\pi}^{0} -\sqrt{2}\sin\frac{x}{2}\cos nx\,dx + \frac{1}{\pi}\int_0^{\pi} \sqrt{2}\sin\frac{x}{2}\cos nx\,dx$$

$$= \frac{2\sqrt{2}}{\pi}\int_0^{\pi} \sin\frac{x}{2}\cos nx\,dx$$

$$= \frac{\sqrt{2}}{\pi}\int_0^{\pi}\left[\sin\left(n+\frac{1}{2}\right)x - \sin\left(n-\frac{1}{2}\right)x\right]dx$$

$$= \frac{\sqrt{2}}{\pi}\left[-\frac{\cos\left(n+\frac{1}{2}\right)x}{\left(n+\frac{1}{2}\right)} + \frac{\cos\left(n-\frac{1}{2}\right)x}{\left(n-\frac{1}{2}\right)}\right]_0^{\pi}$$

$$= \frac{\sqrt{2}}{\pi}\left(\frac{1}{n+\frac{1}{2}} - \frac{1}{n-\frac{1}{2}}\right) = \frac{\sqrt{2}}{\pi}\cdot\frac{-1}{n^2-\frac{1}{4}}$$

$$= \frac{-4\sqrt{2}}{\pi}\frac{1}{(2n-1)(2n+1)} \quad (4)$$

$$b_n = \frac{1}{\pi}\int_{-\pi}^{\pi} f(x)\sin nx\,dx$$

$$= \frac{1}{\pi}\int_{-\pi}^{0} -\sqrt{2}\sin\frac{x}{2}\sin nx\,dx + \frac{1}{\pi}\int_0^{\pi} \sqrt{2}\sin\frac{x}{2}\sin nx\,dx$$

$$= \frac{1}{\pi}\int_{\pi}^{0} -\sqrt{2}\sin\frac{(-x)}{2}\sin n(-x)d(-x)$$

$$+ \frac{1}{\pi}\int_0^{\pi} \sqrt{2}\sin\frac{x}{2}\sin nx\,dx$$

$$= \frac{1}{\pi}\int_0^{\pi} -\sqrt{2}\sin\frac{x}{2}\sin nx\,dx + \frac{1}{\pi}\int_0^{\pi} \sqrt{2}\sin\frac{x}{2}\sin nx\,dx$$

$$= 0 \quad (5)$$

Substituting these values in (2) we obtain

$$f(x) = \sqrt{1-\cos x} = \frac{2\sqrt{2}}{\pi} - \frac{4\sqrt{2}}{\pi}\sum_{n=1}^{\infty}\frac{\cos nx}{(2n-1)(2n+1)}$$

$$\text{in } (-\pi,\ \pi) \quad (6)$$

Note 1 The Fourier series expansions for $\sqrt{1-\cos x}$ in $(0, 2\pi)$ and in $(-\pi, \pi)$ are the same.

Example 8.10

Find the Fourier series for $f(x)$ in $(0, 2\pi)$ where

$$f(x) = \begin{bmatrix} 1, & 0 < x \le \pi \\ 2-\dfrac{x}{\pi}, & \pi < x \le 2\pi \end{bmatrix}$$

Solution The Fourier series for the function

$$f(x) = \begin{bmatrix} 1, & 0 < x \le \pi \\ 2-\dfrac{x}{\pi}, & \pi < x \le 2\pi \end{bmatrix} \quad (1)$$

is given by

$$f(x) = \frac{a_0}{2} + \sum_{n=1}^{\infty} a_n\cos nx + \sum_{n=1}^{\infty} b_n\sin nx \quad (2)$$

where

$$a_0 = \frac{1}{\pi}\int_0^{2\pi} f(x)\,dx = \frac{1}{\pi}\int_0^{\pi} 1\cdot dx + \frac{1}{\pi}\int_{\pi}^{2\pi}\left(2-\frac{1}{\pi}x\right)dx$$

$$= \frac{1}{\pi}(x)_0^{\pi} + \frac{1}{\pi}\left(2x - \frac{1}{2\pi}x^2\right)_{\pi}^{2\pi}$$

$$= \frac{1}{\pi}\cdot\pi + \frac{1}{\pi}\left[2(2\pi-\pi) - \frac{1}{2\pi}(4\pi^2-\pi^2)\right]$$

$$= 1 + 2 - \frac{3}{2} = \frac{3}{2} \quad (3)$$

$$a_n = \frac{1}{\pi}\int_0^{2\pi} f(x)\cos nx\,dx$$

$$= \frac{1}{\pi}\int_0^{\pi} 1\cdot\cos nx\,dx + \frac{1}{\pi}\int_{\pi}^{2\pi}\left(2-\frac{1}{\pi}x\right)\cos nx\,dx$$

$$= \frac{1}{\pi}\left(\frac{\sin nx}{n}\right)_0^{\pi} + \frac{1}{\pi}\left[\left(2-\frac{x}{\pi}\right)\left(\frac{\sin nx}{n}\right)\right.$$

$$\left. + \frac{1}{\pi}\left(\frac{-\cos n\pi}{n^2}\right)\right]_{\pi}^{2\pi}$$

$$= \frac{-1}{\pi^2}\left[\frac{1-(-1)^n}{n^2}\right] = \begin{bmatrix} 0, & \text{if } n \text{ is even} \\ \dfrac{-2}{n^2\pi^2}, & \text{if } n \text{ is odd} \end{bmatrix} \quad (4)$$

$$\because\ t_1 = t_2 = 0 \text{ at } \pi,\ 2\pi,\ 0$$

$$\Rightarrow\ a_{2m-1} = -\frac{2}{\pi^2}\frac{1}{(2m-1)^2}$$

$$b_n = \frac{1}{\pi}\int_0^{2\pi} f(x)\sin nx\,dx$$

$$= \int_0^{\pi} 1\cdot \sin nx\,dx + \frac{1}{\pi}\int_{\pi}^{2\pi}\left(2-\frac{1}{\pi}x\right)\sin nx\,dx$$

$$= \frac{1}{\pi}\left(\frac{-\cos nx}{n}\right)_0^{\pi} + \frac{1}{\pi}\left[\left(2-\frac{x}{\pi}\right)\left(\frac{-\cos nx}{n}\right)\right.$$

$$\left. + \frac{1}{\pi}\left(\frac{-\sin nx}{n^2}\right)\right]_{\pi}^{2\pi}$$

$$= \frac{-1}{\pi n}[(\cos n\pi - 1)] + \frac{1}{\pi}\left[0 + 1\cdot\frac{\cos n\pi}{n} - \frac{1}{\pi n^2}\cdot 0\right]$$

$$= \frac{1-(-1)^n}{\pi n} + \frac{(-1)^n}{\pi n} = \frac{1}{\pi n} \qquad (5)$$

$$\Rightarrow \quad b_m = \frac{1}{\pi m}$$

Substituting these values in (2) we obtain

$$f(x) = \frac{3}{4} - \frac{2}{\pi^2}\sum_{m=1}^{\infty}\frac{\cos(2m-1)x}{(2m-1)^2} + \frac{1}{\pi}\sum_{m=1}^{\infty}\frac{\sin mx}{m} \qquad (6)$$

Example 8.11

Find the Fourier series of the function $f(x) = \begin{bmatrix} x, & 0 \le x \le \pi \\ 2\pi - x, & \pi \le x \le 2\pi \end{bmatrix}$ and deduce that $\frac{1}{1^2} + \frac{1}{3^2} + \frac{1}{5^2} + \cdots = \frac{\pi^2}{8}$. [JNTU 2003, 1997 S]

Solution The Fourier series for the function

$$f(x) = \begin{bmatrix} x, & 0 \le x \le \pi \\ 2\pi - x, & \pi \le x \le 2\pi \end{bmatrix} \qquad (1)$$

is given by

$$f(x) = \frac{a_0}{2} + \sum_{n=1}^{\infty} a_n \cos nx + \sum_{n=1}^{\infty} b_n \sin nx \qquad (2)$$

where

$$a_0 = \frac{1}{\pi}\int_0^{2\pi} f(x)\,dx = \frac{1}{\pi}\int_0^{\pi} x\,dx + \frac{1}{\pi}\int_{\pi}^{2\pi}(2\pi - x)\,dx$$

$$= \frac{1}{\pi}\left(\frac{x^2}{2}\right)_0^{\pi} + \frac{1}{\pi}\left(2\pi x - \frac{x^2}{2}\right)_{\pi}^{2\pi}$$

$$= \frac{1}{\pi}\cdot\frac{\pi^2}{2} + \frac{1}{\pi}\left[2\pi(2\pi - \pi) - \frac{4\pi^2 - \pi^2}{2}\right] = \pi \qquad (3)$$

$$a_n = \frac{1}{\pi}\int_0^{2\pi} f(x)\cos nx\,dx$$

$$= \frac{1}{\pi}\int_0^{\pi} x\cos nx\,dx + \frac{1}{\pi}\int_{\pi}^{2\pi}(2\pi - x)\cos nx\,dx$$

$$= \frac{1}{\pi}\left[x\cdot\frac{\sin nx}{n} - 1\cdot\frac{-\cos nx}{n^2}\right]_0^{\pi}$$

$$+ \frac{1}{\pi}\left[(2\pi - x)\cdot\frac{\sin nx}{n} - (-1)\cdot\left(\frac{-\cos nx}{n^2}\right)\right]_{\pi}^{2\pi}$$

$$= \frac{1}{\pi}\left[0 + \frac{1}{n^2}(\cos n\pi - \cos 0)\right]$$

$$+ \frac{1}{\pi}\left[0 - \frac{1}{n^2}(\cos 2n\pi - \cos n\pi)\right]$$

$$= \frac{2}{\pi n^2}\left[(-1)^n - 1\right] \qquad (4)$$

$$b_n = \frac{1}{\pi}\int_0^{2\pi} f(x)\sin nx\,dx$$

$$= \frac{1}{\pi}\int_0^{\pi} x\sin nx\,dx + \frac{1}{\pi}\int_{\pi}^{2\pi}(2\pi - x)\sin nx\,dx$$

$$= \frac{1}{\pi}\left[x\left(-\frac{\cos nx}{n}\right) - 1\cdot\left(-\frac{\sin nx}{n^2}\right)\right]_0^{\pi}$$

$$+ \frac{1}{\pi}\left[(2\pi - x)\left(-\frac{\cos nx}{n}\right) + 1\left(-\frac{\sin nx}{n^2}\right)\right]_{\pi}^{2\pi}$$

$$= \frac{1}{\pi}\left[-\frac{\pi\cos n\pi - 0}{n} + 0\right] + \frac{1}{\pi}\left[\left(\frac{0 + \pi\cos n\pi}{n}\right) + 0\right]$$

$$= 0 \qquad (5)$$

Substituting these values in (2) we obtain

$$f(x) = \frac{\pi}{2} + \frac{2}{\pi}\sum\frac{1}{n^2}[(-1)^n - 1]\cos nx$$

$$= \frac{\pi}{2} - \frac{4}{\pi}\sum_{m=0}^{\infty}\frac{\cos(2m+1)x}{(2m+1)^2} \qquad (6)$$

Deduction

$x = \pi$ is a point of continuity for $f(x)$ since $f(\pi-) = f(\pi+) = \pi$ and the series converges to $f(\pi)$. Putting $x = \pi$ in (6) we obtain

$$\pi = \frac{\pi}{2} + \frac{4}{\pi}\left(\frac{1}{1^2} + \frac{1}{3^2} + \frac{1}{5^2} + \frac{1}{7^2} + \cdots\right)$$

$$\Rightarrow \quad \frac{\pi^2}{8} = \frac{1}{1^2} + \frac{1}{3^2} + \frac{1}{5^2} + \cdots \quad (7)$$

Example 8.12

Find the Fourier series for the function

$$f(x) = \begin{cases} 0, & \text{for } -\pi < x < 0 \\ \sin x, & \text{for } 0 < x < \pi \end{cases}$$

[JNTU 1994S 1996, 2001 S]

or

Show that $f(x) = \dfrac{1}{\pi} + \dfrac{\sin x}{2} - \dfrac{2}{\pi}\sum_{n=1}^{\infty}\dfrac{\cos 2mx}{4m^2 - 1}$

Hence deduce the results

$$\frac{1}{1.3} - \frac{1}{3.5} + \frac{1}{5.7} - \cdots = \frac{1}{4}(\pi - 2);$$

$$\frac{1}{1.3} - \frac{1}{3.5} + \frac{1}{5.7} - \cdots = \frac{1}{2}$$

Solution The given function $f(x)$ is defined by

$$f(x) = \begin{cases} 0, & \text{for } -\pi < x < 0 \\ \sin x, & \text{for } 0 < x < \pi \end{cases} \quad (1)$$

The Fourier series for the function is given by

$$f(x) = \frac{a_0}{2} + \sum_{n=1}^{\infty} a_n \cos nx + \sum_{n=1}^{\infty} b_n \sin nx \quad (2)$$

where

$$a_0 = \frac{1}{\pi}\int_{-\pi}^{\pi} f(x)\,dx = \frac{1}{\pi}\int_{-\pi}^{0} 0\,dx + \frac{1}{\pi}\int_0^{\pi} \sin x\,dx$$

$$= 0 + \frac{1}{\pi}(-\cos x)_0^{\pi} = \frac{1}{\pi}(-\cos\pi + \cos 0) = \frac{2}{\pi} \quad (3)$$

$$a_n = \frac{1}{\pi}\int_{-\pi}^{\pi} f(x)\cos nx\,dx$$

$$= \frac{1}{\pi}\int_{-\pi}^{0} 0\cos nx\,dx + \frac{1}{\pi}\int_0^{\pi} \sin x \cos nx\,dx$$

$$= 0 + \frac{1}{2\pi}\int_0^{\pi}\left[\sin(n+1)x - \sin(n-1)x\right]dx$$

$$= \frac{1}{2\pi}\left[-\frac{\cos(n+1)x}{n+1} + \frac{\cos(n-1)x}{n-1}\right]_0^{\pi} \quad (n \neq 1)$$

$$= \frac{1}{2\pi}\left[-\frac{\cos(n+1)\pi - \cos 0}{n+1} + \frac{\cos(n-1)\pi - \cos 0}{n-1}\right]$$

$$= \frac{1}{2\pi}\left[\frac{-(-1)^{n+1} + 1}{n+1} + \frac{(-1)^{n-1} - 1}{n-1}\right]$$

$\cos n\pi = (-1)^n; \cos 0 = 1$

$$= \begin{cases} 0, & \text{if } n \text{ is odd } (n \neq 1) \\ -\dfrac{2}{\pi(n^2-1)}, & \text{if } n \text{ is even} \end{cases} \quad \because (-1)^{n+1} = (-1)^{n-1}$$

$$= \begin{cases} +1, & (n, \text{ odd}) \\ -1, & (n, \text{ even} = 2m) \end{cases} \quad \Rightarrow \quad a_{2m} = -\frac{2}{\pi}\frac{1}{(4m^2-1)} \quad (4)$$

***n* = 1**

$$a_1 = \frac{1}{\pi}\int_{-\pi}^{\pi} f(x)\cos x\,dx$$

$$= \frac{1}{\pi}\int_{-\pi}^{0} 0\cdot\cos x\,dx + \frac{1}{\pi}\int_0^{\pi}\sin x\cos x\,dx$$

$$= \frac{1}{2\pi}\int_0^{\pi}\sin 2x\,dx = \frac{1}{2\pi}\left(-\frac{\cos 2x}{2}\right)_0^{\pi}$$

$$= \frac{-1}{4\pi}(\cos 2\pi - \cos 0) = 0 \quad (5)$$

$(\because \cos 2\pi = \cos 0 = 1)$

$$b_n = \frac{1}{\pi}\int_{-\pi}^{\pi} f(x)\sin nx\,dx$$

$$= \frac{1}{\pi}\int_{-\pi}^{0} 0\cdot\sin nx\,dx + \frac{1}{\pi}\int_0^{\pi}\sin x\sin nx\,dx$$

$$= 0 + \frac{1}{2\pi}\int_0^{\pi}[\cos(n-1)x - \cos(n+1)x]\,dx$$

$$= \frac{1}{2\pi}\left[\frac{\sin(n-1)x}{n-1} - \frac{\sin(n+1)x}{n+1}\right]_0^{\pi} = 0 \quad (n \neq 1) (6)$$

$\because \sin nx = 0$ at $x = \pi, 0$

$n = 1$

$$b_1 = \frac{1}{\pi}\int_{-\pi}^{\pi} f(x)\sin x\,dx$$
$$= \frac{1}{\pi}\int_{-\pi}^{0} 0\cdot\sin x\,dx + \frac{1}{\pi}\int_{0}^{\pi}\sin x\sin x\,dx$$
$$= \frac{1}{2\pi}\int_{0}^{\pi}(1-\cos 2x)\,dx = \frac{1}{2\pi}\left(x - \frac{\sin 2x}{2}\right)_0^{\pi}$$
$$= \frac{1}{2\pi}\left[(\pi - 0) - \frac{1}{2}(\sin 2\pi - \sin 0)\right]$$
$$= \frac{1}{2} \quad (\because \sin 2\pi = \sin 0 = 0) \qquad (7)$$

Substituting these values in (2) we have

$$f(x) = \frac{1}{\pi} - \frac{2}{\pi}\sum_{m=1}^{\infty}\frac{\cos 2mx}{4m^2-1} + \frac{1}{2}\sin x \qquad (8)$$

$x = \pi/2$ is a point of continuity and the series converges to $f(\pi/2) = \sin(\pi/2) = 1$

Putting $x = \pi/2$ in (8) we obtain

$$1 = \frac{1}{\pi} + \frac{1}{2}\times 1 - \frac{2}{\pi}\sum_{m=1}^{\infty}\frac{\cos m\pi}{(2m-1)(2m+1)}$$
$$\Rightarrow \quad \frac{1}{2} - \frac{1}{\pi} = -\frac{2}{\pi}\sum_{m=1}^{\infty}\frac{(-1)^m}{(2m-1)(2m+1)}$$
$$\because \cos m\pi = (-1)^m$$

Multiplying both sides by $\pi/2$ we obtain

$$\frac{\pi}{4} - \frac{1}{2} = \frac{1}{1.3} - \frac{1}{3.5} + \frac{1}{5.7} - \frac{1}{7.9} + \cdots \qquad (9)$$

Again, $x = 0$ is a point of continuity and the series converges to $f(0) = 0$. Putting $x = 0$ in (8) we obtain

$$0 = \frac{1}{\pi} - \frac{2}{\pi}\sum_{m=1}^{\infty}\frac{1}{(2m-1)(2m+1)} \qquad (10)$$
$$\Rightarrow \quad \frac{1}{2} = \frac{1}{1.3} + \frac{1}{3.5} + \frac{1}{5.7} + \cdots$$

EXERCISE 8.1

Find the Fourier series expansions of the following functions in the intervals given against them.

$f(x)$, Interval

1. $\begin{cases} 5 & (0, \pi) \\ -5 & (\pi, 2\pi) \end{cases}$

   Ans: $\frac{20}{\pi}\sum_{n\ odd}\frac{\sin nx}{n}$

2. $\begin{cases} \cos x & (0, \pi) \\ -\cos x & (-\pi, 0) \end{cases}$

   Ans: $\frac{8}{\pi}\sum_{m=1}^{\infty}\frac{\sin 2mx}{(2m-1)(2m+1)}$

3. $\cosh ax \quad (-\pi, \pi)$

   Ans: $\frac{2a}{\pi}\sinh a\pi\left[\frac{1}{2a^2} + \sum_{n=1}^{\infty}\frac{(-1)^n}{a^2+n^2}\cos nx\right]$

4. $x - x^2 \quad (-\pi, \pi)$

   Ans: $-\frac{\pi^2}{3} + 4\sum_{n=1}^{\infty}\frac{\cos nx}{n^2} + 2\sum_{n=1}^{\infty}\frac{(-1)^n\sin nx}{n}$

5. $\begin{cases} -x & (-\pi, 0) \\ x & (0, \pi) \end{cases}$

   Deduce that $\frac{\pi^2}{8} = \sum_{n=1}^{\infty}\frac{1}{(2n-1)^2}$

   Ans: $\frac{\pi}{2} - \frac{4}{\pi}\sum_{n=1}^{\infty}\frac{\cos(2n-1)x}{(2n-1)^2}$

6. $\begin{cases} x & \left(-\frac{\pi}{2}, \frac{\pi}{2}\right) \\ \pi - x & \left(\frac{\pi}{2}, \frac{3\pi}{2}\right) \end{cases}$

   Find the value of $1 + \frac{1}{3^2} + \frac{1}{5^2} + \cdots$

   Ans: $\frac{4}{\pi}\left(\sum_{n=1}^{\infty}\frac{(-1)^n\sin(2n-1)x}{(2n-1)^2}\right)$ Sum $= \frac{\pi^2}{8}$

7. $x + x^2 \quad (-\pi, \pi)$

   Show that $\sum\frac{1}{n^2} = \frac{\pi^2}{6}$

   Ans: $\frac{\pi^2}{3} + 4\sum_{n=1}^{\infty}\frac{(-1)^n}{n}\cos nx + 2\sum_{n=1}^{\infty}\frac{(-1)^{n+1}}{n}\sin nx$

8. $\begin{cases} \cos \alpha x & (0,\ 2\pi) \\ & \alpha \notin Z \end{cases}$

Deduce that π cosec $\alpha\pi$

$$= \frac{1}{\alpha} + \sum_{n=1}^{\infty} (-1)^n \left(\frac{1}{\alpha+n} + \frac{1}{\alpha-n} \right)$$

Ans: $\dfrac{\sin 2\alpha\pi}{2\alpha\pi} + \dfrac{\alpha \sin 2\alpha\pi}{\pi} \displaystyle\sum_{n=1}^{\infty} \frac{\cos nx}{\alpha^2 - n^2} - \dfrac{(1-\cos 2\alpha\pi)}{\pi} \displaystyle\sum_{n=1}^{\infty} \frac{n \sin nx}{\alpha^2 - n^2}$

9. $\begin{cases} 1 & (0,\ \pi) \\ 2 - x/\pi & (\pi,\ 2\pi) \end{cases}$

Ans: $\dfrac{3}{4} - \dfrac{2}{\pi^2} \left[\displaystyle\sum_{n=1}^{\infty} \frac{\cos(2n-1)x}{(2n-1)^2} \right] + \dfrac{1}{\pi} \left[\displaystyle\sum_{n=1}^{\infty} \frac{\sin nx}{n} \right]$

10. $\begin{cases} 0, & (-\pi,\ 0) \\ \dfrac{\pi x}{4}, & (0,\ \pi) \end{cases}$

Ans: $\dfrac{\pi}{16} - \dfrac{1}{2} \left(\cos x + \dfrac{1}{3^2} \cos 3x + \dfrac{1}{5^2} \cos 5x + \cdots \right) + \dfrac{\pi}{4} \left(\sin x - \dfrac{1}{2} \sin 2x + \dfrac{1}{3} \sin 3x - \cdots \right)$

8.8 Change of Interval: Fourier Series in Interval $(\alpha, \alpha + 2l)$

So far we have considered expansion of a function $f(x)$ defined on an interval of length 2π in a Fourier series of period 2π. In most engineering and physical applications, the period is not 2π but is of arbitrary length $2l$. By a suitable transformation we can obtain formulas for Fourier series expansion of a function in an interval $(\alpha, \alpha + 2l)$.

Let $F(z)$ be defined in an interval $(\beta, \beta + 2\pi)$ of length 2π.

Then the Fourier series of $F(z)$ is given by

$$F(z) = \frac{a_0}{2} + \sum_{n=1}^{\infty} (a_n \cos nz + b_n \sin nz) \quad (8.27)$$

where the Fourier constants are given by Euler's formulas

$$(a_0,\ a_n,\ b_n) = \frac{1}{\pi} \int_{\beta}^{\beta+2\pi} F(z)(1, \cos nz, \sin nz)\, dz \quad (8.28)$$

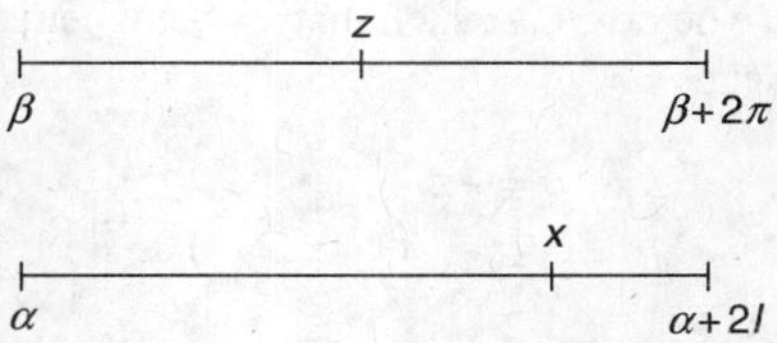

If we change the interval from $(\beta, \beta + 2\pi)$ of length 2π to an interval $(\alpha, \alpha + 2l)$ of length $2l$ then we have to substitute

$$\frac{\text{variable}}{\text{length of the interval}} = \frac{z}{2\pi} = \frac{x}{2l}$$

$$\Rightarrow \quad z = \frac{\pi x}{l},\ dz = \frac{\pi}{l} dx \quad (8.29)$$

Also,

$$z = \beta \quad \Rightarrow \quad x = \frac{l}{\pi} \beta = \alpha \text{ (say)}$$

$$z = \beta + 2\pi \quad \Rightarrow \quad x \frac{l}{\pi} (\beta + 2\pi) = \frac{l\beta}{\pi} + 2l = \alpha + 2l$$

Now, the Fourier series of $F(z) = F(\pi x/l) = f(x)$ (say) is given by

$$f(x) = \frac{a_0}{2} + \sum_{n=1}^{\infty} a_n \cos \frac{n\pi x}{l} + \sum_{n=1}^{\infty} b_n \sin \frac{n\pi x}{l} \quad (8.30)$$

where the Fourier coefficients a_0, a_n and b_n are given by Euler's formulas

$$(a_0,\ a_n,\ b_n) = \frac{1}{l} \int_{\alpha}^{\alpha+2l} f(x) \left(1,\ \cos \frac{n\pi x}{l},\ \sin \frac{n\pi x}{l} \right) dx \quad (8.31)$$

Note 1 Putting $x = 0$ in (8.31) we have formulas for $f(x)$ defined in $(0, 2l)$ and putting $\alpha = -l$ in (8.31) we have formulas for $f(x)$ defined in $(-l, l)$.

8.8.1 Fourier Series in any Arbitrary Interval (a, b)

Let z and x be related by

$$z = lx + m \quad (8.32)$$

We determine x such that $z = -\pi$ when $x = a$ and $z = \pi$ when $x = b$.

$$\therefore \quad \left.\begin{aligned} la + m &= -\pi \\ lb + m &= \pi \end{aligned}\right\} \quad \therefore\ l = \frac{2\pi}{b-a} \tag{8.33}$$

$$m = \pi - bl = \pi - \frac{2\pi b}{b-a}$$

$$\text{or} \quad m = -\frac{b+a}{b-a}\pi$$

The Fourier series of $F(z')$ in $(-\pi, \pi)$ is given by

$$F(z') = \frac{a_0}{2} + \sum_{n=1}^{\infty} a_n \cos nz' + \sum_{n=1}^{\infty} b_n \sin nz' \tag{8.34}$$

where

$$z' = \frac{2\pi x}{b-a}$$

$$(a_0, a_n, b_n) = \frac{1}{\pi}\int_{-\pi}^{\pi} F(z')(1, \cos nz', \sin nz')\, dz' \tag{8.35}$$

Now let $F(z') = F\left(\frac{2\pi x}{b-a}\right) = f(x)$. Then the Fourier series of $f(x)$ in (a, b) is given by

$$f(x) = \frac{a_0}{2} + \sum_{n=1}^{\infty} a_n \cos\frac{2n\pi x}{b-a} + \sum_{n=1}^{\infty} b_n \sin\frac{2n\pi x}{b-a} \tag{8.36}$$

where

$$(a_0, a_n, b_n) = \frac{2}{b-a}\int_a^b f(x)\left(1, \cos\frac{2n\pi x}{b-a}, \sin\frac{2n\pi x}{b-a}\right) dx \tag{8.37}$$

Example 8.13

Find the Fourier series expansion for $f(x) = \pi - x$ in $0 < x < \pi$.

Solution Here $a = 0$, $b = \pi$, $\frac{b-a}{2} = \frac{\pi - 0}{2} = \frac{\pi}{2}$

The Fourier series is

$$f(x) = \pi - x = \frac{a_0}{2} + \sum_{n=1}^{\infty} a_n \cos 2nx + \sum_{n=1}^{\infty} b_n \sin 2nx \tag{1}$$

where $a_0 = \frac{2}{\pi}\int_0^{\pi} (\pi - x)\, dx = \frac{2}{\pi}\left(\pi x - \frac{x^2}{2}\right)_0^{\pi} = \pi$ (2)

$$a_n = \frac{2}{\pi}\int_0^{\pi} (\pi - x)\cos 2nx\, dx \tag{3}$$

$$= \frac{2}{\pi}\cdot\left[(\pi - x)\left(\frac{\sin 2nx}{2n}\right) + \left(-\frac{\cos 2nx}{4n^2}\right)\right]_0^{\pi} = 0$$

$$b_n = \frac{2}{\pi}\int_0^{\pi} (\pi - x)\sin 2nx\, dx \tag{4}$$

$$= \frac{2}{\pi}\left[(\pi - x)\left(-\frac{\cos 2nx}{2n}\right) + \left(-\frac{\sin 2nx}{4n^2}\right)\right]_0^{\pi}$$

$$= \frac{2}{\pi}\frac{\pi}{2n} = \frac{1}{n}$$

$$\therefore \quad \pi - x = \frac{\pi}{2} + \frac{\sin 2x}{1} + \frac{\sin 4x}{2} + \frac{\sin 6x}{3} + \cdots \tag{5}$$

Example 8.14

Find the Fourier series expansion of $f(x) = e^{ax}$ in $(-l, l)$.

Solution The Fourier series expansion of

$$f(x) = e^{ax} \text{ in } (-l, l) \tag{1}$$

is given by

$$e^{ax} = \frac{a_0}{2} + \sum_{n=1}^{\infty} a_n \cos\frac{n\pi x}{l} + \sum_{n=1}^{\infty} b_n \sin\frac{n\pi x}{l} \tag{2}$$

where

$$a_0 = \frac{1}{l}\int_{-l}^{l} e^{ax}\, dx = \frac{1}{l}\left(\frac{e^{ax}}{a}\right)_{-l}^{l}$$

$$= \frac{1}{al}(e^{al} - e^{-al}) = \frac{2}{al}\sinh al \tag{3}$$

$$a_n = \frac{1}{l}\int_{-l}^{l} e^{ax}\cos\frac{n\pi x}{l}\, dx$$

$$= \frac{1}{l}\left[\frac{e^{ax}}{a^2 + (\frac{n\pi}{l})^2}\left(a\cos\frac{n\pi x}{l} + \frac{n\pi}{l}\sin\frac{n\pi x}{l}\right)\right]_{-l}^{l}$$

$$= \frac{l}{l^2a^2+n^2\pi^2}\left[e^{al}\left(a\cos n\pi + \frac{n\pi}{l}\sin n\pi\right) - e^{-al}\left(a\cos n\pi - \frac{n\pi}{l}\sin n\pi\right)\right]$$

$$= \frac{la(-1)^n}{l^2a^2+n^2\pi^2} 2\sinh al \qquad \left| \begin{array}{l}\cos n\pi = (-1)^n \\ \text{and } \sin n\pi = 0\end{array}\right. \quad (4)$$

$$b_n = \frac{1}{l}\int_{-l}^{l} e^{ax}\sin\frac{n\pi x}{l}\,dx$$

$$= \frac{1}{l}\left[\frac{e^{ax}}{a^2+(\frac{n\pi}{l})^2}\left(a\sin\frac{n\pi x}{l} - \frac{n\pi}{l}\cos\frac{n\pi x}{l}\right)\right]_{-l}^{l}$$

$$= \frac{l}{l^2a^2+n^2\pi^2}\left[e^{al}\left(a\sin n\pi - \frac{n\pi}{l}\cos n\pi\right) - e^{-al}\left(-a\sin n\pi - \frac{n\pi}{l}\cos n\pi\right)\right]$$

$$= \frac{l}{l^2a^2+n^2\pi^2}\left(-\frac{n\pi}{l}(-1)^n\right)(e^{al}-e^{-al})$$

$$= \frac{n\pi(-1)^{n+1}}{l^2a^2+n^2\pi^2} 2\sinh al \qquad \left| \begin{array}{l}\cos n\pi = (-1)^n \\ \text{and } \sin n\pi = 0\end{array}\right. \quad (5)$$

The Fourier series expansion is given by

$$e^{ax} = 2\sinh al\left[\frac{1}{2al} + la\sum_{n=1}^{\infty}\frac{(-1)^n}{l^2a^2+n^2\pi^2}\cos\frac{n\pi x}{l} + \pi\sum_{n=1}^{\infty}\frac{(-1)^{n-1}n}{l^2+n^2\pi^2}\sin\frac{n\pi x}{l}\right] \quad (6)$$

Deductions

Putting $a = 1$, $a = -1$ in turn we obtain

$$e^{x} = 2\sinh l\left[\frac{1}{2l} + l\sum_{n=1}^{\infty}\frac{(-1)^n}{l^2+n^2\pi^2}\cos\frac{n\pi x}{l} + \pi\sum_{n=1}^{\infty}\frac{(-1)^{n-1}n}{l^2+n^2\pi^2}\sin\frac{n\pi x}{l}\right] \quad (7)$$

$$e^{-x} = 2\sinh l\left[\frac{1}{2l} + l\sum_{n=1}^{\infty}\frac{(-1)^n}{l^2+n^2\pi^2}\cos\frac{n\pi x}{l} - \pi\sum_{n=1}^{\infty}\frac{(-1)^{n-1}n}{l^2+n^2\pi^2}\sin\frac{n\pi x}{l}\right] \quad (8)$$

$$\frac{(7)+(8)}{2}: \quad \cosh x = 2\sinh l\left[\frac{1}{2l} + l\sum_{n=1}^{\infty}\frac{(-1)^n}{l^2+n^2\pi^2}\cos\frac{n\pi x}{l}\right] \quad (9)$$

$$\frac{(7)-(8)}{2}: \quad \sinh x = 2\pi\sinh l\left[\sum_{n=1}^{\infty}\frac{(-1)^{n-1}n}{l^2+n^2\pi^2}\sin\frac{n\pi x}{l}\right] \quad (10)$$

Example 8.15

Find the Fourier series expansion for

$$f(x) = ax + bx^2 \text{ in } (-l, l)$$

Solution The Fourier series expansion for

$$f(x) = ax + bx^2 \text{ in } (-l, l) \quad (1)$$

is given by

$$f(x) = ax + bx^2 = \frac{a_0}{2} + \sum_{n=1}^{\infty} a_n\cos\frac{n\pi x}{l} + \sum_{n=1}^{\infty} b_n\sin\frac{n\pi x}{l} \quad (2)$$

where

$$a_0 = \frac{1}{l}\int_{-l}^{l}(ax+bx^2)\,dx = \frac{2b}{l}\int_0^l x^2\,dx$$

$$= \frac{2b}{l}\left(\frac{x^3}{3}\right)_0^l = \frac{2bl^2}{3}$$

$$\because \quad \int_{-a}^{a} f(x)\,dx = \begin{cases} 0, & \text{if } f \text{ is odd} \\ 2\int_0^a f(x)\,dx, & \text{if } f \text{ is even}\end{cases} \quad (3)$$

$$a_n = \frac{1}{l}\int_{-l}^{l}(ax+bx^2)\cos\frac{n\pi x}{l}\,dx$$

$$= 0 + \frac{2b}{l}\int_0^l x^2\cos\frac{n\pi x}{l}\,dx \qquad \left(x\cos\frac{n\pi x}{l} \text{ is odd}\right)$$

$$= \frac{2b}{l}\left[x^2\frac{l}{n\pi}\sin\frac{n\pi x}{l} - 2x\frac{l^2}{n^2\pi^2}\left(-\cos\frac{n\pi x}{l}\right) + 2\frac{l^3}{n^3\pi^3}\left(-\sin\frac{n\pi x}{l}\right)\right]_0^l \quad (4)$$

$$= \frac{4bl^2}{n^2\pi^2}(-1)^n \qquad \left\{ \begin{array}{l} t_1 = t_3 = 0 \\ \because \sin\frac{n\pi x}{l} = 0 \text{ at } x = l,\ 0 \\ t_2 = 0 \text{ at } x = 0 \end{array} \right.$$

$$b_n = \frac{1}{l}\int_{-l}^{l}(ax+bx^2)\sin\frac{n\pi x}{l}\,dx$$

$$= \frac{2a}{l}\int_0^l x\sin\frac{n\pi x}{l}\,dx + 0 \qquad \left(x^2\sin\frac{n\pi x}{l} \text{ is odd}\right)$$

$$= \frac{2a}{l}\left[x\cdot\frac{l}{n\pi}\left(-\cos\frac{n\pi x}{l}\right) - 1\cdot\frac{l^2}{n^2\pi^2}\left(-\sin\frac{n\pi x}{l}\right)\right]_0^l$$

$$= \frac{2a}{l}\left[l\cdot\frac{l}{n\pi}(-\cos n\pi) - 0 + 0\right]$$

$$= \frac{2al}{\pi}(-1)^{n-1}, \qquad \left[\sin\frac{n\pi x}{l} = 0 \text{ at } l,\ 0\right] \quad (5)$$

$$\therefore \quad ax+bx^2 = \frac{bl^2}{3} - \frac{4bl^2}{\pi^2}\sum_{n=1}^{\infty}\frac{(-1)^{n-1}}{n^2}\cos\frac{n\pi x}{l} + \frac{2al}{\pi}\sum_{n=1}^{\infty}\frac{(-1)^{n-1}}{n}\sin\frac{n\pi x}{l} \quad (6)$$

Deductions

$a = 1, b = -1$

$$x - x^2 = \frac{-l^3}{3} + \frac{4l^2}{\pi^2}\sum_{n=1}^{\infty}\frac{(-1)^{n-1}}{n^2}\cos\frac{n\pi x}{l} + \frac{2l}{\pi}\sum_{n=1}^{\infty}\frac{(-1)^{n-1}}{n}\sin\frac{n\pi x}{l} \quad (7)$$

$a = 1, b = 1$

$$x + x^2 = \frac{l^3}{3} - \frac{4l^2}{\pi^2}\sum\frac{(-1)^{n-1}}{n^2}\cos\frac{n\pi x}{l} + \frac{2l}{\pi}\sum_{n=1}^{\infty}\frac{(-1)^{n-1}}{n}\sin\frac{n\pi x}{l} \quad (8)$$

Choosing $l = 1$ in the above results (6), (7) and (8) we get Fourier series expansions in the interval (–1, 1) for the functions $ax + bx^2$, $x - x^2$, $x + x^2$, respectively.

Example 8.16

A sinusoidal voltage $E \sin wt$ is passed through a half-wave rectifier which clips the regative portion of the wave. Develop a Fourier series for the resulting periodic function.

$$U(t) = \left\{ \begin{array}{ll} 0, & \text{when } -\frac{T}{2} < t < 0 \\ E\sin wt, & \text{when } 0 < t < \frac{T}{2} \quad \text{where } T = \frac{2\pi}{w} \end{array} \right\}$$

Solution The Fourier series for the function

$$U(t) = \left\{ \begin{array}{ll} 0, & \text{when } -\frac{T}{2} < t < 0 \\ E\sin wt, & \text{when } 0 < t < \frac{T}{2} \quad \text{where } T = \frac{2\pi}{w} \end{array} \right\} \quad (1)$$

is given by

$$U(t) = \frac{a_0}{2} + \sum_{n=1}^{\infty} a_n\cos\frac{2n\pi t}{T} + \sum_{n=1}^{\infty} b_n\sin\frac{2n\pi t}{T} \quad (2)$$

$$\text{or} \quad U(t) = \frac{a_0}{2} + \sum_{n=1}^{\infty} a_n\cos nwt + \sum_{n=1}^{\infty} b_n\sin nwt \quad (3)$$

where

$$a_0 = \frac{2}{T}\int_{-T/2}^{T/2} U(t)\,dt = \frac{2}{T}\left[\int_{-T/2}^{0} 0\cdot dt + \int_0^{T/2} E\sin wt\,dt\right]$$

$$= \frac{2}{T}\left[0 + E\left(-\frac{\cos wt}{w}\right)\right]_0^{T/2}$$

$$= \frac{-2E}{wT}(\cos\pi - 1) = \frac{2E}{\pi} \qquad (\because wT = 2\pi) \quad (4)$$

$$a_n = \frac{2}{T}\int_{-T/2}^{T/2} U(t)\cos nwt\,dt = \frac{2}{T}\int_0^{T/2} E\sin wt\cos nwt\,dt$$

$$= \frac{E}{T}\int_0^{T/2}[\sin(n+1)wt - \sin(n-1)wt]\,dt$$

$$= \frac{E}{T}\left[\frac{-\cos(n+1)wt}{(n+1)w} + \frac{\cos(n-1)wt}{(n-1)w}\right]_0^{T/2}, \qquad (n \neq 1)$$

$$= \frac{E}{wT}\left[\frac{-\cos(n+1)\pi + 1}{(n+1)} + \frac{\cos(n-1)\pi - 1}{(n-1)}\right]$$

$$= \frac{E}{wT}\left[\frac{-(-1)^{n+1} + 1}{n+1} + \frac{(-1)^{n+1} - 1}{n-1}\right]$$

$(n \neq 1)$

$$= \begin{cases} 0, & \text{if } n \text{ is odd} \\ \dfrac{2E}{2\pi}\left(\dfrac{1}{n+1} - \dfrac{1}{n-1}\right) = \dfrac{-2E}{\pi(n^2-1)}, & \text{if } n \text{ is even} \end{cases} \quad (5)$$

$n = 1$

$$a_1 = \frac{2}{T}\int_0^{T/2} E \sin wt \cos wt \, dt = \frac{E}{T}\int_0^{T/2} \sin 2wt \, dt$$

$$= \frac{E}{2Tw}(-\cos 2wt)_0^{T/2} = \frac{E}{2wT}(\cos wT - \cos 0)$$

$$= \frac{E}{4\pi}(\cos 2\pi - \cos 0) = 0 \quad (6)$$

$$b_n = \frac{2}{T}\int_0^{T/2} E \sin wt \sin nwt \, dt$$

$$= \frac{E}{T}\int_0^{T/2} [\cos(n-1)wt - \cos(n+1)wt] \, dt$$

$$= \frac{E}{Tw}\left[\frac{\sin(n-1)wt}{(n-1)} - \frac{\sin(n+1)wt}{n+1}\right]_0^{T/2} = 0,$$

$(n \neq 1)$ (7)

$n = 1$

$$b_1 = \frac{2}{T}\int_0^{T/2} E \sin wt \sin wt \, dt = \frac{E}{T}\int_0^{T/2} (1 - \cos 2wt) \, dt$$

$$= \frac{E}{T}\left(t - \frac{\sin 2wt}{2w}\right)_0^{T/2} = \frac{E}{2} \quad (8)$$

$(\because \quad \sin wT = \sin 2\pi = \sin 0 = 0)$

Substituting the values of the constants in (2) we obtain

$$U(t) = \frac{E}{\pi} + \frac{E}{2}\sin wt - \frac{2E}{\pi}\sum_{m=1}^{\infty}\frac{\cos(2m)wt}{(2m-1)(2m+1)} \quad (9)$$

EXERCISE 8.2

1. Find the Fourier series of $f(x) = \begin{cases} d - x, 0 < x < d \\ 0, d < x < 2d \end{cases}$

   Ans: $f(x) = \dfrac{l}{4} + \dfrac{2l}{\pi^2}\displaystyle\sum_{n=1}^{\infty}\dfrac{\cos(2n-1)\pi x}{(2n-1)^2} + \dfrac{l}{\pi}\displaystyle\sum_{n=1}^{\infty}\dfrac{\sin\frac{n\pi x}{l}}{n}$

2. Obtain the Fourier series expansion of $f(x) = \dfrac{\pi - x}{2}$ in $0 < x < 2$.

   Ans: $\dfrac{\pi - x}{2} = \dfrac{\pi - 1}{2} + \displaystyle\sum_{n=1}^{\infty}\dfrac{\sin n\pi x}{n\pi}$

3. Find the Fourier series for $f(x) = e^x$ in $(-l, l)$.

   Ans: $e^x = \sinh l\left[\dfrac{1}{l} + 2l\displaystyle\sum_{n=1}^{\infty}\dfrac{(-1)^n \cos\frac{n\pi x}{l}}{l^2 + n^2\pi^2}\right.$

   $\left. + 2\pi\displaystyle\sum_{n=1}^{\infty}\dfrac{(-1)^{n-1} n \sin\frac{n\pi x}{l}}{l^2 + n^2\pi^2}\right]$

4. Expand $f(x) = \begin{cases} \pi x, & 0 < x < 1 \\ 0, & 1 < x < 2 \end{cases}$ into a Fourier series

   Ans: $f(x) = \dfrac{\pi}{4} - \dfrac{2}{\pi}\displaystyle\sum_{n=1}^{\infty}\dfrac{\cos(2n-1)\pi x}{(2n-1)^2} + \displaystyle\sum_{n=1}^{\infty}\dfrac{(-1)^{n-1}\sin n\pi x}{n}$

5. Find the Fourier series of $f(x) = \begin{cases} x, -1 < x < 0 \\ x + 2, 0 < x < 1 \end{cases}$ and deduce that $1 - \dfrac{1}{3} + \dfrac{1}{5} - \dfrac{1}{7} + \cdots = \dfrac{\pi}{4}$.

   Ans: $f(x) = 1 + \dfrac{2}{\pi}\displaystyle\sum_{n=1}^{\infty}\dfrac{(1 - (-1^n))}{n}\sin n\pi x$

6. Expand $f(x) = \begin{cases} 0 & -l < x < 0 \\ k & 0 < x < l \end{cases}$ as a Fourier series.

   Ans: $f(x) = \dfrac{k}{2} + \dfrac{2k}{\pi}\displaystyle\sum_{n=1}^{\infty}\dfrac{\sin(2n-1)\pi\frac{x}{l}}{2n-1}$

7. Find the Fourier series of $f(x) = \begin{cases} l - x, & 0 < x \le l \\ 0, & l \le x < 2l \end{cases}$

   Ans:

   $f(x) = \dfrac{1}{4} + \left(\dfrac{2l}{\pi^2}\right)\displaystyle\sum_{n=1}^{\infty}\dfrac{\cos(2n-1)\pi\frac{x}{l}}{(2n-1)^2} + \left(\dfrac{l}{\pi}\right)\displaystyle\sum_{n=1}^{\infty}\dfrac{\sin n\pi\frac{x}{l}}{n}$

8. Obtain the Fourier series expansion for $f(x) = \begin{cases} 0, & -8 < x < 0 \\ 4, & 0 < x < 4 \\ 0, & 4 < x < 8 \end{cases}$

   Ans: $f(x) = 1 + \displaystyle\sum\dfrac{(-1)^n}{(2n-1)}\cos\dfrac{(2n-1)\pi x}{8}$

   $-\dfrac{4}{\pi}\displaystyle\sum_{n=1}^{\infty}\dfrac{\sin\frac{(2n-1)\pi x}{8}}{(2n-1)}$

8.9 Fourier Series Expansions of Even and Odd Functions in $(-l, l)$

We now work out some examples on Fourier series expansions of even and odd functions in $(-l, l)$ and $(-\pi, \pi)$ by applying the properties discussed in sections 8.3 and 8.6 on pp 8.2–8.5.

Example 8.17

Find the Fourier series of $f(x) = x$ in $(-\pi, \pi)$.

Solution The function is defined in $S = (-\pi, \pi)$ by

$$f(x) = x \quad (1)$$

Here (i) $x \in S \Rightarrow -x \in S$ and (ii) $f(-x) = -x = -f(x)$.

$\therefore f(x)$ is an odd function and hence $a_0 = a_n = 0$ for all x.

The Fourier series of the function is given by

$$x = \sum_{n=1}^{\infty} b_n \sin nx \quad (2)$$

where

$$b_n = \frac{2}{\pi}\int_0^{\pi} x \sin nx\, dx = \frac{2}{\pi}\left[x\left(-\frac{\cos nx}{n}\right) - 1\left(-\frac{\sin nx}{n^2}\right)\right]_0^{\pi}$$

$$= \frac{2}{\pi}\left(\frac{-\pi\cos n\pi}{n}\right) = \frac{2(-1)^{n+1}}{n} \quad (3)$$

$$\begin{bmatrix} t_2 = 0 & \text{at } x = \pi, 0 \\ t_1 = 0 & \text{at } x = 0 \end{bmatrix}$$

Substituting the value in (2) we obtain

$$x = 2\sum_{n=1}^{\infty} \frac{(-1)^{n+1}}{n} \sin nx \quad (4)$$

Deduction

$x = \pi/2$ is a point of continuity of $f(x) = x$ and the series converges to

$$f\left(\frac{\pi}{2}\right) = \frac{\pi}{2} \text{ at } x = \frac{\pi}{2} \qquad \sin\frac{n\pi}{2} = \begin{bmatrix} 0 & \text{if } n \text{ is even} \\ (-1)^{n-1} & \text{if } n \text{ is odd} \\ & \text{and } n = 2m-1 \end{bmatrix}$$

$$\therefore \frac{\pi}{2} = 2\sum_{m=1}^{\infty} \frac{(-1)^{m-1}}{(2m-1)} \Rightarrow \frac{\pi}{4} = \frac{1}{1} - \frac{1}{3} + \frac{1}{5} + \frac{1}{7} + \cdots \quad (5)$$

Example 8.18

Find the Fourier series for $f(x) = \sin x$ in $-\pi < x < \pi$. [JNTU 2002]

Solution The function is defined in $S = (-\pi, \pi)$ by

$$f(x) = \sin x \quad (1)$$

Here

(i) $x \in S \Rightarrow -x \in S$ and (ii) $f(-x) = \sin(-x) = -\sin x = -f(x) \quad \forall x \in S$.

$\therefore$ f is an odd function in S and hence $a_0 = a_n = 0$ for all n.

The Fourier series of the function is given by

$$\sin x = \sum_{n=1}^{\infty} b_n \sin nx \quad (2)$$

where

$$b_n = \frac{2}{\pi}\int_0^{\pi} \sin x \sin nx\, dx$$

$$= \frac{1}{\pi}\int_0^{\pi} [\cos(n-1)x - \cos(n+1)x]\, dx \quad (3)$$

$$= \frac{1}{\pi}\left[\frac{\sin(n-1)x}{n-1} - \frac{\sin(n+1)x}{n+1}\right]_0^{\pi} = 0, \qquad n \neq 1 \quad (4)$$

$\because \sin p = 0$ for $p = (n+1)\pi, (n-1)\pi$ or 0

$n = 1$

$$b_1 = \frac{2}{\pi}\int_0^{\pi} \sin x \sin x\, dx = \frac{1}{\pi}\int_0^{\pi} (1 - \cos 2x)\, dx$$

$$= \frac{1}{\pi}\left(x - \frac{\sin 2x}{2}\right)_0^{\pi} = \frac{1}{\pi}\left[(\pi - 0) - \frac{1}{2}(\sin 2\pi - 0)\right] = 1$$

Hence, the required Fourier series for $f(x)$ is

$$f(x) = \sin x$$

Note 1 There is only one term in the Fourier series in this case.

Example 8.19

Find the Fourier series for $f(x) = \begin{cases} 1 + \dfrac{2x}{\pi} & (-\pi, 0) \\ 1 - \dfrac{2x}{\pi} & (0, \pi) \end{cases}$

Solution Consider the function defined in $S = (-\pi, \pi)$ by

$$f(x) = a - b|x| \quad (1)$$

where a, b are constants. Here

(i) $x \in S \Rightarrow -x \in S$ and (ii) $f(-x) = a - b|-x| = a - b|x| = f(x)$

Hence, $f(x)$ is an even function for which the Fourier series is given by

$$f(x) = a - b|x| = \frac{a_0}{2} + \sum_{n=1}^{\infty} a_n \cos nx \quad (b_n = 0 \text{ for all } n) \quad (2)$$

where

$$a_0 = \frac{2}{\pi}\int_0^{\pi} f(x)\,dx = \frac{2}{\pi}\int_0^{\pi}(a - b|x|)\,dx = \frac{2}{\pi}\int_0^{\pi}(a - bx)\,dx$$

$$= \frac{2}{\pi}\left(ax - b\frac{x^2}{2}\right)_0^{\pi} = \frac{2}{\pi}\left(a\pi - b\frac{\pi^2}{2}\right) = 2\left(a - \frac{1}{2}b\pi\right)$$

$$a_n = \frac{2}{\pi}\int_0^{\pi} f(x)\cos nx\,dx = \frac{2}{\pi}\int_0^{\pi}(a - bx)\cos nx\,dx$$

$$= \frac{2}{\pi}\left[(a - bx)\left(\frac{\sin nx}{n}\right) + b\left(-\frac{\cos nx}{n^2}\right)\right]_0^{\pi} \quad (3)$$

$$= \frac{-2b}{\pi}\left(\frac{\cos n\pi - \cos 0}{n^2}\right)$$

$$= -\frac{2b}{\pi}\left(\frac{(-1)^n - 1}{n^2}\right) \quad (4)$$

$$= \begin{cases} 0, & \text{if } n \text{ is even} \\ \dfrac{4b}{\pi(2m-1)^2}, & \text{if } n \text{ is odd, } n = 2m-1 \\ & (\sin nx = 0 \text{ at } x = \pi, 0) \end{cases}$$

Substituting the values in (2) we have

$$a - b|x| = a - \frac{1}{2}b\pi + \frac{4b}{\pi}\sum_{m=1}^{\infty}\frac{\cos(2m-1)x}{(2m-1)^2} \quad (5)$$

Deductions

(i) $a = 1, b = \frac{2}{\pi}$; $1 - \frac{2}{\pi}|x| = \frac{8}{\pi^2}\sum_{n=1}^{\infty}\frac{\cos(2n-1)x}{(2n-1)^2}$ (6)

(ii) $a = 1,\ b = 1$; $1 - |x| = 1 - \frac{\pi}{2} + \frac{4}{\pi}\sum_{n=1}^{\infty}\frac{\cos(2n-1)x}{(2n-1)^2}$ (7)

$x = 0$ point of continuity for $f(x)$ and the series converges to $f(0) = 1$

Hence, from (6) for $x = 0$

$$1 = \frac{8}{\pi^2}\sum\frac{1}{(2n-1)^2} \quad (8)$$

$$\Rightarrow \sum\frac{1}{(2n-1)^2} = \frac{1}{1^2} + \frac{1}{3^2} + \frac{1}{5^2} + \cdots = \frac{\pi^2}{8}$$

Example 8.20

Find the Fourier series expansion for

$$f(x) = \begin{cases} -k, & -\pi < x < 0 \\ k, & 0 < x < \pi \end{cases} \quad \text{where } k > 0;$$

Deduce that $\frac{\pi}{4} = \sum_{n=1}^{\infty}\frac{(-1)^{n-1}}{2n-1}$

[JNTU 1995, 2001, 2002]

Solution The function $f(x) = \begin{cases} -k, & -\pi < x < 0 \\ k, & 0 < x < \pi \end{cases}$ (1)

is defined in the set $S = (-\pi, 0) \cup (0, \pi)$.

$\therefore$ We find that (i) $x \in S \Rightarrow -x \in S$ and (ii) $f(x) = k$ for $0 < x < \pi$.

$$\Rightarrow \left.\begin{aligned} -k &= -f(x) \\ &= f(-x) \end{aligned}\right\} \text{ for } 0 > -x > -\pi$$

$\therefore$ f is an odd function in S.

Also, from Figure 8.1 (p 8.26) we see that the graph of the function is symmetrical with respect to the origin and hence f is odd in S; consequently, the Fourier series consists of only sines and it is given by

$$f(x) = \sum_{n=1}^{\infty} b_n \sin nx \quad (2)$$

where

$$b_n = \frac{2}{\pi}\int_0^{\pi} k\sin nx\,dx = \frac{2k}{\pi}\left(-\frac{\cos nx}{n}\right)_0^{\pi}$$

$$= -\frac{2k}{n}(\cos n\pi - \cos 0)$$

$$= \frac{-2k}{\pi n}((-1)^n - 1) = \begin{cases} 0, & \text{if } n \text{ is even} \\ \dfrac{4k}{\pi}, & \text{if } n \text{ is odd} \end{cases} \quad (3)$$

$$\therefore \quad f(x) = \frac{4k}{\pi}\sum_{n=1}^{\infty}\frac{\sin(2n-1)x}{(2n-1)} \tag{4}$$

Deduction

f is continuous at $x = \pi/2$. The series (4) converges to $f(\pi/2) = k$ at $x = \pi/2$.

$$\therefore \quad k = \frac{4k}{\pi}\left(1 - \frac{1}{3} + \frac{1}{5} + \frac{1}{7} + \frac{1}{9} - \frac{1}{11} + \cdots\right)$$

or $$\frac{\pi}{4} = 1 - \frac{1}{3} + \frac{1}{5} - \frac{1}{7} + \cdots \tag{5}$$

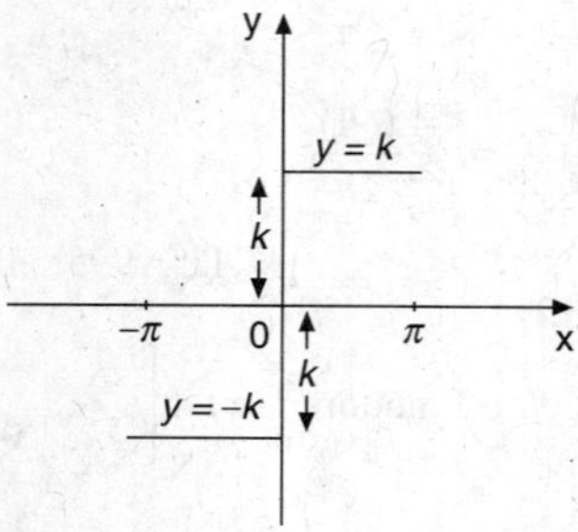

Figure 8.4

Graph of $f(x) = \begin{cases} -k \text{ in } (-\pi, 0) \\ k \text{ in } (0, \pi) \end{cases}$

which is an odd function.

EXERCISE 8.3

Obtain Fourier series expansions with respect to the following even/odd functions in the indicated intervals $(-l, l)$.

1. $f(x) = x^2 \ (-l, l)$ [JNTU 2002]

 Ans: $x^2 = \dfrac{l^2}{3} - \dfrac{4l^2}{\pi^2}\sum_{n=1}^{\infty}\dfrac{(-1)^{n-1}\cos\dfrac{n\pi x}{l}}{n^2}$

2. $f(x) = \sin 3x, \ (-l, l)$

 Ans: $\sin 3x = 2\pi \sin 3l \sum_{n=1}^{\infty}\left(\dfrac{(-1)^{n-1}}{n^2\pi^2 - 9l^2}\right)\sin\dfrac{n\pi x}{l}$

3. Show that sin

 $px = \dfrac{2\sin p\pi}{\pi}\sum_{n=1}^{\infty}\dfrac{(-1)^{n-1} n \sin nx}{n^2 - p^2}$ in $(-\pi, \pi)$ $(p \notin Z)$

4. $f(x) = 4 - x^2$ in $(-2, 2)$

 [JNTU Dec. 2002(4), 2003 S (4), 2003 S (4), 2005 S(3)]

 Ans: $4 - x^2 = \dfrac{8}{3} + \dfrac{16}{\pi^2}\sum_{n=1}^{\infty}\dfrac{(-1)^{n-1}\cos\dfrac{n\pi x}{l}}{n^2}$

5. $f(x) = x^2 - 2$ in $(-2, 2)$

 Ans: $x^2 - 2 = -\dfrac{2}{3} - \dfrac{16}{\pi^2}\left(\cos\dfrac{\pi x}{2} - \dfrac{1}{2^2}\cos\pi x + \dfrac{1}{3^2}\cos\dfrac{3\pi x}{2} - \dfrac{1}{4^2}\cos 2\pi x + \cdots\right)$

6. $f(x) = \cosh x$ in $(-\pi, \pi)$

 Ans: $\cos hx = \dfrac{2\sin h\pi}{\pi}\left(\dfrac{1}{2} + \sum_{n=1}^{\infty}\dfrac{((-1)^n - 1)}{n^2 + 1}\cos nx\right)$

7. $f(x) = |\sin x|$ in $(-\pi, \pi)$

 Ans: $|\sin x| = \dfrac{2}{\pi} - \dfrac{4}{\pi}\left(\dfrac{\cos 2x}{1.3} + \dfrac{\cos 4x}{3.5} + \dfrac{\cos 6x}{5.7} + \cdots\right)$

8. $f(x) = \begin{cases} 1 + \dfrac{2x}{\pi}, & -\pi \le x \le 0 \\ 1 - \dfrac{2x}{\pi}, & 0 \le x \le \pi \end{cases}$

 Ans: $f(x) = \dfrac{8}{\pi^2}\left(\sum_{n=1}^{\infty}\dfrac{\cos(2n-1)x}{(2n-1)^2}\right)$

9. $f(x) = \begin{cases} 0, & -2 \le x < -1 \\ k, & -1 < x < 1 \\ 0, & 1 < x < 2 \end{cases}$

 Ans: $f(x) = \dfrac{k}{2} + \dfrac{2k}{\pi}\left(\cos\dfrac{\pi x}{2} - \dfrac{1}{3}\cos\dfrac{3\pi x}{2} + \dfrac{1}{5}\cos\dfrac{5\pi x}{2} - \dfrac{1}{7}\cos\dfrac{7\pi x}{2} + \cdots\right)$

10. $f(x) = \begin{cases} \dfrac{\pi}{2} + x, & -\pi < x < 0 \\ \dfrac{\pi}{2} - x, & 0 < x < \pi \end{cases}$

 Deduce that $\dfrac{1}{1^2} + \dfrac{1}{3^2} + \dfrac{1}{5^2} + \dfrac{1}{7^2} + \cdots = \dfrac{\pi^2}{8}$.

 Ans: $f(x) = \dfrac{4}{\pi}\sum_{n \text{ odd}}\dfrac{\cos nx}{n^2}$

8.10 Half-Range Fourier Sine/Cosine Series: Odd and Even Periodic Continuations

In most engineering and physical applications we may require Fourier series expansion for a function defined in the interval (0, l) only. Such a series is known as half-range Fourier series. Since $f(x)$ is now neither periodic nor classifiable as even or odd we have the option of obtaining Fourier sine series (FSS) only or Fourier cosine series (FCS) only in the interval (0, l), as the case may be. To obtain a Fourier sine series for $f(x)$ we construct a new function $g(x)$ such that

1. $g(x) = f(x)$ in the interval (0, l)
2. $g(x)$ is an odd function in the interval ($-l$, l) and is periodic with period $2l$

Such a function $g(x)$ is called an *odd periodic continuation* (*or extension*) of $f(x)$ (Figure 8.2). The FSS for $g(x)$ in ($-l$, l) is given by

$$g(x) = \sum_{n=1}^{\infty} b_n \frac{\sin n\pi x}{l}$$

where

$$b_n = \frac{2}{l}\int_0^l g(x)\sin\frac{n\pi x}{l}dx = \frac{2}{l}\int_0^l f(x)\sin\frac{n\pi x}{l}dx$$

with $a_0 = a_n = 0$ for all n.

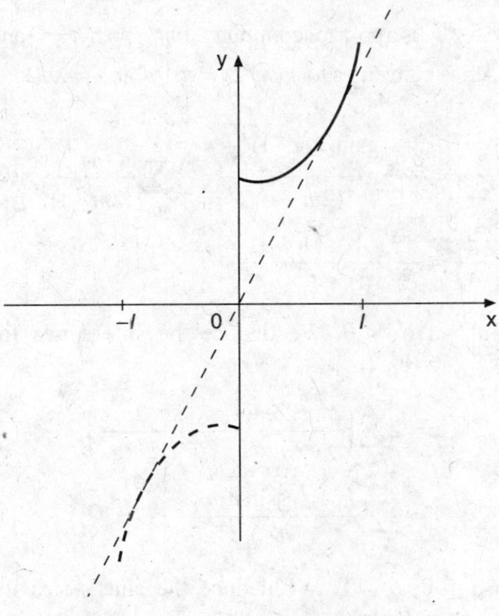

Figure 8.5 Odd periodic continuation

Similarly to obtain the half-range FCS for $f(x)$ we define $g(x)$ such that

1. $g(x) = f(x)$ in (0, l)
2. $g(x)$ is an even function in the interval ($-l$, l) and is periodic with period $2l$

Such a function $g(x)$ is called an *even periodic continuation* (*or extension*) of $f(x)$ (Figure 8.2).

The FCS for $g(x)$ in ($-l$, l) is given by

$$g(x) = \frac{a_0}{2} + \sum_{n=1}^{\infty} a_n \cos\frac{n\pi x}{l}$$

where

$$a_0 = \frac{2}{l}\int_0^l f(x)\,dx, \quad \text{and}$$

$$a_n = \frac{2}{l}\int_0^l f(x)\cos\frac{n\pi x}{l}dx \text{ and } b_n = 0 \text{ for all } n$$

Note 1 The above expansions for $f(x)$ are valid only in (0, l).

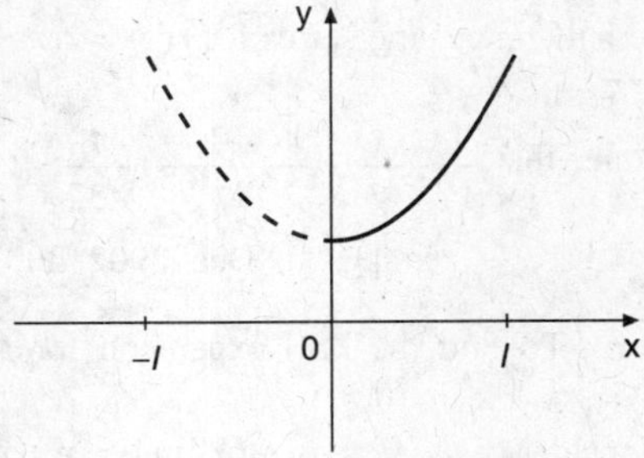

Figure 8.6 Even periodic continuation

Example 8.21

Find a FSS for $f(x) = a + bx$ in $0 < x < l$ where a and b are constants.

Solution Here the function is defined by

$$f(x) = a + bx \quad \text{in } 0 < x < l \tag{1}$$

and we are required to find Fourier half-range sine series for (1), which is given by

$$f(x) = a + bx = \sum_{n=1}^{\infty} b_n \sin\frac{n\pi x}{l} \tag{2}$$

where

$$b_n = \frac{2}{l}\int_0^l f(x)\sin\frac{n\pi x}{l}dx = \frac{2}{l}\int_0^l (a+bx)\sin\frac{n\pi x}{l}dx$$

$$= \frac{2}{l}\cdot\left[(a+bx)\left(-\frac{l}{n\pi}\cos\frac{n\pi x}{l}\right) - b\left(-\frac{l^2}{n^2\pi^2}\sin\frac{n\pi x}{l}\right)\right]_0^l$$

$$t_2 = 0 \text{ at } x = l, 0$$

$$= \frac{-2}{l}\cdot\frac{l}{n\pi}[(a+bl)(\cos n\pi) - a] + 0$$

$$= \frac{2}{n\pi}\left[a(1-(-1)^n) - bl(-1)^n\right] \quad (3)$$

Substituting these values in (2) we obtain

$$a + bx = \frac{4a}{\pi}\sum_{m=1}^{\infty}\frac{\sin[(2m-1)\pi x/l]}{(2m-1)} + \frac{2bl}{\pi}\sum_{n=1}^{\infty}\frac{(-1)^{n-1}\sin\left(\frac{n\pi x}{l}\right)}{n} \quad (4)$$

Example 8.22

Find the half-range sine series for $f(x) = x(\pi - x)$ in $0 < x < \pi$

Deduce that $\frac{1}{1^3} - \frac{1}{3^3} + \frac{1}{5^3} - \frac{1}{7^3} + \cdots \frac{\pi^2}{32}$

[JNTU Dec. 2002, 2003 (2)]

Solution To find the FSS expansion for $f(x) = x(\pi - x) = \pi x - x^2$ in $(0, \pi)$

we consider $f(x) = ax + bx^2$ in $0 < x < \pi$ (1)

The required series is

$$f(x) = ax + bx^2 = \sum_{n=1}^{\infty} b_n \sin nx \quad (2)$$

where

$$b_n = \frac{2}{\pi}\int_0^\pi f(x)\sin nx\, dx = \frac{2}{\pi}\int_0^\pi (ax+bx^2)\sin nx\, dx$$

$$= \frac{2}{x}\left[(ax+bx^2)\left(-\frac{\cos nx}{n}\right) - (a+2bx)\left(-\frac{\sin nx}{n^2}\right) + 2b\left(\frac{\cos nx}{n^3}\right)\right]_0^\pi$$

$$= \frac{2}{\pi}\left[\frac{(a\pi + b\pi^2)}{n}(-1)^{n+1} + \frac{2b}{n^3}((-1)^n - 1)\right]$$

$\sin nx = 0$ at $x = \pi, 0$

$t_1 = 0$ at $x = 0$

$\because \cos n\pi = (-1)^n, \cos 0 = 1$

$$b_n = \begin{cases} \dfrac{-2(a+b\pi)}{n} + 0, & \text{if } n \text{ is even} \\ \dfrac{2(a+b\pi)}{n} - \dfrac{8b}{\pi n^3}, & \text{if } n \text{ is odd} \end{cases} \quad (3)$$

Substituting these values in (2) we obtain

$$ax + bx^2 = 2(a+b\pi)\sum_{m=1}^{\infty}\left(\frac{\sin(2m-1)x}{(2m-1)} - \frac{\sin 2mx}{2m}\right) - \frac{8b}{\pi}\sum_{m=1}^{\infty}\frac{\sin(2m-1)x}{(2m-1)^3} \quad (4)$$

Deduction

Putting

(i) $a = \pi$, $b = -1$. We obtain from (4)

$$x(\pi - x) = \frac{8}{\pi}\sum_{m=1}^{\infty}\frac{\sin(2m-1)x}{(2m-1)^3} \quad (5)$$

$x = \pi/2$ is a point of continuity for $f(x) = x(\pi - x)$ and the series converges to $f(\pi/2) = \pi^2/4$ at $x = \pi/2$

$$\therefore \frac{\pi^2}{4} = \frac{8}{\pi}\sum_{m=1}^{\infty}\frac{\sin(2m-1)\frac{\pi}{2}}{(2m-1)^3} = \frac{8}{\pi}\sum_{m=1}^{\infty}\frac{(-1)^{m-1}}{(2m-1)^3} \quad (6)$$

$$\Rightarrow \frac{\pi^3}{32} = \frac{1}{1^3} - \frac{1}{3^3} + \frac{1}{5^3} - \frac{1}{7^3} + \cdots$$

(ii) $a = 1$, $b = 0$. We deduce the sine series for $f(x) = x$ in $(0, \pi)$

$$x = 2\sum_{m=1}^{\infty}\left(\frac{\sin(2m-1)x}{2m-1} - \frac{\sin 2mx}{2m}\right) \quad (7)$$

$$= 2\sum_{n=1}^{\infty}\frac{(-1)^{n-1}\sin nx}{n}$$

(iii) $a = 0$, $b = 1$. We deduce the sine series for $f(x) = x^2$ in $(0, \pi)$

$$x^2 = 2\pi\sum_{n=1}^{\infty}\frac{(-1)^n \sin nx}{n} - \frac{1}{\pi}\sum_{n=1}^{\infty}\frac{\sin(2n-1)x}{(2n-1)^3} \quad (8)$$

Example 8.23
Obtain the half-range FSS for $f(x) = e^x$ in $0 < x < 1$. [JNTU 2003, 2003(2)]

Solution The half-range FSS of e^x is $e^x =$

$$e^x = \sum_{n=1}^{\infty} b_n \sin n\pi x \quad (\because l = 1) \tag{1}$$

where

$$b_n = 2\int_0^1 e^x \sin n\pi x\, dx$$

$$= \left[\frac{2e^x}{n^2\pi^2+1}(1\cdot \sin n\pi x - n\pi \cos n\pi x)\right]_0^1$$

$$= \frac{2e}{n^2\pi^2+1}(0 - n\pi \cos n\pi) - \frac{2.1}{n^2\pi^2+1}(0 - n\pi\cdot 1)$$

$$= \frac{2n\pi}{n^2\pi^2+1}(1-(-1)^n e) \tag{2}$$

The FSS for $f(x) = e^x$ in (0, 1)

$$e^x = \pi\left[\frac{2(1+e)}{\pi^2+1}\sin \pi x + \frac{2(1-e)}{4\pi^2+1}\sin 2\pi x \right. \tag{3}$$

$$\left. + \frac{3(1+e)}{9\pi^2+1}\sin 3\pi x + \cdots\right]$$

Example 8.24
Find the Fourier half-range sine series for $f(x) = x$ in $(0, \pi)$. [JNTU 2003 (4)]

Solution The half-range FSS for

$$f(x) = x \text{ in } (0, \pi) \tag{1}$$

is given by
$$f(x) = x = \sum_{n=1}^{\infty} b_n \sin nx \tag{2}$$

where

$$b_n = \frac{2}{\pi}\int_0^{\pi} f(x)\sin nx\, dx = \frac{2}{\pi}\int_0^{\pi} x \sin nx\, dx$$

$$= \frac{2}{\pi}\left[x\left(-\frac{\cos x}{n}\right) - 1\left(-\frac{\sin nx}{n^2}\right)\right]_0^{\pi} \qquad \begin{matrix} t_2 = 0 & \text{at } x = \pi, 0 \\ t_1 = 0 & \text{at } x = 0 \end{matrix}$$

$$= -\frac{2\pi}{\pi n}\cos n\pi = -2\cdot\frac{(-1)^n}{n} \tag{3}$$

Substituting for b_n in (2) we obtain

$$x = 2\sum_{n=1}^{\infty}\frac{(-1)^{n-1}}{n}\sin nx \tag{4}$$

Example 8.25
Find the half-range sine series for $f(t) = t - t^2$ in $0 < t < 1$. [JNTU May/June 2002]

Solution The half-range FSS for the function

$$f(t) = t - t^2 \quad \text{in} \quad 0 < t < 1 \tag{1}$$

is given by

$$f(t) = \sum b_n \sin n\pi t \quad (\because \; l = 1 \text{ here}) \tag{2}$$

where

$$b_n = \frac{2}{1}\int_0^1 f(t)\sin n\pi t\, dt = 2\int_0^1 (t-t^2)\sin n\pi t\, dt$$

$$= 2\left[(t-t^2)\left(-\frac{\cos n\pi t}{n\pi}\right) - (1-2t)\left(-\frac{\sin n\pi t}{n^2\pi^2}\right)\right.$$

$$\left. + (-2)\left(\frac{\cos n\pi t}{n^3\pi^3}\right)\right]_0^1$$

$$= -\frac{4}{n^3\pi^3}(\cos n\pi - 1) = -\frac{4}{n^3\pi^3}((-1)^n - 1)$$

$$= \begin{cases} 0, \text{ if } n \text{ is even} \\ \dfrac{8}{n^3\pi^3}, \text{ if } n \text{ is odd} \end{cases}$$

Hence, the FSS is

$$t - t^2 = \frac{8}{\pi^3}\sum_{n=1}^{\infty}\frac{\sin(2n-1)\pi t}{(2n-1)^3} \tag{3}$$

Example 8.26
Expand $\cos \pi x$ in (0, 1) as an FSS (or) show that

$$\cos \pi x = \frac{8}{\pi}\sum_{n=1}^{\infty}\frac{n}{4n^2-1}\sin 2n\pi x \text{ in } (0,\ 1)$$

[JNTU 2004 S(3)]

Solution The FSS of $\cos \pi x$ is given by

$$\cos \pi x = \sum_{n=1}^{\infty} b_n \sin n\pi x \quad (\because \; l = 1) \tag{1}$$

where

$$b_n = 2\int_0^1 \cos \pi x \cdot \sin n\pi x \, dx$$

$$= \int_0^1 [\sin(n+1)\pi x + \sin(n-1)\pi x] \, dx$$

$$= \frac{-1}{\pi}\left[\frac{\cos(n+1)\pi x}{n+1} + \frac{\sin(n-1)\pi x}{n-1}\right]_0^1, \ (n \neq 1)$$

$$= \frac{-1}{\pi}\left[\frac{(-1)^{n+1}-1}{n+1} + \frac{(-1)^{n-1}-1}{n-1}\right]$$

$$= \begin{cases} 0, & \text{if } n \text{ is odd} \\ \dfrac{4n}{\pi(n^2-1)}, & \text{if } n \text{ is even.} \end{cases} \quad (2)$$

***n* = 1**

$$b_1 = \int_0^1 2\sin \pi x \cos \pi x \, dx = \frac{1}{\pi}(\sin^2 \pi x)_0^1 = 0$$

∴ The FSS for cos πx is

$$\cos \pi x = \frac{8}{\pi}\sum_{m=1}^{\infty} \frac{m \sin 2mx}{4m^2-1} \quad (3)$$

Example 8.27

Prove that the function $f(x) = x$ can be expanded in a series of cosines in

$$0 < x < \pi \text{ as } x = \frac{\pi}{2} - \frac{4}{\pi}\left(\frac{\cos x}{1^2} + \frac{\cos 3x}{3^2} + \frac{\cos 5x}{5^2} + \cdots\right)$$

[JNTU 2003 S]

Hence, deduce that $\dfrac{\pi^2}{8} = \dfrac{1}{1^2} + \dfrac{1}{3^2} + \dfrac{1}{5^2} + \cdots$

Solution Here we are required to find Fourier series in the positive interval (0, π) for $f(x) = x$ in terms of cosines. This means that we have to find half-range FCS for

$$f(x) = x \quad (1)$$

and this is given by $f(x) = \dfrac{a_0}{2} + \sum_{n=1}^{\infty} a_n \cos nx \quad (2)$

where

$$a_0 = \frac{2}{\pi}\int_0^{\pi} f(x)\, dx = \frac{2}{\pi}\int_0^{\pi} x\, dx = \frac{2}{\pi}\left(\frac{x^2}{2}\right)_0^{\pi} = \pi \quad (3)$$

$$a_n = \frac{2}{\pi}\int_0^{\pi} f(x)\cos nx\, dx = \frac{2}{\pi}\int_0^{\pi} x\cos nx\, dx$$

$$= \frac{2}{\pi}\left[x\left(\frac{\sin nx}{n}\right) - 1\left(-\frac{\cos nx}{n^2}\right)\right]_0^{\pi}$$

$$= \frac{2}{\pi}\left[0 + \frac{1}{n^2}(\cos nx - \cos 0)\right]$$

$$= \frac{2}{\pi n^2}\left[(-1)^n - 1\right] \quad \because \quad \sin \pi = \sin 0 = 0$$

$$= \begin{cases} 0, & \text{if } n \text{ is even} \\ \dfrac{-4}{\pi(2m-1)^2}, & \text{if } n \text{ is odd and } n = 2m-1 \end{cases}$$

(4)

Substituting these values in (2) we obtain

$$x = \frac{\pi}{2} - \frac{4}{\pi}\sum_{m=1}^{\infty} \frac{\cos(2m-1)x}{(2m-1)^2} \quad (5)$$

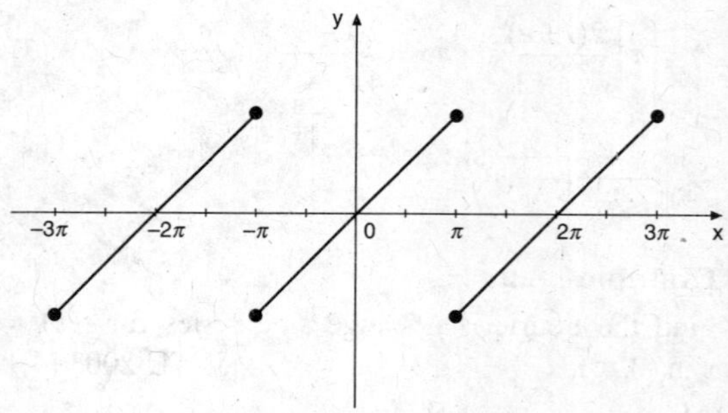

Figure 8.7 $f(x) = x$ defined in $(0, \pi)$ and its extensions

Deductions

$x = 0$ is a point of discontinuity for $f(x)$;

$$f(0) = \frac{f(0-) + f(0+)}{2} = 0$$

The series converges to 0 at $x = 0$

$$\therefore \quad 0 = \frac{\pi}{2} - \frac{4}{\pi}\left(\frac{1}{1^2} + \frac{1}{3^2} + \frac{1}{5^2} + \cdots\right)$$

$$\Rightarrow \quad \frac{1}{1^2} + \frac{1}{3^2} + \frac{1}{5^2} + \cdots = \frac{\pi^2}{8} \quad (6)$$

Example 8.28

Find the half-range cosine series for $f(x) = x(2 - x)$ in $0 \le x < 2$ and hence find the sum of the series. [JNTU 2002, 2003, 2004 (3)]

Solution The half-range FCS for

$$f(x) = x(2 - x) \quad \text{in } 0 \le x < 2 \tag{1}$$

is given by

$$x(2-x) = \frac{a_0}{2} + \sum_{n=1}^{\infty} a_n \cos\frac{n\pi x}{2} \qquad (\because\ l = 2) \tag{2}$$

where

$$a_0 = \frac{2}{2}\int_0^2 f(x)\,dx = \int_0^2 (2x - x^2)\,dx = \left(x^2 - \frac{1}{3}x^3\right)_0^2 \tag{3}$$

$$= 4 - \frac{8}{3} = \frac{4}{3}$$

$$a_n = \frac{2}{2}\int_0^2 f(x)\cos\frac{n\pi x}{2}\,dx = \int_0^2 (2x - x^2)\cos\frac{n\pi x}{2}\,dx$$

$$= \left[(2x - x^2)\left(\frac{2}{n\pi}\right)\cdot\sin\frac{n\pi x}{2} - (2 - 2x)\cdot \frac{4}{n^2\pi^2}\left(-\cos\frac{n\pi x}{2}\right) + (-2)\frac{8}{n^3\pi^3}\left(-\sin\frac{n\pi x}{2}\right)\right]_0^2$$

$$t_1 = t_3 = 0 \quad \text{at } x = 2, 0$$

$$= \frac{8}{n^2\pi^2}(-\cos n\pi - \cos 0)$$

$$= -\frac{8}{n^2\pi^2}((-1^n) + 1)\begin{bmatrix} 0, & \text{if } n \text{ is odd} \\ -\dfrac{16}{n^2\pi^2}, & \text{if } n \text{ is even} \end{bmatrix} \tag{4}$$

Substituting these values in (2) we obtain

$$2x - x^2 = \frac{2}{3} - \frac{4}{\pi^2}\sum_{m=1}^{\infty}\frac{\cos m\pi x}{m^2} \tag{5}$$

$$= \frac{2}{3} - \frac{4}{\pi^2}\left(\cos\pi x + \frac{1}{2^2}\cos 2\pi x + \frac{1}{3^2}\cos 3\pi x + \cdots\right)$$

Deduction

Putting $x = 1$ in (5) we have

$$\frac{1}{1^2} - \frac{1}{2^2} + \frac{1}{3^2} - \frac{1}{4^2} + \frac{1}{5^2} - \frac{1}{6^2} + \cdots = \frac{\pi^2}{12} \tag{6}$$

Example 8. 29

For the function defined by the graph below, find the half-range FCS. [JNTU 2003 S]

Solution The graph is composed of two line segments OP and PB in intervals $\left(0, \frac{\pi}{2}\right)$ and $\left(\frac{\pi}{2}, \pi\right)$, respectively, their slopes being +1 and −1.

Equation of $OP : y = x \quad \left(0, \frac{\pi}{2}\right)$

Equation of $PB : y = -(x - \pi) \quad \left(\frac{\pi}{2}, \pi\right)$

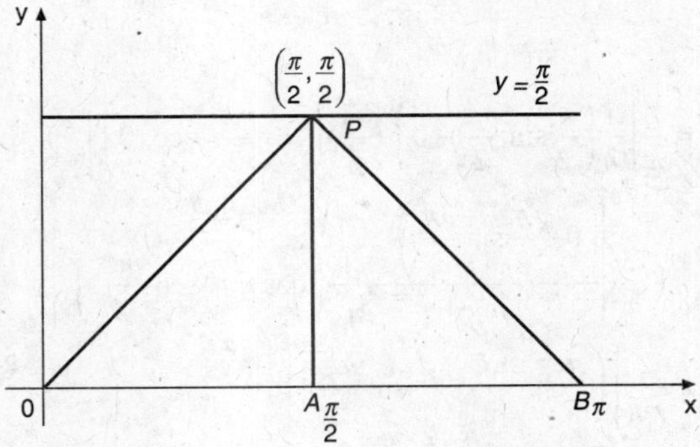

Figure. 8.8 $f(x) = \begin{cases} x \text{ in } (0, \pi/2) \\ \pi - x \text{ in } (\pi/2, \pi) \end{cases}$ (EX.8.29)

So, the function $f(x)$ is defined by

$$f(x) = \begin{cases} x, & 0 < x < \pi/2 \\ \pi - x, & \pi/2 < x < \pi \end{cases} \tag{1}$$

The half-range FCS is

$$f(x) = \frac{1}{2}a_0 + \sum_{n=1} a_n \cos nx \tag{2}$$

where

$$a_0 = \frac{2}{\pi}\int_0^{\pi} f(x)\,dx = \frac{2}{\pi}\int_0^{\pi/2} x\,dx + \frac{2}{\pi}\int_{\pi/2}^{\pi}(\pi - x)\,dx$$

$$= \frac{2}{\pi}\left(\frac{x^2}{2}\right)_0^{\pi/2} + \frac{2}{\pi}\left(\pi x - \frac{x^2}{2}\right)_{\pi/2}^{\pi}$$

$$= \frac{2}{\pi}\cdot\frac{\pi^2}{8} + \frac{2}{\pi}\left[\pi(\pi - \pi/2) - \frac{1}{2}\left(\pi^2 - \frac{\pi^2}{4}\right)\right] = \frac{\pi}{2} \quad (3)$$

$$a_n = \frac{2}{\pi}\int_0^{\pi} f(x)\cos nx\, dx$$

$$= \frac{2}{\pi}\int_0^{\pi/2} x\cos nx\, dx + \frac{2}{\pi}\int_{\pi/2}^{\pi} (\pi - x)\cos nx\, dx$$

$$= \frac{2}{\pi}\left[x\left(\frac{\sin nx}{n}\right) - 1\cdot\left(-\frac{\cos nx}{n^2}\right)\right]_0^{\pi/2}$$

$$+ \frac{2}{\pi}\left[(\pi - x)\left(\frac{\sin nx}{n}\right) - (-1)\cdot\left(-\frac{\cos nx}{n^2}\right)\right]_{\pi/2}^{\pi}$$

$$= \frac{2}{\pi}\left[\frac{1}{n}\left(\frac{\pi}{2}\sin\frac{n\pi}{2} - 0\right) + \frac{1}{n^2}\left(\cos\frac{n\pi}{2} - \cos 0\right)\right]$$

$$+ \frac{2}{\pi}\left[\frac{\left(0 - \frac{\pi}{2}\sin\frac{n\pi}{2}\right)}{n} - \frac{1}{n^2}\left(\cos n\pi - \cos\frac{n\pi}{2}\right)\right]$$

$$= \frac{2}{\pi n}\left[\left(\frac{\pi}{2}\sin\frac{n\pi}{2} - \frac{\pi}{2}\sin\frac{n\pi}{2}\right)\right]$$

$$+ \frac{2}{\pi n^2}\left[2\cos\frac{n\pi}{2} - \cos 0 - \cos n\pi\right]$$

$$= 0 + \frac{2}{\pi n^2}\left[2\cos\frac{n\pi}{2} - (1 + (-1)^n)\right]$$

$$= \begin{cases} 0, & \text{if } n \text{ is odd} \\ \dfrac{4}{\pi(2m)^2}(\cos m\pi - 1), & \text{if } n \text{ is even and } n = 2m \end{cases} \quad (4)$$

$$a_1 = a_3 = a_5 = \cdots = 0 \text{ and } a_{2m} = \frac{1}{\pi m^2}(\cos m\pi - 1)$$

$$= \begin{cases} 0, & \text{if } m \text{ is even} \\ \dfrac{-2}{\pi(2k-1)^2}, & \text{if } m \text{ is odd, } m = 2k - 1 \end{cases} \quad (5)$$

$$\left.\begin{aligned} a_{2(2k)} &= a_{4k} = 0 \\ a_{2(2k-1)} &= a_{4k-2} = -\frac{2}{\pi(2k-1)^2} \end{aligned}\right\} k \in N \quad (6)$$

$$\therefore \quad a_2 = -\frac{2}{\pi},\ a_6 = -\frac{2}{\pi}\cdot\frac{1}{3^2},\ a_{10} = -\frac{2}{\pi}\cdot\frac{1}{5^2}, \ldots$$

$$k = 1, 2, 3, 4, \ldots$$

Substituting these values in (2) we obtain

$$f(x) = \frac{\pi}{4} - \frac{2}{\pi}\left(\frac{\cos 2x}{1^2} + \frac{\cos 6x}{3^2} + \frac{\cos 10x}{5^2} + \cdots\right) \quad (7)$$

Example 8.30

Find the FCS for $f(x) = x \sin x$ in $(0, \pi)$ and show that

$$1 + \frac{2}{1.3} - \frac{2}{3.5} + \frac{2}{5.7} - \frac{2}{7.9} + \cdots = \frac{\pi}{2}. \quad \text{[JNTU 2002 (2)]}$$

Solution Here we have to find half-range cosine series for

$$f(x) = x\sin x \quad \text{in } (0, \pi) \quad (1)$$

The required Fourier half-range cosine series is

$$f(x) = x\sin x = \frac{a_0}{2} + \sum_{n=1}^{\infty} a_n \cos nx \quad (2)$$

where

$$a_0 = \frac{2}{\pi}\int_0^{\pi} f(x)\, dx = \frac{2}{\pi}\int_0^{\pi} x\sin x\, dx$$

$$= \frac{2}{\pi}[x(-\cos x) - 1\cdot(-\sin x)]_0^{\pi} = \frac{-2}{\pi}\pi\cos\pi = 2$$

$$(\because \ \cos\pi = -1) \quad \begin{cases} \sin x = 0 & \text{at } x = \pi, 0 \\ t_1 = 0 & \text{at } x = 0 \end{cases} \quad (3)$$

$$a_n = \frac{2}{\pi}\int_0^{\pi} f(x)\cos nx\, dx = \frac{2}{\pi}\int_0^{\pi} x\sin x\cos nx\, dx$$

$$= \frac{1}{\pi}\int_0^{\pi} x[\sin(n+1)x - \sin(n-1)x]\, dx$$

$$= \frac{1}{\pi}\left[x\left\{\frac{-\cos(n+1)x}{n+1} + \frac{\cos(n-1)x}{n-1}\right\}\right.$$

$$\left. + 1\left\{\frac{\sin(n+1)x}{(n+1)^2} - \frac{\sin(n-1)x}{(n-1)^2}\right\}\right]_0^{\pi}, \quad (n \neq 1)$$

$$=\frac{-(-1)^{n+1}}{n+1}+\frac{(-1)^{n-1}}{n-1}=(-1)^{n-1}\left(\frac{-1}{n+1}+\frac{1}{n-1}\right)$$

$$\begin{Bmatrix}\sin(n+1)x=\sin(n-1)x=0\\ at\ x=\pi,0\\ t_1=0\ at\ x=0\end{Bmatrix}$$

$$=\frac{2(-1)^{n-1}}{(n^2-1)}(n\neq 1);$$

$$a_2=\frac{-2}{1.3},\quad a_3=\frac{2}{2.4},\quad a_4=\frac{-2}{3.5},\quad a_5=\frac{2}{4.6}.$$

$$a_1=\frac{2}{\pi}\int_0^{\pi}x\sin x\cos x\,dx=\frac{1}{\pi}\int_0^{\pi}x\sin 2x\,dx$$

$$=\frac{1}{\pi}\left[x\left(-\frac{\cos 2x}{2}\right)-1\cdot\left(-\frac{\sin 2x}{2^2}\right)\right]_0^{\pi}=-\frac{1}{2}\qquad(4)$$

$$=\begin{Bmatrix}\sin 2x=0 & at\ x=\pi,0\\ x(\cdot)=0 & at\ x=0\end{Bmatrix}$$

Substituting these values in (2) we obtain

$$x\sin x=1-\frac{1}{2}\cos x-\frac{2}{1.3}\cos 2x+\frac{2}{2.4}\cos 3x$$
$$-\frac{2}{3.5}\cos 4x+\cdots\qquad(5)$$

Deduction

$x=\pi/2$ is a point of continuity for $f(x)=x\sin x$ and the series converges to $f(\pi/2)=\pi/2$ when $x=\pi/2$

$$\frac{\pi}{2}=1+\frac{1}{2}(0)-\frac{2}{1.3}(-1)+\frac{2}{2.4}(0)+\frac{2}{3.5}(1)+\cdots$$

$$=1+\frac{2}{1.3}-\frac{2}{3.5}+\frac{2}{5.7}-\frac{2}{7.9}+\cdots\qquad(6)$$

EXERCISE 8.4

(I) Obain half-range FSS for 1 to 5

1. $f(x)=x$ in (0, 2).

Ans: $f(x)=x=\frac{4}{\pi}\sum_{n=1}^{\infty}\frac{(-1)^{n-1}\sin\frac{n\pi x}{2}}{n}$

2. $f(x)=x$ in $(0,\pi)$.

Ans: $f(x)=x=\pi\sum_{n=1}^{\infty}\frac{(-1)^{n-1}\sin nx}{n}$

3. $f(x)=\begin{cases}\frac{2x}{l}, & 0\le x\le\frac{l}{2}\\ \frac{2(l-x)}{l}, & \frac{l}{2}\le x\le l\end{cases}$

Ans: $f(x)=\frac{8}{\pi^2}\sum_{n=1}^{\infty}\frac{\sin\frac{n\pi}{2}}{n^2}\sin\frac{n\pi x}{l}$

4. $f(x)=\begin{cases}\sin x, & \left(0,\frac{\pi}{4}\right)\\ \cos x, & \left(\frac{\pi}{4},\frac{\pi}{2}\right)\end{cases}$ [JNTU 1999]

Ans: $f(x)=\frac{4\sqrt{2}}{\pi}\sum_{n=0}^{\infty}\frac{(-1)^{n+1}\sin(4n+2)x}{(4n+1)(4n+3)}$

5. $f(x)=\begin{cases}\frac{1}{4}-x, & 0<x<\frac{1}{2}\\ x-\frac{3}{4}, & \frac{1}{2}<x<1\end{cases}$ [JNTU 2003 (1)]

Ans: $f(x)=\frac{1}{\pi^2}\left[(\pi-4)\sin\pi x+\frac{3\pi+4}{3^2}\sin 3\pi x+\frac{5\pi-4}{5^2}\sin 5\pi x-\cdots\right]$

(II) Obtain half-range FCS for 6 to 10

6. $f(x)=x$ in $0\le x\le\pi$. Deduce $\frac{1}{1^2}+\frac{1}{3^2}+\frac{1}{5^2}+\cdots=\frac{\pi^2}{8}$. [JNTU 2003 S]

Ans: $f(x)=x=\frac{\pi}{4}-\frac{4}{\pi}\sum_{n=1}^{\infty}\frac{\cos(2x-1)x}{(2n-1)^2}$.

7. $f(x)=\sin\frac{\pi x}{l}$, $(0<x<1$. [JNTU 2004 S(4)]

Ans: $f(x)=\sin\frac{\pi x}{l}=\frac{2}{\pi}-\frac{4}{\pi}\left(\frac{\cos\frac{2\pi x}{l}}{1.3}+\frac{\cos\frac{4\pi x}{l}}{3.5}+\cdots\right)$

8. $f(x)=x$ in (0, 2).

Ans: $f(x)=x=1-\frac{8}{\pi^2}\sum_{n=1}^{\infty}\frac{\cos\left(\frac{2n-1}{2}\pi x\right)}{(2n-1)^2}$

9. $f(x)=\begin{cases} kx, & 0\le x\le \frac{1}{2} \\ k(l-x), & \frac{l}{2}\le x\le l \end{cases}$

Deduce the value of $\frac{1}{1^2}+\frac{1}{3^2}+\frac{1}{5^2}+\frac{1}{7^2}+\cdots$

[JNTU 2005 S(3)]

Ans: $f(x)=\frac{kl}{4}-\frac{2kl}{\pi^2}\sum_{n\ odd}^{\infty}\frac{\cos 2n\pi\frac{x}{l}}{n^2}\ \frac{\pi^2}{8}$

10. $f(x)=\begin{cases} 1, & 0<x<\frac{3}{2} \\ 1-x, & \frac{3}{2}<x<3 \end{cases}$

Ans: $f(x)=-\frac{1}{8}+\frac{3}{\pi^2}\left[(\pi+2)\cos\frac{\pi x}{3}-\cos\frac{2\pi x}{2}\right.$
$\left.+\frac{2-3\pi}{9}\cos\frac{3\pi x}{3}-\cdots\right].$

8.11 Root Mean Square (RMS) Value of a Function

In many engineering applications, one has to deal with the concept of root mean square (RMS) value or effective value of a function.

The RMS value of a function $y = f(x)$ over a given interval (a, b) is defined by

$$\bar{y}=\left(\frac{1}{b-a}\int_a^b y^2\,dx\right)^{1/2} \tag{8.38}$$

If the interval is $(-l, l)$ then $\bar{y}=\left(\frac{1}{2l}\int_{-l}^{l}[f(x)]^2\,dx\right)^{1/2}$

$$\bar{y}^2=\frac{1}{2l}\int_{-l}^{l}[f(x)]^2\,dx \tag{8.39}$$

8.11.1 Parseval's Formula

Theorem 8.1

If the Fourier series for $f(x)$ converges uniformly in $(-l, l)$ then

$$\int_{-l}^{l}[f(x)]^2\,dx=l\left[\frac{1}{2}a_0^2+\sum_{n=1}^{\infty}(a_n^2+b_n^2)\right] \tag{8.40}$$

Proof Let the Fourier series for $f(x)$ in $(-l, l)$ be

$$f(x)=\frac{1}{2}a_0+\sum_{n=1}^{\infty}a_n\cos\frac{n\pi x}{l}+\sum_{n=1}^{\infty}b_n\sin\frac{n\pi x}{l} \tag{8.41}$$

Multiplying both sides of (8.41) by $f(x)$ and integrating term by term from $-l$ to l we obtain

$$\int_{-l}^{l}[f(x)]^2\,dx=\frac{a_0}{2}\int_{-l}^{l}f(x)\,dx+\sum_{n=1}^{\infty}a_n\int_{-l}^{l}f(x)\cos\frac{n\pi x}{l}dx$$
$$+\sum_{n=1}^{\infty}b_n\int_{-l}^{l}f(x)\sin\frac{n\pi x}{l}dx \tag{8.42}$$

Also (a_0, a_n, b_n)

$$=\frac{1}{l}\int_{-l}^{l}f(x)\left(1, \cos\frac{n\pi x}{l}, \sin\frac{n\pi x}{l}\right)dx$$

$$\Rightarrow\ \int_{-l}^{l}f(x)\left(1, \cos\frac{n\pi x}{l}, \sin\frac{n\pi x}{l}\right)dx=(la_0, la_n, lb_n) \tag{8.43}$$

Substituting (8.42) we have

$$\int_{-l}^{l}[f(x)]^2\,dx=l\left[\frac{1}{2}a_0^2+\sum_{n=1}^{\infty}(a_n^2+b_n^2)\right] \tag{8.44}$$

which is known as Parseval's formula.

Related Formulae

1. If $f(x)$ is an even function in $(-l, l)$, (8.43) becomes

$$\frac{2}{l}\int_0^l[f(x)]^2\,dx=\frac{1}{2}a_0^2+\sum_{n=1}^{\infty}a_n^2,\qquad b_n=0 \tag{8.45}$$

2. If $f(x)$ is an odd function in $(-l, l)$, (8.43) becomes

$$\frac{2}{l}\int_0^l[f(x)]^2\,dx=\sum_{n=1}^{\infty}b_n^2,\ a_0=a_n=0 \tag{8.46}$$

3. If the half-range cosine series for $f(x)$ in $(0, l)$ is $f(x)=\frac{1}{2}a_0+\sum_{n=1}^{\infty}a_n\cos\frac{n\pi x}{l}$ then

$$\int_0^l [f(x)]^2\,dx = \frac{1}{2}l\left[\frac{1}{2}a_0^2 + a_1^2 + a_2^2 + a_3^2 + \cdots\right] \tag{8.47}$$

4. If the half-range sine series for $f(x)$ in $(0, l)$ is $f(x) = \sum_{n=1}^{\infty} b_n \sin\frac{n\pi x}{l}$ then

$$\int_0^l [f(x)]^2\,dx = \frac{1}{2}l[b_1^2 + b_2^2 + b_3^2 + \cdots] \tag{8.48}$$

Example 8.31

Find the Fourier expansion of $f(x)=x^2$ in $-\pi<x<\pi$. Using the series find the RMS value of $f(x)$ in the interval. Deduce $\frac{1}{1^4}+\frac{1}{2^4}+\frac{1}{3^4}+\cdots=\frac{\pi^4}{90}$.

Solution Since $f(x) = x^2$ is even in $(-\pi, \pi)$ the Fourier series is given by

$$f(x) = x^2 = \frac{a_0}{2} + \sum_{n=1}^{\infty} a_n \cos nx, \quad b_n = 0 \tag{1}$$

where

$$a_0 = \frac{2}{\pi}\int_0^\pi x^2\,dx = \frac{2}{\pi}\left(\frac{x^3}{3}\right)_0^\pi = \frac{2}{3}\cdot\pi^2 \tag{2}$$

$$a_n = \frac{2}{\pi}\int_0^\pi x^2 \cos nx\,dx$$

$$= \frac{2}{\pi}\left[x^2\left(\frac{\sin nx}{n}\right) - 2x\left(-\frac{\cos nx}{n^2}\right) + 2\left(-\frac{\sin nx}{n^2}\right)\right]_0^\pi$$

$$= \frac{2}{\pi}\cdot 2\pi\frac{\cos n\pi}{n^2} = 4\frac{(-1)^n}{n^2} \tag{3}$$

$t_1 = t_3 = 0$ at $x = \pi, 0$ $\quad t_2 = 0$ at $x = 0$

RMS value of

$$f(x) = \left[\frac{1}{2\pi}\int_{-\pi}^{\pi}[f(x)]^2\,dx\right]^{1/2} = \frac{1}{2\pi}\cdot\int_{-\pi}^{\pi} x^4\,dx \tag{4}$$

$$= \sqrt{\frac{1}{2\pi}2\cdot\left(\frac{x^5}{5}\right)_0^\pi} = \frac{\pi^2}{\sqrt{5}}$$

By RMS value in series

$$\int_{-\pi}^{\pi}[f(x)]^2\,dx = 2\pi\left[\frac{1}{4}\left(\frac{2\pi^2}{3}\right)^2 + \frac{1}{2}\sum\left(\frac{4(-1)^n}{n^2}\right)^2\right]$$

$$\Rightarrow \frac{4x^4}{45} = 8\sum\frac{1}{n^4} \Rightarrow \frac{1}{1^4}+\frac{1}{2^4}+\frac{1}{3^4}+\cdots=\frac{\pi^4}{90} \tag{5}$$

Example 8.32

Find (a) the half-range cosine series and (b) sine series for $f(x) = x$ in $0 < x < l$.

Also, deduce the value of $\frac{1}{1^4}+\frac{1}{3^4}+\frac{1}{5^4}\cdots$

[JNTU 2001].

Solution

(a) The half-range cosine series is given by

$$f(x) = x = \frac{1}{2}a_0 + \sum_{n=0}^{\infty} a_n \cos\frac{n\pi x}{l} \tag{1}$$

where

$$a_0 = \frac{2}{l}\int_0^l f(x)\,dx = \frac{2}{l}\int_0^l x\,dx = \frac{2}{l}\left(\frac{x^2}{2}\right)_0^l = l \tag{2}$$

$$a_n = \frac{2}{l}\int_0^l f(x)\cos\frac{n\pi x}{l}\,dx = \frac{2}{l}\int_0^l x\cos\frac{n\pi x}{l}\,dx$$

$$= \frac{2}{l}\left[x\left(\frac{1}{n\pi}\sin\frac{n\pi x}{l}\right) - 1\cdot\left(-\frac{l^2}{n^2\pi^2}\cos\frac{n\pi x}{l}\right)\right]_0^l$$

$$= \frac{2}{l}\left[\frac{l^2}{n^2\pi^2}\right](\cos n\pi - \cos 0) = \frac{2l}{n^2\pi^2}((-1)^n - 1)$$

$$= \begin{cases} 0 & \text{if } n \text{ is even} \\ \dfrac{-4l}{n^2\pi^2} & \text{if } n \text{ is odd} \end{cases} \tag{3}$$

$$\therefore\; x = \frac{l}{2} - \frac{4l}{\pi^2}\sum_{n\,\text{odd}}\frac{\cos\frac{n\pi x}{l}}{n^2} \tag{4}$$

(b) The half-range sine series is given by

$$f(x) = x = \sum_{n=0}^{\infty} b_n \sin\frac{n\pi x}{l} \tag{1}$$

where

$$b_n = \frac{2}{l}\int_0^l f(x)\sin\frac{n\pi x}{l}\,dx = \frac{2}{l}\int_0^l x\sin\frac{n\pi x}{l}\,dx \quad (2)$$

$$= \frac{2}{l}\Bigg[\underbrace{x\left(-\frac{l}{n\pi}\cos\frac{n\pi x}{l}\right)}_{=0 \text{ at } x=0} \underbrace{-1\cdot\left(\frac{l^2}{n^2\pi^2}\right)\left(-\sin\frac{n\pi x}{l}\right)}_{=0 \text{ at } x=l,0}\Bigg]_0^l$$

$$= \frac{2}{l}l\left(-\frac{l}{n\pi}\right)\cos n\pi = -\frac{2l}{n\pi}(-1)^n$$

$$\therefore\; x = \frac{2l}{\pi}\sum_{n=1}^{\infty}\frac{(-1)^{n-1}}{n}\sin\frac{n\pi x}{l} \quad (3)$$

Deduction

8.11.2 By Parseval's Related Formulas

(1) $$\int_0^l [f(x)]^2\,dx = \int_0^l x^2\,dx = l\left[\frac{l^2}{4} + \frac{1}{2}\sum_{n=1,3,5\ldots}\frac{16l^2}{n^4\pi^4}\right]$$

$$\frac{l^3}{3} = l^3\left[\frac{1}{4} + \frac{1}{2}\sum_{n=1,3,5}\frac{16}{n^4\pi^4}\right] \Rightarrow \frac{\pi^4}{12} = 8\sum_{n \text{ odd}}\frac{1}{n^4} \quad (4)$$

$$\frac{1}{1^4} + \frac{1}{3^4} + \frac{1}{5^4} + \cdots = \frac{\pi^4}{96}$$

(2) $$\int_0^l x^2\,dx = l\left[\frac{1}{2}\sum\frac{4l^2}{n^2\pi^2}\right] \Rightarrow \frac{\pi^2}{6} = \sum_{n=1}^{\infty}\frac{1}{n^2}. \quad (5)$$

EXERCISE 8.5

1. If $f(x) = \frac{1}{2}a_0 + \sum_{n=1}^{\infty} a_n\cos\frac{n\pi x}{l}$ $(0 < x < 1)$ show that

$$\int_0^l (f(x))^2\,dx = \frac{1}{2}l\left(\frac{1}{2}a_0^2 + a_1^2 + a_2^2 + a_3^2 + \cdots\right)$$

2. If $f(x) = \frac{1}{2}a_0 + \sum a_n\cos\frac{n\pi x}{l} + \sum b_n\sin\frac{n\pi x}{l}$ in $(0 < x < 2l)$ prove that

$$\int_0^{2l}(f(x))^2\,dx = 2l\left[\frac{1}{4}a_0^2 + \frac{1}{2}\sum(a_n^2 + b_n^2)\right]$$

3. Using the Fourier cosine series for $f(x) = x(\pi - x)$ in $0 < x < \pi$ show that

$$\frac{1}{1^4} + \frac{1}{2^4} + \frac{1}{3^4} + \cdots = \frac{\pi^4}{90}$$

4. If $f(x) = \sum b_n\sin\frac{n\pi x}{l}$ in $0 < x < l$ where

$b_n = \frac{2}{l}\int_0^l f(x)\sin\frac{n\pi x}{l}$ then show that

$$\int_0^l (f(x))^2\,dx = \frac{1}{2}l\sum_{n=1}^{\infty} b_n^2.$$

Numerical Series Arising From Certain Fourier Series

1. $\frac{\pi}{4} = 1 - \frac{1}{3} + \frac{1}{5} - \frac{1}{7} + \cdots = \sum_{n=1}^{\infty}\frac{(-1)^{n-1}}{n-1}$ (Gregory series)

2. $\frac{\pi^2}{8} = 1 + \frac{1}{3^2} + \frac{1}{5^2} + \frac{1}{7^2} + \cdots = \sum_{n=1}^{\infty}\frac{1}{(2n-1)^2}$

3. $\frac{\pi - 2}{4} = \frac{1}{1.3} - \frac{1}{3.5} + \frac{1}{5.7} - \frac{1}{7.9} + \cdots$

$$= \sum_{n=1}^{\infty}\frac{(-1)^{n-1}}{(4n^2-1)} = \sum_{n=1}^{\infty}\frac{(-1)^{n-1}}{(2n-1)(2n+1)}$$

4. $\frac{\pi^2}{6} = \frac{1}{1^2} + \frac{1}{2^2} + \frac{1}{3^2} + \frac{1}{4^2} + \cdots = \sum_{n=1}^{\infty}\frac{1}{n^2}$

5. $\frac{\pi^2}{12} = \frac{1}{1^2} - \frac{1}{2^2} + \frac{1}{3^2} - \frac{1}{4^2} + \frac{1}{5^2} - \frac{1}{6^2} + \cdots = \sum_{n=1}^{\infty}\frac{(-1)^{n-1}}{n^2}$

6. $\frac{1}{2} = \frac{1}{1.3} + \frac{1}{3.5} + \frac{1}{5.7} + \frac{1}{7.9} + \cdots = \sum_{n=1}^{\infty}\frac{1}{(2n-1)(2n+1)}$

$$S_n = \frac{1}{2}\left[\left(\frac{1}{1} - \frac{1}{3}\right) + \left(\frac{1}{2} - \frac{1}{5}\right) + \left(\frac{1}{5} - \frac{1}{7}\right) + \cdots + \left(\frac{1}{2n-1} - \frac{1}{2n+1}\right)\right]$$

$$= \frac{1}{2}\left(1 - \frac{1}{2n+1}\right) \to \frac{1}{2} \text{ as } n \to \infty$$

(Except the first and the last terms all other terms get cancelled)

From (3) and (6) we obtain

7. $\frac{\pi}{8} = \frac{1}{1.3} + \frac{1}{5.7} + \frac{1}{9.11} + \cdots = \sum_{n=1}^{\infty}\frac{1}{(4n-3)(4n-1)}.$

Fourier Integral Transforms

9

9.1 Introduction

Integral transforms are useful in solving initial and boundary value problems and in evaluating certain integrals. Laplace and Fourier transforms are two important transforms which are widely used in engineering and physical applications. They are used in the solution of conduction of heat, vibration of strings, oscillations of elastic beams, transmission lines, *etc.*

Here we define Fourier transform together with Fourier sine and cosine transforms, their inverses and study their properties and consider evaluation of certain integrals.

9.2 Integral Transforms

A linear integral transform or simply an integral transform of a function $f(x)$ is defined by

$$T\{f(x)\} = \int_a^b f(x)K(s,x)\,dx \qquad (9.1)$$

where $K(s, x)$, called the Kernel of the transformation, is a known function. For different choices of the Kernel $K(s, x)$ and the limits a and b, we obtain different transforms.

9.2.1 Laplace Transform

When $K(s,x) = e^{-sx}$ $(s > 0)$, $a = 0$ and $b = \infty$, we obtain the Laplace transform of $f(x)$. It is defined by

$$F(s) = L\{f(x)\} = \int_0^\infty f(x)e^{-sx}\,dx \quad (s > 0) \qquad (9.2)$$

9.2.2 Fourier Transform

When $K(s, x) = e^{isx}$, $a = -\infty$ and $b = \infty$, we obtain the Fourier transform of $f(x)$. It is defined by

$$F(s) = F\{f(x)\} = \int_{-\infty}^{\infty} f(x)e^{isx}\,dx \qquad (9.3)$$

9.3 Fourier Integral Theorem

Let $f(x)$ be a function satisfying the following Dirichlet conditions in every interval $(-l, l)$ $(l > 0)$ however large:

1. $f(x)$ is periodic; $f(x)$ and its integrals are single-valued and finite (bounded).
2. $f(x)$ has at most a finite number of discontinuities in any one period.
3. $f(x)$ has at most a finite number of maxima or minima.

Then the Fourier series expansion of $f(x)$ in $(-l, l)$ is given by

$$f(x) = \frac{a_0}{2} + \sum_{n=1}^{\infty} a_n \cos\frac{n\pi x}{l} + \sum_{n=1}^{\infty} b_n \sin\frac{n\pi x}{l} \qquad (9.4)$$

where

$$(a_0, a_n, b_n) = \frac{1}{l}\int_{-l}^{l} f(t)\left(1, \cos\frac{n\pi t}{l}, \sin\frac{n\pi t}{l}\right) dt \qquad (9.5)$$

Substituting for a_0, a_n, b_n from (9.5), equation (9.4) becomes

$$\begin{aligned} f(x) &= \frac{1}{2l}\int_{-l}^{l} f(t)dt + \frac{1}{l}\sum_{n=1}^{\infty}\int_{-l}^{l} \cos\frac{n\pi t}{l}\cos\frac{n\pi x}{l} f(t)\,dt \\ &\quad + \frac{1}{l}\sum_{n=1}^{\infty}\int_{-l}^{l} \sin\frac{n\pi t}{l}\sin\frac{n\pi x}{l} f(t)\,dt \\ &= \frac{1}{2l}\int_{-l}^{l} f(t)dt + \frac{1}{l}\sum_{n=1}^{\infty}\int_{-l}^{l} \cos\frac{n\pi(t-x)}{l} f(t)\,dt \end{aligned} \qquad (9.6)$$

If we further assume that $f(x)$ is absolutely integrable for all x

i.e., $$\int_{-\infty}^{\infty}|f(x)|\,dx$$

is convergent, then

$$\left.\begin{aligned}&\frac{1}{2l}\int_{-l}^{l}f(t)\,dt\to 0 && \text{as } l\to\infty\\ &\left|\frac{1}{2l}\int_{-l}^{l}f(t)\,dt\right|\le\frac{1}{2l}\int_{-l}^{l}|f(t)|\,dt\to 0 && \text{as } l\to\infty\end{aligned}\right]\quad(9.7)$$

Now (9.6) becomes

$$f(x)=\frac{1}{l}\sum_{n=0}^{\infty}\int_{-l}^{l}\cos\frac{n\pi(t-x)}{l}f(t)\,dt$$

Putting $\frac{\pi}{l}=\delta s$ or $\frac{1}{l}=\frac{\delta s}{\pi}$ we have, on taking limit as $n\to\infty$ or $\delta s\to 0$

$$f(x)=\frac{1}{\pi}\lim_{n\to\infty}\sum_{n=0}^{\infty}\delta s\int_{-l}^{l}\cos[n\delta s(t-x)]f(t)\,dt$$

$$=\frac{1}{\pi}\int_{0}^{\infty}\left[\int_{-\infty}^{\infty}\cos[s(t-x)f(t)\,dt]\right]\quad(9.8)$$

$$\text{as }\lim_{\substack{n\to\infty\\ \text{or}\\ \delta s\to 0}}\phi(n\delta s)=\int_{0}^{\infty}\phi(s)\,ds$$

which is called the Fourier Integral (FI) of $f(x)$.

Note 1 A rigorous proof of the above theorem is beyond the scope of this book.

Note 2 Fourier integral representation for a function $f(x)$ is useful in solving certain differential and integral equations.

9.3.1 Fourier Sine and Cosine Integrals (FSI's and FCI's)

Writing the expansion for cosine in (9.8) we have

$$f(x)=\frac{1}{\pi}\int_{0}^{\infty}\int_{-\infty}^{\infty}(\cos st\cos sx+\sin st\sin sx)f(t)\,dt\,ds$$

$$=\frac{1}{\pi}\int_{0}^{\infty}\cos sx\left(\int_{-\infty}^{\infty}\cos st\,f(t)\,dt\right)ds$$

$$+\frac{1}{\pi}\int_{0}^{\infty}\sin sx\left(\int_{-\infty}^{\infty}\sin st\,f(t)\,dt\right)ds\quad(9.9)$$

When $f(t)$ is an odd function in $(-\infty,\infty)$ then the integral in the first brackets in (9.9) is zero since the integrand $\cos st\,f(t)$ is an odd function in $(-\infty,\infty)$ and so we obtain

$$f(x)=\frac{2}{\pi}\int_{0}^{\infty}\sin sx\left(\int_{0}^{\infty}\sin st\,f(t)\,dt\right)ds\quad(9.10)$$

which is called the *Fourier Sine Integral (FSI)* of $f(x)$.

When $f(t)$ is an even function in $(-\infty,\infty)$ then the integral in the second brackets in (9.9) is zero since the integrand $\sin st\,f(t)$ is an odd function in $(-\infty,\infty)$ and so we obtain

$$f(x)=\frac{2}{\pi}\int_{0}^{\infty}\cos sx\left(\int_{0}^{\infty}\cos st\,f(t)\,dt\right)ds\quad(9.11)$$

which is called the *Fourier Cosine Integral (FCI)* of $f(x)$.

9.4 Fourier Integral in Complex Form

$\cos s(t-x)$ is an even function of s in $(-\infty,\infty)$ and so we can write (9.8) in the form

$$f(x)=\frac{1}{2\pi}\int_{-\infty}^{\infty}\int_{-\infty}^{\infty}\cos[s(t-x)]f(t)\,dt\,ds\quad(9.12)$$

Also, $\sin s(t-x)$ is an odd function of s in $(-\infty,\infty)$. So,we can write

$$0=\frac{1}{2\pi}\int_{-\infty}^{\infty}\int_{-\infty}^{\infty}\sin[s(t-x)]f(t)\,dt\,ds\quad(9.13)$$

Now, multiplying (9.13) by 'i' and adding it to (9.12) we obtain

$$f(x)=\frac{1}{2\pi}\int_{-\infty}^{\infty}\int_{-\infty}^{\infty}\{\cos[s(t-x)]$$
$$+\,i\sin[s(t-x)]\}f(t)\,dt\,ds$$

Using Euler's formula $e^{i\theta} = \cos\theta + i\sin\theta$ we can write the above integral as

$$f(x) = \frac{1}{2\pi}\int_{-\infty}^{\infty}\int_{-\infty}^{\infty} f(t)e^{is(t-x)}\,dt\,ds \qquad (9.14)$$

which is called *Fourier Integral of f(x) in complex form (FICF)*

9.4.1 Fourier Integral Representation of a Function

By (9.9) a function $f(x)$ may be represented by a FI as

$$f(x) = \frac{1}{\pi}\int_0^{\infty}[A(s)\cos sx + B(s)\sin sx]\,ds \qquad (9.15)$$

where

$$A(s) = \int_{-\infty}^{\infty} f(t)\cos st\,dt \qquad (9.16)$$

and

$$B(s) = \int_{-\infty}^{\infty} f(t)\sin st\,dt \qquad (9.17)$$

If $f(x)$ is an odd function in $(-\infty, \infty)$ then

$$f(x) = \frac{1}{\pi}\int_0^{\infty} B(s)\sin sx\,ds \qquad (9.18)$$

where

$$B(s) = 2\int_0^{\infty} f(t)\sin st\,dt \qquad (9.19)$$

and if $f(x)$ is an even function in $(-\infty, \infty)$ then

$$f(x) = \frac{1}{\pi}\int_0^{\infty} A(s)\cos sx\,ds \qquad (9.20)$$

where

$$A(s) = 2\int_0^{\infty} f(t)\cos st\,dt \qquad (9.21)$$

9.5 Fourier Transform of $f(x)$

Writing the exponential function $e^{ix(t-x)}$ as a product of two exponential functions $e^{ix(t-x)} = e^{ist}\cdot e^{-isx}$ the FI (9.14) may be put in the following form

$$f(x) = \frac{1}{2\pi}\int_{-\infty}^{\infty} e^{-isx}\left(\int_{-\infty}^{\infty} f(t)e^{ist}\,dt\right)ds \qquad (9.22)$$

The expression inside the brackets is a function of s. Denoting it by $F(s)$ and replacing t by x, we have the Fourier transform (FT) of $f(x)$. Then the inverse Fourier transform (IFT) of $F(s)$ is given by (9.22).

Fourier Transform

The Fourier transformation of $f(x)$, denoted by $F\{f(x)\}$, is defined by

$$F\{f(x)\} = F(s) = \int_{-\infty}^{\infty} f(x)e^{isx}\,dx \qquad (9.23)$$

Inverse Fourier Transform

The inverse FT of $F(s)$, denoted by $F^{-1}\{F(s)\}$ is defined by

$$F^{-1}\{F(s)\} = f(x) = \frac{1}{2\pi}\int_{-\infty}^{\infty} f(s)e^{-isx}\,ds \qquad (9.24)$$

Thus, the function $F(s)$ defined by (9.23) is called the *Fourier Transform* of $f(x)$ and is denoted by

$$F\{f(x)\} = F(s) \qquad (9.25)$$

and the function $f(x)$ given by (9.24) is called the Inverse Fourier transform of $F(s)$ and is denoted by

$$F^{-1}\{F(s)\} = f(x) \qquad (9.26)$$

The process of obtaining the FT $F\{f(x)\} = F(s)$ from a given function $f(x)$ is called the FT method or simply *Fourier Transform.*

Existence of Fourier Transform

The following conditions are sufficient for the existence of the FT of a function $f(x)$:

1. $f(x)$ is piecewise continuous on every finite interval.
2. $f(x)$ is absolutely integrable for all x.

9.5.1 Fourier Sine Transform (FST) and Fourier Cosine Transform (FCT)

The FSI in (9.10) can be written by replacing t by x as

$$f(x) = \int_0^{\infty}\sin sx\left(\int_0^{\infty} f(x)\sin sx\,dx\right)ds \qquad (9.27)$$

The expression in the brackets in (9.27) is a function of s denoted by $F_s(s)$. Writing

$$F_s\{f(x)\} = F_s(s) = \int_0^\infty f(x)\sin sx\, dx \qquad (9.28)$$

equation (9.20) becomes

$$F_s^{-1}\{F_s(s)\} = f(x) = \frac{2}{\pi}\int_0^\infty F_s(s)\sin sx\, ds \qquad (9.29)$$

The function defined by (9.28) is called the Fourier sine transform (FST) of $f(x)$ and that defined by (9.29) is called the inverse Fourier sine transform of $F_s(s)$.

The Fourier cosine integral (9.11) can be written by replacing t by x as

$$f(x) = \int_0^\infty \cos sx\left(\int_0^\infty f(x)\cos sx\, dx\right)ds \qquad (9.30)$$

The expression in the brackets (9.30) is a function of s denoted by $F_c(s)$. Writing

$$F_c\{f(x)\} = F_c(s) = \int_0^\infty f(x)\cos sx\, dx \qquad (9.31)$$

$$F_c^{-1}\{F_c(s)\} = f(x) = \frac{2}{\pi}\int_0^\infty F_c(s)\cos sx\, dx \qquad (9.32)$$

The functions $F_c(s)$ and $F_c^{-1}\{F_c(s)\} = f(x)$ are, respectively, called the Fourier cosine transform and inverse Fourier cosine transform, respectively.

9.6 Finite Fourier Sine Transform and Finite Fourier Cosine Transform (FFCT)

The finite Fourier sine transform of $f(x)$ in $0 < x < l$ is defined by

$$F_s(n) = \int_0^l f(x)\sin\frac{n\pi x}{l}\, dx \qquad (9.33)$$

where n is an integer.

The function $f(x)$ is then called the inverse finite Fourier sine transform of $F_s(n)$ and is given by

$$f(x) = \frac{2}{l}\sum_{n=1}^{\infty} F_s(n)\sin\frac{n\pi x}{l} \qquad (9.34)$$

Similarly the finite Fourier cosine transform of $f(x)$ in $0 < x < l$ is defined by

$$F_c(n) = \int_0^l f(x)\cos\frac{n\pi x}{l}\, dx \qquad (9.35)$$

where n is an integer.

The function $f(x)$ is then called the inverse finite Fourier cosine transform of $F_c(n)$ and is given by

$$f(x) = \frac{1}{l}F_c(0) + \frac{2}{l}\sum_{n=1}^{\infty} F_c(n)\cos\frac{n\pi x}{l} \qquad (9.36)$$

Note 1 The finite Fourier sine transform is useful for solving problems with boundary condition of heat distribution (say) on two parallel boundaries while the finite Fourier cosine transform is useful for solving problems in which the velocity distribution (say) normal to two parallel boundaries are prescribed.

9.6.1 FT, FST and FCT Alternative definitions

The Fourier transform, FST and FCT are alternatively defined as follows:

Definition 1

FT $\quad F(s) = \frac{1}{2\pi}\int_{-\infty}^{\infty} f(x)e^{isx}\, dx,$

IFT $\quad f(x) = \int_{-\infty}^{\infty} F(s)e^{-isx}\, ds \qquad (9.37)$

FST $\quad F_s(s) = \frac{2}{\pi}\int_0^\infty f(x)\sin sx\, dx,$

IFST $\quad f(x) = \int_0^\infty F_s(s)\sin sx\, ds \qquad (9.38)$

FCT $\quad F_c(s) = \frac{2}{\pi}\int_0^\infty f(x)\cos sx\, dx,$

IFCT $\quad f(x) = \int_0^\infty F_c(s)\cos sx\, ds \qquad (9.39)$

Definition 2

FT $\quad F(s) = \frac{1}{\sqrt{2\pi}}\int_{-\infty}^{\infty} f(x)e^{isx}\, dx,$

IFT $$f(x)=\frac{1}{\sqrt{2\pi}}\int_{-\infty}^{\infty}F(s)e^{-isx}\,ds \qquad (9.40)$$

FST $$F_s(s)=\sqrt{\frac{2}{\pi}}\int_0^{\infty}f(x)\sin sx\,dx$$

IFST $$f(x)=\sqrt{\frac{2}{\pi}}\int_0^{\infty}F_s(s)\sin sx\,ds \qquad (9.41)$$

FCT $$F_c(s)=\sqrt{\frac{2}{\pi}}\int_0^{\infty}f(x)\cos sx\,dx$$

IFCT $$f(x)=\sqrt{\frac{2}{\pi}}\int_0^{\infty}F_c(s)\cos sx\,ds \qquad (9.42)$$

Note 1 Each of these notations has its advantages and disadvantages. So, one can adopt any one of these definitions. If both transformation and inverse transformation are involved completing one cycle there will not be any difference in the results. If only one of these is involved the results will be different. The student, while answering a question, should understand which notation is used and then answer, clearly stating the notation he or she is using.

9.7 Convolution Theorem for Fourier Transforms

9.7.1 Convolution

Definition: The convolution of two functions $f(x)$ and $g(x)$ over the interval $(-\infty, \infty)$ denoted by $f * g$ is defined by

$$h(x) = f * g = \int_{-\infty}^{\infty} f(u)g(x-u)\,dx \qquad (9.43)$$

9.7.2 Convolution Theorem

Theorem 9.1 If $F(s) = F\{f(x)\}$ and $G(s) = F\{g(x)\}$ are the Fourier transforms of $f(x)$ and $g(x)$ then the Fourier transform of the convolution of $f(x)$ and $g(x)$ is the product of their transforms

$$F\{f(x) * g(x)\} = F\{f(x)\}\cdot F\{g(x)\} \qquad (9.44)$$

Proof By the definition of Fourier transforms, we have

$$F\{f(x)*g(x)\} = F\left\{\int_{-\infty}^{\infty} f(u)g(x-u)\,du\right\} = \int_{-\infty}^{\infty}\left[\int_{-\infty}^{\infty} f(u)g(x-u)du\right]e^{isx}\,dx$$

$$= \int_{-\infty}^{\infty} f(u)\left[\int_{-\infty}^{\infty} g(x-u)e^{isx}dx\right]du, \quad \text{Changing the order of integration}$$

$$= \int_{-\infty}^{\infty} f(u)\left[\int_{-\infty}^{\infty} e^{is(x-u)}g(x-u)d(x-u)\right]e^{isu}\,du \qquad \because\ dx = d(x-u),\ e^{isx} = e^{is(x-u)}\cdot e^{isu}$$

$$= \int_{-\infty}^{\infty} f(u)e^{isu}\left[\int_{-\infty}^{\infty} g(t)e^{ist}\,dt\right]du, \quad \text{where } t = x-u, dt = d(x-u)$$

$$= \int_{-\infty}^{\infty} f(u)\cdot e^{isu}\,du\cdot G(s) = F(s)\cdot G(s) = F\{f(x)\}\cdot F\{g(x)\}$$

9.7.3 Relation between Laplace and Fourier Transforms

If $$f(t) = \begin{cases} g(t)e^{-xt}, & t > 0 \\ 0, & t < 0 \end{cases} \qquad (9.45)$$

then $F\{f(t)\} = L\{g(t)\}$

Proof

$$F\{f(t)\} = \int_{-\infty}^{\infty} f(t)\cdot e^{ist}\,dt$$

$$= \int_{-\infty}^{0} 0\cdot dt + \int_0^{\infty} g(t)\cdot e^{-xt}\cdot e^{ist}\,dt$$

$$= \int_{-\infty}^{\infty} g(t)e^{-pt}\,dt \qquad \begin{array}{r} p = x - is \text{ and} \\ Re\ p = Re\,(x - is) \end{array}$$

$$= L\{g(t)\}$$

Thus, the Fourier transform of a function $f(t)$ defined by (9.45) is the Laplace transformation of $g(t)$.

9.8 Properties of Fourier Transform

9.8.1. Linearity Property

Theorem 9.2. If $F(s)$ and $G(s)$ are the Fourier transforms of $f(x)$ and $g(x)$, respectively, then

$$F\{af(x) + bg(x)\} = aF(s) + bG(s)$$
$$= aF\{f(x)\} + aF\{g(x)\}$$

where a and b are any constants.

Proof By definition we have

$$F(s) = \int_{-\infty}^{\infty} f(x)e^{isx}\,dx \quad \text{and}$$

$$G(s) = \int_{-\infty}^{\infty} g(x)e^{isx}\,dx$$

$$F\{af(x)+bg(x)\} = \int_{-\infty}^{\infty}\{af(x)+bg(x)\}e^{isx}\,dx$$
$$= a\int_{-\infty}^{\infty} f(x)e^{isx}\,dx + b\int_{-\infty}^{\infty} g(x)e^{isx}\,dx$$
$$= aF(s)+bG(s)$$
$$= aF\{f(x)\}+bF\{g(x)\} \qquad (9.46)$$

9.8.2. Change of Scale Property or Damping Rule

Theorem 9.3 If $F(s)$ is the complex Fourier transform of $f(x)$ then

$$F\{f(ax)\} = \frac{1}{a}F(s/a) \quad (a \neq 0) \qquad (9.47)$$

Proof Since $F(s)$ is the complex Fourier transform of $f(x)$ we have

$$F\{f(x)\} = F(s) = \int_{-\infty}^{\infty} f(x)e^{isx}\,dx$$

So, the complex Fourier transform of $f(ax)$ is given by

$$F\{f(ax)\} = \int_{-\infty}^{\infty} f(ax)e^{isx}\,dx = \frac{1}{a}\int_{-\infty}^{\infty} f(u)e^{i(s/a)u}\,du$$

$$= \frac{1}{a}F(s/a) \qquad \left| \begin{array}{l} ax = u, \quad adx = du \\ \text{The limits are same} \end{array} \right.$$

Corollary If $F_s(s)$ and $F_c(s)$ are the Fourier sine transform and Fourier cosine transform of $f(x)$, respectively, then $F_s(f(ax)) = (1/a)F_s(s/a)$ and $F_c(f(ax)) = (1/a)F_c(s/a)$.

9.8.3. Shifting Property

Theorem 9.4 If $F(s)$ is the complex Fourier transform of $f(x)$ then

$$F\{f(x-a)\} = e^{isa}F\{f(x)\} = e^{isa}F(s) \qquad (9.48)$$

Proof It is given that $F(s) = \int_{-\infty}^{\infty} f(x)e^{isx}\,dx$

By definition

$$F\{f(x-a)\} = \int_{-\infty}^{\infty} f(x-a)e^{isx}\,dx$$
$$= e^{isa}\int_{-\infty}^{\infty} f(u)e^{isu}\,du = e^{isa}F(s),$$

where $u = x - a$ and $du = dx$.

9.8.4. Modulation Theorem

Theorem 9.5 If $F(s)$ is the complex Fourier transform of $f(x)$ then

(a) $F\{f(x)\cos ax\} = \frac{1}{2}[F(s+a)+F(s-a)]$ and

(b) $F\{f(x)\sin ax\} = \frac{1}{2}[F(s+a)-F(s-a)]$. (9.49)

Proof It is given that $F(s) = \int_{-\infty}^{\infty} f(x)e^{isx}\,dx$

(a) By definition

$$F\{f(x)\cos ax\} = \int_{-\infty}^{\infty} f(x)\cos ax\, e^{isx}\,dx$$
$$= \int_{-\infty}^{\infty} f(x)\left(\frac{e^{iax}+e^{-iax}}{2}\right)e^{isx}\,dx$$
$$= \frac{1}{2}\left[\int_{-\infty}^{\infty} f(x)e^{ix(s+a)}\,dx + \int_{-\infty}^{\infty} f(x)e^{ix(s-a)}\,dx\right]$$
$$= \frac{1}{2}[F(s+a)+F(s-a)]$$

(b) can be proved similarly.

Corollary If $F_s(s)$ and $F_c(s)$ are the Fourier sine transform and the Fourier cosine transform of $f(x)$, respectively, then

(a) $F_s\{f(x)\cos ax\} = \frac{1}{2}[F_s(s+a)+F_s(s-a)]$

(b) $F_c\{f(x)\sin ax\} = \frac{1}{2}[F_s(s+a)-F_s(s-a)]$

(c) $F_s\{f(x)\sin ax\} = \frac{1}{2}[F_c(s-a)-F_c(s+a)]$

(9.50)

Example 9.1

Using Fourier integral show that

$$e^{-ax} = \frac{2a}{\pi}\int_{-\infty}^{\infty}\frac{\cos\lambda x}{\lambda^2+a^2}d\lambda, \quad a>0, \quad x>0$$

Solution The Fourier cosine integral is given by

$$f(x) = \frac{2}{\pi}\int_0^{\infty}\cos\lambda x\int_0^{\infty}f(t)\cos\lambda t\,dt\,d\lambda \quad (1)$$

Let $f(t) = e^{-at}$ so that $f(x) = e^{-ax}$. Substituting in the above integral, we have

$$e^{-ax} = \frac{2}{\pi}\int_0^{\infty}\cos\lambda x\int_0^{\infty}\cos\lambda x\left(\int_0^{\infty}e^{-at}\cos\lambda t\,dt\right)d\lambda$$

$$= \frac{2}{\pi}\int_0^{\infty}\cos\lambda x\left[\frac{e^{-at}}{a^2+\lambda^2}(-a\cos\lambda t+b\sin\lambda t)\right]_0^{\infty}d\lambda$$

$$= \frac{2}{\pi}\int_0^{\infty}\cos\lambda x\left[0-\frac{e^{-a0}}{a^2+\lambda^2}(-a\cdot 1+b\cdot 0)\right]d\lambda$$

$$= \frac{2a}{\pi}\int_0^{\infty}\frac{\cos\lambda x}{a^2+\lambda^2}d\lambda$$

$$\Rightarrow \int_0^{\infty}\frac{\cos\lambda x}{a^2+\lambda^2}d\lambda = \frac{\pi e^{-ax}}{2a} \quad (2)$$

Example 9.2

Prove that $\frac{\pi e^{-x}}{2} = \int_0^{\infty}\frac{\cos\lambda x}{\lambda^2+1}d\lambda.$

Solution Putting $a = 1$ in Example 9.1, we obtain

$$\int_0^{\infty}\frac{\cos\lambda x}{a^2+\lambda^2}d\lambda = \frac{\pi e^{-x}}{2}$$

Example 9.3

Express $f(x) = \begin{bmatrix} 0, & -\infty<x<-1 \\ 1, & -1<x<1 \\ 0, & 1<x<\infty \end{bmatrix}$ as a Fourier integral and hence evaluate $\int_0^{\infty}\frac{\sin s\cos sx}{s}ds$

Solution The Fourier integral of $f(x)$ is given by

$$f(x) = \frac{1}{\pi}\int_0^{\infty}\int_{-\infty}^{\infty}f(t)\cos s(t-x)\,dt\,ds \quad (1)$$

The funtion $f(t)$ is defined by

$$f(t) = \begin{bmatrix} 0, & \text{in } (-\infty,1) \\ 1, & \text{in } (-1,1) \\ 0, & \text{in } (1,\infty) \end{bmatrix} \quad (2)$$

∴ We obtain

$$f(x) = \frac{1}{\pi}\int_0^{\infty}\int_{-1}^{1}[\cos s(t-x)\,dt]ds$$

$$= \frac{1}{\pi}\int_0^{\infty}\left[\frac{\sin s(t-x)}{s}\right]_{-1}^{1}ds$$

$$= \frac{1}{\pi}\int_0^{\infty}\frac{1}{s}[\sin s(1-x)-\sin s(-1-x)]ds$$

$$= \frac{2}{\pi}\int_0^{\infty}\frac{\sin s\cos sx}{s}ds \quad (3)$$

which is the Fourier integral of $f(x)$. From (3) we obtain

$$\int_0^{\infty}\frac{\sin s\cos sx}{s}ds = \frac{\pi}{2}f(x) = \begin{bmatrix} \frac{\pi}{2}, & |x|\le 1 \\ 0, & |x|\ge 1 \end{bmatrix} \text{ by (2)}$$

At $x = \pm 1$, $f(x)$ is discontinuous and the value of the integral is

$$\frac{1}{2}[f(1-0)+(1+0)] = \frac{1}{2}\left(\frac{\pi}{2}+0\right) = \frac{\pi}{4} \quad (4)$$

Example 9.4

Prove that $\int_0^{\infty}\frac{\sin x}{x} = \frac{\pi}{2}$

Solution From Example 9.3

$$\int_0^\infty \frac{\sin s \cos sx}{s} ds = \frac{\pi}{2}, \qquad \text{for } |x| < 1$$

Putting $x = 0$ we have

$$\int_0^\infty \frac{\sin x}{x} dx = \frac{\pi}{2}$$

where we have replaced the dummy variable s by x.

Example 9.5

Using the Fourier integral representation, show that

$$\int_0^\infty \frac{\cos sx + s \sin sx}{1+s^2} ds = \begin{cases} 0, & x<0 \\ \frac{\pi}{2}, & x=0 \\ \pi e^{-s}, & x>0. \end{cases}$$

Solution Let $f(x)$ be defined by

$$f(x) = \begin{cases} 0, & (x<0) \\ \frac{1}{2}, & (x=0) \\ e^{-x}, & (x>0) \end{cases} \qquad (1)$$

We now find the Fourier integral representation of $f(x)$ in the exponential form, which is given by

$$f(x) = \frac{1}{2\pi}\int_{-\infty}^{\infty}\left[\int_{-\infty}^{\infty} f(t)e^{ist}\,dt\right]e^{-isx}\,ds \qquad (2)$$

$$\text{Let} \quad I = \int_{-\infty}^{\infty} f(t)e^{ist}dt = \int_{-\infty}^{0} 0\cdot e^{ist}dt + \int_0^\infty e^{-t}\cdot e^{ist}dt = \left[\frac{e^{-t(1-is)}}{is-1}\right]_0^\infty = \frac{0-1}{is-1} = \frac{1+is}{1+s^2}$$

$$\therefore \quad f(x) = \frac{1}{2\pi}\int_{-\infty}^{\infty}\frac{1+is}{1+s^2}e^{isx}ds = \frac{1}{2\pi}\int_{-\infty}^{\infty}\frac{(1+is)(\cos sx - i\sin sx)}{1+s^2}ds$$

$$= \frac{1}{2\pi}\int_{-\infty}^{\infty}\frac{\cos sx + s\sin sx}{1+s^2}ds + 0, \qquad \begin{cases}\text{The second integral} = 0 \\ \text{since the integrand is odd in}(-\infty,\infty)\end{cases}$$

$$= \frac{1}{\pi}\int_0^\infty \frac{\cos sx + s\sin sx}{1+s^2}ds, \quad \text{since the integrand is even in}(-\infty,\infty)$$

$$= e^{-x}, \qquad \text{for } x>0 \quad \text{by definition of } f(x) \qquad (3)$$

The function $f(x)$ is discontinuous at $x = 0$ and its value is

$$f(0) = \frac{1}{2}\int_{-\infty}^{\infty}[f(0+0)+f(0-0)] = \frac{1}{2}(1+0) = \frac{1}{2}$$

$$\therefore \quad \frac{1}{\pi}\int_0^\infty \frac{\cos sx + s\sin sx}{1+s^2}ds = \frac{1}{2}$$

$$\Rightarrow \quad \int_0^\infty \frac{\cos sx + s\sin sx}{1+s^2}ds = \frac{\pi}{2}, \quad \text{for } x=0$$

$$= \pi e^{-x}, \quad \text{for } x>0.$$

(4)

Example 9.6

Express $f(x) = \begin{cases} 1, & \text{for } |x| \le a \\ 0, & \text{for } |x| > a \end{cases}$ as a Fourier integral.

Evaluate

(a) $\int_0^\infty \frac{\sin x}{x} dx$ (b) $\int_{-\infty}^{\infty} \frac{\sin as \cos sx}{s} ds$

[JNTU 2003(2), 2004(3)]

Solution The Fourier integral of $f(x)$is

$$f(x) = \frac{1}{\pi}\int_0^\infty \int_{-\infty}^{\infty} f(t)\cos s(t-x)\,dt\,ds$$

$$= \frac{1}{\pi}\int_0^\infty \left[\int_{-a}^{a} \cos s(t-x)dt\right]ds = \frac{1}{\pi}\int_0^\infty \frac{\sin s(t-x)}{s}\Bigg|_{-a}^{a} ds$$

$$= \frac{1}{\pi}\int_0^\infty \frac{\sin s(a-x)+\sin s(a+x)}{s}ds$$

$$= \frac{2}{\pi}\int_0^\infty \frac{\sin sa\cos sx}{s}ds$$

$$\Rightarrow \quad \int_0^\infty \frac{\sin sa \cos sx}{s}ds = \frac{\pi}{2}f(x) = \begin{cases} \frac{\pi}{2}.1, & |x|\le a \\ 0, & |x|>a \end{cases}$$

For $|x| = a \quad \Leftrightarrow \quad x = \pm a$, which are points of discontinuity the value of the integral is

$$\frac{\pi}{2}\frac{f(a+0)+f(a-0)}{2} = \frac{\pi}{2}\cdot\left(\frac{1+0}{2}\right) = \frac{\pi}{4}$$

Choosing $x = 0$ and $a = 1$ we obtain $\int_0^\infty \frac{\sin x}{x}dx = \frac{\pi}{2}$ where we have replaced the dummy variable s by x.

Example 9.7

Express the function $f(x) = \begin{cases} 1, & \text{for } 0 \le x \le \pi \\ 0, & \text{for } x > \pi \end{cases}$ as a Fourier sine integral.

Hence, evaluate $\int_0^\infty \frac{1-\cos\pi}{s}\sin sx\, ds$

Solution The Fourier sine integral of $f(x)$ is given by

$$f(x) = \frac{2}{\pi}\int_0^\infty \sin sx\left(\int_0^\infty f(t)\sin st\, dt\right)ds$$

$$= \frac{2}{\pi}\int_0^\infty \sin sx\left(\int_0^\infty 1\cdot \sin st\, dt\right)ds$$

$$= \frac{2}{\pi}\int_0^\infty \sin sx\left(\frac{-\cos st}{s}\right)_0^\pi ds$$

$$= \frac{2}{\pi}\int_0^\infty \sin sx\left(\frac{-\cos\pi + \cos 0}{s}\right)ds$$

$$= \frac{2}{\pi}\int_0^\infty \frac{\sin sx(1-\cos s\pi)}{s}ds$$

$$\Rightarrow \quad \int_0^\infty \frac{\sin sx(1-\cos s\pi)}{s}ds = \frac{\pi}{2}f(x) = \begin{cases} \frac{\pi}{2}, & 0 \le x < \pi \\ 0, & x > \pi \end{cases}$$

At $x = \pi$ which is a point of discontinuity for $f(x)$ the value of the integral is

$$\frac{\pi}{2}\cdot\left(\frac{f(\pi-0)+f(\pi+0)}{2}\right) = \frac{\pi}{2}\left(\frac{1+0}{2}\right) = \frac{\pi}{4}$$

Example 9.8

Find the Fourier transform of

$$f(x) = \begin{cases} 1-x^2, & |x|\le 1 \\ 0, & |x|>1 \end{cases}$$

Hence, evaluate $\int_0^\infty \frac{x\cos x - \sin x}{x^3}\cos\frac{x}{2}dx$

[JNTU 2002, 2004*S*]

Solution The Fourier transform of $f(x)$ is given by

$$F\{f(x)\} = F(s) = \int_{-\infty}^{\infty} f(x)e^{isx}dx$$

$$= \int_{-\infty}^{-1} 0\cdot e^{isx}dx + \int_{-1}^{1}(1-x^2)e^{isx}dx + \int_1^\infty 0\cdot e^{isx}dx$$

$$= \left[(1-x^2)\frac{e^{isx}}{is} + 2x\cdot\frac{e^{isx}}{(is)^2} - 2\frac{e^{isx}}{(is)^3}\right]_{-1}^{1}$$

$$= 0 + 2\cdot\frac{1\cdot e^{is} - (-1)e^{-is}}{-s^2} + 2\cdot\frac{e^{is}-e^{-is}}{is^3}$$

$$= -\frac{4}{s^2}\cos s + \frac{4}{s^3}\sin s = -\frac{4}{s^3}(s\cos s - \sin s) \quad (1)$$

By the inverse Fourier transform, we have

$$F^{-1}\{F(s)\} = f(x) = \frac{1}{2\pi}\int_{-\infty}^{\infty} F(s)e^{-isx}ds$$

$$= \frac{1}{2\pi}\int_{-\infty}^{\infty} -\frac{4}{s^3}(s\cos s - \sin s)e^{-isx}ds$$

$$= \begin{cases} 1-x^2, & |x|\le 1 \\ 0, & |x|>1 \end{cases} \quad (2)$$

Putting $x = 1/2$ we obtain from the above equation (2)

$$-\frac{2}{\pi}\int_{-\infty}^{\infty}\frac{(s\cos s-\sin s)}{s^3}e^{\frac{-is}{2}}ds = 1-\left(\frac{1}{2}\right)^2 = \frac{3}{4}$$

$$\Rightarrow \int_{-\infty}^{\infty}\frac{s\cos s-\sin s}{s^3}\left(\cos\frac{s}{2}-i\sin\frac{s}{2}\right)ds = -\frac{3\pi}{8} \quad (3)$$

Equating the real parts on either side

$$\int_{-\infty}^{\infty}\frac{s\cos s-\sin s}{s^3}\left(\cos\frac{s}{2}\right)ds = -\frac{3\pi}{8}$$

$$\Rightarrow \int_0^{\infty}\frac{x\cos x-\sin x}{x^3}\cos\frac{x}{2}dx = -\frac{3\pi}{16}, \quad (4)$$

since the integrand is even in $(-\infty, \infty)$ and we have replaced the dummy variable s by x.

Example 9.9

Find the Fourier transform of

$$f(x) = \begin{cases} 1, & \text{for } |x| < 1 \\ 0, & \text{for } |x| > 1 \end{cases}$$

Hence evaluate $\int_0^{\infty}\frac{\sin x}{x}dx$

Solution The Fourier transform of $f(x)$ is given by

$$F\{f(x)\} = F(s) = \int_{-\infty}^{\infty} f(x)e^{isx}dx = \int_{-1}^{1} 1\cdot e^{isx}dx$$

$$= \left(\frac{e^{isx}}{is}\right)_{-1}^{1} = \frac{e^{is}-e^{-is}}{is} = \frac{2}{s}\left(\frac{e^{is}-e^{-is}}{2i}\right)$$

$$= \frac{2}{s}\sin s \qquad (s \neq 0) \qquad (1)$$

We have $\lim_{s\to 0}\frac{\sin s}{s} = 1$ and hence $F\{f(x)\} = F(s) = 2$ when $s = 0$ (2)

By the inverse Fourier transform we have

$$F^{-1}\{F(s)\} = \frac{1}{2\pi}\int_{-\infty}^{\infty} F(s)e^{-isx}ds$$

$$= \frac{1}{\pi}\int_{-\infty}^{\infty}\frac{\sin s}{s}e^{-isx}ds$$

$$= f(x) = \begin{cases} 1, & \text{for} |x| < 1 \\ 0, & \text{for} |x| > 1 \end{cases} \qquad (3)$$

Putting $x = 0$ we get

$$\frac{1}{\pi}\int_{-\infty}^{\infty}\frac{\sin s}{s}ds = 1 \quad \text{or} \quad \int_{-\infty}^{\infty}\frac{\sin s}{s}ds = \pi \qquad (4)$$

Since the integrand is an even function in $(-\infty, \infty)$ we have

$$\int_0^{\infty}\frac{\sin x}{x}dx = \frac{\pi}{2} \qquad (5)$$

where we have replaced the dummy variables s by x.

Example 9.10

Find the Fourier (a) cosine and (b) sine transform of $f(x) = e^{-ax}$ $(x > 0, a > 0)$.

Deduce the values of

$$\int_0^{\infty}\frac{\cos sx}{a^2+s^2}ds \quad \text{and} \quad \int_0^{\infty}\frac{s\sin sx}{a^2+s^2}ds$$

Solution

(a) The Fourier cosine transform of $f(x)$ is given by

$$F_c\{f(x)\} = F_c(s) = \int_0^{\infty} f(x)\cos sx\,dx \qquad (1)$$

$$\Rightarrow \quad F_c\{e^{-ax}\} = F_c(s) = \int_0^{\infty} e^{-ax}\cos sx\,dx$$

$$= \left[\frac{e^{-ax}}{a^2+s^2}(-a\cos sx + s\sin sx)\right]_0^{\infty}$$

$$= 0 - \frac{e^0}{a^2+s^2}(-a.1+0) = \frac{a}{a^2+s^2} \qquad (2)$$

The inverse Fourier cosine transform of $F_c(s)$ is

$$F_c^{-1}\{F_c(s)\} = f(x) = e^{-ax} = \frac{2}{\pi}\int_0^{\infty} F_c(s)\cos sx\,ds$$

$$= \frac{2a}{\pi}\int_0^{\infty}\frac{\cos sx\,ds}{a^2+s^2}$$

$$\Rightarrow \int_0^{\infty}\frac{\cos sx}{a^2+s^2}ds = \frac{\pi e^{-ax}}{2a} \qquad (3)$$

(b) The Fourier sine transform of $f(x)$ is given by

$$F_s\{f(x)\} = F_s(s) = \int_0^{\infty} f(x)\sin sx\,dx \qquad (4)$$

$$F_s\{e^{-ax}\} = F_s(s) = \int_0^{\infty} e^{-ax}\sin sx\,dx \qquad (5)$$

$$=\left[\frac{e^{-ax}}{a^2+s^2}(-a\sin sx - s\cos sx)\right]_0^{\infty}$$

$$=0-\frac{1}{a^2+b^2}(-a.0-s.1)$$

$$=\frac{s}{a^2+s^2} \quad (6)$$

The inverse Fourier sine transform of $F_s(s)$ is

$$F_c^{-1}\{F_s(s)\}=f(x)=e^{-ax}=\frac{2}{\pi}\int_0^{\infty}F_s(s)\sin sx\,ds \quad (7)$$

$$=\frac{2}{\pi}\int_0^{\infty}\frac{s}{a^2+s^2}\sin sx\,dx$$

$$\Rightarrow \quad \int_0^{\infty}\frac{s}{a^2+s^2}\sin sx\,dx=\frac{\pi}{2}e^{-ax} \quad (8)$$

Corollary Putting $a = 0$ in the above result we obtain

$$\int_0^{\infty}\frac{\sin sx}{s}ds=\frac{\pi}{2} \quad (9)$$

Example 9.11

Find the Fourier sine transform of $e^{-|x|}$. Hence, show that

$$\int_0^{\infty}\frac{x\sin ax}{1+x^2}dx=\frac{\pi e^{-a}}{2} \quad (a>0)$$

Solution The variable x is positive in $(0, \infty)$ so that $|x| = x$ for $(0, \infty)$.

The Fourier sine transform of $f(x) = e^{-|x|} = e^{-x}$ is given by

$$F_s\{f(x)\}=F_s\{e^{-x}\}=F_s(s)$$

$$=\int_0^{\infty}f(x)\sin sx\,dx=\int_0^{\infty}e^{-x}\sin sx\,dx$$

$$=\left[\frac{e^{-x}}{1+s^2}(-\sin sx - s\cos sx)\right]_0^{\infty}$$

$$=0+\frac{e^0}{1+s^2}(0+s\cos 0)=\frac{s}{1+s^2} \quad (1)$$

By the inverse Fourier sine transform of $F_s(s)$ we have

$$F^{-1}\{F_s(s)\}=f(x)=\frac{2}{\pi}\int_0^{\infty}F_s(s)\sin sx\,ds$$

$$\Rightarrow \quad e^{-x}=\frac{2}{\pi}\int_0^{\infty}\frac{s}{1+s^2}\sin sx\,ds \quad (2)$$

If we replace 'x' by 'a' we obtain

$$e^{-a}=\frac{2}{\pi}\int_0\frac{s}{1+s^2}\sin sa\,ds=\frac{2}{\pi}\int_0\frac{x\sin ax}{1+x^2}dx \quad (3)$$

where we have replaced the dummy variable s by x.

$$\text{Hence} \quad \int_0^{\infty}\frac{x\sin ax}{1+x^2}dx=\frac{\pi e^{-a}}{2} \quad (4)$$

Example 9.12

Find the Fourier sine transform of e^{-ax}/x

Solution By definition, the Fourier sine transform of $f(x)$ is

$$F_s\{f(x)\}=\{F_s(s)\}=\int_0^{\infty}f(x)\sin sx\,dx \quad (1)$$

$$f(x)=\frac{e^{-ax}}{x} \quad \Rightarrow \quad F_s(s)=\int_0^{\infty}\frac{e^{-ax}}{x}\sin sx\,dx \quad (2)$$

Differentiating both sides with respect to 's'

$$\frac{d}{ds}[F_s(s)]=\int_0^{\infty}\frac{e^{-ax}}{x}\frac{\partial}{\partial s}(\sin sx)dx=\int_0^{\infty}e^{-ax}\cos sx\,dx$$

$$=\left[\frac{e^{-ax}}{a^2+s^2}(-a\cos sx+s\sin sx)\right]_0^{\infty}$$

$$=0-\frac{e^0}{a^2+s^2}(-a.1+0)=\frac{a}{a^2+s^2} \quad (3)$$

Integrating now with respect to 's' we obtain

$$F_s(s)=a\int\frac{1}{a^2+s^2}ds=\tan^{-1}\frac{s}{a}+c$$

When $s=0, F_s(0)=0 \Rightarrow 0=\tan^{-1}0+c \Rightarrow c=0$

Hence, $F_s(s)=\tan^{-1}(s/a)$. (4)

Example 9.13

Find the Fourier cosine transform of

$$f(x) = \begin{cases} x, & \text{for } 0 < x < 1 \\ 2-x, & \text{for } 1 < x < 2 \\ 0, & \text{for } x > 2 \end{cases}$$

[JNTU 2002, 2003*S*, 2004*S*]

Solution The Fourier cosine transform of $f(x)$ is given by

$$F_c\{f(x)\} = F_c(s) = \int_0^\infty f(x)\cos sx\,dx$$

$$= \int_0^1 x\cos sx\,dx + \int_1^2 (2-x)\cos sx\,dx + \int_2^\infty 0.dx$$

$$= \left[x\frac{\sin sx}{s} - \left(\frac{-\cos sx}{s^2}\right)\right]_0^1$$

$$+\left[(2-x)\frac{\sin sx}{s} - (-1)\left(\frac{-\cos sx}{s^2}\right)\right]_1^2$$

$$= \frac{1\cdot\sin s - 0}{s} + \frac{1}{s^2}(\cos s - \cos 0)$$

$$+\frac{(0-\sin s)}{s} - \frac{(\cos 2s - \cos s)}{s^2}$$

$$= \frac{2\cos s - \cos 2s - 1}{s^2}$$

Example 9.14

Find the Fourier cosine transform of $f(x) = 1/(1+x^2)$. Hence, derive the Fourier sine transform of

$$\phi(x) = \frac{x}{1+x^2}$$

Solution By definition, the Fourier cosine transform of $f(x)$ is given by

$$I = F_c\{f(x)\} = F_c(s) = \int_0^\infty f(x)\cos sx\,dx$$

$$= \int_0^\infty \frac{\cos sx}{1+x^2}dx \qquad (1)$$

Differentiating both sides of (1) with respect to 's'

$$\frac{d}{ds}F_c(s) = \int_0^\infty \frac{1}{1+x^2}\frac{\partial}{\partial s}(\cos sx)dx = -\int_0^\infty \frac{x\sin sx}{1+x^2}dx$$

$$= -\int_0^\infty \frac{x^2\sin sx}{x(1+x^2)}dx,$$ inserting x in the numerator and denominator

$$= -\int_0^\infty \frac{\left[(1+x^2)-1\right]\sin sx\,dx}{x(1+x^2)}$$

$$= -\int_0^\infty \frac{1}{x}\sin sx dx + \int_0^\infty \frac{\sin sx}{x(1+x^2)}dx$$

$$= -\frac{\pi}{2} + \int_0^\infty \frac{\sin sx}{x(1+x^2)}dx \quad \because \int_0^\infty \frac{\sin sx}{s}dx = \frac{\pi}{2} \qquad (2)$$

Differentiating both sides of (2) with respect to s again

$$\frac{d^2}{ds^2}F_c(s) = \int_0^\infty \frac{1}{x(1+x^2)}\frac{\partial}{\partial s}(\sin sx)\,dx$$

$$= \int_0^\infty \frac{\cos sx}{1+x^2}dx = F_c(s), \text{ by (1)}$$

$$\Rightarrow (D^2-1)F_c(s) = 0 \qquad (3)$$

The general solution of (3) is

$$F_c(s) = c_1e^s + c_2e^{-s} \qquad (4)$$

$$\frac{d}{ds}(F_c(s)) = c_1e^s - c_2e^{-s} \qquad (5)$$

When $s = 0$, (1) and (4) give

$$F_c(0) = c_1 + c_2 = \int_0^\infty \frac{dx}{1+x^2} = \frac{\pi}{2} \qquad (6)$$

Also $$F_c(0) = c_1 - c_2 = -\frac{\pi}{2}, \text{ by (5)} \qquad (7)$$

Solving (6) and (7) we get

$$c_1 = 0,\ c_2 = \pi/2 \qquad (8)$$

From (1) and (4) we have

$$F_c\{f(x)\} = I = \frac{\pi}{2}e^{-s} \qquad (9)$$

Now

$$F_s\{\phi(x)\} = F_s\left\{\frac{x}{1+x^2}\right\} = \int_0^\infty \frac{x}{1+x^2}\sin sx\,dx$$

$$= \frac{d}{ds}F_c(s) = -\frac{\pi}{2}e^{-s} \qquad (10)$$

Example 9.15

Find the Fourier sine and cosine transform of $e^{-ax}(a > 0)$ and deduce the inverse formula.

[JNTU 2002, 2004]

Solution The Fourier sine transform of $f(x) = e^{-ax}$ is

$$F_s\{e^{-ax}\} = F_s(s) = \int_0^\infty e^{-ax} \sin sx\, dx$$

$$= \left[\frac{e^{-ax}}{a^2+s^2}(-a\sin sx - s\cos sx)\right]_0^\infty$$

$$= 0 + \frac{s}{a^2+s^2} = \frac{s}{a^2+s^2} \quad (1)$$

The inverse Fourier sine transform for $F_s(s) = \dfrac{s}{a^2+b^2}$ is

$$F_s^{-1}\{F_s(s)\} = f(x) = e^{-ax} = \frac{2}{\pi}\int_0^\infty \frac{s}{a^2+s^2}\sin sx\, ds \quad (2)$$

The Fourier cosine transform of $f(x) = e^{-ax}$ is

$$F_c\{f(x)\} = F_c\{e^{-ax}\} = F_c(s) = \int_0^\infty e^{-ax}\cos sx\, dx$$

$$= \left[\frac{e^{-ax}}{a^2+s^2}(-a\cos sx + s\sin sx)\right]_0^\infty$$

$$= 0 + \frac{1}{a^2+s^2}(a.1+0) = \frac{a}{a^2+s^2} \quad (3)$$

The inversion formula for

$$F_c(s) = \frac{a}{a^2+s^2}$$

is $F_c^{-1}\{F_c(s)\} = f(x) = \dfrac{2a}{\pi}\displaystyle\int_0^\infty \frac{\cos sx}{a^2+s^2}\, ds = e^{-ax}$ (4)

Example 9.16

Solve the intergral equation

$$\int_0^\infty f(x)\cos sx\, dx = \begin{cases} 1-s, & 0 \le s \le 1 \\ 0, & s > 1 \end{cases}$$

Hence evaluate $\displaystyle\int_0^\infty \frac{\sin^2 x}{x^2}\, dx$

Solution By definition, the Fourier cosine transform of $f(x)$ is

$$F_c\{f(x)\} = F_c(s) = \int_0^\infty f(x)\cos sx\, ds$$

$$= \begin{cases} 1-s, & 0 \le s \le 1 \\ 0, & s > 1 \end{cases} \quad (1)$$

By the inversion formula we have

$$F_c^{-1}\{F_c(s)\} = f(x) = \frac{2}{\pi}\int_0^\pi F_c(s)\cos sx\, ds$$

$$= \frac{2}{\pi}\int_0^1 (1-s)\cos sx \quad ds$$

$$= \frac{2}{\pi}\left[(1-s)\frac{\sin sx}{x} - (-1)\frac{-\cos sx}{x^2}\right]_0^1$$

$$= \frac{2}{\pi}\left[\left(\frac{0-0}{x}\right) - \frac{\cos x - 1}{x^2}\right] = \frac{2}{\pi}\frac{1-\cos x}{x^2}$$

Now

$$F_c(s) = \int_0^\infty f(x)\cos sx\, dx = \frac{2}{\pi}\int_0^\infty \frac{1-\cos x}{x^2}\cos sx\, dx \quad (2)$$

From (1) and (2) we obtain

$$\frac{2}{\pi}\int_0^\infty \frac{1-\cos x}{x^2}\cos sx\, dx = \begin{cases} 1-s, & 0 \le s \le 1 \\ 0, & s > 1 \end{cases} \quad (3)$$

Taking the limit as $s \to 0$ we obtain

$$\frac{2}{\pi}\int_0^\infty \frac{1-\cos x}{x^2}\, dx = \frac{4}{\pi}\int_0^\infty \frac{\sin^2 x/2}{x^2}\, dx = 1 \qquad \text{Put } \frac{x}{2} = t,\ dx = 2\,dt$$

$$\Rightarrow \frac{4}{\pi}\int_0^\infty \frac{\sin^2 t}{4t^2}2dt = \frac{2}{\pi}\int_0^\infty \frac{\sin^2 t}{t^2}\, dt = 1$$

$$\Rightarrow \int_0^\infty \frac{\sin^2 x}{x^2}\, dx = \frac{\pi}{2} \quad (4)$$

where we have replaced the dummy variable t by x.

Example 9.17

Show that the Fourier transform of $e^{-x^2/2}$ is reciprocal.

[JNTU 2002, 2004S]

Solution Here we have to use definition 2 for the Fourier transform, *viz.*

$$F\{f(x)\} = F(s) = \frac{1}{\sqrt{2\pi}}\int_{-\infty}^{\infty} f(x)e^{isx}\,dx$$

$$F\{e^{-x^2/2}\} = F(s) = \frac{1}{\sqrt{2\pi}}\int_{-\infty}^{\infty} e^{-x^2/2}e^{isx}\,dx$$

$$= \frac{1}{\sqrt{2\pi}}\int_{-\infty}^{\infty} e^{-\frac{1}{2}(x^2-2isx)}\,dx$$

$$x^2 - 2isx + (is)^2 - (is)^2 - (is)^2 = (x-is)^2 + s^2$$

$$= \frac{1}{\sqrt{2\pi}}e^{-s^2/2}\int_{-\infty}^{\infty} e^{-\left(\frac{x-is}{\sqrt{2}}\right)^2}\,dx$$

Put $\frac{x-is}{\sqrt{2}} = u \Rightarrow \frac{1}{\sqrt{2}}dx = du$;

limits are $\begin{cases} x=-\infty \Rightarrow u=-\infty \\ x=\infty \Rightarrow u=\infty \end{cases}$

Hence $F\{e^{-x^2/2}\} = \frac{e^{-s^2/2}}{\sqrt{2\pi}}\cdot\sqrt{2}\int_{-\infty}^{\infty} e^{-u^2/2}\,du$

$$= \frac{e^{-s^2/2}}{\sqrt{2\pi}}, \sqrt{2\pi} = e^{-s^2/2}$$

$$\because \int_{-\infty}^{\infty} e^{-u^2/2}\,du = \sqrt{\pi}$$

Here the function $f(x)$ and the Fourier transform $F(s)$ are the same. Hence, $e^{-x^2/2}$ is self-reciprocal.

Note 1 If we use the other definitions there will be difference of a constant factor.

Example 9.18

Find the Fourier sine transform of $(1/x)(x^2+a^2)$

Hence, deduce the Fourier cosine transform of $\frac{1}{x(x^2+a^2)}$. [JNTU 2001, 2002, 2003]

Solution The Fourier sine transform of $f(x)$ is given by

$$F_s\{f(x)\} = F_s(s) = \int_0^{\infty} f(x)\sin sx\,dx$$

$$F_s\left\{\frac{1}{x(x^2+a^2)}\right\} = F_s(s) = \int_0^{\infty}\frac{\sin sx}{x(x^2+a^2)}dx$$

Now, let $I(s) = \int_0^{\infty}\frac{\sin sx}{x(x^2+a^2)}dx$ (1)

Differentiating both sides of (1) with respect to 's'

$$\frac{dI(s)}{ds} = \int_0^{\infty}\frac{1}{x(x^2+a^2)}\frac{\partial}{\partial s}(\sin sx)dx = \int_0^{\infty}\frac{\cos sx}{x^2+a^2}dx \quad (2)$$

Again differentiating both sides of (2) with respect to 's'

$$\frac{d^2I(s)}{ds^2} = \int_0^{\infty}\frac{1}{x^2+a^2}\frac{\partial}{\partial s}(\cos sx)\,dx = -\int_0^{\infty}\frac{x\sin sx}{x^2+a^2}dx$$

$$= -\int_0^{\infty}\frac{x^2\sin sx}{x(x^2+a^2)}dx, \quad \text{(inserting } x \text{ in the numerator and denominator)}$$

$$= -\int_0^{\infty}\frac{\left[(x^2+a^2)-a^2\right]\sin sx}{x(x^2+a^2)}dx, \quad \text{(adding and subtracting } a^2\text{)}$$

$$= -\int_0^{\infty}\frac{\sin sx}{x}dx + a^2\int_0^{\infty}\frac{\sin sx}{x(x^2+a^2)}dx = -\frac{\pi}{2} + a^2 I$$

$\left[\because \int_0^{\infty}\frac{\sin sx}{x}dx = \frac{\pi}{2}\right.$ corollary of Example (9.10) on page 10

$$\Rightarrow (D^2 - a^2)I = -\frac{\pi}{2}, \quad \text{which is a second-order linear differential equation} \quad (3)$$

The general solution of (3) comprises two parts: the complementary function (CF) and the particular intergal (PI).

$$PI = \frac{1}{D^2 - a^2}\left(-\frac{\pi}{2}\right) = \frac{\pi}{2a^2}\frac{1}{\left(1-\frac{D^2}{a^2}\right)}1$$

$$= \frac{\pi}{2a^2}\left(1-\frac{D^2}{a^2}\right)^{-1} 1 = \frac{\pi}{2a^2}\left(1+\frac{D^2}{a^2}+\cdots\right)1 = \frac{\pi}{2a^2}$$

$CF = Ae^{as} + Be^{-as}$

∴ The general solution is

$$I(s) = Ae^{as} + Be^{-as} + \frac{\pi}{2a^2} \qquad (4)$$

$$\frac{dI(s)}{ds} = aAe^{as} - aBe^{-as} \qquad (5)$$

Putting $s = 0$ in (1) and (4) we obtain

$$I(0) = A + B + \frac{\pi}{2a^2} = 0 \quad \Rightarrow \quad A + B = -\frac{\pi}{2a^2} \qquad (6)$$

and similarly from (2) and (5) we obtain

$$\left.\frac{dI}{ds}\right|_{s=0} = aA - aB = \int_0^\infty \frac{dx}{x^2+a^2} = \left[\frac{1}{a}\tan^{-1}\frac{x}{a}\right]_0^\infty$$

$$= \frac{\pi}{2a} \quad \Rightarrow \quad A - B = \frac{\pi}{2a^2} \qquad (7)$$

Solving (6) and (7) we obtain

$$A = 0, \quad B = -\frac{\pi}{2a^2} \qquad (8)$$

Substituting these values in (4) we have

$$I(s) = \frac{\pi}{2a^2}(1 - e^{-as}), \qquad (9)$$

which is the required Fourier sine transform of

$$f(x) = \frac{1}{x(x^2+a^2)}$$

Now we deduce the Fourier cosine transform of

$$g(x) = \frac{1}{x^2+a^2}$$

The Fourier cosine transform of $g(x) = \dfrac{1}{x^2+a^2}$ is given by

$$F_c\{g(x)\} = G_c(s) \;=\; \int_0^\infty g(x)\cos sx\, dx$$

$$\Rightarrow \quad F_c\left\{\frac{1}{x^2+a^2}\right\} = G_c(s) \;=\; \int_0^\infty \frac{1}{x^2+a^2}\cos sx\, dx$$

$$= \frac{dI(s)}{ds} = \frac{\pi}{2a^2}(0 + ae^{-as})$$

$$= \frac{\pi}{2a}e^{-as}$$

∴ The Fourier cosine transform of $g(x) = 1/(x^2+a^2)$ is $G_c(s) = (\pi/2a)\,e^{-as}$. (10)

EXERCISE 9.1

1. Find the Fourier transform of $e^{-x^2/2}$, $(-\infty < x < \infty)$.

   Ans: $\dfrac{\sqrt{\pi}}{2}e^{-s^2/4}$

2. Find the Fourier transform of

   (a) $f(x) = \begin{cases} 1, & |x| < a \\ 0, & |x| > a \end{cases}$ (b) $f(x) = \begin{cases} x^2, & |x| < a \\ 0, & |x| > a \end{cases}$

   Ans: (a) $\dfrac{2\sin as}{s}$ (b) $[(a^2s^2 - 2)\sin ax + 2as\cos as]$

3. Express $f(x) = \begin{cases} 1, & |x| \le 1 \\ 0, & |x| > 1 \end{cases}$ as a Fourier integral

   Hence, evaluate $\displaystyle\int_0^\infty \frac{\sin s\cos sx}{s}ds$

   Ans: $\dfrac{2}{\pi}\displaystyle\int_0^\infty \frac{\sin s\cos sx}{s}ds, \quad I = \begin{cases} \frac{\pi}{4}, & |x| \times 1 \\ \frac{\pi}{4}, & |x| = 1 \\ 0, & |x| > 1 \end{cases}$

4. Using Fourier integral show that

   $$\int_0^\infty \frac{\sin\lambda\pi\sin\lambda x}{1-\lambda^2} = \begin{cases} \frac{\pi}{2}\sin x, & (0 \le x \le \pi) \\ 0, & (x > \pi) \end{cases}$$

5. Using Fourier integral show that

   $$\int_0^\infty \frac{\lambda\sin\lambda x}{a^2+\lambda^2}d\lambda = \frac{\pi}{2}e^{-ax}, \quad (x > 0, a > 0)$$

6. Using Fourier intergral representation show that

   (a) $\displaystyle\int_0^\infty \frac{s\sin sx}{1+s^2}ds = \frac{\pi}{2}e^{-x}(x > 0)$

(b) $\int_0^\infty \frac{\cos sx}{1+s^2} ds = \frac{\pi}{2} e^{-x} (x \geq 0)$

(c) $\int_0^\infty \frac{\sin s \cos sx}{s} ds = \frac{\pi}{2} \ (0 < x < 1)$

7. Prove that $\int_0^\infty \frac{\lambda \sin \lambda x}{(\lambda^2 + a^2)(\lambda^2 + b^2)} d\lambda = \frac{\pi}{2}\left(\frac{e^{-ax} - e^{-bx}}{b^2 - a^2}\right)$,

$a > 0, b > 0$.

[**Hint:** Use FSI with $f(t) = e^{-at} - e^{-bt}$.]

8. Find the Fourier sine transform of

(a) $f(x) = \begin{cases} \sin x, & 0 < x < a \\ 0, & x > a \end{cases}$

(b) $f(x) = \begin{cases} x, & 0 < x < 1 \\ 2 - x, & 1 < x < 2 \\ 0, & x > 2 \end{cases}$

Ans: (a) $\frac{1}{2}\left(\frac{\sin a(1-s)}{1-s} - \frac{\sin a(1+s)}{1+s}\right)$

9. Find the Fourier sine transform of

$f(x) = \frac{1}{x(x^2 + a^2)}$

Ans: $\frac{\pi}{2a^2}(1 - e^{-as})$

10. Find the Finite fourier sine and cosine transforms of $f(x) = 2x, \quad 0 < x < 4$

Ans: $F_s(n) = \frac{-32(-1)^n}{n\pi}; F_c(n) = 32\frac{(-1)^n - 1}{n^2\pi^2}$

11. Solve $\int_0^\infty f(x) \cos sx \, dx = e^{-s}$

Ans: $f(x) = \frac{2}{\pi} \cdot \frac{1}{1+x^2}$

12. Solve $\int_0^\infty f(x) \sin sx \, dx = \begin{cases} 1, & 0 \leq s < 1 \\ 2, & 1 \leq s \leq 2 \\ 0, & s > 2 \end{cases}$

Ans: $f(x) = (2 + 2 \cos x - 4 \cos 2x)/\pi x$

13. Find the Fourier cosine transform of

(a) $e^{-ax} \cos ax$ (b) $e^{-ax} \sin ax$

Ans: (a) $2a\left(\frac{2a^2 + b^2}{s^4 + 4a^4}\right)$ (b) $\frac{a(2a^2 - s^2)}{s^2 + 4a^4}$

14. Find the Fourier cosine transform of

$f(x) = \begin{cases} x, & 0 < x < 1 \\ 2, & 1 < x < 2 \\ 0, & x > 2 \end{cases}$

Ans: $\frac{2(1 - \cos s)\cos s}{s^2}$

9.9 Parseval's Identity for Fourier Transforms

Theorem 9.6 If $F(s)$ and $G(s)$ are the Fourier transforms of $f(x)$ and $g(x)$, respectively, then

(a) $$\frac{1}{2\pi}\int_{-\infty}^{\infty} F(s)\bar{G}(s) ds = \int_{-\infty}^{\infty} f(x)\bar{g}(x)\, dx \quad (9.51)$$

(b) $$\frac{1}{2\pi}\int_{-\infty}^{\infty} [F(s)]^2 ds = \int_{-\infty}^{\infty} |f(x)|^2\, dx \quad (9.52)$$

Where $\bar{g}$ denotes the complex conjugate of g.

Proof Let $g(x)$ be the inverse Fourier transform of $G(s)$ so that

$$g(x) = \frac{1}{2\pi}\int_{-\infty}^{\infty} G(s)e^{-isx} dx$$

(a) Taking conjugates on both sides

$$\bar{g}(x) = \frac{1}{2\pi}\int_{-\infty}^{\infty} \bar{G}(s)e^{isx} ds$$

$$\therefore \quad \int_{-\infty}^{\infty} f(x)\bar{g}(x)\, dx = \int_{-\infty}^{\infty} f(x)\left(\frac{1}{2\pi}\int_{-\infty}^{\infty} \bar{G}(s)e^{isx} ds\right) dx$$

$$= \frac{1}{2\pi}\int_{-\infty}^{\infty} \bar{G}(s)\left(\int_{-\infty}^{\infty} f(x)e^{isx} dx\right) ds$$

$$= \frac{1}{2\pi}\int_{-\infty}^{\infty} \bar{G}(s) f(s)\, ds,$$

(Changing the order of integration)

by the definition of Fourier transform

(b) Setting $g(x) = f(x)$ for all x we obtain

$$\frac{1}{2\pi}\int_{-\infty}^{\infty} F(s)\bar{F}(s)\, ds = \int_{-\infty}^{\infty} f(x)\bar{f}(x)\, dx$$

$$\Rightarrow \quad \frac{1}{2\pi}\int_{-\infty}^{\infty}|F(s)|^2\,ds = \int_{-\infty}^{\infty}|f(x)|^2\,dx$$

9.10 Parseval's Identities for Fourier Sine and Cosine Transforms

Theorem 9.7 Similarly, we can obtain the following results.

1. $\frac{2}{\pi}\int_0^{\infty} F_s(s)G_s(s)\,ds = \int_0^{\infty} f(x)g(x)\,dx$ (9.53)

2. $\frac{2}{\pi}\int_0^{\infty} F_c(s)G_c(s)\,ds = \int_0^{\infty} f(x)g(x)\,dx$ (9.54)

3. $\frac{2}{\pi}\int_0^{\infty} |F_s(s)|^2\,ds = \int_0^{\infty} |f(x)|^2\,dx$ (9.55)

4. $\frac{2}{\pi}\int_0^{\infty} |F_c(s)|^2\,ds = \int_0^{\infty} |f(x)|^2\,dx$ (9.56)

Example 9.19

If $f(x) = \begin{cases} 1, & |x|<1 \\ 0, & |x|>1 \end{cases}$ find the Fourier transform.

Using Parseval's identity prove that

$$\int_0^{\infty}\left(\frac{\sin t}{t}\right)^2 dt = \frac{\pi}{2}. \qquad \text{[JNTU 2003 } S \text{ (1)]}$$

Solution The Fourier transform of $f(x)$ is given by

$$F\{f(s)\} = F(s) = \int_{-\infty}^{\infty} f(x)e^{isx}\,dx = \int_{-1}^{1} e^{isx}\,dx = \left(\frac{e^{isx}}{is}\right)_{-1}^{1}$$

$$= \frac{e^{is} - e^{-is}}{is} = \frac{2}{s}\sin s \qquad (1)$$

Parsevel's identity for Fourier transform is

$$\int_{-\infty}^{\infty}|f(x)|^2\,dx = \frac{1}{2\pi}\int_{-\infty}^{\infty}|F(s)|^2\,ds$$

$$(x)_{-1}^{1} = 2 = \frac{4}{2\pi}\int_{-\infty}^{\infty}\left|\frac{\sin s}{s}\right|^2 ds = \frac{4}{2\pi}\cdot 2\int_0^{\infty}\left|\frac{\sin s}{s}\right|^2 ds$$

$$\Rightarrow \int_0^{\infty}\left|\frac{\sin s}{s}\right|^2 ds = \frac{\pi}{2} \quad \Rightarrow \quad \int_0^{\infty}\left(\frac{\sin t}{t}\right)^2 dt = \frac{\pi}{2},$$

replacing the dummy variable s by t (2)

Example 9.20

Using Parseval's identities prove that

$$\int_0^{\infty}\left(\frac{1-\cos x}{x}\right)^2 dx = \frac{\pi}{2}$$

Solution The Fourier sine transform of $f(x) = \begin{cases} 1, & 0<x<1 \\ 0, & x>1 \end{cases}$ is given by

$$F_s\{f(x)\} = F_s(s) = \int_0^{\infty} f(x)\sin sx\,dx$$

$$= \int_0^{1} 1\sin sx\,dx + \int_1^{\infty} 0\sin sx\,dx$$

$$= \left(-\frac{\cos sx}{s}\right)_0^1 + 0 = \left(\frac{1-\cos s}{s}\right) \qquad (1)$$

By Parseval's identity for Fourier sine transform we have

$$\frac{2}{\pi}\int_0^{\infty}|F_s(s)|^2\,ds = \int_0^{\infty}|f(x)|^2\,dx$$

$$\Rightarrow \quad \frac{2}{\pi}\int_0^{\infty}|F_s(s)|^2\,ds = \int_0^{1}1^2\,dx = 1$$

$$\Rightarrow \quad \int_0^{\infty}\left(\frac{1-\cos s}{s}\right)^2 ds = \frac{\pi}{2}$$

$$\Rightarrow \quad \int_0^{\infty}\left(\frac{1-\cos x}{x}\right)^2 dx = \frac{\pi}{2} \qquad (2)$$

where we have replaced the dummy variable s by x.

Example 9.21

Evaluate $\int_0^{\infty}\frac{dx}{(x^2+a^2)(x^2+b^2)}$ using Parseval's identity

Solution The Fourier cosine transform of $f(x) = e^{-ax}$ is given by

$$F_c\{f(x)\} = F_c\{e^{-ax}\} = F_c(s) = \frac{a}{s^2+a^2} \qquad (1)$$

and the Fourier cosine transform of $g(x) = e^{-bx}$ is given by

$$F_c\{g(x)\} = F_c\{e^{-bx}\} = F_c(s) = \frac{b}{s^2+b^2} \qquad (2)$$

Parseval's identity for Fourier cosine transforms is

$$\frac{2}{\pi}\int_0^\infty F_c(s)G_c(s)\,ds = \int_0^\infty f(x)g(x)\,dx$$

$$\frac{2}{\pi}\int_0^\infty \frac{a}{s^2+a^2}\cdot\frac{b}{s^2+b^2}\,ds = \int_0^\infty e^{-ax}e^{-bx}\,dx$$

$$= \int_0^\infty e^{-(a+b)x}\,dx = \left[\frac{e^{-(a+b)x}}{-(a+b)}\right]_0^\infty$$

$$= \frac{0-1}{-(a+b)} = \frac{1}{a+b}$$

$$\Rightarrow \int_0^\infty \frac{ds}{(x^2+a^2)(x^2+b^2)} = \frac{\pi}{2ab(a+b)} \quad (3)$$

Example 9.22

Evaluate $\int_0^\infty \frac{dx}{(a^2+x^2)^2}$ using Parseval's identity.

Solution The Fourier cosine transform of $f(x) = e^{-ax}$ is given by

$$F_c\{f(x)\} = F_c\{e^{-ax}\} = F_c(s) = \frac{a}{s^2+a^2} \quad (1)$$

Parseval's identity is

$$\frac{2}{\pi}\int_0^\infty |F_c(s)|^2\,ds = \int_0^\infty |F_c(x)|^2\,dx$$

$$\Rightarrow \frac{2}{\pi}\int_0^\infty \left(\frac{a}{s^2+a^2}\right)^2 ds = \int_0^\infty |e^{-ax}|^2\,dx = \int_0^\infty e^{-2ax}\,dx \quad (2)$$

$$\frac{2}{\pi}a^2\int_0^\infty \frac{ds}{s^2+a^2} = \int_0^\infty e^{-2ax}\,dx = -\frac{1}{2a}(e^{-2ax})_0^\infty$$

$$= -\frac{1}{2a}(0-1) = \frac{1}{2a} \quad (3)$$

$$\Rightarrow \int_0^\infty \frac{ds}{s^2+a^2} = \frac{\pi}{4a^3} \quad (4)$$

EXERCISE 9.2

Using Parseval's identities prove the following:

1. $\int_0^\infty \frac{(x\cos x - \sin x)^2}{x^6}\,dx = \frac{\pi}{15}$
2. $\int_0^\infty \frac{dx}{(x^2+9)(x^2+4)} = \frac{\pi}{60}$
3. $\int_0^\infty \frac{x^2}{(x^2+1)^2}\,dx = \frac{\pi}{4}$
4. Find the Fourier transform of $f(x) = \begin{cases} 1-|x|, & |x|<1 \\ 0, & |x|>1 \end{cases}$ [JNTU 2002S]

   Hence, deduce that $\int_0^\infty \left(\frac{\sin x}{x}\right)^4 dx = \frac{\pi}{3}$
5. Give that $f(x) = \begin{cases} a-|x|, & |x|<a \\ 0, & |x|>a \end{cases}$
6. Evaluate $\int_0^\infty \frac{\sin^4 x}{x^2}\,dx$

   Prove that $\int_0^\infty \left(\frac{\sin x}{x}\right)^2 dx = \frac{\pi}{2}$

Ans: $\frac{\pi}{2}$

Partial Differential Equations

10

10.1 Introduction

The reader is familiar with ordinary differential equations. These equations involve functions of a single variable only and their derivatives. In many problems that arise in geometry, physics, population dynamics, social sciences, medicine and engineering, one has to deal with equations containing functions of more than one variable and partial derivatives with respect to these independent variables. Such equations are called partial differential equations. Thus, a partial differential equation is an equation of the form

$$f\left(\frac{\partial z}{\partial t}, \frac{\partial z}{\partial x}, \frac{\partial z}{\partial y}, \ldots, \frac{\partial^2 z}{\partial x^2}, \ldots, \frac{\partial^2 z}{\partial t \partial x}, \ldots, z, x, y, t\right) = 0 \tag{10.1}$$

containing independent variables $t, x, y, \ldots$ an unknown function $z = z(x, y, \ldots, t)$ and partial derivatives $\frac{\partial z}{\partial t}, \frac{\partial z}{\partial x}, \frac{\partial z}{\partial y}, \ldots, \frac{\partial^2 z}{\partial x^2}, \ldots, \frac{\partial^2 z}{\partial t \partial x}, \ldots,$ with respect to these variables $t, x, y, \ldots$

10.2 Order, Linearity and Homogeneity of a Partial Differential Equation

10.2.1 Order

The order of a partial differential equation is the order of the highest derivative appearing in it.

10.2.2 Linearity

As in the case of an ordinary differential equation, we say that a partial differential equation is linear if it is of the first degree in the dependent variable (the unknown function) and its partial derivatives and are not multiplied together.

10.2.3 Homogeneity

A linear partial differential equation is called homogeneous if it contains no term free from the unknown function and its derivatives; otherwise, it is called a nonhomogeneous equation.

The following examples of partial differential equations with their order and nature noted against each of them are meant to illustrate the points explained above.

Partial differential equation	Order	Linear/nonlinear, homogeneous/nonhomogeneous
1. $\frac{\partial z}{\partial x} + \frac{\partial z}{\partial y} = z$	One	Linear, homogeneous
2. $x\frac{\partial z}{\partial x} + y\frac{\partial z}{\partial y} = 2z$	One	Linear, homogeneous
3. $\frac{\partial^2 z}{\partial x^2} + \frac{\partial^2 z}{\partial y^2} = 0$ (Laplace's Equation)	Two	Linear, homogeneous
4. $\left(\frac{\partial z}{\partial x}\right)\left(\frac{\partial z}{\partial y}\right) = 4xy$	One	Nonlinear, nonhomogeneous

(Continued)

Partial differential equation	Order	Linear/nonlinear, homogeneous/nonhomogeneous
5. $\frac{\partial z}{\partial t} = c^2 \frac{\partial^2 z}{\partial x^2}$ (Heat conduction equation)	Two	Linear, homogeneous
6. $\frac{\partial^2 z}{\partial t^2} = c^2 \frac{\partial^2 z}{\partial x^2}$ (Wave equation)	Two	Linear, homogeneous
7. $\left(\frac{\partial^2 z}{\partial x^2}\right)^2 = xy$	Two	Nonlinear, nonhomogeneous
8. $\left(\frac{\partial^2 z}{\partial x \partial y}\right)^4 = \left(\frac{\partial z}{\partial x}\right)^3$	Two	Nonlinear, homogeneous

10.3 Origin of Partial Differential Equation

1. Consider the equation

$$x^2 + y^2 + (z - c)^2 = a^2 \quad (10.2)$$

where c and a are arbitrary constants. It represents the set of all spheres with their centres on the z-axis.

Differentiating with respect to x and y we get

$$x + p(z-c) = 0, \quad y + q(z-c) = 0; \quad (10.3), (10.4)$$

$$\left(p = \frac{\partial z}{\partial x}, \quad q = \frac{\partial z}{\partial y}\right)$$

Eliminating the arbitrary constant c from (10.3) and (10.4) we obtain the first-order linear partial differential equation

$$yp - xq = 0 \quad (10.5)$$

which characterizes the set of all spheres with centres on the z-axis.

2. Consider again the equation

$$x^2 + y^2 = (z - c)^2 \tan^2 \alpha \quad (10.6)$$

where c and α are arbitrary constants. Equation (10.6) represents the set of all right circular cones whose axes coincide with the z-axis.

Differentiating (10.6) partially with respect to x and y we get

$$p(z - c)\tan^2 \alpha = x, \quad q(z - c)\tan^2 \alpha = y \quad (10.7), (10.8)$$

respectively. Eliminating the constants c and α between these equations we again obtain equation (10.5).

3. The spheres and cones are surfaces of revolution which have the line OZ as their axis of symmetry. Now, consider the equation

$$z = f(x^2 + y^2) \quad (10.9)$$

where f is an arbitrary function. Equation (10.9) characterizes surfaces of revolution having z-axis as the axis of symmetry.

Differentiating equation (10.9) partially with respect to x and y we get

$$p = 2xf'(u), \quad q = 2yf'(u) \quad (10.10), (10.11)$$

where $u = x^2 + y^2$ and $f'(u) = df/du$. Eliminating f between equations (10.10) and (10.11), we again obtain the first-order linear partial differential equation (10.5).

Formation of PDE Ordinary differential equations are formed by eliminating arbitrary constants only, whereas partial differential equations are formed by eliminating (a) arbitrary constants or (b) arbitrary functions.

We know that the order of an ordinary differential equation is equal to the number of arbitrary constants to be eliminated from a relation.

In the case of partial differential equations, if the number of arbitrary constants to be eliminated is equal to the number of independent variables involved in a relation, one obtains a *first*-order partial differential equation, and if the number of arbitrary constants to be eliminated is more than the number of independent variables, then one obtains a higher-order partial differential equation.

Further, if one arbitrary function is to be eliminated from a relation, then a first-order partial differential equation is obtained; and if two arbitrary functions are to be eliminated, then a second-order partial differential equation is obtained and so on.

A first-order partial differential equation is of the form

$$f(x, y, z, p, q) = 0 \tag{10.12}$$

where x, y are independent variables, z is a dependent variable and $p = \partial z/\partial x$ and $q = \partial z/\partial y$.

10.4 Formation of Partial Differential Equation by Elimination of Two Arbitrary Constants

Consider a relation of the type

$$F(x, y, z, a, b) = 0 \tag{10.13}$$

where a and b are arbitrary constants.

Differentiating (10.13) partially with respect to x and y we have

$$\frac{\partial F}{\partial x} + p\frac{\partial F}{\partial z} = 0 \quad \text{and} \quad \frac{\partial F}{\partial y} + q\frac{\partial F}{\partial z} = 0 \qquad (10.14), (10.15)$$

We can now eliminate the two arbitrary constants a and b between the equations (10.13)–(10.15) and obtain a first order partial differential equation of the form

$$f(x, y, z, p, q) = 0 \tag{10.16}$$

We now work out a few examples to show how partial differential equations are formed by eliminating arbitrary constants.

Example 10.1

Show by eliminating the arbitrary constants a and b from

$$(x-a)^2 + (y-b)^2 + z^2 = 1 \tag{1}$$

the partial differential equation

$$z^2(p^2 + q^2 + 1) = 1 \tag{2}$$

is obtained, which is nonlinear.

Solution Differentiating (1) partially with respect to x and y

$$(x-a) + zp = 0, \ (y-b) + zq = 0 \qquad (3), (4)$$

Eliminating a and b from (1), (3) and (4) we get

$$z^2(p^2 + q^2 + 1) = 1$$

which is nonlinear.

Examples

Form the partial differential equation by eliminating the arbitrary constants a and b from each of the relations given below:

Example 10.2

$$z = ax + by$$

Solution We have

$$z = ax + by \tag{1}$$

Differentiating (1) partially with respect to x and y we have

$$\frac{\partial z}{\partial x} = p = a, \quad \frac{\partial z}{\partial y} = q = b \qquad (2), (3)$$

Eliminating a and b from equations (1)–(3) we obtain the first-order partial differential equation

$$z = px + qy \tag{4}$$

Example 10.3

$$z = (x^2 + a^2)(y^2 + b^2)$$

Solution We have

$$z = (x^2 + a^2)(y^2 + b^2) \tag{1}$$

Differentiating (1) partially with respect to x and y

$$\frac{\partial z}{\partial x} = p = 2x(y^2 + b^2) \ \Rightarrow \ y^2 + b^2 = \frac{p}{2x} \tag{2}$$

$$\frac{\partial z}{\partial y} = q = 2y(x^2 + a^2) \ \Rightarrow \ x^2 + a^2 = \frac{q}{2y} \tag{3}$$

Eliminating a and b from (1)–(3) we get

$$z = \frac{q}{2y} \cdot \frac{p}{2x} \quad \text{or} \quad pq = 4xyz \qquad (4)$$

Example 10.4

$$z = \frac{x^2}{a^2} + \frac{y^2}{b^2}$$

Solution We have

$$z = \frac{x^2}{a^2} + \frac{y^2}{b^2} \qquad (1)$$

Differentiating (1) partially with respect to x and y

$$\frac{\partial z}{\partial x} = p = \frac{2x}{a^2}, \quad \frac{\partial z}{\partial y} = q = \frac{2y}{b^2} \qquad (2), (3)$$

Eliminating a and b from (1)–(3) we get

$$z = \frac{x}{2}p + \frac{y}{2}q \quad \Rightarrow \quad px + qy = 2z \qquad (4)$$

Example 10.5

$$z = (x + a)(y + b)$$

Solution We have

$$z = (x + a)(y + b) \qquad (1)$$

Differentiating (1) partially with respect to x and y

$$\frac{\partial z}{\partial x} = p = y + b, \quad \frac{\partial z}{\partial y} = q = x + a \qquad (2), (3)$$

Eliminating a and b from (1)–(3) we have

$$z = pq \qquad (4)$$

Example 10.6

$$2z = (ax + y)^2 + b$$

Solution We have

$$2z = (ax + y)^2 + b \qquad (1)$$

Differentiating (1) partially with respect to x and y

$$2p = 2a(ax + y), \quad 2q = 2(ax + y) \qquad (2), (3)$$

Eliminating a and b from (1)–(3) we get

$$px + qy = q^2 \qquad (4)$$

which is nonlinear.

Example 10.7

$$ax^2 + by^2 + z^2 = 1$$

Solution We have

$$ax^2 + by^2 + z^2 = 1 \qquad (1)$$

Differentiating (1) partially with respect to x and y

$$zp + ax = 0, \quad zq + by = 0 \qquad (2), (3)$$

Multiplying (2) by x and (3) by y and adding

$$z(px + qy) + (ax^2 + by^2) = 0$$

$$\Rightarrow \quad z(px + qy) = -(ax^2 + by^2) = -(1 - z^2), \quad \text{by (1)}$$

$$\Rightarrow \quad z(px + qy) = z^2 - 1 \qquad (4)$$

EXERCISE 10.1

Form the partial differential equation in each of the following cases, by eliminating the arbitrary constants a and b (Questions 1–10):

1. $z = ax^2 + by^2$.

   Ans: $2z = px + qy$

2. $(x - a)^2 + (y - b)^2 = z^2 \cot^2 \alpha$.

   Ans: $p^2 + q^2 = \tan^2 \alpha$

3. $z = ax + by + a^2 + b^2$.

   Ans: $z = px + qy + p^2 + q^2$

4. $z = axy + b$.

   Ans: $px = qy$

5. $z = \frac{1}{2}\left(\sqrt{x + a} + \sqrt{y - a} + b\right)$. [JNTU 2003]

   Ans: $16(p^2 + q^2)(x + y) = 1$

6. $u = a(x + y) + b(x - y) + abz + c$.

   Ans: $u_x^2 - u_y^2 = 4u_z$

7. $z = xy + y\sqrt{x^2 + a^2} + b$.

   Ans: $px + qy = pq$

8. $z = axe^y + \frac{1}{2}a^2e^{2y} + b$.

   Ans: $q = px + p^2$

9. $z = ax + by + \left(\frac{a}{b}\right) - b.$ [JNTU 2003]

Ans: $z = px + qy + \left(\frac{p}{q}\right) - q$

10. $z = a\log\left[\frac{b(y-1)}{(1-x)}\right].$ [JNTU 2003]

Ans: $px + qy = p + q$

11. Form the partial differential equation of all spheres of radius a with their centres on the x–y plane.

Ans: $z^2(p^2 + q^2 + 1) = a^2$

12. Form the partial differential equation of all planes through the origin.

Ans: $z = px + qy$

10.5 Formation of Partial Differential Equations by Elimination of Arbitrary Functions

(a) Elimination of one arbitrary function of the form $z = f(u)$ where $u = u(x, y, z)$

Let

$$z = f(u) \tag{10.17}$$

where $f(u)$ is an arbitrary function of u where $u = u(x, y, z)$ a known function of x, y and z.

Differentiating (10.17) partially with respect to x and y

$$p = f'\cdot\left(\frac{\partial u}{\partial x} + \frac{\partial u}{\partial z}p\right), \quad q = f'\cdot\left(\frac{\partial u}{\partial y} + \frac{\partial u}{\partial z}q\right) \quad (10.18), (10.19)$$

where $'$ on f denotes differentiation with respect to the argument u. Eliminating f from equations (10.18)–(10.19) we obtain a first-order pàrtial differential equation.

(b) Elimination of two arbitrary functions of the form $z = f(x)\cdot g(y)$

Let

$$z = f(x)\cdot g(y) \tag{10.20}$$

where f is a function of x and g is a function of y alone.

Differentiating (10.20) partially with respect to x and y, respectively, we have

$$\frac{\partial z}{\partial x} = p = f'g, \quad \frac{\partial z}{\partial y} = q = fg', \quad \frac{\partial^2 z}{\partial x\partial y} = s = f'g'$$

$$pq = f'g\cdot fg' = f'g'\cdot fg = sz \quad \text{or} \quad pq = sz$$

where

$$s = \frac{\partial^2 z}{\partial x\partial y} \tag{10.21}$$

(c) Formation of PDE by elimination of arbitrary function of the form $F(u, v) = 0$

Consider a relation between x, y and z of the type

$$F(u, v) = 0 \tag{10.22}$$

where u and v are known functions of x, y and z; and F is an arbitrary function of u and v. Also, z is a function of x and y.

Differentiating (10.22) by chain rule, with respect to x and y we obtain

$$\frac{\partial F}{\partial u}\left(\frac{\partial u}{\partial x} + \frac{\partial u}{\partial z}p\right) + \frac{\partial F}{\partial v}\left(\frac{\partial v}{\partial x} + \frac{\partial v}{\partial z}p\right) = 0 \tag{10.23}$$

$$\frac{\partial F}{\partial u}\left(\frac{\partial u}{\partial y} + \frac{\partial u}{\partial z}q\right) + \frac{\partial F}{\partial v}\left(\frac{\partial v}{\partial y} + \frac{\partial v}{\partial z}q\right) = 0 \tag{10.24}$$

Eliminating F between (10.23) and (10.24) we get

$$\begin{vmatrix} u_x + u_z p & v_x + v_z p \\ u_y + u_z q & v_y + u_z q \end{vmatrix} = 0 \tag{10.25}$$

where $u_x = \partial u/\partial x$, $u_y = \partial u/\partial y$ and so on.

Expanding the determinant in (10.25), collecting the terms and simplifying we have

$$Pp + Qq = R \tag{10.26}$$

where $P = \frac{\partial(u,v)}{\partial(y,z)}$, $Q = \frac{\partial(u,v)}{\partial(z,x)}$, $R = \frac{\partial(u,v)}{\partial(x,y)}$

which are functions of x, y and z and do not contain p and q. This equation is called Lagrange's[1] linear equation. In fact, it is a quasi-linear equation since the dependent variable may be present in P, Q and R.

If P and Q are independent of z and R is linear in z the equation (10.26) is called linear.

[1] LAGRANGE, Joseph-Louis (1736–1813) great French analyst, algebraist, number theorist, probabilist, physicist, and astronomer, contributed especially to the calculas of variations, analytic mechanics and astronomer.

Equation (10.26) is of the type given by equation (10.12), which need not be linear, as Example 10.1 above will illustrate.

Examples

Form the partial differential equation by eliminating the arbitrary function f from each of the following relations [Examples 10.8–10.15 except (Example 10.13) in which functions f and g have to be eliminated.]

Example 10.8

$$z = x f(x+y)$$

Solution We have

$$z = xf(x+y) \quad (1)$$

Differentiating (1) partially with respect to x and y

$$p = 1f + xf', \quad q = 0 + xf' \quad (2), (3)$$

Eliminating f between (1)–(3) we have

$$p = \frac{z}{x} + q \;\Rightarrow\; z = (p - q)x \quad (4)$$

which is the required partial differential equation.

Example 10.9

$$z = x - y + f(xy)$$

Solution We have

$$z = x - y + f(xy) \quad (1)$$

Differentiating (1) partially with respect to x and y

$$p = 1 + f' \cdot y, \quad q = -1 + f' \cdot x \quad (2), (3)$$

Eliminating f' between (1) and (2) we get the partial differential equation

$$px - qy = x + y \quad (4)$$

Example 10.10

$$z = e^{ax-by} f(ax+by)$$

Solution We have

$$z = e^{ax-by} f(ax+by) \quad (1)$$

Differentiating (1) partially with respect to x and y

$$p = ae^{ax-by} f + e^{ax-by} f' \cdot a,$$

$$q = -be^{ax-by} f + e^{ax-by} f' \cdot b \quad (2), (3)$$

Eliminating f between (1)–(3) we have

$$bp - aq = 2abz \quad (4)$$

Example 10.11

$$z = x^n f\left(\frac{y}{x}\right)$$

Solution We have

$$z = x^n f\left(\frac{y}{x}\right) \quad (1)$$

Differentiating (1) partially with respect to x and y

$$p = nx^{n-1} f + x^n f' \cdot \left(\frac{-y}{x^2}\right), \quad q = x^n f' \cdot \frac{1}{x} \quad (2), (3)$$

We have from (2) and (3)

$$px + qy = nx^n f - yx^{n-1} f' + yx^{n-1} f' = nz,$$

$$\Rightarrow \quad px + qy = nz, \quad \text{by (1)} \quad (4)$$

Example 10.12

$$yz + zx + xy = f\left(\frac{z}{x+y}\right). \quad \text{[JNTU 2005 S]}$$

Solution We have

$$yz + zx + xy = f\left(\frac{z}{x+y}\right) \quad (1)$$

Differentiating (1) with respect to x and y

$$y + z + (x+y)p = f' \cdot \frac{(x+y)p - z}{(x+y)^2},$$

$$z + x + (x+y)q = f' \cdot \frac{(x+y)q - z}{(x+y)^2} \quad (2), (3)$$

Eliminating f' between (2) and (3) we have

$$\frac{y + z + (x+y)p}{x + z + (x+y)q} = \frac{(x+y)p - z}{(x+y)q - z} = \frac{y + 2z}{x + 2z}\left(= \frac{\text{diff.of nrs.}}{\text{diff.of drs.}}\right)$$

$$(x+y)[p(x+2z) - q(y+2z)] = z[x + 2z - (y+2z)]$$

$$\Rightarrow \quad px - qy + 2z(p - q) = z(x-y)/(x+y)$$

Example 10.13

$$z = f(x+at) + g(x-at). \quad \text{[Osmania 1999]}$$

Solution We have

$$z = f(x+at) + g(x-at) \quad (1)$$

Differentiating (1) partially with respect to x and t twice

$$\frac{\partial z}{\partial x}=f'+g', \quad \frac{\partial z}{\partial t}=af'-ag', \tag{2}$$

$$\frac{\partial^2 z}{\partial x^2}=f''+g'', \quad \frac{\partial^2 z}{\partial t^2}=a^2 f''+a^2 g'' \tag{3}$$

Eliminating the arbitrary functions from (3) we have

$$\frac{\partial^2 z}{\partial t^2}=a^2\frac{\partial^2 z}{\partial x^2} \tag{4}$$

which is a partial differential equation of the second order.

Example 10.14

$$z=y^2+2f\left(\frac{1}{x}+\log y\right) \quad \text{[Madras 2000]}$$

Solution We have

$$z=y^2+2f\left(\frac{1}{x}+\log y\right) \tag{1}$$

Differentiating (1) with respect to x and y

$$p=0+2f'\cdot\left(-\frac{1}{x^2}\right), \quad q=2y+2f'\cdot\frac{1}{y} \quad (2), (3)$$

Eliminating f' between (2) and (3) we get

$$px^2+qy=2y^2 \tag{4}$$

which is the required partial differential equation.

Example 10.15

$$f(xy+z^2, x+y+z)=0$$

[Madras 1995, Kerala 1990 S]

Solution We have

$$f(u,v)=0 \tag{1}$$

where

$$u=xy+z^2, \quad v=x+y+z \tag{2}$$

We know that the partial differential equation is

$$\frac{\partial(u,v)}{\partial(y,z)}p+\frac{\partial(u,v)}{\partial(z,x)}q=\frac{\partial(u,v)}{\partial(x,y)}$$

where

$$\frac{\partial(u,v)}{\partial(y,z)}=\begin{vmatrix} x & 2z \\ 1 & 1\end{vmatrix}=x-2z;$$

$$\frac{\partial(u,v)}{\partial(z,x)}=\begin{vmatrix} 2z & y \\ 1 & 1\end{vmatrix}=2z-y;$$

$$\frac{\partial(u,v)}{\partial(x,y)}=\begin{vmatrix} y & x \\ 1 & 1\end{vmatrix}=y-x \tag{3}$$

Hence, the required partial differential equation is

$$(x-2z)p-(y-2z)q=y-x \tag{4}$$

EXERCISE 10.2

Form the partial differential equations by eliminating the arbitrary functions from the following relations:

1. $z=f\left(\frac{xy}{z}\right)$. [Andhra 1991]

   Ans: $px=qy$

2. $xyz=f(x+y+z)$. [Osmania 1995, Calicut 1994]

   Ans: $x(y-z)p+y(z-x)q=z(x-y)$

3. $x+y+z=f(x^2+y^2+z^2)$. [Madras 2000, Ranchi 1990]

   Ans: $(y-z)p+(z-x)q=x-y$

4. $f(x^2+y^2, z-xy)=0$. [Madurai 1990]

   Ans: $py-qx=y^2-x^2$

5. $z=f(x^2-y^2)$. [Madras 1991]

   Ans: $py+qx=0$

6. $z=yf(x)+xg(y)$. [Madras 1993, Karnataka 1993, Madurai 1998]

   Ans: $xy\frac{\partial^2 z}{\partial x\partial y}=x\frac{\partial z}{\partial x}+y\frac{\partial z}{\partial y}-z$

7. $z=f(y)+\phi(x+y)$. [JNTU 2005 S (1)]

   Ans: $\frac{\partial^2 z}{\partial x\partial y}=\frac{\partial^2 z}{\partial x^2}$

8. $z=xy+f(x^2+y^2)$. [JNTU 2005 S (2)]

   Ans: $py-qx=y^2-x^2$

10.6 Classification of First-Order Partial Differential Equations

10.6.1 Linear Equation

Let $z=z(x, y)$ be a function of two variables x and y. A first-order partial differential equation is said to be linear if it is linear in p, q and z, *i.e.* if it is of the form

$$P(x, y)p + Q(x, y)q = R(x, y)z + S(x, y)$$

where

$$p = \frac{\partial z}{\partial x} \quad \text{and} \quad q = \frac{\partial z}{\partial y} \tag{10.27}$$

E.g.

$$\left.\begin{aligned} yp - xq &= xyz + x \\ y^2 p - xyq &= x(z - 2y) \end{aligned}\right\} \tag{10.28}$$

10.6.2 Semi-Linear Equation

A first-order partial differential equation is said to be semi-linear if it is linear in p and q and the coefficients of p and q are functions of x and y alone, *i.e.* if it is of the form

$$P(x, y)p + Q(x, y)q = R(x, y, z) \tag{10.29}$$

E.g.

$$\left.\begin{aligned} &1.\quad e^x p - yxq = xz^2 \\ &2.\quad \sqrt{x}p + \sqrt{y}q = \sqrt{z} \end{aligned}\right\} \tag{10.30}$$

10.6.3 Quasi-Linear Equation

A first-order partial differential equation is said to be quasi-linear if it is linear in p and q, *i.e.* if it is of the form

$$P(x, y, z)p + Q(x, y, z)q = R(x, y, z) \tag{10.31}$$

E.g.

$$\left.\begin{aligned} &1.\quad (x^2 - yz)p + (y^2 - zx)q = z^2 - xy \\ &2.\quad (x^2 + y^2)p - xyq = z^3x + y^2 \end{aligned}\right\} \tag{10.32}$$

Note that every semi-linear partial differential equation is quasi-linear.

10.6.4 Nonlinear Equation

A first-order partial differential equation is said to be nonlinear if it does not fall into any one of the above types.

10.7 Classification of Solutions of First-Order Partial Differential Equation

Consider a first-order partial differential equation of the form

$$f(x, y, z, p, q) = 0 \tag{10.33}$$

Let

$$z = z(x, y) \tag{10.34}$$

be a continuously differentiable function of x and y in a region D.

Compute $p = \partial z/\partial x$ and $q = \partial z/\partial y$ from (10.34) and substitute in (10.33). If this reduces (10.33) into an identity in x and y then (10.34) defines a solution of partial differential equation (4.33).

Equation (10.34) defines a surface in three-dimensional space. It is called an integral surface of partial differential equation (10.33).

There are different types of solutions of equation (10.33).

10.7.1 Complete Integral

A two-parameter family of solutions

$$z = F(x, y, z, a, b) \tag{10.35}$$

is called a complete integral (or complete solution) of (10.33) if in the region D the matrix

$$M = \begin{bmatrix} \dfrac{\partial F}{\partial a} & \dfrac{\partial^2 F}{\partial a \partial x} & \dfrac{\partial^2 F}{\partial a \partial y} \\ \dfrac{\partial F}{\partial b} & \dfrac{\partial^2 F}{\partial b \partial x} & \dfrac{\partial^2 F}{\partial b \partial y} \end{bmatrix} \tag{10.36}$$

is of rank two.

10.7.2 General Integral

Let a and b be related and suppose

$$b = \phi(a) \tag{10.37}$$

Substituting this in (10.35) we get

$$z = F(x, y, a, \phi(a)) \tag{10.38}$$

which is a one-parameter family of solutions of (10.33).

This is a sub-family of the two-parameter family given by (10.35). The envelope of (10.38), if it exists, is obtained by eliminating a between (10.38) and

$$\frac{\partial F}{\partial a} + \frac{\partial F}{\partial b}\phi'(a) = 0 \tag{10.39}$$

If (10.39) can be solved for a then

$$a = a(x, y) \tag{10.40}$$

Substituting for a in (10.38) we obtain an integral surface as

$$z = F(x, y, a(x, y), \phi(a(x, y))) \tag{10.41}$$

If ϕ is arbitrary then (10.41) is called a general integral (or general solution) of the partial differential equation (10.33).

10.7.3 Particular Integral

When a particular function ϕ is used in (10.41) we obtain a particular integral or particular solution of the partial differential equation (10.33).

10.7.4 Singular Integral

In some cases we find another integral which satisfies the partial differential equation (10.33) but is not a particular integral of (10.33). If it exists, it is obtained by eliminating a and b from

$$z = F(x, y, a, b), \quad \frac{\partial F}{\partial a} = 0 \quad \text{and} \quad \frac{\partial F}{\partial b} = 0 \qquad (10.42), (10.43), (10.44)$$

and it is the envelope of the two-parameter family of surfaces

$$z = F(x, y, a, b) \qquad (10.45)$$

and it is called a singular integral (or a singular solution).

In a sense, a general integral provides a much broader class of solutions of the partial differential equation than does a complete integral.

However, it is possible to derive a general integral when a complete integral is known.

Note that for the partial differential equation

$$z = px + qy \qquad (10.46)$$

the relation

$$z = ax + by \qquad (10.47)$$

is a complete integral while the relation

$$z = yf\left(\frac{y}{x}\right) \qquad (10.48)$$

is a general integral.

10.8 Equations Solvable by Direct Integration

We now consider partial differential equations which can be solved by direct integration. While carrying out integration with respect to a variable the other variable is held fixed. So, in place of constant of integration we have to add an arbitrary function of the variable held fixed.

Example 10.16

Solve: $\dfrac{\partial^2 z}{\partial x^2} = xy^2$.

Solution The given partial differential equation is

$$\frac{\partial^2 z}{\partial x^2} = xy^2 \qquad (1)$$

Integrating once partially with respect to x we get

$$\frac{\partial z}{\partial x} = \frac{x^2}{2} \cdot y^2 + f(y) \qquad (2)$$

and integrating (2) partially with respect to x we get the solution as

$$z = \frac{x^3}{6} \cdot y^2 + xf(y) + g(y) \qquad (3)$$

where f and g are arbitrary functions of y.

Example 10.17

Solve: $\dfrac{\partial^2 z}{\partial x \partial y} = \sin x \tan y$.

Solution The given partial differential equation is

$$\frac{\partial^2 z}{\partial x \partial y} = \sin x \tan y \qquad (1)$$

Integrating (1) partially with respect to x

$$\frac{\partial z}{\partial y} = -\cos x \tan y + f(y) \qquad (2)$$

where $f(y)$ is an arbitrary function of y. Integrating (2) again partially with respect to y

$$z = -\cos x \cdot \log \sec y + g(y) + h(x) \qquad (3)$$

where $g(y) = \int f(y)\, dy$ and $h(x)$ are arbitrary functions of y and x, respectively.

Example 10.18

Solve: $\dfrac{\partial^2 z}{\partial y \partial x} = 2xe^y$.

Solution The given partial differential equation is

$$\frac{\partial^2 z}{\partial y \partial x} = 2xe^y \qquad (1)$$

Integrating (1) partially with respect to y we get

$$\frac{\partial z}{\partial x} = 2xe^y + f(x) \quad (2)$$

where $f(x)$ is an arbitrary function of x. Integrating (2) again partially with respect to x we get

$$z = x^2e^y + \int f(x)\,dx + g(y) \quad (3)$$

which is the solution of (1). We can write (3) as

$$z = x^2e^y + h(x) + g(y) \quad (4)$$

where $h(x)$ and $g(y)$ are arbitrary functions of x and y, respectively.

Example 10.19

Solve: $\dfrac{\partial^3 z}{\partial x^2 \partial y} = \cos(2x+3y)$.

Solution The given partial differential equation is

$$\frac{\partial^3 z}{\partial x^2 \partial y} = \cos(2x+3y) \quad (1)$$

Integrating (1) partially with respect to x

$$\frac{\partial^2 z}{\partial x \partial y} = \frac{1}{2}\sin(2x+3y) + f(y) \quad (2)$$

where $f(y)$ is an arbitrary function of y. Integrating (2) partially with respect to x again

$$\frac{\partial z}{\partial y} = -\frac{1}{4}\cdot\cos(2x+3y) + xf(y) + g(y) \quad (3)$$

where g is an arbitrary function of y. Integrating (3) partially with respect to y,

$$z = -\frac{1}{12}\sin(2x+3y) + x\int f(y)\,dy + \int g(y)\,dy + h(x)$$

which can be put in the form

$$z = f_1(x) + f_2(y) + xf_3(y) - \frac{1}{12}\sin(2x+3y) \quad (4)$$

where $f_1(x)$ is an arbitrary function of x and $f_2(y)$ and $f_3(y)$ are arbitrary functions of y.

Example 10.20

Solve: $\dfrac{\partial^2 z}{\partial x^2} - a^2 z = 0$. If $\dfrac{\partial z}{\partial x} = a\sin y$ and $\dfrac{\partial z}{\partial y} = 0$ when $x = 0$.

Solution The given partial differential equation is

$$\frac{\partial^2 z}{\partial x^2} - a^2 z = 0 \quad (1)$$

We have to solve (1) under the conditions:

When $x = 0$, $\dfrac{\partial z}{\partial x} = a\sin y$ and $\dfrac{\partial z}{\partial y} = 0$ (2), (3)

If we treat z as a function of x alone, the solution of (1) is

$$z = A\sinh x + B\cosh x \quad (4)$$

where A and B are constants. Since z is a function of y also we can take z as

$$z = A(y)\sinh x + B(y)\cosh x$$

Now $\left.\dfrac{\partial z}{\partial x}\right|_{x=0} = A(y)\cosh x + B(y)\sinh x\big|_{x=0}$

$\Rightarrow \quad A.1 + B.0 = a\sin y$

$\left.\dfrac{\partial z}{\partial y}\right|_{x=0} = A'(y)\sin x + B'(y)\cosh x\big|_{x=0}$

$\Rightarrow \quad A'.0 + B'.1 = 0$

$\therefore \quad A(y) = a\sin y$ and $B(y) = $ constant.

Hence $\quad z = (a\sin y)\sinh x + b\cosh x$,

where b is an arbitrary constant.

EXERCISE 10.3

1. Solve $\partial^2 z/\partial x\partial y = \sin x \sin y$ given that $\partial z/\partial y = -2\sin y$ when $x = 0$ and $z = 0$ when y is an odd multiple of $\pi/2$. [Madras 1994 S, Mysore 1999 S]

   Ans: $z = (1 + \cos x)\cos y$

2. Solve $\partial^3 z/\partial x^2\partial y + 18xy^2 + \sin(2x - y) = 0$.

   Ans: $z = \dfrac{1}{4}\cos(2x - y) - x^3y^3 + xf(y) + g(y) + h(x)$

3. Solve $\partial^2 z/\partial x^2 + z = 0$ given that when $x = 0$, $z = e^y$ and $\partial z/\partial x = 1$. [Mysore 1987 S, Madras 1993, Karnataka 1994]

4. Solve $\partial^2 u / \partial x \partial t = e^{-t} \cos x$. [Madurai 1988, Mysore 1987]

Ans: $u = -e^{-t} \sin x + \phi(x) + \psi(t)$

5. Solve $\left(\partial^2 z / \partial x \partial y\right) = \exp(x+y)$.

Ans: $z = e^x \cdot e^y + \int f(y)dy + \phi(x)$

6. Solve $\partial^2 z / \partial y^2 = \sin(xy)$.

Ans: $z = -\dfrac{\sin xy}{x^2} + y\phi(x) + \psi(x)$

10.9 Quasi-Linear Equations of First Order

Quasi-linear partial differential equations of first order can be written as

$$Pp + Qq = R \tag{10.49}$$

where P, Q, R are functions of x, y and z. They do not involve p or q. Equation (10.49) is called Lagrange's equation. Here, linear means that p and q appear to the first degree only. This is in contrast to the situation in ordinary differential equations where z must also be of first degree.

Note that partial differential equation

$$x\frac{\partial z}{\partial x} + y\frac{\partial z}{\partial y} = z^2 + x^2 \tag{10.50}$$

is linear while the ordinary differential equation

$$x\frac{dz}{dx} = z^2 + x^2 \tag{10.51}$$

is nonlinear.

Theorem 10.1 The general solution of the quasi-linear partial differential equation

$$Pp + Qq = R \tag{10.52}$$

is

$$F(u, v) = 0 \tag{10.53}$$

where F is an arbitrary function and $u(x, y, z) = c_1$ and $v(x, y, z) = c_2$ form a solution of the equations

$$\frac{dx}{P} = \frac{dy}{Q} = \frac{dz}{R} \tag{10.54}$$

Proof If

$$u(x, y, z) = c_1, \quad v(x, y, z) = c_2 \tag{10.55}$$

satisfy equations (10.54) then

$$u_x\,dx + u_y\,dy + u_z\,dz = 0, \tag{10.56}$$

$$\text{and } v_x\,dx + v_y\,dy + v_z\,dz = 0 \tag{10.57}$$

must be compatible with (10.54) so that we must have

$$Pu_x + Qu_y + Ru_z = 0, \quad Pv_x + Qv_y + Rv_z = 0 \qquad (10.58), (10.59)$$

From (10.58) and (10.59) we have

$$\frac{P}{\dfrac{\partial(u,v)}{\partial(y,z)}} = \frac{Q}{\dfrac{\partial(u,v)}{\partial(z,x)}} = \frac{R}{\dfrac{\partial(u,v)}{\partial(x,y)}} \tag{10.60}$$

Now, differentiating (10.53) with respect to x and y

$$\frac{\partial F}{\partial u}\left(\frac{\partial u}{\partial x} + \frac{\partial u}{\partial z}p\right) + \frac{\partial F}{\partial v}\left(\frac{\partial v}{\partial x} + \frac{\partial v}{\partial z}p\right) = 0 \tag{10.61}$$

$$\frac{\partial F}{\partial u}\left(\frac{\partial u}{\partial y} + \frac{\partial u}{\partial z}q\right) + \frac{\partial F}{\partial v}\left(\frac{\partial v}{\partial y} + \frac{\partial v}{\partial z}q\right) = 0 \tag{10.62}$$

Eliminating $\partial F/\partial u$ and $\partial F/\partial v$ from (10.61)–(10.62) we get

$$\begin{vmatrix} \dfrac{\partial u}{\partial x} + \dfrac{\partial u}{\partial z}p & \dfrac{\partial v}{\partial x} + \dfrac{\partial v}{\partial z}p \\ \dfrac{\partial u}{\partial y} + \dfrac{\partial u}{\partial z}q & \dfrac{\partial v}{\partial y} + \dfrac{\partial v}{\partial z}q \end{vmatrix} = 0, \quad \text{or}$$

$$\frac{\partial(u,v)}{\partial(y,z)}p + \frac{\partial(u,v)}{\partial(z,x)}q = \frac{\partial(u,v)}{\partial(x,y)}$$

$$\text{or } Pp + Qq = R \tag{10.63}$$

where $P = \dfrac{\partial(u,v)}{\partial(y,z)}$, $Q = \dfrac{\partial(u,v)}{\partial(z,x)}$, $R = \dfrac{\partial(u,v)}{\partial(x,y)}$ are functions of x, y, z.

Substituting from (10.60) into (10.63) we see that (10.53) is a solution of (10.49) if u and v are given by (10.55).

10.10 Solution of Linear, Semi-Linear and Quasi-Linear Equations

If the equation is linear or semi-linear or can be written in these forms then we can solve them as follows.

10.10.1 All the Variables are Separable

Example 10.21

Solve: $zx^2p + zy^2q = 1 - z^2$.

Solution The equation can be thrown into the form

$$x^2p + y^2q = \frac{1-z^2}{z} \tag{1}$$

which is a semi-linear equation.

Lagrange's auxiliary equations are

$$\frac{dx}{x^2} = \frac{dy}{y^2} = \frac{zdz}{1-z^2} \tag{2}$$

Integrating the first and last equations we have

$$\frac{1}{x} - \frac{1}{y} = c_1, \quad \frac{1}{y} = \frac{1}{2}\log(1-z^2) + \log c_2$$

or $\dfrac{(e^{1/y})}{\sqrt{1-z^2}} = c_2$, respectively (3), (4)

General solution is

$$F\left(\frac{1}{x} - \frac{1}{y}, \frac{(e^{1/y})}{\sqrt{1-z^2}}\right) = 0 \tag{5}$$

10.10.2 Two Variables are Separable

Example 10.22

Solve: $mp - lq = z\tan(lx + my)$.

Solution This is a semi-linear equation.

Lagrange's auxiliary equations are

$$\frac{dx}{m} = \frac{dy}{-l} = \frac{dz}{z\tan(lx+my)} \tag{1}$$

From the first equation we have

$$ldx + mdy = 0 \quad \Rightarrow \quad lx + my = c_1 \tag{2}$$

Using this in the last equation

$$\frac{dy}{-l} = \frac{dz}{z\tan c_1} \quad (\text{since } lx + my = c_1)$$

$$\Rightarrow \quad \tan c_1 dy + \frac{ldz}{z} = 0 \tag{3), (4}$$

Integrating we have

$$y\tan(lx + my) + l\log z = \log c_2$$

$$\text{or} \quad z^l \cdot e^{y\tan(lx+my)} = c_2 \tag{5}$$

General solution is

$$F(lx + my, z^l \cdot e^{y\tan(lx+my)}) = 0 \tag{6}$$

Example 10.23

Solve: $yp - xq = xyz + xy$.

Solution The given equation is linear. Lagrange's auxiliary equations are

$$\frac{dx}{y} = \frac{dy}{-x} = \frac{dz}{xyz + xy} \tag{1}$$

From the first equation we have

$$xdx + ydy = 0 \tag{2}$$

Integrating we get

$$x^2 + y^2 = c_1 \tag{3}$$

From the last equation

$$\frac{dy}{-1} = \frac{dz}{y(z+1)} \quad \Rightarrow \quad ydy + \frac{dz}{z+1} = 0 \tag{4}$$

Integrating,

$$\frac{y^2}{2} + \log(z+1) = \log c_2 \quad \Rightarrow \quad (z+1)e^{y^2/2} = c_2 \tag{5}$$

General solution is

$$F\left(x^2 + y^2, (z+1)e^{y^2/2}\right) = 0 \tag{6}$$

10.10.3 Method of Multipliers

In addition to the above methods, we can apply the following method called **method of multipliers**.

Example 10.24

Solve: $x^2(y - z)p + y^2(z - x)q = z^2(x - y)$.

Solution Lagrange's auxiliary equations are

$$\frac{dx}{x^2(y-z)} = \frac{dy}{y^2(z-x)} = \frac{dz}{z^2(x-y)} \tag{1}$$

Using multipliers $\left(\frac{1}{x}, \frac{1}{y}, \frac{1}{z}\right)$ and $\left(\frac{1}{x^2}, \frac{1}{y^2}, \frac{1}{z^2}\right)$ in turn we get

$$\text{each ratio} = \frac{\frac{1}{x}dx + \frac{1}{y}dy + \frac{1}{z}dz}{x(y-z) + y(z-x) + z(x-y) = 0}$$

$$\Rightarrow \quad \frac{dx}{x} + \frac{dy}{y} + \frac{dz}{z} = 0 \tag{2}$$

$$\Rightarrow \quad \log x + \log y + \log z = \log c_1$$

$$\Rightarrow \quad xyz = c_1 \qquad (3)$$

$$\text{each ratio} = \frac{\frac{1}{x^2}dx + \frac{1}{y^2}dy + \frac{1}{z^2}dz}{(y-z)+(z-x)+(x-y)=0}$$

$$\Rightarrow \quad \frac{1}{x^2}dx + \frac{1}{y^2}dy + \frac{1}{z^2}dz = 0 \qquad (4)$$

$$\Rightarrow \quad \frac{1}{x}+\frac{1}{y}+\frac{1}{z} = c_2 \qquad (5)$$

General solution is

$$F\left(xyz, \ \frac{1}{x}+\frac{1}{y}+\frac{1}{z}\right) = 0 \qquad (6)$$

Example 10.25

Solve: $zxp + zyq = 1 + z^2$.

Solution Lagrange's auxiliary equations are

$$\frac{dx}{zx} = \frac{dy}{zy} = \frac{dz}{1+z^2} \qquad (1)$$

$$\Rightarrow \quad \frac{dx}{x} = \frac{dy}{y} = \frac{zdz}{1+z^2}$$

Taking the first two ratios,

$$\frac{dx}{x} = \frac{dy}{y} \quad \Rightarrow \quad \log x = \log y + \text{const., on integration}$$

$$\Rightarrow \quad \frac{x}{y} = c_1 \qquad (2)$$

Taking the last two ratios,

$$\frac{dy}{y} = \frac{zdz}{1+z^2} \quad \Rightarrow \quad \log y = \frac{1}{2}\log(1+z^2) + \text{constant (on integration)}$$

$$\Rightarrow \quad \frac{y}{\sqrt{1+z^2}} = c_2 \qquad (3)$$

General solution is $F\left(\dfrac{x}{y}, \ \dfrac{y}{\sqrt{1+z^2}}\right) = 0 \quad (4)$

Example 10.26

Solve $yp + xq = (x^2 - y^2 + z^2)\,x$.

Solution Lagrange's auxiliary equations are

$$\frac{dx}{y} = \frac{dy}{x} = \frac{dz}{(x^2 - y^2 + z^2)x} \qquad (1)$$

From the first two ratios we get

$$x\,dx - y\,dy = 0 \quad \Rightarrow \quad x^2 - y^2 = a^2$$

(a^2 is an arbitrary constant)

Considering the last two ratios we have

$$dy = \frac{dz}{a^2 + z^2} \quad \because \quad x^2 - y^2 = a^2 \qquad (2)$$

Integrating,

$$y = \frac{1}{a}\tan^{-1}\frac{z}{a} + b = \frac{1}{\sqrt{x^2-y^2}}\tan^{-1}\frac{z}{\sqrt{x^2-y^2}} + b$$

(b is an arbitrary constant) (3)

General solution is

$$F\left(x^2 - y^2, \ y - \frac{1}{\sqrt{x^2-y^2}}\tan^{-1}\frac{z}{\sqrt{x^2-y^2}}\right) = 0 \quad (4)$$

Example 10.27

Solve $yp + xq = (x + y)z$.

Solution Lagrange's auxiliary equations are

$$\frac{dx}{y} = \frac{dy}{x} = \frac{dz}{(x+y)z} \qquad (1)$$

From the first two ratios

$$x\,dx - y\,dy = 0 \quad \Rightarrow \quad x^2 - y^2 = c_1 \qquad (2)$$

Again, each ratio

$$= \frac{dx + dy}{y + x} = \frac{dz}{(x+y)z} \quad \Rightarrow \quad d(x+y) = \frac{dz}{z},$$

cancelling $(x + y)$

On integration we have

$$x + y - \log z = c_2 \qquad (3)$$

General solution is

$$F(x^2 - y^2, \ x + y - \log z) = 0 \qquad (4)$$

Example 10.28

Solve $xp - yq = xyz$.

Solution Lagrange's auxiliary equations are

$$\frac{dx}{x} = \frac{dy}{-y} = \frac{dz}{xyz} \quad (1)$$

From the first two ratios

$$\frac{dx}{x} = \frac{dy}{-y} \Rightarrow x\,dy + y\,dx = 0 \Rightarrow xy = c_1 \quad (2)$$

From the last two ratios

$$\frac{dy}{-y} = \frac{dz}{c_1 z} \quad \because \; xy = c_1$$

Integrating

$$\frac{1}{c_1}\log z + \log y = \text{constant} \Rightarrow \frac{1}{xy}\log z + \log y = c_2 \quad (3)$$

General solution is

$$F\left(xy, \frac{1}{xy}\log z + \log y\right) = 0 \quad (4)$$

Example 10.29

Solve $zp + yq = x$

Solution Lagrange's auxiliary equations are

$$\frac{dx}{z} = \frac{dy}{y} = \frac{dz}{x} \quad (1)$$

From the first and last ratios

$$\frac{dx}{z} = \frac{dz}{x} \Rightarrow x\,dx - z\,dz = 0$$

Integrating we get

$$x^2 - z^2 = c_1 \quad (2)$$

Each of the ratios (1) $= \dfrac{dy}{y} = \dfrac{dx + dy + dz}{z + y + x}$

Integrating,

$$\log y = \log(x + y + z) + \log c_2 \Rightarrow \frac{y}{(x + y + z)} = c_2 \quad (3)$$

General solution is

$$F\left(x^2 - z^2, \frac{y}{(x + y + z)}\right) = 0 \quad (4)$$

Example 10.30

Solve $(y - z)p + (z - x)q = (x - y)$.

Solution Lagrange's auxiliary equations are

$$\frac{dx}{y - z} = \frac{dy}{z - x} = \frac{dz}{x - y} \quad (1)$$

Taking (1, 1, 1) as Lagrange's multipliers, each ratio

$$= \frac{dx + dy + dz}{(y - z) + (z - x) + (x - y) = 0}$$

$$\Rightarrow d(x + y + z) = 0 \Rightarrow x + y + z = c_1 \quad (2)$$

Taking (x, y, z) as Lagrange's multipliers, each ratio

$$= \frac{x\,dx + y\,dy + z\,dz}{x(y - z) + y(z - x) + z(x - y) = 0}$$

$$\Rightarrow d(x^2 + y^2 + z^2) = 0 \Rightarrow x^2 + y^2 + z^2 = c_2 \quad (3)$$

General solution is

$$F\,(x + y + z, x^2 + y^2 + z^2) = 0 \quad (4)$$

Example 10.31

Solve $(\sec x)p + (\sin x - y \sec x \tan x)q = (a^2 - z^2)$.

Solution Lagrange's auxiliary equations are

$$\frac{dx}{\sec x} = \frac{dy}{\sin x - y \sec x \tan x} = \frac{dz}{a^2 - z^2} \quad (1)$$

From the first and last ratios

$$\cos x\,dx = \frac{dz}{a^2 - z^2} \text{ or } (\cos x)\,dx = \frac{dz}{a^2 - z^2}$$

On integration, we get

$$\sin x = \frac{1}{2a}\log\frac{a + z}{a - z} + \text{constant},$$

or $$2a(\sin x) - \log\left[\frac{(a + z)}{(a - z)}\right] = c_1 \quad (2)$$

From the first two ratios

$$\sec x \frac{dy}{dx} = \sin x - y \sec x \tan x$$

or $\quad \sec x\dfrac{dy}{dx}+y\sec x\tan x=\sin x$

or $\quad d(y\sec x)-\sin x\,dx=0$

Integrating we have

$$y\sec x+\cos x=c_2 \tag{3}$$

General solution is

$$F\left(2a(\sin x)-\log\frac{a+z}{a-z},\ y\sec x+\cos x\right)=0 \tag{4}$$

Example 10.32

Solve $x(1-xy)p-y(1+xy)q=z(1-xy)$.

Solution Lagrange's auxiliary equations are

$$\frac{dx}{x(1-xy)}=\frac{dy}{-y(1+xy)}=\frac{dz}{z(1-xy)} \tag{1}$$

From the first and last ratios $\dfrac{dx}{x}=\dfrac{dz}{z}$, cancelling $(1-xy)$.

Integrating

$$\log x=\log z+\log c_1 \ \Rightarrow\ \frac{x}{z}=c_1 \tag{2}$$

On cross-multiplication the first two ratios yield

$$-y\,dx-xy^2\,dx=x\,dy-x^2y\,dy$$

$\Rightarrow\quad x^2y\,dy-xy^2\,dx=y\,dx+x\,dy$, on transposition

$\Rightarrow\quad \dfrac{dy}{y}-\dfrac{dx}{x}=\dfrac{d(xy)}{(xy)^2}$, multiplying by $\dfrac{1}{(xy)^2}$

On integration we get

$$\log y-\log x=-\frac{1}{xy}+\text{constant}$$

$$\frac{1}{xy}+\log\left(\frac{y}{x}\right)=c_2 \tag{3}$$

General solution is

$$F\left(\frac{x}{z},\ \frac{1}{xy}+\log\left(\frac{y}{x}\right)\right)=0 \tag{4}$$

Example 10.33

Solve $(x^2-y^2-z^2)p+2xyq=2xz$.

[Bangalore 1990, Gorakhpur 1991, Andhra 1989]

Solution Lagrange's auxiliary equations are

$$\frac{dx}{x^2-y^2-z^2}=\frac{dy}{2xy}=\frac{dz}{2xz} \tag{1}$$

Considering the last two ratios, we have $\dfrac{dy}{y}=\dfrac{dz}{z}$. On integration, we get

$$\log y=\log z+\text{constant} \ \Rightarrow\ \frac{y}{z}=c_1 \tag{2}$$

Using x, y, z as multipliers, we get that each ratio $=\dfrac{x\,dx+y\,dy+z\,dz}{x(x^2+y^2+z^2)}=\dfrac{dz}{2xz}$

or $\quad \dfrac{d(x^2+y^2+z^2)}{(x^2+y^2+z^2)}=\dfrac{dz}{z}$, on multiplication by $2x$.

This gives, on integration,

$$\log(x^2+y^2+z^2)-\log z=\log c_2$$

or
$$\frac{(x^2+y^2+z^2)}{z}=c_2 \tag{3}$$

From (2) and (3), the general solution is

$$F\left(\frac{y}{z},\ \frac{(x^2+y^2+z^2)}{z}\right)=0 \tag{4}$$

Example 10.34

Solve $(z^2-2yz-y^2)p+(xy+zx)q=xy-zx$.

[Bangalore 1990, Madras 1997 S]

Solution Lagrange's auxiliary equations are

$$\frac{dx}{z^2-2yz-y^2}=\frac{dy}{xy+zx}=\frac{dz}{xy-zx} \tag{1}$$

Using x, y, z as multipliers, each ratio

$$=\frac{x\,dx+y\,dy+z\,dz}{xz^2-2xyz-xy^2+xy^2+xyz+xyz-z^2x=0}$$

$$\Rightarrow\quad x^2+y^2+z^2=c_1 \tag{2}$$

Considering the last two ratios, we have

$$\frac{dy}{y+z}=\frac{dz}{y-z}=\frac{dy-dz}{2z}\Rightarrow 2z\,dz=(y-z)\,d(y-z)$$

On integrating and simplifying

$$y^2 - 2yz - z^2 = c_2 \quad (3)$$

The general solution

$$F(x^2 + y^2 + z^2, y^2 - 2yz - z^2) = 0 \quad (4)$$

Example 10.35
Solve $x^n p + y^n p = z^n$.

Solution Lagrange's auxiliary equations

$$\frac{dx}{x^n} = \frac{dy}{y^n} = \frac{dz}{z^n} \quad (1)$$

On integration we get two solutions

$$x^{-n+1} - y^{-n+1} = c_1, x^{-n+1} - z^{-n+1} = c_2 \ (n \neq 1), \quad (2), (3)$$

$$\log x - \log y = \log c_1, \log x - \log z = \log c_2 \ (n = 1)$$

$$\Rightarrow \quad \frac{x}{y} = c_1, \frac{x}{z} = c_2 \quad (4), (5)$$

General solution when $n \neq 1$ is $F(x^{-n+1} - y^{-n+1}, x^{-n+1} - z^{-n+1}) = 0$ and (6)

when $n \neq 1$ is $\quad F\left(\frac{x}{y}, \frac{x}{z}\right) = 0 \quad (7)$

Example 10.36
Solve $x(y - z)p + y(z - x)q = z(x - y)$. [JNTU 2002, 2003]

Solution Lagrange's auxiliary equations are

$$\frac{dx}{x(y-z)} = \frac{dy}{y(z-x)} = \frac{dz}{z(x-y)} \quad (1)$$

Now, each ratio

$$= \frac{\frac{1}{x}dx}{y-z} = \frac{\frac{1}{y}dy}{z-x} = \frac{\frac{1}{z}dz}{x-y} = \frac{\frac{dx}{x} + \frac{dy}{y} + \frac{dz}{z}}{0} \Rightarrow xyz = c_1 \quad (2)$$

Also each ratio $\quad = \dfrac{dx + dy + dz}{0} \quad (3)$

$$\Rightarrow \quad x + y + z = c_2$$

General solution is

$$F(xyz, x + y + z) = 0 \quad (4)$$

Example 10.37
Solve $(x^2 - yz)p + (y^2 - zx)q = z^2 - xy$. [AMIE 1997, Madras 1994 S, 1998; Bhopal 1991, Karnataka 1990, Ranchi 1996]

Solution Langrange's auxiliary equations are

$$\frac{dx}{x^2 - yz} = \frac{dy}{y^2 - zx} = \frac{dz}{z^2 - xy} \quad (1)$$

$$\text{Each ratio} = \frac{x\,dx + y\,dy + z\,dz}{x^3 + y^3 + z^3 - 3xyz}$$

$$= \frac{dx + dy + dz}{x^2 + y^2 + z^2 - xy - yz - zx}$$

$$\Rightarrow \quad \frac{x\,dx + y\,dy + z\,dz}{x + y + z} = \frac{dx + dy + dz}{1} \quad (2)$$

$$\because \quad x^3 + y^3 + z^3 - 3xyz = (x + y + z)(x^2 + y^2 + z^2 - xy - yz - zx)$$

$$\Rightarrow \quad d(x^2 + y^2 + z^2) = 2(x + y + z)d(x + y + z)$$

$$\Rightarrow \quad (x + y + z)^2 - (x^2 + y^2 + z^2) = \text{constant}$$

$$\Rightarrow \quad xy + yz + zx = c_1 \quad (3)$$

$$\text{Each ratio} = \frac{dx - dy}{(x^2 - yz) - (y^2 - zx)}$$

$$= \frac{dx - dz}{(x^2 - yz) - (z^2 - xy)}$$

$$= \frac{d(x-y)}{(x-y)(x+y+z)} = \frac{d(x-z)}{(x-z)(x+y+z)}$$

$$\Rightarrow \quad \frac{d(x-y)}{x-y} = \frac{d(x-y)}{x-z} \quad \Rightarrow \quad \frac{x-y}{x-z} = c_2 \quad (4)$$

From (3) and (4) we can write the general solution as

$$F\left(xy + yz + zx, \frac{x-y}{x-z}\right) = 0 \quad (5)$$

Example 10.38
Solve $(y - z)p + (x - y)q = z - x$. [Punjab 1987 S]

Solution Lagrange's auxiliary equations are

$$\frac{dx}{y-z} = \frac{dy}{x-y} = \frac{dz}{z-x} \quad (1)$$

$$\text{Each ratio} = \frac{dx + dy + dz}{(y-z) + (x-y) + (z-x) = 0}$$

$$\Rightarrow \quad d(x + y + z) = 0 \quad \Rightarrow \quad x + y + z = c_1 \quad (2)$$

$$\text{Each ratio} = \frac{-x\,dx}{-x(y-z)} = \frac{z\,dy + y\,dz}{z(x-y)+y(z-x)}$$

$$\frac{dx}{y-z} = \frac{d(yz)}{-x(y-z)} \Rightarrow \frac{x^2}{2} + yz = c_2 \quad (3)$$

From (2) and (3) we write the general solution as

$$F\left(x+y+z, \frac{x^2}{2}+yz\right) = 0 \quad (4)$$

Example 10.39

Solve $xp + (2x - y)q = (1 - z)$.

Solution Lagrange's auxiliary equations are

$$\frac{dx}{x} = \frac{dy}{2x-y} = \frac{dz}{1-z} \quad (1)$$

From the first and last ratios $\dfrac{dx}{x} = \dfrac{dz}{1-z}$

Integrating we get

$$\log x = -\log(1-z) + \log c_1$$

$$\Rightarrow \quad x(1-z) = c_1 \quad (2)$$

From the first two ratios $\dfrac{dx}{x} = \dfrac{dy}{2x-y}$

We have, on cross-multiplication,

$$2x\,dx - y\,dx = x\,dy \quad \text{or} \quad 2x\,dx = x\,dy + y\,dx$$

$$\text{or} \quad d(x^2) - d(xy) = 0$$

Integrating

$$x^2 - xy = c_2 \quad (3)$$

General solution is

$$F(x(1-z),\ x^2 - xy) = 0 \quad (4)$$

Example 10.40

Solve $(x + 2y^2)p + yq = -\cos z$.

Solution Lagrange's auxiliary equations are

$$\frac{dx}{x+2y^2} = \frac{dy}{y} = -\sec z\,dz \quad (1)$$

From the last two ratios $\dfrac{dy}{y} = \dfrac{dz}{-\cos z}$

Integrating we get

$$\log y = -\log(\sec z + \tan z) + \log c_1$$

$$\text{or} \quad y(\sec z + \tan z) = c_1 \quad (2)$$

From the first two ratios $\dfrac{dx}{x+2y^2} = \dfrac{dy}{y}$

On cross-multiplication and transposition

$$y\,dx - x\,dy = 2y^2\,dy.$$

$$\frac{y\,dx - x\,dy}{y^2} = 2\,dy \Rightarrow d\left(\frac{x}{y}\right) = 2\,dy$$

Integrating, we get

$$\frac{x}{y} = 2y + c_2 \quad (3)$$

General solution is

$$F\left(y(\sec z + \tan z), \frac{x}{y} - 2y\right) = 0 \quad (4)$$

EXERCISE 10.4

1. $px + qy = z$.

   Ans: $F\left(\frac{x}{y}, \frac{x}{z}\right) = 0$

2. $px^2 + qy^2 = z^2$.

   Ans: $F\left(\frac{1}{x} - \frac{1}{y}, \frac{1}{x} - \frac{1}{z}\right) = 0$

3. $\frac{p}{x} + \frac{q}{y} = \frac{1}{z}$.

   Ans: $F(x^2 - y^2, x^2 - z^2) = 0$

4. $\frac{y^2 z}{x} p + xzq = y^2$ or $y^2 zp + x^2 zq = xy^2$. [Madras 1995, 1997]

   Ans: $F(x^3 - y^3, x^2 - z^2) = 0$

5. $z(xp - yq) = y^2 - x^2$.

   Ans: $F(x^2 + y^2 + z^2, xy) = 0$

6. $px(z - 2y^2) = (z - qy)(z - y^2 - 2x^3)$.

   Ans: $F\left(\frac{y}{z}, y^2 - z\right) = 0$

7. $x(x + y)p - y(x + y)q = -(x - y)(2x + 2y + z)$.

   Ans: $F(xy, (x + y + z)(x + y)) = 0$

8. $x^2(y - z)p + y^2(z - x)q = z^2(x - y)$. [Madurai 1990]

   Ans: $F(xyz, (x^{-1} + y^{-1} + z^{-1})) = 0$

9. $x(y^2 - z^2)p + y(z^2 - x^2)q = z(x^2 - y^2)$.

Ans: $F(xyz,\ (x^2 + y^2 + z^2)) = 0$

10. $(mz - ny)p + (nx - lz)q = ly - mx$.

[AMIE 1990, Madras 1994 S]

Ans: $F(lx + my + nx,\ x^2 + y^2 + z^2) = 0$

11. $(y - zx)p + (x + yz)q = x^2 + y^2$.

Ans: $F(x^2 - y^2 + z^2,\ xy - z) = 0$

12. $(b - c)a^{-1}yzp + (c - a)b^{-1}zxq = (a - b)c^{-1}xy$.

Ans: $F(ax^2 + by^2 + cz^2,\ a^2x^2 + b^2y^2 + c^2z^2) = 0$

13. $(y^2 + z^2 - x^2)p - 2xyq = -2zx$.

Ans: $F\left(\dfrac{y}{z},\ \dfrac{(x^2 + y^2 + z^2)}{y}\right) = 0$

14. $p \tan x + q \tan y = \tan z$.

[Andhra 1990, Kerala 1987 S]

Ans: $F\left(\dfrac{\sin x}{\sin y},\ \dfrac{\sin x}{\sin z}\right) = 0$

15. $xe^y p + (1 - e^y)q = ze^y$.

Ans: $F\left(\dfrac{x}{z},\ (x + z)(1 - e^y)\right) = 0$

10.11 Nonlinear Equations of First Order

The most general partial differential equation of the first order in two independent variables is of the form

$$f(x, y, z, p, q) = 0 \tag{10.64}$$

which may not be linear.

We do not propose to study any general method of integration of equation (10.64) but consider only four standard forms which admit integration by very short processes. Also, many equations can be reduced to one or the other of these four forms.

The general integral, the singular integral and the complete integral must be indicated in each case. Otherwise, the equation is not considered fully solved.

Standard Form I: pq-equation: Equation of the form $f(p, q) = 0$ where x, y, z do not occur explicitly.

To solve such an equation we put $p = a$ and find the value of q in terms of a, *i.e.* $q = f(a)$. Then we substitute these values in

$$dz = p\, dx + q\, dy \tag{10.65}$$

so that we have $dz = a\, dx + f(a)\, dy$.

Integrating, we get the complete integral as

$$z = ax + \int f(a)\, dy + c \quad \text{or} \quad z = ax + f(a)y + c \tag{10.66}$$

where a and c are arbitrary constants.

Note 1 We can put $q = a$ instead of $p = a$ and proceed to obtain the complete integral.

Example 10.41

Solve $pq = k$

Solution The complete integral is

$$z = ax + \frac{k}{a}y + c \tag{1}$$

where a and c are arbitrary constants

The general integral is obtained by eliminating a between the equations

$$z = ax + \frac{k}{a}y + \phi(a)_1 \tag{2}$$

where c has been replaced by $\phi(a)$ and

$$0 = x - \frac{k}{a^2}y + \phi'(a) \tag{3}$$

obtained by differentiating with respect to a.

The *singular integral*, if it exists, is determined from the equation (1)

$$z = ax + \frac{k}{a}y + c$$

and

$$0 = x - \frac{k}{a^2}, \quad 0 = 1 \tag{4), (5}$$

obtained from (1) on differentiating it with respect to a and c, respectively.

The inconsistency of the last equation shows that singular integral does not exist in this case.

Example 10.42

Solve $pq = x^\alpha y^\beta z^\gamma$.

Solution The given equation is

$$\frac{\partial z}{\partial x}\frac{\partial z}{\partial y} = x^\alpha y^\beta z^\gamma \tag{1}$$

We can write (1) as

$$\frac{z^{-\gamma/2}}{x^\alpha}\frac{\partial z}{\partial x} \cdot \frac{z^{-\gamma/2}}{y^\beta}\frac{\partial z}{\partial y} = 1 \tag{2}$$

Case (i) $\alpha \neq 1, \beta \neq -1, \gamma \neq -2$

Put $z^{-\gamma/2} dz = dZ \Rightarrow Z = \dfrac{z^{-(\gamma/2)+1}}{1-\frac{\gamma}{2}}, \quad \gamma \neq 2$

$$x^{\alpha} dx = dX \Rightarrow X = \frac{x^{\alpha+1}}{\alpha+1}, \quad \alpha \neq -1,$$

$$y^{\beta} dy = dY \Rightarrow Y = \frac{y^{\beta+1}}{\beta+1}, \quad \beta \neq -1$$

Equation (2) now becomes,

$$\frac{\partial Z}{\partial X} \cdot \frac{\partial Z}{\partial Y} = 1 \quad (3)$$

whose complete integral is

$$Z = aX + bY + c \quad \text{where} \quad ab = 1$$

or
$$\frac{z^{-(\gamma/2)+1}}{1-\frac{\gamma}{2}} = \frac{a}{\alpha+1} x^{\alpha+1} + \frac{1}{a} \cdot \frac{1}{\beta+1} y^{\beta+1} + c \quad (4)$$

Case (ii) $\alpha = -1, \beta = -1, \gamma = -2$, we have

$$\frac{\partial z}{\partial x} \cdot \frac{\partial z}{\partial y} = \frac{1}{x} \cdot \frac{1}{y} \cdot \frac{1}{z^2}, \quad \Rightarrow \quad \frac{z\partial z}{x^{-1}\partial x} \cdot \frac{z\partial z}{y^{-1}\partial y} = 1$$

Put $z\,dz = dZ \Rightarrow Z = \frac{1}{2} z^2$

$$x^{-1} dx = dX \Rightarrow X = \log x$$

$$y^{-1} dy = dY \Rightarrow Y = \log y$$

The CI is

$$Z = aX + bY + c \quad \text{where} \quad ab = 1 \quad \text{or} \quad b = \frac{1}{a}$$

$$\Rightarrow \frac{1}{2} z^2 = a \log x + \frac{1}{a} \log y + c \quad (5)$$

To find the General Integral

Case (i) Writing $c = \phi(a)$ in the complete integral at (4) we have

$$\frac{z^{-(\gamma/2)+1}}{1-\frac{\gamma}{2}} = \frac{a}{\alpha+1} x^{\alpha+1} + \frac{1}{a} \frac{1}{\beta+1} y^{\beta+1} + \phi(a) \quad (6)$$

Differentiating (6) with respect to a we get

$$0 = \frac{1}{\alpha+1} x^{\alpha+1} - \frac{1}{a^2} \frac{1}{\beta+1} y^{\beta+1} + \phi'(a) \quad (7)$$

The general integral is the eliminant of a between (6) and (7).

Case (ii) Writing $c = \phi(a)$ in the complete Integral at (5)

$$\frac{1}{2} z^2 = a \log x + \frac{1}{a} \log x + \frac{1}{a} \log y + \phi(a) \quad (8)$$

Differentiating (8) with respect to a

$$0 = \log x - \frac{1}{a^2} \log y + \phi'(a) \quad (9)$$

Eliminant of a between (8) and (9) is the general integral.

To Find the Singular Integral

Case (i) The singular integral, if it exists, is obtained from the following equations:

$$\frac{z^{-(\gamma/2)+1}}{1-\frac{\gamma}{2}} = \frac{a}{\alpha+1} x^{\alpha+1} + \frac{1}{a} \frac{1}{\beta+1} y^{\beta+1} + c \quad (10)$$

Differentiating partially with respect to a and c, respectively, we get

$$0 = \frac{1}{\alpha+1} x^{\alpha+1} - \frac{1}{a^2} \frac{1}{\beta+1} y^{\beta+1} \quad 0 = 1 \quad (11), (12)$$

The inconsistency in the last equation shows that the singular integral does not exist in this case.

Case (ii) We can show that in this case also singular integral does not exist.

Example 10.43

Find a complete integral of

$$p + q = pq \quad (1)$$

Solution Put $p = a$, then $q = \dfrac{a}{a-1}$ and the equation $dz = p\,dx + q\,dy$ becomes

$$dz = a\,dx + \frac{a}{a-1} dy \quad (2)$$

whose solution is

$$z = ax + \frac{a}{a-1} y + c \quad (3)$$

which is the required complete integral.

Standard Form II: *zpq*-equation: Equation of the form $f(z, p, q) = 0$ where the independent variables do not appear explicitly.

In this case put

$$q = ap \text{ (or } p = aq) \tag{10.67}$$

The given equation becomes $f(z, p, ap) = 0$. Solving for p we have

$$p = \phi(a, z) \tag{10.68}$$

Substituting in $dz = pdx + qdy$

we have

$$dz = p\,dx + ap\,dy = pd(x + ay) \tag{10.69}$$

$$\Rightarrow \quad d\,(x + ay) = \frac{dz}{\phi(a, z)}$$

$$\Rightarrow \quad x + ay + c = \int \frac{dz}{\phi(a, z)} \tag{10.70}$$

where c is a arbitrary constant

Example 10.44

Solve $p^2 = z^2(1 - pq)$. (1)

Solution Put $q = ap$ in the given partial differential equation (1) so that we have

$$p^2 = z^2(1 - ap^2) \tag{2}$$

$$p^2(1 + az^2) = z^2 \quad \text{or} \quad \frac{1}{p^2} = \frac{1 + az^2}{z^2} \tag{3}$$

Now, $dz = p\,dx + q\,dy$

$$= p\,dx + ap\,dy = p\,d(x + ay)$$

$$x + ay + c = \int \frac{dz}{p} = \int \frac{\sqrt{1 + az^2}}{z} dz = \int \frac{d\theta}{\sin\theta\cos^2\theta}$$

$$= \int \frac{\cos^2\theta + \sin^2\theta}{\sin\theta\cos^2\theta} d\theta = \int (\sec\theta + \sec\theta\tan\theta) d\theta$$

$$= \log(\operatorname{cosec}\theta - \cot\theta) + c$$

$$= \log\left[\left(\sqrt{1 + az^2} - 1\right)\Big/\sqrt{az}\right] + \sqrt{1 + az^2}$$

put $az^2 = \tan^2\theta$

$$\Rightarrow z = \frac{1}{\sqrt{a}}\tan\theta$$

$$dz = \frac{1}{\sqrt{a}}\sec^2\theta\, d\theta = \frac{1}{\sqrt{a}}\frac{1}{\cos^2\theta} d\theta \tag{4}$$

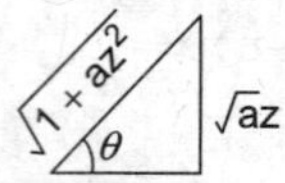

Figure 10.1

$$\therefore \frac{\sqrt{1 + az^2}}{z} = \frac{\sec\theta}{\frac{1}{\sqrt{a}}\tan\theta} = \frac{\sqrt{a}}{\sin\theta}$$

where c is an arbitrary constant.

Example 10.45

Solve $zpq = p^{1/2} + q^{1/2}$. (1)

Solution Putting $q = ap$ in (1) we get

$$zap^2 = p^{1/2}(1 + a^{1/2}) \Rightarrow \quad \frac{1}{p^{3/2}} = \frac{a}{1 + \sqrt{a}} \cdot z \tag{2}$$

Now, $dz = p\,dx + q\,dy = p\,d(x + ay)$ (3)

$$d(x + ay) = \frac{dz}{p} = \left(\frac{a}{1 + \sqrt{a}}\right)^{2/3} z^{2/3}\,dz$$

on integration we get

$$x + ay + c = \left(\frac{a}{1 + \sqrt{a}}\right)^{2/3} \frac{3}{5} z^{5/3} \tag{4}$$

where c is an arbitrary constant.

Standard Form III: Separable equation $f(x, p) = g(y, q)$.

We can take $f(x, p) = a$ and $g(y, q) = a$ (10.71)

Solving for p and q we get

$$p = f_1(a, x), \quad q = g_1(a, y) \qquad (10.72), (10.73)$$

Substituting in $dz = p\,dx + q\,dy = f_1(a, x)\,dx + g_1(a, y)\,dy$ (10.74)

Integrating we get the CI as

$$z = \int f_1(a, x)\,dx + \int g_1(a, y)\,dy + c \tag{10.75}$$

Example 10.46

Solve $p + q + x + y$

Solution We can write the equation as

$$p - x = y - q = a \tag{1}$$

so that

$$p = x + a, \; q = y - a \tag{2}$$

Substituting in

$$dz = p\,dx + q\,dy \tag{3}$$

we have

$$dz = (x + a)\,dx + (y - a)\,dy \qquad (4)$$

Integrating after multiplying throughout by 2, the complete integral is

$$2z = (x + a)^2 + (y - a)^2 + c \qquad (5)$$

where c is an arbitrary constant.

Example 10.47

Solve $zpq = xy$.

Solution The equation can be written as

$$\left(\frac{z^{1/2}\partial z}{\partial x}\right)\left(\frac{z^{1/2}\partial z}{\partial y}\right) = xy \qquad (1)$$

Put $\quad z^{1/2}dz = dZ \Rightarrow Z = \frac{2}{3}z^{3/2} \qquad (2)$

$$\frac{\left(\frac{\partial Z}{\partial x}\right)}{x} = \frac{y}{\left(\frac{\partial Z}{\partial y}\right)} = a \qquad (3)$$

$$\therefore \quad \frac{\partial Z}{\partial x} = ax, \quad \frac{\partial Z}{\partial y} = \frac{y}{a} \qquad (4)$$

Substituting in the equation

$$dZ = \left(\frac{\partial Z}{\partial x}\right)dx + \left(\frac{\partial Z}{\partial y}\right)dy, \qquad (5)$$

We have $dZ = ax\,dx + \frac{y}{a}dy$, which on integrating yields,

$$\frac{2}{3}z^{3/2} = \frac{a}{2}x^2 + \frac{y^2}{2a} + c \qquad (6)$$

where c is an arbitrary constant.

The general integral and singular integral (if it exists) is obtained as explained in 10.7.

Standard Form IV: Clairaut's[2] Equation $z = px + qy + f(p, q)$.

A partial differential equation of the type

$$z = px + qy + f(p, q) \qquad (10.76)$$

is called a Clairaut's equation.

Complete Integral A complete integral of an equation of type (10.76) is obtained by replacing p and q by arbitrary constants a and b, respectively. Thus, a complete integral of equation (10.76) is

$$z = ax + by + f(a, b) \qquad (10.77)$$

One can readily verify that (10.77) is a solution of equation (10.76)

Also, writing (10.77) as

$$z = F(x, y, a, b) = ax + by + f(a, b) \qquad (10.78)$$

we observe that the matrix

$$M = \begin{bmatrix} \frac{\partial F}{\partial a} & \frac{\partial^2 F}{\partial x \partial a} & \frac{\partial^2 F}{\partial y \partial a} \\ \frac{\partial F}{\partial b} & \frac{\partial^2 F}{\partial x \partial b} & \frac{\partial^2 F}{\partial y \partial b} \end{bmatrix} = \begin{bmatrix} x + \frac{\partial f}{\partial a} & 1 & 0 \\ y + \frac{\partial f}{\partial b} & 0 & 1 \end{bmatrix} \qquad (10.79)$$

is of rank two. Hence, (10.78) is indeed a complete integral of equation.

General integral and singular integral (if it exists) are obtained as explained in 10.7.

Singular Integral The complete integral is

$$z = ax + by + ab \qquad (10.80)$$

Differentiating (10.80) partially with respect to a and b we get

$$0 = x + b \qquad (10.81)$$

$$0 = y + b \qquad (10.82)$$

respectively. Eliminating a and b between (10.81), (10.82) we have singular solution of (10.76) as

$$z + xy = 0$$

Examples

Find complete integral in each of the following cases:

Example 10.48

$$z = px + qy - 2\sqrt{pq} \qquad \text{[JNTU 2003]}$$

Solution The given partial differential equation

$$z = px + qy - 2\sqrt{pq} \qquad (1)$$

is an equation of Clairaut's type.

The complete integral of (1) is

---

[2] CLAIRAUT, Alexis Claude (1713–1765), was a French analyst, differential geometer and astronomer.

$$z = ax + by - 2\sqrt{ab} \qquad (2)$$

where a and b are arbitrary constants.

Example 10.49

$$z = px + qy + \frac{p^4 + q^4}{pq} \qquad \text{[JNTU 2003]}$$

Solution The given partial differential equation

$$z = px + qy + \frac{p^4 + q^4}{pq} \qquad (1)$$

is an equation of Clairaut's type.

The complete integral of this equation is

$$z = ax + by + \frac{a^4 + b^4}{ab} \qquad (2)$$

Example 10.50

$$2q(z - px - qy) = 1 + q^2 \qquad \text{[JNTU 2005 S]}$$

The given partial differential equation can be written as

$$z = px + qy + \frac{1 + q^2}{2q} \qquad (1)$$

which is an equation of Clairaut's type.

The complete integral of this equation is

$$z = ax + by + \frac{1 + b^2}{2b} \qquad (2)$$

Example 10.51

Find the singular integral in Example 10.48.

Solution The complete integral is

$$z = ax + by - 2\sqrt{ab} \qquad (1)$$

Differentiating this with respect to a and b, respectively, we get

$$0 = x - \sqrt{\frac{b}{a}}, \quad 0 = y - \sqrt{\frac{a}{b}} \qquad (2), (3)$$

Eliminating a and b from these equations we get the singular integral as

$$xy = 1 \qquad (4)$$

EXERCISE 10.5

1. $q^2 = e^{-p/\alpha}$.

   Ans: $z = ax + e^{-a/2a}y + c$

2. $p + q = pq$.

   Ans: $z = ax + \frac{a}{a-1}y + c$

3. $p^2 + q^2 = 1$. [Osmania 2000]

   Ans: $z = ax + \sqrt{1 - a^2}y + c$

4. $x^2p^2 + y^2q^2 = z$. [Osmania 2000 s, Madras 1998]

   $\left[\textbf{Hint: } \left(\frac{z^{-1}\partial z}{x^{-1}\partial x}\right)^2 + \left(\frac{z^{-1}\partial z}{y^{-1}\partial y}\right)^2 = 1.\right]$

   Ans: $\log z = a \log x + \sqrt{1 - a^2} \cdot \log y + c$

5. $z = pq$.

   Ans: $(x + ay + c)^2 = 4az$

6. $zq = p^2$.

   Ans: $z = ce^{ax} + a^2y$

7. $p(1 + q)qz$. [Kerala 1990 S, Punjab 1986]

   Ans: $\log(az - 1) = (x + ay + c)$

8. $z = p^2 + q^2$. [Calicut 1994, Andhra 1990, Madurai 1990]

   Ans: $4z(1 + a^2) = (x + ay + c)^2$

9. $z(p^2 - q^2) = x - y$. [Madras 1991, Calicut 1991, Karnataka 1990]

   Ans: $z^{3/2} = (x + a)^{3/2} + (y + a)^{3/2} + c$

10. $q = xyp^2$.

    Ans: $z = 2\sqrt{ax} + \frac{1}{2}ay^2 + c$

11. $yp + xq + pq = 0$. [Marathwada 1993]

    Ans: $2z = ax^2 - \frac{a}{a+1}y^2 + c$

12. $p + q = \sin x + \sin y$. [Madras 1993, Karnataka 1993]

    Ans: $z = a(x - y) - (\cos x + \cos y) + c$

13. $z = px + qy + p^2 + q^2$.

    Ans: $z = ax + by + a^2 + b^2$

14. $z = px + qy + \lambda\sqrt{1 + p^2 + q^2}$.

    Ans: $z = ax + by + \lambda\sqrt{1 + a^2 + b^2}$

15. $z = px + qy + \sin(p + q)$.

    Ans: $z = ax + by + \sin(a + b)$

10.12 Euler's Method of Separation of Variables

When we model scientific, engineering, biotechnological and other processes there arise initial and boundary value problems involving partial differential equations. The general solutions of these partial differential equations are in the form of arbitrary functions which are not suitable for determining the exact solutions satisfying the given initial and boundary conditions.

The method of separation of variables, due to Euler[3] is a simple, yet powerful technique in breaking up a partial differential equation into an ordinary differential equation, which can be easily solved using the known methods.

For a partial differential equation in the function $u = u(x, y)$ where x and y are independent variables, we assume that the solution is separable, that is,

$$u(x, y) = X(x)Y(y) \tag{10.83}$$

where $X(x)$ is a function of x alone and $Y(y)$ is a function of y alone. Substituting in the given partial differential equation, separating the variables and assuming each side must be equal to the same constant λ (say), we obtain two ordinary differential equations, whose solution gives the solution of the problem. The method is best illustrated through examples.

Example 10.52

Solve $\dfrac{\partial u}{\partial x} = 2\dfrac{\partial u}{\partial t} + u$ by the method of separation of variables given that $u(x, 0) = 6e^{-3x}$.

Solution We assume the solution to be

$$u(x, t) = X(x)T(t) \tag{1}$$

where $X(x)$ is a function of x alone and $T(t)$ is a function of t alone. We have

$$\frac{\partial u}{\partial x} = X'T \quad \text{and} \quad \frac{\partial u}{\partial t} = XT' \qquad (2), (3)$$

where $'$ denotes differentiation with respect to the argument.

Substituting in the given partial differential equation, we get

$$X'T = 2XT' + XT \quad \text{or} \quad (X' - X)T = 2XT' \tag{4}$$

Dividing both sides of (4) by $2XT$ we get

$$\frac{X' - X}{2X} = \frac{T'}{T} = \lambda \quad \text{(say)} \qquad (5), (6)$$

$$\Rightarrow \quad X' - X - 2\lambda X = 0$$

$$\text{or} \quad \frac{X'}{X} = 1 + 2\lambda, \quad \text{and} \quad \frac{T'}{T} = \lambda$$

Solving (5) we get

$$\log X = (1 + 2\lambda)x + \log c_1 \quad \text{or} \quad X = c_1 e^{(1+2\lambda)x} \tag{7}$$

and solving (6) we get

$$\log T = \lambda t + \log c_2 \quad \text{or} \quad T = c_2 e^{\lambda t} \tag{8}$$

Combining (7) and (8) we may write the solution as

$$u(x,t) = XT = c_1 c_2 e^{(1+2\lambda)x} e^{\lambda t} \tag{9}$$

Now $\quad 6e^{-3x} = u(x, 0) = c_1 c_2 e^{(1+2\lambda)x}$

$$\Rightarrow \quad c_1 c_2 = 6 \quad \text{and} \quad 1 + 2\lambda = -3$$

The latter equation gives $\lambda = -2$

Substituting in the general solution we get

$$u(x, y) = 6e^{-3x} e^{-2t} = 6e^{-(3x+2t)}$$

which is the required solution.

Example 10.53

Solve by the method of separation of variables $4\dfrac{\partial u}{\partial x} + \dfrac{\partial u}{\partial y} = 3u$ given that $u = 3e^{-y} - 3e^{-5y}$ when $x = 0$.

Solution Let

$$u = X(x)Y(y) \tag{1}$$

where $X(x)$ is a function of x alone and $Y(y)$ is a function of y alone. Calculating the derivatives and substituting in (1) we have

$$4X'Y + XY' - 3XY = 0$$

$$\Rightarrow \quad 4\frac{X'}{X} - 3 = -\frac{Y'}{Y} = \lambda \quad \text{(say)}$$

---
[3] Euler, Leonhard (1707–1783), was a gifted Swiss mathematician and the most prolific mathematician, who contibuted to almost all branches of mathematics even after becoming totally blind in 1771. He studied under John Bernoulli and became a professor of mathematics at St. Petersburg, Russia. He is the first modern mathematical universalist.

Thus, we have two ordinary differential equations

$$4X' - 3X = \lambda X \quad \text{and} \quad Y' + \lambda Y = 0 \qquad (2), (3)$$

Taking trial solution of the form e^{mx}, e^{my} for the above equations we obtain auxiliary equations

$$4m - 3 - \lambda = 0 \Rightarrow m = \frac{3+\lambda}{4}, \text{ and } m = -k \qquad (4), (5)$$

respectively.

The solution for X is of the form

$$X = c_1 e^{((3+\lambda)/4)\,x} \qquad (6)$$

and the solution for Y is of the form

$$Y = c_2 e^{-\lambda y} \qquad (7)$$

The solution for u can be taken as

$$u(x, y) = c_1 e^{((3+\lambda_1)/4)x} \cdot e^{-\lambda_1 y} + c_2 e^{((3+\lambda_2)/4)x} \cdot e^{-\lambda_2 y} + \cdots \qquad (8)$$

where we have used the principle of superposition of solutions since the equation is linear.

Putting $x = 0$ in (8) we get

$$u(0, y) = 3e^{-y} - 3e^{-5y} = c_1 e^{((3+\lambda_1)/4)0} \cdot e^{-\lambda_1 y} + c_2 e^{((3+\lambda_2)/4)0} e^{-\lambda_2 y}$$

and $c_3 = c_4 = \cdots = 0$

Equating like terms on both sides

$$3e^{-y} - 3e^{-5y} = c_1 e^{-\lambda_1 y} + c_2 e^{-\lambda_2 y}$$

$$\Rightarrow \quad c_1 = 3, c_2 = -3, \quad \lambda_1 = 1, \quad \lambda_2 = 5$$

and $\qquad c_3 = c_4 = \cdots = 0$

The required solution is

$$u(x, y) = 3e^{x-y} - 3e^{2x-5y}$$

Example 10.54

Solve $\dfrac{\partial^2 u}{\partial x^2} = \dfrac{\partial u}{\partial y} + 2u$ subject to conditions. $u = 0$, $\dfrac{\partial u}{\partial x} = 1 + e^{-3y}$ when $x = 0$.

Solution Let $u = X(x)Y(y)$ be a solution of $\dfrac{\partial^2 u}{\partial x^2} = \dfrac{\partial u}{\partial y} + 2u$. (1)

Substituting in the equation (1)

$$X''Y = XY' + 2XY \quad \Rightarrow \quad \frac{X''}{X} - 2 = \frac{Y'}{Y} = k \quad \text{(say)} \qquad (2)$$

We obtain the ordinary differential equations $X'' - (2 + k)X = 0$, $Y'' - kY = 0$, where k is a constant.

Taking the solutions as $X = e^{mx}$, $Y = e^{my}$ we get (3), (4)

Auxillary equations $m^2 - (2 + k) = 0$, $m - k = 0$ (5), (6)

$\therefore$ The solution is of the form $(e^{\sqrt{2+kx}}, e^{-\sqrt{2+kx}})\, e^{ky}$

If we take the solution as

$$u = (c_1 e^{\sqrt{2+kx}} + c_2 e^{-\sqrt{2+kx}})\, e^{ky}$$

The condition $x = 0 \quad \Rightarrow \quad u = 0$ which gives

$$c_1 + c_2 = 0, \quad (e^{ky} \neq 0), \quad \text{or } c_2 = -c_1 \qquad (7)$$

Since $e^{\alpha x} - e^{-\alpha x} = 2 \sinh \alpha x$, and we have to satisfy another condition by the principle of superposition of solutions, we may take

$$u(x, y) = c_1 \sinh(\sqrt{2+k_1}x)\, e^{k_1 y} + c_2 \sinh(\sqrt{2+k_2}x)\, e^{k_2 y} + \cdots \qquad (8)$$

$$\left.\frac{\partial u}{\partial x}\right|_{x=0} = (\sqrt{2+k_1}c_1 \cosh(\sqrt{2+k_1}x)\, e^{k_1 y} + \sqrt{2+k_2}c_2 \cosh(\sqrt{2+k_2})\, e^{k_2 y} + \cdots)_{x=0} \qquad (9)$$

$$= (\sqrt{2+k_1}c_1.1.e^{k_1 y} + \sqrt{2+k_2}c_2.1.e^{k_2 y} + c_3 \cdots + c_4 \cdots) = 1 + e^{-3y} \qquad (10)$$

$$\Rightarrow \quad k_1 = 0, \quad c_1 = \frac{1}{\sqrt{2}}, \quad k_2 = -3,$$

$$c_2 = \frac{1}{i}, \quad c_3 = c_4 = \cdots = 0 \qquad (11)$$

Since $\sqrt{2+k_2} \cdot c_2 = 1$ for $k_2 = -3 \Rightarrow c_2 = \dfrac{1}{i}$.

The second solution is

$$c_2 \sinh\sqrt{2+k_2}xe^{k_2 y} = \frac{1}{i}\frac{e^{ix} - e^{-ix}}{2}.e^{-3y} \sin xe^{-3y}$$

$$\therefore \quad u(x, y) = \frac{1}{\sqrt{2}} \sinh \sqrt{2}x + \sin x e^{-3y} \quad (12)$$

Example 10.55

Solve

$$\frac{\partial^2 u}{\partial x^2} - 2\frac{\partial u}{\partial x} + \frac{\partial u}{\partial y} = 0 \quad (1)$$

Solution

$$\text{Let } u = X(x) \cdot Y(y) \quad (2)$$

where X is a function of x alone and Y is a function of y alone, be a solution of equation (1). Calculating the derivatives and substituting in (1) we have

$$X''Y - 2X'Y + XY' = 0 \quad \Rightarrow \quad \frac{X''}{X} - 2\frac{Y'}{Y} = k(\text{say})$$

Thus, we obtain two ordinary differential equations

$$X'' - 2X' - kX = 0 \quad \text{and} \quad Y' + kY = 0 \quad (3), (4)$$

where k is a constant. Taking trial solutions, $X = e^{mx}$, $Y = e^{my}$, the auxiliary equations are

$$m^2 - 2mk = 0, \quad m + k = 0$$

$$\Rightarrow \quad m = \frac{2 \pm \sqrt{4 + 4k}}{2} = 1 \pm \sqrt{1 + k}, \quad m = -k$$

$$X = (c_1 e^{\sqrt{1+kx}} + c_2 e^{-\sqrt{1+kx}})\, e^x, \qquad Y = c_3 e^{-ky}$$

$$u(x, y) = (Ae^{\sqrt{1+kx}} + Be^{-\sqrt{1+kx}})\, e^{x-ky}$$

where A and B are two arbitrary constants, consistent with the order of the differential equation.

EXERCISE 10.6

Solve the following equations by separation of variables:

1. $u_{xt} = e^{-t} \cos x$ with $u(x, 0) = 0$ and $\dfrac{\partial u(0,t)}{\partial t} = 0$.

   Ans: $u(x,t) = \sin x - e^{-t} \sin x$

2. $4\dfrac{\partial u}{\partial x} + \dfrac{\partial u}{\partial y} = 3u$ with $u = e^{-5y}$ when $x = 0$. [JNTU 2004 S]

   Ans: $u = e^{2x-5y}$

3. $px^3 + qx^2 = 0 \left(p = \dfrac{\partial z}{\partial x},\ q = \dfrac{\partial}{\partial y} \right)$. [JNTU 2004 S]

   Ans: $z = ce^{k(x^3/3 - y^4/4)}$

4. $2\dfrac{\partial^2 u}{\partial x} - \dfrac{\partial u}{\partial y} = 0$.

   Ans: $u(x, y) = Ae^{\sqrt{k}x} + Be^{-\sqrt{k}x}$

5. $3\dfrac{\partial u}{\partial x} + 2\dfrac{\partial u}{\partial y} = 0, \quad u(x,0) = 4e^{-x}$.

   Ans: $u(x, y) = 4e^{-(2x-3y)/2}$

6. Find a solution of $\dfrac{\partial^2 u}{\partial x^2} = \dfrac{\partial u}{\partial y} + 2u$ in the form $u = X(x)Y(y)$. Solve the equation subject to the conditions $u = 0$ and $\dfrac{\partial u}{\partial x} = 1 + e^{-3y}$ when $x = 0$ for all values of y. [Andhra 2000, Nagpur 1997]

   Ans: $u = \dfrac{1}{\sqrt{2}} \sinh \sqrt{2}x + e^{-3y} \sin x$

10.13 Classification of Second-Order Partial Differential Equations

10.13.1 Introduction

Many physical and engineering applications such as fluid flow, heat transfer and wave motion involve second-order partial differential equations and hence we take up a study of these equations and their solution by separation of variables method.

The general second-order partial differential equation may be written in the form

$$L[u] = A\frac{\partial^2 u}{\partial x^2} + B\frac{\partial^2 u}{\partial x \partial y} + C\frac{\partial^2 u}{\partial y^2} + H\left(x, y, u, \frac{\partial u}{\partial x}, \frac{\partial u}{\partial y} \right) = 0 \quad (10.84)$$

Equation (10.84) is called *semilinear* if A, B and C are functions of the independent variables x and y only. On the other hand, if A, B and C are functions of x, y, u, $\partial u / \partial x$ and $\partial u / \partial y$ then (10.84) is called *quasi-linear*. If A, B and C are functions of x and y and H is a linear function of u, $\partial u / \partial x$ and $\partial u / \partial y$ then equation (10.84) is called *linear*. The general second-order linear partial differential equation in two independent variables x and y may be written as

$$A(x,y)\frac{\partial^2 u}{\partial x^2}+B(x,y)\frac{\partial^2 u}{\partial x\partial y}+C(x,y)\frac{\partial^2 u}{\partial y^2}+D(x,y)\frac{\partial u}{\partial x}$$
$$+E(x,y)\frac{\partial u}{\partial y}+F(x,y)u+G(x,y)=0 \qquad (10.85)$$

If $G \neq 0$ then equation (10.85) is called nonhomogeneous and if $G = 0$ then it is called homogeneous.

10.13.2 Classification of Equations

We call the quantity $\Delta = B^2 - 4AC$ the discriminant and classify equation (10.85) as hyperbolic, parabolic or elliptic according as $\Delta > 0$, $= 0$ or < 0. The following are well-known examples of these three types.

Hyperbolic Type

1. One-dimensional wave equation

$$\frac{\partial^2 u}{\partial t^2} = a^2\frac{\partial^2 u}{\partial x^2} \qquad (10.86)$$

Here $A = a^2$, $B = 0$, $C = -1$ and $\Delta = B^2 - 4AC = 4a^2 > 0$.

Parabolic Type

2. One-dimensional heat-flow equation

$$\frac{\partial u}{\partial t} = a^2\frac{\partial^2 u}{\partial x^2} \qquad (10.87)$$

Here $A = a^2$, $B = C = 0$ and $\Delta = B^2 - 4AC = 0$.

Elliptic Type

3. (a) Two-dimensional Laplace equation

$$\nabla^2 u = \frac{\partial^2 u}{\partial x^2}+\frac{\partial^2 u}{\partial y^2} = 0 \quad \left(\nabla^2 = \frac{\partial^2}{\partial x^2}+\frac{\partial^2}{\partial y^2}\right) \qquad (10.88)$$

Here $A = 1$, $B = 0$, $C = 1$ and $\Delta = B^2 - 4AC = -4 < 0$.

(b) Poisson's equation

$$\nabla^2 u = \frac{\partial^2 u}{\partial x^2}+\frac{\partial^2 u}{\partial y^2} = f(x,y) \qquad (10.89)$$

Equations (10.86)–(10.88) are homogeneous while equation (10.89) is nonhomogeneous.

10.13.3 Initial and Boundary Value Problems and their Solution

The unique solution corresponding to a particular physical problem is obtained by use of additional information arising from the physical situation. If this information is given on the boundary as *boundary conditions*, the resulting problem is called a boundary value problem (BVP). If this information is given at one instant as *initial conditions*, the resulting problem is called an initial value problem (IVP).

The principle of superposition of solutions is applicable as long as the equation is linear and homogeneous. That is, if u_n is a solution for each n then $\Sigma_{n=1}^{\infty} u_n$, which is a linear combination of the solutions $\{u_n\}$ is also a solution of the equation.

The hyperbolic and parabolic types of equations are either initial value problems or initial and boundary value problems, whereas the elliptic-type equation is always a boundary value problem. The boundary conditions may be one of the following three types:

1. *Dirichlet problem (First boundary value problem)*
   The solution is prescribed along the boundary.
2. *Neumann problem (Second boundary value problem)*
   The derivative of the solution is prescribed along the boundary.
3. Mixed problem (Third boundary value problem)
   The solution and its derivative are prescribed along the boundary.

Any of the above conditions is called homogeneous if it is a zero-condition and nonhomogeneous if it is a nonzero condition.

10.13.4 Solution of One-dimensional Heat Equation (or diffusion equation)

Consider a long and thin wire, rod or bar OA of length l and of constant cross-section and homogeneous heat-conducting material. Let the bar be placed along the x-axis with one end O coinciding

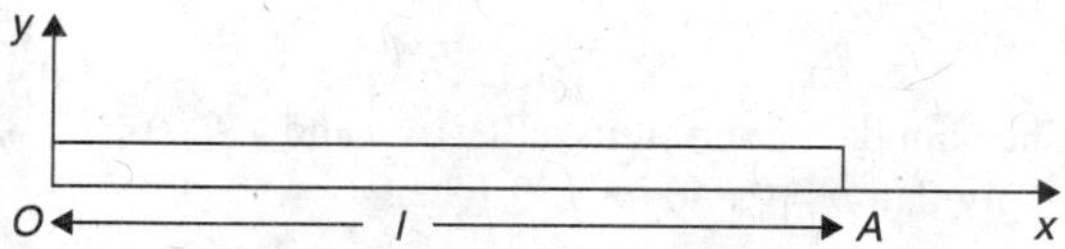

Figure 10.2 Heat conduction along a bar

with the origin and the other end A at a distance l from O (Figure 10.2).
Suppose that the lateral surface of the bar is perfectly insulated, so that heat flow is along the x-direction only. Therefore, the temperature u of the bar depends on x and t only. The initial boundary value problem consists of one-dimensional heat equation:

$$\frac{\partial u}{\partial t} = a^2 \frac{\partial^2 u}{\partial x^2} \quad (10.90)$$

where a^2 is the thermal diffusivity. The boundary conditions at the ends O and A are

$$u(0, t) = 0 = u(l, t) \quad (10.91, 10.92)$$

respectively. The initial temperature distribution in the bar is

$$u(x, 0) = f(x) \quad (10.93)$$

where $f(x)$ is a given function of x. The solution by the method of separation of variables reduces the initial boundary value problem (IBVP) to that of two ordinary differential equations.

Assume that the solution $u(x, t)$ is separable

i.e., $$u(x, t) = X(x)\, T(t) \quad (10.94)$$

where $X(x)$ is a function of x only and $T(t)$ is a function of t only.
Differentiating (10.94) with respect to t and x we get

$$\frac{\partial u}{\partial t} = XT', \quad \frac{\partial^2 u}{\partial x^2} = X'' T \quad (10.95)$$

where $'$ denotes differentiation with respect to the corresponding independent variable. Substituting into (10.90) we have

$$XT' = a^2 X'' T \quad \Rightarrow \quad \frac{X''}{X} = \frac{T'}{a^2 T} \quad (10.96)$$

Since the L.H. member is a function of x only and the R.H. member is a function of t only, both sides must be equal to the same constant λ, say. So, we obtain two ordinary differential equations

$$X'' - \lambda X = 0, \qquad T' - a^2 \lambda T = 0 \quad (10.97, 10.98)$$

Three cases arise: *Case 1.* $\lambda > 0$; *Case 2.* $\lambda = 0$; *Case 3.* $\lambda < 0$.

Case 1: $\lambda > 0$. The general solution is

$$u(x,t) = X(x)\, T(t) = (C_1 e^{\sqrt{\lambda}x} + C_2 e^{-\sqrt{\lambda}x})(C_3 e^{a^2 \lambda t}) \quad (10.99)$$

Equation (10.99) shows that the solution has unbounded temperature for large t due to exponential growth, which is not physically possible.

Case 2: $\lambda = 0$. The general solution is

$$u(x, t) = X(x)\, T(t) = (ax + b) \quad (10.100)$$

which is independent of time. This is also not possible.

Case 3: $\lambda < 0$. We write $\lambda = -p^2$ where p is real. The general solution in this case is

$$u(x,t) = X(x)T(t) = (A\cos px + B\sin px)(C\, e^{-a^2 p^2 t}) \quad (10.101)$$

The boundary condtion $u(0, t) = 0$ implies that $A = 0$ so that we have

$$u(x,t) = BC\, e^{-a^2 p^2 t} \sin px \quad (10.102)$$

The boundary condition $u(l, t) = 0$ requires that

$$\sin pl = 0$$

which holds if

$$pl = n\pi \Rightarrow p = p_n = \frac{n\pi}{l} (n = 1, 2, 3, \ldots) \quad (10.103)$$

The constant p_n is an eigenvalue and the function $\sin p_n x$ is an eigenfunction. Now, we can write the solution, using the principle of superposition, as

$$u(x,t) = \sum_{n=1}^{\infty} u_n(x,t) = \sum_{n=1}^{\infty} b_n \sin p_n x \;\; e^{-a^2 p_n^2 t}$$

$$\left(p_n = \frac{n\pi}{l}, \quad n = 1, 2, 3, \ldots \right) \quad (10.104)$$

The initial conditon (10.93) will be satisfied at $t = 0$ if

$$u(x,0) = f(x) = \sum_{n=1}^{\infty} b_n \sin p_n x \quad (10.105)$$

that is, if $f(x)$ can be expanded in a convergent half-range Fourier series in $(0, l)$. The b_n are given by

$$b_n = \frac{2}{l}\int_0^l f(x)\ \sin p_n x\ dx \qquad (10.106)$$

Note 1 Solution of partial differential equations by separation of variables method cannot be applied in all cases. It is only a certain special set of boundary conditions that allows us to separate the variables.

Example 10.56

A long copper rod with insulated lateral surface has its left end maintained at a temperature of 0°C and its right end at $x = 2$m maintained at 100°C. Find the temperature $u(x,t)$ if the initial condition

$$u(x,0) = t(x) = \begin{cases} 100x & 0 < x < 1 \\ 100 & 1 < x < 2 \end{cases}$$

Solution We have to solve the partial differential equation for heat conduction

$$\frac{\partial u}{\partial t} = a^2 \frac{\partial^2 u}{\partial x^2} \qquad (1)$$

under the boundary conditions

$$u(0,t) = 0 \quad \text{and} \quad u(l,t) = u(2,t) = 100 \qquad (2), (3)$$

and the initial condition

$$u(x,0) = f(x) = \begin{Bmatrix} 100x & 0 < x < 1 \\ 100 & 1 < x < 2 \end{Bmatrix} \qquad (4)$$

Assuming seperation of variables in the form

$$u(x,t) = X(x)\ T(t) \qquad (5)$$

we get

$$\frac{X''}{X} = \frac{T'}{a^2 T} = \lambda \qquad (6)$$

where λ is the separation constant. In this problem the eigenvalue $\lambda = 0$ is important. The solution for $\lambda = 0$ is

$$X(x) = ax + b, \quad T(t) = c \qquad (7)$$

$$u(x,t) = c(ax + b) \qquad (8)$$

The boundary conditions $u(0, t) = 0$ and $u(2, t) = 100$ imply that $b = 0$ and $ac = 50$. Then

$$u(x,t) = 50x \qquad (9)$$

Now taking up the case of exponential decay of temperature namely $\lambda = -p^2$ where p is real we have the general solution as

$$u(x,t) = (A\cos px + B\sin px)\ Ce^{-a^2p^2t} \qquad (10)$$

Superimposing the above two solutions we obtain a more general solution

$$u(x,t) = 50x + (A\cos px + B\sin px)\ Ce^{-a^2p^2t} \qquad (11)$$

The condition $u(0, t) = 0$ requires that $A = 0$ and the condition $u(l, t) = u(2, t) = 100$ demands that

$$100 = 50 \times 2 + BC\sin pl\ e^{-a^2p^2t} \quad \Rightarrow \quad \sin pl = 0$$

This will be satisfied if

$$pl = n\pi \quad \Rightarrow p = p_n = \frac{n\pi}{l} = \frac{n\pi}{2} \qquad (12)$$

$$n = 1, 2, 3, \ldots \qquad (\because l = 2)$$

The solution, by the principle of superposition of solutions, is

$$u(x,t) = \sum_{n=1}^{\infty} u_n(x,t) = 50x + \sum b_n \sin p_n x e^{-a^2p_n^2t} \qquad (13)$$

This must satisfy the initial condition (4) and hence

$$f(x) = 50x + \sum_{n=1}^{\infty} b_n \sin\frac{n\pi x}{2} \qquad (14)$$

Expanding $[f(x) - 50x]$ in a half-range Fourier sine series in $[0, 2]$ we get the solution. The Fourier coefficients b_n are given by

$$b_n = \frac{2}{l}\int_0^l (f(x) - 50x)\sin\frac{n\pi x}{l}\,dx = \frac{2}{2}\int_0^1 (100x - 50x)\sin\frac{n\pi x}{2}\,dx + \frac{2}{2}\int_1^2 (100 - 50x)\sin\frac{n\pi x}{2}\,dx$$

$$= 50\left[-\frac{2x}{n\pi}\cos\frac{n\pi x}{2}+\frac{4}{n^2\pi^2}\sin\frac{n\pi x}{2}\right]_0^1+\left[(100-50x)\frac{2}{n\pi}\left(-\cos\frac{n\pi x}{2}\right)+50\left(\frac{2}{n\pi}\right)^2\left(-\sin\frac{n\pi x}{2}\right)\right]_1^2$$

$$= -\frac{100}{n\pi}\cos\frac{n\pi}{2}+\frac{200}{n^2\pi^2}\sin\frac{n\pi}{2}+50\times\left(\frac{2}{n\pi}\right)^2\sin\frac{n\pi}{2}=\frac{400}{\pi^2 n^2}\sin\frac{n\pi}{2} \quad (15)$$

Finally, the solution is

$$u(x,t)=50x+\frac{400}{\pi^2}\sum_{n=1}^{\infty}\frac{1}{n^2}\sin\frac{n\pi}{2}\cdot\sin\frac{n\pi x}{2}e^{-a^2n^2\pi^2t/4} \quad (16)$$

Example 10.57

An insulated rod of length l has its ends A and B maintained at 0°C and 100°C, respectively, until steady-state conditions prevail. If B is suddenly reduced to 0°C and maintained at 0°C, find the temperature at a distance x from A at time t.

[JNTU 2003 (Set 4)]

Solution Let $u(x, t)$ be the temperature at time t at a distance x from A. The equation for the conduction of heat is

$$\frac{\partial u}{\partial t}=a^2\frac{\partial^2 u}{\partial x^2} \quad (1)$$

where 'a^2' is the diffusivity of the material of the rod.

In the steady state when u depends only on x we get from (1):

$$\frac{d^2u}{dx^2}=0 \quad (2)$$

whose general solution is

$$u(x)=ax+b \quad (3)$$

Boundary conditions are $u(0) = 0$ and $u(l) = 100$. So we obtain $b = 0$ and $a=100/l$. This gives $u(x) = (100/l)x$ at time $t = 0$. Thus, we have the initial condition

$$u(x,0)=\frac{100}{l}x \quad (4)$$

Boundary conditions for unsteady flow are:

$$u(0,t)=0 \quad \text{and} \quad u(l,t)=0 \quad \text{for all } t \quad (5),(6)$$

Now, we have to solve equation (1) under the conditions (4)–(6). A solution of (1) is of the form

$$u(x,t)=(A\cos px+B\sin px)\,e^{-a^2p^2t} \quad (7)$$

$$u(0,t)=0 \Rightarrow Ae^{-a^2p^2t}=0 \Rightarrow A=0 \quad (8)$$

Equation (7) becomes

$$u(x,t)=B\sin px\,e^{-a^2p^2t} \quad (9)$$

$$u(l,t)=0 \Rightarrow B\sin pl\cdot e^{-a^2p^2t}=0 \Rightarrow p=\frac{n\pi}{l}$$

$$(\because B\neq 0) \quad (10)$$

By the principle of superposition of solutions we may write the solution as

$$u(x,t)=\sum_{n=1}^{\infty}B_n\sin p_nx\,e^{-a^2p_n^2t} \quad \text{where } p_n=\frac{n\pi}{l} \quad (11)$$

Imposing the initial condition (4) on the solution (11) we have

$$u(x,0)=\sum_{n=1}^{\infty}B_n\sin p_nx=\frac{100}{l}x, \quad (12)$$

which is the half-range Fourier series expansion in (0, l) for the function $(100/l)x$. Therefore, B_n are given by

$$B_n=\frac{2}{l}\int_0^l\frac{100}{l}x\sin p_nx\,dx$$

$$=\frac{200}{l^2}\left[\left\{x\left(-\frac{\cos p_nx}{p_n}\right)\right\}_0^l-\int_0^l-\frac{\cos p_nx}{p_n}dx\right]$$

$$=\frac{200}{l^2}\left[0-\frac{l\cos p_nl}{p_n}+\frac{1}{p_n^2}(\sin p_nx)_0^l\right] \quad (p_nl=n\pi)$$

$$=\frac{200}{l^2}\cdot\frac{l\cdot(-1)^{n-1}}{\left(\frac{n\pi}{l}\right)}=\frac{200(-1)^{n-1}}{n\pi} \quad (13)$$

The solution for the problem is

$$u(x,t)=\frac{200}{\pi}\sum_{n=1}^{\infty}\frac{(-1)^{n-1}}{n}\sin\frac{n\pi x}{l}e^{-\left(\frac{a^2n^2\pi^2t}{l^2}\right)} \quad (14)$$

Example 10.58

A homogeneous rod of conducting material of length 100 cm has its ends kept at zero temperature, and the temperature is initially

$$u(x,0)=\begin{cases} x & 0\le x\le 50 \\ 100-x & 50\le x\le 100\end{cases}$$

Find the temperature $u(x, t)$ at any time.

[JNTU 2004s (Set 3)]

Solution We have to solve the differential equation for the conduction of heat

$$\frac{\partial u}{\partial t}=a^2\frac{\partial^2 u}{\partial x^2} \tag{1}$$

under the boundary and initial conditions

Boundary conditions:

1. $u(0, t)=0$ for all t (2)
2. $u(100, t)=0$ for all t (3)

Initial condition:

$$u(x,0)=\begin{Bmatrix} x & 0\le x\le 50 \\ 100-x & 50\le x\le 100\end{Bmatrix} \tag{4}$$

A solution of equation (1) may be taken as

$$u(x,t)=(A\cos px+B\sin px)\ e^{-a^2p^2t} \tag{5}$$

$u(0, t)=0 \quad\Rightarrow\quad A=0$ equation (5) becomes

$$u(x,t)=B\sin px\ e^{-a^2p^2t} \tag{6}$$

$$u(100,t)=0\Rightarrow B\sin p100=0$$

$$\Rightarrow 100p=n\pi \quad (\because B\neq 0)$$

$$\Rightarrow p=p_n=\frac{n\pi}{100} \tag{7}$$

By the principle of superposition of solutions we may take the solution as

$$u(x,t)=\sum B_n\sin p_nx\ \ e^{-a^2p_n^2t} \tag{8}$$

Imposition of the initial condition (4) on (8) yield,

$$\sum B_n\sin nx=u(x,0)=\begin{Bmatrix} x & 0\le x\le 50 \\ 100-x & 50\le x\le 100\end{Bmatrix} \tag{9}$$

We now expand $u(x, 0)$ in a half-range Fourier sine series in (0, 100) and determine B_n.

Now, B_n are given by

$$B_n=\frac{2}{l}\int_0^l u(x)\sin\frac{n\pi x}{l}\,dx=\frac{2}{100}\left[\int_0^{50}x\sin p_nx\,dx+\int_{50}^{100}(100-x)\sin p_nx\,dx\right]$$

$$=\frac{2}{100}\left[\left\{x\left(-\frac{\cos p_nx}{p_n}\right)-1\left(-\frac{\sin p_nx}{p_n^2}\right)\right\}_0^{50}+\left\{(100-x)\left(-\frac{\cos p_nx}{p_n}\right)+\left(-\frac{\sin p_nx}{p_n^2}\right)\right\}_{50}^{100}\right]$$

$=0$ at 0, 50 $\qquad =0$ at 0 $\qquad =0$ at 100, 50 $\qquad =0$ at 100

$$\left[\begin{array}{l}\cos p_nx=\cos\dfrac{n\pi x}{100}=\begin{cases}1 & \text{at } x=0\\ 0 & \text{at } x=50\\ (-1)^n & \text{at } x=100\end{cases}\\[2ex] \sin p_nx=\sin\dfrac{n\pi x}{100}=\begin{cases}0 & \text{at } x=0,100\\ \sin\dfrac{n\pi}{2} & \text{at } x=50\end{cases}\end{array}\right.$$

$$=\frac{2}{100}\left(\frac{\sin\frac{n\pi}{2}}{p_n^2}+\frac{\sin\frac{n\pi}{2}}{p_n^2}\right)=\frac{4}{100}\cdot\frac{100^2}{\pi^2n^2}\cdot\sin\frac{n\pi}{2}=\frac{400}{\pi^2n^2}\sin\frac{n\pi}{2}$$

$$=\begin{cases}0, & \text{if } n \text{ is even}\\ \dfrac{400}{\pi^2}\dfrac{(-1)^n}{n^2} & \text{if } n \text{ is odd}\end{cases}$$

Finally, the required solution of the problem is

$$u(x,t)=\frac{400}{\pi^2}\sum_{n=1}^{\infty}\frac{(-1^n)}{(2n-1)^2}\sin\frac{(2n-1)\pi x}{100}$$
$$\exp\left(-\left(\frac{2n-1}{100}\right)^2\pi^2c^2t\right)$$

Example 10.59

Find the temperature $u(x, t)$ in a homogeneous bar of heat conducting material of length l cm with its ends kept at zero temperature and initial temperature given by $\alpha x(l-x)/l^2$.

Solution The initial boundary value problem consists of the following:

1. Partial differential equation for conduction of heat:

$$\frac{\partial u}{\partial t}=a^2\frac{\partial^2 u}{\partial x^2} \qquad (1)$$

2. Boundary conditions:

$$u(0, t)=0, \quad u(l, t)=0 \quad \text{for all } t \qquad (2), (3)$$

3. Initial condition:

$$u(x, 0)=\alpha x(l-x)/l^2, \quad 0<x<l \qquad (4)$$

A general solution of (1) is

$$u(x,t)=(A\cos px+B\sin px)\,e^{-a^2p^2t} \qquad (5)$$

Boundary condition (2) is satisfied if we set $A = 0$ and boundary condition (3) is satisfied if

$$u(l, t)=B\sin pl\,e^{-a^2p^2t}=0\Rightarrow pl=n\pi$$
$$\Rightarrow p=p_n=\frac{n\pi}{l} \quad (\because B\neq 0) \qquad (6)$$

By the principle of superposition of solutions we may write

$$u(x, t)=\sum_{n=1}^{\infty}B_n\sin p_nx\quad e^{-a^2p^2t} \quad (p_n=n\pi/l) \qquad (7)$$

By the imposition of the initial condition (4) on (7) we get

$$u(x, 0)=\frac{\alpha x(l-x)}{l^2}=\sum_{n=1}^{\infty}B_n\sin p_nx$$

which is the half-range Fourier sine series in $(0, l)$ for $f(x)=\alpha x(l-x)/l^2$. The constants B_n are given by

$$B_n=\frac{2}{l}\int_0^l f(x)\sin p_nx\,dx=\frac{2}{l}\int_0^l\frac{\alpha x(l-x)}{l^2}\sin p_nx\,dx$$
$$=\frac{2\alpha}{l^3}\int_0^l(lx-x^2)\sin p_nx\,dx$$
$$=\frac{2\alpha}{l^3}\left[(lx-x^2)\left(\frac{-\cos p_nx}{p_n}\right)\right.$$
$$\left.-(l-2x)\left(\frac{-\sin p_nx}{p_n^2}\right)+(-2)\frac{\cos p_nx}{p_n^3}\right]_0^l$$
$$=\frac{2\alpha}{l^3}\left[0+0-\frac{2l^3}{n^3\pi^3}((-1)^n-1)\right]=\frac{8\alpha}{n^3\pi^3}, \quad (n:\text{odd})$$

Hence the temperature distribution in the bar is given by

$$u(x,t)=\frac{8\alpha}{\pi^3}\sum_{n=1}^{\infty}\frac{1}{(2n-1)^3}\sin\frac{(2n-1)\pi x}{l}$$
$$\exp\left(-\frac{(2n-1)^2}{l^2}\pi^2a^2t\right)$$

Example 10.60

Find the temperature in a thin metal rod of length l with both ends insulated and with initial temperature in the rod $\sin(\pi x/l)$.

Solution The initial boundary value problem consists of the following

1. Partial differential equation for conduction of heat

$$\frac{\partial u}{\partial t}=a^2\frac{\partial^2 u}{\partial x^2} \qquad (1)$$

2. Boundary conditions: (Insulation at both ends)

$$\left.\frac{\partial u}{\partial x}\right|_{x=0}=0; \qquad \left.\frac{\partial u}{\partial x}\right|_{x=l}=0 \qquad (2), (3)$$

3. Initial condition:

$$u(x,0)=\sin\left(\frac{\pi x}{l}\right) \qquad (4)$$

The general solution of equation (1) is of the form

$$u(x,t)=\frac{A_0}{2}+(A\cos px+B\sin px)\,e^{-a^2p^2t} \qquad (5)$$

(Note: we have added the constant $A_0/2$ since in this case we have half-range Fourier *cosine* series expansion)

Differentiating (5) we have

$$\frac{\partial u}{\partial x} = (-pA \sin px + pB \cos px)\, e^{-a^2 p^2 t} \qquad (6)$$

Boundary condition (2) is satisfied if we set $B = 0$. Also, boundary condition (3) is satisfied if

$$\sin pl = 0 \quad (\because A \neq 0) \Rightarrow p = p_n = \frac{n\pi}{l} \qquad (7)$$

Therefore, by the principle of superposition, the solution may be taken as

$$u(x,t) = \frac{A_0}{2} + \sum_{n=1}^{\infty} A_n \cos p_n x \cdot e^{-a^2 p_n^2 t} \qquad (8)$$

Here

$$A_0 = \frac{2}{l}\int_0^l f(x)\,dx = \frac{2}{l}\int_0^l \sin\frac{\pi x}{l}\,dx = \frac{2}{l}\left(-\frac{\cos\frac{\pi x}{l}}{\left(\frac{\pi}{l}\right)}\right)_0^l$$

$$= \frac{4}{l}\cdot\frac{l}{\pi} = \frac{4}{\pi} \qquad (9)$$

$$A_1 = \frac{2}{l}\int_0^l f(x)\cos\frac{\pi x}{l}\,dx = \frac{2}{l}\int_0^l \sin\frac{\pi x}{l}\cos\frac{\pi x}{l}\,dx$$

$$= \frac{1}{\pi}\left(\sin^2\frac{\pi x}{l}\right)_0^l = 0 \qquad (10)$$

$n \neq 1$

$$A_n = \frac{2}{l}\int_0^l f(x)\cos\frac{n\pi x}{l}\,dx = \frac{2}{l}\int_0^l \sin\frac{\pi x}{l}\cos\frac{n\pi x}{l}\,dx$$

$$= \frac{1}{l}\int_0^l \left[\sin\frac{(n+1)\pi x}{l} - \sin\frac{(n-1)\pi x}{l}\right]dx$$

$$= \frac{1}{l}\left[\frac{-l}{(n+1)\pi}\cdot\cos\frac{(n+1)\pi x}{l} + \frac{l}{(n-1)\pi}\cdot\cos(n-1)\frac{\pi x}{l}\right]_0^l$$

$$= \frac{1}{\pi}\left[\frac{1}{n+1}\left(1-(-1)^{n+1}\right) - \frac{1}{n-1}\left(1-(-1)^{n-1}\right)\right]$$

$$= \frac{2}{\pi}\left(\frac{1}{n+1} - \frac{1}{n-1}\right) = -\frac{4}{\pi(n^2-1)}$$

$$= -\frac{4}{\pi(4n^2-1)} \quad (n \text{ even}) \qquad (11)$$

The temperature distribution in the rod is

$$u(x,t) = \frac{2}{\pi} - \frac{4}{\pi}\sum_{n=1}^{\infty}\frac{1}{4n^2-1}\cos\left(\frac{2n\pi x}{l}\right)$$
$$\exp\left(-\frac{4n^2 a^2 \pi^2}{l^2}t\right) \qquad (12)$$

Example 10.61

A homogeneous rod of conduction material of length l has its ends kept at zero temperature. The temperature at the centre is T and it falls uniformly to zero at the two ends. Find the temperature at any time t.

Solution The initial boundary value problem consists of solving the partial differential equation for heat conduction:

$$\frac{\partial u}{\partial t} = c^2\frac{\partial^2 u}{\partial x^2} \qquad (1)$$

under the boundary conditions

$$u(0,t) = 0 = u(l,t) \qquad (2), (3)$$

and the initial condition that $u(x,0)$ falls uniformly to zero at the ends.

To find this condition we have to solve the steady-state equation: $d^2u/dx^2 = 0$ whose general solution is

$$u(x) = ax + b \qquad (4)$$

If $C = l/2$ is the centre of the rod then for the portion AC of the rod we have

$$\left.\begin{array}{l}\text{at } A: u(0) = a.0 + b = 0 \Rightarrow b = 0 \\ \text{and at } C: u\left(\frac{l}{2}\right) = a\frac{l}{2} = T \Rightarrow a = \frac{2T}{l}\end{array}\right\}\text{so that:}$$

$$u(x) = \frac{2T}{l}x, \quad 0 \le x \le \frac{l}{2} \qquad (5)$$

For the portion CB of the rod we have

$$\left.\begin{array}{l}\text{at } B: u(l) = al + b = 0 \Rightarrow b = -al \\ \text{and at } C: u(l/2) = T = a\frac{l}{2} - al = -\frac{1}{2}al \Rightarrow a = -\frac{2T}{l}\end{array}\right\}$$

$$\text{so that } u(x) = -\frac{2T}{l}x + 2T \quad \frac{l}{2} \le x \le l \qquad (6)$$

Consequently the initial condition is

$$u(x,0)=\begin{cases}\dfrac{2T}{l}x, & 0\le x\le\dfrac{l}{2}\\[2mm] \dfrac{2T}{l}(l-x), & \dfrac{l}{2}\le x\le l.\end{cases} \quad (7)$$

By the method of separation of variables the solution of (1) may be put in the form

$$u(x,t)=(A\cos px+B\sin px)\,e^{-c^2p^2t} \quad (8)$$

Boundary condition (2) is satisfied if we set $A = 0$, and boundary conditon (3) is satisfied if we set $\sin pl = 0$

$$\Rightarrow \quad p=p_n=n\pi/l \quad (9)$$

Thus, by the principle of superposition of solutions, we may write the general solution as

$$u(x,t)=\sum_{n=1}^{\infty} B_n \sin p_n x.e^{-c^2p_n^2t} \quad (10)$$

Imposing the initial condition (7) we must have

$$u(x,0)=\sum_{n=1}^{\infty} B_n \sin p_n x=\begin{cases}\dfrac{2T}{l}x, & 0\le x\le\dfrac{l}{2}\\[2mm] \dfrac{2T}{l}(l-x), & \dfrac{l}{2}\le x\le l\end{cases} \quad (11)$$

Equation (11) is a half-range Fourier sine series expansion in $(0,l)$ for the function $u(x,0)$ and so the constants B_n are given by

$$B_n=\frac{2}{l}\int_0^l u(x)\sin p_n x\,dx=\frac{2}{l}\int_0^{l/2}\frac{2T}{l}x\sin p_n x\,dx$$

$$+\frac{2}{l}\int_{l/2}^{l}\frac{2T}{l}(l-x)\sin p_n x\,dx \quad \left(p_n=\frac{n\pi}{l}\right)$$

$$=\frac{4T}{l^2}\left[x\left(\frac{-\cos p_n x}{p_n}\right)-1\left(\frac{-\sin p_n x}{p_n^2}\right)\right]_0^{l/2}$$

$$+\frac{4T}{l^2}\left[(l-x)\left(-\frac{\cos p_n x}{p_n}\right)+1\cdot\left(-\frac{\sin p_n x}{p_n^2}\right)\right]_{l/2}^{l}$$

$$=\frac{4T}{l^2}\cdot\frac{(-1)^n}{n^2\pi^2}l^2+\frac{4T}{l^2}\cdot\frac{(-1)^n}{n^2\pi^2}\cdot l^2=\frac{8T}{\pi^2}\cdot\frac{(-1)^n}{n^2} \quad (13)$$

$$\sin p_n l=\sin n\pi=\sin 0=0$$

$$\sin\frac{p_n l}{2}=\sin\frac{n\pi}{2}=(-1)^n$$

$$\cos p_n l=\cos n\pi=(-1)^n$$

$$\cos\frac{p_n l}{2}=\cos\frac{n\pi}{2}=0$$

$$\cos 0=1$$

Finally, the solution of the problem is

$$u(x,t)=\frac{8T}{\pi^2}\sum_{n=1}^{\infty}\frac{(-1)^n}{n^2}\sin\frac{n\pi x}{l}\exp\left(-\frac{n^2\pi^2c^2t}{l^2}\right) \quad (14)$$

EXERCISE 10.7

1. Solve $\dfrac{\partial\theta}{\partial t}=k\dfrac{\partial^2\theta}{\partial x^2}$ such that

   (a) θ is finite as $t\to\infty$.

   (b) $\partial\theta/\partial x=0$ when $x=0$ and $\theta=0$ when $x=l$ for all t.

   (c) $\theta=\theta_0$ when $t=0$ for all x in $0<x<l$.

   Ans: $\theta=\dfrac{4\theta_0}{\pi}\Big[\cos p_1 x.e^{-p_1^2kt}-\dfrac{1}{3}\cos p_3 x.e^{-p_3^2kt}+\dfrac{1}{5}\cos p_5 x.e^{-p_5^2kt}-\ldots\Big]\left(p_n=\dfrac{n\pi}{l}\right)$

2. A bar 10 cm long with insulated sides has its ends A and B maintained at temperatures 50°C and 100°C, respectively, until steady-state conditions prevail. The temperature at A is suddenly raised to 90°C and at the same time that at B is lowered to 60°C. Find the temperature distribution in the bar at time t.

   [Mysore 1997, Warangal 1996]

   Ans: $u=(x,t)=90-3x-\dfrac{80}{\pi}\sum_{n=1}^{\infty}\dfrac{1}{n}\sin\dfrac{n\pi x}{5}\cdot\exp(-n^2\pi^2c^2t/25)$

3. Solve $\partial u/\partial t=\partial^2u/\partial x^2$ with boundary conditions $u(x,0)=3\sin n\pi x$ $u(0,t)=0$, $u(l,t)=0$ when $0<x<1$, $t>0$.

   [Osmania 1995, Kerala 1990]

   Ans: $u(x,t)=3\sum_{n=1}^{\infty}\sin n\pi x\,\exp(-n^2\pi^2t)$

4. The ends A and B of a rod 20m long have temperatures at 30°C and 80°C, respectively, until steady-state conditions prevail. The temperatures of the ends are changed to 40°C and 60°C, respectively. Find the temperature distribution in the rod at time t.

   [Kerala 1995, Madras 1991]

Ans: $u(x,t)=40+x-\dfrac{20}{\pi}\sum_{n=1}^{\infty}\dfrac{2(-1)^n+1}{n}$

$$\sin\frac{n\pi x}{20}\exp\left(\frac{-c^2n^2\pi^2t}{400}\right)$$

5. Find the solution of one-dimensional heat equation $\partial u/\partial t=c^2(\partial^2u/\partial x^2)$ under the boundary conditions $u(0, t)=u(l, t)=0$ and the initial conditions $u(x, 0)=x$, $0<x<l$, l being the length of the rod.

Ans: $u(x,t)=\dfrac{2l}{\pi}\sum\dfrac{(-1)^{n-1}}{n}\sin p_nx\exp(-p_n^2c^2t)$

$(p_n=n\pi/l)$

6. Solve $\partial u/\partial t=c^2(\partial^2u/\partial x^2)$ subject to the conditions:

(a) $u(0, t)=0$

(b) $\dfrac{\partial u}{\partial t}(l,t)=0$

(c) $u(x,0)=\dfrac{u_0x}{l}$, $0\le x\le l$.

Ans: $u(x,t)=\dfrac{\partial u_0}{x^2}\sum_{n=1}^{\infty}\dfrac{(-1)^{n-1}}{(2n-1)}\sin\left(\dfrac{2n-1}{2l}\pi x\right)$

$$\exp\left(-\left(\frac{2n-1}{2l}\right)^2\pi^2c^2t\right).$$

7. The ends A and B of a rod 30cm long have their temperatures kept at 20°C and 80°C until steady-state conditions prevail. The temperature of the end B is suddenly reduced to 60°C and kept so while the end A is raised to 40°C. Find the temperature distribution of the rod at time t.

Ans: $u(x,t)=40+\dfrac{2x}{3}-\dfrac{40}{\pi}\sum_{n=1}^{\infty}\dfrac{1}{n}\sin\dfrac{n\pi x}{15}$

$$\exp(-n^2\pi^2c^2t/225)$$

8. Find the solution of $\partial u/\partial t=c^2(\partial^2u/\partial x^2)$ under the conditions

(a) $u(0, t)=0=u(l, t)$ for all t.

(b) $u(x,0)=2\sin\dfrac{n\pi x}{l}$ for all x.

Ans: $u(x,t)=2\sum_{n=1}^{\infty}\sin\dfrac{n\pi x}{l}\exp(-n^2\pi^2c^2t/l)$.

10.13.5 One-dimensional Wave Equation

Consider the vibrations of an elastic string placed along the x-axis, stretched to length l between two fixed points $x=0$ and $x=l$. First we consider the problem when there is an initial displacement but no initial velocity (string released from rest). Next we consider motion of a string with an initial velocity but no initial displacement (string given an initial blow, but from its horizontal stretched position). Finally, we consider the case of both initial velocity and initial displacement.

10. 13.6 Vibrating String with Zero Initial Velocity

Consider an elastic string of length l, fastened at its ends on the x-axis at $x=0$ and $x=l$. The string is displaced, then released from rest to vibrate in the xy-plane. We want to find the displacement function $y(x, t)$, whose graph is a curve in the xy-plane showing the shape of the string at time t. The boundary value problem for the displacement function $y(x, t)$ consists in the solution of the partial differential equation

$$\frac{\partial^2y}{\partial t^2}=a^2\frac{\partial^2y}{\partial x^2}\quad\text{for } 0<x<l,\ t>0 \qquad (10.107)$$

under the boundary conditions:

$$y(0,t)=y(l, t)=0\quad\text{for } t\ge 0 \qquad (10.108, 10.109)$$

and the initial conditions:

$$\left.\begin{aligned}&y(x,0)=f(x)\\&\frac{\partial y}{\partial t}(x,0)=0\end{aligned}\right\}\quad\text{for } 0\le x\le l \qquad \begin{aligned}(10.110)\\(10.111)\end{aligned}$$

The graph of $f(x)$ is the position of the string before release.

The separation of variables method consists of attempting a solution of the form $y(x,t)=X(t)\,T(t)$ where $X(x)$ is a function of x only and $T(t)$ is a function of t only. Substituting into the wave equation we obtain

$$XT''=a^2X''T \qquad (10.112)$$

where $'$ denotes differentiation with respect to the respective independent variable. Then

$$\frac{X''}{X}=\frac{T''}{a^2T} \qquad (10.113)$$

The left-hand member is a function of x only and the right-hand member is a function of t only. The equality is possible only if both the quantities are equal to the same constant, which we take as $-\lambda^2 < 0$ for convenience. This is called the separation constant. (Taking the separation constant as 0 or positive real number leads to trivial solutions.)

We now have

$$\frac{X''}{X} = \frac{T''}{a^2 T} = -\lambda^2 \quad (10.114)$$

we obtain two ordinary differential equations

$$X'' + \lambda^2 X = 0 \quad \text{and} \quad T'' + a^2\lambda^2 T = 0 \quad (10.115, 10.116)$$

The boundary conditions $y(0, t) = 0$, and $y(l, t) = 0$ yield $X(0) = 0$ and $X(l) = 0$, respectively. The general solutions of equations are

$$X(x) = A\cos\lambda x + B\sin\lambda x \quad (10.117)$$

$$T(t) = C\cos a\lambda t + D\sin a\lambda t \quad (10.118)$$

The condition $X(0) = 0$ implies that $A = 0$ and the condition $X(l) = 0$ imples that $\sin\lambda l = 0$ ($\because B \neq 0$) Therefore, $\lambda l = n\pi$ or $\lambda = \lambda_n = n\pi/l$, which are the eigenvalues of the problem. The corresponding eigenfunctions are

$$X_n(x) = B_n \sin\lambda_n x \quad (n = 1, 2, 3, \ldots) \quad (10.119)$$

The initial condition (the string is released from rest)

$$\frac{\partial y}{\partial t}(x,0) = 0 \quad \text{gives} \quad X(x)T'(0) = 0 \;\Rightarrow T'(0) = 0 \quad (10.120)$$

Hence

$$-a\lambda_n(C\sin a\lambda_n t - D\cos a\lambda_n t)\big|_{t=0} = 0 \;\Rightarrow\; D = 0$$

Therefore, we obtain

$$T(t) = T_n(t) = C_n \cos a\lambda_n t \quad (n = 1, 2, 3, \ldots) \quad (10.121)$$

Now we take the solutions for the problem as

$$y_n(x, t) = C_n \sin\lambda_n x \cos\lambda_n at$$

$$\left(\lambda_n = \frac{n\pi}{l};\, n = 1, 2, 3, \ldots\right) \quad (10.122)$$

Each of these functions satisfies the wave equation, both boundary conditions and the initial condition $\partial y/\partial t(x,0) = 0$. We need to satisfy the condition $y(x, 0) = f(x)$. This is achieved by an infinite superposition of solutions in the form

$$y(x, t) = \sum_{n=1}^{\infty} y_n(x, t) = \sum_{n=1}^{\infty} C_n \sin\lambda_n x \cos\lambda_n at$$

$$\left(\lambda_n = \frac{n\pi}{l};\, n = 1, 2, 3, \ldots\right) \quad (10.123)$$

we must choose the C_n's to satisfy

$$y(x, 0) = \sum_{n=1}^{\infty} y_n(x, 0) = \sum_{n=1}^{\infty} C_n \sin\lambda_n x = f(x) \quad (10.124)$$

This series is the half-range Fourier sine series of $f(x)$ in $[0, l]$. The Fourier constants are given by

$$C_n = \frac{2}{l}\int_0^l f(x)\sin\frac{n\pi x}{l}\,dx \quad (10.125)$$

Example 10.62

A string AB of length l is fastened at both ends A and B. At a distance 'b' from the end A, the string is transeversely displaced to a distance 'd' and released from rest when it is in this position. Find the solution for the initial displacement function and zero initial velocity.

Solution: Let $y(x, t)$ be the displacement of the string. The initial displacement is given by APB.

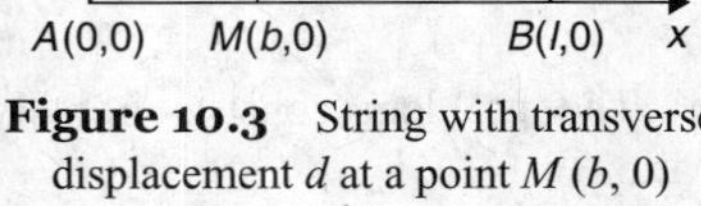

Figure 10.3 String with transverse displacement d at a point $M(b, 0)$

Equation of AP is $y=(d/b)x$ (1)

Equation of PB is $y=\dfrac{d(x-l)}{b-l}$ (2)

The problem is to solve one-dimensional wave equation

$$\frac{\partial^2 y}{\partial t^2}=a^2\frac{\partial^2 y}{\partial x^2} \quad (3)$$

with boundary conditions

$$y(0,t)=0=y(l,t) \quad (4),(5)$$

and initial conditions

$$y(x,0)=f(x)=\begin{cases}\dfrac{d}{b}x & \text{for } 0\le x\le b\\ \dfrac{d(x-l)}{b-l} & \text{for } b\le x\le l\end{cases} \quad (6)$$

The solution is given by

$$y(x,t)=\sum_{n=1}^{\infty}C_n\sin\frac{n\pi x}{l}\cos\frac{n\pi at}{l} \quad (7)$$

where C_n are given by

$$C_n=\frac{2}{l}\int_0^l f(x)\sin\frac{n\pi x}{l}\,dx=\frac{2}{l}\left[\int_0^b\frac{d}{b}x\sin\frac{n\pi x}{l}dx+\int_b^l\frac{d(x-l)}{b-l}\sin\frac{n\pi x}{l}dx\right]$$

$$=\frac{2d}{bl}\left[x\left(\frac{-l}{n\pi}\right)\cos\frac{n\pi x}{l}-\left(\frac{-l^2}{n^2\pi^2}\right)\sin\frac{n\pi x}{l}\right]_0^b+\frac{2d}{l(b-l)}\left[(x-l)\left(\frac{-l}{n\pi}\right)\cos\frac{n\pi x}{l}-\left(\frac{-l^2}{n^2\pi^2}\right)\sin\frac{n\pi x}{l}\right]_b^l$$

$$=\frac{-2dbl}{bl\,n\pi}\cos\frac{n\pi b}{l}+\frac{2dl^2}{bln^2\pi^2}\sin\frac{n\pi b}{l}+\frac{2d}{n\pi}\cos\frac{n\pi b}{l}-\frac{2dl^2}{l(b-l)n^2\pi^2}\sin\frac{n\pi b}{l}$$

$$=\frac{2dl^2}{b(l-b)n^2\pi^2}\sin\frac{n\pi b}{l} \quad (8)$$

Hence, the displacement of the string at any point x and time t is given by

$$y(x,t)=\frac{2dl^2}{b(l-b)\pi^2}\sum_{n=1}^{\infty}\frac{1}{n^2}\sin\frac{n\pi b}{l}\cdot\sin\frac{n\pi x}{l}\cdot\cos\frac{n\pi at}{l} \quad (9)$$

Example 10.63

A string of length l fastened at both ends $A=(0,0)$ and $B=(l,0)$ undergoes initially a transversal displacement given by

$$f(x)=\begin{cases}x & \text{for } 0\le x\le l/2\\ l-x & \text{for } l/2\le x\le l\end{cases}$$

and is released at rest when it is in this position. Find the displacement function $y(x,t)$ for the subsequent motion.

Solution The problem consists of solving the wave equation

$$\frac{\partial^2 y}{\partial t^2}=a^2\frac{\partial^2 y}{\partial x^2} \quad (1)$$

under the boundary conditions

$$y(0,l)=0=y(l,t) \quad (2),(3)$$

and the initial conditions:

Initial displacement:

$$y(x,0)=f(x)=\begin{cases}x & \text{for } 0\le x\le l/2\\ l-x & \text{for } l/2\le x\le l\end{cases} \quad (4)$$

Initial velocity: $\dfrac{\partial y}{\partial t}(x,0)=g(x)=0$ (5)

The solution of equation (1) under the boundary conditions (2) and (3) and the zero initial velocity is

$$y(x,t)=\sum_{n=1}^{\infty}C_n\sin\frac{n\pi x}{l}\cos\frac{n\pi at}{l} \quad (6)$$

where

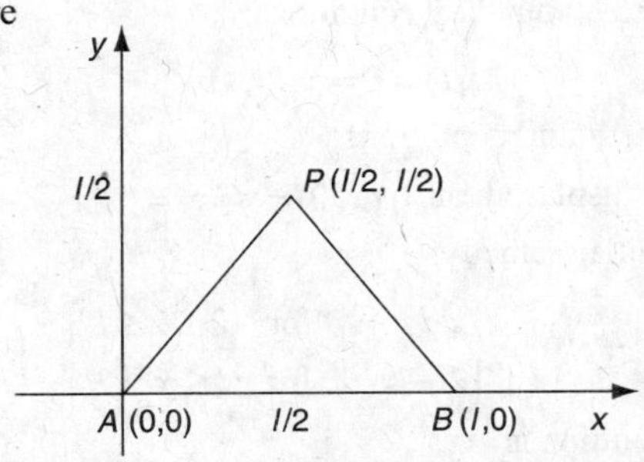

Figure 10.4 String with transverse displacement at the mid point

$$C_n = \frac{2}{l}\int_0^l f(x)\sin\frac{n\pi x}{l}\,dx$$

$$= \frac{2}{l}\int_0^{l/2} x\sin\frac{n\pi x}{l}\,dx + \frac{2}{l}\int_{l/2}^{l}(l-x)\sin\frac{n\pi x}{l}\,dx$$

$$= \frac{2}{l}\left[x\left(\frac{-l}{n\pi}\right)\cos\frac{n\pi x}{l} - \left(\frac{-l^2}{n^2\pi^2}\right)\sin\frac{n\pi x}{l}\right]_0^{l/2}$$

= 0 at 0   = 0 at 0

$$+\frac{2}{l}\left[(l-x)\left(\frac{-l}{n\pi}\right)\cos\frac{n\pi x}{l} - (-1)\left(\frac{-l^2}{n^2\pi^2}\right)\sin\frac{n\pi x}{l}\right]_{l/2}^{l}$$

= 0 at l = 0 at l

$$= \frac{2}{l}\left[\frac{l}{2}\left(-\frac{l}{n\pi}\right)\cos\frac{n\pi}{2} + \frac{l^2}{n^2\pi^2}\sin\frac{n\pi}{2}\right]$$

$$+\frac{2}{l}\left[\frac{l}{2}\cdot\frac{l}{n\pi}\cos\frac{n\pi}{2} + \frac{l^2}{n^2\pi^2}\sin\frac{n\pi}{2}\right]$$

$$= \frac{4l}{n^2\pi^2}\sin\frac{n\pi}{2} = \left\{\begin{array}{ll} 0 & \text{if } n \text{ is even} \\ \dfrac{4l}{n^2\pi^2}(-1)^{(n-1)/2} & \text{if } n \text{ is odd}\end{array}\right\} \tag{7}$$

The solution for the initial displacement given by (4) and zero initial velocity is

$$y(x,t) = \frac{4}{\pi}\sum_{m=1}^{\infty}\frac{(-1)^{m-1}}{(2m-1)^2}\sin\frac{(2m-1)\pi x}{l}\cos\frac{(2m-1)\pi at}{l} \tag{8}$$

10.13.7 Vibrating String with Given Initial Velocity and Zero Initial Displacement

Next, we consider the case when the string is released from its horizontal position with zero initial displacement but with an initial velocity given at x by $g(x)$. The boundary value problem for the displacement function is

$$\frac{\partial^2 y}{\partial t^2} = a^2\frac{\partial^2 y}{\partial x^2} \quad \text{for } 0 \le x \le l,\ t > 0 \tag{10.126}$$

$$y(0,t) = 0 = y(l,t) \quad \text{for } t > 0 \tag{10.127, 10.128}$$

zero initial displacement: $y(x,0) = 0$
prescribed initial velocity: $\dfrac{\partial y}{\partial t}(x,0) = g(x)$
for $0 \le x \le l$ (10.129, 10.130)

By the method of separation of variables we set $y(x, t) = x(t)\,T(t)$ and obtain ordinary differential equations

$$X'' + \lambda^2 X = 0 \quad \text{and} \quad T'' + a^2\lambda^2 T = 0 \tag{10.131, 10.132}$$

The boundary conditions are same as before and hence we obtain eigenvalues

$$\lambda_n^2 = \frac{n^2\pi^2}{l^2} \tag{10.133}$$

and the corresponding eigenfunctions are constant multiples of

$$X_n = \sin\frac{n\pi x}{l} \tag{10.134}$$

with the values of λ as $\lambda = \lambda_n = n\pi/l$ the differential equation for T becomes

$$T'' + a^2\frac{n^2\pi^2}{l^2}T = 0 \tag{10.135}$$

whose general solution is

$T(t) = A\cos(n\pi\, at/l) + B\sin(n\pi\, at/l)$

The initial condition of zero initial displacement gives

$$y\,(x, 0) = 0 = X\,(x)\,T\,(0) \;\Rightarrow\; T\,(0) = 0 \tag{10.136}$$

Since $T(0) = A = 0$ solutions for $T(t)$ are constant multiples of $\sin n\pi\, at/l$. Thus, for $n = 1, 2, 3, \ldots$ we have functions

$$y_n(x,t) = C_n \sin\frac{n\pi x}{l}\sin\frac{n\pi at}{l} \qquad (10.137)$$

Each of these functions satisfies the wave equation, the boundary conditions and the zero initial displacement condition. In order to satisfy the initial velocity condition $(\partial y/\partial t)(x, 0) = g(x)$ we invoke the superposition principle and write

$$y(x, t) = \sum_{n=1}^{\infty} y_n(x, t) = \sum_{n=1}^{\infty} C_n \sin\frac{n\pi x}{l}\sin\frac{n\pi at}{l} \qquad (10.138)$$

We assume that the series admits of term-by-term differentiation. So, we get

$$\frac{\partial y}{\partial t}(x, t) = \sum_{n=1}^{\infty} C_n\left(\frac{n\pi a}{l}\right)\sin\left(\frac{n\pi x}{l}\right)\cos\left(\frac{n\pi at}{l}\right) \qquad (10.139)$$

Now, the initial velocity condition yields

$$\frac{\partial y}{\partial t}(x, 0) = \sum_{n=1}^{\infty} C_n\left(\frac{n\pi a}{l}\right)\sin\frac{n\pi x}{l}\cdot 1 = g(x) \qquad (10.140)$$

This is the half-range Fourier series expansion of $g(x)$ on $[0, l]$. Here, the entire coefficient of $\sin n\pi x/l$ is the Fourier sine coefficient of $g(x)$ on $[0, l]$ so that we have

$$C_n\cdot\frac{n\pi a}{l} = \frac{2}{l}\int_0^l g(x)\sin\frac{n\pi x}{l}\,dx$$

$$\text{or}\quad C_n = \frac{2}{n\pi a}\int_0^l g(x)\sin\frac{n\pi x}{l}\,dx \qquad (10.141)$$

Example 10.64

Find the displacement $y(x, t)$ of a string stretched between two fixed points at a distance $2l$ apart when the string is initially at rest in equilibrium position and the points of the string are given initial velocity $g(x)$, given by

$$g(x) = \begin{cases} x/l & \text{for } 0 \le x \le l \\ (2l - x)/l & \text{for } l \le x \le 2l \end{cases}$$

Solution We have to solve the wave equation

$$\frac{\partial^2 y}{\partial t^2} = a^2\frac{\partial^2 y}{\partial x^2} \qquad (1)$$

under the boundary conditions:

$$y(0, t) = 0 = y(2l, t) \qquad (2), (3)$$

and the initial conditions:

initial displacement: $y(x, 0) = f(x) = 0$ (4)

and initial velocity:

$$g(x, 0) = \begin{Bmatrix} x/l & \text{for } 0 \le x \le l \\ 2l - x/l & \text{for } l \le x \le 2l \end{Bmatrix} \qquad (5)$$

The solution is

$$y(x, t) = \frac{2l}{\pi a}\sum_{n=1}^{\infty}\frac{b_n}{n}\sin\frac{n\pi x}{2l}\sin\frac{n\pi at}{2l} \qquad (6)$$

where

$$b_n = \frac{2}{2l}\int_0^{2l} g(x)\sin\frac{n\pi x}{2l}\,dx$$

$$= \frac{1}{l}\int_0^l \frac{x}{l}\sin\frac{n\pi x}{2l}\,dx + \frac{1}{l}\int_l^{2l}\frac{2l - x}{l}\sin\frac{n\pi x}{2l}\,dx$$

$$= \frac{1}{l}\left[\frac{x}{l}(-1)\left(\frac{2l}{n\pi}\right)\cos\frac{n\pi x}{2l} - \frac{1}{l}(-1)\frac{4l^2}{n^2\pi^2}\sin\frac{n\pi x}{2l}\right]_0^l$$

$$+\frac{1}{l}\left[\frac{2l - x}{l}\cdot\frac{-2l}{n\pi}\cdot\cos\left(\frac{n\pi x}{2l}\right) - \left(\frac{-1}{l}\right)\cdot\frac{-4l^2}{n^2\pi^2}\cdot\sin\frac{n\pi x}{2l}\right]_l^{2l}$$

$$= -\frac{2}{n\pi}\cos\frac{n\pi l}{2l} + \frac{4}{n^2\pi^2}\sin\frac{n\pi l}{2l}$$

$$+\frac{2}{n\pi}\cos\frac{n\pi}{2} + \frac{4}{n^2\pi^2}\sin\frac{n\pi}{2}$$

$$= \frac{8}{n^2\pi^2}\sin\frac{n\pi}{2} \qquad (7)$$

Hence, the displacement function is given by

$$y(x, t) = \frac{2l}{\pi a}\cdot\frac{8}{\pi^2}\sum_{n=1}^{\infty}\frac{1}{n}\cdot\frac{1}{n^2}\sin\frac{n\pi}{2}\cdot\sin\frac{n\pi x}{2l}\sin\frac{n\pi at}{2l}$$

$$\Rightarrow\ y(x, t) = \frac{16l}{\pi^3 a}\sum_{n=1}^{\infty}\frac{1}{n^3}\sin\frac{n\pi}{2}\cdot\sin\frac{n\pi x}{2l}\sin\frac{n\pi at}{2l} \qquad (8)$$

Example 10.65

A tightly stretched string with fixed end points is initially at rest in its equilibrium position, and each of its points is given a velocity v, which is given by

$$v(x)=\begin{cases} Cx & \text{for } 0\le x\le \dfrac{l}{2} \\ C(l-x) & \text{for } \dfrac{l}{2}\le x\le l \end{cases}$$

Find the displacement $y(x, t)$.

[JNTU 1994S, 2001, 2002]

Solution We have to solve the wave equation

$$\frac{\partial^2 y}{\partial t^2}=a^2\frac{\partial^2 y}{\partial x^2} \tag{1}$$

under the boundary conditions:

$$y(0,t)=0=y(l,t) \quad \text{for} \quad t\ge 0 \qquad (2), (3)$$

and the initial conditions:

zero initial displacement:

$$y(x,0)=0 \quad (0\le x\le l) \tag{4}$$

prescribed initial velocity:

$$\frac{\partial y}{\partial t}(x,0)=v(x)=\begin{cases} Cx & \text{for } 0\le x\le \dfrac{l}{2} \\ C(l-x) & \text{for } \dfrac{l}{2}\le x\le l \end{cases} \tag{5}$$

Solving equation (1) by the method of separation of variables under the conditions is given by

$$y(x,t)=\sum_{n=1}^{\infty} y_n(x,t)=\sum_{n=1}^{\infty} C_n \sin\frac{n\pi x}{l}\sin\frac{n\pi at}{l} \tag{6}$$

Differentiating (6) partially with respect to 't' we obtain

$$\frac{\partial y}{\partial t}(x,t)=\sum_{n=1}^{\infty}\frac{n\pi a}{l}C_n\sin\frac{n\pi x}{l}\cos\frac{n\pi at}{l} \tag{7}$$

$$\Rightarrow \quad \frac{\partial y}{\partial t}(x,0)=\sum_{n=1}^{\infty}\left(\frac{n\pi a}{l}C_n\right)\sin\frac{n\pi x}{l}\cdot 1=v(x), \tag{8}$$

This is the half-range Fourier sine series expansion for $v(x)$ in $[0, l]$. Therefore, we have

$$\frac{n\pi a}{l}\cdot C_n=\frac{2}{l}\int_0^l v(x)\sin\frac{n\pi x}{l}\,dx$$

$$=\frac{2}{l}\int_0^{l/2} Cx\sin\frac{n\pi x}{l}\,dx+\frac{2}{l}\int_{l/2}^{l} C(l-x)\sin\frac{n\pi x}{l}\,dx$$

$$=\frac{2C}{l}\cdot\left[x\left(-\frac{l}{n\pi}\right)\cos\frac{n\pi x}{l}-1\cdot\left(-\frac{l^2}{n^2\pi^2}\right)\sin\frac{n\pi x}{l}\right]_0^{l/2}$$

$$= 0 \text{ at } 0 \qquad\qquad = 0 \text{ at } 0$$

$$+\frac{2C}{l}\left[(l-x)\left(-\frac{l}{n\pi}\right)\cos\frac{n\pi x}{l}-(-1)\left(-\frac{l^2}{n^2\pi^2}\right)\sin\frac{n\pi x}{l}\right]_{l/2}^{l}$$

$$= 0 \text{ at } l \qquad\qquad = 0 \text{ at } l$$

$$=\frac{2C}{l}\left[-\frac{l^2}{2n\pi}\cos\frac{n\pi}{2}+\frac{l^2}{n^2\pi^2}\sin\frac{n\pi}{2}\right]$$

$$-\frac{2C}{l}\left[-\frac{l^2}{2n\pi}\cos\frac{n\pi}{2}-\frac{l^2}{n^2\pi^2}\sin\frac{n\pi}{2}\right]$$

$$=\frac{4C}{l}\cdot\frac{l^2}{n^2\pi^2}\sin\frac{n\pi}{2}=\frac{4Cl}{n^2\pi^2}\sin\frac{n\pi}{2}$$

$$\Rightarrow \quad C_n=\frac{l}{n\pi a}\cdot\frac{4Cl}{n^2\pi^2}\sin\frac{n\pi}{2}=\frac{4l^2C}{an^3\pi^3}\sin\frac{n\pi}{2} \tag{9}$$

Hence, the solution of the problem is

$$y(x,t)=\frac{4Cl^2}{a\pi^3}\sum_{n=1}^{\infty}\frac{\sin\frac{n\pi}{2}}{n^3}\sin\frac{n\pi x}{l}\cos\frac{n\pi at}{l} \quad (10)$$

Example 10.66

Find the solution for the above problem when the string is released from its horizontal position with an initial velocity given by $g(x)=x(1+\cos\pi x/l)$.

Deduce the result for $a=1$ and $l=\pi$.

Solution Following the procedure of the above problem, the solution in the present case is obtained as

$$y(x,t)=\sum_{n=1}^{\infty}y_n(x,t)=\sum_{n=1}^{\infty}C_n\sin\frac{n\pi x}{l}\sin\frac{n\pi at}{l} \quad (1)$$

The prescribed initial velocity in this case is

$$\frac{\partial y}{\partial t}(x,0)=g(x)=x\left(1+\cos\frac{\pi x}{l}\right) \quad (2)$$

Differentiating (1) partially with respect to 't' we have

$$\frac{\partial y}{\partial t}(x,t)=\sum_{n=1}^{\infty}\left(\frac{n\pi a}{l}\right)C_n\sin\frac{n\pi x}{l}\cos\frac{n\pi at}{l} \quad (3)$$

Imposing the initial veloctiy condition (2) we obtain

$$\frac{\partial y}{\partial t}(x,0)=\sum_{n=1}^{\infty}\left(\frac{n\pi a}{l}C_n\right)\sin\frac{n\pi x}{l}=g(x)=x+x\cos\frac{\pi x}{l} \quad (4)$$

This is the half-range Fourier series expansion for $g(x)$ in $[0,l]$. Therefore, we have

$$\begin{aligned}\frac{n\pi a}{l}C_n&=\frac{2}{l}\int_0^l g(x)\sin\frac{n\pi x}{l}dx\\&=\frac{2}{l}\int_0^l x\left(1+\cos\frac{\pi x}{l}\right)\sin\frac{n\pi x}{l}dx\\&=\frac{2}{l}\int_0^l x\sin\frac{n\pi x}{l}dx+\frac{2}{l}\int_0^l x\sin\frac{n\pi x}{l}\cos\frac{\pi x}{l}dx\end{aligned} \quad (5)$$

For evaluation of the second intergral, we have to separate the cases.

Case (1) $n=1$ and Case (2) $n\neq 1$

Case 1

$n=1$

$$\begin{aligned}\frac{\pi a}{l}C_1&=\frac{2}{l}\int_0^l x\sin\frac{\pi x}{l}dx+\frac{2}{l}\int_0^l x\sin\frac{\pi x}{l}\cos\frac{\pi x}{l}dx\\&=\frac{2}{l}\int_0^l x\sin\frac{\pi x}{l}dx+\frac{1}{l}\int_0^l x\sin\frac{2\pi x}{l}dx\\&=\frac{2}{l}\left[x\left(-\frac{l}{\pi}\right)\cos\frac{\pi x}{l}-1\cdot\left(-\frac{l^2}{\pi^2}\right)\sin\frac{\pi x}{l}\right]_0^l\end{aligned}$$

$= 0$ at 0 $\qquad = 0$ at l, 0

$$+\frac{1}{l}\left[x\left(-\frac{l}{2\pi}\right)\cos\frac{2\pi x}{l}-\left(-\frac{l^2}{4\pi^2}\right)\sin\frac{2\pi x}{l}\right]_0^l$$

$= 0$ at 0 $\qquad = 0$ at l, 0

$$=\frac{2}{l}\left[-\frac{l^2}{\pi}(-1)\right]+\frac{1}{l}\left[-\frac{l^2}{2\pi}.1\right]=\frac{2l}{\pi}-\frac{l}{2\pi}=\frac{3l}{2\pi}$$

$$\Rightarrow\quad C_1=\frac{3l^2}{2a\pi^2} \quad (6)$$

Case 2

$n\neq 1$

$$\begin{aligned}\frac{n\pi a}{l}C_n&=\frac{2}{l}\int_0^l x\sin\frac{n\pi x}{l}dx+\frac{2}{l}\int_0^l x\sin\frac{n\pi x}{l}\cos\frac{\pi x}{l}dx\\&=\frac{2}{l}\int_0^l x\sin\frac{n\pi x}{l}dx+\frac{1}{l}\int_0^l\left[x\sin\frac{(n+1)\pi x}{l}dx+x\sin\frac{(x-1)\pi x}{l}dx\right]\end{aligned}$$

$$= \frac{2}{l}\left[x\left(-\frac{l}{n\pi}\right)\cdot \cos\frac{n\pi x}{l} - 1\cdot\left(-\frac{l^2}{n^2\pi^2}\right)\sin\frac{n\pi x}{l}\right]_0^l$$

$= 0$ at 0 $\qquad\qquad = 0$ at l, 0

$$+\frac{1}{l}\left[x\left(-\frac{l}{(n+1)\pi}\right)\cos\frac{(n+1)\pi x}{l} - 1\cdot\left(-\frac{l^2}{(n+1)^2\pi^2}\right)\sin\frac{(n+1)\pi x}{l}\right]$$

$= 0$ at 0 $\qquad\qquad = 0$ at l, 0

$$+\frac{1}{l}\left[x\left(-\frac{l}{(n-1)\pi}\right)\cos\frac{(n-1)\pi x}{l} - 1\left(-\frac{l^2}{(n-1)^2\pi^2}\right)\sin\frac{(n-1)\pi x}{l}\right]_0^l$$

$= 0$ at 0 $\qquad\qquad = 0$ at l, 0

$$= \frac{2}{l}\left[-\frac{l^2}{n\pi}\cdot(-1)^n\right] + \frac{1}{l}\left[-\frac{l^2}{(n+1)\pi}\cdot(-1)^{n+1}\right] + \frac{1}{l}\left[-\frac{l^2}{(n-1)\pi}\cdot(-1)^{n-1}\right]$$

$$= -\frac{2l(-1)^n}{n\pi} + \frac{l(-1)^n}{\pi}\left(\frac{1}{n+1}+\frac{1}{n-1}\right) = \frac{2l(-1)^n}{\pi n}\cdot\frac{1}{n^2-1}$$

$$\Rightarrow \quad C_n = \frac{2l^2(-1)^n}{a\pi^2 n^2(n^2-1)} \quad (n \neq 1) \tag{7}$$

Therefore, the solution for $g(x) = x(1 + \cos \pi x/l)$ as the initial velocity function is

$$y(x,t) = \frac{3l^2}{2a\pi^2}\sin\frac{\pi x}{l}\sin\frac{\pi a t}{l} + \frac{2l^2}{a\pi^2}\sum_{n=2}^{\infty}\frac{(-1)^n}{n^2(n^2-1)} \sin\frac{n\pi x}{l}\sin\frac{n\pi a t}{l} \tag{8}$$

By taking $a = 1$ and $l = \pi$ in the above result, we obtain the solution for this particular case as

$$y(x,t) = \frac{3}{2}\sin x \sin t + 2\sum_{n=2}^{\infty}\frac{(-1)^n}{n^2(n^2-1)}\sin nx \sin nt \tag{9}$$

10.13.8 Vibrating String with Initial Displacement and Initial Velocity

Consider the motion of the string with both initial displacement given by $f(x)$ and initial velocity given by $g(x)$. We have to now solve two separate problems, one with initial displacement $f(x)$ and zero initial velocity and the other with zero displacement and initial velocity $g(x)$. Let $y_1(x, t)$ and $y_2(x, t)$ be the respective solutions of the two problems and let

$$y(x,t) = y_1(x,t) + y_2(x,t) \tag{10.142}$$

Then y satisfies the wave equation and the boundary conditions. Further,

$$y(x,0) = y_1(x,0) + y_2(x,0) = f(x) + 0 = f(x) \tag{10.143}$$

and

$$\frac{\partial y}{\partial t}(x,0) = \frac{\partial y_1}{\partial t}(x,0) + \frac{\partial y_2}{\partial t}(x,0) = 0 + g(x) = g(x) \tag{10.144}$$

Thus, $y(x, t)$ is the solution in the case of nonzero initial displacement and velocity functions.

Example 10.67

An elastic string of length l, fastened at its ends on the x-axis at $x = 0$ and $x = \lambda$, is given initial displacement $f(x)$ and initial velocity $g(x)$. Find the displacement function $y(x, t)$ by solving the wave equation

$y_{tt} = a^2 y_{xx}$ under the conditions: $y(0, t) = y(l, t) = 0$, $y(x, 0) = f(x) = x$ for

$$y(x,0) = \begin{cases} x & \text{for } 0 \le x \le l/2 \\ l - x & \text{for } l/2 \le x \le l \end{cases}$$

and

$$y_t(x,0) = g(x) = \left\{ x\left(1 + \cos\frac{\pi x}{l}\right)\right..$$

Solution From equations (10.143) and (10.144), we obtain the solution for the present problem as

$$y(x,t) = \frac{4l}{\pi^2}\sum_{m=1}^{\infty}\frac{(-1)^{m-1}}{(2m-1)^2}\cdot\sin\frac{(2m-1)\pi x}{l}\cdot\cos\frac{(2m-1)\pi at}{l} + \frac{3l^2}{2a\pi^2}\sin\frac{\pi x}{l}\sin\frac{\pi at}{l} + \frac{2l^2}{a\pi^2}\sum_{n=2}^{\infty}\frac{(-1)^n}{n^2(n^2-1)}\sin\frac{n\pi x}{l}\sin\frac{n\pi at}{l} \tag{10.145}$$

EXERCISE 10.8

1. Find the displacement of a string stretched between two fixed points at a distance $2l$ apart when the string is initially at rest in equilibrium position and the points of the string are given an initial velocity $g(x)$, which are given by $\partial y/\partial t(x, 0) = g(x) = bx(l - x)$.

   Ans: $y(x, t) = \dfrac{8bl^3}{a\pi^4}\displaystyle\sum_{n=1}^{\infty}\frac{1}{(2n-1)^4}\sin\frac{(2n-1)\pi x}{l}\sin\frac{(2n-1)\pi at}{l}$.

2. A string of length l is stretched and fastened to two fixed points. Find $y(x, t)$ satisfying the wave equation $y_{tt} = a^2 y_{xx}$ when it is given as:

   (a) Initial displacement $y(x, 0) = f(x) = b \sin \pi x/l$.

   Ans: $y(x,t) = b\sin\dfrac{\pi x}{l}\cos\dfrac{\pi at}{l}$.

   (b) Initial triangular deflection:

$$f(x) = \begin{cases} \dfrac{2Cx}{l} & \text{for } 0 < x < l/2 \\ \dfrac{2C(l-x)}{l} & \text{for } l/2 < x < l \end{cases}$$

   Ans: $y(x, t) = \dfrac{8C}{\pi^2}\displaystyle\sum_{n=1}^{\infty}\frac{1}{n^2}\sin\frac{n\pi}{2}\sin\frac{n\pi x}{l}\cos\frac{n\pi at}{l}$.

   (c) Initial velocity: $g(x) = b\sin^3\dfrac{\pi x}{l}$

   Ans:

$$g(x, t) = \frac{bl}{12a\pi}\left[9\sin\frac{\pi x}{l}\sin\frac{\pi at}{l} - \sin\frac{3\pi x}{l}\sin\frac{3\pi at}{l}\right]$$

   (d) Initial velocity: $g(x) = b\sin\dfrac{3\pi x}{l}\cos\dfrac{2\pi x}{l}$

   Ans:

$$y(x, t) = \frac{bl}{2a\pi}\sin\frac{\pi x}{l}\sin\frac{\pi at}{l} + \frac{bl}{5a\pi}\sin\frac{5\pi x}{l}\sin\frac{5\pi at}{l}$$

10.13.9 Laplace's Equation or Potential Equation or Two-dimensional Steady-state Heat Flow Equation

The two-dimensional heat conduction equation is given by

$$\frac{\partial u}{\partial t} = a^2\,\nabla^2 u, \left(\text{where } \nabla^2 = \frac{\partial^2}{\partial x^2} + \frac{\partial^2}{\partial y^2} \text{ is the Laplacian operator}\right)$$

In the case of steady-state heat flow $\partial u/\partial t = 0$ and equation reduces to

$$\nabla^2 u = \frac{\partial^2 u}{\partial x^2} + \frac{\partial^2 u}{\partial y^2} = 0$$

The solution $u(x,y)$ of the above equation can be obtained by the method of separation of variables in a rectangular region both in the Dirichlet problem as well as in Neumann's problem. A rectangular thin plate with its two faces insulated is considered so that the heat flow is two-dimensional. The boundary conditions are prescribed on the four edges of the plate.

Example 10.68

Solve Laplace's equation

$$\frac{\partial^2 u}{\partial x^2} + \frac{\partial^2 u}{\partial y^2} = 0 \tag{1}$$

in the rectangle; $0 < x < a$, $0 < y < b$ in the xy-plane, with the boundary conditions

$u(x, 0) = 0$ on OA; $u(x, b) = 0$ on BC (2), (3)

$u(0, y) = 0$ on OC; $u(a, y) = f(y)$ on AB (4), (5)

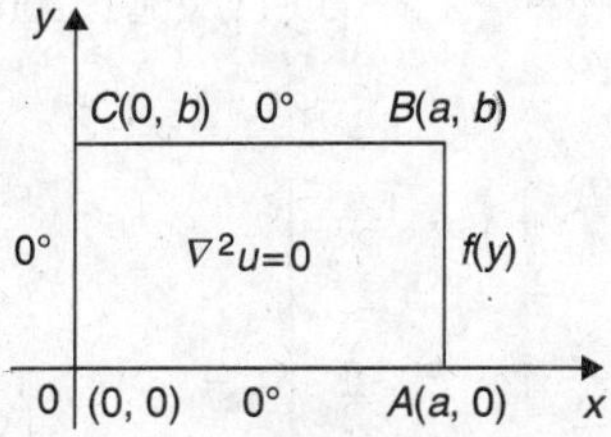

Figure 10.5 Two-dimensional heat flow in a rectangular plate

Solution Let $u(x, y) = X(x)\, Y(y)$ (6)
Substituting in (1) we get

$$\frac{Y''}{Y} = \frac{X''}{X} = -\lambda^2 \tag{7}$$

where the separation constant is taken as negative to get non-trivial solutions. The boundary value problem reduces to solution of the ordinary differential equations

$$Y'' + \lambda^2 Y = 0, \quad X'' - \lambda^2 X = 0 \qquad (8), (9)$$

under the conditons

$$0 = u(x,0) = X(x)Y(0) \Rightarrow Y(0) = 0 \tag{10}$$

$$0 = u(x,b) = X(x)Y(b) \Rightarrow Y(b) = 0 \tag{11}$$

$$0 = u(0,y) = X(0)Y(y) \Rightarrow X(0) = 0 \tag{12}$$

The general solution of equation (8) is

$$Y(y) = A\cos \lambda y + B\sin \lambda y$$

By (10) we get $A = 0$ and by (11) we get $\sin \lambda b = 0$

$$\Rightarrow \lambda_n = \frac{n\pi}{b} (n = 1, 2, 3, \ldots) \tag{13}$$

$\lambda_n = n\pi/b$ are the eigenvalues and the corresponding eigenfunctions are

$$Y_n(y) = \sin\frac{n\pi y}{b} \quad (\text{with } B = 1) \tag{14}$$

Now the general solution of (9) is

$$X_n(x) = C\cos h\frac{n\pi x}{b} + D\sin h\frac{n\pi x}{b} \tag{15}$$

using (12) we get $C = 0$ so that

$$X_n(x) = D\sin\frac{n\pi x}{b}$$

Therefore, the solution of equation (1) satisfying the boundary conditions (2), (3) and (4) is

$$u_n(x, y) = C_n \sinh\frac{n\pi x}{b}\sin\frac{n\pi y}{b}$$

where we have replaced D by C_n.
By the principle of superposition we write the solution as

$$u(x, y) = \sum_{n=1}^{\infty} u_n(x, y) = \sum_{n=1}^{\infty} C_n \sinh\frac{n\pi x}{b}\sin\frac{n\pi y}{b} \tag{16}$$

Lastly we have condition (5) namely $u(a, y) = f(y)$ to be satisfied. This gives

$$u(a, y) = \sum_{n=1}^{\infty} C_n \sinh\frac{n\pi a}{b}\sin\frac{n\pi y}{b} = f(y) \tag{17}$$

This is a half-range Fourier sine series expansion of $f(y)$ in $(0, b)$ and the constants C_n are given by

$$C_n = \sinh\frac{n\pi a}{b} = \frac{2}{b}\int_0^b f(y)\sin\frac{n\pi y}{b}dy$$

or
$$C_n = \frac{2}{b\sinh\frac{n\pi a}{b}}\int_0^b f(y)\sin\frac{n\pi y}{b}dy \tag{18}$$

Thus, this harmonic function $u(x,y)$ satisfying Laplace's equation (1) and the boundary conditions (2)–(5) is given by (16) where the constants C_n are determined by (18) for any specific function $f(y)$.

Example 10.69

A retangular plate is bounded by the lines $x = 0$, $y = 0$, $x = a$ and $y = b$ and the edge temperatures are $u(0, y) = u(x, b) = u(a, y)$ and $u(x, 0) = 5\sin(5\pi x/a) + 3\sin(3\pi x/a)$. Find the steady-state temperature at any point of the plate.

[JNTU 2002, 2003s]

Solution Let $u(x, y)$ be the steady-state temperature at any point $p(x, y)$ of the rectangular plate. We have to solve Laplace's equation

$$\frac{\partial^2 u}{\partial x^2} + \frac{\partial^2 u}{\partial y^2} = 0 \quad \cdots \tag{1}$$

under the boundary conditions

1. $u(0, y) = 0 \quad \text{for} \quad 0 < y < b$ (2)
2. $u(a, y) = 0 \quad \text{for} \quad 0 < y < b$ (3)
3. $u(x, b) = 0 \quad \text{for} \quad 0 < x < a$ (4)
4. $u(x,0) = 5\sin\frac{5\pi x}{a} + 3\sin\frac{3\pi x}{a} \text{ for } 0 < x < a$ (5)

A suitable solution of equation (1), by the method of separation of variables, satisfying boundary conditions (2)–(4) for each $n = 1, 2, 3, \ldots$ is

$$u_n(x, y) = C_n \sin\frac{n\pi x}{a} \sin h\frac{n\pi(y-b)}{a} \quad (6)$$

By the principle of superposition of solution we may write the general solution as

$$u(x, y) = \sum_{n=1}^{\infty} u_n(x, y) = \sum_{n=1}^{\infty} C_n \sin\frac{n\pi x}{a} \sin h\frac{n\pi(y-b)}{a} \quad (7)$$

Imposing condition (5) we get

$$u(x,0) = -\sum_{n=1}^{\infty} C_n \sin\frac{n\pi x}{a} \sin h\frac{n\pi b}{a} = 5\sin\frac{5\pi x}{a} + 3\sin\frac{3\pi x}{a} \quad (8)$$

Equating the coefficients of like terms on either side we get

$$-C_5 \sin h\frac{5\pi b}{a} = 5 \text{ and } -C_3 \sin h\frac{3\pi b}{a} = 3 \text{ and } C_n = 0$$
for all $n \neq 3$ or 5 (9)

Hence, the required solution of steady-state temperature is

$$u(x, y) = \frac{3}{\sin h\frac{3\pi b}{a}} \sin\frac{3\pi x}{a} \sin h\frac{3\pi(b-y)}{a} + \frac{5}{\sin h\frac{5\pi b}{a}} \sin\frac{5\pi x}{a} \sin h\frac{5\pi(b-y)}{a} \quad (10)$$

Example 10.70

An infinitely long plane uniform plate is bounded by two parallel edges and an end at right angles to them. The breadth in π. This end is maintained at a temperature u_0 at all points and the other edges are at zero temperature. Determine the temperature at any point of the plate in the steady state.

[JNTU 2002, 2005 (Set 4)]

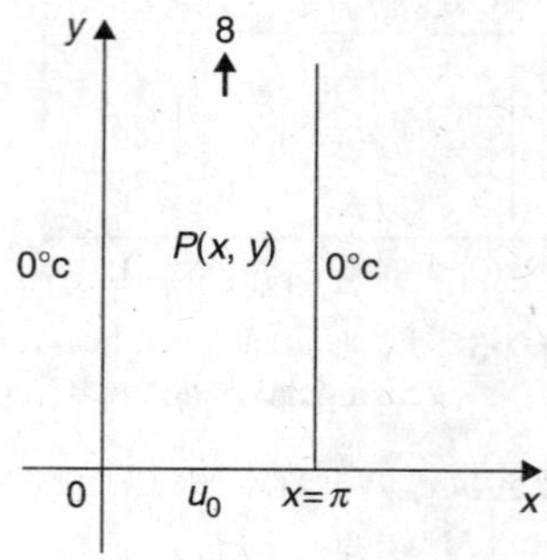

Figure 10.6 Temperature in an infinitely long plate

Solution Let $u(x, y)$ be the temperature at any point $p(x, y)$ of the plate. Then the steady state temperature distribution is given by Laplace's equation

$$\frac{\partial^2 u}{\partial x^2} + \frac{\partial^2 u}{\partial y^2} = 0 \quad \cdots \quad (1)$$

with the boundary conditions

1. $u(0, y) = 0$; 2. $u(\pi, y) = 0$ for all y
3. $u(x, 0) = u_0$; 4. $u(x, \infty) = 0$ for $0 < x < \pi$

The general solution of equation (1) is

$$u(x, y) = (A\cos\lambda x + B\sin\lambda x)(Ce^{\lambda y} + De^{-\lambda y}) \quad (2)$$

By condition (1): $u(0, y)$ we get $A = 0$. Now equation (2) becomes

$$u(x, y) = B\sin\lambda x\,(Ce^{\lambda y} + De^{-\lambda y}) \quad (3)$$

By condition (2): $u(\pi, y) = 0$ we get

$$\sin\lambda\pi = 0 \Rightarrow \lambda = \lambda_n = n \quad (4)$$

so that the solution becomes $u(x, y) = (c_n e^{ny} + D_n e^{-ny})$ $\sin nx$ where BC and BD have been replaced by c_n and d_n, respectively.

By condition (4):

$$u(x, \infty) = 0 \text{ we have } c_n = 0 \text{ for all } n \quad (5)$$

Finally, the general solution satisfying the conditions (1), (2) and (4) is (by the principle of superposition)

$$u(x,y)=\sum_{n=1}^{\infty}u_n(x,y)=\sum_{n=1}^{\infty}d_n \sin nx\, e^{-ny} \qquad (6)$$

Now applying boundary condition (3):

$$u(x,0)=\sum_{n=1}^{\infty}d_n \sin nx = u_0 \qquad (7)$$

Equation (7) is a half-range Fourier sine series expansion for u_0 in $(0, \pi)$ and hence the Fourier constants d_n are given by

$$d_n=\frac{2}{\pi}\int_0^{\pi}u_0 \sin nx\, dx=\frac{-2u_0}{\pi}\left(\frac{\cos nx}{n}\right)_0^{\pi}$$

$$=\frac{-2u_0}{n\pi}(\cos n\pi-1)=\begin{cases}0 & \text{if } n \text{ is even}\\ \dfrac{4u_0}{n\pi} & \text{if } n \text{ is odd}\end{cases} \qquad (8)$$

Now (6) reduces to

$$u(x,y)=\sum_{n=1,3,5}\frac{4u_0}{n\pi}\sin nx\, e^{-ny} \qquad (9)$$

which is the required solution.

Example 10.71

The temperature $u\,(x, y)$ is maintained at 0°C along three edges of a square plate of side 100 cm and the fourth edge is maintained at a constant temperature u_0 until steady-state conditons prevail. Find the temperature at any point (x, y) of the plate and also at the centre of the plate.

[JNTU 2003s (Set 4)]

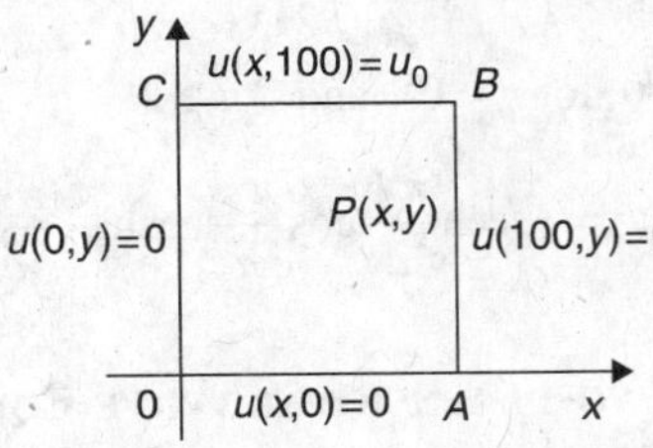

Figure 10.7 Temperature in a square plate

Solution Let the side of the plate be $a = 100$ cm. The temperature function $u\,(x, y)$ satisfies Laplace's equation

$$\frac{\partial^2 u}{\partial x^2}+\frac{\partial^2 u}{\partial y^2}=0 \qquad (1)$$

with the boundary conditons:

1. $u(0,y)=0$ 2. $u(a,y)=0$ for $0\le y\le a$
3. $u(x,0)=0$ 4. $u(x,a)=u_0$ for $0\le x\le a$.

The general solution of equation (1) is

$$u(x,y)=(C_1\cos px + C_2 \sin px)(C_3e^{py}+C_4e^{-py}) \qquad (2)$$

By condition (1) we get $C_1 = 0$ so that equation (2) becomes

$$u(x,y)=C_2 \sin px\,(C_3e^{py}+C_4e^{-py}) \qquad (3)$$

By condition (2) we get $u(a,y)=0=C_2 \sin pa\,(C_3e^{py}+C_4e^{-py}) \Rightarrow p=(n\pi/a)(n=1, 2, 3)$, so that we have

$$u(x,y)=C_2\sin\frac{n\pi x}{a}\left(C_3e^{n\pi y/a}+C_4e^{-n\pi y/a}\right) \qquad (4)$$

Now, condition (3) implies that $C_4 = -C_3$ (5)
The solution may therefore be written as

$$u_n(x,y)=C_n\sin\frac{n\pi x}{a}\sinh\frac{n\pi y}{a} \quad (n=1, 2, 3,\ldots) \qquad (6)$$

for each n. By the principle of superposition we may write

$$u(x,y)=\sum_{n=1}^{\infty}u_n(x,y)=\sum_{n=1}^{\infty}C_n\sin\frac{n\pi x}{a}\sinh\frac{n\pi y}{a} \qquad (7)$$

Imposing the condition (4) we have

$$u(x,a)=u_0=\sum_{n=1}^{\infty}C_n\sinh n\pi,\quad \sin\frac{n\pi x}{a} \qquad (8)$$

Equation (8) is a half-range Fourier sine series expansion of u_0 in $(0, a)$, and the Fourier coefficients are obtained by

$$C_n\sinh n\pi=\frac{2}{a}\int_0^a u_0\sin\frac{n\pi x}{a}\,dx \quad \text{or}$$

$$C_n=\frac{2u_0}{a\sinh n\pi}\int_0^a \sin\frac{n\pi x}{a}\,dx$$

$$\Rightarrow C_n = \frac{2u_0}{a\sinh n\pi}\cdot\left(\frac{-a}{n\pi}\right)\left[\cos\frac{n\pi x}{a}\right]_0^a$$

$$= \frac{2u_0}{a\sinh n\pi}\cdot\frac{-a}{n\pi}\cdot(\cos n\pi - 1)$$

$$= \begin{cases} 0 & \text{if } n \text{ is even} \\ \dfrac{4u_0}{n\pi\sinh n\pi} & \text{if } n \text{ is odd} \end{cases} \quad (9)$$

Substituting for C_n in (7) we have

$$u(x,y) = \sum_{n=1,3,5,\ldots}^{\infty} \frac{4u_0}{n\pi\sinh n\pi}\sin\frac{n\pi x}{a}\sinh\frac{n\pi y}{a}$$

where $a = 100$ cm. (10)

Temperature at the centre = (50, 50) is

$$u(50,50) = \sum_{n=1,3,5,\ldots}^{\infty} \frac{4u_0}{n\pi\sinh n\pi}\sin\frac{n\pi}{2}\sinh\frac{n\pi}{2} \quad (11)$$

Example 10.72

Find the steady-state temperature in a rectangular plate $0 \le x \le a$, $0 \le y \le b$ when the sides $x = 0$, $x = a$ and $y = b$ are insulated while the edge $y = 0$ is kept at temperature $(c/a)\cos \pi x/a$.

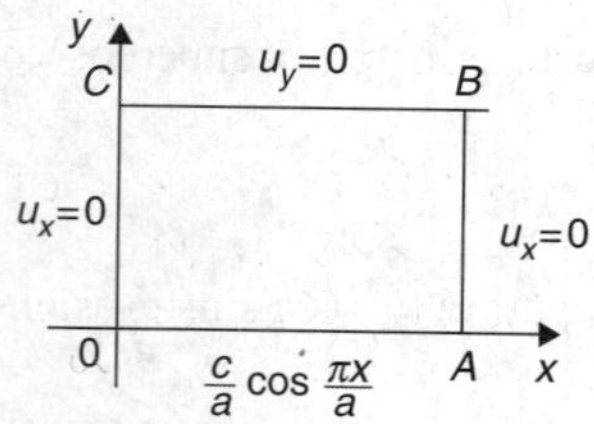

Figure 10.8 Steady state temperature in a rectangular plate

Solution We have to solve Laplace's equation

$$\frac{\partial^2 u}{\partial x^2} + \frac{\partial^2 u}{\partial y^2} = 0 \quad (1)$$

under the boundary conditions

$$u_x\Big|_{x=0} = u_x\Big|_{x=a} = u_y\Big|_{y=b} = 0 \quad (2), (3), (4)$$

and $\quad u(x,0) = f(x) = c\cos\dfrac{\pi x}{a}. \quad (5)$

By separating the variables we get the following two ordinary differential equations:

$$X'' + \lambda^2 X = 0 \quad \text{and} \quad Y'' - \lambda^2 Y = 0 \quad (6), (7)$$

these solutions are

$$X(x) = A\cos\lambda x + B\sin\lambda x;\; Y(y) = Ce^{\lambda y} + De^{-\lambda y} \quad (8), (9)$$

The above boundary conditions imply that

$$X'(0) = (-A\lambda\sin\lambda x + B\lambda\cos\lambda x)_{x=0} = 0 \Rightarrow B = 0$$

$$X'(a) = 0 = -A\lambda\sin\lambda a \Rightarrow \lambda = n\pi/a \quad (n = 0, 1, 2)$$

$$\therefore X_n(x) = A_n\cos\frac{n\pi x}{a} \quad n = 0, 1, 2, \ldots$$

$$Y'(a) = \lambda(Ce^{\lambda b} - De^{-\lambda b}) = 0 \Rightarrow C = \frac{De^{-\lambda b}}{e^{\lambda b}}.$$

Thus $\quad Y(y) = C_n\cosh\dfrac{n\pi(b-y)}{a}.$

Hence, the required solution is

$$u(x,y) = \sum A_n\cos\frac{n\pi x}{a}\cosh\frac{n\pi(b-y)}{a}$$

where

$$A_n = \frac{2}{a\cosh(n\pi b/a)}\int_0^a f(x)\cos\frac{n\pi x}{a}dx$$

$$= \frac{2c}{a\cosh(n\pi b/a)}\int_0^a \cos\frac{\pi x}{a}\cos\frac{n\pi x}{a}$$

$$A_1 = \left(2c/a\cosh\frac{\pi b}{a}\right)\frac{a}{2}$$

$$A_n = 0 \text{ for } n = 0 \text{ and } n \ge 2$$

$$\therefore\; u(x,y) = c\,\text{sech}\frac{\pi b}{a}\cos\frac{\pi x}{a}\cosh\frac{\pi(b-y)}{a}.$$

10.14 Two-dimensional Wave Equation

Vibrations of a rectangular membrane

To solve the problem of a vibrating membrane, We have to determine a solution $u(x, y, t)$ of the two–dimensional wave equation

$$\frac{\partial^2 u}{\partial t^2} = c^2\left(\frac{\partial^2 u}{\partial x^2} + \frac{\partial^2 u}{\partial y^2}\right) \quad (1)$$

$$\left(c^2 = \frac{T}{P}\right)$$

with the boundary conditions

$$u = 0, \tag{2}$$

on the boundary of the membrane for all $t \geq 0$ and two initial conditions

$$u(x, y, o) = f(x, y) \tag{3}$$

and

$$\left.\frac{\partial u}{\partial t}\right|_{t=0} = g(x, y) \tag{4}$$

Where $f(x, y)$ is the given initial displacement and $g(x, y)$ is the given initial velocity.

$u(x, y)$ gives the displacement of the point (x, y) of the membrane from rest ($u = 0$) at time $t = 0$. We note that the conditions (2) – (4) are similar to those for the vibrating string

As a first important case, we consider the rectangular membrane R shown in Fig.

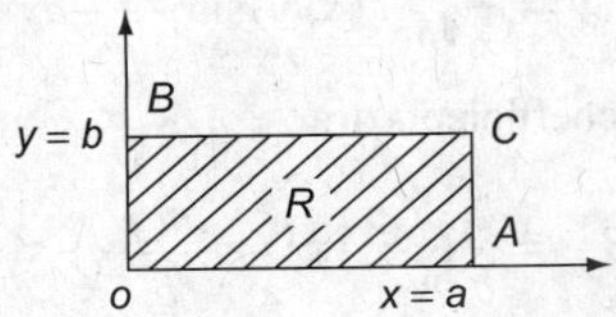

Figure 10.9 Rectangular membrane

By the method of separation of variables we first find solutions of (1) satisfying *b.c.* (2).

Let

$$u(x, y, t) = F(x, y)\, T(t) \tag{5}$$

By substituting this into equation (1) we have

$$F\ddot{T} = c^2(F_{xx}T + F_{yy}T)$$

Where subscripts denote partial derivatives and dots denote derivatives w. r. t. t.

We divide both sides by c^2FT and obtain

$$\frac{\ddot{T}}{c^2T} = \frac{1}{F}(F_{xx} + F_{yy})$$

The LHS depends only on t, while the RHS is independent of t. so, each side must be equal to a constant, $-k^2$ say.Therefore,

$$\frac{\ddot{T}}{c^2T} = \frac{1}{F}(F_{xx} + F_{yy}) = -k^2$$

This gives two equations. For the time function $T(t)$ We have the ordinary differential equation

$$\ddot{T} + \lambda^2 T = 0 \tag{6}$$

where $\lambda^2 = c^2k^2$ and for the amplitude function $F(x, y)$

we have the partial differential equation

$$F_{xx} + F_{yy} + k^2F = 0 \tag{7}$$

This is known as the two-dimensional Helmholtz equation,

To separate this equation we set

$$F(x, y) = \phi(x)\psi(y) \tag{8}$$

Substituting this into (7) we get

$$\phi''\psi + \phi\psi'' + k^2\phi\psi = 0$$

Where prime denotes differention w. r. t the argument

Hermann Von Helmholtz (1821–1894) German physicist known for his basic work in thermodynamics, fluid flow and acousties.

$$\frac{\phi''}{\phi} = -\frac{1}{\psi}\left(\psi'' + v^2\psi\right) = -p^2 \tag{9}$$

Thus we get two ordinary linear differential equations for ϕ and ψ

$$\phi'' + p^2\phi = 0, \psi'' + q^2\psi = 0, q^2 = K^2 - P^2 \tag{10(a), (b)}$$

The general solution of 10 (a), (b) are

$$\phi(x) = A\cos px + B\sin px, \psi(y) = C\cos qy + D\sin qy$$

where A, B, C and D are constants. Since $F(x, y) = \phi(x)\, \psi(y)$ must be zero on the boundary i.e. where $x = 0$, $x = a$, $y = 0$ and $y = b$ therefore we get

$$\phi(0) = 0, \phi(a) = 0, \psi(0) = 0, \psi(b) = 0$$

Now,

$\phi(0)=0 \Rightarrow A=0$ and $\phi(a)=0 \Rightarrow B\sin pa=0$

$$\Rightarrow pa = m\pi$$

$$\Rightarrow p=\frac{m\pi}{a}(m\in z^+)$$

Similarly

$$\psi(0)=0 \Rightarrow C=0 \text{ and } \psi(b)=0 \Rightarrow q=\frac{n\pi}{b}(n\in z^+)$$

We, therefore obtain the solutions

$$\phi_m(x)=\sin\frac{m\pi x}{a} \text{ and } \psi_n(y)=\sin\frac{n\pi y}{b}$$

$$m, n = 1, 2, 3,\ldots$$

$$\therefore F_{mn}(n, y)=\phi_m(x)\phi_m(y)=\sin\frac{m\pi x}{a}\sin\frac{n\pi y}{b} \quad (11)$$

$$m, n = 1, 2, 3,\ldots$$

are the solutions of (7) that are zero. on the boundary of the membrane.

Eigenfunctions and eigenvalues

Since $q^2 = k^2 - p^2$ and $\lambda = ck$ we have

$$\lambda = c\sqrt{p^2+q^2}$$

Hence to $p=\frac{m\pi}{a}$ and $q=\frac{n\pi}{b}$ there corresponds the value

$$\lambda=\lambda_{mn}=c\pi\sqrt{\frac{m^2}{a^2}+\frac{n^2}{b^2}} \quad (12)$$

$$m, n = 1, 2, 3\ldots$$

Now, the general solution of

$$T_{mn}(t)=B_{mn}\cos\lambda_{mn}t+B'_{mn}\sin\lambda_{mn}t$$

Finally the general solution of eq. is

$$u_{mn}(x, y, b)=(B_{mn}\cos\lambda_{mn}t+B'_{mn}\sin\lambda_{mn}t)$$

$$\times\sin\frac{m\pi x}{a}\sin\frac{n\pi y}{b} \quad (13)$$

These functions are called the *eigenfunctions* and the numbers. λ_{mn} are called the *eigenvalues* of the vibrating membrane. The frequency of u_{mn} is $\lambda_{mn}/2\pi$.

Solution of the Entire problem and Double Fourier series

By the principle of superposition of solutions the general solution of the problem may be written as

$$u(x, y, t)=\sum_{m=1}^{\infty}\sum_{n=1}^{\infty}u_{mn}(x, y, t)$$

$$=\sum_{m=1}^{\infty}\sum_{n=1}^{\infty}(B_{mn}\cos\lambda_{mn}t+B'_{mn}\sin\lambda_{mn}t)\sin\frac{m\pi x}{a}\sin\frac{n\pi y}{b} \quad (14)$$

Imposing condition (3) we obtain

$$u(x, y, 0)=\sum_{m=1}^{\infty}\sum_{n=1}^{\infty}B_{mn}\sin\frac{m\pi x}{a}\sin\frac{n\pi y}{b}=f(x, y) \quad (15)$$

This series is called a *double Fourier series.* The constants in (15) are determined by setting

$$K_m(y)=\sum_{n=1}^{\infty}B_{mn}\sin\frac{x\pi y}{b} \quad (16)$$

and writing (15) in the form

$$f(x, y)=\sum_{m=1}^{\infty}K_m(y)\sin\frac{m\pi x}{a} \quad (17)$$

For fixed y this is the fourier since series of $f(x, y)$, considered as a function of x. we see that the Coefficients of this expansion are

$$k_m(y)=\frac{2}{a}\int_0^a f(x, y)\sin\frac{m\pi x}{a}dx \quad (18)$$

So, the coefficients are

$$B_{mn}=\frac{2}{b}\int_0^b k_m(y)\sin\frac{n\pi y}{b}dy \quad (19)$$

From (18) and (19) we obtain the *generalized Euler formula*

$$B_{mn}=\frac{4}{ab}\int_0^b\int_0^a f(x, y)\sin\frac{m\pi x}{a}\sin\frac{n\pi y}{b}dxdy$$

$$m, x = 1, 2, 3,\ldots \quad (20)$$

for the Fourier coefficients of $f(x, y)$ on the double Fourier series (15).

The constants B_{mn} in (14) are now found in terms of $f(x, y)$. To determine the B'_{mn} we differentiate (14) termwise w.r.t. t and use the condition (4); we obtain

$$\left.\frac{\partial u}{\partial t}\right|_{t=0}=\sum_{m=1}^{\infty}\sum_{n=1}^{\infty}B'_{mn}\lambda_{mn}\sin\frac{m\pi x}{a}\sin\frac{n\pi y}{b}=g(x, y) \quad (21)$$

Assuming that $g(x, y)$ can be expanded in this double Fourier series we find the coefficients as

$$B'_{mn} = \frac{4}{ab\lambda_{mn}} \int_0^b \int_0^a g(x, y) \sin \frac{m\pi x}{a} \sin \frac{n\pi y}{b} dx dy \quad (22)$$

$$m, n = 1, 2, 3, \cdots$$

Now, for (14) to satisfy the initial conditions, the coefficients B_{mn} and B'_{mn} must be chosen according to (21) and (22).

The vibration of a membrane are given by the wave equation $\frac{\partial^2 u}{\partial t^2} = c^2 \nabla^2 u$ where $\nabla^2 = \frac{\partial^2 u}{\partial x^2} + \frac{\partial^2}{\partial y^2}$

Example 10.73

Find the deflection $u(x, y, t)$ of a square membrane if $a = b = 1$ and $c = 1$ if the initial velocity is zero and the initial deflection $f(x, y) = A \sin \pi x \sin 2\pi y$ (Coimbatore 1985)

Solution Putting $a = b = 1$ and $f(x, y) = A \sin \pi x \sin 2\pi y$ in eq. (20) we get

$$B_{mn} = 4A \int_0^1 \int_0^1 \sin \pi x \sin 2\pi y \sin m\pi x \sin n\pi y \, dy dx$$

$$= 4A \int_0^1 \sin \pi x \sin m\pi x \, dX \left(\int_0^1 \sin 2\pi y \sin n\pi y \, dy \right.$$

$$= 0 \quad \text{for } m \neq 1$$

$$= 4A\left(\tfrac{1}{2}\right) \int_0^1 \sin 2\pi y \sin n\pi y \, dy, \text{for } m = 1$$

$$\because \int_0^1 \sin \pi x \sin \pi x \, dX = \frac{1}{2}$$

$$\Rightarrow B_{mn} = 2A \int_0^1 \sin 2\pi y \sin n\pi y \, dy = 0 \text{ for } n \neq 2$$

$$= 2A\left(\frac{1}{2}\right) = A \text{ for } n = 2$$

$\therefore B_{12} = A$ Also, from (12)

$$\lambda_{mn} = c\pi \sqrt{\frac{m^2}{a^2} + \frac{n^2}{b^2}}$$

$$\Rightarrow \lambda_{12} = 1 \cdot \pi \sqrt{\frac{12}{12} + \frac{2^2}{1^2}} = \sqrt{5}\pi$$

$$c = 1, a = 1, h = 1$$

$$m = 1, n = 2$$

From eq. (13), the required solution is therefore,

$$u(x, y, t) = A \sin \pi x \sin 2\pi y \cos\left(\sqrt{5}\pi t\right)$$

EXERCISE 10.9

Solve the two-dimensional Laplace's equation. $(\partial^2 u/\partial x^2) + (\partial^2 u/\partial y^2) = 0$ in the region $0 < x < a$, $0 < y < b$ bounded by a metal plate with the following boundary conditions:

1. $u(0, y) = u(x, 0) = x(x, b) = 0$ and $u(a, y) = g(y)$ for $0 < y < b$

   Ans: $u(x, y) = \sum_{n=1}^{\infty} B_n \sinh \frac{n\pi x}{b} \sin \frac{n\pi y}{b}$

   where $B_n = \frac{2}{b} \text{ cosech} \frac{n\pi a}{b} \int_0^h g(y) \sin \frac{n\pi y}{b} dy$

2. Solve problem 1 completely by taking $g(y) = 100$.

   Ans: $u(x, y) = \sum_{n=1,3,5}^{\infty} \frac{400}{n\pi \sinh(n\pi a/b)} \sin h \frac{n\pi x}{b} \sin \frac{n\pi y}{b}$.

3. $u(0, y) = u(a, y) = u(x, b) = 0$ and $u(x, 0) = f(x) = x(a - x),\ 0 < x < a$

   Ans: $u(x, y) = \frac{8a^2}{\pi^2} \sum_{n=1,3,5}^{\infty} \frac{\text{cosech}(m\pi a/b)}{m^3} \sin \frac{m\pi x}{a}$

   $\sinh \frac{m\pi(b - y)}{a}$

4. A square plate has its faces and the edge $y = 0$ insulated. Its edges $x = 0$ and $x = \pi$ are kept at zero temperature and its fourth edge $y = \pi$ is kept at temperature $f(x)$. Find the steady-state temperature at any point of the plate. [JNTU Suppl. 2003]

   Ans: $u(x, y) = \frac{2}{\pi} \sum_{n=1}^{\infty} \frac{\sin nx \cosh ny}{\cosh n\pi} \int_0^{\pi} f(x) \sin nx \, dx$

5. Find the solution of problem 4 if $f(x) = 100$.

   Ans: $u(x, y) = \frac{400}{\pi} \sum_{n=1,3,5\ldots}^{\infty} \frac{\sin nx \cosh ny}{x \cosh n\pi}$.

6. Solve $(\partial^2 u/\partial x^2) + (\partial^2 u/\partial y^2) = 0$ for $0 < x < \pi, 0 < y < \pi$ given that

   $u(0, y) = u(\pi, y) = u(x, \pi) = 0$ and $u(x, 0) = \sin^2 x$.

   Ans: $u(x, y) = \frac{8}{\pi} \sum_{n=1,3,5\ldots}^{\infty} \frac{(e^{ny} - e^{2n\pi} - e^{-ny})}{(n^2 - 2^2)(e^{2n\pi} - 1)} \sin nx$.

7. A tightly stretched unit square membrane starts vibrating from rest and its initial displacement is $A \sin 2\pi x \sin \pi y$. Show that the deflection at any instant is $A \sin 2\pi x \sin \pi y \cdot (\sqrt{5}\pi ct)$. (Calicut 1987 M)

8. Find the vibration $u(x, y, t)$ of a rectangular membrane $(0 < x < a, 0 < y < b)$ whose boundary is fixed, given that it starts from rest and $u(x, y, 0) = kxy(a - x)(b - y)$ (Kurukshetra 1983)

Ans:

$$u(x, y, t) = \sum_{m=1}^{\infty}\sum_{n=1}^{\infty} \sin m\pi x \sin \frac{n\pi y}{2} (A_{mn} \cos \lambda_{mn} t)$$

where $A_{mn} = \dfrac{64a^2b^2h}{\pi^6 m^3 n^3}$ if m, n are odd

$$\left(\lambda_{mn} = c\pi\sqrt{m^2 + \frac{n^2}{4}}\right) = 0, \text{ otherwise}$$

9. Find the vibrations of a rectangular membrane of sides $a = 4$ ft and $b = 2$ ft. if the tension is 12.5 b/ft., the density is 2.5 slugs/ft^2, the initial velocity is zero and the initial displacement is $f(x, y) = \frac{1}{10}(4x - x^2)(2y - y^2)$ ft.

$$u(x, y, t) = 0.42605$$

Ans: $$\sum_{m,\,n\ \text{odd}}\sum \frac{1}{m^3n^3} \cos\left[\frac{5\pi}{4}\sqrt{m^2 + 4n^2}t\right] \sin\frac{m\pi x}{4} \sin\frac{n\pi y}{2}$$

$$= 0.42605\left[\cos\left(\frac{5\sqrt{5}\pi}{4}t\right)\sin\frac{\pi x}{4}\sin\frac{\pi y}{4} + \frac{1}{27}\cos\frac{5\sqrt{37}\pi t}{4}\sin\frac{\pi x}{4}\sin\frac{3\pi y}{2} + \frac{1}{27}\cos\left(\frac{5\sqrt{3}\pi t}{4}\right)\sin\frac{3\pi x}{4}\sin\frac{\pi y}{2} + \frac{1}{729}\cos\left(\frac{5\sqrt{3}\pi t}{4}\right)\sin\frac{3\pi x}{4}\sin\frac{3\pi y}{4} + \cdots\right]$$

Z-Transforms and Solution of Difference Equations

11

11.1 INTRODUCTION

Let A and B be the domain and range sets of specified functions. A transform is a mapping $T : A \to B$ from A into B. Recall that we have earlier studied Laplace and Fourier transforms. These are *integral transforms* by which we mean that these mappings are defined through integrals. Further, these transforms belong to a class of transforms dealing with functions of continuous variables. Now we consider a transformation called the Z-transformation which deals with functions of discrete variables. Indeed the Z-transform is the discrete analogue of the Laplace transform. Consequently, for every operational rule and application of Laplace transform we have an operational rule and application of Z-transform.

The Z-transforms find application in the areas of communication systems, control systems, analysis of signals and linear time-invariant (LTI) systems. Further, we have seen that Laplace transforms are applied to solve initial value problems involving linear differential equations. Now, we will see how Z-transforms are applied to solve difference equations.

11.2 Z-TRANSFORM: DEFINITION

Let $\langle u(n)\rangle$, or $\langle u(1), u(2), \ldots u(n) \ldots\rangle$ or $\langle u_n\rangle = \langle u_0, u_1, \ldots u_n, \ldots\rangle$ be a sequence of real numbers, which is a special function defined on the set of natural numbers $\mathbb{N} = \{0, 1, 2, \ldots, n, \ldots\}$

Suppose that

$$u_0 + u_1 \cdot \frac{1}{z} + u_2 \cdot \frac{1}{z^2} + \cdots + u_n \cdot \frac{1}{z^n} + \cdots$$

where z is a complex number, a series generated by the sequence $\langle u_n\rangle$. Let the series converge for all z in a domain D. Then these values define a complex-valued function $\overline{u}(z)$ on the domain D in

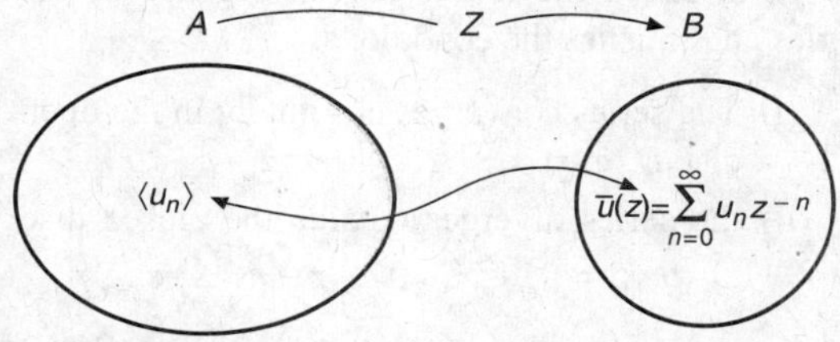

Figure 11.1 Z-transform of sequence $\langle u_n\rangle$ and its inverse transform

B, which is called the Z-transform of $\langle u(n)\rangle$. Then, the Z-transform of $\langle u(n)\rangle$ or $\langle u_n\rangle$ is

$$Z(\langle u_n\rangle) = \sum_{n=0}^{\infty} u(n)\cdot z^{-n} = \overline{u}(z) \qquad (11.1)$$

Note 1 The transform defined by (11.1) is called a one-sided (right-sided) or unilateral Z-transform.

Note 2 The Z-transform defined for all integers $n = 0, \pm 1, \pm 2, \pm 3, \ldots$ is called a two-sided or bilateral Z-transform.

Note 3 One has to show the sequence $\langle u_n\rangle$ as the argument of Z as in (11.1) and not the nth term u_n.

In the following we consider only one-sided or unilateral Z-transforms.

Inverse Z-transform

Let the infinite series on the RHS of (11.1) converge to $\overline{u}(z)$ in a domain D depending on the sequence $\langle u_n\rangle$. Then the mapping is invertible and the inverse Z-transform denoted by $Z^{-1}[\overline{u}(z)]$ is defined by

$$Z^{-1}[\overline{u}(z)] = \langle u(n)\rangle \quad \text{or} \quad \langle u_n\rangle \qquad (11.2)$$

Convergence of the series. An infinite series of the form

$$\sum_{n=0}^{\infty} a_n(z-a)^n = a_0 + a_1(z-a) + a_2(z-a)^2 + \cdots \quad (11.3)$$

where $a, a_0, a_1, a_2, \ldots$ are complex numbers is called a power series. With each series of the form (11.3) we can associate a real number r $(0 < r < \infty)$ defined by

$$\frac{1}{r} = \limsup |a_n|^{1/n}, \quad (0 \le r < \infty) \quad (11.4)$$

It is called the radius of convergence of the series and satisfies the conditions:

(i) The series converges absolutely in the open disc $D(a, r) = \{z \in \mathbb{C} \mid 0 < |z-a| < r\}$.

(ii) the series diverges outside the closed disc $\bar{D}(a, r) = \{z \in \mathbb{C} \mid 0 < |z-a| \le r\}$.

Consequently, the region of convergence (ROC) of one-sided Z-transform

$$\bar{u}(z) = \sum_{n=0}^{\infty} u_n \cdot z^{-n} \quad (11.5)$$

is $|z| > a > 0$, *i.e.* the exterior of a circle with centre at the origin and of radius 'a' ('a' is a positive real number).

Similarly, the ROC of a left-sided Z-transform

$$\bar{u}(z) = \sum_{n=-\infty}^{0} u_n \cdot z^{-n} \quad (11.6)$$

is the region $|z| < b$ and the ROC of a two-sided (or bilateral) Z-transform $\bar{u}(z) = \sum_{n=-\infty}^{0} u_n z^{-n}$ is the annular region $a < |z| < b$ where a and b are some positive real numbers.

11.2.1 Two Special Sequences

Unit Impulse Sequence

The unit impulse sequence $\langle \delta(n) \rangle$ or $\langle \delta_n \rangle$, $n = 0, \pm 1, \pm 2 \ldots$ is defined by (Fig. 11.2)

$$\delta(n) \text{ or } \delta_n = \begin{cases} 1 & \text{for } n = 0 \\ 0 & \text{for } n \ne 0 \end{cases} \quad (11.7)$$

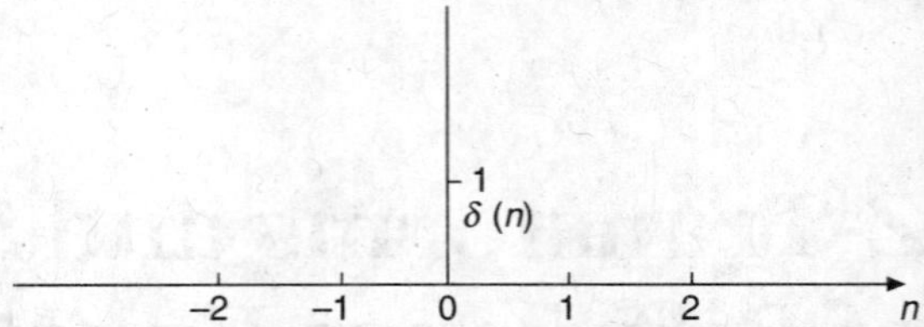

Figure 11.2 Graph of unit impulse sequence

Unit Step Sequence

The unit step sequence $\langle H(n) \rangle$ or $\langle H_n \rangle$, $n = 0, \pm 1, \pm 2, \ldots$ is defined by (Fig. 11.3)

$$H(n) \text{ or } H_n = \begin{cases} 1 & \text{for } n \ge 0 \\ 0 & \text{for } n < 0 \end{cases} \quad (11.8)$$

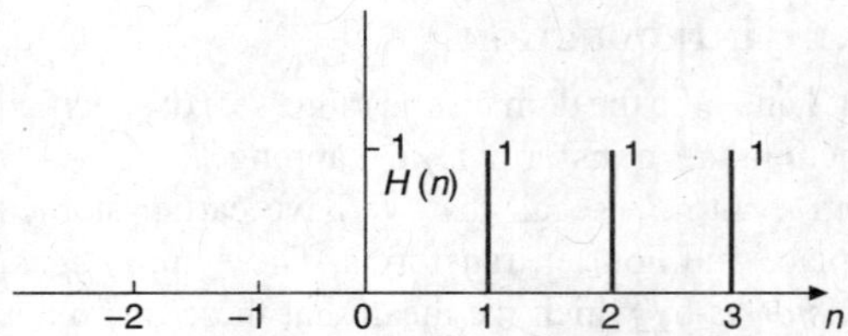

Figure 11.3 Graph of unit step sequence $\langle H_n \rangle$

The relation between the unit impulse sequence and the unit step sequence

We have

$$\delta(n-k) \text{ or } \delta_{n-k} = \begin{cases} 1 & \text{for } n = k \\ 0 & \text{for } n \ne k \end{cases} \quad (11.9)$$

and
$$H(n-k) \text{ or } H_{n-k} = \begin{cases} 1 & \text{for } n \ge k \\ 0 & \text{for } n < k \end{cases} \quad (11.10)$$

$$\therefore \quad H(n) \text{ or } H_n = \sum_{k=-\infty}^{\infty} \delta(k)$$

and
$$\delta(n) = H(n) - H(n-1)$$

Also,

$$f(n) = \sum_{k=-\infty}^{\infty} f(k)\delta(n-k) \quad (11.11)$$

11.2.2 Z-Transforms of Unit Step and Unit Impulse Sequences

Example 11.1

Find the Z-transform of discrete unit step sequence $\langle H_n \rangle$ where

$$H_n = \begin{cases} 1 & \text{for } n \ge 0 \\ 0 & \text{for } n < 0 \end{cases} \quad (11.12)$$

Solution By definition

$$Z(\langle u_n\rangle)=\sum_{n=0}^{\infty}u_n z^{-n}=\sum_{n=0}^{\infty}H_n z^{-n}=\sum_{n=0}^{\infty}1\cdot z^{-n}$$
$$=1+\frac{1}{z}+\frac{1}{z^2}+\cdots+\frac{1}{z^n}+\cdots$$

(an infinite geometric series with common ratio $=1/z$ which converges absolutely for $|z|>1$)

$$=\frac{1}{1-\frac{1}{z}}=\frac{z}{z-1} \qquad (11.13)$$

Example 11.2

Find the *Z*-transform of the unit impulse sequence $\langle\delta(n)\rangle$ where

$$\delta(n) \text{ or } \delta_n=\begin{cases}1 & \text{for } n=0\\ 0 & \text{for } n\neq 0\end{cases} \qquad (11.14)$$

Solution By definition

$$Z(\langle u_n\rangle)=\sum_{n=0}^{\infty}u_n z^{-n} \quad\Rightarrow\quad Z(\langle\delta_n\rangle)$$
$$=\sum_{n=0}^{\infty}\delta_n\cdot z^{-n}=1+0+0+\cdots=1 \qquad (11.15)$$

Example 11.3

Find the *Z*-transform of $\langle u_n|_{n=0}^{4}\rangle$ where $u_0=-3$, $u_1=2$, $u_2=5$, $u_3=0$ and $u_4=12$.

Solution We have

$$Z(\langle u_n\rangle)=\sum_{n=0}^{\infty}u_n z^{-n}=u_0\cdot 1+u_1\cdot z^{-1}+u_2\cdot z^{-2}$$
$$+u_3\cdot z^{-3}+u_4\cdot z^{-4}$$

Since n takes values $n=0,1,2,3,4$ only

$$=(-3)1+2\cdot\frac{1}{z}+5\cdot\frac{1}{z^2}+0\cdot\frac{1}{z^3}+12\cdot\frac{1}{z^4}$$
$$=-3+2\cdot\frac{1}{z}+5\cdot\frac{1}{z^2}+12\cdot\frac{1}{z^4}$$

Example 11.4

Find the *Z*-transform of $u_n=1/2^n, -3\geq n\geq 3$.

Solution We have

$$Z(\langle u_n\rangle)=\sum_{n=-\infty}^{\infty}u_n\cdot z^{-n}$$
$$=u_{-3}\cdot z^{-(-3)}+u_{-2}\cdot z^{-(-2)}$$
$$+u_{-1}\cdot z^{-(-1)}+u_0\cdot z^{0}+u_1\cdot z^{-1}$$
$$+u_2 z^{-2}+u_3\cdot z^{-3},$$

since n takes values $-3,-2,-1,0,1,2,3$ only

$$=8z^3+4z^2+2z+1+\frac{1}{2}\cdot\frac{1}{z}$$
$$+\frac{1}{4}\cdot\frac{1}{z^2}+\frac{1}{8}\cdot\frac{1}{z^3}$$

Example 11.5

Find the *Z*-transform of $u_n=(2/3)^n, n\in\mathbb{N}$.

Solution We have

$$Z(\langle u_n\rangle)=\sum_{n=0}^{\infty}u_n\cdot z^{-n}$$
$$=u_0\cdot 1+u_1\cdot\frac{1}{z}+u_2\cdot\frac{1}{z^2}$$
$$+u_3\cdot\frac{1}{z^3}+\cdots+u_n\cdot\frac{1}{z^n}+\cdots$$
$$=1\cdot 1+\frac{2}{3}\cdot\frac{1}{z}+\left(\frac{2}{3}\right)^2\frac{1}{z^2}$$
$$+\left(\frac{2}{3}\right)^3\frac{1}{z^3}+\cdots+\left(\frac{2}{3}\right)^n\frac{1}{z^n}+\cdots$$
$$=1+\left(\frac{2}{3z}\right)+\left(\frac{2}{3z}\right)^2+\left(\frac{2}{3z}\right)^3$$
$$+\cdots+\left(\frac{2}{3z}\right)^n+\cdots$$

[which is a GS with c.r. $=2/3z$ and converges absolutely for $|2/3z|<1$, *i.e.* $|z|>2/3$]

$$=\frac{1}{1-\left(\frac{2}{3z}\right)}=\frac{3z}{3z-2}$$

11.3 Z-Transforms of Some Standard Functions (Special Sequences)

11.3.1. Unit Constant Sequence

$$u_n = 1 \quad \forall\, n \in \mathbb{N};$$

$$\langle u_n\rangle = \langle 1, 1, 1, 1, 1, \cdots\rangle \tag{11.16}$$

$$Z(\langle u_n\rangle) = Z(\langle 1\rangle) = \sum_{n=0}^{\infty} u_n \cdot z^{-n} = \sum_{n=0}^{\infty} 1 \cdot \frac{1}{z^n}$$

$$= 1 + \frac{1}{z} + \frac{1}{z^2} + \cdots + \frac{1}{z^n} + \cdots$$

$$\begin{bmatrix}\text{infinite GS which converges}\\ \text{absolutely for } |z| > 1\end{bmatrix}$$

$$= \frac{1}{1-\frac{1}{z}} = \frac{z}{z-1} \tag{11.17}$$

11.3.2. Alternating Unit Constant Sequence

$$u_n = (-1)^n \quad \forall\, n \in \mathbb{N};$$

$$\langle u_n\rangle = \langle(-1)^n\rangle = \langle 1, -1, 1, -1, \ldots\rangle \tag{11.18}$$

$$Z(\langle u_n\rangle) = Z(\langle(-1)^n\rangle)$$

$$= \sum_{n=0}^{\infty} u_n \cdot z^{-n} = \sum_{n=0}^{\infty} (-1)^n \frac{1}{z^n}$$

$$= 1 - \frac{1}{z} + \frac{1}{z^2} - \frac{1}{z^3} + \cdots + (-1)^n \frac{1}{z^n} + \cdots,$$

$$\begin{bmatrix}\text{infinite GS which converges}\\ \text{absolutely for } |z| > 1\end{bmatrix}$$

$$= \frac{1}{1-(-\frac{1}{z})} = \frac{z}{z+1} \tag{11.19}$$

11.3.3. Geometric Sequence

$$u_n = a^n \quad \forall\, n \in \mathbb{N};\ a \in \mathbb{R} \text{ or } \mathbb{C};$$

$$\langle u_n\rangle = \langle a^n\rangle = \langle 1, a, a^2, \ldots a_n, \ldots\rangle \tag{11.20}$$

$$Z(\langle u_n\rangle) = Z(\langle a^n\rangle) = \sum_{n=0}^{\infty} u_n \cdot z^{-n} = \sum_{n=0}^{\infty} a^n \cdot \frac{1}{z^n}$$

$$= 1 + \frac{a}{z} + \left(\frac{a}{z}\right)^2 + \cdots + \left(\frac{a}{z}\right)^n + \cdots$$

$$\begin{bmatrix}\text{infinite GS which converges}\\ \text{absolutely for } \left|\frac{a}{z}\right| < 1 (i.e.) |z| > |a|\end{bmatrix}$$

$$= \frac{1}{1-\frac{a}{z}} = \frac{z}{z-a} \tag{11.21}$$

Corollary Putting $a = 1$ and -1 in turn we get results 1 and 2, respsectively.

11.3.4. Natural Number Sequence

$$u_n = n \quad \forall\, n \in \mathbb{N};$$

$$\langle u_n\rangle = \langle n\rangle = \langle 0, 1, 2, \ldots\rangle \tag{11.22}$$

$$Z(\langle u_n\rangle) = Z(\langle n\rangle) = \sum_{n=0}^{\infty} u_n \cdot z^{-n} = \sum_{n=0}^{\infty} n \cdot \frac{1}{z^n}$$

$$= 0 + 1\cdot\frac{1}{z} + 2\cdot\frac{1}{z^2} + \cdots + n\cdot\frac{1}{z^n} + \cdots$$

$$= \frac{1}{z}\left(1 + \frac{2}{z} + \frac{3}{z^2} + \cdots + \frac{n}{z^{n-1}}\right)$$

$$\left|\begin{array}{l}(1-x)^{-2} = 1 + 2x + 3x^2\\ \qquad + \cdots + nx^{n-1} + \cdots\\ \qquad\text{(Binomial Theorem)}\end{array}\right.$$

$$= \frac{1}{z}\left(1-\frac{1}{z}\right)^{-2} = \frac{1}{z}\cdot\frac{z^2}{(z-1)^2} = \frac{z}{(z-1)^2},$$

converges absolutely for $|z| > 1$ (11.23)

11.3.5. Reciprocal Factorial Sequence

$$u_n = \frac{1}{n!} \quad n \in \mathbb{N};$$

$$\langle u_n\rangle = \left\langle\frac{1}{n!}\right\rangle = \left\langle 1, \frac{1}{1!}, \frac{1}{2!}, \ldots, \frac{1}{n!}, \ldots\right\rangle \tag{11.24}$$

$$Z(\langle u_n\rangle) = z\left(\left\langle\frac{1}{n!}\right\rangle\right) = \sum_{n=0}^{\infty} u_n z^{-n}$$

$$= \sum_{n=0}^{\infty} \frac{1}{n!}\cdot\frac{1}{z^n}$$

$$= 1 + \frac{1}{1!}\cdot\frac{1}{z} + \frac{1}{2!}\cdot\frac{1}{z^2} + \cdots + \frac{1}{n!}\cdot\frac{1}{z^n} + \cdots$$

$$= e^{(1/z)} \quad \left(\because\ e^t = 1 + \frac{t}{1!} + \frac{t^2}{2!} + \cdots + \frac{t^n}{n!} + \cdots\right) \tag{11.25}$$

11.3.6. Power-Cum-Reciprocal Factorial Sequence

$$u_n = \frac{a^n}{n!} \quad n \in \mathbb{N};$$

$$\langle u_n \rangle = \left\langle \frac{a^n}{n!} \right\rangle = \left\langle 1, \frac{a}{1!}, \frac{a^2}{2!}, \ldots, \frac{a^n}{n!}, \ldots \right\rangle \quad (11.26)$$

$$Z(\langle u_n \rangle) = z\left(\left\langle \frac{a^n}{n!} \right\rangle\right) = \sum_{n=0}^{\infty} u_n z^{-n}$$

$$= \sum_{n=0}^{\infty} \frac{a^n}{n!}\frac{1}{z^n} = \sum_{n=0}^{\infty} \frac{1}{n!}\left(\frac{a}{z}\right)^n$$

$$= 1 + \frac{1}{1!}\left(\frac{a}{z}\right) + \frac{1}{2!}\left(\frac{a}{z}\right)^2 + \frac{1}{3!}\left(\frac{a}{z}\right)^3$$

$$+ \cdots + \frac{1}{n!}\left(\frac{a}{z}\right)^n + \cdots = e^{(a/z)} \quad (11.27)$$

11.3.7. Binomial Coefficient Sequence

$$u_n = {}^nC_k \; 0 \le k \le n$$

$$\langle u_n \rangle = \langle {}^nC_k \rangle = \langle {}^nC_0, {}^nC_1, {}^nC_2, \ldots, {}^nC_n \rangle \quad (11.28)$$

$$Z(\langle u_k \rangle) = Z(\langle {}^nC_k \rangle) = \sum_{k=0}^{n} u_k z^{-k} = \sum_{k=0}^{n} {}^nC_k z^{-k}$$

$$= {}^nC_0 \cdot 1 + {}^nC_1 \cdot \frac{1}{z} + {}^nC_2 \cdot \frac{1}{z^2}$$

$$+ \cdots + {}^nC_n \cdot \frac{1}{z^n} = \left(1 + \frac{1}{z}\right)^n$$

(Binomial Theorem) (11.29)

11.3.8. Power-Cum-Reciprocal Factorial Sequence with a Multiple

$$u_n = \frac{a^n}{n!}e^{-a}; \quad \langle u_n \rangle = \left\langle \frac{a^n}{n!}e^{-a} \right\rangle$$

$$= \left\langle e^{-a}, \frac{a}{1!}e^{-a}, \frac{a}{2!}e^{-a}, \ldots, \frac{a^n}{n!}e^{-a} \ldots \right\rangle \quad (11.30)$$

$$Z(\langle u_n \rangle) = Z\left(\left\langle \frac{a^n}{n!}e^{-e} \right\rangle\right)$$

$$= \sum_{n=0}^{\infty} u_n z^{-n} = \sum_{n=0}^{\infty} \frac{a^n}{n!}e^{-a}$$

$$= 1 \cdot e^{-a} + \frac{a}{1!}e^{-a} + \frac{a^2}{2!}e^{-a} + \cdots$$

$$+ \frac{a^n}{n!}e^{-a} + \cdots = e^{-a}e^{(a/z)} = e^{a(1/z-1)} \quad (11.31)$$

11.4 Recurrence Formula for the Sequence of a Power of Natural Numbers

Theorem 11.1 Let $u_n = n^p; n, p \in \mathbb{N}, \langle u_n \rangle = \langle n^p \rangle = \langle 0.\, 1^p, 2^p, 3^p \ldots n^p, \ldots \rangle$

If $Z(\langle u_n \rangle) = Z(\langle n^p \rangle)$

$$= \sum_{n=0}^{\infty} u_n z^{-n} = \sum_{n=0}^{\infty} n^p \cdot z^{-n} \quad (11.32)$$

then $Z(\langle n^p \rangle) = z\dfrac{d}{dz}[Z(\langle n^{p-1} \rangle)]$ or

$$-z \cdot Z'(\langle n^{p-1} \rangle) \quad (11.33)$$

Proof Differentiating $Z(\langle n^p \rangle) = \sum_{n=0}^{\infty} n^p z^{-n}$ with respect to 'z' we have

$$Z'(\langle n^p \rangle) = \frac{d}{dz}[Z(\langle n^p \rangle)] = \frac{d}{dz}\left(\sum_{n=0}^{\infty} n^p \cdot z^{-n}\right)$$

$$= \sum_{n=0}^{\infty} n^p \cdot (-n) z^{-n-1} = -z^{-1}\sum_{n=0}^{\infty} n^{p+1} z^{-n} \quad (11.34)$$

Replacing p by $(p-1)$ in (11.34) we get

$$Z'(\langle n^{p-1} \rangle) = \frac{d}{dz}[Z(\langle n^{p-1} \rangle)] = -z^{-1} \cdot \sum_{n=0}^{\infty} n^p \cdot z^{-n}$$

$$= -z^{-1} \cdot Z(\langle n^p \rangle) \quad \text{by (11.32)}$$

$$\Rightarrow \quad Z(\langle n^p \rangle) = -z \cdot Z'(\langle n^{p-1} \rangle)$$

$$= -z\frac{d}{dz}(Z\langle n^{p-1} \rangle) \cdots \quad (11.35)$$

Corollary 1 Putting $p = 1$ in (11.33) we get

$$Z(\langle n \rangle) = -z \cdot \frac{d}{dz}(Z\langle 1 \rangle) = -z \cdot \frac{d}{dz}\left(\frac{z}{z-1}\right)$$

$$= -z \cdot \frac{(z-1)\cdot 1 - z \cdot 1}{(z-1)^2} = \frac{z}{(z-1)^2}$$

Corollary 2 Putting $p = 2$ in (11.33) we get

$$Z(\langle n^2\rangle) = -z\cdot\frac{d}{dz}(Z\langle n\rangle) = -z\cdot\frac{d}{dz}\left[\frac{z}{(z-1)^2}\right]$$

$$= (-z)\cdot\frac{(z-1)^2\cdot 1 - z\cdot 2(z-1)}{(z-1)^4} = \frac{z^2+z}{(z-1)^3} \tag{11.36}$$

For $p = 3$ and 4 we get

Corollary 3 $$Z(\langle n^3\rangle) = \frac{z^3+4z^2+z}{(z-1)^4}; \tag{11.37}$$

and

Corollary 4 $$Z(\langle n^4\rangle) = \frac{z^4+11z^3+11z^2+z}{(z-1)^5} \tag{11.38}$$

11.5 Properties of Z-Transforms

11.5.1. Linearity

Theorem 11.2

(a) *Z* distributes over sum

$$Z(\langle u_n + v_n\rangle) = Z(\langle u_n\rangle + \langle v_n\rangle)$$
$$= Z(\langle u_n\rangle) + Z(\langle v_n\rangle) \tag{11.39}$$

(b) Scalar multiplication

$$Z(k\langle u_n\rangle) = k \quad Z(\langle u_n\rangle) \tag{11.40}$$

(Transform of sum = sum of the transforms)

Proof By definition,

(a) $$Z(\langle u_n\rangle + \langle v_n\rangle) = Z(\langle u_n + v_n\rangle)$$
$$= \sum_{n=0}^{\infty}(u_n+v_n)z^{-n}$$
$$= \sum_{n=0}^{\infty}u_n z^{-n} + \sum_{n=0}^{\infty}v_n z^{-n}$$
$$= Z(\langle u_n\rangle) + Z(\langle v_n\rangle)$$

Scalar multiplication

(b) $Z(k\langle u_n\rangle) = k \quad Z(\langle u_n\rangle)$

(Transform of scalar times a sequence = Scalar times the transform of the sequence)

Proof By definition,

$$Z(k\langle u_n\rangle) = Z(\langle ku_n\rangle) = \sum_{n=0}^{\infty}(ku_n)z^{-n}$$
$$= k\sum_{n=0}^{\infty}u_n z^{-n} = k \quad Z(\langle u_n\rangle)$$

Combining the above two results, the *linearity property* can be stated as

$$Z(k\langle u_n\rangle + l\langle v_n\rangle) = Z(\langle ku_n + lv_n\rangle)$$
$$= kZ(\langle u_n\rangle) + lZ(\langle v_n\rangle)$$

11.5.2. Change of Scale or Damping Rule

Theorem 11.3 If $Z(\langle u_n\rangle) = \bar{u}(z)$ then

$$Z(\langle a^{-n}u_n\rangle) = \bar{u}(az) \tag{11.41}$$

Proof By definition,

$$Z(\langle a^{-n}u_n\rangle) = \sum_{n=0}^{\infty}a^{-n}u_n z^{-n} = \sum_{n=0}^{\infty}u_n(az)^{-n}$$

Corollary $$Z(\langle a^n u_n\rangle) = \bar{u}\left(\frac{z}{a}\right) \tag{11.42}$$

Results Obtained by the Application of Damping Rule

(a) $$Z(\langle na^n\rangle) = \frac{az}{(z-a)^2} \tag{11.43}$$

In $Z(\langle n\rangle) = \dfrac{z}{(z-1)^2}$, replacing z by z/a

we get $$Z(\langle na^n\rangle) = \frac{(z/a)}{[(z/a)-1]^2} = \frac{az}{(z-a)^2}$$

(b) $$Z(\langle n^2a^n\rangle) = \frac{az^2+a^2z}{(z-a)^3} \tag{11.44}$$

Replacing z by $\dfrac{z}{a}$ in $Z(\langle n^2\rangle) = \dfrac{z^2+z}{(z-1)^3}$

we get $$Z(\langle n^2a^n\rangle) = \frac{(z/a)^2+(z/a)}{[(z/a)-1]^3}$$
$$= \frac{a(z^2+az)}{(z-a)^3}$$

Note 1 In the following (11.48) and (11.49) involving cos and sin functions z is a real variable.

(c) (i) $Z(\langle \cos nt \rangle) = \dfrac{z(z-\cos t)}{z^2 - 2z\cos t + 1}$ [JNTU 2001, 2002, S] (11.45)

(ii) $Z(\langle \sin nt \rangle) = \dfrac{z \sin t}{z^2 - 2z\cos t + 1}$ (11.46)

We have $Z(\langle 1 \rangle) = \dfrac{z}{z-1}$ (11.13)

By the damping rule

$$Z(\langle a^n \cdot 1 \rangle) = \frac{(z/a)}{(z/a)-1} = \frac{z}{z-a} \quad (11.47)$$

Now $Z(\langle e^{-\text{int}} \rangle) = Z(\langle (e^{-it})^n \cdot 1 \rangle)$

$$= \frac{z}{z-e^{-it}} \cdot \frac{z-e^{it}}{z-e^{it}}$$

$$= \frac{z[z-(\cos t + i \sin t)]}{z^2 - 2z\cos t + 1} \quad (11.48)$$

$\because$ $e^{it} = \cos t + i \sin t$ (Euler's Theorem) and $\dfrac{e^{it}+e^{it}}{2} = \cos t$.

By Linearity Property

$Z(\langle e^{-int} \rangle) = Z(\langle \cos nt - i \sin nt \rangle)$;

$e^{-int} = \cos nt - i \sin nt$ (De Moivre's Theorem)

$= Z(\langle \cos nt \rangle) - iZ(\langle \sin nt \rangle)$ (11.49)

On equating the real and imaginary parts we obtain from (11.9) and (11.10)

(i) $Z(\langle \cos nt \rangle) = \dfrac{z(z-\cos t)}{z^2 - 2z\cos t + 1}$;

(ii) $Z(\langle \sin nt \rangle) = \dfrac{z \sin t}{z^2 - 2z\cos t + 1}$

Corollary Replacing t by 'it' we get

$$Z(\langle \cosh nt \rangle) = \frac{z(z-\cosh t)}{z^2 - 2z\cosh t + 1};$$

$$Z(\langle \sinh nt \rangle) = \frac{z \sinh t}{z^2 - 2z\cosh t + 1}$$

(d) By the application of the damping rule we obtain, on replacing z by z/a

(i) $Z(\langle a^n \cos nt \rangle) = \dfrac{z(z-a\cos t)}{z^2 - 2az\cos t + a^2}$;

(ii) $Z(\langle a^n \sin nt \rangle) = \dfrac{az \sin t}{z^2 - 2az\cos t + a^2}$

(e) By the application of the damping rule to $Z\left(\left\langle \dfrac{1}{n!} \right\rangle\right) = e^{1/z}$ we obtain $Z\left(\left\langle \dfrac{a^n}{n!} \right\rangle\right) = e^{a/z}$.

11.5.3. Shifting Property

Theorem 11.4

(a) First Shifting Theorem (Shifting u_n to the right)

If $Z(\langle u_n \rangle) = \bar{u}(z)$ then $Z(\langle u_{n-k} \rangle) = z^{-k}\bar{u}(z)$ for $k > 0$.

Proof By definition,

$$Z(\langle u_{n-k} \rangle) = \sum_{n=0}^{\infty} u_{n-k} z^{-n}$$

$$= u_0 z^{-k} + u_1 z^{-(k+1)} + u_2 z^{-(k+2)} + \cdots$$

($\because$ $u_n = 0$ for $n < 0$, i.e. u_{-k}, u_{1-k}, u_{2-k}, u_{-1} are zero)

$$= z^{-k} \sum_{n=0}^{\infty} u_n z^{-n} = z^{-k}\bar{u}(z)$$

Theorem 11.5

(b) Second Shifting Theorem (Shifting u_n to the left)

If $Z(\langle u_n \rangle) = \bar{u}(z)$ then $Z(\langle u_{n+k} \rangle) = z^k\left[\bar{u}(z) - u_0 - \frac{u_1}{z} - \cdots - \frac{u_{k-1}}{z^{k-1}}\right]$

Proof By definition,

$$Z(\langle u_{n+k} \rangle) = \sum_{n=0}^{\infty} u_{n+k} z^{-n} = z^k \sum_{n=0}^{\infty} u_{n+k} z^{-(n+k)}$$

$$= z^k \left[u_k z^{-k} + u_{1+k} z^{-(1+k)} + u_{2+k} z^{-(2+k)} + \cdots \right]$$

$$\begin{bmatrix}\text{adding and subtracting} \\ u_0 + u_1 z^{-1} + u_2 z^{-2} + \cdots + u_{k-1} z^{-(k-1)}\end{bmatrix}$$

$$= z^k \left[\sum_{n=0}^{\infty} u_n z^{-n} - \sum_{n=0}^{k-1} u_n z^{-n} \right]$$

$$= z^k \left[\bar{u}(z) - u_0 - u_1 z^{-1} - u_2 z^{-2} - \cdots - u_{k-1} z^{-(k+1)} \right]$$

Particular results: For $k = 1, 2, 3$ we get

$$Z(\langle u_{n+1} \rangle) = z[\bar{u}(z) - u_0]$$

$$Z(\langle u_{n+2} \rangle) = z^2[\bar{u}(z) - u_0 - u_1 z^{-1}] \tag{11.50}$$

$$Z(\langle u_{n+2} \rangle) = z^3[\bar{u}(z) - u_0 - u_1 z^{-1} - u_2 z^{-2}]$$

11.5.4. Multiplication by *n*

Theorem 11.6 If $Z(\langle u_n \rangle) = \bar{u}(z)$ then

$$Z(\langle nu_n \rangle) = -z \frac{d\,\bar{u}(z)}{dz} \tag{11.51}$$

Proof By definition,

$$Z(\langle nu_n \rangle) = \sum_{n=0}^{\infty} n \cdot u_n z^{-n} = (-z) \sum_{n=0}^{\infty} u_n (-n) z^{-n-1},$$

multiplying and dividing by z

$$= (-z) \sum_{n=0}^{\infty} u_n \left(\frac{d}{dz} \right) (z^{-n})$$

$$\left[\because \frac{d}{dz}(z^{-n}) = -nz^{-n-1} \right]$$

$$= (-z) \sum_{n=0}^{\infty} \left(\frac{d}{dz} \right) (u_n z^{-n})$$

$$[\because u_n \text{ does not contain } z]$$

$$= (-z) \frac{d}{dz} \left(\sum_{n=0}^{\infty} u_n z^{-n} \right) = (-z) \frac{d}{dz} \bar{u}(z) \tag{11.52}$$

assuming the uniform convergence of the series, and that term-wise differentiation is valid.

By the principle of mathematical induction we can prove the general result:

$$Z(\langle n^p u_n \rangle) = (-z)^p \frac{d^p \bar{u}(z)}{dz^p}, \quad p = 0, 1, 2, 3, \cdots \tag{11.53}$$

11.5.5. Division by *n*

Theorem 11.7 If $Z(\langle u_n \rangle) = \bar{u}(z)$ then

$$Z\left(\left\langle \frac{u_n}{n} \right\rangle \right) = -\int_0^z t^{-1} \bar{u}(t)\, dt \tag{11.54}$$

Proof By definition,

$$Z\left(\left\langle \frac{u_n}{n} \right\rangle \right) = \sum \frac{u_n}{n} z^{-n} = -\sum u_n \left(\frac{z^{-n}}{-n} \right),$$

inserting -1 both in the numerator and denominator

$$= -\sum u_n \int_0^z t^{-n-1} dt \quad \text{since} \quad \int_0^z t^{-n-1} dt = \frac{t^{-n}}{-n} \Big|_0^z = \frac{z^{-n}}{-n} \quad \text{for } n \neq 0$$

$$= -\int_0^z \sum u_n \cdot t^{-n-1} dt, \text{ interchanging } \sum \text{ and } \int$$

$$= -\int_0^z t^{-1} \left(\sum u_n t^{-n} \right) dt = -\int_0^z t^{-1} \bar{u}(t)\, dt \tag{11.55}$$

Now, we deal with two theorems—the initial value theorem (IVT) and the final value theorem (FVT)—which determine the values of u_n for $n = 0$ and for $n \to \infty$, respectively, without the complete knowledge of u_n.

11.5.6. Initial Value Theorem

Theorem 11.8 If $Z(\langle u_n \rangle) = \bar{u}(z)$ then

$$u_0 = \lim_{z \to \infty} \bar{u}(z) \tag{11.56}$$

Proof By definition,

$$\bar{u}(z) = Z(\langle u_n \rangle)$$

$$= u_0 + u_1 \cdot \frac{1}{z} + u_2 \cdot \frac{1}{z^2} + \cdots + u_n \cdot \frac{1}{z^n} + \cdots$$

Taking the limit as $z \to \infty$ we get

(a) $\lim_{z \to \infty} \bar{u}(z) = u_0$

(b) $\lim_{z \to \infty} z[\bar{u}(z) - u_0]$

$$= \lim_{z \to \infty} \left[u_1 + u_2 \cdot \frac{1}{z} + u_3 \cdot \frac{1}{z^2} + \cdots \right] = u_1$$

(c) $\lim_{z\to\infty} z^2\left[\bar{u}(z)-u_0-\frac{u_1}{z}\right]$

$$= \lim_{z\to\infty}\left[u_2+\frac{u_3}{z}+\frac{u_4}{z^2}+\cdots\right]=u_2 \quad (11.57)$$

11.5.7. Final Value Theorem

Theorem 11.9 If $Z(\langle u_n\rangle)=\bar{u}(z)$ then

$$\lim_{n\to\infty} u_n = \lim_{z\to 1}(z-1)\bar{u}(z)$$

[JNTU 2002 *S*] (11.58)

Proof By linearity property

$$Z(\langle u_{n+1}-u_n\rangle)=Z(\langle u_{n+1}\rangle)-Z(\langle u_n\rangle)$$
$$= Z[\bar{u}(z)-u_0]-\bar{u}(z),$$

by shifting property

$$=(z-1)\bar{u}(z)-u_0 z$$

$$\Rightarrow\ (z-1)\bar{u}(z)-u_0 z=(\langle u_{n+1}-u_n\rangle)$$
$$=\sum_{n=0}^{\infty}(u_{n+1}-u_n)z^{-n},$$

by definition

Taking the limits as $z\to 1$we have

$$\lim_{z\to 1}[(z-1)\bar{u}(z)]-u_0 1=\sum_{n=0}^{\infty}(u_{n+1}-u_n)\cdot 1$$
$$=\lim_{n\to\infty}\sum_{k=0}^{n}(u_{k+1}-u_k)$$
$$=\lim_{n\to\infty}[(u_1-u_0)+(u_2-u_1)+\cdots+(u_{n+1}-u_n)]$$
$$=\lim_{n\to\infty}(u_{n+1}-u_0)$$
$$=\lim_{n\to\infty}u_{n+1}-u_0$$

Hence

$$\lim_{n\to\infty}u_n=\lim_{z\to 1}[(z-1)\bar{u}(z)] \quad (11.59)$$

11.5.8. Convolution Theorem

Theorem 11.10

$Z^{-1}(\bar{u}(z))=u_n$ and $Z^{-1}(\bar{v}(z))=v_n$ then

$Z^{-1}(\bar{u}(z).\bar{v}(z))=u_n * v_n =$ convolution of u_n and v_n

$$=\left\langle\sum_{m=0}^{n}u_m v_{n-m}\right\rangle \quad (11.60)$$

Proof By definition,

$$\bar{u}(z)\cdot\bar{v}(z)=\left(\sum_{n=0}^{\infty}u_n z^{-n}\right)\left(\sum_{n=0}^{\infty}v_n z^{-n}\right)$$
$$=u_0v_0+(u_0v_1+u_1v_0)z^{-1}+(u_0v_2+u_1v_1+u_2v_0)z^{-2}+\cdots+(u_0v_n+u_1v_{n-1}+u_2v_{n-2}+\cdots+u_mv_{n-m}+\cdots+u_nv_0)z^{-n}+\cdots$$
$$=\sum_{n=0}^{\infty}(u_0v_n+u_1v_{n-1}+u_2v_{n-2}+\cdots+u_mv_{n-m}+\cdots+u_nv_0)z^{-n}$$
$$=Z(\langle u_0v_n+u_1v_{n-1}+u_2v_{n-2}+\cdots+u_mv_{n-m}+\cdots+u_nv_0\rangle),\ \text{by definition}$$
$$=Z\left(\left\langle\sum_{m=0}^{m}u_mv_{n-m}\right\rangle\right)$$

Taking inverse *Z*-transforms of both sides

$$Z^{-1}(\bar{u}(z)\cdot\bar{v}(z))=\left\langle\sum_{m=0}^{n}u_mv_{n-m}\right\rangle \quad (11.60)$$

Example 11.6

Evaluate $Z\left(\left\langle\frac{1}{(n+1)!}\right\rangle\right)$. [JNTU 2001]

Solution By definition of *Z*-transforms

$$Z(\langle u_n\rangle)=\sum_{n=0}^{\infty}u_nz^{-n}$$

Here $\langle u_n\rangle=\left\langle\frac{1}{(n+1)!}\right\rangle$

$$=\left\langle\frac{1}{1!},\frac{1}{2!},\cdots\frac{1}{(n+1)!},\cdots\right\rangle \quad (1)$$

$$\therefore\ Z\left(\left\langle\frac{1}{(n+1)!}\right\rangle\right)=\sum_{n=0}^{\infty}\frac{1}{(n+1)!}z^{-n}=\frac{1}{1!}+\frac{1}{2!}z^{-1}+\frac{1}{3!}z^{-2}+\cdots+\frac{1}{(n+1)!}z^{-n}+\cdots$$
$$=z\left[\frac{1}{1!}z^{-1}+\frac{1}{2!}z^{-2}+\frac{1}{3!}z^{-3}+\cdots+\frac{1}{(n+1)!}z^{-(n+1)}+\cdots\right],$$

(multiplying and dividing by z)

$$=z(e^{1/z}-1) \quad (2)$$

Example 11.7

Find $Z\left(\left\langle \frac{1}{n+1} \right\rangle\right)$.

Solution By definition of Z-transforms

$$Z(\langle u_n \rangle) = \sum_{n=0}^{\infty} u_n z^{-n}$$

Here $\quad \langle u_n \rangle = \left\langle \frac{1}{n+1} \right\rangle = \left\langle 1, \frac{1}{2}, \frac{1}{3}, \cdots, \frac{1}{n+1}, \cdots \right\rangle \quad (1)$

$$Z\left(\left\langle \frac{1}{n+1} \right\rangle\right) = \sum_{n=0}^{\infty} \frac{1}{n+1} z^{-n}$$

$$= 1 + \frac{1}{2} z^{-1} + \frac{1}{3} z^{-2} + \cdots + \frac{1}{n+1} z^{-n} + \cdots$$

$$= z\left(z^{-1} + \frac{1}{2} z^{-2} + \frac{1}{3} z^{-3} + \cdots + \frac{1}{n+1} z^{-n-1} + \cdots \right)$$

(multiplying and dividing by z) (2)

$$= z\left[-\log\left(1 - \frac{1}{z}\right) \right]$$

$$\because \quad -\log(1-t) = t + \frac{t^2}{2} + \cdots + \frac{t^n}{n} + \cdots$$

Table 11.1 Z-Transforms of Some Standard Functions (Special Sequences)

S. No	Sequence $\langle u_n \rangle$	Z-transform: $\bar{u}(z)$
1.	$\langle 1 \rangle$	$\frac{z}{z-1}$
2.	$\langle n \rangle$	$\frac{z}{(z-1)^2}$
3.	$\langle n^2 \rangle$	$(z^2+z)/(z-1)^2$
4.	$\langle n^p \rangle$ $(p \in \mathbb{N})$	$-z\frac{d}{dz}[Z(\langle n^{p-1} \rangle)]$
5.	$\left\langle \frac{1}{n!} \right\rangle$	$e^{1/z}$
6.	$\langle a^n \rangle$	$z/(z-a)$
7.	$\langle nu_n \rangle$	$-z\frac{d}{dz}(\bar{u}(z))$
8.	$\langle a^n u_n \rangle$	$\bar{u}(z/a)$
9.	$\langle H_n \rangle$ (Unit step sequence)	$\frac{z}{z-1}$
10.	$\langle \delta_n \rangle$ (Unit impulse sequence)	1
11.	$\langle u_{n+1} \rangle$	$z(\bar{u}(z) - u_0)$
12.	$\langle u_{n+2} \rangle$	$z^2(\bar{u}(z) - u_0 - u_1 z^{-1})$
13.	$\langle u_{n+3} \rangle$	$z^3(\bar{u}(z) - u_0 - u_1 z^{-1} - u_2 z^{-2})$
14.	$\langle u_{n-k} \rangle$	$z^{-k}\bar{u}(z)$
15.	u_0	$\lim_{z\to\infty} (z-1)\bar{u}(z)$
16.	$\lim_{n\to\infty} u_n$	$\lim_{z\to 1} (z-1)\bar{u}(z)$
17.	$\langle \cos nt \rangle$	$\frac{z(z-\cos t)}{z^2 - 2z\cos t + 1}$
18.	$\langle \sin nt \rangle$	$\frac{z \sin t}{z^2 - 2z\cos t + 1}$

EXERCISE 11.1

Definition

1. Find the *Z*-transform of $\langle u_n \rangle$, $(n = 0, 1, 2, 3, 4)$ where $u_0 = 4, u_1 = -5, u_2 = 0, u_3 = 0$ and $u_4 = 2$.

   Ans: $4 - \frac{5}{z} + \frac{2}{z^4}$

Linearity

2. Find the *Z*-transform of $\langle u_n \rangle$ where $u_n = 6n + 3\sin\frac{n\pi}{4} - 5a^2\ n \in \mathbb{N}$.

3. Using $Z(\langle n^2 \rangle) = \frac{z^2+z}{(z-1)^3}$ show that $Z(\langle (n+1)^2 \rangle) = \frac{z^3+z^2}{(z-1)^3}$.

4. Evaluate $Z(\langle (\cos\theta + i\sin\theta)^n \rangle)$. Hence prove that

   $$Z(\langle \cos n\theta \rangle) = \frac{z(z-\cos\theta)}{z^2 - 2z\cos\theta + 1} \text{ and}$$

   $$Z(\langle \sin n\theta \rangle) = \frac{z\sin\theta}{z^2 - 2z\cos\theta + 1}.$$

   [JNTU 2001, 2002*S*]

Second Shifting Theorem

5. Show that $Z(\langle \sin(n+1)\theta \rangle) = \frac{z^2\sin\theta}{z^2 - 2z\cos\theta + 1}$.

   [JNTU 2002*S*]

Initial Value Theorem and Second Shifting Theorem

6. Find $Z(\langle u_{n+2} \rangle)$ if $Z(\langle u_n \rangle) = \frac{z}{z+1} + \frac{z}{z^2+1}$.

   Ans: $\frac{z(z^2 - z + 2)}{(z-1)(z^2+1)}$

Damping Rule

7. Evaluate $Z(\langle e^{-an}\sin n\theta \rangle)$.

   Ans: $\frac{ze^a\sin\theta}{z^2e^{2a} - 2ze^a\cos s\theta + 1}$

Multiplication by *n*

8. Evaluate $Z(\langle na^n \rangle)$.

   Ans: $\frac{az}{(z-a)^2}$

9. Evaluate $z(\langle na^n \rangle)$ using the definition of *Z*-transform.

   Ans: $\frac{az}{(z-a)^2}$

Recurrence Formula

10. Using the recurrence formula $Z(\langle n^p \rangle) = z(d/dz)[Z(\langle n^{p-1} \rangle)]$, find $Z(\langle n^3 \rangle)$ and $Z(\langle n^4 \rangle)$ given that $Z(\langle n^2 \rangle) = \frac{z^2+2}{(z-1)^3}$.

    Ans: $Z(\langle n^3 \rangle) = \frac{z^3 + 4z^2 + z}{(z-1)^4}$;

    $Z(\langle n^4 \rangle) = \frac{z^4 + 11z^3 + 11z^2 + z}{(z-1)^5}$

11.6 INVERSE *Z*-TRANSFORM

We have already defined the inverse *Z*-transforms of $\overline{u}(z)$, a function of a complex variable *z*, by

$$Z^{-1}(\overline{u}(z)) = \langle u_n \rangle \qquad (11.61)$$

which exists provided the series, $\sum_{n=0}^{\infty} u_n z^{-n}$ converges and $\langle u_n \rangle$ is the sequence generating the series.

We will now consider methods for finding the inverse *Z*-transforms when a function $\overline{u}(z)$ is given.

11.6.1 Methods for Evaluation of Inverse *Z*-Transforms

Method 1: Use of the Table of Transforms

The *Z*-transforms of some special sequences are tabulated. If the given function $\overline{u}(z)$ is one of those tabulated functions or is a linear combination of them we can readily find the inverse transform.

Example 11.8

Find $Z^{-1}\left(\frac{z}{z-1}\right)$.

Solution Since $Z(\langle 1 \rangle) = \frac{z}{z-1}$, the inverse *Z*-transform of $\left(\frac{z}{z-1}\right)$ is the sequence $\langle 1 \rangle = \langle 1, 1, 1, 1, 1, \ldots \rangle$.

Example 11.9

Find $Z^{-1}(\overline{u}(z))$ where

$$\overline{u}(z) = \frac{3z}{(z-1)^2} + 5e^{1/z} + 2z\log\left(z - \frac{1}{z}\right) + \frac{1}{4}\frac{z(z-\cos\theta)}{z^2 - 2z\cos\theta + 1}. \qquad (1)$$

Solution

$$Z^{-1}(\bar{u}(z)) = 3Z^{-1}\left(\frac{z}{(z-1)^2}\right) + 5Z^{-1}(e^{1/z})$$
$$+2Z^{-1}\left[z\log\left(z-\frac{1}{z}\right)\right]$$
$$+\frac{1}{4}Z^{-1}\left(\frac{z(z-\cos\theta)}{z^2-2z\cos\theta+1}\right),$$

by linearity

$$= 3\cdot\langle n\rangle + 5\cdot\left\langle\frac{1}{n!}\right\rangle$$
$$-2\left\langle\frac{1}{n+1}\right\rangle + \frac{1}{4}\langle\cos n\theta\rangle$$

$$\therefore \quad \langle u_n\rangle = \left\langle 3n + \frac{5}{n!} - \frac{2}{n+1} + \frac{1}{4}\cos n\theta\right\rangle, \tag{2}$$

is the inverse Z-transform of the given function.

Method 2: Method of Partial Fractions

If $\bar{u}(z)$ is a rational function of z then put $\bar{u}(z)/z$ into partial fractions by the usual procedure.

Multiply both sides by z. Write the inverse transform of each fraction. By the linearity property, their sum is the inverse transform of $\bar{u}(z)$.

Examples

Find the inverse Z-transforms of each of the following functions (11.10–11.12).

Example 11.10

$$\frac{z+2}{z^2-5z+6}.$$

Solution

Let
$$\bar{u}(z) = \frac{z+2}{z^2-5z+6} = \frac{z+2}{(z-2)(z-3)} \tag{1}$$

Write
$$\frac{\bar{u}(z)}{z} = \frac{z+2}{z(z-2)(z-3)} = \frac{A}{z} + \frac{B}{(z-2)} + \frac{C}{(z-3)} \tag{2}$$

where A, B, C are constants to be determined.

$$\left.\begin{aligned} A &= \left.\frac{z+2}{(z-2)(z-3)}\right|_{z=0} = \frac{2}{(-2)(-3)} = \frac{1}{3} \\ B &= \left.\frac{z+2}{z(z-3)}\right|_{z=2} = \frac{2+2}{2(2-3)} = -2 \\ C &= \left.\frac{z+2}{z(z-2)}\right|_{z=3} = \frac{3+2}{3(3-2)} = \frac{5}{3} \end{aligned}\right\} \tag{3}$$

Multiplying both sides of (2) by z and substituting for A, B, C from (3) we have

$$\bar{u}(z) = \frac{1}{3} + (-2)\cdot\frac{z}{z-2} + \frac{5}{3}\cdot\frac{z}{z-3} \tag{4}$$

Now, we take the inverse Z-transforms of both sides of (4) and obtain

$$\begin{aligned} \langle u_n\rangle &= \frac{1}{3}\langle\delta_{\text{n}}\rangle - 2\langle 2^n\rangle + \frac{5}{3}\langle 3^n\rangle \\ &= \left\langle\frac{1}{3}\delta_n - 2^{n-1} + 5\cdot 3^{n-1}\right\rangle \end{aligned} \tag{5}$$

by linearity property, where δ_n is the unit impulse function.

Example 11.11

$$\frac{3z^2+z}{(5z-1)(5z+2)}$$ [JNTU, 2001]

Solution Let
$$\bar{u}(z) = \frac{3z^2+z}{(5z-1)(5z+2)} \tag{1}$$

write
$$\frac{\bar{u}}{z} = \frac{3z+1}{(5z-1)(5z+2)} = \frac{A}{5z-1} + \frac{B}{5z+2} \tag{2}$$

where A and B are constants to be determined.

$$\left.\begin{aligned} A &= \left.\frac{3z+1}{5z+2}\right|_{z=\frac{1}{5}} = \frac{3(\frac{1}{5})+1}{5\cdot\frac{1}{5}+2} = \frac{8}{15} \\ B &= \left.\frac{3z+1}{5z-2}\right|_{z=-\frac{2}{5}} = \frac{3(-\frac{2}{5})+1}{5(-\frac{2}{5})-1} = \frac{1}{15} \end{aligned}\right\} \tag{3}$$

Multiplying both sides of (2) by z and substituting for A and B from (3) we have

$$\begin{aligned} \bar{u}(z) &= \frac{8}{15}\cdot\frac{z}{5z-1} + \frac{1}{15}\cdot\frac{z}{5z+2} \\ &= \frac{8}{75}\cdot\frac{z}{z-(\frac{1}{5})} + \frac{1}{75}\frac{z}{z-(-\frac{2}{5})}, \end{aligned} \tag{4}$$

putting the fractions in the form $z/(z-a)$ for inversion. Now we apply the inverse *Z*-transforms on both sides and by linearity property we have

$$Z^{-1}(\overline{u}(z)) = \langle u_n \rangle = \frac{8}{75} Z^{-1}\left(\frac{z}{z-\frac{1}{5}}\right) + \frac{1}{75} Z^{-1}\left(\frac{z}{z-\left(-\frac{2}{5}\right)}\right)$$

$$= \frac{8}{75}\left\langle \left(\frac{1}{5}\right)^n \right\rangle + \frac{1}{75}\left\langle \left(-\frac{2}{5}\right)^n \right\rangle$$

$$\langle u_n \rangle = \frac{1}{3}\langle (8-2^n)5^{-n-2} \rangle \quad (5)$$

Example 11.12

$$\frac{2z}{(z-1)(z^2+1)}$$

Solution Let $\overline{u}(z) = \dfrac{2z}{(z-1)(z^2+1)}$ (1)

write $\dfrac{\overline{u}(z)}{2} = \dfrac{2}{(z-1)(z^2+1)} = \dfrac{A}{z-1} + \dfrac{Bz+C}{z^2+1}$ (2)

where *A*, *B* and *C* are constants to be determined.

$$\left.\begin{array}{r} A(z^2+1) + (Bz+C)(z-1) = 2 \\ \text{Put } z = 1, 0 \text{ we get } A = 1, \\ \text{and } A + C(-1) = 2 \Rightarrow C = -1 \\ \text{Equating the coefficient of } z^2 \\ A + B = 0 \Rightarrow B = -1 \end{array}\right] \quad (3)$$

$$\overline{u}(z) = \frac{z}{z-1} - \frac{z^2+z}{z^2+1} = \frac{z}{z-1} - \frac{z^2}{z^2+1} - \frac{z}{z^2+1} \quad (4)$$

Now, we apply the inverse *Z*-transforms to both sides of (4) and by linearity we have

$$\langle u_n \rangle = \langle 1 \rangle - \left\langle \cos\frac{n\pi}{2} \right\rangle - \left\langle \sin\frac{n\pi}{2} \right\rangle$$

$$= \left\langle 1 - \cos\frac{n\pi}{2} - \sin\frac{n\pi}{2} \right\rangle \quad (5)$$

Method 3: Long Division Method

If $\overline{u}(z)$ is a rational function of z and the denominator cannot be factored, then by long division we can write $\overline{u}(z) = \sum_{n=0}^{\infty} u_n z^{-n}$ so that $z^{-1}(\overline{u}(z)) = \langle u_n \rangle$.

This method is simple and straightforward.

Example 11.13

Find $Z^{-1}\left(\dfrac{10z}{z^2-32+2}\right)$.

Let $\overline{u}(z) = \dfrac{10z}{z^2-3z+2} = \dfrac{10z^{-1}}{1-3z^{-1}+2z^{-2}}$ (1)

(multiplying the numerator and denominator by z^{-2})

By long division

$$\begin{array}{r|l} & 10z^{-1} + 30z^{-2} + 70z^{-3} + 150z^{-4} + \cdots \\ 1-3z^{-1}+2z^{-2} & 10z^{-1} \\ & \underline{10z^{-1} - 30z^{-2} + 20z^{-3}} \\ & 30z^{-2} - 20z^{-3} \\ & \underline{30z^{-2} - 90z^{-3} + 60z^{-4}} \\ & 70z^{-3} - 60z^{-4} \\ & \underline{70z^{-3} - 210z^{-4} + 140z^{-5}} \\ & 150z^{-4} - 140z^{-5} \end{array}$$

$$\therefore \quad \overline{u}(z) = 10z^{-1} + 30z^{-2} + 70z^{-3} + 150z^{-4} + \cdots \quad (2)$$

$$= 10\,[z^{-1} + 3z^{-2} + 7z^{-3} + 15z^{-4} + \cdots]$$

$$= 10\,[(1 + z^{-1} + 3\,.\,z^{-2} + 7z^{-3} + 15z^{-4} + \cdots) - 1], \text{ adding and subtracting 1}$$

$$= 10\,[(2^0-1) + (2^1-1)z^{-1} + (2^2-1)z^{-2} + (2^3-1)z^{-3} + (2^4-1)z^{-4} + \cdots]$$

$$= 10\sum_{n=0}^{\infty}(2^n-1)z^{-n} \quad (3)$$

By definition

$$\overline{u}(z) = Z(\langle u_n \rangle) = \sum_{n=0}^{\infty} u_n z^{-n} \quad (4)$$

Comparing (3) and (4) we have the inverse *Z*-transform as

$$\langle u_n \rangle = 10\,\langle 2^n - 1 \rangle \; n = 0, 1, 2, \ldots \quad (5)$$

Method 4: Inverse Integral Method [Contour Integral Method or Method of Residues]

Using the theory of complex variables, it can be shown that the inverse Z-transform of a function $\overline{u}(z)$ is given by

$$\langle u_n \rangle = \frac{1}{2\pi i}\int_c \overline{u}(z) z^{n-1} dz \qquad (11.62)$$

where C is a closed contour in the ROC of $\overline{u}(z)$ enclosing the origin and all the n (say) isolated singularities of the integrand $[\overline{u}(z)zn^{-1}]$. But by the Cauchy's Residue Theorem the integral in (11.62) is equal to $2\pi i\left(\sum_{k=1}^{n} r_k\right)$, where r_k is the residue at the kth isolated singularity of $\overline{u}(z)$. $z = a$ is a singularity of $\overline{u}(z)$ if $\overline{u}(z)$ is not analytic (not differentiable or has no power series expansion) at $z = a$.

Example 11.14

Evaluate $Z^{-1}\left(\frac{z}{z^2+7z+10}\right)$ by contour integral method.

Solution Let $\overline{u}(z) = \frac{z}{z^2+7z+10} = \frac{z}{(z+2)(z+5)}$.

The inverse transform is $\langle u_n \rangle = \frac{1}{2\pi i}\int_c z^{n-1}\overline{u}(z)\,dz = \sum r_k$ (= sum of the residues)

$$= \frac{1}{2\pi i}\int_c z^{n-1}\frac{z}{(z+2)(z+5)}dz$$

$$= \frac{1}{2\pi i}\int_c \frac{z^n}{(z+2)(z+5)}dz$$

where C is a simple closed curve enclosing the singularities at $z = -2$ and -5 (which are the zeros of the denominator).

Singularities are at $z = -2, -5$ which are simple poles (poles of order one)

$$= \operatorname*{Res}_{z=-2}(\overline{u}(z)z^{n-1})$$

$$= \lim_{z\to -2}(z+2)\frac{z^n}{(z+2)(z+5)} = \frac{(-2)^n}{3}$$

$$= \operatorname*{Res}_{z=-5}(\overline{u}(z)z^{n-1})$$

$$= \lim_{z\to -5}(z+5)\frac{z^n}{(z+2)(z+5)} = \frac{(-5)^n}{-3}$$

$\langle u_n \rangle$ = Sum of residues

$$= \left\langle\frac{(-2)^n}{3}\right\rangle + \left\langle\frac{(-5)^n}{-3}\right\rangle \langle u_n \rangle$$

$$= \left\langle\frac{(-1)^n}{3}(2^n - 5^n)\right\rangle$$

Example 11.15

Evaluate $Z^{-1}\left(\frac{z^2+z}{(z-1)(z^2+1)}\right)$ using the method of residues.

Solution Let $\overline{u}(z)\frac{z^2+z}{(z-1)(z^2+1)}$.

The inverse transform is $\langle u_n \rangle = \frac{1}{2\pi i}\int_c z^{n-1}\overline{u}(z)\,dz = \sum r_k$ (= sum of the residues)

$$= \frac{1}{2\pi i}\int_c z^{n-1}\frac{z^2+z}{(z-1)(z^2+1)}dz$$

$$= \frac{1}{2\pi i}\int_c z^n\frac{z+1}{(z-1)(z^2+1)}dz$$

where C is a simple closed curve enclosing the singularities at $z = 1, \pm i$ (which are the zeros of the denominator).

Singularities are at $z = 1, \pm i$ which are simple poles (poles of order one)

$$r_1 = \operatorname*{Res}_{z=1}(\overline{u}(z)\cdot z^{n-1})$$

$$= \lim_{z\to 1}(z-1)\cdot z^n\cdot\frac{z+1}{(z-1)(z^2+1)} = 1$$

$$r_2 = \operatorname*{Res}_{z=i}(\overline{u}(z)\cdot z^{n-1})$$

$$= \lim_{z\to i}(z-i)\cdot z^n\frac{z+1}{(z-1)(z-i)(z+i)}$$

$$= \frac{i^n(1+i)}{(i-1)(2i)} = \frac{i^n(1+i)}{-2(1+i)} = -\frac{1}{2}i^n$$

$$r_3 = \operatorname*{Res}_{z=-i}(\overline{u}(z)z^{n-1})$$

$$= -\frac{1}{2}(-i)^n \text{ (replacing } i \text{ by } -i \text{ in } r_2)$$

The required inverse transform

$$\langle u_n \rangle = \sum r_k = \text{ Sum of the residues}$$

$$= \langle 1 \rangle + \left\langle -\frac{1}{2} i^n \right\rangle + \left\langle -\frac{1}{2} (-i)^n \right\rangle$$

$$= \langle 1 \rangle - \frac{1}{2} \langle i^n + (-i)^n \rangle$$

$$= \langle 1 \rangle - \frac{1}{2} 2 \left\langle \cos \frac{n\pi}{2} \right\rangle = \left\langle 1 - \cos \frac{n\pi}{2} \right\rangle$$

since

$$i^n = \cos \frac{n\pi}{2} + i \sin \frac{n\pi}{2}; \; (-i)^n = \cos \frac{n\pi}{2} - i \sin \frac{n\pi}{2};$$

$$i^n + (-i)^n = 2 \cos \frac{n\pi}{2}$$

Method 5: Power Series Method

We expand $\bar{u}(z)$ in powers of $1/z$ using a known formula and then identify the sequence $\langle u_n \rangle$, which is the inverse *Z*-transform of $\bar{u}(z)$.

Example 11.16

Find $Z^{-1}\left(\log \left(\frac{z}{z+1} \right) \right)$ by power series method.

Solution Let

$$\bar{u}(z) = \log \left(\frac{z}{z+1} \right) = -\log \left(\frac{z+1}{z} \right) = -\log \left(1 + \frac{1}{z} \right)$$

$$= -\frac{1}{z} + \frac{1}{z} \frac{1}{z^2} - \frac{1}{3} \cdot \frac{1}{z^3} + \cdots + \frac{(-1)^n}{n} z^{-n} + \cdots$$

$$\therefore \quad \langle u_n \rangle = Z^{-1}(\bar{u}(z)) = \begin{cases} 0 & \text{for } n = 0 \\ \frac{(-1)^n}{n} & \text{for } n > 0 \end{cases}$$

Example 11.17

Find $Z^{-1}\left(\frac{z^2}{(z-1)^2} \right)$ by power series method.

Solution Let

$$\bar{u}(z) = \frac{z^2}{(z-1)^2} = \left(\frac{z-1}{z} \right)^{-2} = \left(1 - \frac{1}{z} \right)^{-2}$$

$$= 1 + 2 \cdot \frac{1}{z} + 3 \cdot \frac{1}{z^2} + \cdots + (n+1) \frac{1}{z^n} + \cdots$$

$$= u_0 + u_1 \cdot \frac{1}{z} + u_2 \frac{1}{z^2} + \cdots + u_n \frac{1}{z^n} + \cdots$$

$$\therefore \quad z^{-1}\left(\frac{z^2}{(z-1)^2} \right) = \langle u_n \rangle$$

$$= \langle n+1 \rangle \quad n = 0, 1, 2, \cdots$$

Example 11.18

Find $z^{-1}\left(\frac{z^2}{z^2+1} \right)$.

Solution Let

$$\bar{u}(z) = \frac{z^2}{z^2+1} = \frac{1}{1+\frac{1}{z^2}} = \left(1 + \frac{1}{z^2} \right)^{-1}$$

$$= 1 - \frac{1}{z^2} + \frac{1}{z^4} - \frac{1}{z^6} + \cdots + \cdots$$

$$= \sum_{n=0}^{\infty} (-1)^{n/2} z^{-n} \; (n \text{ is even})$$

Here the coefficient sequence is $\langle u_0, u_1, u_2, u_3, \ldots \rangle$

$$= \langle 1, 0, 1, 0, 1, 0, 1, \cdots \rangle = \left\langle \cos \frac{n\pi}{2} \right\rangle$$

$$\Rightarrow \quad \bar{u}(z) = \langle u_n \rangle = \left\langle \cos \frac{n\pi}{2} \right\rangle \quad n = 0, 1, 2, 3, \cdots$$

Method 6: Convolution Method

When the product of two transforms is given we can apply the following convolution theorem and find the inverse transform.

Convolution Theorem

If $z^{-1}(\bar{u}(z)) = \langle u_n \rangle$ and

$z^{-1}(\bar{v}(z)) = \langle v_n \rangle$ then

$$z^{-1}(\bar{u}(z) \cdot \bar{v}(z)) = \sum_{m=0}^{n} u_m v_{n-m}$$

$$= \langle u_n \rangle * \langle v_n \rangle \qquad (11.63)$$

Example 11.19

Find $Z^{-1}\left(\frac{z^2}{(z-a)(z-b)} \right)$ using the convolution theorem.

Solution Let

$$\bar{u}(z)=\frac{z}{z-a} \Rightarrow \langle u_n\rangle = Z^{-1}(\bar{u}(z))$$

$$= Z^{-1}\left(\frac{z}{z-a}\right) = \langle a^n\rangle$$

$$\bar{v}(z)=\frac{z}{z-b} \Rightarrow \langle v_n\rangle = Z^{-1}(\bar{v}(z))$$

$$= Z^{-1}\left(\frac{z}{z-b}\right) = \langle b^n\rangle$$

Then

$$Z^{-1}\left(\frac{z}{z-a}\cdot\frac{z}{z-b}\right) = \langle u_n\rangle * \langle v_n\rangle$$

$$= \langle a^n\rangle * \langle b^n\rangle$$

$$= \sum_{m=0}^{\infty} a^m b^{n-m} = b^n \sum_{m=0}^{n} (a/b)^m,$$

a finite GS with c.r. $= a/b$

$$= b^n \frac{(a/b)^{n+1}-1}{(a/b)-1} = \frac{a^{n+1}-b^{n+1}}{a-b}$$

Example 11.20

Find $Z^{-1}\left(\frac{z^3}{(z-3)(z^2+1)}\right)$ using the convolution theorem.

Solution Let

$$\bar{u}(z)=\frac{z}{z-3} \Rightarrow \langle u_n\rangle = Z^{-1}(\bar{u}(z))$$

$$= Z^{-1}\left(\frac{z}{z-3}\right) = \langle 3^n\rangle$$

$$\bar{v}(z)=\frac{z^2}{z^2+1} \Rightarrow \langle v_n\rangle = Z^{-1}(\bar{v}(z))$$

$$= Z^{-1}\left(\frac{z^2}{z^2+1}\right) = \left\langle \cos\frac{n\pi}{2}\right\rangle$$

$$Z^{-1}\left(\frac{z}{z-3}\cdot\frac{z^2}{z^2+1}\right) = \langle u_n\rangle * \langle v_n\rangle$$

$$= \langle 3^n\rangle * \left\langle \cos\frac{n\pi}{2}\right\rangle$$

$$= \left\langle \sum_{m=0}^{n} 3^m \cdot \cos\frac{(n-m)\pi}{2}\right\rangle$$

EXERCISE 11.2

Partial Fractions Method

1. Find $Z^{-1}\left(\frac{8z^2}{(2z-1)(4z-1)}\right)$.

   Ans: $\langle u_n\rangle = \left\langle \frac{1}{2^{n-1}} - \frac{1}{4^n}\right\rangle$

2. Find $Z^{-1}\left(\frac{z^2-20z}{(z-2)^3(z-4)}\right)$. [JNTU 2002]

   Ans: $\langle u_n\rangle = \langle 2^{n-1} + 2^n \,.\, n^2 - 2^{2n-1}\rangle$

3. Find $Z^{-1}\left(\frac{2z^2+3z}{(z+2)(z-4)}\right)$.

   Ans: $\langle u_n\rangle = \left\langle \frac{1}{6}(-2)^n + \frac{11}{6}4^n\right\rangle$

Long Division Method

4. Find $Z^{-1}\left(\frac{2(z^2-z)}{(z^2+1)^2}\right)$.

   Ans: $\langle u_n\rangle = \left\langle 2n\sin\frac{n\pi}{2}\right\rangle$

5. Find $Z^{-1}\left(\frac{z}{(z+1)^2}\right)$, by division method.

   $\langle u_n\rangle = \langle (-1)^{n-1} n\rangle \; n = 0, 1, 2, \ldots$

   Ans: $\langle u_n\rangle = \langle (-1)^{n-1} n\rangle$

Inverse Integral Method

6. Find

   $$Z^{-1}\left(\frac{z}{z^2-2z+2}\right)$$

   Ans: $\langle u_n\rangle = \left\langle 2^{n/2}\sin\frac{n\pi}{4}\right\rangle, n = 0, 1, 2, \ldots$

7. Find $Z^{-1}\left(\frac{z(z+1)}{(z-1)^3}\right)$ $\langle u_n\rangle = \langle n^2\rangle \; n = 0, 1, 2, \ldots$

   Ans: $\langle u_n\rangle = \langle n^2\rangle$

Power Series Method

8. Find $Z^{-1}\left(\frac{z}{z+1}\right)$ by the power series method

   $$\langle u_n\rangle = \begin{cases} 0 & \text{for } n = 0 \\ (-1)^{n/2}, & n > 0 \end{cases}$$

Ans: $\langle u_n \rangle = \begin{cases} 0 & \text{for } n = 0 \\ \dfrac{(-1)^n}{n} & \text{for } n > 0 \end{cases}$

Convolution Theorem

9. Find the inverse *Z*-transform of $\dfrac{z^2}{(z-2)(z-3)}$ by the convolution theorem method.

   Ans: $\langle u_n \rangle = \langle 3^{n+1} - 2^{n+1} \rangle$

10. Using the Convolution Theorem evaluate $Z^{-1}\left(\dfrac{z^2}{(z-1)(z-3)}\right)$.

    Ans: $\langle u_n \rangle = \dfrac{1}{2}\langle 3^{n+1} - 1 \rangle$

11.7 Application of *Z*-Transforms: Solution of a Difference Equations by *Z*-Transform

11.7.1 Introduction

We have earlier seen that the Laplace transform method is quite useful in solving ordinary linear differential equations. Now we will see how *Z*-transform can be applied in solving linear difference equations.

The following are the steps to be followed in solving difference equations.

1. Apply *Z*-transform on both sides of the difference equation.
2. Solve for $\bar{u}(z)$, using the given initial condition.
3. Split the rational expression into partial fractions.
4. Apply inverse *Z*-transform which gives the solution sequence $\langle u_n \rangle$.

Note 1 The reader may refer to the topic "Finite Differences" in Chapter 6 of the text to know the meaning of the terms and the symbols used below.

11.7.2 Difference Equation

An equation involving differences of an unknown function at one or more values of the argument is called a *difference equation*.

E.g.

$$\Delta y_{n+1} - 2y_n = 3 \tag{11.64}$$

$$\Rightarrow \quad y_{n+2} - y_{n+1} - 2y_n = 3 \qquad \therefore \ \Delta y_{n+1} = y_{n+2} - y_{n+1}$$

$$\Rightarrow \quad (E^2 - E - 2)y_n = 3, \ (E = 1 + \Delta) \tag{11.65}$$

11.7.3 Order of a difference equation

The order of a difference equation is the difference of the largest and smallest arguments divided by the length of the interval.

The order of the equation in Example 1 above is 2.

11.7.4 General Solution (Complete Solution)

The expression for y_n satisfying the given difference equation is called its solution. The solution in which the number of arbitrary constants is equal to the order of the difference equation is called its General (Complete) Solution or Integral.

11.7.5 Particular Solution (Particular Integral)

Any solution obtained from the general solution by giving particular values to the arbitrary constants is called a Particular Solution or Particular Integral.

11.7.6 Linear Difference Equation

A difference equation in which $y_{n+1}, y_{n+2}, \ldots$ are of first degree only and are not multiplied together is called a linear difference equation. If the coefficients in the equation are constants then the equation is called a linear difference equation with constant coefficients. The equation

$$y_{n+m} + a_1 y_{n+m-1} + a_2 y_{n+m-2} + \cdots + a_m y_n = f(n) \tag{11.66}$$

where $a_1, a_2, \ldots, a_n$ are constants is a linear difference equation with constant coefficients.

If $f(n) \neq 0$, equation (11.66) is called a non-homogeneous equation. Otherwise, it is called a homogeneous equation or reduced equation.

11.7.7 Complementary Function and Particular Integral

Consider the reduced linear difference equation

$$y_{n+m} + a_1 y_{n+m-1} + a_2 y_{n+m-2} + \cdots + a_m y_n = 0 \tag{11.67}$$

The Complete (general) Solution of (11.67) is

$$u_n = c_1u_1(n) + c_2u_2(n) + \cdots + c_mu_m(n) \qquad (11.68)$$

where $c_1, c_2, \ldots, c_m$ are constants and $u_1(n), u_2(n), \ldots, u_m(n)$ are m linearly independent solutions of (11.68). The Complete Solution of the nonhomogeneous equation (11.66) is

$$y_n = u_n + v_n \; (CS = CF + PI) \qquad (11.69)$$

where u_n is the Complete Solution of the reduced equation (11.67) and v_n is the Particular Integral of (11.66).

11.8 Method for Solving a Linear Difference Equation with Constant Coefficients

11.8.1(a) Complementary Function

Writing $y_{n+k} = E^k y_n$ $k = 0, 1, 2, \ldots, n$ equations (11.66) and (11.67) can be put in operator notation as

$$\phi(E)y_n \equiv (E^m + a_1E^{m-1} + a_2E^{m-2} + \cdots a_m1)\, y_n = f(n)(NH) \qquad (11.70)$$

$$\phi(E)y_n = (E^m + a_1E^{m-1} + a_2E^{m-2} + \cdots a_m1)\, y_n = 0\, (H) \qquad (11.71)$$

Assuming a solution of (11.71) as

$$y_n = \lambda^n \qquad (11.72)$$

We have $E^k y_n = \lambda^{n+k}$ $(K = 0, 1, 2, \ldots, m)$ (11.73)

and we obtain

$$(\lambda^m + a_1\lambda^{m-1} + a_2\lambda^{m-2} + \cdots + a_m1)\lambda^n = 0 \qquad (11.74)$$

Equation (11.74) will be satisfied if the algebraic equation

$$(\lambda^m + a_1\lambda^{m-1} + a_2\lambda^{m-2} + \cdots + a_m1) = 0 \qquad (11.75)$$

holds. Equation (11.75) is called the auxiliary equation (AE) or subsidiary equation of the difference equation. Since the degree of the equation (11.75) is m it has m roots $\lambda_1, \lambda_2, \ldots, \lambda_m$ (say). Solving the AE (11.75) we can write the Complementary Function (CF) of the difference equation (11.70) as shown in the table below.

11.8.1(b) Particular Integral

Operating equation (11.70) by the inverse operator $1/\phi(E)$ we get the particular integral of equation (11.58) as

$$y_n = \frac{1}{\phi(E)} f(n) \qquad (11.77)$$

11.8.2 Short Methods for Finding the Particular Integral

Rule 1 $f(n) = a^n$ and $\phi(a) \neq 0$ (11.78)

(1) $PI = \dfrac{1}{\phi(E)} a^n = \dfrac{1}{\phi(a)} a^n$ (replace E by 'a')

(2) If $\phi(E) \equiv (E - a)\,\psi(E)$ and $\psi(a) \neq 0$ then

$$PI = \frac{1}{\psi(E)}\frac{1}{(E-a)} a^n = \frac{1}{\psi(a)}(na^{n-1}) \qquad (11.79)$$

(3) If $\phi(E) = (E - a)^2\,\psi(E)$ and $\psi(a) \neq 0$ then

Table 11.2 Table of Complementary Functions Corresponding to the Roots of the Auxiliary Equations

S. No.	Roots of AE	Nature of Roots	Complementary Function (CF)
1.	$\lambda_1, \lambda_2, \ldots, \lambda_m$	Real distinct	$C_1\lambda_1^n + C_2\lambda_2^n + \cdots + C_m\lambda_m^n$
2.	$\lambda_1, \lambda_2, \lambda_3, \ldots, \lambda_m$	$\lambda_i \in R$ Two roots equal, others distinct	$(C_1 + C_2n)\lambda_1^n + C_3\lambda_3^n + \cdots + C_m\lambda_m^n$
3.	$\lambda_1, \lambda_1, \ldots, \lambda_1$ (s roots) $\lambda_{s+1}, \ldots, \lambda_m$	$\lambda_i \in R$ s roots equal, others distinct	$(C_1 + C_2n + C_3n^2 + \cdots + C_sn^{s-1})\lambda_1^n + C_{s+1}{\lambda^n}_{s+1} + \cdots + C_n\lambda_m^n$
4.	$\alpha \pm i\beta$	A pair of complex roots	$r^n(C_1\cos n\theta + C_2\sin n\theta)$ where $r = \sqrt{a^2 + \beta^2}$, $\theta = \tan^{-1}\dfrac{\beta}{\alpha}$ (11.76)

$$PI = \frac{1}{\psi(E)}\frac{1}{(E-a)^2}a^n = \frac{1}{\psi(a)}n(n-1)a^{n-2} \quad (11.80)$$

and so on.

Rule 2 $f(n) = \sin \alpha n$ or $\cos \alpha n$

Since $\sin \alpha n = \dfrac{e^{i\alpha n} - e^{-i\alpha n}}{2i} = \dfrac{1}{2i}[(e^{i\alpha})^n - (e^{-i\alpha})^n]$

and $\cos \alpha n = \dfrac{e^{i\alpha n} + e^{-i\alpha n}}{2} = \dfrac{1}{2}[(e^{i\alpha})^n + (e^{-i\alpha})^n]$

apply Rule 1, using $e^{i\alpha}$ and $e^{-i\alpha}$ in place of 'a'

Rule 3 $f(n) = n_p$

$$P.I. = \frac{1}{\phi(E)}n^p = \frac{1}{\phi(1+\Delta)}n^p = \phi^{-1}(1+\Delta)n^p \quad (11.81)$$

Expand $\phi^{-1}(1+\Delta)$ in increasing powers of Δ up to Δ^p and express n^p in the factorial form and apply on each term of the expansion.

Example 11.21

Form a difference equation from

$$y_n = c_1 3^n + c_2(-1)^n \quad (1)$$

Solution Applying the shift operator E twice

$$Ey_n = y_{n+1} = 3 \cdot C_1 3^n - C_2(-1)^n \quad (2)$$

$$E^2 y_n = y_{n+2} = 9 \cdot C_1 \cdot 3^n + C_2(-1)^n \quad (3)$$

Eliminating C_1 and C_2 from (1)–(3) we get

$$\begin{vmatrix} y_n & 1 & 1 \\ y_{n+1} & 3 & -1 \\ y_{n+2} & 9 & 1 \end{vmatrix} = 0 \quad \begin{aligned} &\Rightarrow 12y_n + 8y_{n+1} - 4y_{n+2} = 0 \\ &\Rightarrow y_{n+2} - 2y_{n+1} - 3y_n = 0 \end{aligned} \quad (4, 5)$$

Example 11.22

Form a difference equation from

$$y_n = a{\cdot}2^n + b{\cdot}3^n \quad (1)$$

Solution Applying the shift operator E twice

$$Ey_n = y_{n+1} = 2a \cdot 2^n + 3b \cdot 3^n \quad (2)$$

$$E^2 y_n = y_{n+2} = 4a \cdot 2^n + 9b \cdot 3^n \quad (3)$$

Eliminating a and b from (1)–(3) we get

$$\begin{vmatrix} y_n & 1 & 1 \\ y_{n+1} & 2 & 3 \\ y_{n+2} & 4 & 9 \end{vmatrix} = 0 \quad \begin{aligned} &\Rightarrow 6y_n - 5y_{n+1} + y_{n+2} = 0 \quad (4) \\ &\Rightarrow y_{n+2} - 5y_{n+1} + 6y_n = 0 \end{aligned}$$

Example 11.23

Solve $y_{n+2} - 2y_{n+1} + y_n = 0$.

Solution In operator notation the equation is

$$(E^2 - 2E + 1) \equiv (E-1)^2 y_n = 0$$

$$A.E. \text{ is } (a-1)^2 = 0 \quad \Rightarrow \quad a = 1, 1$$

$\therefore$ The Required Solution is $y_n = (C_1 + C_2 n)(1)^n$

or $\qquad y_n = C_1 + C_2 n$

Example 11.24

The sequence $\langle F_n \rangle = \langle 0, 1, 1, 2, 3, 5, 8, 13, 21, \ldots \rangle$ is called the Fibonacci sequence. Form the difference equation for the sequence and solve it.

Solution In the given sequence each term beyond the second is the sum of the two terms immediately preceding it. If y_n denotes the nth term then

$$y_n = y_{n-1} + y_{n-2} \; (n > 2) \quad (1)$$

or

$$y_{n+2} = y_{n+1} - y_n = 0 \; (n > 0) \quad (2)$$

We have to solve this under the conditions $y_1 = 0, y_2 = 1$

Equation (2) $\Rightarrow \quad (E^2 - E - 1)y_n = 0$

The Auxiliary Equation is

$$x^2 - x - 1 = 0 \Rightarrow x = \frac{1 \pm \sqrt{5}}{2} \quad (3)$$

$\therefore$ The Complete Solution y_n is

$$y_n = C_1\left(\frac{1+\sqrt{5}}{2}\right)^n + C_2\left(\frac{1-\sqrt{5}}{2}\right)^n \text{ for } n > 0 \quad (4)$$

where C_1 and C_2 are arbitrary constants.

When $n = 1, y_1 = 0$

$$\therefore \quad C_1\left(\frac{1+\sqrt{5}}{2}\right) + C_2\left(\frac{1-\sqrt{5}}{2}\right) = 0 \quad (5)$$

When $n = 2, y_2 = 1$

$$\therefore \quad C_1\left(\frac{1+\sqrt{5}}{2}\right) + C_2\left(\frac{1-\sqrt{5}}{2}\right) = 1 \quad (6)$$

Solving (5) and (6) we get

$$C_1 = \frac{5-\sqrt{5}}{10}, C_2 = \frac{5+\sqrt{5}}{10} \qquad (7)$$

Hence, the solution under the stated conditions

$$y_n = \frac{5-\sqrt{5}}{10}\left(\frac{1+\sqrt{5}}{2}\right)^n + \frac{5-\sqrt{5}}{10}\left(\frac{1-\sqrt{5}}{2}\right)^n \qquad (8)$$

Example 11.25

Solve $y_{n+2} + 2y_{n+1} + y_n = n$ with $y_0 = y_1 = 0$.

Solution Taking Z-transform of both sides, $Z(y_{n+2}) + 2Z(y_{n+1}) + Z(y_n) = Z(n)$ by linearity.

$$\Rightarrow \quad z^2(\overline{y}(z) - y_0 - y_1 z^{-1}) + 2z$$

$$\overline{y}(z) - u_0 + \overline{y}(z) = \frac{z}{(z-1)^2} \qquad (1)$$

Putting $y_0 = y_1 = 0$ and solving for $\overline{y}(z)$

$$\overline{y}(z) = \frac{z}{(z-1)^2(z+1)^2} \quad \text{or} \quad \frac{\overline{y}(z)}{z} = \frac{1}{(z-1)^2(z+1)^2}$$

We have to put the rational function into partial fractions.

We see that $\dfrac{1}{z-1} - \dfrac{1}{z+1} = \dfrac{2}{(z-1)(z+1)}$ (2)

Squaring both sides of (1)

$$\frac{1}{(z-1)^2} + \frac{1}{(z+1)^2} - \frac{2}{(z-1)(z+1)} = \frac{4}{(z-1)^2(z+1)^2}$$

or
$$\frac{1}{(z-1)^2(z+1)^2} = \frac{1}{4}\frac{1}{(z-1)^2} + \frac{1}{4}\frac{1}{(z+1)^2}$$

$$-\frac{1}{4}\frac{1}{z-1} + \frac{1}{4}\frac{1}{z+1}, \text{ by (2)}$$

$$\therefore \quad \overline{y}(z) = \frac{1}{4}\cdot\frac{z}{(z-1)^2} + \frac{1}{4}\cdot\frac{z}{(z+1)^2}$$

$$-\frac{1}{4} - \frac{2}{z-1} + \frac{1}{4}\cdot\frac{z}{z+1} \qquad (3)$$

Taking inverse Z-transform of both sides

$$y_n = \frac{1}{4}[n - n(-1)^n - 1^n + (-1)^n]$$

$$= \frac{1}{4}\cdot(n-1)[1-(-1)^n] \qquad (4)$$

Example 11.26

Using Z-transform, solve the difference equation $u_{n+2} - 4u_{n+1} + 3u_n = 5^n$.

Solution Applying Z-transform we get

$$Z(u_{n+2}) - 4Z(u_{n+1}) + 3Z(u_n) = Z(5^n),$$

by linearity (1)

$$\Rightarrow \quad z^2\left(\overline{u}(z) - u_0 - u_1\frac{1}{z}\right) - 4z(\overline{u}(z) - u_0)$$

$$+3\overline{u}(z) = \frac{z}{z-5}$$

$$\Rightarrow \quad \overline{u}(z)(z^2 - 4z + 3) - u_0(z^2 - 4z)$$

$$-u_1 z = \frac{z}{z-5}$$

$$\Rightarrow \quad \frac{\overline{u}(z)}{z} = \frac{1}{(z-1)(z-3)(z-5)} + \frac{u_0(z-4)+u_1}{(z-1)(z-3)}$$

$$= \frac{A}{z-1} + \frac{B}{z-3} + \frac{C}{z-5} + \frac{D}{z-1} + \frac{E}{z-3} \qquad (2)$$

Where $A = \left.\dfrac{1}{(z-3)(z-5)}\right|_{z=1} = \dfrac{1}{8}$ (3)

$$B = \left.\frac{1}{(z-1)(z-5)}\right|_{z=3} = -\frac{1}{4}$$

$$C = \left.\frac{1}{(z-1)(z-3)}\right|_{z=5} = \frac{1}{8}$$

$$D = \left.\frac{u_0(z-4)+u_1}{z-3}\right|_{z=1} = \frac{-3u_0+u_1}{-2}$$

$$E = \left.\frac{u_0(z-4)+u_1}{z-1}\right|_{z=3} = \frac{-u_0+u_1}{2}$$

Taking the inverse Z-transforms and substituting for A, B, C, D, E,

$$u_n = \frac{1}{8}\cdot 1 - \frac{1}{4}\cdot 3^n + \frac{1}{8}3^n + \frac{1}{8}5^n$$

$$-\frac{u_1 - 3u_0}{2}\cdot 1 + \frac{u_1 - u_0}{2}\cdot 3^n$$

$$= \left(\frac{1}{8} + \frac{3u_0 - u_1}{2}\right) + \left(\frac{u_1 - u_0}{2} - \frac{1}{4}\right)3^n + \frac{1}{8}\cdot 5^n \qquad (4)$$

Example 11.27

Solve $u_{n+2} - 4u_{n+1} + 4u_n = 2^n$.

Solution Writing the equation in operator form

$$(E^2 - 4E + 4)u_n \equiv (E-2)^2 u_n = 2^n \qquad (1)$$

The auxiliary equation $x^2 - 4x + 4 = 0$ has two equal roots $x = 2, 2$

$$CF = (c_1 + c_2 n)2^n \qquad (2)$$

$$PI = \frac{1}{(E-2)^2} 2^n = \frac{n(n-1)}{2!} 2^{n-2} = n(n-1)2^{n-3} \qquad (3)$$

$$\therefore \quad CS \text{ is } u_n = (c_1 + c_2 n)2^n + n(n-1)2^{n-3} \qquad (4)$$

Example 11.28

Solve

$$y_{n+2} - 2y_{n+1} + y_n = n^2 \cdot 2^n \qquad (1)$$

Solution Writing the equation in operator notation $(E^2 - 2E + 1)y_n = n^2 \cdot 2^n$. The auxiliary equation $x^2 - 2x + 1 = 0$ has equal roots $x = 1, 1$

$$CF = c_1 + c_2 n \qquad (2)$$

$$PI = \frac{1}{(E-2)^2} 2^n \cdot n^2$$

$$= 2^n \frac{1}{(2E-1)^2} n^2 = 2^n \frac{1}{(1+2\Delta)^2} n^2.$$

$$= 2^n (1 + 2\Delta)^{-2} [n(n-1) + n]$$

$$= 2^n (1 - 4\Delta + 12\Delta^2 \cdots) [(n)_2 + (n)_1]$$

$$= 2^n [(n)_2 + (n)_1 - 4 \quad 2(n)_1 + 1 + 12 \times 2] \qquad (3)$$

$$= 2^n [(n)_2 - 7(n)_1 + 20] = 2^n(n^2 - 8n + 20)$$

Hence, the complete solution is

$$y_n = c_1 + c_2 n + 2^n (n^2 - 8n + 20) \qquad (4)$$

EXERCISE 11.3

1. Solve $y_{n+2} - 7y_{n+1} + 12y_n = 0$ given that $y_0 = 1, y_1 = 2$.

   Ans: $y_n = 2 . 3^n - 4^n$

2. Solve $y_{n+2} - 5y_{n+1} - 6y_n = 2^n$.

   Ans: $y_n = c_1 6^n + c_2(-1)^n - \frac{1}{3} 2^{n-2}$

3. Solve $y_{n+2} - 5y_{n+1} + 9y_n = 3^n$ given that $y_0 = 0, y_1 = 1$.

   Ans: $y_n = \frac{1}{18}(5n \cdot 3^n + n^2 \cdot 3^n)$

4. Solve $u_{n+2} - 3u_{n+1} + 2u_n = 4^n$ given that $u_0 = 0, u_1 = 1$.

   Ans: $y_n = -\frac{2}{3} + 2^{n-1} + \frac{1}{3} 2^{n-1}$

5. Solve $y_{n+3} - 7y_{n+2} + 14y_{n+1} - 8y_n = 0$.

   Ans: $y_n = c_1 + c_2 2^n + c_3 2^{2n}$

6. Solve $y_{n+2} - 5y_{n+1} + 6y_n = n + 2^n$.

   Ans: $y_n = c_1 \cdot 2^n + c_2 \cdot 3^n + \left(\frac{n+1}{2}\right) - n \cdot 2^{n-1}$

7. Solve $y_{n+2} - 2y_{n+1} + y_n = 3n + 5$ with $y_0 = 1, y_1 = 3$.

   Ans: $y_n = (2n+1) + \frac{1}{2} n(n-1)(n+3)$

8. Solve $y_{n+2} - 4y_{n+1} + 4y_n = n^2 \cdot 2^n$ with $y_0 = 2, y_1 = 3$.

   Ans: $y_n = \left[(n+2) + \frac{n(n-1)^2(n-2)}{48}\right] 2^n$

Special Functions

12

12.1 Introduction

In the study of Mathematics, we have so far dealt with algebraic and transcendental functions $x^n, \sum a_n x^n$, $e^x, \log x, \sin x, \cos x, \sinh x, \cosh x$ and so on, which are called elementary functions.

In solving higher engineering problems, there arises a need to use higher transcendental functions which are of two types:

1. Functions defined by improper integrals and
2. Functions defined by convergent infinite series, which are solutions of linear differential equations.

A study of these functions comprises the theory of special functions and we undertake here a brief study of some of the important properties of the following special functions:

The gamma and the beta functions; and the Legendre's and the Bessel's functions.

12.2 Gamma Function

The gamma function, denoted by $\Gamma(p)$, is defined by the integral

$$\Gamma(p) = \int_0^\infty e^{-t} t^{p-1} dt \qquad (p > 0) \qquad (12.1)$$

for all positive real values of p since the improper integral in Eq. (12.1) is convergent for $p > 0$ ($Re\ p > 0$ if P is complex).

The gamma function is known as Euler's integral of the second kind. Replacing p by $(p+1)$ in Eq. (12.1) and integrating by parts we obtain from Eq. (12.1)

$$\Gamma(p+1) = \int_0^\infty e^{-t} t^p \, dt$$

$$= t^p \left(\frac{e^{-t}}{-1} \right) \Big|_0^\infty + p \int_0^\infty t^{p-1} e^{-t} dt$$

$$= p\Gamma(p), \quad \text{[JNTU 2003S]} \qquad (12.2)$$

$$\left(\because \lim_{t \to \infty} t^p e^{-t} = \lim_{t \to \infty} \frac{t^p}{e^t} = 0; \lim_{t \to 0} \frac{t}{e^t} = 0 \right)$$

which is the functional relation of the gamma function.

12.3 Recurrence Relation or Reduction Formula

For $p = 1$, we get from Eq. (12.1)

$$\Gamma(1) = \int_0^\infty e^{-t} t^0 dt = \frac{e^{-t}}{-1} \Big|_0^\infty = 0 - (-1) = 1$$

If p is a positive integer and $p = 2, 3, 4, \ldots, n$, we have from Eq. (12.2)

$$\Gamma(2) = 1\Gamma(1) = 1 = 1! \qquad \because \Gamma(1) = 1$$

$$\Gamma(3) = 2\Gamma(2) = 2.1! = 2!$$

$$\Gamma(4) = 3\Gamma(3) = 3.2! = 3!$$

$$\ldots \quad \ldots$$

$$\Gamma(n) = (n-1)\Gamma(n-1) = (n-1)(n-2)!$$

$$= (n-1)(n-2)\cdots 2.1 = (n-1)! \qquad (12.3)$$

This is a recurrence relation or reduction formula for the gamma function.

This further shows that the gamma function is a generalisation of the elementary factorial function and so we can call it "the generalised factorial function".

12.3.1 Gamma Function for Negative Non-Integer Values

Rewrite Eq. (12.2) as follows:

$$\Gamma(p)=\frac{\Gamma(p+1)}{p} \qquad (12.4)$$

Letting $p\to 0^+$ and 0^-, we get

$$\Gamma(0^+)=\lim_{p\to 0^+}\frac{\Gamma(1)}{p}=\lim_{p\to 0^+}\frac{1}{p}=+\infty$$

$$\Gamma(0^-)=\lim_{p\to 0^-}\frac{\Gamma(1)}{p}=\lim_{p\to 0^-}\frac{1}{p}=-\infty$$

respectively.

This shows that $\Gamma(0)$ is undefined. It follows from Eq. (12.4) that $\Gamma(-1),\Gamma(-2),\Gamma(-3),\ldots$ are all undefined. By repeated application of Eq. (12.4), we get

$$\Gamma(p)=\frac{\Gamma(p+1)}{p}=\frac{\Gamma(p+2)}{p(p+1)}=\cdots$$

$$=\frac{\Gamma(p+k+1)}{p(p+1)\cdots(p+k)} \qquad (12.5)$$

We can determine the values of the gamma function for any real $p<0$ (except for $p=0,-1,-2,\ldots$).

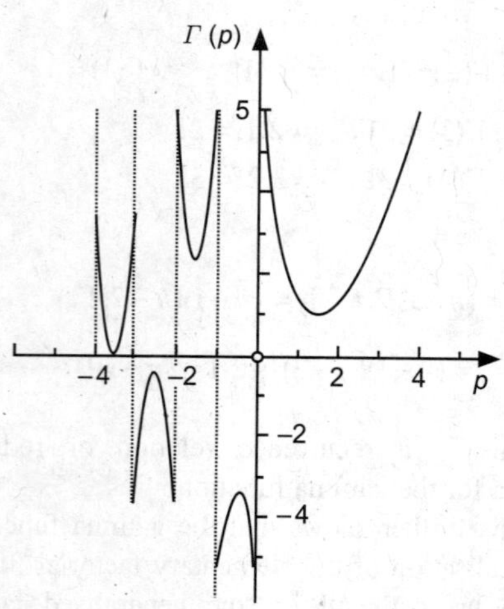

Figure 12.1 Gamma function

The gamma function $\Gamma(p)$ is defined and continuous for any $p>0$, and it is undefined and is discontinuous for $p=0,-1,-2,-3,\ldots$

We may use the relation

$$\Gamma(p)=\frac{\Gamma(p+k+1)}{p(p+1)\cdots(p+k)} \quad (p\neq 0,-1,-2,\ldots) \quad (12.6)$$

for defining the gamma function for negative p $(\neq -1,-2,\ldots)$ choosing for k the smallest integer such that $p+k+1>0$. Then Eq. (12.6) together with Eq. (12.1) gives a definition of $\Gamma(p)$ for all p not equal to zero or a negative integer (Fig. 12.1).

Example 12.1

$$\Gamma(2.5)=(1.5)\Gamma(1.5)=(1.5)(0.5)\Gamma(0.5)$$
$$=\frac{3}{2}\times\frac{1}{2}\times\sqrt{\pi}=\frac{3}{4}\times 1.772$$
$$=3\times 0.443=1.329.$$

Example 12.2

Evaluate $\Gamma(-1.5)$.

Solution We have $p+k+1=-1.5+k+1>0$ or $k>0.5$. Then the least integer value of $k=1$

$$\Gamma(-1.5)=\frac{\Gamma(-1.5+1+1)}{(-1.5)(-0.5)}=\frac{4}{3}\Gamma(0.5)$$
$$=\frac{4}{3}\sqrt{\pi}=\frac{4}{3}\times 1.772$$
$$=\frac{7.088}{3}=2.363.$$

(We will show below $\Gamma(0.5)=\Gamma(1/2)=\sqrt{\pi}=1.772$ approx.).

12.3.2 Some Standard Results

1. Show that $\Gamma(1/2)=\sqrt{\pi}=1.772$. [JNTU 2002]

Solution By definition

$$\Gamma(p)=\int_0^\infty e^{-t}t^{p-1}dt \quad (p>0)$$

Take $p=\frac{1}{2}$

$$\Gamma\left(\frac{1}{2}\right)=\int_0^\infty e^{-t}t^{\frac{1}{2}-1}\,dt=\int_0^\infty e^{-t}t^{-\frac{1}{2}}dt$$

Put $t = x^2$, $dt = 2xdx$
The limits for x are 0 and ∞.

$$\Gamma(1/2) = 2\int_0^\infty e^{-x^2} dx$$

Similarly $\Gamma(1/2) = 2\int_0^\infty e^{-y^2} dy$

Multiplying the two results, we get

$$[\Gamma(1/2)]^2 = 4\int_0^\infty e^{-x^2} dx \int_0^\infty e^{-y^2} dy$$

$$= 4\int_0^\infty \int_0^\infty e^{-(x^2+y^2)} dxdy$$

The double integral in the first quadrant is evaluated by changing to polar coordinates $x = r\cos\theta,\ y = r\sin\theta,\ x^2 + y^2 = r^2$

$$J = \frac{\partial(x,y)}{\partial(r,\theta)} = \begin{vmatrix} \cos\theta & -r\sin\theta \\ \sin\theta & r\cos\theta \end{vmatrix} = r$$

The limits are $\begin{cases} \textit{for } \theta: & 0 \text{ to } \pi/2 \\ \textit{for } r: & 0 \text{ to } \infty \end{cases}$

$$\therefore [\Gamma(1/2)]^2 = 4\int_{r=0}^\infty e^{-r^2} r\,dr \int_0^{\pi/2} d\theta$$

$$= 4\cdot\frac{\pi}{2}\int_0^\infty \left(-\frac{1}{2}\right)\frac{d}{dr}(e^{-r^2})dr$$

$$= \pi$$

$$\Rightarrow \Gamma(1/2) = \sqrt{\pi} = 1.772 \text{ (approx.).}$$

Corollary

1. $\int_0^\infty e^{-x^2} dx = \frac{1}{2}\Gamma\left(\frac{1}{2}\right) = \frac{\sqrt{\pi}}{2} = 0.886$

2. $\int_{-\infty}^0 e^{-x^2} dx = -\int_0^\infty e^{-x^2} dx = \int_0^\infty e^{-(-x)^2} d(-x)$

$$= \frac{\sqrt{\pi}}{2} = 0.886$$

3. $\int_{-\infty}^\infty e^{-x^2} dx = \int_{-\infty}^0 e^{-x^2} dx + \int_0^\infty e^{-x^2} dx$

$$= \frac{\sqrt{\pi}}{2} + \frac{\sqrt{\pi}}{2} = \sqrt{\pi} = 1.772$$

The following results are easily obtained:

1. $\Gamma(3/2) = \frac{1}{2}\Gamma\left(\frac{1}{2}\right) = \frac{\sqrt{\pi}}{2} = \frac{1.772}{2} = 0.886$

2. $\Gamma(-1/2) = \frac{\Gamma\left(-\frac{1}{2}+1\right)}{\left(-\frac{1}{2}\right)} = -2\sqrt{\pi}$

$$= -2\times 1.772 = -3.544$$

3. $\Gamma(-3/2) = \frac{\Gamma\left(-\frac{3}{2}+2\right)}{\left(-\frac{3}{2}\right)\left(-\frac{1}{2}\right)} = \frac{4}{3}\Gamma(1/2)$

$$= \frac{4}{3}\sqrt{\pi} = \frac{4}{3}\times 1.772 = 2.363$$

12.4 VARIOUS INTEGRAL FORMS OF GAMMA FUNCTION

12.4.1 Form I: Integral of Log Function

Prove that $\int_0^1 (\log 1/y)^{p-1} dy = \Gamma(p)$

[JNTU 2003(1)]

Solution Consider $\int_0^1 (\log 1/y)^{p-1} dy$

Put $\log 1/y = x \Leftrightarrow y = e^{-x}, dy = -e^{-x}dx$

When $y \to 1, x \to 0$ and when $y \to 0, x \to \infty$

$$\therefore \int_0^1 (\log 1/y)^{p-1} dy = \int_\infty^0 x^{p-1}(-e^{-x})dx$$

$$= \int_0^\infty e^{-x}x^{p-1} dx = \Gamma(p)$$

Corollary Put $p = \frac{1}{2}$, we get

$$\int_0^1 \left(\log\frac{1}{y}\right)^{-1/2} dy = \int_0^1 \frac{dy}{\sqrt{\log 1/y}}$$

$$= \Gamma(1/2) = \sqrt{\pi}$$

12.4.2 Form II: Exponential Function

Prove that $\int_0^\infty -e^{y^{1/p}} dy = p\Gamma(p)$ or $\Gamma(p+1)$.

[JNTU 2003(4), 2004(1)]

Solution Consider $\int_0^\infty -e^{y^{1/p}} dy$

Put $y^{1/p} = x \;\Rightarrow\; y = x^p, dy = px^{p-1}dx$

When $y = 0, x = 0$ and when $y = \infty, x = \infty$

$$\therefore \int_0^\infty e^{-y^{1/p}} dy = \int_0^\infty e^{-x} px^{p-1} dx = p\int_0^\infty e^{-x} x^{p-1} dx$$

$$= p\Gamma(p) = \Gamma(p+1)$$

Corollary $p = \frac{1}{2}, \int_0^\infty e^{-y^{\frac{1}{1/2}}} dy = \int_0^\infty e^{-y^2} dy$

$$= \frac{1}{2}\Gamma\left(\frac{1}{2}\right) = \frac{\sqrt{\pi}}{2}$$

12.4.3 Form III: Scaling of Variable of Integration

Show that $\int_0^\infty - e^{at} t^{p-1} dt = \frac{\Gamma(p)}{a^p}$

Solution We know that $\Gamma(p) = \int_0^\infty e^{-t} t^{p-1} dt$

Put $t = ax,\ dt = adx$ in the right-side integral

The limits are $t = 0,\ x = 0;\quad t = \infty,\ x = \infty.$

$$\Gamma(p) = \int_0^\infty e^{-ax} (ax)^{p-1} d(ax) = a^p \int_0^\infty e^{-ax} x^{p-1} dx$$

$$\Rightarrow \int_0^\infty e^{-ax} x^{p-1} dx = \frac{\Gamma(p)}{a^p}$$

12.4.4 Form IV: The Product of a Power Function and a Logarithmic Function

Show that $\int_0^1 t^{q-1} (\log 1/t)^{p-1} dt = \frac{\Gamma(p)}{q^p}$

Solution Consider $I = \int_0^1 t^{q-1} (\log 1/t)^{p-1} dt$

Put $\log(1/t) = x$ or $t = e^{-x},\ dt = -e^{-x} dx$

When $t = 0,\ x = \infty;\ t = 1,\ x = 0$

The given integral

$$I = \int_0^\infty (e^{-x})^{q-1} x^{p-1} (-e^{-x}) dx$$

$$= \int_0^\infty e^{-xq} x^{p-1} dx = \frac{\Gamma(p)}{q^p}, \text{ by Form I}$$

Corollary Put $p = m+1,\ q = n+1$

$$\int_0^1 t^n (-\log t)^m dt = (-1)^m \frac{\Gamma(m+1)}{(n+1)^{m+1}}$$

12.4.5 Form V: Product of a Power Function and an Exponential Function

Show that $\int_0^\infty (x^p / p^x) dx = \frac{\Gamma(p+1)}{(\log p)^{p+1}}$

Solution Consider $I = \int_0^\infty (x^p / p^x) dx$

Put $p^x = e^t \Rightarrow x \log p = t,\ dt = (\log p) dx$

When $x = 0,\ t = 0$ and when $x = \infty,\ t = \infty$

$$I = \int_0^\infty (x^p / p^x) dx = \int_0^\infty \left(\frac{t}{\log p}\right)^p \cdot e^{-t} \cdot \frac{dt}{\log p}$$

$$= \frac{1}{(\log p)^{p+1}} \int_0^\infty e^{-t} t^p dt = \frac{\Gamma(p+1)}{(\log p)^{p+1}}$$

Example 12.3

Show that $\int_0^\infty e^{-x^p} x^q dx = \frac{1}{p}\Gamma\left(\frac{q+1}{p}\right)$

Solution Consider $I = \int_0^\infty e^{-x^p} x^q dx$

Put $x^p = t \Rightarrow x = t^{1/p},\ dx = \frac{1}{p} t^{\frac{1}{p}-1} dt$

$$x^q dx = \frac{1}{p} t^{\frac{q+1}{p}-1} dt$$

$$I = \int_0^\infty e^{-x^p} x^q dx = \int_0^\infty e^{-t} t^{\frac{q}{p}} \cdot \frac{1}{p} t^{\frac{1}{p}-1} dt$$

$$= \frac{1}{p} \int_0^\infty e^{-t} t^{\frac{q+1}{p}-1} dt = \frac{1}{p}\Gamma\left(\frac{q+1}{p}\right).$$

The following results are easily obtained:

1. $\int_0^\infty e^{-x^2} x^2 dx = \frac{1}{2}\Gamma\left(\frac{3}{2}\right) = \frac{1}{2} \cdot \frac{1}{2}\Gamma(1/2)$

$$= \frac{\sqrt{\pi}}{4} = \frac{1.772}{4}$$

$$= 0.443\ (p = 2, q = 2)$$

2. $\int_0^\infty e^{-x^2} x dx = \frac{1}{2}\Gamma\left(\frac{1+1}{2}\right) = \frac{1}{2}\ (p = 2, q = 1)$

3. $\int_0^\infty e^{-\sqrt{x}} x dx = 2\Gamma(4) = 2(3!) = 12$

$$\left(p = \frac{1}{2}, q = 1\right)$$

4. $\int_0^\infty (x-1)^2 e^{-\sqrt{x}} dx = \int_0^\infty (x^2 - 2x + 1) e^{-\sqrt{x}} dx$

$= 2[\Gamma(6) - 2\Gamma(4) + 1\Gamma(2)]$

$= 2(5!) - 4(3!) + 2 = 218$

5. $\int_0^\infty e^{-x^{1/3}} dx = 3\Gamma(3) = 6 \quad \left(p = \frac{1}{3}, q = 0\right)$

6. $\int_0^\infty e^{-x^{1/3}} x dx = 3\Gamma\left(\frac{1+1}{1/3}\right) = 3(5!) = 360$

$$\left(p = \frac{1}{3}, q = 1\right)$$

7. $\int_0^\infty (x-1)^2 e^{-x^{1/3}} dx = \int_0^\infty (x^2 - 2x + 1) e^{-x^{1/3}} dx$

$= 3[\Gamma(3\times2+3) - 2\Gamma(3\times1+3) + \Gamma(3\times0+3)]$

$$\left(p = \frac{1}{3}; q = 2, 1, 0 \text{ in turn}\right)$$

$= 3[8! - 2(5!) + 2]$

8. $\int_0^\infty (x-1)^2 e^{-x^3} dx = \frac{1}{3}[1 - 2\Gamma(2/3) + \Gamma(1/3)]$

$$[p = 3; q = 2, 1, 0 \text{ in turn}]$$

9. $\int_0^\infty e^{-x^p} dx = \frac{1}{p}\Gamma\left(\frac{1}{p}\right) \text{ or } \Gamma\left(\frac{1}{p} + 1\right) \quad (q = 0)$

Example 12.4

Evaluate $\int_0^\infty e^{-\sqrt{x}} dx$.

Solution Let $I = \int_0^\infty e^{-\sqrt{x}} dx$

Put $\sqrt{x} = t$ or $x = t^2 \Rightarrow dx = 2t\,dt$

The limits for t are 0 and ∞.

$$\therefore I = \int_0^\infty e^{-t} \cdot 2t\,dt = 2\int_0^\infty e^{-t} \cdot t^{2-1} dt$$

$$= 2\Gamma(2) = 2 \cdot 1\Gamma(1) = 2 \quad \because \Gamma(1) = 1.$$

Example 12.5

Evaluate $\int_0^\infty e^{-\sqrt{x}} x dx$.

Solution Let $I = \int_0^\infty e^{-\sqrt{x}} x dx$

Put $\sqrt{x} = t$ or $x = t^2 \Rightarrow dx = 2tdt$.

The limits for t are 0 and ∞.

$$\therefore I = \int_0^\infty e^{-t} \cdot t^2 \cdot 2t dt = 2\int_0^\infty e^{-t} \cdot t^{4-1} dt$$

$$= 2\Gamma(4) = 2(3!) = 12.$$

Example 12.6

Evaluate $\int_0^\infty (x-1) e^{-\sqrt{x}} dx$.

Solution From Eqs. (12.1) and (12.2), we have

$$\int_0^\infty e^{-\sqrt{x}} dx = 2, \int_0^\infty x e^{-\sqrt{x}} dx = 12$$

$$\therefore \int_0^\infty (x-1) e^{-\sqrt{x}} dx = \int_0^\infty e^{-\sqrt{x}} x dx - \int_0^\infty e^{-\sqrt{x}} dx$$

$$= 12 - 2 = 10.$$

Example 12.7

Evaluate $\int_0^\infty e^{-ax^2} dx$.

Solution Let $I = \int_0^\infty e^{-ax^2} dx$

Put $ax^2 = t \Rightarrow a^2x^2 = at \Rightarrow ax = \sqrt{at}$

$$\Rightarrow dx = \frac{1}{\sqrt{a}} \cdot \frac{1}{2\sqrt{t}} dt.$$

The limits are the same.

$$\therefore I = \int_0^\infty e^{-t} \frac{1}{2\sqrt{at}} dt = \frac{1}{2\sqrt{a}} \int_0^\infty e^{-t} t^{\frac{1}{2}-1} dt$$

$$= \frac{1}{2\sqrt{a}} \Gamma(1/2) = \frac{1}{2\sqrt{a}} \sqrt{\pi} = \frac{1}{2}\sqrt{\frac{\pi}{a}}.$$

Example 12.8

Show that $\int_0^\infty e^{-x^p} dx = \frac{1}{p}\Gamma\left(\frac{1}{p}\right) \text{ or } \Gamma\left(\frac{1}{p} + 1\right)$.

Solution Consider $I = \int_0^\infty e^{-x^p} dx$

Put $x^p = t \Rightarrow x = t^{\frac{1}{p}}, \quad dx = \frac{1}{p} t^{\frac{1}{p}-1} dt$

The limits are the same.

$$\therefore I = \int_0^\infty e^{-x^p} dx = \frac{1}{p} \int_0^\infty e^{-t} t^{\frac{1}{p}-1} dt$$

$$= \frac{1}{p}\Gamma\left(\frac{1}{p}\right) \text{ or } \Gamma\left(\frac{1}{p}+1\right).$$

Example 12.9

Evaluate $\int_0^1 x^4\left(\log\frac{1}{x}\right)^3 dx$.

Solution Put

$x = e^{-t/5} \Rightarrow \log\frac{1}{x} = \frac{t}{5} \quad dx = -\frac{1}{5}e^{-\frac{t}{5}}\,dt$. When

$x = 0, t = \infty;\ \ x = 1, t = 0$

$$\int_0^1 x^4\left(\log\frac{1}{x}\right)^3 dx = -\int_0^\infty e^{-\frac{4}{5}t}\left(\frac{t}{5}\right)^3\frac{1}{5}e^{-\frac{1}{5}t}\,dt$$

$$= \frac{1}{625}\int_0^\infty e^{-t}t^3\,dt = \frac{1}{625}\Gamma(4) = \frac{6}{625}.$$

Example 12.10

If $n \in \mathbb{N}$ then show that

$$2^n\,\Gamma\left(n+\frac{1}{2}\right) = 1\cdot 3\cdot 5\cdots(2n-1)\sqrt{\pi}.$$

Solution We know that $\Gamma(p+1) = p\Gamma(p)$

Taking

$p = \left(n-\frac{1}{2}\right), \left(n-\frac{3}{2}\right), \left(n-\frac{5}{2}\right), \ldots, \left(n-\frac{2n-1}{2}\right)$ in turn

$$\Gamma\left(n+\frac{1}{2}\right) = \Gamma\left(n-\frac{1}{2}+1\right)$$

$$= \left(n-\frac{1}{2}\right)\Gamma\left(n-\frac{1}{2}\right)$$

$$= \left(n-\frac{1}{2}\right)\left(n-\frac{3}{2}\right)\Gamma\left(n-\frac{3}{2}\right)$$

$$= \left(n-\frac{1}{2}\right)\left(n-\frac{3}{2}\right)\left(n-\frac{5}{2}\right)\cdots\frac{1}{2}\Gamma\left(\frac{1}{2}\right)$$

$$= \left(\frac{2n-1}{2}\right)\left(\frac{2n-3}{2}\right)\left(\frac{2n-5}{2}\right)\cdots\frac{5}{2}\cdot\frac{3}{2}\cdot\frac{1}{2}\Gamma\left(\frac{1}{2}\right)$$

$$2^n\Gamma\left(n+\frac{1}{2}\right) = 1\cdot 3\cdot 5\cdot 7\cdots(2n-3)(2n-1)\sqrt{\pi}.$$

Example 12.11

Show that $\int_0^\infty x^4 e^{-x^2}\,dx = \frac{3}{8}\sqrt{\pi}$. [JNTU 2004(3)]

Solution To evaluate the integral put $x^2 = y$

$\Rightarrow x = y^{\frac{1}{2}}, dx = \frac{1}{2\sqrt{y}}dy$

$$\int_0^\infty x^4 e^{-x^2}\,dx = \int_0^\infty e^{-y}\cdot y^2\cdot\frac{1}{2\sqrt{y}}dy$$

$$= \frac{1}{2}\int_0^\infty e^{-y}y^{\frac{5}{2}-1}\,dy = \frac{1}{2}\Gamma(5/2)$$

$$= \frac{1}{2}\cdot\frac{3}{2}\cdot\frac{1}{2}\sqrt{\pi} \quad \because \Gamma(1/2) = \sqrt{\pi}$$

$$= \frac{3}{8}\sqrt{\pi}.$$

EXERCISE 12.1

1. Prove that

$$\int_0^\infty (x-1)^2 e^{-x^3}\,dx = \frac{1}{3}[1 - 2\Gamma(2/3) + \Gamma(1/3)].$$

2. Prove that $\Gamma\left(n+\frac{1}{2}\right) = \frac{(2n)!\sqrt{\pi}}{4^n(n!)}$.

3. Prove that $\int_0^1 x^3(\log x)^3\,dx = -3/128$.

4. Prove that $\int_0^\infty (e^{-st}/\sqrt{t})dt = \sqrt{\pi/s} \quad (s > 0)$.

5. Prove that $\int_0^\infty x^{2n-1}e^{-ax^2}\,dx = \frac{\Gamma(n)}{2a^n}$.

   [**Hint:** Put $ax^2 = t$]

6. Prove that $\int_0^\infty \sqrt{x}e^{-x^2}\,dx = \int_0^\infty \frac{e^{-x^2}}{\sqrt{x}}dx = \frac{\pi}{2\sqrt{2}}$.

7. Prove that $\int_{-\infty}^\infty e^{-a^2x^2}\,dx = \frac{\sqrt{\pi}}{a}$.

8. Prove that $\int_0^1 x^m(\log x)^n\,dx = \frac{(-1)^n(n!)}{(m+1)^{n+1}}$;

   $n \in \mathbb{N}, m > -1$. [JNTU 2001 S, 2004(4)]

9. Prove that $\int_0^\infty x^m e^{-ax^n}\,dx = \frac{1}{na^{(m+1)/n}}\Gamma\left(\frac{m+1}{n}\right)$.

   m, n are positive constants. [JNTU 2004(2)]

12.5 Beta Function

The beta function denoted by $B(p, q)$, a function of two variables p and q, is defined by the integral

$$B(p,q) = \int_0^1 t^{p-1}(1-t)^{q-1}\,dt \ (p > 0, q > (0) \qquad (12.7)$$

for all positive real values of p and q since the improper integral Eq. (12.7) is convergent for $p > 0$, $q > 0$ ($Re\, p > 0$, $Re\, q > 0$ if p and q are complex).

The beta function is known as Euler's integral of the first kind.

12.5.1 Some Standard Results

Symmetry

1. Show that $B(p, q) = B(q, p)$.

Solution Recall the following property of definite integrals

$$\int_0^a f(x)dx = \int_0^a f(a-x)dx$$

$$B(p, q) = \int_0^1 t^{p-1}(1-t)^{q-1}dt$$

$$= \int_0^1 (1-t)^{p-1}[1-(1-t)]^{q-1}dt$$

$$= \int_0^1 t^{q-1}(1-t)^{p-1}dt = B(q, p).$$

12.6 Various Integral Forms of Beta Function

12.6.1 Form I: Beta Function as an Infinite Integral

Show that $B(p, q) = \int_0^\infty \frac{x^{p-1}}{(1+x)^{p+q}}dx$

$$= \int_0^\infty \frac{x^{q-1}}{(1+x)^{p+q}}dx.$$

[JNTU 2002*S*, 2004(4)]

Solution We know that

$$B(p, q) = \int_0^1 t^{p-1}(1-t)^{q-1}dt$$

Put $t = \frac{x}{1+x}$

$$\Rightarrow 1-t = 1-\frac{x}{1+x} = \frac{1}{1+x}$$

$$\Rightarrow x = \frac{1}{1-t} - 1 = \frac{t}{1-t}$$

$$\Rightarrow dt = \frac{(1+x)dx - xdx}{(1+x)^2} = \frac{dx}{(1+x)^2}$$

Limits for x are $t = 0 \Rightarrow x = 0$;

$$t = 1 \Rightarrow x = \infty.$$

$$B(p, q) = \int_0^\infty \left(\frac{x}{1+x}\right)^{p-1} \frac{1}{(1+x)^{q-1}} \frac{dx}{(1+x)^2}$$

$$= \int_0^\infty \frac{x^{p-1}}{(1+x)^{p+q}}dx$$

$$= \int_0^\infty \frac{x^{q-1}}{(1+x)^{p+q}}dx \quad \because B(p, q) = B(q, p).$$

12.6.2 Form II: Beta Function in Symmetric Integral Form

Show that

$$B(p, q) = \int_0^1 \frac{x^{p-1} + x^{q-1}}{(1+x)^{p+q}}dx. \qquad \text{[JNTU 2000]}$$

Solution Form I representation of beta function

$$B(p, q) = \int_0^\infty \frac{x^{p-1}}{(1+x)^{p+q}}dx$$

Splitting the interval of integration 0 to ∞ into 0 to 1 and 1 to ∞, we have

$$B(p, q) = \int_0^1 \frac{x^{p-1}}{(1+x)^{p+q}}dx + \int_1^\infty \frac{x^{p-1}}{(1+x)^{p+q}}dx$$

$$= I_1 + I_2 \quad \text{(say)}$$

Consider the second integral

$$I_2 = \int_1^\infty \frac{x^{p-1}}{(1+x)^{p+q}}dx$$

Put $x = \frac{1}{t} \Rightarrow dx = -\frac{1}{t^2}dt$

The limits are when $x = 1, t = 1$ and when $x = \infty, t = 0$.

$$\therefore \quad I_2 = \int_1^\infty \frac{x^{p-1}}{(1+x)^{p+q}}dx$$

$$= \int_1^0 \frac{1}{t^{p-1}}\left(\frac{t}{1+t}\right)^{p+q}\left(-\frac{1}{t^2}dt\right)$$

$$= \int_0^1 \frac{t^{q-1}}{(1+t)^{p+q}}dt$$

$$= \int_0^1 \frac{x^{q-1}}{(1+x)^{p+q}}dx, \quad \text{replacing } t \text{ by } x$$

$$\therefore \quad B(p, q) = I_1 + I_2$$

$$= \int_0^1 \frac{x^{p-1}}{(1+x)^{p+q}}dx + \int_0^1 \frac{x^{q-1}}{(1+x)^{p+q}}dx$$

$$= \int_0^1 \frac{x^{p-1}+x^{q-1}}{(1+x)^{p+q}}dx$$

12.6.3 Form III: Improper Integral Form

Show that $B(p,q) = a^p b^q \int_0^\infty \frac{x^{p-1}}{(ax+b)^{p+q}}dx.$

Solution Consider

$$I = \int_0^\infty \frac{x^{p-1}}{(ax+b)^{p+q}}dx$$

$$= \frac{1}{b^{p+q}}\int_0^\infty \frac{x^{p-1}dx}{\left(\frac{ax}{b}+1\right)^{p+q}}$$

Put $\frac{ax}{b} = t$ or $x = \frac{b}{a}t$; $dx = \frac{b}{a}dt$

$$\therefore\ I = \frac{1}{b^{p+q}}\int_0^\infty \frac{\left(\frac{b}{a}t\right)^{p-1}}{(t+1)^{p+q}}\frac{b}{a}dt$$

$$= \frac{1}{a^p b^q}\int_0^\infty \frac{t^{p-1}}{(1+t)^{p+q}}dt$$

$$= \frac{1}{a^p b^q}B(p,q), \quad \text{by Form I}$$

12.6.4 Form IV: Integral from 0 to 1 Form

Show that

$$B = (p,q) = (1+a)^p a^q \int_0^1 \frac{x^{p-1}(1-x)^{q-1}}{(x+a)^{p+q}}dx.$$

Solution Beta function is defined by

$$B(p,q) = \int_0^1 t^{p-1}(1-t)^{q-1}dt, \quad (p>0, q>0)$$

Put $t = \frac{(1+a)x}{(x+a)} \Rightarrow \frac{1}{t}dx = \left(\frac{1}{x} - \frac{1}{x+a}\right)dx$

$$= \frac{a}{x(x+a)}dx, \text{ by log differentiation}$$

The limits are $t = 0, x = 0; t = 1, x = 1.$

$$\therefore B(p,q) = \int_0^\infty \frac{(1+a)^{p-1}x^{p-1}}{(x+a)^{p-1}} \cdot \left[1 - \frac{(1+a)x}{x+a}\right]^{q-1} \cdot \frac{(1+a)a dx}{(x+a)^2}$$

$$= (1+a)^p a^q \int_0^1 \frac{x^{p-1}(1-x)^{q-1}}{(x+a)^{p+q}}dx.$$

12.6.5 Form V: Integral from *a* to *b* Form

Show that

$$\int_a^b (x-a)^{p-1}(b-x)^{q-1}dx = (b-a)^{p+q-1}B(p,q)$$

[JNTU 2003(4), 2004(1)]

Solution In the beta function integral

$$B(p,q) = \int_0^1 t^{p-1}(1-t)^{q-1}dt$$

Put $t = \frac{x-a}{b-a} \Rightarrow dt = \frac{dx}{(b-a)}$

The limits are $t = 0, x = a$ and $t = 1, x = b$.

$$\therefore B(p,q) = \int_a^b \frac{(x-a)^{p-1}}{(b-a)^{p-1}} \frac{(b-x)^{q-1}}{(b-a)^{q-1}} \frac{dx}{(b-a)}$$

$$= \frac{1}{(b-a)^{p+q-1}}\int_a^b (x-a)^{p-1}(b-x)^{q-1}dx.$$

Corollary

Put $a = -1, b = 1$; and $a = 0$ we get respectively,

1. $\int_{-1}^1 (x+1)^{p-1}(1-x)^{q-1}dx = 2^{p+q-1}B(p,q).$ [JNTU 2004(3)]
2. $\int_0^b x^{p-1}(b-x)^{q-1}dx = b^{p+q-1}B(p,q)$.

12.6.6 Form VI: Integral of Circular Functions

Show that $B(p,q) = 2\int_0^{\frac{\pi}{2}} \sin^{2p-1}\theta \cos^{2q-1}\theta d\theta$

$p>0, \quad q>0$. [JNTU 2003(2)]

Solution We know that

$$B(p,q) = \int_0^1 t^{p-1}(1-t)^{q-1}dt$$

Put $t=\sin^2\theta;\ dt=2\sin\theta\cos\theta d\theta;\ 1-t=\cos^2\theta.$

The limits are $t=0, \theta=0; t=1, \theta=\frac{\pi}{2}$.

$$B(p,q)=\int_0^{\frac{\pi}{2}}(\sin^2\theta)^{p-1}(\cos^2\theta)^{q-1}(2\sin\theta\cos\theta d\theta)$$

$$=2\int_0^{\frac{\pi}{2}}\sin^{2p-1}\theta\cos^{2q-1}\theta d\theta=2I_{2p-1,2q-1}$$

Corollary

$$I_{p,q}=\int_0^{\frac{\pi}{2}}\sin^p\theta\cos^q\theta d\theta=\frac{1}{2}B\left(\frac{p+1}{2},\frac{q+1}{2}\right)$$

12.6.7 Form VII: Relation of Proportionality

If $p>0, q>0$, then prove that

$$\frac{B(p,q+1)}{q}=\frac{B(p+1,q)}{p}=\frac{B(p,q)}{p+q}.$$

Solution

$$\frac{B(p,q+1)}{q}=\frac{1}{q}B(p,q+1)$$

$$=\frac{1}{q}\int_0^1 t^{p-1}(1-t)^{q+1-1}dt$$

$$=\frac{1}{q}\int_0^1(1-t)^q t^{p-1}dt$$

Evaluating the last integral by parts

$$=\frac{1}{q}(1-t)^q\frac{t^p}{p}\Bigg|_0^1-\frac{1}{q}\int_0^1 q(1-t)^{q-1}(-1)\frac{t^p}{p}dt$$

$$=0+\frac{1}{p}\int_0^1 t^{(p+1)-1}(1-t)^{q-1}dt=\frac{1}{p}B(p+1,q)$$

$$\therefore\ \frac{B(p,q+1)}{q}=\frac{B(p+1,q)}{p}=\frac{\text{sum of numerators}}{\text{sum of denominators}}$$

$$=\frac{B(p,q+1)+B(p+1,q)}{q+p}$$

$$=\frac{1}{p+q}\left[\int_0^1 t^{p-1}(1-t)^q dt+\int_0^1 t^p(1-t)^{q-1}dt\right]$$

$$=\frac{1}{p+q}\int_0^1 t^{p-1}(1-t)^{q-1}[(1-t)+t]dt$$

$$=\frac{1}{p+q}B(p,q)$$

We also have $B(p+1,q)=\frac{p}{p+q}B(p,q)$ and

$$B(p,q+1)=\frac{q}{p+q}B(p,q).$$

Corollary

Show that (i) $B(p,1)=\frac{1}{p}$ (ii) $B(1,q)=\frac{1}{q}$.

Solution

(i) $B(p,1)=\int_0^1 x^{p-1}(1-x)^{1-1}dx$

$$=\int_0^1 x^{p-1}dx=\frac{x^p}{p}\Bigg|_0^1=\frac{1}{p}$$

(ii) $B(1,q)=\int_0^1 x^{1-1}(x)^{q-1}dx$

$$=\int_0^1 x^{q-1}dx=\frac{x^q}{q}\Bigg|_0^1=\frac{1}{q}$$

12.6.8 Form VIII: Beta Function in Explicit Form

Show that

$$B(p,q)=\frac{(q-1)!}{p(p+1)(p+2)\cdots(p+q-2)(p+q-1)}$$

Solution

$$B(p,q)=\int_0^1 x^{p-1}(1-x)^{q-1}dx=\int_0^1(1-x)^{q-1}x^{p-1}dx$$

$$=(1-x)^{q-1}\frac{x^p}{p}\Bigg|_0^1+\frac{q-1}{p}\int_0^1(1-x)^{q-2}x^p\,dx$$

$$=0+\frac{q-1}{p}B(p+1,q-1)$$

$$=\frac{q-1}{p}\cdot\frac{q-2}{p+1}\cdot\frac{q-3}{p+2}\cdots\frac{1}{p+q-2}\cdot B(p+q-1,1)$$

$$=\frac{(q-1)(q-2)(q-3)\cdots 2\cdot 1}{p(p+1)(p+2)\cdots(p+q-2)(p+q-1)}$$

If p, q are positive integers

$$B(p,q)=\frac{(p-1)!(q-1)!}{(p+q-1)!}.$$

12.7 Relation Between Beta and Gamma Functions

Show that $B(p,q)=\dfrac{\Gamma(p)\Gamma(q)}{\Gamma(p+q)}\quad p>0, q>0$

[JNTU 2002, 2003 *S*]

Solution We know that $\dfrac{\Gamma(p)}{z^p}=\int_0^\infty e^{-zx}x^{p-1}dx$

Multiplying both sides by $z^{p+q-1}e^{-z}$ and integrating w.r.t. z from 0 to ∞

$$\Gamma(p)\int_0^\infty z^{q-1}e^{-z}dz=\int_0^\infty\left[\int_0^\infty e^{-z(1+x)}z^{p+q-1}dz\right]x^{p-1}\,dx$$

$$\Gamma(p)\Gamma(q)=\int_0^\infty\frac{\Gamma(p+q)}{(1+x)^{p+q}}x^{p-1}dx$$

$$=\Gamma(p+q)\int_0^\infty\frac{x^{p-1}}{(1+x)^{p+q}}dx$$

$$=\Gamma(p+q)\,B(p,q)$$

or $B(p,q)=\dfrac{\Gamma(p)\Gamma(q)}{\Gamma(p+q)}$

Aliter We know that $\Gamma(p)=\int_0^\infty t^{p-1}e^{-t}\,dt$

Put $t=x^2, dt=2xdx$

The limits are $t=0, x=0;\ t=\infty, x=\infty.$

$$\Gamma(p)=2\int_0^\infty x^{2p-1}e^{-x^2}dx$$

Similarly $\Gamma(q)=2\int_0^\infty y^{2q-1}e^{-y^2}dy$

$$\therefore \Gamma(p)\Gamma(q)=\left(2\int_0^\infty x^{2p-1}e^{-x^2}dx\right)\times\left(2\int_0^\infty y^{2q-1}e^{-y^2}dy\right)$$

$$=4\int_0^\infty\int_0^\infty e^{-(x^2+y^2)}x^{2p-1}y^{2q-1}dxdy$$

Put $x=r\cos\theta,\ y=r\sin\theta\quad dxdy=rd\theta dr$

$$\because J=\frac{\partial(x,y)}{\partial(r,\theta)}=\begin{vmatrix}\cos\theta & -r\sin\theta\\ \sin\theta & r\cos\theta\end{vmatrix}=r;$$

Limits for $r:0$ to ∞ ; $\theta:0$ to $\dfrac{\pi}{2}$.

$$\Gamma(p)\Gamma(q)=4\int_0^{\frac{\pi}{2}}\sin^{2q-1}\theta\cos^{2p-1}\theta d\theta$$

$$\int_0^\infty e^{-r^2}\times r^{2p+2q-1}dr$$

$$=2B(p,q)\int_0^\infty e^{-r^2}r^{2p+2q-1}dr$$

Put $r^2=u;\quad 2rdr=du$

$$=B(p,q)\int_0^\infty e^{-u}u^{p+q-1}du$$

$$=B(p,q)\Gamma(p+q)$$

$$\Rightarrow B(p,q)=\frac{\Gamma(p)\Gamma(q)}{\Gamma(p+q)}.$$

Corollary

$$I_{p,q}=\int_0^{\frac{\pi}{2}}\sin^p\theta\cos^q\theta d\theta=\frac{1}{2}B\left(\frac{p+1}{2},\frac{q+1}{2}\right)$$

$$=\frac{\Gamma\left(\frac{p+1}{2},\frac{q+1}{2}\right)}{2\Gamma\left(\frac{p+q+2}{2}\right)}.$$

12.8 Multiplication Formula

Prove that $\Gamma(p)\Gamma(1-p)=\pi\operatorname{cosec}p\pi\qquad(p\notin\mathbb{Z})$.

Solution We have the following results:

Beta–Gamma relation:

$$\Gamma(p)\Gamma(q)=\Gamma(p+q)B(p,q)\tag{12.8}$$

Infinite integral form of beta function (B Form I):

$$B(p,q)=\int_0^\infty\frac{x^{p-1}}{(1+x)^{p+q}}dx\tag{12.9}$$

By contour integration[1], the value of

$$\int_0^\infty\frac{x^{p-1}}{1+x}dx=\pi\ \operatorname{cosec}\ p\pi\tag{12.10}$$

From Eqs. (12.8) – (12.10) by taking $q=1-p$, we have

$$\Gamma(p)\Gamma(1-p)=B(p,1-p)=\int_0^\infty\frac{x^{p-1}}{1+x}dx$$

$$=\pi\ \operatorname{cosec}\ p\pi\quad(0<p<1)$$

By analytic continuation[1], we get the result

$$\Gamma(p)\Gamma(1-p)=\pi\ \operatorname{cosec}\ p\pi$$

for all non-integral values of p

[1] These are taken for granted and not discussed here.

Corollary

$$p=\frac{1}{2},\quad \left[\Gamma\left(\frac{1}{2}\right)\right]^2=\pi \Rightarrow \Gamma\left(\frac{1}{2}\right)=\sqrt{\pi}$$

12.9 LEGENDRE'S DUPLICATION FORMULA

Prove that

$$\sqrt{\pi}\Gamma(2p)=2^{2p-1}\Gamma(p)\Gamma\left(p+\frac{1}{2}\right).$$

[JNTU 2003 *S*]

Solution We know beta function as integral of circular functions (*B* Form VI)

$$B(p,q)=2\int_0^{\frac{\pi}{2}}\sin^{2p-1}\theta\cos^{2p-1}\theta d\theta$$

Put $q=p$, we get

$$B(p,p)=2\int_0^{\frac{\pi}{2}}\sin^{2p-1}\theta\cos^{2p-1}\theta d\theta$$

$$=2\int_0^{\frac{\pi}{2}}\left(\frac{\sin 2\theta}{2}\right)^{2p-1}d\theta$$

$$=\frac{2}{2^{2p-1}}\int_0^{\frac{\pi}{2}}(\sin 2\theta)^{2p-1}d\theta.$$

Put $2\theta=\phi$, $2d\theta=d\phi$

The limits for ϕ are $0, \pi$.

$$=\frac{1}{2^{2p-1}}\int_0^{\pi}(\sin\phi)^{2p-1}d\phi$$

$$=\frac{1}{2^{p-1}}2\int_0^{\frac{\pi}{2}}\sin^{2p-1}\phi d\phi$$

$$=\frac{1}{2^{2p-1}}B\left(p,\frac{1}{2}\right)$$

$$=\frac{1}{2^{2p-1}}\frac{\Gamma(p)\Gamma\left(\frac{1}{2}\right)}{\Gamma\left(p+\frac{1}{2}\right)}$$

$$=\frac{\sqrt{\pi}}{2^{2p-1}}\frac{\Gamma(p)}{\Gamma\left(p+\frac{1}{2}\right)},$$

by Beta – Gamma relation.

Formula: $\int_0^{2\theta}f(x)dx=2\int_0^a f(x)dx$

if $f(2a-x)=f(x)$. Here $a=\frac{\pi}{2}$

and $f(\pi-\phi)=f(\phi)$.

Example 12.12

Prove that $B(p,q)=B(p+1,q)+B(p,q+1)$.

Solution

$$RHS=B(p+1,q)+B(p,q+1)$$

$$=\int_0^1 x^p(1-x)^{q-1}dx+\int_0^1 x^{p-1}(1-x)^q dx$$

$$=\int_0^1 x^{p-1}(1-x)^{q-1}[x+(1-x)]dx$$

$$=B(p,q).$$

Corollary

$$B(p,q)=B(p+1,q)+B(p,q+1)$$

$$=\frac{\Gamma(p+1)\Gamma(q)}{\Gamma(p+q+1)}+\frac{\Gamma(p)\Gamma(q+1)}{\Gamma(p+q+1)}$$

$$=\frac{p\Gamma(p)\Gamma(q)+q\Gamma(p)\Gamma(q)}{(p+q)\Gamma(p+q)}$$

$$=(p+q)\frac{\Gamma(p)\Gamma(q)}{\Gamma(p+q+1)}\qquad p,q\in\mathbb{N},$$

$$B(p,q)=\frac{(p-1)!(q-1)!}{(p+q-1)!}$$

Example 12.13

Show that $\int_0^1 x^p(1-x^r)^q dx=\frac{1}{r}B\left(\frac{p+1}{r},q+1\right)$.

Solution To evaluate the integral $\int_0^1 x^p(1-x^r)^q dx$,

put $x^r=y\Rightarrow x=y^{\frac{1}{r}}, dx=\frac{1}{r}y^{\frac{1}{r}-1}dy$

The limits are $x=0\Rightarrow y=0;\quad x=1\Rightarrow y=1$.

$$\int_0^1 x^p(1-x^r)^q dx=\int_0^1 y^{\frac{p}{r}}(1-y)^q\frac{1}{r}y^{\frac{1}{r}-1}dy$$

$$= \frac{1}{r}\int_0^1 y^{\frac{p+1}{r}-1}(1-y)^q\,dy$$

$$= \frac{1}{r}B\left(\frac{p+1}{r}, q+1\right).$$

Corollary

1. $\int_0^1 \frac{x\,dx}{\sqrt{1-x^5}} = \frac{1}{5}B\left(\frac{2}{5},\frac{1}{2}\right)$ [JNTU 2001]

$$\left(p=1, q=-\frac{1}{2}, r=5\right)$$

2. $\int_0^1 (1-x^n)^{\frac{1}{n}}\,dx = \frac{1}{n}B\left(\frac{1}{n},\frac{1}{n}+1\right)$

$$= \frac{1}{n}\frac{\Gamma\left(\frac{1}{n}\right)\cdot\frac{1}{n}\Gamma\left(\frac{1}{n}\right)}{\frac{2}{n}\Gamma\left(\frac{2}{n}\right)} = \frac{1}{2n}\frac{\left[\Gamma\left(\frac{1}{n}\right)\right]^2}{\Gamma\left(\frac{2}{n}\right)}$$

$$\left(p=0, r=n, q=\frac{1}{n}\right)$$

3. $\int_0^1 \frac{x^m\,dx}{\sqrt{1-x^n}} = \frac{1}{n}B\left(\frac{m+1}{n},\frac{1}{2}\right)$

$$\left(p=m, r=n, q=-\frac{1}{2}\right)$$

Example 12.14

Prove that if m, n are non-negative integers

$$\int_0^1 (1-x^{\frac{1}{n}})^m\,dx = \frac{m!n!}{(m+n)!}.$$

Solution

$$I = \int_0^1 \left(1-x^{\frac{1}{n}}\right)^m dx$$

Put $x^{\frac{1}{n}} = y \Rightarrow x = y^n,\ dx = ny^{n-1}dy$

The limits are $x = 0 \Rightarrow y = 0$;

$x = 1 \Rightarrow y = 1$.

$$= \int_0^1 (1-y)^m ny^{n-1}dy = n\quad B(n, m+1)$$

$$= n\frac{\Gamma(n)\Gamma(m+1)}{\Gamma(m+n+1)} = \frac{n!m!}{(m+n)!}.$$

Example 12.15

Prove that $\int_0^{\frac{\pi}{2}} \frac{\cos^{2m-1}\theta\sin^{2n-1}\theta\,d\theta}{(a\cos^2\theta+b\sin^2\theta)^{m+n}} = \frac{B(m,n)}{2a^m b^n}$.

Solution Denote the integral by I and put $\tan\theta = t \Rightarrow \sec^2\theta\,d\theta = dt$.

The limits are when $\theta = 0, t = 0; \theta = \frac{\pi}{2}, t = \infty$.

$$I = \int_0^{\frac{\pi}{2}} \frac{\cos^{2m-1}\theta\sin^{2n-1}\theta\,d\theta}{(a\cos^2\theta+b\sin^2\theta)^{m+n}}$$

$$= \int_0^{\frac{\pi}{2}} \frac{\cos^{2m+1}\theta\cdot\sin^{2n-1}\theta\cdot\sec^2\theta\,d\theta}{\cos^{2m+2n}\theta(a+b\tan^2\theta)^{m+n}}$$

$$= \int_0^{\infty} \frac{t^{2n-1}\,dt}{(a+bt^2)^{m+n}}$$

Put $bt^2 = ay$ or $t = \sqrt{\frac{a}{b}}\sqrt{y}$

$$dt = \sqrt{\frac{a}{b}}\frac{1}{2\sqrt{y}}dy$$

$$= \frac{1}{2}\int_0^{\infty} \frac{\left(\frac{a}{b}y\right)^{n-\frac{1}{2}}\left(\frac{a}{b}\right)^{\frac{1}{2}}\cdot y^{-\frac{1}{2}}dy}{a^{m+n}(1+y)^{m+n}}$$

$$= \frac{1}{2a^m b^n}\int_0^{\infty} \frac{y^{n-1}}{(1+y)^{m+n}}dy = \frac{B(m,n)}{2a^m b^n}.$$

Example 12.16

Prove that $\int_0^{\infty} \frac{x^2(1-x^2)dx}{(1+x)^8} = 0$.

Solution

$$\int_0^{\infty} \frac{x^2(1-x^2)dx}{(1+x)^8} = \int_0^{\infty} \frac{x^{3-1}dx}{(1+x)^{5+3}} - \int_0^{\infty} \frac{x^{5-1}dx}{(1+x)^{5+3}}$$

$$= B(3,5) - B(5,3) = 0$$

$$\because B(p,q) = B(q,p).$$

Example 12.17

Evaluate $\int_0^{\infty} \frac{x\,dx}{1+x^6}$.

Solution Put $x^6 = y \Rightarrow x = y^{\frac{1}{6}}$; $dx = \frac{1}{6}y^{-\frac{5}{6}}dy$

The limits are: when $x = 0, y = 0;\ x = \infty, y = \infty.$

$$I = \int_0^\infty \frac{x\,dx}{1+x^6} = \frac{1}{6}\int_0^\infty \frac{y^{\frac{1}{6}}y^{-\frac{5}{6}}dy}{1+y}$$

$$= \frac{1}{6}\int_0^\infty \frac{y^{-\frac{2}{3}}dy}{1+y} = \frac{1}{6}B\left(\frac{1}{3},\frac{2}{3}\right)$$

$$= \frac{1}{6}\frac{\Gamma\left(\frac{1}{3}\right)\Gamma\left(\frac{2}{3}\right)}{\Gamma\left(\frac{1}{3}+\frac{2}{3}\right)} = \frac{1}{6}\frac{\pi}{\sin\frac{\pi}{3}} = \frac{1}{6}\frac{\pi}{\left(\frac{\sqrt{3}}{2}\right)}$$

$$= \frac{\pi}{3\sqrt{3}} \quad (\Gamma(1) = 1).$$

Example 12.18

Express in terms of gamma function $\int_0^{\frac{\pi}{2}}\sqrt{\tan\theta}\,d\theta.$

Solution

$$\int_0^{\frac{\pi}{2}}\sqrt{\tan\theta}\,d\theta = \int_0^{\frac{\pi}{2}}\sin^{\frac{1}{2}}\theta\cos^{-\frac{1}{2}}\theta\,d\theta$$

$$= \frac{\Gamma\left(\frac{\frac{1}{2}+1}{2}\right)\Gamma\left(\frac{-\frac{1}{2}+1}{2}\right)}{2\Gamma\left(\frac{\frac{1}{2}-\frac{1}{2}+2}{2}\right)} = \frac{\Gamma\left(\frac{3}{4}\right)\Gamma\left(\frac{1}{4}\right)}{2}.$$

Example 12.19

Evaluate $\int_0^{\frac{\pi}{2}}\left(\sqrt{\tan\theta}+\sqrt{\sec\theta}\right)d\theta.$

Solution

$$\int_0^{\frac{\pi}{2}}\left(\sqrt{\tan\theta}+\sqrt{\sec\theta}\right)d\theta$$

$$= \int_0^{\frac{\pi}{2}}\sqrt{\tan\theta}\,d\theta + \int_0^{\frac{\pi}{2}}\sqrt{\sec\theta}\,d\theta$$

$$= \int_0^{\frac{\pi}{2}}\sin^{\frac{1}{2}}\theta\cos^{-\frac{1}{2}}\theta\,d\theta + \int_0^{\frac{\pi}{2}}\cos^{-\frac{1}{2}}\theta\,d\theta$$

$$= \frac{1}{2}B\left(\frac{\frac{1}{2}+1}{2},\frac{-\frac{1}{2}+1}{2}\right)+\frac{1}{2}B\left(\frac{1}{2},\frac{-\frac{1}{2}+1}{2}\right)$$

$$= \frac{1}{2}B\left(\frac{3}{4},\frac{1}{4}\right)+\frac{1}{2}B\left(\frac{1}{2},\frac{1}{4}\right)$$

$$= \frac{1}{2}\frac{\Gamma\left(\frac{3}{4}\right)\Gamma\left(\frac{1}{4}\right)}{\Gamma\left(\frac{3}{4}+\frac{1}{4}\right)}+\frac{1}{2}\frac{\Gamma\left(\frac{1}{2}\right)\Gamma\left(\frac{1}{4}\right)}{\Gamma\left(\frac{1}{2}+\frac{1}{4}\right)}$$

$$= \frac{1}{2}\Gamma\left(\frac{3}{4}\right)\Gamma\left(\frac{1}{4}\right)+\frac{\sqrt{\pi}}{2}\frac{\Gamma\left(\frac{1}{4}\right)}{\Gamma\left(\frac{3}{4}\right)}.$$

Example 12.20

Evaluate $\int_0^2 x^4(8-x^3)^{-\frac{1}{3}}dx.$

Solution $I = \int_0^2 x^4(8-x^3)^{-\frac{1}{3}}dx$

Put $x^3 = 8\sin^2\theta \Rightarrow 3x^2dx = 16\sin\theta\cos\theta\,d\theta.$

When $x \to 0, \theta \to 0$ and when $x \to 2, \theta \to \frac{\pi}{2}.$

$$\therefore I = \int_0^{\frac{\pi}{2}}(8\sin^2\theta)^{\frac{2}{3}}(8-8\sin^2\theta)^{-\frac{1}{3}}\cdot\frac{16}{3}\cdot\sin\theta\cos\theta\,d\theta$$

$$= \frac{32}{3}\int_0^{\frac{\pi}{2}}\sin^{\frac{7}{3}}\theta\cos^{\frac{1}{3}}\theta\,d\theta$$

$$= \frac{32}{3}\cdot\frac{1}{2}\cdot B\left(\frac{\frac{7}{3}+1}{2},\frac{\frac{1}{3}+1}{2}\right)$$

$$= \frac{16}{3}B\left(\frac{5}{3},\frac{2}{3}\right).$$

Example 12.21

Show that $\int_0^1\frac{x^2dx}{\sqrt{1-x^4}}\cdot\int_0^1\frac{dx}{\sqrt{1-x^4}} = \frac{\pi}{4}.$

[JNTU 2003S]

Solution In the LHS integrals,

put $x^2 = \sin\theta \Rightarrow x = \sin^{\frac{1}{2}}\theta,$

$$dx = \frac{1}{2}\sin^{-\frac{1}{2}}\theta\cos\theta d\theta.$$

The limits are $x = 0 \Rightarrow \theta = 0;\ x = 1 \Rightarrow \theta = \frac{\pi}{2}$.

$$\therefore \int_0^1 \frac{x^2\,dx}{\sqrt{1-x^4}} \cdot \int_0^1 \frac{dx}{\sqrt{1-x^4}}$$

$$= \int_0^{\frac{\pi}{2}} \frac{\sin\theta \cdot \frac{1}{2}\sin^{-\frac{1}{2}}\theta\cos\theta d\theta}{\sqrt{(1-\sin^2\theta)}}$$

$$\cdot \int_0^{\frac{\pi}{2}} \frac{\frac{1}{2}\sin^{-\frac{1}{2}}\theta\cos\theta d\theta}{\sqrt{(1-\sin^2\theta)}}$$

$$= \frac{1}{2}\int_0^{\frac{\pi}{2}} \sin^{\frac{1}{2}}\theta d\theta \cdot \frac{1}{2}\int_0^{\frac{\pi}{2}} \sin^{-\frac{1}{2}}\theta d\theta$$

$$= \frac{1}{4}, \frac{1}{2}B\left(\frac{\frac{1}{2}+1}{2}, \frac{1}{2}\right) \cdot \frac{1}{2}B\left(\frac{-\frac{1}{2}+1}{2}, \frac{1}{2}\right)$$

$$= \frac{1}{16} \frac{\Gamma\left(\frac{3}{4}\right)\Gamma\left(\frac{1}{2}\right)}{\Gamma\left(\frac{5}{4}\right)} \cdot \frac{\Gamma\left(\frac{1}{4}\right)\Gamma\left(\frac{1}{2}\right)}{\Gamma\left(\frac{3}{4}\right)}$$

$$= \frac{\pi}{16} \cdot \frac{\Gamma\left(\frac{1}{4}\right)}{\frac{1}{4}\Gamma\left(\frac{1}{4}\right)} = \frac{\pi}{4}.$$

Example 12.22

Prove that $\int_0^{\frac{\pi}{2}} \sin^2\theta\cos^4\theta d\theta = \frac{\pi}{32}$. [JNTU 2003(3)]

Solution We know that

$$\int_0^{\frac{\pi}{2}} \sin^p\theta\cos^q\theta d\theta = \frac{1}{2}B\left(\frac{p+1}{2}, \frac{q+1}{2}\right)$$

$$= \frac{\Gamma\left(\frac{p+1}{2}\right)\Gamma\left(\frac{q+1}{2}\right)}{2\Gamma\left(\frac{p+q+2}{2}\right)}$$

$p = 2, \quad q = 4$

$$\therefore \int_0^{\frac{\pi}{2}} \sin^2\theta\cos^4\theta d\theta = \frac{\Gamma\left(\frac{2+1}{2}\right)\Gamma\left(\frac{4+1}{2}\right)}{2\Gamma\left(\frac{2+4+2}{2}\right)}$$

$$= \frac{\Gamma\left(\frac{3}{2}\right)\Gamma\left(\frac{5}{2}\right)}{2\Gamma(4)}$$

$$= \frac{\frac{1}{2}\Gamma\left(\frac{1}{2}\right)\frac{3}{2}\frac{1}{2}\Gamma\left(\frac{1}{2}\right)}{2 \cdot 6} = \frac{\pi}{32}.$$

$$\left(\Gamma(4) = 3! = 6;\ \Gamma\left(\frac{1}{2}\right) = \sqrt{\pi}\right)$$

Example 12.23

Show that $\int_0^\infty \sqrt{x}e^{-x^3}\,dx = \frac{\sqrt{\pi}}{3}$. [JNTU 2003(3)]

Solution Put $x^3 = t \Rightarrow x = t^{\frac{1}{3}}, dx = \frac{1}{3}t^{-\frac{2}{3}}dt$.

The given integral

$$= \int_0^\infty t^{\frac{1}{6}}e^{-t}\frac{1}{3}t^{-\frac{2}{3}}dt = \frac{1}{3}\int_0^\infty t^{-\frac{1}{2}}e^{-t}dt$$

$$= \frac{1}{3}\int_0^\infty e^{-t}t^{\frac{1}{2}-1}\,dt = \frac{1}{3}\Gamma\left(\frac{1}{2}\right) = \frac{\sqrt{\pi}}{3}$$

Example 12.24

Show that $\iint_D x^{m-1}y^{n-1}\,dx\,dy = \frac{\Gamma(m)\Gamma(n)}{\Gamma(m+n+1)}a^{m+n}$

where D is the domain $x \geq 0, y \geq 0$ and $x + y \leq a$.

Solution Let $x = aX$ and $y = aY$, then

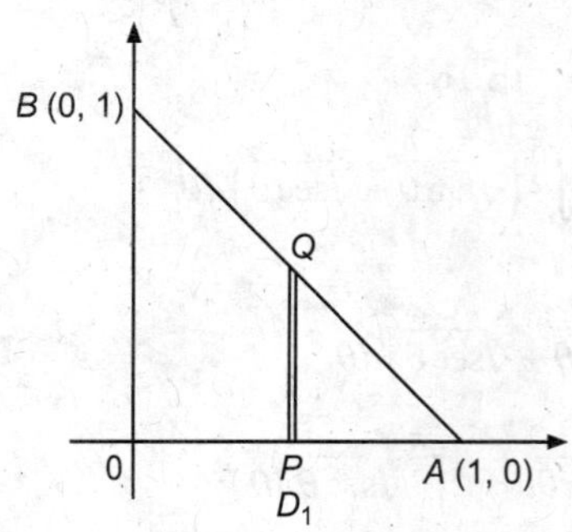

Figure 12.2 Evaluation of Integral by use of Gamma functions

$$\iint_D x^{m-1} y^{n-1}\, dx\, dy$$

$$= \iint_{D1} (aX)^{m-1} (aY)^{n-1} a^2\, dX\, dY,$$

where D_1 is the domain $X \geq 0$, $Y \geq 0$ and $X + Y \leq 1$ (Fig. 12.2)

$$= a^{m+n} \int_0^1 \int_0^{1-X} X^{m-1} Y^{n-1}\, dX\, dY$$

$$= a^{m+n} \int_0^1 X^{m-1} \left(\frac{Y^n}{n} \right)_0^{1-X} dX$$

$$= \frac{a^{m+n}}{n} \int_0^1 X^{m-1} (1-X)^n\, dX$$

$$= \frac{a^{m+n}}{n} B(m, n+1)$$

$$= \frac{a^{m+n}}{n} \cdot \frac{\Gamma(m)\Gamma(n+1)}{\Gamma(m+n+1)}$$

$$= a^{m+n} \frac{\Gamma(m)\Gamma(n)}{\Gamma(m+n+1)}.$$

12.9.1 Dirichlet's Integral

Example 12.25

Establish Dirichlet's integral

$$\iiint_V x^{l-1} y^{m-1} z^{n-1}\, dx dy dz = \frac{\Gamma(l)\Gamma(m)\Gamma(n)}{\Gamma(l+m+n+1)}$$

where V is the region $x \geq 0, y \geq 0, z \geq 0$ and $x + y + z \leq 1$.

Solution Let $y + z = 1 - x = a$, then $z \leq a - y$

$$\iiint_V x^{l-1} y^{m-1} z^{n-1}\, dx dy dz$$

$$= \int_0^1 \int_0^{1-x} \int_0^{1-x-y} x^{l-1} y^{m-1} z^{n-1}\, dx dy dz$$

$$= \int_0^1 x^{l-1} \left[\int_0^a \int_0^{a-y} (y^{m-1} z^{n-1}\, dz\, dy) \right] dx$$

$$= \int_0^1 x^{l-1} \frac{\Gamma(m)\Gamma(n)}{\Gamma(m+n+1)} a^{m+n}\, dx, \quad \text{(Ex. 1.24)}$$

$$= \frac{\Gamma(m)\Gamma(n)}{\Gamma(m+n+1)} \int_0^1 x^{l-1} (1-x)^{m+n}\, dx$$

$$\because\ a = (1-x)$$

$$= \frac{\Gamma(m)\Gamma(n)}{\Gamma(m+n+1)} B(l, m+n+1)$$

$$= \frac{\Gamma(m)\Gamma(n)}{\Gamma(m+n+1)} \frac{\Gamma(l)\Gamma(m+n+1)}{\Gamma(l+m+n+1)}$$

$$= \frac{\Gamma(l)\Gamma(m)\Gamma(n)}{\Gamma(l+m+n+1)}.$$

Example 12.26

Evaluate $\iiint_V xyz\, dx dy dz$ taken over the volume V of the tetrahedron given by $x \geq 0, y \geq 0, z \geq 0$ and $x + y + z \leq 1$. [JNTU 2003 *S* (3)]

Solution Let $y + z \leq (1-x) = h$

The triple integral

$$\iiint_V xyz\, dx dy dz = \int_0^1 \int_0^{1-x} \int_0^{1-x-y} xyz\, dx dy dz$$

$$= \int_0^1 x \left[\int_0^h y \left(\int_0^{h-y} z\, dz \right) dy \right] dx$$

$$= \int_0^1 x \left[\int_0^1 (hY) \int_0^{1-Y} (hZ\, dZ) h\, dY \right] dx$$

where $\frac{y}{h} = Y$, $\frac{z}{h} = Z$

$$= \int_0^1 xh^4 \left[\int_0^1 Y \left(\int_0^{1-Y} Z\, dZ \right) dY \right] dx$$

$$= \int_0^1 xh^4 \left[\int_0^1 Y \left(\frac{Z^2}{2} \right)_0^{1-Y} dY \right] dx$$

$$= \frac{1}{2} \int_0^1 xh^4 \left[\int_0^1 Y(1-Y)^2\, dY \right] dx$$

$$= \frac{1}{2} \int_0^1 xh^4 \left(\int_0^1 Y^{2-1} (1-Y)^{3-1}\, dY \right) dx$$

$$= \frac{1}{2} \int_0^1 xh^4\, B(2,3)\, dx$$

$$= \frac{\Gamma(2)\Gamma(3)}{2\Gamma(5)} \int_0^1 x(1-x)^4\, dx \qquad (\because h = 1-x)$$

$$= \frac{1!2!}{2(4!)} B(2,5) = \frac{1}{24} \cdot \frac{\Gamma(2)\Gamma(5)}{\Gamma(7)}$$

$$= \frac{1}{24} \cdot \frac{1!4!}{6!} = \frac{1}{720}.$$

Aliter: Put $l = m = n = 2$ in Dirichlet's integral, in the last example.

EXERCISE 12.2

1. Show that $\int_0^1 x^7(1-x)^8\,dx = \dfrac{7!8!}{16!}$.

2. Prove that $\int_0^1 \dfrac{x^2\,dx}{\sqrt{1-x^4}} \cdot \int_0^1 \dfrac{dx}{\sqrt{1+x^4}} = \dfrac{\pi}{4\sqrt{2}}$.

   [**Hint:** Put $x^2 = \sin\theta, x^2 = \tan\theta$, respectively to evaluate the two integrals]

3. Prove that

$$\int_0^{\frac{\pi}{2}} \tan^p\theta\,d\theta = \int_0^{\frac{\pi}{2}} \cot^p\theta\,d\theta = \frac{1}{2}B\left(\frac{1+p}{2}, \frac{1-p}{2}\right)$$

$$= \frac{\pi}{2}\sec\frac{p\pi}{2}\ (-1 < p < 1).$$

4. Prove that

$$B(p, 1-p) = \Gamma(p)\Gamma(1-p) = \int_0^\infty \frac{t^{p-1}}{1+t}dt$$

$= \pi\,\text{cosec}\,p\pi$. Deduce that

(a) $B\left(\dfrac{1}{3}, \dfrac{2}{3}\right) = \Gamma\left(\dfrac{1}{3}\right)\Gamma\left(\dfrac{2}{3}\right) = \dfrac{2\pi}{\sqrt{3}}$

(b) $B\left(\dfrac{1}{4}, \dfrac{3}{4}\right) = \Gamma\left(\dfrac{1}{4}\right)\Gamma\left(\dfrac{3}{4}\right) = \sqrt{2}\pi.$

5. Prove that $\int_0^\infty \dfrac{x^{p-1} - x^{q-1}}{(1+x)^{p+q}}dx = 0.$

6. Prove that

$$\int_0^{\frac{\pi}{2}} \sin^p\theta\,d\theta = \int_0^{\frac{\pi}{2}} \cos^p\theta\,d\theta$$

$$= \frac{1}{2}B\left(\frac{p+1}{2}, \frac{1}{2}\right) = \sqrt{\pi}\frac{\Gamma\left(\frac{p+1}{2}\right)}{2\Gamma\left(\frac{p}{2}+1\right)}.$$

[JNTU 2003]

7. Prove that

$$\int_0^\infty \frac{x^a}{(1+x^b)^c}dx = \frac{1}{b}B\left(\frac{a+1}{b}, \frac{cb-a-1}{b}\right)$$

$$= \frac{1}{b}\frac{\Gamma\left(\frac{a+1}{b}\right)\Gamma\left(\frac{cb-a-1}{b}\right)}{\Gamma(c)}$$

$[a > -1, b > 0, bc > a+1.]$

8. Prove that $\int_a^b (x-a)^p(b-x)^q\,dx$

$$= (b-a)^{p+q-1}B(p+1, q+1).$$

   [**Hint:** Put $x = a + (b-a)t$]

9. Prove that $\int_0^2 x^2(2-x)^{\frac{7}{2}}dx = 2^{\frac{13}{2}}B\left(3, \dfrac{9}{2}\right)$.

10. Prove that

(a) $\int_0^\infty \dfrac{x^8(1-x^6)dx}{(1+x)^{24}} = 0$

(b) $\int_0^\infty \dfrac{x^2(1-x^2)dx}{(1+x)^8} = 0.$

11. Prove that $\int_0^\infty \dfrac{x^3(1+x^5)dx}{(1+x)^{13}} = \dfrac{1}{990}$.

12. Prove that

(a) $\Gamma\left(\dfrac{1}{2}+x\right)\Gamma\left(\dfrac{1}{2}-x\right) = \pi\sec\pi x$

(b) $\Gamma(x)\Gamma(-x) = -\dfrac{\pi\,\text{cosec}\,\pi x}{x}$.

13. Prove that $\int_0^1 \dfrac{dx}{\sqrt{1-x^n}} = \dfrac{\sqrt{\pi}}{n}\dfrac{\Gamma\left(\frac{1}{n}\right)}{\Gamma\left(\frac{1}{n}+\frac{1}{2}\right)}$.

[JNTU 2003]

12.10 LEGENDRE FUNCTIONS

12.10.1 Introduction

If a homogeneous linear differential equation of the form $L(y) = (\Sigma a_r D^r)y = 0$ has constant coefficients a_r, it can be solved by elementary methods, and the functions involved in solutions are elementary functions such as $x^n, e^x, \log x, \sin x$ *etc.* However, if such an equation has variable coefficients and functions of the independent variable x, it has to be solved by other methods. Legendre's equation and Bessel's equation are two very important equations of this type. We take up below, solution of these equations by application of two standard methods of solution:

1. The power series method and
2. The Frobenius[2] method, which is an extension of the power series method.

12.10.2 Power Series Method of Solution of Linear Differential Equations

The power series method, which gives solutions in the form of power series, is the standard method for solving ordinary linear differential equations with variable coefficients.

Power series

An infinite series of the form $\Sigma_{m=0}^{\infty} a_m (x-x_0)^m = a_0 + a_1(x-x_0) + a_2(x-x_0)^2 + \cdots$ where $a_0, a_1, a_2, \ldots$ are real constants is called a power series. These constants are called the coefficients of the series. x_0 is a constant and is called the centre of the series, and x is a real variable.

For the choice $x_0 = 0$, we obtain the power series

$$\sum_{m=0}^{\infty} a_m x^m = a_o + a_1 x + a_2 x^2 + \cdots \tag{12.11}$$

with the origin as its centre.

Convergence interval

If there exists a positive real number R such that for all x in $I = (x-R, x+R)$ the series Eq. (12.11) is convergent, then I is called the interval of convergence for the series Eq. (12.11) and R is called the radius of convergence. R is defined by

$$R = (\lim_{n\to\infty} a_n^{\frac{1}{n}})^{-1} \text{ or } R = \lim_{n\to\infty} \frac{a_n}{a_{n+1}}.$$

Real analytic functions

A real function $f(x)$ is called analytic at $x = x_0$ if it can be represented by a power series in powers of $(x - x_0)$ with radius of convergence $R > 0$.

[2] Georg Frobenius (1849–1917), German mathematician, also made important contributions to the theory of matrices and groups.

12.10.3 Existence of Series Solutions: Method of Frobenius

The condition for the existence of power series solutions of the type

$$\sum_{n=0}^{\infty} a_n (x-x_0)^n \tag{12.12}$$

for a second order differential equation with variable coefficients, written in the standard form

$$y'' + p(x)y' + q(x)y = 0 \tag{12.13}$$

is that the coefficient functions $p(x)$ and $q(x)$ must be analytic at $x = x_0$ (equivalently expansible in Taylor's series or differentiable any number of times). In the case of Bessel's equation

$$y'' + \frac{1}{x}y' + \left(\frac{x^2 - p^2}{x^2}\right)y = 0 \tag{12.14}$$

the above condition is not satisfied since $\frac{1}{x}$ and $\frac{x^2 - p^2}{x^2}$ are not analytic at $x = 0$ and the power series method fails. The method of Frobenius which is an extension of the power series method applies to an important class of equations of the form

$$y'' + \frac{p(x)}{x}y' + \frac{q(x)}{x^2}y = 0 \tag{12.15}$$

where the functions $p(x)$ and $q(x)$ are analytic at $x = 0$, or equivalently, $p(x)$ and $q(x)$ are expansible as power series. It gives at least one solution that can be represented as

$$y(x) = x^m \sum_{r=0}^{\infty} a_r x^r = x^m (a_0 + a_1 x + a_2 x^2 + \cdots) \tag{12.16}$$

where the exponent m may be any real or complex number and m is chosen so that $a_0 \neq 0$.

Eq. (12.15) also has a second solution, which is linearly independent of the above solution in the form Eq. (12.16) with a different m and different coefficients or a logarithmic term.

Bessel's equation Eq. (12.14) is of the form Eq. (12.15) with $p(x) = 1$ and $q(x) = x^2 - n^2$ analytic at $x = 0$ so that the method of Frobenius applies.

Regular and singular points

The point x_0 at which the coefficients p and q are analytic is called a regular point of Eq. (12.13). Then the power series method can be applied for its solution.

In the case of Legendre's equation

$$(1-x^2)y''-2xy'+n(n+1)y=0$$

which can be put in the standard form

$y''-\dfrac{2x}{1-x^2}y'+\dfrac{n(n+1)}{1-x^2}y=0\ (|x|\neq 1)$ the functions $p(x)=-\dfrac{2x}{1-x^2}$ and $q(x)=\dfrac{n(n+1)}{1-x^2}$ are expressible as a power series about $x=0$ and so they are analytic at $x=0$. Hence the equation can be solved by the power series method.

If x_0 is not regular, it is called a singular point. In respect of Bessel's Eq. (12.14) the point $x=0$ is not regular and is therefore a singular point of Eq. (12.14). But $xp(x)=1$ and $x^2q(x)=x^2-p^2$ are analytic at $x=0$ and hence we can apply Frobenius method for the solution of Eq. (12.14).

The series solution of $p(x)y''+q(x)y'+r(x)y=0$ by Frobenius method consists of the following steps:

1. Assume that y given by eq. (12.16) is a solution of the equation.
2. Compute y' and y'' and substitute for y, y' and y'' in the equation.
3. Equate to zero the coefficient of the lowest power of x; it gives a quadratic in m, which is known as the indicial equation, giving two values for m.
4. Equate to zero the coefficients of the other powers of x and find the values of $a_1, a_2, a_3, \cdots$ in terms of a_0.

12.10.4 Legendre Functions

The power series method of solution can be applied to find the solution of the Legendre's[3] differential equation

$$(1-x^2)y''-2xy'+n(n+1)y=0$$

$$\text{or}\quad [(1-x^2)y']'+n(n+1)y=0 \tag{12.17}$$

which arises in boundary value problems having spherical symmetry. Here n is a real constant. We can put Eq. (12.17) in the standard form

$$y''-\frac{2x}{1-x^2}y'+\frac{n(n+1)}{1-x^2}y=0 \tag{12.18}$$

by dividing it by the coefficient $(1-x^2)$ of y'' in Eq. (12.17). Now, the coefficients of y' and y in Eq. (12.18) are

$$-\frac{2x}{1-x^2}\quad\text{and}\quad\frac{n(n+1)}{1-x^2}$$

respectively, and these functions are analytic at $x=0$ (*i.e.*, derivable any number of times). Hence we may apply the power series method for its solution.

Let us assume a power series solution of Eq. (12.17) in the form

$$y(x)=\sum_{r=0}^{\infty}a_r x^r \tag{12.19}$$

Differentiating Eq. (12.18) w.r.t. 'x' twice, we have

$$y'(x)=\sum_{r=1}^{\infty}ra_r x^{r-1},$$

$$y''(x)=\sum_{r=2}^{\infty}r(r-1)a_r x^{r-2} \tag{12.20}$$

Substituting the expressions for y, y' and y'' from Eqs. (12.19) and (12.20) in Eq. (12.17)

$$(1-x^2)\sum_{r=2}^{\infty}r(r-1)a_r x^{r-2}-2x\sum_{r=1}^{\infty}ra_r x^{r-1}$$

$$+n(n+1)\sum_{r=0}^{\infty}a_r x^r=0$$

$$\Rightarrow \sum_{r=2}^{\infty}r(r-1)a_r x^{r-2}-\sum_{r=2}^{\infty}r(r-1)a_r x^r$$

$$-2\sum_{r=1}^{\infty}ra_r x^r+n(n+1)\sum_{r=0}^{\infty}a_r x^r=0 \tag{12.21}$$

[3] Adrien Marie Legendre (1752–1833), French mathematician, who became a professor in Paris in 1775 made important contributions to special functions, elliptic integrals, number theory and the calculus of variations. His book Elements de Geometrie (1794) became very famous and had 12 editions in less than 30 years.

By writing each series in the expanded form and arranging each power of x in a column, we have

$$2\cdot 1\, a_2 + 3\cdot 2\, a_3 x + 4\cdot 3\, a_4 x^2 + \cdots +$$

$$(r+2)(r+1)\, a_{r+2} x^r + \cdots - 2\cdot 1 a_2 x^2 - \cdots -$$

$$r(r-1)a_r \cdot x^r - \cdots - 2\cdot 1 a_1 x - 2\cdot 2 a_2 x^2 - \cdots$$

$$-2r\cdot a_r \cdot x^r - \cdots n(n+1)a_0 + n(n+1)a_1 x$$

$$+n(n+1)a_2 x^2 + \cdots + n(n+1)a_r \cdot x^r + \cdots = 0$$

Since Eq. (12.19) is a solution of Eq. (12.17), this must be an identity in x. So, the sum of the coefficients of each power of x must be zero. This implies that

$$2a_2 + n(n+1)a_0 = 0 \quad \text{for coefficients of } x^0 \quad (12.22)$$

$$6a_3 + [-2 + n(n+1)]a_1 = 0 \quad \text{for coefficients of } x^1 \quad (12.23)$$

In general, when $r = 2, 3, \cdots$

$$(r+2)(r+1)a_{r+2} + [-r(r-1) - 2r$$

$$+n(n+1)]a_r = 0 \quad (12.24)$$

Simplifying the expression in square brackets, we obtain $(n-r)(n-r+1)$ so that

$$a_{r+2} = -\frac{(n-r)(n+r+1)}{(r+2)(r+1)} a_r \quad r = 0, 1, 2, \ldots \quad (12.25)$$

This is called a *recurrence relation* or *recursion formula*. It gives all the coefficients starting from a_2 onwards in terms of a_0 and a_1, which are considered as arbitrary constants. Thus we have

$$a_2 = -\frac{n(n+1)}{2!} a_0,$$

$$a_3 = -\frac{(n-1)(n+2)}{3!} a_1,$$

$$a_4 = -\frac{(n-2)(n+3)}{4\cdot 3} a_2$$

$$= \frac{(n-2)n(n+1)(n+3)}{4!} a_0,$$

$$a_5 = -\frac{(n-3)(n+4)}{5\cdot 4} a_3$$

$$= \frac{(n-3)(n-1)(n+2)(n+4)}{5!} a_1, \text{ etc.} \quad (12.26)$$

Substituting these coefficients in Eq. (12.19), we obtain

$$y(x) = a_0 y_1(x) + a_1 y_2(x) \quad 12.27)$$

where

$$y_1(x) = 1 - \frac{n(n+1)}{2!} x^2 + \frac{(n-2)n(n+1)(n+3)}{4!} x^4 - \cdots \quad (12.28)$$

$$\text{and } y_2(x) = x - \frac{(n-1)(n+2)}{3!} x^3 + \frac{(n-3)(n-1)(n+2)(n+4)}{5!} x^5 - \cdots \quad (12.29)$$

The two series converge for $|x| < 1$, if they are non-terminating. Since y_1 contains only the even powers of x and y_2 contains only the odd powers of x, the ratio y_1/y_2 is not a constant so that y_1 and y_2 are linearly independent. Hence Eq. (12.27) is a general solution of Eq. (12.17) in the interval $-1 < x < 1$.

12.10.5 Legendre Polynomials $P_n(x)$

If the parameter n in Legendre's Eq. (12.17) is a non-negative integer n, then the right-hand side of Eq. (12.25) is zero when $r = n$. This implies that $a_{n+2} = a_{n+4} = \cdots = 0$. Hence, if n is even, $y_1(x)$ reduces to a polynomial of degree n, while $y_2(x)$ remains an infinite series.

Similarly, if n is odd, $y_2(x)$ reduces to a polynomial of degree n, while $y_1(x)$ remains an infinite series.

In either case, the terminating series solution (*i.e.*, the polynomial solution) of the Legendre's equation, multiplied by some constants are called Legendre polynomials or zonal harmonics of order n and are denoted by $P_n(x)$. The series which is non-terminating is known as Legendre's function of the second kind and is denoted by $Q_n(x)$. Thus, for a non-negative integer n the general solution of Legendre's Eq. (12.17) is

$$y(x) = AP_n(x) + BQ_n(x) \quad (12.30)$$

where $P_n(x)$ is a polynomial of degree n and $Q_n(x)$ is a non-terminating series, which is unbounded at $x = \pm 1$.

From Eq. (12.25), we have

$$a_r = \frac{-(r+2)(r+1)}{(n-r)(n+r+1)} a_{r+2} \quad \text{for } r \le (n-2) \quad (12.31)$$

Now, all the non-vanishing coefficients may be expressed in terms of the coefficient a_n of the highest power of x of the polynomial. The coefficient a_n which is still arbitrary may be chosen as $a_n = 1$ when $n = 0$ and

$$a_n = \frac{(2n)!}{2^n (n!)^2} = \frac{1\cdot 3\cdot 5\cdots(2n-1)}{n!} \qquad (12.32)$$

for $n = 1, 2, \ldots$

For this choice of a_n, all these polynomials will have the value 1 when $x = 1$, *i.e.*, $P_n(1) = 1$. We obtain from Eq. (12.31) and Eq. (12.32),

$$a_{n-2} = -\frac{n(n-1)}{2(2n-1)} a_n = -\frac{n(n-1)}{2(2n-1)} \cdot \frac{(2n)!}{2^n (n!)^2}$$

$$= -\frac{n(n-1)2n(2n-1)(2n-2)!}{2(2n-1)2^n n(n-1)! n(n-1)(n-2)!}$$

$$\text{i.e., } a_{n-2} = -\frac{(2n-2)!}{2^n (n-1)!(n-2)!}$$

$$\text{Similarly, } a_{n-4} = -\frac{(n-2)(n-3)}{4(2n-3)} a_{n-2}$$

$$= \frac{(2n-4)!}{2^n 2!(n-2)!(n-4)!}$$

and, in general, when $(n-2r) \geq 0$

$$a_{n-2r} = (-1)^r \frac{(2n-2r)!}{2^n r!(n-r)!(n-2r)!} \qquad (12.33)$$

The resulting solution of Legendre's differential equation Eq. (12.17) is called the Legendre polynomial of degree n, denoted by $P_n(x)$ and is given by

$$P_n(x) = \sum_{r=0}^{M} \frac{(-1)^r (2n-2r)!}{2^n r!(n-r)!(n-2r)!} x^{2n-2r} \qquad (12.34)$$

where $M = n/2$ or $(n-1)/2$ according as n is even or odd so that M is an integer.

Legendre's equation Eq. (12.17) can also be solved as a series of descending powers of x, assuming a solution of Eq. (12.17) in the form $y = \sum_{r=0}^{\infty} a_r x^{m-r}$

Differentiating w.r.t. x twice, we obtain

$$y' = \sum_{r=0}^{\infty} (m-r) a_r x^{m-r-1},$$

$$y'' = \sum_{r=0}^{\infty} (m-r)(m-r-1) a_r x^{m-r-2}$$

Substituting these expressions in Eq. (12.17)

$$(1-x^2)\sum_{r=0}^{\infty} a_r (m-r)(m-r-1) x^{m-r-2}$$

$$-2x\sum_{r=0}^{\infty}(m-r)x^{m-r-1} + n(n+1)\sum_{r=0}^{\infty} a_r x^{m-r} = 0$$

$$\Rightarrow \sum_{r=0}^{\infty} a_r [(m-r)(m-r-1)x^{m-r-2} + \{n(n+1)$$

$$-(m-r)(m-r-1) - 2(m-r)\}x^{m-r}] = 0$$

$$\Rightarrow \sum_{r=0}^{\infty} a_r [(m-r)(m-r-1)x^{m-r-2} + \{n(n+1)$$

$$-(m-r)(m-r+1)\}x^{m-r}] = 0$$

$$\Rightarrow \sum_{r=0}^{\infty} a_r [(m-r)(m-r-1)x^{m-r-2} + \{n^2 - (m-r)^2$$

$$+n - (m-r)\}x^{m-r}] = 0$$

$$\Rightarrow \sum_{r=0}^{\infty} a_r [(m-r)(m-r-1)x^{m-r-2} + (n-m+r)$$

$$(n+m-r+1)x^{m-r}] = 0$$

Equating to zero the coefficient of the highest power of x, *i.e.*, x^m, we get $a_0(n-m)(n+m+1) = 0$ which is obtained by putting $r = 0$ in the coefficient of x^{m-r} in (12.19). This implies that $m = n$ or $m = -n-1$ ($\because$ $a_0 \neq 0$)

Equating now to zero the coefficient of the next power of x, *i.e.*, x^{m-1}, we get $a_1(n-m+1)(n+m) = 0$ which is obtained by putting $r = 1$ in the coefficient of x^{m-r-1} in Eq. (12.35). This implies that $a_1 = 0$ $\because$ $m \neq -n$ and $m \neq n+1$

To obtain the recurrence relation, we equate the coefficient of x^{m-r} to zero:

$$(m-r+2)(m-r+1)a_{r-2} - (m-r-n)$$

$$(m-r+n+1)a_r = 0$$

$$\Rightarrow \quad a_r = -\frac{(m-r+2)(m-r+1)}{(m-r+n)(m-r+n+1)} a_{r-2}$$

$$r = 2, 3, 4, \ldots$$

From this relation, we see that
$a_1 = a_3 = a_5 = \cdots = a_{2r-1} = 0.$

Case (i) $m = n$, We have

$$a_r = \frac{(n-r+2)(n-r+1)}{r(2n-r+1)}a_{r-2}$$

Putting $r = 2, 4, 6, \ldots$, we have

$$a_2 = -\frac{n(n-1)}{2(2n-1)}a_0$$

$$a_4 = -\frac{(n-2)(n-3)}{4(2n-3)}a_2$$

$$= \frac{n(n-1)(n-2)(n-3)}{2\cdot 4\cdot (2n-1)(2n-3)}a_0$$

$$a_6 = -\frac{(n-4)(n-5)}{6(2n-5)}a_4$$

$$= -\frac{n(n-1)(n-2)(n-3)(n-4)(n-5)}{2\cdot 4\cdot 6\cdot (2n-1)(2n-3)(2n-5)}a_0, etc.$$

We obtain a solution of Eq. (12.17) as

$$y_1 = a_0\left[x^n - \frac{n(n-1)}{2(2n-1)}x^{n-2} + \frac{n(n-1)(n-2)(n-3)}{2\cdot 4\cdot (2n-1)(2n-3)}x^{n-4} - \cdots\right]$$

If we choose a_0 as $a_0 = \dfrac{1\cdot 3\cdot 5\cdots(2n-1)}{n!}$ the first solution y_1 is obtained and is called Legendre's polynomial and is denoted by $P_n(x)$. This is called the Legendre's function of the first kind of degree n.

Case (ii) $m = -n-1$, In this case,

$$a_r = \frac{(n+r-1)(n+r)}{r(2n+r+1)}a_{r-2}$$

Putting $r = 2, 4, 6\cdots$

$$a_2 = \frac{(n+1)(n+2)}{2(2n+3)}a_0$$

$$a_4 = \frac{(n+3)(n+4)}{4(2n+5)}a_2$$

$$= \frac{(n+1)(n+2)(n+3)(n+4)}{2\cdot 4\cdot (2n+3)(2n+5)}a_0, \cdots$$

$$\therefore\ y_2 = \sum_{r=0}^{\infty} a_r x^{-n-1-r}$$

$$= a_0\left[x^{-n-1} + \frac{(n+1)(n+2)}{2(2n+3)}x^{-n-3} + \frac{(n+1)(n+2)(n+3)(n+4)}{2\cdot 4\cdot (2n+3)(2n+5)}x^{-n-5} + \cdots\right]$$

Choosing $a_0 = \dfrac{n!}{1\cdot 3\cdot 5\cdots(2n+1)}$ the second solution y_2 of Legendre's Eq. (12.17) is obtained and it is called Legendre's function of the second kind of degree n and is denoted by $Q_n(x)$. The general solution of Legendre's Eq. (12.17) is $y = AP_n(x) + BQ_n(x)$, where A and B are arbitrary constants.

The first few Legendre polynomials are shown in Fig. 12.3

$P_0(x) = 1 \quad (n = 0, M = 0)$

$P_1(x) = x \quad (n = 1, M = 0)$

$P_2(x) = \frac{1}{2}(3x^2 - 1) \quad (n = 2, M = 1, r = 0 \text{ to } 1)$

$P_3(x) = \frac{1}{2}(5x^3 - 3x) \quad (n = 3, M = 1, r = 0 \text{ to } 1)$

$$P_4(x) = \frac{1}{8}(35x^4 - 30x^2 + 3)$$

$(n = 4, M = 2, r = 0 \text{ to } 2)$

$$P_5(x) = \frac{1}{8}(63x^5 - 70x^3 + 15x)$$

$(n = 5, M = 2, r = 0 \text{ to } 2)$

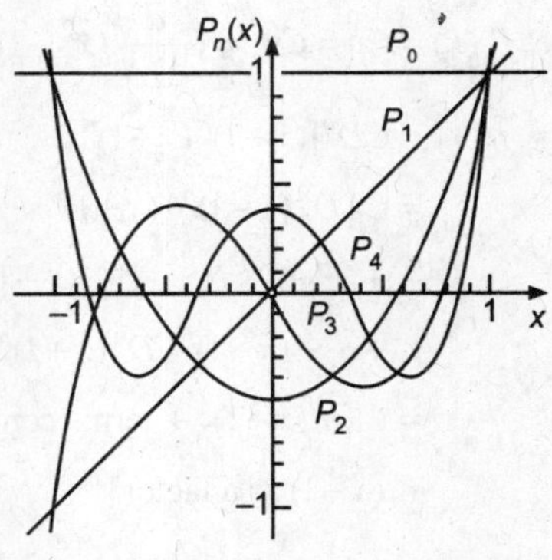

Figure 12.3 Legendre polynomials

$$P_6(x)=\frac{1}{16}[231x^6-315x^4+105x^2-5]$$

$(n=6, M=3, r=0 \text{ to } 3)$ (12.35)

Rodrigues'[4] Formula for Legendre Polynomials $P_n(x)$

Example 12.27

Show that $P_n(x)=\frac{1}{n!2^n}\frac{d^n}{dx^n}[(x^2-1)^n]\,(n\in\mathbb{N})$ (12.1)

Solution Let $y=(x^2-1)^n$. Then

$$y_1=\frac{dy}{dx}=n(x^2-1)^{n-1}2x$$

$$\Rightarrow -(1-x^2)\frac{dy}{dx}=2nx(x^2-1)^n=2nxy$$

$$\Rightarrow (1-x^2)\frac{dy}{dx}+2nxy=0 \quad (12.2)$$

Differentiating Eq. (2) $(n+1)$ times using Leibnitz's theorem, we have

$$(1-x^2)y_{n+2}+(n+1)(-2x)y_{n+1}-(n+1)ny_n$$
$$+2n[xy_{n+1}+(n+1)\cdot 1\cdot y_n]=0$$

or $(1-x^2)\frac{d^2v}{dx^2}-2x\frac{dv}{dx}+n(n+1)v=0$ (where $v=y_n$)

which is Legendre's differential equation having $P_n(x)$ and $Q_n(x)$ as solutions.

Since $v=y_n=\frac{d^n}{dx^n}[(x^2-1)^n]$ contains positive powers of x only, v must be a constant multiple of $P_n(x)$.

$$\text{Let } P_n(x)=Cv=Cy_n=C\frac{d^n}{dx^n}(x^2-1)^n$$
$$=CD^n[(x-1)^n(x+1)^n]$$
$$=C[D^n(x-1)^n\cdot(x+1)^n$$
$$+{}^nC_1\cdot D^{n-1}(x-1)^n\cdot(x+1)^{n-1}$$
$$+\cdots+(x-1)^n\cdot D^n(x+1)^n]$$
$$=C[n!(x+1)^n+\text{terms containing}$$
$$(x-1)\text{ as a factor}]$$

[4] Olinde Rodrigues (1794–1851) was a French mathematician and economist.

Putting $x=1$, we get

$$1=P_n(1)=C(n!2^n)\Rightarrow C=\frac{1}{n!2^n}$$

$$\therefore\ P_n(x)=\frac{1}{n!2^n}\frac{d^n}{dx^n}[(x^2-1)^n]$$

Note

1. The first few Legendre polynomials obtained earlier can also be obtained using Rodrigues' formula.
2. Any polynomial $f(x)$ of degree n can be expressed as a linear combination of Legendre polynomials

   $P_n(x)$ as $f(x)=\sum_{m=0}^{n}C_mP_m(x)$

12.10.6 Generating Function for Legendre Polynomials $P_n(x)$

Generating function: Let $<f_n(x)>$ be a sequence of functions. A function $w(t, x)$ is called a generating function for the functions $f_n(x)$ if $w(t, x)=\sum_{n=-\infty}^{\infty}f_n(x)t^n$.

Example 12.28

Show that $(1-2xt+t^2)^{-1/2}=\sum_{n=0}^{\infty}t^nP_n(x)$.

Solution We have, by binomial theorem,

$$(1-2xt+t^2)^{-1/2}=(1-u)^{-\frac{1}{2}}$$
$$=1+\frac{1}{2}u+\frac{1\cdot 3}{2\cdot 4}u^2+\frac{1\cdot 3\cdot 5}{2\cdot 4\cdot 6}u^3+\cdots$$
$$+\frac{1\cdot 3\cdot 5\cdots(2m-1)}{2\cdot 4\cdot 6\cdots(2m)}u^m+\cdots$$

where $u=2xt-t^2$.

$$=1+\sum_{m=1}^{\infty}\frac{1\cdot 3\cdot 5\cdots(2m-1)}{2\cdot 4\cdot 6\cdots(2m)}(2xt-t^2)^m. \quad (12.1)$$

Now, $u^m=t^m(2x-t)^m$

$$=t^m[(2x)^m-m(2x)^{m-1}+\frac{m(m-1)}{2!}$$
$$\cdot(2x)^{m-2}\cdot t^2+\cdots+(-1)^mt^m]$$

Substituting in Eq. (1), we get

$$(1-2xt+t^2)^{-1/2} = 1+\sum_{m=1}^{\infty}\frac{1\cdot 3\cdot 5\cdots(2m-1)}{2\cdot 4\cdot 6\cdots(2m)}[(2x)^m t^m - m(2x)^{m-1}\cdot t^{m+1} + \frac{m(m-1)}{2!}\cdot(2x)^{m-2}t^{m+2}+\cdots+(-1)^m t^{2m}. \quad (12.2)$$

Thus, the coefficient of t^n in the expansion of $(1-2xt+t^2)^{-\frac{1}{2}}$ is

$$\frac{1\cdot 3\cdots(2n-1)}{2\cdot 4\cdots(2n)}(2x)^n - \frac{1\cdot 3\cdots(2n-3)}{2\cdot 4\cdots(2n-2)}(n-1)(2x)^{n-2}$$

$$+\frac{1\cdot 3\cdots(2n-5)}{2\cdot 4\cdots(2n-4)}\frac{(n-2)(n-3)}{2!}(2x)^{n-4}+\cdots$$

$$=\frac{1\cdot 3\cdot 5\cdots(2n-1)}{n!}\left[x^n - \frac{n(n-1)}{2\cdot(2n-1)}x^{n-2} + \frac{n(n-1)(n-2)(n-3)}{2\cdot 4\cdot(2n-1)(2n-3)}x^{n-4}-\cdots\right] = P_n(x)$$

$$\therefore\ (1-2xt+t^2)^{-1/2} = \sum_{n=0}^{\infty} t^n P_n(x). \quad (12.3)$$

12.10.7 Recurrence Relations of Legendre Functions

Example 12.29

Show that the following recurrence relations are satisfied by the Legendre polynomials $P_n(x)$:

$\text{RR1}: (2n+1)xP_n(x) = (n+1)P_{n+1}(x)+nP_{n-1}(x)$

$\text{RR2}: nP_n(x) = xP_n'(x) - P_{n-1}'(x)$

$\text{RR3}: (2n+1)P_n(x) = P_{n+1}'(x) - P_{n-1}'(x)$

$\text{RR4}: P_n'(x) = xP_{n-1}'(x)+nP_{n-1}(x)$

$\text{RR5}: (1-x^2)P_n'(x) = n[P_{n-1}(x) - xP_n(x)]$

$\text{RR6}: (1-x^2)P_n'(x) = (n+1)[xP_n(x) - P_{n+1}(x)]$

Solution

To prove RR1

We know that $(1-2xt+t^2)^{-1/2} = \sum_{n=0}^{\infty} t^n P_n(x)$. (12.1)

Differentiating Eq. (1) partially w.r.t. t, we get

$$-\frac{1}{2}(1-2xt+t^2)^{-3/2}(-2x+2t) = \sum_{n=1}^{\infty} nt^{n-1}P_n(x).$$

Multiplying this by $(1-2xt+t^2)$ and using Eq. (1), we get

$$(x-t)\sum_{n=0}^{\infty} t^n P_n(x) = (1-2xt+t^2)\sum_{n=1}^{\infty} nt^{n-1}P_n(x)$$

Equating the coefficients of t^n

$$xP_n(x) - P_{n-1}(x) = (n+1)P_{n+1}(x) - 2nxP_n(x) + (n-1)P_{n-1}(x)$$

$$\Rightarrow (2n+1)xP_n(x) = (n+1)P_{n+1}(x)+nP_{n-1}(x)$$

To prove RR2

$$nP_n(x) = xP_n'(x) - P_{n-1}'(x)$$

Differentiating Eq. (1) partially w.r.t. t and x, we get respectively

$$(x-t)(1-2xt+t^2)^{-3/2} = \sum_{n=1}^{\infty} nt^{n-1}P_n(x)$$

and $$t(1-2xt+t^2)^{-3/2} = \sum_{n=0}^{\infty} t^n P_n'(x).$$

From these we have, on multiplying the first by t and the second by $(x-t)$,

$$t(x-t)(1-2xt+t^2)^{-3/2} = \sum_{n=1}^{\infty} nt^n P_n(x) = (x-t)\sum_{n=0}^{\infty} t^n P_n'(x)$$

Equating the coefficients of t^n

$$nP_n(x) = xP_n'(x) - P_{n-1}'(x)$$

To prove RR3

$$(2n+1)P_n(x) = P_{n+1}'(x) - P_{n-1}'(x)$$

Differentiating RR1: $(2n+1)xP_n(x) = (n+1)\,P_{n+1}(x)+nP_{n-1}(x)$ w.r.t. x we have

$$(2n+1)xP_n'(n) + (2n+1)P_n(x) = (n+1)P_{n+1}'(x) + nP_{n-1}'(x)$$

RR2 is $xP_n'(x) = nP_n(x) + P_{n-1}'(x)$

$$\therefore\ (2n+1)[nP_n(x)+P_{n-1}'(x)] + (2n+1)P_n(x)$$

$$= (n+1)P'_{n+1}(x) + nP'_{n-1}(x).$$

$$\Rightarrow (2n+1)(n+1)P_n(x) = (n+1)P'_{n+1}(x) -(n+1)P'_{n-1}(x).$$

$$\Rightarrow (2n+1)P_n(x) = P'_{n+1} - P'_{n-1}(x),$$

cancelling $(n+1) \neq 0$.

To prove RR4

$$P'_n(x) = xP'_{n-1}(x) + nP_{n-1}(x)$$

$$\text{RR3}: (2n+1)P_n(x) = P'_{n+1}(x) - P'_{n-1}(x)$$

$$\text{RR2}: nP_n(x) = xP_{n'}(x) - P'_{n-1}(x)$$

Subtracting we get $(n+1)P_n(x) = P'_{n+1}(x) -xP_{n'}(x)$ Replace $(n+1)$ by n

$$nP_{n-1}(x) = P_{n'}(x) - xP'_{n-1}(x)$$

or $$P_{n'}(x) = nP_{n-1}(x) + xP'_{n-1}(x).$$

To prove RR5

$$(1-x^2)P'_n(x) = n[P_{n-1}(x) - xP_n(x)]$$

$$\text{RR4}: P_{n'}(x) - xP'_{n-1}(x) = nP_{n-1}(x)$$

$$\text{RR2}: xP'_n(x) - P'_{n-1}(x) = nP_n(x)$$

Multiplying the last equation by x and subtracting it from the previous one

$$(1-x^2)P'_n(x) = n[P_{n-1}(x) - xP_n(x)].$$

To prove RR6

$$(1-x^2)P'_n(x) = (n+1)[xP_n(x) - P_{n+1}(x)]$$

Rewriting RR1 as

$$(n+1)xP_n(x) + nxP_n(x) = (n+1)P_{n+1}(x) + nP_{n-1}(x)$$

$$\Rightarrow (n+1)[xP_n(x) - P_{n+1}(x)] = n[P_{n-1}(x) - xP_n(x)]$$

$$= (1-x^2)P'_n(x) \quad \text{by RR 5}$$

$$\Rightarrow (1-x^2)P'_n(x) = (n+1)[xP_n(x) - P_{n+1}(x)].$$

Note

The above recurrence relations can also be proved by using Rodrigues' formula.

Example 12.30

(i) $P_n(1) = 1$; (ii) $P_n(-x) = (-1)^n P_n(x)$; (iii) $P_n(-1)^n$.

Solution We know that

$$\sum_{n=0}^{\infty} t^n P_n(x) = (1-2xt+t^2)^{-\frac{1}{2}}. \tag{1.1}$$

(i) Put $x = 1$ in Eq. (1)

$$\sum_{n=0}^{\infty} t^n P_n(1) = (1-2t+t^2)^{-\frac{1}{2}} = [(1-t)^2]^{-\frac{1}{2}}$$

$$= (1-t)^{-1} = 1 + t + t^2 + t^3 + \cdots$$

$$= \sum_{n=0}^{\infty} t^n \cdot 1$$

Equating the coefficients of t^n on both sides

$$P_n(1) = 1$$

(ii) Replacing t by $-t$ and x by $-x$ in Eq. (1), we get

$$\sum_{n=0}^{\infty} (-1)^n t^n P_n(-x) = [1 - 2(-x)(-t) + (-t)^2]^{-\frac{1}{2}}$$

$$= (1-2xt+t^2)^{-\frac{1}{2}} = \sum_{n=0}^{\infty} t^n P_n(x).$$

Equating the coefficients of t^n on both sides

$$(-1)^n P_n(-x) = P_n(x) \quad \text{or} \quad P_n(-x) = (-1)^n P_n(x).$$

(iii) Put $x = 1$ in the above result. Then,

$$P_n(-1) = (-1)^n. \qquad \because P_n(1) = 1.$$

12.10.8 Orthogonality of Functions

A set of functions $f_1, f_2, f_3, \ldots$ defined on some interval $I = \{x \in R \mid a \leq x \leq b\}$ is said to be *orthogonal* on I with respect to a *weightfunction* $p(x) > 0$ if

$$\int_a^b p(x) f_m(x) f_n(x)\,dx = 0 \qquad \text{for } m \neq n \tag{12.36}$$

The *norm* $\| f_n \|$ of f_n is defined by

$$\| f_n \| = \sqrt{\int_a^b p(x) f_n^2(x)\,dx} \tag{12.37}$$

The functions are called *orthonormal* on I if they are orthogonal on I and all have a norm equal to 1.

In respect of functions with $p(x)=1,$ we simply say 'orthogonal'. Thus, functions $f_1, f_2, f_3, \ldots$ are *orthogonal* on some interval I if

$$\int_a^b f_m(x) f_n(x) dx = 0 \qquad \text{for } m \neq n \quad (12.38)$$

The *norm* $\| f_n \|$ of f_n is then simply given by

$$\| f_n \| = \sqrt{\int_a^b f_n^2(x) dx} \qquad (12.39)$$

The functions are called *orthonormai* on I if they are orthogonal on I and all of them are of norm equal to 1.

12.10.9 Orthogonality of Legendre Polynomials $P_n(x)$

Example 12.31

Show that the Legendre Polynomials $P_0(x), P_1(x), P_2(x), \ldots$ are orthogonal on $I = [-1, 1]$. That is,

$$\int_{-1}^{1} P_m(x) P_n(x) dx = \begin{cases} 0 & \text{for } m \neq n \\ \dfrac{2}{2n+1} & \text{for } m = n \end{cases} \qquad (12.1)$$

Solution

Case (i) $m \neq n$

Let $u = P_m(x)$ and $v = P_n(x)$ be the solutions of the Legendre's equation

$$(1-x^2)y'' - 2xy' + p(p+1)y = 0$$

so that we have

$$(1-x^2)u'' - 2xu' + m(m+1)u = 0 \qquad (12.2)$$

$$(1-x^2)v'' - 2xv' + n(n+1)v = 0 \qquad (12.3)$$

Multiplying Eq. (2) by v and Eq. (3) by u and subtracting, we have

$$(1-x^2)(u''v - uv'') - 2x(u'v - uv') + [m(m+1)$$
$$-n(n+1)]uv = 0$$

$$(1-x^2)(u'v - uv')' + (1-x^2)'(u'v - uv')$$
$$+(m-n)(m+n+1)uv = 0$$

$$[(1-x^2)(u'v - uv')]' + (m-n)(m+n+1)uv = 0$$

On transposing, we have

$$-(m-n)(m+n+1)uv = [(1-x^2)(u'v - uv')]'$$

Integrating both sides w.r.t. x between -1 and 1, we get

$$-(m-n)(m+n+1)\int_{-1}^{1} uv dx = (1-x^2)(u'v - uv')\Big|_{-1}^{1}$$
$$= 0 \quad m \neq n$$

$$\Rightarrow \int_{-1}^{1} uv dx = 0 \text{ or}$$

$$\int_{-1}^{1} P_m(x) P_n(x) dx = 0 \; (m \neq n). \qquad (12.4)$$

Case (ii) $m = n$. We know that

$$\sum t^n P_n(x) = P_0(x) + tP_1(x) + t^2 P_2(x) + \cdots$$
$$= (1 - 2xt + t^2)^{-\frac{1}{2}} \qquad (12.5)$$

Squaring both sides we have

$$P_0^2(x) + t^2 P_1^2(x) + t^4 P_2^2(x) + \cdots + 2[tP_0(x)P_1(x)$$
$$+t^2(P_0(x)P_2(x)) + t^3(P_0(x)P_3(x)$$
$$+P_1(x)P_2(x)) + \cdots] = (1 - 2xt + t^2)^{-1} \qquad (12.6)$$

$$\sum_{n=0}^{\infty} t^{2n} P_n^2(x) + 2 \sum_{m=1(m \neq n)}^{\infty} t^{m+n} P_m(x) P_n(x)$$
$$= (1 - 2xt + t^2)^{-1} \qquad (12.7)$$

Integrating both sides w.r.t. x between -1 and 1, we get

$$\sum_{n=0}^{\infty} t^{2n} \int_{-1}^{1} P_n^2(x) dx + 0 = \int_{-1}^{1} \frac{dx}{1 - 2xt + t^2}, \qquad (12.8)$$

by Eq. (4).

$$= \left[-\frac{1}{2t} \log(1 - 2xt + t^2) \right]_{-1}^{1}$$

$$= -\frac{1}{2t} [\log(1-t)^2 - \log(1+t)^2]$$

$$= -\frac{1}{t} [\log(1-t) - \log(1+t)]$$

$$=\frac{1}{t}\left[t+\frac{t^2}{2}+\frac{t^3}{3}+\cdots+\left(t-\frac{t^2}{2}+\frac{t^3}{3}-\cdots\right)\right]$$

$$=2\left[1+\frac{t^2}{3}+\frac{t^4}{5}+\cdots+\frac{t^{2n}}{2n+1}+\cdots\right]$$

$$=2\sum_{n=0}^{\infty}\frac{t^{2n}}{2n+1} \qquad (12.9)$$

Equating the coefficients of t^{2n} on both sides, we get

$$\int_{-1}^{1}P_n^2(x)dx=\frac{2}{2n+1}. \qquad (12.10)$$

Example 12.32

Prove that

$$P'_{n+1}(x)+P'_n(x)=P_0(x)+3P_1(x)+5P_2(x)+\cdots+(2n+1)P_n(x)$$

$$=\sum_{m=0}^{n}(2m+1)P_m(x) \qquad (12.1)$$

Solution RR3 $\Rightarrow (2n+1)P_n=P'_{n+1}-P'_{n-1}$

Replacing n by $1, 2, \ldots (n-1), n$

$$3P_1=P'_2-P'_0;\quad 5P_2=P'_3-P'_1;\quad 7P_3=P'_4-P'_2$$

$$\cdots\quad\cdots$$

$$(2n-1)P_{n-1}=P'_n-P'_{n-2}$$

$$(2n+1)P_n=P'_{n+1}-P'_{n-1}$$

Adding we get

$$3P_1+5P_2+7P_3+\cdots+(2n-1)P_{n-1}+(2n+1)P_n$$

$$=-P'_0-P'_1+P'_n+P'_{n+1}$$

$$\Rightarrow P_{n+1}'(x)+P'_n(x)=\sum_{m=0}^{n}(2m+1)P_m(x).$$

$(\because\ -P_{1'}=-P_0 \text{ and } P_{0'}=0)$

12.10.10 Betrami's Result

Example 12.33

Prove that

$$(2n+1)(x^2-1)P'_n(x)=n(n+1)[P_{n+1}(x)-P_{n-1}(x)]$$

Solution

$$\text{RR5}: (1-x^2)P_{n'}(x)=n[P_{n-1}(x)-xP_n(x)] \quad (12.1)$$

$$\text{RR6}: (1-x^2)P_{n'}(x)=(n+1)[xP_n(x)-P_{n+1}(x)] \quad (12.2)$$

Multiplying Eq. (1) by $(n+1)$, Eq. (2) by n and adding

$$[(n+1)+n](1-x^2)P_{n'}(x)$$

$$=n(n+1)[P_{n-1}(x)-xP_n(x)+xP_n(x)-P_{n+1}(x)]$$

$$\Rightarrow (2n+1)(x^2-1)P_{n'}(x)=n(n+1)[P_{n+1}(x)-P_{n-1}(x)].$$

12.10.11 Christoffel's Expansion

Example 12.34

Prove that $P_{n'}=(2n-1)P_{n-1}+(2n-5)P_{n-3}+(2n-9)P_{n-5}+\cdots+3P_1$ or P_0 according as n is even or odd.

Solution RR3: $(2n+1)P_n=P'_{n+1}-P'_{n-1}$

$$\Rightarrow P'_{n+1}=(2n+1)P_n+P'_{n-1}$$

Replacing n by $(n-1)$, we have

$$P'_n=(2n-1)P_{n-1}+P'_{n-2}-5pt \quad (12.1)$$

Replacing n by $(n-2), (n-4), (n-6), \ldots$ and finally by 2 (if n is even) and 3 (if n is odd).

$$P'_{n-2}=(2n-5)P_{n-3}+P'_{n-4} \quad (12.2)$$

$$P'_{n-4}=(2n-9)P_{n-5}+P'_{n-6} \quad (12.3)$$

$$\cdots\quad\cdots$$

$$P'_2=3P_1+P_{0'} \quad (n \text{ even}) \quad (12.4)$$

$$P_{3'}=5P_2+P' \quad (n \text{ odd}) \quad (12.5)$$

Adding these

$$P_{n'}=(2n-1)P_{n-1}+(2n-5)P_{n-3}+(2n-9)P_{n-5}+\cdots+3P_1 \quad (n \text{ even}) \because\ P_{0'}=0$$

$$=(2n-1)P_{n-1}+(2n-5)P_{n-3}+(2n-9)P_{n-5}+\cdots+P_{1'} \quad (n \text{ odd}).$$

12.10.12 Christoffel's Summation Formula

Example 12.35

Prove that $\sum_{m=0}^{n}(2m+1)P_m(x)P_m(y)$

$$= (n+1)\times\frac{[P_{n+1}(x)P_n(y)-P_{n+1}(y)P_n(x)]}{(x-y)}\ (x\neq y)$$

Solution

$$\text{RR1}\Rightarrow (2m+1)xP_m(x)=(m+1)P_{m+1}(x)+mP_{m-1}(x) \quad (12.1)$$

$$(2m+1)yP_m(y)=(m+1)P_{m+1}(y)+mP_{m-1}(y) \quad (12.2)$$

(1)$\cdot P_m(y)-$(2)$\cdot P_m(x)$ yields

$$(2m+1)(x-y)P_m(x)P_m(y)$$

$$=(m+1)\times[P_{m+1}(x)P_m(y)-P_{m+1}(y)P_m(x)]$$
$$-m[P_{m-1}(y)P_m(x)-P_{m-1}(x)P_m(y)] \quad (12.3)$$

Substituting $m=0,1,2,\ldots n$ in Eq. (3)

$$1\cdot(x-y)P_0(x)P_0(y)=1[P_1(x)P_0(y)-P_1(y)P_0(x)] \quad (12.4)$$

$$3\cdot(x-y)P_1(x)P_1(y)$$
$$=2\cdot[P_2(x)P_1(y)-P_2(y)P_1(x)]$$
$$-1\cdot[P_1(x)P_0(y)-P_1(y)P_0(x)] \quad (12.5)$$

$$5\cdot(x-y)P_2(x)P_2(y)$$
$$=3\cdot[P_3(x)P_2(y)-P_3(y)P_2(x)]$$
$$-2\cdot[P_2(x)P_1(y)-P_2(y)P_1(x)] \quad (12.6)$$

… …

$$(2n-1)(x-y)P_{n-1}(x)P_{n-1}(y)$$
$$=n[P_n(x)P_{n-1}(y)-P_n(y)P_{n-1}(x)]-$$
$$(n-1)[P_{n-1}(x)P_{n-2}(y)-P_{n-1}(y)P_{n-2}(x)] \quad (12.7)$$

$$(2n+1)(x-y)P_n(x)P_n(y)$$
$$=(n+1)\times[P_{n+1}(x)P_n(y)-P_{n+1}(y)P_n(x)]$$
$$-n[P_n(x)P_{n-1}(y)-P_n(y)P_{n-1}(x)] \quad (12.8)$$

Adding Eqs. (4) – (8)

$$(x-y)\sum_{m=0}^{n}(2m+1)P_m(x)P_m(y)$$

$$=(n+1)\times[P_{n+1}(x)P_n(y)-P_{n+1}(y)P_n(x)]$$

$$\Rightarrow\sum_{m=0}^{n}(2m+1)P_m(x)P_m(y)$$

$$=(n+1)\times\frac{[P_{n+1}(x)P_n(y)-P_{n+1}(y)P_n(x)]}{(x-y)}\ (x\neq y).$$

12.10.13 Laplace's First Integral for $P_n(x)$

Example 12.36

Show that when $n\in\mathbb{N}$,

$$P_n(x)=\frac{1}{\pi}\int_0^{\pi}(x\pm\sqrt{x^2-1}\cos\theta)^n\,d\theta$$

Solution We know that

$$\int_0^{\pi}\frac{d\theta}{a\pm b\cos\theta}=\frac{\pi}{\sqrt{a^2-b^2}}\quad(a^2>b^2)\quad(12.1)$$

Let $a=1-tx$ and $b=t\sqrt{x^2-1}$. We have $a^2-b^2=1-2tx+t^2$ so that Eq. (1) becomes

$$\int_0^{\pi}(1-tx\pm t\sqrt{x^2-1}\cos\theta)^{-1}d\theta=\pi(1-2tx+t^2)^{-1/2}$$

$$=\pi\sum_{n=0}^{\infty}t^nP_n(x)$$

[$\because (1-2xt+t^2)^{-\frac{1}{2}}$ is the generating function for $P_n(x)$].

$$\Rightarrow\pi\sum_{n=0}^{\infty}t^nP_n(x)=\int_0^{\pi}(1-tu)^{-1}d\theta$$

[where $u=x\pm\sqrt{x^2-1}\cos\theta$]

$$=\int_0^{\pi}\sum_{n=0}^{\infty}(tu)^n\,d\theta$$

$$=\sum_{n=0}^{\pi}t^n\int_0^{\pi}(x\pm\sqrt{x^2-1}\cos\theta)^n\,d\theta$$

Equating the coefficients of t^n,

$$\pi P_n(x)=\int_0^{\pi}(x\pm\sqrt{x^2-1}\cos\theta)^n\,d\theta$$

$$\text{or } P_n(x)=\frac{1}{\pi}\int_0^{\pi}(x\pm\sqrt{x^2-1}\cos\theta)^n\,d\theta.$$

Example 12.37

Show that

$$P_n(\cos\phi) = \frac{1}{\pi}\int_0^\pi (\cos\phi + i\sin\phi\cos\theta)^n\, d\theta$$

Solution Laplace's first integral for $P_n(x)$ is

$$P_n(x) = \frac{1}{\pi}\int_0^\pi (x \pm \sqrt{x^2-1}\cos\theta)^n\, d\theta$$

Put $x = \cos\phi$ then

$$\sqrt{x^2-1} = \sqrt{\cos^2\phi - 1} = \sqrt{-1}\sqrt{1-\cos^2\phi} = i\sin\phi$$

$$\therefore\quad P_n(\cos\phi) = \frac{1}{\pi}\int_0^\pi (\cos\phi + i\sin\phi\cos\theta)d\theta.$$

12.10.14 Laplace's Second Integral for $P_n(x)$

Example 12.38

Prove that

$$P_n(x) = \frac{1}{\pi}\int_0^\pi (x \pm \sqrt{x^2-1}\cos\theta)^{-n-1}\, d\theta \ (n \in \mathbb{N})$$

Solution We know that

$$\int_0^\pi \frac{d\theta}{a \pm b\cos\theta} = \frac{\pi}{\sqrt{a^2-b^2}} \qquad (a^2 > b^2)$$

Let $a = tx - 1$ and $b = t\sqrt{x^2-1}$. Then

$$a^2 - b^2 = 1 - 2xt + t^2 = t^2\left(\frac{1}{t^2} - 2x\frac{1}{t} + 1\right)$$

$$\therefore\quad \frac{\pi}{t}\left(1 - 2x\frac{1}{t} + \frac{1}{t^2}\right)^{-1/2}$$

$$= \int_0^\pi (-1 + tx \pm t\sqrt{x^2-1}\cos\theta)^{-1} d\theta$$

$$\Rightarrow \frac{\pi}{t}\sum_{n=0}^{\infty}\left(\frac{1}{t}\right)^n P_n(x) = \int_0^\pi (u-1)^{-1} d\theta,$$

where $u = t[x \pm \sqrt{x^2-1}\cos\theta]$

$$= \int_0^\pi \frac{1}{u}\left(1 - \frac{1}{u}\right)^{-1} d\theta = \int_0^\pi \frac{1}{u}\left(1 + \frac{1}{u} + \frac{1}{u^2} + \cdots\right) d\theta$$

$$= \int_0^\pi \left(\sum_{n=0}^{\infty} \frac{1}{u^{n+1}}\right) d\theta$$

$$= \sum_{n=0}^{\infty}\int_0^\pi \frac{d\theta}{t^{n+1}(x \pm \sqrt{x^2-1}\cos\theta)^{n+1}}$$

Equating the coefficients of $\frac{1}{t^{n+1}}$ on both sides

$$\pi P_n(x) = \int_0^\pi \frac{d\theta}{(x \pm \sqrt{x^2-1}\cos\theta)^{n+1}}$$

$$\text{or } P_n(x) = \frac{1}{\pi}\int_0^\pi \frac{d\theta}{(x \pm \sqrt{x^2-1}\cos\theta)^{n+1}}$$

$$\because\quad P_{-(n+1)} = P_n(x)$$

12.10.15 Expansion of $f(x)$ in a Series of Legendre Polynomials

Example 12.39

Let $f(x)$ be expressible in a series of Legendre polynomials. Thus

$$f(x) = \sum_{n=0}^{\infty} C_n P_n(x)$$

$$= C_0P_0(x) + C_1P_1(x) + \cdots + C_nP_n(x) + \cdots \quad (12.1)$$

where C_n are constants to be determined.

Then multiplying both sides by $P_n(x)$ and integrating w.r.t. x from -1 to 1, we get

$$\int_{-1}^{1} f(x)P_n(x)dx = \int_{-1}^{1}[C_0P_0(x) + C_1P_1(x) + \cdots + C_nP_n(x) + \cdots]P_n(x)dx$$

$$= C_n\int_{-1}^{1} P_n^2(x)dx = C_n \cdot \frac{2}{2n+1}$$

$$\Rightarrow C_n = \frac{2n+1}{2}\int_{-1}^{1} f(x)P_n(x)dx$$

$$\because \int_{-1}^{1} P_m(x)P_n(x)dx = 0 \qquad \text{if } m \neq n$$

$$\int_{-1}^{1} P_n^2(x)dx = \frac{2}{2n+1} \qquad \text{if } m = n.$$

Example 12.40

Show that $P_0(x) = 1, P_1(x) = x, P_2(x) = \frac{1}{2}(3x^2-1)$ and hence express $2x^2 - 4x + 2$ in terms of Legendre polynomials. [JNTU 2003 *S*]

Solution From Rodrigues' formula

$$P_n(x) = \frac{1}{2^n\, n!}\frac{d^n}{dx^n}(x^2-1)^n$$

Taking $n = 0, 1, 2$ we have

$$P_0(x) = \frac{d^0}{dx^0}(x^2-1)^0 = 1,$$

$$P_1(x) = \frac{1}{2^1 1!}\frac{d}{dx}(x^2-1) = x,$$

$$P_2(x) = \frac{1}{2^2 2!}\frac{d^2}{dx^2}(x^2-1)^2$$

$$\Rightarrow P_2(x) = \frac{1}{8}\frac{d}{dx}[2(x^2-1)\cdot 2x]$$

$$= \frac{1}{2}\frac{d}{dx}(x^3-x) = \frac{1}{2}(3x^2-1)$$

$$\Rightarrow x^2 = \frac{1}{3}[2P_2(x)+1]$$

$$= \frac{1}{3}[2P_2(x)+P_0(x)] \quad \because P_0(x)=1$$

$$\therefore 2x^2-4x+2 = \frac{2}{3}[2P_2(x)+P_0(x)]-4P_1(x)+2P_0(x)$$

$$= \frac{4}{3}[P_2(x)-3P_1(x)+2P_0(x)].$$

Example 12.41

Express x^3+2x^2-x-3 in terms if Legendre polynomials. [JNTU 2004 *S* (Set 1)]

Solution We have $P_0(x)=1, P_1(x)=x,$

$$P_2(x) = \frac{1}{2}(3x^2-1),\ P_3(x) = \frac{1}{2}(5x^3-3x)$$

$$\Rightarrow x^3 = \frac{2}{5}P_3(x)+\frac{3}{5}x = \frac{2}{5}P_3(x)+\frac{3}{5}P_1(x)$$

$$x^2 = \frac{1}{3}[2P_2(x)+P_0(x)] \quad x = P_1(x)$$

$$\therefore x^3+2x^2-x-3 = \frac{2}{5}P_3(x)+\frac{3}{5}P_1(x)+\frac{4}{3}P_2(x)$$

$$+\frac{2}{3}P_0(x)-P_1(x)-3P_0(x)$$

$$= \frac{2}{5}P_3(x)+\frac{4}{3}P_2(x)-\frac{2}{5}P_1(x)-\frac{7}{3}P_0(x).$$

Example 12.42

Using Rodrigues' formula prove that

$$\int_{-1}^{1} x^m P_n(x)dx = 0 \qquad \text{if } m<n.$$

Solution Rodrigues' formula for $P_n(x)$ is

$$P_n(x) = \frac{1}{2^n n!}\frac{d^n}{dx^n}(x^2-1)^n$$

$$\Rightarrow \int_{-1}^{1} x^m P_n(x)dx = \frac{1}{2^n n!}\int_{-1}^{1} x^m \frac{d^n}{dx^n}(x^2-1)^n dx$$

$$= \frac{1}{2^n n!}\int_{-1}^{1} x^m D^n (x^2-1)^n dx$$

where $D^n = \frac{d^n}{dx^n}$

$$= \frac{1}{2^n n!}\int_{-1}^{1} x^m D[D^{n-1}(x^2-1)^n]dx$$

$$= \frac{1}{2^n n!}[x^m D^{n-1}(x^2-1)^n]_{-1}^{1}$$

$$-\int_{-1}^{1} mx^{m-1}D^{n-1}(x^2-1)^n dx$$

(by integration by parts).

$$= 0 - \frac{m}{2^n n!}\int_{-1}^{1} x^{m-1}D^{n-1}(x^2-1)^n dx$$

($\because D^{n-1}(x^2-1)^n$ contains (x^2-1) as a factor)

Proceeding similarly, integrating by parts $(n-1)$ times, we get

$$\int_{-1}^{1} x^m P_n(x)dx = \frac{(-1)^n m}{2^n n!}\int_{-1}^{1}(x^2-1)^n D^{n-1}(x^{m-1})dx = 0$$

$$[\because \ D^{m-1}(x^{m-1}) = 0 \qquad \text{if } m<n]$$

$$\Rightarrow \int_{-1}^{1} x^m P_n(x)dx = 0 \qquad \text{if } m<n.$$

Example 12.43

Prove that $(1-2xt+t^2)^{-1/2}$ is a solution of the

equation $t\dfrac{\partial^2(tu)}{\partial t^2}+\dfrac{\partial}{\partial x}\left[(1-x^2)\dfrac{\partial u}{\partial x}\right]=0.$

Solution We know that

$$(1-2xt+t^2)^{-1/2} = \sum_{n=0}^{\infty} t^n P_n(x).$$

Denoting each side by u, we have

$$tu = \sum_{n=0}^{\infty} t^{n+1}P_n.$$

$$\Rightarrow t\frac{\partial^2(tu)}{\partial t^2} = \sum_{n=0}^{\infty} n(n+1)t^n P_n \text{ and } \frac{\partial u}{\partial x} = \sum_{n=0}^{\infty} t^n P_{n'} \quad (12.1)$$

$$\therefore \quad \frac{\partial}{\partial x}\left[(1-x^2)\frac{\partial u}{\partial x}\right] = \frac{\partial}{\partial x}\left[(1-x^2)\sum_{n=0}^{\infty} t^n P_{n'}\right]$$

$$= (1-x^2)\sum_{n=0}^{\infty} t^n P_{n''} - 2x\sum_{n=0}^{\infty} t^n P_{n'} \qquad (12.2)$$

From Eqs. (1) and (2) we have

$$t\frac{\partial^2(tu)}{\partial t^2}+\frac{\partial}{\partial x}\left[(1-x^2)\frac{\partial u}{\partial x}\right]$$

$$=\sum_{n=0}^{\infty}n(n+1)t^nP_n+(1-x^2)$$

$$\cdot\sum_{n=0}^{\infty}t^nP_{n''}-2x\sum_{n=0}^{\infty}t^nP_{n'}$$

$$=\sum_{n=0}^{\infty}t^n[n(n+1)P_n-2xP_{n'}+(1-x^2)P_{n''}]=0,$$

since P_n satisfies Legendre's equation:

$$(1-x^2)y''-\ 2xy'+n(n+1)y=0.$$

Example 12.44

Prove that $\dfrac{1+t}{t\sqrt{1-2xt+t^2}}-\dfrac{1}{t}=\sum_{n=0}^{\infty}[P_n(x)$ $+P_{n+1}(x)]t^n$

[JNTU 2005S (Set 4)]

Solution Since

$$(1-2xt+t^2)^{-\frac{1}{2}}=\sum_{n=0}^{\infty}P_n(x)t^n \quad (12.1)$$

$$\therefore\ \frac{1+t}{t\sqrt{1-2xt+t^2}}-\frac{1}{t}$$

$$=\frac{1}{t}(1-2xt+t^2)^{-\frac{1}{2}}+(1-2xt+t^2)^{-\frac{1}{2}}-\frac{1}{t}$$

$$=\frac{1}{t}\sum_{n=0}^{\infty}P_n(x)t^n+\sum_{n=0}^{\infty}P_n(x)t^n-\frac{1}{t} \quad \text{by Eq. (1)}$$

$$=\frac{1}{t}\left(P_0(x)+\sum_{n=1}^{\infty}P_nt^n\right)+\sum_{n=0}^{\infty}P_n(x)t^n-\frac{1}{t}$$

$$=\frac{1}{t}+\frac{1}{t}\sum_{n=1}^{\infty}P_n(x)t^n+\sum_{n=0}^{\infty}P_n(x)t^n-\frac{1}{t} \quad \because P_0(x)=1$$

$$=\sum_{n=1}^{\infty}P_n(x)t^{n-1}+\sum_{n=0}^{\infty}P_n(x)t^n$$

$$=\sum_{n=0}^{\infty}P_{n+1}(x)t^n+\sum_{n=0}^{\infty}P_n(x)t^n,$$

replacing n by $(n+1)$ in the first summation

$$=\sum_{n=0}^{\infty}[P_n(x)+P_{n+1}(x)]t^n.$$

Example 12.45

Expand $f(x)$ in a series of Legendre polynomials if

$$f(x)=\begin{cases}x, & 0<x<1\\ 0, & -1<x<0\end{cases}$$

Solution Let $f(x)=C_0P_0(x)+C_1P_1(x)+\cdots+C_nP_n(x)+\cdots$. Then

$$C_n=\left(\frac{2n+1}{2}\right)\int_{-1}^{1}f(x)P_n(x)dx$$

$$=\left(\frac{2n+1}{2}\right)\left[\int_{-1}^{0}0\cdot P_n(x)dx+\int_{0}^{1}xP_n(x)dx\right]$$

$$=\left(\frac{2n+1}{2}\right)\int_{0}^{1}xP_n(x)dx$$

$$n=0,\ C_0=\frac{1}{2}\int_0^1 x\,dx=\frac{1}{4}$$

$$n=1,\ C_1=\frac{3}{2}\int_0^1 x^2dx=\frac{1}{2}$$

$$n=2,\ C_2=\frac{5}{2}\int_0^1\frac{x}{2}(3x^2-1)dx=\frac{5}{16} \quad \text{and so on.}$$

EXERCISE 12.3

1. Show that (a) $P_{n'}(1)=\dfrac{1}{2}n(n+1)$;

   (b) $P_{n'}(-1)=(-1)^{n-1}\dfrac{n}{2}(n+1)$.

   [**Hint:** Use generating functions or $(1-x^2)P_{n''}-2xP_{n'}+n(n+1)P_n=0$]

2. Show that $\int_{-1}^{1}P_n(x)dx=\begin{cases}0 & n\neq 0\\ 2 & n=0\end{cases}$.

   [**Hint :** Use orthogonal property.]

3. Show that $x^5=\dfrac{1}{63}[8P_5(x)+28P_3(x)+27P_1(x)]$.

4. Show that (a) $P_{2n}(0)=\dfrac{(-1)^n(2n)!}{2^{2n}(n!)^2}$;

   (b) $P_{2n+1}(0)=0$. [JNTU 2006 (Set 2)]

5. Show that

$$\int_{-1}^{1}(1-2xt+t^2)^{-1/2}P_n(x)dx=\frac{2t^n}{2t+1}.$$

6. Show that

$$\int_0^{\pi}P_n(\cos\phi)\cos n\phi d\phi=\frac{1\cdot 3\cdot 5\cdots(2n-1)}{2\cdot 4\cdot 6\cdots 2n}.$$

7. Show that $(P_{n+1}-P_{n-1})/_{(2n+1)}=\int P_n dx+c.$

8. Show that $\int_{-1}^{1}xP_n(x)P_{n-1}(x)dx=\frac{2n}{4n^2-1}.$

[**Hint :** Use recurrence relation RR1 and orthogonal property.]

9. Show that

$$\int_{-1}^{1}x^2P_{n+1}(x)P_{n-1}(x)dx=\frac{2n(n+1)}{(2n-1)(2n+1)(2n+3)}.$$

[JNTU 2003 *S* (Set 4)]

[**Hint :** Use RR1]

10. Show that $2P_3(x)+3P_1(x)=5x^3$.

[JNTU 2006 (Set 3)]

12.11 Bessel Functions

12.11.1 Introduction

Bessel functions are solutions of Bessel's differential equation which arises in the solution of Laplace's equation in cylindrical coordinates. It occurs in many boundary value problems arising in electrical fields, mechanical vibrations and heat conduction.

12.11.2 Bessel Functions

The second order linear differential equation

$$x^2y''+xy'+(x^2-p^2)y=0$$

$$\text{or}\quad \frac{1}{x}\frac{d}{dx}\left(x\frac{dy}{dx}\right)+\left(1-\frac{p^2}{x^2}\right)y=0 \tag{12.40}$$

is called Bessel's[5] equation of order p. Its particular solutions are called Bessel functions of order p. p is a real (or complex) constant. Here we assume p to be real non-negative number.

As mentioned earlier, Eq. (12.40) can be solved by the method of Frobenius. Writing Eq. (12.40) in the standard form we have

$$y''+\frac{1}{x}y'+\left(1-\frac{p^2}{x^2}\right)y=0 \tag{12.41}$$

We assume a solution of Eq. (2) in the form

$$y(x)=\sum_{r=0}^{\infty}a_rx^{m+r}$$

$$=x^m(a_0+a_1x+\cdots+a_rx^r+\cdots) \tag{12.42}$$

Differentiating Eq. (12.42) w.r.t. x twice, we get

$$y'(x)=\sum_{r=0}^{\infty}(m+r)a_rx^{m+r-1} \tag{12.43}$$

$$y''(x)=\sum_{r=0}^{\infty}(m+r)(m+r-1)a_rx^{m+r-2} \tag{12.44}$$

Substituting Eqs. (12.42)--(12.44) into Eq. (12.40) we obtain

$$\sum[(m+r)(m+r-1)+(m+r)-p^2]a_rx^{m+r}$$

$$+\sum a_rx^{m+r+2}=0. \tag{12.45}$$

Equation (12.45) must be identically satisfied. This implies that the coefficient of each power of x must be zero.

The lowest power of x is x^m obtained for $r=0$. Equating to zero the coefficient of x^m, we get the indicial equation

$$m^2-p^2=0\Rightarrow m=p \text{ or } -p \because a_0\neq 0 \tag{12.46}$$

Equating to zero the coefficient of x^{m+1}, we obtain

$$[(m+1)^2-p^2]a_1=0\Rightarrow a_1=0\because (m+1)^2\neq p^2$$

Generally equating to zero the coefficient of x^{m+r+2}, we get the coefficient recurrence relation

$$[(m+r+2)^2-p^2]a_{r+2}+a_r=0$$

$$\Rightarrow a_{r+2}=-\frac{a_r}{(m-p+r+2)(m+p+r+2)} \tag{12.47}$$

Putting $r=1,3,5,\ldots$ we see that the odd coefficients vanish, *i.e.*, $a_{2r-1}=0$ for all $r=1,2,3,\ldots$. Putting $r=0,2,4,\ldots$ we get

[5] Named after the German mathematician and astronomer Friedrich Wilhelm Bessel (1784–1846) whose paper on Bessel functions dated 1824 appeared in 1826. He started out as an apprentice of a trade company studying astronomy on his own in his spare time, later became an assistant at a small private observatory and finally became director of new Konigsberg observatory.

$$a_2 = -\frac{a_0}{(m-p+2)(m+p+2)},$$

$$a_4 = -\frac{a_2}{(m-p+4)(m+p+4)}$$

$$= \frac{a_0}{(m-p+2)(m+p+2)(m-p+4)\times(m+p+4)},$$

$$a_6 = -\frac{a_4}{(m-p+6)(m+p+6)}$$

$$= \frac{-a_0}{(m-p+2)(m+p+2)(m-p+4)(m+p+4)(m-p+6)(m+p+6)} \text{etc.}$$

A solution of Eq. (12.40) is

$$y = a_0 x^m \left[1 - \frac{x^2}{[(m+2)^2-p^2]} + \frac{x^4}{[(m+2)^2-p^2][(m+4)^2-p^2]} - \frac{x^6}{[(m+2)^2-p^2][(m+4)^2-p^2][(m+6)^2-p^2]} + \cdots\right] \quad (12.48)$$

We get different types of solutions depending on the values of p.

12.11.3 Bessel Functions of Non-Integral Order p: $J_p(x)$ and $J_{-p}(x)$

Case (i) p is not an integer.

In this case, we get two linearly independent solutions for $m = p$ and $m = -p$. For $m = p$, we have

$$y_1 = a_0 x^p \left[1 - \frac{x^2}{2(2p+2)} + \frac{x^4}{2\cdot 4(2p+2)(2p+4)} - \frac{x^6}{2\cdot 4\cdot 6(2p+2)(2p+4)(2p+6)} + \cdots\right]$$

$$= a_0 x^p \left[1 + (-1)\frac{x^2}{2(1!)(p+1)} + (-1)^2 \frac{x^4}{2^4(2!)(p+1)(p+2)} + (-1)^3 \times \frac{x^6}{2^6(3!)(p+1)(p+2)(p+3)} + \cdots\right]$$

$$= a_0 x^p \sum_{r=0}^{\infty} \frac{(-1)^r}{2^{2r}(r!)(p+1)(p+2)\cdots(p+r)} x^{2r}$$

$$= a_0 x^p \sum_{r=0}^{\infty} \frac{(-1)^r \Gamma(p+1) x^{2r}}{2^{2r}\Gamma(r+1)\Gamma(p+r+1)}$$

Since a_0 is arbitrary we choose it as $a_0 = \dfrac{1}{2^p\,\Gamma(p+1)}$ and denote the resulting solution by $J_p(x)$. This is called Bessel function of the first kind of order p. Thus

$$J_p(x) = \sum_{r=0}^{\infty} \frac{(-1)^r}{r!\,\Gamma(p+r+1)} \left(\frac{x}{2}\right)^{p+2r} \quad (12.49)$$

A second linearly independent solution corresponding to $m = -p$ is

$$J_{-p}(x) = \sum_{r=0}^{\infty} \frac{(-1)^r}{r!\,\Gamma(-p+r+1)} \left(\frac{x}{2}\right)^{-p+2r} \quad (12.50)$$

which is also called Bessel function of the first kind of order $-p$. Both the series converge for all x as can be seen by D' Alembert's ratio test.

When p is not an integer, the complete solution of Eq. (12.40) is

$$y(x) = AJ_p(x) + BJ_{-p}(x) \quad (12.51)$$

where A and B are arbitrary constants.

12.11.4 Bessel Functions of Order Zero and One: $J_0(x)$, $J_1(x)$

Case (ii) $p = 0$

Putting $p = 0$ in Eq. (12.40) we obtain Bessel's equation of order zero as

$$xy'' + y' + xy = 0 \quad \text{or} \quad (xy')' + xy = 0 \quad (12.52)$$

Its solution obtained from Eq. (12.48) by putting $p = 0$ and taking $a_0 = 1$ is

$$y = x^m \left[1 - \frac{x^2}{(m+2)^2} + \frac{x^4}{(m+2)^2(m+4)^2}\right.$$

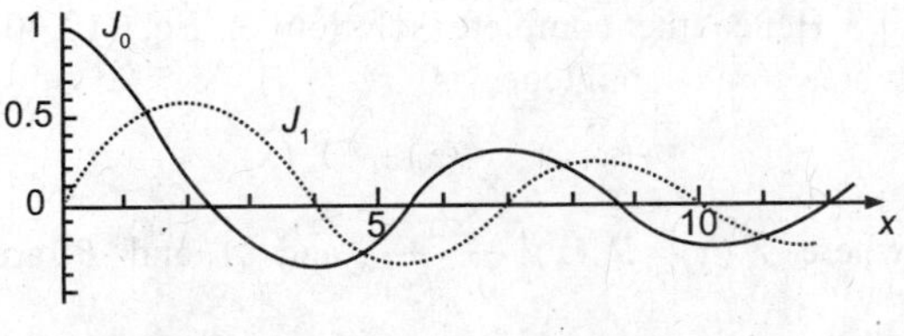

Figure 12.4 Bessel functions of the first kind

$$-\frac{x^6}{(m+2)(m+4)(m+6)}+\cdots\Bigg] \quad (12.53)$$

which is the solution of Eq. (12.52) if $m=0$.

Thus, the first solution of Eq. (12.52) is

$$J_0(x)=\sum_{r=0}^{\infty}\frac{(-1)^r}{(r!)}\left(\frac{x}{2}\right)^{2r}$$

$$=1-\frac{x^2}{2^2(1!)^2}+\frac{x^4}{2^4(2!)^2}-\frac{x^6}{2^6(3!)^2}+\cdots \quad (12.54)$$

which is called Bessel's function of the first kind of order zero, and appears to be similar to the cosine function.

For $p=1$, we obtain the Bessel function of order 1

$$J_1(x)=\sum_{r=0}^{\infty}\frac{(-1)^r x^{2r+1}}{2^{2r+1}r!(r+1)!}$$

$$=\frac{x}{2}-\frac{x^3}{2^3 1!2!}+\frac{x^5}{2^5\cdot 2!3!}-\frac{x^7}{2^7\cdot 3!4!}+\cdots \quad (12.55)$$

This looks similar to the sine function. The height of the waves decreases with increasing x.

Dividing Bessel's Eq. (12.40) by x^2 we can put it in the standard form $y''+\frac{1}{x}y+\left(1-\frac{n^2}{x^2}\right)y=0$

where p has been replaced by n. The term $\frac{n^2}{x^2}$ is 0 for $n=0$. Further, this term as well as the term $\frac{1}{x}y'$ are small in absolute value for large x so that the Bessel's equation comes closer to $y''+y=0$ whose solutions are sine and cosine functions. Also, $\frac{1}{x}y$ acts as a damping term which is partly responsible for the decrease in height (see Fig. 12.4).

For large x, we can derive the result that

$$J_n(x)=\sqrt{\frac{2}{\pi x}}\cos\left(x-\frac{2n+1}{4}\pi\right).$$

12.11.5 Bessel Function of Second Kind of Order Zero $Y_0(x)$

Differentiating Eq. (12.53) partially w.r.t. m, we get

$$\frac{\partial y}{\partial m}=x^m\log x\left[1-\frac{x^2}{(m+2)^2}+\frac{x^4}{(m+2)^2(m+4)^2}-\frac{x^6}{(m+2)^2(m+4)^2(m+6)^2}+\cdots\right]$$

$$+x^m\left[\frac{x^2}{(m+2)^2}\cdot\frac{2}{m+2}-\frac{x^4}{(m+2)^2(m+4)^2}\times\left\{\frac{2}{m+2}+\frac{2}{m+4}\right\}+\cdots\right]$$

A second, independent solution of Eq. (12.52) denoted by $Y_0(x)$ is given by $\left(\frac{\partial y}{\partial m}\right)_{m=0}$

$$Y_0(x)=J_0(x)\log x\left[\left(\frac{x}{2}\right)^2-\frac{1}{(2!)^2}\left(1+\frac{1}{2}\right)\left(\frac{x}{2}\right)^4+\frac{1}{(3!)^2}\left(1+\frac{1}{2}+\frac{1}{3}\right)\left(\frac{x}{2}\right)^6-\cdots\right]$$

$$=J_0(x)\log x+\sum_{r=1}^{\infty}\frac{(-1)^{r+1}}{(r!)^2}\times\left(1+\frac{1}{2}+\frac{1}{3}+\cdots+\frac{1}{r}\right)\left(\frac{x}{2}\right)^{2r} \quad (12.56)$$

$Y_0(x)$ is called the Bessel function of the second kind of order zero or Neumann function. So, the complete solution of Bessel Eq. (12.52) of order zero is

$$y(x)=AJ_0(x)+BY_0(x) \quad (12.57)$$

where A and B are arbitrary constants.

12.11.6 Bessel Functions of Integral Order: Linear Dependence of $J_n(x)$ and $J_{-n}(x)$

Case (iii) $p=n$ (an integer):

To prove that $J_{-n}(x)=(-1)^n J_n(x)$.

Taking $p=-n$, a positive integer so that $-n$ is a negative integer, we obtain from Eq. (12.50)

$$J_{-n}(x)=\sum_{r=0}^{\infty}\frac{(-1)^r}{r!\Gamma(-n+r+1)}\left(\frac{x}{2}\right)^{-n+2r}$$

$$=(-1)^n\sum_{(r-n)=0}^{\infty}\frac{(-1)^{r-n}}{(r-n+n)!\Gamma(r-n+1)}\left(\frac{x}{2}\right)^{2(r-n)+n}$$

$\because\ \Gamma(r-n+1)=\infty$ for $r<n$

$$=(-1)^n\sum_{s=0}^{\infty}\frac{(-1)^s}{(s+n)!\Gamma(s+1)}\left(\frac{x}{2}\right)^{2s+n}$$

where $s=r-n\geq 0$

$$=(-1)^n J_n(x). \qquad (12.58)$$

This proves that when p is an integer Bessel functions $J_p(x)$ and $J_{-p}(x)$ are linearly dependent.

12.11.7 Bessel Functions of the Second Kind of Order n: $Y_n(x)$: Determination of Second Solution $Y_n(x)$ by the Method of Variation of Parameters

Since $J_n(x)$ and $J_{-n}(x)$ are linearly dependent when n is an integer, the second solution of Eq. (12.40) is obtained by the following method.

Let $y=uv$, where $v=J_n(x)$, be a solution of Eq. (12.40). Then

$$y'=u'v+uv', \qquad y''=u''v+2u'v'+uv''$$

Substituting in Eq. (12.40), we get

$$x^2[u''v+2u'v'+uv'']+x[u'v+uv']+(x^2-n^2)uv=0$$

$$\Rightarrow x^2[u''v+2u'v']+xu'v=0.$$

The coefficient of u is zero since $v=J_n(x)$ is a solution of Eq. (12.40).

$\Rightarrow \dfrac{u''}{u'}+2\dfrac{v'}{v}+\dfrac{1}{x}=0$, on dividing each term by $x^2u'v$

$\Rightarrow (\log u')'+2(\log v)'+(\log x)'=0$

$\Rightarrow \log(xu'\cdot v^2)=\log B$, on integration w.r.t. x

$\Rightarrow u(x)=B\displaystyle\int\frac{dx}{xJ_n^2(x)}+A$, on dropping logs,

after dividing by xv^2 and integrating.

Hence the complete solution of Eq. (12.40) when $p=n$, an integer, is

$$y(x)=AJ_n(x)+BY_n(x)$$

where $Y_n(x)=J_n(x)\displaystyle\int\frac{dx}{xJ_n^2(x)}$, and A and B are arbitrary constants.

12.11.8 Generating Functions for Bessel Functions

Let $<f_n>$ where $n\in\mathbb{Z}$ be a sequence of functions f_n. A function $w(t,x)$ such that $w(t,x)=\sum_{n=-\infty}^{\infty}f_n(x)$ is called a generating function of the functions f_n as already given at Section 12.10.6 (p. 21).

Generating function for Bessel functions $J_n(x)$ of integral order n.

Example 12.46

Prove that $$e^{\frac{x}{2}\left(t-\frac{1}{t}\right)}=\sum_{n=-\infty}^{\infty}t^nJ_n(x) \qquad (12.1)$$

Solution

$$e^{\frac{x}{2}\left(t-\frac{1}{t}\right)}=e^{\frac{x}{2}t}\cdot e^{-\frac{x}{2t}}$$

$$=\left[1+\frac{1}{1!}\left(\frac{x}{2}\right)t+\frac{1}{2!}\left(\frac{x}{2}\right)^2t^2+\cdots+\frac{1}{n!}\left(\frac{x}{2}\right)^nt^n+\cdots\right]$$

$$\times\left[1-\frac{1}{1!}\left(\frac{x}{2}\right)\frac{1}{t}+\frac{1}{2!}\left(\frac{x}{2}\right)^2\frac{1}{t^2}-\cdots+(-1)^n\frac{1}{n!}\left(\frac{x}{2}\right)^n\frac{1}{t^n}+\cdots\right]$$

In the product on the RHS

$$\text{Coeff. of } t^0=1-\frac{1}{(1!)^2}\left(\frac{x}{2}\right)^2+\frac{1}{(2!)^2}\left(\frac{x}{2}\right)^4-\frac{1}{(3!)^2}\left(\frac{x}{2}\right)^6-\cdots+(-1)\frac{n!}{(n!)^2}\left(\frac{x}{2}\right)^{2n}+\cdots$$

$$= \sum \frac{(-1)^r}{(r!)^2}\left(\frac{x}{2}\right)^{2n} = J_0(x) \qquad (12.2)$$

$$\text{Coeff. of } t^n = \frac{1}{n!}\left(\frac{x}{2}\right)^n - \frac{1}{(n+1)!1!}\left(\frac{x}{2}\right)^{n+2} + \frac{1}{(n+2)!2!}\left(\frac{x}{2}\right)^{n+4} - \cdots$$

$$= \sum \frac{(-1)^r}{r!(n+r)!}\left(\frac{x}{2}\right)^{n+2r} = J_n(x) \quad (12.3)$$

$$\text{Coeff. of } t^{-n} = (-1)^n\left[\frac{1}{n!}\left(\frac{x}{2}\right)^n - \frac{1}{(n+1)!1!}\left(\frac{x}{2}\right)^{n+2} + \frac{1}{(n+2)!2!}\left(\frac{x}{2}\right)^{n+4} - \cdots\right]$$

$$= (-1)^n \sum \frac{(-1)^r}{r!(n+r)!}\left(\frac{x}{2}\right)^{n+2r}$$

$$= (-1)^n J_n(x) = J_{-n}(x) \qquad (12.4)$$

The result follows from Eqs. (2), (3) and (4).

12.11.9 Recurrence Relations of Bessel Functions

Example 12.47

The following recurrence relations connect Bessel functions of different orders and are very useful in solving problems involving Bessel functions. These relations are true for general p but we prove them taking $p = n$.

$$\text{RR1}: \frac{d}{dx}(x^n J_n(x)) = x^n J_{n-1}(x)$$

$$\text{RR2}: \frac{d}{dx}(x^{-n} J_n(x)) = -x^{-n} J_{n+1}(x)$$

$$\text{RR3}: J_{n'}(x) + \frac{n}{x} J_n(x) = J_{n-1}(x)$$

$$\text{RR4}: J_{n'}(x) - \frac{n}{x} J_n(x) = J_{n+1}(x)$$

$$\text{RR5}: 2J_{n'}(x) = J_{n-1}(x) - J_{n+1}(x)$$

$$\text{RR6}: \frac{2n}{x} J_n(x) = J_{n-1}(x) + J_{n+1}(x)$$

Solution

To prove RR1

$$\frac{d}{dx}(x^n J_n(x)) = x^n J_{n-1}(x)$$

$$\text{We know that } J_n(x) = \sum_{r=0}^{\infty} \frac{(-1)^r}{r!\Gamma(n+r+1)}\left(\frac{x}{2}\right)^{n+2r}$$

$$\Rightarrow \quad x^n J_n(x) = \sum_{r=0}^{\infty} \frac{(-1)^r x^{2n+2r}}{2^{n+2r} r!(n+r)\Gamma(n+r)}$$

$$\Rightarrow \frac{d}{dx}[x^n J_n(x)] = \sum_{r=0}^{\infty} \frac{(-1)^r x^{2n+2r-1}}{2^{n+2r-1} r!\Gamma(n+r)}$$

$$= x^n J_{n-1}(x).$$

To prove RR2

$$\frac{d}{dx}(x^{-n} J_n(x)) = -x^{-n} J_{n+1}(x)$$

$$\text{We know that } J_n(x) = \sum_{r=0}^{\infty} \frac{(-1)^r}{r!\Gamma(n+r+1)}\left(\frac{x}{2}\right)^{n+2r}$$

$$\Rightarrow x^{-n} J_n(x) = \sum_{r=0}^{\infty} \frac{(-1)^r x^{2r}}{2^{n+2r} r!\Gamma(n+r+1)}$$

$$\Rightarrow \frac{d}{dx}[x^{-n} \cdot J_n(x)] = -\sum_{r=1}^{\infty} \frac{(-1)^{r-1} x^{2r-1}}{2^{n+2r-1}(r-1)!\,\Gamma(n+1+r)}$$

$$= -x^{-n} \sum_{s=0}^{\infty} \frac{(-1)^s}{s!\Gamma(n+1+s+1)} \times \left(\frac{x}{2}\right)^{n+1+2s}$$

$$= -x^{-n} J_{n+1}(x),$$

replacing $r-1$ by s.

To prove RR3

$$J_n'(x) + \frac{n}{x} J_n(x) = J_{n-1}(x)$$

$$RR1 \quad \Rightarrow \frac{d}{dx}(x^n J_n(x)) = x^n J_{n-1}(x)$$

$$\Rightarrow x^n J_{n'}(x) + n x^{n-1} J_n(x) = x^n J_{n-1}(x)$$

$$\Rightarrow J_{n'}(x)+\frac{n}{x}J_n(x)=J_{n-1}(x),$$

on dividing by x^n.

To prove RR4

$$J_{n'}(x)-\frac{n}{x}J_n(x)=-J_{n+1}(x)$$

RR2 $\Rightarrow \frac{d}{dx}(x^{-n}J_n(x))=-x^{-n}J_{n+1}(x)$

$$\Rightarrow x^{-n}J_{n'}(x)-nx^{-n-1}J_n(x)=-x^{-n}J_{n+1}(x)$$

$$\Rightarrow J_{n'}(x)-\frac{n}{x}J_n(x)=-J_{n+1}(x),$$

on dividing by x^{-n}.

To prove RR5

$$2J_{n'}(x)=J_{n-1}(x)-J_{n+1}(x)$$

Adding RR3 $\Rightarrow J_{n'}(x)+\frac{n}{x}J_n(x)=J_{n-1}(x)$

and RR4 $\Rightarrow J_{n'}(x)-\frac{n}{x}J_n(x)=-J_{n+1}(x)$

We get $2J_{n'}(x)=J_{n-1}(x)-J_{n+1}(x)$.

To prove RR6

$$2\frac{n}{x}J_n(x)=J_{n-1}(x)+J_{n+1}(x)$$

We have RR3 $\Rightarrow J_{n'}(x)+\frac{n}{x}J_n(x)=J_{n-1}(x)$

RR4 $\Rightarrow J_{n'}(x)-\frac{n}{x}J_n(x)=-J_{n+1}(x)$

Subtracting, we get

$$2\frac{n}{x}J_n(x)=J_{n-1}(x)+J_{n+1}(x).$$

12.11.10 Bessel's Functions of Half- Integral Order

Bessel function J_p of orders $p=n\pm\frac{1}{2}$ $(n=0,1,2,3,\ldots)$ are elementary functions. They can be expressed in terms of sines and cosines and powers of x.

Example 12.48

Prove that

(i) $J_{1/2}(x)=\sqrt{\frac{2}{\pi x}}\sin x$; (ii) $J_{-1/2}(x)=\sqrt{\frac{2}{\pi x}}\cos x$;

(iii) $J_{1/2}^2(x)+J_{-1/2}^2(x)=\frac{2}{\pi x}$

Solution (i) We know that

$$J_p(x)=\frac{x^p}{2^p\Gamma(p+1)}\left[1-\frac{x^2}{2\cdot 2(p+1)}+\frac{x^4}{2\cdot 4\cdot 2^2(p+1)(p+2)}-\cdots\right]$$

Putting $p=\frac{1}{2}$

$$J_{\frac{1}{2}}(x)=\frac{x^{1/2}}{2^{1/2}\Gamma(\frac{1}{2}+1)}\left[1-\frac{x^2}{2\cdot 2\left(\frac{1}{2}+1\right)}+\frac{x^4}{2\cdot 4\cdot 2^2\left(\frac{1}{2}+1\right)\left(\frac{1}{2}+2\right)}-\cdots\right]$$

$$=\frac{x^{1/2}}{2^{1/2}x\cdot\frac{1}{2}\sqrt{\pi}}\left[x-\frac{x^3}{2\cdot 3}+\frac{x^5}{2\cdot 4\cdot 3\cdot 5}-\cdots\right]$$

$$=\sqrt{\frac{2}{\pi x}}\left[x-\frac{x^3}{3!}+\frac{x^5}{5!}-\cdots\right],$$

$$=\sqrt{\frac{2}{\pi x}}\sin x$$

on multiplying and dividing by x, and noting

$$\Gamma\left(\frac{1}{2}+1\right)=\frac{1}{2}\Gamma\left(\frac{1}{2}\right)=\frac{1}{2}\sqrt{\pi}$$

(ii) We know that

$$J_p(x)=\frac{x^p}{2^p\Gamma(p+1)}\left[1-\frac{x^2}{2\cdot 2(p+1)}+\frac{x^4}{2\cdot 4\cdot 2^2(p+1)(p+2)}-\cdots\right]$$

Putting $p=-\frac{1}{2}$

$$J_{-\frac{1}{2}}(x)=\frac{x^{-1/2}}{2^{-1/2}\Gamma(-\frac{1}{2}+1)}\left[1-\frac{x^2}{2\cdot 2(-\frac{1}{2}+1)}+\frac{x^4}{2\cdot 4\cdot 2^2\left(-\frac{1}{2}+1\right)\left(-\frac{1}{2}+2\right)}-\cdots\right]$$

$$=\sqrt{\frac{2}{\pi x}}\left[1-\frac{x^2}{2!}+\frac{x^4}{4!}-\cdots\right]$$

$$=\sqrt{\frac{2}{\pi x}}\cos x \because \Gamma\left(\frac{1}{2}\right)=\sqrt{\pi}$$

(iii) Squaring and adding the above results we have

$$J_{\frac{1}{2}}^2(x)+J_{-\frac{1}{2}}^2(x)=\frac{2}{\pi x}(\sin^2 x+\cos^2 x)=\frac{2}{\pi x}.$$

Example 12.49

Prove that (i) $J_{3/2}(x)=\sqrt{\frac{2}{\pi x}}\left(\frac{\sin x}{x}-\cos x\right)$

(ii) $J_{-3/2}(x)=-\sqrt{\frac{2}{\pi x}}\left(\sin x+\frac{\cos x}{x}\right)$

Solution

(i) RR6: $\frac{2p}{x}J_p(x)=J_{p-1}(x)+J_{p+1}(x)$, where p is taken in place of n.

Taking $p=\frac{1}{2}$ we have

$$J_{3/2}(x)=\frac{1}{x}J_{\frac{1}{2}}(x)-J_{-\frac{1}{2}}(x)=\sqrt{\frac{2}{\pi x}}\left(\frac{\sin x}{x}-\cos x\right)$$

(ii) Taking $p=-\frac{1}{2}$ in

RR6: $\frac{2p}{x}J_p(x)=J_{p-1}(x)+J_{p+1}(x)$, we have

$$\frac{2\left(-\frac{1}{2}\right)}{x}J_{-\frac{1}{2}}(x)=J_{-3/2}(x)+J_{1/2}(x)$$

or $J_{-3/2}(x)=-\frac{1}{x}J_{-\frac{1}{2}}(x)-J_{\frac{1}{2}}(x)$

$$=-\frac{1}{x}\cdot\sqrt{\frac{2}{\pi x}}\cos x-\sqrt{\frac{2}{\pi x}}\sin x$$

$$=-\sqrt{\frac{2}{\pi x}}\left(\frac{\cos x}{x}+\sin x\right).$$

12.11.11 Differential Equation Reducible to Bessel's Equation

The differential equation

$$x^2\frac{d^2y}{dx^2}+x\frac{dy}{dx}+(\lambda^2x^2-p^2)y=0$$

where λ is a parameter, can be written as

$$(\lambda x)^2\frac{d^2y}{d(\lambda x)^2}+(\lambda x)\frac{dy}{d(\lambda x)}+[(\lambda x)^2-p^2]y=0$$

$$\text{or } t^2\frac{d^2y}{dt^2}+t\frac{dy}{dt}+(t^2-p^2)y=0 \qquad (t=\lambda x)$$

which is Bessel's equation. When p is a non-integer the general solution is

$$y=C_1J_p(t)+C_2J_{-p}(t)=C_1J_p(\lambda x)+C_2J_{-p}(\lambda x)$$

and when $p=n$ is an integer the general solution is

$$y=C_1J_n(t)+C_2Y_n(t)=C_1J_n(\lambda x)+C_2Y_n(\lambda x).$$

12.11.12 Orthogonality

Definition of orthogonality of functions

A set of functions $f_1, f_2,\ldots$ defined on some interval $I=[a,b,]$ is said to be orthogonal on I with respect to a *weight function* $w(x)>0$ if

$$\int_a^b w(x)f_m(x)f_n(x)dx=0 \qquad \text{for } m\neq n$$

The norm $\|f_m\|$ of f_m is defined by

$$\|f_m\|=\sqrt{\int_a^b w(x)f_m^2(x)dx}$$

The functions are called orthonormal of I if they are orthogonal on I and all have norm equal to unity.

For 'orthogonal w.r.t. $w(x)=1$' we simply say orthogonal on I. Thus, functions $f_1, f_2,\ldots$ are orthogonal on some interval I if

$$\int_a^b f_m(x)f_n(x)dx = 0 \quad \text{for } m \neq n.$$

The norm $\| f_m \|$ of f_m is then simply defined by

$$\| f_m \| = \sqrt{\int_a^b f_m^2(x)dx}$$

and the functions are called *orthonormal* on $a \le x \le b$ if they are orthogonal there and all have norm equal to unity.

Orthogonal set, orthonormal set

Example 12.51

Prove that the functions $f_m(x) = \sin mx$ $(m = 1, 2, 3, \ldots)$ form an orthogonal set on $-\pi \le x \le \pi$.

Solution Here $a = -\pi$, $b = \pi$, $f_m(x) = \sin mx$ and we have

$$\int_a^b f_m(x)f_n(x)dx = \int_{-\pi}^{\pi} \sin m \sin nx dx$$

$$= \frac{1}{2}\int_{-\pi}^{\pi} \cos(m-n)xdx - \frac{1}{2}\int_{-\pi}^{\pi} \cos(m+n)xdx$$

$$= \frac{1}{2}\cdot\frac{\sin(m-n)x}{m-n}\bigg|_{-\pi}^{\pi} - \frac{1}{2}\cdot\frac{\sin(m+n)x}{m+n}\bigg|_{-\pi}^{\pi} = 0$$

since the functions vanish at both the limits.

Also, $\| f_m \|^2 = \int_{-\pi}^{\pi} \sin^2 mxdx = \frac{1}{2}\int_{-\pi}^{\pi}(1-\cos 2mx)dx$

$$= \frac{1}{2}\cdot 2\pi - 0 = \pi$$

$\therefore$ The norm $= \sqrt{\pi}$; hence the orthonormal set is

$$\frac{\sin x}{\sqrt{\pi}}, \frac{\sin 2x}{\sqrt{\pi}}, \frac{\sin 3x}{\sqrt{\pi}}, \ldots$$

Orthogonality of Bessel functions

Example 12.50

Prove that, for each fixed non-negative integer n,

$$\int_0^a xJ_n(\alpha x)J_n(\beta x)dx = \begin{cases} 0, & \text{if } \alpha \neq \beta \\ \dfrac{a^2}{2}J_{n+1}^2(a\alpha), & \text{if } \alpha = \beta \end{cases}$$

where α and β are roots of $J_n(ax) = 0$.

Solution Let $u(x) = J_n(\alpha x)$ and $v(x) = J_n(\beta x)$ be the solutions of the equations

$$x^2u'' + xu' + (\alpha^2x^2 - n^2)u = 0 \qquad (12.1)$$

$$x^2v'' + xv' + (\beta^2x^2 - n^2)v = 0 \qquad (12.2)$$

respectively. Hence $u'(x) = \alpha J_{n'}(\alpha x), v'(x) = \beta J_{n'}(\beta x)$.

Now multiplying Eq. (1) by $\frac{v}{x}$ and Eq. (2) by $\frac{u}{x}$ and subtracting,

$$x(u''v - uv'') + (u'v - uv') + (\alpha^2 - \beta^2)xuv = 0$$

$$\Rightarrow \frac{d}{dx}[x(u'v - uv')] = (\beta^2 - \alpha^2)xuv \qquad (12.3)$$

Integrating both sides of Eq. (3) w.r.t. x from $x = 0$ to a

$$(\beta^2 - \alpha^2)\int_0^a xuvdx = [x(u'v - uv')]_0^a$$

$$= a[u'(a)v(a) - u(a)v'(a)] \qquad (12.4)$$

This implies that

$$(\beta^2 - \alpha^2)\int_0^a xJ_n(\alpha x)J_n(\beta x)dx$$

$$= a[\alpha J_{n'}(\alpha a)J_n(\beta a) - \beta J_{n'}(\beta a)J_n(\alpha a)]$$

$$\Rightarrow \int_0^a xJ_n(\alpha x)J_n(\beta x)dx$$

$$= \frac{a}{\beta^2 - \alpha^2}[\alpha J_{n'}(\alpha a)J_n(\beta a) - \beta J_{n'}(\beta a)J_n(\alpha a)] \qquad (12.5)$$

Case (i) $\alpha \neq \beta$

That is, α, β are two district roots of $J_n(ax)$ so that $J_n(a\alpha) = J_n(a\beta) = 0$.

Therefore, we obtain from Eq. (5)

$$\int_0^a xJ_n(\alpha x)J_n(\beta x)dx = 0$$

which is the orthogonality relation for Bessel functions.

Case (ii) $\alpha = \beta$

In this case, the RHS of Eq. (5) assumes the indeterminate form $\frac{0}{0}$. Hence we apply L' Hospital's Rule by differentiating w.r.t. β and evaluating the limits.

$$\lim_{\beta \to \alpha}\int_0^a xJ_n(\alpha x)J_n(\beta x)dx$$

$$= \lim_{\beta\to\alpha}\left(\frac{a}{\beta^2-\alpha^2}\right)[\alpha J_{n'}(a\alpha)J_n(a\beta)-0],$$

$$\because J_n(a\alpha)=0$$

$$= \lim_{\beta\to\alpha}\frac{a}{2\beta}[\alpha J_{n'}(a\alpha).aJ_{n'}(a\beta)]$$

$$= \frac{a^2}{2}[J_{n'}(a\alpha)]^2 = \frac{a^2}{2}J_{n+1}^2(a\alpha).$$

By recurrence relation RR4:

$$J_{n'}(x)=\frac{n}{x}J_n(x)-J_{n+1}(x)$$

$$\Rightarrow J_{n'}(a\alpha)=\frac{n}{a\alpha}J_n(a\alpha)-J_{n+1}(a\alpha)$$

$$=0-J_{n+1}(a\alpha)=-J_{n+1}(a\alpha),$$

$$(\because J_n(a\alpha)=0)$$

If we put $x=a\alpha$ in the recurrence relation RR6, we have

$$J_{n-1}(a\alpha)+J_{n+1}(a\alpha)=\frac{2n}{a\alpha}J_n(a\alpha)=0$$

Thus for $\alpha=\beta$

$$\int_0^a xJ_n(\alpha x)J_n(\beta x)dx=\frac{a^2}{2}J_{n+1}(a\alpha)=\frac{a^2}{2}J_{n-1}(a\alpha)$$

12.11.13 Integrals of Bessel Functions

Example 12.52

Prove that $\int x^{-p}J_{p+1}(x)dx=-x^{-p}J_p(x)+c.$

Solution Recurrence relation RR1 is

$$\frac{d}{dx}(x^pJ_p(x))=x^pJ_{p-1}(x)$$

Integrating both sides we have

$$\int x^pJ_{p-1}(x)dx=x^pJ_p(x)+c$$

For $p=1$, we get

$$\int xJ_0(x)dx=xJ_1(x)+c$$

Recurrence relation RR2 is

$$\frac{d}{dx}(x^{-p}J_p(x))=-x^{-p}J_{p+1}(x)$$

Integrating both sides, we have

$$\int x^{-p}J_{p+1}(x)dx=-x^{-p}J_p(x)+c$$

For $p=0$ we get

$$\int J_1(x)dx=-J_0(x)+c.$$

In general, $\int x^mJ_n(x)dx$ for m and n integer with $m+n\geq 0$ can be integrated by parts completely if $m+n$ is odd.

But when $m+n$ is even the integral depends on the residual integral $\int J_0(x)dx$ which has been tabulated.

Integrating,

$$J_{p'}(x)=\frac{1}{2}[J_{p-1}(x)-J_{p+1}(x)]$$

We get $2J_p(x)=\int J_{p-1}(x)dx-\int J_{p+1}(x)dx$

or $\int J_{p+1}(x)dx=\int J_{p-1}(x)dx-2J_p(x).$

12.11.14 Expansion of Sine and Cosine in Terms of Bessel Functions

Example 12.53

Show that (i) $\cos x=J_0-2J_2+2J_4-\cdots$

(ii) $\sin x=2(J_1-J_3+J_5-\cdots).$

Solution We know that

$$e^{\frac{x}{2}\left(t-\frac{1}{t}\right)}=\sum_{n=-\infty}^{\infty}t^nJ_n(x)$$

$$=\sum_{-\infty}^{-1}t^nJ_n(x)+J_0(x)+\sum_{1}^{\infty}t^nJ_n(x)$$

$$=J_0(x)+\sum_{n=1}^{\infty}[t^{-n}J_{-n}(x)+t^nJ_n(x)]$$

$$=J_0(x)+\left(t-\frac{1}{t}\right)J_1(x)+\left(t^2+\frac{1}{t^2}\right)J_2(x)+\cdots$$

$$=J_0(x)+\sum_{n=1}^{\infty}(t^n+(-1)^nt^{-n})J_n(x)$$

Put $t=e^{i\theta}$

$$e^{x(e^{i\theta}-e^{-i\theta})/2}=J_0+(e^{i\theta}-e^{-i\theta})J_1+(e^{2i\theta}+e^{-2i\theta})J_2+\cdots$$

$$\Rightarrow e^{ix\sin\theta}=J_0+(2i\sin\theta)J_1+(2\cos 2\theta)J_2+\cdots$$

Separating the real and imaginary parts we have

$$\cos(x\sin\theta) = J_0 + 2\cos 2\theta J_2 + 2\cos 4\theta J_4 + \cdots$$

$$\sin(x\sin\theta) = 2\sin\theta J_1 + 2\sin 3\theta J_3 + 2\sin 5\theta J_5 + \cdots$$

Putting $\theta = \pi/2$, we get

$$\cos x = J_0 - 2J_2 + 2J_4 + \cdots$$

$$\sin x = 2(J_1 - J_3 + J_5 - \cdots).$$

Example 12.54

Find series expansions for $J_0(x)$ and $J_1(x)$.

Solution We know that

$$J_n(x) = \sum_{r=0}^{\infty} \frac{(-1)^r}{r!(n+r)!}\left(\frac{x}{2}\right)^{n+2r}$$

Putting $n = 0$ we get

$$J_0(x) = \sum_{r=0}^{\infty} \frac{(-1)^r}{(r!)^2}\left(\frac{x}{2}\right)^{2r}$$

$$= 1 - \frac{1}{(1!)^2}\left(\frac{x}{2}\right)^2 + \frac{1}{(2!)^2}\left(\frac{x}{2}\right)^4 - \frac{1}{(3!)^2}\left(\frac{x}{2}\right)^6 + \cdots$$

Putting $n = 1$ we get

$$J_1(x) = \sum_{r=0}^{\infty} \frac{(-1)^r}{r!(r+1)!}\left(\frac{x}{2}\right)^{2r+1}$$

$$= \frac{x}{2}\left[1 - \frac{1}{1!2!}\left(\frac{x}{2}\right)^2 + \frac{1}{2!3!}\left(\frac{x}{2}\right)^4 - \frac{1}{3!4!}\left(\frac{x}{2}\right)^6 + \cdots\right].$$

Example 12.55

Express $J_5(x)$ in terms of $J_0(x)$ and $J_1(x)$.

Solution

$$\text{RR6:}\quad \frac{2n}{x}J_n(x) = J_{n-1}(x) + J_{n+1}(x)$$

$$\Rightarrow \quad J_{n+1}(x) = \frac{2n}{x}J_n(x) - J_{n-1}(x) \qquad (12.1)$$

Putting $n = 1, 2, 3, 4$ we get

$$J_2 = \frac{2}{x}J_1 - J_0 \qquad (12.2)$$

$$J_3 = \frac{4}{x}J_2 - J_1 \qquad (12.3)$$

$$J_4 = \frac{6}{x}J_3 - J_2 \qquad (12.4)$$

$$J_5 = \frac{8}{x}J_4 - J_3 \qquad (12.5)$$

where the argument x has been omitted for convenience. Substituting from Eq. (2), Eq. (3) becomes

$$J_3 = \frac{4}{x}\left(\frac{2}{x}J_1 - J_0\right) - J_1 = \left(\frac{8}{x^2} - 1\right)J_1 - \frac{4}{x}J_0 \qquad (12.6)$$

Substituting Eqs. (6) and (2) into Eq. (4)

$$J_4 = \frac{6}{x}\left[\left(\frac{8}{x^2} - 1\right)J_1 - \frac{4}{x}J_0\right] - \left[\frac{2}{x}J_1 - J_0\right]$$

$$= \left(\frac{48}{x^3} - \frac{8}{x}\right)J_1 + \left(1 - \frac{24}{x^2}\right)J_0 \qquad (12.7)$$

Substituting Eqs. (7) and (6) into Eq. (5),

$$J_5 = \frac{8}{x}\left[\left(\frac{48}{x^3} - \frac{8}{x}\right)J_1 + \left(1 - \frac{24}{x^2}\right)J_0\right] - \left[\left(\frac{8}{x^2} - 1\right)J_1 - \frac{4}{x}J_0\right]$$

$$\Rightarrow \quad J_5 = \left(\frac{384}{x^4} - \frac{72}{x^2} - 1\right)J_1 + \left(\frac{12}{x} - \frac{192}{x^3}\right)J_0.$$

Example 12.56

Express $J_{5/2}(x)$ in terms of sine and cosine functions.

Solution RR6: $\frac{2n}{x}J_n(x) = J_{n-1}(x) + J_{n+1}(x)$

Put $n = \frac{3}{2}$ we get $\frac{3}{x}J_{3/2}(x) = J_{\frac{1}{2}}(x) + J_{5/2}(x)$

$$\Rightarrow J_{5/2}(x) = \frac{3}{x}J_{3/2}(x) - J_{1/2}(x)$$

$$= \frac{3}{x}\cdot\sqrt{\frac{2}{\pi x}}\left(\frac{\sin x}{x} - \cos x\right) - \sqrt{\frac{2}{\pi x}}\sin x$$

$$= \sqrt{\frac{2}{\pi x}}\left[\left(\frac{3 - x^2}{x^2}\right)\sin x - \frac{3}{x}\cos x\right]$$

$$J_{\frac{1}{2}}(x) = \sqrt{\frac{2}{\pi x}}\sin x; J_{-\frac{1}{2}}(x) = \sqrt{\frac{2}{\pi x}}\cos x$$

$$J_{3/2}(x) = \sqrt{\frac{2}{\pi x}}\left(\frac{\sin x}{x} - \cos x\right).$$

Example 12.57

Prove that

(i) $J_{-n}(x) = (-1)^n J_n(x)$

(ii) $J_n(-x) = (-1)^n J_n(x)$

(iii) $J_n(x)$ is an even or odd function of x according as n is even or odd.

Solution

(i) We know that for any non-negative real p,

$$J_p(x) = \sum_{r=0}^{\infty} \frac{(-1)^r}{r!\Gamma(p+r+1)}\left(\frac{x}{2}\right)^{p+2r} \quad (12.1)$$

Let $p = -n$ where n is a positive integer. Then

$$J_{-n}(x) = \sum_{r=0}^{\infty} \frac{(-1)^r}{r!\Gamma(-n+r+1)}\left(\frac{x}{2}\right)^{-n+2r} \quad (12.2)$$

Since $\lim \Gamma(t) = \infty$, as $t \to 0$ or a negative integer, each term in the above summation is zero as long as

$$-n+r+1 \le 0 \quad (12.3)$$

i.e., as long as $r \le (n-1)$. So, the summation starts with $r = n$,

$$J_{-n}(x) = \sum_{r=n}^{\infty} \frac{(-1)^r}{r!\Gamma(-n+r+1)}\left(\frac{x}{2}\right)^{-n+2r}$$

$$= (-1)^n \sum_{s=0}^{\infty} \frac{(-1)^s}{(n+s)!s!}\left(\frac{x}{2}\right)^{n+2s}$$

where $s = r - n$ or $r = n + s$

$$= (-1)^n J_n(x). \quad (12.4)$$

(ii) Let n be a non-negative integer.

Replacing x by $-x$, we have

$$J_n(-x) = \sum_{r=0}^{\infty} \frac{(-1)^r}{r!\Gamma(n+r+1)}\left(\frac{-x}{2}\right)^{n+2r}$$

$$= (-1)^n \sum_{r=0}^{\infty} \frac{(-1)^r}{r!\Gamma(n+r+1)}\left(\frac{x}{2}\right)^{n+2r},$$

$$\because (-1)^{n+2r} = (-1)^n(-1)^{2r} = (-1)^n$$

$$\Rightarrow J_n(-x) = (-1)^n J_n(x). \quad (12.5)$$

Let n be a negative integer and $n = -m$ where m is a positive integer. Then

$$J_n(x) = J_{-m}(x) = (-1)^m J_m(x) \quad (12.6)$$

Replacing x by $-x$, we have

$$J_n(-x) = (-1)^m J_m(-x) = (-1)^m$$

$$\cdot(-1)^m J_m(x) \text{ by Eq. (5)} \because m > 0$$

$$= (-1)^m J_{-m}(x) \qquad \text{by Eq.(6)}$$

$$= (-1)^{2m}(-1)^{-m} J_{-m}(x)$$

$$= (-1)^{-m} J_{-m}(x) \qquad \because (-1)^{2m} = 1$$

$$= (-1)^n J_n(x) \qquad \because -m = n$$

(iii) We have from (ii) $J_n(-x) = (-1)^n J_n(x)$

$\Rightarrow J_n(-x) = J_n(x)$ if n is even $\because (-1)^n = 1$

and $J_n(-x) = -J_n(x)$ if n is odd $\because (-1)^n = -1$.

Thus J_n is even or odd according as n is even or odd.

Example 12.58

Prove that

(a) $\frac{d}{dx}(J_n^2 + J_{n+1}^2) = 2\left(\frac{n}{x}J_n^2 - \frac{n+1}{x}J_{n+1}^2\right)$

or $J_n J_{n'} + J_{n+1}J'_{n+1} = \frac{1}{x}[nJ_n^2 - (n+1)J_{n+1}]$

(b) $\frac{d}{dx}(xJ_n J_{n+1}) = 2\left(\frac{n}{x}J_n^2 - \frac{n+1}{x}J_{n+1}^2\right)$

Solution

$$\text{(a)} \quad \frac{d}{dx}(J_n^2 + J_{n+1}^2) = 2J_n J_{n'} + 2J_{n+1}J_{n+1}{}' \quad (12.1)$$

$$\text{Now, RR4}: J_{n'} = \frac{n}{x}J_n - J_{n+1} \quad (12.2)$$

$$\text{RR3}: J_{n'} = -\frac{n}{x}J_n + J_{n-1}$$

$$\Rightarrow J'_{n+1} = -\frac{n+1}{x}J_{n+1} + J_n \quad (12.3)$$

replacing n by $(n+1)$.

Substituting for J'_n and J'_{n+1} into Eq. (1)

$$\frac{d}{dx}(J_n^2 + J_{n+1}^2) = 2J_n\left(\frac{n}{x}J_n - J_{n+1}\right)$$

$$+2J_{n+1}\left(-\frac{n+1}{x}J_{n+1}+J_n\right)$$

$$=2\left(\frac{n}{x}J_n^2-\frac{n+1}{x}J_{n+1}^2\right).$$

(b) $$\frac{d}{dx}(xJ_nJ_{n+1})=J_nJ_{n+1}+x(J_{n+1}J_n'+J_nJ_{n+1}') \quad (12.1)$$

$$\text{RR4: } J_{n'}=\frac{n}{x}J_n-J_{n+1}; \quad (12.2)$$

$$\text{RR3: } J_{n+1}'=-\frac{n+1}{x}J_{n+1}+J_n. \quad (12.3)$$

Substituting in Eq. (1), we get

$$\frac{d}{dx}(xJ_nJ_{n+1})=J_nJ_{n+1}+x\left[\left(\frac{n}{x}J_nJ_{n+1}-J_{n+1}^2\right.\right.$$

$$\left.-\frac{n+1}{x}J_nJ_{n+1}+J_n^2\right]$$

$$=x(J_n^2-J_{n+1}^2)$$

Example 12.59

Prove that $J_0^2+2(J_1^2+J_2^2+\cdots)=1$. Hence show that $|J_0|\le 1$ and $|J_n|\le 1/2$ for $n\ge 1$.

Solution We have

$$\frac{d}{dx}(J_n^2+J_{n+1}^2)=2\left(\frac{n}{x}J_n^2-\frac{n+1}{x}J_{n+1}^2\right)$$

Putting $n=0,1,2,3,\ldots$

$$\frac{d}{dx}(J_0^2+J_1^2)=2\left(0-\frac{1}{x}J_1^2\right) \quad \because \lim_{n\to 0}J_n=0$$

$$\frac{d}{dx}(J_1^2+J_2^2)=2\left(\frac{1}{x}J_1^2-\frac{2}{x}J_2^2\right)$$

$$\frac{d}{dx}(J_2^2+J_3^2)=2\left(\frac{2}{x}J_2^2-\frac{3}{x}J_3^2\right)$$

… …

Adding these we get

$$\frac{d}{dx}[J_0^2+2(J_1^2+J_2^2+\cdots)]=0$$

Integrating we obtain

$$J_0^2+2(J_1^2+J_2^2+\cdots)=c$$

where $c=1$ since $[J_0^2]_{x=0}=J_0^2(0)=1$ and $J_n(0)=0$ for $n\ge 1$ $\therefore$ $J_0^2+2(J_1^2+J_2^0+\cdots)=1$

Since each term is non-negative

$J_0^2\le 1$ or $|J_0|^2\le 1$ and $2|J_n^2|=2|J_n|^2\le 1$

$\Rightarrow |J_n|\le 1/2$ for $n\ge 1$.

EXERCISE 12.4

Prove the following:

1. $J_3(x)=\dfrac{8-x^2}{x^2}J_1(x)-\dfrac{4}{x}J_0(x).$
2. $J_4(x)=\left(\dfrac{48}{x^3}-\dfrac{8}{x}\right)J_1(x)+\left(1-\dfrac{24}{x^2}\right)J_0(x).$
3. $J_0''(x)=\dfrac{1}{2}(J_2(x)-J_0(x)).$
4. $J_2(x)=J_0''(x)-\dfrac{1}{x}J_0'(x).$
5. $\dfrac{J_2(x)}{J_1(x)}=\dfrac{1}{x}-\dfrac{J_0''(x)}{J_0'(x)}.$
6. $\dfrac{d}{dx}(xJ_1(x))=xJ_0(x).$
7. $\dfrac{d}{dx}(x^nJ_n(ax))=ax^nJ_{n-1}(ax).$
8. $J_{n-1}=\dfrac{2}{x}[nJ_n-(n+2)J_{n+2}+\cdots].$

   Deduce that $\dfrac{x}{2}J_n=(n+1)J_{n+1}-(n+3)J_{n+3}$
   $-(n+5)J_{n+5}-\cdots.$
9. $J_pJ_{-p}'-J_{-p}J_p'=-\dfrac{2\sin p\pi}{\pi x}.$

   Deduce that $\left(\dfrac{J_{-p}}{J_p}\right)=-\dfrac{2\sin p\pi}{\pi xJ_p^2}.$
10. (a) $\int J_0J_1dx=\dfrac{1}{2}J_0^2$

    (b) $\int xJ_0(\lambda x)dx=\dfrac{a}{\lambda}J_1(a\lambda).$
11. $\pi J_n(\dot{x})=\int_0^{\pi}\cos(x\sin\theta-x\theta)d\theta.$
12. $\int J_3(x)dx=-J_2(x)-\dfrac{2}{x}J_1(x).$

[JNTU 2003 (Set 2)]

CHEBYSHEV POLYNOMIALS

One of the important differential equations that gives rise to special functions is the Chebyshev differential equation

$$(1-x^2)\,y''-xy'+n^2\,y=0$$

where n is a positive integer. The singularities of Eq. (1) are $x = 1$. If we seek a power series solution of Eq. (1) about $x = 0$ of the form

$$y(x)=\sum_{m=0}^{\infty} c_m\,x^m \qquad (2)$$

then this series solution is convergent in $|x|<1$ since the distance between $x = 0$ and the nearest singularity is 1.

Differentiating (2) w.r.t. x twice we obtain

$$y'(x)=\sum_{m=1}^{\infty} mc_m\,x^{m-1} \text{ and } y''(x)$$

$$=\sum_{m=2}^{\infty} m(m-1)c_m\,x^{m-2} \qquad (3);\ (4)$$

Substituting (2) – (4) in Eq. (1) we obtain

$$\sum_{m=2}^{\infty} m(m-1)c_m\,x^{m-2}-\sum_{m=2}^{\infty} m(m-1)c_m\,x^m$$

$$-\sum_{m=1}^{\infty} m\,c_m\,x^m+n^2\sum_{m=0}^{\infty} c_m\,x^m=0$$

or

$$2c_2+6c_3x+\sum_{m=4}^{\infty} m(m-1)c_m\,x^{m-2}-$$

$$\sum_{m=2}^{\infty} m(m-1)c_m\,x^m-c_1x$$

$$-\sum_{m=2} m\,c_m\,x^m+n^2(c_0+c_1x)+$$

$$n^2\sum_{m=2}^{\infty} c_m\,x^m=0$$

Substituting $m-2=t$ in the first sum, we get

$$(2c_2+n^2c_0)+(6c_3-c_1+n^2c_1)x+$$

$$\sum_{t=2}^{\infty}(t+2)(t+1)c_{t+2}\,x^{t+2}$$

$$-\sum_{m=2}^{\infty}[m(m-1)+m-n^2]c_m\,x^m=0$$

Since t is a dummy variable, we can combine the third and fourth terms of this equation. Equating the coefficients of various powers of x to zero, we obtain

$$2c_2+n^2\,c_0=0,\ 6c_3-c_1+n^2\,c_1=0$$

and

$$(m+2)(m+1)c_{m+2}-(m^2-n^2)c_m=0 \quad m=2,3,\ldots$$

$$\therefore c_2=-\frac{n^2}{2!}c_0,\ c_3=\frac{1-n^2}{3!}c_1=-\frac{n^2-1}{3!}c_1$$

$$c_{m+2}=-\frac{n^2-m^2}{(m+2)(m+1)}c_m,\ m\ge 2 \qquad (2)$$

we have

$$c_4=-\frac{(n^2-2^2)}{4\cdot 3}c_2=\frac{1}{4!}n^2(n^2-2^2)c_0$$

$$c_5=-\frac{(n^2-3^2)}{5\cdot 4}c_3=\frac{1}{5!}(n^2-1^2)(n^2-3^2)c_1,\ldots$$

Substituting in the power series solution, we obtain

$$y(x)=c_0\left[1-\frac{1}{2!}n^2x^2+\frac{1}{4!}(n^2-2^2)n^2x^4+\cdots\right]$$

$$+c_1\left[x-\frac{1}{3!}(x^2-1^2)x^3+\right.$$

$$\left.\frac{1}{5!}(x^2-3^2)(x^2-1^2)x^5+\cdots\right]$$

$$=c_0\,y_0\,(x)+c_1\,y_1\,(x)\ldots \qquad (3)$$

where

$$y_0(x)=1-\frac{1}{2!}n^2x^2+\frac{1}{4!}(x^2-2^2)n^2x^4\ldots$$

$$y_1(x)=x-\frac{1}{3!}(x^2-1^2)x^3+$$

$$\frac{1}{5!}(n^2-3^2)(n^2-1^2)x^5\ldots$$

The series $y_0\,(x)$, $y_1\,(x)$ converge for $|x|<1$. Now $y_0\,(x)$ contains even powers of x only and $y_1\,(x)$

contains odd powers of x only. Hence, the two solutions $y_0(x)$, $y_1(x)$ are linearly independent solutions of the chebyshev differential equation (1).

As n takes the value zero and even positive integral values we obtain for

$$n=0:\quad y_0(x)=1,$$

$$n=2:\quad y_0(x)=1-2x^2,$$

$$n=4:\quad y_0(x)=1-8x^2+8x^4,\ldots$$

Hence, $y_0(x)$ reduces to an even degree polynomial as n takes even positive integral values, whereas $y_1(x)$ remains an infinite series.

As n takes odd positive integral values, we get for

$$n=1:\quad y_1(x)=x$$

$$n=3:\quad y_1(x)=\frac{1}{3}(3x-4x^3)$$

$$n=5:\quad y_1(x)=\frac{1}{5}(5x-20x^3+16x^5),\ldots$$

Hence, $y_1(x)$ reduces to an odd degree polynomial as n takes odd positive integral values, whereas $y_0(x)$ remains an infinite series. These polynomial give rise to an important class of polynomials called *chebyshev polynomials*.

Changing the independent variable x by the substitution $x = \cos\theta$ we have

$$\frac{dy}{dx}=\frac{dy}{d\theta}\cdot\frac{d\theta}{dx}=-\frac{1}{\sin\theta}\frac{dy}{d\theta}.$$

$$\frac{d^2y}{dx^2}=\frac{d}{d\theta}\left(-\frac{1}{\sin\theta}\frac{dy}{d\theta}\right)\frac{d\theta}{dx}$$

$$=\left[-\frac{1}{\sin\theta}\frac{d^2y}{d\theta^2}+\frac{\cos\theta}{\sin^2\theta}\frac{dy}{d\theta}\right]\left(-\frac{1}{\sin\theta}\right)$$

$$=\frac{1}{\sin^2\theta}\frac{d^2y}{d\theta^2}-\frac{\cos\theta}{\sin^3\theta}\frac{dy}{d\theta}$$

Substituting in the differential eq. (1) we obtain

$$\sin^2\theta\left(\frac{1}{\sin^2\theta}\frac{d^2y}{d\theta^2}-\frac{\cos\theta}{\sin^3\theta}\frac{dy}{d\theta}\right)-$$

$$\cos\theta\left(-\frac{1}{\sin\theta}\frac{dy}{d\theta}\right)+n^2y=0$$

$$\Rightarrow \frac{d^2y}{d\theta^2}+n^2y=0 \qquad (4)$$

The general solution of the differential Eq. (4) is

$$y(\theta)=A\cos n\theta+B\sin n\theta$$

∴ The solution of eq. (1) can be written as

$$y(x)=A\cos(n\cos^{-1}x)+B\sin(n\cos^{-1}x) \qquad (5)$$

Thus, $\cos(n\cos^{-1}x)$ and $\sin(n\cos^{-1}x)$ are two linearly independent solutions of eq. (1). Denoting the first solution by $T_n(x)$ we have

$$T_n(x)=\cos(n\cos^{-1}x) \quad -1\le x\le 1 \qquad (6)$$

which is called the *chebyshev polynomial of first kind*.

Denoting the second linearly independent solution by

$$U_n(x)=\sin(n\cos^{-1}x) - 1\le x\le 1 \qquad (7)$$

which is called the *chebyshev polynomial of second kind*.

Chebyshev polynomials of First kind

The Chebyshev polynomials of first kind are given by

$$T_n(x)=\cos n\theta=\cos(n\cos^{-1}x),\ -1\le x\le 1 \quad (8)$$

we note that $|T_n(x)|\le 1$.

Recurrence relation for Chebyshev polynomials $T_n(x)$

The Chebyshev polynomials $T_n(x)$ satisfy the following recurrence relation

$$T_{n+1}(x)-2x\,T_n(x)+T_{n-1}(x)=0 \qquad (9)$$

We have

$$T_{n+1}(x)=\cos\left[(n+1)\cos^{-1}x\right]=\cos(n+1)\theta$$

$$T_{n-1}(x)=\cos\left[(n-1)\cos^{-1}x\right]=\cos(n-1)\theta$$

where $\cos^{-1}x=\theta$. Hence

$$T_{n+1}(x)=\cos(n\theta+\theta)=\cos n\theta\cos\theta-\sin n\theta\sin\theta$$

and $T_{n-1}(x)=\cos(n\theta-\theta)=\cos n\theta\cos\theta+\sin n\theta\sin\theta$

Adding we obtain

$$T_{n+1}(x)+T_{n-1}(x)=2\cos n\theta\cos\theta=2x\,T_n(x)$$

$$\Rightarrow T_{n+1}(x)-2x\,T_n(x)+T_{n-1}(x)=0$$

From Eq. (8) we have

$T_0(x) = \cos 0 = 1,\ T_1(x) = \cos(\cos^{-1} x) = x$

Using the recurrence relation (Eq. (9)) we have

$$\begin{aligned}
T_2(x) &= 2xT_1(x) - T_0(x)\\
&= 2x^2 - 1\\
T_3(x) &= 2xT_2(x) - T_1(x)\\
&= 2x(2x^2-1) - x\\
&= 4x^3 - 3x\\
T_4(x) &= 2xT_3(x) - T_2(x)\\
&= 2x(4x^3-3x) - (2x^2-1)\\
&= 8x^4 - 8x^2 + 1\\
T_5(x) &= 2xT_4(x) - T_3(x)\\
&= 2x(8x^4 - 8x^2 + 1) - (4x^3 - 3x).\\
&= 16x^5 - 20x^3 + 5x \qquad (10)
\end{aligned}$$

$\therefore$ $T_n(x)$ is a polynomial of degree n. If n is even $T_n(x)$ is an even degree polynomial and if n is odd $T_n(x)$ is an odd degree polynomial.

To express x^n in terms of Chebyshev polynomials

From Eqns. (10) we obtain

$$1 = T_0,\quad x = T_1,\quad x^2 = \frac{1}{2}T_0 + \frac{1}{2}T_2$$

$$\begin{aligned}
x^3 &= \frac{1}{4}(3T_1 + T_3)\\
x^4 &= \frac{1}{8}\left[T_4 + 8\times\frac{1}{2}(T_0+T_2) - T_0\right]\\
&= \frac{1}{8}(3T_0 + 4T_2 + T_4)\\
x^5 &= \frac{1}{16}\left[T_5 + 20\times\frac{1}{4}(3T_1+T_3) - 5T_1\right]\\
&= \frac{1}{16}(10T_1 + 5T_3 + T_5) \qquad (11)
\end{aligned}$$

The relationship between $T_n'(x)$ and $T_n(x)$ is given by the recurrence relation

$$(1-x^2)T_n'(x) = -nxT_n(x) + nT_{n-1}(x) \qquad (12)$$

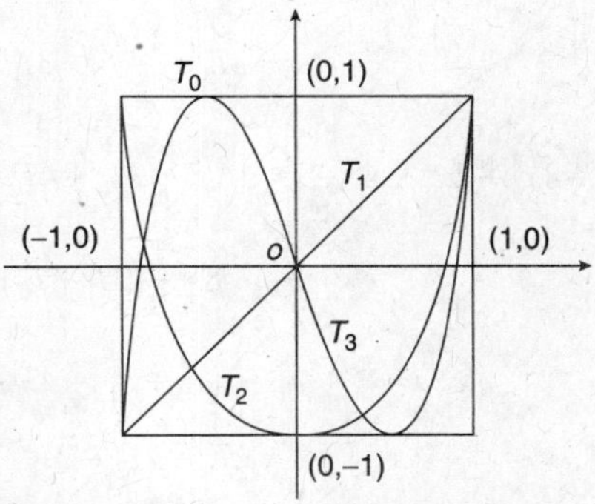

Figure 12.5 Chebyshev polynomials

Putting $x = \cos\theta$ we get

$$\begin{aligned}
\text{LHS} &= (1-\cos^2\theta)\frac{d}{d\theta}\cos n\theta\frac{d\theta}{dx}\\
&= \sin^2\theta(-n\sin n\theta)\left(-\frac{1}{\sin\theta}\right)\\
&= n\sin\theta\sin n\theta\\
\text{RHS} &= n\cos(n-1)\theta - n\cos\theta\cos n\theta\\
&= n[\cos n\theta\cos\theta + \sin n\theta\sin\theta -\\
&\qquad \cos\theta\cos n\theta]\\
&= n\sin\theta\sin n\theta
\end{aligned}$$

Since LHS = RHS Eq. (1) is proved.

Example 12.60 Write the following polynomials in terms of chebyshev polynomials of first kind.

(a) $2x^2 - 5x + 7$ (b) $8x^3 + 11x^2 - 3x + 4$

(c) $16x^4 - 8x^3 - 2x^2 + 4x - 3$

Solution We have

$$1 = T_0,\ x = T_1,\ x^2 = \frac{1}{2}(T_0+T_2),\ x^3 = \frac{1}{4}(3T_1+T_3)$$

$$x^4 = \frac{1}{8}(3T_0 + 4T_2 + T_4)$$

$$\begin{aligned}
2x^2 - 5x + 7 &= 2\times\frac{1}{2}(T_0+T_2) - 5T_1 + 7T_0\\
&= T_2 - 5T_1 + 8T_0
\end{aligned}$$

$$\begin{aligned}
8x^3 + 11x^2 - 3x + 4 &= 8\times\frac{1}{4}(3T_1+T_3) +\\
&\quad \frac{11}{2}(T_0+T_2) - 3T_1 + 4T_0
\end{aligned}$$

$$= 2T_3 + \frac{11}{2}T_2 + 3T_1 + \frac{19}{2}T_0$$

$$16x^4 - 8x^3 - 2x^2 + 4x - 3 = 16 \times \frac{1}{8}(3T_0 + 4T_2 + T_4)$$

$$- 8 \times \frac{1}{4}(3T_1 + T_0) - 2 \times$$

$$\frac{1}{2}(T_0 + T_2) + 4T_1 - 3T_0$$

$$= 2T_4 + 7T_2 - 2T_1.$$

Example 12.61 Find the expression for

(a) $T_3'(x)$ in terms of $T_3(x)$ and $T_2(x)$; and

(b) $T_5'(x)$ in terms of $T_5(x)$ and $T_4(x)$.

Solution Consider the Recurrence for much

$$(1-x^2)T_n'(x) = -nxT_n(x) + nT_{n-1}(x)$$

(a) Taking $n = 3$ we have

$$(1-x^2)T_3'(x) = -3xT_3(x) + 3T_2(x)$$

$$\Rightarrow T_3'(x) = \frac{3}{1-x^2}[T_2(x) - xT_3(x)]$$

(b) Taking $n = 5$ we have

$$(1-x^2)T_5'(x) = -5xT_5(x) + 5T_4(x)$$

$$\Rightarrow T_5'(x) = \frac{5}{1-x^2}[T_4(x) - xT_5(x)]$$

Zeros of $T_n(x)$

Equating $T_n(x)$ to zero we obtain

$$T_n(x) = \cos(n\cos^{-1}x) = 0 = \cos(2k+1)\frac{\pi}{2}$$

$$k = 0, 1, 2, \ldots (n-1).$$

$$\Rightarrow n\cos^{-1}x = (2k+1)\frac{\pi}{2} \Rightarrow x = \cos(2k+1)\frac{\pi}{2n} \quad (13)$$

$$k = 0, 1, 2, \ldots (n-1).$$

The n simple zeros of $T_n(x)$ are given by (13).

Turning Points (Extreme values) of $T_n(x)$

Putting

$$T_n'(x) = 0 \Rightarrow T_n'(x) = n\frac{\sin(n\cos^{-1}x)}{\sqrt{1-x^2}} = 0$$

$$\Rightarrow \sin(n\cos^{-1}x) = 0 = \sin k\pi$$

$$\Rightarrow n\cos^{-1}x = k\pi \Rightarrow x = \cos\frac{k\pi}{n} \quad (14)$$

$$k = 1, 2, \ldots (n-1).$$

$T_n(x)$ attains its relative maximum or minimum at $(n-1)$ points given by eq. (14). At these points we have

$$T_n(x_k) = \cos^{-1}(n\cos^{-1}x_k) = \cos k\pi = (-1)^k$$

Also, at the end points of the interval $[-1, 1]$ we have

$$T_n(-1) = \cos n\pi = (-1)^n \text{ and } T_n(1) = \cos(0) = 1.$$

$\therefore$ $T_n(x)$ attains its maximum and minimum values at the $(n+1)$ points

$$x_k = \cos\left(\frac{k\pi}{n}\right) \quad k = 0, 1, 2, \ldots n. \quad (15)$$

Leading Coefficient of $T_n(x)$

We have

$$T_n(x) = \cos n\theta = \text{Re}(e^{in\theta}) = \text{Re}(\cos\theta + i\sin\theta)^n$$

$$= \text{Re}\Big[\cos^n\theta + {}^nc_1\cos^{n-1}\theta(i\sin\theta) + {}^nc_2\cos^{n-2}\theta(i\sin\theta)^2 + \cdots\Big]$$

expanding by Binomial Theorem

$$= \cos^n\theta - {}^nc_2\cos^{n-2}\theta\sin^2\theta + {}^nc_4\cos^{n-4}\theta\sin^4\theta - \cdots$$

$$= x^n - {}^nc_2x^{n-2}(1-x^2) + {}^nc_4x^{n-4}(1-x^2)^2 - \cdots$$

$$= ({}^nc_0 + {}^nc_2 + {}^nc_4 + \cdots)x^n +$$

terms containing x to lower degree

$$= 2^{n-1}x^n + \cdots \quad (16)$$

This shows that $T_n(x)$ is a polynomial of degree n and its leading coefficient is 2^{x-1}.

Generating Function of Chebyshev Polynomials $T_n(x)$

Prove that

$$\frac{1-xt}{1-2xt+t^2} = \sum_{n=0}^{\infty} T_n(x)t^n \tag{17}$$

Proof: The function on the LHS of (17), which is a function of two variables x and t is called the generating function of chebyshev polynomials $T_n(x)$. We have

$$\text{RHS} = \sum_{n=0}^{\infty} T_n(x)t^n = \sum_{n=0}^{\infty}(0)n\theta\cdot t^n = \sum_{n=0}^{\infty}\text{Re}\, e^{in\theta}\cdot t^n$$

$$= \text{Re}\sum_{n=0}^{\infty} e^{in\theta}\cdot t^n = \text{Re}\sum_{n=0}^{\infty}(e^{i\theta t})^n = \text{Re}\left(\frac{1}{1-e^{i\theta t}}\right)$$

$$= \text{Re}\frac{1}{1-\cos\theta t - i\sin\theta t}$$

$$= \text{Re}\frac{1}{(1-xt)-it\sqrt{i-x^2}}[\because \cos\theta = x]$$

$$= \text{Re}\frac{(1-xt)+it\sqrt{1-x^2}}{(1-xt)^2+t^2(1-x^2)}$$

$$= \frac{(1-xt)}{1-2xt+t^2} = \text{LHS}.$$

Hence the result (17) is proved.

Alternate expansion for $T_n(x)$

Writing $(1-2xt+t^2)^{-1} = [1-t(2x-t)]^{-1}$ and expanding it by using Binomial Theorem we get

$$(1-2xt+t^2)^{-1} = [1-t(2x-t)]^{-1}$$
$$= 1 + t(2x-t) + t^2(2x-t)^2$$
$$+\cdots t^{n-r}(2x-t)^{n-r}+\cdots$$

Now, the term containing t^n in the expansion of the product $t^{n-r}(2x-t)^{n-r}$ is obtained from

$$t^{n-r}\left[{}^{n-r}c_r(-t)^r(2x)^{n-2r}\right]$$
$$= (-1)^r\, {}^{n-r}c_r(2x)^{n-2r}t^n$$

$\therefore$ The coefficient of t^n from all the terms is t_n where

$$b_n = \sum_{r=0}^{\infty}(-1)^r\, {}^{n-r}c_r(2x)^{n-2r}$$

Also, the term containing t^{n-1} in the expansion of the product $t^{n-r}(2x-t)^{n-r}$ is obtained from

$$t^{n-r}\left[{}^{n-r}c_{r-1}(-t)^{r-1}(2x)^{n-2r+1}\right]$$
$$= (-1)^{n-1}\, {}^{n-r}c_{r-1}(2x)^{n-2r+1}t^{n-1}$$

$\therefore$ The coefficient of t^{n-1} from all the terms is b_{n-1} where

$$b_{n-1} = \sum_{r=1}^{\infty}(-1)^{r-1}\, {}^{n-r}c_{r-1}(2x)^{n-2r+1}$$

Now, the LHS of Eq. (17) yields

$$(1-xt)(b_0+b_1t+\cdots+b_{n-1}t^{n-1}+b_nt^n+\cdots)$$
$$= (b_1-b_0x)t+\cdots+(b_n-x\cdot b_{n-1})t^n+\cdots$$

Thus, we obtain the coefficient of t^n as

$$b_n - xb_{n-1} = \sum_{r=0}^{\infty}(-1)^r\, {}^{n-r}c'_r(2x)^{n-2r}$$
$$-\sum_{r=1}^{\infty}(-1)^{r-1}\, {}^{n-r}c_{r-1}2^{n-2r+1}x^{n-2r+2}$$

In the second sum we put $r-1 = s$ so that we have, replacing the dummy variable s by r:

$$b_n - xb_{n-1} = \sum_{r=0}^{\infty}(-1)^r\, {}^{n-r}c_r(2x)^{n-2r}$$
$$-\frac{1}{2}\sum_{r=0}^{\infty}(-1)^r\, {}^{n-r-1}c_r\cdot(2x)^{n-2r}$$

$$= \frac{1}{2}\left[(2\,{}^nc_0 - {}^{n-1}c_0)(2x)^n - (2\,{}^{n-1}c_1 - {}^{n-2}c_1)\right.$$
$$\left.(2x)^{n-2}+\cdots\cdots+2({}^{n-r}c_r - {}^{n-r-1}c_r)(2x)^{n-2r}+\cdots\right]$$

Consequently

$$T_n(x) = \frac{1}{2}\left[(2x)^n - (2\,{}^{n-1}c_1 - {}^{n-2}c_1)(2x)^{n-2}\right.$$
$$\left.+(2\,{}^{n-2}c_2 - {}^{n-3}c_2)(2x)^{n-4} - \cdots\right]$$

We can also obtain the expression for x^n in terms of the chebyshev polynomials as follows:

$$x^n = (\cos\theta)^n = \frac{1}{2^n}\left(e^{i\theta}+e^{-i\theta}\right)^n$$

$$= \frac{1}{2^n}\left[e^{in\theta} + {}^nc_1e^{i(n-1)\theta}e^{-i\theta} + {}^nc_2e^{i(n-2)\theta}e^{-2i\theta}+\cdots\right.$$
$$\left.+{}^nc_{n-1}e^{i\theta}\cdot e^{-i(n-1)\theta}+e^{-in\theta}\right]$$

$$= \frac{1}{2^{n-1}}\left[\left(\frac{e^{in\theta}+e^{-in\theta}}{2}\right)+{}^nc_1\left(\frac{e^{i(n-2)\theta}+e^{-i(n-2)\theta}}{2}\right)\right.$$

$$\left.+\cdots+{}^nc_r\left(\frac{e^{i(n-2r)\theta}+e^{-i(n-2r)\theta}}{2}\right)+\cdots+\cdots\right]$$

Collecting the equidistant terms from both ends

$$= \frac{1}{2^{n-1}}\left[\cos n\theta+{}^nc_1\cos(n-2)\theta+\right.$$

$$\left.\cdots+{}^nc_r\cos(n-2r)\theta+\cdots\right]$$

$$= \frac{1}{2^{x-1}}\left[T_n(x)+{}^nc_1T_{n-2}(x)+\right.$$

$$\left.\cdots+{}^nc_r+T_{n-2r}(x)+\cdots\right]$$

If n is even, the last term is $\frac{1}{2}\cdot{}^nc_{\frac{n}{2}}T_0(x)$.

Integration of $T_n(x)$

Consider

$$\int T_n(x)dX = \int \cos n\theta\cdot d(\cos\theta) = -\int \cos n\theta\sin\theta d\theta$$

$$= -\frac{1}{2}\int[\sin(n+1)\theta-\sin(n-1)\theta]d\theta$$

$$= \frac{1}{2}\left[\frac{\cos(n+1)\theta}{n+1}-\frac{\cos(n-1)\theta}{n-1}\right] \quad n>1$$

$$= \frac{1}{2}\left[\frac{1}{n+1}\cdot T_{n+1}(x)-\frac{1}{n-1}\cdot T_{n-1}(x)\right]$$

The above result does not hold good for $n = 0$ and 1.

For these values of n we have

$$\int T_0(x)dx = \int 1\cdot dX = x = T_1(x)$$

$$\int T_1(x)dX = \int x\,dX = \frac{x^2}{2} = \frac{\cos^2\theta}{2}$$

$$= \frac{1}{4}(2\cos^2\theta) = \frac{1}{4}(1+\cos 2\theta)$$

$$= \frac{1}{4}(T_0+T_2).$$

Orthogonality of Chebyshev Polynomials $T_n(x)$

Prove that

$$\int_{-1}^{1}\frac{T_m(x)T_n(x)}{\sqrt{1-x^2}}dX = \begin{cases}0, & \text{if } m\neq n\\ \pi, & \text{if } m=n=0\\ \frac{\pi}{2}, & \text{if } m=n\neq 0\end{cases}$$

Solution Case (i) $m = n = 0$. In this case, we get (since $T_0 = 1$)

$$I = \int_{-1}^{1}\frac{1}{\sqrt{1-x^2}}dX = \sin^{-1}x\Big|_{-1}^{1}$$

$$= \sin^{-1}1-\sin^{-1}(-1) = \frac{\pi}{2}-\left(-\frac{\pi}{2}\right) = \pi$$

Case (ii) $m = n \neq 0$. In this case, we get have to evaluate

$$I = \int_{-1}^{1}\frac{T_n^2(x)}{\sqrt{1-x^2}}dX = \int_{-1}^{1}\frac{\cos^2(n\cos^{-1}x)}{\sqrt{1-x^2}}dX$$

Put $\cos^{-1}x = \theta$ or $x = \cos\theta \Rightarrow dX = -\sin^{\theta}d\theta$ and the limits $x = -1$ and $x = 1$ become $\theta = \pi$ and $\theta = 0$ respectively.

$$\therefore I = \int_{\pi}^{0}\frac{\cos^2 x\theta}{\sqrt{1-\cos^2\theta}}(-\sin\theta)d\theta$$

$$= -\frac{1}{2}\int_{\pi}^{0}(1+\cos 2\theta)d\theta$$

$$= \frac{1}{2}\int_{0}^{\pi}(1+\cos 2\theta)d\theta$$

$$= \frac{1}{2}\left(\theta+\frac{1}{2}\sin 2\theta\right)_0^{\pi} = \frac{\pi}{2}$$

Case (iii) $m \neq n$. Since $T_m(x) = \cos(m\cos^{-1}x)$ and $T_n(x) = \cos(n\cos^{-1}x)$ we have no evaluate

$$I = \int_{-1}^{1}\frac{\cos(m\cos^{-1}x)\cos(n\cos^{-1}x)}{\sqrt{1-x^2}}dX$$

Put $\cos^{-1}x = \theta \Rightarrow x = \cos\theta,\ dX$

$= -\sin\theta\, d\theta, = \sqrt{1-x^2} = \sin\theta$ and the limits $x =$ -1 and $x = 1$ become $\theta = \pi$ and $\theta = 0$ respectively.

$$\therefore I = -\int_{\pi}^{0} \frac{\cos m\theta \cos n\theta}{\sin\theta}(\sin\theta)d\theta$$

$$= +\frac{1}{2}\int_{0}^{\pi}[\cos(m+n)\theta + \cos(m-n)\theta]$$

$$= \left[\frac{\sin(m+n)\theta}{m+n} + \frac{\sin(m-n)\theta}{m-n}\right]_0^{\pi} \quad m \neq n$$

$$= 0$$

Consequently the chebyshev polynomials $T_n(x)$ are orthogonal w.r.t. the weight function

$$w(x) = \frac{1}{\sqrt{1-x^2}} \text{ on } [-1, 1].$$

Chebyshev series

Let $f(x)$ be a continuous function, having continuous derivatives on the internal $[-1, 1]$. Now, $f(x)$ can be uniquely written as an infinite series

$$f(x) = \sum_{r=0}^{\infty} a_r T_r(x) = a_0 T_0(x) + a_1 T_1(x) + a_2 T_2(x) + a_3 T_3(x) + \cdots \quad (1)$$

which converges uniformly in $[-1, 1]$. Multiplying both sides of the equation by $T_m(x)/\sqrt{1-x^2}$ and integrating with respect to x over $[-1, 1]$ and using the orthogonal property of chebyshev polynomials we get

$$\int_{-1}^{1} f(x)\frac{T_m(x)}{\sqrt{1-x^2}}dX = \sum_{n=0}^{\infty} a_n \int_{-1}^{1}\frac{T_m(x)T_n(x)}{\sqrt{1-x^2}}dx$$

$$= \begin{cases} \pi a_0 \text{ for } m = n = 0 \\ \frac{\pi}{2}c_n \text{ for } m = n \neq 0. \end{cases} \quad (2)$$

Hence

$$c_0 = \frac{1}{\pi}\int_{-1}^{1}\frac{f(x)}{\sqrt{1-x^2}}dX,\ c_m = \frac{2}{\pi}\int_{-1}^{1}\frac{f(x)T_m(x)}{\sqrt{1-x^2}}dx \quad (3)$$

$$m = 1, 2, 3, \ldots$$

If we write the series given in Eq. (1) as

$$f(x) = \frac{1}{2}c_0 T_0(x) + c_1 T_1(x) + c_2 T_2(x) + \cdots$$

then using Eq. (3) we can write

$$c_m = \frac{2}{\pi}\int_{-1}^{1}\frac{f(x)T_m(x)}{\sqrt{1-x^2}}dx \quad (4)$$

$$m = 0, 1, 2, \ldots$$

This series may also be written as

$$f(x) = \sum_{n=0}^{\infty}{}' c_n T_n(x)$$

where Σ' means that the coefficient of the first term $T_0(x)$ is multiplied by $\frac{1}{2}$.

Example 12.62 Expand $f(x) = x^3 + x$, $-1 \leq x \leq 1$ in a Chebyshev series.

Solution We have

$$f(x) = x^3 + x$$

Write $f(x) = c_0 T_0(x) + c_1 T_1(x) + c_2 T_2(x) + \cdots$

we obtain

$$c_0 = \frac{1}{\pi}\int_{-1}^{1}\frac{f(x)}{\sqrt{1-x^2}}dX$$

$$= \frac{1}{\pi}\int_{-1}^{1}\frac{x^3+x}{\sqrt{1-x^2}}dX = 0 \quad \text{(odd function)}$$

$$c_m = \frac{2}{\pi}\int_{-1}^{1}\frac{(x^3+x)T_m(x)}{\sqrt{1-x^2}}dx = 0 \quad \text{for } m > 3$$

$$c_1 = \frac{2}{\pi}\int_{-1}^{1}\frac{(x^3+x)T_1(x)}{\sqrt{1-x^2}}dx = \frac{2}{\pi}\int_{-1}^{1}\frac{x^4+x^2}{\sqrt{1-x^2}}dX$$

$$= \frac{4}{\pi}\int_{0}^{1}\frac{x^4+x^2}{\sqrt{1-x^2}}dX = \frac{4}{\pi}\int_{0}^{\frac{\pi}{2}}\frac{\cos^4\theta + \cos^2\theta}{\sin\theta}\sin\theta d\theta$$

$$= \frac{4}{\pi}\left(\frac{3}{4}\cdot\frac{1}{2}\cdot\frac{\pi}{2} + \frac{1}{2}\cdot\frac{\pi}{2}\right) = \frac{7}{4}$$

$$c_2 = \frac{2}{\pi}\int_{-1}^{1}\frac{(x^3+x)}{\sqrt{1-x^2}}(2x^2-1)dX = 0 \quad \text{(odd function)}$$

$$c_3 = \frac{2}{\pi}\int_{-1}^{1}\frac{(x^3+x)(4x^3-3x)}{\sqrt{1-x^2}}$$

$$= \frac{4}{\pi}\int_0^{\frac{\pi}{2}}(\cos^3\theta+\cos\theta)(4\cos^3\theta-3\cos\theta)d\theta$$

$$= \frac{4}{\pi}\int_0^{\frac{\pi}{2}}(4\cos^6\theta+\cos^4\theta-3\cos^2\theta)d\theta$$

$$= \frac{4}{\pi}\left[4\left(\frac{5}{6}\cdot\frac{3}{4}\cdot\frac{1}{2}\cdot\frac{\pi}{2}\right)+\frac{3}{4}\cdot\frac{1}{2}\cdot\frac{\pi}{2}-3\left(\frac{1}{2}\cdot\frac{\pi}{2}\right)\right]=\frac{1}{4}$$

Hence $f(x)=\frac{1}{4}(T_3+7T_1)$

EXERCISE 12.5

Prove the following (1 – 7)

1. $T_n(1)=1$
2. $T_n(-1)=(-1)^n$
3. $T_n(-x)=(-1)^n\,T_n(x)$
4. $T_{2n}(0)=(-1)^n$
5. $T_{2n+1}(0)=0$
6. $\int_{-1}^{1}\frac{T_n(x)}{\sqrt{1-x^2}}dX=0$
7. $\int_{-1}^{1}\frac{x^m T_n(x)}{\sqrt{1-x^2}}dX=0 \quad (m<n)$
8. Show that $\frac{1-t^2}{1-2xt+t^2}=T_0(x)+2\sum_{n=1}^{\infty}T_n(x)t^n$

Functions of a Complex Variable

13

13.1 Introduction

Many engineering problems can be solved using Complex Variable Theory. Some of the elementary problems dealing with applications to electric circuits or mechanical vibrating systems need only elementary knowledge of complex analysis. Advanced problems in heat conduction, fluid flow and electrostatics require knowledge of complex analysis.

Equations such as $x^2 = -1$ have no solutions in real variables. By defining $i = \sqrt{-1}$ as a complex unit, the system of real numbers can be extended to the complex number system.

13.2 Complex Numbers—Complex Plane

A complex number z is an ordered pair of real numbers x and y written as $z = (x, y)$. Here x is called the real part and y the imaginary part of z. We write

$$x = Re\, z,\ y = Im\, z$$

Equality of two complex numbers

Two complex numbers $z_1 = (x_1, y_1)$ and $z_2 = (x_2, y_2)$ are equal if and only if their real parts are equal and their imaginary parts are equal, *i.e.*,

$$x_1 = x_2, y_1 = y_2$$

Imaginary unit

(0, 1) is called the *imaginary unit* and is denoted by *i*.

$$i = (0, 1) \tag{13.1}$$

Addition and multiplication

Addition of two complex numbers $z_1 = (x_1, y_1)$ and $z_2 = (x_2, y_2)$ is defined by

$$\begin{aligned} z_1 + z_2 &= (x_1, y_1) + (x_2, y_2) \\ &= (x_1 + x_2, y_1 + y_2) \end{aligned} \tag{13.2}$$

Multiplication is defined by

$$\begin{aligned} z_1 \cdot z_2 &= (x_1, y_1)(x_2, y_2) \\ &= (x_1x_2 - y_1y_2, x_1y_2 + x_2y_1) \end{aligned} \tag{13.3}$$

In particular, these two definitions imply that

$$(x_1, 0) + (x_2, 0) = (x_1 + x_2, 0) \text{ and}$$
$$(x_1, 0)\,(x_2, 0) = (x_1x_2, 0)$$

as for real numbers x_1 and x_2. Hence the complex numbers *extend* the reals. We can write

$$(x, 0) = x, \quad (0, y) = iy, \quad (x, y \in \mathbb{R}) \tag{13.4}$$

$$\begin{aligned} \because iy &= (0, 1)(y, 0) \\ &= (0 \cdot y - 1 \cdot 0, 0 \cdot 0 + 1 \cdot y) \\ &= (0, y). \end{aligned}$$

Since $(x, y) = (x, 0) + (0, y) = x + iy$ it has great practical advantage to write $z = x + iy$. If $x = 0$ then $z = iy$ and is called a pure imaginary number. Also Eqs. (13.1) and (13.3) give

$$i^2 = i \cdot i = (0, 1)(0, 1) = (-1, 0) = -1 \tag{13.5}$$

By the standard notation, addition may be done as follows:

$$(x_1 + iy_1) + (x_2 + iy_2) = (x_1 + x_2) + i(y_1 + y_2)$$

Multiplication can be done in the usual way noting that $i^2 = -1$

$$(x_1 + iy_1)(x_2 + iy_2) = (x_1x_2 - y_1y_2) + i(x_1y_2 + x_2y_1)$$

Real and imaginary parts

Let $z_1 = 3 - 4i$; $z_2 = -5 + i$.

Then $Re\, z_1 = 3$ $Im\, z_1 = -4$

$Re\, z_2 = -5$ $Im\, z_2 = 1$

Sum and product

$$z_1 + z_2 = (3-4i) + (-5+i) = -2-3i$$
$$z_1 z_2 = (3-4i)(-5+i)$$
$$= (-15+4) + i(3+20) = -11+23i$$

Subtraction and division

Subtraction and division are inverse operations of addition and multiplication.

Let $z_1 = (x_1, y_1)$ and $z_2 = (x_2, y_2)$. The difference $z_1 - z_2$ is the complex number $z = (x, y)$ such that

$$z + z_2 = z_1 \Rightarrow (x, y) + (x_2, y_2) = (x_1, y_1)$$
$$\Rightarrow (x + x_2, y + y_2) = (x_1, y_1)$$
$$\Rightarrow x + x_2 = x_1, y + y_2 = y_1$$
$$\Rightarrow (x, y) = (x_1 - x_2, y_1 - y_2).$$
$$(x+iy) + (x_2 + iy_2) = (x_1 + iy_1)$$
$$\Rightarrow x + x_2 = x_1, y + y_2 = y_1$$
$$\Rightarrow x = x_1 - x_2, \quad y = y_1 - y_2$$
$$\Rightarrow z = z_1 - z_2$$
$$= (x_1 - x_2) + i(y_1 - y_2).$$

The *quotient* z_1/z_2 ($z_2 \neq 0$) is the complex number z such that $zz_2 = z_1$

$$(x+iy)(x_2 + iy_2) = (x_1 + iy_1)$$
$$\Rightarrow xx_2 - yy_2 = x_1, xy_2 + yx_2 = y_1$$
$$xx_2 - yy_2 = x_1 \mid x_2 \mid y_2$$
$$xy_2 + yx_2 = y_1 \mid y_2 \mid x_2$$

$$x(x_2^2 + y_2^2) = x_1 x_2 + y_1 y_2$$
$$y(x_2^2 + y_2^2) = x_2 y_1 - x_1 y_2$$

Solving $z = (x, y)$

$$= \frac{x_1 x_2 + y_1 y_2}{x_2^2 + y_2^2} + i \frac{x_2 y_1 - x_1 y_2}{x_2^2 + y_2^2}$$

In practice, we do as follows:

$$\frac{z_1}{z_2} = \frac{x_1 + iy_1}{x_2 + iy_2} \cdot \frac{x_2 - iy_2}{x_2 - iy_2}$$
$$= \frac{x_1 x_2 + y_1 y_2}{x_2^2 + y_2^2} + i \frac{x_2 y_1 - x_1 y_2}{x_2^2 + y_2^2}$$

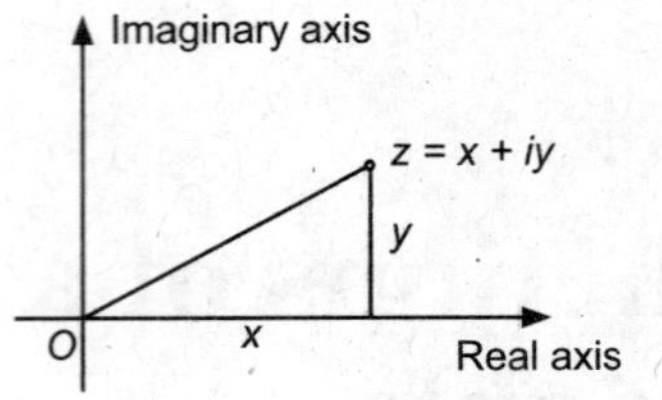

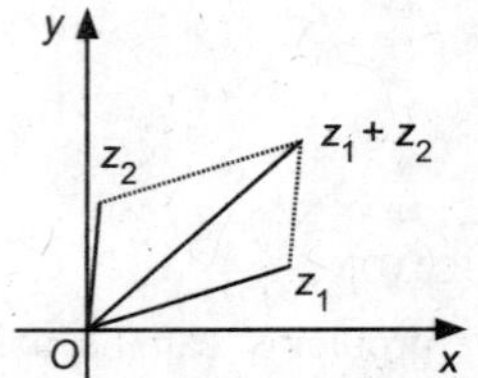

Addition of complex numbers

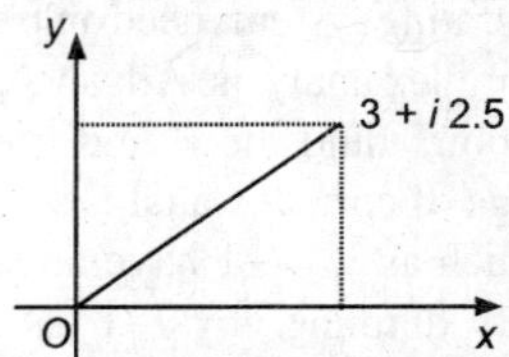

The number 3 + *i* 2.5 in the complex plane

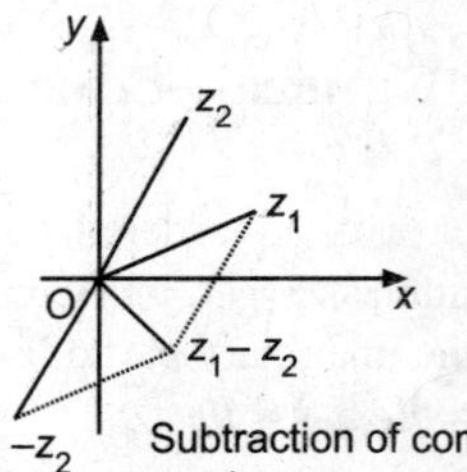

Subtraction of complex numbers

Figure 13.1

Difference and quotient

Let $z_1 = 3 - 2i$ and $z_2 = 4 + i$

$$z = z_1 - z_2 = (3-2i) - (4+i) = -1-3i$$
$$\frac{z_1}{z_2} = \frac{3-2i}{4+i}\frac{4-i}{4-i} = \frac{12+(-2)(-1)}{4^2+1^2}$$
$$+ i \cdot \frac{-8-3}{4^2+1^2} = \frac{14}{17} + i\frac{(-11)}{17}$$

Complex numbers satisfy

1. closure
2. commutative

3. associative
4. existence of identity and
5. existence of inverse

laws w.r.t the operations of addition and multiplication and also the distributive laws as real numbers and form a field.

$\mathbb{Q}, \mathbb{R}, \mathbb{C}$ are fields; $\mathbb{Q}, \mathbb{R}$ satisfy order relation '<'; $\mathbb{R}$ is complete ordered field. No such order relation exists in $\mathbb{C}$

There is one-one correspondence between the set of reals $\mathbb{R}$ and points in a line, and the set of complex numbers $\mathbb{C}$ and the points in a plane.

Geometrical Representation of Complex Numbers

Complex plane or Argand plane

Choose two perpendicular lines, the horizontal line called the *real axis* and the vertical line called the *imaginary axis*. Thus we have a *cartesian coordinate system.* Plot a given complex number $z=(x, y)=x+iy$ as a point P with coordinates x and y.

Addition and subtraction can be visualized as shown above (Fig. 13.1)

Complex conjugate numbers

The complex conjugate $\bar{z}$ of a complex number $z = x + iy$ is defined by

$$\bar{z} = x - iy$$

Replace i by $-i$ in the complex number to get its conjugate. It is obtained geometrically by reflecting the point z in the real axis.

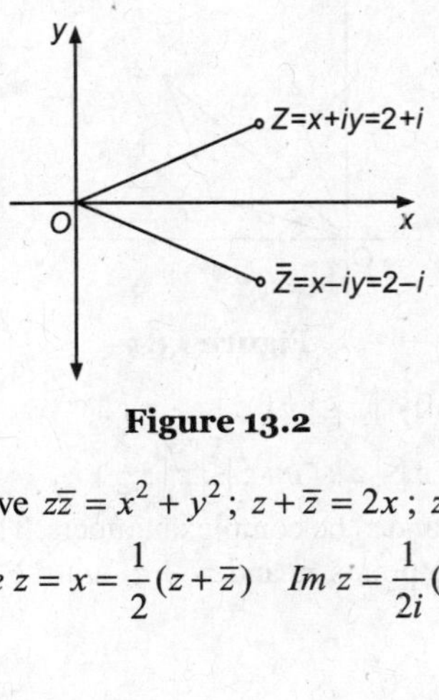

Figure 13.2

We have $z\bar{z} = x^2 + y^2$; $z + \bar{z} = 2x$; $z - \bar{z} = 2iy$

$$\therefore \quad Re\, z = x = \frac{1}{2}(z+\bar{z}) \quad Im\, z = \frac{1}{2i}(z-\bar{z})$$

If z is real, $z = x$, then $\bar{z} = z$. Conversely if $\bar{z} = z$ then z is real and $z = x$.

1. $\overline{(z_1 + z_2)} = \bar{z}_1 + \bar{z}_2$

Conjugate of sum = Sum of conjugates

2. $\overline{(z_1 - z_2)} = \bar{z}_1 - \bar{z}_2$

Conjugate of difference = Difference of conjugates

3. $\overline{(z_1 z_2)} = \bar{z}_1 \cdot \bar{z}_2$

Conjugate of product = Product of conjugates

4. $\overline{(z_1 / z_2)} = \bar{z}_1 / \bar{z}_2$

Conjugate of quotient = Quotient of conjugates

Example 13.1

Let $z_1 = 2 - 3i$, $z_2 = 4 + 5i$

$$Im\, z_1 = \frac{1}{2i}(z_1 - \bar{z}_1) = \frac{1}{2i}[(2-3i)-(2+3i)] = 3$$

$$\overline{(z_1 z_2)} = \overline{(2-3i)(4+5i)} = \overline{(23-2i)} = 23 + 2i$$

$$\bar{z}_1 \cdot \bar{z}_2 = (2+3i)(4-5i) = 23 + 2i$$

EXERCISE 13.1

1. Powers of the imaginary unit:
   Show that $i^2 = -1, i^3 = -i, i^4 = 1, i^5 = i, \cdots$ and $\frac{1}{i} = -i, \frac{1}{i^2} = -1, \frac{1}{i^3} = i, \cdots; \frac{1+i}{1-i} = i.$
2. Multiplication by i is geometrically a counter-clockwise rotation through $\pi/2$ (90°). Verify this by plotting z and iz and the angle of rotation for $z = 4 + 2i$, $z = -1 + i$, $z = 5 - 2i$.

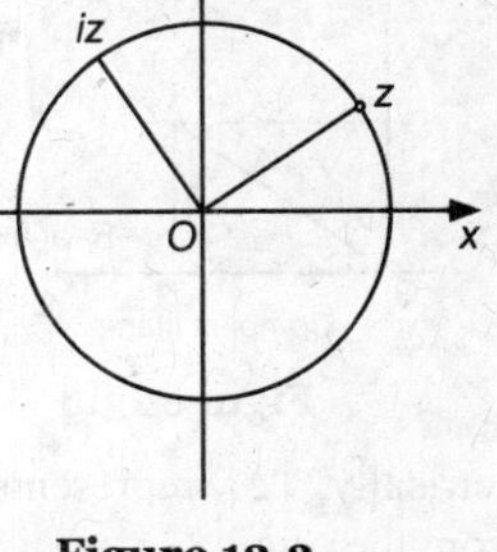

Figure 13.3

3. Let $z_1 = -2 + 3i$, $z_2 = 1 + i$. Find

(a) $z_1 z_2 = (-2+3i)(1+i)$

$= (-2-3) + i(3-2) = -5 + i$

(b) $\dfrac{1}{z_1} = \dfrac{1}{1+i} = \dfrac{1-i}{1^2+1^2} = \dfrac{1}{2} - \dfrac{1}{2}i$

(c) $\dfrac{z_2}{z_1} = \dfrac{1+i}{-2+3i} = \dfrac{(1+i)(-2-3i)}{(-2)^2+3^2} = \dfrac{1}{13} - \dfrac{5}{13}i$

(d) $Re(z_1)^2 = Re(5-12i) = 5 \quad (Re\, z_1)^2 = (-2)^2 = 4$

$z = x + iy$

(e) $Im\left(\dfrac{1}{z}\right) = Im\dfrac{1}{x+iy} = Im\dfrac{x-iy}{x^2+y^2} = -\dfrac{y}{x^2+y^2}$

(f) $(1+i)^2 = 2i$; $Re(1+i)^2 = 0$; $(1+i)^4 = (2i)^2 = -4$

(g) $\left|\dfrac{z+i}{1-iz}\right| = \left|\dfrac{x+i(y+1)}{1-i(x+iy)}\right| = \dfrac{\sqrt{x^2+(y+1)^2}}{\sqrt{(1+4)^2+x^2}} = 1$

Polar Form of Complex Numbers, Powers and Roots

If we put

$x = r\cos\theta$, $y = r\sin\theta$
$z = x + iy$ takes the polar form
$z = r(\cos\theta + i\sin\theta)$ or $re^{i\theta}$.

r is called the *absolute value* or *modulus* of z and is denoted by $|z|$. $|z| = r = \sqrt{x^2+y^2} = \sqrt{z\bar{z}}$

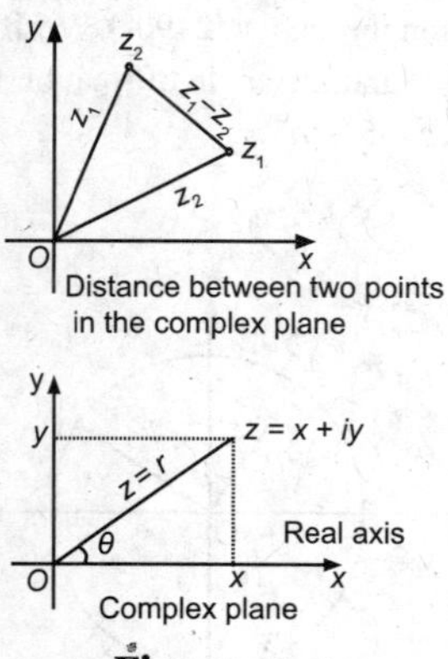

Figure 13.4

Geometrically, $|z|$ represents the distance of the point from the origin.

The inequality $|z_1| > |z_2|$ means that the point z_1 is farther away from the origin than the point z_2, and $|z_1 - z_2|$ is the distance between the points z_1 and z_2.

θ is called the argument or amplitude of z denoted by arg z or amp z.

$$\theta = \arg z = \tan^{-1}\frac{y}{x}$$

Geometrically, θ is the directed angle from the positive x-axis to OP. For $z = 0$, this angle θ is undefined. For $z \neq 0$ it is determined only up to an integer multiple of 2π. The value of θ that lies in the interval $-\pi < \theta \leq \pi$ is called the principal value of the argument of $z \neq 0$ and is denoted by Arg z with capital A. Thus $-\pi <$ Arg $z \leq \pi$.

Example 13.2

$$z = 1 + i = \sqrt{2}\left(\cos\frac{p}{4} + i\sin\frac{p}{4}\right).$$

$$\therefore |z| = \sqrt{2} \quad \arg z = \frac{1}{4}p \pm 2np \quad n = 0, 1, 2, \ldots$$

$$\text{Arg } z = \frac{p}{4} \text{ (principal value).}$$

Since for any complex number $z = x + iy$, $|z| = \sqrt{x^2+y^2} \geq |x|$

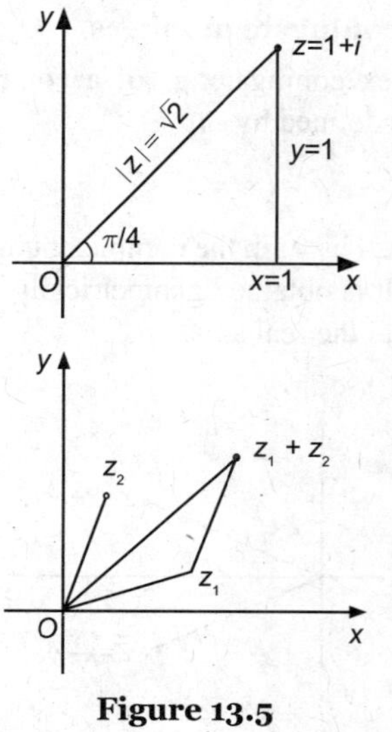

Figure 13.5

Similarly $|z| \geq |y|$

$\therefore |Re\, z| \leq |z|$; $|Im\, z| \leq |z|$.

Let z_1 and z_2 be complex numbers. Then the origin and the points z_1 and $z_1 + z_2$ are the vertices of a triangle whose sides are $|z_1|, |z_2|$ and $|z_1+z_2|$. We obtain *triangle inequality*

$$|z_1+z_2| \le |z_1|+|z_2|.$$

Also, we have

$$||z_1|-|z_2|| \le |z_1-z_2|$$

(The difference of two sides of a triangle cannot exceed the third side – Fig. 13.6.)

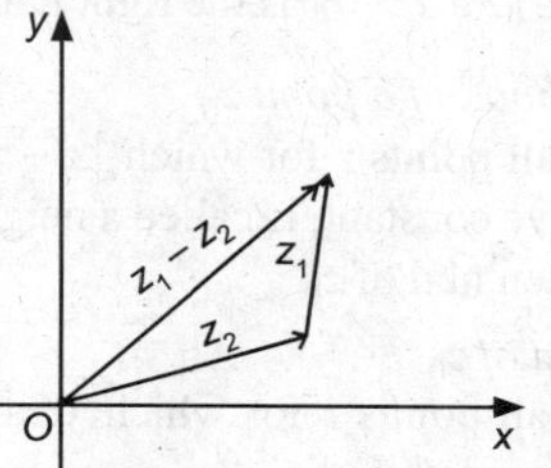

Figure 13.6

Multiplication and Division in Polar Form

Let $z_1 = r_1(\cos\theta_1 + i\sin\theta_1)$,

$$z_2 = r_2(\cos\theta_2 + i\sin\theta_2)$$

$$z_1 z_2 = r_1 r_2\left[(\cos\theta_1\cos\theta_2 - \sin\theta_1\sin\theta_2)\right.$$
$$\left. + i(\sin\theta_1\cos\theta_2 + \cos\theta_1\sin\theta_2)\right]$$
$$= r_1 r_2 \operatorname{cis}(\theta_1+\theta_2)$$

$$|z_1 z_2| = |z_1||z_2|$$

$\arg(z_1 z_2) = \arg z_1 + \arg z_2$, up to multiple of 2π

$$z = \frac{z_1}{z_2} \Leftrightarrow zz_2 = z_1 \therefore |zz_2| = |z||z_2| = |z_1|$$

$\arg(zz_2) = \arg z + \arg z_2 = \arg z_1$

$$\therefore \left|\frac{z_1}{z_2}\right| = \frac{|z_1|}{|z_2|} \quad (z_2 \ne 0)$$

$\arg\left(\dfrac{z_1}{z_2}\right) = \arg z_1 - \arg z_2$, up to multiple of 2π

$$\frac{z_1}{z_2} = \frac{r_1}{r_2}\operatorname{cis}(\theta_1-\theta_2)$$

Example 13.3

$$z_1 = -2+2i \quad z_2 = 3i$$

$$z_1 z_2 = -6-6i; \quad \frac{z_1}{z_2} = \frac{2}{3}+\frac{2}{3}i$$

$$\operatorname{Arg} z_1 = \frac{3\pi}{4} \quad \operatorname{Arg} z_2 = \frac{\pi}{2}$$

$$|z_1 z_2| = 6\sqrt{2} = 3\sqrt{1} = |z_1||z_2|$$

$$\operatorname{Arg} z_1 z_2 = \frac{-3\pi}{4} = \operatorname{Arg} z_1 - \operatorname{Arg} z_2 - 2\pi$$
$$= \frac{\pi}{4} - \operatorname{Arg} z_1 - \operatorname{Arg} z_2$$

Division

The quotient $z = z_1/z_2$ is the number z satisfying $zz_2 = z_1$.

Hence $|zz_2| = |z||z_2| = |z_1|$. $\arg zz_2 = \arg z + \arg z_2 = \arg z_1$. This gives $|z_1/z_2| = |z_1|/|z_2|\ (z_2 \ne 0)$ and $\arg(z_1/z_2) = \arg z_1 - \arg z_2$.

Integer Powers: DeMoivre's Formula

$$z^n = r^n(\cos n\theta + i\sin n\theta) \qquad n = 0, 1, 2, \ldots$$

DeMoivre's formula

$$(\cos\theta + i\sin\theta)^n = \cos n\theta + i\sin n\theta$$

Put $n = 2$

$$\cos^2\theta - \sin^2\theta + 2i\sin\theta\cos\theta = \cos 2\theta + i\sin 2\theta$$

$\Rightarrow \cos 2\theta = \cos^2\theta - \sin^2\theta$ and $\sin 2\theta = 2\sin\theta\cos\theta$

Roots

$$z = w^n \ (n = 1, 2, 3, \ldots) \quad w = \sqrt[n]{z} \quad n\text{th root of } z$$

Remark

This symbol is multi-valued (n-valued) in contrast to the usual convention in *real calculus*.

$$z = r(\cos\theta + i\sin\theta) \quad w = R(\cos\phi + i\sin\phi)$$

$$w^n = R^n(\cos n\phi \sin n\phi) = z = r(\cos\theta + i\sin\theta)$$

By equating absolute values $R^n = r \Rightarrow R = \sqrt[n]{r}$ root is real positive and unique.

By equating arguments $n\phi = \theta + 2k\pi$

$$\text{thus } \phi = \frac{\theta}{n} + \frac{2k\pi}{n} \quad k = 0, 1, 2, \ldots (n-1).$$

For these values of k, we get n *distinct* values of w; these values repeat themselves for higher values of k.

Consequently, $\sqrt[n]{z}$ for $z \ne 0$ has the n district values

$$\sqrt[n]{z} = \sqrt[n]{r}\left(\cos\frac{\theta+2k\pi}{n} + i\sin\frac{\theta+2k\pi}{n}\right) \qquad (13.6)$$

where $k = 0, 1, \cdots (n-1)$. These n values, lie on a circle of radius $\sqrt[n]{r}$ with centre at the origin and constitute the vertices of a regular polygon of n sides.

The value of $\sqrt[n]{z}$ obtained by taking the principal value of Arg z and $k = 0$ in Eq. (13.6) is called the principal value of $w = \sqrt[n]{z}$. In particular, taking $z = 1$ we have $|z| = r = 1$ and Arg $z = 0$. Then Eq. (13.6) gives

$$\sqrt[n]{r} = \cos\frac{2k\pi}{n} + i\sin\frac{2k\pi}{n} \quad k = 0, 1, \ldots (n-1) \qquad (13.7)$$

These n values are called the nth *roots of unity*. They lie on the circle of radius 1 and centre 0, briefly called the *unit circle* (see Fig. 13.7).

If w denotes the value corresponding to $k = 1$ in Eq. (13.7) the n values of $\sqrt[n]{1}$ can be written as

$$1, w, w^2, \cdots w^{n-1}$$

Similarly if w_1 is any nth root of an arbitrary complex number z then the n values of $\sqrt[n]{z}$ in Eq. (13.6) are

$$w_1, w_1 w, w_1 w^2, \cdots w_1 w^{n-1}$$

because multiplying w_1 by w^k corresponds to increasing the argument of w_1 by $2k\pi/n$.

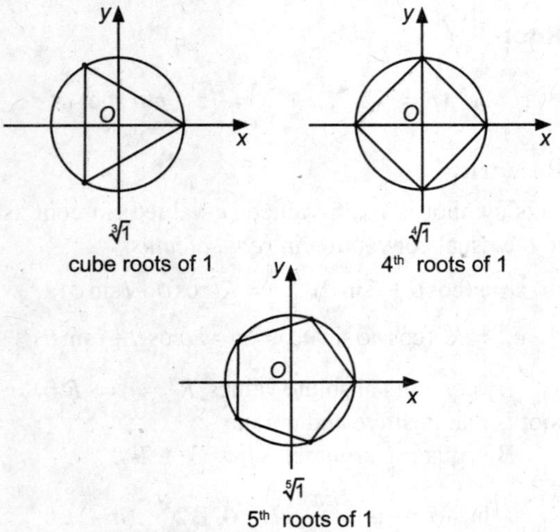

Figure 13.7

Definitions of Certain Terms

Distance between two points

If 'z' and 'a' are points in the complex plane $\mathbb{C}$ the distance between them is $d(z, a) = |z - a|$

Set of points

By a set of points we mean a collection of points in the complex plane $\mathbb{C}$.

Example 13.4

Circle: $S_1 = \{z \in \mathbb{C} \mid |iz - a| = \rho\}$ is called a circle with centre 'a' and radius. ρ.

Right half-plane

$S_2 = \{z \in \mathbb{C} \mid Re\, z > 0\}$ is the right half-plane.

Neighbourhood of a point z_0

The set of all points z for which $|z - z_0| < \delta$ where δ is a positive constant, is called a neighbourhood of z_0 and written nbd of z_0.

Deleted nbd of z_0

The set of all points z for which $0 < |z - z_0| < \delta$ is called a deleted nbd of z_0.

Open circular disc

The set of all z for which $|z - z_0| < \rho$ is called an open circular disc.

Closed circular disc

The set of all z for which $|z - z_0| \le \rho$ is called a closed circular disc.

Open annulus or circular ring

The set of all points z such that $\rho_1 < |z - z_0| < \rho_2$ $(\rho_1 < \rho_2)$ is called an open annulus.

Closed annulus or circular ring

The set of all points z such that $\rho_1 \le |z - z_0| \le \rho_2$ $(\rho_1 < \rho_2)$ is called a closed annulus.

Interior point of a set *S*

A point z_0 is called an interior point of a set S if there is a nbd of z_0 containing only the points of S.

Open set

A set S is called open if every point of S has a nbd every point of which belongs to S.

Example 13.5

1. The points inside a circle.
2. The points inside a square.
3. Right half-plane $Re\, z = x > 0$.

Open half-plane

Open upper half-plane: The set of points z such that $Im\, z = y > 0$.

Open lower half-plane: The set of points z such that $Im\, z = y > 0$.

Closed set

A set S is said to be closed if its complement S' is open.

Example 13.6

The set of points on and inside the unit circle $|z| = 1$.

Bounded set

A set S is called bounded if all its points lie within a circle of sufficiently large radius.

Example 13.7

Points in a rectangle, ellipse or polygon.

Note

Points on a straight line or in an infinite strip do not form a bounded set.

Connected Set

A set S in which any pair of points can be joined by a polygonal arc entirely lying in S, is called a connected set.

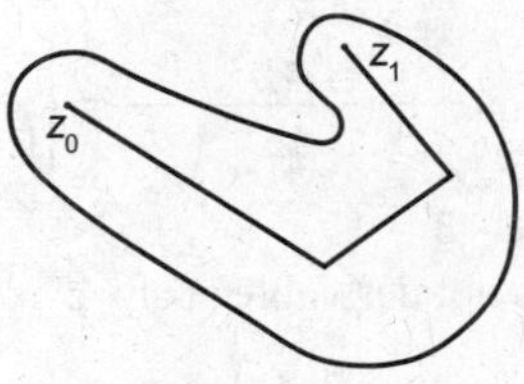

Figure 13.8

Domain

An open connected set is called a domain.

Example 13.8

Interior of a circle or a square or a closed polygon.

Boundary point of a set S

A point every nbd of which contains both points in S and points not in S.

Example 13.9

Points on the two circles C_1 and C_2 of an annulus.

Note

If S is open, no boundary point belongs to S. If S is closed, every boundary point belongs to S. $|z| \leq 1$: *closed*, bounded set.

$Im\, z^2 = xy > 0$: open, unbounded set.

$\rho_1 < |z - z_0 < \rho_2$: open, bounded, unconnected set.

Region

A set of points consisting of a domain and some or all of its boundary points.

13.2.1 Complex Function

Let S be a set of complex numbers. A function f defined on S is a rule that assigns to every z in S a complex number w, called the value of f at z. We write

$$w = f(z)$$

z varies in S and is called a *complex variable* and S is called the *domain of definition of f*.

$$z = x + iy,\ w = u(x, y) + iv(x, y)$$

Example 13.10

$w = f(z) = z^2 + 2z$

Let $z = x + iy$; $z^2 = (x + iy)^2 = x^2 - y^2 + 2ixy$;

$2z = 2x + i2y$

$\therefore w = u(x, y) + iv(x, y) = x^2 - y^2 + 2x + i(2xy + 2y)$

$Re\, w = u(x, y) = x^2 - y^2 + 2x;\quad Im\, w = 2xy + 2y$

13.2.2 Limit of a Function

Let $f(z)$ be a function defined in some deleted nbd of z_0 and l be a complex number. If for every $\epsilon > 0$ there corresponds a $\delta > 0$ such that

$$0 < |z - z_0| < \delta \Rightarrow |f(z) - l| < \epsilon$$

then we say that $f(z)$ tends to l as z tends to z_0. We write $\lim\limits_{z \to z_0} f(z) = l$.

13.2.3 Continuity at z_0

Let $f(z)$ be defined in a nbd of z_0. If $\lim\limits_{z \to z_0} f(z) = f(z_0)$ then we say that f is continuous at z_0.

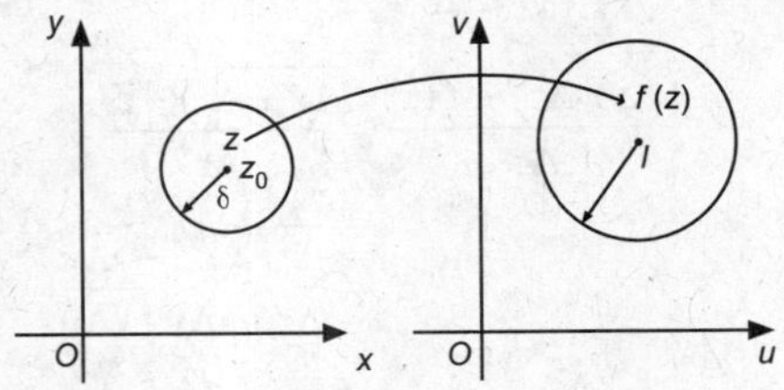

Figure 13.9

13.2.4 Differentiability

A complex function $f(z)$ is said to be differentiable at a point 'z_0', if the limit

$$f'(z_0) = \lim_{\Delta z \to z_0} \frac{f(z_0 + \Delta z) - f(z_0)}{\Delta z}$$

exists. The limit $f'(z_0)$ is called the derivative of $f(z)$ at z_0. Putting $\Delta z = z - z_0$ we can write the limit as follows:

$$f'(z_0) = \lim_{z \to z_0} \frac{f(z) - f(z_0)}{z - z_0}$$

Remark

In real calculus, there are two directions – right-side and left-side – to approach a point x_0. The $\lim_{x \to x_0} f(x)$ exists ⇔

(1) right-side limit $\lim_{x \to x_0+} f(x)$ exists

(2) left-side limit $\lim_{x \to x_0-} f(x)$ exists

(3) both are equal.

But in the case of complex numbers the situation is different. One can approach a point z_0 in the complex plane from any of the infinite number of directions. The above limit must exist and be the same when we approach z_0 in any of these directions. Thus the condition of differentiation is stringent in complex variables. Rules of differentiation are the same as in real calculus. All the complex-valued functions whose real-valued counterparts are differentiable are differentiable.

It is important to note that many simple functions do not have derivatives at any point.

Example 13.11

$\bar{z}$ is not differentiable at any point z.

Solution Let $f(z) = \bar{z} = x - iy$. Writing $\Delta z = \Delta x + i\Delta y$

$$\frac{f(z+\Delta z) - f(z)}{\Delta z} = \frac{\overline{(z+\Delta z)} - \bar{z}}{\Delta z}$$

$$= \frac{\overline{\Delta z}}{\Delta z} = \frac{\Delta x - i\Delta y}{\Delta x + i\Delta y} \qquad (1)$$

If $\Delta y = 0$ this is $+1$ and if $\Delta x = 0$ this is -1. Thus Eq. (1) approaches $+1$ along path I but -1 along path II. Hence the limit of Eq. (1) as $\Delta z \to 0$ does not exist at any point z.

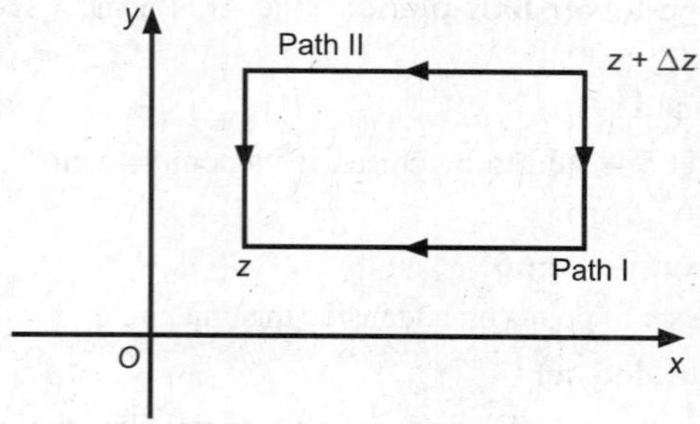

Figure 13.10

Example 13.12

Show that $f(z) = Re\, z = x$ is not differentiable at any z.

Solution

$$\frac{f(z+\Delta z) - f(z)}{\Delta z} = \frac{Re(z + \Delta z - z)}{\Delta x + i\Delta y}$$

$$= \frac{x + \Delta x - x}{\Delta x + i\Delta y}$$

$$\Rightarrow \lim_{\Delta z \to 0} \frac{f(z+\Delta z) - f(z)}{\Delta z} = \begin{cases} 1 & \text{if } \Delta y = 0 \\ 0 & \text{if } \Delta y \neq 0 \end{cases}$$

Example 13.13

$f(z) = |z|^2$ is not derivable at any z except at $z = 0$.

Solution

$$\frac{f(z+\Delta z) - f(z)}{\Delta z} = \frac{z\bar{z} + z\overline{\Delta z} + \bar{z}\Delta z + \Delta z\overline{\Delta z} - z\bar{z}}{\Delta z}$$

Case (i) $\Delta y \to 0, \Delta x \to 0$

$$\lim_{\Delta z \to 0} \frac{f(z+\Delta z) - f(z)}{\Delta z} = \lim_{\Delta x \to 0} \frac{z\Delta x + \Delta x \cdot \bar{z} + \Delta x \cdot \Delta x}{\Delta x}$$

$$= z + \bar{z}$$

Case (ii) $\Delta x \to 0, \Delta y \to 0$

$$\lim_{\Delta z \to 0} \frac{f(z+\Delta z) - f(z)}{\Delta z}$$

$$= \lim_{\Delta y \to 0} \frac{z(-i\Delta y) + i\Delta y\bar{z} + (i\Delta y)(-i\Delta y)}{i\Delta y}$$

$$= -z + \bar{z}$$

$f'(z)$ exists $\Leftrightarrow z + \bar{z} = -z + \bar{z} \Rightarrow 2z = 0 \Rightarrow z = 0$

$f'(z)$ does not exist though $f(z) = |z|^2$ is continuous at all z (for $z \neq 0$).

13.2.5 Analytic Functions: Definition of Analyticity

A function $f(z)$ is said to be analytic in a domain D if $f(z)$ is defined and differentiable at all points of D.

The function $f(z)$ is said to be analytic at a point $z = z_0$ in D if $f(z)$ is analytic in a nbd of z_0 in D.

By an *analytic function* we mean a function that is analytic in *some* domain.

'Regular' and 'holomorphic' are other terms used for analytic functions. The fact that a function $f(z)$ is holomorphic in a domain D is expressed as

$$f \in H(D).$$

13.2.6 [1]Cauchy–[2]Riemann Equations

The Cauchy–Riemann equations (C.R.Es) provide us a test for the analyticity of a complex function

$$w = f(z) = u(x, y) + iv(x, y)$$

f is analytic in a domain $D \Leftrightarrow$ (i) the first partial derivatives of u, v exist and are continuous in D, and (ii) u, v satisfy Cauchy–Riemann equations $u_x = v_y$, $u_y = -v_x$ in D.

Example 13.14

$f(z) = z^2 = x^2 - y^2 + i2xy$ is analytic for all z

$u = x^2 - y^2$, $v = 2xy$ satisfy CREs $u_x = 2x$,

$u_y = -2y$, $v_x = 2y$, $v_y = 2x \therefore u_x = v_y\ u_y = -v_u$

13.2.7 Cauchy–Riemann Equations in Cartesian Coordinates

Theorem 13.1 If $f(z) = u(x, y) + iv(x, y)$ is differentiable at z then at this point the first order partial derivatives of u and v exist and satisfy the CREs

$$u_x = v_y; \quad u_y = -v_x \tag{13.8}$$

Proof: By hypothesis f is differentiable at z. So

$$f'(z) = \lim_{\Delta z \to 0} \frac{f(z + \Delta z) - f(z)}{\Delta z} \text{ exists.} \tag{13.9}$$

Hence the limit value along Paths I and II must be the same.

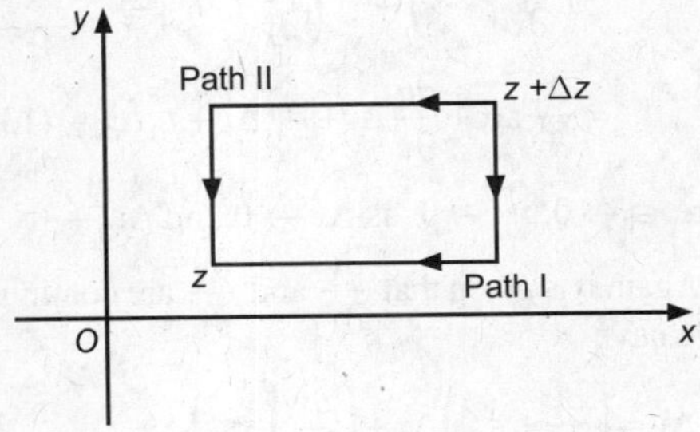

Figure 13.11

$$z = x + iy, \quad \Delta z = \Delta x + i\Delta y$$

$$f'(z) = \lim_{\Delta z \to 0} \frac{u(x + \Delta x, y + \Delta y) - u(x, y)}{\Delta x + i\Delta y} + i \lim_{\Delta z \to 0} \frac{v(x + \Delta x, y + \Delta y) - v(x, y)}{\Delta x + i\Delta y} \tag{13.10}$$

Choose Path I. We let $\Delta y \to 0$ first and then $\Delta x \to 0$. After $\Delta y = 0$, $\Delta z = \Delta x$, then Eq. (13.10) becomes

$$f'(z) = u_x + iv_x \tag{13.11}$$

Similarly along Path II, we let $\Delta x \to 0$ first and then $\Delta y \to 0$. After $\Delta x = 0$, $\Delta z = i\Delta y$, then Eq. (13.10) becomes

$$f'(z) = -iu_y + v_y \qquad \because \frac{1}{i} = -i \tag{13.12}$$

$\therefore$ Existence of $f'(z)$ implies the existence of u_x, v_x, u_y, v_y in Eqs. (13.11) and (13.12).

Equating the real and imaginary parts of Eqs. (13.11) and (13.12) we get the CREs.

Theorem 13.2 If two real-valued functions $u(x, y)$, $v(x, y)$ of two real variables x and y have continuous

---

[1] Cauchy, Augustin Louis (1789–1857) great French analyst, applied mathematician and group theorist. After Euler, the most prolific mathematician in history.

He worked on wave theory, elasticity and group theory, introduced modern rigour into calculus, founded the theory of functions of a complex variable, inaugurated modern era in differential equations with his existence theorems.

[2] Riemann, Georg Friedrich Bernhard (1826–1866) Great innovative German mathematician who made fundamental contributions to geometry and the theory of analytic functions of a complex variable and also to number theory, potential theory, topology and mathematical physics. Riemannian geometry provides the foundation for modern relativity theory.

first partial derivatives that satisfy CREs in some domain D then the complex function $f(z) = u(x, y) + iv(x, y)$ is analytic in D.

Proof: It is given that $\frac{\partial u}{\partial x}$ and $\frac{\partial u}{\partial y}$ are continuous. We have,

$$\Delta u = u(x+\Delta x, y+\Delta y) - u(x, y)$$
$$= [u(x+\Delta x, y+\Delta y) - u(x, y+\Delta y)] + [u(x, y+\Delta y) - u(x, y)]$$
$$= \left(\frac{\partial u}{\partial x} + \epsilon_1\right)\Delta x + \left(\frac{\partial u}{\partial y} + \eta_1\right)\Delta y$$
$$= \frac{\partial u}{\partial x}\Delta x + \frac{\partial u}{\partial y}\Delta y + \epsilon_1 \Delta x + \eta_1 \Delta y \quad (13.13)$$

where $\epsilon_1 \to 0$, $\eta_1 \to 0$ as $\Delta x \to 0$ and $\Delta y \to 0$

Again it is given that $\frac{\partial v}{\partial x}$ and $\frac{\partial v}{\partial y}$ are continuous. So we have,

$$\Delta v = \left(\frac{\partial v}{\partial x} + \epsilon_2\right)\Delta x + \left(\frac{\partial v}{\partial y} + \eta_2\right)\Delta y$$
$$= \frac{\partial v}{\partial x}\Delta x + \frac{\partial v}{\partial y}\Delta y + \epsilon_2 \Delta x + \eta_2 \Delta y \quad (13.14)$$

where $\epsilon_2 \to 0$, $\eta_2 \to 0$ as $\Delta x \to 0$ and $\Delta y \to 0$

$$\therefore \Delta w = \Delta u + i\Delta v$$
$$= \left(\frac{\partial u}{\partial x} + i\frac{\partial v}{\partial x}\right)\Delta x + \left(\frac{\partial u}{\partial y} + i\frac{\partial v}{\partial y}\right)\Delta y + \epsilon \Delta x + \eta \Delta y \quad (13.15)$$

where $\epsilon = \epsilon_1 + i\epsilon_2 \to 0$, $\eta = \eta_1 + i\eta_2 \to 0$ as $\Delta x \to 0$ and $\Delta y \to 0$

Using CREs, we can write Eq. (13.15) as

$$\Delta w = \left(\frac{\partial u}{\partial x} + i\frac{\partial v}{\partial x}\right)\Delta x + \left(-\frac{\partial v}{\partial x} + i\frac{\partial u}{\partial x}\right)\Delta y + \epsilon \Delta x + \eta \Delta y$$
$$= \left(\frac{\partial u}{\partial x} + i\frac{\partial v}{\partial x}\right)(\Delta x + i\Delta y) + \epsilon \Delta x + \eta \Delta y$$

Dividing both sides by $\Delta z = \Delta x + i\Delta y$ and taking limit as $\Delta z \to 0$

$$\frac{dw}{dz} = f'(z) = \lim_{\Delta z \to 0} \frac{\Delta w}{\Delta z} = \frac{\partial u}{\partial x} + i\frac{\partial v}{\partial x}$$

So, the derivative exists and is unique.

$$\therefore f(z) \in H(D).$$

Alternative forms of CREs

Theorem 13.3 (Complex form of CREs). Let $f(z) = u(x, y) + iv(x, y)$ be differentiable in a domain D. Then the CREs $u_x = v_y$ and $u_y = -v_x$ can be put in a complex form as

$$f_x = -if_y.$$

Proof: Let

$$f(z) = u(x, y) + iv(x, y). \text{ Then,}$$
$$f_x = u_x + iv_x, \qquad f_y = u_y + iv_y$$
$$\text{Now, } f_x = -if_y$$
$$\Leftrightarrow u_x + iv_x = -i(u_y + iv_y)$$
$$\Leftrightarrow u_x = v_y \text{ and } u_y = -v_x$$

13.2.8 Cauchy–Riemann Equations in Polar Coordinates

Theorem 13.4 (CREs in polar coordinates). If $f(z) = u(r, \theta) + iv(r, \theta)$ is differentiable at $z = re^{i\theta}$ then $\frac{\partial u}{\partial r} = \frac{1}{r}\frac{\partial v}{\partial \theta}$ and $\frac{\partial v}{\partial r} = -\frac{1}{r}\frac{\partial u}{\partial \theta}$.

Also, $f'(z) = \frac{r}{z}\left(\frac{\partial u}{\partial r} + i\frac{\partial v}{\partial r}\right)$.

Proof: We know that $x = r\cos\theta$, $y = r\sin\theta$

$$\therefore \frac{\partial u}{\partial r} = \frac{\partial u}{\partial x}\frac{\partial x}{\partial r} + \frac{\partial u}{\partial y}\frac{\partial y}{\partial r}$$
$$= \frac{\partial u}{\partial x}\cos\theta + \frac{\partial u}{\partial y}\sin\theta$$

$$\text{and } \frac{\partial v}{\partial \theta} = \frac{\partial v}{\partial x}\frac{\partial x}{\partial \theta} + \frac{\partial v}{\partial y}\frac{\partial y}{\partial \theta}$$
$$= \frac{\partial v}{\partial x}(-r\sin\theta) + \frac{\partial v}{\partial y}(r\cos\theta)$$
$$\Rightarrow \frac{1}{r}\frac{\partial v}{\partial \theta} = -\frac{\partial v}{\partial x}\sin\theta + \frac{\partial v}{\partial y}\cos\theta$$
$$= \frac{\partial u}{\partial y}\sin\theta + \frac{\partial u}{\partial x}\cos\theta$$
$$= \frac{\partial u}{\partial r} \quad \text{by CREs}$$

Thus $\frac{\partial u}{\partial r} = \frac{1}{r}\frac{\partial v}{\partial \theta}$.

Similarly, we can prove that $\frac{\partial v}{\partial r} = -\frac{1}{r}\frac{\partial u}{\partial \theta}$ by CREs.

Now, $r\left(\frac{\partial u}{\partial r}+i\frac{\partial v}{\partial r}\right)$

$$=r\left[\left(\frac{\partial u}{\partial x}\frac{\partial x}{\partial r}+\frac{\partial u}{\partial y}\frac{\partial y}{\partial r}\right)+i\left(\frac{\partial v}{\partial x}\frac{\partial x}{\partial r}+\frac{\partial v}{\partial y}\frac{\partial y}{\partial r}\right)\right]$$

$$=r\left[\left(\frac{\partial u}{\partial x}\cos\theta+\frac{\partial u}{\partial y}\sin\theta\right)+i\left(\frac{\partial v}{\partial x}\cos\theta+\frac{\partial v}{\partial y}\sin\theta\right)\right]$$

$$=r\cos\theta\left(\frac{\partial u}{\partial x}+i\frac{\partial v}{\partial x}\right)+r\sin\theta\left(\frac{\partial u}{\partial y}+i\frac{\partial v}{\partial y}\right)$$

$$=xf'(z)+iyf'(z)=zf'(z)$$

$$\text{or } f'(z)=\frac{r}{z}\left(\frac{\partial u}{\partial r}+i\frac{\partial v}{\partial r}\right).$$

CRES in Polar Coordinates [Alternative Method]

In polar coordinates, CREs are

$$u_r=\frac{1}{r}v_\theta,\ v_r=-\frac{1}{r}u_\theta.$$

Let $x = r\cos\theta$, $y = r\sin\theta$. Then $z = x + iy = r\cos\theta + ir\sin\theta = re^{i\theta}$.

So, $u+iv=f(z)=f(re^{i\theta})$

Differentiating partially w.r.t r and θ, we get

$$\frac{\partial u}{\partial r}+i\frac{\partial u}{\partial r}=f'(re^{i\theta})\cdot e^{i\theta}$$

$$\frac{\partial u}{\partial \theta}+i\frac{\partial v}{\partial \theta}=f'(re^{i\theta})ire^{i\theta}$$

$$=ir\left(\frac{\partial u}{\partial r}+i\frac{\partial v}{\partial r}\right)$$

Equating the real and imaginary parts

$$\frac{\partial u}{\partial r}=\frac{1}{r}\frac{\partial v}{\partial \theta},\qquad \frac{\partial v}{\partial r}=-\frac{1}{r}\frac{\partial u}{\partial \theta}.$$

Note

1. If $f(z) = u(x, y) + iv(x, y)$ is analytic in a domain D then $Re\, f = u = u(x, y)$ and $\text{Im}\, f = v = v(x, y)$ have partial derivatives u_x, u_y, v_x and v_y and satisfy the partial differential equations $u_x = v_y$ and $u_y = -v_x$ called the Cauchy–Riemann Equations.

2. The derivative $f'(z)$ can be calculated using either of the two formulas:

$$f'(z) = u_x + iv_x \text{ or } f'(z) = -iu_y + v_y.$$

3. Let $f(z) = u + iv$. Then the Cauchy–Riemann Equations $u_x = v_y$ and $u_y = -v_x$ are only necessary conditions for analyticity (differentiability) of f but not sufficient.

4. Let $f(z) = u + iv$. Then a sufficient condition for analyticity of f is that u and v have continuous partial derivatives and also satisfy CREs $u_x = v_y$ and $u_y = -v_x$.

Properties of analytic functions

1. The sum, the difference, the product and the quotient of analytic functions are analytic. $f, g \in H(D) \Rightarrow f \pm g, fg, f/g\ (g(z) \neq 0) \in H(D)$.

2. An analytic function of an analytic function is analytic.

3. An entire function of an entire function is entire. (A function which is analytic in the entire complex plane is called an entire function, *e.g.* e^z, polynomial, sin z, cos z etc.)

4. $f \in H(D) \Rightarrow f$ is differentiable at all $z \in D$. $\Rightarrow f$ is continuous at all $z \in D$.

5. If f is analytic at a point $z = a$ then f is differentiable at $z = a$. However, the converse is not true.

Example 13.15

We have proved that $f(z)=|z|^2$ is differentiable only at $z = 0$ and nowhere else. This function is *not analytic* at $z = 0$ since according to the definition of analyticity at a point, f must be differentiable not only at $z = 0$ but at every point of some nbd of 0.

Example 13.16

If $f(z) = z^2$ is analytic for all z, then the CREs are satisfied.

For $f(z)=\bar{z}=x-iy$ we have $u = x$, $v = -y$

CRE(2) $u_y=0=-v_x$ is satisfied

CRE(1) $u_x=1\neq v_y=-1$ is not satisfied.

$\therefore \bar{z}$ is not analytic.

An analytic function of constant absolute value is constant

Example 13.17

Show that if $f(z)$ is analytic in D and $|f(z)| = k$, a constant in D then $f(z)$ is constant in D.

Solution By assumption $u^2 + v^2 = k^2$.

By differentiating partially w.r.t x and y, we get

$$uu_x + vv_x = 0, uu_y + vv_y = 0$$

$$\Rightarrow \text{(a) } uu_x - vu_y = 0 \quad \text{(b) } uu_y + vu_x = 0 \qquad (1)$$

$$\because v_x = -u_y, \quad v_y = u_x \text{ by CREs.}$$

To eliminate u_y, multiply 1(a) by u and 1(b) by v and add. We get

$$(u^2 + v^2)u_x = 0.$$

Similarly to eliminate u_x, multiply 1(a) by $-v$ and 1(b) by u and add. We get

$$(u^2 + v^2)u_y = 0.$$

If $k^2 = u^2 + v^2 = 0$, then $u = v = 0 \Rightarrow f = 0$.

If $k \neq 0$ then $u_x = u_y = 0 \Rightarrow v_x = v_y = 0$, by CREs.

$\therefore u =$ constant, $v =$ constant $\Rightarrow f(z)$ is constant of D.

Example 13.18 Show that

(a) an analytic function with constant real part is constant; and

(b) an analytic function with constant imaginary part is constant.

Solution Let $w = u + iv$ be an analytic function. Since Cauchy Riemann Equations are satisfied, we have

$$\frac{\partial u}{\partial x} = \frac{\partial v}{\partial y} \text{ and } \frac{\partial u}{\partial y} = -\frac{\partial v}{\partial x} \qquad (1)$$

(a) Real part u is constant $\Rightarrow u = k_1$ say

$$\therefore \frac{\partial u}{\partial x} = 0 \text{ and } \frac{\partial u}{\partial y} = 0$$

From eq. (1) we have

$$\frac{\partial v}{\partial y} = 0 \text{ and } \frac{\partial v}{\partial x} = 0$$

Thus, V is independent of x and y so that $v = k_2$ a constant

$$\therefore w = u + iv = k_1 + ik_2 \text{ (constant)}$$

Hence the analytic function is constant.

(b) we can similarly prove that an analytic function with constant imaginary part is constant.

Limits, continuity, differentiability

Example 13.19

$$f(z) = \begin{cases} \dfrac{z^2 + 3iz - 2}{z + i} & \text{for } z \neq -i \\ a & \text{for } z = i \end{cases}$$

Find 'a' so that f is continuous.

Solution $f(z)$ is continuous everywhere except at $z = -i$.

Path I:

$$\lim_{z \to -i} f(z) = \lim_{\substack{x \to 0 \\ y \to -1}} \frac{(x+iy)^2 + 3i(x+iy) - 2}{x + iy + i}$$

$$= \lim_{y \to -1} \frac{-y^2 - 3y - 2}{i(y+1)}$$

$$= \lim_{y \to -1} \frac{-(y+1)(y+2)}{i(y+1)}$$

$$= \lim_{y \to -1} \frac{-(y+2)}{i} = -\frac{1}{i} = i$$

Path II:

$$\lim_{z \to -i} f(z) = \lim_{\substack{y \to -1 \\ x \to 0}} \frac{(x+iy)^2 + 3i(x+iy) - 2}{x + iy + i}$$

$$= \lim_{x \to 0} \frac{2(x-i) + 3i}{1} = i,$$

by L' Hospital's rule

Thus $\lim_{z \to -i} f(z) = i = a = f(-i)$. Hence f is continuous if $a = i$, but not otherwise.

Example 13.20

Show that $f(z) = \dfrac{\bar{z}}{z}$ is not continuous at $z = 0$.

Solution Let $z = x + iy$.

Path I:

Along the x-axis:

$$y = 0, \ \lim_{z \to 0} \frac{\bar{z}}{z} = \lim_{y \to 0} \frac{x - iy}{x + iy} = 1$$

Path II:
Along the y-axis:

$$x = 0,\ \lim_{z\to 0}\frac{\bar{z}}{z} = \lim_{\substack{x=0\\ y\to 0}}\frac{x-iy}{x+iy} = -1$$

$\lim_{z\to 0}\frac{\bar{z}}{z}$ does not exist. Hence f is not continuous at $z = 0$.

Example 13.21

Show that $f(z) = \bar{z}$ is continuous at all z.

Solution We have $|f(z) - f(z_0)| = |\bar{z} - \bar{z}_0|$. For any given $\in$ choose $\in\ = \delta$. We have $|z - z_0| < \delta$ $\Rightarrow |f(z) - f(z_0)| < \in \Rightarrow f(z)$ is continuous at z_0.

Since z_0 is arbitrary $f(z)$ is continuous for all z.

Example 13.22

Show that $f(z) = |z|$ is continuous at all z.

Solution We have

$|f(z) - f(z_0)| = ||z| - |z_0|| \le |z - z_0|$.

For any $\in\ > 0$ choose $\in\ = \delta$ then
$|z - z_0| < \delta \Rightarrow |f(z) - f(z_0)| < \in \Rightarrow f$ is continuous at z_0.

Since z_0 is arbitrary f is continuous at all z.

Example 13.23

Show that $\lim_{z\to 0}\frac{x^2y}{x^4+y^2}$ does not exist though the function approaches the same limit along every straight line through the origin.

Solution

Path I:

$$\lim_{\substack{x\to 0\\ y\to 0}}\frac{x^2y}{x^4+y^2} = \lim_{y\to 0} 0 = 0$$

Path II:

$$\lim_{\substack{y\to 0\\ x\to 0}}\frac{x^2y}{x^4+y^2} = \lim_{x\to 0} 0 = 0$$

Path III:
Along any straight line $y = mx$ through (0, 0),

$$\lim_{\substack{y=mx\\ x\to 0}}\frac{x^2y}{x^4+y^2} = \lim_{x\to 0}\frac{mx^3}{x^4+m^2x^2}$$

$$= \lim_{x\to 0}\frac{mx}{x^2+m^2} = 0$$

Path IV:
Along the parabolic path $y = mx^2$

$$\lim_{\substack{y=mx^2\\ x\to 0}}\frac{x^2y}{x^4+y^2} = \lim_{x\to 0}\frac{mx^4}{x^4+m^2x^4} = \lim_{x\to 0}\frac{m}{1+m^2}$$

$$= \frac{m}{1+m^2} \neq 0 \text{ if } m \neq 0$$

Along Path IV the limit is dependent on m.

Example 13.24

$f(z) = \frac{z}{|z|}$ $(z \neq 0)$ and $f(0) = 0$. Is $f(z)$ continuous at $z = 0$?

Solution

$$\lim_{z\to 0} f(z) = \lim_{z\to 0}\frac{x+iy}{\sqrt{x^2+y^2}};$$

Path I:

$$\lim_{\substack{x\to 0\\ y\to 0}}\frac{x+iy}{\sqrt{x^2+y^2}} = i$$

Path II:

$$\lim_{\substack{y\to 0\\ x\to 0}}\frac{x+iy}{\sqrt{x^2+y^2}} = 1$$

Since the limit does not exist f is discontinuous at $z = 0$.

Example 13.25

Show that every differentiable function is continuous.

Solution Let $f(z)$ be differentiable at 'a'. Then

$$f'(a) = \lim_{z\to a}\frac{f(z)-f(a)}{z-a} \text{ exists.}$$

$\therefore f(a)$ is well-defined. Consider

$$\lim_{z\to a}[f(z) - f(a)] = \lim_{z\to a}\frac{f(z)-f(a)}{(z-a)}(z-a)$$

$$= \lim_{z\to a}\frac{f(z)-f(a)}{(z-a)}\lim_{z\to a}(z-a)$$

$$= f'(a)\lim_{z\to a}(z-a) = 0$$

$$\Rightarrow \lim_{z\to a} f(z) = \lim_{z\to a} f(a) = f(a)$$

Hence f is continuous at $z = a$.

Remark

The converse of the above result is not true. That is, a function may be continuous but not differentiable.

Counter-examples:

The following functions are continuous at points specified against them but they are not differentiable at those points.

(i) $f(z)=\bar{z}$ at any point
(ii) $g(z)=|z|^2$ at zero
(iii) $h(z)=Re\, z$ at any point.

Example 13.26

Find the regions of analyticity of the functions

(i) $f(z)=e^z$ (ii) $f(z)=(x-y)^2+2i(x+y)$

(iii) $f(z)=\dfrac{z-2i}{z^2+1}$.

Solution

(i) $f(z)=e^z=e^{x+iy}=e^x(\cos y+\sin y)=u+iv$ (*say*)

$\Rightarrow u=e^x\cos y,\qquad v=e^x\sin y$

$\Rightarrow u_x=e^x\cos y,\qquad u_y=-e^x\sin y,$

$v_x=e^x\sin y,\qquad v_y=e^x\cos y$

$\Rightarrow u_x=v_y;\qquad u_y=-v_x$

u_x, u_y, v_x, v_y, are continuous. CREs are satisfied for all x and y.

Hence e^z is analytic throughout the complex plane and it is an entire function.

(ii) $u=(x-y)^2;\qquad v=2(x+y)$

$u_x=2(x-y);\qquad u_y=-2(x-y);$

$v_x=2, v_y=2$

$u_x=v_y\Rightarrow x-y=1$

also $u_y=-v_x\Rightarrow x-y=1$

CREs are satisfied only along the straight line $x-y=1$.

$f'(z)=u_x+iv_x=2(x-y)+2i=2+2i$ exists along $x-y=1$ and not through any region (nbd) R. Hence $f(z)$ is nowhere analytic.

(iii) $f(z)=\dfrac{z-2i}{z^2+1}$ is the quotient of polynomials $z-2i$ and z^2+1 which are analytic throughout $\mathbb{C}$. So, $f(z)$ is analytic everywhere except at points where $z^2+1=0$, *i.e.*, $z=\pm i$.

Example 13.27

Show that z^2 is analytic for all z.

Solution

$f(z)=u+iv=z^2=(x+iy)^2=(x^2-y^2)+i(2xy)$

$\Rightarrow u=x^2-y^2,\quad v=2xy$

$\Rightarrow u_x=2x;\quad v_y=2x;\quad u_y=-2y;\quad v_x=2y$

$\Rightarrow u_x=v_y,\quad u_y=-v_x$

u_x, u_y, v_x, v_y, are continuous and CREs are satisfied; $f(z)$ is analytic in $\mathbb{C}$.

Example 13.28

Show that $f(z)=z+2\bar{z}$ is not analytic anywhere in the complex plane. [JNTU 2000, 2003(2)]

Solution Let $z=x+iy$ and

$$f(z)=u+iv=z+2\bar{z}=(x+iy)+2(x-iy)$$

$$\Rightarrow u=3x, v=-y$$

$$\Rightarrow u_x=3, v_x=0, u_y=0, v_y=1$$

CREs $u_x\neq v_y, u_y=-v_u$ are not satisfied.

Hence $f(z)$ is not analytic anywhere.

Example 13.29

Show that $\left(\dfrac{\partial^2}{\partial x^2}+\dfrac{\partial^2}{\partial y^2}\right)\log|f'(z)|=0$ where $f(z)$ is an analytic function. [JNTU 2002]

Solution

Since $x=\dfrac{1}{2}(z+\bar{z}),\ y=\dfrac{1}{2i}(z-\bar{z})=-\dfrac{i}{2}(z-\bar{z})$, we have

$$\frac{\partial}{\partial z}=\frac{\partial}{\partial x}\left(\frac{\partial x}{\partial z}\right)+\frac{\partial}{\partial y}\left(\frac{\partial y}{\partial z}\right)=\frac{1}{2}\left(\frac{\partial}{\partial x}-i\frac{\partial}{\partial y}\right);$$

Similarly, $\dfrac{\partial}{\partial \bar{z}}=\dfrac{1}{2}\left(\dfrac{\partial}{\partial x}+i\dfrac{\partial}{\partial y}\right)$

Now $\dfrac{\partial^2}{\partial z\partial \bar{z}}=\dfrac{1}{4}\left(\dfrac{\partial^2}{\partial x^2}+\dfrac{\partial^2}{\partial y^2}\right)$

$$\therefore \left(\frac{\partial^2}{\partial x^2}+\frac{\partial^2}{\partial y^2}\right)(\log|f'(z)|)$$

$$=4\frac{\partial^2}{\partial z\partial \bar{z}}\left(\frac{1}{2}\log|f'(z)|^2\right)$$

$$=2\frac{\partial^2}{\partial z\partial \bar{z}}\log[f'(z)f'(\bar{z})]$$

$[\because |z|^2=z\bar{z}]$

$$=2\frac{\partial^2}{\partial z\partial \bar{z}}(\log f'(z)+\log f'(\bar{z}))$$

$$=2\left[\frac{\partial}{\partial \bar{z}}\frac{f''(z)}{f'(z)}+\frac{\partial}{\partial z}\frac{f''(\bar{z})}{f'(\bar{z})}\right]$$

$$= 2(0+0) = 0.$$

Since $f(z)$ is analytic $f'(z)$ is analytic. $f'(\bar{z})$ is also analytic

$$\therefore \frac{\partial f'(z)}{\partial \bar{z}} = 0; \quad \frac{\partial f'(\bar{z})}{\partial z} = 0.$$

Example 13.30

If $f(z) = u + iv$ is an analytic function of z prove that

$$\left(\frac{\partial^2}{\partial x^2} + \frac{\partial^2}{\partial y^2}\right)|f(z)|^p = p^2 |f(z)|^{p-2} |f'(z)|^2.$$

[JNTU 2003S (4)]

Solution We know that

$$\frac{\partial^2}{\partial x^2} + \frac{\partial^2}{\partial y^2} = 4\frac{\partial^2}{\partial z \partial \bar{z}}$$

and $|f(z)|^2 = u^2 + v^2 = f(z)f(\bar{z})$.

$$\therefore \left(\frac{\partial^2}{\partial x^2} + \frac{\partial^2}{\partial y^2}\right)|f(z)|^p = 4\frac{\partial^2}{\partial z \partial \bar{z}}|f(z)|^p$$

$$= 4\frac{\partial^2}{\partial z \partial \bar{z}}[f(z)f(\bar{z})]$$

$$= 4\frac{\partial^2}{\partial z \partial \bar{z}}[(f(z))^{p/2}(f(\bar{z}))^{p/2}]$$

Differentiate first w.r.t $\bar{z}$ taking $f(z)$ as constant and then w.r.t z taking $f(\bar{z})$ as constant; we have

$$\left(\frac{\partial^2}{\partial x^2} + \frac{\partial^2}{\partial y^2}\right)|f(z)|^p$$

$$= 4\frac{\partial}{\partial z}\left[(f(z))^{p/2}\frac{p}{2}(f(\bar{z}))^{\frac{p}{2}-1} f'(\bar{z})\right]$$

$$= 4\left[\frac{p}{2}(f(z))^{\frac{p}{2}-1} f'(z) \cdot \frac{p}{2} f(\bar{z})^{\frac{p}{2}-1} f'(\bar{z})\right]$$

$$= 4 \cdot \frac{p^2}{4}[f(z)f(\bar{z})]^{\frac{p-2}{2}} (f'(z)f'(\bar{z}))$$

$$= p^2(|f(z)|^2)^{\frac{p-2}{2}} |f'(z)|^2$$

$$= p^2 |f(z)|^{p-2} |f'(z)|^2.$$

Example 13.31

Find all values of k such that $f(z) = e^x$ (cos ky + i sin ky) is analytic. [JNTU 2003(4)]

Solution

Let $f(z) = u + iv = e^x(\cos ky + i \sin ky)$

$$\Rightarrow u = e^x \cos ky, \qquad v = e^x \sin ky$$

$$\Rightarrow u_x = e^x \cos ky, \qquad u_y = -ke^x \sin ky$$

$$v_x = e^x \sin ky, \qquad v_y = ke^x \cos ky$$

CREs. $u_x = v_y \Rightarrow k = 1, \quad u_y = -v_x \Rightarrow k = 1$

$f(z)$ is analytic $\Leftrightarrow k = 1$

Example 13.32

Prove that $\left(\frac{\partial^2}{\partial x^2} + \frac{\partial^2}{\partial y^2}\right)|Re f(z)|^2 = 2|f'(z)|^2$ where $w = f(z)$ is analytic.

[JNTU 2004(4), 2004S, 2005]

Solution

Let $f(z) = u + iv \Rightarrow Re\, f(z) = u$.

Also $f'(z) = u_x + iv_x$ so that $|f'(z)| = \sqrt{u_x^2 + v_x^2}$

Squaring $|f'(z)|^2 = u_x^2 + v_x^2$ (1)

Now $\frac{\partial}{\partial x}(u^2) = 2u\frac{\partial u}{\partial x} \Rightarrow \frac{\partial^2}{\partial x^2}(u^2) = 2[u_x^2 + uu_{xx}]$ (2)

Similarly $\frac{\partial^2}{\partial y^2}(u^2) = 2\left[u_y^2 + uu_{yy}\right]$ (3)

Now Eqs. (2) + (3) gives

$$\nabla^2(u^2) = \left(\frac{\partial^2}{\partial x^2} + \frac{\partial^2}{\partial y^2}\right)u^2$$

$$= 2[u_x^2 + u_y^2 + u(u_{xx} + u_{yy})] \quad (4)$$

We know that $Re\, f(z) = u$ satisfies the Laplace's equation so that

$$\nabla^2 u = \frac{\partial^2 u}{\partial x^2} + \frac{\partial^2 u}{\partial y^2} = 0 \quad (5)$$

From Eqs. (4) and (5)

$$\nabla^2(u^2) = 2(u_x^2 + u_y^2)$$

$$= 2|f'(z)|^2 \text{ by Eq. (1)}$$

$$\Rightarrow \left(\frac{\partial^2}{\partial x^2} + \frac{\partial^2}{\partial y^2}\right)|Re\ f(z)|^2 = 2|f'(z)|^2$$

Example 13.33

Determine whether the function $2xy + i(x^2 - y^2)$ is analytic. [JNTU 2003(4)]

Solution

Let $f(z) = u + iv = 2xy + i(x^2 - y^2)$

$$u = 2xy; \quad v = x^2 - y^2$$

$$\frac{\partial u}{\partial x} = 2y; \quad \frac{\partial u}{\partial y} = 2x;$$

$$\frac{\partial v}{\partial x} = 2x; \quad \frac{\partial v}{\partial y} = -2y$$

CREs are $u_x = v_y$ and $u_y = -v_x$ which are clearly not satisfied. Hence the given function is not analytic.

Example 13.34

Show that $f(z) = xy^2(x + iy)/(x^2 + y^4)$ $(z \neq 0)$ and $f(0) = 0$ is not analytic at $z = 0$ although CREs are satisfied at the origin.

Solution

$$\frac{f(z)-f(0)}{z-0} = \frac{f(z)-0}{z} = \frac{xy^2(x+iy)}{(x+iy)(x^2+y^4)}$$

$$= \frac{xy^2}{x^2+y^4}$$

Clearly, $\lim\limits_{\substack{x\to 0\\ y\to 0}} \dfrac{xy^2}{x^2+y^4} = \lim\limits_{\substack{y\to 0\\ x\to 0}} \dfrac{xy^2}{x^2+y^4} = 0$

Along the path $y = mx$,

$$\lim_{z\to 0}\frac{f(z)-f(0)}{z-0} = \lim_{x\to 0}\frac{x\cdot m^2x^2}{x^2+m^4x^4}$$

$$= \lim_{x\to 0}\frac{m^2x}{1+m^4} = 0$$

Also, along the path $x = my^2$,

$$\lim_{z\to 0}\frac{f(z)-f(0)}{z-0} = \lim_{y\to 0}\frac{my^2\cdot y^2}{m^2y^4+y^4}$$

$$= \lim_{y\to 0}\frac{m}{m^2+1} \neq 0$$

Since the limit value depends on m and is different for different paths, the limit does not exist. Hence $f(z)$ is not differentiable at $z = 0$. So, $f(z)$ is not analytic at $z = 0$.

We now prove that CREs are not satisfied at $z = 0$.

Let $f(z) = u + iv = \dfrac{xy^2(x+iy)}{x^2+y^4}$

$$\Rightarrow u = \frac{x^2y^2}{x^2+y^4},\ v = \frac{xy^3}{x^2+y^4} \text{ for } z \neq 0$$

Also $u(0, 0) = 0$ and $v(0, 0) = 0$

$\because f(z) = 0$ at $z = 0$

$$\frac{\partial u}{\partial x} = \lim_{x\to 0}\frac{u(x,0)-u(0,0)}{x} = \lim_{x\to 0}\frac{0-0}{x} = 0$$

$$\frac{\partial u}{\partial y} = \lim_{y\to 0}\frac{u(0,y)-u(0,0)}{y} = \lim_{y\to 0}\frac{0-0}{y} = 0$$

$$\frac{\partial v}{\partial x} = \lim_{x\to 0}\frac{v(x,0)-v(0,0)}{x} = \lim_{x\to 0}\frac{0-0}{x} = 0$$

$$\frac{\partial v}{\partial y} = \lim_{y\to 0}\frac{v(0,y)-v(0,0)}{y} = \lim_{y\to 0}\frac{0-0}{y} = 0$$

The CREs are satisfied at the origin $z = 0$. Consequently, $f(z)$ is not analytic at (0, 0) though CREs are satisfied at $z = 0$.

Example 13.35

Show that the function $f(z) = \sqrt{|xy|}$ is not analytic at the origin, although Cauchy–Riemann equations are satisfied at the origin.

Solution Let $f(z) = u(x, y) + iv(x, y) = \sqrt{|xy|}$ so that $u(x, y) = \sqrt{|xy|}$ and $v(x, y) = 0$. We compute the partial derivatives $\dfrac{\partial u}{\partial x}, \dfrac{\partial u}{\partial y}, \dfrac{\partial v}{\partial x}, \dfrac{\partial v}{\partial y}$ at the origin.

$$\frac{\partial u}{\partial x} = \lim_{x\to 0}\frac{u(x,0)-u(0,0)}{x} = \lim_{x\to 0}\frac{0-0}{x} = 0$$

$$\frac{\partial u}{\partial y} = \lim_{y\to 0}\frac{u(0,y)-u(0,0)}{y} = \lim_{y\to 0}\frac{0-0}{y} = 0$$

$$\frac{\partial v}{\partial x} = \lim_{x\to 0}\frac{0-0}{x} = 0$$

$$\frac{\partial v}{\partial y} = \lim_{y\to 0}\frac{0-0}{y} = 0$$

Hence, Cauchy–Riemann equations $\dfrac{\partial u}{\partial x} = \dfrac{\partial v}{\partial y}$, $\dfrac{\partial u}{\partial y} = -\dfrac{\partial v}{\partial x}$ are satisfied at the origin.

We now find the derivative at the origin.

$$f'(0) = \lim_{z\to 0}\frac{f(z)-f(0)}{z}$$

$$= \lim_{z\to 0}\frac{\sqrt{|xy|}-0}{x+iy}$$

The limit must exist along any path leading to the origin. If we allow $z = (x, y) \to (0, 0)$ along the straight line $y = mx$, we obtain

$$f'(0) = \lim_{x\to 0} \frac{\sqrt{|mx^2|}}{(1+im)x} = \frac{\sqrt{|m|}}{1+im},$$

which depends on m and hence not unique. Hence $f'(0)$ does not exist.

Note

In this problem, $\frac{\partial u}{\partial x}, \frac{\partial u}{\partial y}, \frac{\partial v}{\partial x}, \frac{\partial v}{\partial y}$ all exist at $(x, y) = (0, 0)$ and also satisfy Cauchy–Riemann equations at the origin. Yet $f'(0)$ does not exist because the partial derivatives are not continuous at the origin.

Example 13.36

Prove that the function $f(z)$ defined by

$$f(z) = \begin{cases} \dfrac{x^3(1+i) - y^3(1-i)}{x^2+y^2}, & (z \neq 0) \\ 0, & (z = 0) \end{cases}$$

is continuous and the Cauchy–Riemann equations are satisfied at the origin but $f'(0)$ does not exist.

[VTU 2001, JNTU 2000, Punjab 1999]

Solution

$$\lim_{z\to 0} f(z) = \lim_{\substack{x\to 0\\ y\to 0}} \frac{x^3(1+i) - y^3(1-i)}{x^2+y^2}$$

$$= \lim_{y\to 0}[-y(1-i)] = 0$$

$$\lim_{z\to 0} f(z) = \lim_{\substack{y\to 0\\ x\to 0}} \frac{x^3(1+i) - y^3(1-i)}{x^2+y^2}$$

$$= \lim_{x\to 0} x(1+i) = 0$$

Also, $f(0) = 0$ by hypothesis so that $\lim_{z\to 0} f(z) = f(0)$, when $x \to 0, y \to 0$ and $y \to 0, x \to 0$.

Now, let x and y both tend to 0 simultaneously along the path $y = mx$

$$\lim_{z\to 0} f(z) = \lim_{\substack{y=mx\\ x\to 0}} \frac{x^3(1+i) - m^3x^3(1-i)}{(1+m^2)x^2}$$

$$= \lim_{x\to 0} \frac{x[1+i-m^3(1-i)]}{1+m^2} = 0$$

Thus, irrespective the manner in which $z \to 0$ we have $\lim_{z\to 0} f(z) = 0 = f(0)$, which implies that the function is continuous at $z = 0$.

$$f(z) = u(x, y) + iv(x, y)$$

$$= \frac{x^3 - y^3}{x^2+y^2} + i\frac{x^3+y^3}{x^2+y^2}.$$

Also,

$$\left.\begin{matrix} u(0, 0) \\ v(0, 0) \end{matrix}\right\} f(0) = 0.$$

Now, we complete partial derivatives for u, v at the origin

$$\frac{\partial u}{\partial x} = \lim_{x\to 0} \frac{u(x, 0) - u(0, 0)}{x} = \lim_{x\to 0} \frac{x}{x} = 1$$

$$\frac{\partial u}{\partial y} = \lim_{y\to 0} \frac{u(0, y) - u(0, 0)}{y} = \lim_{y\to 0} -\frac{y}{y} = -1$$

$$\frac{\partial v}{\partial x} = \lim_{x\to 0} \frac{v(x, 0) - v(0, 0)}{x} = \lim_{x\to 0} \frac{x}{x} = 1$$

$$\frac{\partial v}{\partial y} = \lim_{y\to 0} \frac{v(0, y) - v(0, 0)}{y} = \lim_{y\to 0} \frac{y}{y} = 1$$

Clearly, Cauchy–Riemann equations $\frac{\partial u}{\partial x} = \frac{\partial v}{\partial y}$, $\frac{\partial u}{\partial y} = -\frac{\partial v}{\partial x}$ are satisfied at the origin.

Also,

$$f'(0) = \lim_{z\to 0} \frac{f(z) - f(0)}{z} = \lim_{z\to 0} \frac{f(z)}{z}$$

$$= \lim_{z\to 0} \frac{(x^3 - y^3) + i(x^3 + y^3)}{(x^2+y^2)(x+iy)}$$

$$= \frac{1 - m^3 + i(1+m^3)}{(1+m^2)(1+im)}, \text{ along the path } y = mx$$

$f'(0)$ depends on m and is not unique. Therefore, $f'(z)$ does not exist at the origin.

$f(z)$ is not analytic at the origin even though it is continuous and satisfies Cauchy–Riemann equations at $z = 0$ because the partial derivatives are not continuous at the origin.

13.2.9 Orthogonal Trajectories

The curves $u(x, y) = c_1$ and $v(x, y) = c_2$ where $f = u + iv \in H(D)$ represents two families of curves in the xy-plane which are mutually orthogonal.

Suppose a member of

$$u(x, y) = c_1 \qquad (13.16)$$

cuts a member of

$$v(x, y) = c_2 \qquad (13.17)$$

at a point $P(x_0, y_0)$.

By implicit differentiation of $u(x, y) = c_1$ we get $\frac{\partial u}{\partial x} + \frac{\partial u}{\partial y}\frac{dy}{dx} = 0 \Rightarrow m_1 =$ slope of curves Eq. (1) at $P = \left(\frac{dy}{dx}\right)_P = -\left(\frac{u_x}{u_y}\right)_P$.

Similarly, differentiation of $v(x, y) = c_2$ gives $m_2 =$ slope of curves (2) at $P = \left(\frac{dy}{dx}\right)_P = -\left(\frac{v_x}{v_y}\right)_P$.

The product of the slopes of the curves at the intersection point P $m_1 \cdot m_2 = \left(-\frac{u_x}{u_y}\right)\cdot\left(-\frac{v_x}{v_y}\right)_P = -1$

$\because u_x = v_y, u_y = -v_x$ by CREs.

Hence the family of curves $u(x, y) = c_1$ and the family $v(x, y) = c_2$ cut each other orthogonally.

Example 13.37

Find the orthogonal trajectories of

$$u(x, y) = x^3 y - xy^3 = c \qquad (1)$$

Solution The curves given by $v(x, y) =$ constant will be the required orthogonal trajectories if $f(z) = u + iv$ is analytic.

Differentiating (1) partially w.r.t. x and y

$$u_x = 3x^2 y - y^3; \qquad u_y = x^3 - 3xy^2;$$

$\therefore v_y = u_x = 3x^2 y - y^3$, by Cauchy Riemann Eqs.

Integrating w.r.t y, we get $v = \frac{3}{2}x^2 y^2 - \frac{1}{4}y^4 + h(x)$

Differentiating w.r.t x, we get

$$v_x = 3xy^2 + h'(x) = -u_y = -x^3 + 3xy^2$$

$$\Rightarrow h(x) = -\frac{1}{4}x^4 + c_1,$$

where c_1 is an arbitrary constant. The required orthogonal trajectory is

$$v(x, y) = \frac{3}{2}x^2 y^2 - \frac{1}{4}(x^4 + y^4) = k$$

Example 13.38

Prove that $u(x, y) = x^3 - 3xy^2 = c_1$ and $v(x, y) = 3x^2 y - y^3 = c_2$ cut each other orthogonally.

[JNTU 2003(3)]

Solution Let a member of the family

$$u(x, y) = x^3 - 3xy^2 = c_1 \qquad (1)$$

intersect a member of the family

$$v(x, y) = 3x^2 y - y^3 = c_2 \text{ at } P = (x_0, y_0) \qquad (2)$$

Differentiating Eqs. (1) and (2), we get

$$3x^2 - 3y^2 - 6xyy^1 = 0 \Rightarrow m_1 = (y')_P = \frac{x_0^2 - y_0^2}{2x_0 y_0} \qquad (3)$$

$$2xy + x^2 y' - x^2 y' = 0 \Rightarrow m_2 = (y')_P = -\frac{2x_0 y_0}{x_0^2 - y_0^2} \qquad (4)$$

Clearly the product of their slopes at $P = -1$ and hence the problem.

Example 13.39

Find the orthogonal trajectories of the family of the curves $r^2 \cos 2\theta = c$.

Solution Let $u(r, \theta) = r^2 \cos 2\theta = c$

$$\text{Then } \frac{\partial u}{\partial r} = 2r\cos 2\theta; \qquad \frac{\partial u}{\partial \theta} = -2r^2 \sin 2\theta$$

$$\text{CREs } \frac{\partial v}{\partial \theta} = r\frac{\partial u}{\partial \theta} \Rightarrow \frac{\partial v}{\partial \theta} = 2r^2 \cos 2\theta$$

on integrating $v(r, \theta) = r^2 \sin 2\theta + c_1$

Differentiating v w.r.t r, we get

$$\frac{\partial v}{\partial r} = 2r\sin 2\theta + \frac{\partial c_1}{\partial r}$$

$$= -\frac{1}{r}\frac{dr}{d\theta} = -\frac{1}{r}(-2r^2 \sin 2\theta) = 2r\sin 2\theta$$

$$\text{Also } \frac{dc_1}{dr} = 0 \Rightarrow c_1 = \text{constant}$$

$\therefore$ The orthogonal trajectories of curves $u(r, \theta) = r^2 \cos 2\theta = c$ are the curves $v(r, \theta) = r^2 \sin 2\theta = c$.

EXERCISE 13.2

1. Verify CREs for the functions $f(z) =$
   (a) z^2, (b) $iz+\bar{z}$, (c) $\sin z$.
2. Prove that the following functions are nowhere differentiable:
   (a) $f(z)=|z|$, (b) $f(z)=Re\,\bar{z}$,
   (c) $f(z)=2x+ixy^2$.
3. Prove that the following functions are differentiable at every point:
   (a) $f(z)=x^2-y^2-2xy+i(x^2-y^2+2xy)$
   (b) $f(z)=2x-3y+i(3x+2y)$
   (c) $f(z)=iz+2$.
4. Prove that $f(z)=\cos x(\cosh y+a\sinh y)+i\sin x(\cosh y+b\sinh y)$ is analytic if $a=b=-1$.
5. Determine constants a, b and c such that the following functions are differentiable at every point:
   (a) $f(z)=x+ay+i(bx+cy)$
   (b) $f(z)=e^x\cos ay+i\,e^x\sin(y+b)+c$.
6. Prove that the functions $f(z)$ and $\overline{f}(z)$ are simultaneously analytic.

13.3 Laplace's Equation: Harmonic and Conjugate Harmonic Functions

The real part $u(x, y)$ and the imaginary part $v(x, y)$ of an analytic function $f(z) = u(x, y) + iv(x, y)$ satisfy the most important differential equation in physics: Laplace's equation $\nabla^2 F=\dfrac{\partial^2 F}{\partial x^2}+\dfrac{\partial^2 F}{\partial y^2}=0$

(∇ is called nabla or del) (*viz.*) $\nabla^2 u=u_{xx}+u_{yy}=0$ in D $\nabla^2 v=v_{xx}+v_{yy}=0$ in D

Equation $\nabla^2 F = 0$ occurs in gravitation, electrostatics, fluid flow, heat conduction and so on.

Theorem 13.5 (Laplace's equation) If $f(z) = u(x, y) + iv(x, y)$ is analytic in a domain D then u and v satisfy Laplace's equation

$$\nabla^2 u=u_{xx}+u_{yy}=0, \tag{13.18}$$

$$\nabla^2 u=v_{xx}+v_{yy}=0 \tag{13.19}$$

in D and have continuous second partial derivatives in D.

Proof: Differentiating $u_x = v_y$ w.r.t x and $u_y = -v_x$ w.r.t y, we have

$$u_{xx}=v_{yx} \quad\text{and}\quad u_{yy}=-v_{xy} \tag{13.20}$$

The derivative of an analytic function is itself analytic $\Rightarrow u$, v have continuous partial derivatives of all orders.

In particular, $v_{yx} = v_{xy}$.

Adding Eq. (13.20) we obtain Eq. (13.18). Equation (13.19) is similarly proved.

Solutions of Laplace's equation having *continuous* second-order partial derivatives are called *harmonic functions* and their theory is called *Potential Theory*.

Hence the real and imaginary parts of an analytic function are harmonic functions.

If two harmonic functions u and v satisfy CREs in a domain D they are the real and imaginary parts of an analytic function f in D. v is said to be a conjugate harmonic function of u in D and vice versa.

Let $f(z) = u + iv$ be an analytic function whose real part u is known. We can find the imaginary part v and also the function $f(z)$ by following either of the two methods:

Method 1: Exact Differential Method using CREs
u is a harmonic function

$$\Rightarrow \nabla^2 u=\frac{\partial^2 u}{\partial x^2}+\frac{\partial^2 u}{\partial y^2}=0\ u,\ v \text{ satisfy CREs} \tag{13.21}$$

$$\frac{\partial u}{\partial x}=\frac{\partial v}{\partial y},\quad \frac{\partial u}{\partial y}=-\frac{\partial v}{\partial x} \tag{13.22}$$

$$\text{Let } \frac{\partial v}{\partial x}dx+\frac{\partial v}{\partial y}dy=Mdx+Ndy$$

where

$$M=\frac{\partial v}{\partial x}=-\frac{\partial u}{\partial y};\ N=\frac{\partial v}{\partial y}=\frac{\partial u}{\partial x} \text{ by Eq. (13.22)}$$

$$\text{Now } \frac{\partial M}{\partial y}-\frac{\partial N}{\partial x}=-\frac{\partial^2 u}{\partial y^2}-\frac{\partial^2 u}{\partial x^2}$$

$$=-\nabla^2 u=0 \text{ by Eq. (13.21)}$$

$$\Rightarrow \frac{\partial M}{\partial y}=\frac{\partial N}{\partial x}\Rightarrow Mdx+Ndy=dv \tag{13.23}$$

is an exact differential equation.

Hence integrating Eq. (13.23) we can find v. Having known u and v, $f(z) = u + iv$ can be determined.

Method 2: **Milne–Thomson's method**

Since $z = x + iy, \bar{z} = x - iy$ we have

$$x = \frac{1}{2}(z + \bar{z}),\ y = \frac{1}{2i}(z - \bar{z})$$

$$\therefore f(z) = u(x, y) + iv(x, y)$$

$$= u\left(\frac{z+\bar{z}}{2}, \frac{z-\bar{z}}{2i}\right) + iv\left(\frac{z+\bar{z}}{2}, \frac{z-\bar{z}}{2i}\right) \quad (13.24)$$

Considering Eq. (13.24) as a formal identity in z and $\bar{z}$ and putting $\bar{z} = z$, we get

$$f(z) = u(z, 0) + iv(z, 0) \quad (13.25)$$

which is obtained from $f(z) = u(x, y) + iv(x, y)$ by replacing x by z and y by 0.

Now $f'(z) = \frac{\partial u}{\partial x} + i\frac{\partial v}{\partial x} = \frac{\partial u}{\partial x} - i\frac{\partial u}{\partial y}$

($\because u_x = v_y,\ \ u_y = -v_x$ by CREs)

If we set $\frac{\partial u}{\partial x} = \Phi_1(x, y)$ and $\frac{\partial u}{\partial y} = \Phi_2(x, y)$ then,

$$f'(z) = \Phi_1(x, y) - i\Phi_2(x, y) \quad (13.26)$$

Replacing x by z and y by 0 in Eq. (13.26), we obtain

$$f'(z) = \Phi_1(z, 0) - i\Phi_2(z, 0)$$

$$\Rightarrow f(z) = [\Phi_1(z, 0) - i\Phi_2(z, 0)] + c_1$$

where c_1 is a complex constant.

Similarly if $v(x, y)$ is given, we can find u such that $u + iv$ is analytic.

$$f'(z) = \frac{\partial u}{\partial x} + i\frac{\partial v}{\partial x} = \frac{\partial v}{\partial y} + i\frac{\partial v}{\partial x} \quad \text{by CREs}$$

$$= \Psi_1(x, y) + i\Psi_2(x, y)$$

where $\frac{\partial v}{\partial y} = \Psi_1(x, y)$ and $\frac{\partial v}{\partial x} = \Psi_2(x, y)$

$$= \Psi_1(z, 0) + i\Psi_2(z, 0),$$

replacing x by z and y by 0.

Integrating we get

$$f(z) = \int [\Psi_1(z, 0) + i\Psi_2(z, 0)]dz + c_2$$

where c_2 is a complex constant.

Example 13.40

Verify that $u = x^2 - y^2 - y$ is harmonic in C and find a conjugate harmonic function v of u.

Solution $\nabla^2 u = 0$ by direct calculation. Now $u_x = 2x,\ u_y = -2y - 1$. Hence, because of CREs a conjugate v of u must satisfy the following:

$$v_y = u_x = 2x; \quad (1)$$

$$v_x = -u_y = 2_y + 1 \quad (2)$$

Integrating Eq. (1) w.r.t y and differentiating the result w.r.t x, we get

$$v = 2xy + h(x),$$

$$v_x = 2y + \frac{dh}{dx}$$

Comparison with Eq. (2) shows that $dh/dx = 1$ $\Rightarrow h(x) = x + c$. Hence $v = 2xy + x + c$ (c any real constant) is the most general conjugate harmonic of the given function u. The corresponding analytic function is

$$f(z) = u + iv = x^2 - y^2 - y + i(2xy + x + c)$$

$$= z^2 + iz + ic$$

The conjugate of a harmonic function is uniquely determined up to an arbitrary real additive constant.

Harmonic Functions

Example 13.41

If $u(x, y)$ and $v(x, y)$ are harmonic functions in a region R, prove that the function $\left(\frac{\partial u}{\partial y} - \frac{\partial v}{\partial x}\right) + i\left(\frac{\partial u}{\partial x} + \frac{\partial v}{\partial y}\right)$ is an analytic function.

[JNTU 2003*S*(1)]

Solution u and v are harmonic functions

$$\Rightarrow \frac{\partial^2 u}{\partial x^2} + \frac{\partial^2 u}{\partial y^2} = 0, \qquad \frac{\partial^2 v}{\partial x^2} + \frac{\partial^2 v}{\partial y^2} = 0 \quad (1)$$

$$\Rightarrow \frac{\partial^2 u}{\partial y^2} = -\frac{\partial^2 u}{\partial x^2}, \qquad \frac{\partial^2 v}{\partial y^2} = -\frac{\partial^2 v}{\partial x^2} \quad (2)$$

$$\text{Let} \quad U = \frac{\partial u}{\partial y} - \frac{\partial v}{\partial x}, \qquad V = \frac{\partial u}{\partial x} + \frac{\partial v}{\partial y} \quad (3), (4)$$

Differentiating Eqs. (3) and (4) partially w.r.t x and again w.r.t y we get

$$\frac{\partial U}{\partial x} = \frac{\partial^2 u}{\partial x \partial y} - \frac{\partial^2 v}{\partial x^2} \text{ and } \frac{\partial V}{\partial x} = \frac{\partial^2 u}{\partial x^2} + \frac{\partial^2 v}{\partial x \partial y} \quad (5), (6)$$

$$\frac{\partial U}{\partial y} = \frac{\partial^2 u}{\partial y^2} - \frac{\partial^2 v}{\partial y \partial x} \text{ and } \frac{\partial V}{\partial y} = \frac{\partial^2 u}{\partial y \partial x} + \frac{\partial^2 v}{\partial y^2} \quad (7), (8)$$

Now $\dfrac{\partial U}{\partial x}=\dfrac{\partial^2 u}{\partial x\partial y}+\dfrac{\partial^2 v}{\partial y^2}$ by Eq. (2)

$$=\frac{\partial}{\partial y}\left(\frac{\partial u}{\partial x}+\frac{\partial v}{\partial y}\right)=\frac{\partial V}{\partial y}$$

$$\frac{\partial U}{\partial y}=-\frac{\partial^2 u}{\partial x^2}-\frac{\partial^2 v}{\partial y\partial x} \quad \text{by Eq. (1)}$$

$$=-\frac{\partial}{\partial x}\left(\frac{\partial u}{\partial x}+\frac{\partial v}{\partial y}\right)=-\frac{\partial V}{\partial x}$$

$$\therefore \frac{\partial U}{\partial x}=\frac{\partial V}{\partial y},\quad \frac{\partial U}{\partial y}=-\frac{\partial V}{\partial x}.$$

Thus U, V satisfy CREs and their partial derivatives U_x, U_y, V_x, V_y are continuous.

Hence $U+iV=\left(\dfrac{\partial v}{\partial y}-\dfrac{\partial v}{\partial x}\right)+i\left(\dfrac{\partial u}{\partial x}+\dfrac{\partial v}{\partial y}\right)$ is an analytic function.

13.3.1 Harmonic and Conjugate Harmonic Functions

Example 13.42

Show that $u=2\log(x^2+y^2)$ is harmonic and find its harmonic conjugate. [JNTU 2006(1)]

Solution $u=2\log(x^2+y^2)$

$$\Rightarrow \frac{\partial u}{\partial x}=\frac{2}{x^2+y^2}\cdot 2x=\frac{4x}{x^2+y^2};\ \frac{\partial u}{\partial y}=\frac{4y}{x^2+y^2}$$

$$\frac{\partial^2 u}{\partial x^2}=\frac{(x^2+y^2)4-4x\cdot 2x}{(x^2+y^2)^2}=\frac{4(y^2-x^2)}{(x^2+y^2)^2}$$

$$\frac{\partial^2 u}{\partial y^2}=\frac{(x^2+y^2)4-4y\cdot 2y}{(x^2+y^2)^2}=\frac{-4(y^2-x^2)}{(x^2+y^2)^2}$$

Adding we get $\nabla^2 u=\dfrac{\partial^2 u}{\partial x^2}+\dfrac{\partial^2 u}{\partial y^2}=0$

$\Rightarrow u$ is an harmonic function.

If $v(x, y)$ is the harmonic conjugate of $u(x, y)$ then

$$dv=\frac{\partial v}{\partial x}dx+\frac{\partial v}{\partial y}dy=-\frac{\partial u}{\partial y}dx+\frac{\partial u}{\partial y}dy$$

(using CREs)

$$=\frac{-4y}{x^2+y^2}dx+\frac{4x}{x^2+y^2}dy$$

$$=\frac{4\left(\dfrac{xdy-ydx}{x^2}\right)}{\dfrac{x^2+y^2}{x^2}}=4\frac{d\left(\dfrac{y}{x}\right)}{1+\left(\dfrac{y}{x}\right)^2}$$

Integrating we get $v=4\tan^{-1}\dfrac{y}{x}+c.$

Example 13.43

Show that if u and v are conjugate harmonic functions then the product uv is a harmonic function.

Solution u, v are conjugate harmonic functions

$$\Rightarrow u_{xx}+u_{yy}=0,\quad v_{xx}+v_{yy}=0 \qquad (1), (2)$$

$$u_x=v_y,\quad u_y=-v_x \qquad (3), (4)$$

If $\Phi=uv$ then $\Phi_x=uv_x+vu_x$,

$$\Phi_{xx}=uv_{xx}+2u_xv_x+vu_{xx}$$

and $\Phi_{yy}=uv_{yy}+2u_yv_y+vu_{yy}$

$$=uv_{yy}-2v_xu_x+vu_{yy}, \text{ by CREs}$$

Now $\Phi_{xx}+\Phi_{yy}=u(v_{xx}+v_{yy})+v(u_{xx}+u_{yy})=0.$

Hence $\Phi=uv$ is a harmonic function.

Example 13.44

Show that $v(x, y)=-\sin x\sinh y$ is harmonic. Find the harmonic conjugate of v.

Solution Differentiating v partially w.r.t x and y, we get

$$v_x=-\cos x\sinh y,\quad v_{xx}=\sin x\sinh y$$

$$v_y=-\sin x\cosh y,\quad v_{yy}=-\sin x\sinh y$$

$$\nabla^2 v=v_{xx}+v_{yy}=0 \quad \therefore v \text{ is harmonic.}$$

To find u (harmonic conjugate of v), by CREs,

$$u_x=v_y=-\sin x\cosh y, \qquad (1)$$

$$u_y=-v_x=\cos x\sinh y \qquad (2)$$

Integrating Eq. (1) partially w.r.t x, we get

$$u(x, y)=\cos x\cosh y+h(y) \qquad (3)$$

where $h(y)$ is an arbitrary function of y.

Differentiating Eq. (3) partially w.r.t y, we get

$$u_y = \cos x \sinh y + h'(y) = -v_x = \cos x \sinh y$$
$$\Rightarrow h'(y) = 0 \Rightarrow h(y) = \text{constant}$$
$$\therefore u(x, y) = \cos x \cosh y + c$$

The required analytic function is

$$f(z) = \cos x \cosh y + c + i(-\sin x \sinh y)$$
$$= \cos(x + iy) + c = \cos z + c.$$

Example 13.45

Prove that $u = 2x - x^3 + 3xy^2$ is harmonic and find its harmonic conjugate. Find also the corresponding analytic function.

Solution Let $u = 2x - x^3 + 3xy^2$

$$\Rightarrow u_x = 2 - 3x^2 + 3y^2, \quad u_{xx} = -6x,$$
$$u_y = 6xy, \quad u_{yy} = 6x$$
$$\nabla^2 u = \frac{\partial^2 u}{\partial x^2} + \frac{\partial^2 u}{\partial y^2} = 0 \Rightarrow u \text{ is harmonic}$$

Let v be a harmonic conjugate of u so that $f(z) = u + iv$ is analytic.

By CREs we have $v_y = u_x = 2 - 3x^2 + 3y^2$

Integrating w.r.t y, we get

$$v = 2y - 3x^2 y + y^3 + A(x) \tag{1}$$

where $A(x)$ is an arbitrary function of x.

Differentiating Eq. (1) partially w.r.t x, we get $v_x = -6xy + A'(x) = -u_y = -6xy \Rightarrow A'(x) = 0$ so that $A(x) = c$ a pure complex constant

$$\Rightarrow v = 2y - 3x^2 y + y^3 + c$$

Now

$$f(z) = (2x - x^3 + 3xy^2) + i(2y - 3x^2 y + y^3) + ic$$
$$= 2(x + iy) - [(x^3 - 3xy^2) + i(3x^2 y - y^3)] + ic$$
$$= 2z - z^3 + ic.$$

Example 13.46

Show that $v(x, y) = -\sin x \sinh y$ is harmonic. Find the harmonic conjugate of v.

Solution Differentiating $v(x, y)$ partially w.r.t x and y, we get

$$v_x = -\cos x \sinh y, \quad v_{xx} = \sin x \sinh y \tag{1}$$
$$v_y = -\sin x \cosh y, \quad v_{yy} = -\sin x \sinh y \tag{2}$$

$\therefore$ v is a harmonic function in D

To find u, the harmonic conjugate of v:

$$\text{By CREs } u_x = v_y = -\sin x \cosh y, \tag{3(a)}$$
$$u_y = -v_x = \cos x \sinh y \tag{3(b)}$$

Integrating 3(a) partially w.r.t x, we get

$$u = \cos x \cosh y + h(y) \tag{4}$$

where $h(y)$ is an arbitrary function of y.

Now differentiating Eq. (4) w.r.t y, we get

$$u_y = \cos x \sinh y + h'(y)$$
$$= -v_x = \cos x \sinh y, \quad \text{by Eq. 3(b)}$$
$$\Rightarrow h'(y) = 0 \Rightarrow h(y) = c \text{ (const.)}$$

The required analytic function is

$$f(z) = u(x, y) + iv(x, y)$$
$$= \cos x \cosh y - i \sin x \sinh y + c$$
$$= \cos z + c$$

Milne–Thomson's method

Example 13.47

Find the analytic function $f(z) = u + iv$ where $u(x, y) = e^x(x \cos y - y \sin y) + 2\sin x \sinh y + x^3 - 3xy^2 + y$.

Solution Differentiating $u(x, y)$ partially w.r.t x and y, we get

$$u_x = e^x(x \cos y - y \sin y + \cos y) + 2\cos x \sinh y + 3x^2 - 3y^2$$
$$u_y = e^x(-x \sin y - \sin y - y \cos y) + 2\sin x \cosh y - 6xy + 1$$

We know that $f'(z) = u_x + iv_x = u_x - iu_y$,

By CREs $v_x = -u_y$

Replace x by z and y by 0 to obtain $f'(z)$

$$\therefore f'(z) = e^z(z \cdot 1 - 0 + 1) + 0 + 3z^2 - 0 - i[e^z(0 - 0 - 0) + 2\sin z \cdot 1 - 6z \cdot 0 + 1]$$
$$= e^z(z + 1) + 3z^2 - 2i \sin z - i$$

Integrating, we obtain

$$f(z) = ze^z + z^3 + 2i \cos z - iz + c.$$

Example 13.48

Show that $u(x, y) = \sin x \cosh y + 2\cos x \sinh y + x^2 - y^2 + 4xy$ is harmonic. Find the analytic function $f(z)$ with u as its real part.

Solution Differentiating u partially w.r.t x and y twice, we get

$$u_x = \cos x \cosh y - 2\sin x \sinh y + 2x + 4y$$
$$u_{xx} = -\sin x \cosh y - 2\cos x \sinh y + 2$$
$$u_y = \sin x \sinh y + 2\cos x \cosh y - 2y + 4x$$
$$u_{yy} = \sin x \cosh y + 2\cos x \sinh y - 2$$
$$\Rightarrow \nabla^2 u = \frac{\partial^2 u}{\partial x^2} + \frac{\partial^2 u}{\partial y^2} = 0$$

Hence u is harmonic.

Let $\Phi_1(x, y) = u_x$ and $\Phi_2(x, y) = u_y$

so that $\Phi_1(z, 0) = \cos z \cosh 0 - 2\sin z \sinh 0 + 2z$

$$= \cos z + 2z$$

and $\Phi_2(z, 0) = \sin z \sinh 0 + 2\cos z \cosh 0 - 2\times 0 + 4z$

$$= 2\cos z + 4z$$

$$\therefore f(z) = \int [\Phi_1(z, 0) - i\Phi_2(z, 0)]dz$$
$$= \int [\cos z + 2z - i(2\cos z + 4z)]dz$$
$$= \sin z + z^2 - 2i\sin z - 2iz^2 + c$$

by Milne–Thomson's method, where c is an arbitrary complex constant

Example 13.49

If $u(x, y) = \sin 2x/(\cosh 2y + \cos 2x)$ is the real part of analytic function $f(z) = u(x, y) + iv(x, y)$. Find $f(z)$.

Solution We can easily verify that $u_{xx} + u_{yy} = 0$ and hence u is harmonic. Differentiating u partially w.r.t x and y, we get

$$u_x = \frac{(\cosh 2y + \cos 2x)2\cos 2x + 2\sin^2 2x}{(\cosh^2 y + \cos 2x)^2}$$
$$= \frac{2\cosh 2y \cos 2x + 2}{(\cosh 2y + \cos 2x)^2}$$
$$u_y = \frac{-2\sin 2x \sinh 2y}{(\cosh 2y + \cos 2x)^2}$$

Let $\Phi_1(x, y) = u_x$ and $\Phi_2(x, y) = u_y$

$$\therefore \Phi_1(z, 0) = \frac{2\cos 2z \cosh 0 + 2}{(\cosh 0 + \cos 2z)^2}$$
$$= \frac{2}{1 + \cos 2z} = \sec^2 z$$

and $\Phi_2(z, 0) = 0$

$$f(z) = \int [\Phi_1(z, 0) - i\Phi_2(z, 0)]dz$$
$$= \int \sec^2 z dz = \tan z + c.$$

Example 13.50

Find the analytic function $f(z) = u + iv$ if

$$u + v = \frac{\sin 2x}{\cosh 2y - \cos 2x}.$$

Solution

$$u + v = \frac{\sin 2x}{\cosh 2y - \cos 2x} \quad (1)$$
$$\Rightarrow u_x + v_x = \frac{2(\cosh 2y - \cos 2x)\cos 2x - 2\sin^2 x}{(\cosh 2y - \cos 2x)^2} \quad (2)$$
$$u_y + v_y = \frac{-2\sin 2x \sinh 2y}{(\cosh 2y - \cos 2x)^2} \quad (3)$$

Since $f(z) = u + iv$ is an analytic function

$$u_x = v_y,\ u_y = -v_x \text{(CRES)}$$
$$(2) \Rightarrow u_x - u_y = \frac{-2(\cosh 2y - \cos 2x)\cos 2x - 2\sin^2 2x}{(\cosh 2y - \cos 2x)^2}$$

Replacing x by z and y by 0,

$$u_x(z, 0) - u_y(z, 0) = \frac{2(1 - \cos 2z)\cos 2z - 2\sin^2 2z}{(1 - \cos 2z)^2}$$
$$= \frac{2\cos 2z - 2(\cos^2 2z + \sin^2 2z)}{(1 - \cos 2z)^2}$$
$$= \frac{-2(1 - \cos 2z)}{(1 - \cos 2z)^2} = -\text{cosec}^2 z \quad (4)$$

$$\text{Eq. (3)} \Rightarrow u_y + u_x = \frac{-2\sin 2x \sinh 2y}{(\cosh 2y - \cos 2x)^2}$$

Replacing x by z and y by 0,

$$u_y(z, 0) + u_x(z, 0) = 0 \tag{5}$$

Adding Eqs. (4) and (5), we get

$$2u_x(z, 0) = -\text{cosec}^2 z$$

$$\text{or } u_x(z, 0) = -\frac{1}{2}\text{cosec}^2 z \tag{6}$$

Subtracting Eqs. (4) from (5), we get

$$2u_y(z, 0) = \text{cosec}^2 z$$

$$\text{or } \quad u_y(z, 0) = \frac{1}{2}\text{cosec}^2 z \tag{7}$$

Now $f(z) = u(z, 0) + iv(z, 0)$

$$\Rightarrow f'(z) = u_x(z, 0) + iv_x(z, 0)$$

$$\Rightarrow f'(z) = u_x(z, 0) - iv_y(z, 0)$$

$$= -\frac{1}{2}(1+i)\text{cosec}^2 z \text{ by Eqs. (6) and (7).}$$

Integrating w.r.t z, we obtain

$$f(z) = \frac{1}{2}(1+i)\cot z + c$$

where c is an arbitrary complex constant.

Example 13.51

If $u + v = (x - y)(x^2 + 4xy + y^2)$ find the analytic function $f(z) = u + iv$.

Solution

$$f(z) = u + iv$$

$$\Rightarrow (1+i)f(z) = (u - v) + i(u + v)$$

$$= U + iV \quad \text{(say)}$$

Now $V = u + v = (x - y)(x^2 + 4xy + y^2)$

$$\Rightarrow \left.\begin{aligned} V_x &= (x-y)(2x+4y) + x^2 \\ &\quad +4xy + y^2 = \Psi_2(x, y) \\ V_y &= (x-y)(4x+2y) \\ &\quad -(x^2+4xy+y^2) = \Psi_1(x, y) \end{aligned}\right\} \text{(say)}$$

$$\Rightarrow \Psi_1(z, 0) = 3z^2, \Psi_2(z, 0) = 3z^2$$

By Milne–Thomson's method,

$$(1+i)f(z) = \int[\Psi_1(z, 0) + i\Psi_2(z, 0)]dz$$

$$= (1+i)\int 3z^2 dz = (1+i)z^3 + c$$

$$\Rightarrow f(z) = z^3 + c$$

where c is an arbitrary complex constant.

Example 13.52

If $v(x, y) = x^4 - 6x^2y^2 + y^4$ find $f(z) = u(x, y) + iv(x, y)$ such that $f(z)$ is analytic.

Solution Differentiating v partially w.r.t x and y twice, we get

$$v_x = 4x^3 - 12xy^2 = -u_y \tag{1}$$

$$v_y = -12x^2y + 4y^3 = u_x \tag{2}$$

$$v_{xx} = 12x^2 - 12y^2 \tag{3}$$

$$v_{yy} = -12x^2 + 12y^2 \tag{4}$$

$$\Rightarrow \nabla^2 v = 0 \Rightarrow v \text{ is harmonic}$$

Integrating Eq. (2) w.r.t x, we get

$$u = -4x^3y + 4xy^3 + A(y) \tag{5}$$

where $A(y)$ is an arbitrary function of y. Differentiating w.r.t to y, we get

$$u_y = -4x^3 + 12xy^2 + A'(y) = -v_x$$

$$= -4x^3 + 12xy^2 \Rightarrow A'(y) = 0$$

$$A(y) = c = \text{constant}$$

$$\therefore u(x, y) = -4x^3y + 4xy^3 + c$$

Hence

$$f(z) = (-4x^3y + 4xy^3 + c) + i(x^4 - 6x^2y^2 + y^4)$$

$$= i[(x^4 - 6x^2y^2 + y^4) + i(4x^3y - 4xy^3)] + c$$

$$= i(x + iy)^4 + c = iz^4 + c$$

Milne–Thomson's method:

Let $\Psi_1(x, y) = v_y = -12x^2y + 4y^3$

and $\Psi_2(x, y) = v_x = 4x^3 - 12xy^2$

$$\Psi_1(z, 0) = 0 \quad \text{and} \quad \Psi_2(z, 0) = 4z^3$$

$$\therefore f(z) = \int[\Psi_1(z, 0) + i\Psi_2(z, 0)]dz$$

$$= i\int 4z^3 dz$$

$$f(z) = iz^4 + c$$

where c is an arbitrary complex constant.

Example 13.53

Show that $u = \sin x \cosh y + 2\cos x \sinh y + x^2 - y^2 + 4xy$ satisfies the Laplace equation. Find the corresponding analytic function. [JNTU 2008]

Solution It is given that

$$u(x, y) = \sin x \cosh y + 2\cos x \sinh y + x^2 - y^2 + 4xy \quad (1)$$

Differentiating Eq. (1) twice partially w.r.t x and y, respectively, we get

$$u_x = \cos x \cosh y - 2\sin x \sinh y + 2x + 4y \quad (2)$$

$$u_{xx} = -\sin x \cosh y - 2\cos x \sinh y + 2 \quad (3)$$

$$u_y = \sin x \sinh y + 2\cos x \cosh y - 2y + 4x \quad (4)$$

$$u_{yy} = \sin x \cosh y + 2\cos x \sinh y - 2 \quad (5)$$

Adding Eqs. (3) and (5), we obtain

$$\nabla^2 u = u_{xx} + u_{yy} = 0 \quad (6)$$

u satisfies Laplace's equation and hence u is a harmonic function.

$$\text{Also, } f'(z) = \frac{\partial u}{\partial x} + i\frac{\partial v}{\partial x} = \frac{\partial u}{\partial x} - i\frac{\partial u}{\partial y}$$

$$= \Phi_1(x, y) - i\Phi_2(x, y) \quad \text{(say)}$$

Replace x by z and y by 0 (Milne–Thomson's method)

$$\Rightarrow \cos x \cosh y - 2\sin x \sinh y + 2x + 4y - i(\sin x \sinh y + 2\cos x \cosh y - 2y + 4x)$$

$$= \Phi_1(z, 0) - i\Phi_2(z, 0) = \Phi(z) \quad \text{(say)}$$

$$= (\cos z \cdot 1 - 0 + 2z + 0) - i(\sin z \cdot 0 + 2\cos z \cdot 1 - 2 \times 0 + 4z)$$

$$\text{or } \Phi(z) = (1 - 2i)\cos z + 2(1 - 2i)z$$

Integrating w.r.t z,

$$f(z) = \int\Phi(z)dz + c = (1 - 2i)\sin z + (1 - 2i)z^2 + c.$$

Example 13.54

Find the analytic function whose imaginary part is $e^{-x}(x \cos y + y \sin y)$. [JNTU 1998S]

Solution Let $f(z) = u + iv$ be the analytic function. Then $v = e^{-x}(x \cos y + y \sin y)$

$$\text{We know that } f'(z) = u_x + iv_x = v_y + iv_x$$

$$= \Phi_1 + i\Phi_2 = \Phi(z) \quad \text{say}$$

Since $f(z) = u + iv \in H(D)$,

$$u_x = v_y, u_y = -v_x \quad \text{(CREs)}$$

$$\text{Now } v_x = e^{-x}(-x\cos y - y\sin y + \cos y)$$

$$= \Phi_2(z) = e^{-z}(-z + 1)$$

$$v_y = e^{-x}(-x\sin y + \sin y + y\cos y)$$

$$= \Phi_1(z) = e^{-z}(0)$$

$$\Phi(z) = \Phi_1(z, 0) + i\Phi_2(z, 0) = i(1 - z)e^{-z}$$

Integrating w.r.t z, we get

$$f(z) = \int\Phi(z)dz + c = ize^{-z} + c.$$

Example 13.55

Find the analytic function whose real part is $u = e^x[(x^2 - y^2)\cos y - 2xy\sin y]$. [JNTU 2008]

Solution

$$u = e^x[(x^2 - y^2)\cos y - 2xy\sin y]$$

$$\Rightarrow u_x = e^x[(x^2 - y^2)\cos y - 2xy\sin y + 2x\cos y - 2y\sin y] = \Phi_1(x, y)$$

$$u_y = e^x[-2y\cos y - (x^2 - y^2)\sin y - 2x\sin y - 2xy\cos y] = \Phi_2(x, y)$$

$$\therefore f'(z) = u_x + iv_x = u_x - iu_y$$

$$= \Phi_1(z, 0) - i\Phi_2(z, 0)$$

$$= e^z(z^2 + 2z) - e^z(0) = e^z(z^2 + 2z)$$

Integrating, $f(z) = z^2 e^z + c$

Example 13.56

If $f(z) = u + iv \in H(D)$ then prove that

(i) $\nabla^2 |f(z)|^2 = 4|f'(z)|^2$. [JNTU 1993]

(ii) $\nabla^2 u^p = p(p-1)u^{p-2}|f'(z)|^2$. [JNTU 1997S]

Solution Since $f(z) = v + iv$ is analytic in a domain D, the functions $u = u(x, y)$ and $v = v(x, y)$ satisfy Cauchy–Riemann equations in D

$$u_x = v_y \qquad u_y = -v_x \tag{1}$$

Further, u and v are harmonic functions in D which implies that u, v satisfy Laplace's equation $\nabla^2 \Phi = 0$, *i.e.*, $\nabla^2 u = 0$, $\nabla^2 u = 0$ in D

$$\left(\nabla^2 = \frac{\partial^2}{\partial x^2} + \frac{\partial^2}{\partial y^2}\right) \tag{2}$$

(i) $f(z) = u + iv \Rightarrow |f(z)|^2 = u^2 + v^2$

We calculate partial derivatives w.r.t x and y of $|f(z)|^2 = u^2 + v^2$

$$\frac{\partial}{\partial x}|f(z)|^2 = \frac{\partial}{\partial x}(u^2 + v^2) = 2uu_x + 2vv_x$$

$$\frac{\partial^2}{\partial x^2}|f(z)|^2 = \frac{\partial^2}{\partial x^2}(u^2 + v^2) = \frac{\partial}{\partial x}(2uu_x + 2vv_x)$$

$$= 2u_x^2 + 2uu_{xx} + 2v_x^2 + 2vv_{xx}$$

Similarly,

$$\frac{\partial^2}{\partial y^2}|f(z)|^2 = 2u_y^2 + 2uu_{yy} + 2v_y^2 + 2vv_{yy} \tag{3}$$

Adding the two results (3) and (4)

$$\left(\frac{\partial^2}{\partial x^2} + \frac{\partial^2}{\partial y^2}\right)|f(z)|^2 = 2(u_x^2 + v_x^2 + u_y^2 + v_y^2)$$

$$+ 2u\nabla^2 u + 2v\nabla^2 v \tag{4}$$

$$= 2(u_x^2 + v_x^2 + u_y^2 + v_y^2), \text{ by Eq. (2)}$$

$$= 4(u_x^2 + u_y^2), \text{ by Eq. (1)}$$

$$= 4|u_x - iu_y|^2$$

$$= 4|f'(z)|^2 \tag{5}$$

(ii) Differentiating u^p partially w.r.t x twice, we get

$$\frac{\partial}{\partial x}u^p = pu^{p-1}\frac{\partial u}{\partial x};$$

$$\frac{\partial^2}{\partial x^2}u^p = p(p-1)u^{p-2}\left(\frac{\partial u}{\partial x}\right)^2 + pu^p + \frac{\partial^2 u}{\partial x^2} \tag{6}$$

Similarly, we obtain

$$\frac{\partial^2}{\partial y^2}u^p = p(p-1)u^{p-2}\left(\frac{\partial u}{\partial y}\right)^2 + pu^{p-1}\frac{\partial^2 u}{\partial y^2} \tag{7}$$

Adding Eqs. (6) and (7), we obtain

$$\left(\frac{\partial^2}{\partial x^2} + \frac{\partial^2}{\partial y^2}\right)u^p$$

$$= p(p-1)u^{p-2}\left[\left(\frac{\partial u}{\partial x}\right)^2 + \left(\frac{\partial u}{\partial y}\right)^2\right]$$

$$+ pu^{p-1}\nabla^2 u$$

$$= p(p-1)u^{p-2}(u_x^2 + u_y^2) \because \nabla_u^2 = 0$$

$$= p(p-1)u^{p-2}|u_x - iu_y|^2$$

$$= p(p-1)u^{p-2}|f'(z)|^2.$$

EXERCISE 13.3

1. Prove that the following functions are harmonic. Find their harmonic conjugates.

(a) $u = x^2 - y^2 - y$,

Ans: $v = 2xy + x + c$.

(b) $v = x^2 - y^2 + \dfrac{x}{x^2 + y^2}$,

Ans: $u = -2xy + \dfrac{x}{x^2 + y^2} + c$.

(c) $u = 3xy^2 - x^3$,

Ans: $v = y^3 - 3x^2 y + c$

(d) $v = y^2 - x^2$,

Ans: $u = 2xy + c$

2. Find the analytic function $f(z) = u + iv$ in question 1.

Ans: (a) $z^2 + iz + c$

(b) $i\left(z^2 + \dfrac{1}{z}\right) + c$

(c) $-z^3 + ic$

(d) $-iz^2 + c$

3. Find the analytic function $f(z) = v + iv$ such that $u - v = \dfrac{e^y - \cos x + \sin x}{\cosh y - \cos x}$ and $f(\pi/2) = (3 - i)/2$.

   Ans: $f(z) = \cot(z/2) + (1 - i)/2$

4. Find the analytic function $f(z) = u + iv$ such that $u - v = \dfrac{\cos x + \sin x - e^{-y}}{2\cos x - e^y - e^{-y}}$ and $f(\pi/2) = 0$.

   Ans: $f(z) = -\dfrac{1}{2}\cot(z/2) + \dfrac{1}{2}\pi$

5. Find the analytic function $f(z) = u + iv$ such that $u + v = \dfrac{2\sin 2x}{e^{2y} - e^{-2y} - 2\cos 2x}$.

   Ans: $f(z) = \dfrac{-i}{1+i}\cot z + \dfrac{c}{1+i}$

6. If $f(z) = u + iv$ is an analytic function of $z = x - iy$ and $u - v = c^x(\cos y - \sin y)$ find $f(z)$ in terms of z. [JNTU 1995S]

   Ans: $f(z) = e^z + c(1 + i)$

7. Show that $u(x, y) = \sin x \cosh y + 2\cos x \sinh y + x^2 - y^2 + 4xy$ is harmonic. Find an analytic function $f(z)$ with u as its real point.

   Ans: $f(z) = \sin z + z^2 - 2i \sin z - 2iz^2 + c$

8. Find the real part of the analytic function whose imaginary part is e^{-x} $[2xy \cos y + (y^2 - x^2)\sin y]$. Find the analytic function.

   Ans: $u = e^{-x}[(x^2 - y^2)\cos y + 2xy \sin y]$

   $f(z) = e^{-z} \cdot z^2$

9. Prove that $u + iv$ is analytic if $u = \Phi_y - \Psi_x$ and $v = \Phi_x + \Psi_y$ where Φ and Ψ are harmonic function.

10. If $f(z)$ is analytic, prove that

    $$\left(\frac{\partial^2}{\partial x^2} + \frac{\partial^2}{\partial y^2}\right)|f(z)|^2 = 4|f'(z)|^2.$$

11. If $f(z) = u + iv$ is analytic prove that

    $$\left(\frac{\partial^2}{\partial x^2} + \frac{\partial^2}{\partial y^2}\right)\log|f(z)| = 0 \text{ and } \nabla^2 (\text{amp } f(z)) = 0.$$

12. From Laplace's equation for $u(x, y)$ prove that $\dfrac{\partial^2 u}{\partial z \partial \bar{z}} = 0$.

13. Show that $f(x, y) = x^3 y - xy^3 + xy + x + y$ can be the imaginary part of an analytic function of $z = x + iy$.

14. Find the analytic function $f(z) = u(r, \theta) + iv(r, \theta)$ such that

    (a) $u(r, \theta) = -r^3 \sin 3\theta$

    (b) $v(r, \theta) = \left(r - \dfrac{1}{r}\right)\sin\theta \ (r \neq 0)$.

    Ans: (a) $iz^3 + ic$

    (b) $\left(r + \dfrac{1}{r}\right)\cos\theta + \left(r - \dfrac{1}{r}\right)\sin\theta + c$

15. If $f(z) = u + iv = \dfrac{1}{z}$ show that the curves $u(x, y) = c_1$ and $v(x, y) = c_2$ intersect orthogonally.

Elementary Functions

14

14.1 Introduction

The reader is familiar with elementary functions studied in real calculus. One can define similar functions involving a complex variable $z = x + iy$ which reduce to the real-valued functions when we put $y = 0$ so that $z = x$ (real). Complex functions have a wide range of applications. Most of the properties of real functions, such as continuity and differentiability, hold in the case of complex functions. Some of them have interesting properties, not apparent in the case of real functions.

In this chapter, we define the elementary functions of a complex variable and study their properties. In the following, $z = x + iy$ and $w = u + iv$ in Cartesian coordinates, and $z = re^{i\theta}$ and $w = Re^{i\phi}$ in polars.

14.2 Elementary Functions of a Complex Variable

Complex functions $w = f(z)$ defined by the following rules are examples of elementary complex functions.

Power function: $f(z) = z^n \quad (n \in \mathbb{N}, z \in \mathbb{C})$

Polynomial function: $f(z) = a_0 + a_1 z + a_2 z^2 + \cdots + a_n z^n$ denoted by $P_n(z)$ $(a_r \in \mathbb{C}, a_n \neq 0)$ is a polynomial of degree n.

Rational function: $f(z) = P_n(z) / Q_m(z)$ $(Q_m \neq 0)$ where P_n and Q_m are polynomials.

Algebraic function: A function $w = f(z)$ which is a solution of the polynomial equation

$$\sum_{r=0}^{n} P_r(z) w^r \equiv P_0(z) + P_1(z) w + \cdots + P_{n-1} w^{n-1} + P_n(z) w^n = 0$$

where $P_n(z) \neq 0$, $P_r(z)$ $(r = 0, 1, \ldots, n)$ are polynomials in z and $n \in \mathbb{N}$ is called an algebraic function.

Polynomials and rational functions are special cases of algebraic functions.

Exponential, logarithmic, trigonometric and hyperbolic functions are other elementary functions.

14.2.1 Exponential Function

The exponential function of a complex variable $z = x + iy$ denoted by e^z or $\exp(z)$ is defined in terms of the real functions e^x, $\cos y$ and $\sin y$ as

$$e^z = e^{x+iy} = e^x \cdot e^{iy} = e^x(\cos y + i \sin y) \qquad (14.1)$$

This definition is motivated by the fact that e^z is a natural extension of the real exponential function e^x and reduces to it when we put $y = 0$ in Eq. (3.1).

Properties

1. If we write $w = f(z) = u + iv = f(x + iy) = e^x (\cos y + i \sin y)$, then

$$u = e^x \cos y \quad \text{and} \quad v = e^x \sin y \qquad (14.2)$$

Differentiating these partially w.r.t. x and y respectively, we get

$$\left.\begin{array}{ll} u_x = e^x \cos y & v_y = e^x \cos y \\ u_y = -e^x \sin y & v_x = e^x \sin y \end{array}\right\} \qquad (14.3)$$

From Eq. (14.3) it is clear that $u_x = v_y$ and $u_y = -v_x$; thus the Cauchy–Riemann equations (CREs) are satisfied for all x and y.

$\therefore e^z$ is analytic for all z. That is, e^z is an entire function. Indeed,

$$f'(z) = \frac{d}{dz} e^z = u_x + iv_x = e^x \cos y + ie^x \sin y$$
$$= e^x (\cos y + i \sin y)$$
$$= e^x \cdot e^{iy} = e^z \tag{14.4}$$

2. $f(z) = e^z \neq 0$ for any z and hence e^z has no zeros in $\mathbb{C}$. Also, $e^z = Re^{i\phi} \Rightarrow R = e^x > 0$ and $y = \phi$ and $|e^{iy}| = 1$. Hence $|e^z| = e^x \neq 0$.

3. Now, if we set $z = iy\ (x = 0)$ in Eq. (3.1) we obtain the Euler formula

$$e^{iy} = \cos y + i \sin y \tag{14.5}$$

Also, the polar form of the complex number $z = r(\cos\theta + i\sin\theta)$ may be written as $z = re^{i\theta}$. Taking $y = 2\pi$, we obtain

$$e^{2\pi i} = \cos 2\pi + i \sin 2\pi$$
$$= 1 + i \cdot 0 = 1 \tag{14.6}$$

Again, taking $y = \frac{\pi}{2}, \pi, -\frac{\pi}{2}$ and $-\pi$ in turn, we obtain the following results:

$$e^{\frac{\pi i}{2}} = \cos\frac{\pi}{2} + i\sin\frac{\pi}{2} = 0 + i \cdot 1 = i$$
$$e^{\pi i} = \cos\pi + i\sin\pi = -1 + i \cdot 0 = -1$$
$$e^{\frac{-\pi}{2}i} = \cos\left(-\frac{\pi}{2}\right) + i\sin\left(-\frac{\pi}{2}\right)$$
$$= 0 + (-1) \cdot i = -i$$
$$e^{-\pi i} = \cos(-\pi) + i\sin(-\pi)$$
$$= -1 + 0 \cdot i = -1$$

Also, $e^{2n\pi i} = \cos 2n\pi + i \sin 2n\pi = 1 + i \cdot 0 = 1$ for all $n \in \mathbb{Z}$.

From the definition of e^z, we have

$$e^{z_1 + z_2} = e^{z_1} \cdot e^{z_2} \quad z_1, z_2 \in \mathbb{C} \tag{14.7}$$

Let $z_1 = x_1 + iy_1$ and $z_2 = x_2 + iy_2$

$$\text{RHS} = e^{z_1} \cdot e^{z_2} = e^{x_1 + iy_1} \cdot e^{x_2 + iy_2}$$
$$= e^{x_1}(\cos y_1 + i \sin y_1) \cdot e^{x_2}(\cos y_2 + i \sin y_2)$$
$$= e^{x_1} \cdot e^{x_2}(\cos y_1 + i\sin y_1)(\cos y_2 + i \sin y_2)$$
$$= e^{x_1 + x_2}[(\cos y_1 \cos y_2 - \sin y_1 \sin y_2)$$
$$+ i(\sin y_1 \cos y_2 + \cos y_1 \sin y_2)]$$
$$= e^{x_1 + x_2}[\cos(y_1 + y_2) + i \sin(y_1 + y_2)]$$
$$= e^{x_1 + x_2} \cdot e^{i(y_1 + y_2)} = e^{(x_1 + iy_1) + (x_2 + iy_2)}$$
$$= e^{z_1 + z_2}$$
$$= \text{LHS}$$

Similarly we can prove that

$$e^{z_1 - z_2} = e^{z_1} \cdot e^{-z_2} = \frac{e^{z_1}}{e^{z_2}} \quad (\because e^{z_2} \neq 0) \tag{14.8}$$

$$(e^z)^n = e^{nz} \text{ for any } n \in \mathbb{Z} \text{ (by induction)} \tag{14.9}$$

From Eq. (3.5), we have

$$|e^{iy}| = |\cos y + i \sin y|$$
$$= \sqrt{(\cos y)^2 + (\sin y)^2} = 1 \tag{14.10}$$

That is, for purely imaginary exponents the exponential function has an absolute value of unity.

Also, from Eqs. (3.1) and (3.11)

$$|e^z| = |e^x||e^{iy}| = e^x \cdot 1 = e^x \tag{14.11}$$

Hence arg $z = y \pm 2n\pi, n = 0, 1, 2, 3, \ldots$ since the result $|e^z| = e^x$ shows that Eq. (3.1) is actually e^z in polar form.

From $|e^z| = e^x \neq 0$ in Eq. (3.12), we see that

$$e^z \neq 0 \qquad \text{for all } z \in \mathbb{C} \tag{14.12}$$

Thus, the exponential function of e^z is an entire function which never vanishes. This is in contrast to non-constant polynomials which are also entire but always have a zero as proved by the fundamental theorem of algebra.

Periodicity of e^z

We have $e^z = e^x(\cos y + i \sin y)$. Since cos and sin functions are periodic with period 2π, *i.e.*,

$$\cos(y + 2\pi) = \cos y \text{ and } \sin(y + 2\pi) = \sin y$$

we note that

$$e^{z + 2\pi i} = e^{x + iy + 2\pi i} = e^{x + i(y + 2\pi)}$$
$$= e^x \cdot e^{i(y + 2\pi)}$$
$$= e^x[\cos(y + 2\pi) + i \sin(y + 2\pi)]$$
$$= e^x(\cos y + i \sin y) = e^x \cdot e^{iy} = e^{x + iy} = e^z$$

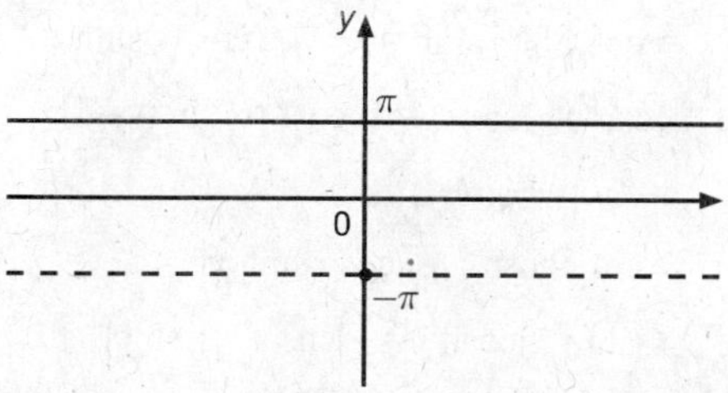

Figure 14.1

Thus we find that the exponential function e^z is a periodic function with period $2\pi i$ (an imaginary number).

Consider an infinite horizontal strip of width 2π (Fig. 14.1),

$$-\pi < y < \pi.$$

In the z-plane, we see that all values that the mapping $w = f(z) = e^z$ can assume, are assumed in the above horizontal strip once. This infinite strip is called a fundamental region of e^z.

In geometrical terms, e^z maps a fundamental region bijectively onto the entire complex plane.

14.2.2 Trigonometric Functions

We are familiar with the Euler formulas

$$e^{ix} = \cos x + i\sin x, \tag{14.13}$$

$$e^{-ix} = \cos x - i\sin x \tag{14.14}$$

Adding and subtracting Eqs. (3.13) and (3.14), we obtain

$$\cos x = \frac{e^{ix} + e^{-ix}}{2}, \tag{14.15}$$

$$\sin x = \frac{e^{ix} - e^{-ix}}{2i} \tag{14.16}$$

where x is real.

On the same lines, we define $\cos z$ and $\sin z$ for complex values $z = x + iy$ by

$$\cos z = \frac{e^{iz} + e^{-iz}}{2}, \tag{14.17}$$

$$\sin z = \frac{e^{iz} - e^{-iz}}{2i} \tag{14.18}$$

which reduce to Eqs. (14.15) and (14.16) when we put $y = 0$.

Further, as in calculus, we define

$$\tan z = \frac{\sin z}{\cos z} \tag{14.19}$$

$$\sec z = \frac{1}{\cos z} \text{ if } \cos z \neq 0, \tag{14.20}$$

$$\cot z = \frac{\cos z}{\sin z}, \tag{14.21}$$

$$\operatorname{cosec} z = \frac{1}{\sin z}, \text{ if } \sin z \neq 0 \tag{14.22}$$

Using these definitions one can easily prove that all identities which hold for real x are true for complex z.

Since e^z is entire $\cos z$ and $\sin z$ are entire. Tan z and $\sec z$ are not entire; they are analytic at all z except at the points where $\cos z = 0$. Also, $\cot z$ and $\operatorname{cosec} z$ are analytic at all points except where $\sin z = 0$.

The derivations of $\cos z$ and $\sin z$:

$$\frac{d}{dz}(\cos z) = \frac{d}{dz}\left(\frac{e^{iz} + e^{-iz}}{2}\right) = -i\left(\frac{e^{iz} - e^{-iz}}{2}\right)$$

$$= -\left(\frac{e^{iz} - e^{-iz}}{2i}\right) = -\sin z \tag{14.23}$$

$$\frac{d}{dz}(\sin z) = \frac{d}{dz}\left(\frac{e^{iz} - e^{-iz}}{2i}\right)$$

$$= i\left(\frac{e^{iz} + e^{-iz}}{2i}\right)$$

$$= \frac{e^{iz} + e^{-iz}}{2} = \cos z .2 \tag{14.24}$$

By the quotient rule, we have

$$\frac{d}{dz}(\tan z) = \frac{d}{dz}\left(\frac{\sin z}{\cos z}\right)$$

$$= \frac{\cos z(\cos z) - \sin z(-\sin z)}{\cos^2 z}$$

$$= \frac{1}{\cos^2 z} = \sec^2 z \tag{14.25}$$

$$\frac{d}{dz}(\cot z) = \frac{d}{dz}\left(\frac{\cos z}{\sin z}\right)$$

$$= \frac{\sin z(-\sin z) - \cos z(\cos z)}{\sin^2 z}$$

$$= \frac{-1}{\sin^2 z} = -\operatorname{cosec}^2 z \tag{14.26}$$

Euler's formula is valid for complex variable

$$e^{iz} = \cos z + i\sin z \quad \text{for all } z \in \mathbb{C} \qquad (14.27)$$

We now find real and imaginary parts of $\cos z$ and $\sin z$, and also the absolute values and periodicity of these functions.

Example 14.1

Show that

$$\cos z = \cos x\cosh y - i\sin x\sinh y \qquad (1)$$

$$\sin z = \sin x\cosh y + i\cos x\sinh y \qquad (2)$$

and

$$\left.\begin{aligned}|\cos z|^2 &= \cos^2 x + \sinh^2 y\\ |\sin z|^2 &= \sin^2 x + \sinh^2 y\end{aligned}\right\}. \qquad (3)$$

Solution We have,

$$\cos z = \frac{e^{iz} + e^{-iz}}{2}$$
$$= \frac{1}{2}[e^{i(x+iy)} + e^{-i(x+iy)}]$$
$$= \frac{1}{2}[e^{ix}\cdot e^{-y} + e^{-ix}\cdot e^{y}]$$
$$= \frac{1}{2}e^{-y}(\cos x + i\sin x) + \frac{1}{2}e^{y}(\cos x - i\sin x)$$
$$= \frac{1}{2}(e^{y} + e^{-y})\cos x - i\frac{1}{2}(e^{y} - e^{-y})\sin x$$
$$= \cosh y\cdot\cos x - i\sinh y\sin x$$
$$= \cos x\cosh y - i\sin x\sinh y$$

$$\therefore \cosh y = \frac{e^{y} + e^{-y}}{2}, \quad \sinh y = \frac{e^{y} - e^{-y}}{2}$$

Again,

$$\sin z = \frac{e^{iz} - e^{-iz}}{2i} = \frac{1}{2i}\left[e^{i(x+iy)} - e^{-i(x+iy)}\right]$$
$$= \frac{-i}{2}\left[e^{ix}\cdot e^{-y} - e^{-ix}\cdot e^{y}\right]$$
$$= \frac{-i}{2}e^{-y}(\cos x + i\sin x) + \frac{i}{2}e^{y}(\cos x - i\sin x)$$
$$= \sin x\left(\frac{e^{y} + e^{-y}}{2}\right) + i\cdot\cos x\left(\frac{e^{y} - e^{-y}}{2}\right)$$
$$= \sin x\cosh y + i\cos x\sinh y$$

$$|\cos z|^2 = |\cos x\cosh y - i\sin x\sinh y|^2$$
$$= |\cos x|^2|\cosh y|^2 + |\sin x|^2|\sinh y|^2$$
$$= \cos^2 x(1 + \sinh^2 y) + (1 - \cos^2 x)\sinh^2 y$$
$$[\because \cosh^2 y - \sinh^2 = 1, \cos^2 x + \sin^2 x = 1]$$
$$= \cos^2 x + \sinh^2 y \qquad (4)$$

$$|\sin z|^2 = |\sin x\cosh y + i\cos x\sinh y|^2$$
$$= |\sin x|^2|\cosh y|^2 + |\cos x|^2|\sinh y|^2$$
$$= \sin^2 x(1 + \sinh^2 y) + (1 - \sin^2 x)\sinh^2 y$$
$$= \sin^2 x + \sinh^2 y \qquad (5)$$

$\cos z$ and $\sin z$ are periodic with period 2π

$$\cos(z + 2\pi) = \cos(\overline{x + 2\pi} + iy)$$
$$= \cos(x + 2\pi)\cosh y - i\sin(x + 2\pi)\sinh y$$
$$= \cos x\cosh y - i\sin x\sinh y$$
$$= \cos z \qquad (6)$$

$$\sin(z + 2\pi) = \sin(\overline{x + 2\pi} + iy)$$
$$= \sin(x + 2\pi)\cosh y + i\cos(x + 2\pi)\sinh y$$
$$= \sin x\cosh y + i\cos x\sinh y$$
$$= \sin z \qquad (7)$$

These results show that $\cos z$ and $\sin z$ are periodic with period 2π. Sec z and cosec z which are reciprocals of $\cos z$ and $\sin z$, respectively, are also periodic with period 2π. Tan z and $\cot z$ are periodic with period π.

$$\tan(z + \pi)$$
$$= \frac{\sin(z + \pi)}{\cos(z + \pi)}$$
$$= \frac{\sin(\overline{x + \pi} + iy)}{\cos(\overline{x + \pi} + iy)}$$
$$= \frac{\sin(x + \pi)\cosh y + i\cos(x + \pi)\sinh y}{\cos(x + \pi)\cosh y - i\sin(x + \pi)\sinh y}$$
$$= \frac{-\sin x\cosh y - i\cos x\sinh y}{-\cos x\cosh y + i\sin x\sinh y}$$
$$= \frac{\sin x\cosh y + i\cos x\sinh y}{\cos x\cosh y - i\sin x\sinh y}$$
$$= \frac{\sin z}{\cos z} = \tan z$$

$$\begin{aligned}
&\cot(z+\pi)\\
&=\frac{\cos(z+\pi)}{\sin(z+\pi)}\\
&=\frac{\cos(x+\pi+iy)}{\sin(x+\pi+iy)}\\
&=\frac{\cos(x+\pi)\cosh y - i\sin(x+\pi)\sinh y}{\sin(x+\pi)\cosh y + i\cos(x+\pi)\sinh y}\\
&=\frac{-\cos x\cosh y + i\sin x\sinh y}{-\sin x\cosh y - i\cos x\sinh y}\\
&\times\frac{\cos x\cosh y - i\sin x\sinh y}{\sin x\cosh y + i\cos x\sinh y}\\
&=\frac{\cos z}{\sin z}=\cot z
\end{aligned}$$

From the formulas (4) and (5), we see that there is an essential difference between real and complex cosine and sine functions.

We know that $|\cos x|\le 1$ and $|\sin x|\le 1$ which show that real cosine and sine functions are bounded whereas their complex counterparts are no longer bounded. They approach infinity in absolute value as $y\to\infty$ since $\sinh y=\dfrac{e^y-e^{-y}}{2}$ which is present in $|\cos z|$ and $|\sin z|$ [formulas (4) and (5)] tend to ∞ as $y\to\infty$.

Zeros of cos z and sin z

We have

$$|\cos z|^2=\cos^2 x+\sinh^2 y,$$

$$|\sin z|^2=\sin^2 x+\sinh^2 y$$

$$|\cos z|=0\Leftrightarrow \cos x=0, \sinh y=0$$

$$\Leftrightarrow x=(2n+1)\frac{\pi}{2},\quad y=0\quad n\in Z$$

$$|\sin z|=0\Leftrightarrow \sin x=0,\ \sinh y=0$$

$$\Leftrightarrow x=n\pi,\quad y=0\quad n\in Z$$

The zeros of $\cos z$ and $\sin z$ are

$$z=z_n=(x_n,0)=\left(n\frac{\pi}{2},0\right)$$

according as n is odd or even integer.

Thus the zeros of $\cos z$ and $\sin z$ are those of real cosine and sine functions, respectively.

General formulas for the real trigonometric functions continue to hold for complex values. Thus

$$\cos(z_1+z_2)=\cos z_1\cos z_2-\sin z_1\sin z_2$$
$$\cos(z_1-z_2)=\cos z_1\cos z_2+\sin z_1\sin z_2$$
$$\sin(z_1+z_2)=\sin z_1\cos z_2+\cos z_1\sin z_2$$
$$\sin(z_1-z_2)=\sin z_1\cos z_2-\cos z_1\sin z_2$$
$$\cos^2 z+\sin^2 z=1$$

Solution of trigonometric equations

Example 14.2

Solve $\cos z=5$

Solution $\cos z=5$

$$\Rightarrow \frac{(e^{iz}+e^{-iz})}{2}=5$$

$$\Rightarrow e^{iz}+e^{-iz}=10$$

$$\Rightarrow t^2-10t+1=0$$

where $t=e^{iz}=e^{-y+ix}$

Solving the quadratic in *t*, we have

$$\begin{aligned}
t&=e^{-y+ix}\\
&=5\pm\sqrt{24}\\
&=5\pm 2\sqrt{6}\\
&=9.899 \text{ or } 0.101
\end{aligned}$$

$$\Rightarrow e^{-y}=9.899 \text{ or } 0.101;\quad e^{ix}=1$$

$$y=\pm 2.292, x=2n\pi; z=2n\pi\pm 2.292i$$

14.2.3 Hyperbolic Functions

The complex hyperbolic cosine and sine are defined by

$$\cosh z=\frac{e^z+e^{-z}}{2},\quad \sinh z=\frac{e^z-e^{-z}}{2} \tag{14.28}$$

Observe that these formulas are the same as their counterparts for real variables. These have derivatives

$$(\cosh z)'=\sinh z$$

$$(\sinh z)'=\cosh z\quad \text{for all } z \tag{14.29}$$

as in calculus. So, they are entire functions.

The other hyperbolic functions are defined by

$$\tanh z = \frac{\sinh z}{\cosh z}, \quad \coth z = \frac{\cosh z}{\sinh z}$$

$$\text{sech}\, z = \frac{1}{\cosh z}, \quad \text{cosec}h\, z = \frac{1}{\sinh z} \qquad (14.30)$$

Relation between complex trigonometric and hyperbolic functions

Replacing z by iz in Eq. (3.28), we get

$$\cosh(iz) = \frac{e^{iz} + e^{-iz}}{2} = \cos z$$

$$\sinh(iz) = \frac{e^{iz} - e^{-iz}}{2} = i \sin z \qquad (14.31)$$

Further, from trigonometric formulas

$$\cos z = \frac{e^{iz} + e^{-iz}}{2}, \quad \sin z = \frac{e^{iz} - e^{-iz}}{2i}$$

we get

$$\cos(iz) = \frac{e^{-z} + e^{z}}{2} = \cosh z$$

$$\sin(iz) = \frac{e^{-z} - e^{z}}{2i}$$

$$= i \cdot \frac{e^{z} - e^{-z}}{2}$$

$$= i \sinh z \qquad (14.32)$$

In real calculus, trigonometric and hyperbolic functions are *not related* but their complex analogues are *related*. This shows the advantage of working in complex variables. The theory of complex variables gives a more unified and deeper understanding of special functions. Table 14.1 gives a comparison of complex trigonometric and hyperbolic functions.

Formulas for hyperbolic functions

Prove that

1. (i) $\cosh z = \cosh x \cos y + i \sinh x \sin y$
   (ii) $\sinh z = \sinh x \cos y + i \cosh x \sin y$
2. (i) $\cosh(z_1 + z_2)$
   $= \cosh z_1 \cosh z_2 + \sinh \dot{z}_1 \sinh z_2$
   (ii) $\sinh(z_1 + z_2)$
   $= \sinh z_1 \cosh z_2 + \cosh z_1 \sinh z_2$
3. (i) $\cosh^2 z - \sinh^2 z = 1$
   (ii) $\cosh^2 z + \sinh^2 z = \cosh 2z \qquad (14.33)$

Solution

1. (i)
$$\cosh z = \frac{e^z + e^{-z}}{2}$$
$$= \frac{1}{2}\left[e^{x+iy} + e^{-x-iy}\right]$$
$$= \frac{1}{2} e^x (\cos y + i \sin y) + \frac{1}{2} e^{-x} (\cos y - i \sin y)$$
$$= \left(\frac{e^x + e^{-x}}{2}\right) \cos y + i\left(\frac{e^x - e^{-x}}{2}\right) \sin y$$
$$= \cosh x \cos y + i \sinh x \sin y$$

(ii)
$$\sinh z = \frac{e^z - e^{-z}}{2}$$
$$= \frac{1}{2}\left[e^{x+iy} - e^{-x-iy}\right]$$
$$= \frac{1}{2} e^x (\cos y + i \sin y) - \frac{1}{2} e^{-x} (\cos y - i \sin y)$$
$$= \left(\frac{e^x - e^{-x}}{2}\right) \cos y + i\left(\frac{e^x + e^{-x}}{2}\right) \sin y$$
$$= \sinh x \cos y + i \cosh x \sin y$$

2.(i)
$$\text{RHS} = \frac{e^{z_1} + e^{-z_1}}{2} \cdot \frac{e^{z_2} + e^{-z_2}}{2} + \frac{e^{z_1} - e^{-z_1}}{2} \cdot \frac{e^{z_2} - e^{-z_2}}{2}$$
$$= \frac{1}{4}\Big[\left(e^{z_1+z_2} + e^{-(z_1+z_2)} + e^{z_1-z_2} + e^{-(z_1-z_2)}\right) + \left(e^{z_1+z_2} + e^{(z_1+z_2)} - e^{z_1-z_2} - e^{-(z_1-z_2)}\right)\Big]$$
$$= \frac{1}{2}\left[e^{z_1 z_2} + e^{-(z_1+z_2)}\right]$$
$$= \cosh(z_1 + z_2) = \text{LHS}$$

Table 14.1 Relation between Complex Trigonometric and Hyperbolic Functions

$\cos z = \dfrac{e^{iz}+e^{-iz}}{2} = \cosh(iz)$	$\cosh t = \dfrac{e^{t}+e^{-t}}{2} = \dfrac{e^{-t}+e^{t}}{2}$
	$= \dfrac{e^{i(it)}+e^{-i(it)}}{2} = \cos(it)$
$\sin z = \dfrac{e^{iz}-e^{-iz}}{2i} = \dfrac{1}{i}\sinh(iz)$	$\sinh t = \dfrac{e^{t}-e^{-t}}{2} = -\dfrac{e^{-t}-e^{t}}{2}$
	$= \dfrac{1}{i}\dfrac{e^{i(it)}-e^{-i(it)}}{2i} = \dfrac{1}{i}\sin(it)$
$\tan z = \dfrac{\sin z}{\cos z} = \dfrac{1}{i}\tanh(iz)(\cos z \neq 0)$	$\tanh t = \dfrac{1}{i}\tan(it)$
$\cot z = \dfrac{\cos z}{\sin z} = i\coth(iz)(\sin z \neq 0)$	$\coth t = i\cot(it)$
$\sec z = \dfrac{1}{\cos z} = \operatorname{sech}(iz)(\cos z \neq 0)$	$\operatorname{sech} t = \sec(it)$
$\operatorname{cosec} z = \dfrac{1}{\sin z} = i\operatorname{cosech}(iz)(\sin z \neq 0)$	$\operatorname{cosech} t = i\operatorname{cosec}(\text{it})$

(ii) $$\text{RHS} = \frac{e^{z_1}-e^{-z_1}}{2}\cdot\frac{e^{z_2}+e^{-z_2}}{2} + \frac{e^{z_1}+e^{-z_1}}{2}\cdot\frac{e^{z_2}-e^{-z_2}}{2}$$

$$= \frac{1}{4}\Big[e^{z_1+z_2} - e^{-(z_1+z_2)} + e^{z_1-z_2} - e^{-(z_1-z_2)} + e^{z_1+z_2} - e^{-(z_1+z_2)} - \left(e^{z_1-z_2} - e^{-(z_1-z_2)}\right)\Big]$$

$$= \frac{e^{z_1+z_2} - e^{-(z_1+z_2)}}{2}$$

$$= \sinh(z_1+z_2) = \text{LHS}$$

3. (i) $$\text{LHS} = \left(\frac{e^{z}+e^{-z}}{2}\right)^2 - \left(\frac{e^{2}-e^{-z}}{2}\right)^2$$

$$= 4\cdot\frac{e^{z}}{2}\cdot\frac{e^{-z}}{2} = 1 = \text{RHS}$$

$$\because (a+b)^2 - (a-b)^2 = 4ab$$

(ii) $$\text{LHS} = \left(\frac{e^{2}+e^{-z}}{2}\right)^2 + \left(\frac{e^{2}-e^{-z}}{2}\right)^2$$

$$= 2\left(\frac{e^{2z}+e^{-2z}}{4}\right)$$

$$= \cosh 2z = \text{RHS}$$

$$\because\ (a+b)^2 + (a+b)^2 = 2(a^2+b^2)$$

14.2.4 Logarithm

We now introduce complex logarithm. This is more complicated than its real counterpart, which is a special case of the complex logarithm.

The natural logarithm of $z = x+iy$ is defined as the inverse of the exponential function and is denoted by $\ln z$ or $\log z$.

$$w = \ln z$$

is defined for $z \neq 0$ by

$$e^{w} = z \tag{14.34}$$

Since $e^{w} \neq 0$ for all w, $z = 0$ is not possible. Setting $w = u+iv$ and $z = re^{i\theta}$, we get

$$e^{w} = e^{u+iv} = re^{i\theta} \tag{14.35}$$

Table 14.2

Natural Log	Principal Value
$\ln 1 = 0, \quad \pm 2\pi i, \pm 4\pi i, \ldots$	$\text{Ln } 1 = 0$
$\ln 8 = 2.079442 \pm 2n\pi i$	$\text{Ln } 8 = 2.079442$
$\ln(-1) = \pm\pi i, \pm 3\pi i, \pm 5\pi i, \ldots$	$\text{Ln } (-1) = \pi i$
$\ln(-8) = 2.079442 \pm (2n+1)\pi i,$	$\text{Ln } (-8) = 2.079442 + \pi i$
$\ln i = \frac{\pi i}{2}, -\frac{3}{2}\pi i, \frac{5}{2}\pi i, \ldots$	$\text{Ln } i = \frac{\pi i}{2}$
$\ln 8i = 2.079442 + \frac{\pi i}{2} \pm 2n\pi i$	$\text{Ln } 8i = 2.079442 \pm \frac{\pi i}{2}$
$\ln(-8i) = 2.079442 - \frac{\pi i}{2} \pm 2n\pi i$	$\text{Ln } (-8i) = 2.079442 - \frac{\pi i}{2}$
$\ln(6-8i) = \ln 10 + i \arg(6-8i)$	$\text{Ln } (6-8i) = 2.302585 - 0.927295i$
$= 2.302585 - 0.927295i + 2n\pi i$	

Now we note that e^w has the absolute value e^u and the argument v. Equating the absolute values and the arguments on either side, we have

$$e^u = r \quad \text{and} \quad v = \theta \qquad (14.36)$$

$e^u = r \Rightarrow u = \ln r$, the real logarithm of the positive number $r = |z|$. Hence, $w = u + iv = \ln z$ is given by

$$1. \ln z = \ln r + i\theta,\ r = |z| > 0,\ \theta = \arg z \qquad (14.37)$$

Note

Since the argument of z is determined only up to integer multiples of 2π, the complex natural logarithm $\ln z\ (z \neq 0)$ is infinitely many-valued.

The value of $\ln z$ corresponding to the principal value $\arg z$ such that $-\pi < \arg z \leq \pi$ is denoted by $\text{Ln } z$ (note the capital L) and is called the principal value of $\ln z$. Thus

$$2. \text{Ln } z = \ln |z| + i \arg z \quad (z \neq 0) \qquad (14.38)$$

This $\ln z$ is single-valued and it is therefore a function in the usual sense.

Since the other values of $\ln z$ differ by the integer multiples of 2π, they are given by

$$3. \ln z = \ln z \pm 2n\pi i \quad n = 1, 2, 3, \ldots \qquad (14.39)$$

They all have the same real part and their imaginary parts differ by integer multiples of 2π. These points can be seen from the graph of $\text{Ln } (6-8i)$ shown for some values in Fig. 14.2.

If z is positive real, then $\text{Arg } z = 0$ and $\text{Ln } z$ becomes identical with the real natural logarithm of real calculus.

If z is negative real (for which the natural log of calculus is not defined) we have $\text{Arg } z = \pi$ and

$$\text{Ln } z = \ln |z| + \pi i \qquad (14.40)$$

Values of natural logarithm and its principal values for some specified numerical values are centered in Table 14.2.

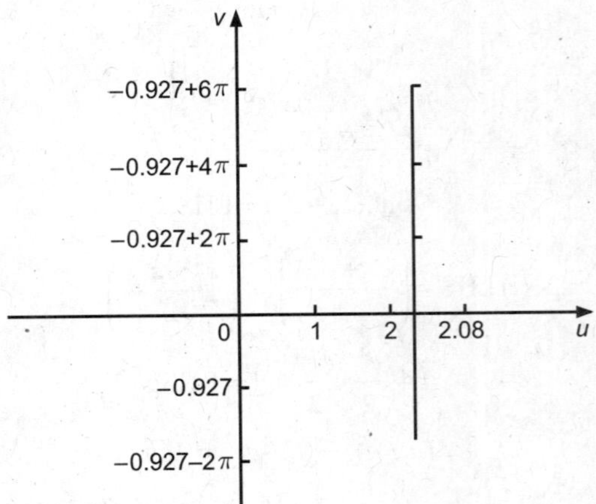

Some values of Ln (6–8*i*)

Figure 14.2

The following familiar laws for real natural logarithm continues to hold for complex values:

4. (i) $\ln(z_1 z_2) = \ln z_1 + \ln z_2$
(ii) $\ln(z_1/z_2) = \ln z_1 - \ln z_2$ (14.41)

Caution: These relations are to be understood in the sense that the cosh value of one side is also contained in the value of the another side.

Illustration of the functional relations

Let

$$z_1 = z_2 = e^{\pi i} = -1$$

Taking the principal values

$$\ln z_1 = \text{Ln } z_2 = \pi i$$

Result 4(i) holds provided we write $\ln z_1 z_2 = \ln 1 = 2\pi$. It is not true for the principal value $\ln z_1 z_2 = \ln 1 = 0$

From Equation (14.34),

$$\ln z = \ln r + i\theta \quad r = |z| > 0, \theta = \arg z$$

we have

$$e^{\ln r} = r$$

for positive real r and so we obtain

$$e^{\ln z} = z$$

as expected, but since $\arg(e^z) = y \pm 2n\pi$ is multi-valued so is

$$\ln e^z = z \pm 2n\pi i \quad n = 0,1,2,\ldots$$

For every fixed non-negative integer n formula

$$\ln z = \ln z \pm 2n\pi i \quad n = 1,2,3,\ldots$$

defines a function. We prove that each of these functions, in particular, the principal value Ln z is analytic except at $z = 0$ and on the negative real axis (where the imaginary part of such a fraction is not even continuous but has a jump of magnitude 2π). We do this by proving that

$$(\ln z)' = \frac{1}{z}$$

In (1), we have

$$\ln z = u + iv$$

where

$$u = \ln r = \ln |z| = \frac{1}{2}\ln(x^2 + y^2)$$

$$v = \arg z = \text{arc tan}\frac{y}{x} + C$$

where C is a constant which is a multiple of $2n\pi$.

We calculate the partial derivatives of u and v and observe that they satisfy the CREs:

$$u_x = \frac{x}{x^2 + y^2};$$

$$v_y = \frac{1}{1+\left(\frac{y}{x}\right)^2} \cdot \frac{1}{x} = \frac{x}{x^2 + y^2};$$

$$u_y = \frac{y}{x^2 + y^2};$$

$$v_x = -\frac{1}{1+\left(\frac{y}{x}\right)^2} \cdot \left(-\frac{y}{x^2}\right)$$

$$= -\frac{y}{x^2 + y^2}$$

so that

$$u_x = v_y, \quad u_y = -v_x$$

The formula $f'(z) = u_x + iv_x$ gives

$$(\ln z)' = u_x + iv_x = \frac{x}{x^2 + y^2} + i\left(-\frac{y}{x^2 + y^2}\right)$$

$$= \frac{x - iy}{x^2 + y^2}$$

$$= \frac{\bar{z}}{z\bar{z}} = \frac{1}{z}$$

14.2.5 General powers of z: z^α ($\alpha \in$ C)

The general power of a complex number z, *i.e.*, z^α ($\alpha \in \mathbb{C}$) is defined by the equation

$$z^\alpha = e^{\alpha(\ln z)} \quad \text{or} \quad \exp(\alpha \ln z) \quad (z \neq 0)$$

Since $\ln z$ is multi-valued z^α is, in general, multi-valued.

Case 1 If $\alpha = k$ ($k \in \mathbb{N}$), then z^k is single-valued and is identical with the usual nth power of z. Indeed,

$$z^k = \exp(k \ln z) = \exp[k \ln r + ik(\theta + 2n\pi)]$$

is the single-valued kth power of z, which is analytic.

When $\alpha = -k$ ($k \in \mathbb{N}$), then $z^{-k} = \dfrac{1}{z^k}$ is single-valued.

$$z^{-k} = \exp(-k \ln z) = \exp[-k \ln r - ik(\theta + 2n\pi)].$$

Case 2 If $\alpha = \dfrac{1}{k}$ ($k \in \mathbb{N}$), then

$$z^{\frac{1}{k}} = \exp\left(\frac{1}{k}\ln z\right) = \exp\left[\frac{1}{k}\ln r + \frac{i}{k}(\theta + 2n\pi)\right]$$

which is one of the k distinct values corresponding to $n = 0,1,2,\ldots(k-1)$ and it is analytic.

Case 3 If $\alpha = \frac{p}{q}$ where $p, q \in \mathbb{N}$ and p and q have no common factors except 1, then

$$z^{\alpha} = z^{\frac{p}{q}} = \exp\left[\left(\frac{p}{q}\ln r\right) + i\frac{p}{q}(\theta + 2n\pi)\right]$$

which is one of the q distinct values corresponding to $q = 0,1,2,\ldots(q-1)$ and it is analytic.

Case 4 If α is real and irrational or complex, then z^{α} is infinitely many-valued. All the index laws hold for complex variables also as in real calculus.

Real and imaginary parts of $(a+ib)^{(x+iy)}$

By definition

$$\begin{aligned}(a+ib)^{(x+iy)} &= e^{(x+iy)\ln(a+ib)} \\ &= e^{(x+iy)[\ln r + i(\theta+2n\pi)]} \\ &= (a^2+b^2)^{\frac{x}{2}} e^{-y(\theta+2n\pi)} \\ &\quad \times \text{cis}\left[\frac{y}{2}\ln(a^2+b^2) + x(\theta+2n\pi)\right]\end{aligned}$$

where $r = \sqrt{a^2+b^2}$; $\theta = \tan^{-1}\frac{b}{a}$; $n = 0, \pm1, \pm2, \ldots$.

The principal value of the expression is obtained for $n = 0$.

14.2.6 Inverse Trigonometric (Circular) Functions and Inverse Hyperbolic Functions

We have discussed above the trigonometric (circular) and hyperbolic functions of a complex variable, having defined them through the complex exponential function. Inverse trigonometric and inverse hyperbolic functions are now defined through the complex logarithm. All these functions are multi-valued functions. But when specific values of the square roots and logarithms are used these functions become single-valued.

$$w = \cos^{-1} z$$

Inverse cosine function is obtained by inverting the equation $z = \cos w$. Thus we have $w = \cos^{-1} z$. But

$$\cos w = \frac{e^{iw} + e^{-iw}}{2} = z$$

so that $e^{iw} + e^{-iw} - 2z = 0$.

Substituting $t = e^{iw}$, we obtain the quadratic

$$t^2 - 2zt + 1 = 0 \quad (t = e^{iw})$$

Solving, we get

$$t = e^{iw} = z \pm \sqrt{z^2 - 1}$$

$$\Rightarrow \cos^{-1} z = w = \frac{1}{i}\ln(z + \sqrt{z^2-1})$$

$$= -\frac{1}{i}\ln(z - \sqrt{z^2-1})$$

$$\because \quad z - \sqrt{z^2-1} = \frac{1}{z + \sqrt{z^2-1}}$$

$$w = \sin^{-1} z$$

Inverse sine function is obtained by inverting the equation $z = \sin w$. Thus we have $w = \sin^{-1} z$. But

$$\sin w = \frac{e^{iw} - e^{-iw}}{2i} = z$$

so that $e^{iw} - e^{-iw} - 2iz = 0$.

Substituting $t = e^{iw}$, we obtain the quadratic

$$t^2 - 2izt - 1 = 0 \quad (t = e^{iw})$$

Solving, we get

$$t = e^{iw} = iz \pm \sqrt{1 - z^2}$$

$$\Rightarrow \sin^{-1} z = w = \frac{1}{i}\ln\left(iz \pm \sqrt{1-z^2}\right)$$

Similarly, we can solve for the other inverse trigonometric functions. Inverse trigonometric functions in terms of logarithmic function are given below, in which we have omitted the additive constant $2k\pi$ $(k = 0, \pm1, \pm2, \pm3, \ldots)$.

Inverse trigonometric (circular) functions

$$\sin^{-1}z = \frac{1}{i}\ln(iz + \sqrt{1-z^2})$$

$$\operatorname{cosec}^{-1}z = \frac{1}{i}\ln\left(\frac{i+\sqrt{z^2-1}}{z}\right)$$

$$= \frac{1}{i}\ln\left(i\frac{1}{z} + \sqrt{1-\frac{1}{z^2}}\right)$$

$$= \sin^{-1}\left(\frac{1}{z}\right)$$

$$\cos^{-1}z = \frac{1}{i}\ln(z + \sqrt{z^2-1})$$

$$\sec^{-1}z = \frac{1}{i}\ln\left(\frac{1+\sqrt{1-z^2}}{z}\right)$$

$$= \frac{1}{i}\ln\left(\frac{1}{z} + \sqrt{\frac{1}{z^2}-1}\right)$$

$$= \cos^{-1}\left(\frac{1}{z}\right)$$

$$\tan^{-1}z = \frac{1}{2i}\ln\frac{1+iz}{1-iz}$$

$$\cot^{-1}z = \frac{1}{2i}\ln\frac{z+i}{z-i}$$

$$= \frac{1}{2i}\ln\frac{1+\frac{i}{z}}{1-\frac{i}{z}} = \tan^{-1}\left(\frac{i}{z}\right)$$

Derivatives of inverse trigonometric (circular) functions

$$\frac{d}{dz}(\sin^{-1}z) = \frac{1}{\sqrt{1-z^2}}$$

$$\frac{d}{dz}(\operatorname{cosec}^{-1}z) = -\frac{1}{z\sqrt{z^2-1}}$$

$$\frac{d}{dz}(\cos^{-1}z) = \frac{-1}{\sqrt{1-z^2}}$$

$$\frac{d}{dz}(\sec^{-1}z) = \frac{1}{z\sqrt{z^2-1}}$$

$$\frac{d}{dz}(\tan^{-1}z) = \frac{1}{1+z^2}$$

$$\frac{d}{dz}(\cot^{-1}z) = \frac{-1}{1+z^2}$$

Note

The above differentiation formulas are the same as in real calculus. We now give below the inverse hyperbolic functions and their derivatives.

Inverse hyperbolic functions

$z = \sinh w \Rightarrow w = \sinh^{-1}z$. Similarly, other inverse hyperbolic functions are defined.

Inverse hyperbolic functions are multi-valued. They are expressed in terms of natural logarithms where an additive constant $2k\pi$ $(k = 0, \pm1, \pm2, \ldots)$ to be added is omitted.

$$\sinh^{-1}z = \ln\left(z + \sqrt{z^2+1}\right);$$

$$\operatorname{cosech}^{-1}z = \ln\left(\frac{1+\sqrt{z^2+1}}{z}\right)$$

$$= \sinh^{-1}\left(\frac{1}{z}\right)$$

$$\cosh^{-1}z = \ln(z + \sqrt{z^2-1});$$

$$\operatorname{sech}^{-1}z = \ln\left(\frac{1+\sqrt{1-z^2}}{z}\right)$$

$$= \cosh^{-1}\left(\frac{1}{z}\right)$$

$$\tanh^{-1}z = \frac{1}{2}\ln\left(\frac{1+z}{1-z}\right);$$

$$\coth^{-1}z = \frac{1}{2}\ln\left(\frac{z+1}{z-1}\right) = \tanh^{-1}\left(\frac{1}{z}\right)$$

Derivatives of inverse hyperbolic functions

$$\frac{d}{dz}(\sinh^{-1}z) = \frac{1}{\sqrt{1+z^2}}$$

$$\frac{d}{dz}(\operatorname{cosech}^{-1}z) = \frac{-1}{z\sqrt{z^2+1}}$$

$$\frac{d}{dz}(\cosh^{-1}z) = \frac{1}{\sqrt{z^2-1}}$$

$$\frac{d}{dz}(\operatorname{sech}^{-1}z) = \frac{-1}{z\sqrt{1-z^2}}$$

$$\frac{d}{dz}(\tanh^{-1}z) = \frac{1}{1-z^2}$$

$$\frac{d}{dz}(\coth^{-1}z) = \frac{1}{1-z^2}$$

Example 14.3

Find all solutions of the equation $e^z = 2+i$.

Solution $z = x+iy$. Taking modulus on both sides of $e^z = 2+i$

$$|e^z| = e^x = \sqrt{2^2+1^2} = \sqrt{5}$$

$$\Rightarrow x = \frac{1}{2}\ln 5 = 0.8045$$

$$e^x = \sqrt{5}, e^x \cos y = 2 \quad \text{and} \quad e^x \sin y = 1$$

$$\Rightarrow \cos y = \frac{2\sqrt{5}}{5} = 0.8944,$$

$$\sin y = \frac{\sqrt{5}}{5} = 0.4472$$

$$\Rightarrow y = 0.4324$$

The given equation has infinitely many solutions $z = 0.8045 + i0.4324 \pm 2k\pi i \quad k = 0,1,2,\ldots$.

Example 14.4

Find the real and imaginary parts of e^{z^3}.

Solution

$$z = x+iy \Rightarrow z^3 = (x+iy)^3$$

$$= x^3 + 3x^2(iy) + 3x(iy)^2 + (iy)^3$$

$$= (x^3 - 3xy^2) + i(3x^2y - y^3)$$

$$\exp(z^3) = e^{(x^3-3xy^2)}[\cos(3x^2y - y^3) + i\sin(3x^2y - y^3)]$$

$$Re[\exp(z^3)] = e^{(x^3-3xy^2)} \cdot \cos(3x^2y - y^3)$$

$$Im[\exp(z^3)] = e^{(x^3-3xy^2)} \cdot \sin(3x^2y - y^3).$$

Example 14.5

If $\sin(A+iB) = x+iy$, prove that (a)

(i) $$\frac{x^2}{(\cosh^2 B)} + \frac{y^2}{(\sinh^2 B)} = 1$$

(ii) $$\frac{x^2}{(\sin^2 A)} - \frac{y^2}{(\cos^2 A)} = 1.$$

Solution We know that,

$$\sin(A+iB) = \sin A\cos(iB) + \cos A\sin(iB)$$

$$= \sin A\cosh B + i\cos A\sinh B$$

$$= x+iy \text{ (given)}$$

$$\Rightarrow x = \sin A\cosh B \quad \text{and}$$

$$y = \cos A\sinh B$$

$$\Rightarrow \left.\begin{aligned} \frac{x}{\cosh B} &= \sin A, \\ \frac{y}{\sinh B} &= \cos A \end{aligned}\right\} \quad (1)$$

Squaring and adding,

$$\frac{x^2}{\cosh^2 B} + \frac{y^2}{\sinh^2 B} = 1 \quad (\sin^2 A + \cos^2 A = 1)$$

From Eq. (1), we also have

$$\frac{x}{\sin A} = \cosh B \quad \text{and} \quad \frac{y}{\cos A} = \sinh B$$

Squaring and subtracting,

$$\frac{x^2}{\sin^2 A} - \frac{y^2}{\cos^2 A} = 1$$

Example 14.6

Find all the roots of $\sin z = 2$.

[JNTU 1999, 2006 (Set 4)]

Solution

$$\sin z = 2 \Rightarrow \frac{e^{iz} - e^{-iz}}{2i} = 2$$

$$\Rightarrow e^{iz} - \frac{1}{e^{iz}} = 4i$$

$$\Rightarrow (e^{iz})^2 - 4i(e^{iz}) - 1 = 0$$

$$\Rightarrow t^2 - 4it - 1 = 0 \quad (t = e^{iz})$$

Solving the quadratic in t, we have

$$t = e^{iz} = (2 \pm \sqrt{3})i$$

$$\Rightarrow z = \frac{1}{i}[\ln(2+\sqrt{3}) + \ln i \pm 2k\pi i];$$

$$k = 0,1,2,\ldots$$

$$z = \frac{1}{i}[\ln(2+\sqrt{3}) + i\frac{\pi}{2} \pm 2k\pi i];$$

$$\ln i = \ln e^{i\frac{\pi}{2}} = i\frac{\pi}{2}$$

$$z = 2k\pi + \frac{\pi}{2} - i\ln(2+\sqrt{3}); \quad \frac{1}{i} = -i.$$

Example 14.7

Find the general values i^i and $\ln i^i$.

Solution We have $i = e^{\ln i}$. Taking ith power on both sides,

$$i^i = e^{i\ln i} = e^{i(\ln i \pm 2k\pi i)}$$

$$= e^{\mp 2k\pi} \cdot e^{i\ln i}$$

$$= e^{\mp 2k\pi} \cdot e^{\frac{-\pi}{2}} \quad [\because \ln i = \ln e^{i\frac{\pi}{2}} = i\frac{\pi}{2}]$$

or $\quad i^i = e^{(2k-\frac{1}{2})\pi}$ where $k = 0, \pm 1, \pm 2, \ldots$

Example 14.8

Separate the following functions into real and imaginary parts (i) $\cot z$; (ii) $\operatorname{cosec} z$.

Solution

(i) $\cot z = \cot(x+iy)$

$$= \frac{\cos(x+iy)}{\sin(x+iy)}$$

$$= \frac{2\cos(x+iy)\sin(x-iy)}{2\sin(x+iy)\sin(x-iy)}$$

(Introduce $2\sin(x-iy)$ in the numerator and denominator to make the denominator a real function.)

$$= \frac{\sin 2x - \sin 2iy}{\cos 2iy - \cos 2x}$$

$$= \frac{\sin 2x - i\sinh 2y}{\cosh 2y - \cos 2x}, \quad [\sin(it) = i\sinh t]$$

$$= \frac{\sin 2x}{\cosh 2y - \cos 2x} + i\left(\frac{-\sinh 2y}{\cosh 2y - \cos 2x}\right)$$

$[\because \cos(it) = \cosh t]$

(ii) $\operatorname{cosec} z = \operatorname{cosec}(x+iy)$

$$= \frac{1}{\sin(x+iy)}$$

$$= \frac{2\sin(x-iy)}{2\sin(x+iy)\sin(x-iy)}$$

(Introduce $2\sin(x-iy)$ in the numerator and denominator to make the denominator a real function)

$$= \frac{2[\sin x\cos(iy) - \cos x\sin(iy)}{\cos 2iy - \cos 2x}$$

$$= \frac{2\sin x\cosh y - 2i\cos x\sinh y}{\cosh 2y - \cos 2x}$$

$$= \frac{2\sin x\cosh y}{\cosh 2y - \cos 2x} + i\left(\frac{-2\cos x\sinh y}{\cosh 2y - \cos 2x}\right).$$

Example 14.9

Separate the real and imaginary parts of the functions (i) $\tanh z$; (ii) $\operatorname{cosec} z$.

Solution

(i) $\tanh z = \tanh(x+iy)$

$$= \frac{1}{i}\tan[i(x+iy)]; \quad [\because \tan(it) = i\tanh t]$$

$$= i\tan(ix-y) = i\tan(y-ix)$$

$$= i\frac{\sin(y-ix)}{\cos(y-ix)}$$

$$= i\cdot\frac{2\sin(y-ix)\cos(y+ix)}{2\cos(y-ix)\cos(y+ix)}$$

(Introduce $2\cos(y+ix)$ in the numerator and denominator to make the denominator a real function.)

$$= i\cdot\frac{\sin 2y - \sin 2ix}{\cos 2y + \cos 2ix}$$

$$= i\cdot\frac{\sin 2y - i\sinh 2x}{\cosh 2x + \cos 2y},$$

$[\because \sin it = i\sinh t, \cos it = \cosh t]$

$$= \frac{\sinh 2x}{\cosh 2x + \cos 2y} + i\left(\frac{\sin 2y}{\cosh 2x + \cos 2y}\right)$$

(ii) $\operatorname{cosec} z = \dfrac{1}{\sinh(x+iy)}$

$$= \frac{1}{\frac{1}{i}\sin i(x+iy)}$$

$$= \frac{1}{\sin(ix-y)}$$

$$= \frac{i\cdot 2\sin(ix+y)}{2\sin(ix+y)\sin(ix-y)}$$

$$= \frac{i\,2(\sin(ix)\cos y + \cos(ix)\sin y)}{\cos 2y - \cos(2ix)}$$

$$= i\cdot\frac{2(i\sinh x\cos y + \cosh x\sin y)}{\cos 2y - \cosh 2x}$$

$$= \frac{-2\sinh x\cos y}{\cos 2y - \cosh 2x} + i\frac{2\cosh x\sin y}{\cos 2y - \cosh 2x}$$

$$= \frac{2\sinh x \cos y}{\cosh 2x - \cos 2y}$$

$$+ i\left(\frac{-2\cosh x \sin y}{\cosh 2x - \cos 2y}\right).$$

Example 14.10

If $\tan\left(\frac{\pi}{6} + i\alpha\right) = x + iy$, prove that

$x^2 + y^2 + \frac{2}{\sqrt{3}}x = 1$. [JNTU 2004 (Set 2)]

Solution

$$\tan\left(\frac{\pi}{6} + i\alpha\right) = x + iy \qquad (1)$$

$$\tan\left(\frac{\pi}{6} - i\alpha\right) = x - iy \qquad (2)$$

$$\therefore 2\cdot\frac{\pi}{6} = \left(\frac{\pi}{6} + i\alpha\right) + \left(\frac{\pi}{6} - i\alpha\right)$$

$$\tan\left(\frac{2\pi}{6}\right) = \tan\frac{\pi}{3}$$

$$= \tan\left[\left(\frac{\pi}{6} + i\alpha\right) + \left(\frac{\pi}{6} - i\alpha\right)\right] \qquad (3)$$

$$= \frac{\tan\left(\frac{\pi}{6} + i\alpha\right) + \tan\left(\frac{\pi}{6} - i\alpha\right)}{1 - \tan\left(\frac{\pi}{6} + i\alpha\right)\tan\left(\frac{\pi}{6} - i\alpha\right)}$$

$$\Rightarrow \sqrt{3} = \frac{(x+iy) + (x-iy)}{1 - (x+iy)(x-iy)}$$

$$= \frac{2x}{1 - x^2 - y^2}, \qquad (4)$$

$$\Rightarrow 1 - x^2 - y^2 = \frac{2}{\sqrt{3}}x \quad \text{or}$$

$$x^2 + y^2 + \frac{2}{\sqrt{3}}x = 1.$$

Example 3.11

If $\tan(x+iy) = A + iB$ show that $A^2 + B^2 + 2A\cot 2x = 1$.

Solution

$$A = Re\tan(x+iy)$$

$$= \frac{\sin 2x}{\cos 2x + \cosh 2y},$$

$$B = Im\tan(x+iy)$$

$$= \frac{\sinh 2y}{\cos 2x + \cosh 2y} \quad A^2 + B^2 + 2A\cot 2x$$

$$= \frac{\sin^2 2x + \sinh^2 2y}{(\cos 2x + \cosh 2y)^2}$$

$$+ 2\cot 2x \cdot \frac{\sin 2x}{(\cos 2x + \cosh 2y)}$$

$$= \frac{\sin^2 2x + \sinh^2 2y + 2\cos 2x(\cos 2x + \cosh 2y)}{(\cos 2x + \cosh 2y)^2}$$

$$= \frac{1 + \sinh^2 2y + 2\cos 2x\cosh 2y + \cos^2 2x}{(\cos 2x + \cosh 2y)^2}$$

$$= \frac{\cosh^2 2y + 2\cos 2x\cosh 2y + \cos^2 2x}{(\cos 2x + \cosh 2y)^2}$$

$$= \frac{(\cosh 2y + \cos 2x)^2}{(\cos 2x + \cosh 2y)^2} = 1.$$

Example 14.12

Find the general and principal values of
i) $\log(1 + i\sqrt{3})$; (ii) $\log(-i)$, (iii) $\log(-1)$.

[JNTU 2003, 2005]

Solution

(i) Put $1 = r\cos\theta$, $\sqrt{3} = r\sin\theta$ so that

$r = \sqrt{1^2 + (\sqrt{3})^2} = 2$ and $\tan\theta = \sqrt{3} \Rightarrow \theta = \frac{\pi}{3}$.

$$\log(1 + i\sqrt{3}) = \log r + i\theta = \log 2 + i\left(\frac{\pi}{3}\right)$$

The general value is

$$\log(1 + i\sqrt{3}) = \log 2 + i\left(\frac{\pi}{3} \pm 2k\pi\right),$$

$$k = 0, 1, 2, 3, \ldots$$

The principal argument (i.e., the value of θ which lies between $-\pi$ and π) is $\frac{\pi}{3}$.

So, the principal value is $\log 2 + i(\frac{\pi}{3})$ which is obtained by putting $k = 0$ in the general value.

(ii) Put $0 = r\cos\theta$, and $-1 = r\sin\theta$ so that $r = 1$ and $\theta = -\frac{\pi}{2}$

The general value is

$$\log(-1) = \log r + i\theta + 2k\pi i$$

$$= \log 1 + i\left(-\frac{\pi}{2}\right) + 2k\pi i$$

Principal value is $\log 1 - i\frac{\pi}{2}$ $\quad (\because -\pi \le \frac{\pi}{2} \le \pi)$

(iii) $z = x + iy = -1$

$$\Rightarrow x = -1,\ y = 0,\ r = \sqrt{x^2 + y^2} = 1,$$

$$\cos\theta = \frac{x}{r} = -1,\quad \sin\theta = \frac{y}{r} = 0 \Rightarrow \theta = \pi$$

The general value is

$$\therefore \log(-1) = \log 1 + i(\pi \pm 2k\pi)$$

The principal value is

$$\log(-1) = \log 1 + i\pi.$$

Example 14.13

Determine the values of $(1+i)^i$.

Solution By definition,

$$(1+i)^i = \exp[i\ln(1+i)]$$

Expressing $(1+i)$ in polar form, we have

$$x = 1, y = 1, r = \sqrt{x^2 + y^2} = \sqrt{1^2 + 1^2} = \sqrt{2}$$

$$\cos\theta = \frac{1}{\sqrt{2}}, \sin\theta = \frac{1}{\sqrt{2}} \Rightarrow \theta = \frac{\pi}{4} \pm 2n\pi$$

$$\therefore 1 + i = \sqrt{2}\exp\left[i\left(\frac{\pi}{4} \pm 2n\pi\right)\right] n = 0,1,2,\ldots$$

$$\ln z = \ln(x+iy) = \frac{1}{2}\log\sqrt{x^2+y^2} + i\left(\tan^{-1}\frac{b}{a} \pm 2n\pi\right)$$

$$\Rightarrow \ln(1+i) = \frac{1}{2}\log 2 + i\left(\frac{\pi}{4} \pm 2n\pi\right)$$

$$\Rightarrow i\ln(1+i) = \frac{i}{2}\log 2 - \left(\frac{\pi}{4} \pm 2n\pi\right)$$

$$\Rightarrow (1+i)^i = \exp[i\ln(1+i)]$$

$$= \exp\left(\frac{1}{2}\log 2\right)\exp\left[-\frac{\pi}{4} \pm 2n\pi\right]$$

$$= \sqrt{2}\exp\left(-\frac{\pi}{4} \pm 2n\pi\right);$$

$$n = 0,1,2,\ldots.$$

Example 14.14

Find all the values of z which satisfy (i) $e^z = -2$; (ii) $e^z = 1 + i$. [JNTU 2006 (Set 4)]

Solution

(i) $e^z = -2$;

$$\Rightarrow |e^z| = e^x = \sqrt{(2)^2 + 0^2} = 2 \Rightarrow x = \ln 2$$

$$e^z = -2 \Rightarrow e^x(\cos y + i\sin y) = -2 + 0\cot i$$

$$\Rightarrow e^x\cos y = -2,\quad e^x\sin y = 0$$

Since $e^x = 2$, $\cos y = -1$ and $\sin y = 0 \Rightarrow y = \pi$

Hence $z = x + iy = \ln 2 + i\pi \pm 2k\pi i$,

$$k = 0,1,2,\ldots$$

(ii) $e^z = 1 + i \Rightarrow |e^z| = e^x = \sqrt{1^2 + 1^2} = \sqrt{2} \Rightarrow$

$x = \ln\sqrt{2}$

$$e^z = 1 + i$$

$$\Rightarrow e^{x+iy} = 1 + i$$

$$\Rightarrow e^x(\cos y + i\sin y) = 1 + i$$

$$\Rightarrow e^x\cos y = 1 \text{ and } e^x\sin y = 1$$

$$\text{But } e^x = \sqrt{2}$$

$$\Rightarrow \cos y = \frac{1}{\sqrt{2}}, \sin y = \frac{1}{\sqrt{2}}$$

$$\Rightarrow y = \frac{\pi}{4}$$

Hence $z = \ln\sqrt{2} + i\frac{\pi}{4} \pm 2k\pi i,\ k = 0,1,2,\ldots$

Example 14.15

Determine all the values of $(x+iy)^{a+ib}$ and in particular those of $(1-i)^{1+i}$.

Solution Let $w = u + iv = (x+iy)^{a+ib}$. Applying logarithms, we have

$$\ln w = \ln(u+iv) = (a+ib)\ln(x+iy)$$

$$= (a+ib)\ln(re^{(i\theta)})$$

$$= (a+ib)(\ln r + i\theta);$$

$$r = \sqrt{x^2+y^2}, \theta = \tan^{-1}\frac{y}{x}$$

$$= (a\ln r - b\theta) + i(b\ln r + a\theta)$$

$$\Rightarrow w = u + iv = e^{a\ln r - b\theta} \cdot e^{i(b\ln r + a\theta)}$$
$$= e^{a\ln r - b\theta}[\cos(b\ln r + a\theta) + i\sin(b\ln r + a\theta)]$$
$$\Rightarrow u = e^{a\ln r - b\theta}\cos(b\ln r + a\theta),$$

where $r = \sqrt{x^2 + y^2}$, $v = e^{a\ln r - b\theta}\sin(b\ln r + a\theta)$

and $\theta = \tan^{-1}\dfrac{y}{x}$.

(For general solution, replace θ by $\theta + 2k\pi$, $k = 0, \pm 1, \pm 2, \ldots$)

For $(1-i)^{1+i}$, choose $x = 1,\ y = -1$;
$r = \sqrt{1^2 + (-1)^2} = \sqrt{2},\ \theta = \tan^{-1} = -\pi/4$
$a = 1, b = 1$.

The general value is $\dfrac{-\pi}{4} + 2k\pi$

$$(1-i)^{1+i}$$
$$= e^{\ln\sqrt{2} - \frac{\pi}{4} + 2k\pi} \cdot \operatorname{cis}\left[\ln\sqrt{2} + \left(-\frac{\pi}{4} + 2k\pi\right)\right]$$
$$= \sqrt{2}\exp\left(\frac{-\pi}{4} + 2k\pi\right)$$
$$\times \operatorname{cis}\left[\ln\sqrt{2} + \left(\frac{-\pi}{4} + 2k\pi\right)\right]$$

For $k = 0$, we obtain the principal value of the given complex number.

Example 14.16

Find all the principal values of $(1+i\sqrt{3})^{(1+i\sqrt{3})}$.

[JNTU 2005 *S* (Set 2)]

Solution In Example 3.15, choose $x = a = 1$, $y = b = \sqrt{3}$.

$$r = \sqrt{1^2 + (\sqrt{3})^2} = 2, \theta = \frac{\pi}{3}$$
$$(1+i\sqrt{3})^{(1+i\sqrt{3})} = e^{\log 2 - \sqrt{3}\cdot\frac{\pi}{3}}$$
$$\times \operatorname{cis}\left(\sqrt{3}\log 2 + \frac{\pi}{3} \pm 2k\pi\right)$$

For $k = 0$, we obtain the principal values as

$$2e^{\frac{-\pi}{\sqrt{3}}}\operatorname{cis}\left(\sqrt{3}\log 2 + \frac{\pi}{3}\right).$$

Example 14.17

Find all the principal values of $\left(\dfrac{\sqrt{3}}{2} + i\dfrac{1}{\sqrt{2}}\right)^{(1+i\sqrt{3})}$.

[JNTU 2005 *S* (Set 4)]

Solution In Example 3.15, choose $x = \sqrt{\dfrac{3}{2}}$, $y = \dfrac{1}{\sqrt{2}}$, $a = 1, b = \sqrt{3}$

$$r = \sqrt{x^2 + y^2}$$
$$= \sqrt{\left(\frac{\sqrt{3}}{2}\right)^2 + \left(\frac{1}{\sqrt{2}}\right)^2}$$
$$= \sqrt{\frac{3}{4} + \frac{1}{2}} = \frac{\sqrt{5}}{2};$$
$$\theta = \tan^{-1}\frac{\frac{1}{\sqrt{2}}}{\frac{\sqrt{3}}{2}} = \tan^{-1}\frac{\sqrt{2}}{3}$$
$$\left(\frac{\sqrt{3}}{2} + i\frac{1}{\sqrt{2}}\right)^{(1+i\sqrt{3})}$$
$$= e^{(1+i\sqrt{3})\left(\log\frac{\sqrt{5}}{2} + i\tan^{-1}\frac{\sqrt{2}}{3}\right)}$$
$$= e^{\log\frac{\sqrt{5}}{2} - \sqrt{3}\tan^{-1}\frac{\sqrt{2}}{3} \pm 2k\pi}$$
$$\times \operatorname{cis}\left(\sqrt{3}\log\frac{\sqrt{5}}{2} + \tan^{-1}\frac{\sqrt{2}}{3} \pm 2k\pi\right)$$

The principal value for $k = 0$ is

$$\frac{\sqrt{5}}{2}e^{-\sqrt{3}\tan^{-1}\sqrt{2/3}}$$
$$\cdot \operatorname{cis}\left(\sqrt{3}\log\frac{\sqrt{5}}{2} + \tan^{-1}\sqrt{2/3}\right).$$

EXERCISE 14.1

1. If $e^x = \sec\theta + \tan\theta$, prove that $\cosh x = \sec\theta$. **[Hint:** $e^{-x} = \sec\theta - \tan\theta$]
2. Show that $\log(1 + i\tan\theta) = \log\sec\theta + i\theta$.

3. Separate the following functions into real and imaginary parts:

(a) $\tan(x+iy)$.

Ans: $\dfrac{\sin 2x}{\cos 2x+\cosh 2y}+i\dfrac{\sinh 2y}{\cos 2x+\cosh 2y}$

(b) $\sec(x+iy)$.

Ans: $\dfrac{2\cos x\cosh y}{\cos 2x+\cosh 2y}+i\dfrac{2\sin x\sinh y}{\cos 2x+\cosh 2y}$

(c) $\coth(x+iy)$.

Ans: $\dfrac{\sinh 2x}{\cosh 2x-\cos 2y}+i\dfrac{-\sin 2y}{\cosh 2x-\cos 2y}$

(d) $\tan^{-1}(x+iy)$.

Ans: $\dfrac{1}{2}\tan^{-1}\left(\dfrac{2x}{1-x^2-y^2}\right)$

$+\dfrac{i}{2}\tanh^{-1}\left(\dfrac{2y}{1+x^2+y^2}\right)$

4. If $u+iv=\cosh(x+iy)$, prove that

(a) $\dfrac{u^2}{\cosh^2 x}+\dfrac{v^2}{\sinh^2 x}=1$;

(b) $\dfrac{u^2}{\cos^2 y}-\dfrac{v^2}{\sin^2 v}=1$.

5. If $(\cos\theta+i\phi)=\cos\alpha+i\sin\alpha$, prove that $\sin^2\theta=\pm\sin\alpha$.

6. If $\sin(x+iy)=r(\cos\alpha+i\sin\alpha)$, prove that

$$y=\frac{1}{2}\log\left[\frac{\cos(x-\alpha)}{\cos(x+\alpha)}\right].$$

7. Prove that $\tan\left[i\log\left(\dfrac{a-ib}{a+ib}\right)\right]=\dfrac{2ab}{a^2-b^2}$.

8. Prove that the real part of the principal value of $i\log(1+i)$ is equal to $e^{\frac{\pi^2}{8}}\cdot\cos\left(\dfrac{\pi}{4}\log 2\right)$.

9. Find all the roots of $\sin z=2$.

Ans: $z=2k\pi+\dfrac{\pi}{2}-i\log(2+\sqrt{3})$

10. Find all the values of i^{-2i}.

Ans: $e^{(4k+1)\pi}$ $k=0\pm1\pm2,\ldots$ Pr. V. e^{π}

11. Determine all the values of $(1+i)^i$.

Ans: $e^{i\ln\sqrt{2}+i\left(\frac{\pi}{4}\pm 2k\pi\right)}$

12. Prove that $\tan^{-1}z=\dfrac{i}{2}\ln\left(\dfrac{i+z}{i-z}\right)$.

13. Find all the roots of $\tanh z+2=0$.

Ans: $z=-\dfrac{1}{2}\log 3+i\left(2k+\dfrac{1}{2}\right)\pi;\ k=0,1,2,\ldots$

14. Determine all the roots of $\sinh z=i$.

Ans: $z=\left(2k+\dfrac{1}{2}\right)\pi i;\ k=0,1,2,\ldots$

Complex Integration

15

15.1 Introduction

Complex integration is useful in calculating certain real integrals that cannot be evaluated by the usual methods as, for instance, in finding the inverse Laplace, Fourier or Z-transforms in applications. It also enables us to establish some basic properties of analytic functions such as the existence of higher order derivatives, in the case of analytic functions.

In this chapter, we define and explain the concept of complex integral, state and prove Cauchy's integral formula. We also prove Cauchy's generalised integral formula, which shows that if a function is analyticin a domain D, it possesses the derivatives of all orders in D.

15.2 Basic Concepts

The concept of line integral is a natural generalisation of [1]Riemann's integral of real calculus defined by

$$\int_a^b f(x)dx = \lim_{\substack{n\to\infty \\ |\Delta x|\to 0}} \sum_{r=1}^{n} f(C_r)\Delta x_r \qquad (15.1)$$

where C_r is any point in the rth subinterval $I_r = [x_{r-1}, x_r]$ of length $\Delta x_r = x_r - x_{r-1}$ of the interval $[a, b]$. Here we integrate $f(x)$ from $x = a$ to $x = b$. In a line integral, we integrate a given function along a curve in the plane. Hence it is actually a curve integral but it has become a standard practice to call it a line integral.

[1] Georg Friedrich Bernhard Riemann (1826–1866), great innovative German mathematician who made fundamental contributions to geometry and the theory of analytic functions of a complex variable to number theory, potential theory, topology and mathematical physics. His Riemannian geometry provides the foundation for modern relativity theory.

Before we define complex line integral and state some of its properties we define certain terms.

15.2.1 Simple Curve

A curve is said to be simple if it does not cut or touch itself, *i.e.*, $z(t_1) \neq z(t_2)$ for $t_1 \neq t_2$.

But $z(a) = z(b)$ is allowed, in which case the curve is said to be closed.

15.2.2 Closed Curve

A curve whose initial and terminal points coincide is called a closed curve or path.

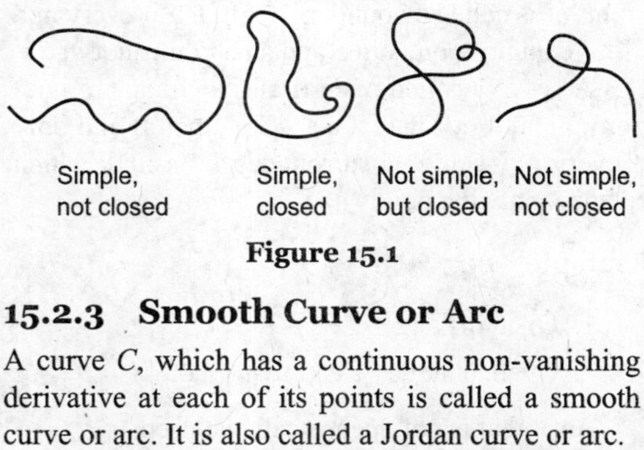

Figure 15.1

15.2.3 Smooth Curve or Arc

A curve C, which has a continuous non-vanishing derivative at each of its points is called a smooth curve or arc. It is also called a Jordan curve or arc.

Geometrically, it means that C has a unique and continuously turning tangent at each point of C.

Example 15.1

Circles and ellipses are closed smooth curves while the parabola and the hyperbola are open smooth curves.

15.2.4 Contour

The curve obtained by joining finitely many smooth arcs is called a contour. If the integration is along a contour we call it a contour integration.

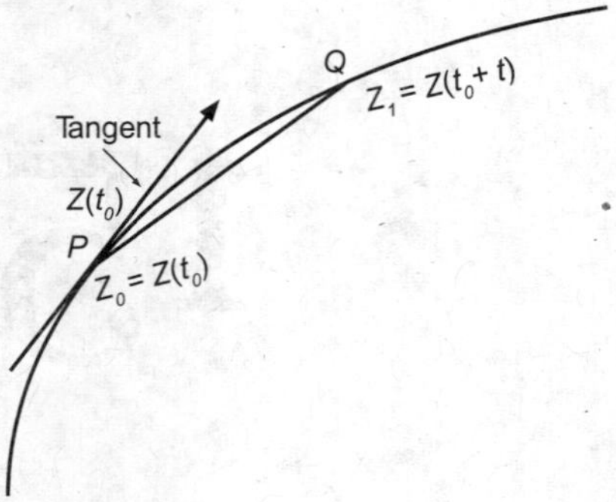

Figure 15.2

Example 15.2

Closed contours: Circle, triangle, square, rectangle, closed polygon, etc. Open contours: Arms of an angle, open polygon, etc.

The set of all complex numbers denoted by $\mathbb{C}$ is called the complex plane. A complex number z is also called a point of $\mathbb{C}$. By a *neighbourhood* of a point 'a' we mean the set of points z such that $|z-a| < \delta$.

Let S be a subset of $\mathbb{C}$. A point 'a' is called an *interior* point of S if there is a neighbourhood of 'a' (nbd of a) such that every point in it belongs to S. A point 'b' is called a boundary point of S if every nbd of 'b' contains points in S and also points not in S.

A set S is called open if all its points are interior points of S. That is, a set S is open if and only if every point of S is surrounded by a nbd, which contains only the points of S.

Example 15.3

1. Open disk $\{z \in \mathbb{C} : |z-a| < r\}$,
2. Open annulus $\{z \in \mathbb{C} : r_1 < |z-a| < r_2\}$

A set that is not open is called closed.

Example 15.4

1. Closed disk: $\{z \in \mathbb{C} : |z-a| \leq r\}$,
2. Closed annulus: $\{z \in \mathbb{C} : r_1 \leq |z-a| \leq r_2\}$

An open set 'S' is called connected if any two points of 'S' can be joined by a broken line entirely lying in 'S'.

An open connected set is called a *domain*.. A domain together with none, some or all of its boundary points is called a *region*.

15.2.5 Simply-Connected Domain

A domain D is called simply-connected if it is such that every simple closed path in D encloses only the points of D.

Example 15.5

Open disk, open ellipse or the interior of any simple closed curve.

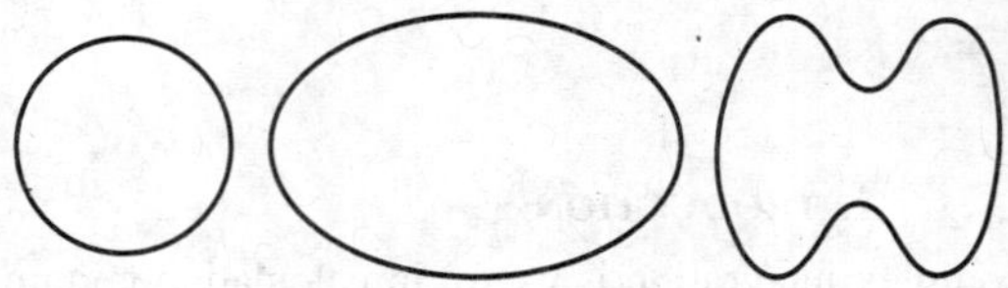

Figure 15.3

Simple closed curves are also called contours and the integrals over them are called contour integrals, as already mentioned.

15.2.6 Multiply-Connected Domain

A domain D is called multiply-connected if it is not simply-connected.

Annulus $r_1 < |z-a| < r$; punctured disk

$0 < |z| < 1$ (disk without centre).

15.3 Complex Line Integral

We know that there are two types of integrals: (1) definite integrals and (2) indefinite integrals or anti-derivatives.

An indefinite integral is a function whose derivative is a given analytic function in a region. By interchanging the columns in the table of functions and their derivatives we get the table of functions and their indefinite integrals.

$f(z)\left[=\int g(z)dx\right]$	$g(z)[=f'(z)]$
$\dfrac{z^{n+1}}{n+1}(n \neq -1)$	z^n
$\text{Ln}\, z$	$\dfrac{1}{z}$
e^z	e^z
$\sin z$	$\cos z$
$\tan z$	$\sec^2 z$, and so on.

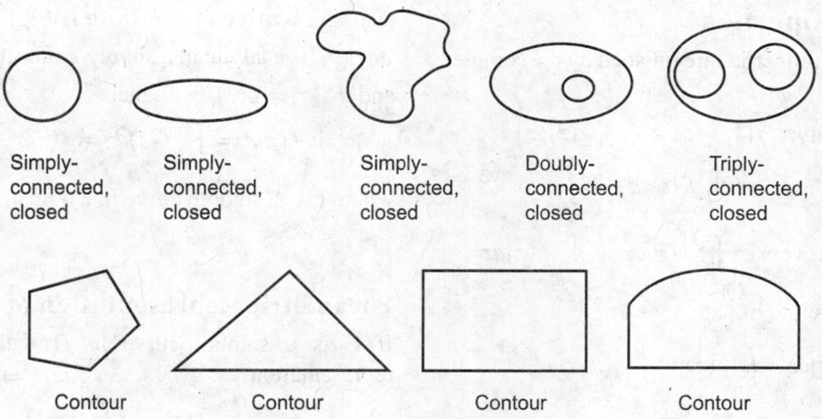

Figure 15.4

Complex definite integrals are called complex line integrals and are denoted by

$$\int_C f(z)dz$$

where the integrand and $f(z)$ is integrated over a given curve C in the complex plane. The curve C is called the path of integration. C may be represented by a parameter t as

$$z(t) = x(t) + iy(t) \quad (a \le t \le b) \tag{15.2}$$

The direction in which t increases is considered to be the positive sense on C and this gives an orientation to C.

The point A corresponding to $t = a$ is called the initial point on C and the point B corresponding to $t = b$ is called the terminal point on C.

For example, parametric equations of a straight line are

$$x = (1-t)x_1 + tx_2, \quad y = (1-t)y_1 + ty_2$$

and those of a circle $|z-a| = r$ are

$$x = x_1 + r\cos t, \quad y = y_1 + r\sin t.$$

15.3.1 Definition of the Complex Line Integral

Let C be a smooth curve in $\mathbb{C}$. Suppose $f(z)$ is a continuous function given at each point of C. We subdivide the interval $a \le t \le b$ by points

$$a = t_0, t_1, \ldots, t_{n-1}, t_n = b \tag{15.3}$$

where $t_0 < t_1 < t_2 < \cdots < t_{n-1} < t$. To this subdivision there corresponds a subdivision of C, indicated by points

$$z_0, z_1, \ldots, z_{n-1}, z_n \tag{15.4}$$

where $z_i = z(t_i)$. Let ζ_i be an arbitrary point between z_{i-1} and z_i. We form the sum

$$S_n = \sum_{i=1}^{n} f(z_i)\Delta z_i \quad \text{where } \Delta z_i = z_i - z_{i-1} \tag{15.5}$$

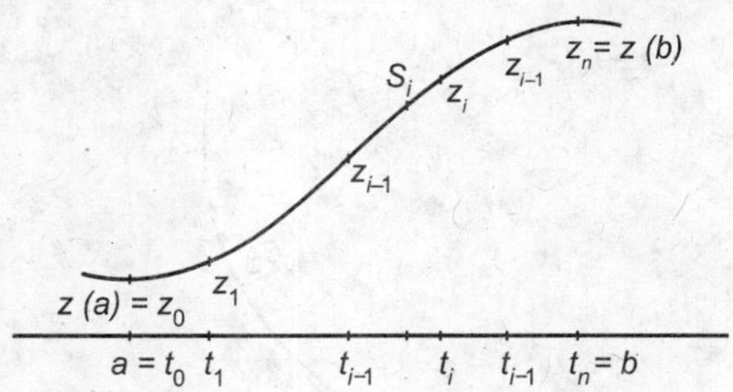

Figure 15.5

The limit of the sequence $\{S_n\}$ as $n \to \infty$ and $\max|\Delta z_i| \to 0$, if it exists, is called the line integral of $f(z)$ over C.

We write it as

$$\int_C f(z)dz.$$

Actually, it is a curve integral but it is standard practice to call it a line integral.

15.3.2 Properties

The following properties are satisfied by the complex line integral:

1. Linearity: $\int_C [k_1 f_1(z)dz + k_2 f_2(z)dz]$
$$= k_1 \int_C f_1(z)dz + k_2 \int_C f_2(z)dz$$
2. Sense reversal: $\int_{z_1}^{z_2} f(z)dz = -\int_{z_2}^{z_1} f(z)dz$
3. Partitioning of path:
$$\int_C f(z)dz = \int_{C_1} f(z)dz + \int_{C_2} f(z)dz$$

15.3.3 Relation Between Real and Complex Line Integrals

Let $w = f(z)$ be a complex-valued function defined in a region R. Suppose $z = x+iy$ and $w = u(x,y)+iv(x,y)$. Then the complex line integral $\int_C f(z)dz$ over C in R can be expressed as a sum of two line integrals of real functions as follows:

$$\int_C f(z)dz = \int_C (u+iv)(dx+idy)$$
$$= \int_C (udx - vdy) + i\int_C (vdx+udy)$$

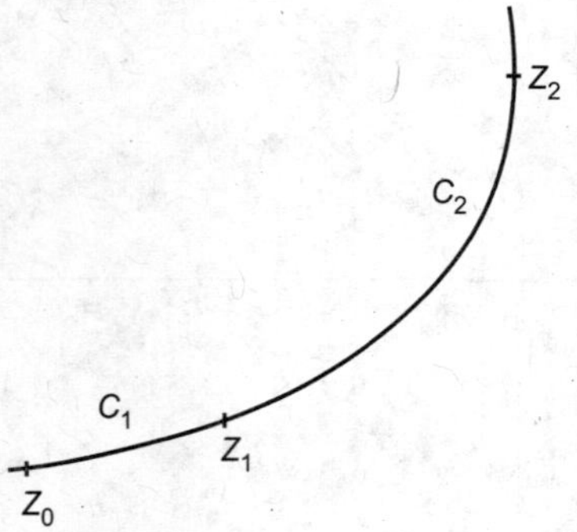

Figure 15.6

15.3.4 Evaluation of Complex Line Integral

A complex line integral can be integrated adopting one of the following two methods:

Antiderivative (Indefinite Integral) Method

If $f(z)$ is analytic in a simply-connected domain D and $F'(z) = f(z)$ in D, then

$$\int_C f(z)dz = \int_{z_1}^{z_2} f(z)dz = F(z_2) - F(z_1)$$

where C is a smooth curve in a domain D and $z_1, z_2 \in D$.

Parametric Representation Method

If C is a smooth curve in D with parametric representation $z = z(t),\ a \le t \le b$, then

$$\int_C f(z)dz = \int_a^b f(z(t)) \cdot \frac{dz}{dt} \cdot dt.$$

Thus the integral is converted into an ordinary integral in t and then evaluated.

15.3.5 Analytic Functions: Path Independence

Example 15.6

Evaluate $\int_0^{2+i} z^2 dz$ along (i) the real axis from $x = 0$ to 2 and then the vertical from $y = 0$ to $2+i$; and (ii) the imaginary axis from $y = 0$ to i and then the horizontal line from $x = 0$ to $2+i$.

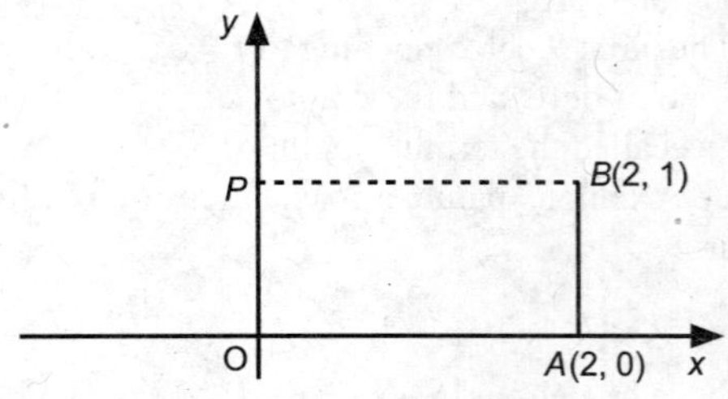

Figure 15.7

Solution Let $\int_0^{2+i} z^2 dz = I_1 + I_2$

We have $z^2 = (x+iy)^2 = x^2 - y^2 + 2ixy$

$$I_1 = \int_{OA} z^2 dz; \quad I_2 = \int_{AB} z^2 dz$$

(i) $dz = d(x+iy) = dx + idy$

Along OA: $y = 0,\ dy = 0;$ x varies from 0 to 2

Along AB: $x = 2$, $z = 2 + iy, dz = idy$, $z^2 = 4 - y^2 + 4iy$; y varies from 0 to 1

$$I_1 = \int_{OA} z^2 dz = \int_0^2 x^2 dx = \left[\frac{x^3}{3}\right]_0^2 = \frac{8}{3}$$

$$I_2 = \int_{AB} z^2 dz = \int_{y=0}^1 (4 - y^2 + 4iy) i dy$$

$$= i\left[4y - \frac{1}{3}y^3 + \frac{4}{2}iy^2\right]_0^1 = i(4 - \frac{1}{3} + 2i)$$

$$= -2 + \frac{11}{3}i$$

$$\therefore \int_0^{2+i} z^2 dz = I_1 + I_2 = \frac{8}{3} - 2 + \frac{11}{3}i = \frac{2}{3} + \frac{11}{3}i$$

$$I_1 = \int_{OP} z^2 dz; \quad I_2 = \int_{PB} z^2 dz$$

(ii) Along OP: $x = 0$, $z = x + iy$, $dz = idy$,

$z^2 = (iy)^2 = -y^2$; y varies from 0 to 1

$$I_1 = \int_{OP} z^2 dz = \int_{y=0}^1 - y^2 i dy = \left[-i\frac{y^3}{3}\right]_0^1 = -\frac{i}{3}$$

Along PB: $y = 1$, $z = x + i, dz = dx$; x varies from 0 to 2

$$I_2 = \int_{PB} z^2 dz = \int_0^2 (x+i)^2 dx = \int_0^2 (x^2 - 1 + 2ix) dx$$

$$= \left[\frac{x^3}{3} - x + ix^2\right]_0^2 = \frac{8}{3} - 2 + i4 = \frac{2}{3} + 4i$$

$$\therefore \int_C z^2 dz = I_1 + I_2 = -\frac{i}{3} + \frac{2}{3} + 4i = \frac{2}{3} + \frac{11}{3}i.$$

Note
The line integral is independent of the path of integration C when $f(z)$ is analytic.

Example 15.7
Evaluate $\int_C (z^2 + 3z + 2) dz$ where C is the arc of the cycloid $x = a(\theta + \sin\theta)$, $y = a(1 - \cos\theta)$ between the points $(0,0)$ and $(a\pi, 2a)$.

Solution The integrand $f(z) = z^2 + 3z + 2$ is an entire function (analytic in the entire complex plane). So, its integral along any curve connecting two fixed points will be the same, whatever the path may be.

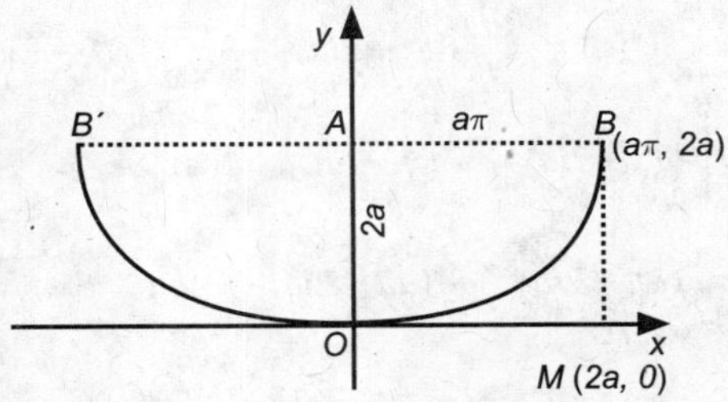

Figure 15.8

We can evaluate the integral by the indefinite integral method. If $F(z) = \frac{1}{3}z^3 + \frac{3}{2}z^2 + 2z$, then $F'(z) = f(z)$ so that

$$\int_C f(z) dz = \int_0^{(a\pi + 2ia)} F'(z) dz = F(z)\Big|_0^{(a\pi+2ia)}$$

$$= \frac{1}{3}(a\pi + 2ia)^3 + \frac{3}{2}(a\pi + 2ia)^2$$

$$+ 2(a\pi + 2ia)$$

If we want to evaluate the integral by representation of path we may choose the simplest path consisting of straight line segments
(i) OA joining $O = (0,0)$ and $A = (0, 2a)$ and
(ii) AB joining $A = (0, 2a)$ and $B = (a\pi, 2a)$

$$\therefore \int_C (z^2 + 3z + 2) dz = \int_{OA} f(z) dz + \int_{AB} f(z) dz$$

$$= I_1 + I_2$$

$$= \frac{1}{3}(a\pi + 2ia)^3 + \frac{3}{2}(a\pi + 2ia)^2$$

$$+ 2(a\pi + 2ia)$$

Along OA: $x = 0$, because $z = iy$, $dz = idy$; y varies from 0 to $2a$

$$I_1 = \int_{OA} f(z) dz = \int_0^{2a} [(iy)^2 + 3(iy) + 2] d(iy)$$

$$= i\left[\frac{-y^3}{3} + 3i\frac{y^2}{2} + 2y\right]_0^{2a}$$

$$= -\frac{1}{3}i(2a)^3 - \frac{3}{2}(2a)^2 + 2i(2a)$$

Along AB: $z = x + 2ia$, $dz = dx$; $y = 2a$; x varies from 0 to $a\pi$

$$I_2 = \int_{AB} f(z)dz = \int_0^{a\pi}[(x+2ia)^2 + 3(x+2ia)+2]dx$$

$$= \left[\frac{1}{3}(x+2ia)^3 + \frac{3}{2}(x+2ia)^2 + 2(x)\right]_0^{a\pi}$$

$$= \frac{1}{3}(a\pi+2ia)^3 + \frac{3}{2}(a\pi+2ia)^2 + 2a\pi$$

$$+\frac{1}{3}i(2a)^3 + \frac{3}{2}(2a)^2.$$

Example 15.8

Evaluate $\int_{(0,0)}^{(1,3)}[3x^2 ydx + (x^3 - 3y^2)]dy$ along the curve (i) $y = 3x$ and (ii) $y = 3x^2$. [JNTU 2003S (Set 3)]

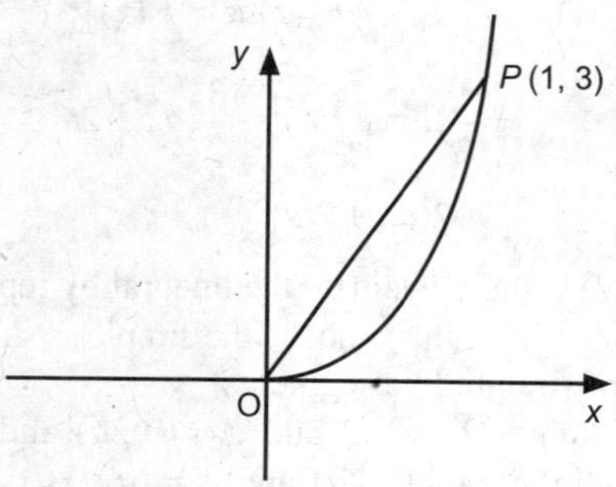

Figure 15.9

Solution Treating x as a parameter we can convert the integral in terms of x.

(i) $y = 3x, dy = 3dx$. Substituting for y and dy,

$$\int_{(0,0)}^{(1,3)}[3x^2 ydx + (x^3 - 3y^2)dy]$$

$$= \int_{x=0}^{1}[3x^2 \cdot 3xdx + (x^3 - 3\cdot 9x^2)3dx]$$

$$= \int_0^1 (12x^3 - 81x^2)dx$$

$$= 3x^4 - \frac{81}{3}x^3 \Big|_0^1 = 3 - 27 = -24$$

(ii) $y = 3x^2$; $dy = 6xdx$. Substituting for y and dy,

$$\int_{(0,0)}^{(1,3)}[3x^2 ydx + (x^3 - 3y^2)dy]$$

$$= \int_{x=0}^{1}[3x^2 \cdot 3x^2 \cdot dx + (x^3 - 3\cdot 9x^4)6x\,dx]$$

$$= \int_0^1 (15x^4 - 162x^5)dx$$

$$= (3x^5 - 27x^6)\Big|_0^1 = 3 - 27 = -24.$$

Note

The value of the integral along both paths is the same. We know from vector integral calculus that a line integral $\int_C \bar{F}(\bar{r})\cdot d\bar{r} = \int_C (F_1 dx + F_2 dy)$ with continuous F_1, F_2 in a domain D is independent of path in D if $\bar{F} = (F_1, F_2)$ is the gradient of a scalar function Φ in D such that $\bar{F} = \nabla\Phi$ or, in components,

$$F_1 = \frac{\partial\Phi}{\partial x}, \qquad F_2 = \frac{\partial\Phi}{\partial y}.$$

Here $F_1 = \frac{\partial\Phi}{\partial x} = 3x^2 y;\quad F_2 = \frac{\partial\Phi}{\partial y} x^3 - 3y^2$

$$\Rightarrow \quad \Phi(x,y) = x^3 y - y^3 = C.$$

15.3.6 Non-Analytic Functions: Path Dependence

Example 15.9

Evaluate $\int_0^{1+i}(x - y + ix^2)dz$ along the line $z = 0$ to $z = 1+i$. [JNTU 1993]

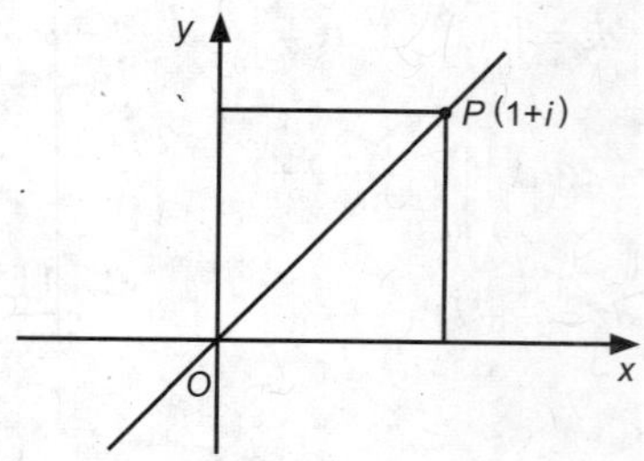

Figure 15.10

Solution Equation of OP is $y = x$

Let $z = x + iy = x + ix = (1+i)x$

$dz = (1+i)dx$; x varies from 0 to 1

$$\therefore \quad \int_0^{1+i}(x - y + ix^2)dz = (1+i)\int_0^1 ix^2 dx$$

$$= (1+i)i\left[\frac{x^3}{3}\right]_0^1 = \frac{i-1}{3}.$$

Example 15.10

Evaluate $\int_{z=1-i}^{2+i}(2x+iy+1)dz$ along the path

(i) $x=t+1,\ y=2t^2-1$ and

(ii) $x=t+1,\ y=2t-1$. [JNTU 1999S, 2005 (Set 1)]

Solution

(i) $x=t+1,\ y=2t^2-1$ are parametric equations of the parabola $y+1=2(x-1)^2$

$z=x+iy=(t+1)+i(2t^2-1),\ dz=(1+4it)dt;$

t varies from 0 to 1

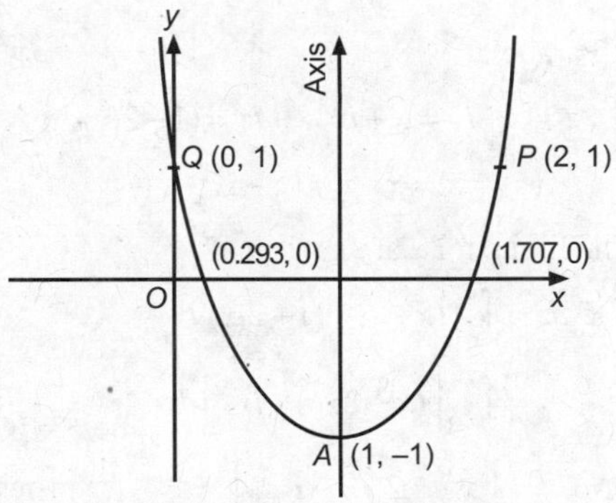

Figure 15.11

$$\int_{z=1-i}^{2+i}(2x+iy+1)dz$$

$$=\int_{t=0}^{1}[2t+2+1+i(2t^2-1)](1+i4t)dt$$

$$=\int_0^1[(2t+3+4t-8t^3)+i(8t^2+12t+2t^2-1)]dt$$

$$=\left[(3t+3t^2-2t^4)+i\left(\frac{10t^3}{3}+6t^2-t\right)\right]_0^1$$

$$=(3+3-2)+i\left(\frac{10}{3}+6-1\right)=4+\frac{25}{3}i$$

(ii) $x=t+1,\ y=2t-1$ are the parametric equations of the straight line $y=2x-3$. The limits for t are 0 and 1.

$$\int_{z=1-i}^{2+i}(2x+iy+1)dz$$

$$=\int_{t=0}^{1}\left[(2t+3)+i(2t-1)\right](dt+i2dt)$$

$$=(1+2i)\left[t^2+3t+i(t^2-t)\right]_0^1$$

$$=(1+2i)4=4+8i.$$

Example 15.11

Integrate $f(z)=x^2+ixy$ from $A(1,1)$ to $B(2,8)$ along the straight line AB.

Solution Equation of AB is $y=7x-6$. Let $z=x+iy,\ dz=dx+idy=1+7idx;$ x varies from 1 to 2

$$\int_{AB}(x^2+ixy)dz$$

$$=\int_{x=1}^{2}[x^2+i(7x^2-6x)](1+7i)dx$$

$$=(1+7i)\left[\frac{x^3}{3}+i\left(\frac{7}{3}x^3-3x^2\right)\right]_1^2$$

$$=(1+7i)\left[\frac{2^3-1}{3}+i\left\{\frac{7}{3}(2^3-1)-3(2^2-1)\right\}\right]$$

$$=(1+7i)\left[\frac{7}{3}+i\left(\frac{49}{3}-9\right)\right]=(1+7i)\left(\frac{7+22i}{3}\right)$$

$$=\frac{7-154+i(22+49)}{3}=\frac{-147+i71}{3}.$$

Example 15.12

Evaluate $\int_C \bar{z}\,dz$ where C is the upper half of the unit circle $|z|=1$ taken in the clockwise sense.

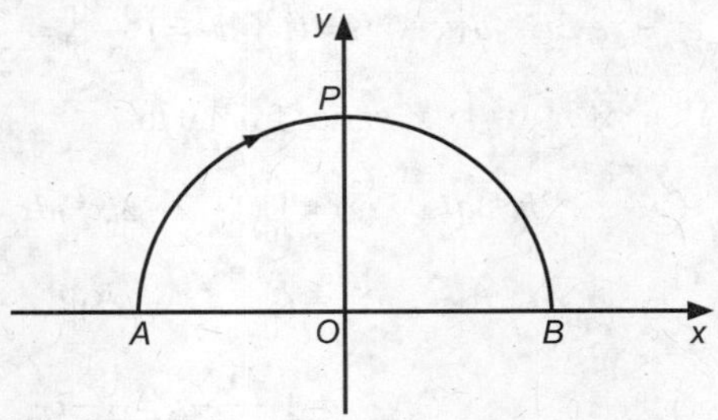

Figure 15.12

Solution The upper half of the unit circle $|z|=1$ taken in the clockwise sense is given by $z=e^{it}\ (\pi\le t\le 0),\quad dz=ie^{it}\,dt$ so that $\bar{z}=e^{-it}$

$$\therefore\quad \int_C \bar{z}\,dz=\int_\pi^0 e^{-it}ie^{it}\,dt=it\Big|_\pi^0=-i\pi.$$

Example 15.13

Evaluate $\int_c f(z)dz$ where $f(z) = y - x - i3x^2$ and (i) C is the straight line segment from 0 to $1+i$ and also (ii) when C consists of line segments OA from $z = 0$ to $z = i$ and AB from $z = i$ to $z = 1 + i$.

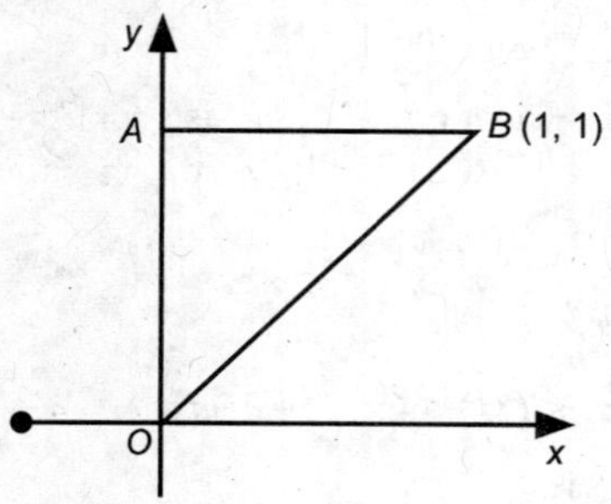

Figure 15.13

Solution

(i) If $O = (0, 0)$ and $B = (1, 1)$ then the equation of OB is $y = x$ and x varies from 0 to 1 (Fig. 15.13).

$$z = x + iy \Rightarrow dz = (1+i)dx \quad \because \quad y = x$$

$$\int_C f(z)dz = \int_C (y - x - i3x^2)dz = -3i\int_0^1 x^2(1+i)dx$$

$$= (3-3i)\frac{x^3}{3}\Big|_0^1 = (1-i)$$

(ii) Along OA: $x = 0$; y varies from 0 to 1

$$\int_{OA}(y - x - 3ix^2)d(x+iy) = i\int_0^1 y\,dy = i\frac{y^2}{2}\Big|_0^1 = \frac{i}{2}$$

Along AB: $y = 1$; x varies from 0 to 1

$$\int_{AB}(y - x - 3ix^2)d(x+iy) = \int_0^1 (1 - x - 3ix^2)dx$$

$$= (x - \frac{x^2}{2} - ix^3)\Big|_0^1$$

$$= 1 - \frac{1}{2} - i = \frac{1}{2} - i$$

$$\therefore \quad \int_C f(z)dz = \frac{i}{2} + \frac{1}{2} - i = \frac{1-i}{2}.$$

Example 15.14

Evaluate $\int_C \bar{z}\,dz$ where C is (i) the line segment joining the points (1, 1) and (2, 4) and (ii) the curve $x = t, y = t^2$ joining the points (1, 1) and (2, 4).

Solution

(i) C: Straight line

$$y = 3x - 2, z = x + iy = x + i(3x-2), dz = (1+3i)dx$$

$$f(z) = \bar{z} = x - iy = x - i(3x-2).$$

The limits for x are $x = 1$ to 2.

$$\int_C \bar{z}\,dz = \int_1^2 [(1-3i)x + 2i](1+3i)dx$$

$$= \left[10\cdot\frac{x^2}{2} + (2i-6)x\right]_1^2$$

$$= 5(2^2 - 1) + (2i-6)(2-1)$$

$$= 15 - 6 + 2i = 9 + 2i$$

(ii) C: Parabola

$$y = x^2, z = x + iy = x + ix^2, \quad dz = (1+2ix)dx$$

$$f(z) = \bar{z} = x - iy = x - ix^2.$$

The limits for x are $x = 1$ to 2

$$\int_C \bar{z}\,dz = \int_{x=1}^2 (x - ix^2)(1+2ix)dx$$

$$= \int_{x=1}^2 \left[(x + 2x^3) + ix^2\right]dx$$

$$= \left[\frac{x^2}{2} + \frac{2}{4}x^4 + \frac{i}{3}x^3\right]_1^2$$

$$= \frac{1}{2}(2^2 - 1) + \frac{1}{2}(2^4 - 1) + \frac{i}{3}(2^3 - 1)$$

$$= \frac{3}{2} + \frac{15}{2} + \frac{7i}{3} = 9 + \frac{7i}{3}.$$

Example 15.15

Evaluate $\int_0^{1+i}(x^2 - iy)dz$ along the paths (i) $y = x$ and (ii) $y = x^2$.

Solution

(i) Equation of OB is $y = x, dy = dx$, x varies from 0 to 1 (Fig. 15.14)

$$\int_0^{1+i}(x^2 - iy)dz = \int_{(0,0)}^{(1,1)}(x^2 - iy)(dx + idy)$$

$$= \int_{x=0}^1 (x^2 - ix)(dx + idx)$$

$$= (1+i)\int_{x=0}^1 (x^2 - ix)dx$$

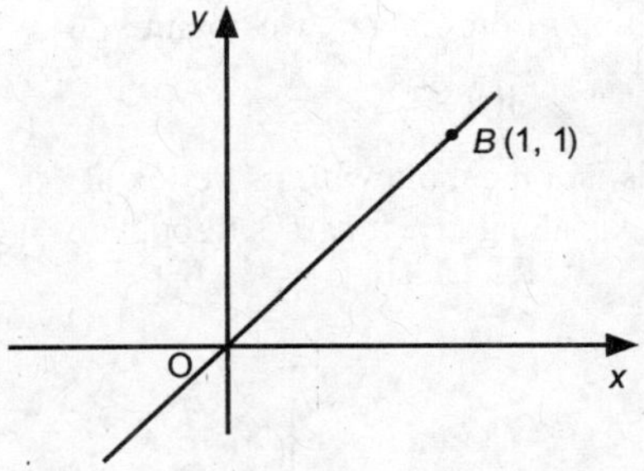

Figure 15.14

$$= (1+i)\left[\frac{x^3}{3} - i\frac{x^2}{2}\right]_0^1 = (1+i)\left(\frac{1}{3} - \frac{i}{2}\right)$$

$$= \frac{1}{3} + \frac{1}{2} + i\left(\frac{1}{3} - \frac{1}{2}\right) = \frac{5}{6} - \frac{i}{6}$$

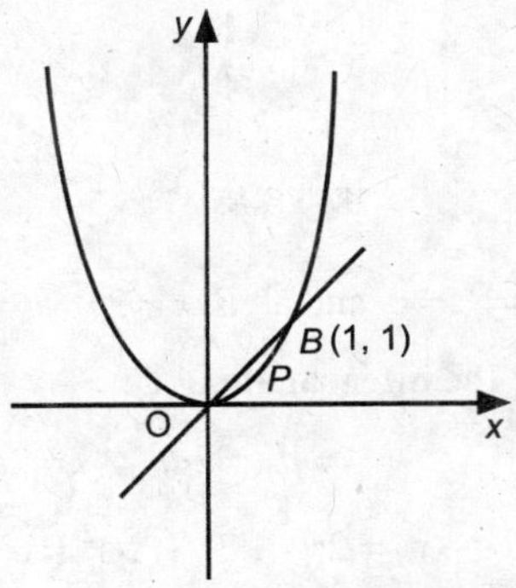

Figure 15.15

(ii) Equation of OPB is $y = x^2$, $dy = 2x dx$; x varies from 0 to 1 (Fig. 15.15)

$$\int_0^{1+i} (x^2 - iy)dz = \int_{(0,0)}^{(1,1)} (x^2 - ix^2)(dx + i2x dx)$$

$$= (1-i)\int_{x=0}^{1} (x^2 dx + i2x^3 dx)$$

$$= (1-i)\left[\frac{x^3}{3} + i\frac{2}{4}x^4\right]_0^1$$

$$= (1-i)\left(\frac{1}{3} + i\frac{1}{2}\right)$$

$$= \left(\frac{1}{3} + \frac{1}{2}\right) + i\left(\frac{1}{2} - \frac{1}{3}\right) = \frac{5}{6} + \frac{i}{6}.$$

Example 15.16

Integrate $f(z) = x^2 + ixy$ from $A(1,1)$ to $B(2,8)$ along (i) the straight line AB and (ii) the curve $C: x = t$, $y = t^3$.

Solution

(i) Equation of straight line AB is $\frac{y-1}{8-1} = \frac{x-1}{2-1}$

$\Rightarrow y = 7x - 6, dy = 7dx$. The limits for x are $x = 1$ to 2

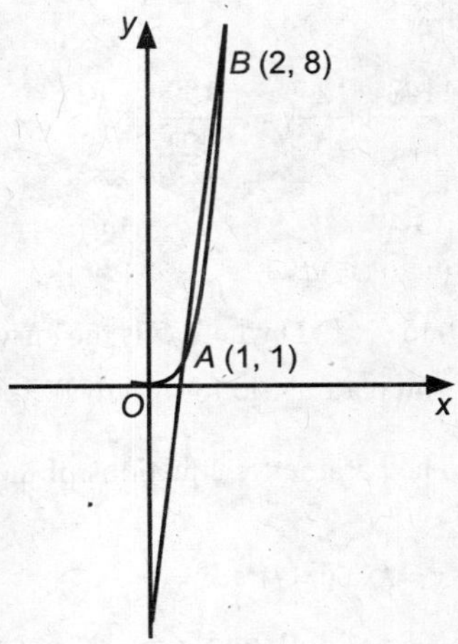

Figure 15.16

$$\int_{AB} f(z)dz = \int_{AB} (x^2 + ixy)d(x + iy)$$

$$= \int_1^2 [x^2 + ix(7x-6)](1+7i)dx$$

$$= (7i+1)\int_1^2 [(7i+1)x^2 - 6ix]dx$$

$$= (14i - 48)\left[\frac{x^3}{3}\right]_1^2 + (42 - 6i)\left[\frac{x^2}{2}\right]_1^2$$

$$= (14i - 48)\frac{7}{3} + (42 - 6i)\frac{3}{2}$$

$$= \frac{98}{3}i - 112 + 63 - 9i = -49 + \frac{71}{3}i$$

(ii) $C: x = t, y = t^3$. The limits for t are $t = 1$ to 2

$$\int_C f(z)dz = \int_{(1,1)}^{(2,8)} (x^2 + ixy)d(x + iy)$$

$$= \int_{t=1}^{2} (t^2 + it^4)d(t + it^3)$$

$$= \int_{t=1}^{2} (t^2 + it^4)(1 + 3it^2)dt$$

$$= \int_{t=1}^{2} [(t^2 - 3t^6) + 4it^4]dt$$

$$= \left[\frac{1}{3}t^3 - \frac{3}{7}t^7 + i\frac{4}{5}t^5\right]_1^2$$

$$= \frac{2^3 - 1}{3} - \frac{3(2^7 - 1)}{7} + \frac{4i}{5}(2^5 - 1)$$

$$= \frac{7}{3} - \frac{3.127}{7} + \frac{4i}{5}\cdot 31$$

$$= \frac{49 - 1143}{21} + \frac{124}{5}i = -\frac{1094}{21} + \frac{124}{5}i.$$

Example 15.17

Evaluate $\int_C (y^2 + z^2)dx + (z^2 + x^2)dy + (x^2 + y^2)dz$ from (0, 0, 0) to (1, 1, 1) where C is the curve $x = t$, $y = t^2, z = t^3$ in parametric form. [JNTU 2004(4)]

Solution The parametric equations of the path of integration C are

$$x = t, \quad y = t^2 \quad z = t^3 \tag{1}$$

The ends $A = (0,0,0)$ and $B = (1,1,1)$ correspond to $t = 0$ and $t = 1$, $dx = dt, dy = 2tdt, dz = 3t^2 dt$.

$$\therefore \int_C (y^2 + z^2)dx + (z^2 + x^2)dy + (x^2 + y^2)dz$$

$$= \int_0^1 (t^4 + t^6)dt + (t^6 + t^2)2tdt + (t^2 + t^4)3t^2 dt$$

$$= \int_0^1 (2t^7 + 4t^6 + 4t^4 + 2t^3)dt$$

$$= \left[\frac{2}{8}t^8 + \frac{4}{7}t^7 + \frac{4}{5}t^5 + \frac{2}{4}t^4\right]_0^1$$

$$= \frac{1}{4} + \frac{4}{7} + \frac{4}{5} + \frac{1}{2} = \frac{297}{140}.$$

Example 15.18

$\int (2y + x^2)dx + (3x - y)dy$ along the parabola $x = 2t$, $y = t^2 + 3$ joining the points (0, 3) and (2, 4).

Solution Equations of the parabola are

$$x = 2t, \quad y = t^2 + 3 \tag{1}$$

Eliminating t from Eq. (1), we obtain

$$x^2 = 4(y - 3) \tag{2}$$

This is a parabola with its vertex at $A = (0,3)$ and its axis along OY. It passes through $B = (2,4)$.

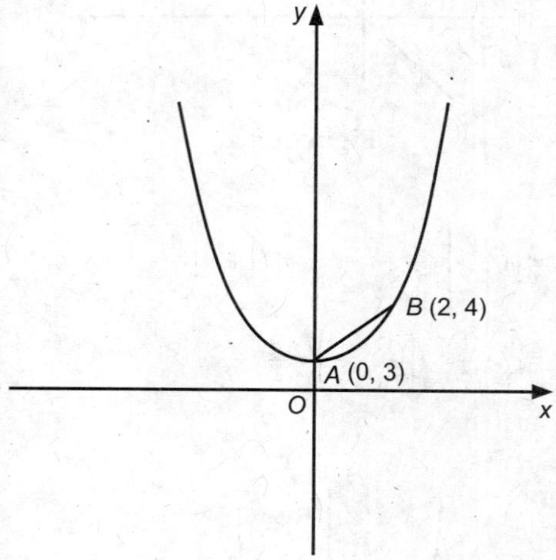

Figure 15.17

At A, $x = 0 \Rightarrow t = 0$ and at B, $x = 2 \Rightarrow t = 1$.

To evaluate the integral along *C*:

$C: x = 2t$, $y = t^2 + 3 \Rightarrow dx = 2dt$, $dy = 2tdt$

$$2y + x^2 = 2(t^2 + 3) + (2t)^2 = 6t^2 + 6$$

$$3x - y = 3(2t) - (t^2 + 3) = 6t - t^2 - 3$$

Now, the given integral is

$$\int_C (2y + x^2)dx + (3x - y)dy$$

$$= \int_0^1 (6t^2 + 6)2dt + (6t - t^2 - 3)2tdt$$

$$= \int_0^1 (12t^2 + 12 + 12t^2 - 2t^3 - 6t)dt$$

$$= \int_0^1 (-2t^3 + 24t^2 - 6t + 12)dt$$

$$= \left[-\frac{2}{4}t^4 + \frac{24}{3}t^3 - \frac{6}{2}t^2 + 12t\right]_0^1$$

$$= -\frac{1}{2} + 8 - 3 + 12 = \frac{33}{2}$$

To evaluate the integral along AB:

Equations of AB are

$$\frac{x}{x-0}=\frac{y}{4-3}=t \Rightarrow x=2t, y=t$$

The limits of integration are $t=0, t=1$

$$\therefore \int_{AB}(2y+x^2)dx+(3x-y)dy$$

$$=\int_0^1(2t+4t^2)2dt+(6t-t)dt$$

$$=\int_0^1(8t^2+9t)dt=\frac{8}{3}t^3+\frac{9}{2}t^2\Big|_0^1$$

$$=\frac{8}{3}+\frac{9}{2}=\frac{16+27}{6}=\frac{43}{6}$$

The value is different because the integrand is not an analytic function of z.

Example 15.19

Evaluate $\int_C(x-2y)dx+(y^2-x^2)dy$ where C is the boundary of the first quadrant of the circle $x^2+y^2=4$. [JNTU 2003(3)]

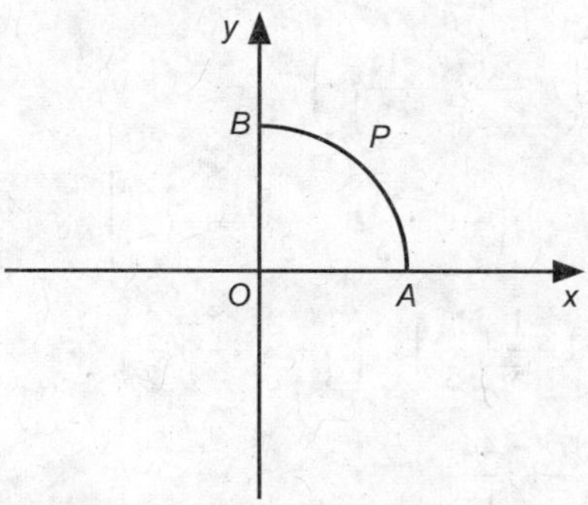

Figure 15.18

Solution

The path of integration is $C: x^2+y^2=4$. (1)

Parametric equations of C are

$$x=2\cos t, \quad y=2\sin t \ (15.2)$$

$$dx=-2\sin t dt, \quad dy=2\cos t dt$$

$$\int_{APB}f(z)\,dz$$

$$=\int_0^{\pi/2}[(2\cos t-4\sin t)(-2\sin t\,dt)$$

$$+4(\sin^2 t-\cos^2 t)2\cos t\,dt]$$

$$=\int_0^{\pi/2}[-4\sin t\cos t dt+8\sin^2 t\,d$$

$$+8(2\sin^2 t-1)\cos t\,dt]$$

$$=\int_0^{\pi/2}[-4\sin t\,d(\sin t)+4dt-4\cos 2t\,dt$$

$$+16\sin^2 t\,d(\sin t)-8\,d(\sin t)$$

$$=\left[-2\sin^2 t+4t-2\sin 2t+\frac{16}{3}\sin^3 t-8\sin t\right]_0^{\pi/2}$$

$$=-2\cdot 1+4\cdot\frac{\pi}{2}-2\cdot 0+\frac{16}{3}\cdot 1-8\cdot 1$$

$$=2\pi-\frac{14}{3}.$$

Example 15.20

Evaluate $\int_{(0,0)}^{(1,1)}(3x^2+4xy+ix^2)dz$ along $y=x^2$.

[JNTU 2005S (Set 3)]

Solution Put $x=t$, $y=t^2$, $z=t+it^2$, and $dz=(1+2it)dt$ where t varies from 0 to 1. The given integral becomes

$$\int_{t=0}^1(3t^2+4t\cdot t^2+it^2)(1+2it)dt$$

$$=\int_0^1(3t^2+4t^3+it^2)(1+2it)dt$$

$$=\int_0^1[(3t^2+2t^3)+it^2+6t^3+3t^4]dt$$

$$=\left[\left(t^3+\frac{1}{2}t^4\right)+i\left(\frac{1}{3}t^3+\frac{3}{2}t^4+\frac{8}{5}t^5\right)\right]_0^1$$

$$=1+\frac{1}{2}+i\left(\frac{1}{3}+\frac{3}{2}+\frac{8}{5}\right)=\frac{3}{2}+\frac{103}{30}i.$$

Example 15.21

Evaluate $\int_{1-i}^{2+i}(2x+iy+1)dz$ along the straight line joining 1 − i and 2 + i [JNTU 2006 (Set 1)]

Solution Let the given points be $A=(1,-1)$ and $B=(2,1)$.

Equation of AB is $\dfrac{y+1}{1-(-1)} = \dfrac{x-1}{2-1} = t$ (say)

$\Rightarrow x = t+1,\ y = 2t-1.$

$z = x+iy = (t+1)+i(2t-1);\ dz = dt + 2ydt;$

t varies from $t = 0$ to 1

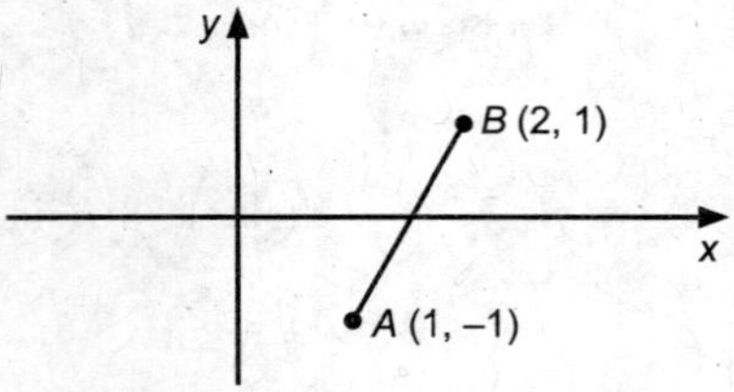

Figure 15.19

$$\therefore \int_{1-i}^{2+i}(2x+iy+1)dz$$
$$= \int_0^1[(2t+3)+i(2t-1)](2i+1)dt$$
$$= \int_0^1[(2t+3-4t+2)+i(2t-1+4t+6)]dt$$
$$= \int_0^1[(-2t+5)+i(6t+5)]dt$$
$$= [-t^2+5t+i(3t^2+5t)]_0^1$$
$$= -1+5+i(3+5) = 4+8i.$$

Example 15.22

Evaluate $\int_0^{1+1}(x-y^2+ix^3)dz$ along the real axis from $z = 0$ to 1. [JNTU 2005*S* (Set 4)]

Solution Let $z = x+iy \Rightarrow dz = dx+idy$. Along the x-axis, $y = 0$ and $dz = dx$; x varies from $x = 0$ to 1.

$$\int_0^{1+i}(x-y^2+ix^3)dz = \int_0^1(x-0+ix^3)dx$$
$$= \left[\frac{x^2}{2}+\frac{i}{4}x^4\right]_0^1 = \frac{1}{2}+\frac{i}{4}.$$

Example 15.23

Evaluate $\int_{(0,0)}^{(1,1)}\left[3x^2+5y+i(x^2-y^2)\right]dz$ along $y^2 = x$. [JNTU 2006 (Set 4)]

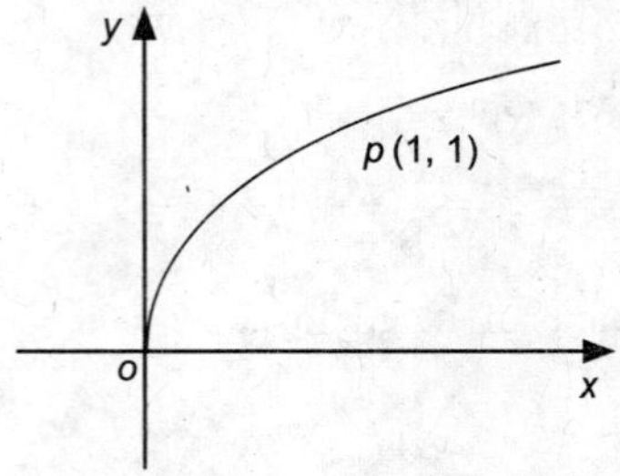

Figure 15.20

Solution Put $y = t$, $x = t^2$, $z = t^2+it \Rightarrow dz = (2t+i)dt$; t varies from 0 to 1.

$$I = \int_{(0,0)}^{(1,1)}[3x^2+5y+i(x^2-y^2)]dz$$
$$= \int_0^1[3t^4+5t+i(t^4-t^2)(2t+1)dt$$
$$= \int_0^1[(6t^5+10t^2-t^4+t^2) + i(2t^5-2t^3+3t^4+5t)]dt$$
$$= \int_0^1[(6t^5-t^4+11t^2) + i(2t^5+3t^4-2t^3+5t)]dt$$
$$= \left[\left(t^6-\frac{1}{5}t^5+\frac{11}{3}t^3\right) + i\left(\frac{1}{3}t^6+\frac{3}{5}t^5-\frac{1}{2}t^4+\frac{5}{2}t^2\right)\right]_0^1$$
$$= \left(1-\frac{1}{5}+\frac{11}{3}\right)+i\left(\frac{1}{3}+\frac{3}{5}-\frac{1}{2}+\frac{5}{2}\right)$$
$$= \frac{67}{15}+i\frac{44}{15}.$$

Example 15.24

Evaluate $\int_C(y+2xy)dx+(x^2-2xy)dt$ where C is the boundary of the region enclosed by $y = x^2$ and $x = y^2$. [JNTU 2006 (Set 4)]

Solution C consists of two curves C_1 defined by $y = x^2$ and C_2 defined by $y^2 = x$.

Integral along C_1

We take $x = t$, $y = t^2$, $dx = dt$, $dy = 2tdt$; t varies from 0 to 1.

$$\int_{C_1}(y^2+2xy)dx+(x^2-2xy)dy$$

$$=\int_0^1(t^4+2t^3)dt+(t^2-2t^3)2tdt$$

$$=\int_0^1(-3t^4+4t^3)dt$$

$$=\left[-\frac{3}{5}t^5+t^4\right]_0^1$$

$$=-\frac{3}{5}+1=\frac{2}{5}$$

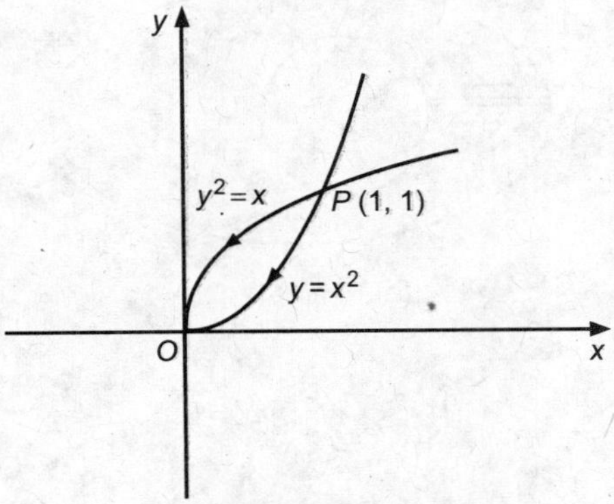

Figure 15.21

Integral along C_2

We take $x = t^2$, $y = t$, $dx = 2tdt$, $dy = dt$; t varies from 1 to 0.

$$\int_{C_2}(y^2+2xy)dx+(x^2-2xy)dy$$

$$=\int_1^0(t^2+2t^3)2tdt+(t^4-2t^3)dt$$

$$=\int_1^0 5t^4dt=[t^5]_1^0=-1$$

$$\therefore\quad \int_C(y^2+2xy)dx+(x^2-2xy)dy=\int_{c_1}+\int_{c_2}$$

$$=\frac{2}{5}-1=\frac{-3}{5}.$$

15.4 CAUCHY–GOURSAT THEOREM

[2]Cauchy's integral theorem (C.I.T.) or Cauchy's theorem

Theorem 15.1 If $f(z) \in H(D)$ where D is a simply-connected domain, then

$$\int_C f(z)dz = 0$$

for every simple, closed curve C in D.

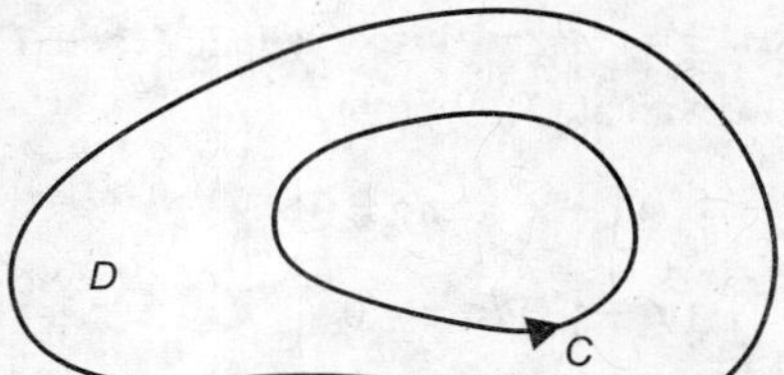

Figure 15.22

Proof: Let $f(z) = u(x, y) + iv(x, y)$, $dz = dx + idy$ so that

$$\int_C f(z)dz = \int_C (u+iv)(dx+idy)$$

$$=\int_C(udx-vdy)+i\int_C(vdx+udy) \quad (15.6)$$

$$=I_1+iI_2 \text{ (say)} \quad (15.7)$$

Let us assume that $f'(z)$ is continuous in D Therefore, partial derivatives u_x, u_y, v_x and v_y are continuous in D containing C and satisfy the Cauchy–Riemann equations

$$u_x = v_y, \qquad u_y = -v_x \quad (15.8)$$

Hence we can apply Green's theorem in plane to Eq. (15.7).

---

[2] Cauchy, Augustin Louis (1789–1857), great French analyst, applied mathematician and group theorist. After Euler, the most prolific mathematician in history. He worked on wave, elasticity and group theory; introduced modern rigour into calculus, founded the theory of functions of a complex variable; inaugurated modern era in differential equations with his existence theorems.

[3]Green's Theorem

If F_1 and F_2 are continuous functions of x and y having continuous partial derivatives $\frac{\partial F_1}{\partial y}$ and $\frac{\partial F_2}{\partial x}$ in a region R bounded by a simple closed curve C, then

$$\oint_C F_1 dx + F_2 dy = \int_R\int\left(\frac{\partial F_2}{\partial x} - \frac{\partial F_1}{\partial y}\right)dxdy$$

where C is traversed in the counter-clockwise direction.

Take $F_1 = u, F_2 = -v$ to evaluate I_1 and $F_1 = v$, $F_2 = u$ to evaluate I_2. We obtain

$$\left.\begin{aligned} I_1 &= -\iint_D (u_y + v_x)dxdy = 0 \\ I_2 &= \iint_D (u_x - v_y)dxdy = 0 \end{aligned}\right\} \quad \text{by Eq. (15.8)}$$

$$\therefore \quad \int_C f(z)dz = 0.$$

15.4.1 Alternative Statement of Cauchy's Theorem

If $f(z)$ is analytic at all points inside and on a simple closed curve C, then $\int_C f(z)dz = 0$.

Note

1. Cauchy's integral theorem forms the basis for the theory of complex integration.
2. An additional condition that $f'(z)$ is continuous in D is assumed to apply Green's theorem in the proof. French mathematician E. Goursat (1858–1936) proved the theorem in 1900 without this assumption and this theorem is known as Cauchy–Goursat theorem. His proof is somewhat complicated and hence omitted here.
3. The Cauchy–Riemann equations are precisely the conditions for two real integrals in Eq. (15.7) above to be independent of path. Hence the line integral of an analytic function $f(z)$ in D is independent of the path joining any two points in D.

[3] Green, George (1793–1841), English analyst and applied mathematician.

15.5 Cauchy's Theorem for Multiply-Connected Domain

Theorem 15.2 If $f(z)$ is analytic in a doubly-connected domain D bounded by two simple closed curves C_1 and C_2, then

$$\int_{C_1} f(z)dz = \int_{C_2} f(z)dz.$$

Proof: Select two points A and B on C_1 and C_2 respectively and join them by an arc AB. This converts the doubly-connected domain D into a simply connected one bounded by $APABQBA$. By Cauchy's integral theorem,

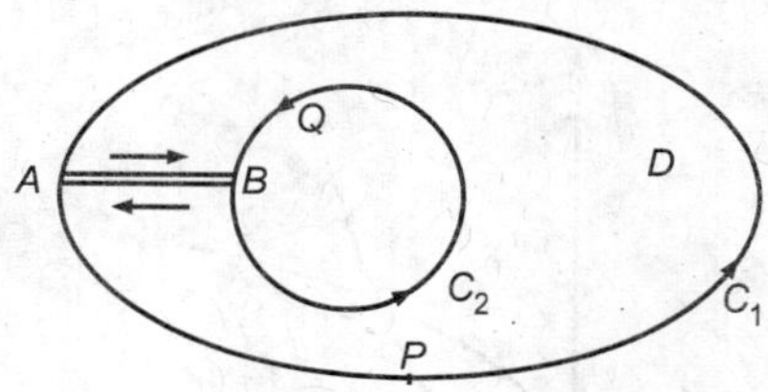

Figure 15.23

$$\int_{APABQBA} f(z)dz = 0$$

$$\Rightarrow \int_{APA} + \int_{AB} + \int_{BQB} + \int_{BA} f(z)dz = 0$$

$$\Rightarrow \int_{C_1} f(z)dz = \int_{C_2} f(z)dz$$

$$\because \quad \int_{BA} f(z)dz = -\int_{AB} f(z)dz$$

$$\text{and } \int_{BQB} f(z)dz = -\int_{APA} f(z)dz.$$

Note

We can extend the result to p-ply connected domains, $p \geqslant 3$.

Alternative Statement

Let C be a simple closed curve. Let C_j $(j = 1, 2, \cdots n)$ be a finite number of simple closed curves lying in the interior of C such that the interiors of C_j's are disjoint. Suppose D is the closed region consisting of all points within and on C except the points interior to each C_j. Let B be the entire oriented boundary of D consisting of C and all C_j's described in such a direction that all the points of D are to the left of B. If f be a function analytic in D, then $\int_B f(z)dz = 0$.

Proof: Let L_1 be a polygonal path joining a point of C to a point of C_1; L_2 a polygonal path joining a point of C_1 to a point of $C_2, \cdots L_j$, a polygonal path joining a point of C_{j-1} to a point of C_j $(j = 1, 2, \cdots n)$ such that no two L_j's cross each other (Fig. 15.24).

This divides the region D into two simply-connected regions D_1 and D_2. If B_1 and B_2 are the boundaries of D_1 and D_2 respectively, then by Cauchy's theorem for simply-connected region

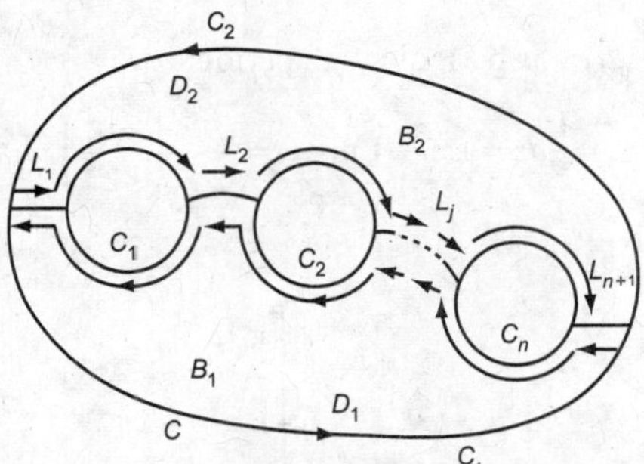

Curves $C_1 \cup C_2 = C$
Curves $B_1 \cup B_2 = B$,
which consist of curves
C_j and lines L_j
(Domains $D_1 \cup D_2 = D$)

Figure 15.24

$$\int_{B_1} f(z)dz = 0 \quad \text{and} \quad \int_{B_2} f(z)dz = 0$$

Further, $\int_{B_1} f(z)dz + \int_{B_2} f(z)dz = \int_B f(z)dz$ since the integrals along L_j are taken twice in opposite directions and thus get cancelled out.

$$\therefore \quad \int_B f(z)dz = 0.$$

Remark

We note that $B = C - C_1 - C_2 - \cdots C_n$ and hence the above theorem can also be written as

$$\int_C f(z)dz = \int_{C_1} + \int_{C_2} + \cdots + \int_{C_n} f(z)dz$$

In particular, if C is a simple closed curve and C_0 is another simple closed curve lying in the interior of C and $f \in H(D)$, where D consists of all points inside and on C excluding those points which are interior to C_0 then $\int_C f(z)dz = \int_{C_0} f(z)dz$.

15.5.1 Basic Result: Integral of $\frac{1}{z}$ Around $|z| = 1$ (Unit Circle)

Prove that $\oint_C \frac{1}{z} dz = 2\pi i$ where C is the unit circle $|z| = 1$ traced out in the counter-clockwise direction.

Solution Let $z(t) = \cos t + \sin t = e^{it}$

$\dot{z}(t) = \frac{dz}{dt} = ie^{it}$. As z traverses around the unit circle, t varies from $t = 0$ to $t = 2\pi$

$$\oint_C \frac{dz}{z} = \int_0^{2\pi} e^{-it} \cdot ie^{it} dt = 2\pi i.$$

15.5.2 Integral of Integer Power of $(z - a)$ Around Circle of Radius r

Prove that $\int_C (z-a)^m \, dz = \begin{cases} 0 & m \neq -1 \\ 2\pi i & m = -1 \end{cases}$

where C is any simple closed path enclosing 'a'.

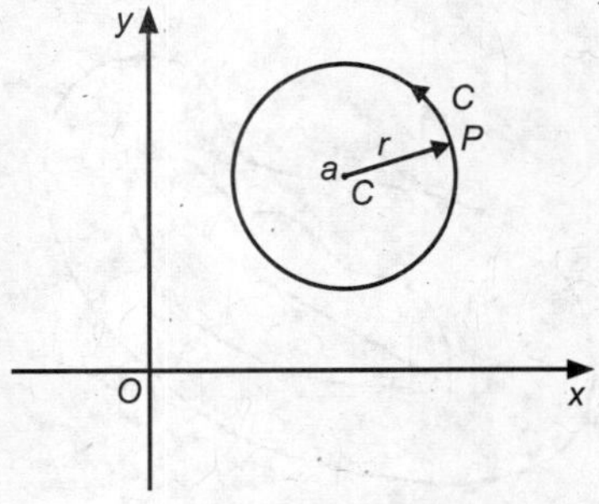

Figure 15.25

Solution Integrate $f(z) = (z-a)^m, m \in \mathbb{Z}$ and 'a' a constant counter-clockwise around the circle C of radius r and centre 'a'.

$$z(t) = a + r(\cos t + i\sin t) = a + re^{it},\ 0 \le t \le 2\pi$$

$$z - a = re^{it} \Rightarrow (z-a)^m = r^m e^{imt}, \quad dz = ire^{it}dt$$

(i) $m \neq -1$

$$\oint_C (z-a)^m dz = \int_0^{2\pi} r^m e^{imt} ire^{it} dt$$

$$= ir^{m+1}\int_0^{2\pi} e^{i(m+1)t} dt \qquad (1)$$

(ii) $m = -1$

$$= \frac{r^{(m+1)}}{m+1}(e^{(m+1)2\pi i} - 1) = 0$$

$$\oint (z-a)^{-1} dz = \frac{1}{r}\int_0^{2\pi} e^{-it} \cdot ire^{it} dt$$

$$= i\oint_0 dt = 2\pi i \qquad (2)$$

By the principle of continuous deformation of path C, the result holds for all simple closed paths C containing a circle of radius r $(0 < r < \infty)$ and centre 'a'.

15.5.3 Evaluation of Complex Line Integral—Method 1

Indefinite integral of an analytic function theorem:

Theorem 15.3 If $f(z)$ is analytic in a simply-connected domain D, there exists an analytic function $F(z)$ such that $F'(z) = f(z)$ in D for all paths joining z_0 and z_1 in D.

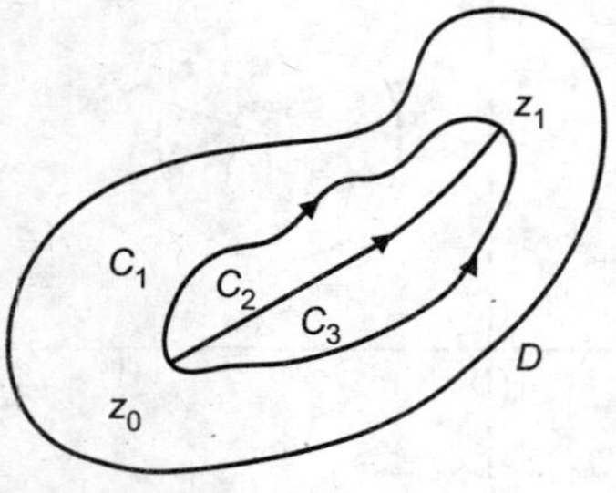

Figure 15.26

Proof:

$$\int_{z_0}^{z_1} f(z)dz = F(z_1) - F(z_0) \therefore \quad F'(z) = f(z).$$

Note that we can write z_0 and z_1 instead of path C since we get the same value for all C joining z_0 to z_1.

Example 15.25

1. $\displaystyle\int_0^{1+i} z^2 dz = \frac{1}{3}z^3 \Big|_0^{1+i} = \frac{1}{3}(1+i)^3 - 0 = -\frac{2}{3} + \frac{2}{3}i.$

2. $\displaystyle\int_0^{\frac{\pi}{2}i} e^z dz = e^z \Big|_0^{\frac{\pi i}{2}} = e^{\frac{\pi i}{2}} - e^0$

$$= \cos\frac{\pi}{2} + i\sin\frac{\pi}{2} - 1 = i - 1.$$

3. $\displaystyle\int_{-\pi i}^{\pi i} \cos z dz = \sin z \Big|_{-\pi i}^{\pi i} = 2\sin \pi i = 2i\sinh \pi.$

4. $\displaystyle\int_{8+\pi i}^{8-3\pi i} e^{z/2} dz = 2e^{z/2}\Big|_{8+\pi i}^{8-3\pi i} = 2(e^{4-\frac{3}{2}\pi i} - e^{4+\frac{\pi i}{2}})$

$$2e^4(e^{\frac{\pi i}{2}} - e^{\frac{\pi i}{2}}) = 0$$

$\therefore e^z$ is periodic with period $2\pi i$.

5. $\displaystyle\int_{-i}^{i} \frac{1}{z} dz = \text{Ln}\, i - \text{Ln}(-i) = \frac{i\pi}{2} - \left(\frac{-i\pi}{2}\right) = i\pi.$

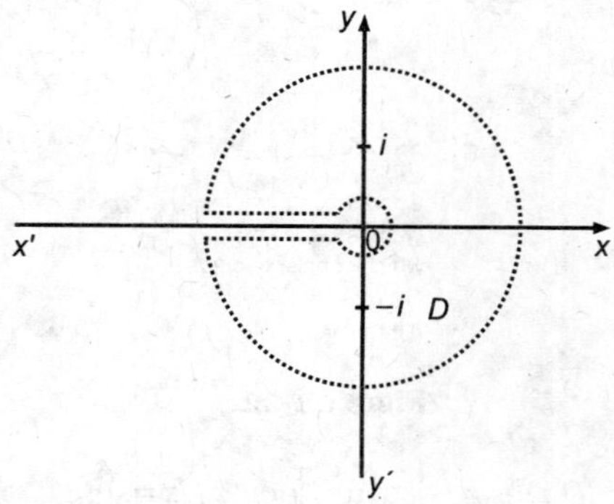

Figure 15.27

Here D is the complex plane without 0 and the negative real axis OX', where Ln z is not analytic and is clearly a simply-connected domain.

Method 2

Use of representation of path

Theorem 15.4 Let C be a piece-wise smooth path represented by $z(t) = x(t) + iy(t)$ $(a \le t \le b)$.

Suppose $f(z)$ is continuous on C. Then

$$\int_C f(z)dz = \int_a^b f[z(t)]\cdot\frac{dt}{dt}\cdot dt$$
$$= \int_a^b (u+iv)\left(\frac{dx}{dt}+i\frac{dy}{dt}\right)dt$$
$$= \int_C (udx - vdy) + i\int_C (udy + vdx)$$

where $f(z) = u + iv; u = u(x, y), v = v(x, y)$.

Steps:

1. Represent the path C in the form $z(t) = x(t) + iy(t) \;\; (a \le t \le b)$.
2. Compute the derivative $\dot{z}(t) = \dfrac{dz(t)}{dt}$.
3. Substitute $z(t)$ for every z in $f(z)$. Hence $x(t)$ for x and $y(t)$ for y.
4. Integrate $f[z(t)]\dot{z}(t)$ over t from a to b.

15.5.4 Value of Line Integral: Independence of Path

We have observed that the value of a line integral of a given function $f(z)$ from a point z_1 to another point z_2 is, in general, dependent upon the path C of integration and not merely on z_1 and z_2.

We call an integral of $f(z)$ *independent of path in a domain D* if, for every z_1 and z_2 in D its value depends only on z_1 and z_2 and not on the choice of path C in D.

Theorem 15.5 If $f(z)$ is analytic in a simply-connected domain D, then the integral of $f(z)$ is independent of the path in D.

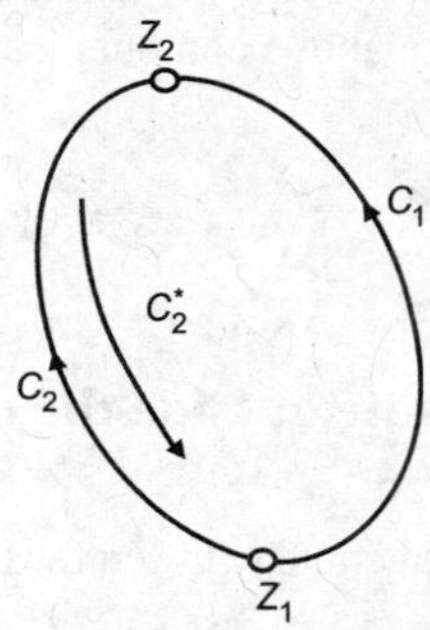

Figure 15.28

Proof: Let z_1 and z_2 be any points in D. Consider two paths C_1, C_2 in D from z_1 to z_2. If C_2^* denotes the path from z_2 to z_1, then $C = C_1 + C_2^*$ is a closed path in D. By C.I.T,

$$\int_{C_1+C_2^*} f(z)dz = 0$$
$$\Rightarrow \int_{C_1} f(z)dz = -\int_{C_2^*} f(z)dz = \int_{C_2} f(z)dz$$

This proves the theorem for paths having only the two end points in common. We can extend the theorem to the case of paths with more common points.

15.5.5 Integral of Non-Analytic Function: Dependence on Path of Integration

The following example illustrates that the complex line integral of a non-analytic function depends not only on the end points z_1 and z_2, but also on the path of integration joining z_1 and z_2.

Example 15.26

Evaluate $\int_C f(z)dz = \int_C Re\, z\, dz$ along the two given paths: (i) Along C^*, the straight line joining $O(0,0)$ and $P(1,2)$ and (ii) along C_1, the straight line joining $O(0,0)$ and $A(1,0)$ and then along C_2, the straight line joining $A(1,0)$ and $P(1,2)$.

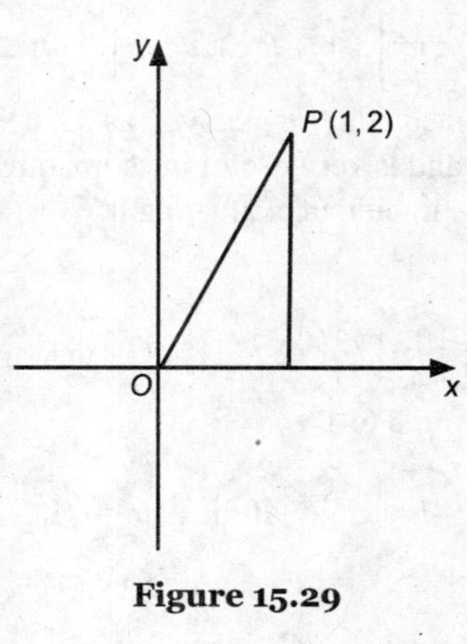

Figure 15.29

Solution

(i) Equation of OP is $\dfrac{y}{2} = \dfrac{x}{1} = t(\text{say}) \Rightarrow x = t, y = 2t$.

$$z(t) = x(t) + iy(t) = t + 2it, \quad \frac{dz}{dt} = 1 + 2i.$$

Along OP: t varies from $t = 0$ to 1; $Re\, z = t$

$$\int_C Re\, z\, dz = \int_0^1 t(1+2i)dt = (1+2i)\frac{t^2}{2}\Big|_0^1 = \frac{1}{2} + i$$

(ii) C consists of $C_1 : z(t) = x(t) + io = t$ where t varies from 0 to 1 and $dz = dt; C_2 : z(t) = x(t) + iy(t) = 1 + it$ where t varies from 0 to 2 and $dz = idt$

$$\int_C Re\, z\, dz = \int_{C_1} Re\, z\, dz + \int_{C_2} Re\, z\, dz$$

$$= \int_0^1 tdt + \int_0^2 idt$$

$$= \frac{1}{2}t^2\Big|_0^1 + it\Big|_0^2 = \frac{1}{2} + 2i$$

Clearly, the two results are different.

15.5.6 Bound for the Absolute Value of an Integral (ML-Inequality)

We say that a function $f(z)$ defined on a simply-connected domain D is bounded if there exists a positive real number M such that $|f(z)| \leq M$ for all z in D.

Suppose $f(z)$ is defined on a simply-connected domain D and C is any path of integration in C. Then $f(z)$ is bounded by M on C. If L is the length of the path C we have

$$\left|\int_C f(z)dz\right| \leq \int_C |f(z)|\, dz \leq M\int_C dz \leq M$$

This inequality is known as Cauchy's ML-inequality and is very useful in estimating the value of an integral without evaluating it.

Example 15.27

Show that $\left|\int_C z^2 dz\right| \leqslant 2\sqrt{2}$ if C is the straight line segment from 0 to $1 + i$.

Solution L =length of the path $= \sqrt{1^2 + 1^2} = \sqrt{2}$

$$|f(z)| = |z^2| \leq |z|^2 \leq (\sqrt{2})^2 =$$

Hence by Cauchy's ML-inequality

$$\left|\int_C z^2 dz\right| \leq 2\sqrt{2}.$$

15.5.7 Verification and Application of C.I.T.

Functions with no singularities (Entire functions)

The following functions are entire functions, *i.e.*, they have no singularities.
Polynomial function:

$$P_n(z) = a_0 + a_1 z + \cdots + a_n z^n \,(a_n \neq 0)$$

Exponential function: e^z
sine and cosine functions: $\sin z$, $\cos z$
sinh and cosh functions: $\sinh z$, $\cosh z$

Integrals of these functions along any simple closed path C is zero. In particular,

(i) $\oint_C z^n dz = 0 \quad n = 0,1,2,\ldots.$

(ii) $\oint_C e^z dz = 0 \quad \oint_C \sin z dz = 0 \quad \oint_C \cos z dz$

(iii) $\oint_C \sinh dz = 0 \quad \oint_C \cosh z dz = 0$

Example 15.28

Evaluate $I = \oint_C (5z^4 - 3z^2 + 6)dz$ where C is

(a) the unit circle $|z| = 1$ and
(b) the square $ABCD$ with $A = (0,0)$, $B = (1,0)$, $C = (1,1)$ and $D = (0,1)$
(c) The parabola $y = x^2 (C_1)$ and parabola $x = y^2 (C_2)$ joining (0, 0) and (1, 1).

Solution The integral $f(z) = 5z^4 - 3z^2 + 6$ is a polynomial in z. So, $f(z)$ is an entire function. Also, each of the above contours is a simple closed path. By C.I.T.,

$$\oint_C f(z)dz = \int_C (5z^4 - 3z^2 + 6)dz = 0.$$

Singularities lie outside the contour

Example 15.29

Let C be the unit circle with centre at 0, *i.e.*, $|z| = 1$.

(i) $\oint_c \sec z dz = 0$ and (ii) $\oint \tan z dz = 0$.

Solution Sec z and $\tan z$ are analytic everywhere except at $z = \pm\frac{\pi}{2}, \pm\frac{3\pi}{2}, \ldots$ lying outside $|z| = 1$ since the singularities of $\sec z$ and $\tan z$ are the zeros of $\cos z$.

Example 15.30

$$\oint_c \frac{dz}{z^2+2} = 0.$$

Solution $\frac{1}{z^2+2}$ is analytic except at $z = \pm\sqrt{2}i$.

Singularities of $\frac{1}{z^2+2}$ are the zeros of $z^2+2=0$ and these points lie outside C; so, no singularities lie inside C.

Example 15.31

$$\oint \frac{dz}{z^2+4} = 0.$$

Solution $\frac{1}{z^2+4}$ is analytic except at $z = \pm zi$, which lie outside C; so, no singularities inside C.

Example 15.32

$$\oint \frac{dx}{z^2+3} = 0.$$

Solution $\frac{1}{z^2+3}$ is analytic except at $z = \pm\sqrt{3}i$, which lie outside C; so, no singularities lie inside C.

Example 15.33

$$\oint_C \frac{dz}{z^2+4} \text{ where } C: |z-2| = 2 \Leftrightarrow x^2+y^2-4x = 0.$$

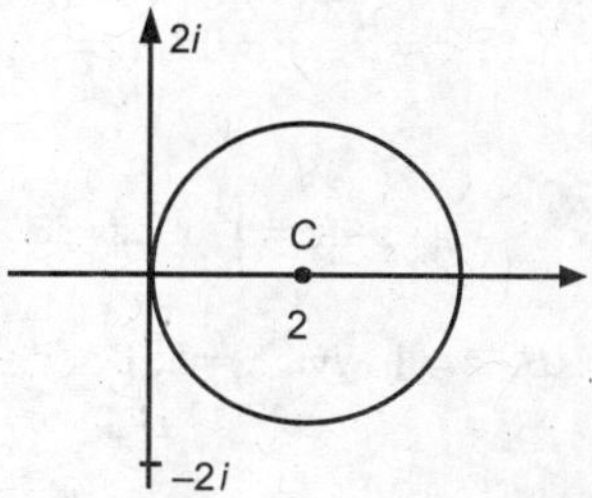

Figure 15.30

Solution The singularities of $\frac{1}{z^2+4}$ are at $z^2+4=0$, *i.e.*, $z = \pm 2i$ which lie outside C.

Hence C.I.T. applies and the integral is zero.

C.I.T. is not applicable in the following example:

Example 15.34

$$\oint \frac{dz}{z^2+4} \text{ where } C: |z-2| = 3 \Leftrightarrow x^2+y^2-4x-5 = 0.$$

Solution The singularities of $\frac{1}{z^2+4}$ are at $z^2+4=0$; *i.e.*, $z = \pm 2i$, *i.e.*, $A = (0,2)$ and $B = (0,-2)$, centre $= C = (2,0)$, radius $r = 3$.

The distance between C and A is

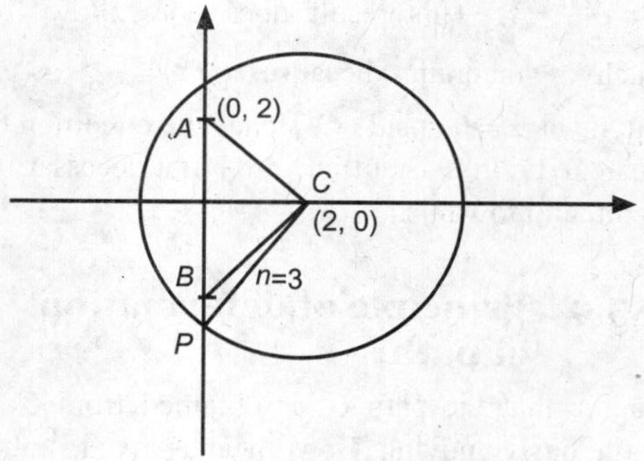

Figure 15.31

$$CA = \sqrt{2^2+2^2} = 2\sqrt{2} = 2.828 < \text{radius } r\ (=3)$$

$\therefore$ The singularities lie inside C. Hence C.I.T. is not applicable.

15.5.8 Non-Analytic Functions

There exist functions $f(z)$ and simple closed paths C such that $\oint_C f(z)dz = 0$.

For example, let $f(z) = \bar{z}$ and $C: |z| = 1$.

If $z(t) = e^{it}, dz = ie^{it}dt$, then t varies from 0 to 2π

$$\oint_C f(z)dz = \oint_C \bar{z}dz = \oint_0 e^{-it}.ie^{it}\,dt = i\int_0^{2\pi} dt = 2\pi i$$

This does not contradict Cauchy's theorem because $f(z) = \bar{z}$ is not analytic for any z, and hence the condition of Cauchy's theorem that $f(z) = \bar{z}$ should be analytic is not satisfied.

The condition of analyticity of Cauchy's theorem is only a sufficient condition but not necessary:

Example 15.35

Show that the condition that $f(z)$ must be analytic for $\oint_C f(z)dz = 0$ is only a sufficient condition but not necessary.

Solution Let $f(z) = 1/z^2$ and C: $|z| = 1$, the unit circle

$$\oint f(z)dz = \oint_C \frac{1}{z^2}dz = \oint_0 e^{-2it} ie^{it} dt$$
$$= i\int_0^{2\pi} e^{-it} dt = 0$$

$z(t) = e^{it}$, $dz = ie^{it}$; t varies from 0 to 2π since $e^0 = e^{2\pi i} = 1$. This result does not follow from Cauchy's theorem, because $f(z) = \frac{1}{z^2}$ is not analytic at $z = 0$ inside C. Hence the condition that f be analytic in D is sufficient but not necessary for the integral to vanish.

15.5.9 Principle of deformation of path:

Imagine that the path C_2 is obtained from C_1 by continuously moving C_1 with ends fixed, until it coincides with C_2. As long as our path of deformation always contains only points at which $f(z)$ is analytic, the integral retains the same value. This is called the *principle of deformation of path*.

Example 15.36

Show that $\int_C \frac{xdx - ydy}{\sqrt{x^2 + y^2}}$ is independent of the path of integration which does not pass through the origin.

[JNTU 2003 (Set 1)]

Solution Let $f(z) = \frac{z}{|z|} \neq 0$. Then $f(z)$ is analytic throughout the complex plane except at $z = 0 \Leftrightarrow (x, y) = (0,0)$, the origin. If C is any simple closed path not containing the origin as an interior point or a boundary point, then by Cauchy's integral theorem

$$\int_C f(z)dz = \int_C \frac{z}{|z|}dz = \int_C \frac{(x+iy)(dx+idy)}{\sqrt{x^2+y^2}} = 0$$

$$\Rightarrow \int_C \frac{xdx - ydy}{\sqrt{x^2+y^2}} + i\int_C \frac{xdy + ydx}{\sqrt{x^2+y^2}} = 0$$

$$\Rightarrow \int_C \frac{xdx - ydy}{\sqrt{x^2+y^2}} = 0 \text{ (considering the real part)}$$

This implies that the integral is independent of the path of integration.

Example 15.37

Verify Cauchy's theorem for the function $f(z) = z^3$ when the integral is taken around the boundary of the rectangle with vertices $-1, 1, 1+i$ and $-1+i$.

Solution By C.I.T. $\int_C z^3 dz = 0$. z^3 is analytic in D and C is a piece-wise smooth simple closed curve.

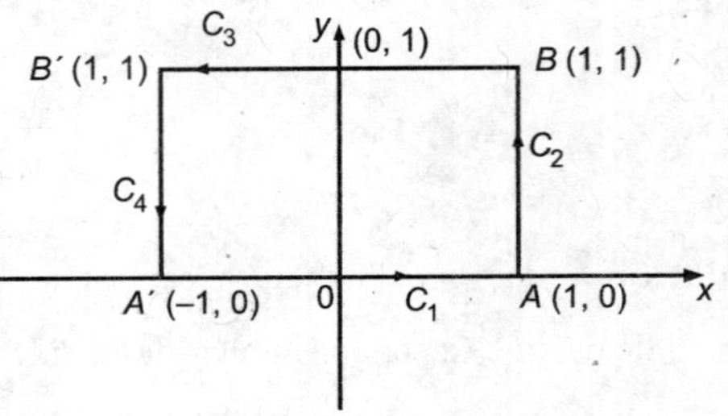

Figure 15.32

The integral $\int_C f(z)dz = \int_C z^3 dz$ had to be evaluated along the contour C, which is the boundary of the rectangle $A'ABB'$ consisting of curves (straight line segments) C_1, C_2, C_3 and C_4. Hence

$$\int_C z^3 dz = \int_{C_1} z^3 dz + \int_{C_2} z^3 dz + \int_{C_3} z^3 dz + \int_{C_4} z^3 dz \quad (1)$$

Along C_1: $A'A$; $z = x$, $y = 0$, $dy = 0$, $-1 \le x \le 1$, $dz = idy$

$$\int_{C_1} z^3 dz = \int_{-1}^{1} x^3 dz = 0 \quad (2)$$

Along C_2: AB; $z = 1 + iy$, $x = 1$, $0 \le y \le 1$, $dx = 0$, $dz = idy$

$$\int_{C_2} z^3 dz = \int_0^1 (1+iy)^3 (idy)$$
$$= i\int_0^1 (1 + 3iy - 3y^2 - iy^3)dy$$
$$= i\left[y + \frac{3}{2}iy^2 - y^3 - \frac{i}{4}y^4\right]_0^1$$

$$= i\left(1+\frac{3}{2}i-1-\frac{1}{4}i\right) = -5/4 \quad (3)$$

Along C_3: BB'; $z = x+i$, x varies from 1 to -1, $dy = 0$, $dz - dx$

$$\int_{C_3} z^3 dz = \int_1^{-1}(x+i)^3 dx$$

$$= \int_1^{-1}(x^3+3ix^2-3x-i)dx$$

$$= 0+2i+0-2i = 0 \quad (4)$$

Along C_4: $B'A'$, $z = -1+iy$, y varies from 1 to 0, $dx = 0$, $dz = idy$

$$\int_{C_4} z^3 dz = \int_1^{0}(-1+iy)^3 idy$$

$$= i\int_1^{0}(-1+3iy+3y^2-iy^3)dy$$

$$= i\left[-y+\frac{3}{2}iy^2+y^3-\frac{i}{4}y^4\right]_1^0$$

$$= i\left(1-\frac{3}{2}i-1+\frac{i}{4}\right) = \frac{5}{4} \quad (5)$$

Substituting into Eq. (1) from Eqs. (2) – (5), we get $\int_C z^3 dz = 0$. Thus C.I.T. is verified.

Example 15.38

Verify C.I.T. for $f(z) = z^2$ taken over the boundary of the square with vertices $\pm 1, \pm i$ in the counter-clockwise direction.

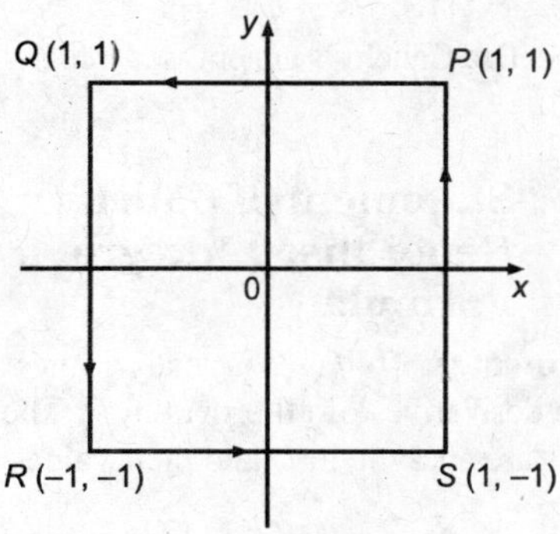

Figure 15.33

Solution By C.I.T., $\oint_C z^2 dz = 0$

$\because$ z^2 is analytic everywhere and C is a piece-wise smooth, simple and closed curve.

Also,

$$\oint_C z^2 dz = \int_{PQ} + \int_{QR} + \int_{RS} + \int_{SP} z^2 dz \quad (1)$$

Along PQ: $y = 1$, x varies from 1 to -1, $z = x+i$, $dz = dx$

$$\int_{PQ} z^2 dz = \int_1^{-1}(x+i)^2 dx = \int_1^{-1}(x^2+2ix-1)dx$$

$$= \int_1^{-1}(x^2-1)dx = \left[\frac{1}{3}x^3 - x\right]_1^{-1}$$

$$= 2\left(-\frac{1}{3}+1\right) = \frac{4}{3} \quad (2)$$

Along QR: $x = -1$, y varies from 1 to -1, $z = -1+iy$, $dz = idy$

$$\int_{QR} z^2 dz = \int_1^{-1}(-1+iy)^2 dy$$

$$= i\int_1^{-1}(1-2iy-y^2)dy$$

$$= i\left[y-\frac{1}{3}y^3\right]_1^{-1} = 2i\left(-1+\frac{1}{3}\right) = -\frac{4}{3}i \quad (3)$$

Along RS: $y = -1$, $-1 \le y \le 1$, $z = x - i$, $dz = dx$

$$\int_{RS} z^2 dz = \int_{-1}^{1}(x-i)^2 dx \quad (4)$$

$$= \int_{-1}^{1}(x^2-2ix-1)dx$$

$$= 2\left[\frac{1}{3}x^3 - x\right]_0^1 = 2\left(\frac{1}{3}-1\right) = -\frac{4}{3} \quad (5)$$

Along SP: $x = 1$, $-1 \le y \le 1$, $z = 1+iy$, $dz = idy$

$$\int_{SP} z^2 dz = \int_{-1}^{1}(1+iy)^2 idy$$

$$= i\int_{-1}^{1}[1+2iy+(-y^2)]dy = 2i\left(y-\frac{1}{3}y^3\right)$$

$$= 2i\left(1-\frac{1}{3}\right) = \frac{4}{3}i \quad (6)$$

Adding Eqs. (2) and (3), we get $\oint_C z^2 dz = 0$. Then C.I.T. is verified.

15.6 Cauchy's Integral Formula (C.I.F.) or Cauchy's Formula Theorem

Theorem 15.6 Let $f(z)$ be analytic in a simply-connected domain D. Then for any point 'a' in D and for any simple closed path C in D that encloses 'a', we have

$$\oint_C \frac{f(z)}{z-a}dz = 2\pi i f(a) \qquad (15.9)$$

the integration being taken in the counter-clockwise direction.

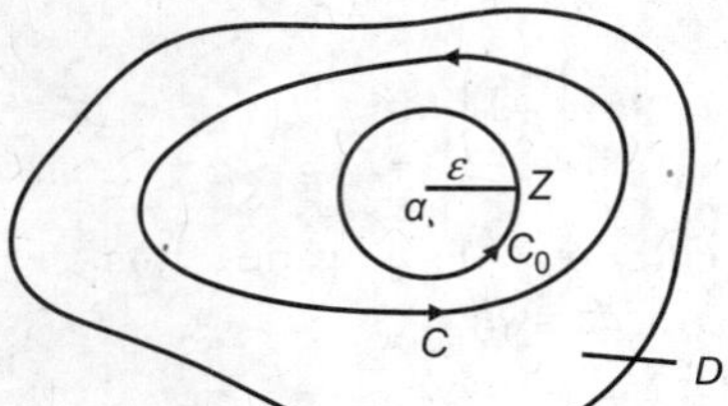

Figure 15.34

Proof: Enclose 'a' in a small circle C_0 of radius ε. Then the function $\frac{f(z)}{z-a}$ is analytic at all points in the simply-connected domain D except at $z = a$. By the extended Cauchy's theorem (Note 4 under C.I.T.)

$$\oint_C \frac{f(z)}{z-a}dz = \oint_{C_0} \frac{f(z)}{z-a}dz$$

$$= \int_0^{2\pi} \frac{f(a+\varepsilon e^{it})}{\varepsilon e^{it}} i\varepsilon e^{it} dt$$

$$= if(a)\int_0^{2\pi} dt, \text{ as } \varepsilon \to 0$$

$$= 2\pi i f(a) \qquad (15.10)$$

On C_0 : $z = a + \varepsilon e^{it}$, $0 \le t \le 2\pi$, $dz = i\varepsilon e^{it} dt$. As the moving point $P(t)$ traces the circle C_0 once in the counter-clockwise direction, t varies from 0 to 2π.

Alternative statement: Let $f(z)$ be analytic inside and on a simple closed curve C. If 'a' is any point within C then $\oint_C \frac{f(z)}{z-a}dz = 2\pi i f(a)$.

Note

Dividing both sides of Eq. (15.9) by $2\pi i$ we can represent $f(a)$ by contour integral. Thus,

$$f(a) = \frac{1}{2\pi i}\oint_c \frac{f(z)}{z-a}dz \qquad (15.11)$$

Note

1. Cauchy's integral formula is an important consequence of Cauchy's theorem. It establishes a relation between the value of an analytic function at an interior point 'a' and the value of the function on the boundary.

2. Cauchy's integral formula is a useful tool for evaluating integrals, finding derivatives of an analytic function $f(z)$ at some point $z = a$ and in developing $f(z)$ in a Taylor's series.

$$f(z) = \sum_{n=0}^{\infty} \frac{f^{(n)}(a)}{n!}(z-a)^n$$

about a point $z = a$ where

$$f^{(n)}(a) = \frac{n!}{2\pi i}\oint_C \frac{f(z)}{(z-a)^{n+1}}dz \quad n = 1, 2, 3, \ldots.$$

15.6.1 Derivatives of Analytic Function (Cauchy's Generalised Integral Formula)

Formally differentiating Cauchy's integral formula under the integral sign w.r.t 'a' we obtain,

$$f'(a) = \frac{1}{2\pi i}\oint_C \frac{(-1)f(z)}{(z-a)}dz = \frac{1}{2\pi i}\oint_C \frac{f(z)}{(z-a)^2}dz$$

and after n differentiations,

$$f^{(n)}(a) = \frac{n!}{2\pi i}\oint_C \frac{f(z)}{(z-a)^{n+1}}dz \quad n = 1, 2, 3, \ldots$$

which is called Cauchy's generalised integral formula (G.C.I.F.).

15.6.2 Statement of Cauchy's Generalised Integral Formula

Theorem 15.7 If $f(z)$ is analytic in a domain D, it has derivatives of all orders in D. The derived functions are analytic in D and their values are

$$f'(a) = \frac{a}{2\pi i}\oint_C \frac{f(x)}{(z-a)^2}dz$$

and $f^{(n)}(a)=\frac{n!}{2\pi i}\oint_C \frac{f(z)}{(z-a)^{n+1}}dz,\quad n=1,2,3,\ldots$

where C is any simple closed path entirely lying in D which encloses $z=a$.

Corollary Let C be a simple closed curve enclosing the origin then $\frac{a^n}{n!}=\frac{1}{2\pi i}\int_C \frac{e^{az}}{z^{n+1}}dz.$

Proof: Cauchy's generalised integral formula is

$$\frac{f^{(n)}(a)}{n!}=\frac{1}{2\pi i}\oint_C \frac{f(z)}{(z-a)^{n+1}}dz$$

Put $a=0,\ f(z)=e^{az}\Rightarrow f^{(n)}(z)=a^n e^{az}$

$$\therefore\quad f^{(n)}(0)=\frac{n!}{2\pi i}\int_C \frac{e^{az}}{z^{n+1}}dz$$

$$\Rightarrow \frac{a^n}{n!}=\frac{1}{2\pi i}\int_C \frac{e^{az}}{z^{n+1}}dz$$

15.7 Morera's Theorem (Converse of Cauchy's Theorem)

Theorem 15.8 If $f(z)$ is continuous in a simply-connected domain D and $\int_C f(z)dz=0$ around every simple, closed curve C in D, then $f(z)\in H(D)$.

Proof: Let z_0 be a fixed point and z be a variable point in D. By hypothesis, $\int_C f(z)dz=0$ around every simple, closed path C in D. This implies that $\int_{z_0}^{z} f(z)dz$ is independent of the path joining z_0 and z; it depends only on z. Thus the integral defines a function in z, *i.e.*, $\int_{z_0}^{z} f(z)\,dz=F(z)$

Let $F(z)=U+iV$ and $f(z)=u+iv$ so that

$$U+iV=\int_{z_0}^{z}(u+iv)(dx+idy)$$

$$=\int_{z_0}^{z}(udx-vdy)+i\int_{z_0}^{z}(vdx+udy)$$

$$\Rightarrow\quad U=\int_{z_0}^{z}(udx-vdy)\quad V=\int_{z_0}^{z}(vdx+udy)$$

Differentiating U and V w.r.t x and y, we get

$$U_x=u,\quad U_y=-v,\quad V_x=v,\qquad V_y=u.$$

Since $f(z)$ is continuous by hypothesis these partial derivatives are continuous and satisfy CREs.

$$U_x=V_y,\qquad U_y=-V_x$$

$\therefore F(z)$ is analytic. Also,

$$F'(z)=\frac{\partial U}{\partial x}+i\frac{\partial V}{\partial x}=u+iv=f(z).$$

By Cauchy's generalised integral formula,

$F(z)\in H(D)\ \Rightarrow F'(z), F''(z)=f'(z)$ exist in D.
$\Rightarrow f(z)\in H(D)$.

15.8 Cauchy's Inequality

Theorem 15.9 If $f(z)$ is analytic within and on the circle $C:|z-a|=r$, then $|f^{(n)}(a)|\le\frac{n!\,M}{r^n}$ where $|f(z)|\le M$.

Proof: By Cauchy's generalised integral formula,

$$f^{(n)}(a)=\frac{n!}{2\pi i}\oint_C\frac{f(z)dz}{(z-a)^{n+1}}\qquad n=1,2,\ldots$$

Taking absolute values on both sides

$$|f^{(n)}(a)|=\left|\frac{n!}{2\pi i}\right|\left|\oint_C\frac{f(z)}{(z-a)^{n+1}}dz\right|$$

$$\le\left|\frac{n}{2\pi i}\right|\left|\int_C\frac{f(z)}{(z-a)^{n+1}}dz\right|$$

$$\le\frac{n}{2\pi}\frac{M}{r^{(n+1)}}\cdot 2\pi r\le\frac{n!M}{r^n}.$$

Example 15.39

Evaluate $\int_C\frac{e^z+\sin z}{z-\frac{\pi i}{2}}dz$ where (i) $C:|z|=1$ and (ii) $C:|z|=2$.

Solution The only singularity of the integrand $z=\frac{\pi i}{2}=(1.5708\ldots)i$ lies outside $|z|=1$ but lies inside $|z|=2$. Thus the integrand and $\frac{e^z+\sin z}{z-\frac{\pi i}{2}}$ is analytic inside $|z|=1$. By Cauchy's integral theorem we have

(i) $\int_C\frac{e^z+\sin z}{z-\frac{\pi}{2}i}dz=0$

(ii) By Cauchy's integral formula

$$\int_C \frac{f(z)}{(z-a)^{n+1}} dz = \frac{2\pi i}{n!} f^{(n)}(a), \text{ we have}$$

$$\int_C \frac{e^2 + \sin z}{z - \dfrac{\pi i}{2}} dz = \frac{2\pi i}{0!}\left(i + \sinh\frac{\pi}{2}\right)$$

$$\left[\begin{array}{c} a = \dfrac{\pi i}{2}, n = 0, f(z) = e^z + \sin z \\ f(a) = f\left(\dfrac{\pi i}{2}\right) = e^{\frac{\pi i}{2}} + \sin\dfrac{\pi i}{2} = i + \sinh\dfrac{\pi}{2} \end{array}\right].$$

Example 15.40

Evaluate $\oint_C \frac{\tan z}{z^2 - 1} dz$ around the circle $|z| = 3/2$.

Solution $\tan z$ is not analytic at $z = \pm\pi/2$, $\pm 3\pi/2, \ldots$ but all these points lie outside C. $\frac{1}{z^2-1} = \frac{1}{2}\left(\frac{1}{z-1} - \frac{1}{z+1}\right)$ has singularities at $z = 1$ and $z = -1$, respectively.

$$\oint_C \frac{\tan z}{z^2-1} dz = \frac{1}{2}\int_C \frac{\tan z}{z-1} dz - \frac{1}{2}\oint_C \frac{\tan z}{z+1} dz$$

$$= \frac{1}{2} 2\pi i[\tan 1 - \tan(-1)] = 2\pi i \tan 1.$$

Example 15.41

Evaluate $\int_C \frac{z^2 - z - 1}{z(z-1)^2} dz$ where $C: |z - \frac{1}{2}| = 1$ using Cauchy's integral formula.

Solution The given curve is

$$C: |z - \frac{1}{2}| = 1 \Leftrightarrow \left(x - \frac{1}{2}\right)^2 + y^2 = 1^2,$$

which is a circle with centre at $\left(\frac{1}{2}, 0\right)$ and radius = 1. The integral has singularities at $z = 0$ and $z = i$ of which the former lies inside C and the latter lies outside C. We apply Cauchy's generalised integral formula

$$\oint_C \frac{f(z)}{(z-a)^{n+1}} dz = \frac{2\pi i}{n!} f^{(n)}(a) \qquad (1)$$

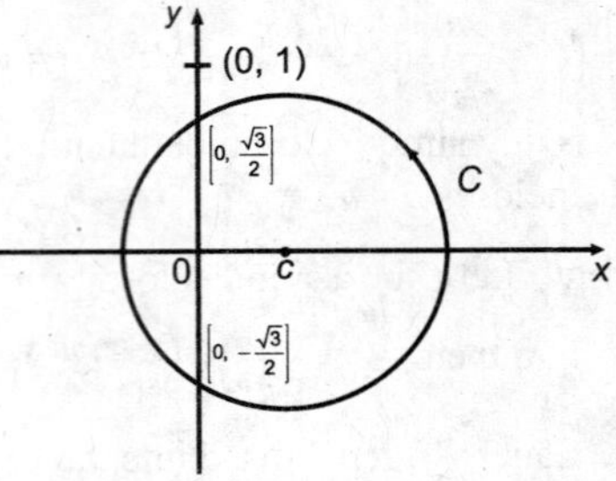

Figure 15.35

$$\text{Here } a = 0, n = 0, f(z) = \frac{z^2 - z - 1}{(z-i)^2}$$

$$f(a) = f(0) = \frac{-1}{(-i)^2} = 1$$

$$\therefore \int_C \frac{(z^2 - z - 1)/(z-i)^2}{z} dz = \frac{2\pi i}{1} f(0) = 2\pi i.$$

Example 15.42

Evaluate $\int_C \frac{z^3 - 2z + 1}{(z-i)^2} dz$ where C is the circle $|z| = 2$, using Cauchy's integral formula.

[JNTU 2003 (Set 4)]

Solution The given curve is $|z| = 2 \Leftrightarrow x^2 + y^2 = 2^2$ which is a circle of radius 2 with its centre at the origin. The integrand has a singularity at $z = i$, which lies inside C. We apply Cauchy's generalised integral formula

$$\int_C \frac{f(z)dz}{(z-a)^{n+1}} = \frac{2\pi i}{n!} f^{(n)}(a) \qquad (1)$$

$$\text{Here } a = i, n = 1, f(z) = z^3 - 2^2 + 1$$

$$\Rightarrow f'(z) = 3z^2 - 2; \quad f'(i) = 3i^2 - 2 = -5$$

$$\therefore \int_C \frac{z^3 - 2z + 1}{(z-i)^2} dz = \frac{2\pi i}{1!} f'(i) = 2\pi i(-5) = -10\pi i.$$

Example 15.43

Evaluate $\int_C \frac{z^3 - \sin 3z}{(z - \pi/2)^3} dz$ with $C: |z| = 2$ using Cauchy's integral formula. [JNTU 2005 *S* (Set 5)]

Solution The given curve is $|z| = 2 \Leftrightarrow x^2 + y^2 = 2^2$, which is a circle of radius 2 with its centre at the origin. The integrand has a unique singularity (a pole of order 3) at $z = \pi/2 = 1.5708\ldots < 2$ and hence lies inside C.

Let $f(z) = z^3 - \sin 3z$. We apply Cauchy's integral formula $\int_C \frac{f(z)dz}{(z-a)^{n+1}} = \frac{2\pi i}{n!} f^{(n)}(a)$.

Here $a = \pi/2, n = 2$ and $f''(z) = 6z + 9\sin 3z$

$$\Rightarrow f''\left(\frac{\pi}{2}\right) = 6\cdot\frac{\pi}{2} + 9\sin 3\left(\frac{\pi}{2}\right)$$

$$= 3\pi - 9 = 3(\pi - 3)$$

$$\int_C \frac{z^3 - \sin 3z}{(z-\frac{\pi}{2})^3} dz = \frac{2\pi i}{2} f''\left(\frac{\pi}{2}\right) = 3\pi i(\pi - 3).$$

Example 15.44

Evaluate $\int_C \frac{ze^z dz}{(z+a)^3}$ where c is any simple closed curve enclosing the point $z = -a$, using Cauchy's integral formula. [JNTU 2004, 2004 *S* (Set 1)]

Solution The given curve is any simple closed path enclosing the only singular point $z = -a$ of the integrand. We apply Cauchy's generalised integral formula

$$\int \frac{f(z)}{(z-z_0)^{n+1}} dz = \frac{2\pi i}{n!} f^{(n)}(z_0) \qquad (1)$$

Here $z_0 = -a, n = 2, f(z) = ze^2$

$$f''(z) = (e^z)''z + 2(e^z)'(z)' = (z+2)e^z$$

$$f''(-a) = (2-a)e^{-a}$$

Substituting in Eq. (1), we get

$$\int_c \frac{ze^z}{(z+a)^3} dz = \frac{2\pi i}{2} f''(z_0) = \pi i(2-a)e^{-a}.$$

Example 15.45

Evaluate $\int_C \frac{z^3 e^{-z} dz}{(z-1)^3}$ using Cauchy's theorem where C is the circle $|z-1| = 1/2$.

[JNTU 2003, 2006 (Set 4)]

Solution The equation of the given curve is

$$C : |z-1| = 1/2 \Leftrightarrow (x-1)^2 + (y-0)^2 = \left(\frac{1}{2}\right)^2$$

It is a circle. Its centre $= (1,0)$; radius $r = \frac{1}{2}$.

The integrand is $\frac{z^3}{(z-i)^3 e^z}$ and it has a unique

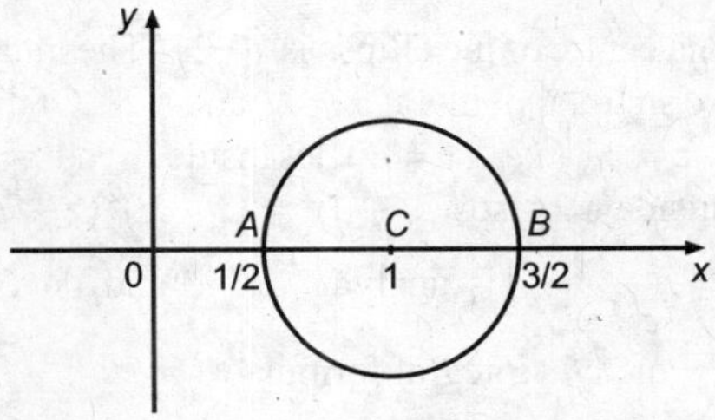

Figure 15.36

singularity at $z = 1$ which lies inside C. Let $f(z) = z^3 e^{-z}$, which is analytic at all points inside C. We apply Cauchy's generalised integral formula

$$f^{(n)}(a) = \frac{n!}{2\pi i} \int_C \frac{f(z)}{(z-a)^{n+1}} dz$$

Here we have $a = 1, n = 2$,

$$f''(2) = (z^3)'' e^{-2} + 2(z^3)'(te^{-2})' + z^3(e^{-z})''$$

$$= [6z + 6z^2(-1) + z^3]e^{-z}$$

by Leibnitiz's theorem

$$\Rightarrow f''(a) = f''(1) = (6\cdot 1 - 6\cdot 1^2 + 1^3)e^{-2} = e^{-1}$$

$$\therefore \int_C \frac{z^3 e^{-z}}{(z-1)^3} dz = \frac{2\pi i}{2} f''(1) = \pi i e^{-1}.$$

Example 15.46

Evaluate $\int_C \frac{dz}{(z^2+4)^2}$ where C is the circumference of the ellipse $x^2 + 4(y-2)^2 = 4$. [JNTU 2003 (Set 1)]

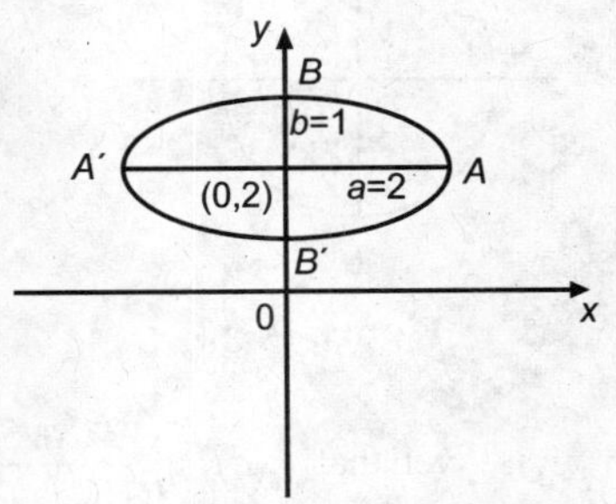

Figure 15.37

Solution The given curve is the ellipse whose equation is

$$\frac{(x-0)^2}{2^2} + \frac{(y-2)^2}{1^2} = 1 \qquad (1)$$

The centre of the ellipse is $(0,2)$. The integrand has two singular points at $z=2i$ and $-2i$. Of the two points $z=2i$, (*i.e.*, $y=2$) lies inside C and $z=-2i$ lies outside C. Also $(z^2+4)^2=(z+2i)^2(z-2i)^2$.

$f(z)=\dfrac{1}{(z+2i)^2}$ is analytic inside and on C. We apply Cauchy's integral formula

$$f^{(n)}(a)=\frac{n!}{2\pi i}\oint_C \frac{f(z)}{(z-a)^{n+1}} \tag{2}$$

Here $f'(z)=\dfrac{-2}{(z+2i)^3}$; $a=2i$ and $n=1$

Substituting these in Eq. (1), we get

$$\int_C \frac{\frac{1}{(z+2i)^2}dz}{(z-2i)^2}=2\pi i\left[\frac{d}{dz}\left(\frac{1}{(z+2i)^2}\right)\right]_{z=2i}$$

$$=2\pi i\left[\frac{-2}{(z+2i)^3}\right]_{z=2i}=\frac{-4\pi i}{(4i)^3}=\frac{\pi}{16}.$$

Example 15.47

If $f(a)=\int_c \dfrac{3z^2+7z+1}{z-a}dz$ where C is $|z|=2$ find (i) $f(3)$, (ii) $f(1)$, (iii) $f(1-i)$, (iv) $f''(1-i)$.

[JNTU 1998]

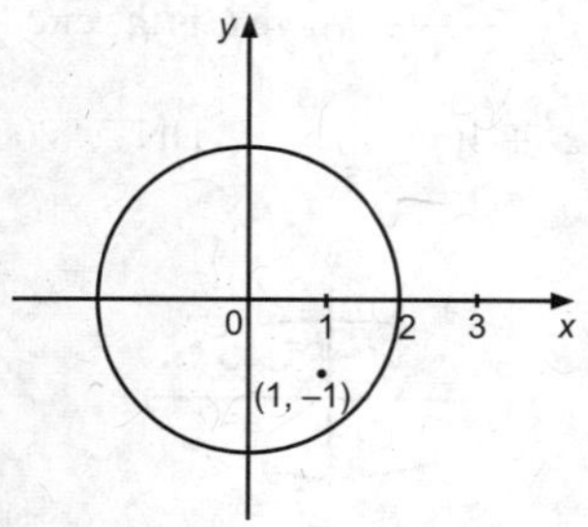

Figure 15.38

Solution It is given that $f(a)=\int_C \dfrac{3z^2+7z+1}{z-a}dz$

where C is the circle $|z|=2 \Leftrightarrow x^2+y^2=2^2$

(i) $\therefore\ f(3)=\int_C \dfrac{3z^2+7z+1}{z-3}dz$

Since $z=3$ lies outside the circle $C:|z|=2$, the integrand $\dfrac{3z^2+7z+1}{z-3}$ is analytic within and on C.

$\therefore$ By Cauchy's theorem,

$$f(3)=\int_C \frac{3z^2+7z+1}{z-3}dz=0$$

(ii) $f(1)=\int_C \dfrac{3z^2+7z+1}{z-1}dz$

Since $z=1$ lies inside the circle $C:|z|=2$, the integrand $\dfrac{3z^2+7z+1}{z-1}$ has a singularity at $z=1$ and is of the form $\dfrac{g(z)}{z-1}$ where $g(z)=3z^2+7z+1$ is analytic everywhere.

$\therefore$ By Cauchy's integral theorem,

$$f(1)=\int_C \frac{3z^2+7z+1}{z-1}dz=2\pi i\cdot g(1)$$

$$=2\pi i[3z^2+7z+1]_{z=1}=22\pi i$$

(iii) $f(1-i)=\int_C \dfrac{3z^2+7z+1}{z-(1-i)}dz$

Since $|1-i|=\sqrt{2}<2$, the point $1-i$ lies inside the circle C: $|z|=2$, the integrand $\dfrac{3z^2+7z+1}{z-(1-i)}$ has a singularity at $z=1-\mathrm{i}$ and is of the form $\dfrac{g(z)}{z-(1-i)}$ where $g(z)=3z^2+7z+1$ is analytic everywhere. By Cauchy's integral theorem,

$$f(1-i)=\int_C \frac{3z^2+7z+1}{z-(1-i)}dz=2\pi i g(1-i)$$

$$=2\pi i\,[3z^2+7z+1]_{z=1-i}$$

$$=2\pi i\,[(3z+7)z+1]_{z=1-i}$$

$$=2\pi i\,[(10-3i)(1-i)+1]$$

$$=2\pi i\,[7-13i+1]=2\pi(13+8i)$$

(iv) Differentiating the given equation twice w.r.t 'a', we get

$$f''(a)=2\int_C \frac{3z^2+7z+1}{(z-a)^3}dz$$

$$\therefore\ f''(1-i)=2\int_C \frac{3z^2+7z+1}{[z-(1-i)]^{2+1}}dz$$

$$=2\cdot\frac{2\pi i}{2!}\cdot g''(1-i)=12\pi i$$

$$\begin{bmatrix} g(z) & =3z^2+7z+1 & \\ g''(z) & =6, & g''(1-i)=6\end{bmatrix}.$$

Example 15.48

Evaluate $\int_C \frac{ze^z}{z^2-1}dz$ where C is the circle $|z|=2$.

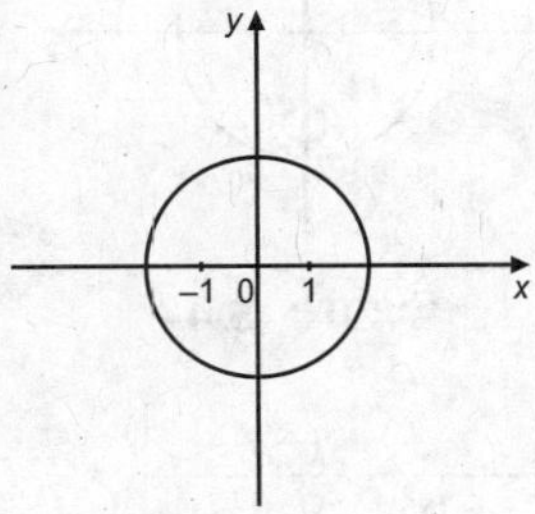

Figure 15.39

Solution The singularities of the integrand $\frac{ze^z}{z^2-1}$ are at $z=-1$ and 1, which lie inside the circle $|z|=2$.

Now, $\frac{z}{z^2-1}=\frac{1}{2}\left(\frac{1}{z-1}+\frac{1}{z+1}\right)$

$$\therefore \int_C \frac{ze^z}{z^2-1}dz=\frac{1}{2}\int_C\frac{e^z}{z-1}dz+\frac{1}{2}\int_C\frac{e^z}{z+1}dz$$

$$=\frac{1}{2}\cdot 2\pi i\cdot[e^z]_{z=1}+\frac{1}{2}\cdot\pi i[e^z]_{z=-1}$$

$$=2\pi i\frac{(e^1+e^{-1})}{2}=2\pi i\cdot\cosh 1.$$

(by Cauchy's integral theorem)

Example 15.49

Evaluate $\int_C\frac{e^{2z}}{(z-1)(z-2)}dz$ where C is the circle $|z|=3$. [JNTU 1999 *S*, 2001]

Solution

$$\frac{1}{(z-1)(z-2)}=\frac{1}{z-2}-\frac{1}{z-1}$$

$$\therefore \int_C\frac{e^{2z}}{(z-i)(z-2)}dz=\int_C\frac{e^{2z}dz}{z-2}-\int_C\frac{e^{2z}dz}{z-1}$$

$$=2\pi i[e^{2z}]_{z=2}-2\pi i[e^{2z}]_{z=1},$$

$$=2\pi i(e^4-e^2)$$

(by Cauchy's integral theorem)

$$=4\pi ie^3\sinh 1 \text{ or } 4\pi e^3\sin i$$

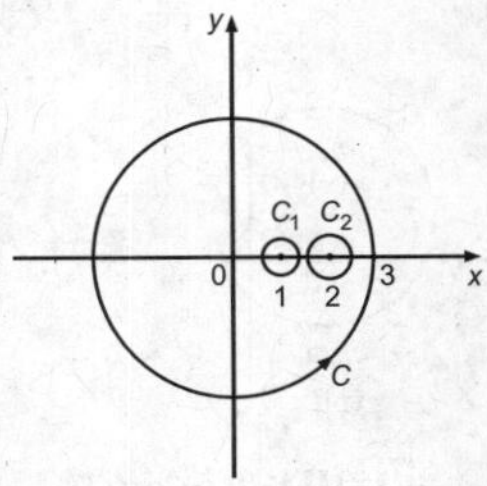

Figure 15.40

Aliter:

$$\int_c\frac{e^{2z}dz}{(z-1)(z-2)}=\int_{c_1}\frac{\left(\frac{e^{2z}}{z-2}\right)}{z-1}dz+\int_{c_2}\frac{\left(\frac{e^{2z}}{z-2}\right)}{z-2}dz$$

$$=2\pi i\left[\frac{e^{2z}}{z-2}\right]_{z=1}+2\pi i\left[\frac{e^{2z}}{z-1}\right]_{z=2}$$

$$=2\pi i\left[\frac{e^{2\cdot 1}}{1-2}\right]+2\pi i\left[\frac{e^{2\cdot 2}}{2-1}\right]$$

$$=2\pi i(e^4-e^2)=4\pi ie^3\sinh 1 \text{ or } 4\pi e^3\sin i.$$

Example 15.50

Evaluate $\int_C\frac{z+4}{z^2+2z+5}dz$ where C is the circle $|z|=1$. [JNTU 1998]

Solution $z^2+2z+5=(z+1)^2-(2i)^2$
$=(z+1+2i)(z+1-2i)$

Hence the singularities of the integrand are at $-1-2i,-1+2i$, *i.e.*, at $P_1=(-1,-2)$ and $P_2=(-1,2)$, $C_1:|z|=1$; $|z|=1\Rightarrow x^2+y^2=1$

Let $f(x,y)=x^2+y^2-1$.

Then $f(-1,2)$ and $f(-1,-2)$ are positive so that the points $(-1,2)$ and $(-1,-2)$ lie outside C. The integrand is analytic at all points inside C.

$\therefore \int_C\frac{z+4}{z^2+2z+5}dz=0$, by Cauchy's integral theorem.

Example 15.51

Evaluate $\int_C\frac{z+4}{z^2+2z+5}dz$ where C is the circle (i) $|z+1-i|=2$ and (ii) $|z+1+i|=2$.

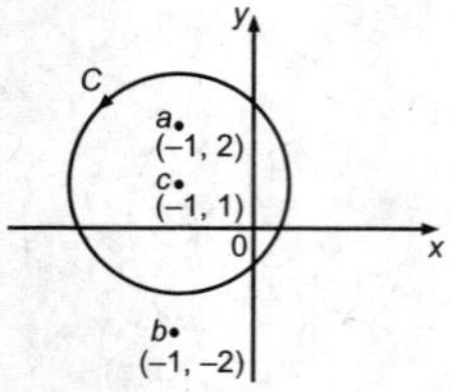

Figure 15.41

Solution Since $z^2+2z+5=(z+1)^2+2^2=(z+1+2i)(z+1-2i)$, the integrand has singularities at $z=a=-1+2i,\ z=b=-1-2i$.

(i) $C: |z+1-i|=2 \Leftrightarrow (x+1)^2+(y-1)^2=2^2$.

The singularity $a=(-1,2)$ lies inside C and the singularity $b=(-1,-2)$ lies outside C.

$$\therefore \int_C \frac{\left(\dfrac{z+4}{z-b}\right)dz}{z-a}=2\pi i\left[\frac{z+4}{z-b}\right]_{z=a}=2\pi i\left(\frac{a+4}{z-b}\right)$$

$$=2\pi i\frac{3+2i}{4i}=\frac{\pi}{2}(3+2i)$$

(ii) $C: |z+1+i|=2 \Leftrightarrow (x+1)^2+(y+1)^2=2^2$.

The singularity $a=(-1,2)$ lies outside C and the singularity $b=(-1,-2)$ lies inside C.

$$\therefore \int_C \frac{\left(\dfrac{z+4}{z-a}\right)dz}{z-b}=2\pi i\left[\frac{z+4}{z-a}\right]_{z=b}=2\pi i\left(\frac{b+4}{b-a}\right)$$

$$=2\pi i\left(\frac{3-2i}{-4i}\right)=-\frac{\pi}{2}(3-2i).$$

Example 15.52

Evaluate $\oint_C \dfrac{e^z}{(z^2+\pi^2)^2}dz$ where C is $|z|=4$ using Cauchy's integral theorem.

[JNTU 1999, 2004, 2006 (Set 1)]

Solution The singularities of the integrand are the zeros of $(z^2+\pi^2)^2=0 \Leftrightarrow z=\pm\pi i$.

$$\therefore \int_C \frac{e^z}{(z^2+\pi^2)^2}dz=\int_{C_1}\frac{\left(\dfrac{e^z}{(z+\pi i)^2}\right)}{(z-\pi i)^2}dz+\int_{C_2}\frac{\left(\dfrac{e^z}{(z-\pi i)^2}\right)}{(z+\pi i)^2}dz$$

$$=\frac{2\pi i}{1!}\left[\frac{d}{dz}\frac{e^z}{(z-\pi i)^2}\right]_{z=-\pi i}+\frac{2\pi i}{1!}\left[\frac{d}{dz}\frac{e^z}{(z+\pi i)^2}\right]_{z=\pi i},$$

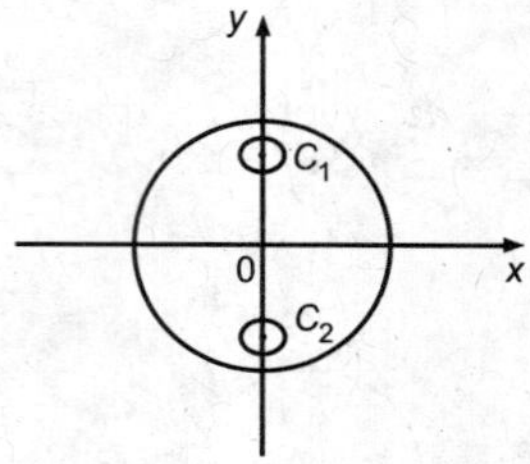

Figure 15.42

$$=2\pi i\left[\frac{e^z}{(z-\pi i)^2}+\frac{(-2)e^z}{(z-\pi i)^3}\right]_{z=-\pi i}+2\pi i\left[\frac{e^z}{(z+\pi i)^2}+\frac{(-2)e^z}{(z+\pi i)^3}\right]_{z=-\pi i}$$

(by generalised Cauchy's integral formula)

$$=2\pi i\left[\frac{e^{-\pi i}}{(-2\pi i)^2}-\frac{2e^{-\pi i}}{(-2\pi i)^3}\right]+2\pi i\left[\frac{e^{\pi i}}{(2\pi i)^2}-\frac{2e^{\pi i}}{(2\pi i)^3}\right]$$

$$=2\pi i\left(\frac{e^{\pi i}+e^{-\pi i}}{-4\pi^2}\right)+\frac{2\pi i}{4\pi^3 i}\left(e^{\pi i}-e^{-\pi i}\right)$$

$$=-\frac{i}{\pi}\cos\pi+\frac{i}{\pi^2}\sin\pi=\frac{i}{\pi}$$

$$\frac{e^{\pi i}+e^{-\pi i}}{2}=\cos\pi=1$$

$$\frac{e^{\pi i}-e^{-\pi i}}{2i}=\sin\pi=0.$$

Example 15.53

Evaluate $\int_C \dfrac{z-1}{(z+1)^2(z-2)}dz$ where C is $|z-i|=2$.

[JNTU 1998 *S*]

Solution C is the circle $|z-i|=2 \Leftrightarrow x^2+(y-1)^2=2^2$. The singularities of the integrand are at $z=a=-1$ and $z=b=2$ out of which the former lies inside C while the latter lies outside C.

$$\int_C \frac{z-1}{(z+1)^2(z-2)}dz=\int_C \frac{\left(\dfrac{z-1}{z-2}\right)}{(z+1)^2}dz$$

$$=\frac{2\pi i}{1!}\left[\frac{d}{dz}\left(\frac{z-1}{z-2}\right)\right]_{z=-1}$$

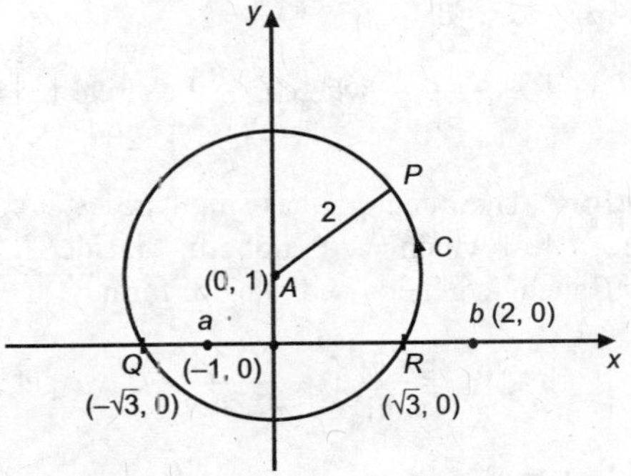

Figure 15.43

$$= 2\pi i\left[\frac{(z-2)1-(z-1)\cdot 1}{(z-2)^2}\right]_{z=-1}$$

$$= 2\pi i\left(\frac{-3+2}{9}\right) = -\frac{2\pi i}{9}.$$

Example 15.54

Let C be the circle $|z|=3$ described in the anti-clockwise direction. Let

$$g(a) = \int_C \frac{2z^2-z-2}{z-a}\,dz \quad (|a| \neq 3)$$

show that $g(a) = 8\pi i$. Find $g(a)$ if $|a|>3$.

Solution $\dfrac{2z^2-z-2}{z-a}$ is analytic for all z except for $z=a$, which may be (i) inside C, (ii) on C, or (iii) outside C.

(i) $z=a$ is inside C if $|a|<3$ $\dfrac{2z^2-z-2}{z-a}$ is analytic for all z except $z=a$.

Taking $f(z) = 2z^2-z-2$,

$$g(a) = \int_C \frac{2z^2-z-2}{z-a}\,dz = \int_C \frac{f(z)}{z-a}\,dz$$

$= 2\pi i f(a)$ (by Cauchy's integral formula)

$$= 2\pi i(2\cdot a^2 - a - 2)$$

$$\therefore\quad g(2) = 2\pi i(2\cdot 2^2 - 2 - 2) = 8\pi i$$

(ii) $z=a$ is outside C if $|a|>3$. $\dfrac{2z^2-z-2}{z-a}$ is analytic for all z inside and on C.

$\therefore \displaystyle\int_C \frac{2z^2-z-2}{z-a}\,dz = 0$ by Cauchy's integral theorem.

Example 15.55

Evaluate $\displaystyle\int_C \frac{\log z\,dz}{(z-1)^3}$ where C: $|z-1| = \dfrac{1}{2}$ using Cauchy's integral Formula. [JNTU 2003 (Set 4)]

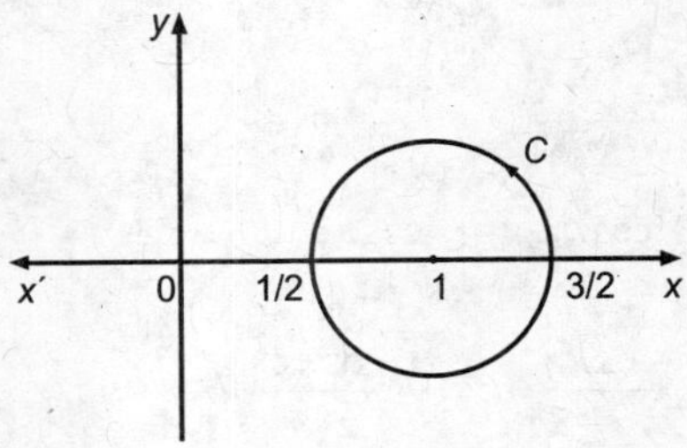

Figure 15.44

Solution The function $f(z) = \log z$ is analytic inside C ($\log z$ is not analytic along OX' including the origin) and the integrand has a singular point at $z=1$, which lies inside C. We apply Cauchy's generalised integral formula

$$\int_C \frac{f(z)}{(z-a)^{n+1}}\,dz = \frac{2\pi i}{n!} f^{(n)}(a) \qquad (1)$$

Here $a=1$, $n=2$ and $f(z) = \log z$

$$\Rightarrow f''(z) = -\frac{1}{z^2},\ f''(1) = -\frac{1}{1^2} = -1$$

$$\therefore\quad \int_C \frac{\log z}{(z-1)^3}\,dz = \frac{2\pi i}{2} f''(1) = -\pi i.$$

Example 15.56

Evaluate $\displaystyle\int_C \frac{\cos \pi z^2}{(z-1)(z-2)^3}\,dz$ where C is $|z|=3$ by using Cauchy's integral formula.

Solution The integrand has singularities at $z=1$ and $z=2$ both of which lie inside C. We put $\dfrac{1}{(z-1)^3(z-1)}$ into partial fractions. We know that $\dfrac{1}{(z-2)(z-1)} = \dfrac{1}{z-2} - \dfrac{1}{z-1}$. Multiplying both sides by $\dfrac{1}{z-2}$ twice, we get

$$\frac{1}{(z-1)^2(z-1)} = \frac{1}{(z-2)^2} - \frac{1}{(z-2)(z-1)}$$

$$= \frac{1}{(z-2)^2} - \frac{1}{z-2} + \frac{1}{z-1}$$

$$\frac{1}{(z-1)^3(z-1)}=\frac{1}{(z-2)^3}-\frac{1}{(z-2)^2}+\frac{1}{z-2}-\frac{1}{z-1}$$

By Cauchy's integral formula,

$$\int_C \frac{f(z)}{(z-a)^{n+1}}dz=\frac{2\pi i}{n!}f^{(n)}(a);$$

$$\int_c \frac{\cos \pi z^2 dz}{(z-1)(z-2)^3}$$

$$=\int_C \frac{\cos\pi z^2}{(z-2)^3}-\int_C \frac{\cos\pi z^2}{(z-2)^2}dz$$

$$+\int_C \frac{\cos\pi z^2}{z-2}dz-\int_C \frac{\cos\pi z^2}{z-1}dz$$

$$=2\pi i\left[\frac{1}{2!}f''(2)-\frac{1}{1!}f'(2)+\frac{1}{0!}f(2)-f(1)\right]$$

$$=2\pi i[-8\pi^2-0+1-1]=-16\pi^3 i.$$

We have,

$$f(z)=\cos\pi z^2 f'(z)=-2z\pi\sin\pi z^2$$

$$f''(z)=-2\pi\sin\pi z^2-4\pi^2 z^2\cos\pi z^2$$

$$f''(2)=-2\pi\sin 4\pi-16\pi^2\cos 4\pi=0-16\pi^2$$

$$f'(2)=-4\pi\sin 4\pi=0$$

$$f(2)=\cos 4\pi=1$$

$$f(1)=\cos\pi=-1.$$

Example 15.57

Evaluate $\int_C\left[\frac{e^z}{z^3}+\frac{z^4}{(z+i)^2}\right]dz$ where $C{:}\ |z|=2$, using Cauchy's integral formula.

[JNTU 2004, 2004 *S* (Set 1)]

Solution The given curve C is $|z|=2 \Leftrightarrow x^2+y^2=2^2$ which is a circle with radius = 2 and centre at the origin. We apply Cauchy's generalised integral formula

$$\int_C \frac{f(z)}{(z-a)^{n+1}}dz=\frac{2\pi i}{n!}f^{(n)}(a) \qquad (1)$$

The integrands have singularities at $z = 0$ and $z=-i$ inside C. In the first case, $a=0, n=2, f(z)=e^2$ and $f''(0)=1$ and in the second case $a = i$, $n=1, f(z)=z^4, f'(z)=4z^3, f'(-i)=4(-i)^3=4i$

$$\therefore \int_C\left[\frac{e^z}{z^3}+\frac{z^4}{(z+i)^2}\right]dz=2\pi i\left(\frac{f''(0)}{2}+\frac{f'(-i)}{1}\right)$$

$$=\pi i\cdot 1+2\pi i\cdot 4i=\pi(i-8).$$

Example 15.58

Evaluate $\int_C \frac{dz}{z^8(z+4)}$ where C is the circle $|z|=2$.

[JNTU 2004 (Set 2)]

Solution The integrand has singularities at $z = 0$ and $z = -4$ of which $z = 0$ alone lies inside C. We apply Cauchy's generalised integral formula

$$f^{(n)}(a)=\frac{2\pi i}{n!}\oint_C \frac{f(z)}{(z-a)^{n+1}}dz, \text{ we get}$$

$$\int_C \frac{\left(\frac{1}{z+4}\right)dz}{z^8}=\frac{2\pi i}{7!}(-1)\cdot 7!\cdot\frac{1}{4^8}=-\frac{\pi i}{2^{15}}.$$

By taking $f(z)=\frac{1}{z+4}$ and

$$a=0, n=7, f^{(n)}(z)=(-1)^n n!(z+4)^{-n-1}$$

$$f^{(7)}(0)=[(-1)^7 !(z+4)^{-8}]_{z=0}=(-1)7!\frac{1}{4^8}.$$

Example 15.59

Evaluate $\int_C \frac{\cos z-\sin z}{(z+i)^3}dz$ with $C{:}\ |z|=2$ using Cauchy's integral formula.

[JNTU 2005 *S* (Set 1)]

Solution The integrand has a singular point at $z=-i$ which lies inside $C{:}\ |z|=2$. Let $f(z)=\cos z-\sin z$ then $f(z)$ is analytic inside and on C. We apply Cauchy's integral formula

$$\int_C \frac{f(z)}{(z-a)^{n+1}}dz=\frac{2\pi i}{n!}f^{(n)}(a) \qquad (1)$$

Here $a=-i, n=2, f(z)=\cos z-\sin z$

$$f''(z)=-\cos z+\sin z$$

$$f''(-i)=\cos(-i)+\sin(-i)=-\cos i-\sin i$$

$$=-\frac{(1+i)e+(1-i)e^{-1}}{2}$$

$$\therefore \int_C \frac{\cos z-\sin z}{(z+i)^3}dz=\frac{2\pi i}{2}\left[-\frac{(1+1)e+(1-i)e^{-1}}{2}\right]$$

$$=-\frac{\pi i}{2}\left[(1+i)e+(1-i)e^{-1}\right].$$

Example 15.60

Evaluate $\int_C \frac{dz}{z^3(z+4)}$ where C is $|z| = 2$ using Cauchy's integral formula.

[JNTU 2006 (Set 1)]

Solution The integrand has singularities at $z = 0$ and $z = -4$ of which only $z = 0$ lies inside C. We apply Cauchy's integral formula

$$\int_C \frac{f(z)dz}{(z-a)^{n+1}} = \frac{2\pi i}{n!} f^{(n)}(a) \qquad (1)$$

Here $a = 0, n = 2;\ f(z) = \frac{1}{z+4}$

$$\Rightarrow f''(z) = \frac{(-1)(-2)}{(z+4)^3},\ f''(0) = \frac{2}{4^3} = \frac{1}{32}$$

$$\therefore \int_C \frac{\left(\frac{1}{z+4}\right)dz}{z^3} = \frac{2\pi i}{2!} f''(0) = \frac{\pi i}{32}.$$

EXERCISE 15.1

1. Evaluate $\oint_C \frac{e^z}{z-2} dz$. (a) $C : |z| =$ [illegible] and (b) $C : |z| = 3$
   Ans: (a) 0, (b) $2\pi i e^2$
2. Evaluate $\oint_C \frac{e^{iz}}{Z^3} dz$ if $C : |z| = 1$.
   Ans: $-\pi i$
3. Evaluate $\oint_c f(z)dz$ around $C : |z - i| = 1$ if
   (a) $f(z) = \frac{1}{z^2+4}$, (b) $f(z) = \frac{1}{(z^2+4)^2}$.
   [JNTU 2000]
   Ans: (a) $\pi/2$, (b) $\pi/16$
4. If $f(a) = \int_C \frac{z^2 - z + 5}{z - a} dz$ where C is the ellipse $\frac{x^2}{4} + \frac{y^2}{9} = 1$. Evaluate (a) $f(1)$, (b) $f(i)$.
   Ans: (a) $f(1) = 10\pi i$, (b) $2\pi i(4 - i)$
5. Using the formula for the derivatives of an analytic functions prove that $\oint_C \frac{\sin^6 z}{(z-\pi/6)^3} dz = \frac{21\pi i}{16}$ where C is given by $|z| = 1$.
6. Evaluate $\oint_C \frac{\sin \pi z^2 + \cos \pi z^2}{(z-1)(z-2)} dz$ where C is the circle $|z| = 3$.
   Ans: $4\pi i$
7. Evaluate $\oint_C \frac{\sin z}{z^2 e^z} dz$ where C is the circle $|z| = 1$.
   Ans: $2\pi i$
8. Evaluate $\int_C \frac{e^{2z}}{(z-1)(z-2)} dz$ where C is the circle $|z| = 3$ [JNTU 2001]
   Ans: $2\pi i(e^4 - e^2)$.
9. Evaluate $\int_C \frac{z^3 - \sin 3z}{(z - \pi/2)^3} dz$ where C is the circle $|z| = 2$. [JNTU 2003 *S*, 2005 *S*, 2007 *S*, 2008 *S* (Set 3)]
   Ans: $3\pi i(\pi - 3)$
10. Evaluate $\int_C \frac{z}{z^2+1} dz$ where C is the curve $\left|z + \frac{1}{z}\right| = 2$. [JNTU 1996]
    Ans: $2\pi i$
11. Using Cauchy's integral formula evaluate $\int_C \frac{z^4}{(z+1)(z-i)^2} dz$ where C is the ellipse $9x^2 + 4y^2 = 36$. [JNTU 2004 *S* (Set 2)]
    Ans: $4\pi(1 - i)$
12. Prove that $\int_C \frac{dz}{z} = \pm \pi i$ according as C is the lower or upper semicircle of $|z| = 1$.
13. Evaluate $\int_{3+4i}^{4-3i} (6z^2 + 8iz)dz$ along any path joining $3 + 4i$ and $4 - 3i$.
    Ans: $238 - 266i$
14. Evaluate $\int_C \frac{z^2 e^z}{(z-1)^3} dz$ where C is $|z| = 2$.
    Ans: $14\pi i e^2$
15. If $F(a) = \int_C \frac{4z^2 + z + 5}{z - a} dz$ where C is the ellipse $x = 2\cos\theta, y = 3\sin\theta\ (0 \le \theta \le 2\pi)$.
    Find (a) $F(4)$, (b) $F(i), F'(-1) F''(-i)$.
    Ans: (a) 0, (b) $2\pi(i-1) - 14\pi i, 16\pi i$

16. Evaluate $\int_C \frac{z^2+1}{z^2-1}dz$ where C is the circle $|z-1|=1$.

Ans: $2\pi i$

17. Evaluate $\int_C \frac{z}{z^2+1}dz$ where C is the circle $|z+i|=1$.

Ans: πi

18. Evaluate $\int_C \frac{e^{2z}}{(z+1)^4}dz$ where C is $|z|=2$.

Ans: $8\pi i/3e^2$

19. If $F(a)=\int_C \frac{3z^2+7z+1}{z-a}dz$ where C is the circle $|z|=2$, find the values of $F(1), F'(-1)$. and $F''(1-i)$. [JNTU 1994, 1998]

Ans: $22\pi i, 2\pi i, 12\pi i$

20. Evaluate $\int_C \frac{z+1}{z^3-2z^2}dz$ where C is the unit circle $|z|=1$.

Ans: $-\frac{3}{2}\pi i$

Complex Power Series

16

16.1 Introduction

In calculus, we have studied [1]Taylor's and [2]Maclaurin's power series expansions in the forms $\sum a_n(z-a)^n$ and $\sum a_n z^n$ in respect of elementary functions. We now extend these to functions of complex variables. Further, we study here another type of expansion known as [3]Laurent's series expansion which involves both positive as well as negative powers of $(z - a)$. These expansions help us in the study of the properties of functions, in numerical computations and in the evaluation of certain real integrals.

16.2 Sequences and Series

Let S be a set of complex numbers. We can form a complex sequence by choosing from S a first number u_1, a second number u_2, and so on. Here $u_1, u_2, \ldots u_n, \ldots$ are called the first, the second, … the n th term of the sequence.

Thus, we can regard a sequence as a set of ordered pairs of numbers (n, u_n) whose first coordinate n is a natural number and whose second coordinate, the term u_n, is a complex number. Since this correspondence is single-valued, sequences can be defined as functions.

[1] Taylor, Brook (1685–1731), English analyst, geometer and philosopher. He has published books on perspective and calculus of finite differences.

[2] Maclaurin, Colin (1698–1746), Scottish mathematician and physicist.

[3] Laurent, Paul Matthieu Hermann (1841–1908), French analyst. He is best known for the Laurent series that generalises the Taylor's series.

Complex Sequence

Definition Complex sequence is a function whose domain of definition is a set of natural numbers and whose range is a set of complex numbers. Thus, a function $u: N \to S$ where $S \subseteq \mathbb{C}$ is called a complex sequence.

A sequence is denoted by

$$< u_1, u_2, \ldots u_n, \ldots > \quad \text{or} \quad < u_n |_{n=1}^{\infty} >$$

Convergent Sequence

Definition Let l be a complex number. A sequence $< u_n >$ is said to converge to the number l if, for every $\varepsilon > 0$ there exists a positive integer m such that $| s_n - l | < \varepsilon$ for all $n \geq m$. We write $\lim_{n\to\infty} s_n = l$.

A sequence which does not converge is said to diverge.

Infinite Complex Series

An infinite complex series is merely a special kind of an important sequence. Form a new sequence $< s_n >$ as follows:

$$s_1 = u_1;$$
$$s_2 = u_1 + u_2;$$
$$s_3 = u_1 + u_2 + u_3;$$
$$\vdots$$
$$s_n = u_1 + u_2 + \cdots + u_n = \sum_{k=1}^{n} u_k \, (n = 1, 2, 3, \ldots)$$

Definition The ordered pair of sequences $(<u_n>, <s_n>)$ is called an infinite series. sn is called the nth partial sum of the series. The series is said to converge or diverge according as <sn>

converges or diverges. The series is symbolically written as

$$u_1 + u_2 + \ldots + u_n + \cdots \quad \text{or} \quad \sum_{k=1}^{n} u_k \ (n = 1, 2, 3, \ldots)$$

16.3 Power Series

An infinite series of the form

$$\sum_{n=0}^{\infty} a_n (z-a)^n = a_0 + a_1(z-a) + a_2(z-a)^2 + \cdots$$

is called a power series about $z = a$. If we choose $a = 0$,we get

$$\sum_{n=0}^{\infty} a_n z^n = a_0 + a_1 z + a_2 z^2 + \cdots.$$

Power series about the origin

Here $a_0, a_1, a_2, \ldots$ are complex (or real) constants and are called the coefficients of the series and 'a' is called the centre of the series.

Region of convergence (R.O.C)

A set of all points z for which the series converges is called the region of convergence.

Convergence of power series in a disk

As regards the region of convergence of a power series, three distinct cases arise:

1. The series converges only at $z = a$.
2. The series converges for all z.
3. The series converges for $|z-a| < R$ (i.e., inside a circular disk $|z-a| = R$) and diverges for $|z-a| > R$ (i.e., outside it). Here R is known as the radius of convergence and $|z-a| = R$, the circle of convergence.

Note

A series may converge or diverge at the points on the circle of convergence.

16.4 Series of Complex Functions

Consider a series of complex functions

$$f_1(z) + f_2(z) + \ldots + f_n(z) + \cdots \tag{16.1}$$

$$\text{Let } \lim_{n\to\infty} \left| \frac{f_{n+1}(z)}{f_n(z)} \right| = |g(z)|; \tag{16.2}$$

$$\text{Let } \lim_{n\to\infty} |f_n(z)|^{\frac{1}{n}} = |g(z)| \tag{16.3}$$

The series at Eq. (16.1) converges absolutely for the values of z for which $0 < |g(z)| < 1$ and diverges for the values of z for which $|g(z)| > 1$ by D'Alembert's Ratio test or Cauchy's root test as the case may be.

If the values of z are such that $|g(z)| = 1$ no conclusion is possible. The series may be convergent or divergent.

Consider the power series

$$\sum a_n z^n \tag{16.4}$$

Let $\lim_{n\to\infty} a_n^{\frac{1}{n}} = \frac{1}{R}$ so that $\lim_{n\to\infty} \left| a_n z^n \right|^{\frac{1}{n}} = \frac{|z|}{R}$.

For $|z| < R$, the series (16.4) is absolutely convergent and for $|z| > R$ the series (16.4) is divergent.

Here R is the radius of convergence and $|z| = R$, the circle of convergence.

Example 16.1

1. $\sum \frac{z^n}{n!}$; here $a_n = \frac{1}{n!}$;

$$R = \lim_{n\to\infty} \frac{a_n}{a_{n+!}} = \lim \frac{(n+1)!}{n!} = \lim(n+1) = \infty$$

The series converges absolutely over the entire complex plane.

2. $\sum \left(1+\frac{1}{n}\right)^{n^2} z^n$; here $a_n = \left(1+\frac{1}{n}\right)^{n^2}$;

$$\frac{1}{R} = \lim_{n\to\infty} a_n^{\frac{1}{n}} = \lim_{n\to\infty} \left(1+\frac{1}{n}\right)^n = e \Rightarrow R = \frac{1}{e}.$$

3. $\sum \frac{2^n}{2^n+1}$; here $a_n = \frac{1}{2^n+1}$;

$$\frac{1}{R} = \lim_{n\to\infty} \frac{a_{n+1}}{a_n} = \lim_{n\to\infty} \frac{2^n+1}{2^{n+1}+1} = \frac{1}{2} \Rightarrow R = 2.$$

16.5 Uniform Convergence of a Series of Functions

Definition A series $\sum_{m=0}^{\infty} f_m(z)$ with sum $s(z)$ is called uniformly convergent in a region G if, for every $\varepsilon > 0$ we can find an $N = N(\varepsilon)$, not depending on z, such that

$|s(z) - s_n(z)| < \varepsilon$ for all $n > N(\varepsilon)$ and for all $z \in G$

We can show that the geometric series $\sum z^n$ is

(a) uniformly convergent in any closed disk $|z| \le r < 1$; but

(b) not uniformly convergent in its whole disk of convergence $|z| < 1$.

Relation between the behaviour of real and complex series

Let

$$(a_1 + ib_1) + (a_2 + ib_2) + \ldots + (a_n + ib_n) + \cdots \quad (16.5)$$

be a series of complex terms. Here the a's and the b's are real numbers. If the real series $\sum a_n$ and $\sum b_n$ converge to the sums A and B respectively, then the series (16.1) converges to the sum $A + iB$.

If the series (16.1) is convergent, then $\lim_{n\to\infty}(a_n + ib_n) = 0$. This is a necessary condition for convergence. If the condition fails to hold, then the series is divergent.

16.6 Weierstrass's M-Test

The series

$$f_1(z) + f_2(z) + \cdots + f_n(z) + \cdots \quad (16.6)$$

is uniformly convergent in a region R if there is a convergent series of positive constants $\sum M_n$ such that $|f_n(z)| \le M_n$ for all z in R.

A uniformly convergent series of continuous complex functions is continuous and can be differentiated or integrated term by term.

Term-by-term differentiation or integration of a non-uniformly convergent series of functions is not permissible—it leads to wrong results.

16.7 Taylor's Theorem (Taylor Series)

Theorem 16.1 Let D be a simply-connected domain and $f(z) \in H(D)$. Suppose C is a circle in D with centre 'a'. Then, for any z inside C

$$f(z) = f(a) + \frac{f'(a)}{1!}(z-a) + \frac{f''(a)}{2!}(z-a)^2 + \ldots + \frac{f^{(n)}(a)}{n!}(z-a)^n + \cdots \quad (16.7)$$

or $f(z) = \sum_{n=0}^{\infty} a_n (z-a)^n$ where $a_n = \dfrac{f^{(n)}(a)}{n!}$

Proof: Let z be any point inside C. Draw a circle C_1 with centre 'a' enclosing z. Let t be a point on C_1. Then

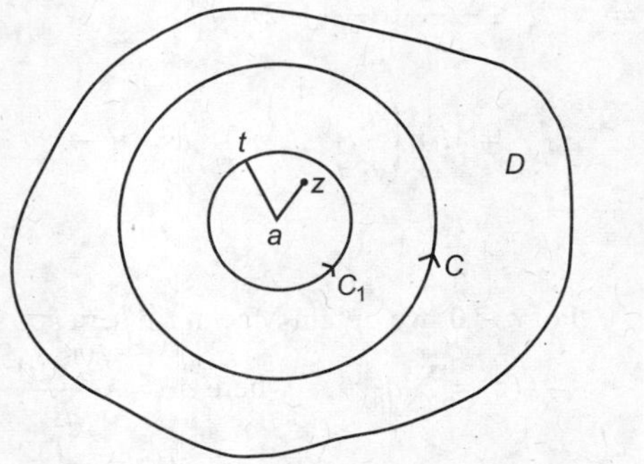

Figure 16.1

$$\frac{1}{t-z} = \frac{1}{(t-a)-(z-a)} = \frac{1}{t-a}\left(1 - \frac{z-a}{t-a}\right)^{-1}$$

$$= \frac{1}{t-a}\left[1 + \left(\frac{z-a}{t-a}\right) + \left(\frac{z-a}{t-a}\right)^2 + \cdots\right] \quad (16.8)$$

Since $\left|\dfrac{z-a}{t-a}\right| < 1$ the series converges uniformly.

Multiplying Eq. (16.8) by $\dfrac{1}{2\pi i} f(t)$ and integrating each term over C_1 we get

$$\frac{1}{2\pi i}\int_{C_1} \frac{f(t)}{t-z}\,dt = \frac{1}{2\pi i}\int_{C_1} \frac{f(t)}{t-a}\,dt + \frac{z-a}{2\pi i}\int_{C_1} \frac{f(t)}{(t-a)^2}\,dt + \cdots + \frac{(z-a)^n}{2\pi i}\int_{C_1} \frac{f(t)}{(t-a)^{n+1}}\,dt + \cdots$$

$$\Rightarrow f(z) = \sum_{n=0}^{\infty} a_n (z-a)^n, \quad (16.9)$$

where $a_n = \dfrac{f^{(n)}(a)}{n!}$ by Cauchy's integral formula.

Remainder after *n* terms

$$R_n = \frac{(z-a)^{n+1}}{2\pi i}\int_{C_1}\frac{f(t)}{(t-z)(t-a)^{n+1}}dt$$

$$|z-a| = r, \quad |t-a| = r_1, \quad |t-z| \geq r_1 - r$$

$$\left|\frac{f(t)}{t-z}\right| \leq M \quad (f(t) \text{ is analytic and hence bounded})$$

$$\Rightarrow \quad |R_n| \leq \frac{r^{n+1}}{2\pi}\cdot\frac{1}{r_1^{n+1}}\cdot M\,2\pi r_1$$

$$= M\,r_1\left(\frac{r}{r_1}\right)^{n+1} \to 0 \quad \text{as} \quad n\to\infty$$

$$\because \frac{r}{r_1} < 1$$

For $a = 0$, we obtain Maclaurin series

$$f(z) = \sum_{n=0}^{\infty} a_n z^n, \quad \text{where } a_n = \frac{f^{(n)}(0)}{n!}$$

$$= f(0) + \frac{f'(0)}{1!}z + \frac{f''(0)}{2!}z^2 + \cdots.$$

16.7.1 Important Special Taylor Series

1. Geometric series:

$$f(z) = \frac{1}{1-z}$$

$$\Rightarrow f^{(n)}(z) = \frac{n!}{(1-z)^{n+1}}; \qquad f^{(n)}(0) = n!$$

$$\frac{1}{1-z} = \sum z^n$$

$$= 1 + z + z^2 + \cdots |z| < 1 \qquad (16.10)$$

$$1 + \alpha + \alpha^2 + \cdots + \alpha^n = \frac{1-\alpha^{n+1}}{1-\alpha} = \frac{1}{1-\alpha} - \frac{\alpha^{n+1}}{1-\alpha}$$

$$\Rightarrow \frac{1}{1-\alpha} = 1 + \alpha + \alpha^2 + \cdots + \alpha^n = \frac{\alpha^{n+1}}{1-\alpha} \quad (\alpha \neq 1)$$

$$\alpha = \frac{z-a}{t-a} \Rightarrow \frac{1}{t-z} = \frac{1}{(t-a)(1-\frac{z-a}{ta})}$$

$$= \frac{1}{t-a}\left[1 + \left(\frac{z-a}{t-a}\right) + \left(\frac{z-a}{t-a}\right)^2 + \cdots + \left(\frac{z-a}{t-a}\right)^n\right] + \left(\frac{z-a}{t-a}\right)^{n+1}\cdot\frac{1}{t-z}$$

$f(z)$ has singularity at $z = 1$, which lies on the circle of convergence $|z| = 1$.

2. Exponential function e^z analytic for all z:

$$e^z = \sum \frac{z^n}{n!}$$

$$= 1 + \frac{z}{1!} + \frac{z^2}{2!} + \cdots + \frac{z^n}{n!} + \cdots \quad |z| < \infty$$

Put $z = iy, \quad e^{iy} = \cos y + i\sin y$

$$e^{iy} = \sum\frac{(iy)^n}{n!} = \sum\frac{(-1)^k y^{2k}}{(2k)!} + i\sum\frac{(-1)^k y^{2k+1}}{(2k+1)!}$$

3. Trigonometric and hyperbolic functions:

$$\cos z = \sum_{n=0}^{\infty}\frac{(-1)^n z^{2n}}{(2n)!}$$

$$= 1 - \frac{z^2}{2!} + \frac{z^4}{4!} - \cdots + \cdots \quad |z| < \infty$$

$$\sin z = \sum_{n=0}^{\infty}\frac{(-1)^n z^{2n+1}}{(2n+1)!}$$

$$= z - \frac{z^3}{3!} + \frac{z^5}{5!} - \cdots \quad |z| < \infty$$

$$\tan z = z + \frac{z^3}{3} + \frac{2z^5}{15} + \cdots \quad |z| < \frac{\pi}{2}$$

$$\sec z = 1 + \frac{z^2}{2} + \frac{5z^4}{24} + \cdots \quad |z| < \frac{\pi}{2}$$

$$\cosh z = \sum_{n=0}^{\infty}\frac{z^{2n}}{(2n)!}$$

$$= 1 + \frac{z^2}{2!} + \frac{z^4}{4!} + \cdots \quad |z| < \infty$$

$$\sinh z = \sum_{n=0}^{\infty}\frac{z^{2n+1}}{(2n+1)!}$$

$$= z + \frac{z^3}{3!} + \frac{z^5}{5!} + \cdots \quad |z| < \infty$$

4. Logarithm:

$$\text{Ln}(1+z) = z - \frac{z^2}{2} + \frac{z^3}{3} - \cdots \quad |z| < 1$$

$$-\text{Ln}(1-z) = z + \frac{z^2}{2} + \frac{z^3}{3} + \cdots \quad |z| < 1$$

$$\frac{1}{2}\text{Ln}\frac{1+z}{1-z}=z+\frac{z^3}{3}+\cdots;\quad |z|<1$$

Binomial Series

$$\frac{1}{1-z}=\sum_{n=0}^{\infty}z^n=1+z+z^2+\cdots;\quad |z|<1$$

$$\frac{1}{1+z}=\sum_{n=0}^{\infty}(-1)^n z^n=1-z+z^2-\cdots;\quad |z|<1$$

$$\frac{1}{(1-z)^m}=\sum_{m=0}^{\infty}\frac{(m+n-1)!}{(m-1)!n!}z^n$$

(m, non-negative integer) $|z|<1$

$$\frac{1}{(1+z)^m}=\sum_{n=0}^{\infty}\binom{-m}{n}z^n$$

Remark

Eventhough we can directly apply the formulas of Taylor or Laurent (Sec. 16.8 below) for obtaining expansions of given functions we can frequently obtain the expansions by making use of standard expansions of elementary functions, sometimes by a change of variable as the following worked-out examples will illustrate.

Example 16.2

Find the Taylor series expansions of $f(z)=e^z$ about $z=a$. Determine the region of convergence.

Solution Taylor series of $f(z)$ is

$$f(z)=\sum_{n=0}^{\infty}\frac{f^{(n)}(a)}{n!}(z-a)^n$$

$$f(z)=e^z\Rightarrow f^{(n)}(z)=e^z,\quad f^{(n)}(a)=e^a$$

$$\therefore\ e^z=\sum_{n=0}^{\infty}\frac{e^a}{n!}(z-a)^n=e^a\sum_{n=0}^{\infty}\frac{(z-a)^n}{n!}$$

Alternatively

$$e^z=e^{(z-a)+a}=e^a\cdot e^{z-a}$$

$$=e^a\sum_{n=0}^{\infty}\frac{(z-a)^n}{n!}\quad \text{valid for } |z|<\infty$$

Example 16.3

Expand $f(z)=\dfrac{z-1}{z+1}$ as a Taylor's series (i) about $z=0$ and (ii) about $z=1$.

Solution

(i) $$f(z)=\frac{z-1}{z+1}=\frac{(z+1)-2}{z+1}$$
$$=1-\frac{2}{z+1}$$
$$=1-2(1+z)^{-1}$$
$$=1-2[1-z+z^2-z^3+\cdots]$$
$$=-1+2z-2z^2+2z^3-\cdots$$
(Binomial theorem)

The region of convergence is $|z|<1$.

(ii) $$f(z)=\frac{z-1}{z+1}=\frac{z-1}{2+(z-1)}$$
$$=\frac{z-1}{2}\left(\frac{1}{1+\frac{z-1}{2}}\right)$$
$$=\frac{z-1}{2}\left(1+\frac{z-1}{2}\right)^{-1}$$
$$=\frac{z-1}{2}\left[1+\frac{z-1}{2}+\left(\frac{z-1}{2}\right)^2-\left(\frac{z-1}{2}\right)^3+\cdots\right]$$
(Binomial theorem)
$$=\frac{z-1}{2}-\frac{(z-1)^2}{2^2}-\frac{(z-1)^3}{2^3}-\cdots$$

valid for $\left|\dfrac{z-1}{2}\right|<1$

i.e. valid in circular disk $|z-1|<2$.

Example 16.4

Show that

(i) $\dfrac{1}{z^2}=1+\sum_{n=1}^{\infty}(n+1)(z+1)^n$ when $|z+1|<1$

(ii) $\dfrac{1}{z^2}=\dfrac{1}{4}+\dfrac{1}{4}\sum_{n=1}^{\infty}(-1)^n(n+1)\left(\dfrac{z-2}{2}\right)^n$

when $|z-2|<2$.

Solution

(i) $$\frac{1}{z^2}=\frac{1}{[1-(z+1)]^2}=[1-(z+1)]^{-2}$$

$$=1+2(z+1)+3(z+1)^2+4(z+1)^3+\cdots \quad \text{if } |z+1|<1$$

$$=1+\sum_{n=0}^{\infty}(n+1)(z+1)^n \quad \text{where } |z+1|<1$$

(ii) $$\frac{1}{z^2}=\frac{1}{[(z-2)+2]^2}=\frac{1}{4}\left(1+\frac{z-2}{2}\right)^{-2}$$

$$=\frac{1}{4}\left[1-2\left(\frac{z-2}{2}\right)+3\left(\frac{z-2}{2}\right)^3-\dots\right] \quad \text{if } \left|\frac{z-2}{2}\right|<1$$

$$=\frac{1}{4}-\frac{1}{4}\cdot 2\left(\frac{z-2}{2}\right)+\frac{1}{4}\cdot 3\left(\frac{z-2}{2}\right)^2-\dots$$

$$=\frac{1}{4}+\frac{1}{4}\sum_{n=1}^{\infty}(-1)^n(n+1)\left(\frac{z-2}{2}\right)^n.$$

Example 16.5

Expand the following function about the given points $f(z)=\dfrac{k}{az+b}$ about z_0.

Solution We have

$$f(z)=\frac{k}{az-az_0+(az_0+b)},$$

(adding and subtracting az_0)

$$=\frac{1}{az_0+b}\left[\frac{k}{1+\dfrac{a(z-z_0)}{az_0+b}}\right]=\frac{k}{q}\left[\frac{1}{1+p(z-z_0)}\right]$$

where $q=az_0+b$, $p=\dfrac{a}{q}$

$$=\frac{k}{q}\sum_{n=0}^{\infty}(-1)^n p^n(z-z_0)^n, \text{ (by binomial theorem)}$$

if $|p(z-z_0)|<1$ or $|z-z_0|<\dfrac{1}{p}=\dfrac{q}{a}$

$$=z_0+\frac{b}{a}$$

$$\Rightarrow f(z)=\frac{k}{az_0+b}\sum_{n=0}^{\infty}(-1)^n\left(\frac{1}{az_0+b}\right)^n(z-z_0)^n.$$

Example 16.6

Expand $f(z)=\dfrac{3}{5-2z}$ about $z_0=1-i$ by Taylor series.

Solution In Ex. 16.5 taking $a=-2$, $b=5$, $k=3$ and $z_0=1-i$

We have $q=az_0+b=-2(1-i)+5=3+2i$

$$p=\frac{a}{q}=\frac{-2}{3+2i}\cdot\frac{3-2i}{3-2i}=\frac{-6+4i}{13}$$

$$\therefore \quad \frac{3}{5-2z}=\frac{9-6i}{13}\sum_{n=0}^{\infty}(-1)^n\frac{(-6+4i)^n}{13^n}(z-1+i)^n.$$

Example 16.7

Expand $f(z)=\dfrac{1}{z^2+z-2}$ about $z=-1$ by Taylor series.

Solution

$$f(z)=\frac{1}{z^2+z-2}=\frac{1}{3}\frac{1}{z-1}-\frac{1}{3}\frac{1}{z+2},$$

putting into partial fractions

About $z=-1$ means in powers of $z+1$

$$\therefore \quad f(z)=\frac{1}{3}\frac{1}{z-1}-\frac{1}{3}\frac{1}{z+2}$$

$$=\frac{1}{3}\frac{1}{(z+1)-2}-\frac{1}{3}\frac{1}{(z+1)+1}$$

$$=\frac{1}{3}\cdot\frac{1}{-2\left[1-\dfrac{z+1}{2}\right]}-\frac{1}{3}\cdot\frac{1}{1+(z+1)}$$

$$=\frac{1}{3}\cdot\frac{1}{-2}\sum_{n=0}^{\infty}\left(\frac{z+1}{2}\right)^n-\frac{1}{3}\cdot\sum_{n=0}^{\infty}(-1)^n(z+1)^n,$$

by binomial theorem.

$$=-\frac{1}{6}\sum_{n=0}^{\infty}\left(\frac{z+1}{2}\right)^n-\frac{1}{3}\sum_{n=0}^{\infty}(-1)^n(z+1)^n$$

valid for $|z+1|<1$.

Example 16.8

Expand $f(z)=e^{1+z}$ in powers of $(z-1)$.

[JNTU 2003, 2004 (Set 4)]

Solution We have

$$e^{1+z}=e^2\cdot e^{z-1}=e^z\left[1+\frac{z-1}{1!}+\frac{(z-1)^2}{2!}+\dots\right]$$

Aliter

By direct application of Taylor's theorem

$$f(z)=\sum_{n=0}^{\infty}(z-a)^n\frac{f^{(n)}(a)}{n!}$$

with $a=1, f(z)=e^{1+z}$ we get

$$e^{1+z}=\sum_{n=0}^{\infty}(z-1)^n\frac{e^{1+1}}{n!}=e^2\sum_{n=0}^{\infty}\frac{(z-1)^n}{n!}$$

valid for all z.

Example 16.9

Expand ze^{2z} in a Taylor's series about $z=-1$ and determine the region of convergence.

Solution

$$ze^{2z}=\frac{ze^{2(z+1)}}{e^2}=\frac{1}{e^2}\left[(z+1)e^{2(z+1)}-e^{2(z+1)}\right]$$

$$=\frac{1}{e^2}\left\{(z+1)\left[1+\frac{2(z+1)}{1!}+\frac{4(z+1)^2}{2!}+\cdots\right]\right.$$

$$\left.-\left[1+\frac{2(z+1)}{1!}+\frac{4(z+1)^2}{2!}+\cdots\right]\right\}$$

$$=\frac{1}{e^2}\left\{(z+1)+\frac{2(z+1)^2}{1!}+\frac{2^2(z+1)^3}{2!}+\cdots\right.$$

$$\left.-\left[1+\frac{2(z+1)}{1!}+\frac{2^2(z+1)^2}{2!}+\cdots\right]\right\}$$

$$=\frac{1}{e^2}\left[-1+\left(1-\frac{2}{1!}\right)(z+1)+\left(\frac{2}{1!}-\frac{2^2}{2!}\right)(z+1)^2\right.$$

$$\left.+\left(\frac{2^2}{2!}-\frac{2^3}{3!}\right)(z+1)^3+\cdots\right]$$

valid for all z in the complex plane.

Example 16.10

Expand $f(z)=\dfrac{z^2-4}{z^2+5z+4}$ valid for $|z|<1$.

[JNTU 1994*S*]

Solution Putting $f(z)$ into partial fractions, we have

$$f(z)=1-\frac{1}{z+1}-\frac{4}{z+4}=1-(1+z)^{-1}-\left(1+\frac{z}{4}\right)^{-1}$$

$$=1-(1-z+z^2-z^3+\cdots)$$

$$-\left(1-\frac{z}{4}+\frac{z^2}{16}-\frac{z^3}{64}+\cdots\right)$$

by binomial theorem if $|z|<1$

$$=-1+\frac{5}{4}z-\frac{17}{16}z^2+\frac{65}{64}z^3-\cdots \text{ valid for } |z|<1$$

$$z^2+5z+4=(z+1)(z+4)$$

$$\text{Let } \frac{z^2-4}{(z+1)(z+4)}=1+\frac{A}{z+1}+\frac{B}{z+4}$$

$$\Rightarrow (z+1)(z+4)+A(z+4)+B(z+1)=z^2-4$$

Put $z=-1$

$$A(-1+4)=(-1)^2-4\Rightarrow A=-1$$

$$z=-4$$

$$B(-4+1)=(-4)^2-4\Rightarrow B=-4.$$

Example 16.11

Expand $f(z)=\sin z$ about $z=\dfrac{\pi}{4}$.

Solution Taylor's series of $f(z)$ about $z=a$ is

$$f(z)=f(a)+(z-a)f'(a)+\frac{(z-a)^2}{2!}f''(a)+\cdots \quad (16.1)$$

Here $a=\dfrac{\pi}{4}$; $f(z)=\sin z$; $f\left(\dfrac{\pi}{4}\right)=\dfrac{1}{\sqrt{2}}$

$f'(z)=\cos z$; $f''(z)=-\sin z$; $f'''(z)=-\cos z$; *etc.*

$$f'\left(\frac{\pi}{4}\right)=\frac{1}{\sqrt{2}};\ f''\left(\frac{\pi}{4}\right)=-\frac{1}{\sqrt{2}};\ f'''\left(\frac{\pi}{4}\right)=-\frac{1}{\sqrt{2}}$$

Putting these values in Eq. (1)

$$\sin z=\frac{1}{\sqrt{2}}+\left(z-\frac{\pi}{4}\right)\frac{1}{\sqrt{2}}+\frac{\left(z-\frac{\pi}{4}\right)^2}{2!}\left(-\frac{1}{\sqrt{2}}\right)$$

$$+\frac{\left(z-\frac{\pi}{4}\right)^3}{3!}\left(-\frac{1}{\sqrt{2}}\right)+\cdots \text{ valid for all } z.$$

Example 16.12

Expand $f(z)=\cos z$ about $z=\dfrac{\pi}{2}$.

Solution

$$\text{Put } t=z-\frac{\pi}{2}\Rightarrow \cos z=\cot\left(t+\frac{\pi}{2}\right)=-\sin t$$

$$=-\left[t-\frac{1}{3!}t^3+\frac{1}{5!}t^5-\cdots\right]$$

$$=-\left(z-\frac{\pi}{2}\right)+\frac{1}{3!}\left(z-\frac{\pi}{2}\right)^3-\frac{1}{5!}\left(z-\frac{\pi}{2}\right)^5+\cdots$$

valid for all z.

Example 16.13

Expand $f(z)=\sin\frac{1}{z-1}$ valid in $1<|z|<\infty$.

Solution Put $t=\frac{1}{z-1}\Rightarrow z-1=\frac{1}{t}$

$z=1\Rightarrow t=\infty$ and $z=\infty\Rightarrow t=0$

But $\sin t=t-\frac{1}{3!}t^3+\frac{1}{5!}t^5-\cdots$

$$\Rightarrow \sin\left(\frac{1}{z-1}\right)=\frac{1}{z-1}-\frac{1}{3!}\frac{1}{(z-1)^3}+\frac{1}{5!}\frac{1}{(z-1)^5}-\cdots.$$

Example 16.14

Find Taylor's expansion for $f(z)=\frac{1}{(1+z)^2}$ with the centre at $-i$. [JNTU 1998*S*]

Solution By Taylor's theorem

$$f(z)=f(a)+(z-a)f'(a)+\frac{(z-a)^2}{2!}f''(a)+\cdots$$
$$+\frac{(z-a)^n}{n!}f^{(n)}(a)+\cdots$$

Here $a=-i$ and $f(z)=\frac{1}{(1+z)^2}$

$$\Rightarrow f^{(n)}(z)=(-1)^n\frac{(n+1)!}{(1+z)^{n+2}}$$

$$f(-i)=\frac{1}{(1-i)^2}=\frac{i}{2};$$

$$f^{(n)}(-i)=\frac{(-1)^n(n+1)!}{(1-i)^{n+2}}$$

$$\therefore \quad \frac{1}{(1+z)^2}=\frac{i}{2}+\sum_{n=1}^{\infty}\frac{(z+i)^n}{n!}\frac{(-1)^n(n+1)!}{(1-i)^{n+2}}$$
$$=\frac{i}{2}+\sum_{n=1}^{\infty}(-1)^n(n+1)\frac{(z+i)^n}{(1-i)^{n+2}}$$
$$=\frac{i}{2}+\sum_{n=1}^{\infty}(-1)^n(n+1)\frac{(z+i)^n}{(1-i)^n}\frac{1}{(1-i)^2}$$
$$=\frac{i}{2}+\sum_{n=1}^{\infty}(-1)^n(n+1)\frac{(z+i)^n}{(1-i)^n(-2i)}$$
$$=\frac{i}{2}+\frac{i}{2}\sum_{n=1}^{\infty}(-1)^n(n+1)\frac{(z+i)^n}{(1-i)^n}.$$

Example 16.15

Obtain the Taylor's series to represent the function $\frac{z^2-1}{(z+2)(z+3)}$ in the region $|z|<2$.

[JNTU 2005*S* (Set 4), 2006 (Set 3)]

Solution

Let $f(z)=\frac{z^2-1}{(z+2)(z+3)}=1+\frac{3}{z+2}-\frac{8}{z+3}.$

Resolving into partial fractions

$$=1+\frac{3}{2}\left(1+\frac{z}{2}\right)^{-1}-\frac{8}{3}\left(1+\frac{z}{3}\right)^{-1}$$
$$=1+\frac{3}{2}\left(1-\frac{z}{2}+\frac{z^2}{4}-\frac{z^3}{8}+\cdots\right)$$
$$-\frac{8}{3}\left(1-\frac{z}{3}+\frac{z^2}{9}-\frac{z^3}{27}+\cdots\right).$$

By binomial theorem,

$$=1+\frac{3}{2}\sum_{n=0}^{\infty}\frac{(-1)^n}{2^n}z^n-\frac{8}{3}\sum_{n=0}^{\infty}\frac{(-1)^n}{3^n}z^n$$
$$=1+\sum_{n=0}^{\infty}(-1)^n\left[\frac{3}{2^{n+!}}-\frac{8}{3^{n+1}}\right]z^n,$$

valid for $|z|<2$.

Example 16.16

Expand $\log(1-z)$ when $|z|<1$ using Taylor's series. [JNTU 2003*S* (Set 1)]

Solution

Let $f(z)=\log(1-z)$ $\quad f(0)=\log 1=0$

$$f'(z)=\frac{-1}{1-z}\quad f'(0)=\frac{-1}{1-0}=-1$$

$$f''(z)=\frac{-1}{(1-z)^2}\quad f''(0)=\frac{-1}{(1-0)^2}=-1$$

$$f'''(z)=\frac{-2}{(1-z)^3}\quad f'''(0)=\frac{-2}{(1-0)^3}=-2\ etc.$$

Taylor's series is

$$f(z)=f(0)+\frac{zf'(0)}{1!}+\frac{z^2}{2!}f''(0)+\frac{z^3}{3!}f'''(0)+\cdots$$
$$=0-z-\frac{z^2}{2!}-\frac{2z^2}{3!}-\cdots=-\left(z+\frac{z^2}{2}+\frac{z^3}{3}+\cdots\right)$$

Example 16.17
Find Taylor's expansion of $f(z)=\dfrac{2z^3+1}{z^2+z}$ about $z=1$. [JNTU 2006 (Set 1)]

Solution Resolving $f(z)$ into partial fractions, we have

$$f(z)=2z-2+\frac{2z+1}{z(z+1)} \quad \text{(by division)}$$

$$=2z-2+\frac{1}{z}+\frac{1}{z+1}$$

$$\Rightarrow f^{(n)}(z)=(-1)^n n!\left[\frac{1}{z^{n+1}}+\frac{1}{(z+1)^{n+1}}\right] \quad \text{for } n\ge 2$$

$$f^{(n)}(1)=(-1)^n n!\left(1+\frac{1}{2^{n+1}}\right)$$

By Taylor's theorem,

$$f(z)=f(1)+\sum_{n=1}^{\infty}\frac{(z-1)^n}{n!}f^{(n)}(1)$$

$$=\frac{3}{2}+\sum_{n=1}^{\infty}\frac{(z-1)^n}{n!}(-1)^n n!\left(1+\frac{1}{2^{n+1}}\right)$$

$$=\frac{3}{2}+\sum_{n=1}^{\infty}(-1)^n\left(1+\frac{1}{2^{n+1}}\right)(z-1)^n.$$

16.8 LAURENT'S SERIES

Theorem 16.2 If $f(z)$ is analytic in the closed annular domain D bounded by the two concentric circles C_1 and C_2 of radii r_1 and r_2 $(r_1>r_2)$ and with centre 'a', then for all z in D

$$f(z)=\sum_{n=-\infty}^{\infty}a_n(z-a)^n \tag{16.10}$$

where $$a_n=\frac{1}{2\pi i}\int_C\frac{f(t)}{(t-a)^{n+1}}dt \tag{16.11}$$

over a simple closed path C in the annulus described in the counter-clockwise direction.

Proof: By Cauchy's integral formula,

$$f(z)=\frac{1}{2\pi i}\int_{C_1}\frac{f(t)}{t-z}dt-\frac{1}{2\pi i}\int_{C_2}\frac{f(t)}{t-z}dt$$

$$=I_1-I_2 \text{ (say)} \tag{16.12}$$

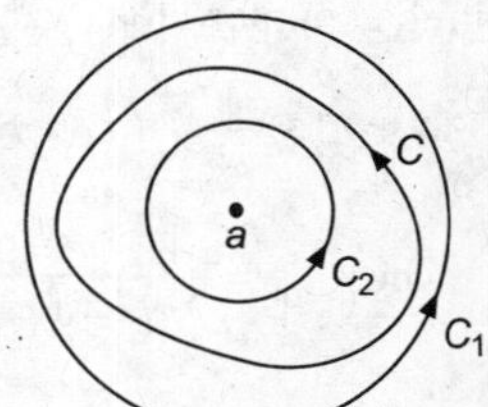

Figure 16.2

Since $\left|\dfrac{z-a}{t-a}\right|<1$ for any t on C_1, we can replace $\dfrac{1}{t-z}$ in I_1 by

$$\frac{1}{t-a}\left[1+\left(\frac{z-a}{t-a}\right)+\left(\frac{z-a}{t-a}\right)^2+\cdots+\left(\frac{z-a}{t-a}\right)^n+\cdots\right]$$

Then performing term-by-term integration, we get

$$I_1=\frac{1}{2\pi i}\int_{C_1}\frac{f(t)}{t-z}dt=\sum_{n=0}^{\infty}\frac{(z-a)^n}{2\pi i}\int_{C_1}\frac{f(t)}{(t-a)^{n+1}}dt$$

$$=\sum_{n=0}^{\infty}a_n(z-a)^n \tag{16.13}$$

where $a_n=\dfrac{1}{2\pi i}\displaystyle\int_C\frac{f(t)}{(t-a)^{n+1}}dt$

and C is a simple closed curve in D, taken in the counter-clockwise direction.

Since $\left|\dfrac{z-a}{t-a}\right|>1$ for any t on C_2, we can replace $\dfrac{1}{t-z}$ in I_2 by

$$-\frac{1}{z-a}\left[1+\left(\frac{t-a}{z-a}\right)+\left(\frac{t-a}{z-a}\right)^2+\cdots+\left(\frac{t-a}{z-a}\right)^n+\cdots\right]$$

Then performing term-by-term integration, we get

$$I_2=-\frac{1}{2\pi i}\int_{C_2}\frac{f(t)}{t-z}dt$$

$$=-\sum_{n=1}^{\infty}\frac{(z-a)^{-n}}{2\pi i}\int_{C_2}\frac{f(t)}{(t-a)^{-n+1}}dt$$

$$=\sum_{n=1}^{\infty}\frac{a_{-n}}{(z-a)^n} \text{ where } a_{-n}=\frac{1}{2\pi i}\int_C\frac{f(t)}{(t-a)^{-n+1}}dt$$

C being taken in the counter-clockwise direction.

$$\therefore \quad f(z) = \sum_{n=-\infty}^{\infty} a_n (z-a)^n$$

$$\text{where } a_n = \frac{1}{2\pi i}\int_C \frac{f(t)dt}{(t-a)^{n+1}}.$$

Remark

If $f(z)$ is analytic at a point $z = a$, then we obtain a unique Taylor's series expansion for $f(z)$ about $z = a$. We can also determine its region of convergence, which is unique.

This is not the case in respect of Laurent's series expansion of function $f(z)$ about a point $z = a$. Different Laurent's series expansions in powers of $(z - a)$ can be obtained for a function $f(z)$ whose regions of convergence are different. This point can be illustrated through the following example:

Example 16.18

Expand $f(z) = \dfrac{1}{(z-1)(z-2)(z-4)} = \dfrac{1}{3}\dfrac{1}{z-1} - \dfrac{1}{2}\dfrac{1}{z-2} + \dfrac{1}{6}\dfrac{1}{z-4}$, putting into partial fraction; $f(z)$ has singular points at $z = 1$, 2 and 4.

I Taylor's series expansions

1. *With centre at* $z = 0$: Maclaurin's series expansion of f(z) about $z = 0$ in $|z| < 1$.

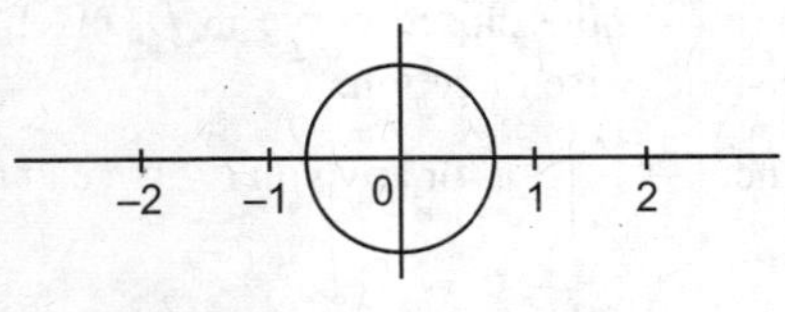

Figure 16.3

$$f(z) = -\frac{1}{3}(1-z)^{-1} - \frac{1}{2}\left(\frac{1}{-2}\right)\left(1-\frac{z}{2}\right)^{-1} + \frac{1}{6}\left(\frac{1}{-4}\right)\left(1-\frac{z}{4}\right)^{-1}$$

$$= -\frac{1}{3}\sum_{n=0}^{\infty} z^n + \frac{1}{4}\sum_{n=0}^{\infty}\left(\frac{z}{2}\right)^n - \frac{1}{24}\sum_{n=0}^{\infty}\left(\frac{z}{4}\right)^n$$

$$\text{for } |z| < 1$$

$$\left(\because \ |z| < 1 \Rightarrow \left|\frac{z}{2}\right| < 1 \Rightarrow \left|\frac{z}{4}\right| < 1\right)$$

2. *With centre at* $z = \frac{3}{2}$: Taylor's series expansion of f(z) about $z = \frac{3}{2}$ in $|z - \frac{3}{2}| < \frac{1}{2}$.

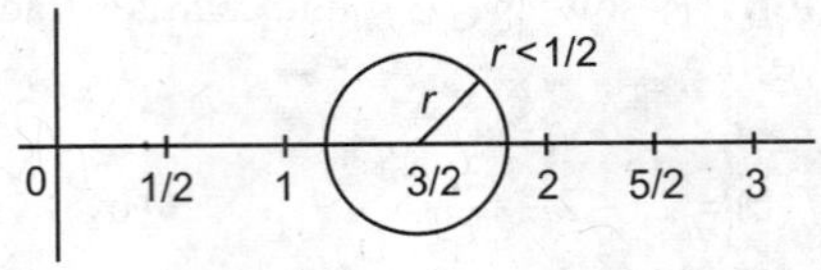

Figure 16.4

$$f(z) = \frac{1}{3}\frac{1}{z-1} - \frac{1}{2}\frac{1}{z-2} + \frac{1}{6}\frac{1}{z-4}.$$

Put $z - \frac{3}{2} = t \Rightarrow z = t + \frac{3}{2}$

$$= \frac{1}{3}\frac{1}{t+\frac{1}{2}} - \frac{1}{2}\frac{1}{t-\frac{1}{2}} + \frac{1}{6}\frac{1}{t-\frac{5}{2}}$$

$$= \frac{2}{3}(1+2t)^{-1} + (1-2t)^{-1} - \frac{1}{15}\left(1-\frac{2t}{5}\right)^{-1}$$

$$\text{for } |t| = \left|z-\frac{3}{2}\right| < \frac{1}{2}$$

$$= \frac{2}{3}\sum_{n=0}^{\infty}(-1)^n(2t)^n + \sum_{n=0}^{\infty}(2t)^n - \frac{1}{15}\sum_{n=0}^{\infty}\left(\frac{2t}{5}\right)^n$$

$$= \sum_{n=0}^{\infty} 2^n\left[(-1)^n \cdot \frac{2}{3} + 1 - \frac{1}{15}\cdot\frac{1}{5^n}\right]\left(z-\frac{3}{2}\right)^n$$

3. *With centre at* $z = 3$: Taylor's series expansion of $f(z)$ about $z = 3$ in $|z - 3| < 1$.

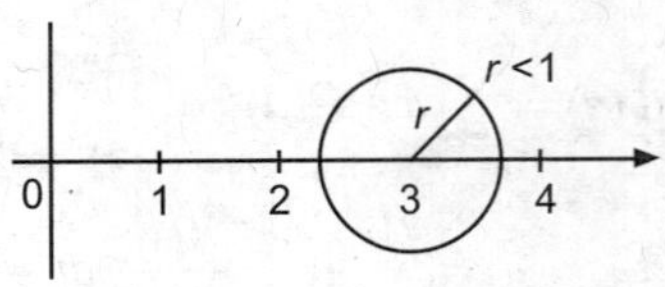

Figure 16.5

$$f(z) = \frac{1}{3}\frac{1}{z-1} - \frac{1}{2}\frac{1}{z-2} + \frac{1}{6}\frac{1}{z-4}$$

Put $z - 3 = t \Rightarrow z = t + 3$

$$= \frac{1}{3}\frac{1}{t+2} - \frac{1}{2}\frac{1}{t+1} + \frac{1}{6}\frac{1}{t-1}$$

$$= \frac{1}{3}\cdot\frac{1}{2}\left(1+\frac{t}{2}\right)^{-1} - \frac{1}{2}(1+t)^{-1} - \frac{1}{6}(1-t)^{-1}$$

$$= \sum_{n=0}^{\infty}\left[\frac{1}{6}\cdot(-1)^n \frac{1}{2^n} - \frac{1}{2}(-1)^n - \frac{1}{6}\right](z-3)^n$$

for $|z-3|<1$

II Laurent's series expansions

1. *With centre* $z = 0$: Laurent's series expansion valid for $1<|z|<2$.

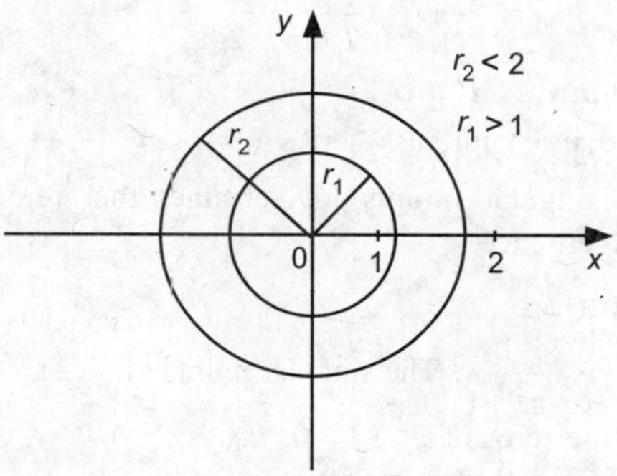

Figure 16.6

$$f(z) = \frac{1}{3}\frac{1}{z-1} - \frac{1}{2}\frac{1}{z-2} + \frac{1}{6}\frac{1}{z-4}$$

$$= \frac{1}{3z}\left(1-\frac{1}{z}\right)^{-1} + \frac{1}{4}\left(1-\frac{z}{2}\right)^{-1} - \frac{1}{24}\left(1-\frac{z}{4}\right)^{-1}$$

$$= \frac{1}{3z}\sum_{n=0}^{\infty}\frac{1}{z^n} + \frac{1}{4}\sum_{n=0}^{\infty}\left(\frac{z}{2}\right)^n - \frac{1}{24}\sum_{n=0}^{\infty}\left(\frac{z}{4}\right)^n$$

$$= \frac{1}{32}\sum_{n=0}^{\infty}\frac{1}{z^n} + \sum_{n=0}^{\infty}\left(\frac{1}{2^{n+2}} - \frac{1}{3}\frac{1}{2^{2n+3}}\right)z^n$$

valid for $1<|z|<2$

2. *With centre* $z = 0$: Laurent's series expansion valid for $2<|z|<4$.

$$f(z) = \frac{1}{3}\frac{1}{z-1} - \frac{1}{2}\frac{1}{z-2} + \frac{1}{6}\frac{1}{z-4}$$

$$= \frac{1}{3z}\left(1-\frac{1}{z}\right)^{-1} - \frac{1}{2z}\left(1-\frac{2}{z}\right)^{-1} + \frac{1}{6(-4)}\left(1-\frac{z}{4}\right)^{-1}$$

$$= \frac{1}{3}\sum_{n=0}^{\infty}\frac{1}{z^{n+1}} - \frac{1}{2}\sum 2^n\cdot\frac{1}{z^{n+1}} - \frac{1}{3}\sum\frac{1}{2^{2n+3}}z^n.$$

valid for $2<|z|<4$

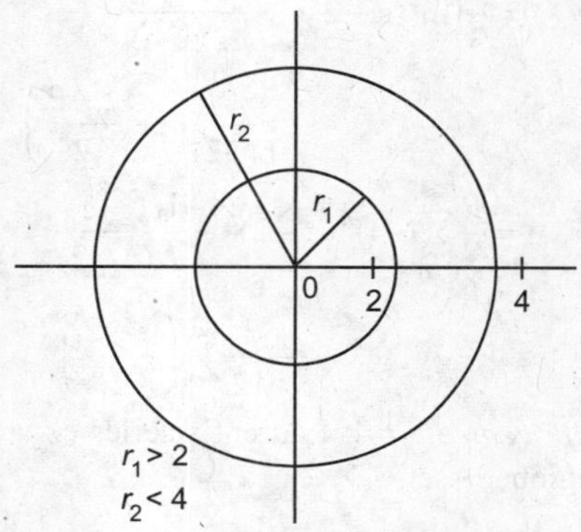

Figure 16.7

3. *With centre* $z = 1$: Laurent's series expansion in the region $0<|z-1|<1$.

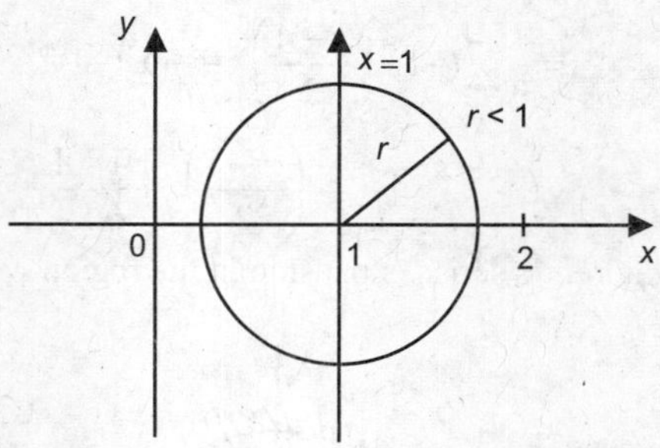

Figure 16.8

$$f(z) = \frac{1}{3}\frac{1}{z-1} - \frac{1}{2}\frac{1}{(z-1)-1} + \frac{1}{6}\frac{1}{(z-1)-3}$$

$$= \frac{1}{3}\frac{1}{z-1} - \frac{1}{2}[1-(z-1)]^{-1} + \frac{1}{6(-3)}\left(1-\frac{z-1}{3}\right)^{-1}$$

$$= \frac{1}{3}\frac{1}{z-1} + \frac{1}{2}\sum(z-1)^n - \frac{1}{18}\sum\left(\frac{z-1}{3}\right)^n$$

4. *With centre z* = 2: Laurent's series expansion in the region $0<|z-2|<1$.

$$f(z)=\frac{1}{3}\frac{1}{(z-2)+1}-\frac{1}{2}\frac{1}{z-2}+\frac{1}{6}\frac{1}{(z-2)-2}$$

$$=\frac{1}{3}[1+(z-2)]^{-1}-\frac{1}{2}\frac{1}{z-2}+\frac{1}{6(-2)}\left(1-\frac{z-2}{2}\right)^{-1}$$

$$=\frac{1}{3}\sum_{n=0}^{\infty}(-1)^n(z-2)^n-\frac{1}{2}\frac{1}{(z-2)}-\frac{1}{12}\sum_{n=0}^{\infty}\left(\frac{z-2}{2}\right)^n$$

5. *With centre z* = 4: Laurent's series expansion in the region $0<|z-4|<2$.

$$f(z)=\frac{1}{3}\frac{1}{(z-4)+3}-\frac{1}{2}\frac{1}{(z-4)+2}+\frac{1}{6}\frac{1}{z-4}$$

$$=\frac{1}{3(3)}\left(1+\frac{z-4}{3}\right)^{-1}-\frac{1}{2(2)}\left(1+\frac{z+4}{2}\right)^{-1}+\frac{1}{6}\frac{1}{z-4}$$

$$=\frac{1}{9}\sum(-1)^n\left(\frac{z-4}{3}\right)^n-\frac{1}{4}\sum(-1)^n\left(\frac{z-4}{2}\right)^n+\frac{1}{6}\frac{1}{z-4}$$

6. Laurent's series expansion in the region $|z|>4$:

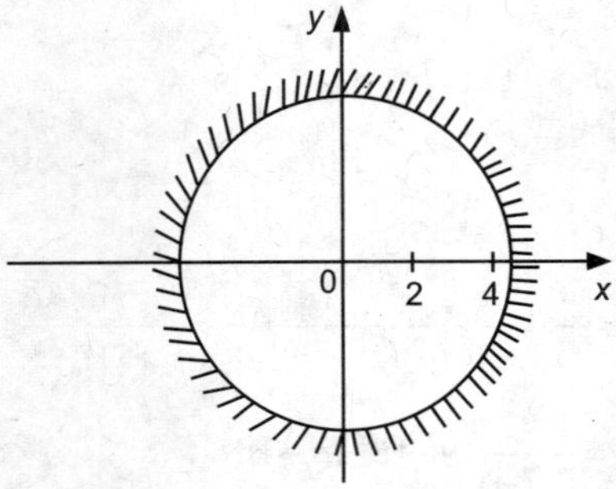

Figure 16.9

$$f(z)=\frac{1}{3z}\frac{1}{1-\frac{1}{z}}-\frac{1}{2z}\frac{1}{1-\frac{2}{z}}+\frac{1}{6z}\frac{1}{1-\frac{4}{z}}$$

$$=\frac{1}{3z}\left(1-\frac{1}{z}\right)^{-1}-\frac{1}{2z}\left(1-\frac{2}{z}\right)^{-1}+\frac{1}{6z}\left(1-\frac{4}{z}\right)^{-1}$$

$$=\frac{1}{3z}\sum_{n=0}^{\infty}\frac{1}{z^n}-\frac{1}{2z}\sum_{n=0}^{\infty}\left(\frac{2}{z}\right)^n+\frac{1}{6z}\sum_{n=0}^{\infty}\left(\frac{4}{z}\right)^n.$$

Example 16.19

Find the Laurent's series of $f(z)=\frac{e^z}{z^2}$ for $0<|z|<\infty$

Solution

$$f(z)=\frac{e^z}{z^2}=\frac{1}{z^2}\left(1+\frac{z}{1!}+\frac{z^2}{2!}+\frac{z^3}{3!}+\frac{z^4}{4!}+\cdots\right)$$

$$=\frac{1}{z^2}+\frac{1}{z}+\frac{1}{2}+\frac{z}{6}+\frac{z^2}{24}+\cdots.$$

Example 16.20

Find the Laurent's series of $f(z)=\frac{1}{z^2+1}$ about its singular points. Determine the region of convergence.

Solution

$f(z)=\frac{1}{z^2+1}$. The singular points of $f(z)=\frac{1}{z^2+1}$ are the zeros of $z^2+1=0$, *i.e.*, $z=\pm i$.

(i) Laurent's series about $z=i$

$$f(z)=\frac{1}{z^2+1}=\frac{1}{z-i}\frac{1}{z+i}=\frac{1}{z-i}\cdot\frac{1}{2i+(z-i)}$$

$$=\frac{1}{z-i}\cdot\frac{1}{2i}\left(1+\frac{z-i}{2i}\right),$$

$$=\frac{1}{zi(z-i)}\sum_{n=0}^{\infty}(-1)^n\left(\frac{z-i}{2i}\right)^n,$$

if $|z-i|<|2i|=2$.

$$=\sum_{n=0}^{\infty}(-1)^n\frac{(z-i)^{n-1}}{(2i)^{n+1}},\text{ if } |z-i|<2.$$

(ii) Laurent's series about $z=-i$

$$f(z)=\frac{1}{z^2+1}=\frac{1}{z+i}\cdot\frac{1}{z-i}=\frac{1}{z+i}\frac{1}{z+i-2i}$$

$$=\frac{1}{z+i}\frac{1}{-2i}\left(1-\frac{z+i}{2i}\right)^{-1}$$

$$=\frac{1}{-2i(z+i)}\sum_{n=0}^{\infty}\left(\frac{z+i}{2i}\right)^n,$$

if $\left|\frac{z+i}{2i}\right|<1$ or $|z+i|<2$

$$= -\sum_{n=0}^{\infty} \frac{(z+i)^{n-1}}{(2i)^{n+1}}, \quad \text{if } |z+i| < 2.$$

Example 16.21

Find the Laurent's series expansion of $\dfrac{1}{(z+1)(z+2)}$ for the range $0 < |z+1| < 2$.

[JNTU 1998S]

Solution

$$f(z) = \frac{1}{(z+1)(z+2)} = \frac{1}{z+1} - \frac{1}{z+2} = \frac{1}{t} - \frac{1}{t+1}$$

where $t = z+1$.

Case (i) Expansion for $0 < |z+1| < 1$ *i.e.*, $0 < |t| < 1$

$$\frac{1}{t} - \frac{1}{t+1} = \frac{1}{t} - (1+t)^{-1}$$

$$= \frac{1}{t} - (1 - t + t^2 - t^3 + \ldots),$$

by binomial thorem.

$$\Rightarrow f(z) = \frac{1}{z+1} - 1 + (z+1) - (z+1)^2 + (z+1)^3 - \cdots \text{ for } 0 < |z+1| < 1$$

Case (ii) Expansion for $1 < |z+1| < 2$ *i.e.*, $1 < |t| < 2$

$$\frac{1}{t} - \frac{1}{t+1} = \frac{1}{t} - \frac{1}{t}\left(1 + \frac{1}{t}\right)^{-1}$$

$$= \frac{1}{t} - \frac{1}{t}\left(1 - \frac{1}{t} + \frac{1}{t^2} - \frac{1}{t^3} + \cdots\right)$$

$$= \frac{1}{t^2} - \frac{1}{t^3} + \frac{1}{t^4} - \cdots$$

$$\Rightarrow f(z) = \frac{1}{(z+1)^2} - \frac{1}{(z+1)^3} + \frac{1}{(z+1)^4} - \cdots$$

for $1 < |z+1| < 2$

Combining the two results, we have Laurent's series expansion for $0 < |z+1| < 2$

$$\frac{1}{(z+1)(z+2)} = \cdots + \frac{1}{(z+1)^4} - \frac{1}{(z+1)^3} + \frac{1}{(z+1)^2} + \frac{1}{z+1} - 1 + (z+1) - (z+1)^2 + (z+1)^3 - \cdots.$$

Example 16.22

Obtain the Taylor's or Laurent's series which represents the function $\dfrac{1}{(1+z^2)(z+2)}$ when (i) $|z| < 1$, (ii) $1 < |z| < 2$ and (iii) $|z| > 2$.

Solution

$$\text{Let } f(z) = \frac{1}{(1+z^2)(z+2)} = \frac{A}{z+2} + \frac{Bz+C}{z^2+1}$$

A, B, C are found to be $A = \frac{1}{5}, \quad B = -\frac{1}{5}, \quad C = \frac{2}{5}.$

(i) $|z| < 1$

$$f(z) = \frac{1}{5}\cdot\frac{1}{2}\left(1 + \frac{z}{2}\right)^{-1} - \frac{1}{5}(z-2)(1+z^2)^{-1}$$

$$= \frac{1}{10}\left(1 - \frac{z}{2} + \frac{z^2}{2^2} - \cdots\right) - \frac{1}{5}(z-2)(1 - z^2 + z^4 - z^6 + \cdots)$$

This is the Taylor's series expansion of $f(z)$ valid for $|z| < 1$ since it contains non-negative powers of z.

(ii) $1 < |z| < 2$

$$f(z) = \frac{1}{5}\cdot\frac{1}{2}\left(1 + \frac{z}{2}\right)^{-1} - \frac{z-2}{5.z^2}\left(1 + \frac{1}{z^2}\right)^{-1}$$

$$= \frac{1}{10}\left(1 - \frac{z}{2} + \frac{z^2}{2^2} - \cdots\right) - \frac{z-2}{5\cdot z^2}\left(1 - \frac{1}{z^2} + \frac{1}{z^4} - \cdots\right)$$

This is the Laurent's series expansion of $f(z)$ valid for $1 < |z| < 2$ since it contains both positive and negative powers of z.

(iii) $|z| > 2$.

$$f(z) = \frac{1}{5z}\left(1 + \frac{2}{z}\right)^{-1} - \frac{1}{5}\cdot\frac{z-2}{z^2}\left(1 + \frac{1}{z^2}\right)^{-1}$$

$$= \frac{1}{5z}\left(1 - \frac{2}{z} + \left(\frac{2}{z}\right)^2 - \left(\frac{2}{z}\right)^3 + \cdots\right) - \frac{1}{5z^2}(z-2)\left(1 - \frac{1}{z^2} + \frac{1}{z^4} - \frac{1}{z^6} + \cdots\right)$$

This is again the Laurent's series expansion of $f(z)$ valid for $|z| > 2$, since it contains terms with negative powers of z.

Example 16.23

Find the Laurent's expansion $f(z)=\dfrac{7z-2}{(z+1)z(z-2)}$ in $1<|z+1|<3$. [JNTU 2003S]

Solution Put $z+1=t \Rightarrow z=t-1, z-2=t-3$

$$f(z)=\frac{7(t-1)-2}{t(t-1)(t-3)}=\frac{7t-9}{t(t-1)(t-3)}$$

$$=\frac{A}{t}+\frac{B}{t-1}+\frac{C}{t-3}=\frac{-3}{t}-\frac{1}{t\left(1-\frac{1}{t}\right)}-\frac{2}{3\left(1-\frac{t}{3}\right)}$$

$$=-\frac{3}{t}-\frac{1}{t}\left(1-\frac{1}{t}\right)^{-1}-\frac{2}{3}\left(1-\frac{t}{3}\right)^{-1}$$

$$=-\frac{3}{t}+\frac{1}{t}\left(1+\frac{1}{t}+\frac{1}{t^2}+\cdots\right)$$

$$-\frac{2}{3}\left(1+\frac{t}{3}+\frac{t^2}{9}+\frac{t^3}{27}+\cdots\right)$$

$$=-\frac{2}{z+1}+\frac{1}{(z+1)^2}+\frac{1}{(z+1)^3}$$

$$-\frac{2}{3}\left[1+\frac{1}{3}(z+1)+\frac{1}{3^2}(z+1)^2+\cdots\right]$$

valid for $1<|z+1|, |z+1|>3$.

This is the required Laurent's series expansion for $f(z)$ in $1<|z+1|<3$.

Example 16.24

Obtain the Laurent's expansion for

$f(z)=\dfrac{1}{(z+2)(1+z)^2}$ in (i) $|z|<2$ and (ii) $|1+z|>1$. [JNTU 2003S]

Solution It is easily seen that

$$\frac{1}{(1+z)(2+z)}=\frac{1}{1+z}-\frac{1}{z+2}.$$

Multiplying both sides by $\dfrac{1}{z+1}$ and using this result, we obtain

$$\frac{1}{(1+z)^2(2+z)}=\frac{1}{(1+z)^2}-\frac{1}{(1+z)(2+z)}$$

$$=\frac{1}{(1+z)^2}-\frac{1}{1+z}+\frac{1}{2+z}$$

(i) $|z|<2$

$$f(z)=\frac{1}{(1+z)^2}-\frac{1}{1+z}+\frac{1}{z+2}$$

$$=(1+z)^{-2}-(1+z)^{-1}+\frac{1}{2}\left(1+\frac{z}{2}\right)^{-1}$$

$$=(1-2z+3z^2-4z^3+\cdots)-(1-z+z^2$$

$$-z^3+\cdots)+\frac{1}{2}\left(1-\frac{z}{2}+\frac{z^2}{4}-\frac{z^3}{8}+\cdots\right)$$

$$=\frac{1}{2}-\frac{5}{4}z+\frac{17}{8}z^2-\frac{49}{16}z^3+\cdots$$

(ii) $|1+z|>1$. Put $1+z=t$

$$f(z)=\frac{1}{(1+z)^2}-\frac{1}{1+z}+\frac{1}{z+2}$$

$$=\frac{1}{t^2}-\frac{1}{t}+\frac{1}{t+1}=\frac{1}{t^2}-\frac{1}{t}+\frac{1}{t}\left(1+\frac{1}{t}\right)^{-1}$$

$$=\frac{1}{t^2}-\frac{1}{t}+\frac{1}{t}\left(1-\frac{1}{t}+\frac{1}{t^2}-\frac{1}{t^3}+\cdots\right)$$

$$=\frac{1}{t^2}-\frac{1}{t}+\frac{1}{t}-\frac{1}{t^2}+\frac{1}{t^3}-\frac{1}{t^4}+\cdots$$

$$=\frac{1}{(z+1)^3}-\frac{1}{(z+1)^4}+\frac{1}{(z+1)^5}-\frac{1}{(z+1)^6}+\cdots.$$

Example 16.25

Find the Laurent's series expansion for $f(z)=\dfrac{1}{z^2(1-z)}$ and find the region of convergence. [JNTU 2006 (Set 1)]

Solution The singular points of $f(z)=\dfrac{1}{z^2(1-z)}$ are $z=0$ and $z=1$.

$$f(z)=\frac{1}{z^2}(1-z)^{-1}=\frac{1}{z^2}\sum z^n=\frac{1}{z^2}(1+z+z^2+\cdots)$$

$$=\frac{1}{z^2}+\frac{1}{z}+1+z+z^2+\cdots,\quad \text{if } 0<|z|<1$$

Also,

$$f(z)=\frac{1}{z^2}\frac{1}{(-z)(1-\frac{1}{z})}=\frac{1}{-z^3}\sum\left(\frac{1}{z}\right)^n$$

$$=-\frac{1}{z^3}\left(1+\frac{1}{z}+\frac{1}{z^2}+\frac{1}{z^3}+\cdots\right)$$

$$=-\left(\frac{1}{z^3}+\frac{1}{z^4}+\frac{1}{z^5}+\cdots\right)$$

if $z \neq 0$ and $\left|\frac{1}{z}\right| < 1$ *i.e.*, if $|z| > 1$.

Example 16.26

Expand $f(z) = \frac{z}{z^2+1}$ for $|z-3| > 2$. [JNTU 2000]

Solution The singular points of

$f(z) = \frac{1}{2}\left(\frac{1}{z+i} + \frac{1}{z-i}\right)$ are $z = i$ and $-i$.

$$f(z) = \frac{1}{2}\left(\frac{1}{z+i} + \frac{1}{z-i}\right)$$

$$= \frac{1}{2}\left(\frac{1}{t+3+i} + \frac{1}{t+3-i}\right),$$

(Putting $z-3 = t \Rightarrow z = t+3$)

$$= \frac{1}{2t}\left(\frac{1}{1+\frac{3+i}{t}} + \frac{1}{1+\frac{3-i}{t}}\right)$$

$$= \frac{1}{2t}\left[\left(1+\frac{3+i}{t}\right)^{-1} + \left(1+\frac{3-i}{t}\right)^{-1}\right]$$

$$= \frac{1}{2t}\sum_{n=0}^{\infty}(-1)^n[(3+i)^n + (3-i)^n]\frac{1}{t^n}$$

$$= \frac{1}{2(z-3)}\sum_{n=0}^{\infty}(-1)^n[(3+i)^n + (3-i)^n]\frac{1}{(z-3)^n}$$

valid for $|3 \pm i| < |z-3|$ or $|z-3| > \sqrt{10}$.

Example 16.27

Find the Laurent's series expansion of the function $f(z) = \frac{z^2-6z-1}{(z-1)(z-3)(z+2)}$ in the region $3 < |z+2| < 5$ [JNTU 2001, 2005*S* (Set 1)]

Solution Let

$$f(z) = \frac{z^2-6z-1}{(z-1)(z-3)(z+2)}$$

$$= \frac{A}{z-1} + \frac{B}{z-3} + \frac{C}{z+2}$$

where $A = \frac{1-6-1}{(1-3)(1+2)} = 1,$

$$B = \frac{3^2-6\cdot 3-1}{(3-1)(3+2)} = 1,$$

$$C = \frac{(-2)^2-6(-2)-1}{(-2-1)(-2-3)} = 1$$

Put $z+2 = t \Rightarrow z = t-2$

Now

$$f(z) = \frac{1}{t-3} - \frac{1}{t-5} + \frac{1}{t}$$

$$= \frac{1}{t\left(1-\frac{3}{t}\right)} - \frac{1}{-5\left(1-\frac{t}{5}\right)} + \frac{1}{t}$$

$$= \frac{1}{t} + \frac{1}{t}\sum_{n=0}^{\infty}\left(\frac{3}{t}\right)^n + \frac{1}{5}\sum_{n=0}^{\infty}\left(\frac{t}{5}\right)^n$$

$$= \frac{1}{z+2} + \sum_{n=0}^{\infty}3^n\frac{1}{(z+2)^{n+1}} + \sum_{n=0}^{\infty}\frac{1}{5^{n+1}}(z+2)^n$$

valid for $3 < |z+2| < 5$.

Example 16.28

Expand $f(z) = \frac{z+3}{z(z^2-z-2)}$ in powers of z if (i) $|z| < 1$ (ii) $1 < |z| < 2$ and (iii) $|z| > 2$.

[JNTU 2001*S*]

Solution Let

$$f(z) = \frac{z+3}{z(z+1)(z-2)} = \frac{A}{z} + \frac{B}{z+1} + \frac{C}{z-2}$$

$\because \quad z^2 - z - 2 = (z+1)(z-2)$

We find $A = -3/2$, $B = \frac{-1+3}{-1(-1-2)} = \frac{2}{3},$

$$C = \frac{2+3}{2(2+1)} = \frac{5}{6}$$

(i) $|z| < 1$

$$f(z) = \frac{A}{z} + B(1+z)^{-1} - \frac{C}{2}\left(1-\frac{z}{2}\right)^{-1}$$

$$= -\frac{3}{2}\cdot\frac{1}{z} + \frac{2}{3}\sum_{n=0}^{\infty}(-1)^n z^n - \frac{5}{12}\sum_{n=0}^{\infty}\left(\frac{z}{2}\right)^n$$

$$= -\frac{3}{2}\cdot\frac{1}{z} + \frac{2}{3}(1-z+z^2-z^3+\cdots)$$

$$- \frac{5}{12}\left(1+\frac{z}{2}+\frac{z^2}{4}+\frac{z^3}{8}+\cdots\right)$$

$$= -\frac{3}{2}\cdot\frac{1}{z} + \frac{1}{4} - \frac{7}{8}z + \frac{9}{16}z^2 - \frac{23}{32}z^3 + \cdots$$

(ii) $1 < |z| < 2$

$$f(z) = -\frac{3}{2}\cdot\frac{1}{z} + \frac{2}{3z}\left(1+\frac{1}{z}\right)^{-1} - \frac{5}{6\times 2}\left(1-\frac{z}{2}\right)^{-1}$$

$$= -\frac{3}{2}\cdot\frac{1}{z} + \frac{2}{3}\cdot\frac{1}{z}\sum(-1)^n\frac{1}{z^n} - \frac{5}{12}\sum\left(\frac{z}{2}\right)^n$$

$$= -\frac{3}{2}\cdot\frac{1}{z}+\frac{2}{3z}\left(1-\frac{1}{z}+\frac{1}{z^2}-\frac{1}{z^3}+\cdots\right)$$

$$-\frac{5}{12}\left(1+\frac{z}{2}+\frac{z^2}{2^2}+\frac{z^3}{2^3}+\cdots\right)$$

$$= -\frac{5}{12}-\frac{5}{6}\cdot\frac{1}{z}+\frac{2}{3}\left(-\frac{1}{z^2}+\frac{1}{z^3}-\cdots\right)$$

$$-\frac{5}{12}\left(\frac{z}{2}+\frac{z^2}{4}+\frac{z^3}{8}+\cdots\right)$$

(iii) $|z|>2$

$$f(z) = -\frac{3}{2}\cdot\frac{1}{z}+\frac{2}{3z}\left(1+\frac{1}{z}\right)^{-1}+\frac{5}{6z}\left(1-\frac{2}{z}\right)^{-1}$$

$$= -\frac{3}{2}\cdot\frac{1}{z}+\frac{2}{3z}\left(1-\frac{1}{z}+\frac{1}{z^2}-\frac{1}{z^3}+\cdots\right)$$

$$+\frac{5}{6z}\left(1+\frac{2}{z}+\frac{4}{z^2}+\frac{8}{z^3}+\cdots\right)$$

$$= \frac{1}{z^2}+\frac{4}{z^3}+\frac{6}{z^4}+\cdots.$$

Example 16.29

Expand $f(z)=\dfrac{e^{2z}}{(z-1)^3}$ as Laurent's series about the singular point $z = 1$. [JNTU 2002]

Solution

$$f(z)=\frac{e^{2z}}{(z-1)^3}=\frac{e^{2t}\cdot e^2}{t^3},$$

Put $z-1=t \Rightarrow z=t+1$

$$= e^2\cdot\frac{1}{t^3}\left[1+(2t)+\frac{(2t)^2}{2!}+\frac{(2t)^3}{3!}+\cdots\right]$$

$$= e^2\cdot\sum_{n=0}^{\infty}\frac{2^n}{n!}t^{n-3} = e^2\sum_{n=0}^{\infty}\frac{2^n}{n!}(z-1)^{n-3}, z\neq 1$$

$$= e^2\sum_{n=0}^{\infty}\frac{2^n}{n!}(z-1)^{n-3}, \quad \text{if } |z-1|>0.$$

Example 16.30

Obtain all the Laurent's series of the function $f(z)=\dfrac{7z-2}{(z+1)z(z-2)}$ about $z_0 = -1$.

[JNTU 2003S (Set 4), 2004S, 2008 (Set 2)]

Solution $f(z)$ has singular points at $z = 0, -1$ and 2.

Let $f(z)=\dfrac{7z-2}{(z+1)z(z-2)}=\dfrac{A}{z}+\dfrac{B}{z+1}+\dfrac{C}{z-2}$

$$=\frac{1}{z}-\frac{3}{z+1}+\frac{2}{z-2}$$

where $A=\dfrac{7(0)-2}{(0+1)(0-2)}=1,$

$$B=\frac{7(-1)-2}{(-1)(-1-2)}=-3,$$

$$C=\frac{7(2)-2}{(2+1)2}=2$$

Three types of Laurent's series expansions are possible about $z_0=-1$ in the regions (i) $0<|z+1|<1$, (ii) $1<|z+1|<3$ and (iii) $|z+1|>3$.

(i) In the annular region $0<|z+1|<1$

$$f(z)=\frac{1}{z}-\frac{3}{z+1}+\frac{2}{z-2}$$

$$=\frac{-1}{1-(z+1)}-\frac{3}{z+1}+\frac{2}{(z+1)-3}$$

$$=\frac{-1}{1-(z+1)}-\frac{3}{z+1}+\frac{2}{-3\left(1-\frac{z+1}{3}\right)}$$

$$=-\frac{3}{z+1}-\sum_{n=0}^{\infty}(z+1)^n-\frac{2}{3}\sum\left(\frac{2+1}{3}\right)^n$$

$$=-\frac{3}{z+1}-[1+(z+1)+(z+1)^2+\cdots]$$

$$-\frac{2}{3}\left[1+\frac{z+1}{3}+\frac{(z+1)^2}{9}+\cdots\right]$$

$$=-\frac{3}{z+1}-\frac{5}{3}-\frac{11}{9}(z+1)-\frac{29}{27}(z+1)^2$$

$$-\frac{83}{81}(z+1)^3-\cdots.$$

(ii) In the annular region $1<|z+1|<3$

$$f(z)=\frac{1}{z}-\frac{3}{z+1}+\frac{2}{z-2}$$

$$=\frac{1}{(z+1)-1}-\frac{3}{z+1}+\frac{2}{(z+1)-3}$$

$$=\frac{1}{z+1}\left(1-\frac{1}{z+1}\right)^{-1}-\frac{3}{z+1}-\frac{2}{3}\left(1-\frac{z+1}{3}\right)^{-1}$$

$$=-\frac{3}{z+1}+\frac{1}{z+1}\left(1+\frac{1}{z+1}+\frac{1}{(z+1)^2}+\cdots\right)$$

$$-\frac{2}{3}\left(1+\frac{z+1}{3}+\frac{(z+1)^2}{9}+\frac{(z+1)^3}{27}+\cdots\right)$$

$$= -\frac{2}{z+1} + \frac{1}{(z+1)^2} + \frac{1}{(z+1)^3} + \cdots$$
$$-\frac{2}{3}\left(1 + \frac{z+1}{3} + \frac{(z+1)^2}{9} + \frac{(z+1)^3}{27} + \cdots\right)$$

(iii) In the region $|z+1| > 3$

$$f(z) = \frac{1}{z} - \frac{3}{z+1} + \frac{2}{z-2}$$
$$= \frac{1}{(z+1)-1} - \frac{3}{z+1} + \frac{2}{(z+1)-3}$$
$$= \frac{1}{z+1}\left(1 - \frac{1}{z+1}\right)^{-1}$$
$$-\frac{3}{z+1} + \frac{2}{z+1}\left(1 - \frac{3}{z+1}\right)^{-1}$$
$$= -\frac{3}{z+1} + \frac{1}{z+1}\left[1 + \frac{1}{z+1} + \frac{1}{(z+1)^2} + \cdots\right]$$
$$+\frac{2}{z+1}\left[1 + \frac{3}{z+1} + \frac{9}{(z+1)^2} + \cdots\right]$$
$$= \frac{7}{(z+1)^2} + \frac{19}{(z+1)^3} + \frac{33}{(z+1)^4} + \cdots$$

Example 16.31

Find the Laurent's series of $\dfrac{z^2-1}{(z+2)(z+3)}$ for $|z| > 3$

[JNTU 2004, 2006 (Set 4)]

Solution Let

$$f(z) = \frac{z^2-1}{(z+2)(z+3)} = 1 + \frac{A}{z+2} + \frac{B}{z+3}$$
$$\Rightarrow z^2 + 5z + 6 + A(z+3) + B(z+2) = z^2 - 1$$
$$\Rightarrow A(z+3) + B(z+2) = -7 - 5z$$
$$z = -2 \Rightarrow A = 3; z = -3 \Rightarrow -B = 8$$
$$f(z) = 1 + \frac{3}{z+2} - \frac{8}{z+3}$$
$$= 1 + \frac{3}{z}\left(1 + \frac{2}{z}\right)^{-1} - \frac{8}{z}\left(1 + \frac{3}{z}\right)^{-1}$$
$$= 1 + \frac{3}{z}\left(1 - \frac{2}{z} + \frac{4}{z^2} - \frac{8}{z^3} + \cdots\right)$$
$$-\frac{8}{z}\left(1 - \frac{3}{z} + \frac{9}{z^2} - \frac{27}{z^3} + \cdots\right), \text{ if } |z| > 3$$
$$= 1 - \frac{5}{z} + \frac{18}{z^2} - \frac{60}{z^3} + \cdots \text{ valid for } |z| > 3.$$

Example 16.32

Find the Laurent's series of the function $f(z) = \dfrac{z}{(z+1)(z+2)}$ about $z = -2$. [JNTU 2005*S* (3)]

Solution

$$f(z) = \frac{z}{(z+1)(z+2)} = \frac{t-2}{t(t-1)} = \frac{2}{t} - \frac{1}{t-1}$$
$$[\text{Put } z+2 = t \Rightarrow z = t-2]$$
$$= \frac{2}{t} + (1-t)^{-1} = \frac{2}{t} + (1 + t + t^2 + \cdots)$$
$$= \frac{2}{z+2} + 1 + (z+2) + (z+2)^2 + \cdots.$$

Example 16.33

Find the Laurent's expansion of $\dfrac{1}{z^2-4z+3}$ for $1 < |z| < 3$.

[JNTU 2006(Set 4)]

Solution

$$f(z) = \frac{1}{z^2-4z+3} = \frac{1}{(z-1)(z-3)}$$
$$= \frac{1}{2}\left(\frac{1}{z-3} - \frac{1}{z-1}\right)$$
$$= \frac{1}{2\times 3}\frac{1}{(1-z/3)} + \frac{1}{2z}\frac{1}{(1-1/z)}$$
$$= -\frac{1}{6}\left(1 - \frac{z}{3}\right)^{-1} + \frac{1}{2z}\left(1 - \frac{1}{z}\right)^{-1}$$
$$= -\frac{1}{6}\left(1 + \frac{z}{3} + \frac{z^2}{9} + \frac{z^3}{27} + \cdots\right)$$
$$+\frac{1}{2z}(1 + z^{-1} + z^{-2} + z^{-3} + \cdots)$$
$$= -\frac{1}{6}\left(1 + \frac{1}{3}z + \frac{1}{9}z^2 + \frac{1}{27}z^3 + \cdots\right)$$
$$+\frac{1}{2}\left(\frac{1}{z} + \frac{1}{z^2} + \frac{1}{z^3} + \cdots\right),$$

valid for $1 < |z| < 3$.

Example 16.34

Expand $f(z) = \dfrac{1}{z^2-3z+2}$ in the region (i) $0 < |z-1| < 1$, (ii) $1 < |z| < 2$

[JNTU 2003*S*, 2004*S*, 2006 (Set 2)]

Solution The singular points of $f(z) = \dfrac{1}{z-2} - \dfrac{1}{z-1}$ are $z = 1$ and 2.

(i) $f(z)=\frac{1}{z-2}-\frac{1}{z-1}=\frac{1}{t-1}-\frac{1}{t}$

[Put $z-1=\Rightarrow z=t+1$]

$$=-\frac{1}{t}-\frac{1}{1-t}=-\frac{1}{t}-(1+t+t^2+\cdots)$$

$$=-\frac{1}{z-1}-[1+(z+1)+(z-1)^2+\cdots]$$

if $0<|z-1|<1$

(ii) $f(z)=\frac{1}{z-2}-\frac{1}{z-1}$

$$=\frac{1}{-2}\frac{1}{(1-z/2)}-\frac{1}{z(1-1/z)}$$

$$=-\frac{1}{2}(1-z/2)^{-1}-\frac{1}{z}\left(1-\frac{1}{z}\right)^{-1}$$

$$=-\frac{1}{2}\left(1+\frac{z}{2}+\frac{z^2}{4}+\frac{z^3}{8}+\cdots\right)$$

$$-\frac{1}{z}\left(1+\frac{1}{z}+\frac{1}{z^2}+\frac{1}{z^3}+\cdots\right)$$

$$=-\frac{1}{2}\left(1+\frac{1}{2}z+\frac{1}{4}z^2+\frac{1}{8}z^3+\cdots\right)$$

$$-\left(\frac{1}{z}+\frac{1}{z^2}+\frac{1}{z^3}+\cdots\right)$$

valid for $1<|z|<2$.

16.9 Higher Derivatives of Analytic Functions

We will prove in the following that an analytic function has derivatives of all orders. It follows, in particular, that the derivative of an analytic function is again an analytic function. This is not the case with real functions.

Consider a function $f(z)$ which is analytic in a region D. Let $z \in D$ and suppose C is any circle with centre z such that the circle and its interior are contained in D. By Cauchy's integral formula, we have

$$f(z)=\frac{1}{2\pi i}\int_C \frac{f(t)}{t-z}dt.$$

We will proceed to prove that

$$f'(z)=\frac{1}{2\pi i}\int_C \frac{f(t)}{(t-z)^2}dt$$

and, in general,

$$f^{(n)}(z)=\frac{n!}{2\pi i}\int_C \frac{f(t)}{(t-z)^{n+1}}dt$$

Theorem 16.3 Let f be analytic inside and on a simple closed curve C. Let z be any point inside C. Then $f'(z)=\frac{1}{2\pi i}\int_C \frac{f(t)}{(t-z)^2}dt$

Proof: We have, by Cauchy's integral formula,

$$f(z)=\frac{1}{2\pi i}\int_C \frac{f(t)}{t-z}dt \Rightarrow \frac{f(z+h)-f(z)}{h}$$

$$=\frac{1}{h}\left[\frac{1}{2\pi i}\int_C f(t)\left(\frac{1}{t-z-h}-\frac{1}{t-z}\right)dt\right] \quad (16.14)$$

$$=\frac{1}{2\pi i}\int_C \frac{f(t)dt}{(t-z-h)(t-z)}$$

$$\Rightarrow \frac{f(z+h)-f(z)}{h}-\frac{1}{2\pi i}\int_C \frac{f(t)}{(t-z)}dt$$

$$=\frac{1}{2\pi i}\int_C \frac{f(t)}{t-z}\left(\frac{1}{t-z-h}-\frac{1}{t-z}\right)dt$$

$$=\frac{h}{2\pi i}\int_C \frac{f(t)dt}{(t-z-h)(t-z)^2} \quad (16.15)$$

Let M be the maximum value of $|f(t)|$ on C. Let $L = 2\pi R$ be the length of C and d be the shortest distance from z to any point on C. Then, for any point t on C we have

$$|t-z|\ge d \text{ and } |t-z-h|\ge|t-z|-|h|\ge d-|h|$$

$$\Rightarrow \left|\frac{f(t)}{(t-z-h)(t-z)^2}\right|\le\frac{M}{d^2(d-|h|)}$$

Therefore, we have from Eq. (16.14)

$$\left|\frac{f(z+h)-f(z)}{h}-\frac{1}{2\pi i}\int_C \frac{f(t)dt}{(t-z)^2}\right|$$

$$=\frac{|h|}{2\pi}\left|\int_C \frac{f(t)dt}{(t-z-h)(t-z)^2}\right|$$

$$\le\frac{|h|}{2\pi}\cdot\frac{2\pi MR}{d^2(d-|h|)}$$

$$=\frac{|h|MR}{d^2(d-|h|)}\to 0 \quad \text{as} \quad h\to 0$$

$$\Rightarrow \lim_{h\to 0}\left|\frac{f(z+h)-f(z)}{h}-\frac{1}{2\pi i}\int_C \frac{f(t)dt}{(t-z)^2}\right|=0$$

$$\Rightarrow f'(z)=\frac{1}{2\pi i}\int_C \frac{f(t)dt}{(t-z)^2}.$$

Remark

By using induction on n, we can prove that for any positive integer n we have

$$f^{(n)}(z)=\frac{n!}{2\pi i}\int_C \frac{f(t)}{(t-z)^{n+1}}dt.$$

EXERCISE 16.1

1. Expand $f(z) = \sin z$ in a Taylor's series about $z=\pi/4$.

Ans: $\sin z=\frac{1}{\sqrt{2}}+\left(z-\frac{\pi}{4}\right)\frac{1}{\sqrt{2}}+\frac{(z-\pi/4)^2}{2!}\left(-\frac{1}{\sqrt{2}}\right)+\frac{(z-\pi/4)^3}{3!}\left(-\frac{1}{\sqrt{2}}\right)+\cdots$

2. Show that $\frac{1}{z^2}=\sum_{n=0}^{\infty}(n+1)(z+1)^n$ if $|z+1|<1$.

3. Find the Taylor's expansion of $f(z)=\frac{2z^3+1}{z^2+z}$ about the point $z=i$.

Ans: $\frac{2z^3+1}{z^2+z}=\frac{i-3}{z}+(z-i)\left(3+\frac{i}{2}\right)+\cdots$

4. Find the Taylor's expansion for $f(z)=\frac{1}{(1+z)^2}$ with centre at $-i$.

Ans: $\frac{1}{(1+z)^2}=\frac{i}{2}+\frac{i}{2}\sum_{n=0}^{\infty}(z+i)^n(-1)^n\frac{(n+1)^n}{(1-i)^n}$

5. Expand $\cos z$ in Taylor's series about $z=\pi/2$.

Ans: $\cos z=-\left(z-\frac{\pi}{2}\right)+\frac{1}{3!}-\left(2-\frac{\pi}{2}\right)^3-\frac{1}{5!}(z-\pi/2)^5+\cdots$

6. Expand $\frac{z^2-1}{(z+2)(z+3)}$ in the region

(a) $|z|<2$ and (b) $2<|z|<3$.

Ans: (a) $1+\frac{3}{2}\sum_{n=0}^{\infty}(-1)^n\frac{z^n}{2^n}-\frac{8}{3}\sum_{n=0}^{\infty}(-1)^n\frac{z^n}{3^n}$

(b) $1+\frac{3}{2}\sum_{n=0}^{\infty}(-1)^n\frac{2^n}{z^n}-\frac{8}{3}\sum_{n=0}^{\infty}(-1)^n\frac{z^n}{3^n}$

7. Expand $f(z)=\frac{1}{(z-1)(z-2)}$ in a Taylor or Laurent series valid in

(a) $|z|<1$ and (b) $1<|z|<2$.

Ans: (a) $-\frac{1}{2}\left(1+\frac{z}{2}+\frac{z^2}{4}+\frac{z^3}{8}+\cdots\right)+(1+z+z^2+\cdots)$

(Taylor's series for $|z|<1$)

(b) $-\frac{1}{2}\left(1+\frac{z}{2}+\frac{z^2}{4}+\frac{z^3}{8}+\cdots\right)-\frac{1}{2}\left(1+\frac{1}{z}+\frac{1}{z^2}+\frac{1}{z^3}+\cdots\right)$

(Laurent's series for $1<|z|<2$)

8. Expand $\frac{z^3-4}{z^2+5z+4}$ in Laurent's series valid in

(a) $|z|>4$ (b) $1<|z|<4$.

Ans: (a) $1-\frac{1}{z}+\frac{1}{z^2}-\frac{1}{z^3}+\cdots-\frac{4}{z}+\frac{16}{z^2}-\frac{64}{z^3}+\cdots$ for $|z|>4$

(b) $1-\frac{1}{z}+\frac{1}{z^2}-\frac{1}{z^3}+\cdots-1+\frac{z}{4}-\frac{z^2}{16}+\cdots$ for $1<|z|<4$

9. If $0<|z-1|<2$, express $f(z)=\frac{z}{(z-1)(z-3)}$ in a series of positive and negative powers of $(z-1)$.

Ans: $-\frac{1}{2(z-1)}-\frac{3}{4}\sum_{n=0}^{\infty}\left(1-\frac{1}{z}\right)^n$

10. Expand $f(z)=\frac{z}{(z+1)(z+2)}$ in Laurent's series about $z=-2$.

Ans: $\frac{2}{z+2}+1+(z+2)+(z+2)^2+\cdots$

11. Expand $\frac{1}{(z-a)(z-b)}$ for (a) $a<|z|<b$ and (b) $|z|>b\ (z<b)$.

Ans: (a) $\frac{1}{a-b}\left[\frac{1}{b}\left(1+\frac{z}{b}+\frac{z^2}{b^2}+\cdots\right)+\frac{1}{z}\left(1+\frac{a}{z}+\frac{a^2}{z^2}+\cdots\right)\right]$

(b) $\frac{1}{b-a}\left[\frac{1}{z}\left(1+\frac{b}{z}+\frac{b^2}{z^2}+\cdots\right)+\frac{1}{z}\left(1+\frac{a}{z}+\frac{a^2}{z^2}+\cdots\right)\right]$

12. Find the Laurent's series of

$f(z)=\frac{z^2-6z-1}{(z-1)(z-3)(z+2)}$ in the region

$3<|z+2|<5.$

Ans: $\frac{2}{z+2}+\frac{3}{(z+2)^2}+\frac{3^2}{(z+2)^3}+\cdots+\frac{1}{5}\left[1+\frac{z+2}{5}+\frac{(z+2)^2}{5^2}+\frac{(z+2)^3}{5^3}+\cdots\right]$

Calculus of Residues

17

17.1 EVALUATION OF REAL INTEGRALS

17.1.1 Introduction

It is interesting to note that the residue theorem provides us with an elegant and simple method for evaluation of certain classes of real integrals, which cannot be evaluated by the usual methods of real calculus. Evaluation of a real definite integral is done by converting it into a contour integral. In order to do this one has to identify

1. an appropriate closed contour C and
2. a suitable complex function to be chosen as an integrand.

We classify the real integrals to be evaluated under different types, identify the closed contour C, the complex function to be integrated in each case and evaluate the integrals. But at first we will briefly study about zeros, singularities and their types, the theory of residues and Cauchy's residue theorem.

17.1.2 Zeros and Singularities

Zero of an analytic function: A zero of an analytic function $f(z)$ is a value a of z such that $f(a) = 0$.
Order of a zero: If an analytic function $f(z)$ can be expressed in the form $f(z) = (z - a)^m\, \phi(z)$ where $\phi(z)$ is analytic and $\phi(a) \neq 0$ then $z = a$ is called a zero of $f(z)$ of order m. If $m = 1$ it is called a simple zero.

Example 17.1

1. $f(z) = z^3 - z^2 - z + 1 = (z-1)^2(z+1)$;
   $z = -1$ is a simple zero and $z = 1$ is a zero of order 2 of f.
2. $f(z) = \dfrac{1}{z-2}$; $z = \infty$ is a simple zero of f.
3. $f(z) = \sin z$; $z = 0, \pm\pi, \pm 2\pi, \ldots$ are simple zeros of f.
4. $f(z) = e^z$ has no zero.

17.1.3 Types of Singularities

1. Isolated singularities: A point $z = a$ is called an isolated singularity of an analytic function $f(z)$ if (a) $f(z)$ is not analytic at $z = a$ and if $f(z)$ is analytic in some deleted nbd of $z = a$, *i.e.*, there exists a nbd of 'a' which contains no other singularity of $f(z)$.

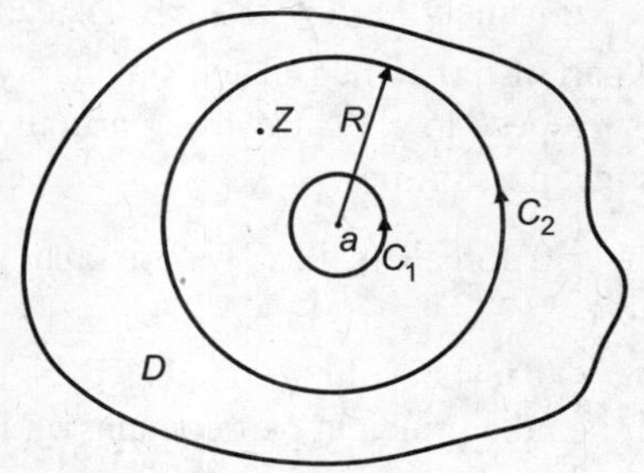

Figure 17.1

Example 17.2

1. $f(z) = \dfrac{1}{z^2 - 1}$; $z = \pm 1$ are two isolated singularities of $f(z)$.
2. $f(z) = \cot z$; $z = 0, \pm\pi, \pm 2\pi, \ldots, \pm n\pi$ are an infinite number of isolated singularities of $f(z)$.

2. Non-isolated singularities: A singularity of an analytic function $f(z)$ that is not isolated is called a non-isolated singularity.

Example 17.3

Consider $f(z)=\cot\frac{\pi}{z}$. The function is not analytic when $\sin\frac{\pi}{z}=0$, *i.e.*, when $\frac{\pi}{z}=n\pi, n=\pm 1,\pm 2,\ldots$.

In other words, $z=\pm 1,\pm\frac{1}{2},\pm\frac{1}{3},\ldots$ are all isolated singularities as there are no singularities in the nbd of any of them. $z = 0$ is a non-isolated singularity because in a δ-nbd of 0, we find more singular points for large n that lie in it.

Types of singularities Let $f(z) \in H(D - a)$, *i.e.*, $f(z)$ is analytic within a domain D except at the point $z = a$, which is an isolated singular point. Draw a circle C_1, with centre a and radius as small as we please. Draw another large concentric circle C_2 of radius R lying wholly within D. $f(z)$ is analytic in the annular region bounded by circles C_1 and C_2. If z is any point inside this region, by Laurent's theorem,

$$f(z)=\sum_{n=0}^{\infty}a_n(z-a)^n+\sum_{n=1}^{\infty}b_n(z-a)^{-n}$$
$$\text{for } 0<|z-a|<\delta$$

The second part of the series with negative powers of $(z - a)$ namely $\sum_{n=1}^{\infty}b_n(z-a)^{-n}$ is called the principal part of $f(z)$ at the isolated singularity $z = a$.

With respect to the principal part, there are three distinct possibilities:

(i) bn = 0 for all n, i.e., the principal part is absent.

(ii) bn ≠ 0 for an infinite number of values of n, i.e., the principal part contains an infinite number of terms.

(iii) The principal part contains a finite number of terms.

These three cases give rise to three kinds of singularities.

1. Removable singularity: In this case, $f(z)=\sum_{n=0}^{\infty}(z-a)^n$ (Taylor series). The singularity can be removed by defining $f(z)$ at $z = a$ in such a way that it is analytic at $z = a$.

Example 17.4

Let $f(z)=\frac{\sin z}{z}=\frac{1}{z}\left(z-\frac{z^3}{3!}+\cdots\right)=1-\frac{z^2}{3!}+\frac{z^4}{5!}+\cdots$

If we set $f(z)=\frac{\sin z}{z}=1$ at $z=0$, $\lim_{z\to a}f(z)$ exists and is a finite quantity.

2. Essential singularity: In this case, there are an infinite number of terms in the principal part which contains terms with negative powers of $(z-a)$. The series $\sum_{n=1}^{\infty}b_n(z-a)^{-n}=b_1(z-a)^{-1}+b_2\quad(z-a)^{-2}+\cdots$ contains non-vanishing coefficients b_n for an infinite number of values of n.

Example 17.5

$f(z)=e^{1/z^2}=1+\frac{1}{z^2}+\frac{1}{2!z^4}+\frac{1}{3!z^6}+\cdots$ has an infinite number of negative powers of z (though $b_{2n-1}=0$ for all n).

$z = 0$ is an essential singularity. In this case, $\lim_{z\to a}f(z)$ does not exist.

3. Pole: In this case, the principal part contains a finite number of terms. If in the series $\sum b_n(z-a)^{-n}, b_n=0$ for $n\geq m$ then

$$\lim_{z\to a}(z-a)^m f(z)=l\neq 0 \text{ exists}$$

and in this case $z = a$ is called a pole of order m of $f(z)$. If $m = 1$, it is called a simple pole.

Example 17.6

$$f(z)=\frac{e^z}{z^2}=\frac{1}{z^2}\left(1+\frac{z}{1!}+\frac{z^2}{2!}+\frac{z^2}{2!}+\frac{z^3}{3!}+\frac{z^4}{4!}+\cdots\right)$$
$$=\frac{1}{z^2}+\frac{1}{z}+\frac{1}{2}+\frac{1}{6}z+\cdots$$

$\lim_{z\to 0} z^2 f(z)=1$ (finite)

Hence $z = 0$ is a pole of order 2 (double pole) of $f(z)$.

Singularity at infinity

Put $z=\frac{1}{t}$ in $f(z)$ and obtain $f\left(\frac{1}{t}\right)=F(t)$, say.

Then the nature of singularity at ∞ is defined to be the same as that of $F(t)$ at $t = 0$.

$f(z) = z^3$ has a pole of order 3 at $z = \infty$ since $f\left(\frac{1}{t}\right)=\frac{1}{t^3}$ has a pole of order 3 at $t = 0$.

$f(z) = e^z$ has an essential singularity at $z = \infty$ since $f\left(\frac{1}{t}\right)=e^{\frac{1}{t}}$ has an essential singularity at $t = 0$.

$f(z) = \text{cosec}\ \pi z$ has a non-isolated singularity at $z = \infty$ since $f\left(\frac{1}{t}\right) = \text{cosec}\frac{\pi}{t} = \frac{1}{\sin \pi t}$ has a non-isolated singularity at $t = 0$.

The theory of residues deals with the evaluation of integrals of the form

$$I = \oint_C f(z)dz \qquad (17.1)$$

taken round a simple closed path C.

If $f(z)$ is analytic everywhere on C and inside C, then $I = 0$ by Cauchy's integral theorem (CIT).

If $f(z)$ is analytic everywhere on C and inside C except for an isolated singularity at $z = a$ inside C, then $f(z)$ can be expanded in a Laurent series

$$f(z) = \sum_{n=0}^{\infty} a_n (z-a)^n + \frac{b_1}{z-a} + \frac{b_2}{(z-a)^2} + \cdots \qquad (17.2)$$

which converges in $0 < |z-a| < R$. Here the constants b_n are given by

$$b_n = \frac{1}{2\pi i}\oint_C (z-a)^{n-1} f(z)dz. \qquad (17.3)$$

(Cauchy's generalised integral formula)

Formula (17.3) can be used for determining the constants b_n by evaluating the integral, which is more difficult. Instead we can use the formula for evaluating the integral since we can obtain Laurent series expansion for a given function in several ways. Taking $n = 1$ in Eq. (17.3), we have (Cauchy's integral formula)

$$b_1 = \frac{1}{2\pi i}\oint_C f(z)dz \qquad (17.4)$$

$$\text{or } \oint_C f(z)dz = 2\pi i b_1 \qquad (17.5)$$

This formula enables us to evaluate the integral on the LHS of Eq. (17.5) when we know b_1, which can be found through one or the other of the methods of Laurent's series expansion. The coefficient b_1 is called the residue of $f(z)$ at $z = a$ and we denote it by

$$b_1 = \text{Res}_{z=a} f(z) \text{ or } b_1 = \text{Res}\ (f : z = a). \qquad (17.6)$$

17.1.4 Formulas for Residues at Poles

Residue at simple pole

(i) If $z = a$ is a simple pole of $f(z)$, we have by Laurent's theorem

$$f(z) = \frac{b_1}{z-a} + a_0 + a_1(z-a) + a_n(z-a)^z + \cdots$$

$$0 < |z-a| < R$$

$$\Rightarrow (z-a)f(z) = b_1 + (z-a)\sum_{n=0}^{\infty} a_n (z-a)^n$$

$$b_1 \neq 0$$

$$\Rightarrow b_1 = \lim_{z\to a}[(z-a)f(z)] \qquad \text{(Formula I)}$$

(ii) Let $f(z) = \frac{p(z)}{q(z)}$ p, q are analytic, $p(a) \neq 0$, $q(z)$ has simple zero at $z = a$.

By Taylor's theorem,

$$q(z) = (z-a)q'(a) + \frac{(z-a)^2}{2!}q''(a) + \cdots$$

$$\text{Res}_{z=a} f(z) = \lim_{z\to a}(z-a)\frac{p(z)}{q'(z)}$$

$$= \lim_{z\to a}\frac{(z-a)p(z)}{(z-a)\left[q'(a) + \frac{q''(a)^{(z-a)}}{2!} + \cdots\right]}$$

$$= \frac{p(a)}{q'(a)}$$

$$\text{Res}_{z=a} f(z) = \text{Res}_{z\to a}\frac{p(z)}{q'(z)}$$

$$= \frac{p(a)}{q'(a)}. \qquad \text{(Formula II)}$$

Example 17.7

Find the residue at a simple pole $z = i$ for $f(z) = \frac{9z+i}{z(z^2+1)}$.

Solution

$$\text{Res}_{z=i} f(z) = \text{Res}_{z=i}\left[\frac{9z+i}{z(z^2+1)}\right]$$

$$= \lim_{z\to i}\frac{(z-i)(9z+i)}{(z-i)z(z+i)} = \lim_{z\to i}\frac{9z+i}{z(z+i)}$$

$$= \frac{10i}{2i^2} = -5i \qquad \text{(Formula I)}$$

Example 17.8
Find the residues at simple poles $z = 0, 1$ for
$$f(z) = \frac{4z-3}{z(z-1)}.$$

Solution

$$\text{Res}_{z=0}\left[\frac{4z-3}{z(z-1)}\right] = \left[\frac{4z-3}{2z-1}\right]_{z=0}$$
$$= 3 \quad \text{(Formula II)}$$

$$\text{Res}_{z=1}\left[\frac{4z-3}{z(z-1)}\right] = \left[\frac{4z-3}{2z-1}\right]_{z=1}$$
$$= 1 \quad \text{(Formula II)}$$

Example 17.9
Find the residues of $f(z) = \dfrac{z}{z^2+1}$ at poles. [JNTU 1995*S*]

Solution $f(z)$ has two simple poles at $z = \pm i$

$$\text{Res}_{z=i} f(z) = \lim_{z\to i}(z-i)f(z)$$
$$= \lim_{z\to i}(z-i)\frac{z}{(z-i)(z+i)}$$
$$= \lim_{z\to i}\frac{z}{z+i} = \frac{i}{2i} = \frac{1}{2}$$
$$\text{Res}_{z=-i} f(z) = \lim_{z\to -i}(z+i)f(z)$$
$$= \lim_{z\to -i}(z+i)\frac{z}{(z+i)(z-i)}$$
$$= \lim_{z\to -i}\frac{z}{z-i} = \frac{-i}{-2i} = \frac{1}{2}.$$

Example 17.10
Find the poles and the corresponding residues of
$$f(z) = \frac{2z+1}{z^2-z-2}.$$ [JNTU 1994*S*]

Solution $f(z)$ has simple poles at $z = -1, 2$

$$\text{Res}_{z\to -1} f(z) = \lim_{z\to -1}(z+1)f(z)$$
$$= \lim_{z\to -1}(z+1)\frac{2z+1}{(z+1)(z-2)}$$
$$= \lim_{z\to -1}\frac{2z+1}{z-2} = \frac{2(-1)+1}{-1-2} = \frac{1}{3}$$

$$\text{Res}_{z\to +2} f(z) = \lim_{z\to 2}(z-2)f(z)$$
$$= \lim_{z\to 2}(z-2)\frac{2z+1}{(z+1)(z-2)}$$
$$= \lim_{z\to 2}\frac{2z+1}{z+1}$$
$$= \frac{2(2)+1}{2+1} = \frac{5}{3}.$$

Residue at a pole of order *m*

Let $f(z)$ have a pole of order $m > 1$ at $z = a$. Then Laurent's series for $f(z)$ is

$$f(z) = \frac{b_m}{(z-a)^m} + \frac{b_{m-1}}{(z-a)^{m-1}} + \cdots + \frac{b_1}{(z-a)}$$
$$+ \sum_{n=0}^{\infty} a_n(z-a)^n \quad (b_m \neq 0) \tag{17.7}$$

Multiplying Eq. (17.7) by $(z-a)^m$, we get

$$(z-a)^m f(z) = b_m + b_{m-1}(z-a) + \cdots + b_1$$
$$(z-a)^{m-1} + \sum_{n=0}^{\infty} a_n(z-a)^{n+m} \tag{17.8}$$

Differentiating both sides of Eq. (17.8) $(m-1)$ times w.r.t z, we have

$$\frac{d^{m-1}}{dz^{m-1}}[(z-a)^m f(z)]$$
$$= (m-1)!b_1 + \sum \frac{(m+n)!}{(n+1)!}(z-a)^{n+1}$$

Taking limits as $z \to a$, we get,

$$b_1 = \frac{1}{(m-1)!}\lim_{z\to a}\frac{d^{m-1}}{dz^{m-1}}[(z-a)^m f(z)].$$
(Formula III) (17.9)

Corollary $m = 1$ (simple pole)

$$b_1 = \lim_{z\to a}[(z-a)f(z)].$$

Note

If $z = a$ is an essential singularity, then the above formulas fail so that the residue is obtained by expanding $f(z)$ about $z = a$ in the Laurent's series and identifying the coefficient b_1 of $(z-a)^{-1}$.

17.1.5 Cauchy's Residue Theorem

Theorem 17.1 Let $f(z)$ be analytic inside and on a simple closed path C except at a finite number of singularities at z_j ($j = 1, 2, \ldots, m$) inside C. Then

$$\oint_C f(z)dz = 2\pi i \sum_{j=1}^{m} \text{Res}_{z=z_j} f(z) \quad (17.10)$$

Proof: Enclose each z_j in a circle C_j with radius so small that these circles and closed path C are all separated and do not overlap. Then $f(z)$ is analytic in the multiply-connected domain D bounded by these circles $C_1, C_2, \ldots, C_m$ and C.

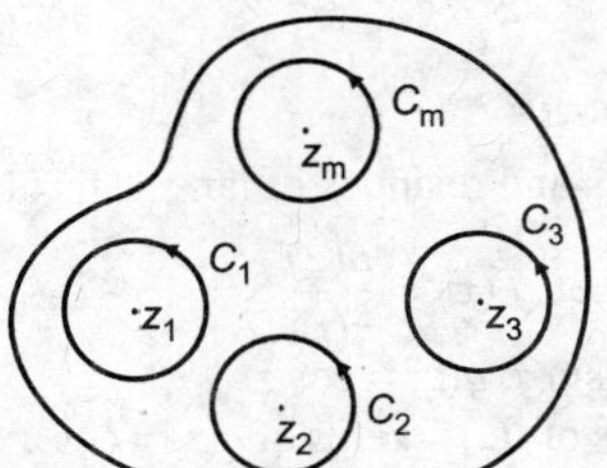

Figure 17.2 Cauchy's Residue Theorem

By Cauchy's integral theorem (CIT)

$$\oint_C f(z)dz + \oint_{\Sigma C_j} f(z)dz = 0 \quad (17.11)$$

where C is taken in the counter-clockwise direction and C_j in the clockwise direction. Transposing and changing the direction of each C_j counter-clockwise and using the definition of residue

$$\oint_C f(z)dz = \oint_{\Sigma C_j} f(z)dz$$

$$= 2\pi i \sum_{j=1}^{m} \text{Res}_{z=z_j} f(z). \quad (17.12)$$

Note

This theorem is important in several applications. Using the theorem we can evaluate the complex integral and real integrals. First, we consider evaluation of some complex integrals and then we take up the evaluation of real integrals.

Example 17.11

Evaluate $I = \oint_C \frac{3-2z}{z^2 - z} dz$ counter-clockwise around any simple closed path C such that

(i) the singularities 0 and 1 lie inside C,

(ii) the singularity 0 lies inside and 1 lies outside C,

(iii) the singularity 1 lies inside and 0 lies outside C,

(iv) the singularities 0 and 1 lie outside C.

Solution The integrand $f(z) = \frac{3-2z}{z(z-1)}$ has simple poles at $z = 0$ and $z = 1$.

$$\text{Res}_{z=0} f(z) = \lim_{z\to 0}\left[z\left(\frac{3-2z}{z(z-1)}\right)\right]$$

$$= \lim_{z\to 0}\frac{3-2z}{z-1} = -3$$

$$\text{Res}_{z=1} f(z) = \lim_{z\to 1}\left[(z-1)\frac{3-2z}{z(z-1)}\right]$$

$$= \lim_{z\to 1}\frac{3-2z}{z} = 1$$

(i) $I = 2\pi i \sum_{\text{Res}_{z=z_j}} f(z) = 2\pi i(-3+1) = -4\pi i$

(ii) $I = 2\pi i \sum_{\text{Res}_{z=z_j}} f(z) = 2\pi i(-3+0) = -6\pi i$

(iii) $I = 2\pi i \sum_{\text{Res}_{z=z_j}} f(z) = 2\pi i(0+1) = 2\pi i$

(iv) $I = 2\pi i \sum_{\text{Res}_{z=z_j}} f(z) = 2\pi i(0) = 0.$

Example 17.12

Evaluate $I = \oint_C \left(\frac{ze^{z\pi}}{z^4 - 16} + ze^{\frac{\pi}{z^2}}\right)dz$ where C is the ellipse $16x^2 + y^2 = 16$.

Solution Let $f_1(z) = \frac{ze^{z\pi}}{z^4 - 16} = \frac{p(z)}{q(z)}$ (say)

It has simple poles at $z = \pm 2i, \pm 2$ of which $\pm 2i$ lie inside C and ± 2 lie outside C

$$\text{Res}_{z=2i} f_1(z) = \lim_{z\to 2i}\frac{p(z)}{q'(z)} = \left.\frac{ze^{\pi z}}{4z^3}\right|_{z=2i}$$

$$= \frac{e^{2i\pi}}{4(2i)^2} = -\frac{\cos 2\pi + \sin 2\pi}{8} = -\frac{1}{8}$$

$$\text{Res}_{z=-2i} f_1(z) = \lim_{z\to -2i} \frac{p(z)}{q'(z)} = \left.\frac{e^{\pi z}}{4z^2}\right|_{z=-2i}$$

$$= \frac{e^{-2i\pi}}{4(-2i)^2} = -\frac{1}{8}$$

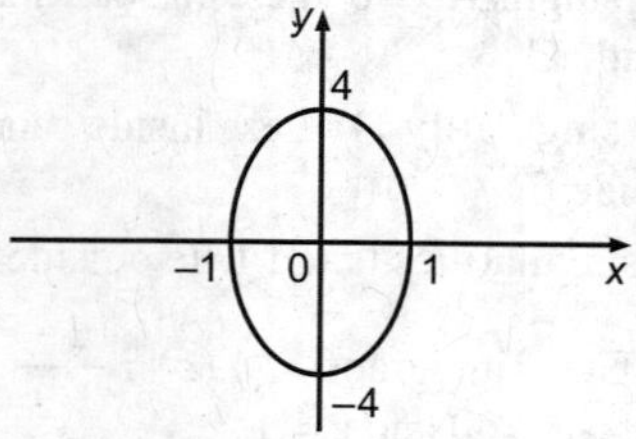

Figure 17.3

$f_2(z) = ze^{\frac{\pi}{z}}$ has an essential singularity at 0 with residue $\dfrac{\pi^2}{2}$

$$\because \quad ze^{\frac{\pi}{z}} = z\left[1+\frac{\pi}{z}+\frac{\pi^2}{2!z^2}+\cdots\right] = z+\pi+\frac{\pi^2}{2!z}+\cdots$$

$$\therefore \quad I = 2\pi i\left(-\frac{1}{8}-\frac{1}{8}+\frac{\pi^2}{2}\right) = \pi i\left(\pi^2-\frac{1}{2}\right).$$

Example 17.13

Find the poles of $f(z) = \dfrac{z^2}{(z-1)^2(z+2)}$ and the residue at each pole. [JNTU 1999, 2003*S*]

Solution $z_1 = 1$ is a pole of order 2 and $z_2 = -2$ is a simple pole of $f(z)$.

$$\text{Res}_{z=1} f(z) = \lim_{z\to 1} \frac{1}{1!}\frac{d}{dz}[(z-1)^2 f(z)]$$

$$= \lim_{z\to 1}\left(\frac{z^2}{z+2}\right) = \lim_{z\to 1}\frac{(z+2)2z - z^2\cdot 1}{(z+2)^2}$$

$$= \lim_{z\to 1}\frac{z^2+4z}{(z+2)^2} = \frac{5}{9}$$

$$\text{Res}_{z=-2} f(z) = \lim_{z\to -2}(z+2)f(z) = \lim_{z\to -2}\frac{z^2}{(z-1)^2} = \frac{4}{9}.$$

Example 17.14

Find the residue of $\dfrac{z^2}{z^4-1}$ at singular points that lie inside the circle $|z|=2$. [JNTU 1998*S*]

Solution Let $f(z) = \dfrac{z^2}{z^4-1} = \dfrac{p(z)}{q(z)}$ (say)

The singularities of

$$f(z) = \frac{z^2}{z^4-1} = \frac{z^2}{(z-1)(z+1)(z-i)(z+i)}$$

are simple poles at $z = 1, -1, i, -i$ all of which lie inside the circle $|z|=2$ whose centre is at the origin and radius = 2.

$$\text{Res}_{z=a} f(z) = \frac{p(a)}{q'(a)} = \frac{1}{4a} \Rightarrow \text{Res}_{z=1} f(z) = \frac{1}{4};$$

$$\text{Res}_{z=-1} f(z) = \frac{1}{4(-1)} = -\frac{1}{4}$$

$$\text{Res}_{z=i} f(z) = \frac{1}{4i}; \quad \text{Res}_{z=-i} f(z) = \frac{1}{4(-i)} = -\frac{1}{4i}$$

Example 17.15

Find the poles and residues of tan *z*. [JNTU 2003*S*]

Solution Let $f(z) = \dfrac{p(z)}{q(z)}$ where $p(z) = \sin z$, $q(z) = \cos z$.

The poles of $f(z)$ = zeros of $q(z) = \cos z$ which are

$$z_n = (2n+1)\frac{\pi}{2}, \; n \in \mathbb{Z}$$

$$\text{Res}_{z=z_n} f(z) = \frac{p(z_n)}{q'(z_n)} = \frac{\sin(2n+1)\frac{\pi}{2}}{-\sin(2n+1)\frac{\pi}{2}} = -1.$$

Example 17.16

Calculate the residue at $z = 0$ of

$f(z) = \dfrac{1+e^z}{z\cos z + \sin z}$. [JNTU 2003*S*]

Solution

$\text{Res}_{z=0} f(z)$

$$= \lim_{z\to 0}(z-0)f(z) = \lim_{z\to 0}\frac{z(1+e^z)}{z\cos z+\sin z} \quad \left(\frac{0}{0} \text{ form}\right)$$

$$= \lim_{z\to 0}\frac{ze^z + (1+e^z).1}{\cos z + z(-\sin z)+\cos z}$$

$$= \frac{0+1+1}{1+1} = 1 \quad \text{by } L \text{ Hospital's rule.}$$

Example 17.17

Find the poles of $f(z) = \dfrac{1+z}{(1-z)^2}$ and the residue at each pole. [JNTU 2000*S*, 2003]

Solution $z = 1$ is a pole of order 2 of $f(z)$

$$\text{Res}_{z=1} f(z) = \text{Res}_{z=1} \frac{1+z}{(1-z)^2}$$

$$= \frac{1}{1!} \lim_{z\to 1} \frac{d}{dz}\left[(1-z)^2 \frac{1+z}{(1-z)^2}\right] = 1.$$

Example 17.18

Find the poles of $f(z) = \dfrac{z^2-2z}{(z+1)^2(z^2+1)}$ and the residues at these poles. [JNTU 2000*S*, 2003]

Solution $z = -1$ is a pole of order 2 and $z = \pm i$ are simple poles of $f(z)$.

$$\text{Res}_{z=i} f(z) = \lim_{z\to i} \frac{(z-i)(z^2-2z)}{(z-i)(z+i)(z+1)^2}$$

$$= \frac{i^2-2i}{2i(i+1)^2} = \frac{1+2i}{4}$$

$$\text{Res}_{z=-i} f(z) = \lim_{z\to -i} \frac{(z+i)(z^2-3z)}{(z-i)(z+i)(z+1)^2}$$

$$= -\frac{1}{2i}\frac{-1+2i}{(1-i)^2} = -\frac{1}{2i}\frac{-1+2i}{-2i} = \frac{1-2i}{4}$$

$$\text{Res}_{z=-i} f(z) = \frac{1}{1!}\lim_{z\to -1} \frac{d}{dz}\left(\frac{z^2-2z}{z^2+1}\right)$$

$$= \left.\frac{(z^2+1)2z-(z^2-2z)2z}{(z^2+1)^2}\right|_{z=-1}$$

$$= \left.\frac{2z(2z+1)}{(z^2+1)^2}\right|_{z=-1} = \frac{-2(-2+1)}{\left[(-1)^2+1\right]^2} = \frac{1}{2}.$$

EXERCISE 17.1

1. Find the poles of the following functions and the residue at each poles:

(a) $\dfrac{z+1}{z^2(z-2)}$. [JNTU 2004, 2004*S* (1), 2006(2)]

Ans: Pole $z = 0$ (double), Res: $-\dfrac{3}{4}$;

$z = 2$ (simple), Res: $\dfrac{3}{4}$

(b) $\dfrac{z^2-2z}{(z+1)^2(z^2+1)}$. [JNTU 2000*S*, 2003]

Ans: $z = -1$ (double), Res: $\dfrac{1}{2}$;

$z = \pm 1$ (simple), Res: $\dfrac{1 \pm 2i}{4}$

(c) $\dfrac{1-e^{2z}}{z^4}$. [JNTU 2000, 2003*S*]

Ans: $z = 0$ (order 4), Res: $-\dfrac{4}{3}$.

(d) $\cot z$.

Ans: $z = n\pi$ (simple), $n \in \mathbb{Z}$, Res: 1

(e) $\dfrac{2z+1}{z^2-z-2}$.

Ans: $z = -1, 2$ (simple), Res: $\dfrac{1}{3}, \dfrac{5}{3}$

(f) $\dfrac{ze^z}{(z-1)^3}$. [JNTU 1996, 2003]

Ans: $z = 1$ (order 3), Res: $\dfrac{3e}{2}$

(g) $\tan z$ [JNTU 2003*S*]

Ans: $z = (2n+1)\dfrac{\pi}{2}$, $n \in \mathbb{Z}$, Res: -1

(h) $\dfrac{1}{1-\cos z}$.

Ans: $z = 2n\pi$, $n \in \mathbb{Z}$, Res: 1

2. Find the residue of $\dfrac{1}{(z^2+1)^3}$ at $z = i$.

Ans: $\dfrac{3}{16i}$

3. Find the sum of the residues of $f(z) = \dfrac{\sin z}{z\cos z}$.

Ans: At 0 its poles inside the circle $|z| = 2$

4. Find the singularities and their nature in respect of the following functions pole of order 2:

(a) $\dfrac{e^{3z}}{(z-1)^2}$.

Ans: $z = 3$ is an essential singularity

(b) $(2z-3)\sin\left(\dfrac{1}{z-3}\right)$.

Ans: $z = 0$ is a removable singularity

(c) $\dfrac{1-\cos z}{z^3}$.

Ans: $z = 3$ is an essential singularity

(d) $\exp\left[\dfrac{(z)}{(z-3)}\right]$.

Ans: $z = \infty$ is a removable singularity

(e) $\tanh z$.

Ans: $z = n\pi i$ $(n \in \mathbb{Z})$ are simple poles

(f) $\dfrac{1-e^{2z}}{z^4}$.

Ans: $z = 0$ is a pole of order 3

5. Evaluate the following integrals:

(a) $\int_C \frac{4-3z}{z^2-z}dz \quad C:|z|=2.$

Ans: $-6\pi i$

(b) $\int_C \frac{z-3}{z^2+2z+5}dz \quad C:|z+1-i|=2.$

Ans: $\pi(i-2)$

(c) $\int_C \tan z\, dz \quad C:|z|=2.$

Ans: $-4\pi i$

(d) $\int_C \frac{dz}{(z^2+4)^2} \quad C:|z-i|=2.$

Ans: $\frac{\pi}{16}$

(e) $\int_C \frac{3z^2+z-1}{(z^2-1)(z-3)}dz \quad C:|z|=2.$

Ans: $-\frac{5\pi i}{4}$

(f) $\int_C \frac{e^z dz}{(z-1)(z+2)} \quad C:|z-1|=1.$

Ans: $\frac{2i\pi e}{3}$

(g) $\int_C \frac{\sin \pi z^2+\cos \pi z^2}{(z-1)^2(z-2)}dz \quad C:|z|=3.$

Ans: $4\pi i(\pi+1)$

(h) $\int_C \frac{3\cos z}{2i-3z}dz \quad C:|z|=1.$

Ans: $-2\pi \cosh\left(\frac{2}{3}\right)$

(i) $\int_C \frac{(z-3)dz}{(z^2+2z+5)} \quad C:|z+1+i|=2.$

[JNTU 2003(2), 2004(2), 2004*S*, 2006(2)]

Ans: $\pi(2+i)$

(j) $\int_C \frac{(z^2-2z)dz}{(z+1)^2(z^2+1)} \quad C:|z|=10.$

Ans: 0

17.1.6 Type I: Integrals of the Type $\int_0^{2\pi} F(\cos\theta, \sin\theta)d\theta$

Type I : Integrals of rational functions of cos θ and sin θ (integration around the unit circle $C:|z|=1$)

Consider integrals of the type

$$I=\int_0^{2\pi} F(\cos\theta, \sin\theta)d\theta \tag{17.13}$$

where F is a real, rational function of cos θ and sin θ, which is finite on $0 \le \theta \le 2\pi$.

To convert Eq. (17.13) into a contour integral, we transform F into a complex function by putting

$$e^{i\theta}=z \Rightarrow e^{-i\theta}=\frac{1}{z}.$$

Differentiating $e^{i\theta}=z$ w.r.t. θ we get

$$ie^{i\theta}d\theta = izd\theta = dz \quad \text{or} \quad d\theta=\frac{dz}{iz}.$$

$$\text{Also } \cos\theta = \frac{e^{i\theta}+e^{-i\theta}}{2}=\frac{z+\frac{1}{z}}{2}=\frac{z^2+1}{2z}$$

$$\sin\theta = \frac{e^{i\theta}-e^{-i\theta}}{2i}=\frac{z-\frac{1}{z}}{2i}=\frac{z^2-1}{2iz}$$

$$\therefore \text{ Also, } F(\cos\theta, \sin\theta)=F\left(\frac{z^2+1}{2z}, \frac{z^2-1}{2iz}\right) = f(z) \text{ (say)}$$

Now the integral becomes

$$I=\int_C f(z)\frac{dz}{iz}.$$

As θ varies from 0 to 2π the complex variable z traverses the unit circle $|z|=1$ once, in the counter-clockwise sense. Recalling the following property of definite integrals

$$\begin{cases}\int_0^{2a} F(x)dx = 2\int_0^a F(x)dx, \text{ if } F(2a-x)=F(x)\\ \quad = 0, \quad \text{if } F(2a-x)=-F(x)\end{cases}$$

we have

$$\int_0^{\pi} F(\cos\theta, \sin\theta)d\theta = \frac{1}{2}\int_0^{2\pi} F(\cos\theta, \sin\theta)d\theta$$

$$\because F(2\pi-\theta)=F(\theta)$$

Example 17.19

Show that $\int_0^{2\pi} \frac{d\theta}{5+4\cos\theta}=\frac{2\pi}{3}$.

Solution To evaluate the integral $I=\int_0^{2\pi} \frac{d\theta}{5+4\cos\theta}$, we have to convert it into a contour integral over the unit circle $C:|z|=1$.

Put $e^{i\theta}=z, \quad d\theta=\frac{dz}{iz}$

$$\cos\theta = \frac{z^2+1}{2z}, \quad \sin\theta = \frac{z^2-1}{2iz}$$

$$\therefore I = \int_C \frac{1}{5+4\left(\frac{z^2+1}{2z}\right)}\frac{dz}{iz} = \frac{1}{i}\int_C \frac{dz}{2z^2+5z+2}$$

$$= \frac{1}{2i}\int_C \frac{dz}{\left(z+\frac{1}{2}\right)(z+2)}$$

There are two singularities for the integrand $f(z) = \frac{1}{\left(z+\frac{1}{2}\right)(z+2)}$ at $z = -\frac{1}{2}$ and -2; $z = -\frac{1}{2}$ is a simple pole that lies in C.

$$\therefore \text{Res}_{z=-\frac{1}{2}} f(z) = \lim_{z\to -\frac{1}{2}}\left(z+\frac{1}{2}\right)f(z)$$

$$= \lim_{z\to -\frac{1}{2}}\left(z+\frac{1}{2}\right)\frac{1}{\left(z+\frac{1}{2}\right)(z+2)} = \frac{2}{3}$$

By Cauchy's residue theorem, we have

$$\int_0^{2\pi}\frac{d\theta}{5+4\cos\theta} = 2\pi i\sum \text{Res} f(z) = 2\pi i\frac{1}{2i}\frac{2}{3} = \frac{2\pi}{3}.$$

Example 17.20

Evaluate $\int_0^{2\pi}\frac{d\theta}{2+\cos\theta}$. [JNTU 2003]

Solution To evaluate the integral $I = \int_0^{2\pi}\frac{d\theta}{2+\cos\theta}$, we have to convert it into a contour integral over the unit circle $C: |z| = 1$.

Put $e^{i\theta} = z \Rightarrow d\theta = \frac{dz}{iz}$ and $\cos\theta = \frac{z^2+1}{2z}$

$$\therefore I = \int_0^{2\pi}\frac{2d\theta}{4+2\cos\theta} = \int_C \frac{2}{4+\frac{z^2+1}{z}}\frac{dz}{iz}$$

$$= \frac{2}{i}\int_C \frac{dz}{z^2+4z+1}$$

$$= \frac{2}{i}\times 2\pi i \quad \text{(sum of residues at poles of } f(z))$$

$$= \frac{2}{i}\frac{2\pi i}{2\sqrt{3}} = \frac{2\pi}{\sqrt{3}}$$

$$= z^2+4z+1 = (z+2)^2-(\sqrt{3})^2$$

$$= (z+2-\sqrt{3})(z+2+\sqrt{3})$$

$$= (z-\alpha)(z-\beta) \quad \text{(say)}$$

$\alpha = -2+\sqrt{3}$ lies inside C

while $\beta = -2-\sqrt{3}$ lies outside C.

$$\therefore \text{Res}_{z=\alpha} f(z) = \lim_{z\to\alpha}(z-\alpha)\frac{1}{(z-\alpha)(z-\beta)}$$

$$= \frac{1}{\alpha-\beta} = \frac{1}{2\sqrt{3}}.$$

Example 17.21

Show that $\int_0^{2\pi}\frac{d\theta}{a+b\cos\theta} = \frac{2\pi}{\sqrt{a^2-b^2}}$ $(a > |b| > 0)$.

Solution To evaluate the integral $I = \int_0^{2\pi}\frac{d\theta}{a+b\cos\theta}$, we have to convert it into a contour integral over the unit circle $C: |z| = 1$.

Put $e^{i\theta} = z$, $d\theta = \frac{dz}{iz}$,

$$\cos\theta = \frac{e^{i\theta}+e^{-i\theta}}{2} = \frac{z+z^{-1}}{2} = \frac{z^2+1}{2z}$$

$$I = \int_C \frac{1}{a+b\left(\frac{z^2+1}{2z}\right)}\frac{dz}{iz} = \frac{2}{i}\int_C \frac{dz}{bz^2+2az+b}$$

$$= 2\pi i\frac{2}{i}\sum \text{Res} f(z) \tag{1}$$

by Cauchy's residue theorem.

The singularities of the integrand $f(z) = \frac{1}{bz^2+2az+b}$ which are the roots of $bz^2 + 2az + b = 0$ are α and β where $\alpha+\beta = -\frac{2a}{b}$; $\alpha\beta = 1$;

$$(\alpha-\beta)^2 = (\alpha+\beta)^2 - 4\alpha\beta = \frac{4a^2}{b^2} - 4$$

$$= \frac{4(a^2-b^2)}{b^2}$$

$$\therefore \alpha-\beta = \frac{2}{b}\sqrt{a^2-b^2}$$

Since $a > |b| > 0, |\beta| > 1$ but $|\alpha\beta| = 1$ we have $|\alpha| < 1$. The only singularity that lies inside C is the simple pole at $z = \alpha$.

$$\text{Res}_{z=\alpha} f(z) = \lim_{z\to\alpha}(z-\alpha)\frac{1}{b(z-\alpha)(z-\beta)}$$

$$= \frac{1}{b}\frac{1}{\alpha-\beta} = \frac{1}{2\sqrt{a^2-b^2}} \tag{2}$$

Substituting in Eq. (1), we have

$$I=\int_0^{2\pi}\frac{d\theta}{a+b\cos\theta}=\frac{2\pi}{\sqrt{a^2-b^2}}.$$

Note

We can evaluate $I=\int_0^{2\pi}\frac{d\theta}{a+b\sin\theta}$ similarly and show that $I=\frac{2\pi}{\sqrt{a^2-b^2}}\left(\sin\theta=\frac{z^2-1}{2iz}\right).$

We can evaluate

$$I=\int_0^{\pi}\frac{d\theta}{a+b\cos\theta}\quad (a>|b|>0)$$

$$=\frac{1}{2}\int_0^{2\pi}\frac{d\theta}{a+b\cos\theta}$$

$$=\frac{1}{2}\frac{2\pi}{\sqrt{a^2-b^2}}=\frac{\pi}{\sqrt{a^2-b^2}}$$

Particular cases

1. $a=\pi, b=1\Rightarrow\int_0^{\pi}\frac{d\theta}{\pi+\cos\theta}=\frac{\pi}{\sqrt{\pi^2-1}}$ [JNTU 1998]
2. $a=3, b=2\Rightarrow\int_0^{\pi}\frac{d\theta}{3+2\cos\theta}=\frac{\pi}{\sqrt{5}}$
3. $a=2, b=-1\Rightarrow\int_0^{\pi}\frac{d\theta}{2-\sin\theta}=\frac{\pi}{\sqrt{3}}$

Example 17.22

Prove that $\int_0^{2\pi}\frac{d\theta}{(a+b\cos\theta)^2}=\frac{2\pi a}{(a^2-b^2)^{\frac{3}{2}}}$ $(a>b>0)$. [JNTU 2003*S*]

Solution We have proved that

$$\int_0^{2\pi}\frac{d\theta}{a+b\cos\theta}=\frac{2\pi}{\sqrt{a^2-b^2}}$$

Differentiating w.r.t *a* under the integral sign, we get

$$\int_0^{2\pi}\frac{\partial}{\partial a}\left(\frac{1}{a+b\cos\theta}\right)d\theta=-\int_0^{2\pi}\frac{d\theta}{(a+b\cos\theta)^2}$$

$$=2\pi\left(-\frac{1}{2}\right)(a^2-b^2)^{-\frac{3}{2}}2a$$

$$=\frac{-2\pi a}{(a^2-b^2)^{\frac{3}{2}}}$$

$$\Rightarrow\int_0^{2\pi}\frac{d\theta}{(a+b\cos\theta)^2}=\frac{2\pi a}{(a^2-b^2)^{\frac{3}{2}}}$$

Example 17.23

Evaluate $\int_0^{2\pi}\frac{d\theta}{1-2p\cos\theta+p^2}$ $(0<p<1)$. [JNTU 2002, 2003]

Solution To evaluate the integral $I=\int_0^{2\pi}\frac{d\theta}{1-2p\cos\theta+p^2}$, we have to convert it into a contour integral over the unit circle $C:|z|=1$.

Put $e^{i\theta}=z,\quad d\theta=\frac{dz}{iz},$

$$\cos\theta=\frac{e^{i\theta}+e^{-i\theta}}{2}=\frac{z+z^{-1}}{2}=\frac{z^2+1}{2z}$$

$$I=\int_C\frac{1}{1-2p\left(\frac{z^2+1}{2z}\right)+p^2}\frac{dz}{iz}$$

$$=\frac{-1}{pi}\int_C\frac{dz}{z^2-\left(p+\frac{1}{p}\right)z+1}$$

$$=\frac{-1}{pi}\int_C\frac{dz}{(z-p)\left(z-\frac{1}{p}\right)}$$

The singularities of the integrand $f(z)=\frac{1}{(z-p)\left(z-\frac{1}{p}\right)}$ are simple poles at $z=p$ and $\frac{1}{p}$ of which $z=p$ lies inside $C:|z|=1$ $(\because p<1)$.

$$\therefore I=-\frac{1}{pi}\cdot 2\pi i\cdot \text{Res}_{z=p}f(z)$$

$$=-\frac{2\pi}{p}\lim_{z\to p}(z-p)\frac{1}{(z-p)\left(z-\frac{1}{p}\right)}$$

$$=-\frac{2\pi}{p}\frac{1}{\left(p-\frac{1}{p}\right)}=\frac{2\pi}{1-p^2}.$$

Example 17.24

Evaluate $\int_0^{2\pi}\frac{\cos^2 3\theta}{5-4\cos 2\theta}d\theta$.

Solution To evaluate the integral

$$I=\int_0^{2\pi}\frac{\cos^2 3\theta}{5-4\cos 2\theta}d\theta=\frac{1}{2}\int_0^{2\pi}\frac{1+\cos 6\theta}{5-4\cos 2\theta}d\theta,$$

we have to convert it into a contour integral over the unit circle $C:|z|=1$.

Putting $e^{i\theta} = z, \quad d\theta = \dfrac{dz}{iz}$,

$$\cos 2\theta = \frac{e^{2i\theta} + e^{-2i\theta}}{2} = \frac{z^2 + z^{-2}}{2} = \frac{z^4+1}{2z^2}$$

$$1+\cos 6\theta = Re(1+e^{6i\theta})$$

$$= Re(1+z^6) \quad \text{(Note this step.)}$$

$$\therefore \quad I = \frac{1}{2} Re \int_C \frac{1+z^6}{5-2\left(z^2+\frac{1}{z^2}\right)} \frac{dz}{iz}$$

$$= -\frac{1}{2\cdot 2} \mathrm{Re} \frac{1}{i} \int_C \frac{z(1+z^6)}{(z^2-2)\left(z^2-\frac{1}{2}\right)} dz$$

The singularities of the integrand $f(z) = \dfrac{z(1+z^6)}{(z^2-2)\left(z^2-\frac{1}{2}\right)}$ are simple poles at $z = \pm\dfrac{1}{\sqrt{2}}$ (inside C) and $\pm\sqrt{2}$ (outside C).

$$R_1 = \operatorname*{Res}_{z=\frac{1}{\sqrt{2}}} f(z)$$

$$= \lim_{z\to\frac{1}{\sqrt{2}}} \frac{\left(z-\frac{1}{\sqrt{2}}\right) z(1+z^6)}{\left(z-\frac{1}{\sqrt{2}}\right)\left(z+\frac{1}{\sqrt{2}}\right)(z^2-2)}$$

$$= \frac{\frac{1}{\sqrt{2}}\left(1+\frac{1}{8}\right)}{\left(\frac{1}{\sqrt{2}}+\frac{1}{\sqrt{2}}\right)\left(\frac{1}{2}-2\right)} = \frac{\frac{9}{8}}{-3} = -\frac{3}{8}$$

$$R_2 = \operatorname*{Res}_{z=-\frac{1}{\sqrt{2}}} f(z)$$

$$= \lim_{z\to-\frac{1}{\sqrt{2}}} \frac{\left(z+\frac{1}{\sqrt{2}}\right) z(1+z^6)}{\left(z+\frac{1}{\sqrt{2}}\right)\left(z-\frac{1}{\sqrt{2}}\right)(z^2-2)}$$

$$= \frac{-\frac{1}{\sqrt{2}}\left(1+\frac{1}{8}\right)}{\left(-\frac{1}{\sqrt{2}}-\frac{1}{\sqrt{2}}\right)\left(\frac{1}{2}-2\right)} = \frac{\frac{9}{8}}{-3} = -\frac{3}{8}$$

$$\therefore \quad I = -\frac{1}{4} \mathrm{Re} \frac{1}{i} 2\pi i \left(-\frac{3}{8}-\frac{3}{8}\right) = \frac{3\pi}{8}.$$

Aliter

$$I = \int_0^{2\pi} \frac{\cos^2 3\theta}{5-4\cos 2\theta} d\theta = \frac{1}{2} \int_0^{2\pi} \frac{1+\cos 6\theta}{5-4\cos 2\theta} d\theta;$$

Putting $2\theta = \phi, d\theta = \dfrac{1}{2} d\phi$ and changing the limits for ϕ as $\phi = 0$ to $\phi = 4\pi$

$$= \frac{1}{4} \int_0^{4\pi} \frac{1+\cos 3\phi}{5-4\cos\phi} d\phi$$

$$= \frac{1}{4}\cdot 2 \int_0^{2\pi} \frac{1+\cos 3\phi}{5-4\cos\phi} d\phi \quad \text{property of definite integrals sec.6.1.6 } P\,6-8$$

$$= \frac{1}{2} \int_0^{2\pi} \frac{1+\cos 3\phi}{5-4\cos\phi} d\phi$$

We convert the integral into a contour integral around the unit circle $c: |z| = 1$ by putting $e^{i\phi} = z \Rightarrow d\phi = \dfrac{dz}{iz}, \cos\phi = \dfrac{z^2+1}{2z}$ and $1+\cos 3\phi = \mathrm{Re}\left(1+e^{3i\phi}\right)$ $= \mathrm{Re}\left(1+z^3\right)$

$$\therefore I = \frac{1}{2} \mathrm{Re} \int_c \frac{1+z^3}{5-4\left(\frac{z^2+1}{2z}\right)} \frac{dz}{iz}$$

$$= \mathrm{Re} \frac{1}{-4i} \int_c \frac{1+z^3}{z^2 - \frac{5}{2}z + 1} dz$$

$$= \mathrm{Re}\left(-\frac{1}{4i}\right) \int_c \frac{1+z^3}{\left(z-\frac{1}{2}\right)(z-2)} dz$$

$$= \mathrm{Re}\left(-\frac{1}{4i}\right) \times 2\pi i \times \frac{1+\frac{1}{2^3}}{\frac{1}{2}-2} = \frac{3\pi}{8}$$

The integrand $f(z) = \dfrac{1+z^3}{\left(z-\frac{1}{2}\right)(z-2)}$ has simple poles at $z = \dfrac{1}{2}$ (lying inside c) and $z = 2$ (lying outside c) and

$$\operatorname*{Res}_{z=\frac{1}{2}} f(z) = \lim_{z\to\frac{1}{2}} \left(z-\frac{1}{2}\right) \frac{1+z^3}{\left(z-\frac{1}{2}\right)(z-2)} = \frac{1+\frac{1}{2^3}}{\frac{1}{2}-2}$$

$$= \frac{\frac{9}{8}}{\frac{-3}{2}} = \frac{-3}{4}.$$

Example 17.25

Show that $\displaystyle\int_0^{2\pi} \frac{\sin^2\theta}{a+b\cos\theta} d\theta = \frac{2\pi}{b^2}\left(a-\sqrt{a^2-b^2}\right)$ $(a > b > 0)$. [JNTU 2000, 2003]

Solution We convert the integral $I=\int_0^{2\pi}\frac{\sin^2\theta}{a+b\cos\theta}d\theta$ into a contour integral taken round the unit circle $C:|z|=1$ by putting

$$e^{i\theta}=z\Rightarrow d\theta=\frac{dz}{iz};\quad (0\le\theta\le 2\pi)$$

$$\sin^2\theta=\frac{1}{2}(1-\cos 2\theta)$$

$$=\frac{1}{2}\text{Re}(1-z^2)\quad \text{(Note this step.)}$$

$$a+b\cos\theta=a+\frac{b}{2z}(z^2+1)$$

$$=\frac{b}{2z}\left(z^2+\frac{2a}{b}z+1\right)$$

$$I=Re\int_C\frac{(1-z^2)dz}{ib\left(z^2+\frac{2a}{b}z+1\right)}$$

The integrand $g(z)=\dfrac{(1-z^2)}{ib\left(z^2+\frac{2a}{b}z+1\right)}$ has simple poles at α,β [where $\alpha+\beta=-\frac{2a}{b}$, $\alpha\beta=1$ and $\alpha-\beta=\frac{2}{b}\sqrt{a^2-b^2}$] of which α lies inside C and β lies outside C since $\because|\alpha|=\frac{1}{|\beta|}<1$.

$$\text{Res}_{z=\alpha}\,g(z)=\lim_{z\to\alpha}(z-\alpha)\frac{2(1-z^2)}{ib(z-\alpha)(z-\beta)}$$

$$=\frac{1-\alpha^2}{ib(\alpha-\beta)}=\frac{\alpha\beta-\alpha^2}{ib(\alpha-\beta)}\quad\because \alpha\beta=1\ \text{(Note this step.)}$$

$$=-\frac{\alpha}{ib}=\frac{1}{ib}\left(\frac{a-\sqrt{a^2-b^2}}{b}\right)$$

$$I=Re\,2\pi i\frac{1}{ib^2}(a-\sqrt{a^2-b^2})$$

$$=\frac{2\pi}{b^2}(a-\sqrt{a^2-b^2})\quad \text{[JNTU 1995S].}$$

Corollary

$$\int_0^{2\pi}\frac{\sin^2\theta}{5-4\cos\theta}d\theta=\frac{\pi}{4}$$

Example 17.26

Show that $I=\int_0^{2\pi}\frac{\cos 3\theta}{5-4\cos\theta}d\theta=\frac{\pi}{12}$.

[JNTU 2003S]

Solution Convert the integral into a contour integral taken round the unit circle $C:|z|=1$ by putting $z=e^{i\theta}$ $(0\le\theta\le 2\pi)$ and $d\theta=\frac{dz}{iz}$

$$\cos 3\theta=Re\,z^3$$

$$5-4\cos\theta=5-4\left(\frac{z^2+1}{2z}\right)=-\frac{2}{z}\left(z^2-\frac{5}{2}z+1\right)$$

$$=-\frac{2}{z}(z-2)\left(z-\frac{1}{2}\right).$$

Now, $I=\int_0^{2\pi}\frac{\cos 3\theta}{5-4\cos\theta}d\theta=Re\int_C f(z)dz$

where $f(z)=\dfrac{z^3}{-2i(z-2)\left(z-\frac{1}{2}\right)}$ has simple poles at $z=2$ (outside C) and $z=\frac{1}{2}$ (inside C)

$$\text{Res}_{z=\frac{1}{2}}f(z)=\lim_{z\to\frac{1}{2}}\frac{\left(z-\frac{1}{2}\right)z^3}{-2i(z-2)\left(z-\frac{1}{2}\right)}$$

$$=\frac{\frac{1}{8}}{-2i\left(\frac{1}{2}-2\right)}=\frac{1}{24i}$$

$$\therefore\quad I=Re\int_C f(z)dz$$

$$=2\pi i\frac{1}{24i}=\frac{\pi}{12}.$$

Example 17.27

Evaluate $I=\int_0^{2\pi}\frac{\sin^2\theta}{5-4\cos\theta}$. [JNTU 1995S]

Solution Convert the integral into a contour integral taken round the unit circle $C:|z|=1$ by putting

$$z=e^{i\theta}\ (0\le\theta\le 2\pi)\quad\text{and}\quad d\theta=\frac{dz}{iz}$$

$$\sin^2\theta=\frac{1}{2}(1-\cos 2\theta)=\frac{1}{2}Re(1-z^2)$$

$$5-4\cos\theta=5-4\left(\frac{z^2+1}{2z}\right)=-\frac{2}{z}\left(z^2-\frac{5}{2}z+1\right)$$

$$=-\frac{2}{z}\left[(z-2)\left(z-\frac{1}{2}\right)\right]$$

We have to evaluate the integral

$$I = -\frac{1}{4} Re \frac{1}{i}\int_C f(z)dz$$

$$= -\frac{1}{4} Re \frac{1}{i}\int_C \frac{1-z^2}{(z-2)\left(z-\frac{1}{2}\right)} dz$$

Here $f(z) = \frac{1-z^2}{(z-2)\left(z-\frac{1}{2}\right)}$ has simple poles at $z = 2$ (which lies outside C) and $z = \frac{1}{2}$ (which lies inside C)

$$\text{Res}_{z=\frac{1}{2}} f(z) = \left.\frac{1-z^2}{z-2}\right|_{z=\frac{1}{2}} = \frac{1-\left(\frac{1}{2}\right)^2}{\frac{1}{2}-2}$$

$$= -\frac{3}{4}\cdot\frac{2}{3} = -\frac{1}{2}$$

$$\therefore I = -\frac{1}{4}\text{Re}\frac{1}{i}2\pi i\left(-\frac{1}{2}\right) = \frac{\pi}{4}$$

Example 17.28

Show that $I = \int_0^{\pi} \frac{\cos 2\theta}{1-2a\cos\theta + a^2} d\theta = \frac{\pi a^2}{1-a^2}$; $(a^2 < 1)$.

Solution Let $f(0) = \frac{\cos 2\theta}{1-2a\cos\theta + a^2}$

$$\therefore\ I = \int_0^{\pi} f(\theta)d\theta = \frac{1}{2}\int_0^{2\pi} f(\theta)d\theta$$

$$[f(2\pi - \theta) = f(\theta)]$$

We convert this into a contour integral taken round the unit circle $C: |z| = 1$ by putting

$$z = e^{i\theta} (0 \le \theta \le 2\pi) \text{ and } d\theta = \frac{dz}{iz}$$

$$\cos\theta = \frac{z^2+1}{2z}$$

$$\cos 2\theta = Re\, e^{2i\theta} = Re\, z^2$$

$$1 - 2a\cos\theta + a^2 = -a\left[2\cos\theta - \left(a + \frac{1}{a}\right)\right]$$

$$= -\frac{a}{z}\left[z^2 - \left(a+\frac{1}{a}\right)z + 1\right]$$

$$= -\frac{a}{z}\left[(z-a)\left(z-\frac{1}{a}\right)\right]$$

The integrand as a function of z is

$$g(z) = \frac{z^2}{-ai(z-a)\left(z-\frac{1}{a}\right)}.$$

Its singularities are simple poles at $z = a$ and $\frac{1}{a}$ of which $z = a$ alone lies inside C $\because |a| < 1$

$$\text{Res}_{z=a} g(z) = \lim_{z\to a} \frac{(z-a)z^2}{-ai(z-a)\left(z-\frac{1}{a}\right)}$$

$$= \frac{a^2}{-ai\left(a-\frac{1}{a}\right)} = \frac{a^2}{i(1-a^2)}$$

$$\therefore\ I = \frac{1}{2}\ Re\ 2\pi i \sum \text{Res } g(z)$$

$$= \frac{1}{2} 2\pi i \frac{a^2}{i(1-a^2)}$$

$$= \frac{\pi a^2}{1-a^2}.$$

Example 17.29

Evaluate $\int_0^{2\pi} \frac{d\theta}{3 - 2\cos\theta + \sin\theta}$.

Solution To evaluate the integral, $I = \int_0^{2\pi}\frac{d\theta}{3-2\cos\theta+\sin\theta}$, we have to convert it into a contour integral over the unit circle $C: |z| = 1$.

Put $e^{i\theta} = z$, $d\theta = \frac{dz}{iz}$,

$$\cos\theta = \frac{z^2+1}{2z}, \quad \sin\theta = \frac{z^2-1}{2iz}$$

$$\therefore \int_C \frac{2}{6 - 2\left(\frac{z^2+1}{z}\right) + \left(\frac{z^2-1}{iz}\right)}\frac{dz}{iz}$$

$$= 2\frac{dz}{6iz - 2i(z^2+1) + z^2 - 1}$$

$$= 2\int_C \frac{dz}{(1-2i)z^2 + 6iz - (1+2i)}$$

$$= 2\pi i \sum \text{Res} f(z)$$

where $f(z) = [(1-2i)z^2 + 6iz - (1+2i)]^{-1}$

$f(z)$ has simple poles at α, β which are the roots of $(1-2i)z^2 + 6iz - (1+2i)$ so that

$$[f(z)]^{-1} = [(1-2i)(z-\alpha)(z-\beta)]^{-1}$$

$$\therefore \alpha, \beta = \frac{-6i \pm \sqrt{(-6i)^2 + 4(1+2i)(1-2i)}}{2(1-2i)}$$

$$= \frac{-3i \pm \sqrt{-9+5}}{1-2i} = \frac{(-i, -5i)}{1-2i}$$

$$\because |\alpha\beta| = 1 \text{ and } |\beta| > 1$$

$$\therefore \alpha = \frac{-i}{1-2i} \text{ lies inside } C$$

$$\text{Res}_{z=\alpha} f(z) = \lim_{z\to\alpha}(z-\alpha)\frac{1}{(1-2i)(z-\alpha)(z-\beta)}$$

$$= \frac{1}{1-2i}\frac{1}{\alpha-\beta} = \frac{1}{1-2i}\frac{1-2i}{4i} = \frac{1}{4i}$$

$$\therefore I = 2\cdot 2\pi i \cdot \frac{1}{4i} = \pi.$$

Example 17.30

Show that $\int_0^{2\pi} e^{\cos\theta}\cos(n\theta - \sin\theta)\, d\theta = \dfrac{2\pi}{n!}$ where n is a positive integer.

Solution Let

$$I = \int_0^{2\pi} e^{\cos\theta}[\cos(n\theta - \sin\theta) - i\sin(n\theta - \sin\theta)]d\theta \quad (1)$$

$$= \int_0^{2\pi} e^{\cos\theta} e^{-i(n\theta - \sin\theta)} d\theta$$

$$= \int_0^{2\pi} e^{\cos\theta + i\sin\theta} \cdot e^{-in\theta} d\theta$$

We convert this integral into a contour integral over unit circle $C: |z| = 1$ by putting $e^{i\theta} = z,\ d\theta = \dfrac{dz}{iz}$.

Now we have

$$I = \int_C e^z \cdot z^{-n}\frac{dz}{iz} = \frac{1}{i}\int_C f(z)dz$$

where $f(z) = e^z z^{-n-1}$.

$f(z)$ has a pole of order $(n+1)$ at $z = 0$

$\text{Res}_{z=0} f(z) =$ coeff. of $\dfrac{1}{z}$ in the Laurent's series expansion of $f(z)$

$$= \frac{1}{z^{n+1}}\sum_{n=0}^{\infty}\frac{1}{n!}z^n = \frac{1}{n!}$$

Equating the real and imaginary parts in Eq. (1), we get

$$\int_0^{2\pi} e^{\cos\theta}\cos(n\theta - \sin\theta)d\theta = \frac{2\pi}{n!}$$

$$\int_0^{2\pi} e^{\cos\theta}\sin(n\theta - \sin\theta)d\theta = 0 \quad \therefore I = \frac{2\pi}{n!}.$$

EXERCISE 17.2

1. Evaluate the following integrals:

(a) $\displaystyle\int_0^{\pi}\frac{d\theta}{\pi + \cos\theta}$. [JNTU 1998]

Ans: $\dfrac{\pi}{\sqrt{\pi^2 - 1}}$

(b) $\displaystyle\int_0^{2\pi}\frac{\cos\theta}{3+\sin\theta}d\theta$. [JNTU 1996, 2003*S*]

Ans: 0

(c) $\displaystyle\int_0^{\pi}\frac{\cos 2\theta d\theta}{1 - 2a\cos\theta + a^2}$ $(a^2 < 1)$.

Ans: $\dfrac{\pi a^2}{(1-a^2)}$

(d) $\displaystyle\int_0^{\pi}\frac{a d\theta}{a^2 + \sin^2\theta}$ $(a > 0)$.

Ans: $\dfrac{\pi}{\sqrt{1+a^2}}$

(e) $\displaystyle\int_0^{2\pi}\frac{d\theta}{4\cos^2\theta + \sin^2\theta}$.

Ans: π

(f) $\displaystyle\int_0^{2\pi}\frac{d\theta}{5 - 3\cos\theta}$. [JNTU 1997, 2003*S*]

Ans: $\dfrac{\pi}{2}$

(g) $\displaystyle\int_0^{2\pi}\frac{\cos^2 3\theta}{1 - 2a\cos 2\theta + a^2}$ $0 < a < 1$.

Ans: $\dfrac{\pi(1 - a + a^2)}{(1-a)}$

(h) $\displaystyle\int_0^{2\pi}\frac{\sin^2\theta d\theta}{a + b\cos\theta}$ $a > b > 0$.

Ans: $\dfrac{2\pi}{b^2}(a - \sqrt{a^2 - b^2})$

2. Show that $\int_0^{2\pi} \frac{\cos 3\theta}{5-4\cos\theta} d\theta = \frac{\pi}{12}$.

[JNTU 2003S]

3. Evaluate $\int_0^{2\pi} \frac{\sin^2\theta}{5-4\cos\theta} d\theta$.

[JNTU 1995S]

Ans: $\pi/4$

4. Show that $\int_{-\pi}^{\pi} \frac{d\theta}{5+4\sin\theta} = \frac{2\pi}{3}$.

[JNTU 1994S]

5. Prove that $\int_0^{2\pi} \frac{d\theta}{(a+b\cos\theta)^2} = \frac{2\pi a}{(a^2-b^2)^{3/2}}$ $(a > b > 0)$.

[JNTU 2003S]

6. Show that $\int_0^{2\pi} \frac{d\theta}{1-2p\sin\theta+p^2} = \frac{2\pi}{1-p^2}$ $(|p|<1)$.

7. Show that $\int_0^{2\pi} \frac{\cos 2\theta}{5+4\cos\theta} d\theta = \frac{\pi}{6}$.

8. Evaluate $\int_0^{2\pi} \frac{d\theta}{(5-3\sin\theta)^2}$.

Ans: $\frac{5\pi}{32}$

9. Prove that $\int_0^{2\pi} \frac{d\theta}{(5-4\cos\theta)^2} = \frac{10\pi}{27}$.

10. Prove that $\int_0^{2\pi} \frac{d\theta}{13+5\cos\theta} = \frac{\pi}{6}$.

11. Prove that $\int_0^{2\pi} \frac{d\theta}{17-8\cos\theta} = \frac{2\pi}{15}$.

12. Prove that $\int_0^{\pi} \frac{1+2\cos\theta}{5+4\cos\theta} d\theta = 0$.

17.1.7 Type II: Integral of the Type $\int_{-\infty}^{\infty} f(x)dx$

Improper integral of rational functions of the type $\int_{-\infty}^{\infty} f(x)dx$ (Integration around a semicircle S in the upper half-plane bounded by its diameter L along the x-axis).

We first explain the meaning of an improper integral.

Improper integral

An integral of the type

$$\int_{-\infty}^{\infty} f(x)dx \tag{17.14}$$

where the interval of integration is not finite is called an improper integral of the first kind.

By definition

$$\int_{-\infty}^{\infty} f(x)dx = \lim_{a\to-\infty}\int_a^0 f(x)dx + \lim_{b\to\infty}\int_0^b f(x)dx \tag{17.15}$$

If both the limits exist, then we may combine the two limits and write

$$\int_{-\infty}^{\infty} f(x)dx = \lim_{R\to\infty}\int_{-R}^{R} f(x)dx \tag{17.16}$$

We assume that $f(x)$ is a real, rational function of the form

$$f(x) = \frac{p(x)}{q(x)}$$

where $p(x)$ and $q(x)$ are polynomials and that deg $q(x) \geq$ deg $p(x) + 2$ and $q(x) \neq 0$ for all real x, *i.e.*, $f(z)$ has no singularities on the real axis.

To evaluate integrals of the type Eq. (17.14), we consider contour integral $\int_C f(z)dz$ around a simple closed path C consisting of a semicircle S in the upper half-plane and the line segment $L = [-R, R]$ along the x-axis. Since $f(x)$ is rational, $f(z)$ has finitely many poles. The radius R of the semicircle is taken sufficiently large so that C encloses all the poles of $f(z)$ that lie in the upper half-plane.

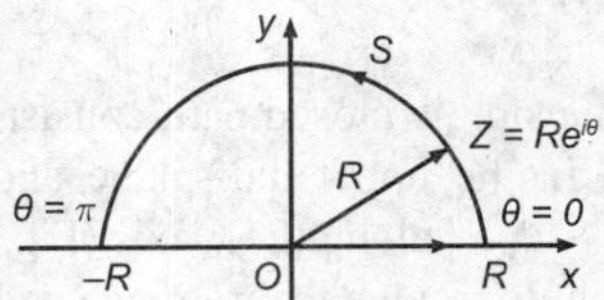

Figure 17.4

By Cauchy's residue theorem, we have

$$\begin{aligned}\int_C f(z)dz &= \int_S f(z)dz + \int_{-R}^{R} f(x)dx \\ &= 2\pi i \sum \text{Res}\, f(z) \end{aligned} \tag{17.17}$$

where the sum consists of the residues of $f(z)$ at all the poles of $f(z)$ in the upper half-plane. From Eq. (17.17), we obtain

$$\int_{-R}^{R} f(x)dx = 2\pi i \sum \text{Res} f(z) - \int_S f(z)dz \qquad (17.18)$$

To prove that $\int_S f(z)dz \to 0$ as $R \to \infty$

Consider $\int_S f(z)dz$ where S is a fixed semicircle with radius R = constant. Put $z = Re^{i\theta}$ where θ varies from 0 to π as z ranges along the semicircle S.

$$\text{Now } |f(z)| = \frac{|p(z)|}{|q(z)|} < \frac{K}{|z|^2} \quad (|z| = R > R_0)$$

since deg $q(z) \geq$ deg $p(z) + 2$, for sufficiently large K and R.

By the ML-inequality, we have

$$\left|\int_S f(z)dz\right| < \frac{K}{R^2}\pi R = \frac{K\pi}{R} \quad (R > R_0)$$

Hence as $R \to \infty$ the value of the integral over S approaches zero from Eqs. (17.16) and (17.18), we get $\int_{-\infty}^{\infty} f(x)dx = 2\pi i \sum \text{Res} f(z)$ where the sum consists of the residues at all poles of $f(z)$ in the upper half-plane.

Example 17.31

Evaluate $\int_{-\infty}^{\infty} \frac{dx}{x^2+1}$.

Solution To evaluate the integral

$I = \int_{-\infty}^{\infty} \frac{dx}{x^2+1}$, we consider the contour integral

$$\int_C f(z)dz = \int_C \frac{dz}{z^2+1} \qquad (1)$$

where C is the simple closed path consisting of the semicircle S in the upper half-plane and the line segment $L = [-R, R]$ along the x-axis (Fig. 17.4).

By Cauchy's residue theorem, we have

$$\int_C \frac{dz}{z^2+1} = \int_S \frac{dz}{z^2+1} + \int_{-R}^{R} \frac{dx}{x^2+1}$$

$$= 2\pi i \sum \text{Res} f(z) \qquad (2)$$

where the sum $\sum \text{Res} f(z)$ indicates the sum of the residues of $f(z)$ at all the poles that lie in the upper half-plane.

$$\text{Here } f(z) = \frac{1}{z^2+1} \qquad (3)$$

which has simple poles at $z = \pm i$. Of these $z = i$ lies in the upper half-plane and $z = -i$ lies outside the closed contour C and is not of interest to us.

$$\text{Res}_{z=i} \frac{1}{z^2+1} = \lim_{z \to i} (z-i)\frac{1}{(z-i)(z+i)}$$

$$= \frac{1}{i+i} = \frac{1}{2i} \qquad (4)$$

Estimation of $\int_S f(z)dz$:

$$f(z) = \frac{1}{z^2+1}$$

On S, $z = Re^{i\theta}$ $(0 \leq \theta \leq \pi)$

$\because |z^2+1| > R^2 - 1$ (for $R > 1$);

$\therefore \frac{1}{|z^2+1|} < \frac{1}{R^2-1}$ (for $R > 1$)

By Cauchy's ML-inequality, we have

$$\left|\int_S f(z)dz\right| = \left|\int_S \frac{1}{z^2+1}dz\right|$$

$$\leq \frac{\pi R}{R^2-1} \to 0 \qquad (5)$$

as $R \to \infty$

As $R \to \infty$ we get from Eqs. (2), (4) and (5)

$$\int_{-\infty}^{\infty} \frac{dx}{x^2+1} = 2\pi i \frac{1}{2i} = \pi.$$

Remark

1. $\int_0^{\infty} \frac{dx}{x^2+1} = \frac{1}{2}\int_{-\infty}^{\infty} \frac{dx}{x^2+1} = \frac{\pi}{2}$

   since $f(x) = \frac{1}{x^2+1}$ is an even function.

2. $\int_{-\infty}^{\infty} \frac{dx}{x^2+1} = \int_{-\infty}^{0} \frac{dx}{x^2+1} + \int_0^{\infty} \frac{dx}{x^2+1}$

$$= \int_0^{\infty} \frac{dx}{x^2+1} + \int_0^{\infty} \frac{dx}{x^2+1}$$

$$= 2 \cdot \frac{\pi}{2} = \pi$$

$$\because \int_{-\infty}^{0} \frac{dx}{x^2+1} = \int_0^{\infty} \frac{1}{\frac{1}{y^2}+1} \frac{1}{y^2}dy = \int_0^{\infty} \frac{dy}{y^2+1}$$

Put $x = -\frac{1}{y}$, $dx = \frac{1}{y^2}dy$. The limits are $x = -\infty \Rightarrow y = 0$; $x = 0 \Rightarrow y = \infty$.

The above integral can easily be evaluated by the methods of calculus. In fact

$$\int_0^\infty \frac{dx}{1+x^2} = \tan^{-1}x\Big|_0^\infty = \frac{\pi}{2} - 0 = \frac{\pi}{2}$$

This example has been worked out here to illustrate the method of evaluation of real integrals by the residue theory and compare the result with that obtained by the integral calculus method.

Example 17.32

Evaluate $\int_0^\infty \frac{dx}{x^2+a^2}$.

Solution

$$\int_0^\infty \frac{dx}{x^2+a^2} = \frac{1}{2}\int_{-\infty}^\infty \frac{dx}{x^2+a^2} \quad (1)$$

We consider

$$\int_C f(z)dz = \int_C \frac{dz}{z^2+a^2}$$

$$= \int_S \frac{dz}{z^2+a^2} + \int_L \frac{dx}{x^2+a^2} = I_1 + I_2 \quad (2)$$

$S: z = Re^{i\theta} \quad (0 \le \theta \le \pi), \quad dz = iRe^{i\theta}d\theta$

S is the semicircle in the upper half-plane and L: $[-R, R]$, the line segment along the x-axis.

$$z = x,\ dz = dx;$$

$$f(z) = \frac{1}{z^2+a^2} = f(x) = \frac{1}{x^2+a^2}$$

$$|z^2+a^2| \ge |z|^2 - |a|^2 = R^2 - |a|^2$$

$$\frac{1}{|z^2+a^2|} \le \frac{1}{R^2-|a|^2}$$

To estimate $I = \int_S \frac{dz}{z^2+a^2}$,

$$|I_1| = \left|\int_S \frac{dz}{z^2+a^2}\right| \le \frac{\pi R}{R^2-|a|^2} \to 0 \text{ as } R \to \infty \quad (3)$$

The integral $\int_L \frac{dz}{z^2+a^2}$ on the x-axis: $z = x$, $dz = dx$

$$I_2 = \int_C \frac{dz}{z^2+a^2} = \int_{-R}^R \frac{dx}{x^2+a^2}$$

$$\to \int_{-\infty}^\infty \frac{dx}{x^2+a^2} \quad \text{as } R \to \infty \quad (4)$$

The integrand $f(z) = \frac{1}{z^2+a^2}$ has simple poles at $z = \pm ia$ of which $z = ia$ alone lies in the upper half-plane.

$$\text{Res}_{z=ia} \frac{1}{z^2+a^2} = \lim_{z\to ia}(z-ia)\frac{1}{(z-ia)(z+ia)}$$

$$= \frac{1}{ia+ia} = \frac{1}{zia}$$

$$\int_C \frac{dz}{z^2+a^2} = 2\pi i.\frac{1}{2ia} = \frac{\pi}{a}$$

$$\Rightarrow \int_0^\infty \frac{dx}{x^2+a^2} = \frac{\pi}{2a} \quad \text{by Eqs. (2), (3) and (4)}$$

Note

From integral calculus, we know that

$$\int_0^\infty \frac{dx}{x^2+a^2} = \frac{1}{a}\tan^{-1}\frac{x}{a}\Big|_0^\infty = \frac{\pi}{2a}$$

Example 17.33

Evaluate $I = \int_0^\infty \frac{dx}{(x^2+a^2)^2}$.

Solution Let C be the closed contour consisting of (i) Semicircle S: $z = Re^{i\theta}$ in the upper half-plane $dz = iRe^{i\theta}\,d\theta$, θ varies from 0 to π.

$$f(z) = f(Re^{i\theta}) \Rightarrow \frac{1}{(z^2+a^2)^2} = \frac{1}{(R^2e^{2i\theta}+a^2)^2}$$

(ii) Line segment L: $[-R, R]$ on the x-axis

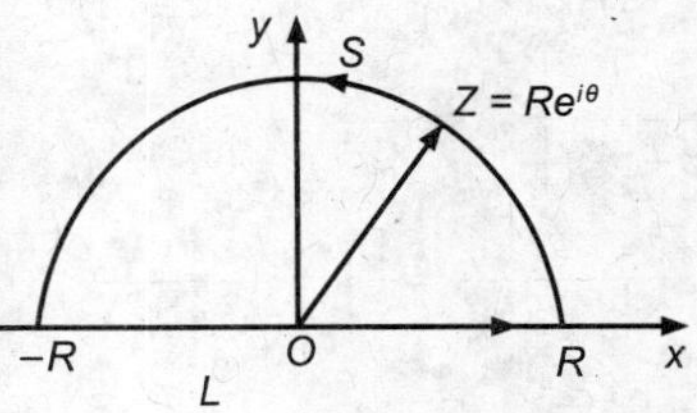

Figure 17.5

$$z = x, \quad dz = dx \text{ and } f(z) = f(x) = \frac{1}{(x^2+a^2)^2}$$

$$\int_C f(z)dz = \int_S f(Re^{i\theta})iRe^{i\theta}d\theta + \int_L f(x)dx$$

$$= I_1 + I_2 \quad \text{(say)}$$

where $f(z) = \frac{1}{(z^2+a^2)^2}$.

On S: Estimation of I_1, $z = Re^{i\theta}$; $dz = iRe^{i\theta}\,d\theta$, θ varies from 0 to π, *i.e.*, $0 \le \theta \le \pi$

$$|I_1| = \left|\int_S f(z)dz\right| = \left|\int_0^{\pi} f(R\ e^{i\theta})iR\ e^{i\theta}\,d\theta\right|$$

$$\leq \int_0^{\pi} |f(Re^{i\theta})|\,Rd\theta$$

$$< \frac{\pi R}{(R^2-a^2)^2} \to 0 \text{ as } R\to\infty$$

$$\Rightarrow \quad I_1 = 0 \quad \text{as } R\to\infty$$

$$f(z) = \frac{1}{(z^2+a^2)^2}$$

$$|z^2+a^2| > ||z|^2 - |a|^2| = R^2 - a^2$$

$$\frac{1}{(z^2+a^2)^2} < \frac{1}{(R^2-a^2)^2}$$

I_2: On L, $z = x$, $dz = dx$; x varies from $-R$ to R.

$$\lim_{R\to\infty}\int_{-R}^{R}\frac{dx}{(x^2+a^2)^2} = \int_C f(z)dz$$

$$= 2\pi i\sum_{k=1}^{n}\text{Res}_{z=z_k} f(z).$$

$f(z) = \dfrac{1}{(x^2+a^2)^2}$ has singularities at $z = \pm ia$ which are double poles, of which only $z = ia$ lies in the upper half-plane.

$$\text{Res}_{z=ia} f(z)$$

$$= \frac{1}{(2-1)!}\lim_{z\to ia}\frac{d}{dz}\left[(z-ai)^2\frac{1}{(z-ai)^2(z+ai)^2}\right]$$

$$= \left.\frac{-2}{(z+ia)^3}\right|_{z=ia} = \frac{-1}{4a^3}$$

$$\therefore I = \int_0^{\infty}\frac{dx}{(x^2+a^2)^2} = \frac{1}{2}\int_{-\infty}^{\infty}\frac{dx}{(x^2+a^2)^2}$$

$$= \frac{1}{2}\cdot 2\pi i\sum \text{Res}\, f(z) = \pi i\left(\frac{-i}{4a^3}\right) = \frac{\pi}{4a^3}.$$

Note

We have proved that $\displaystyle\int_0^{\infty}\frac{dx}{x^2+a^2} = \frac{\pi}{2a}$

Differentiating both sides partially w.r.t the parameter 'a', we get

$$-\frac{\pi}{2a^2} = \frac{\partial}{\partial a}\left(\int_0^{\infty}\frac{dx}{x^2+a^2}\right) = \int_0^{\infty}\frac{-2a\ dx}{(x^2+a^2)^2}$$

$$\Rightarrow \int_0^{\infty}\frac{dx}{(x^2+a^2)^2} = \frac{\pi}{4a^3}.$$

Example 17.34

Show that $\displaystyle\int_0^{\infty}\frac{x^2dx}{(x^2+a^2)(x^2+b^2)} = \frac{\pi}{2(a+b)}$.

Solution Take $f(z) = \dfrac{z^2}{(z^2+a^2)(z^2+b^2)}$ and consider $I = \int_C f(z)dz$ where C comprises of

(i) S: $z = Re^{i\theta}$, θ varies in $[0, \pi]$,
(ii) the semicircle in the upper half-plane,
(iii) L: $[-R, R]$, the line segment along the x-axis on which $z = x$, $dz = x$ and $f(z) = f(x)$.

$f(z)$ has simple poles at $z = \pm ia, \pm ib$ of which $z = ia, ib$ alone lie in the upper half-plane.

$$\text{Res}_{z=ia} f(z) = \lim_{z\to ia}(z-ia)\frac{z^2}{(z-ia)(z+ia)(z^2+b^2)}$$

$$= \frac{-a^2}{zia(-a^2+b^2)} = \frac{a}{2i(a^2-b^2)}$$

$$\text{Res}_{z=ib} f(z) = \lim_{z\to ib}(z-ib)\frac{z^2}{(z-ib)(z+ib)(z^2+a^2)}$$

$$= \frac{-b^2}{zib(-b^2+a^2)} = \frac{b}{2i(a^2-b^2)}$$

$$\text{Sum of residues} = \frac{1}{2i(a+b)}$$

$$I = 2\pi i\frac{1}{2i(a+b)}$$

$$= \frac{\pi}{a+b} = I_1 + I_2 \qquad (1)$$

$$\text{where } |I_1| = \left|\int_S f(z)dz\right| \leq \frac{R^2\cdot\pi R}{(R^2-a^2)(R^2-b^2)}$$

$$\to 0 \text{ as } R\to\infty \qquad (2)$$

$$\because |z^2+a^2| \leq R^2 - |a|^2 \qquad (3)$$

$$\Rightarrow \frac{1}{|z^2+a^2|} > \frac{1}{R^2-|a|^2},\quad \textit{etc.}$$

$$I_2 = \int_C f(z)dz = \int_{-R}^{R}\frac{x^2dx}{(x^2+a^2)(x^2+b^2)}$$

$$\Rightarrow \int_{-\infty}^{\infty}\frac{x^2dx}{(x^2+a^2)(x^2+b^2)} \text{ as } R\to\infty \qquad (4)$$

$$\text{Hence } \int_0^{\infty}\frac{x^2dx}{(x^2+a^2)(x^2+b^2)}$$

$$= \frac{1}{2}\int_{-\infty}^{\infty} \frac{dx}{(x^2+a^2)(x^2+b^2)}$$

$$= \frac{\pi}{2(a+b)}$$

Note

Resolving $\dfrac{x^2}{(x^2+a^2)(x^2+b^2)}$ into partial fractions

$$\frac{x^2}{(x^2+a^2)(x^2+b^2)} = \frac{1}{a^2-b^2}\left(a^2 \cdot \frac{1}{x^2+a^2} - b^2 \frac{1}{x^2+b^2}\right)$$

we get

$$\int_0^{\infty} \frac{x^2 dx}{(x^2+a^2)(x^2+b^2)} = \frac{1}{a^2-b^2}\left(a^2 \cdot \frac{\pi}{2a} - b^2 \frac{\pi}{2b}\right)$$

$$= \frac{\pi}{2(a+b)} \qquad (1)$$

$$\because \int_0^{\infty} \frac{dx}{x^2+a^2} = \frac{\pi}{2a}, \quad etc.$$

Resolving $\dfrac{1}{(x^2+a^2)(x^2+b^2)}$ into partial fractions,

$$\frac{1}{(x^2+a^2)(x^2+b^2)} = \frac{1}{b^2-a^2}\left(\frac{1}{x^2+a^2} - \frac{1}{x^2+b^2}\right).$$

We get

$$\int_0^{\infty} \frac{dx}{(x^2+a^2)(x^2+b^2)} = \frac{1}{b^2-a^2}\left(\frac{\pi}{2a} - \frac{\pi}{2b}\right) \qquad (2)$$

$$= \frac{\pi}{2ab(a+b)};$$

$$\because \int_0^{\infty} \frac{dx}{x^2+a^2} = \frac{\pi}{2a}, \quad etc.$$

Corollary

1. $\displaystyle\int_0^{\infty} \frac{x^2 dx}{(x^2+a^2)^2} = \frac{\pi}{4a}$ Put $b = a$ in Eq. (1)
2. $\displaystyle\int_0^{\infty} \frac{dx}{(x^2+a^2)^2} = \frac{\pi}{4a^3}$ Put $b = a$ in Eq. (2)

Example 17.35

Evaluate $I = \displaystyle\int_0^{\infty} \frac{x^2+px+q}{(x^2+a^2)(x^2+b^2)} dx \quad a > 0, b > 0.$

Solution Consider

$$\int_C f(z)dz = \int_S f(z)dz + \int_L f(z)dz$$

where $f(z) = \dfrac{z^2+pz+q}{(z^2+a^2)(z^2+b^2)}$, the integral being taken around C which consists of

(i) Semicircle $S : z = Re^{i\theta}, dz = iR\, e^{i\theta}\, d\theta$ in the upper half-plane and θ varies from 0 to π as z traces the semicircle S.

(ii) Line segment L: $[-R, R]$ of the x-axis on which $z = x$, $dz = dx$ and x varies from $-R$ to R as z traces the line segment and $R \to \infty$.

Estimation of the integral on S

$$\left|\int_S f(z)dz\right| \le \int_0^{\pi} \left|f(Re^{i\theta})\right| R d\theta$$

$$\le \pi R \frac{R^2+pR+q}{(R^2-a^2)(R^2-b^2)}$$

$$= \frac{1}{R} \to 0 \text{ as } R \to \infty$$

$$\because (z^2+pz+q) \le |z|^2 + p|z| + q = R^2 + pR + q$$

$$|z^2+a^2| \ge ||z|^2 - |a|^2|, \quad |z^2+b^2| \ge ||z|^2 - |b|^2|$$

$$\Rightarrow \frac{1}{(z^2+a^2)(z^2+b^2)} \le \frac{1}{(R^2-a^2)(R^2-b^2)}$$

Evaluation of the integral on L

The function $f(z) = \dfrac{z^2+pz+q}{(z^2+a^2)(z^2+b^2)}$ has simple poles at $z = \pm ia, \pm ib$ of which $z = ia, ib$ alone lie in the upper half-plane.

$$\text{Res}_{z=ia} f(z) = \left.\frac{z^2+pz+q}{(z+ia)(z^2+b^2)}\right|_{z=ia}$$

$$= \frac{-a^2+pia+q}{2ia(b^2-a^2)};$$

$$\text{Res}_{z=ib} f(z) = \left.\frac{z^2+pz+q}{(z+ib)(z^2+a^2)}\right|_{z=ib}$$

$$= \frac{-b^2+pib+q}{2ib(a^2-b^2)}$$

$$\text{Sum of residues } = \frac{1}{zi(a^2-b^2)}\left[a-b+q\left(\frac{1}{b}-\frac{1}{a}\right)\right]$$

$$= \frac{1}{2i(a+b)}\left(1+\frac{q}{ab}\right)$$

We know that $\int_0^\infty f(x)dx = \frac{1}{2}\int_{-\infty}^\infty f(x)dx$ if $f(x)$ is even, and equal to 0 if $f(x)$ is odd

$$\left(\text{as } R\to\infty, \left|\int_S f(z)dz\right| \to 0\right.$$

$$\left.\text{and } \int_L f(z)dz = \int_{-\infty}^\infty \frac{x^2+px+q}{(x^2+a^2)(x^2+b^2)}dx\right).$$

By Cauchy's residue theorem, we have

$$\therefore I = \frac{1}{2}\int_C f(z)dz = \frac{1}{2}2\pi i.\frac{1}{2i(a+b)}$$

$$= \frac{\pi}{2(a+b)}\left(1+\frac{q}{ab}\right)$$

(p-term is missing because it is an odd function).

Particular cases

(i) $p = 0, q = 3, a = 1, b = 2$

$$\int_{-\infty}^\infty \frac{x^2+3}{(x^2+1^2)(x^2+2^2)}dx = \frac{\pi}{(1+2)}\left(1+\frac{3}{1x^2}\right)$$

$$= \frac{5\pi}{6}$$

$$\Rightarrow \int_0^\infty \frac{x^3+3}{(x^2+1^2)(x^2+2^2)}dx = \frac{5\pi}{12}$$

(ii) $p = -1, q = 2, a = 1, b = 2$

$$\int_{-\infty}^\infty \frac{x^2-x+2}{x^4+10x^2+a}dx = \int_{-\infty}^\infty \frac{x^2-x+2}{(x^2+1)(x^2+3^3)}dx$$

$$= \frac{\pi}{1+3}\left(1+\frac{2}{1x3}\right) = \frac{5\pi}{12}$$

$$\Rightarrow \int_0^\infty \frac{x^2-x+2}{(x^2+1^2)(x^2+2^2)}dx = \frac{5\pi}{24}$$

(iii) $p = 0 = q$

$$\int_{-\infty}^\infty \frac{x^2}{(x^2+a^2)(x^2+b^2)}dx = \frac{\pi}{(a+b)}$$

$$\Rightarrow \int_0^\infty \frac{x^2}{(x^2+a^2)(x^2+b^2)}dx = \frac{\pi}{2(a+b}$$

(iv) $a = 1, b = 2, p = 0, q = 0$

$$\int_{-\infty}^\infty \frac{x^2dx}{(x^2+1)(x^2+2^2)} = \frac{\pi}{1+2}\left(1+\frac{0}{1x^2}\right)$$

$$= \frac{\pi}{3}$$

$$\Rightarrow \int_0^\infty \frac{x^2dx}{(x^2+1)(x^2+2^2)} = \frac{\pi}{6}$$

(v) $$\int_{-\infty}^\infty \frac{x^2dx}{(x^2+3^2)(x^2+2^2)} = \frac{\pi}{3+2}\left(1+\frac{0}{2x^2}\right)$$

$$= \frac{\pi}{5}$$

$$\Rightarrow \int_0^\infty \frac{x^2dx}{(x^2+3^2)(x^2+2^2)} = \frac{\pi}{10}$$

Remark

From Example 17.32 we have

$$I = \int_0^\infty \frac{dx}{x^2+a^2} = \frac{\pi}{2a} \text{ and } J = \int_0^\infty \frac{dx}{x^2+b^2} = \frac{\pi}{2b}$$

$$\therefore \int_0^\infty \frac{x^2dx}{(x^2+a^2)(x^2+b^2)}$$

$$= \frac{1}{a^2-b^2}\int_0^\infty\left(\frac{a^2}{x^2+a^2} - \frac{b^2}{x^2+b^2}\right)dx$$

$$= \frac{1}{a^2-b^2}\left(a^2\cdot\frac{\pi}{2a} - b^2\cdot\frac{\pi}{2b}\right) = \frac{\pi}{2(a+b)}$$

$$\int_0^\infty \frac{dx}{(x^2+a^2)(x^2+b^2)}$$

$$= \frac{1}{a^2-b^2}\int_0^\infty\left(\frac{1}{x^2+b^2} - \frac{1}{x^2+a^2}\right)dx$$

$$= \frac{1}{a^2-b^2}\left(\frac{\pi}{2b} - \frac{\pi}{2a}\right) = \frac{\pi}{2ab(a+b)}$$

Choosing $b = a$, we get

$$\int_0^\infty \frac{x^2dx}{(x^2+a^2)^2} = \frac{\pi}{4a}; \quad \int_0^\infty \frac{dx}{(x^2+a^2)^2} = \frac{\pi}{4a^3}.$$

Example 17.36

Show that $I = \int_{-\infty}^\infty \frac{dx}{x^4+1} = \frac{\pi}{\sqrt{2}}$.

Solution Let $f(z) = \frac{1}{z^4+1}$ and c be the contour consisting of

(i) Semicircle S: $z = Re^{i\theta}$, $dz = iRe^{i\theta}\,d\theta$ in the upper half-plane where θ ranges from 0 to π as z traces S and

(ii) Line segment L: $[-R, R]$ along the x-axis where $z = x$, $dz = dx$ and x ranges from $-R$ to R and $R\to\infty$.

$$\int_C \frac{1}{z^4+1}dz = \int_S \frac{1}{z^4+1}dz + \int_L \frac{1}{z^4+1}dz \quad (1)$$

Estimation of the integral on *S*

$$\left|\int_S \frac{1}{z^4+1}dz\right| \le \int_S^{\pi} |f(Re^{i\theta})||iRe^{i\theta}|\,d\theta$$

$$\le \frac{\pi R}{R^4-1} \to 0 \text{ as } R\to\infty$$

$$\because |z^4+1| \ge |z|^4 - 1 = R^4-1$$

$$\Rightarrow \left|\frac{1}{z^4+1}\right| < \frac{1}{R^4-1}$$

Evaluation of the integral on *L*

As $R\to\infty$ we have from Eq. (1)

$$I = \int_{-\infty}^{\infty}\frac{dx}{x^4+1} = \int_C \frac{dz}{z^4+1} \quad (1)$$

The singularities of $f(z)$ are the zeros of

$$x^4+1=0 \Rightarrow z^4 = -1 = e^{\pi i}e^{2k\pi i}$$

$$\Rightarrow z = e^{\frac{1}{4}(2k+1)\pi i} \quad (k=0,1,2,3)$$

$$\text{or } z_1 = e^{\frac{\pi i}{4}},\ z_2 = e^{\frac{3\pi i}{4}},$$

$$z_3 = e^{\frac{5\pi i}{4}} = e^{\frac{-3\pi i}{4}},\ z_4 = e^{\frac{7\pi i}{4}} = e^{\frac{-\pi i}{4}}$$

of which z_1 and z_2 alone lie in the upper half-plane.

$$\text{Res}_{z=z_1} f(z) = \left.\frac{1}{4z^3}\right|_{z=e^{\frac{\pi i}{4}}} = \frac{1}{4}e^{\frac{-3\pi i}{4}};$$

$$\text{Res}_{z=z_2} f(z) = \left.\frac{1}{4z^3}\right|_{z=e^{\frac{-9\pi i}{4}}} = \frac{1}{4}e^{-\frac{\pi i}{4}}$$

By Cauchy's residue theorem, as $R\to\infty$, we have from Eq. (2)

$$I = \int_{-\infty}^{\infty}\frac{dx}{x^4+1} = 2\pi i\cdot\frac{1}{4}e^{-\frac{\pi i}{4}}\left(1+e^{-\frac{2\pi i}{4}}\right)$$

$$= \frac{\pi i}{2}\cdot\frac{1-i}{\sqrt{2}}(1-i) = \frac{\pi i(-2i)}{2\sqrt{2}} = \frac{\pi}{\sqrt{2}}.$$

Corollary $\displaystyle\int_0^{\infty}\frac{dx}{x^4+1} = \frac{\pi}{2\sqrt{2}}$

Example 17.37

Show that $I = \displaystyle\int_0^{\infty}\frac{dx}{x^6+1}$.

Solution Let $f(z) = \dfrac{1}{z^6+1}$ and C be the contour consisting of

(i) Semicircle S: $z = Re^{i\theta}$; $dz = iRe^{i\theta}\,d\theta$ and θ ranges from 0 to π as z traces S and

(ii) Line segment L: $[-R, R]$ on the x-axis where $z = x$, $dz = dx$ and x ranges from $-R$ to R and finally $R\to\infty$.

Estimation of the integral on *S*

$$\int_C \frac{dz}{z^6+1} = \int_S \frac{dz}{z^6+1} + \int_C \frac{dz}{z^6+1} \quad (1)$$

$$\left|\int_S \frac{dz}{z^4+1}\right| \le \int_0^{\pi}|f(Re^{i\theta})|R\,d\theta$$

$$\le \frac{\pi R}{R^6-1} \to 0 \text{ as } R\to\infty \quad (2)$$

$$\because\ |z^6+1| \ge |z|^6 - 1 = R^6-1$$

Evaluation of the integral on *L*

$$\int_L \frac{dx}{x^6+1} = \int_{-R}^{R}\frac{dx}{z^6+1}$$

$$\to \int_{-\infty}^{\infty}\frac{dx}{x^6+1} \text{ as } R\to\infty$$

By Cauchy's residue theorem, as $R\to\infty$, we have from Eqs. (1) and (2)

$$\int_{-\infty}^{\infty}\frac{dx}{x^6+1} = \int_C \frac{dz}{z^6+1} = 2\pi i\sum \text{Res}\, f(z)$$

The singularities of $f(z) = \dfrac{1}{z^6+1}$ are at

$$z^6 = e^{\pi i}\cdot e^{2k\pi i} \Rightarrow z = e^{\frac{(2k+1)\pi i}{6}} \quad (k=0,1,2,3,4,5)$$

of which $z_1 = e^{\frac{\pi i}{6}}$, $z_2 = e^{\frac{3\pi i}{6}}$, $z_3 = e^{\frac{5\pi i}{6}}$ lie in the upper half-plane.

$$\text{Res}_{z=z_1} f = \left.\frac{1}{6z^5}\right|_{z=e^{\frac{\pi i}{6}}} = \frac{1}{6}e^{\frac{-5\pi i}{6}} = \frac{1}{6}e^{\frac{3\pi i}{6}}e^{-\frac{2\pi i}{6}}$$

$$= \frac{1}{6}e^{-\frac{\pi i}{2}}e^{\frac{\pi i}{3}} = -\frac{1}{6}e^{-\frac{\pi i}{2}}$$

$$\text{Res}_{z=z_2} f = \left.\frac{1}{6z^5}\right|_{z=e^{\frac{3\pi i}{6}}} = \frac{1}{6}e^{\frac{-15\pi i}{6}} = \frac{1}{6}e^{-\frac{3\pi i}{6}}e^{-\frac{12\pi i}{6}}$$

$$= \frac{1}{6}e^{-\frac{\pi i}{2}}e^{-2\pi i} = -\frac{i}{6}$$

$$\text{Res}_{z=z_3} f = \left.\frac{1}{6z^5}\right|_{z=e^{\frac{5\pi i}{6}}} = \frac{1}{6}e^{\frac{-25\pi i}{6}} = \frac{1}{6}e^{-\frac{27\pi i}{6}}e^{\frac{2\pi i}{6}}$$

$$= \frac{1}{6}e^{-\frac{\pi i}{2}}e^{\frac{\pi i}{3}} = -\frac{i}{6}e^{\frac{\pi i}{3}}$$

By Cauchy's residue theorem, we have from Eqs. (1) and (2) (as $R \to \infty$)

$$\int_C \frac{dz}{z^6+1} = \int_{-\infty}^{\infty} \frac{dx}{x^6+1} = 2\pi i \sum \text{Res} f$$

$$\left[\because \left|\int_S f(z)dz\right| \to 0 \text{ as } R \to \infty\right]$$

$$= 2\pi i \cdot \frac{1}{6} \cdot e^{-\frac{\pi i}{2}}\left(e^{-\frac{\pi i}{2}} + 1 + e^{\frac{\pi i}{3}}\right)$$

$$= \frac{\pi i}{3}(-i)\left(1 + 2\cos\frac{\pi}{3}\right)$$

$$= \frac{\pi}{3}\left(1 + 2x\frac{1}{2}\right) = \frac{2\pi}{3}$$

$$\Rightarrow \int_0^{\infty} \frac{dx}{x^6+1} = \frac{1}{2}\int_{-\infty}^{\infty} \frac{dx}{x^6+1} = \frac{1}{2}\frac{2\pi}{3} = \frac{\pi}{3}.$$

Example 17.38

Evaluate $\int_{-\infty}^{\infty} \frac{x^2}{1+x^4} dx.$

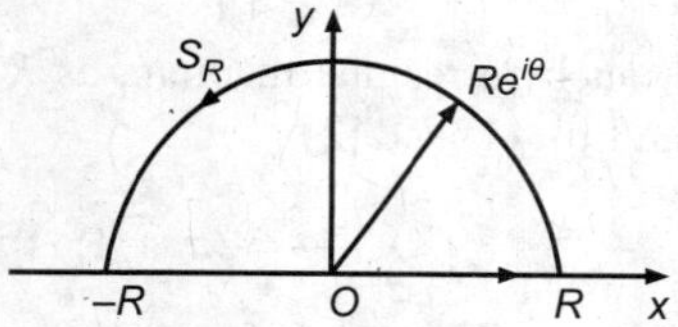

Figure 17.6

Solution We have to evaluate

$$\int_{-\infty}^{\infty} \frac{x^2}{1+x^4} dx \tag{1}$$

$$\text{Let } f(z) = \frac{z^2}{1+z^4} \tag{2}$$

which is the rational function P/Q where $f(z) = z^2$ and $Q(z) = 1 + z^4$. Since deg Q – deg $P = 2$, therefore

$$\lim_{R\to\infty} \int_{S_R} f = 0$$

where S_R is the semicircle of radius R in the upper half-plane, *i.e.*,

$$\lim_{R\to\infty} \int_0^{\pi} f(Re^{i\theta})\text{Re}^{i\theta}\, d\theta = 0 \tag{3}$$

The poles of the integrand are the zeros of $z^4 + 1$, *i.e.*, the solutions of $z^4 = -1$, which are $e^{\pi i \frac{x}{4}}$ for $k = 1, -1, 3, -3$.

Since $Q'(z) = 4z^3 \neq 0$ unless $z = 0$, we conclude that all the poles of f are simple. The two poles in the upper half-plane are $z_1 = e^{\frac{\pi i}{4}}$ and $z_2 = e^{\frac{3\pi i}{4}}$

$$\left.\begin{aligned} \text{Res}(f:z_1) &= \frac{P(z_1)}{Q'(z_1)} = \frac{z_1^2}{4z_1^3} = \frac{1}{4}\frac{1}{z_1} = \frac{1}{4}e^{-\frac{\pi i}{4}} \\ \text{Res}(f:z_2) &= \frac{P(z_2)}{Q'(z_2)} = \frac{z_2^2}{4z_2^3} = \frac{1}{4}\frac{1}{z_2} = \frac{1}{4}e^{-\frac{3\pi i}{4}} \end{aligned}\right\} \tag{4}$$

By Cauchy's residue theorem, we have

$$\int_C f(z)dz = \int_{-R}^{R} \frac{x^2}{1+x^4} dx + \int_{S_R} \frac{z^2}{z^4+1} dz$$

$$= 2\pi i \sum_{k=1}^{2} \text{Res}(f:z_k) \tag{5}$$

$$= 2\pi i\left(\frac{1}{4}e^{-\frac{\pi i}{4}} + \frac{1}{4}e^{-\frac{3\pi i}{4}}\right) = \frac{\pi}{2}\left[e^{\frac{\pi i}{2}}\left(e^{-\frac{\pi i}{4}} + e^{-\frac{3\pi i}{4}}\right)\right]$$

$$= \frac{\pi}{2}\left(e^{\frac{\pi i}{4}} + e^{-\frac{\pi i}{4}}\right) = \pi\cos\frac{\pi}{4} = \frac{\pi}{\sqrt{2}}$$

Letting $R \to \infty$ and using Eq. (3), we obtain $\int_{-\infty}^{\infty} \frac{x^2}{1+x^4} dx = \frac{\pi}{\sqrt{2}}$. That the integral $\int_{S_R} f(z)dz$ on the semicircle S_R tends to 0 as $R \to \infty$ can be seen independently as follows:

On S_R: $z = Re^{i\theta}$, $|z_1 + z_2| \geq |z_1| - |z_2|$

$$\because |z^4+1| = |R^4 e^{4i\theta} + 1| \geq |R^4 e^{4i\theta}| - 1 = R^4 - 1$$

$$|f| = \left|\frac{z^2}{1+z^4}\right| \leq \frac{R^2}{R^4-1}$$

$$\therefore \left|\int_{S_R} f(z)dz\right| = \left|\int_0^{\pi} f(R e^{i\theta}) iR e^{i\theta} d\theta\right|$$

$$\leq \int_0^{\pi} |f| R d\theta \leq \frac{R^2}{R^4-1}\pi$$

$$R \to 0 \text{ as } R \to \infty.$$

Example 17.39

Show that $\int_{-\infty}^{\infty} \frac{x^2 dx}{(x^2+1)^2(x^2+2x^2+2)} = \frac{7\pi}{50}.$

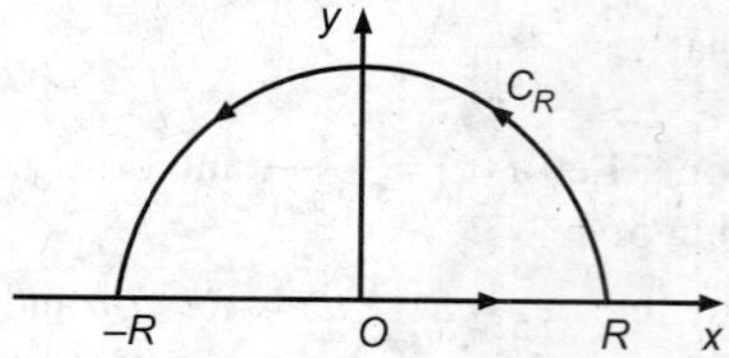

Figure 17.7

Solution We have to prove

$$\int_{-\infty}^{\infty}\frac{x^2dx}{(x^2+1)^2(x^2+2x^2+2)}=\frac{7\pi}{50} \quad (1)$$

$$\text{Let } f(z)=\frac{z^2}{(z^2+1)^2(z^2+2z+2)}$$

Consider

$$\int_C f(z)dz=\int_{-R}^{R}\frac{x^2dx}{(x^2+1)^2(x^2+2x+2)}+\int_C f(z)dz \quad (2)$$

where C is the contour shown Fig. 17.7.

The singularities of the integrand inside C are the pole at $z = i$ of order 2 and the pole at $z = -1 + i$ of order 1.

Res $(f: i)$

$$=\lim_{z\to i}\frac{d}{dz}\left[(z-i)^2\frac{z^2}{(z+i)^2(z-i)^2(z^2+2z+2)}\right]$$

$$=\lim_{z\to i}\frac{(z^2+2z+2)(2z)[(z+i)-z]+(2z+2)}{(z+i)^3(z^2+2z+2)^2}$$

$$=\frac{-2(1+2i)+(-2+2i-2+2i)}{-8i(1+2i)^2}$$

$$=\frac{(-2-4i-4+4i)i(1-2i)^2}{8\cdot 5^2}$$

$$=\frac{(-6)i(1-4-4i)}{200}=\frac{(3i)(3+4i)}{100}$$

$$=\frac{(9i-12)}{100} \quad (3)$$

Res $(f: -1+i)$

$$=\lim_{z\to -1+i}(z+1-i)\frac{z^2}{(z^2+1)^2(z+1-i)(z+1+i)}$$

$$=\frac{(-1+i)^2}{[(-1+i)^2+1]^2(2i)}=\frac{1-1-2i}{(-2i+1)^2 2i}$$

$$=\frac{-1(1+2i)^2}{5^2}=\frac{-1(1-4+4i)}{25}=\frac{3-4i}{25} \quad (4)$$

By residue theorem,

$$\int_C f(z)dz=\int_{-R}^{R}\frac{x^2}{(x^2+1)^2(x^2+2x+2)}dx+\int_{C_R} f(z)dz$$

$$=2\pi i\left[\frac{9i-12}{100}+\frac{4(3-4i)}{4(25)}\right]$$

$$=\frac{7\pi}{50} \quad (5)$$

$$\because\left|\int_{C_R}\frac{z^2}{(z^2+1)^2(z^2+2z+2)}dz\right|$$

$$\le \pi R\frac{R^2}{(R^2-1)^2(R^2-2R+2)}\to 0 \text{ as } R\to\infty$$

We have from Eq. (5), by letting $R\to\infty$

$$\int_{-\infty}^{\infty}\frac{x^2}{(x^2+1)^2(x^2+2x+2)}dx=\frac{7\pi}{50}.$$

17.1.8 Type II (a): Improper Integrals Involving Trigonometric Functions

Improper integrals involving trigonometric functions (Fourier integrals) of the type

$$\int_{-\infty}^{\infty} f(x)\begin{Bmatrix}\cos mx\\ \sin mx\end{Bmatrix}dx$$

$$\text{i.e., } \int_{-\infty}^{\infty} f(x)\cos mx\,dx \text{ and } \int_{-\infty}^{\infty} f(x)\sin mx\,dx, \quad (17.19)$$

occurring in connection with Fourier integrals may be evaluated by considering the contour integral

$$\oint_C f(z)e^{imz}dz \qquad (m>0) \qquad (17.20)$$

over the closed contour C consisting of
(i) Semicircle S: $z = Re^{i\theta}$, $dz = iRe^{i\theta}\,d\theta$; θ ranging from 0 to π as z traces S in the upper half-plane and
(ii) Line segment L: $[-R, R]$, $z = x$, $dz = dx$; x ranges from $-R$ to R where R is large enough to enclose all poles of the integrand in the upper half-plane and ultimately R is allowed to tend to ∞.

We obtain, by Cauchy's residue theorem

$$\int_C f(z)e^{imz}dz=\int_S f(Re^{i\theta})\exp(imRe^{i\theta})iRe^{i\theta}d\theta+\int_L f(x)e^{imx}dx$$

$$\Rightarrow \lim_{R\to\infty}\int_{-R}^{R} f(x)e^{imx}dx=\int_{-\infty}^{\infty} f(x)e^{imx}dx$$

$$=2\pi i\sum \text{Res}(f(z)e^{imz}) \quad (m>0) \qquad (17.21)$$

where the summation extends to the residues of $f(z)$ e^{imz} at all its poles in the upper half-plane, provided the integral over S namely

$$\left|\int_S f(Re^{i\theta})e^{im(Re^{i\theta})}iRe^{i\theta}d\theta\right| \to 0 \text{ as } R \to \infty \quad (17.22)$$

Now $m > 0$ and S is in the upper half-plane so that $y \geq 0$. Since $z = x + iy$ and

$$|e^{imz}| = |e^{imx+iy}| = |e^{imx}||e^{-my}| = e^{-my} < 1$$

$(\because m > 0, y \geq 0 \text{ and } |e^{imx}| = 1)$

We have $|f(z)e^{imz}| = |f(z)||e^{imz}| \leq |f(z)|$ (17.23)

$$(m > 0, y \geq 0)$$

If $f(z)$ is a rational function $f(z) = \dfrac{p(z)}{q(z)}$ where p, q are polynomials in z such that $(\deg q) \geq (\deg p) + 2$ then $f(z) \sim \dfrac{1}{R^2}$ for large z

$$\left|\int_S f(z)e^{imz}dz\right| \leq \frac{\pi}{R} \quad \text{as } R \to \infty \quad (17.24)$$

Now equating the real and imaginary parts on both sides of Eq. (17.21), we have

$$\int_{-\infty}^{\infty} f(x)\cos mx dx$$
$$= -2\pi \sum Im \text{ Res } [f(z)e^{imz}] (m > 0) \quad (17.25)$$

$$\int_{-\infty}^{\infty} f(x)\sin mx dx$$
$$= 2\pi \sum Re \text{Res } [f(z)e^{imz}] (m > 0) \quad (17.26)$$

Example 17.40

Show that (i) $\displaystyle\int_{-\infty}^{\infty} \frac{\cos mx}{x^2 + a^2} dx = \frac{\pi}{a} e^{-ma}$

(ii) $\displaystyle\int_{-\infty}^{\infty} \frac{\sin mx}{x^2 + a^2} dx = 0.$

Solution Let $f(z) = \dfrac{e^{imz}}{z^2 + a^2}$. We have to evaluate $\displaystyle\int_C \frac{e^{imz}}{z^2 + a^2} dz$ over the closed contour C consisting of

(i) Semicircle S in the upper half-plane and $z = Re^{i\theta}, (0 \leq \theta \leq \pi), dz = iRe^{i\theta}d\theta$ and

(ii) Line segment L along the x-axis = $[-R, R]$ where $z = x$ $dz = dx$, x, varies from $-R$ to R and $R \to \infty$, ultimately.

The integrand has simple poles at $z = \pm ia$ of which $z = ia$ alone lies in the upper half-plane.

The other singularity $z = -ia$ which lies in the lower half-plane is of no interest for us. By Cauchy's residue theorem, we have

$$\int_C f(z)e^{imz}dz = \int_C \frac{e^{imz}}{z^2 + a^2} dz$$
$$= \int_S \frac{e^{imz}}{z^2 + a^2} dz + \int_L \frac{e^{imx}}{x^2 + a^2} dx$$
$$= 2\pi i \sum \text{Res } [f(z)e^{imz}]$$

Now $$\text{Res}_{z=ia}[f(z)e^{imz}] = \text{Res}_{z=ia}\left(\frac{e^{imz}}{z^2 + a^2}\right)$$
$$= \frac{e^{imz}}{z + ia}\Big|_{z=ia} = \frac{e^{-ma}}{2ia}$$

$$\left|\int_S \frac{e^{imz}}{z^2 + a^2} dz\right| \leq \frac{|e^{imz}||e^{-my}}{R^2 - a^2}\pi R$$
$$= \frac{\pi R e^{-my}}{R^2 - a^2}$$
$$\sim \frac{1}{R} \to 0 \quad \text{as} \quad R \to \infty$$

$\therefore$ We have (as $R \to \infty$)

$$\int_{-\infty}^{\infty} \frac{e^{imx}}{x^2 + a^2} dx = 2\pi i . \frac{e^{-ma}}{2ia} = \frac{\pi}{a} e^{-ma}$$

Equating the real and imaginary parts on both sides, we get the results.

Also, $\displaystyle\int_0^{\infty} \frac{\cos mx}{x^2 + a^2} dx = \frac{\pi}{2a} e^{-ma}$ (since the integrand is an even function)

Corollary Replace m by a and a by 1, we get

$$\int_{-\infty}^{\infty} \frac{\cos ax}{x^2 + 1} dx = \pi e^{-a}.$$

Example 17.41

Evaluate $\displaystyle\int_{-\infty}^{\infty} \frac{\cos mx dx}{(x^2 + a^2)(x^2 + b^2)}$ $(a \neq b)$.

Solution Consider $\displaystyle\int_C f(z)e^{imz}dz = \int_S + \int_L$ where C is the closed contour consisting of

(i) Semicircle S in the upper half-plane: $z = Re^{i\theta}$; $dz = iRe^{i\theta}\,d\theta$; θ varying from 0 to π as z traces S.
(ii) Line segment L along the x-axis $[-R, R]$ where $z = x$, $dz = dx$ and x varies from $-R$ to R.

Estimate of the integral on S

Since $|e^{imz}| = |e^{im(x+iy)}| = |e^{imx-my}| = e^{-my}$

$$\leq |(m > 0, y \geq 0)|$$

$$\therefore \left|\int_S \frac{e^{imz}\,dz}{(z^2+a^2)(z^2+b^2)}\right| \leq \frac{\pi R}{(R^2-a^2)(R^2-b^2)}$$

$$\sim \frac{1}{R^3} \to 0 \text{ as } R \to \infty$$

By Cauchy's residue theorem, as $R \to \infty$, Eq. (1) gives

$$\int_{-\infty}^{\infty} \frac{e^{ima}\,dx}{(x^2+a^2)(x^2+b^2)} = \int_C \frac{e^{imz}\,dz}{(z^2+a^2)(z^2+b^2)}$$
$$= 2\pi i \sum \text{Res}\,[f(z)e^{imz}]$$

where the summation extends over residues at all poles of $f(z)e^{imz}$ in the upper half-plane. The integrand has simple poles at $z = \pm ia, \pm ib$ of which $z = ia, ib$ alone lie in the upper half-plane.

$$R_1 = \text{Res}_{z=ia}\,(f(z)e^{imz}) = \left.\frac{e^{imz}}{(z+ia)(z^2+b^2)}\right|_{z=ia}$$

$$= \frac{e^{-ma}}{zia(b^2-a^2)}$$

$$R_2 = \text{Res}_{z=ib}\,(f(z)e^{imz}) = \left.\frac{e^{imz}}{(z+ib)(z^2+a^2)}\right|_{z=ib}$$

$$= \frac{e^{-mb}}{2ib(a^2-b^2)}$$

Sum of residues $= R_1 + R_2$

$$= \frac{i}{2(a^2-b^2)}\left(\frac{e^{-ma}}{a} - \frac{e^{-mb}}{b}\right)$$

$$\int_{-\infty}^{\infty} \frac{\cos mx dx}{(x^2+a^2)(x^2+b^2)} = \frac{-\pi}{a^2-b^2}\left(\frac{e^{-ma}}{a} - \frac{e^{-mb}}{b}\right)$$

$$= \frac{\pi}{a^2-b^2}\left(\frac{e^{-mb}}{b} - \frac{e^{-ma}}{a}\right)$$

Corollary $\displaystyle\int_0^{\infty} \frac{\cos mx}{(x^2+a^2)(x^2+b^2)}\,dx$

$$= \frac{\pi}{2(a^2-b^2)}\left(\frac{e^{-mb}}{b} - \frac{e^{-ma}}{a}\right)$$

17.1.9 Jordan's Lemma

In the foregoing section, we have considered the case of the integrand being a rational function of the form $\dfrac{p(z)}{q(z)}$ where (deg q) ≥ (deg p)+2. There arise situations where this condition is not satisfied. We need a sharper inequality to evaluate such integrals. This is provided by what is known as Jordan's Lemma, which we take up now.

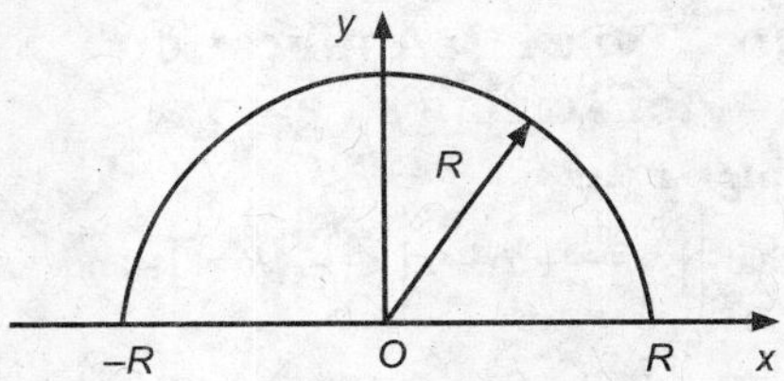

Figure 17.8

Jordan's Lemma: If S is the semicircle in the upper half-plane and $J = \int_S e^{iz}\,dz$, then $|J| \leq \pi(1-e^{-R}) < \pi$.

Proof: On S, we have

$$z = Re^{i\theta}\ (0 \leq \theta \leq \pi) \text{ and } |e^{iz}| = e^{-y} = e^{-R\sin\theta}$$

$$\because |e^{ix}| = 1,\ y = R\sin\theta. \text{ Also } |dz| = Rd\theta$$

$$|J| = \left|\int_S e^{iz}\,dz\right| \leq \int_S |e^{iz}|\,|dz| = \int_0^{\pi} e^{-R\sin\theta} Rd\theta$$

$$= 2\int_0^{\frac{\pi}{2}} e^{-R\sin\theta}\,d\theta \quad \because f(\pi-\theta) = f(\theta)$$

$$\leq 2R\int_0^{\pi/2} e^{-\frac{2R}{\pi}\theta}\,d\theta = 2R\left[e^{-\frac{2R\theta}{\pi}}\left(-\frac{\pi}{2R}\right)\right]_0^{\pi/2}$$

$$= \pi(1-e^{-R}) < \pi \qquad (1)$$

where we have used (2) below
Slope of $OP \geq$ slope of OA

$$\therefore \frac{PM}{OM} \geq \frac{AN}{ON} \Rightarrow \sin(\pi-\theta) = \sin\theta$$

$$0 \leq \frac{\sin\theta}{\theta} \leq \frac{1}{\pi/2} = \frac{2}{\pi}$$

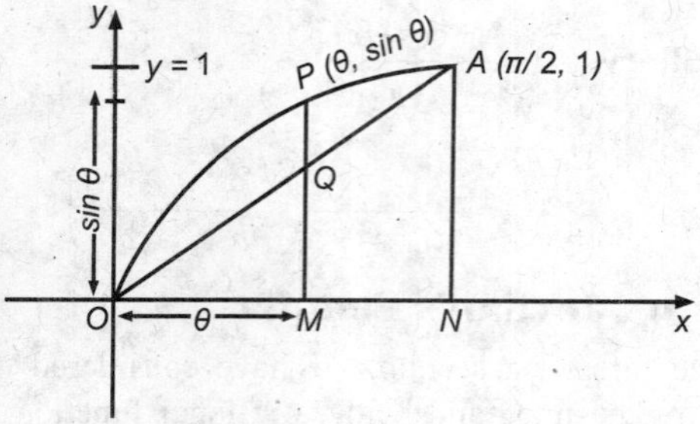

Figure 17.9

$$\Rightarrow \sin\theta \geq \frac{2\theta}{\pi}\ (0 \leq \theta \leq \pi/2) \quad (2)$$

$$\Rightarrow \frac{\sin\theta}{\theta} \geq \frac{1}{\pi/2}$$

17.1.10 Type III: Application of Jordan's Lemma

Example 17.42

Show that $\left|\int_{-R}^{R}\frac{\sin x}{x}dx - \pi\right| < \frac{\pi}{R}$ $(R > 0)$ and deduce that $\int_0^{\infty}\frac{\sin x}{x}dx = \frac{\pi}{2}$.

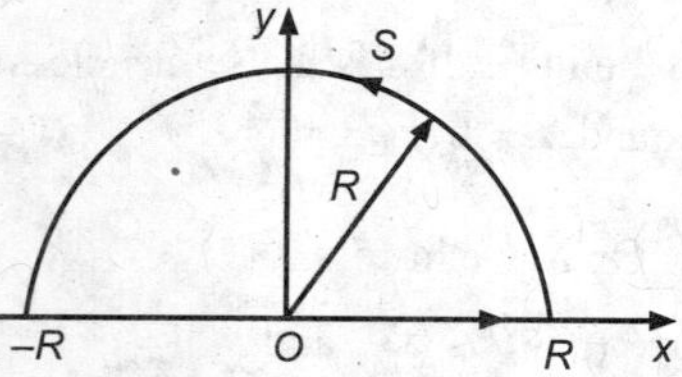

Figure 17.10

Solution Let $f(z) = \frac{e^{iz}-1}{z}$ and consider $\int_C f(z)dz$ where C is the closed contour consisting of the line segment [–R, R] and the semicircle S in the upper half-plane with the line segment [–R, R] as diameter.

$$\text{Now, } \frac{e^{iz}-1}{z} = \frac{1}{z}\left[\left(1+iz+\frac{i^2z^2}{2!}+\frac{i^3z^3}{3!}+\cdots\right)-1\right]$$

$$= i - \frac{1}{2}z - \frac{1}{6}z^2 + \cdots \quad (1)$$

$$\lim_{z\to 0}\frac{e^{iz}-1}{z} = i \quad \text{(finite number).}$$

So, the integrand has a removable singularity at $z = 0$ and the function can be so defined that it is analytic inside C. By Cauchy's theorem, we have

$$\int_C \frac{e^{iz}-1}{z}dz = 0 \Rightarrow \int_{-R}^{R}\frac{e^{ix}-1}{x}dx + \int_S \frac{e^{iz}-1}{z}dz = 0 \quad (2)$$

since there is no singularity inside C.

$$\Rightarrow \int_{-R}^{R}\left(\frac{\cos x - 1}{x} + i\frac{\sin x}{x}\right)dx - \int_S \frac{dz}{z} = \int_S \frac{e^{iz}}{z}dz$$

$$\text{On } S,\ z = Re^{i\theta};\ dz = iRe^{i\theta}d\theta\ (0 \leq \theta \leq \pi) \quad (3)$$

$$\int_S \frac{dz}{z} = \int_0^{\pi}\frac{1}{Re^{i\theta}}i\,\mathrm{Re}^{i\theta}\,d\theta = i\pi \quad (4)$$

Also,

$$\left|\int_S \frac{e^{iz}}{z}dz\right| \leq \int_S \frac{|e^{ix}|\,|e^{-y}|\,|dz|}{|z|} = \frac{1}{R}\int_0^{\pi}e^{-R\sin\theta}\cdot Rd\theta$$

$$= 2\int_0^{\pi/2}e^{-R\sin\theta}d\theta \because f(\pi-\theta) = f(\theta)$$

$$\leq 2\int_0^{\pi/2}e^{\frac{-2R\theta}{\pi}}d\theta = 2e^{-\frac{2R}{\pi}\theta}\left(-\frac{\pi}{2R}\right)\Bigg|_0^{\pi/2}$$

$$\because \sin\theta \geq \frac{2\theta}{\pi}\ (0 \leq \theta \leq \pi/2)$$

$$= \frac{\pi}{R}(1-e^{-R}) < \frac{\pi}{R}, \quad (5)$$

by Jordan's lemma.

Substituting (4) and (5) into (3) we get

$$\left|\left(\int_{-R}^{R}\frac{\cos x - 1}{x}dx\right) + i\left(\int_{-R}^{R}\frac{\sin x}{x}dx - \pi\right)\right|$$

$$\leq \left|-\int_S \frac{e^{iz}}{z}dz\right| < \frac{\pi}{R} \quad (6)$$

$$\Rightarrow \left|\int_{-R}^{R}\frac{\sin x}{x}dx - \pi\right| < \frac{\pi}{R} \quad (7)$$

Taking limits as $R \to \infty$, we get

$$\int_{-\infty}^{\infty}\frac{\sin x}{x}dx = \pi$$

$$\Rightarrow \int_0^{\infty}\frac{\sin x}{x}dx = \pi/2, \quad (8)$$

since the integrand $\frac{\sin x}{x}$ is an even function.

Example 17.43

Show that $\int_0^{\infty}\frac{x\sin mx\,dx}{(x^2+a^2)} = \frac{\pi}{2}e^{-ma}$. [JNTU 2003S]

Solution We integrate $f(z) = \frac{ze^{imz}}{z^2+a^2}$ over the closed contour C shown in Fig. 17.11.

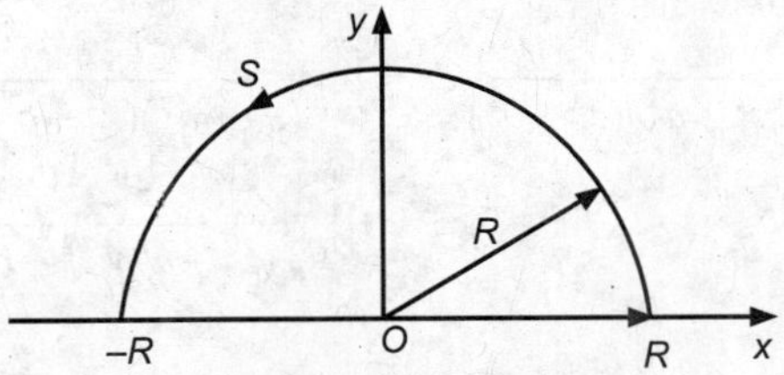

Figure 17.11

It consists of the line segment $L = [-R, R]$ along the real axis and the semicircle S: $z = Re^{i\theta}$ $(0 \le \theta \le \pi)$ in the upper half-plane.

By Cauchy's residue theorem, we have

$$\int_C f(z)dz = \int_{-R}^{R} f(x)dx + \int_S f(z)dz = 2\pi i \sum \text{Res } f(z) \quad (1)$$

The singularities of $f(z) = \dfrac{ze^{imz}}{z^2+a^2}$ are simple poles at $z = \pm ia$ of which $z = ia$ alone lies in the upper half-plane.

$$\text{Res}_{z=ia} f(z) = \lim_{z \to ia} \frac{(z-ia)ze^{imz}}{(z-ia)(z+ia)}$$

$$= \frac{iae^{-ma}}{2ia} = \frac{1}{2}e^{-ma} \quad (2)$$

Now on S $z = Re^{i\theta}$, $dz = iRe^{i\theta}d\theta$;

$$|e^{imz}| = |e^{imx}|e^{-my} = e^{-mR\sin\theta}$$

$$\text{and } \left|\frac{z}{z^2+a^2}\right| \le \frac{R}{R^2-a^2} \quad (3)$$

$$\left|\int_S f(z)dz\right| \le \frac{R^2}{R^2-a^2} \cdot \int_0^{\pi} e^{-mR\sin\theta} d\theta$$

$$= \frac{2R^2}{R^2-a^2} \int_0^{\pi/2} e^{-mR\sin\theta} d\theta$$

$$\because f(\pi-\theta) = f(0)$$

$$\le \frac{2R^2}{R^2-a^2} \int_0^{\pi/2} e^{-\frac{2mR\theta}{\pi}} d\theta$$

$$\because \sin\theta \ge \frac{2\theta}{\pi} \quad (0 \le \theta \le \pi/2)$$

$$= \frac{2R^2}{R^2-a^2} e^{-\frac{2mR}{\pi}\theta} \left(-\frac{\pi}{2mR}\right)\Bigg|_0^{\pi/2}$$

$$= \frac{2R^2}{R^2-a^2} \cdot \frac{\pi}{2mR}(1-e^{-mR})$$

$$= \left(\frac{\pi R}{R^2-a^2}\right)\left(\frac{1-e^{-mR}}{m}\right) \to 0 \text{ as } R \to \infty \quad (4)$$

Letting $R \to \infty$, we obtain from Eqs. (1) – (4)

$$\int_{-\infty}^{\infty} \frac{xe^{imx}}{x^2+a^2} dx = \pi i e^{-ma} \quad (5)$$

Equating the real parts, we have from eq (5)

$$\int_{-\infty}^{\infty} \frac{x\cos mx}{x^2+a^2} dx = 0 \quad (6)$$

This is to be expected since the integrand is an odd function. Again, equating the imaginary parts

$$\int_{-\infty}^{\infty} \frac{x\sin mx}{x^2+a^2} dx = \pi e^{-ma} \quad (7)$$

$$\Rightarrow \int_0^{\infty} \frac{x\sin mx}{x^2+a^2} dx = \frac{\pi}{2}e^{-ma} \quad (8)$$

since the integrand is an even function.

Example 17.44

Evaluate $\displaystyle\int_{-\infty}^{\infty} \frac{x\sin \pi x}{x^2+2x+5} dx.$

Solution We have to evaluate

$$\int_{-\infty}^{\infty} \frac{x\sin \pi x}{x^2+2x+5} dx \quad (1)$$

Consider $\displaystyle\int_C \frac{ze^{i\pi z}}{z^2+2z+5} dz$

$$= \int_{-R}^{R} \frac{xe^{i\pi x}}{x^2+2x+5} dx + \int_{C_R} f(z)dz \quad (2)$$

where C is the part of real axis from $-R$ to $+R$ and the semicircle in the upper half-plane, as shown in Fig. 17.12.

The integrand has simple poles at $z = -1 \pm zi$ but only $z_1 = -1 \pm zi$ lies inside C.

$$\text{Res}(f: -1+zi) = \lim_{z \to z_1} \left\{(z+1-2i)\frac{ze^{i\pi z}}{z^2+2z+5}\right\}$$

$$= (-1+2i)\frac{e^{-i\pi-2\pi}}{4i} \quad (3)$$

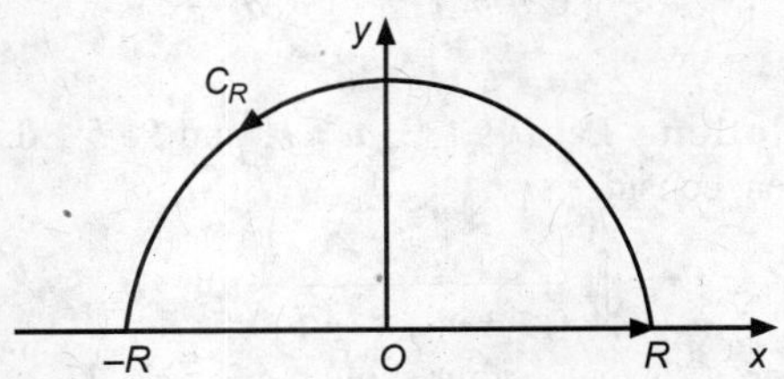

Figure 17.12

By residue theorem,

$$\int_{-R}^{R}\frac{xe^{i\pi x}}{x^2+2x+5}dx+\int_{C_R}f(z)dz=\frac{\pi}{2}(1-2i)e^{-2\pi}$$

$$\Rightarrow \int_{-R}^{R}\frac{x\cos\pi x}{x^2+2x+5}dx+i\int_{-R}^{R}\frac{x\sin\pi x}{x^2+2x+5}dx$$
$$+\int_{-C_R}f(z)dz$$
$$=\frac{\pi}{2}e^{-2\pi}-i\pi e^{-2\pi} \qquad (4)$$

Now $\left|\int_{C_R}f(z)dz\right|$

$$\leq\frac{R}{|R^2-2R-5|}\int_0^{\pi/2}\left|e^{-\pi R\sin\theta}iRe^{i\theta}\right|d\theta$$
$$\because \left|z^2+2z+5\right|\geq R^2-2R-5$$
$$\leq\frac{2R^2}{|R^2-2R-5|}\int_0^{\pi/2}e^{-\pi R\frac{2\theta}{\pi}}d\theta$$
$$\because \sin\theta\geq\frac{2\theta}{\pi}\quad\text{for}\quad 0\leq\theta\leq\frac{\pi}{2}$$
$$=\frac{2R^2}{|R^2-2R-5|}\left(-\frac{1}{2R}\right)[e^{-2R\theta}]_0^{\pi/2}$$
$$=\frac{R}{R^2-2R-5}(1-e^{-\pi R})\to 0 \text{ as } R\to\infty \qquad (5)$$

Allowing $R\to\infty$ we get from Eq. (4), on using Eq. (5), and separating the real and imaginary parts

$$\int_{-\infty}^{\infty}\frac{x\cos\pi x}{x^2+2x+5}dx=\frac{\pi}{2}e^{-2\pi}$$
$$\int_{-\infty}^{\infty}\frac{x\sin\pi x}{x^2+2x+5}dx=-\frac{\pi}{2}e^{-2\pi}.$$

Example 17.45

Show that

$$\int_{-\infty}^{\infty}\frac{x^3e^{ix}}{(x^2+a^2)(x^2+b^2)}dx=\pi i\left(\frac{a^2e^{-a}}{a^2-b^2}+\frac{b^2e^{-b}}{b^2-a^2}\right).$$

Solution Let $a,b\in\mathbb{C}$, $a\neq b$, and $Re\, b>0$. Then, consider

$$\int_C f(z)dz=\int_C\frac{z^{3i}z}{(z^2+a^2)(z^2+b^2)}dz \qquad (1)$$

where C is the contour shown in Fig. 17.13.

$$\int_C\frac{z^3e^{iz}}{(z^2+a^2)(z^2+b^2)}dz=\int_{-R}^{R}\frac{x^3e^{ix}}{(x^2+a^2)(x^2+b^2)}dx$$
$$+\int_{C_R}\frac{z^3e^{iz}}{(z^2+a^2)(z^2+b^2)}dz \qquad (3)$$

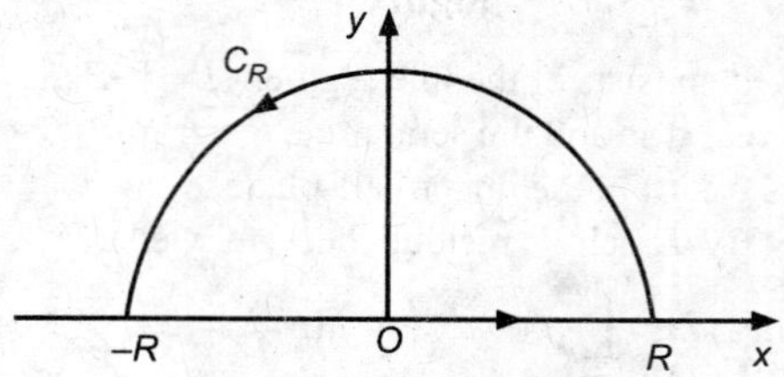

Figure 17.13

The only isolated singularities of the integrand in the upper half-plane are at $z=ia$, $z=ib$.

$$\text{Res}(f:ia)=\left[\frac{z^3e^{iz}}{4z^3+2(a^2+b^2)z}\right]$$
$$=\left[\frac{z^2e^{iz}}{4z^2+2(a^2+b^2)}\right]_{z=ia}$$
$$=\frac{-a^2e^{-a}}{-4a^2+2(a^2+b^2)}$$
$$=\frac{a^2e^{-a}}{2(a^2-b^2)} \qquad (4)$$

$$\text{Res}\,(f:ib)=\frac{-b^2e^{-b}}{-4b^2+2(a^2+b^2)}$$
$$=\frac{b^2e^{-b}}{2(b^2-a^2)} \qquad (5)$$

By Cauchy's residue theorem, we have

$$\int_{-R}^{R}\frac{x^3e^{ix}}{(x^2+a^2)(x^2+b^2)}dx+\int_{C_R}f(z)dz$$
$$=2\pi\sum\text{Res}(f:a_k)$$
$$=\pi i\left(\frac{a^2e^{-a}}{a^2-b^2}+\frac{b^2e^{-b}}{b^2-a^2}\right) \qquad (6)$$

We have $\left|\frac{z^3}{(z^2+a^2)(z^2+b^2)}\right|\leq\frac{M}{R}\,|z|=R\geq R_0$ (7)

$$\therefore \int_{C_R} f(z)dz = \int_{C_R} \frac{M}{R}|e^{iz}||dz| < \frac{M}{R}$$

$\to 0$ as $R \to \infty$ (by Jordan's lemma) (8)

$\therefore$ By letting $R \to \infty$, we have from Eqs. (6) and (8)

$$\int_{-\infty}^{\infty} \frac{x^3 e^{ix}}{(x^2+a^2)(x^2+b^2)} dx = \pi i \left(\frac{a^2 e^{-a}}{a^2-b^2} + \frac{b^2 e^{-b}}{b^2-a^2} \right).$$

17.1.11 Type IV: Poles on the Real Axis (Indentation)

Theorem 17.2 (Simple pole on the real axis.) If $f(z)$ has a simple pole at $z = a$ on the real axis, then

$$\lim_{r\to 0} \int_{C_r} f(z)dz = \pi i\, [\text{Res}_{z=a} f(z)] \quad (17.27)$$

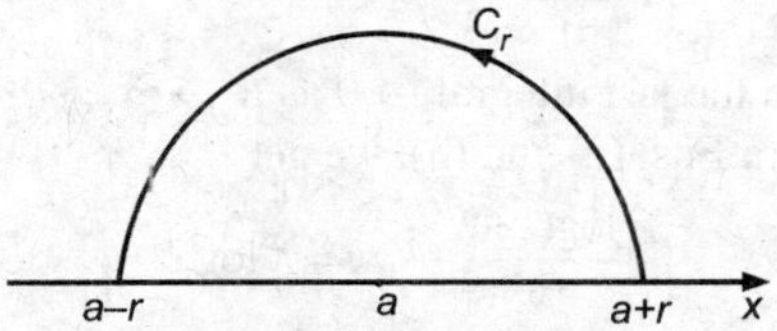

Figure 17.14

Proof: Since $f(z)$ has a simple pole at $z = a$, $f(z)$ has Laurent series

$$f(z) = \frac{b_1}{(z-a)} + g(z), \text{ and } b_1 = \text{Res}_{z=a} f(z) \quad (17.28)$$

where $g(z)$ is analytic on the semicircle of integration (Fig. 17.14)

$$C_r : z = a + re^{i\theta} \qquad 0 \le \theta \le \pi \qquad (17.29)$$

and for all z between C_r and the x-axis

$$\int_{C_r} f(z)dz = b_1 \int_0^{\pi} \frac{1}{re^{i\theta}} i\, r\, e^{i\theta} d\theta + \int_{c_r} g(z)dz$$

$$= b_1 \cdot \pi i$$

since $\left| \int_{C_r} g(z)dz \right| \le M\pi r \to 0,$ as $r \to 0$ by the ML-inequality if $|g(z)| \le M$.

Corollary If b_k $(k = 1, 2, \ldots, n)$ are n simple poles of $f(z)$ on the real axis and C_{r_k} are small semicircles with centre at b_k in the upper half-plane, then

$$\int_{\sum C_{r_k}} f(z)dz = \pi i \sum_{k=1}^{n} \text{Res}_{z=b_k} f(z) = -K$$

where the integral $\int_{C_{r_k}} f(z)dz$ is taken in the **clockwise** direction.

Formula for finding the principal value of an improper integral (the case of *n* simple poles on the real axis)

The principal value (Pr. V.) of improper integral $\int_{-\infty}^{\infty} f(x)dx$ is the limit $\lim_{R\to\infty} \int_{-R}^{R} f(x)dx$ if the limit existsandthePr.V.oftheimproperintegral $\int_{a-r}^{a+r} f(x)dx$, where $x = a$ is a pole is the limit $\lim_{r\to 0} \int_{a-r}^{a+r} f(x)dx$ if the limit exists, where $f(x) = \frac{p(x)}{q(x)}$ is a rational function with (deg q) $\ge$ (deg p) + 2.

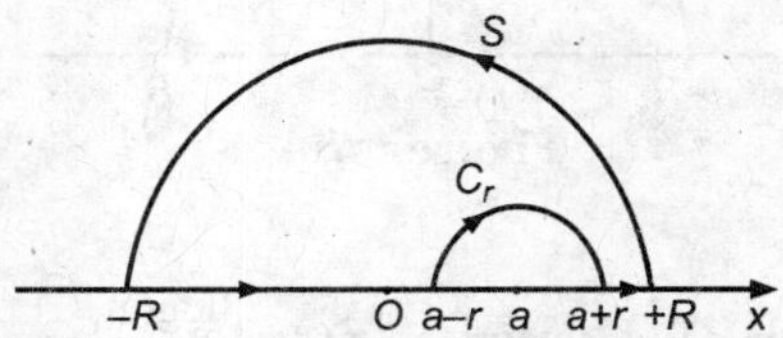

Figure 17.15 The case of n simple poles on the real axis

For sufficiently large R, the integral over the closed contour C consisting of the semicircles S and C_r and the line segments $[-R, a - r]$ and $[a + r, R]$ along the x-axis (Fig. 17.15) has the value J and

$$J = \int_C f(z)dz = 2\pi i \sum_{j=1}^{m} \text{Res}_{z=aj} f(z) \qquad (17.30)$$

where $z = a_j$ $(j = 1, 2, \cdots m)$ are singularities of $f(z)$ in the upper half-plane. Now we have

$$\int_S f(z)dz \to 0 \text{ as } R \to \infty \qquad (17.31)$$

$$\because (\deg q) \ge (\deg p) + 2$$

As $r \to 0$ the integral over C_r, taken in the clockwise direction,

$$\int_{C_r} f(z)dz \to K = -\pi i \sum_{k=1}^{n} \text{Res}_{z=b_k} f(z), \quad (17.32)$$

for each b_k (by Theorem 17.2). It follows that

$$\text{Pr. V.} \int_{-\infty}^{\infty} f(x)dx = 2\pi i \sum_{j=1}^{m} \text{Res}_{z=a_j} f(z) + \pi i \sum_{k=1}^{n} \text{Res}_{z=b_k} f(x) \quad (17.33)$$

where the first sum extends over all poles in the upper half-plane and the second over all poles (assumed to be simple) on the real axis.

Example 17.46

Prove that $\int_0^{\infty} \frac{\log(1+x^2)}{1+x^2} dx = \pi \log 2.$

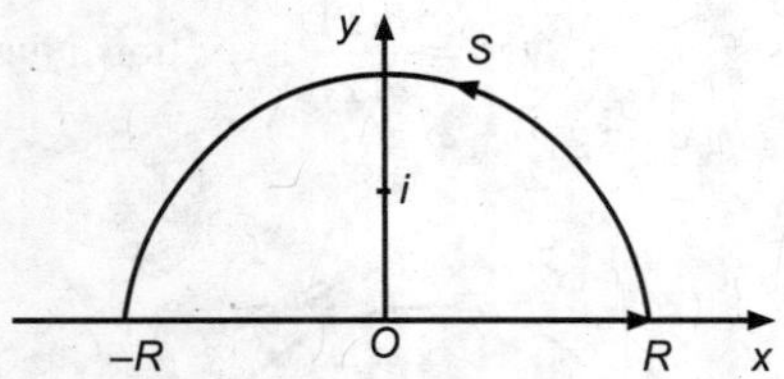

Figure 17.16

Solution We integrate $f(z) = \frac{\log(z+i)}{z^2+1}$ over the closed contour C consisting of the line segment $L = [-R, R]$ along the real axis and the semicircle $S: z = Re^{i\theta}$ $(0 \le \theta \le \pi)$ in the upper half-plane.

The only pole of $f(z)$ inside C is the simple pole at $z = i$.

$$\text{Res}_{z=i} f(z) = \lim_{z \to i} \frac{(z-i)\log(z+i)}{(z-i)(z+i)} = \frac{\log 2i}{2i} = \frac{1}{2i}\left(\log 2 + \frac{\pi i}{2}\right) \quad (1)$$

$$\because \log 2i = \log 2 + \log e^{i\pi/2} = \log 2 + \frac{i\pi}{2} \quad \text{(Pr. V.)}$$

By residue theorem,

$$\int_C \frac{\log(z+i)}{z^2+1} dz = 2\pi i \cdot \frac{1}{2i}\left(\log 2 + \frac{i\pi}{2}\right) = \pi \log 2 + \frac{i\pi^2}{2}, \text{[by Eq. (1)]} \quad (2)$$

$$\text{LHS} = \int_{-R}^{0} \frac{\log(x+i)}{x^2+1} dx + \int_0^R \frac{\log(x+i)}{x^2+1} dx + \int_S \frac{\log(z+i)}{z^2+1} dz$$

$$= \int_0^R \frac{\log(-x+i)}{(-x)^2+1} dx + \int_0^R \frac{\log(x+i)}{x^2+1} dx + \int_S \frac{\log(z+i)}{z^2+1} dz$$

$$= \int_0^R \frac{\log(-x+i)(x+i)}{x^2+1} dx + 0,$$

$$\text{if } \int_S \frac{\log(z+i)}{z^2+1} dz \to 0 \text{ as } R \to \infty$$

$$= \int_0^R \frac{\log(1+x^2)+i\pi}{1+x^2} dx$$

$$= \int_0^R \frac{\log(1+x^2)}{1+x^2} + i\pi \cdot \frac{\pi}{2}$$

provided the last integral $\to 0$ as $R \to \infty$.

From Eqs. (2) and (3), we get

$$\int_0^{\infty} \frac{\log(1+x^2)}{1+x^2} dx = \pi \log 2.$$

On S, $z = Re^{i\theta}$, $dz = iRe^{i\theta} d\theta$ $(0 \le \theta \le \pi)$

$$\left| \int_S \frac{\log(z+i)}{z^2+1} dz \right| \le \frac{\log(R+1)}{R^2-1} \pi R$$

$$\to 0 \quad \text{as} \quad R \to \infty$$

$$\because \frac{\log|x|}{|x|} \to 0 \text{ as } |x| \to \infty.$$

Example 17.47

Find the principal value

$$\text{Pr. V.} \int_{-\infty}^{\infty} \frac{dx}{(x^2-3x+2)(x^2+1)}.$$

Solution Let $f(z) = \frac{1}{(z^2-3z+2)(z^2+1)}$ and C be the closed contour shown in Fig. 17.17.

$f(z)$ has simple poles on the real axis at $z = 1$ and $z = 2$ and a simple pole at $z = i$ inside C (in the upper half-plane). The other singularity of $f(z)$ at $z = -i$ is in the lower half-plane and is thus outside C.

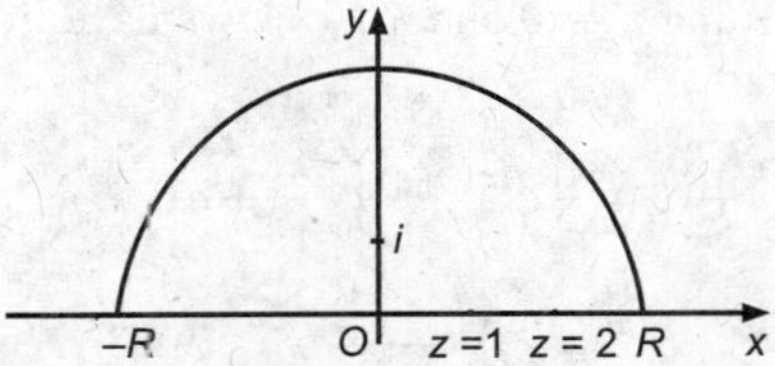

Figure 17.17

$$\text{Res}_{z=1}\quad f(z)=\left.\frac{1}{(z-2)(z^2+1)}\right|_{z=1}=-\frac{1}{2}$$

$$\text{Res}_{z=2}\quad f(z)=\left.\frac{1}{(z-1)(z^2+1)}\right|_{z=2}=\frac{1}{5}$$

$$\text{Res}_{z=i}\quad f(z)=\left.\frac{1}{(z^2-3z+2)(z+i)}\right|_{z=i}$$

$$=\frac{1}{(1-3i)2i}=\frac{1}{6+2i}\cdot\frac{3-i}{3-i}$$

$$=\frac{3-i}{20}$$

Pr. V. $\displaystyle\int_{-\infty}^{\infty}\frac{dx}{(x^2-3x+2)(x^2+1)}$

$$=2\pi i\left(\frac{3-i}{20}\right)+\pi i\left(-\frac{1}{2}+\frac{1}{5}\right)=\pi/10.$$

Example 17.48

Show that $\displaystyle\int_0^{\infty}\frac{\log^2 x}{1+x^2}dx=\frac{\pi^3}{8},\ \int_0^{\infty}\frac{\log x}{1+x^2}dx=0.$

Solution We will prove that

$$\int_0^{\infty}\frac{\log^2 x}{1+x^2}dx=\frac{\pi^3}{8} \tag{1}$$

Let $I=\displaystyle\int_C\frac{\log^2 z}{1+z^2}dz$ where C is the indented semicircle of Fig. 17.18 with inner radius $\varepsilon<1$ and outer radius $R>1$. Let $\log z = Log\, z$, the principal value on C. Then the integrand is analytic except for a simple pole at $z=i$ inside C and hence by Cauchy's residue theorem,

$$I=2\pi i\frac{\log^2 i}{2i}=\pi\left(\frac{i\pi}{2}\right)^2=\frac{-\pi^3}{4} \tag{2}$$

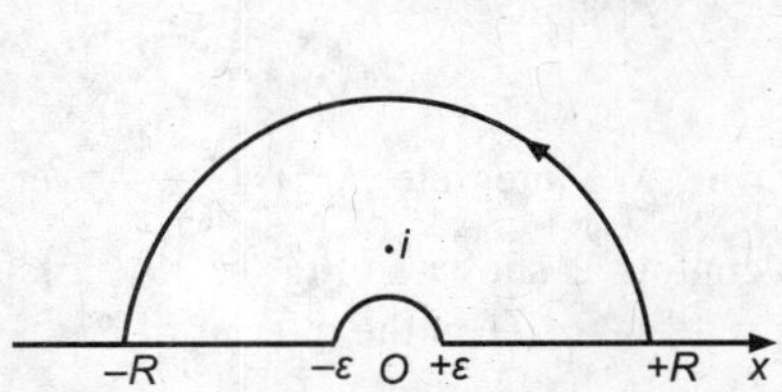

Figure 17.18

on the negative real axis $z=re^{i\pi}$. Hence

$$I=\int_{\in}^{R}\frac{\log^2 x}{1+x^2}dx+\int_{R}^{\in}\frac{(Log\ re^{i}\pi)^2 e^{i\pi}dx}{1+r^2}+J_1+J_2 \tag{3}$$

where J_1 and J_2 are the integrals over the parts of C where $|z|=R$ and $|z|=\in$, respectively. Since $\text{Log}\ re^{i\pi}=\text{Log}\ r+i\pi$ we have

$$(\text{Log}\ re^{i\pi})^2=\text{Log}\ ^2 r+2i\pi\,\text{Log}\ r-\pi^2$$

and setting $r=x$ in the second integral above, we have

$$I=\int_{\in}^{R}\frac{2\log^2 x+2\pi i\log x-\pi^2}{1+x^2}dx+J_1+J_2 \tag{4}$$

On J_1, we have $z=Re^{i\theta}$, $0\le\theta\le\pi$ and so the integrand satisfies

$$\left|\frac{(\log^z)^2}{1+z^2}\right|=\frac{|\log R+i\theta|^2}{|1+z^2|}\le\frac{\log^2 R+\pi^2}{R^2-1}$$

$$\therefore |J_1|\le\pi R\frac{\log^2 R+\pi^2}{R^2-1}$$

Since $(\log R)^2/R\to 0$ as $R\to\infty$. It follows that $J_1\to 0$ as $R\to\infty$. A similar calculation with $\in$ replacing R shows that $J_2\to 0$ as $\in\to 0$. Letting $R\to\infty$ and $\in\to 0$ in Eq. (4) therefore gives

$$\frac{-\pi^3}{4}=\int_0^{\infty}\frac{2\log^2 x+2\pi i\log x}{1+x^2}dx-\pi^2\int_0^{\infty}\frac{dx}{1+x^2}$$

when we recall that $I=-\pi^2/4$ by Eq. (2). The second integral is $\pi/2$ and Eq. (1) follows, by equating real and imaginary parts.

Example 17.49

Show that $\displaystyle\int_0^{\infty}\frac{x^{p-1}}{1+x}dx=\frac{\pi}{\sin p\pi}\quad(0<p<1).$

Solution We integrate $f(z)=\dfrac{z^{p-1}}{1+z}$ over the closed contour C shown in Fig. 17.19, $f(z)$ has a simple pole at $z=-1$ in the interior of C

$$\text{Res}_{z=-1}f(z)=\lim_{z\to -1}\frac{(z+1)z^{p-1}}{(1+z)}=(e^{\pi i})^{p-1} \quad (1)$$

$$=e^{(p-1)\pi i} \quad \because e^{\pi i}=-1$$

By Cauchy's residue theorem, we have

$$\int_C f(z)dz=\int_{AB}f+\int_{C_R}f+\int_{GH}f+\int_{c_r}f$$

$$=2\pi i e^{(p-1)\pi i} \quad (2)$$

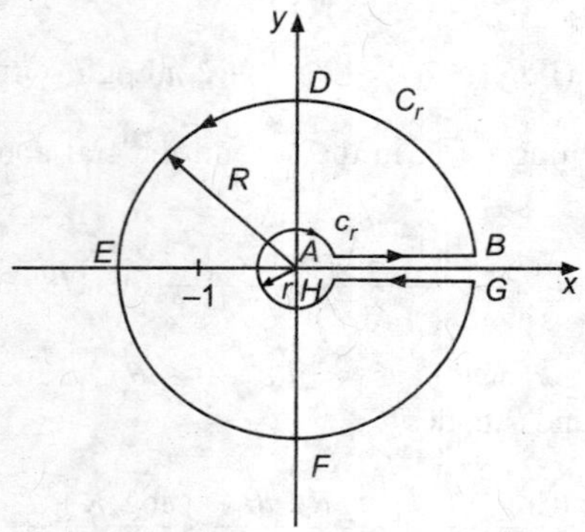

Figure 17.19

Since AB and GH are arbitrarily close to the x-axis and $z=xe^{2\pi i}$ along GH, we have

$$\int_r^R \frac{x^{p-1}}{1+x}dx+\int_0^{2\pi}\frac{(\text{Re}^{i\theta})^{p-1}i\,\text{Re}^{i\theta}}{1+\text{Re}^{i\theta}}d\theta$$
$$+\int_R^r\frac{(xe^{2\pi i})^{p-1}}{1+xe^{2\pi i}}dx+\int_{2\pi}^0\frac{(re^{i\theta})^{p-1}ire^{i\theta}}{1+re^{i\theta}}d\theta$$
$$=2\pi i e^{(p-1)\pi i} \quad (3)$$

Now the absolute values of the second and fourth integrals

$$\left.\begin{aligned}\left|\int_0^{2\pi}\frac{R^p e^{ip\theta}id\theta}{1+\text{Re}^{i\theta}}\right| &\le \frac{2\pi R^p}{R-1}\to 0 \text{ as } R\to\infty\\ \left|\int_{2\pi}^0\frac{r^p e^{ip\theta}id\theta}{1+re^{i\theta}}\right| &\le \frac{2\pi R^p}{1-r}\to 0 \text{ as } r\to 0\end{aligned}\right\} \quad (4)$$

Letting $r\to 0$ and $R\to\infty$, we have from Eqs. (3) and (4)

$$\int_0^\infty \frac{x^{p-1}}{1+x}dx-e^{2\pi 1(p-1)}\int_0^\infty\frac{x^{p-1}}{1+x}dx$$
$$=(1-e^{2\pi i(p-1)})\int_0^\infty\frac{x^{p-1}}{1+x}dx$$
$$=2\pi i e^{(p-1)\pi i}$$
$$\Rightarrow \int_0^\infty\frac{x^{p-1}}{1+x}dx=\frac{2\pi i e^{(p-1)\pi i}}{1-e^{2\pi i(p-1)}}$$
$$=\frac{2\pi i}{e^{p\pi i}-e^{-p\pi i}}=\frac{\pi}{\sin p\pi}$$
$$e^{2\pi i(p-1)}=e^{2\pi p i}\cdot e^{-2\pi i}=e^{2\pi p i} \quad \because e^{-2\pi i}=1$$
$$e^{(p-1)\pi i}=e^{p\pi i}\cdot e^{-\pi i}=-e^{p\pi i} \quad \because e^{-\pi i}=-1.$$

Integration around rectangular/sectorial contours

Example 17.50

By integrating e^{-z^2} around the boundary of the rectangle with vertices $-R$, R, $R+ib$, $-R+ib$ letting $R\to\infty$ and using $\int_0^\infty e^{-z^2}dz=\sqrt{\pi}/2$ show that $\int_0^\infty e^{-x^2}\cos 2bx dx=\frac{\sqrt{\pi}}{2}e^{-b^2}$ (heat conduction problem).

Solution $f(z)=e^{-z^2}$ has no poles inside the contour C (Fig. 17.20). By Cauchy's residue theorem, we have

$$\Rightarrow \int_C e^{-z^2}dz=0$$
$$\Rightarrow \int_{-R}^R e^{-x^2}dx+\int_0^b e^{-(R+iy)^2}idy+\int_R^{-R}e^{-(x+ib)^2}dx$$
$$+\int_b^0 e^{-(-R+iy)^2}idy=0$$

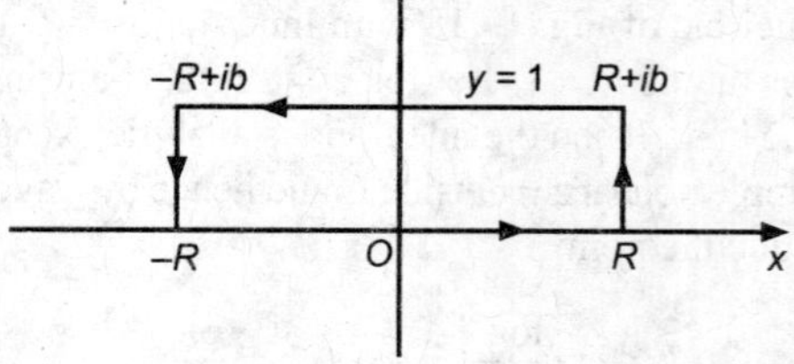

Figure 17.20

As $R \to \infty$

$$I_1 = \int_{-\infty}^{\infty} e^{-x^2} dx = 2\int_0^{\infty} e^{-x^2} dx = \sqrt{\pi} \text{ (given)}$$

$$I_2 = \int_0^b e^{-R^2-2iRy+y^2}\, idy$$

$$= ie^{-R^2}\int_0^b e^{-2iRy+y^2} dy \to 0 \quad \text{as} \quad R \to \infty$$

$$I_3 = \int_{\infty}^{-\infty} e^{x^2+b^2} \cdot e^{-2ibx} dx$$

$$= -e^{b^2}\int_{-\infty}^{\infty} e^{-x^2}(\cos 2bx - i\sin 2bx)dx$$

$$I_4 = \int_b^0 e^{-R^2+iRy+y^2}\, idy$$

$$= ie^{-R^2}\int_b^0 e^{2iRy+y^2} dy \to 0 \quad \text{as} \quad R \to \infty$$

(interchanging limits and changing the sign). Equating the real parts

$$\therefore \sqrt{\pi} - e^{b^2}\int_{-\infty}^{\infty} e^{-x^2}\cos 2bx \; dx = 0$$

$$\Rightarrow \int_{-\infty}^{\infty} e^{-x^2}\cos 2bx \; dx = \sqrt{\pi}e^{-b^2}.$$

Example 17.51

Prove that $\displaystyle\int_0^{\infty} \frac{\cosh ax}{\cosh x} dx = \frac{\pi}{2\cos(\pi a/2)}$ $(|a|<1)$.

Solution Let $f(z) = \dfrac{e^{az}}{\cosh z}$ and consider

$$\int_C f(z)dz \tag{1}$$

where C is the rectangle with vertices at $-R$, R, $R+\pi i$, $-R+\pi i$ as shown in Fig. 17.21.

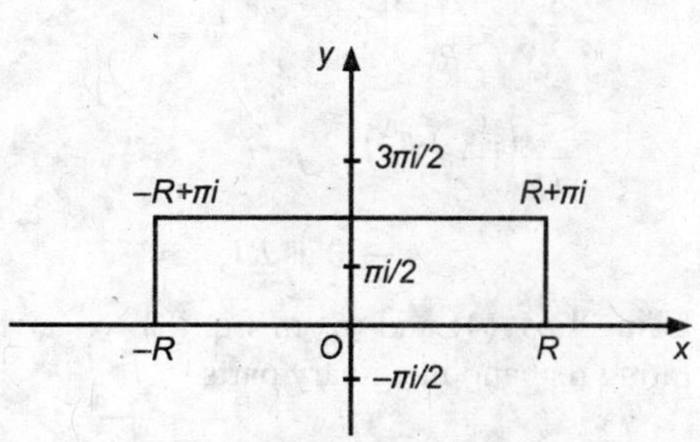

Figure 17.21

The poles are simple and occur where $\cosh z = 0$ *i.e.*, $z = \left(n+\frac{1}{2}\right)\pi i$, $n = 0, \pm 1, \pm 2, \ldots$. The only pole inside C is $\pi i/2$.

$$\text{Res}(f:\pi i/2) = \lim_{z\to \pi i/2}(z-\pi i/2)\frac{e^{az}}{\cosh z} = \frac{e^{a\pi i/2}}{\sinh \pi i/2}$$

$$= \frac{e^{a\pi i/2}}{i\sin \pi/2} = -ie^{a\pi i/2} \tag{2}$$

By residue theorem, we have

$$\int_C f(z)dz = 2\pi i(-ie^{a\pi i/2}) = 2\pi e^{a\pi/2}$$

$$\Rightarrow \int_{-R}^{R} \frac{e^{ax}}{\cosh x}dx + \int_0^{\pi}\frac{e^{a(R+iy)}}{\cosh(R+iy)}idy$$

$$+ \int_R^{-R}\frac{e^{a(x+\pi i)}}{\cosh(x+\pi i)}dx + \int_{\pi}^{0}\frac{e^{a(-R+iy)}}{\cosh(-R+iy)}idy$$

$$= 2\pi e^{a\pi i/2} \tag{3}$$

Denoting the integrals on the LHS of Eq. (3) by I_1, I_2, I_3 and I_4, we find that I_2 and $I_4 \to 0$ as $R \to \infty$ as shown below:

$$\because |\cosh(R+iy)| = \left|\frac{e^{R+iy}+e^{-R-iy}}{2}\right|$$

$$\geq \frac{1}{2}(|e^{R+iy}| - |e^{-R-iy}|)$$

$$= \frac{1}{2}(e^R - e^{-R}) \geq \frac{1}{4}e^R \tag{4}$$

$$\therefore |I_2| = \left|\int_0^{\pi}\frac{e^{a(R+iy)}}{\cosh(R+iy)}idy\right|$$

$$\leq \int_0^{\pi}\frac{e^{aR}}{\frac{1}{4}e^R}dy = 4\pi e^{(a-1)R}$$

$$\to 0 \text{ as } \quad R\to\infty \quad \text{if } |a|<1 \tag{5}$$

Similarly, we can prove that the fourth integral I_4 also approaches 0 as $R \to \infty$

$$\text{Also } \cosh(x+\pi i) = \frac{e^x\cdot e^{\pi i} + e^{-x}\cdot e^{-\pi i}}{2}$$

$$= -\frac{e^x + e^{-x}}{2} = -\cosh x$$

Now let $r \to \infty$ so, I_1, $I_3 \to 0$. In I_3 replace $\cosh(x + \pi i)$ by $-\cosh x$ and interchange its limits. Therefore, we get from Eq. (3)

$$I_1 + I_3 = \int_{-\infty}^{\infty} \frac{e^{ax}}{\cosh x} dx + e^{a\pi i} \int_{-\infty}^{\infty} \frac{e^{ax}}{\cosh x} dx \tag{6}$$

$$= 2\pi e^{a\pi i/2}$$

$$\Rightarrow \int_{-\infty}^{\infty} \frac{e^{ax}}{\cosh x} dx = \frac{2\pi e^{a\pi i/2}}{1+e^{a\pi i}} = \frac{2\pi}{e^{a\pi i/2} + e^{-a\pi i/2}}$$

$$= \frac{\pi}{\cos \pi a/2} \tag{7}$$

$$\text{Now LHS} = \int_{-\infty}^{0} \frac{e^{ax}}{\cosh x} dx + \int_{0}^{\infty} \frac{e^{ax}}{\cosh x} dx$$

$$= \int_{0}^{\infty} \frac{e^{-ax}}{\cosh - x} dx + \int_{0}^{\infty} \frac{e^{ax}}{\cosh x} dx$$

$$= 2\int_{0}^{\infty} \frac{\cosh ax}{\cosh x}$$

$$\Rightarrow \int_{0}^{\infty} \frac{\cosh ax}{\cosh x} dx = \frac{\pi}{2\cos a\pi/2} \tag{8}$$

Miscellaneous definite integrals

Example 17.52

Prove that $\int_0^{\infty} \sin x^2 dx = \int_0^{\infty} \cos x^2 dx = \frac{1}{2}\sqrt{\pi/2}$.

Solution Let C be the contour as shown in Fig. 17.22. By Cauchy's theorem, we have

$$\int_C e^{iz^2} dz = 0$$

$$\Rightarrow \int_{OA} e^{iz^2} dx + \int_{\widehat{AB}} e^{iz^2} dz + \int_{BO} e^{iz^2} dz = 0 \tag{1}$$

On OA, $z = x$ from $x = 0$ to $x = R$

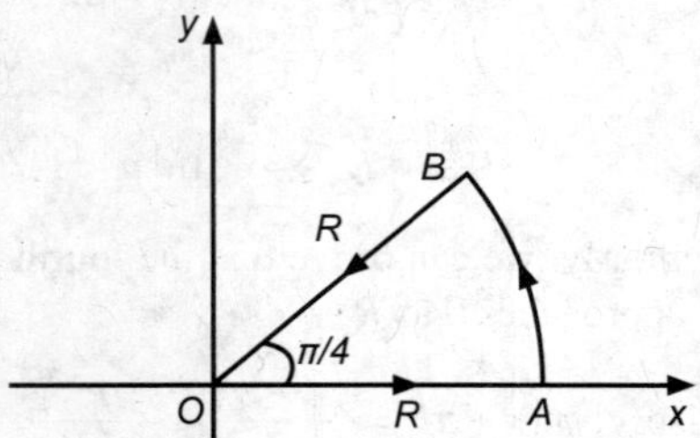

Figure 17.22

On $\widehat{AB}$ $z = Re^{i\theta}$ from $\theta = 0$ to $\theta = \pi/4$ and

On BA, $z = re^{\pi/4}$ from $r = R$ to $r = 0$

Hence from Eq. (1)

$$\int_0^R (\cos x^2 + i\sin x^2)dx + \int_0^{\pi/4} e^{iR^2(\cos 2\theta + i\sin 2\theta)} iRe^{i\theta} d\theta$$

$$+ \int_R^0 e^{ir} e^{i\pi/2} e^{\pi i/4} dx = 0 \tag{2}$$

$$\Rightarrow \int_0^R (\cos x^2 + i\sin x^2)dx$$

$$= e^{\pi i/4} \int_0^R e^{-r^2} dr - \int_0^{\pi/4} e^{iR^2 \cos 2\theta} e^{-R^2 \sin 2\theta} iRe^{i\theta} d\theta \tag{3}$$

As $R \to \infty$,

$$\int_0^R e^{-r^2} dr \to \int_0^{\infty} e^{-r^2} dr = \frac{\Gamma\left(\frac{1}{2}\right)}{2}$$

$$= \frac{1}{2}\sqrt{\pi}$$

$$\therefore e^{\pi i/4} \int_0^{\infty} e^{-r^2} dr = \frac{1}{2}\sqrt{\frac{\pi}{2}} + \frac{i}{2}\sqrt{\frac{\pi}{2}} \tag{4}$$

$$\text{Also, } \left| \int_0^{\pi/4} e^{iR^2 \cos 2\theta} e^{-R^2 \sin 2\theta} iRe^{i\theta} d\theta \right|$$

$$\le R\int_0^{\pi/4} e^{-R^2 \sin 2\theta} d\theta \quad \because \left|e^{iR^2 \cos 2\theta}\right| = \left|e^{i\theta}\right| = 1$$

$$= \frac{R}{2}\int_0^{\pi/2} e^{-R^2 \sin\phi} d\phi, \qquad 2\theta = \phi$$

$$\le \frac{R}{2}\int_0^{\pi/2} e^{-2R^2 \phi\pi} d\phi$$

$$\left[\therefore \sin\phi \ge \frac{2\phi}{\pi} \quad \text{for} \quad 0 \le \phi \le \pi/2\right]$$

$$= \frac{R}{2}\left(-\frac{\pi}{2R^2}\right)\left[e^{-2R^2\phi/\pi}\right]_0^{\pi/2}$$

$$= \frac{\pi}{4R}\left(1 - e^{-R^2}\right) \tag{5}$$

$$\to 0 \text{ as } R \to \infty$$

Using Eqs. (4) and (5) in Eq. (3) we have on separating real and imaginary parts

$$\int_0^{\infty} \cos x^2 dx = \int_0^{\infty} \sin x^2 dx = \frac{1}{2}\sqrt{\frac{\pi}{2}} \tag{6}$$

EXERCISE 17.3

Evaluate the following integrals:

1. $\int_0^\infty \frac{dx}{x^4+a^4}$. [JNTU 2003S (4)]

Ans: $\frac{\pi}{2\sqrt{2}a^3}$

2. $\int_0^\infty \frac{dx}{1+x^2}$. [JNTU 2004S, 2005S, 2006 S (3)]

Ans: $\int_0^\infty \frac{x \sin mx}{x^4+16} dx$

3. $\int_0^\infty \frac{x^2 dx}{(x^3+1)(x^2+4)}$.

Ans: $\frac{\pi}{6}$

4. $\int_0^\infty \frac{\cos x dx}{(1+x^2)^2}$. [JNTU 2003(3)]

Ans: $\frac{\pi}{2e}$

5. $\int_0^\infty \frac{x \sin mx}{x^4+16} dx$. [JNTU 2005 S (1)]

Ans: $\frac{\pi}{2a^2} e^{-\frac{am}{\sqrt{2}}} \sin\left(\frac{am}{\sqrt{2}}\right)$

6. $\int_0^\infty \frac{\cos mn}{(a^2+x^2)^2} dx$.

Ans: $\frac{\pi}{4a^3}(1-ma)e^{-ma} \quad m>0, a>0$

7. $\int_0^\infty \frac{\sin mx}{x} dx \ (m>0)$. [JNTU 2001]

Ans: $\frac{\pi}{2}$

8. $\int_{-\infty}^\infty \frac{e^{ax}}{e^x+1} dx$. [JNTU 2001]

Ans: $\frac{\pi}{\sin a\pi}$

9. $\int_0^\infty \frac{x^2 dx}{x^4+x^2+1}$.

Ans: $\frac{\pi}{2\sqrt{3}}$

10. $\int_0^\infty \frac{x^2 dx}{x^6+1}$.

Ans: $\frac{\pi}{6}$

11. $\int_{-\infty}^\infty \frac{x^2 dx}{(x^2+1)(x^2+4)}$.

[JNTU 2001 S, 2003 S, 2004 S]

Ans: $\frac{\pi}{3}$

12. $\int_{-\infty}^\infty \frac{\sin x dx}{x^2+4x+5}$.

Ans: $\frac{-\pi \sin 2}{e}$

13. $\int_{-\infty}^\infty \frac{\cos x dx}{(x^2+a^2)(x^2+b^2)} \quad (a>b>0)$.

[JNTU 2004 S (2)]

Ans: $\frac{\pi}{(a^2-b^2)}\left(\frac{e^{-b}}{b}-\frac{e^{-a}}{a}\right)$

14. $\int_0^\infty \sin x^2 dx = \int_0^\infty \cos x^2 dx$.

Ans: $\frac{\sqrt{\pi}}{2\sqrt{2}}$

15. $\int_0^\infty \frac{x \sin x}{x^2+a^2} \quad (a>0)$.

Ans: $\frac{\pi}{2}e^{-a}$

Argument Principle and Rouche's Theorem

18

18.1 Introduction

In this chapter, we state and prove the argument principle and Rouche's theorem, which are consequences of Cauchy's residue theorem.

We shall also prove the fundamental theorem of algebra, which we have been using without proof since the beginning of the study of algebra.

The argument principle helps us find the number of zeros or poles of a function in a given region and is useful in determining the stability criteria of linear systems.

The argument principle applies to a class of complex functions called meromorphic functions defined in a domain D whose only singularities in D are poles and so we define meromorphic functions.

18.2 Meromorphic Function

Definition A function f of the complex variable z in a domain D is said to be meromorphic in D if it is analytic in D except for poles, *i.e.*, the only singularities of f are poles in D.

Examples

1. Rational function of the form $\dfrac{p(z)}{q(z)}$ where p and q are polynomials such that q is not a constant function. $\dfrac{(z^2+1)}{(z^2-1)(z^2-5)}$; $\dfrac{z}{(z-1)(z+2)}$
2. Trigonometric functions: $\tan z$, $\cot z$, $\sec z$ and cosec z
3. Hyperbolic functions: $\tanh z$, $\coth z$, $\operatorname{sech} z$ and $\operatorname{cosech} z$

18.3 Argument Principle (Repeated Single Pole/Zero)

Theorem 18.1 Let $f(z)$ be meromorphic within and on a simple closed curve C except for a pole $z = a$ of order (multiplicity) p inside C. Suppose that $f(z) \neq 0$ on C but has a zero $z = b$ of order n inside C. Then

$$\frac{1}{2\pi i}\int_C \frac{f'(z)}{f(z)}dz = n - p. \tag{18.1}$$

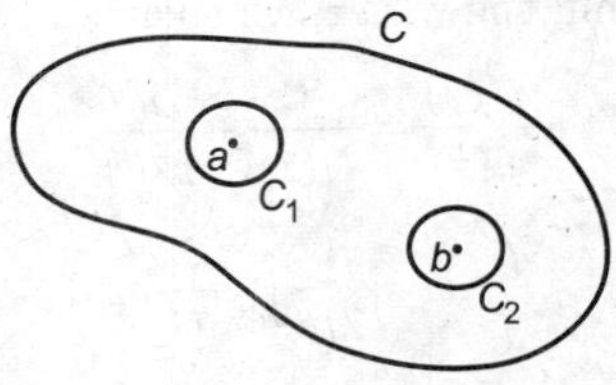

Figure 18.1

Proof: If we enclose $z = a$ and $z = b$ in small non-overlapping circles C_1 and C_2, respectively, then

$$\frac{1}{2\pi i}\int_C \frac{f'(z)}{f(z)}dz = \frac{1}{2\pi i}\int_{C_1} \frac{f'}{f}dz + \frac{1}{2\pi i}\int_{C_2} \frac{f'}{f}dz \tag{18.2}$$

$f(z)$ has a pole of order p at $z = a$. So, we can write

$$f(z) = \frac{F(z)}{(z-a)^p} \tag{18.3}$$

where $F(z)$ is analytic and non-vanishing on and inside C_1. Taking logs on both sides of Eq. (18.3) and differentiating w.r.t z, we obtain

$$\frac{f'(z)}{f(z)} = \frac{F'(z)}{F(z)} - \frac{p}{z-a} \tag{18.4}$$

$$\Rightarrow \frac{1}{2\pi i}\int_{C_1}\frac{f'}{f}dz = \frac{1}{2\pi i}\int_{C_1}\frac{F'}{F}dz$$

$$-\frac{p}{2\pi i}\int_{C_1}\frac{dz}{z-a} \quad (18.5)$$

Now, $\frac{F'}{F}$ is analytic in C_1 since F and hence F' are analytic in C_1, and by Cauchy's integral theorem

$$\int_{C_1}\frac{F'}{F}dz = 0. \quad \text{Also} \quad \int_{C_1}\frac{dz}{z-a} = 2\pi i$$

Therefore, Eq. (18.5) reduces to

$$\frac{1}{2\pi i}\int_{C_1}\frac{f'}{f}dz = 0 - \frac{p}{2\pi i}2\pi i = -p \quad (18.6)$$

Now $f(z)$ has a zero of order n at $z = b$. So, we can write

$$f(z) = (z-b)^n G(z) \quad (18.7)$$

where $G(z)$ is analytic and non-vanishing on and inside C_1. Taking logs on both sides of Eq. (18.7) and differentiating w.r.t z, we obtain

$$\frac{f'(z)}{f(z)} = \frac{G'(z)}{G(z)} + \frac{n}{z-b}$$

$$\Rightarrow \frac{1}{2\pi i}\int_{C_2}\frac{f'}{f}dz = \frac{1}{2\pi i}\int_{C_2}\frac{G'}{G}dz$$

$$+\frac{n}{2\pi i}\int_{C_2}\frac{dz}{z-b} \quad (18.8)$$

By a similar argument as above, we obtain

$$\frac{1}{2\pi i}\int_{C_2}\frac{f'}{f}dz = 0 + \frac{n}{2\pi i}2\pi i = n \quad (18.9)$$

Substituting Eqs. (18.6) and (18.9) in Eq. (18.2), we get

$$\frac{1}{2\pi i}\int_{C}\frac{f'(z)}{f(z)}dz = n - p. \quad (18.10)$$

18.4 Generalised Argument Theorem

Theorem 18.2 Let $f(z)$ be meromorphic within and on a simple closed curve C except for a finite number of poles $a_1, a_2, \ldots, a_j$ repeated $p_1, p_2, \ldots, p_j$ times respectively inside C, and $f(z) \neq 0$ on C, but has a finite number of zeros $b_1, b_2, \ldots, b_k$ repeated $n_1, n_2, \ldots, n_k$ times respectively inside C. Then

$$\frac{1}{2\pi i}\int_{C}\frac{f'(z)}{f(z)}dz = N - P$$

where $N = \sum_{r=1}^{k} n_r$ and $P = \sum_{r=1}^{j} p_r$ are the total number of zeros and poles respectively of $f(z)$ inside C, counted according to their multiplicities.

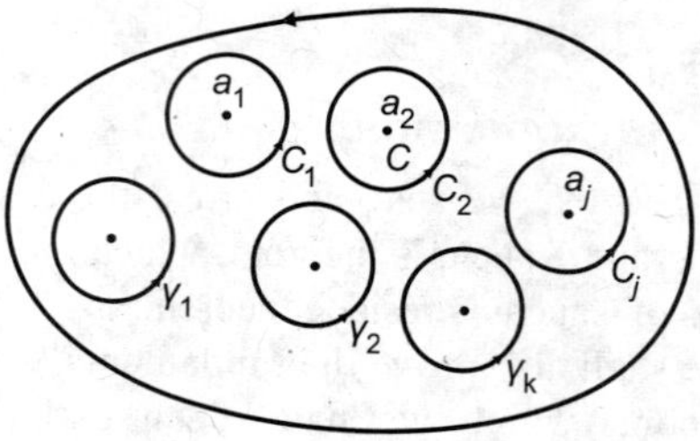

Figure 18.2

Proof: If we enclose a_r $(r = 1, 2, \ldots, j)$ and b_r $(r = 1, 2, \ldots, k)$ in small non-overlapping circles C_r and γ_r respectively, and apply the above argument theorem, we obtain

$$\frac{1}{2\pi i}\int_{C}\frac{f'}{f}dz = \frac{1}{2\pi i}\sum_{r=1}^{k}\int_{\gamma_r}\frac{f'}{f}dz$$

$$+\frac{1}{2\pi i}\sum_{r=1}^{j}\int_{C_r}\frac{f'}{f}dz$$

$$= \sum_{r=1}^{k} n_r - \sum_{r=1}^{j} p_r = N - P.$$

Principle of Argument (Angle)

By argument theorem,

$$N - P = \frac{1}{2\pi i}\int_{C}\frac{f'(z)}{f(z)}dz$$

$$= \frac{1}{2\pi i}\int_{C} d[\ln f(z)] = -\frac{1}{2\pi i}$$

[Variation of $\ln f(z)$ in going completely once around C]

$$= \frac{1}{2\pi i}\text{Var}\,[\ln|f(z)| + i\arg f(z)]$$

$$= \frac{1}{2\pi}\text{Var}\,[\arg f(z)] = \frac{1}{2\pi}$$

(change in the angle $f(z)$ as C traversed completely once)

$$= \frac{1}{2\pi} \Delta_C (\arg f(z))$$

$\because \ln|f|$ is the same at the beginning and at the end of one full circuit around C.

Corollary If $f(z)$ is analytic everywhere so that $p = 0$, then

N = Number of zeros of $f(z)$ inside C

$$= \frac{1}{2\pi} \text{Var}[\arg f(z)] = \frac{1}{2\pi} \Delta_c [\arg f(z)]$$

as z traverses the closed curve C

Geometrically, this means that the number of zeros of $f(z)$ is equal to the number of times the locus C^* of $w = f(z)$ encircles the origin.

Note

If $w = 0$ and C^* does not pass through the origin, then N = net variation of $\arg f(z)$ is zero.

The following theorem due to Rouche gives a simple procedure to find the number of zeros of an analytic function in a given domain.

18.5 [1]ROUCHE'S THEOREM

Theorem 18.3 Let $f(z)$ and $g(z)$ be analytic within and on a simple closed curve C. If $|g(z)| < |f(z)|$ at every point z on C, then $f(z)$ and $f(z) + g(z)$ both have the same number of zeros inside C.

Proof: First, we observe that $f(z)$ and $f(z) + g(z)$ have no zeros on C.

For, if $f(a) = 0$ for some point $z = a$ on C then $g(a) = 0$ on C since $g(a) < f(a)$. This implies that $g(a) = f(a) = 0$ on C, which contradicts the hypothesis that $|g(z)| < |f(a)|$ for all z on C.

Also, if $f(a) + g(a) = 0$ for some point $z = a$ on C then $|f(a)| = |-g(a)| = |g(a)|$, which is again a contradiction.

Therefore, $f(z)$ and $f(z) + g(z)$ have no zeros on C. Thus, the meromorphic function

$$F(z) = \frac{f(z) + g(z)}{f(z)} = 1 + \frac{g(z)}{f(z)} \tag{18.11}$$

[1] Rouche, Engene (1832–1910), French algebraist, analyst geometer and probabilist.

has no zeros or poles on C. By the principle of the argument for F

$$N - P = \frac{1}{2\pi} \arg f(z)|_C \tag{18.12}$$

Also,

$$|F(z) - 1| = \left|\frac{g(z)}{f(z)}\right| < 1 \tag{18.13}$$

for all z on C. Thus, the point $w = F(z)$ lies inside the circle of unit radius with center at 1, as shown in Fig. 18.3

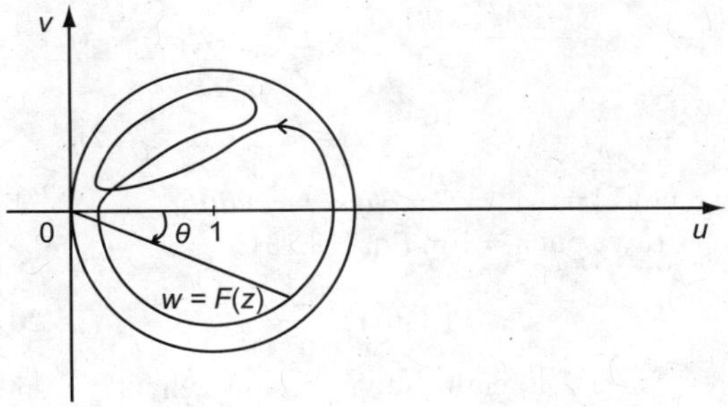

Figure 18.3

Now, it follows that

$$-\frac{\pi}{2} < \arg w < \frac{\pi}{2} \quad \text{or} \quad |\arg w| < \frac{\pi}{2} \tag{18.14}$$

for z on C.

In general,

$$\arg w = \arg F(z) = 2\pi k \tag{18.15}$$

(k : an integer)

If $k \neq 0$, then Eq. (4) is violated. Hence, $k = 0$.

$$\therefore \arg w = \arg F(z) = 0 \tag{18.16}$$

Eq. (2) now gives

$$N = P \tag{18.17}$$

Since the poles of F are the zeros of f and the zeros of F are the zeros of $f(z) + g(z)$, Eq. (7) implies that the number of zeros of $f(z)$ and $f(z) + g(z)$ are equal.

18.6 [2]LIOUVILLE THEOREM

Theorem 18.4 If $f(z)$ is entire and $|f(z)|$ is bounded for all z, then $f(z)$ is constant.

Proof: By Cauchy's generalised integral formula,

$$f^{(n)}(a)=\frac{n!}{2\pi i}\int_C \frac{f(z)}{(z-a)^{n+1}}dz \ (n=1,2,3,\ldots) \tag{18.18}$$

where C is a circle with centre at 'a' and radius 'r'.

Since $|f(z)|$ is bounded, $|f(z)|\le M$. Applying ML-inequality to Eq. (18.12),

$$|f^{(n)}(a)|\le \frac{n!}{2\pi}\left|\int_C \frac{f(z)}{(z-a)^{n+1}}dz\right| \tag{18.19}$$

$$\le \frac{n!}{2\pi}\frac{M(2\pi r)}{r^{n+1}} \quad \because \ C:|z-a|=r,$$

$$\Rightarrow \ |f^{(n)}(a)|\le \frac{n!M}{r^n} \quad L=\text{length of } C=2\pi r$$

which is known as *Cauchy's inequality*.

If we put $n = 1$ in Eq. (18.13)

$$|f'(a)|\le \frac{M}{r} \tag{18.20}$$

Since f is entire (analytic throughout C) and $|f|$ is bounded for all z Eq. (18.14) is true for any r. Letting $r\to\infty$ we have $|f'(a)|\to 0$.

Since 'a' is arbitrary $f'(z)\to 0$ for all z and hence $f(z)$ is constant.

18.7 FUNDAMENTAL THEOREM OF ALGEBRA

Theorem 18.5 Every polynomial of degree n in the complex plane has at least one zero.

Proof: (*By Liouville's theorem*) Let $p(z)$ be a polynomial in z of degree $n>1$, given by

$$p(z)=a_0+a_1z+a_2z^2+\cdots+a_nz^n \tag{18.21}$$

$$(\text{where } a_n\neq 0)$$

Suppose that $p(z)$ has no zero. Then $p(z)\neq 0$ for all z. This implies that $f(z)=\dfrac{1}{p(z)}$ is analytic everywhere. Also, $|f(z)|\to 0$ as $|z|\to\infty$ so that $|f(z)|$ is bounded for all z. By Liouville's theorem, a function which is analytic everywhere and bounded must be constant. This leads to contradiction since $p(z)$ is not constant. Hence, $p(z)$ is zero for at least one value of z.

[2] Liouville, Joseph (1809—1882) French analyst and geometer. First to prove existence of transcendental numbers.

Alternative form of the theorem: Every polynomial of degree n in the complex plane has n zeros, counting the multiplicities.vs10

Proof: (*By Rouche's theorem*) Let $p(z)=a_0+a_1z+a_2z^2+\cdots+a_nz^n$ $(a_n\neq 0)$ be the polynomial. Choose $f(z)=a_nz^n$ and $g(z)=a_1+a_1z+\cdots+a_{n-1}z^{n-1}$. Let C be a circle with centre at the origin and radius $R>1$. Then on C

$$\left|\frac{g(z)}{f(z)}\right|=\frac{|a_0+a_1z+a_2z^2+\cdots+a_{n-1}z^{n-1}|}{|a_nz^n|}$$

$$\le \frac{|a_0|+|a_1|R+|a_2|R^2+\cdots+|a_{n-1}|R^{n-1}}{|a_n|R^n}$$

$$\le \frac{|a_0|R^{n-1}+|a_1|R^{n-1}+|a_2|R^{n-1}+\cdots+|a_{n-1}|R^{n-1}}{|a_n|R^n},$$

replacing each power of R by R^{n-1}, which is larger for $n>1$

$$\le \frac{|a_0|+|a_1|+|a_2|+\cdots+|a_n|}{|a_n|R}$$

For any specified values of the coefficients $a_0,a_1,a_2,\ldots,a_n$, we can choose R such that

$$0<\left|\frac{g(z)}{f(z)}\right|_{z \text{ on } C}=\left|\frac{g(z)}{f(z)}\right|_{|z|=R}$$

$$<1 \ \Rightarrow |g(z)|<|f(z)|$$

By Rouche's theorem, the total number of zeros of the polynomial $p(z)=f(z)+g(z)$ in the circle C is the same as the number of zeros of $f(z)=a_nz^n$ in C.

But $f(z)=a_nz^n$ has n zeros (n-fold zero) all located at $z = 0$. Hence $p(z)=f(z)+g(z)$ has n zeros in C. Since R is arbitrary, this is true for the entire complex plane.

18.8 MAXIMUM MODULUS THEOREM FOR ANALYTIC FUNCTIONS

Theorem 18.6 Let $f(z)$ be a non-constant, analytic and bounded function in a domain D and also on its boundary. Then its absolute value $|f(z)|$ cannot have a maximum at an interior point of D. Consequently, the maximum of $|f(z)|$ is attained on the boundary of D. Similarly, if $f(z)\neq 0$ the minimum of $|f(z)|$ is also attained on the boundary of D.

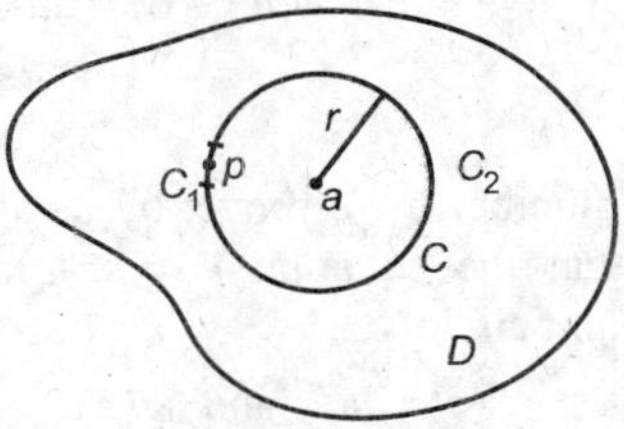

Figure 18.4

Proof: We prove the theorem by the method of contradiction. Assume that $|f(z)|$ has a maximum M at an interior point 'a' of D, *i.e.*, $f(a)=M$ and show that it leads to contradiction. Since $f(z)$ is not constant, $|f(z)|$ is also not constant. Describe a circle C of radius r with centre 'a' such that the interior of C is in D and $|f(z)|<M$ at some point P of C.

Let C_1 be an arc of C containing the point P (Fig. 18.4) Analyticity of $f(z)$ implies continuity of $f(z)$ within and on C. Therefore, $|f(z)|<M$ on C_1 containing the point P. Thus if $\varepsilon>0$ then $|f(z)|\leq M-\varepsilon$ for all z on C_1.

Let the length of C_1 be l_1 so that the complementary arc C_2 of C has the length $(2\pi r-l_1)$ since $|z-a|=r$. Now, we have by Cauchy's integral formula

$$|f(a)|=\frac{1}{2\pi}\left|\int_C\frac{f(z)dz}{z-a}\right|$$

$$\leq\frac{1}{2\pi}\left|\int_{C_1}\frac{f(z)}{z-a}dz\right|+\frac{1}{2\pi}\left|\int_{C_2}\frac{f(z)}{z-a}dz\right|$$

$$\leq\frac{1}{2\pi}\frac{(m-\varepsilon)l_1}{r}+\frac{1}{2\pi}\frac{M(2\pi r-l_1)}{r},$$

by ML - inequality

$$\leq M-\left(\frac{\varepsilon l_1}{2\pi^r}\right)<M$$

which is a contradiction. Hence our assumption is false. This proves the first part of the theorem.

To prove the last part, we observe that $f(z)\neq 0$ in D implies that $\frac{1}{f(z)}$ is analytic in D.

By the above result, $\frac{1}{|f(z)|}$ attains the maximum on the boundary of D so that $|f(z)|$ attains its minimum on the boundary of D. Hence the theorem.

Maximum principle for harmonic functions

Let D be a simply-connected, bounded domain with C as its boundary curve. Then, if $\phi(x,y)$ is a non-constant harmonic function in a domain containing domain D and curve C then $\phi(x,y)$ has neither a maximum nor a minimum in D.

Consequently, the maximum and the minimum of $\phi(x,y)$ occur only on the boundary C of D.

Example 18.1

Evaluate $\int_C\frac{f'(z)}{f(z)}dz$ using the argument principle, where C is a simple closed curve and $f(z)=z^5-8z^3i+2z-3+5i$.

Solution By the fundamental theorem of algebra

$$f(z)=z^5-8z^3i+2z-3+5i$$

being a polynomial and hence analytic everywhere of degree 5 has 5 zeros and has no poles in the complex plane, *i.e.*, $N=5, P=0$

$\therefore$ By the argument principle

$$\int_C\frac{f'(z)}{f(z)}dz=2\pi i(N-P)=2\pi i(5-0)=10\pi i.$$

Example 18.2

Evaluate $\int_C\frac{f'(z)}{f(z)}dz$ using the argument principle, where C is the simple closed curve $|z|=2$ and $f(z)=\frac{z^2-1}{(z^2+z)^2}$.

Solution $f(z)=\frac{z^2-1}{(z^2+z)^2}$ has two zeros of order 1 at $z=\pm1$ and two poles each of order 2 at $z=0$ and -1; otherwise, it is analytic inside $C:|z|=2$ (circle of radius 2 and centre 0).

$\therefore$ By the argument principle

$$\int_C\frac{f'(z)}{f(z)}dz=2\pi i(N-P)=2\pi i(2\cdot1-2\cdot2)=-4\pi i.$$

Example 18.3

Prove that all the roots of the equation $z^7-5z^3+12=0$ lie between the circles $|z|=1$ and $|z|=2$.

Solution Let $f(z)=12$ and $g(z)=z^7-5z^3$. Both $f(z)$ and $g(z)$ are analytic within and on $C_1:|z|=1$.

$$\text{Now } \left|\frac{g(z)}{f(z)}\right|=\left|\frac{z^7-5z^3}{12}\right|\leq\frac{|z|^7+5|z|^3}{12}$$

$$=\frac{1+5}{12}=\frac{1}{2}<1 \text{ on } C_1:|z|=1$$

By Rouche's theorem, both $f(z)$ and $f(z)+g(z)$ have the same number of zeros inside C_1. But $f(z)=12$ has no zeros inside C_1. Thus, $f(z)+g(z)=z^7-5z^3+12$ has no zeros inside C_1.

Next, let $f(z)=z^7$ and $g(z)=-5z^3+12$. We observe that $f(z)$ and $g(z)$ are analytic within and on $C_2:|z|=2$.

$$\text{Now } \left|\frac{g(z)}{f(z)}\right|=\left|\frac{12-5z^3}{z^7}\right|\leq\frac{12+5|z|^3}{|z|^7}$$

$$=\frac{12+5\cdot 2^3}{2^7}=\frac{52}{128}=\frac{13}{32}<1,$$

$$\text{on } C_2:|z|=2$$

By Rouche's theorem, both $f(z)$ and $f(z)+g(z)$ have the same number of zeros inside C_2. But $f(z)=z^7$ has 7 zeros at $z=0$ inside C_2. Consequently, $f(z)+g(z)=z^7-5z^3+12$ also has 7 zeros inside C_2.

Therefore, all the 7 zeros of z^7-5z^3+12 lie between the two circles $C_1:|z|=1$ and $C_2:|z|=2$.

Example 18.4

Evaluate $\int_C \frac{f'(z)}{f(z)}dz$ where $f(z)=\tan\pi z$ and $C:|z|=\pi$.

Solution Since $\tan\pi z=\frac{\sin\pi z}{\cos\pi z}$, the zeros of $\tan\pi z$ are those of $\sin\pi z$, which are $z=\pm n;\ n=0,1,2,\ldots$ and the poles of $\tan\pi z$ are the zeros of $\cos\pi z$, which are $z=\pm\frac{(2n+1)}{2};\ n=0,1,2,\ldots$

The zeros of $\tan\pi z$ lying inside C are $z=0,\pm1,\pm2,\pm3.\Rightarrow N=$ and vthe poles of $\tan\pi z$ lying inside C are $z=\pm\frac{1}{2},\pm\frac{3}{2},\pm\frac{5}{2}\Rightarrow P=6$. Now the radius of C is $\pi=3.1416$ (approximately).

∴ By argument principle,

$$\int_C\frac{f'(z)}{f(z)}dz=2\pi i(7-6)=2\pi i.$$

Example 18.5

If the real number $a>e$ prove, by using Rouche's theorem, that the equation $e^z=az^n$ has n roots inside the unit circle. [JNTU 2001S]

Solution Let $f(z)=az^n$ and $g(z)=-e^z$ so that $f(z)+g(z)=az^n-e^z=0$ is the given equation. Let C denote the unit circle $|z|=1$.

Note that $|f(z)|=|a||z^n|=|a|=a\quad \because|z|=1$.

$$\text{Now } |g(z)|=|-e^z|=|e^z|$$

$$=\left|1+z+\frac{z^2}{2!}+\frac{z^3}{3!}+\cdots\right|$$

$$\leq 1+|z|+\frac{|z|^2}{2}+\cdots$$

$$=1+1+\frac{1}{2!}+\frac{1}{3!}+\cdots=e$$

$$\Rightarrow \left|\frac{g(z)}{f(z)}\right|\leq\frac{e}{a}<1\quad\because a>e$$

∴ By Rouche's theorem, $f(z)+g(z)$ and $f(z)$ have the same number of zeros inside C. But $f(z)=az^n$ has n zeros inside C. So, $f(z)+g(z)=az^n-e^z$ has n zeros inside C.

Example 18.6

Applying Rouche's theorem, find the number of zeros of e^z-4z^n+1 inside the unit circle $|z|=1$.

Solution Let $f(z)=-4z^n$ and $g(z)=e^z+1$ and C be the curve $|z|=1$. On $C:|z|=1$ so that $|f(z)|=4$ and $|g(z)|=|e^z|+1=e+1<|f(z)|$.

∴ $f(z)+g(z)=e^2-4z^n+1$ has the same number of zeros inside the unit circle $C:|z|=1$ as $f(z)$, which has n zeros.

Hence e^z-4z^n+1 has n zeros inside $C:|z|=1$.

Example 18.7

Use Rouche's theorem to show that the equation $z^5+15z+1=0$ has one root in the disk $|z|=\frac{3}{2}$ and four roots in the annulus $\frac{3}{2}<|z|<2$.

[JNTU 2004(2)]

Solution First, we show that all the five roots lie within the circle $C:|z|=2$. Next, we show that one root lies inside the circle $C_1:|z|=\frac{3}{2}$ and the remaining four roots lie inside the annulus $A:\frac{3}{2}<|z|<2$.

Let $f(z)=z^5$ and $g(z)=15z+1$, $f(z)$ and $g(z)$ are polynomials and are hence analytic everywhere and, in particular, inside and on $C:|z|=2$.

$$\text{Now, } \frac{|g(z)|}{|f(z)|}=\frac{|15z+1|}{|z^5|}<\frac{15|z|+1}{|z|^5}$$

$$=\frac{15\cdot 2+1}{2^5}=\frac{31}{32}<1, \text{ on } C:|z|=2$$

By Rouche's theorem, both $f(z)+g(z)=z^5+15z+1$ and $f(z)=z^5$ have the same number of zeros inside $C:|z|=2$. But $f(z)=z^5$ has 5 zeros inside $C:|z|=2$ so that $z^5+15z+1$ has all the 5 zeros inside the circle $C:|z|=2$. Next consider the circle $C_1:|z|=\frac{3}{2}$.

Let $F(z)=15z$ and $G(z)=z^5+1$. F and G are analytic everywhere and, in particular, within and on the circle $C_1:|z|=\frac{3}{2}$.

$$\text{Consider } \frac{|G(z)|}{|F(z)|}=\frac{|z^5+1|}{|15z|}<\frac{|z|^5+1}{15|z|}$$

$$=\frac{\left(\frac{3}{2}\right)^5+1}{15\left(\frac{3}{2}\right)}=\frac{55}{144}<1$$

$$\text{on } C:|z|=\frac{3}{2}$$

$\therefore$ By Rouche's theorem, both $F(z)+G(z)=15z+z^5+1$ and $F(z)=15z$ have the same number of zeros inside the circle $|z|=\frac{3}{2}$. But $f(z)=15z$ has only one zero $(z=0)$ inside $|z|=\frac{3}{2}$. Therefore, the remaining four zeros must lie inside the annulus $A:\frac{3}{2}<|z|<2$.

Example 18.8

Show that the polynomial z^5+z^3+2z+3 has just one zero in the first quadrant of the complex plane.

[JNTU 2004 (Sets 1,4)]

Solution Let

$$f(z)=u+iv=z^5+z^3+2z+3 \tag{1}$$

and $z=Re^{i\theta}\left(0\le\theta\le\frac{\pi}{2}\right)$ where $R\to\infty$

Define the first quadrant as $OABO$ and let C be the boundary of this quadrant (Fig. 18.5). $f(z)$

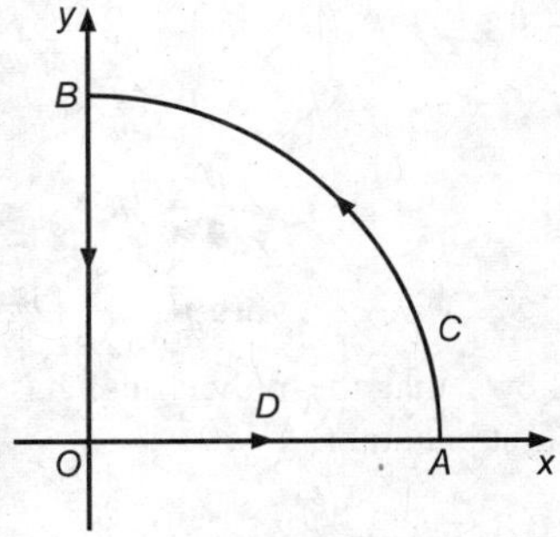

Figure 18.5

is analytic at all points except at $z=\infty$. So, it is analytic within and on C.

Along OA: $z=x$ and x varies from 0 to ∞ so that Eq. (1) becomes

$$u+iv=x^5+x^3+2x+3$$

$$\arg f=\tan^{-1}\frac{v}{u}=\tan^{-1}0$$

$$=0 \text{ for all } x>0\ (\because v=0)$$

$$\therefore \Delta_{OA}(\arg f)=\text{Var}(\arg f)=0$$

Along $\hat{AB}$: $z=Re^{i\theta}\quad 0\le\theta\le\frac{\pi}{2}, R\to\infty$

$$f=R^5e^{5i\theta}\left(1+\frac{1}{R^2e^{2i\theta}}+\frac{2}{R^4e^{4i\theta}}+\frac{3}{R^5e^{5i\theta}}\right)$$

$$\to R^5e^{5i\theta}, \text{ as } R\to\infty$$

$$\arg f=5\theta$$

$$\therefore \Delta_{AB}(\arg f)=\text{Var}(\arg f)$$

$$=(5\theta)_0^{\frac{\pi}{2}}=5\left(\frac{\pi}{2}-0\right)=\frac{5\pi}{2}$$

Along BO: $z=iy$ and y varies from ∞ to 0. Equation (1) takes the form $f=u+iv=iy^5-iy^3+2iy+3$

$$\Delta_{BO}(\arg f) = \tan^{-1}\frac{v}{u}$$

$$= \left[\tan^{-1}\left(\frac{y^5 - y^3 + 2y}{3}\right)\right]_{\infty}^{0}$$

$$= \tan^{-1}0 - \tan^{-1}\infty = -\frac{\pi}{2}$$

Hence $\Delta_C(\arg f) = \Delta_{OA}\arg f + \Delta_{AB}\arg f$

$$+\Delta_{BO}\arg f$$

$$= 0 + \frac{5\pi}{2} - \frac{\pi}{2} = 2\pi$$

$$\therefore N = \frac{1}{2\pi}\Delta_C \arg f = \frac{1}{2\pi}(2\pi) = 1$$

This shows that the polynomial has just one zero in the first quadrant.

Example 18.9

Show that the equation $z^4 + 4(1+i)z + 1 = 0$ has one root in each quadrant. [JNTU 2004 (Sets 1,4)]

Solution Let $f(z) = u + iv = z^4 + 4(1+i)z + 1 = 0$ and $z = Re^{i\theta}$, $0 \le \theta \le \frac{\pi}{2}$ where $R \to \infty$ define the first quadrant $OABO$ and C denotes the complete boundary of the quadrant $OABO$ (Fig. 18.6).

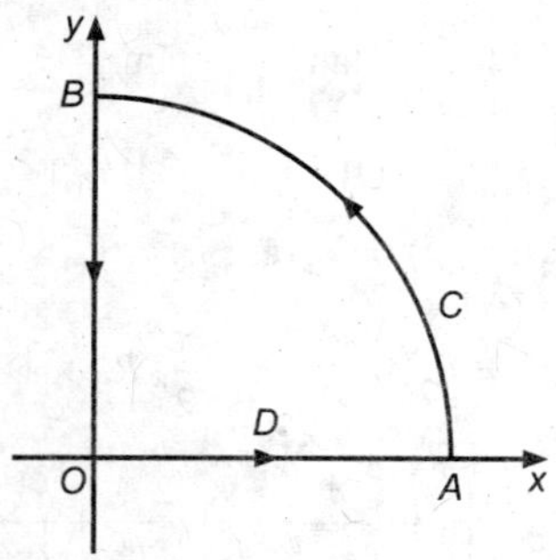

Figure 18.6

Along OA: $z = x$ and x varies from 0 to ∞ and Eq. (1) becomes

$$f = u + iv = x^4 + 4(1+i)x + 1$$

$$= x^4 + 4x + 1 + 4ix$$

$$\Delta_{OA}\arg f = \tan^{-1}\frac{u}{a} = \left[\tan^{-1}\left(\frac{4x}{x^4 + 4x + 1}\right)\right]_0^{\infty}$$

$$= \tan^{-1}0 - \tan^{-1}0 = 0$$

Along AB: $z = Re^{i\theta}$ $0 \le \theta \le \frac{\pi}{2}$, $R \to \infty$

$$\arg f = R^4 e^{4i\theta}\left[1 + \frac{4(1+i)}{R^3 e^{3i\theta}} + \frac{1}{R^4 e^{4i0}}\right]$$

$$\to R^4 e^{4i\theta} \text{ as } R \to \infty$$

$$\therefore \arg f = 4\theta$$

$$\Delta_{AB}\arg f = [4\theta]_0^{\frac{\pi}{2}} = 4\left(\frac{\pi}{2} - 0\right) = 2\pi$$

Along BO: $z = iy$ and y varies from ∞ to 0 so that Eq. (1) becomes

$$f = u + iv = y^4 + 4(1+i)i4 + 1$$

$$= (y^4 - 4y + 1) + i(4y)$$

$$\Delta_{BO}\arg f = \left[\tan^{-1}\left(\frac{4y}{y^4 - 4y + 1}\right)\right]_{\infty}^{0}$$

$$= \tan^{-1}0 - \tan^{-1}0 = 0$$

Hence $\Delta_C \arg f = \Delta_{OA}\arg f + \Delta_{AB}\arg f$

$$+\Delta_{BO}\arg f$$

$$= 0 + 2\pi + 0 = 2\pi$$

$$\Rightarrow \quad N = \frac{1}{2\pi}\Delta_C \arg f$$

$$= \frac{1}{2\pi}(2\pi) = 1$$

This implies that the given equation has one root in the first quadrant. The conjugate complex root lies in the fourth quadrant. The equation being fourth degree has 4 roots; the other roots lie in each of the second and third quadrants.

Example 18.10

Prove that one root of the equation $z^4 + z^3 + 1 = 0$ lies in the first quadrant. [JNTU 2004 (Set 3)]

Solution Let

$$-5ptf(z) = u + iv$$

$$= z^4 + z^3 + 1 = 0 \quad (1)$$

We first prove that the equation has no real root.

Putting $z = x$, we get $f(x) = x^4 + x^3 + 1$ which does not become zero for any positive real number x. So, $f(z) = 0$ has no positive real root.

Let $z=-x$ where $x>0$

Now, $f(-x)=x^4-x^3+1$

$$=x^2(x-1)+1>0 \quad \text{if } x>1$$

$$=x^4+(1-x^3)$$

$$=x^4+(1-x)(1+x+x^2)>0$$

$$\text{if } \quad 0<x<1$$

Thus $f(-x)>0$ if $x>1$, $f(-x)>0$ if $0<x<1$, which is a contradiction.

Hence $f(z)=0$ has no negative real root also.

$\therefore f(z)=0$ has no real root.

We also prove that the equation has no purely imaginary root.

Put $z=iy$; $f(z)=y^4-iy^3+1=0 \Rightarrow y^4+1=0$

$y^3=0$ (equating the real and imaginary parts zero).

These equations are not compatible and hence $f(z)=0$ has no purely imaginary roots.

We now find the number of roots in the first quadrant. Let $z=Re^{i\theta}\left(0\le\theta\le\dfrac{\pi}{2}\right) R\to\infty$ define the first quadrant bounded by the straight line segments OA and BO along the x-and y-axes and the part of circular $\overset{\frown}{AB}$ between these lines. $f(z)$ is analytic within and on C, consisting of OA, BO and $\overset{\frown}{AB}$.

(i) Along OA: $z=x$ and x varies from 0 to ∞

$$\therefore f=u+iv=x^4+x^3+1$$

$$\Delta_{OA}\arg f=\left[\tan^{-1}\frac{v}{u}\right]_0^{\infty}=\tan^{-1}0=0$$

(ii) Along AB: On this circular arc $z=Re^{i\theta}$,

$$0\le\theta\le\frac{\pi}{2}$$

$$f(z)=R^4e^{4i\theta}\left[1+\frac{1}{Re^{i\theta}}+\frac{1}{R^4e^{4i\theta}}\right]$$

$$\to R^4e^{4i\theta} \text{ as } R\to\infty$$

$$\arg f=\tan^{-1}\frac{v}{u}=\tan^{-1}\left(\frac{R^4\sin 4\theta}{R^4\cos 4\theta}\right)$$

$$=\tan^{-1}(\tan 4\theta)=4\theta$$

$$\Delta_{AB}\arg f=[4\theta]_0^{\frac{\pi}{2}}=4\left(\frac{\pi}{2}-0\right)=2\pi$$

(iii) Along BO: $z=iy$ and y varies from ∞ to 0

$$f(z)=u+iv=y^4-iy^3+1$$

$$\Rightarrow \arg f=\tan^{-1}\frac{v}{u}=\left[\tan^{-1}\left(\frac{-y^3}{(y^4+1)}\right)\right]$$

$$=-\tan^{-1}\frac{y^3}{y^4+1}$$

$$\Delta_{BO}\arg f=-\left[\tan^{-1}\frac{y^3}{y^4+1}\right]_{\infty}^{0}$$

$$=-(\tan^{-1}0-\tan^{-1}0)=0$$

$$\therefore \Delta_C\arg f=\Delta_{OA}\arg f+\Delta_{AB}\arg f+\Delta_{BO}\arg f$$

$$=0+2\pi+0=2\pi$$

$$\Rightarrow N=\frac{1}{2\pi}\Delta_C\arg f$$

$$=\frac{1}{2\pi}(2\pi)=1$$

By the argument principle, the equation has only one complex root in the first quadrant.

EXERCISE 18.1

1. Locate the quadrant in which the roots of the equation $z^4+z^3+4z^2+2z+3=0$ are situated. [JNTU 2003S]

   Ans: 2 roots in Q II and 2 in Q IV

2. Find the number of roots of the equation $2z^5-6z^2+z+1=0$ in the region $1\le|z|<2$.

   [**Hint:** Take $f(z)=-6z^2; g(z)=2z^5+z+1$ on $C_1:|z|=1$ and $f(z)=2z^5; g(z)=-6z^2+z+1$ on $C_2:|z|=2$.]

   Ans: 3

3. Use Rouche's theorem to show that the equation $z^5+15z+1=0$ has one root in the disk $|z|>\frac{3}{2}$ and four roots in the annulus $\frac{3}{2}<|z|<2$.

   [**Hint:** Take $f(z)=z^5; g(z)=15z+1$ on $C_1:|z|=2$ and $f(z)=15z; g(z)=z^5+1$ on $C_2:|z|=3/2$.]

   [JNTU 2004, Nov 2008S, Nov 2008 Set 1]

4. Show that the equation $z^3 + |z|i = 0$ has a root in each of the first, second and fourth quadrants.

5. Evaluate $\int_C \frac{f'(z)}{f(z)} dz$ where $f(z) = z^5 - 3iz^2 + 2z - 1 + 1$ and C is a simple closed curve.

Ans: $10\pi i$

6. Apply Rouche's theorem to determine the number of zeros of $z^9 - 2z^6 + z^2 - 8z - 2$ that lie within the circle $C : |z| = 1$.

[**Hint:** Take $f(z) = -8z$,

$$g(z) = z^9 - 2z^6 + z^2 - 2]$$

Ans: One zero

7. By Rouche's theorem determine the number of zeros of $z^4 - 5z + 1$ in the region $1 < |z| < 2$.
[**Hint:** Take $f(z) = -5z, g(z) = z^4 + 1$ on $C_1 : |z| = 1$ and $f(z) = z^4; g(z) = -5z + 1$ on $C_2 : |z| = 2$.]

Ans: Three

8. Evaluate $\int_C \frac{f'(z)}{f(z)} dz$ where $f(z) = \sin \pi z$ and L is the circle $|z| = \pi$.

Ans: $14\pi i$

9. Use Rouche's theorem to find the number of zeros of the polynomial $z^{10} - 6z^7 + 3z^3 + 1$ if $|z| < 1$.

[**Hint:** Take $f(z) = -6z^7, g(z) = z^{10} + 3z^3 + 1$]

Ans: Seven zeros

Conformal Mapping

19

19.1 Introduction

Let $y = f(x)$ be a function in the xy-plane. The pair (x, y) of real numbers x and y can be plotted with the help of two co-ordinate axes Ox and Oy. We thus obtain a graph of the function. It gives a visual representation of the function. This helps in giving a better understanding of the properties of the function.

Now, consider a complex-valued function $w = f(x)$ where $z = (x, y)$ and $w = (u, v)$. The graphical representation as in the case of real-valued functions is not possible in this case. However, if we have two planes, one for each pair of numbers, and the mapping assigns to each point $z = (x, y)$ in the z-plane a corresponding point $w = (u, v)$ in the w-plane, this modified version of the geometrical representation provides a better understanding of the problems. It has several practical applications in potential theory, theory of elasticity, hydrodynamics and other applied fields.

19.1.1 Mapping $f: z \to w$

Consider the complex function $w = f(z)$ where $z = x + iy$ or (x, y) and $w = u + iv$ or (u, v).

To represent this function graphically, we need two planes. One plane is the z-plane in which we plot the values of $z = (x, y)$ and the other is the w-plane in which we plot the corresponding values of $w = (u, v)$. Thus a given function f assigns to each z in its domain of definition D a corresponding point $w = f(z)$ in the w-plane (Fig. 19.1).

We say that f defines a mapping of D onto a range of values of $w = f(z)$. If z is any point in D the point $w = f(z)$ is called the image of z under f. For the points of a curve C in D, the image points form the image curve C^* of C in the w-plane.

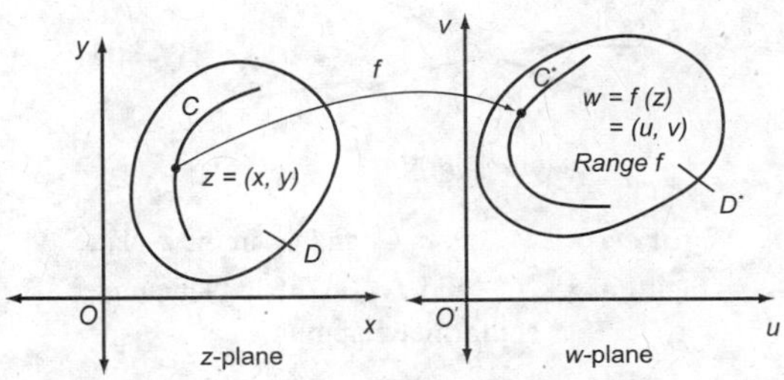

Figure 19.1

Instead of using the phrase "the mapping by a function $w = f(z)$" we briefly say "the mapping (or transformation) $w = f(z)$".

19.1.2 Conformal Mapping

A mapping under which the angles between any two oriented curves are preserved both in magnitude and sense is called a *conformal mapping*. If a mapping preserves angles in magnitude only and not in sense then it is called an *isogonal mapping*.

Example 19.1

$w = f(z) = \bar{z}$. Here $|w| = |\bar{z}| = |z|$ but $\arg w = \arg(\bar{z}) = -\arg(z)$.

Figure 19.2 shows the angle of intersection α of the curves C_1 and C_2 defined as the angle between the oriented tangents to the curves at the point $P(z_0)$ of intersection of C_1 and C_2.

We know that a curve in the xy-plane has parametric representation.

For example, the equation of the circle $x^2 + y^2 = 1$ can be represented by $x = \cos t$, $y = \sin t$ $(0 \le t \le 2\pi)$

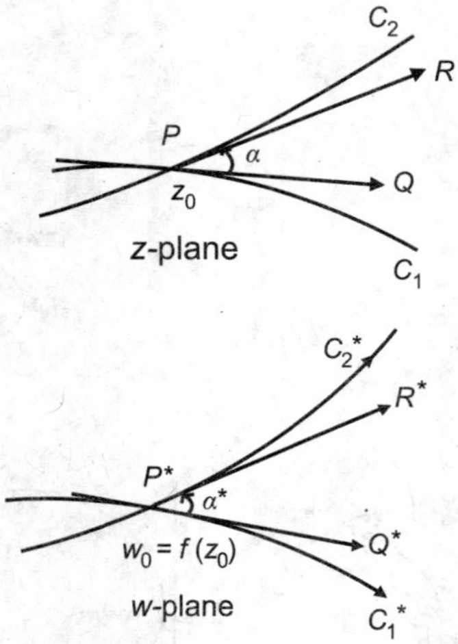

Figure 19.2 Curves C_1 and C_2 in the z-plane and their images C_1^* and C_2^* in the w-plane under a conformal mapping

and the equation of the parabola $y^2 = 4ax$ can be represented by $x = at^2, y = 2at\ (-\infty < t < \infty)$.

In the complex plane, a curve can be represented by $z(t) = x(t) + iy(t)\ (a \le t \le b)$. We assume that C is smooth. That is, at each point $z(t)$ of C, $\dot{z}(t)$ exists and is not zero. The positive sense of C is that in which t is increasing, *i.e.*, $\frac{dz}{dt} = \dot{z}(t) > 0$.

Conformal mapping preserves orthogonality of curves. That is, if $\phi\,(x, y) = C_1$, $\psi\,(x, y) = C_2$ are mutually orthogonal in the z-plane their image curves in the w-plane given by $\phi^*(u,v) = C_1^*$ and $\psi^*(u,v) = C_2^*$ are also mutually orthogonal.

19.2 Conformal Mapping: Conditions for Conformality

The conditions under which the transformation $w = f(z)$ is conformal are stated and proved in the following theorem.

Theorem 19.1 If $f(z)$ is analytic and $f'(z) \neq 0$ in a region R of the z-plane, then the mapping $w = f(z)$ is conformal at all points of R.

Proof: Let $P(z)$ be a point in the region R of the z-plane and $P^*(w)$ the corresponding point in the corresponding region G of the w-plane. Suppose that as P moves on a curve C and P' moves on the corresponding curve C' in G. Let $Q(z + \delta z)$ and $Q'(z + \delta z)$ be the neighbouring points on C and C', respectively, so that $\overrightarrow{PQ} = \delta z$ and $\overrightarrow{P'Q'} = \delta w$.

Then δz is a complex number whose modulus r is the length PQ and amplitude θ is the angle which PQ makes with the x-axis. $\delta z = re^{i\theta}$ and $\delta w = r'e^{i\theta'}$, $\therefore \frac{\delta w}{\delta z} = \frac{r'}{r} e^{i(\theta'-\theta)}$, where r' is the modulus and θ' the amplitude of δw of the tangent at P on C, and the tangent at P' on C makes angles α and α', respectively, with the real axes x and u. Then as $\delta z \to 0$, $\theta \to \alpha$ and $\theta' \to \alpha'$. Hence,

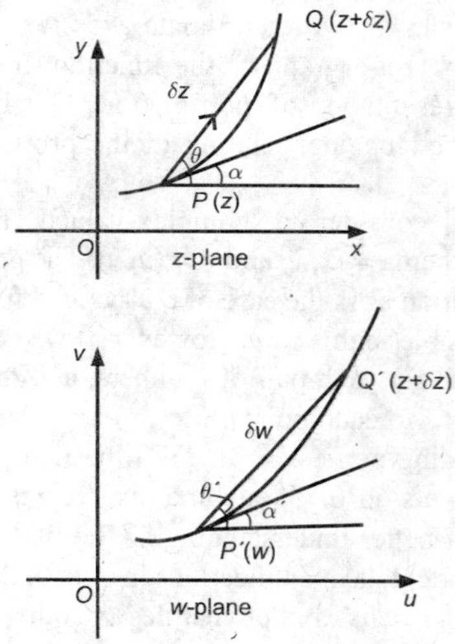

Figure 19.3

$$f'(z) = \frac{dw}{dz} = \lim_{\delta z \to 0} \frac{\delta w}{\delta z} = \lim_{\delta z \to 0} \left(\frac{r'}{r}\right) e^{i(\theta'-\theta)} \qquad (19.1)$$

$$f'(z) \neq 0 \Rightarrow f'(z) = \rho e^{i\phi}$$

so that $\rho = |f'(z)|$ and $\phi = \text{amp}\, f'(z)$

$$\therefore \rho e^{i\phi} = \lim_{\delta z \to 0} \left(\frac{r'}{r}\right) e^{i(\theta'-\theta)}$$

so that $\rho = \lim_{\delta z \to 0} \frac{r'}{r}$ (19.2)

and $\phi = \lim_{\delta z \to 0} (\theta' - \theta) = \alpha' - \alpha.$

If C_1 and C_1' be another pair of curves through P and P', respectively, in the z- and w- planes and thetangents at these points make angles β and β' with the real axis then $\phi = \beta' - \beta \Rightarrow \alpha' - \alpha = \beta' - \beta$ or $\beta - \alpha = \beta' - \alpha' = \gamma$.

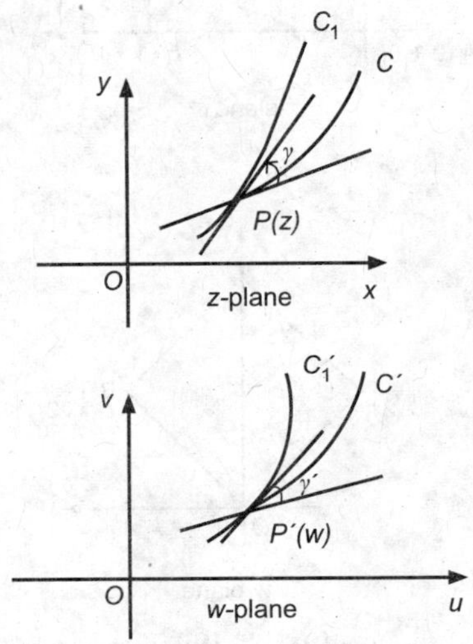

Figure 19.4

The angle between the curves before and after the mapping is preserved in magnitude and sense. Hence the mapping by the analytic function $w = f(z)$ is conformal at each point where $f'(z) \neq 0$.

Note

1. A point at which $f'(z) = 0$ is called a critical point of the transformation.
2. We have obtained $\rho = \lim_{\delta z \to 0} \frac{r'}{r}$. It follows that under the conformal transformation w=f(z), the lengths of arc through P are magnified in the ratio p: 1 where $\rho = |f'(z)|$.
3. A harmonic function remains harmonic under a conformal transformation.

Example 19.2

For the conformal transformation $w = z^2$ show that

(i) the coefficient of magnification at $z = \frac{1}{2} + \frac{i}{2}$ is $\sqrt{2}$ and

(ii) the angle of rotation at $z = \frac{1}{2} + \frac{i}{2}$ is $\frac{\pi}{4}$.

Solution $w = f(z) = z^2 \Rightarrow f'(z) = 2z$ and $f'\left(\frac{1}{2} + \frac{1}{2}i\right) = 1 + i.$

(i) The coefficient of magnification at $z = \frac{1}{2} + \frac{i}{2}$ is $\left|f'\left(\frac{1}{2} + \frac{i}{2}\right)\right| = \sqrt{1^2 + 1^2} = \sqrt{2}$.

(ii) The angle of rotation at $z = \frac{1}{2} + \frac{i}{2}$ is amp $f'\left(\frac{1}{2} + \frac{i}{2}\right) = \tan^{-1}\frac{\frac{1}{2}}{\frac{1}{2}} = \frac{\pi}{4}$.

19.3 Conformal Mapping by Elementary Functions

19.3.1 General Linear Transformation

Consider the mapping

$$w = f(z) = az + b \tag{19.3}$$

($a \neq 0$, b are arbitrary complex constants.)

It maps conformally the extended complex z-plane onto the extended w-plane since Eq. (19.3) is analytic and $f'(z) = a \neq 0$ for any z. If $a = 0$, then Eq. (19.3) reduces to a constant function.

Special cases of linear transformation

Identity transformation

The mapping

$$w = z \tag{19.4}$$

obtained from Eq. (19.3) for $a = 1$, $b = 0$ maps a point z onto itself.

Translation transformation

The mapping

$$w = z + b \tag{19.5}$$

obtained from Eq. (19.3) for $a = 1$ translates or shifts z through a distance $|b|$ in the direction of b.

More specifically, if $z = x + iy$, $b = p + iq$, $w = u + iv$, then the transformation becomes

$$u + iv = (x + iy) + (p + iq)$$
$$= (x + p) + i(y + q)$$

so that $u = x + p$ and $v = y + q$.

Thus the transformation is a mere translation of the axes and it preserves the shape and size of the figure in the z-plane.

For example, the rectangle $ABCD$ in the z-plane is transformed to the rectangle $A'B'C'D'$ in the w-plane under the transformation $w = z + (2 + i)$.

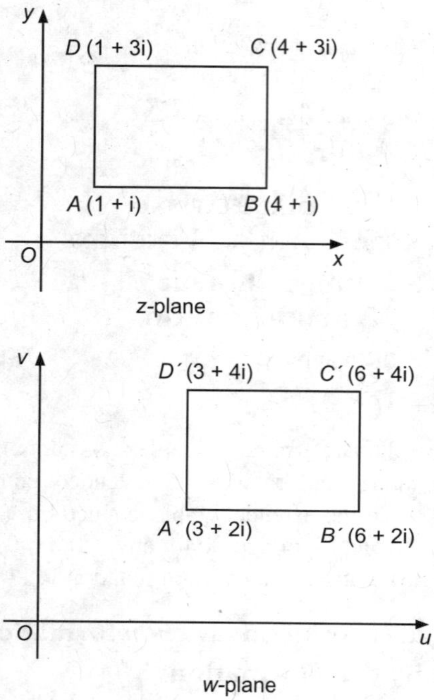

Figure 19.5

Rotation transformation

The mapping

$$w = e^{i\alpha} z \qquad (19.6)$$

obtained from Eq. (19.3) for $a = e^{i\alpha}$ and $b = 0$ rotates a radius vector of a point z through a scalar angle α counter-clockwise if $\alpha > 0$ and clockwise if $\alpha < 0$.

For example, the mapping $w = e^{i\frac{\pi}{4}} z$ or $w = iz$ rotates the square $ABCD$ in the z-plane into the square $A'B'C'D'$, which is rotated through an angle $\frac{\pi}{4}$.

$$A(0, 0) \leftrightarrow A'(0, 0) \quad B(1, 0) \leftrightarrow B'\left(\frac{1}{\sqrt{2}}, \frac{1}{\sqrt{2}}\right)$$

$$C(1, 1) \leftrightarrow C'\left(0, \frac{1}{\sqrt{2}}\right) \quad D(0, 1) \leftrightarrow D'\left(\frac{-1}{\sqrt{2}}, \frac{1}{\sqrt{2}}\right)$$

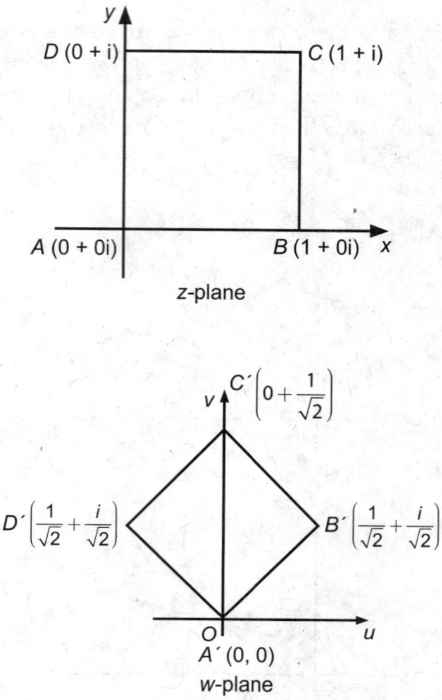

Figure 19.6

Stretching, scaling or magnification

The mapping

$$w = az \qquad (19.7)$$

obtained from Eq. (19.3) for $b = 0$ stretches (magnifies) the radius vector by a factor a if a is real and >1 and contracts if a is real and $0 < a < 1$.

19.3.2 Inversion Transformation

Consider the mapping

$$w = \frac{1}{z} \qquad (19.8)$$

Put $z = re^{i\theta}$ and $w = Re^{i\phi}$ then the transformation becomes

$$Re^{i\phi} = \frac{1}{r}e^{-i\theta} \Rightarrow R = \frac{1}{r} \quad \text{and} \quad \phi = -\theta$$

Thus under the transformation $w = \frac{1}{z}$ a point $P(r, \theta)$ in the z-plane is mapped into the point $P'\left(\frac{1}{r}, -\theta\right)$.

Imagine that the w-plane is superposed on to the z-plane. If P is (r, θ) and P_1 is $\left(\frac{1}{r}, \theta\right)$, then $OP_1 = \frac{1}{r} = \frac{1}{OP} \Rightarrow OP \cdot OP_1 = 1$ so that P_1 is the inverse of P w.r.t the unit circle $|z| = 1$.

Note

The inverse of a point P w.r.t the circle having its centre at 0 and radius r is defined as the point Q on OP such that $OP \cdot OQ = r^2$.

The reflection P' of P in the real axis represents $w = \frac{1}{z}$. Thus the transformation $w = \frac{1}{z}$ is an inversion of z w.r.t the unit circle $|z| = 1$ followed by reflection of the inverse in the real axis.

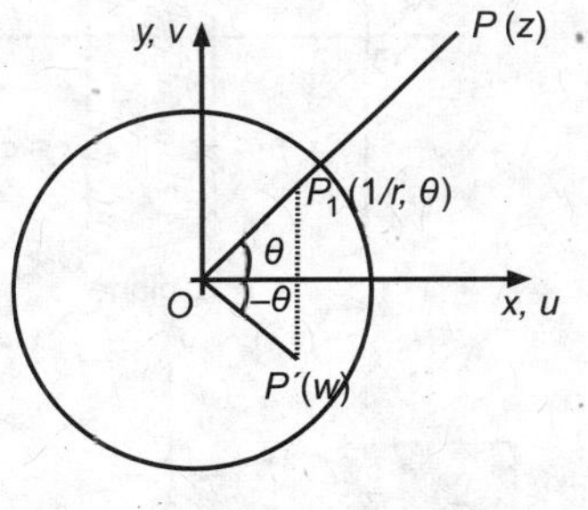

Figure 19.7

Clearly, the transformation $w = \frac{1}{z}$ maps the interior of the unit circle $|z| = 1$ into the exterior of the unit circle $|w| = 1$ and the exterior of $|z| = 1$ into the interior of $|w| = 1$.

However, the origin $z = 0$ is mapped to the point $w = \infty$, called the point at infinity.

We may discuss the transformation $w = \frac{1}{z}$ by changing it to cartesian coordinates. For this, we put

$$z = x + iy \text{ and } w = u + iv$$

$$w = \frac{1}{z} \Rightarrow u + iv = \frac{1}{x + iy}$$

$$= \frac{x - iy}{(x + iy)(x - iy)} = \frac{x - iy}{x^2 + y^2}$$

$$\Rightarrow u = \frac{x}{x^2 + y^2} \text{ and } v = \frac{-y}{x^2 + y^2}$$

$$\text{Also, } z = \frac{1}{w} \Rightarrow x + iy = \frac{1}{u + iv}$$

$$= \frac{u - iv}{(u + iv)(u - iv)} = \frac{u - iv}{u^2 + v^2}$$

$$\Rightarrow x = \frac{u}{u^2 + v^2} \text{ and } y = \frac{-v}{u^2 + v^2}$$

Now consider the second degree general equation with equal coefficients for x^2 and y^2 *viz.*

$$a(x^2 + y^2) + bx + cy + d = 0 \tag{19.9}$$

where a, b, c and d are real constants. If $a \neq 0$ the above equation represents a circle and if $a = 0$ it represents a straight line, substituting for x and y in terms of u and v we get,

$$a\left[\frac{u^2}{(u^2 + v^2)^2} + \frac{v^2}{(u^2 + v^2)^2}\right]$$

$$+ b\left[\frac{u}{u^2 + v^2}\right] - c\left[\frac{v}{u^2 + v^2}\right] + d = 0$$

$$\text{or } d(u^2 + v^2) + bu - cv + a = 0, \tag{19.10}$$

on multiplication throughout by $(u^2 + v^2)$.

The following cases arise

1. $a \neq 0$, $d \neq 0$. Equations (19.9) and (19.10) show that *circles* not passing through the origin are mapped into *circles* not passing through the origin.
2. $a \neq 0$, $d = 0$. Equations (19.9) and (19.10) show that *circles* passing through the origin are mapped onto *straight lines* not passing through the origin.
3. $a = 0$, $d \neq 0$. Equations (19.9) and (19.10) show that *straight lines* passing through the origin are mapped onto *circles* passing through the origin.
4. $a = 0$, $d = 0$. Equations (19.9) and (19.10) show that *straight lines* passing through the origin are mapped onto *straight lines* passing through the origin.

Thus, the mapping $w = \frac{1}{z}$ transforms circles and straight lines into circles and straight lines.

Since a straight line can be considered as a circle with infinite radius, we conclude that under the mapping $w = \frac{1}{z}$ circles are mapped into circles in the w-plane.

19.4 Some Special Transformations

19.4.1 Transformation $w = z^2$

Polar coordinates

Changing the equation into polar coordinates by putting $z = re^{i\theta}$ and $w = Re^{i\phi}$

we have $w = z^2$

$$\Rightarrow Re^{i\phi} = (re^{i\theta})^2 = r^2 e^{2i\theta}$$

$$\Rightarrow R(\cos\phi + i\sin\phi) = r^2(\cos 2\theta + i\sin 2\theta)$$

$$\Rightarrow R = r^2 \text{ and } \phi = 2\theta$$

Hence circles $r = r_0$ (constant) in the z-plane are transformed into circles $R = r_0^{\ 2}$ (constant) in the w-plane.

In particular, the region $1 \le |z| \le \dfrac{3}{2}, \dfrac{\pi}{6} \le \theta \le \dfrac{\pi}{3}$ is mapped onto the region $1 \le |w| \le \dfrac{9}{4}, \dfrac{\pi}{3} \le \phi \le \dfrac{2\pi}{3}$.

Mapping $w = z^2$; lines $|z| = constant$, arg z = *constant* and their images are in the w-plane.

Cartesian coordinates

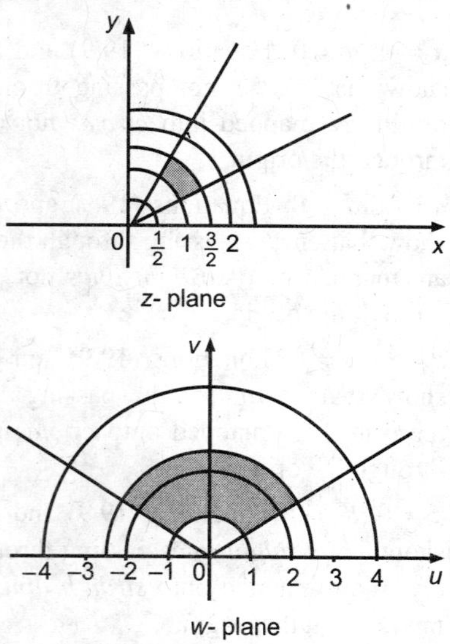

Figure 19.8

Changing the equation $w = z^2$ by putting $z = x + iy$ and $w = u + iv$ where $x, y, u, v \in \mathbb{R}$ we have $w = z^2 \Rightarrow u + iv = (x + iy)^2 = x^2 - y^2 + i(2xy)$. On equating the real and imaginary parts, we get $u = x^2 - y^2$; $v = 2xy$.

Vertical lines

Consider a vertical line $x = c$ (constant) in the z-plane. It is mapped onto the curve

$$u = c^2 - y^2, \ v = 2cy$$

$$\Rightarrow y^2 = c^2 - u, \ 4c^2 y^2 = v^2.$$

Eliminating y between these equations, we get

$$v^2 = 4c^2 (c^2 - u) = -4c^2 (u - c^2)$$

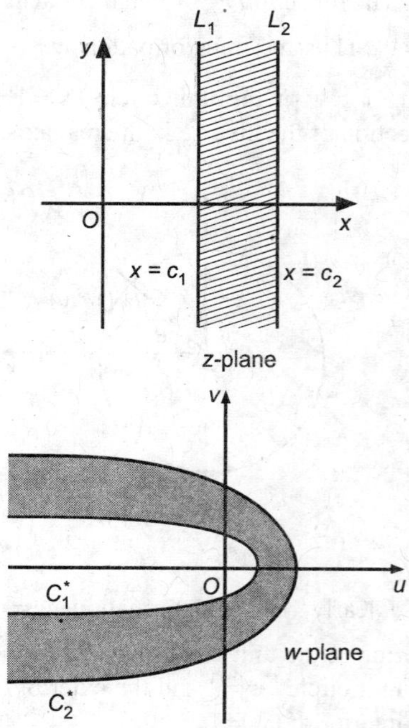

Figure 19.9

These are parabolas with the negative u-axis as their axis and their openings are to the left.

The region between two vertical lines $x = c_1$ and $x = c_2$ ($c_2 > c_1$) is mapped into the region between the parabolas $v^2 = -4c_1^{\ 2}(u - c_1^{\ 2})$ and $v^2 = -4c_2^{\ 2}(u - c_2^{\ 2})$ in the w-plane whose vertices are at $(c_1^{\ 2}, 0)$ and $(c_2^{\ 2}, 0)$, respectively, and their common axis is the negative u-axis.

Horizontal lines

Consider a horizontal line $y = d$ (constant) in the z-plane. It is mapped onto the curve $v^2 = 4d^2 (u + d^2)$, a parabola with its axis along Ou and with its opening to the right.

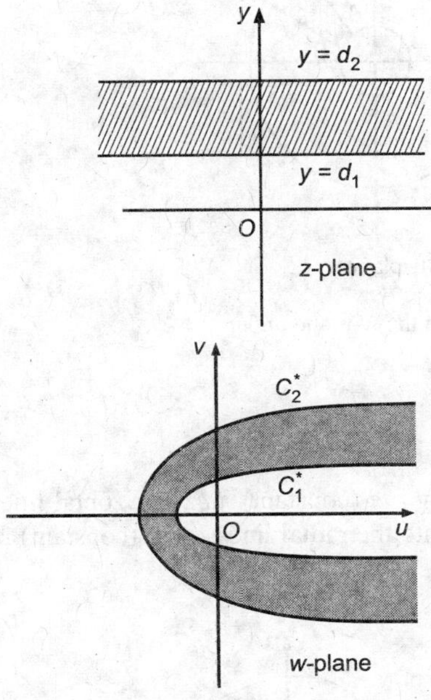

Figure 19.10

$$u = x^2 - d^2, v = 2dx$$

$$\Rightarrow x^2 = u + d^2, 4d^2 x^2 = v^2$$

Eliminating x between these equations, we get

$$v^2 = 4d^2 (u + d^2)$$

The vertex of the parabola is at $(-d^2, 0)$, and the axis is the positive u-axis.

The region between two horizontal lines $y = d_1$ and $y = d_2(d_2 > d_1 \geq 1)$ is mapped into the region between the parabolas $v^2 = 4{d_1}^2(u + {d_1}^2)$ and $v^2 = 4{d_2}^2(u + {d_2}^2)$ in the w-plane.

19.4.2 Transformation $w = z^n$ $(n \in \mathbb{N})$

Differentiating $w = z^n$ w.r.t z, we have

$$\frac{dw}{dz} = nz^{n-1}, \frac{dw}{dz} = 0 \Rightarrow z = 0.$$

So, the mapping is conformal at all points z except at the origin $z = 0$.

Solving $w = z^n$ for z, we have $z = w^{\frac{1}{n}}$ which has n roots. For every non-zero w, this shows that there are n pre-images.

Converting the equation $w = z^n$ into polars by putting $z = re^{i\theta}$ and $w = Re^{i\phi}$, we get $R = r^n$ and $\phi = n\theta$.

A circle $|z| = r$ of radius r and centre at the origin in the z-plane is mapped into a circle of radius r^n and centre at the origin of the w-plane.

The sectorial region $\{z = re^{i\theta}, \alpha \leq \theta \leq \beta\}$ is mapped into the sectorial region $\{w = Re^{i\phi}, n\alpha \leq \phi \leq n\beta\}$ in the w-plane.

In particular, the sector $\left\{z = re^{i\theta}, 0 \leq \theta \leq \frac{\pi}{n}\right\}$ is mapped into the upper half-plane $v \geq 0$ in the w-plane given by

$\{w = Re^{i\phi}, 0 \leq \phi \leq \pi\}$ and the region

$\left\{z = re^{i\theta}, 0 \leq \theta \leq \frac{2\pi}{n}\right\}$ is mapped into the region

$\{w = \text{Re}^{i\phi}, 0 \leq \phi \leq 2\pi\}$, which is the entire w-plane.

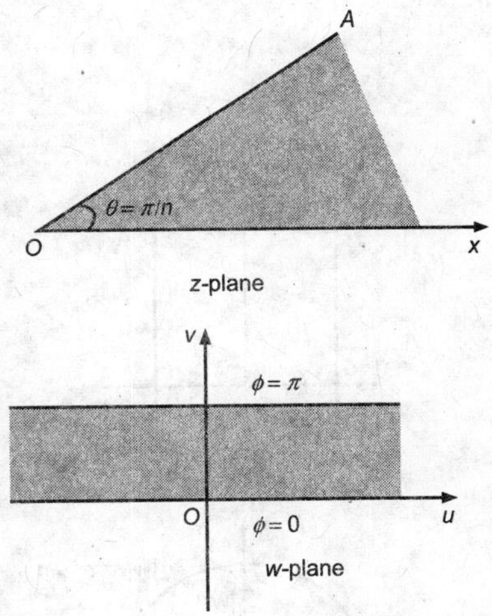

Figure 19.11 Mapping $w = z^n$

19.4.3 Transformation $w = e^z$

Consider the mapping $w = e^z$. Put $z = x + iy$, $x, y \in \mathbb{R}$

$$w = Re^{i\phi} = R(\cos\phi + i\sin\phi)$$

$$\Rightarrow e^z = e^x e^{iy} = e^x(\cos y + i \sin y) \text{ and}$$

$$w = e^z \Rightarrow |w| = R = e^x; \arg w = \phi = y.$$

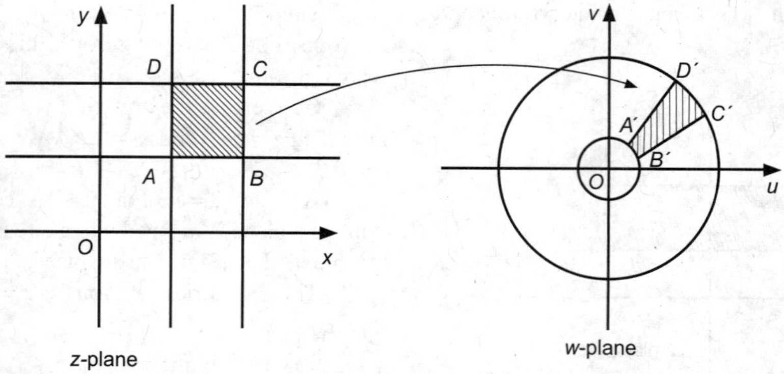

Figure 19.12 Mapping of rectangle into sectorial region in the w-plane under $w = e^z$

Vertical lines in the z-plane

Straight lines $x = c$ (constant), *i.e.*, vertical lines are mapped onto the circles $|w| = R = e^c$ (constant).

If $c > 0$ then the circles in the w-plane have radius greater than 1, if $c < 0$ then the circles have radius less than 1 and if $c = 0$ (y-axis) then the circle is of unit radius.

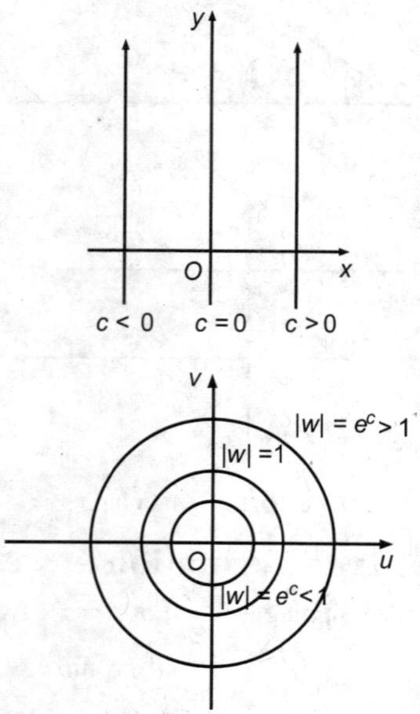

Figure 19.13 Vertical lines mapped into concentric circles under $w = e^z$

Horizontal lines in the z-plane

Straight lines $y = d$ (constant), *i.e.*, horizontal lines are mapped onto the radial lines $\phi = d$ (constant) in the w-plane.

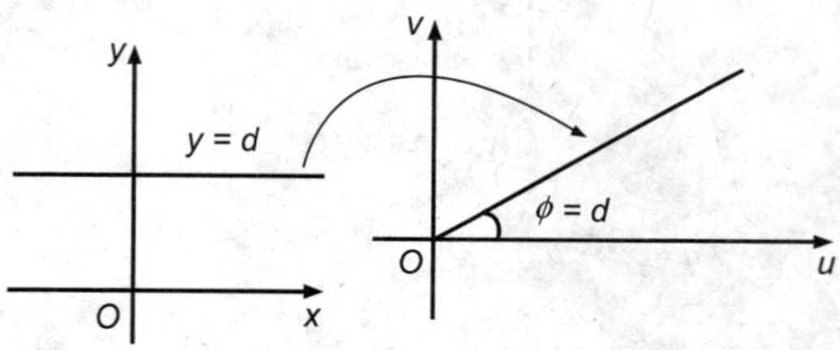

Figure 19.14 Horizontal lines mapped into rays under the mapping $w = e^z$

If $d > 0$ then the ray $\phi = d$ in the w-plane makes a positively oriented angle and if $d < 0$ it makes a negatively oriented angle with the real u-axis.

If $d = 0$ then we have the real axis (x-axis) in the z-plane, and it is mapped onto the real u-axis in the w-plane. The rectangular region $ABCD$ bounded by the straight lines $x = c_1$ and c_2, and $y = d_1$ and d_2 ($c_1 < c_2$, $d_1 < d_2$) in the z-plane is mapped onto the sectorial region $A'B'C'D'$ given by

$$e^{c_1} \le |w| \le e^{c_2} \quad d_1 \le \phi \le d_2 \quad \text{in the } w\text{-plane.}$$

19.4.4 Transformation $w = \sin z$

Consider the mapping

$$w = \sin z \tag{19.11}$$

put $z = x + iy$, $w = u + iv$ where $x, y, u, v \in R$

$$w = u + iv = \sin z = \sin(x + iy)$$

$$= \sin x \cosh y + i \cos x \sinh y \quad (19.12)$$

Equating the real and imaginary parts

$$u = \sin x \cosh y \; v = \cos x \sinh y \quad (19.13)$$

Since these are periodic functions with period 2π therefore Eq. (19.11) is periodic with period 2π. Hence the mapping is not one-to-one if we consider the whole of the xy-plane.

So, we restrict ourselves to the infinite vertical strip D_∞ defined by $\frac{-\pi}{2} \le x \le \frac{\pi}{2}$.

Since $f'(z) = \cos z = 0$ at $z = \pm\frac{\pi}{2}$, the mapping is not conformal at these points.

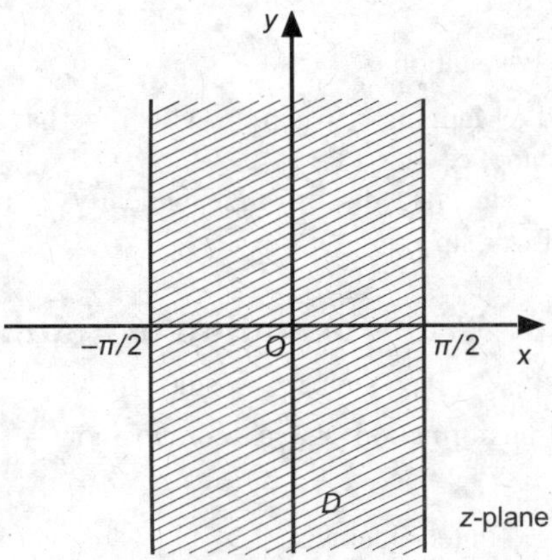

Figure 19.15

Horizontal line $y = 0$ (the x-axis)

For $y = 0$, $\sinh y = 0$; $\cosh y = 1$.

$\therefore$ Equation (19.13) $\Rightarrow u = \sin x$, $v = 0$.

Since $-1 \le \sin x \le 1$, u varies from -1 to 1 $\forall$ x.

Thus the segment $\frac{-\pi}{2} \le x \le \frac{\pi}{2}$ of the x-axis is mapped onto the segment $-1 \le u \le 1$ of the u-axis.

Vertical lines $x = \pm\frac{\pi}{2}$

For $x = \pm\frac{\pi}{2}$, we have $\sin x = \pm 1$, $\cos x = 0$.

Equations (19.13) $\Rightarrow u = \pm \cosh y$, $v = 0$.

Since $\cosh y \ge 1$ and $-\cosh y \le -1$ $\forall y$, the vertical line $x = -\frac{\pi}{2}$ maps onto $u \le -1$, $v = 0$ and the vertical line $x = \frac{\pi}{2}$ maps onto $u \ge 1$, $v = 0$.

Horizontal line segments $y = \pm k$ ($k > 0$), $-\frac{\pi}{2} \le x \le \frac{\pi}{2}$

Let $y = k > 0$.

Equation (19.13) $\Rightarrow$ $u = (\cosh k) \sin x$, $v = (\sinh k) \cos x$ or $\frac{u}{a} = \sin x$, $\frac{v}{b} = \cos x$, where $a = \cosh k$, $b = \sinh k$.

Squaring and adding, we get

$$\frac{u^2}{a^2} + \frac{v^2}{b^2} = 1 \quad (19.14)$$

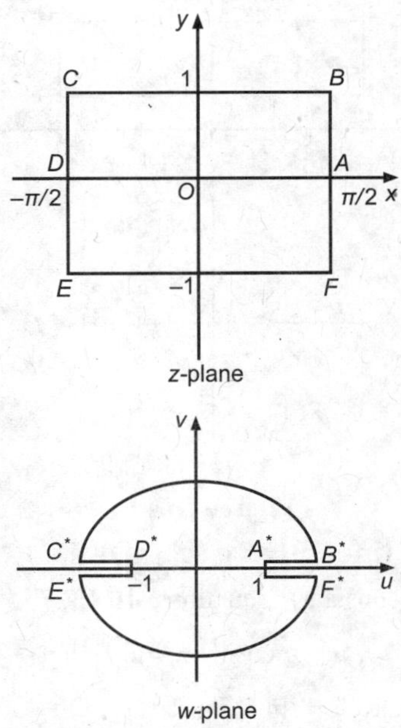

Figure 19.16 Mapping by $w = \sin z$

This equation represents an ellipse. Thus the line segment $y = k > 0\left(\frac{-\pi}{2} \le x \le \frac{\pi}{2}\right)$ is mapped onto the upper half portion of the ellipse showing in Fig. 19.16.

Similarly, the line segment $y = -k$ ($k > 0$), $\frac{-\pi}{2} \le x \le \frac{\pi}{2}$ maps onto the lower half of the ellipse shown in Fig. 19.16.

The focii of the ellipse are at $w = \pm 1$, which are independent of k.

As k varies, we obtain a family of confocal ellipses.

The rectangular region D defined by $\frac{-\pi}{2} < x < \frac{\pi}{2}$, $-k < y < k$ is thus mapped onto the interior of the ellipse (19.14).

Note that the image of the boundary consists of the ellipse (19.14) and the two segments of the x-axis as shown in Fig. 19.16 with $k = 1$.

Image of the rectangle $k_1 < y < k_2, -\pi < x < \pi$

Similarly, the rectangle $-\pi < x < \pi$, $k_1 < y < k_2$ maps onto the elliptic ring with a cut along the negative v-axis.

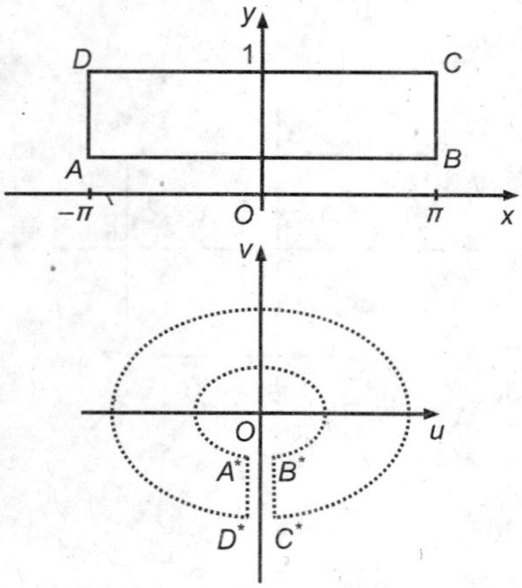

Figure 19.17

Vertical lines $x = \pm c$ $(c > 0)$

Let $x = c$ (constant). Equations (19.13)

$$\Rightarrow u = A \cosh y,\ v = B \sinh y$$

where $A = \sin c$, $B = \cos c$

$$\frac{u}{A} = \cosh y, \quad \frac{v}{B} = \sinh y$$

Squaring and subtracting

$$\frac{u^2}{A^2} - \frac{v^2}{B^2} = 1 \tag{19.15}$$

The equation represents a hyperbola. Thus the vertical lines $x = \pm c$ are mapped onto the confocal hyperbolas (with focii at $w = \pm 1$) on the right and left half-planes, respectively, and these curves cut the ellipses orthogonally.

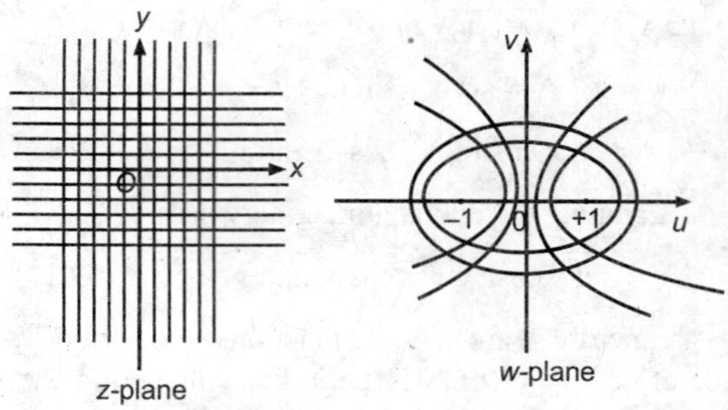

Figure 19.18

19.4.5 Transformation $w = \cos z$

Since $\cos z = \sin\left(z + \frac{\pi}{2}\right)$ the cosine transformation is a composition of the following:

$$w = \cos z = \sin\left(z + \frac{\pi}{2}\right) = \sin z_1, \quad z_1 = z + \frac{\pi}{2}$$

First : Translation $z_1 = z + \frac{\pi}{2}$

Second : Mapping $w = \sin z_1$, which we have discussed above.

We can also discuss independently as in the case of $w = \sin z$.

19.4.6 Transformation $w = \sinh z$

$$w = \sinh z = -i \sin(iz)$$

The transformation is a composition of the following:

(1) Counter-clockwise rotation through $\frac{\pi}{2}(= 90^\circ)$ $\quad z_1 = iz$.

(2) Sine mapping z2 = sin z1.

(3) Clockwise rotation through $\frac{\pi}{2}(= 90^\circ)$ $\quad w = -iz_2$.

19.4.7 Transformation $w = \cosh z$

$$w = \cosh z = \cos(iz)$$

The transformation is a composition of the following:

(1) Rotation through $\frac{\pi}{2}(= 90^\circ)$ in the counter-clockwise direction $z_1 = iz$.

(2) Cosine mapping $w = \cos z_1$.

19.4.8 Logarithm

The natural (Naperian) logarithm of $z = x + iy$ is denoted by $\ln z$ (sometimes also by $\log z$ or $\log e^z$). It is defined as the inverse of the exponential function, *i.e.*, $w = \ln z \Leftrightarrow e^w = z$ for $z \neq 0$.
(Since $e^w \neq 0$, $z = 0$ is impossible)

Put $z = re^{i\theta}$, $w = u + iv$

$$e^w = e^{u+iv} = re^{i\theta}$$
$$|e^w| = e^u \quad \because |e^{iv}| = 1$$
$$\arg e^w = v = 0$$
$$e^u = r \Rightarrow u = \ln r$$

(real lograrithm defined for $r > 0$).
Hence $w = u + iv = \ln z$ is given by

$$\ln z = \ln r + i\theta \quad r = |z| > 0, \theta = \arg z$$

Since the argument of z is determined only up to integer multiples of 2π, the complex natural logarithm $\ln z$ ($z \neq 0$) is infinitely many-valued.

The value of $\ln z$ corresponding to the principal value Arg z is denoted by Ln z and is called the *principal value* of $\ln z$.

Thus

$$\ln z = \ln |z| + i \text{ Arg } z \quad (z \neq 0)$$

For a given $z \neq 0$, Arg z is unique. This implies that $\ln z$ is single-valued and is therefore a function in the usual sense. Since the other values of arg z differ by integer multiples of 2π the *other values* of $\ln z$ are given by

$$\ln z = \text{Ln } z \pm 2n\pi i; \; n = 1, 2, 3 \ldots.$$

They all have the same real part and their imaginary parts differ by integer multiples of 2π. If z is positive real then Arg $z \geq 0$ and $\ln z$ becomes identical with the real natural logarithm of calculus. If z is negative real (natural log of calculus is not defined) then Arg $z = \pi$ and Ln $z = \ln |z| + \pi i$.

Conformal mapping by ln z -- the principle of inverse mapping

The mapping by the inverse $z = f^{-1}(w)$ of $w = f(z)$ is obtained by interchanging the roles of the z-plane and the w-plane in the mapping by $w = f(z)$.

Example 19.3

Mapping by $w = z^2$ and by its inverse, $w = z^2$ is restricted to the right half-plane $x > 0$ and maps this half-plane onto the w-plane without the negative half $u \leq 0$ of the real u-axis because at the origin, the angles are doubled, so that rays θ= Arg z = constant are mapped onto rays ϕ=Arg $w = 2\theta$. Hence its inverse $w = \sqrt{z}$ (with $\sqrt{z} > 0$ for $z = x > 0$) maps the z-plane without the negative half of the x-axis onto the right half-plane $u = Re\, w > 0$.

Natural logarithm

$\ln z$ is the inverse relation of the exponential function $w = e^z$. The latter maps a fundamental strip onto the w-plane without $z = 0$ because $e^z \neq 0$ for every z. Hence by the principle of inverse mapping, the principal value w = Ln z maps the z-plane with $z = 0$ omitted and cut along the real axis, when $\theta = Im(\ln z)$ jumps by 2π onto the horizontal strip $-\pi < v < \pi$ of the plane. Since the mapping $w = \text{Ln } z + 2\pi i$ differs from w = Ln z by the translation $2\pi i$ (vertically upward), this mapping maps the z-plane (cut as before and 0 omitted) onto the strip $-\pi < v \leq 3\pi$. Similarly for each of the many mappings

$$w = \ln z = \ln z \pm 2n\pi i \; (n = 0, 1, 2, \ldots)$$

The corresponding horizontal strips of width 2π images of the z-plane under these mappings together cover the whole of the w-plane without overlapping.

19.4.9 Transformation $w = z + \frac{1}{z}$ (Joukowski[1] Airfoil)

If in the mapping $w = z + \frac{1}{z}$ we put $z = re^{i\theta} = r$ ($\cos\theta + i\sin\theta$) and $w = Re^{i\phi} = R(\cos\phi + i\sin\phi)$ we have

$$w = u + iv = r(\cos\theta + i\sin\theta) + \frac{1}{r}(\cos\theta - i\sin\theta).$$

Equating the real and imaginary parts, we get

$$u = \left(r + \frac{1}{r}\right)\cos\theta, \quad v = \left(r - \frac{1}{r}\right)\sin\theta$$

or $u = a\cos\theta$ and $v = b\sin\theta$

where $a = r + \frac{1}{r}$ and $b = r - \frac{1}{r}$.

Consider the circles $|z| = r = constant$ in the z-plane. Since b must not be equal to zero, we have $r \neq 1$. These non-unit circles are mapped onto ellipses

[1] Joukowski, Nikolai Jegorowitch (1847–1921), Russian applied mathematician and aerodynamicist.

$$\frac{u^2}{a^2}+\frac{v^2}{b^2}=1 \text{ or } \frac{u^2}{\left(r+\frac{1}{r}\right)^2}+\frac{v^2}{\left(r-\frac{1}{r}\right)^2}=1.$$

The unit circle $|z|=r=1$ is mapped onto the line segment $-2 \le u \le 2$ of the u-axis. (real axis), since $a=r+\frac{1}{r}=1+1=2$ and $b=0$ yield $u=2\cos\theta$, $v=0$.

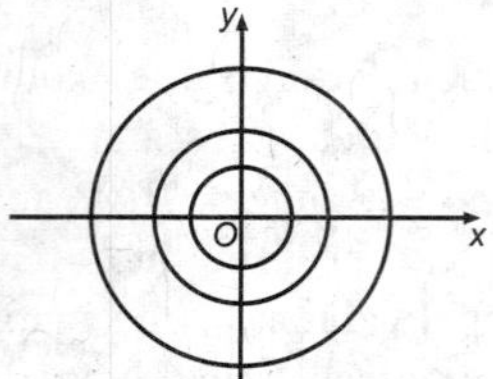

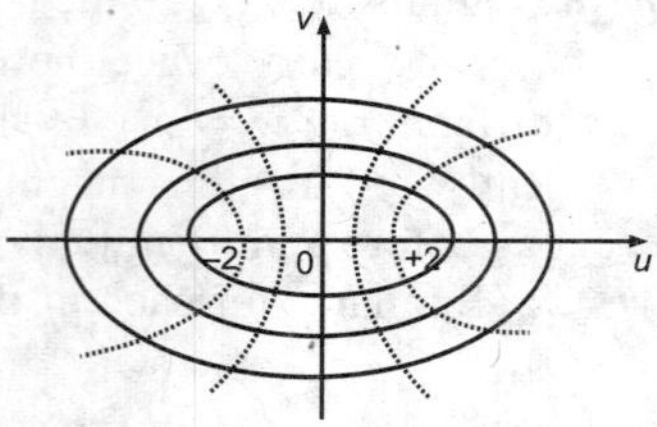

Figure 19.19 Mapping $w=z+1/z$ (Joukowski Airfoil)

Differentiating $w=z+\frac{1}{z}$ w.r.t z, we get

$$\frac{dw}{dz}=1-\frac{1}{z^2}=\frac{(z+1)(z-1)}{z^2}$$

Hence $\frac{dw}{dz}=0 \Rightarrow z=\pm 1$. These are points at which the mapping is not conformal.

Solving the equation $w=z+\frac{1}{z}$ for z, we get $z=\frac{w\pm\sqrt{w^2-4}}{2}$.

So, z is a double-valued function of w. For $r=a$ (constant) as mentioned above, this equation represents an ellipse with focii at

$$\sqrt{\left(r+\frac{1}{r}\right)^2-\left(r-\frac{1}{r}\right)^2}=\sqrt{4}=\pm 2$$

which are independent of r.

The Joukowski's function maps the family of concentric circles $|z|=a$ of the z-plane onto the family of confocal ellipses in the w-plane with focii at $w=\pm 2$.

Case (i) $|z|=1$ (Unit circle) As $r\to 1$, $v=0$, $u=2\cos\theta$ since $|\cos\theta|\le 1$, $-2\le u\le 2$.

This shows that the unit circle $|z|=1$ in the z-plane is mapped into a degenerated ellipse which flattens to the line segment $v=0$, $-2\le u\le 2$ on the real u-axis traversed twice.

Case (ii) As $r\to 0$ the ellipse is transformed into a circle of infinitely large radius. To find the image of the ray $\arg z=\theta_0$ eliminate the parameter n we get

$$\left(\frac{u}{\cos\theta}\right)^2-\left(\frac{v}{\sin\theta}\right)^2=\left(r+\frac{1}{r}\right)^2-\left(r-\frac{1}{r}\right)^2=4$$

$$\text{or} \quad \frac{u^2}{4\cos^2\theta_0}-\frac{v^2}{4\sin^2\theta_0}=1$$

which represents hyperbola with focii at $w=\pm 2$. Thus Joukowski's function defines the transformation of an orthogonal system of polar coordinates in the z-plane into an orthogonal curvilinear system of coordinates, whose coordinate lines are the confocal families of ellipses and hyperbolas in the w-plane.

Example 19.4

Show that the transformation $w=\frac{1}{z}$ maps the hyperbola $x^2-y^2=1$ to the lemniscate.

[JNTU 1985, 1998]

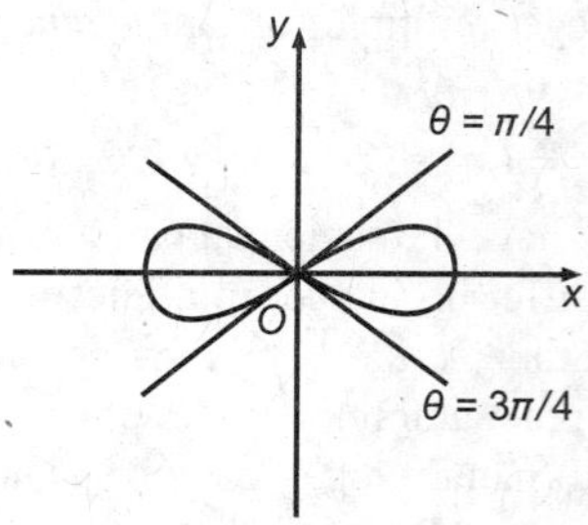

Figure 19.20

Solution The given curve in the z-plane is the hyperbola

$$x^2-y^2=1 \tag{1}$$

The given transformation is

$$w=\frac{1}{z} \tag{2}$$

Let $z=x+iy$, $w=u+iv$.

In polars $z=re^{i\theta}$, $w=Re^{i\phi}$. By Eq. (2) we get

$$Re^{i\phi} = \frac{1}{r}e^{-i\theta} \Rightarrow R = \frac{1}{r}\phi = -\theta$$

Now let us put Eq. (1) in polar form by substituting

$$x = r\cos\theta, y = r\sin\theta$$

Thus $x^2 - y^2 = 1$ gives $r^2 = \dfrac{1}{\cos 2\theta}$

$$\therefore \frac{1}{R^2} = \frac{1}{\cos(-2\phi)} \text{ or } R^2 = \cos 2\phi$$

This is the equation in polar coordinates for the lemniscate.

Example 19.5

Show that the transformation $w = \dfrac{1}{z}$ maps a circle to a circle or to a straight line in the w-plane if the former goes through the origin. [JNTU 2000]

Solution Let $z = x + iy$, $w = u + iv$ where $x, y, u, v \in \mathbb{R}$. Then

$$w = \frac{1}{z} \Rightarrow u + iv = \frac{1}{x+iy} \quad (1)$$

$$= \frac{x-iy}{x^2+y^2} \quad [\because (x+iy)(x-iy) = x^2+y^2]$$

Equating the real and imaginary parts, we get from Eq. (1)

$$u = \frac{x}{x^2+y^2}, \quad v = \frac{-y}{x^2+y^2} \quad (2)$$

Squaring and adding

$$u^2 + v^2 = \frac{x^2+y^2}{(x^2+y^2)^2} = \frac{1}{x^2+y^2} \quad (3)$$

$$x = u\cdot(x^2+y^2) = \frac{u}{u^2+v^2}\, y$$

$$= -v(x^2+y^2) = \frac{-v}{u^2+v^2} \quad (4)$$

The general equation of a circle in the z-plane is

$$A(x^2+y^2) + Bx + Cy + D = 0, \quad (5)$$

[where A, B, C, D are constants]

$$\Rightarrow A\left(\frac{1}{u^2+v^2}\right) + B\left(\frac{u}{u^2+v^2}\right) - C\left(\frac{v}{u^2+v^2}\right) + D = 0$$

$$\Rightarrow D(u^2+v^2) + Bu - Cv + A = 0 \; (D \neq 0) \quad (6)$$

Equation (6) represents a circle in the w-plane. Thus a circle (5) in the z-plane is transformed into a circle (6) in the w-plane. If $D = 0$ then it represents a straight line whose equation is

$$Bu - Cv + A = 0.$$

Example 19.6

Determine the region in the w-plane corresponding to the triangular region bounded by the lines $x = 0$, $y = 0$ and $x + y = 1$ in the z-plane mapped by the transformation $w = e^{i\frac{\pi}{4}z}$.

Solution The transformation is

$$w = u + iv = \left(\cos\frac{\pi}{4} + i\sin\frac{\pi}{4}\right)(x+iy)$$

$$= \frac{1+i}{\sqrt{2}}(x+iy) = \frac{x-y}{\sqrt{2}} + i\frac{x+y}{\sqrt{2}}$$

$$\Rightarrow u = \frac{x-y}{\sqrt{2}} \text{ and } v = \frac{x+y}{\sqrt{2}}$$

The line $x = 0$ maps into

$$u = \frac{-y}{\sqrt{2}} \text{ and } v = \frac{y}{\sqrt{2}} \text{ or } v = -u.$$

The line $y = 0$ maps into

$$u = \frac{x}{\sqrt{2}} \text{ and } v = \frac{x}{\sqrt{2}} \text{ or } v = u.$$

The line $x + y = 1$ maps into $v = \dfrac{1}{\sqrt{2}}$.

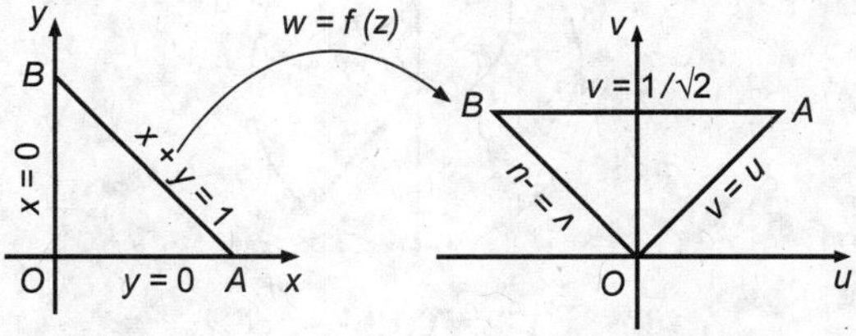

Figure 19.21 The region in the z-plane and the w-plane mapped by the transformation $w = e^{i\frac{\pi}{4}z}$

Example 19.7

Find the image of the infinite ship $\dfrac{1}{4} \leq y \leq \dfrac{1}{2}$ under the transformation $w = \dfrac{1}{z}$ and show the regions graphically.

Solution The transformation is

$$w=\frac{1}{z}\Rightarrow z=\frac{1}{w}$$

$$\Rightarrow x+iy=\frac{1}{u+iv}=\frac{u-iv}{u^2+v^2}$$

$$\Rightarrow x=\frac{u}{u^2+v^2}\quad\text{and}\quad y=\frac{-v}{u^2+v^2}$$

Now, $y=\frac{1}{4}\Rightarrow\frac{-v}{u^2+v^2}=\frac{1}{4}$

$$\Rightarrow u^2+v^2+4v=0$$

$$\Rightarrow u^2+(v+2)^2=2^2$$

Also, $y=\frac{1}{z}\Rightarrow\frac{-v}{u^2+v^2}=\frac{1}{2}$

$$\Rightarrow u^2+v^2+2v=0$$

$$\Rightarrow u^2+(v+1)^2=1$$

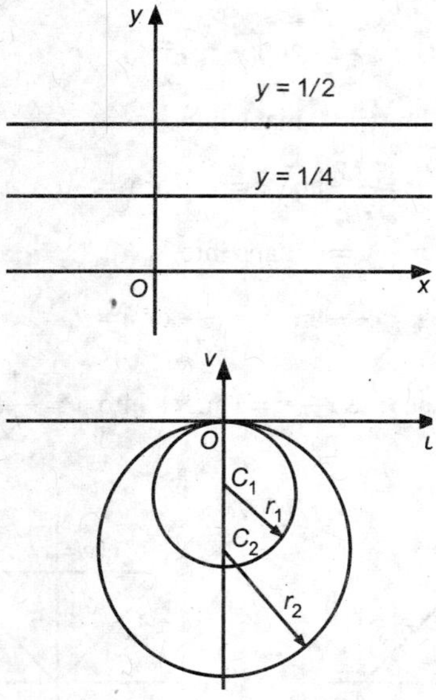

Figure 19.22

$\therefore$ The infinite strip $\frac{1}{4}\le y\le\frac{1}{2}$ is transformed into the region between the two circles:

$u^2+(v+2)^2=2^2$; centre $c_2=(0,-2)$ and radius $r_2=2$ units.

$u^2+(v+1)^2=1^2$; centre $c_1=(0,-1)$ and radius r_1 = 1 units.

Example 19.8

Plot the image of the region $1<|z|<2$ under the transformation $w=2iz+1$. [JNTU 1995]

Solution Let $z=x+iy$, $w=u+iv$ where $x, y, u, v\in\mathbb{R}$.

Now, $w=2iz+1\Rightarrow u+iv=2i(x+iy)+1$

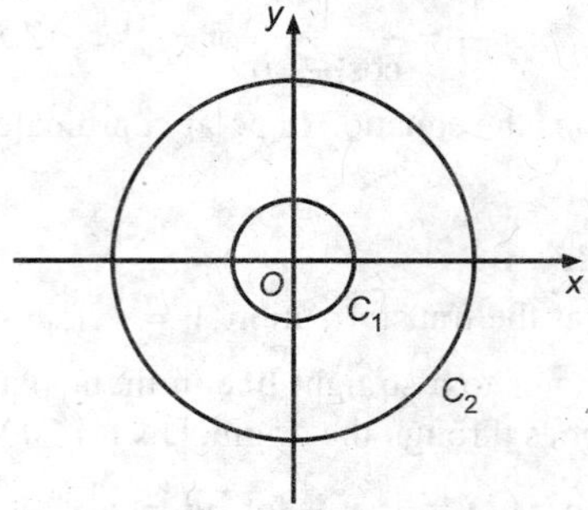

Figure 19.23

Equating the real and imaginary parts, we get $u=-2y+1$ and $v=2x$, *i.e.*, $x=\frac{v}{2}$ and $y=\frac{1-u}{2}$.

The given region $1<|z|<2$ is the annulus bounded by the circles $C_1:|z|=1$ and $C_2:|z|=2$.

$$|z|=1\Rightarrow x^2+y^2=1 \tag{1}$$

$$\Rightarrow (u-1)^2+v^2=1^2 \tag{2}$$

The unit circle C_1: $x^2+y^2=1$ in the z-plane is mapped to the circle $C_1^*:(u-1)^2+v^2=4$ with its centre at $u=1$, $v=0$ and radius = 2 units.

Again

$$|z|=2\Rightarrow x^2+y^2=4 \tag{3}$$

$$\Rightarrow (u-1)^2+v^2=4 \tag{4}$$

The circle C_2: $x^2+y^2=2^2$ with radius = 2 units in the z-plane is mapped to the circle $C_2^*:(u-1)^2+v^2=4^2$ with centre at $u=1$, $v=0$ and radius = 4 units.

Thus, the annular region $1<|z|<2$ is mapped to the annular region $2<|w-1|<4$.

Example 19.9

Find the image of the triangle with vertices i, $1+i$, $1-i$ in the z-plane under the transformation $w=3z+4-2i$. [JNTU 2003]

Solution Let $z = x + iy$ and $w = u + iv$ where x, y, u and $v \in \mathbb{R}$.

$$\therefore u + iv = 3(x + iy) + 4 - 2i$$

Equating the real and imaginary parts, we have

$$u = 3x + 4, \quad (1)$$

$$v = 3y - 2 \quad (2)$$

The coordinates of the vertices of the triangle are

$$z_1 = i = (0, 1),$$

$$z_2 = 1 + i = (1, 1),$$

$$z_3 = 1 - i = (1, -1)$$

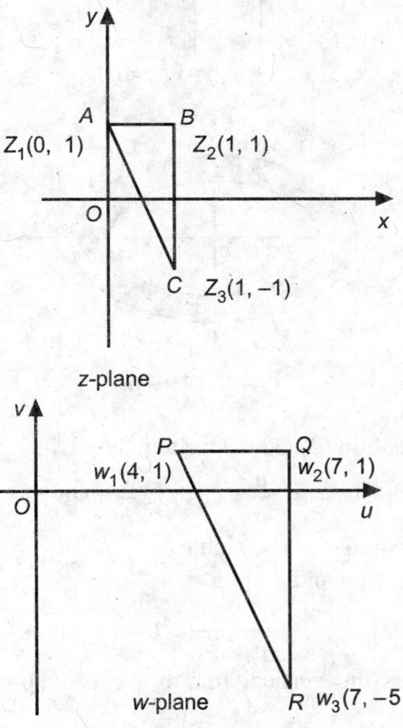

Figure 19.24

The images of the points in the w-plane are

$$(0, 1) \to (4, 1)$$

$$(1, 1) \to (7, 1)$$

$$(1, -1) \to (7, -5)$$

ΔABC in the z-plane is mapped to the triangle ΔPQR in the w-plane.

Example 19.10

Show that the transformation $w = z + \frac{1}{z}$ transforms r = constant in the z-plane into a family of ellipses in the w-plane.

Solution Let

$$z = re^{i\theta} = r(\cos\theta + i\sin\theta) \text{ and } w = u + iv \quad (1)$$

This gives

$$w = u + iv = z + \frac{1}{z} = re^{i\theta} + \frac{1}{r}e^{-i\theta}$$

$$= \left(r + \frac{1}{r}\right)\cos\theta + i\left(r - \frac{1}{r}\right)\sin\theta \quad (2)$$

$$\Rightarrow u = \left(r + \frac{1}{r}\right)\cos\theta \quad v = \left(r - \frac{1}{r}\right)\sin\theta$$

Let $r = k$ (constant) $\neq 1$ so that

$$\frac{u^2}{\left(k + \frac{1}{k}\right)^2} + \frac{v^2}{\left(k - \frac{1}{k}\right)^2} = 1 \quad (3)$$

This equation represents a family of ellipses with semi-major axis $= a = k + \frac{1}{k}$, semi-minor axis $= b = k - \frac{1}{k}$.

If the eccentricity is e, then

$$e^2 = a^2 - b^2 = \left(k + \frac{1}{k}\right)^2 - \left(k - \frac{1}{k}\right)^2 = 4.$$

The focii of the ellipses are at

$$S = (\sqrt{a^2 - b^2}, 0) = (2, 0).$$

The focii of the ellipses are at

$$S' = (-\sqrt{a^2 - b^2}, 0) = (-2, 0).$$

Example 19.11

Under the transformation $w = \frac{1}{z}$ find the image of $|z - 2i| = 2$. [JNTU 2001]

Solution Let

$$z = x + iy, \ w = u + iv \quad \text{where } x, y, u, v \in \mathbb{R}$$

$$w = \frac{1}{z} \Rightarrow u + iv = \frac{1}{x + iy}\,\frac{x - iy}{x - iy} = \frac{x - iy}{x^2 + y^2} \quad (1)$$

Equating the real and imaginary parts, we get

$$u = \frac{x}{x^2 + y^2}, \qquad v = \frac{-y}{x^2 + y^2} \quad (2)$$

Squaring and adding,

$$u^2+v^2=\frac{x^2+y^2}{(x^2+y^2)^2}=\frac{1}{x^2+y^2} \qquad (3)$$

$$\therefore x=u(x^2+y^2)=\frac{u}{u^2+v^2}$$

$$y=-v(x^2+y^2)=\frac{-v}{u^2+v^2} \quad \text{by Eq. (3)} \qquad (4)$$

The given curve in the z-plane is

$$|z-2i|=2 \Rightarrow x^2+(y-2)^2=4$$

$$\Rightarrow x^2+y^2-4y=0$$

which is a circle with centre at (0, 2) and radius = 2.

Substituting for x and y, we have

$$\frac{u^2}{(u^2+v^2)^2}+\frac{v^2}{(u^2+v^2)^2}-\frac{-4v}{u^2+v^2}=0$$

$$\Rightarrow \frac{1}{u^2+v^2}+\frac{4v}{u^2+v^2}=0 \Rightarrow 4v+1=0$$

This is a horizontal straight line in the w-plane.

Example 19.12

If $w=\dfrac{1+iz}{1-iz}$, find the image of $|z|<1$. [JNTU 2002].

Solution

$$w=\frac{1+iz}{1-iz} \Rightarrow w(1-iz)=1+iz$$

$$\Rightarrow iz(1+w)=w-1 \Rightarrow z=\frac{1}{i}\frac{w-1}{w+1}$$

$$\text{Now, } |z|<1 \Rightarrow \left|\frac{1}{i}\frac{w-1}{w+1}\right|<1$$

$$\Rightarrow |w-1|<|i||w+1|=|w+1|$$

Putting $w=u+iv$, we get

$$(u-1)^2+v^2<(u+1)^2+v^2$$

$$\Rightarrow u^2+v^2-2u+1<u^2+v^2+2u+1$$

$$\Rightarrow u>0$$

Thus the image of the interior of the circle $|z|=1$ is mapped onto the right half of the w-plane.

Example 19.13

If $w=\cosh z$, find the mapping of the region $a \le x \le b$ onto the w-plane.

Solution Let $z=x+iy$ and $w=u+iv$ where $x, y, u, v \in \mathbb{R}$

$$\therefore w=\cosh z \Rightarrow u+iv=\cosh(u+iv) \qquad (1)$$

$$=\cosh x\cosh y+i\sinh x\sinh y$$

$$\Rightarrow u=\cosh x\cosh y,\ v=\sinh x\sinh y. \qquad (2)$$

Vertical line $x = c$

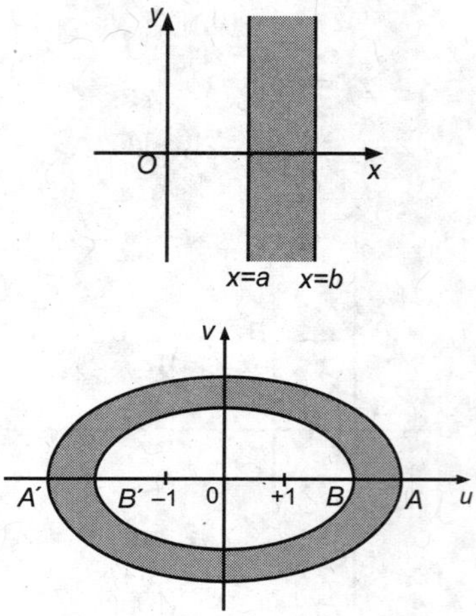

Figure 19.25

Equation (2) can be written as

$$\frac{u}{a}=\cos y, \quad \frac{v}{b}=\sin y$$

where $a=\cosh c$, $b=\sinh c$.

Squaring and adding

$$\frac{u^2}{a^2}+\frac{v^2}{b^2}=1 \qquad (3)$$

Thus the vertical line $x = c$ is mapped onto ellipse (3).

Hence the given vertical strip $a \le x \le b$ in the z-plane is mapped onto the annular region between two coaxial confocal ellipses (3) with $c = a$ and $c = b$ and focii at ± 1.

Example 19.14

Show that the transformation $w=\dfrac{2z+3}{z-4}$ maps the circle $x^2+y^2-4x=0$ into the straight line $4u+3=0$.
[JNTU 2003]

Solution Let

$$z = x + iy, \quad \bar{z} = x - iy$$

$$x^2 + y^2 = (x+iy)(x-iy) = z\bar{z}$$

$$x = \frac{1}{2}(z+\bar{z}), \quad y = \frac{1}{2i}(z-\bar{z})$$

The given transformation is $w = \dfrac{2z+3}{z-4}$

$$\Rightarrow z(w-2) = 4w+3, \text{ or } z = \frac{4w+3}{w-2}.$$

The given equation of the circle is

$$x^2 + y^2 - 4x = 0 \Rightarrow z\bar{z} - 2(z+\bar{z}) = 0$$

$$\Rightarrow \left(\frac{4w+3}{w-2}\right)\left(\frac{4\bar{w}+3}{\bar{w}-2}\right) = 2\left(\frac{4w+3}{w-2}\right) + 2\left(\frac{4\bar{w}+3}{\bar{w}-2}\right)$$

$$\Rightarrow (4w+3)(4\bar{w}+3) = 2(4w+3)(\bar{w}-2) + 2(w-2)(4\bar{w}+3)$$

$$\Rightarrow 16w\bar{w} + 12(w+\bar{w}) + 9 = 2[4w\bar{w} - 8w + 3\bar{w} - 6] + 2[4w\bar{w} - 8\bar{w} + 3w - 6]$$

$$\Rightarrow 22(w+\bar{w}) + 33 = 0$$

$$\Rightarrow 4u + 3 = 0$$

This represents a straight line in the *w*-plane.

Example 19.15
Find the image of the rectangle *R*; $-\pi < x < \pi$, $\frac{1}{2} < y < 1$ under the transformation $w = \sin z$. [JNTU 2003*S*].

Solution Let $z = x + iy$ and $w = u + iv$, then

$$w = u + iv = \sin(x + iy)$$

$$= \sin x \cosh y + i \cos x \sinh y$$

$$\Rightarrow u = \sin x \cosh y, \; v = \cos x \sinh y$$

The horizontal line *FED*: $y = 1$ maps onto the outer ellipse

$$\frac{u^2}{\cosh^2 1} + \frac{v^2}{\sinh^2 1} = 1$$

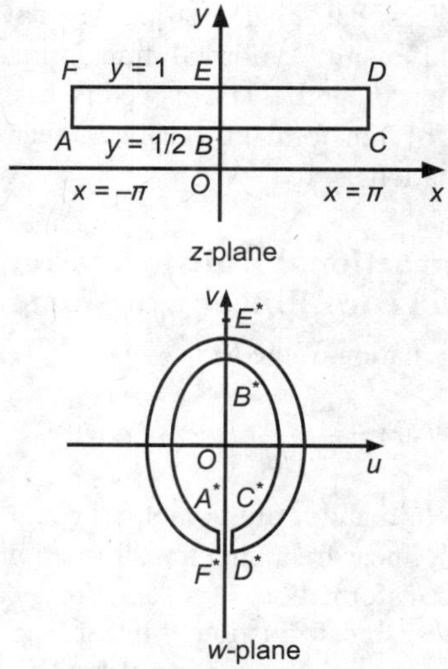

Figure 19.26

and the horizontal line $ABC: y = \frac{1}{2}$ maps onto the inner ellipse

$$\frac{u^2}{\cosh^2\frac{1}{2}} + \frac{v^2}{\sinh^2\frac{1}{2}} = 1$$

Also, the vertical line segment *CD* given by $x = \pi\left(\frac{1}{2} < y < 1\right)$ maps onto C^*D^* given by $u = 0$ and $-\sinh\frac{1}{2} < v < -\sinh 1$.

The vertical line segment A^*F^* given by $x = -\pi$, $\frac{1}{2} < y < 1$ maps onto A^*F^* given by $u = 0$ and $-\sinh\frac{1}{2} < v < -\sinh\frac{1}{2}$.

Thus both C^*D^* and A^*F^* map onto the same line in the *w*-plane given by the cut A^*F^* or C^*D^* along the imaginary axis *Ov*.

19.5 Bilinear or Mobius[2] or Linear Fractional Transformations

Introduction

Conformal mappings are useful in solving boundary value problems. This is done by mapping complicated regions conformally onto standard regions like

[2] Mobius, August Ferdinand (1790--1868) German mathematician, student of Gauss known for his work in the theory of surfaces, projective geometry and mechanics. He showed that the Mobius strip is an example of a non-orientable surface.

circular disks, half-planes, strips. We have already considered many conformal transformations and studied their properties. Here, we discuss an important class of transformations called linear fractional transformations.

Linear fractional transformation (LFT) or Mobius or Bilinear transformation

A transformation of the form

$$w = f(z) = \frac{az+b}{cz+d} \quad (ad - bc \neq 0) \tag{19.16}$$

where a, b, c and d are complex or real constants is called a linear fractional transformation (LFT) or Mobius transformation. It is sometimes called simply as linear transformation or linear map.

Cross-multiplying, we can write Eq. (19.16) in the form

$$Awz + Bz + Cw + D = 0 \tag{19.17}$$

which is linear in z as well as in w. So, it is also called a bilinear transformation.

Differentiating Eq. (19.16) w.r.t z, we have

$$\frac{dw}{dz} = \frac{(cz+d)a-(az+b)c}{(cz+d)^2} = \frac{(ad-bc)}{(cz+d)^2} \tag{19.18}$$

If $(ad - bc) = 0$ then $\frac{dw}{dz} = 0$. So, the transformation is not conformal at any z if $ad - bc = 0$.

Hence we assume that the constants a, b, c, d are such that $ad - bc \neq 0$ in which case we have conformal transformation for all z.

The transformation at Eq. (19.16) is very interesting. First of all, for different choices of the constants a, b, c, d we get the following special cases of LFTs:

$a = 1, c = 0, d = 1$

$$\Rightarrow w = z + b \text{ (translation)} \tag{19.19}$$

$$b = 0, c = 0, d = 1 \Rightarrow w = az \tag{19.20}$$

Simply rotation if $|a| = |e^{i\theta}| = 1$
Magnification or dimination if

$$a = |a| e^{i\theta} (|a| = 1) \tag{19.21}$$

$c = 0$, $d = 1$ $w = az + b$ (linear transformation) (19.22)

$$a = 0, \quad b = 1, \quad c = 1, \quad d = 0$$

$$w = \frac{1}{z} \text{ (inversion)} \tag{19.23}$$

So, the general LFT is composed of all the above types of transformation, which we have already discussed above.

Also, corresponding to each z, Eq. (19.16) gives a unique value of w except when $cz + d = 0$.

(1) $c \neq 0$. In this case, there does not correspond a number w to the value $-\frac{d}{c}$ of z.
To get over this situation, we attach the number '∞' which we call 'the point at infinity'. The complex plane then is called the *extended* complex plane.
(2) $c = 0$. In this case, $a \neq 0$ and $d \neq 0$ in view of the fact that $ad - bc \neq 0$. In this case, $w = \infty$ is the image of the point $z = \infty$.

Inverse mapping of $w = \frac{az+b}{cz+d}$

$$w = \frac{az+b}{cz+d} \Rightarrow w(cz+d) = az + b \text{ or}$$

$$z(wc - a) = b - wd \Rightarrow z = \frac{(-d)w+b}{cw+(-a)} \tag{19.24}$$

This is the inverse of the transformation (19.16)

(i) $c \neq 0. cw - a = 0 \Rightarrow w = \frac{a}{c}$. This point corresponds to $z = \infty$.
(ii) $c = 0$. In view of $ad - bc \neq 0$ we get $a \neq 0$ and $d \neq 0$. Therefore, $w = \infty$ corresponds to $z = \infty$.

Hence the mapping is bijective.

19.6 Fixed Points of the Transformation $w = \frac{az+b}{cz+d}$

A point z is called a fixed point of the transformation if it is such that $f(z) = z$ (not an identity mapping). So, fixed points of Eq. (19.16) are obtained from

$$z = \frac{az+b}{cz+d} \quad or \quad cz^2 - (a-d)z - b = 0 \tag{19.25}$$

This being a quadratic in z has, in general, two values for z. Thus an LFT has two fixed points if it is not an identity.

If an LFT has more than two fixed points, then it must be an identity mapping $w = z$.

An LFT maps the totality of circles and straight lines in the z-plane into the totality of circles and straight lines in the w-plane.

Though there are four constants in Eq. (19.16), on dividing three of them by the fourth non-zero constant we see that the effective constants are three and the mapping can be determined uniquely through three conditions.

Suppose for example $d \neq 0$. Then $w = f(z) = \dfrac{az+b}{cz+d}$ can be written as $w = \dfrac{\left(\frac{a}{d}\right)z + \left(\frac{b}{d}\right)}{\left(\frac{c}{d}\right)z+1}$.

Hence bilinear transformation consists of three independent constants

$$\frac{a}{d}, \ \frac{b}{d}, \ \frac{c}{d}.$$

So, three conditions are required to determine the transformation. That is to say, three distinct points z_1, z_2, z_3 in the z-plane can be mapped into three distinct points w_1, w_2, w_3 in the w-plane.

Cross-ratio of four points

Let z_1, z_2, z_3 and z_4 be four distinct points in the extended complex plane.

Then the cross-ratio of these points denoted by (z_1, z_2, z_3, z_4) is defined as

$$(z_1, z_2, z_3, z_4) = \frac{(z_1 - z_3)(z_2 - z_4)}{(z_1 - z_4)(z_2 - z_3)}$$

if none of z_1, z_2, z_3, z_4 are ∞.

$$= \frac{z_1 - z_3}{z_1 - z_4} \quad \text{if } z_2 \text{ is } \infty$$

$$\left(\because \frac{z_2 - z_4}{z_2 - z_3} = \frac{1 - \left(\frac{z_4}{z_2}\right)}{1 - \left(\frac{z_3}{z_2}\right)} \to 1 \text{ as } z_2 \to \infty \right)$$

$$= \frac{z_2 - z_4}{z_1 - z_4} \quad \text{if } z_3 \text{ is } \infty$$

$$\left(\because \frac{z_1 - z_3}{z_2 - z_3} = \frac{1 - \left(\frac{z_1}{z_3}\right)}{1 - \left(\frac{z_2}{z_3}\right)} \to 1 \text{ as } z_3 \to \infty \right)$$

$$= \frac{z_1 - z_3}{z_2 - z_3} \quad \text{if } z_4 \text{ is } \infty$$

$$\left(\because \frac{z_2 - z_4}{z_1 - z_4} = \frac{1 - \left(\frac{z_2}{z_4}\right)}{1 - \left(\frac{z_1}{z_4}\right)} \to 1 \text{ as } z_4 \to \infty \right)$$

$$= \frac{z_2 - z_4}{z_2 - z_3} \quad \text{if } z_1 \text{ is } \infty$$

$$\left(\because \frac{z_1 - z_3}{z_1 - z_4} = \frac{1 - \left(\frac{z_3}{z_1}\right)}{1 - \left(\frac{z_4}{z_1}\right)} \to 1 \text{ as } z_1 \to \infty \right)$$

Theorem 19.2 Prove that any bilinear transformation is the product of basic transformation namely translation, rotation, magnification or contraction and inversion.

Proof: Let

$$w = f(z) = \frac{az+b}{cz+d} \quad (ad - bc \neq 0) \tag{19.26}$$

Eq. (1) be the given bilinear transformation.

Case (i) Let $c = 0$ so that $ad - bc = ad \neq 0$ this implies that $d \neq 0$.

From Eq. (19.26) we get

$$w = \frac{az+b}{d} = \left(\frac{a}{d}\right)z + \left(\frac{b}{d}\right)$$

Let $f_1(z) = \left(\frac{a}{d}\right)z$ and $f_2(z) = z + \left(\frac{b}{d}\right)$

Then f_1 and f_2 are elementary transformations and

$$(f_1\, f_2)(2) = f_2(f_1(z)) = f_2\left(\frac{a}{d}z\right)$$

$$= \left(\frac{a}{d}\right)z + \left(\frac{b}{d}\right) = f(z).$$

Case (ii) Let $c \neq 0$. Equation (19.26) gives

$$w = f(z) = \frac{az+b}{cz+d} = \frac{a\left(z + \frac{d}{c}\right) + b - \frac{ad}{c}}{c\left(z + \frac{d}{c}\right)}$$

$$= \left(\frac{a}{c}\right) + \frac{b - \frac{ad}{c}}{cz+d}$$

Let $f_1(z) = cz + d, \quad f_2(z) = \dfrac{1}{z},$

$$f_3(z) = \left(b - \frac{ad}{c}\right)z, \quad f_4(z) = \frac{a}{c} + z; \text{ then}$$

$$(f_4\, f_3\, f_2\, f_1)(z) = (f_4\, f_3\, f_2)(cz + d)$$

$$= (f_4\, f_3)\left(\frac{1}{cz+d}\right) = f_4\left(\frac{b - \frac{ad}{c}}{cz+d}\right)$$

$$= \frac{a}{c} + \frac{b - \frac{ad}{c}}{cz+d} = f(z)$$

Thus $f(z)$ is the product of f_1, f_2, f_3 and f_4.
Hence the theorem.

Corollary Each elementary transformation as seen earlier, transforms circles and straight lines into circles and straight lines. It may be noted that a bilinear transformation maps circles and straight lines onto circles and straight lines.

If we observe that straight lines are linearising cases of circles with infinite radius we may conclude that a bilinear transformation maps circles into circles.

Theorem 19.3 Any bilinear transformation preserves the cross-ratio.

Proof: Let

$$w = f(z) = \frac{az+b}{cz+d} (ad - bc \neq 0) \tag{19.27}$$

be the given bilinear transformation.

Let z_1, z_2, z_3 and z_4 be four distinct points of the z-plane and w_1, w_2, w_3 and w_4 be then corresponding images in the w-plane under the transformation (19.27). Suppose all these points are different from ∞. Now, we have to prove that $(z_1, z_2, z_3, z_4) = (w_1, w_2, w_3, w_4)$.

From Eq. (19.27), we have

$$w_i = \frac{az_i + b}{cz_i + d} \quad (i = 1,2,3,4)$$

$$w_1 - w_3 = \frac{az_1 + b}{cz_1 + d} - \frac{az_3 + b}{cz_3 + d}$$

$$= \frac{(ad - bc)(z_1 - z_3)}{(cz_1 + d)(cz_3 + d)} = k_1(z_1 - z_3)$$

where $k_1 = \dfrac{ad - bc}{(cz_1 + d)(cz_3 + d)}$

Similarly, $w_2 - w_4 = k_2(z_2 - z_4)$

where $k_2 = \dfrac{ad - bc}{(cz_2 + d)(cz_4 + d)}$

$$\therefore (w_1 - w_3)(w_2 - w_4) = k_1 k_2 (z_1 - z_3)(z_2 - z_4)$$
$$= k(z_1 - z_3)(z_2 - z_4)$$

where $k = k_1 k_2 = \dfrac{(ad - bc)^2}{\pi_{i=1}^{4}(cz_i + d)}$

Similarly, we can derive the result

$$(w_1 - w_4)(w_2 - w_3) = k(z_1 - z_4)(z_2 - z_3)$$

$$\Rightarrow \frac{(w_1 - w_3)(w_2 - w_4)}{(w_1 - w_4)(w_2 - w_3)} = \frac{(z_1 - z_3)(z_2 - z_4)}{(z_1 - z_4)(z_2 - z_3)}$$

$$\Rightarrow (z_1, z_2, z_3, z_4) = (w_1, w_2, w_3, w_4)$$

Hence the theorem. The proof when one of z_i or w_i is ∞ is similar.

Note
Four distinct points z_i (i = 1, 2, 3, 4) are collinear or concyclic $\Leftrightarrow$ (z_1, z_2, z_3, z_4) is real. Also any bilinear transformation preserves a cross-ratio. Hence it follows that circles and straight lines are mapped onto circles and straight lines. This gives another proof for the corollary of the above theorem.

Theorem 19.4 The transformation $(z_1, z_2, z_3, z_4) = (w_1, w_2, w_3, w_4)$ is a bilinear transformation that maps three distinct points z_1, z_2 and z_3 onto three specified distinct points w_1, w_2 and w_3, respectively.

Proof: Let

$$(z_1, z_2, z_3, z_4) = (w_1, w_2, w_3, w_4) \tag{19.28}$$

$$\Rightarrow \frac{(w - w_1)(w_2 - w_3)}{(w - w_3)(w_1 - w_2)} = \frac{(z - z_1)(z_2 - z_3)}{(z - z_3)(z_1 - z_2)}$$

$$\Rightarrow (w - w_1)(w_2 - w_3)(z - z_3)(z_1 - z_2)$$
$$= (w - w_3)(w_1 - w_2)(z - z_1)(z_2 - z_3) \tag{19.29}$$

$z = z_1 \Rightarrow w = w_1$ and

$z = z_3 \Rightarrow w = w_3$ from *Eq.* (19.29).

Also $z = z_2$

$$\Rightarrow (w-w_1)(w_2-w_3)(z_2-z_3)(z_1-z_2)$$
$$= (w-w_3)(w_1-w_2)(z_2-z_1)(z_2-z_3)$$
$$\Rightarrow (w-w_1)(w_2-w_3) = -(w-w_3)(w_1-w_2).$$

Cancelling out non-zero factors $(z_2 - z_3)$, $(z_1 - z_2)$, we get

$$w(w_2 - w_3 + w_1 - w_2)$$
$$= w_2w_1 - w_3w_2 + w_1w_2 - w_1w_3$$
$$\Rightarrow w(w_1 - w_3) = w_2(w_1 - w_3)$$
$$\Rightarrow w = w_2 \qquad \because w_1 \neq w_3$$

Hence, the required transformation is given by Eq. (19.28).

Example 19.16
Show that the relation $w = \dfrac{5-4z}{4z-2}$ transforms the circle $|z| = 1$ into a circle of unit radius in the w-plane. [JNTU 2003]

Solution Let $z = x + iy$ and $w = u + iv$ so that $\bar{z} = x - iy$, $\bar{w} = u - iv$.

The given transformation is

$$w = \frac{5-4z}{4z-2} \Rightarrow w(4z-2) = 5-4z$$
$$\Rightarrow z(4w+4) = 2w+5$$
$$\Rightarrow z = \frac{2w+5}{4w+4} \text{ and } \bar{z} = \overline{\left(\frac{2w+5}{4(w+1)}\right)} = \frac{2\bar{w}+5}{4(\bar{w}+1)}$$

The given circle is

$$|z| = 1 \Rightarrow z\bar{z} = 1 \Rightarrow \frac{2w+5}{4(w+1)}\frac{2\bar{w}+5}{4(\bar{w}+1)} = 1$$
$$\Rightarrow (2w+5)(2\bar{w}+5) - (4w+4)(4\bar{w}+4) = 0$$
$$\Rightarrow 4w\bar{w} + 10(w+\bar{w}) + 25 - 16w\bar{w} - 16(w+\bar{w}) - 16 = 0$$
$$\Rightarrow 12w\bar{w} + 6(w+\bar{w}) - 9 = 0$$
$$\Rightarrow w\bar{w} + \frac{w+\bar{w}}{2} - \frac{3}{4} = 0 \Rightarrow u^2 + v^2 + u - \frac{3}{4} = 0$$
$$\Rightarrow \left(u + \frac{1}{2}\right)^2 + v^2 = \frac{3}{4} + \frac{1}{4} = 1$$

This is the circle with centre at $\left(-\frac{1}{2}, 0\right)$ and radius = 1 in the w-plane.

Example 19.17
Find the bilinear transformation that maps the points ∞, i, 0 into the points 0, i, ∞, respectively. [JNTU 2003*S*]

Solution By the invariance of cross-ratios, we can write

$$\frac{(z-z_1)(z_2-z_3)}{(z-z_3)(z_2-z_1)} = \frac{(w-w_1)(w_2-w_3)}{(w-w_3)(w_2-w_1)}$$

Here $z_1 = \infty$, $z_2 = i$, $z_3 = 0$; $w_1 = 0$, $w_2 = i$, $w_3 = 0$

$$\therefore \frac{(z-\infty)(i-0)}{(z-0)(i-\infty)} = \frac{(w-0)(i-\infty)}{(w-\infty)(i-0)} \Rightarrow \frac{i}{z} = \frac{w}{i}.$$

Replacing $\dfrac{z-\infty}{i-\infty}$ and $\dfrac{i-\infty}{w-\infty}$ by 1, according to the convention, we get $w = -\dfrac{1}{z}$ as the required bilinear transformation.

Example 19.18
Find the bilinear transformation which maps the points $z = 1, i, -1$ into the points $w = 0, 1, \infty$.

Solution By the invariance of cross ratios, we have

$$\frac{(z-z_1)(z_2-z_3)}{(z-z_3)(z_2-z_1)} = \frac{(w-w_1)(w_2-w_3)}{(w-w_3)(w_2-w_1)}$$
$$\Rightarrow \frac{(z-1)(i+1)}{(z+1)(i-1)} = \frac{(w-0)(1-\infty)}{(w-\infty)(-0)}$$
$$\Rightarrow w = \frac{1}{i}\cdot\frac{z-1}{z+1}$$
$$\because \frac{1-\infty}{w-\infty} = 1 \text{ and } \frac{i+1}{i-1} = \frac{1}{i}.$$

Example 19.19
Show that the transformation $w = \dfrac{2z+3}{z-4}$ maps the circle $x^2 + y^2 - 4x = 0$ into the straight line $4u + 3 = 0$. [JNTU 2003]

Solution Let $z = x + iy$, $w = u + iv$ so that $\bar{z} = x - iy$, $\bar{w} = u - iv$.

Now $x^2 + y^2 = (x+iy)(x-iy) = z\bar{z}$ and

$$x = \frac{1}{2}(z+\bar{z}), \quad y = \frac{1}{2i}(z-\bar{z})$$

The given transformation is,

$$w=\frac{2z+3}{z-4} \Rightarrow w(z-4)=2z+3$$

$$\Rightarrow z(w-2)=4w+3 \Rightarrow z=\frac{4w+3}{w-2}$$

$$\text{Also } \bar{z}=\left(\frac{4\overline{w}+3}{w-2}\right)=\frac{4\overline{w}+3}{\overline{w}-2}.$$

The given equation of the circle is

$$x^2+y^2-4x=0 \Rightarrow z\bar{z}-2(z+\bar{z})=0$$

$$\Rightarrow \left(\frac{4w+3}{w-2}\right)\left(\frac{4\overline{w}+3}{\overline{w}-2}\right)$$

$$-2\left(\frac{4w+3}{w-2}+\frac{4\overline{w}+3}{\overline{w}-2}\right)=0$$

$$\Rightarrow (4w+3)(4\overline{w}+3)-2[(4w+3)(\overline{w}-2)$$
$$+(4\overline{w}+3)(w-2)]=0$$

$$\Rightarrow 16w\overline{w}+12(w+\overline{w})+9-2[4w\overline{w}-8w$$
$$+3\overline{w}-6+4w\overline{w}-8\overline{w}+3w-6]=0$$

$$\Rightarrow 22(w+\overline{w})+33=0$$

$$\Rightarrow 4u+3=0 \quad \because u=\frac{w+\overline{w}}{2}$$

which is the equation of a straight line in the w-plane.

Example 19.20

Determine the bilinear transformation that maps the points (1-2i, 2+i, 2+3i) into the points (2+i, 1+3i, 4). [JNTU 2003, 2004(Set 4), 2004*S* (Set 2)]

Solution Let the required bilinear transformation be

$$\frac{(z-z_1)(z_2-z_3)}{(z-z_3)(z_2-z_1)}=\frac{(w-w_1)(w_2-w_3)}{(w-w_3)(w_2-w_1)} \quad (1)$$

Here

$$\left.\begin{aligned} z_1&=1-2i, \quad z_2=2+i, \quad z_3=2+3i.\\ w_1&=2+i, \quad w_2=1+3i, \quad w_3=4; \end{aligned}\right\} \quad (2)$$

Equation (1) now becomes

$$\frac{(z-1+2i)(-2i)}{(z-2-3i)(1+3i)}=\frac{(w-2-i)(-3+3i)}{(w-4)(-1+2i)}$$

$$\frac{(z-1+2i)}{(z-2-3i)}=\frac{(w-2-i)}{(w-4)}\cdot\frac{(-3+3i)(1+3i)}{(-1+2i)(-2i)}$$

$$=\frac{-3w+6+3i}{w-4} \quad \left(\text{on multiplying by } \frac{1+3i}{-2i}\right)$$

$(w-4)(z-1+2i)=(-3w+6+3i)(z-2-3i)$, on cross-multiplication

$$w(z-1+2i+3z-6-9i)$$
$$=(6z-12-18i)+(4z-4+8i)+(3iz+9-6i)$$

$$w=\frac{z(10+3i)-7-16i}{4z-7-7i}. \quad (3)$$

Example 19.21

Find the LFT that maps the points $-i$, 0, i onto the points -1, i, 1, respectively. [JNTU 1996, 2003*S*]

Solution Let the transformation be

$$\frac{(z-z_1)(z_2-z_3)}{(z-z_3)(z_2-z_1)}=\frac{(w-w_1)(w_2-w_3)}{(w-w_3)(w_2-w_1)} \quad (1)$$

Here $z_1=-i,\ z_2=0,\ z_3=i$
$w_1=-1,\ w_2=i,\ w_3=1$

Substituting in Eq. (1), we get

$$\frac{(z+i)(0-i)}{(z-i)(0+i)}=\frac{(w+1)(i-1)}{(w-1)(i+1)}$$

$$\Rightarrow -\frac{z+i}{z-i}=i\frac{w+1}{w-1} \qquad \left(\because \frac{i-1}{i+1}=i\right)$$

$$\Rightarrow \frac{w+1}{w-1}=\frac{iz-1}{z-i}$$

$$\Rightarrow w=\frac{iz-1+z-i}{iz-1-z+i}$$

(by componendo–dividendo).

$$\Rightarrow w=\frac{(1+i)(z-1)}{(i-1)(z+1)}$$

$$=\frac{1}{i}\frac{z-1}{z+1}=i\frac{1-z}{1+z}.$$

Example 19.22

Find the bilinear transformation that maps the points 2, i, -2 onto 1, i, -1, respectively. [JNTU 1985*S*]

Solution Let the transformation be

$$\frac{(z-z_1)(z_2-z_3)}{(z-z_3)(z_2-z_1)}=\frac{(w-w_1)(w_2-w_3)}{(w-w_3)(w_2-w_1)} \quad (1)$$

Here $z_1=2,\ z_2=i,\ z_3=-2$
$w_1=1,\ w_2=i,\ w_3=-1$

Equation (1) becomes

$$\frac{(z-2)(i+2)}{(z+2)(i-2)}=\frac{(w-1)(i+1)}{(w+1)(i-1)}\frac{i+1}{i-1}$$

$$=\frac{1}{i}\frac{(i+2)^2}{(i-2)(i+2)}=\frac{3+4i}{5}$$

$$\Rightarrow \frac{z-2}{z+2}\frac{3+4i}{5}=\frac{1}{i}\frac{w-1}{w+1}$$

$$\Rightarrow \frac{z-2}{z+2}\frac{-4+3i}{5}=\frac{w-1}{w+1}$$

$$\Rightarrow \frac{w+1}{w-1}=\frac{5z+10}{-4z+8+3iz-6i}$$

By componendo–dividendo,

$$w=\frac{(1+3i)z+18-6i}{(9-3i)z+2+6i}$$

$$=\frac{(1-3i)(1+3i)z+(1-3i)(18-6i)}{(1-3i)(9-3i)z+(1-3i)2(1+3i)}$$

$$=\frac{10z-60i}{-30iz+20}=\frac{z-6i}{-3iz+2}.$$

Example 19.23

Find the LFT which maps –1, 0, 1 onto 0, –1, ∞, respectively. [JNTU 1994*S*]

Solution Let the transformation be

$$\frac{(z-z_1)(z_2-z_3)}{(z-z_3)(z_2-z_1)}=\frac{(w-w_1)(w_2-w_3)}{(w-w_3)(w_2-w_1)} \qquad (1)$$

Here $z_1=-1, z_2=0, z_3=1$

$$w_1=0, w_2=-1, w_3=\infty$$

Substituting in Eq. (1), we get

$$\frac{(z+1)(0-1)}{(z-1)(0+1)}=\frac{(w-0)(-1-\infty)}{(w-\infty)(-1-0)}$$

$$\Rightarrow \frac{z+1}{z-1}=w, \quad \text{replacing } \frac{-1-\infty}{w-\infty} \text{ by } 1.$$

Example 19.24

Find the LFT which maps 0, 1, ∞ onto –1, –*i*, 1, respectively.

Solution Let the transformation be

$$\frac{(z-z_1)(z_2-z_3)}{(z-z_3)(z_2-z_1)}=\frac{(w-w_1)(w_2-w_3)}{(w-w_3)(w_2-w_1)} \qquad (1)$$

Here $z_1=0, z_2=1, z_3=\infty$

$$w_1=-1, w_2=-i, w_3=1$$

Substituting in Eq. (1), we get

$$\frac{(z-0)(1-\infty)}{(z-\infty)(1-0)}=\frac{(w+1)(-i-1)}{(w-1)(-i+1)}\Rightarrow z=\frac{w+1}{w-1}$$

$$\because \frac{i+1}{i-1}=\frac{1}{i};\ \frac{1-\infty}{z-\infty}=1\Rightarrow w=\frac{z+1}{z-1}.$$

Example 19.25

Find the bilinear transformation which maps the points *z* = 1, *i*, –1, respectively, onto *w* = *i*, 0, –*i*. [JNTU 1999*S*, 1997*S*]

Solution Let

$$w=\frac{az+b}{cz+d} \qquad (1)$$

be the required transformation.

When $z=1, w=i$

$$\therefore i=\frac{a\cdot 1+b}{c\cdot 1+d} \quad \text{or} \quad a+b=i(c+d) \qquad (2)$$

When $z=i, w=0$

$$\therefore 0=\frac{a\cdot i+b}{c\cdot i+d} \quad \text{or} \quad ai+b=0 \qquad (3)$$

When $z=-1, w=-i$

$$\therefore -i=\frac{a(-1)+b}{c(-1)+d} \text{ or } -a+b=i(c-d) \qquad (4)$$

From Eqs. (2) and (4), we get $b=ic, a=id$

But $b=-ai$ from Eq. (3)

$$\therefore c=-a=-id=\frac{b}{i} \qquad (5)$$

$$\therefore w=\frac{az-ai}{-az+\frac{a}{i}}=\frac{z-i}{-(z+i)}=\frac{i-z}{i+z}.$$

Example 19.26

Show that $w=\frac{2z+3}{z-4}$ maps the circle $x^2+y^2-4x=0$ onto a straight line $4u+3=0$. [JNTU 2003]

Solution The given transformation is

$$w=\frac{2z+3}{z-4}$$

$$\Rightarrow wz-4w-2z-3=0$$

$$\Rightarrow z(w-2)=4w+3\Rightarrow z=\frac{4w+3}{w-2} \qquad (1)$$

The given equation of the circle in the z-plane is

$$x^2+y^2-4x=0 \Rightarrow z\bar{z}-2(z+\bar{z})=0 \quad (2)$$

(Under the transformation (1), Eq. (2) becomes)

$$\left(\frac{4w+3}{w-2}\right)\left(\frac{4\bar{w}+3}{\bar{w}-2}\right)-2\left(\frac{4w+3}{w-2}+\frac{4\bar{w}+3}{\bar{w}-2}\right)=0$$

$$\Rightarrow (4w+3)(4\bar{w}+3)-2[(4w+3)(\bar{w}-2) + (4\bar{w}+3)+(w-2)]=0$$

$$\Rightarrow 16w\bar{w}+12(w+\bar{w})+9-2[4w\bar{w}-8w+3\bar{w} -6+4\bar{w}w-8\bar{w}+3w-6]=0$$

$$\Rightarrow 22(w+\bar{w})+33=0 \Rightarrow 4u+3=0 \quad (3)$$

which is a straight line in the w-plane.

Example 19.27

If $w=\dfrac{1+iz}{1-iz}$, find the image of $|z|<1$. [JNTU 2002]

Solution The given transformation is

$$w=\frac{1+iz}{1-iz}$$

$$\Rightarrow w-izw=1+iz$$

$$\Rightarrow iz(1+w)=w-1$$

$$\Rightarrow z=\frac{w-1}{i(w+1)}$$

Now $|z|<1 \Rightarrow \left|\dfrac{w-1}{i(w+1)}\right|<1$

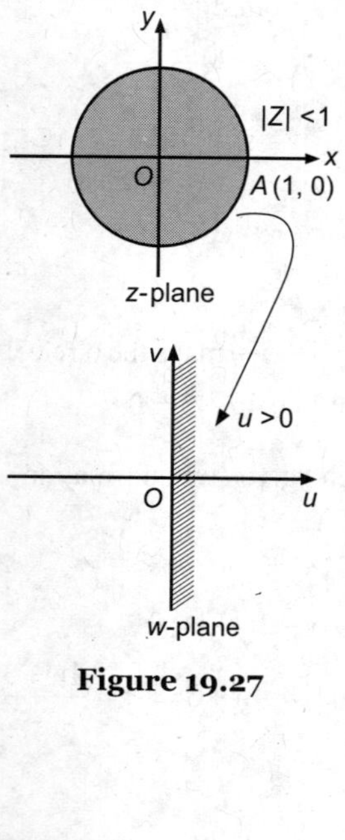

Figure 19.27

$$\Rightarrow |w-1|<|w+1|$$

$$\Rightarrow (u-1)^2+v^2<(u+1)^2+v^2$$

where $w=u+iv$

$$\Rightarrow 4u>0 \text{ or } u>0$$

$\therefore |z|<1$ is mapped into $u>0$, the upper half-plane of the w-plane.

Example 19.28

Find the bilinear transformation which maps the points ∞, i, 0 in the z-plane into -1, $-i$, 1 in the w-plane. [JNTU 2003*S* (2)]

Solution Let the transformation be

$$\frac{(z-z_1)(z_2-z_3)}{(z-z_3)(z_2-z_1)}=\frac{(w-w_1)(w_2-w_3)}{(w-w_3)(w_2-w_1)} \quad (1)$$

Here $\left.\begin{matrix} z_1=\infty, & z_2=i, & z_3=0 \\ w_1=-1, & w_2=-i, & w_3=1 \end{matrix}\right\}$ (2)

Substituting these values in Eq. (1)

$$\frac{(z-\infty)(i-0)}{(z-0)(i-\infty)}=\frac{(w+1)(-i-1)}{(w-1)(-i+1)}$$

$$\Rightarrow \frac{i}{z}=-i\frac{w+1}{w-1}$$

replacing $\dfrac{z-\infty}{i-\infty}$ by 1 and $\dfrac{1+i}{1-i}=i$

$$\Rightarrow \frac{w+1}{w-1}=\frac{1}{-z} \Rightarrow w=\frac{1-z}{1+z} \quad (3)$$

by componendo–bydividendo.

Example 19.29

Find the bilinear transformation that maps the points $(0, 1, \infty)$ in the z-plane onto the points $(-1, -2, -i)$ in the w-plane. [JNTU 2006(4)]

Solution Let the transformation be

$$\frac{(z-z_1)(z_2-z_3)}{(z-z_3)(z_2-z_1)}=\frac{(w-w_1)(w_2-w_3)}{(w-w_3)(w_2-w_1)} \quad (1)$$

Here $\left.\begin{matrix} z_1=0, & z_2=1, & z_3=\infty \\ w_1=-1, & w_2=-2, & w_3=-i \end{matrix}\right\}$ (2)

Substituting these values in Eq. (1)

$$\frac{(z-0)(1-\infty)}{(z-\infty)(1-0)}=\frac{(w+1)(-2+i)}{(w+i)(-2+1)}$$

$$\Rightarrow z=\frac{w+1}{w+i}(2-i),$$

replacing $\frac{1-\infty}{z-\infty}$ by 1

$$\Rightarrow (w+i)z-w(2-i)=2-i$$

$$\Rightarrow w[z-2+i]=2-i-iz$$

$$\Rightarrow w=\frac{2-i-iz}{i-2+z} \qquad (3)$$

which is the required transformation.

Example 19.30

Find the fixed points of the transformation

(i) $w=\frac{2i-6z}{iz-3}$, (ii) $w=\frac{3iz+1}{z+i}$,

(iii) $w=\frac{2z+3}{z+4}$.

Solution The fixed points of the transformation $w=f(z)$ are such that $f(z)=z$.

(i) Putting $w=f(z)=z$, we obtain

$$z=\frac{2i-6z}{iz-3}\Rightarrow iz^2-3z=2i-6z$$

$$\Rightarrow iz^2+3z-2i=0;$$

Solving, we get

$$z=\frac{-3\pm\sqrt{9-4(i)(-2i)}}{2i}=\frac{-3\pm 1}{2i}=i,2i$$

(ii) Putting $w=f(z)=z$, we obtain

$$z=\frac{3iz+1}{z+i}\Rightarrow z^2+iz=3iz+1$$

$$\Rightarrow z^2-21z-1=0$$

$$\Rightarrow (z-i)^2=0\Rightarrow z=i,i$$

(iii) Putting $w=f(z)=z$, we obtain

$$z=\frac{2z+3}{z+4}\Rightarrow z^2+4z=2z+3$$

$$\Rightarrow z^2+2z-3=0\Rightarrow (z+1)^2-2^2=0$$

$$\Rightarrow z+1=\pm 2\Rightarrow z=-1\pm 2=1,-3.$$

Example 19.31

Find the bilinear transformation which maps vertices $(1+i, -i, 2-i)$ of the triangle T of the z-plane into the points $(0, 1, i)$ of the w-plane. [JNTU 2005S (2)]

Solution Let the transformation be

$$\frac{(w-w_1)(w_2-w_3)}{(w-w_3)(w_2-w_1)}=\frac{(z-z_1)(z_2-z_3)}{(z-z_3)(z_2-z_1)} \qquad (1)$$

Here $z_1=1+i$, $z_2=-i$, $z_3=2-i$

$$w_1=0,\ w_2=1,\ w_3=i$$

Substituting these values in Eq. (1), we get

$$\frac{(w-0)(1-i)}{(w-i)(1-0)}=\frac{(z-1-i)(-i-2+i)}{(z-2+i)(-i-1-i)}$$

$$\Rightarrow \frac{w}{w-i}(1-i)=\frac{(z-1-i)}{(z-2+i)}\frac{[+(1+i)(1-i)]}{1+2i}$$

$$=\frac{(1+i)z-2i}{(1+2i)z-(4+3i)}\text{(inverting).}$$

$$\Rightarrow 1-\frac{1}{w}=\frac{(1+2i)z-(4+3i)}{(1+i)z-2i}$$

$$\Rightarrow \frac{i}{w}=\frac{(1+i)z-2i-(1+2i)z+4+3i}{(1+i)z-2i}$$

$$=\frac{-iz+4+i}{(1+i)z-2i}$$

$$\Rightarrow w=i\frac{[(1+i)z-2i]}{(-iz+4+i)}$$

$$=\frac{(-1+i)z+2}{(-i)z+4+i}.$$

Example 19.32

Determine Mobius transformation having α, β as fixed points and mapping 0 to -1.

Solution The quadratic equation with α, β as roots is

$$z^2-(\alpha+\beta)z+\alpha\beta=0$$

For any $\gamma\in\mathbb{C}$, we have

$$z^2(\alpha+\beta)z+\gamma z-\gamma z+\alpha\beta=0$$

Adding and subtracting γz

$$\text{or } z[z-(\alpha+\beta-\gamma)]=\gamma z-\alpha\beta$$

$$\text{or } z=\frac{\gamma z-\alpha\beta}{z-(\alpha+\beta)+\gamma}.$$

Hence the Mobius transformation with fixed points α, β is given by $w = \dfrac{\gamma z - \alpha\beta}{z - (\alpha + \beta) + \gamma}$.

Example 19.33

Find the linear fractional transformation with fixed points 1 and i and mapping 0 to −1.

Solution Take $z = 0$; $w = -1$; $\alpha = 1$, $\beta = i$ in the transformation $w = \dfrac{\gamma z - \alpha\beta}{z - (\alpha + \beta) + \gamma}$, we have

$$-1 = \frac{\gamma \cdot 0 - 1 \cdot i}{0 - (1 + i) + \gamma}$$

$$\Rightarrow -\gamma + 1 + i = -i$$

$$\Rightarrow \gamma = 1 + 2i$$

∴ The required transformation is

$$w = \frac{(2i + 1)z - i}{z + i}.$$

EXERCISE 19.1

1. Find the bilinear transformation that maps z_1, z_2, z_3 onto w_1, w_2, w_3, respectively.

   (a). $z = -1, 0, 1$; $w = 0, i, 3i$.

   Ans: $w = -3i(z + 1)/(z - 3)$

   (b). $z = 0, -i, -1$; $w = i, 1, 0$.

   Ans: $w = -i(z + 1)/(z - 1)$

   (c). $z = \infty, i, 0$; $w = 0, i, \infty$.

   Ans: $w = -\dfrac{1}{z}$

   (d). $z = 1, 0, -1$; $w = i, 1, \infty$.

   Ans: $w = [(-1 + 2i)z + 1]/(z + 1)$

2. Find the fixed points of the transformation.

   (a). $w = (z - 1)/(z + 1)$.

   Ans: $w = \pm i$

   (b). $w = (2z - 5)/(z + 4)$.

   Ans: $-1 \pm 2i$

3. Find the bilinear transformation whose fixed points are

   (a). $i, -i$.

   Ans: $w = \dfrac{cz - 1}{z + c}$

   (b). $1, -1$.

   Ans: $w = \dfrac{cz + 1}{z + c}$

4. Determine the bilinear transformation which maps the points 0, 1, ∞ in the z-plane into −5, −1, 3, respectively, in the w-plane. Find the invariant points.

   Ans: $w = (3z - 5)/(z + 1), z = 1 \pm 2i$

5. Determine the region into which the rectangular region in the z-plane bounded by $x = 0$, $y = 0$, $x = 2$ and $y = 1$ is mapped under the transformation $w = z + (1 - 2i)$.

   Ans: Rectangular bounded by $u = 1$, $u = 3$, $v = -1$, $v = -2$

6. Find the bilinear transformation which maps the points −1, 0, 1 into the points 0, −1, ∞, respectively.

   Ans: $w = \dfrac{z + 1}{z - 1}$

7. Find the bilinear transformation which maps the points (−1, 0, 1) into the points (0, i, 3i). [JNTU 2005]

   Ans: $w = \dfrac{3i(z + 1)}{(3 - z)}$

Question Bank

Multiple Choice Questions

CHAPTER 1 EIGEN VALUES AND EIGEN VECTORS

1. The sum of the characteristic roots of $A = \begin{bmatrix} 0 & 1 \\ 1 & 0 \end{bmatrix}$ is

(a) 1 (b) 2 (c) 0 (d) −1 **Ans:** (c)

2. If the sum and the product of the eigenvalues of $\begin{bmatrix} a+1 & -1 \\ 1 & a-1 \end{bmatrix}$ are s_1 and s_2 respectively then $s_1^2 =$

(a) $4s_2$ (b) $2s_2$
(c) s_2 (d) $s_2/4$ **Ans:** (a)

3. By diagonalisation of $A = \begin{bmatrix} 2 & -2 & -4 \\ -1 & 3 & 4 \\ 1 & -2 & -3 \end{bmatrix}$ we obtain $D = \begin{bmatrix} 0 & 0 & 0 \\ 0 & 1 & 0 \\ 0 & 0 & 1 \end{bmatrix}$. From this we conclude that A is

(a) Symmetric (b) Skew-symmetric
(c) Singular (d) Involutory **Ans:** (c)

4. A is an n-square matrix with tr. $A = 0$ then the coefficient of λ^{n-1} in its characteristic equation is

(a) 1 (b) 0 (c) −1 (d) n **Ans:** (b)

5. If the characteristic equation of matrix A of order 2 is $ax^2 + bx + c = 0$ then the characteristic equation of A^{-1} is

(a) $ax^2 + bx + c = 0$ (b) $bx^2 + cx + a = 0$
(c) $cx^2 + bx + a = 0$ (d) $cx^2 + ax + b = 0$
Ans: (c)

6. If the eigenvalues of A are 3, −1, 7 then the eigenvalues of $(A + I)$ are

(a) 4, 0, 8 (b) 2, −2, 6
(c) 0, 1, 3 (d) 3, −1, 7 **Ans:** (a)

7. The eigenvalues a, b, c of A are in A.P and those of A^2 are in A.P too. Then $b =$

(a) $\dfrac{a^2+2c^2}{a+2c}$ (b) $\dfrac{2a^2+c^2}{2a+c}$
(c) $\dfrac{a+c}{a^2+b^2}$ (d) $\dfrac{a^2+c^2}{a+c}$ **Ans:** (d)

8. If the product of the eigenvalues of A is zero then A is a ________ matrix.

(a) Nonsingular (b) Singular
(c) Involutory (d) Symmetric **Ans:** (b)

9. If the product of the eigenvalues of A is positive then A is a ______ matrix.

(a) Nonsingular (b) Singular
(c) Symmetric (d) Skew-symmetric
Ans: (a)

10. If 2 and 1/2 are the eigenvalues of a 2-square matrix A then those of A^{-1} are

(a) 2, 4 (b) 2, $\frac{1}{2}$ (c) 1, 1 (d) 1, −1
Ans: (b)

11. If the characteristic equation of a 2-square matrix A is $ax^2 + bx + c = 0$ then that of $3A$ is

(a) $ax^3 + 3bx + 9c = 0$
(b) $3ax^2 + bx + c = 0$
(c) $3ax^2 + 9bx + c = 0$
(d) $ax^2 + 3bx + 6c = 0$ **Ans:** (a)

12. If 2 is an eigenvalue of $\begin{bmatrix} 3 & -9 & -12 \\ 1 & a & 4 \\ 0 & 0 & 1 \end{bmatrix}$ then $a =$

(a) 0 (b) 4 (c) −6 (d) −7 **Ans:** (d)

$$\left[\textbf{Hint: } |A - 2I| = 0 \Rightarrow (-1)\begin{vmatrix} 3-2 & -9 \\ 1 & a-2 \end{vmatrix} = 0 \Rightarrow a = -7.\right]$$

13. The latent roots of $\begin{bmatrix} 1 & 1 & -1 \\ 0 & 2 & -1 \\ 0 & 0 & -2 \end{bmatrix}$ are

(a) 1, 0, 0 (b) 1, 2, −2
(c) 0, 0, −2 (d) 1, 1, −1 **Ans:** (b)

14. A is similar to $D = \text{diag}(1, -1, 3)$ then the proper values of A^2 are

(a) 1, −1, 3 (b) 2, −2, 6
(c) 1, 1, 9 (d) $1^2, -1^2, 3^2$ **Ans:** (c)

15. If '0' is a latent root of $\begin{bmatrix} 1 & 2 & 2 \\ 0 & a & 1 \\ 0 & 1 & -1 \end{bmatrix}$ then $a =$

(a) 2 (b) 0 (c) 1 (d) −1 **Ans:** (d)

$$\left[\textbf{Hint: } 0 = \begin{vmatrix} a & 1 \\ 1 & -1 \end{vmatrix} \Rightarrow a = -1.\right]$$

16. The eigenvalues of $\begin{bmatrix} 2 & 1 \\ 1 & 5 \end{bmatrix}$ are the roots of the equation

(a) $x^2 + 7x - 9 = 0$ (b) $x^2 - 7x + 9 = 0$
(c) $x^2 + 7x + 9 = 0$ (d) $x^2 - 7x + 10 = 0$
Ans: (b)

$$\left[\textbf{Hint: } x^2 - (2+5)x + \begin{vmatrix} 2 & 1 \\ 1 & 5 \end{vmatrix} = 0.\right]$$

17. If X_1 and X_2 are the eigenvectors of $A = \begin{bmatrix} 0 & 2 \\ 2 & 0 \end{bmatrix}$ then X_1 and X_2 are

(a) Collinear (b) Orthogonal
(c) Identical (d) Linearly dependent **Ans:** (b)

18. The algebraic multiplicity of the eigenvalue $\lambda = 2$ of the matrix $\begin{bmatrix} 1 & 0 & 2 \\ -1 & 2 & 2 \\ -1 & 0 & 4 \end{bmatrix}$ is

(a) 0 (b) 1 (c) 2 (d) 3 **Ans:** (c)

[**Hint:** Expanding by c_2

$$|A - \lambda I| = (2-\lambda)\begin{vmatrix} 1-\lambda & 2 \\ -1 & 4-\lambda \end{vmatrix} = -(\lambda-2)^2(\lambda-3).]$$

19. If $x^3 - s_1x^2 + s_2x - s_3$ is the characteristic polynomial of $\begin{bmatrix} 1 & 2 & 1 \\ 0 & 3 & 5 \\ 0 & 0 & 4 \end{bmatrix}$ then $(s_1, s_3) =$

(a) (8, −12) (b) (−8, −12)
(c) (−8, 12) (d) (8, 12) **Ans:** (d)

20. If the product of the eigenvalues of $A = \begin{bmatrix} 4 & 2 \\ 6 & 5 \end{bmatrix}$ is divisible by that of the eigenvalues of $B = \begin{bmatrix} 8 & -3 \\ 0 & a \end{bmatrix}$ then $a =$

(a) ±2 (b) 2, 1 (c) ±1 (d) 0, 1 **Ans:** (c)

21. If $A = \begin{bmatrix} -1 & 0 \\ 1 & -4 \end{bmatrix}$ then the eigenvalues of $B = A^T + I$ are

(a) −1, −4 (b) ±1 (c) ±3 (d) 0, −3
Ans: (d)

[**Hint:** $\lambda_1 = -1, \lambda_2 = -4$ are e.v.s of A

$\Rightarrow -1+1 = 0, \; -4+1 = -3$ are those of B]

22. If λ_1, λ_2 are distinct eigenvalues of A then the corresponding eigenvectors X_1, X_2 of A are

(a) Linearly independent (b) Linearly dependent
(c) Orthogonal (d) Orthonormal **Ans:** (a)

23. If $\begin{bmatrix}1\\-1\end{bmatrix}\begin{bmatrix}1\\1\end{bmatrix}$ are the eigenvectors of A then those of A^2 are

(a) $\begin{bmatrix}1\\-1\end{bmatrix}^2, \begin{bmatrix}1\\1\end{bmatrix}^2$ (b) $\begin{bmatrix}1^2\\1^2\end{bmatrix}, \begin{bmatrix}1^2\\(-1)^2\end{bmatrix}$

(c) $\begin{bmatrix}1\\-1\end{bmatrix}, \begin{bmatrix}1\\1\end{bmatrix}$ (d) $\begin{bmatrix}0\\0\end{bmatrix}, \begin{bmatrix}0\\1\end{bmatrix}$ **Ans:** (c)

24. −1, −2 are two eigenvalues of the third-order square matrix A with det $A = 6$. The eigenvalues of $-A$ are

(a) 1, −2, −3 (b) −1, −2, 3
(c) −1, 2, −3 (d) 1, 2, −3 **Ans:** (d)

25. The characteristic equation of second-order matrix A is $ax^2 + bx + c = 0$ then that of $-A$ is

(a) $a + bx + cx^2 = 0$ (b) $ax^2 - bx + c = 0$
(c) $ax^2 + bx - c = 0$ (d) $-ax^2 + bx + c = 0$
Ans: (b)

26. The characteristic equation of a second-order singular matrix A is $ax^2 + bx + c = 0$. Then

(a) $a = 0$ (b) $b = 0$
(c) $c = 0$ (d) $b = c = 0$ **Ans:** (c)

27. If matrix $A = \begin{bmatrix}5 & 1\\0 & 5\end{bmatrix}$ then the latent roots of B, which is similar to A, are

(a) 0, 5 (b) 4, 6 (c) 5, −5 (d) 5, 5
Ans: (d)

28. If the characteristic equation of a third-order matrix A is $a\lambda^3 + b\lambda + c = 0$ then tr. A =

(a) $\frac{-b}{a}$ (b) $\frac{-c}{a}$ (c) 0 (d) 1 **Ans:** (c)

29. If the characteristic equation of a second-order matrix A is $a\lambda^2 + b\lambda + c = 0$ then the product of the eigenvalues of A is

(a) c (b) $\frac{-c}{a}$ (c) $-c$ (d) $\frac{c}{a}$ **Ans:** (d)

30. If a, b, c are the elements of a nonsingular diagonal matrix D then the eigenvalues of D^{-1} are

(a) (1, 1, 1) (b) $(-a, -b, -c)$
(c) (a^{-1}, b^{-1}, c^{-1}) (d) (a, b, c) **Ans:** (c)

31. If A is a nonsingular matrix with λ_i as its eigenvalues then the eigenvalues of Adj A are

(a) $\frac{1}{\lambda_i}$ (b) $\frac{|A|}{\lambda_i}$ (c) λ^2_i (d) λ^{-2}_i **Ans:** (b)

32. If $A = \begin{bmatrix}1 & 2\\2 & -1\end{bmatrix}$ then $A^4 = xI$ where x =

(a) 1 (b) 5 (c) 25 (d) 625 **Ans:** (c)

CHAPTER 2 QUADRATIC FORMS

1. A real quadratic form in n variables is a ______ expression in n variables with ______ coeffcients.

(a) Second degree; real
(b) Second degree; complex
(c) Second degree homogeneous; real
(d) Second degree homogeneous; complex
Ans: (c)

2. The quadratic form in 2 variables among the following

(a) $3x + 5xy - 2y^2$ (b) $8x^2 - 5y + 3y^2$
(c) $x^2 - y^3 + 2xy$ (d) $2x^2 - 4xy + 5y^2$
Ans: (d)

3. The symmetric matrix of the quadratic form $X^T AX = x^2 + 6xy + 5y^2$ is A =

(a) $\begin{bmatrix}1 & 3\\3 & 5\end{bmatrix}$ (b) $\begin{bmatrix}1 & 6\\0 & 5\end{bmatrix}$

(c) $\begin{bmatrix}1 & 2\\3 & 5\end{bmatrix}$ (d) $\begin{bmatrix}5 & 1\\5 & 1\end{bmatrix}$ **Ans:** (a)

4. The symmetric matrix of the quadratic form $X^TAX = 2x_1x_2 + 8x_1x_3 - x_2x_3$ is A =

(a) $\begin{bmatrix}0 & -1 & 2\\-1 & 0 & 3\\2 & 3 & 0\end{bmatrix}$ (b) $\begin{bmatrix}0 & 1 & -4\\1 & 0 & 2\\-4 & 2 & 0\end{bmatrix}$

(c) $\begin{bmatrix} 0 & 1 & 4 \\ 1 & 0 & -\frac{1}{2} \\ 4 & -\frac{1}{2} & 0 \end{bmatrix}$ (d) $\begin{bmatrix} 0 & 2 & 8 \\ 2 & 0 & -1 \\ 8 & -1 & 0 \end{bmatrix}$ **Ans:** (c)

5. The matrix of the quadratic form $x_1^2 + 3x_2^2 + 3x_3^2 - 2x_2x_3$ is $A =$

(a) $\begin{bmatrix} 1 & 0 & 0 \\ 0 & 3 & -1 \\ 0 & -1 & 3 \end{bmatrix}$ (b) $\begin{bmatrix} 1 & -1 & 0 \\ -1 & 3 & 1 \\ 0 & 1 & 3 \end{bmatrix}$

(c) $\begin{bmatrix} 1 & 0 & 0 \\ 0 & 3 & 0 \\ 0 & 0 & 3 \end{bmatrix}$ (d) $\begin{bmatrix} 1 & -1 & 0 \\ -1 & 3 & 0 \\ 0 & 0 & 3 \end{bmatrix}$ **Ans:** (a)

6. The quardratic form $X^T AX$ with $X^T = [X_1, X_2]$ whose matrix $A = \begin{bmatrix} 1 & -1 \\ -1 & 1 \end{bmatrix}$ is

(a) $x_1^2 - 4x_1, x_2 + x_2^2$ (b) $x_1^2 - 2x_1x_2 + x_2^2$

(c) $x_1^2 + 2x_1x_2 + x_3^2$ (d) $x_1^2 + x_2^2$ **Ans:** (b)

7. The quadratic form $X^T AX$ with $X^T = [X_1, X_2, X_3]$ whose matrix $A = \begin{bmatrix} 1 & 0 & -4 \\ 0 & 2 & 0 \\ -4 & 0 & -2 \end{bmatrix}$ is

(a) $2x_1^2 + x_2^2 + 3x_3^2 + 8x_2x_3$

(b) $-2x_1^2 + x_2^2 + 2x_3^2 + 8x_2x_3$

(c) $x_1^2 + x_2^2 - x_3^2 - 2x_1x_3$

(d) $x_1^2 + 2x_2^2 - 2x_3^2 - 8x_1x_3$ **Ans:** (d)

8. If the matrix of the quadratic form $y^2 + 2z^2 + 10xy + 12yz$ is $\begin{bmatrix} 0 & a & 0 \\ a & 1 & b \\ 0 & b & 2 \end{bmatrix}$ then $(a, b) =$

(a) (5, 4) (b) (3, 4) (c) (4, 5) (d) (5, 6)
Ans: (d)

9. The rank of the quadratic form $x^2 + 3y^2 + 3z^2 - 2yz$ is

(a) 1 (b) 2 (c) 3 (d) 0 **Ans:** (c)

10. The nature of the quadratic form $2xy + 2yz + 2zx$ is

(a) Positive definite (b) Negative definite
(c) Indefinite (d) Positive semidefinite
Ans: (c)

11. The index of the quadratic form $x^2 + 3y^2 + 4z^2$ is

(a) 3 (b) 2 (c) 1 (d) 0 **Ans:** (a)

12. The signature of $x^2 + 2xz$ is

(a) 1 (b) 0 (c) 2 (d) 3 **Ans:** (b)

13. The rank of the quadratic form $x^2 + 2yz$ is

(a) 1 (b) 2 (c) 3 (d) 0 **Ans:** (c)

14. The index of a quadratic form is the number of the _____ terms in its canonical form.

(a) Positive (b) Negative
(c) Nonzero (d) Zero **Ans:** (a)

15. The signature of a quadratic form is the number of the _____terms in its canonical form

(a) Positive
(b) Negative
(c) Excess of positive terms over negative
(d) Excess of the nonzero terms over the zero
Ans: (c)

16. The rank of the quadratic form $2x^2 + 4xy + z^2$ is

(a) 1 (b) 0 (c) 2 (d) 3 **Ans:** (d)

17. The column vectors X_1, X_2 and X_3 are pairwise orthogonal if

(a) $X_iX_j^T = 0\ (i \neq j)$ (b) $X_1X_2^T = 0$

(c) $X_2X_3^T = 0$

(d) $X_3X_1^T = 0$ for all $i, j = 1, 2, 3$ **Ans:** (a)

18. If X_1, X_2 and X_3 are three linearly independent eigenvectors of a symmetric matrix A and $\|X_i\| = \alpha_i$ $(i = 1, 2, 3)$ then the modal matrix of $AP =$

(a) $\begin{bmatrix} \frac{X_3}{\alpha_3}, & \frac{X_2}{\alpha_2}, & \frac{X_1}{\alpha_1} \end{bmatrix}$ (b) $\begin{bmatrix} \frac{X_1}{\alpha_1}, & \frac{X_2}{\alpha_2}, & \frac{X_3}{\alpha_3} \end{bmatrix}$

(c) $\begin{bmatrix} \frac{X_2}{\alpha_2}, & \frac{X_3}{\alpha_3}, & \frac{X_1}{\alpha_1} \end{bmatrix}$ (d) $\begin{bmatrix} \frac{X_1}{\alpha_1}, & \frac{X_3}{\alpha_3}, & \frac{X_2}{\alpha_2} \end{bmatrix}$

Ans: (b)

19. The method of reduction of a quadratic form $q = X^T AX$ to its canonical form $Y^T DY$ by applying a linear transformation $X = PY$ where $P = [X_1, X_2, X_3]$, X_1, X_2, X_3 being normalised eigenvectors of A is called ______

(a) Diagonalisation
(b) Orthogonalisation
(c) Lagrange's reduction **Ans:** (b)

20. The method of reduction of a quadratic form $q = X^T AX$ to its canonical form $Y^T DY$ by applying row and column operations simultaneously on $A = I_3AI_3$ to convert A into a diagonal matrix is called ______

(a) Diagonalisation
(b) Orthogonalisation
(c) Lagrange's reduction **Ans:** (a)

21. The method of reduction of the quadratic form $q = X^T AX$ to canonical form by grouping the terms and completing the squares is called ______

(a) Diagonalisation
(b) Orthogonalisation
(c) Lagrange's reduction **Ans:** (c)

22. The normalised modal matrix P corresponding to the symmetric matrix $A = \begin{bmatrix} 1 & 2 \\ 2 & 1 \end{bmatrix}$ is

(a) $\begin{bmatrix} 1 & 1 \\ 1 & -1 \end{bmatrix}$ (b) $\begin{bmatrix} \frac{1}{2} & \frac{1}{2} \\ \frac{1}{2} & -\frac{1}{2} \end{bmatrix}$

(c) $\begin{bmatrix} \frac{1}{\sqrt{3}} & \frac{1}{\sqrt{3}} \\ \frac{1}{\sqrt{3}} & -\frac{1}{\sqrt{3}} \end{bmatrix}$ (d) $\begin{bmatrix} \frac{1}{\sqrt{2}} & \frac{1}{\sqrt{2}} \\ \frac{1}{\sqrt{2}} & -\frac{1}{\sqrt{2}} \end{bmatrix}$ **Ans:** (d)

23. The modal matrix of $A = \begin{bmatrix} 2 & 4 \\ 1 & 5 \end{bmatrix}$ is $P = \begin{bmatrix} -4 & 1 \\ 1 & 1 \end{bmatrix}$. Then the normalised modal matrix is

(a) $\begin{bmatrix} \frac{-4}{\sqrt{17}} & \frac{1}{\sqrt{2}} \\ \frac{1}{\sqrt{17}} & \frac{1}{\sqrt{2}} \end{bmatrix}$ (b) $\begin{bmatrix} \frac{-4}{\sqrt{5}} & \frac{1}{\sqrt{2}} \\ \frac{1}{\sqrt{5}} & \frac{1}{\sqrt{2}} \end{bmatrix}$.

(c) $\begin{bmatrix} -4 & \frac{1}{2} \\ 1 & \frac{1}{2} \end{bmatrix}$ (d) $\begin{bmatrix} \frac{-4}{\sqrt{13}} & \frac{1}{2} \\ \frac{1}{\sqrt{13}} & \frac{1}{2} \end{bmatrix}$ **Ans:** (a)

24. The modal matrix of $A = \begin{bmatrix} 0 & 1 & 0 \\ 1 & 0 & 1 \\ 0 & 1 & 0 \end{bmatrix}$ is $\begin{bmatrix} -1 & 1 & -1 \\ 0 & \sqrt{2} & \sqrt{2} \\ 1 & 1 & -1 \end{bmatrix}$ Then the normalised modal matrix $P =$

(a) $\begin{bmatrix} -\frac{1}{2} & \frac{1}{\sqrt{2}} & -\frac{1}{2} \\ 0 & \frac{1}{\sqrt{2}} & \frac{1}{2} \\ \frac{1}{2} & \frac{1}{\sqrt{2}} & -\frac{1}{2} \end{bmatrix}$ (b) $\begin{bmatrix} -\frac{1}{2} & \frac{1}{2} & -\frac{1}{2} \\ 0 & \frac{1}{2} & \frac{1}{2} \\ \frac{1}{2} & \frac{1}{2} & -\frac{1}{2} \end{bmatrix}$

(c) $\begin{bmatrix} -\frac{1}{\sqrt{2}} & \frac{1}{2} & -\frac{1}{2} \\ 0 & \frac{1}{\sqrt{2}} & \frac{1}{\sqrt{2}} \\ \frac{1}{\sqrt{2}} & \frac{1}{2} & -\frac{1}{2} \end{bmatrix}$ (d) $\begin{bmatrix} -\frac{1}{2} & \frac{1}{2} & -\frac{1}{2} \\ 0 & \frac{\sqrt{2}}{2} & \frac{\sqrt{2}}{2} \\ -\frac{1}{2} & \frac{1}{2} & \frac{-1}{2} \end{bmatrix}$

Ans: (c)

Real quadratic form in canonical form

25. $2y_1^2 + y_2^2 - 3y_2^2$. Its rank is

(a) 1 (b) 3 (c) 2 (d) 0 **Ans:** (b)

Its index is

(a) 1 (b) 0 (c) 2 (d) 3 **Ans:** (c)

Its signature is

(a) 3 (b) 2 (c) 1 (d) 0 **Ans:** (c)

Its nature is

(a) +ve Semidefinite (b) –ve Definite
(c) +ve Definite (d) Indefinite **Ans:** (d)

26. $y_1^2 + 2y_2^2 + 3y_3^2$. Its rank is

(a) 1 (b) 2 (c) 3 (d) 0 **Ans:** (c)

Its index is

(a) 2 (b) 3 (c) 1 (d) 0 **Ans:** (b)

Its signature is

(a) 1 (b) 2 (c) 3 (d) –1 **Ans:** (c)

Its nature is

(a) Positive definite (b) Positive semidefinite
(c) Indefinite (d) Negative definite **Ans:** (a)

27. $y_1^2 - 2y_3^2$. Its rank is

(a) 2 (b) 1 (c) 3 (d) 0 **Ans:** (a)

Its index is

(a) 3 (b) 2 (c) 1 (d) 0 **Ans:** (c)

Its signature is

(a) 2 (b) 0 (c) 2 (d) 1 **Ans:** (b)

Its nature is

(a) Positive definite (b) –ve Definite
(c) Indefinite (d) +ve Semidefinite **Ans:** (c)

28. $3y_1^2 + y_3^2$. Its rank is

(a) 0 (b) 1 (c) 2 (d) 3 **Ans:** (d)

Its index is

(a) 2 (b) 1 (c) 0 (d) 3 **Ans:** (a)

Its signature is

(a) 0 (b) 1 (c) 2 (d) –1 **Ans:** (c)

Its nature is

(a) Positive semidefinite
(b) Positive definite
(c) Negative definite
(d) Negative semidefinite **Ans:** (a)

29. $-2y_1^2 - y_2^2 - y_3^2$; Its rank is

(a) –2 (b) 1 (c) –3 (d) 0 **Ans:** (a)

Its index is

(a) 3 (b) 2 (c) 0 (d) 1 **Ans:** (c)

Its signature is

(a) 2 (b) 1 (c) 3 (d) 0 **Ans:** (c)

Its nature is

(a) Positive definite
(b) Negative definite
(c) Positive semidefinite
(d) Negative semidefinite **Ans:** (b)

30. Quadratic form is $2x_1x_2 + 2x_1x_3 + 2x_2x_3$; Its rank is

(a) 1 (b) 3 (c) 2 (d) 0 **Ans:** (b)

Its index is

(a) 1 (b) 2 (c) 3 (d) 0 **Ans:** (a)

Its signature is

(a) 0 (b) 1 (c) –1 (d) 2 **Ans:** (c)

Its nature is

(a) Positive definite (b) Positive indefinite
(c) Indefinite (d) Positive semidefinite
Ans: (c)

31. $x^2 + y^2 + z^2 + 2yz$; Its rank is

(a) 2 (b) 3 (c) 1 (d) 0 **Ans:** (a)

Its index is

(a) 1 (b) 2 (c) 3 (d) 0 **Ans:** (b)

Its signature is

(a) 2 (b) 1 (c) 0 (d) –1 **Ans:** (a)

Its nature is

(a) Positive definite (b) Negative definite
(c) Positive semidefinite (d) Indefinite
Ans: (c)

CHAPTER 3 SOLUTION OF ALGEBRAIC AND TRANSCENDENTAL EQUATIONS

1. Bisection method is used for _______

(a) Solution of an algebraic or transcendental equation
(b) Differentiation of a function
(c) Integration of a function
(d) Solution of ordinary differential equation
Ans: (a)

2. Regula Falsi method is used for _______

(a) Solution of ordinary differential equation
(b) Differential of a function
(c) Integration of a function
(d) Solution of an algebraic or transcendental equation **Ans:** (d)

3. Fixed point iteration method is used for ______

(a) Differentiation of a function
(b) Integration of a function
(c) Solution of an algebraic or transcendental equation
(d) Solution of ordinary differential equation
Ans: (c)

4. Newton-Raphson method is used for _______

(a) Solution of ordinary differential equation
(b) Solution of an algerbraic or transcendental equation
(c) Differentiation of a function
(d) Integration of a function **Ans:** (b)

5. Secant method is used for ________

(a) Solution of an algebraic or transcendental equation
(b) Solution of an ordinary differential equation
(c) Differentiation of a function
(d) Integration of a function **Ans:** (a)

6. Inverse interpolation can be used for ________

(a) Differentiation of a function
(b) Integration of a function
(c) Solution of an algebraic or transcendental equation
(d) Solution of an ordinary differential equation
Ans: (c)

7. Newton-Raphson method is also called

(a) Method of false position
(b) Method of chord (c) Method of tangents
(d) Method of secants **Ans:** (c)

8. Among the method of solution of equation of the form $f(x) = 0$ the one which is used commonly for its simplicity and great speed is _____ method

(a) Secant (b) Regula Falsi
(c) Bolzano (d) Newton-Raphson
Ans: (d)

9. For _______ method of solution of equations of the form $f(x) = 0$ approximation x_0 is to be very close to the root and $f(x_n) \neq 0$

(a) Bolzano (b) Newton-Raphson
(c) Secant (d) Chord **Ans:** (b)

10. In the bisection method of solution of an equation of the form $f(x)=0$ the covergence of the sequence $\langle x_n \rangle$ of mid points (iterates) to a root of $f(x)=0$ in an interval (a, b) where $f(a)\,f(b) < 0$ is

(a) Assured and very fast
(b) Not assured but very fast
(c) Assured but very slow
(d) Independent on the sequence of point
Ans: (c)

11. The convergence of the fixed point iteration process depends, on the form of the given equation. Which one of the following forms gives convergent sequence at $x = 1$ for $x^3 - x - 1 = 0$

(a) $x = x^3 - 1$ (b) $x = \frac{x+1}{x^2}$

(c) $x = \sqrt{\frac{x+1}{x}}$ (d) None of these **Ans:** (c)

[**Hint:** $f(x) = 0 \pi x = g(x)$ condition for convergence of iterates: $| g(x_n) | < k < |$ which holds for (c).]

12. In the method of False position for solution of an equation of the form $f(x) = 0$ the convergence of the sequence $\langle x_n \rangle$ of iterates to a root of $f(x) = 0$ is

(a) Assured and very fast
(b) Not assured but very fast
(c) Assured but slow
(d) Independent on the sequence of points

Ans: (c)

13. Among the methods of solution of equation of the form $f(x) = 0$, the one which has second order convergence is _______ method.

(a) Newton-Raphson (b) Chords
(c) Secant (d) Bisection **Ans:** (a)

14. Among two methods of solution of equation of the form $f(x) = 0$, the one which has super-linear *i.e.*, (more rapid than linear $|\epsilon_{n-1}| \approx \text{const.} |\epsilon_n|^{1.62}$) convergenceis_______ method.

(a) Bisection (b) Secant
(c) Chord (d) None of these **Ans:** (b)

15. In Newton-Raphson method we approximate the graph of f by suitable

(a) Chords (b) Tangents
(c) Secants (d) None of these **Ans:** (b)

16. Newton's iterative formula for finding a root of $f(x) = 0$ is $x_{n+1} =$

(a) $x_n - \frac{x_n - x_{n-1}}{f(x) - f(x_{n-1})} f(x_n)$

(b) $x_n + \frac{f(x_n)}{f'(x_n)}$ (c) $x_n - \frac{f(x_n)}{f'(x_n)}$

(d) None of these **Ans:** (c)

17. Newton's iterative formula for finding the square root of a positive number N is $x_{n+1} =$

(a) $\frac{1}{2}(x_n - N/x_n)$ (b) $\frac{1}{2}(x_n + N/x_n)$

(c) $\frac{1}{4}(x_n - N/x_n)$ (d) $\frac{1}{4}(x_n + N/x_n)$ **Ans:** (b)

18. Newton's iterative formula for finding the reciprocal of a number $x_{n+1} =$

(a) $x_n (1 - Nx_n)$ (b) $x_n (2 + Nx_n)$
(c) $x_n (2 - Nx_n)$ (d) $x_n (2 - {}^{xn}/_N)$ **Ans:** (c)

19. Newton iterative formula for finding the pth root of a positive number N is $x_{n+1} =$

(a) $\frac{1}{p}[(p-1)x_n - N/(x_n^p)]$

(b) $\frac{1}{p}[(p-1)x_n - N/(x_n^{p-1})]$

(c) $\frac{1}{p}[(p-1)x_n + N/(x_n^p)]$

(d) $\frac{1}{p}[(p-1)x_n + N/(x_n^{p-1})]$ **Ans:** (d)

20. Newton's iterative formula for finding the cube root of a number N is $x_{n+1} =$

(a) $\frac{1}{3}\left[2x_n + \frac{N}{x_n^2}\right]$ (b) $\frac{1}{3}\left[2x_n + \frac{N}{x_n^3}\right]$

(c) $\frac{1}{3}\left[2x_n - \frac{N}{x_n^2}\right]$ (d) $\frac{1}{3}\left[2x_n - \frac{N}{x_n^3}\right]$ **Ans:** (a)

21. The cube root of 24 by Newton's formula, taking $x_0 = 3$ is

(a) 2.889 (b) 2.888 (c) 2.8 (d) 2.7 **Ans:** (a)

22. The square root of 35 by Newton's formula taking $x_0 = 6$ is

(a) 5.806 (b) 5.916
(c) 5.906 (d) 5.816 **Ans:** (b)

23. The Regula Falsi method is related to ________ at a point of the curve

(a) Chord (b) Ordinate
(c) Abscissa (d) Tangent **Ans:** (a)

24. The Newton-Raphson method is related to _______ at a point of the curve.

(a) Chord (b) Ordinate
(c) Abscissa (d) Tangent **Ans:** (d)

CHAPTER 4 INTERPOLATION

1. $\Delta^3 (1-x)(1-2x)(1-3x) =$

(a) –24 (b) –54 (c) –36 (d) –72 **Ans:** (c)

2. $\Delta^n (e^{ax+b}) =$

(a) $(e^2-1)^n e^{ax+b}$ (b) $(e^a-1)^n e^{ax+b}$
(c) $(e^a-1)^{n-1} e^{ax+b}$ (d) $(e-1)^n e^{ax+b}$
Ans: (b)

3. $\Delta(u_x v_x) =$

(a) $u_x(\Delta v_x) + v_{x+1}(\Delta u_x)$
(b) $u_{x-1}(\Delta v_x) + v_x\Delta(u_{x-1})$
(c) $u_{x+1}(\Delta v_x) + v_{x-1}\Delta(\Delta u_x)$
(d) $u_x(\Delta v_x) + v_x(\Delta v_x)$ **Ans:** (a)

4. $\Delta x^{(r)} =$

(a) $r(r-1)x^{(r-2)}$ (b) $rh\, x^{(r-1)}$
(c) $rh\, x^{(r-2)}$ (d) $(r+1)\, h\, x^{(r-1)}$ **Ans:** (b)

5. $\Delta(u_x/v_x) =$

(a) $[v_x(\Delta u_x) - u_x(\Delta v_x)]/v_x v_{x+1}$
(b) $[v_{x-1}(\Delta u_x) - u_{x-1}(\Delta v_x)]/v_{x-1}v_x$
(c) $[v_x(\Delta u_{x-1}) - u_x(\Delta v_{x-1})]/v_x v_{x+1}$
(d) $[v_x(\Delta u_x) + u_x(\Delta v_x)]/v_x v_{x+1}$ **Ans:** (a)

6. $(1+\Delta)(1-\nabla) =$

(a) 0 (b) E^{-1} (c) 1 (d) E **Ans:** (c)

7. $\Delta E^{-1/2} =$

(a) $(E^{1/2}-1)E^{-1/2}$ (b) $E^{1/2} - E^{-1/2}$
(c) $1 - E^{-1}$ (d) ΔE^{-1} **Ans:** (b)

8. $\Delta + \nabla =$

(a) $E - E^{-1}$ (b) $E^{1/2} - E^{-1/2}$
(c) $(E-1)/E$ (d) $\Delta/\nabla - \nabla/\Delta$ **Ans:** (d)

9. $\nabla =$

(a) $(1+\Delta)^{-1}$ (b) $1 - E^{-1}$
(c) $E^{-1} - 1$ (d) $E^{-1} - E$ **Ans:** (b)

10. $\Delta\nabla =$

(a) $\Delta - \nabla$ (b) $\nabla - \Delta$
(c) $\Delta + \nabla$ (d) $E^{-1} - E$ **Ans:** (a)

11. $\delta^2 =$

(a) ΔE^{-1} (b) $E\Delta$
(c) $\Delta^2 E^{-1}$ (d) $E\Delta$ **Ans:** (c)

12. $\Delta - \nabla =$

(a) $\Delta E^{-1/2}$ (b) δ^2 (c) $\Delta E^{1/2}$ (d) δ **Ans:** (b)

13. $\Delta^n\left(\sum_{r=0}^{n} ar\, x^r\right) =$

(a) $a_n h^{n-1} n!$ (b) $a_0 h^n n!$
(c) $a_n h^n n!$ (d) $a_{n-1} h^{n-1}(n-1)!$ **Ans:** (c)

14. The shift operator E and the differential operator D are related by

(a) $E = e^{hD}$ (b) $E + 1 = e^{hD}$
(c) $E - 1 = e^{hD}$ (d) $E^{-1} = e^{hD}$ **Ans:** (a)

15. If the interval of differencing is unity then $\Delta^3(1-x)(1-3x)(1-5x) =$

(a) −60 (b) −90
(c) −50 (d) −75 **Ans:** (b)

$$\left[\textbf{Hint: } \Delta^n\left(\sum_{r=0}^{n} a_n x^r\right) = a_n n!h^n; \text{ here } h = 1.\right]$$

16. If $x^{(r)} = x(x-h)(x-2h)\cdots(x-rh+h)$ is the factorial function then the values of $\Delta^2 x^{(r)}$ and $\Delta^{r+1}x^{(r)}$ are, respectively, ______ and ______.

(a) $r(r-1)h^2x^{(r-2)}$, and 0;
(b) $r(r-1)x^{(r+2)}$, and $n!$;
(c) $r(r-1)x^{(r-2)}$, and 1;
(d) $rx^{(r-1)}$; and 0 **Ans:** (a)

17. If $\Delta^n f(x) = 0$ for $n \geq 5$ then $f(x)$ is a polynomial function of degree

(a) 2 (b) 5 (c) 3 (d) 4 **Ans:** (d)

18. Newton's forward interpolaltion formula is used for interpolating the values of a function f near of the tabulated values of f.

(a) The end (b) The middle
(c) The beginning (d) Any point **Ans:** (c)

19. Newton's backward interpolation formula is used for interpolating the values of a function f near ______ of the tabulated values of f.

(a) The middle (b) The beginning
(c) The end (d) Any point **Ans:** (c)

20. Stirling or Bessel central difference interpolation formula is used for interpolating the values of a function f near the ________ of the tabulated values of f.

(a) Middle (b) Beginning
(c) End (d) Beginning or end **Ans:** (a)

21. Lagrange's interpolation formula is used for interpolating the values of a function f near of the tabulated values of f.

(a) The beginning (b) The end
(c) The middle (d) Any point **Ans:** (d)

22. The missing value of y in the following table

x	100	101	102	103
y	2.0000	2.0043	–	2.0128

(a) 2.0066 (b) 2.0084 (c) 2.0086 (d) 2.0096

Ans: (c)

[**Hint:** Here three values of y are given; the polynomial degree 2 so that $\Delta^3 y_0 = 0 \Rightarrow y_3 - 3y_2 + 3y_1 - y_0 = 0$. Substituting the given values and solving we get the missing $y_2 = 2.0086$.]

23. From the data

x	0	1	2	3
y	7	5	y_2	7

$y_2 =$

(a) 4 (b) 5 (c) 6 (d) 5.5 **Ans:** (b)

[**Hint:** Since 3 values are given the polynomial is of degree 2 so that $\Delta^3 y_0 = 0$.]

$\Rightarrow \quad y_3 - 3y_2 + 3y_1 - y_0 = 0$
$\Rightarrow \quad 3y_2 = 7 + 3 \times 5 - 7 \quad \Rightarrow \quad y_2 = 5$.]

24. From the data

x	0	5	10	15	20	25
y	6	10	$\underline{y_2}$	17	$\underline{y_4}$	31

y_2 and y_4 are, respectively, ______ and ______.

(a) 11.25, 26.25 (b) 14.50, 24.50
(c) 13.25, 22.50 (d) 11.25, 20.50 **Ans:** (c)

25. Compute the missing values,

x	45	50	55	60	65
y	3	$\underline{y_1}$	2	$\underline{y_3}$	−2.4

[**Hint:** $\Delta^3 y_0 = 0 \quad \Rightarrow \quad 3y_1 + y_3 - 9 = 0$, $\Delta_3 y_1 = 0 \Rightarrow y_1 + 3y_2 - 3.6 = 0$ solving $y_1 = 2.925$, $y_3 = 0.225$.]

(a) 2.925, 0.225 (b) 2.725, 0.215
(c) 2.950, 0.250 (d) 2.860, 0.225 **Ans:** (a)

26. Use Lagrange's formula and find $y(1)$ from the data

x	0	2	3
y	3	1	2

(a) 2.0333 (b) 2.6666
(c) 1.6666 (d) 2.3333 **Ans:** (d)

[**Hint:** $y(x)=\dfrac{(x-x_1)(x-x_2)}{(x_0-x_1)(x_0-x_2)}y_0$

$+\dfrac{(x-x_0)(x-x_2)}{(x_1-x_0)(x_1-x_2)}y_1$

$+\dfrac{(x-x_0)(x-x_1)}{(x_2-x_0)(x_2-x_1)}y_2 (x=1)$

$=\dfrac{(1-2)(1-3)}{(0-2)(0-3)}\cdot 3+\dfrac{(1-0)(1-3)}{(2-0)(2-3)}\cdot 1$

$+\dfrac{(1-0)(1-2)}{(3-0)(3-2)}\cdot 2$

$=\dfrac{1}{3}\cdot 3+\dfrac{-2}{-1}1+\dfrac{-1}{3}\cdot 2.$]

27. Using Lagrange's interpolation formula find $y(2)$ from the data

x	0	1	3	4
y	−12	0	6	12

(a) 8 (b) 2 (c) 4 (d) 6 **Ans:** (c)

[**Hint:** $y(x)=\dfrac{(x-x_1)(x-x_2)(x-x_3)}{(x_0-x_1)(x_0-x_2)(x_0-x_3)}y_0$

$+\cdots+\cdots+\cdots \quad \Rightarrow \quad y(2)$

$=\dfrac{(2-1)(2-3)(2-4)}{(0-1)(0-3)(0-4)}(-12)$

$+\dfrac{(2-0)(2-3)(2-4)}{(1-0)(1-3)(1-4)}\cdot 0$

$+\dfrac{(2-0)(2-1)(2-4)}{(3-0)(3-1)(3-4)}6$

$+\dfrac{(2-0)(2-1)(2-3)}{(4-0)(4-1)(4-3)}12$

$=\dfrac{2}{3.4}12+0+\dfrac{2.2}{3.2}6-\dfrac{2}{4.3}12=4.$]

28. Compute x for $y = 1$ by Lagrange's inverse interpolation formula from the data

x	−2	−1	1
y	−7	2	0

(a) 0.1587 (b) 0.2586
(c) 0.2437 (d) 0.1547 **Ans:** (a)

[**Hint:** $x=\dfrac{(y-y_1)(y-y_2)}{(y_0-y_1)(y_0-y_2)}x_0$

$+\cdots+\cdots; x(1)=\dfrac{(1-2)(1-0)}{9.7}(-2)$

$+\dfrac{(1+7)(1-0)}{(2+7)(2-0)}(-1)+\dfrac{(1+7)(1-2)}{(0+7)(0-2)}$

$=\dfrac{1.1}{9.7}\cdot 2-\dfrac{8}{9.2}\cdot 1+\dfrac{8.1}{7.2}\cdot 1$

$=\dfrac{2}{63}-\dfrac{4}{9}+\dfrac{4}{7}=\dfrac{10}{63}=0.1587.$]

29. Compute $x(15)$ by Lagrance's inverse interpolation using the data

x	5	6	9	11
y	12	13	14	16

(a) 10.5 (b) 11.5 (c) 12.30 (d) 13.5
Ans: (b)

[**Hint:** $x(y)=\dfrac{(y-y_1)(y-y_2)(y-y_3)}{(y_0-y_1)(y_0-y_2)(y-y_3)}x_0+\cdots+\cdots+\cdots$

$x(15)=\dfrac{2\cdot 1(-1)}{(-1)\cdot(-2)\cdot(-4)}\times 5+\dfrac{3\cdot 1\cdot(-1)}{1\cdot(-1)\cdot(-3)}\times 6$

$+\dfrac{3\cdot 2\cdot(-1)}{2\cdot 1\cdot(-2)}\times 9+\dfrac{3\cdot 2\cdot 1}{4\cdot 3\cdot 2}11$

$=\dfrac{5}{4}-6+\dfrac{27}{2}+\dfrac{11}{4}=11.5.$]

30. $(\Delta+\nabla)^2=$

(a) $(\nabla-\Delta)^2$ (b) $[(\nabla/\Delta)-(\Delta/\nabla)]^2$
(c) $[(\Delta/\nabla)-(\nabla/\Delta)]^2$ (d) None of these
Ans: (c)

CHAPTER 5 CURVE FITTING

1. Given a set of data points (x_i, y_i) $(i = 0, 1, 2, \ldots n)$ in an interval $I = [x_0, x_n]$ the process of estimating $y = f(x)$ at a point outside I is called

(a) Interpolation (b) Extrapolation
(c) Intrapolation (d) None of these **Ans:** (b)

2. Given a set of data points (x_i, y_i) $(i = 0, 1, 2, \ldots n)$ in an interval $I = [x_0, x_n]$ the process of estimating $y = f(x)$ at a point $\bar{x} \in (x_i, x_{i+1})$ is called

(a) Interpolation (b) Extrapolation
(c) Intrapolation (d) None of these **Ans:** (a)

3. The method of least squares minimizes the _______ of the derivations (errors) of expected values from observed data.

(a) Sum of the squares (b) Sum
(c) Sum of absolute value
(d) Sum of the square roots **Ans:** (a)

4. The equations $\sum y_i = na + b\sum x_i$ and $\sum x_i y_i = a\sum x_i + b\sum x_i^2$ are the normal equation of the _______ fit for the data.

(a) Hyperbolic (b) Parabolic or quadratic
(c) Linear or straight line (d) Polynomial
Ans: (c)

5. The equation $\sum y_i = na + b\sum x_i + c\sum x_i^2$; $\sum x_i y_i = a\sum x_i + b\sum x_i^2 - c\sum x_i^3$ and $\sum x_i^2 y_i = a\sum x_i^2 + b\sum x_i^3 + c\sum x_i^4$ are the normal equation of the _______ fit for the data.

(a) Hyperbolic (b) Parabolic or quadratic
(c) Linear or straight line (d) Polynomial
Ans: (b)

6. The equation $\sum y_i = nA + B\sum x_i$ and $\sum x_i y_i = A\sum x_i + B\sum x_i^2$ where $x = \log x$, $y = \log y$, $A = \log a$ and $B = b$ are the normal equation of the _____ fit for the data.

(a) Hyperbolic (b) Parabolic or quadratic
(c) Geometric or power curve
(d) Polynomial **Ans:** (c)

7. The equation $\sum y_i = nA + B\sum x_i$ and $\sum x_i y_i = A\sum x_i + B\sum x_i^2$ where $y = \log y$, $A = \log a$ and $B = \log = b$ are the normal equation of the _______ fit for the data.

(a) Hyperbolic (b) Parabolic or quadratic
(c) Geometric or power curve
(d) Exponential curve **Ans:** (d)

8. The equation $\sum y_i^* = na + b\sum x_i$ and $\sum x_i y_i^* = a\sum x_i - b\sum x_i^2$ where $y^* = \frac{1}{y}$ are the normal equations of the _______ fit for the data.

(a) Hyperbolic (b) Parabolic
(c) Linear (d) Polynomial **Ans:** (a)

9. The least squares straight line passes through the point $(\overline{x}, \overline{y}) =$

(a) $\left(\sqrt{\sum x_i^2}, \sqrt{\sum y_1^2}\right)$ (b) $\left(\sum x_i, \sum y_i\right)$
(c) $\left(\sum x_i / N_1, \sum y_i / N\right)$ (d) None of these
Ans: (c)

10. The best fitting curve for a set of given data is the one having the least or minimum of the _______ of expected values from the observed data.

(a) Squares of the derivation
(b) Sum of the squares of derivation
(c) Absolute values of the derivation
(d) Absolute sum of the derivation **Ans:** (b)

11. The least squares straight line for the following data

x	1	2	3	4
y	6	4	3	5

is $y = a + bx$ where (a, b) is

(a) (5.5, –0.4) (b) (4.5, 0.4)
(c) (5.5, 0.4) (d) (5.5, 1.4) **Ans:** (a)

[**Hint:** Normal equation $4a + 10b = 18$, $10a + 30b = 43$.]

12. The least squares straight line for the following data

x	0	1	3	6	8
y	1	3	2	5	4

is $y = a + bx$ where $(a, b) =$

(a) (0.176, 1.646) (b) (0.370, 1.646)
(c) (0.376, 1.046) (d) (0.376, 1.646)

[**Hint:** Normal equations are $5a + 18b = 15$; $18a + 110b = 71$.]

CHAPTER 6 NUMERICAL DIFFERENTIATION AND INTEGRATION

1. The following four expressions (Set A) are derivatives, obtained from the interpolation formula mentioned below (Set B)

Set A

(1) $$\left[\Delta y_0 + \left(\frac{2p-1}{2}\right)\Delta^2 y_0 + \left(\frac{3p^2-6p+2}{6}\right)\Delta^3 y_0 + \left(\frac{4p^3-18p^2+22p-6}{24}\right)\Delta^4 y_0 + \cdots\right]$$

(2) $$\left[\Delta^2 y_0 + (p-1)\Delta^3 y_0 + \left(\frac{6p^2-18p+11}{12}\right)\Delta^4 y_0 + \cdots\right]$$

(3) $$\left[\nabla y_n + \left(\frac{2p+1}{2}\right)\nabla^2 y_n + \left(\frac{3p^3+6p+2}{6}\right)\nabla^3 y_n + \cdots\right]$$

(4) $$\left[\nabla^2 y_n + (p+1)\nabla^3 y_n + \left(\frac{6p^2+18p+11}{12}\right)\nabla^4 y_n + \cdots\right]$$

Set B

(1) $hy'(x)$, First-order derivative from Newton's forward formula
(2) $hy'(x)$, First-order derivative from Newton's backward formula
(3) $h^2y''(x)$, Second-order derivative from Newton's forward formula
(4) $h^2y''(x)$, Second-order derivative from Newton's backward formula

The correct pairing of members of Set A and Set B is

(a) (1, 1) (2, 3) (3, 2) (4, 4)
(b) (1, 2) (2, 3) (3, 1) (4, 4)
(c) (1, 1) (2, 4) (3, 2) (4, 3)
(d) (1, 4) (2, 3) (3, 2) (4, 1) **Ans:** (a)

2. The following four expressions (Set A) are derivatives obtained from the interpolation formulas mentioned below (Set B)

Set A

(1) $$\left[\left(\frac{\Delta y_0 + \Delta y_{-1}}{2}\right) + p\Delta^2 y_{-1} + \left(\frac{3p^2-1}{12}\right)(\Delta^3 y_{-1} + \Delta^3 y_{-2}) + \frac{2p^3-p}{12}\Delta^4 y_{-2} + \cdots\right]$$

(2) $$\left[\Delta^2 y_{-1} + \frac{p}{2}(\Delta^3 y_{-1} + \Delta^3 y_{-2}) + \left(\frac{6p^2-1}{12}\right)\Delta^4 y_{-2} + \cdots\right]$$

(3) $$\left[\Delta y_0 + \left(\frac{2p-1}{4}\right)(\Delta^2 y_{-1} + \Delta^2 y_0) + \left(\frac{3p^2-3p+\frac{1}{2}}{6}\right)\Delta^3 y_{-1} + \cdots\right]$$

(4) $$\left[\frac{1}{2}(\Delta^2 y_{-1} + \Delta^2 y_0) + \left(p - \frac{1}{2}\right)\Delta^3 y_{-1} + \cdots\right]$$

Set B

(1) $hy'(x)$, First-order derivative from Stirling's formula
(2) $hy'(x)$, First-order derivative from Bessel's formula
(3) $h^2y''(x)$, Second-order derivative from Stirling's formula
(4) $h^2y''(x)$, Second-order derivative from Bessel's formula

The correct pairing of members of Set A and Set B is

(a) (1, 1) (2, 2) (3, 3) (4, 4)
(b) (1, 1) (2, 3) (3, 2) (4, 4)
(c) (1, 2) (2, 3) (3, 1) (4, 4)
(d) (1, 3) (2, 2) (3, 4) (4, 1) **Ans:** (b)

3. The following four expressions (Set A) are derivatives, obtained from the interpolation formulas mentioned below (Set B) when $p = 0$

Set A

(1) $\left[\Delta y_0 - \frac{1}{2}\Delta^2 y_0 + \frac{1}{3}\Delta^3 y_0 - \cdots\right]$

(2) $\left[\Delta^2 y_0 - \Delta^3 y_0 + \frac{11}{12}\Delta^4 y_0 - \cdots\right]$

(3) $\left[\nabla y_n + \frac{1}{2}\nabla^2 y_n + \frac{1}{3}\nabla^3 y_n + \cdots\right]$

(4) $\left[\nabla^2 y_n + \nabla^3 y_n + \frac{11}{12}\nabla^4 y_n + \cdots\right]$

Set B

(1) $hy'(x)$, First-order derivative from Newton's forward formula
(2) $h^2y''(x)$, Second-order derivative from Newton's forward formula
(3) $hy'(x)$, First-order derivative from Newton's backward formula
(4) $h^2y''(x)$, Second-order derivative from Newton's backward formula

The correct pairing of members of set *A* and set *B* is

(a) (1, 4) (2, 3) (3, 2) (4, 1)
(b) (1, 1) (2, 3) (3, 2) (4, 4)
(c) (1, 1) (2, 2) (3, 3) (4, 4)
(d) (1, 2) (2, 3) (3, 4) (4, 1) **Ans:** (c)

4. The following four expressions (Set A) are derivatives obtained from the interpolation formulas mentioned below (Set B) when $p = 0$

Set A

(1) $\left[\left(\frac{\Delta y_0 + \Delta y_{-1}}{2}\right) - \left(\frac{\Delta^3 y_{-1} + \Delta^3 y_{-2}}{12}\right) + \left(\frac{\Delta^5 y_{-2} + \Delta^5 y_{-1}}{60}\right)\cdots\right]$

(2) $\left[\Delta^2 y_{-1} - \frac{1}{12}\Delta^4 y_{-2} + \frac{1}{90}\Delta^6 y_{-3} + \cdots\right]$

(3) $\left[\Delta y_0 - \left(\frac{\Delta^2 y_{-1} + \Delta^2 y_0}{4}\right) + \left(\frac{\Delta^3 y_{-1}}{12}\right) + \left(\frac{\Delta^4 y_{-2} + \Delta^4 y_{-1}}{24}\right) + \cdots\right]$

(4) $\left[\left(\frac{\Delta^2 y_{-1} + \Delta^2 y_0}{2}\right) - \frac{1}{2}\Delta^3 y_{-1} + \cdots\right]$

Set B

(1) $hy'(x)$, First-order derivative from Stirling's formula
(2) $h^2y''(x)$, Second-order derivative from Stirling's formula
(3) $hy'(x)$, First-order derivative from Bessel's formula
(4) $h^2y''(x)$, Second-order derivative from Bessel's formula

The correct pairing of members of Set A and Set B is

(a) (1, 4) (2, 3) (3, 1) (4, 2)
(b) (1, 4) (2, 3) (3, 2) (4, 1)
(c) (1, 3) (2, 4) (3, 1) (4, 2)
(d) (1, 1) (2, 2) (3, 3) (4, 4) **Ans:** (d)

5. Using the Newton's forward interpolation formula find $y'(1.1)$ if $x_0 = 1.0$, $h = 0.2$, $p = 0.5$, $\Delta y_0 = 0.128$, $\Delta^2 y_0 = 0.288$ and $\Delta^3 y_0 = 0.048$.

(a) 0.61 (b) 0.62 (c) 0.63 (d) 0.64
Ans: (c)

$$\left[\textbf{Hint: } y'(x_0 + h) = \frac{1}{h}\left[\Delta y_0 + \frac{2p-1}{2}\Delta^2 y_0 + \frac{3p^2 - 6p + 2}{6}\Delta^3 y_0\right]\right.$$

$$= \frac{1}{0.2}\left[0.128 + 0 + \frac{3(0.5)^2 - 6(0.5) + 2}{6}(0.048)\right.$$

$$\left. = 0.63.\right]$$

6. Using the data in question 5, find $y''(1.1)$

(a) 6.2 (b) 6.6 (c) 6.4 (d) 6.3 **Ans:** (b)

$$\left[\textbf{Hint:} \quad y''(x_0 + h) = \frac{1}{h^2}\left[\Delta^2 y_0 + (p-1)\Delta^3 y_0\right]\right.$$

$$\left. = \frac{1}{(0.2)^2}[0.288 + (0.5-1)0.048] = 6.6.\right]$$

7. By using Newton's forward differences formula, find $y'(1.2)$ if $x_0 = 1.2$, $p = 0$, $h = 0.2$, $\Delta y_0 = 0.416$, $\Delta^2 y_0 = 0.336$ and $\Delta^3 y_0 = 0.048$

(a) 1.30 (b) 1.32 (c) 1.34 (d) 1.34

Ans: (b)

$$\left[\textbf{Hint:} \quad y'(x_0) = \frac{1}{h}\left[\Delta y_0 - \frac{1}{2}\Delta^2 y_0 + \frac{1}{3}\Delta^3 y_0\right]\right.$$

$$= \frac{1}{0.2}\left[0.416 - \frac{1}{2}(0.336)\right.$$

$$\left.\left. + \frac{1}{3}(0.048)\right] = 1.32.\right]$$

8. Using the data in Question 7, find $y''(1.2)$

(a) 7.2 (b) 7.0 (c) 7.4 (d) 7.6 **Ans:** (a)

$$\left[\textbf{Hint:} \quad y''(x_0) = \frac{1}{h^2}\left[\Delta^2 y_0 - \Delta^3 y_0\right]\right.$$

$$\left. = \frac{1}{(0.2)^2}[0.336 - 0.048] = 7.2.\right]$$

9. Using Stirling's central difference formula find $y'(x)$ at $x = 0.2$ if $p = 0$, $h = 0.1$, $\Delta y_{-1} = 0.10117$, $\Delta y_0 = 0.10318$, $\Delta^3 y_{-2} = 0.00101$ and $\Delta^3 y_{-1} = 0.00104$

(a) 0.101 (b) 0.98 (c) 1.02 (d) 1.01

Ans: (c)

$$\left[\textbf{Hint:} \quad y'(x_0 + h) = \frac{1}{h}\left[\frac{\Delta y_{-1} + \Delta y_0}{2} - \frac{1}{6}\frac{\Delta^3 y_{-2} + \Delta^3 y_{-1}}{2}\right]\right.$$

$$= \frac{1}{0.1}\left[\frac{0.10117 + 0.10318}{2}\right.$$

$$\left.\left. - \frac{1}{12}(0.00101 + 0.00104) = 1.02\right].\right]$$

10. Using Stirling's central difference formula find $y''(x)\ at\ x=0.2$ with the data of Question 9 above.

(a) 0.221 (b) 0.202

(c) 0.101 (d) 0.201 **Ans:** (d)

$$\left[\textbf{Hint:} \quad y''(x_0 + h) = \frac{1}{h^2}\left[\Delta^2 y_{-1} - \frac{1}{12}\Delta^4 y_{-2}\right]\right.$$

$$\left. = \frac{1}{(0.1)^2}\left[0.00201 - \frac{0.00003}{12}\right] = 0.201.\right]$$

11. If $x_0 = 1.0$, $h = 0.2$, $\Delta y_0 = 0.128$, $\Delta^2 y_0 = 0.288$ and $\Delta^3 y_0 = 0.048$ find x for which y is maximum, using Newton's forward formula.

(a) −1.2 (b) 1.2 (c) −1.4 (d) 0.2

Ans: (a)

[**Hint**: Newton's forward interpolation formula

$$y = y_0 + p\Delta y_0 + \frac{p(p-1)}{2}\Delta^2 y_0$$

$$\left. + \frac{p(p-1)(p-2)}{6}\Delta^3 y_0 + \cdots\right]$$

For maximum or minimum of tabulated function

$$\frac{dy}{dp} = 0 \Rightarrow \Delta y_0 + \left(p - \frac{1}{2}\right)\Delta^2 y_0$$

$$+ \left(\frac{p^2}{2} - \frac{p}{2} + \frac{1}{3}\right)\Delta^3 y_0 = 0$$

neglecting $\Delta^n y_0$ for $n \geq 4$.

$$\Rightarrow \quad ap^2 + bp + c = 0 \text{ where } a = \frac{1}{2}\Delta^3 y_0;$$

$$b = \Delta^2 y_0 - \frac{1}{2}\Delta^3 y_0; c = \frac{1}{2}\Delta^2 y_0 + \frac{1}{3}\Delta^3 y_0.$$

Here $a = 0.024, b = 0.288 - \frac{1}{2}(0.048) = 0.264,$

$$c = 0.128 - \frac{1}{2}(0.288) + \frac{1}{3}(0.048) = 0$$

$$ap^2 + bp + c = 0 \quad \Rightarrow \quad p = 0 \text{ or } \quad p = -11$$

$$\Rightarrow \quad x = x_0 + px = 1.0 - 11(0.2) = -1.2$$

12. In Question 11 above find y_{max}

(a) 3.07 (b) 4.87 (c) 3.67 (d) 3.87

Ans: (d)

$$\left[\textbf{Hint: } y = 0 + (-11)(0.128) + \frac{132}{2}(0.288) - \frac{11 \cdot 12 \cdot 13}{6}(0.048) = 3.87.\right]$$

13. Newton–Cote's quadrative formula is $I =$

(a) $nh\left[y_0 + \frac{n}{2}\Delta y_0 + \frac{n(2n-3)}{12}\Delta^2 y_0 + \frac{n(n-2)^2}{24}\Delta^3 y_0 + \cdots\right]$

(b) $h\left[y_0 + \frac{n}{2}\Delta y_0 + \frac{n(n-3)}{12}\Delta^2 y_0 + \frac{n(n-2)^2}{24}\Delta^3 y_0 + \cdots\right]$

(c) $\frac{nh}{2}\left[y_0 + \frac{n}{2}\Delta y_0 + \frac{n(2n-3)}{12}\Delta^2 y_0 + \frac{n(n-2)^2}{24}\Delta^3 y_0 + \cdots\right]$

(d) $nh\left[y_0 + \frac{n}{2}\Delta y_0 + \frac{n(2n-3)}{12}\Delta^2 y_0 + \frac{n(n-2)^2}{24}\Delta^3 y_0 + \cdots\right]$ **Ans:** (a)

14. The composite trapezoidel rule is $\int_{x_0}^{x_n} y\,dx =$

(a) $\frac{h}{2}[y_0 + 2(y_1 + y_2 + \cdots + y_{n-1}) + y_n]$

(b) $h[y_0 + 2(y_1 + y_2 + \cdots + y_{n-1}) + y_n]$

(c) $\frac{h}{2}[y_0 + (y_1 + y_2 + \cdots + y_{n-1}) + y_n]$

(d) $h[y_0 + 3(y_1 + y_2 + \cdots + y_{n-1}) + y_n]$ **Ans:** (a)

15. For applying the Simpson's $\frac{1}{3}$ rule the minimum number of y-values required is

(a) 2 (b) 3 (c) 4 (d) 5 **Ans:** (b)

16. For applying the Simpson's $\frac{3}{8}$ th rule the minimum number of y-values required is

(a) 2 (b) 3 (c) 4 (d) 5 **Ans:** (c)

17. For applying the Weddle's rule the minimum number of y-values required is

(a) 7 (b) 3 (c) 4 (d) 5 **Ans:** (a)

18. The value of $\int_{0.1}^{0.7} dx/x$ (with $h = 0.1$) by Weddle's rule is

(a) 1.964 (b) 1.98 (c) 1.953 (d) 1.944 **Ans:** (c)

19. Evaluate $\int_0^4 e^x dx$ by Simpson's rule given that $e = 2.72$, $e^2 = 7.39$, $e^3 = 20.09$, $e^4 = 54.6$

(a) 53.78 (b) 53.64
(c) 52.47 (d) 53.87 **Ans:** (d)

20. The value of $\int_0^1 dx/x$ by Simpson's $\frac{1}{3}$ rule with $h = \frac{1}{8}$ is

(a) 0.6844 (b) 0.6736
(c) 0.6931 (d) 0.6824 **Ans:** (c)

21. The value of $\int_0^1 dx/(1+x^2)$ by Simpson's $\frac{3}{8}$ rule with $h = \frac{1}{6}$

(a) 0.6932 (b) 0.6848
(c) 0.6947 (d) 0.6984 **Ans:** (a)

22. The value of $\int_0^1 dx/(1+x^2)$ by Weddle's rule is

(a) 0.7854 (b) 0.7654
(c) 0.7846 (d) 0.7884 **Ans:** (a)

23. To apply Simpson's $\frac{1}{3}$ and $\frac{3}{8}$ rules both the interval must be divided into minimum of _______ but interval.

(a) 10 (b) 12 (c) 6 (d) 4 **Ans:** (c)

24. To apply Boole's rule the minimum number of ordinates required is

(a) 5 (b) 6 (c) 7 (d) 8 **Ans:** (a)

CHAPTER 7 NUMBER SOLUTION OF ORDINARY DIFFERENTIAL EQUATIONS

1. Theformula $y_{n+1} = y_n + \frac{h}{1!}y_n' + \frac{h^2}{2!}y_n'' + \frac{h^3}{3!}y_n''' + \cdots (n = 0,1,3,\cdots)$is ______ iteration formula for the solution of $y' f(x, y), y(x_0) = y_0$

(a) Picard's (b) Taylor's (c) Euler's
(d) Euler's modified method **Ans:** (b)

2. The formula $y_{n+1} = y_0 + \int_{x_0}^{x} f(x, y_n)\, dx, n = 0, 1,$ 2, 3,... is _______ iteration formula for the solution of $y' = f(x, y), y(x_0) = y_0$

(a) Picardi's (b) Taylor's (c) Euler's
(d) Euler's modified method **Ans:** (a)

3. The formula $y_{n+1} = y_n + hf(x_n, y_n)$, $n = 0, 1,$ 2, 3, ... is ________ iteration formula for the solution of $y' = f(x, y), y(x_0) = y_0$

(a) Picard's (b) Taylor's (c) Euler's
(d) Euler's modified method **Ans:** (c)

4. The formula

$$y_{n+1}^{(k+1)} = y_n + \frac{h}{2}\left[f(x_n, y_n) + f(x_{n+1}, y_{n+1}^{(k)})\right]$$
$$(k = 0,1,2,\ldots;\ \text{in} = 0,1,2\cdots)$$

is ______ iteration formula for the solution of $y' = f(x, y), y(x_0) = y_0$.

(a) Picard's (b) Taylor's (c) Euler's
(d) Euler's modified method **Ans:** (d)

5. To solve $y' = xy + 1$, $y(0) = 1$ numerically by Taylor's series method, we compute the values of y_0', y_0'', y_0''' and y_0^{iv} as

(a) 1, 1, 2, 3 (b) 1, 0, 2, 3
(c) 1, 1, 0, 3 (d) 1, 0, 1, 2 **Ans:** (a)

6. To solve $y' = x + y$, $y(1) = 0$ numerically by Taylor's series method we compute the values of y_0', y_0'', y_0''' and y_0^{iv} as

(a) 1, 2, 2, 3 (b) 1, 2, 0, 4
(c) 1, 2, 2, 2 (d) 1, 2, 3, 4 **Ans:** (c)

7. In the numerical solution of $y' = f(x, y), y(0) = y_0$ by picard's method the formula for the first approximation y_1 is $y_1 =$

(a) $y_0 - \int_0^x f(x, y_0)\, dx$ (b) $y_0 - \int_{x_0}^x f(x, y_0)\, dx$

(c) $y_0 + \int_{x_0}^x f(x, y_0)\, dx$ (d) $y_0 + \int_0^x f(x, y_0)\, dx$

Ans: (c)

8. In the numerical solution of $y' = xy + 1, y(0) =$ 1 by Picard's method the first approximation is $y_1 =$

(a) $1 + x - \frac{x^2}{2}$ (b) $1 - x + \frac{x^2}{2}$

(c) $1 + x + \frac{1}{2}x^2$ (d) $1 + \frac{x}{2} + \frac{x^2}{3}$ **Ans:** (c)

9. In the numerical solution of $y' = xy + 1, y(0) = 1$ by Picard's method the first approximation for $y_{0.1} =$

(a) 1.100 (b) 1.105
(c) 1.115 (d) 1.110 **Ans:** (b)

10. In the numerical solution of $y' = x - y, y_0 = 1$ by Picard's method the first approximation $y_1 =$

(a) $1+\frac{x}{2}+\frac{x^2}{4}$ (b) $1-x+\frac{1}{2}x^2$

(c) $1+x+\frac{x^2}{2}$ (d) $1+x-\frac{x^2}{2}$ **Ans:** (b)

11. In the numerical solution of $y' = x - y, y_0 = 1$ by Picard's method the first approximation y_1 at $x = 0.2$ is

(a) 0.801 (b) 0.812
(c) 0.820 (d) 0.802 **Ans:** (c)

12. Solving the equation $y' = x + y, y(0) = 1$ are obtains $y(0.1)$ as

(a) 1.2 (b) 0.99 (c) 1.0 (d) 1.1 **Ans:** (d)

13. Solving $dy/dx = 3x^2 + 1;\ y(1) = 2$ the first approximation $y_1 = y(1.5)$ is obtained as

(a) 2 (b) 3 (c) 4 (d) 5 **Ans:** (c)

14. Using Euler's modified method, $y(0.2)$ for $y' = y + e^x, y(0) = 0$ is obtained as

(a) 1.4214 (b) 1.4124
(c) 1.4304 (d) 1.2414 **Ans:** (a)

15. By Euler's modified method $y(0.02)$ is computed for $y' = x^2 + y, y(0) = 1$

(a) 1.0204 (b) 1.0202
(c) 1.2020 (d) 1.0222 **Ans:** (b)

16. In the fourth order Runge–Kutta method for numerical solution of $dy/dx = f(x, y), y(x_0) = y_0$ the formula for the first approximation is $y_1 =$

(a) $\frac{1}{6}(k_1 + k_2 + k_3 + k_4)$

(b) $\frac{1}{6}(k_1 + 2k_2 + 2k_3 + k_4)$

(c) $\frac{1}{6}(k_1 + 2k_2 + 2k_3 + k_4)$

(d) $\frac{1}{6}(k_1 + k_2 + 2k_3 + k_4)$ **Ans:** (c)

17. In Question 16, k_1 and k_2 are given by

(a) $k_1 = hf(x_0, y_0),\ k_2 = hf(x_0 + h, y_0 + k_1)$

(b) $k_1 = hf(x_0, y_0), k_2 = hf\left(x_0 + \frac{1}{2}h, y_0 + \frac{1}{2}k_1\right)$

(c) $k_1 = f(x_0, y_0), k_2 = hf\left(x_0 + \frac{1}{2}h, y_0 + \frac{1}{2}k_1\right)$

(d) $k_1 = hf(x_0, y_0),\ k_2 = hf(x_0 + h, y_0 + k_1)$

Ans: (b)

18. In Question 16, k_3 and k_4 are given by

(a) $k_3 = hf\left(x_0 + h, y_0 + \frac{1}{2}k_2\right)$,

$k_4 = hf\left(x_0 + \frac{1}{2}h, y_0 + \frac{1}{2}k_3\right)$

(b) $k_3 = hf(x_0 + h, y_0 + k_2)$,

$k_4 = hf\left(x_0 + \frac{1}{2}h, y_0 + \frac{1}{2}k_3\right)$

(c) $k_3 = hf\left(x_0 + \frac{1}{2}h, y_0 + k_2\right)$,

$k_4 = hf(x_0 + h, y_0 + k_3)$

(d) $k_3 = hf\left(x_0 + \frac{1}{2}h, y_0 + \frac{1}{2}k_2\right)$,

$k_4 = hf(x_0 + h, y_0 + k_3)$ **Ans:** (d)

19. Milne's predictor formula is

(a) $y_{n+1}^{(p)} = y_{n-1} + \frac{4h}{3}(2y'_{n-2} - y'_{n-1} + 2y'_n)$

(b) $y_{n+1}^{(p)} = y'_{n-1} + \frac{4h}{3}(2y'_{n-2} + y'_{n-1} + 2y'_n)$

(c) $y_{n+1}^{(p)} = y_{n-1} + \frac{2h}{3}(2y'_{n-2} - y'_{n-1} + 2y'_n)$

(d) $y_{n+1}^{(p)} = y_{n-1} + \frac{2h}{3}(2y'_{n-2} + y'_{n-1} + 2y'_n)$

Ans: (a)

20. Milne's corrector formula is

(a) $y_{n+1}^{(c)} = y_{n-1} + \frac{h}{3}(y_{n+1}'^{(p)} + 4y'_n + y'_{n-1})$

(b) $y_{n+1}^{(c)} = y_{n-1} + \frac{h}{3}(y_{n+1}'^{(p)} - 4y'_n + y'_{n-1})$

(c) $y_{n+1}^{(c)} = y_{n-1} + \frac{2h}{3}(y_{n+1}'^{(p)} + 4y_n' + y_{n-1}')$

(d) $y_{n+1}^{(c)} = y_{n-1} + \frac{2h}{3}(y_{n+1}'^{(p)} - 4y_n' + y_{n-1}')$

Ans: (a)

CHAPTER 8 FOURIER SERIES

Classify the following functions defined in the interval mentioned against them as

(a) Even (b) Odd (c) Neither even nor odd
(d) Nonclassifiable and more the correct letter in the brackets.

Function	Interval	
1. $\sin x + \cos x$	$(-\infty, \infty)$	**Ans:** (c)
2. e^x	$(-\infty, \infty)$	**Ans:** (c)
3. $\log x$	$(0, \infty)$	**Ans:** (d)
4. $\log\frac{1+x}{1-x}$	$(-1, 1)$	**Ans:** (b)
5. $\sin x \sinh x$	$(-\infty, \infty)$	**Ans:** (a)
6. $x\log\frac{1+x}{1-x}$	$(-1, 1)$	**Ans:** (a)
7. $\sin^{-1} x + x^2 \log\frac{1+x}{1-x}$	$(-1, 1)$	**Ans:** (b)
8. $\lvert x\rvert$	$(-\infty, \infty)$	**Ans:** (a)

9. The odd function among the following is

(a) $f(x) = \begin{cases} -k, & (-\pi, 0) \\ k, & (0, \pi) \end{cases}$

(b) $f(x) = \begin{cases} -x, & (-\pi, 0) \\ x, & (0, \pi) \end{cases}$

(c) $f(x) = \sqrt{x}(0, \infty)$

(d) $f(x) = \log(-x)\ (-\infty, 0)$ **Ans:** (a)

10. The even function among the following is

(a) $f(x) = \sqrt{x}\quad (0, \infty)$

(b) $f(x) = \log x\quad (0, \infty)$

(c) $f(x) = \begin{cases} -x & (-\pi, 0) \\ x & (0, \pi) \end{cases}$

(d) $f(x) = \begin{cases} -k, & (-\pi, 0) \\ k & (0, \pi) \end{cases}$ **Ans:** (c)

11. The graph of a function is as shown in the figure; then the function is

(a) Even
(b) Odd
(c) Neither even nor odd
(d) Nonclassifiable **Ans:** (b)

12. The graph of a function is as shown in the figure; then the function is

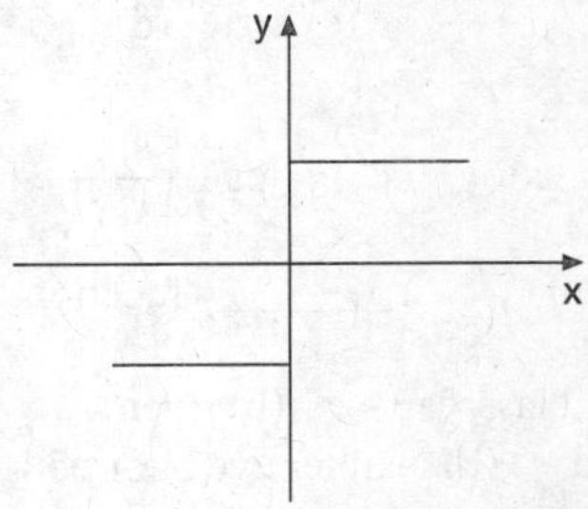

(a) Even
(b) Odd
(c) Neither even nor odd
(d) Nonclassifiable **Ans:** (a)

13. The leading term in the Fourier series expansion of $f(x) = x \cos i$ in $(-\pi, \pi)$ is

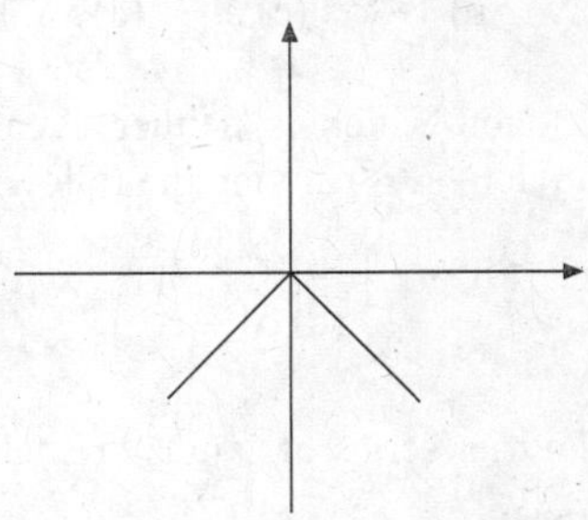

(a) $-\sin x$ (b) $\frac{1}{2}\sin x$

(c) $-\frac{1}{2}\sin x$ (d) $\sin x$ **Ans:** (c)

14. The coeffcients $b_{2n} = 0$ in the Fourier series expansion in $(-\pi, \pi)$ of the function $f(x) =$

(a) $|\sin x|$ (b) $\begin{cases} -k & (-\pi,0) \\ k & (0,\pi) \end{cases}$

(c) x^2 (d) $\cos x$ **Ans:** (b)

15. If $f(x) = \begin{cases} -\pi & (-\pi,0) \\ x & (0,\pi) \end{cases}$ then the value of $f(x)$ at $x = 0$, which is a point of discontinuity, is

(a) $\frac{\pi}{2}$ (b) $-\frac{\pi}{2}$ (c) 0 (d) Not defined **Ans:** (b)

16. If $f(x) = |x|$ in $(-\pi, \pi)$ then the Fourier coeffcient $a_0 =$

(a) π (b) $\frac{\pi}{2}$ (c) 2π (d) $\frac{3\pi}{2}$ **Ans:** (a)

17. The function $f(x) = \log\frac{1+\sin x}{1-\sin x}$ in $(-\infty, \infty)$, $x \neq (2n+1)\frac{\pi}{2}$, is________ function

(a) Nonclassifiable (b) Even
(c) Odd (d) Neither even nor odd **Ans:** (c)

18. The function which is an odd function in $(-\infty, \infty)$ among the following is

(a) $\cos x$ (b) $1+x$ (c) e^{-x} (d) x **Ans:** (d)

19. The function which is an even function in $(-\infty, \infty)$ among the following is

(a) $\cos x$ (b) $1+x$ (c) e^{-x} (d) x **Ans:** (a)

20. The function which is neither even nor odd function in $(-\infty, \infty)$ among the following

(a) $\cos x + \sin x$ (b) $e^x + e^{-x}$
(c) $ex - e^{-x}$ (d) $|x| + x^2$ **Ans:** (a)

21. If the Fourier series of $f(x) = x/\pi$ in $(-\pi, \pi)$ is $\frac{x}{\pi} = \frac{2}{\pi}\sum_{n=1}^{\infty}\frac{(-1)^{n-1}}{n}\sin nx$ then $\frac{1}{1} - \frac{1}{3} + \frac{1}{5} - \frac{1}{7} + \frac{1}{9} - \cdots =$

(a) π (b) $\frac{3\pi}{2}$ (c) $\frac{\pi}{2}$ (d) $\frac{\pi}{4}$ **Ans:** (d)

22. If the Fourier series of $f(x) = \begin{cases} 1+\frac{2\pi}{\pi}, & (-\pi,0) \\ 1-\frac{2\pi}{\pi}, & (\pi,0) \end{cases}$ is $f(x) = \frac{8}{\pi^2}\left(\frac{\cos x}{1^2} + \frac{\cos 3x}{3^2} + \frac{\cos x}{5^2} + \cdots\right)$ then $\frac{1}{1^2} + \frac{1}{3^2} + \frac{1}{5^2} + \cdots =$

(a) $\frac{\pi^2}{12}$ (b) $\frac{\pi^2}{6}$ (c) $\frac{\pi^2}{8}$ (d) $\frac{\pi^2}{4}$ **Ans:** (c)

23. If the Fourier series of $\sin ax$ in $(-\pi, \pi)$ is $\sin ax = \frac{2\sin a\pi}{\pi}\left[\frac{\sin x}{1^2 - a^2} + \frac{2\sin 2x}{2^2 - a^2} + \frac{3\sin 3x}{3^2 - a^2} - \cdots\right]$ where $a \notin z$, then $\frac{1}{1^2 - a^2} - \frac{3}{3^2 - a^2} + \frac{5}{5^2 - a^2} - \cdots =$

(a) $\frac{\pi}{4}\sec\frac{a\pi}{2}$ (b) $\frac{\pi}{2}\sec\frac{a\pi}{2}$

(c) $\frac{\pi}{2}\cos\frac{a\pi}{2}$ (d) $\frac{\pi}{4}\sin\frac{a\pi}{2}$ **Ans:** (a)

24. If the Fourier series of $\sin ax$ in $(-\pi, \pi)$ is $\sin ax = \frac{2\sin a\pi}{\pi}\left[\frac{\sin x}{1^2 - a^2} - \frac{2\sin 3x}{2^2 - a^2} + \frac{3\sin 3x}{3^2 - a^2} - \cdots\right]$ where $a \notin z$, then $\frac{1}{1^2 - (0.1)^2} - \frac{3}{3^2 - (0.1)^2} + \frac{5}{5^2 - (0.1)^2} - \cdots =$

(a) $\frac{\pi}{4}\sec\frac{\pi}{10}$ (b) $\frac{\pi}{2}\sec\frac{\pi}{10}$

(c) $\frac{\pi}{4}\sec\frac{\pi}{20}$ (d) $\frac{\pi}{4}\sin\frac{\pi}{20}$ **Ans:** (c)

25. If $2\pi x - x^2 = \frac{2\pi^2}{3} - \frac{4}{\pi^2}\sum_{n=1}^{\infty}\cos nx$ in $(0, 2\pi)$ then $\frac{1}{1^2} + \frac{1}{2^2} + \frac{1}{3^2} + \cdots =$

(a) $\frac{\pi^2}{16}$ (b) $\frac{\pi^2}{12}$ (c) $\frac{\pi^2}{8}$ (d) $\frac{\pi^2}{6}$

Ans: (d)

26. If $f(x)=\begin{cases}0 & [-\pi,0]\\ x & [0,\pi]\end{cases}$ has Fourier series expansion $f(x)=\frac{\pi}{4}-\frac{2}{\pi}\sum_{n=1}^{\infty}\frac{\cos(2n-1)x}{(2n-1)^2}+\sum_{n=1}^{\infty}\frac{(-1)^{n-1}\sin nx}{n}$ then $1+\frac{1}{3^2}+\frac{1}{5^2}+\frac{1}{7^2}+\cdots=$

(a) $\frac{\pi^2}{16}$ (b) $\frac{\pi^2}{8}$ (c) $\frac{\pi^2}{6}$ (d) $\frac{\pi^2}{12}$ **Ans:** (b)

27. If $x^2=\frac{\pi^2}{3}-4\sum_{n=1}^{\infty}\frac{(-1)^n\cos nx}{n^2}$ in $(-\pi,\pi)$ then $1-\frac{1}{2^2}+\frac{1}{3^2}-\frac{1}{4^2}+\frac{1}{5^2}-\frac{1}{6^2}+\cdots=$

(a) $\frac{\pi^2}{12}$ (b) $\frac{\pi^2}{8}$ (c) $\frac{\pi^2}{6}$ (d) $\frac{\pi^2}{16}$ **Ans:** (a)

28. If $1-\frac{1}{2^2}+\frac{1}{3^2}-\frac{1}{4^2}+\cdots=\frac{\pi^2}{12}$ and $1+\frac{1}{2^2}+\frac{1}{3^2}+\frac{1}{4^2}+\cdots=\frac{\pi^2}{6}$ then $\frac{1}{1^2}+\frac{1}{3^2}+\frac{1}{5^2}+\frac{1}{7^2}+\cdots=$

(a) $\frac{\pi^2}{24}$ (b) $\frac{\pi^2}{16}$ (c) $\frac{\pi^2}{8}$ (d) $\frac{\pi^2}{4}$ **Ans:** (c)

29. If the half-range sine series in $(0,\pi)$ for $f(x)=x(\pi-x)=\frac{8}{\pi}\sum_{n=1}^{\infty}\frac{\sin(2n-1)^x}{(2n-1)^3}$ then $1-\frac{1}{3^3}+\frac{1}{5^3}-\frac{1}{7^3}+\cdots=$

(a) $\frac{\pi^2}{24}$ (b) $\frac{\pi^2}{8}$ (c) $\frac{\pi^2}{16}$ (d) $\frac{\pi^2}{32}$ **Ans:** (d)

30. If the half-range cosine series in $(0,\pi)$ for $f(x)=\sin x=\frac{2}{\pi}-\frac{4}{\pi}\sum_{n=1}^{\infty}\frac{\cos 2nx}{4n^2-1}$ then $\frac{1}{1.3}+\frac{1}{3.5}-\frac{1}{5.7}+\frac{1}{7.9}\cdots=$

(a) $\frac{1}{2}$ (b) $\frac{\pi-2}{2}$ (c) $\frac{\pi+2}{4}$ (d) $\frac{\pi-4}{4}$

Ans: (a)

31. If the half-range sine series of e^x in $(0,\pi)$ is $e^x=\frac{2}{\pi}\left[\frac{1+e^x}{1^2+1}\sin x+\frac{2(1-e^x)}{2^2+1}\sin 2x+\frac{3(1+e^x)}{3^2+1}\sin 3x+\cdots\right]$ then $\frac{1}{1^2+1}-\frac{3}{3^2+1}+\frac{5}{5^2+1}-\cdots=$

(a) $\frac{\pi}{2}\operatorname{sech}\frac{\pi}{2}$ (b) $\frac{\pi}{4}\operatorname{sech}\frac{\pi}{2}$

(c) $\frac{\pi}{4}\cosh\frac{\pi}{2}$ (d) $\frac{\pi}{4}\operatorname{cosech}\frac{\pi}{2}$ **Ans:** (b)

32. If the half-range cosine series in $(0,\pi)$ for $x\sin x=1-\frac{1}{2}\cos x-\frac{2}{1.3}+\cos 2x+\frac{2}{2.4}\cos 3x-\frac{2}{3.5}\cos 4x+\cdots$ then $\frac{1}{1.3}-\frac{1}{3.5}+\frac{1}{5.7}-\frac{1}{7.9}+\cdots=$

(a) $\frac{\pi}{4}-\frac{1}{2}$ (b) $\frac{\pi-1}{4}$ (c) $\frac{\pi-1}{2}$ (d) $\frac{\pi+2}{4}$

Ans: (a)

33. If $\frac{1}{1.3}-\frac{1}{3.5}+\frac{1}{5.7}-\frac{1}{7.9}+\cdots=\frac{\pi}{4}-\frac{1}{2}$ and $\frac{1}{1.3}+\frac{1}{3.5}+\frac{1}{5.7}+\frac{1}{7.9}+\cdots=\frac{1}{2}$ then $\frac{1}{1.3}+\frac{1}{5.7}+\frac{1}{9.11}+\cdots=$

(a) $\frac{\pi}{4}$ (b) $\frac{\pi}{6}$ (c) $\frac{\pi}{8}$ (d) $\frac{\pi}{12}$ **Ans:** (c)

34. If $e^{-ax}=\frac{2\sinh a\pi}{\pi}\left[\left(\frac{1}{2a}-\frac{a\cos x}{1^2+a^2}+\frac{a\cos 2x}{2^2+a^2}-\frac{a\cos 3x}{3^2+a^2}+\cdots\right)-\left(\frac{\sin x}{1^2+a^2}-\frac{2\sin 2x}{2^2+a^2}+\frac{3\sin 3x}{3^2+a^2}+\cdots\right)\right]$ in $(-\pi,\pi)$ then $\frac{1}{2^2+1}-\frac{1}{3^2+1}+\frac{1}{4^2+1}\cdots=$

(a) $\frac{2\pi}{\sinh\pi}$ (b) $\frac{\pi}{\sinh\pi}$

(c) $\frac{\pi}{2\sinh\pi}$ (d) $\frac{\pi}{2}\sinh\pi$ **Ans:** (c)

35. If the Fourier series expansion of $f(x) = x^2$ in $(-l, l)$ is $x^2 = \frac{l^2}{3} - \frac{4l^2}{\pi^2}\sum_{n=1}^{\infty}\frac{(-1)^{n-1}}{n^2}\cos\frac{n\pi x}{l}$ then $\frac{1}{1^2} - \frac{1}{2^2} + \frac{1}{3^2} - \frac{1}{4^2} + \cdots =$

(a) $\frac{\pi^2}{4}$ (b) $\frac{\pi^2}{12}$ (c) $\frac{\pi^2}{6}$ (d) $\frac{\pi^2}{2}$ **Ans:** (b)

36. If the Fourier series expansion of $\cos ax$ in $(-\pi, \pi)$ is $\cos ax = \frac{2a\sin a\pi}{\pi}\left[\frac{1}{2a^2} + \sum_{n=1}^{\infty}\frac{(-1)^{n-1}\cos nx}{n^2 - a^2}\right]$, $(a \notin z)$ then $\frac{1}{2a^2} + \sum_{n=1}^{\infty}\frac{(-1)^{n-1}}{n^2 - a^2} = \ldots$

(a) $\frac{\pi}{a\sin a\pi}$ (b) $\frac{\pi}{2\sin a\pi}$

(c) $\frac{\pi}{2a\sin a\pi}$ (d) $\frac{\pi}{2a\sin 2a\pi}$ **Ans:** (c)

37. If n is an integer and 'a' is not an integer then $\frac{\sin(n+a)\pi}{\sin a\pi} =$

(a) $(-1)^{n-1}$ (b) $(-1)^n$ (c) (-1) (d) 0 **Ans:** (b)

38. If n is an integer and 'a' is not an integer then $\frac{\sin(n-a)\pi}{\sin a\pi} =$

(a) $(-1)^{n-1}$ (b) $(-1)^n$ (c) (-1) (d) 0 **Ans:** (a)

39. If the Fourier series for e^x $(0, 2\pi)$ is given by $e^x = \frac{e^{2\pi} - 1}{\pi}\left[\frac{1}{2} + \sum_{n=1}^{\infty}\frac{\cos nx}{n^2+1} - \sum_{n=1}^{\infty}\frac{n\sin nx}{1+n^2}\right]$ then $\frac{1}{2} + \sum_{n=1}^{\infty}\frac{(-1)^n}{1+n^2} =$

(a) $\frac{\pi}{2}$ sech π (b) π cosech π

(c) $\frac{\pi}{2}$ cosech π (d) π cosech π **Ans:** (c)

[**Hint:** Put $x = \pi$.]

40. If the Fourier series for e^x $(0, 2\pi)$ is given by $e^x = \frac{e^{2\pi} - 1}{\pi}\left[\frac{1}{2} + \sum_{n=1}^{\infty}\frac{\cos nx}{n^2+1} - \sum_{n=1}^{\infty}\frac{n\sin nx}{n^2+1}\right]$ then $\frac{1}{2} - \sum_{n=1}^{\infty}\frac{(-1)^{n-1}}{4n^2+1} =$

(a) $\frac{\pi e^{-\pi/2}}{2\sinh\pi}$ (b) $\frac{e^{-\pi/2}}{2\sinh\pi}$

(c) $\frac{\pi e^{-\pi/2}}{\sinh\pi}$ (d) $\frac{\pi e^{\pi/2}}{2\sinh\pi}$ **Ans:** (a)

[**Hint:** Put $x = \pi/2$ n even.]

41. If the Fourier series of $f(x) = \begin{cases} 0, & [-\pi, 0] \\ \sin x & [0, \pi] \end{cases}$ is $f(x) = \frac{1}{\pi} - \frac{2}{\pi}\sum_{n=1}^{\infty}\frac{\cos 2nx}{4n^2-1} + \frac{1}{2}\sin x$ is then $\frac{1}{1.3} + \frac{1}{5.7} - \frac{1}{9.11} + \cdots =$

(a) $\frac{\pi}{4}$ (b) $\frac{\pi}{2}$ (c) $\frac{3\pi}{8}$ (d) $\frac{\pi}{8}$ **Ans:** (d)

[**Hint:** Put $x = 0$ and $\pi/2$ and obtain, respectively, $\frac{1}{1.3} + \frac{1}{3.5} - \frac{1}{5.7} + \cdots = \frac{1}{2}$ (i) and $\frac{1}{1.3} - \frac{1}{3.5} + \frac{1}{5.7} \cdots = \frac{\pi}{4} - \frac{1}{2}$ (ii)

Then $\frac{1}{1.3} + \frac{1}{5.7} + \frac{1}{9.11} + \cdots = \frac{\text{(i)}+\text{(ii)}}{2} = \frac{\pi}{8}$.]

42. If the Fourier series of $f(x) = \begin{cases} 0, & [-\pi, 0] \\ \sin x, & [0, \pi] \end{cases}$ is $f(x) = \frac{1}{\pi} - \frac{2}{\pi}\sum_{n=1}^{\infty}\frac{\cos 2nx}{4n^2-1} + \frac{1}{2}\sin x$ then $\frac{1}{3.5} - \frac{1}{7.9} + \frac{1}{11.13} + \cdots =$

(a) $\frac{1}{4} - \frac{\pi}{8}$ (b) $\frac{1}{2} + \frac{\pi}{8}$

(c) $\frac{1}{2} - \frac{\pi}{8}$ (d) $\frac{1}{4} + \frac{\pi}{8}$ **Ans:** (c)

[**Hint:** In above question

$\frac{1}{3.5}+\frac{1}{7.9}+\frac{1}{11.13}+\cdots=\frac{(\text{i})-(\text{ii})}{2}=\frac{1}{2}-\frac{\pi}{8}.$]

43. If n is an even number then $\cos n\pi/2 =$

(a) $(-1)^{n/2}$ (b) $(-1)^{(n-1/2)}$ (c) -1 (d) 1 **Ans:** (a)

44. $\sin n\pi =$

(a) $(-1)^n$ (b) -1 (c) 0 (d) 1 **Ans:** (c)

45. The function $f(x)=\begin{cases}\frac{1}{2}(x-x^2), & 0<x<l\\ -\frac{1}{2}(x+x^2), & -l<x<0\end{cases}$ is _______ function is $(-l, l)$.

(a) Nonclassifiable (b) Even
(c) Odd (d) Neither even nor odd **Ans:** (b)

46. If $f(x) = |x|$ in $(-\pi, \pi)$ then the Fourier coeffcient $a_0 =$

(a) π (b) $\frac{\pi}{2}$ (c) $\frac{2}{\pi}$ (d) 2π **Ans:** (a)

47. If $f(x) = 2x$ is represented by Fourier series in $(-\pi, \pi)$ then the value of $a_0 =$

(a) 0 (b) π (c) $\frac{\pi}{2}$ (d) 2π **Ans:** (a)

48. The Euler's integral formula for determining the constants b_n for a function $f(x)$ defined in $(-\pi, \pi)$ is $b_n =$

(a) $\frac{1}{\pi}\int_0^{\pi} f(x)\sin n\pi x\,dx$ (b) $\frac{2}{\pi}\int_0^{\pi} f(x)\,dx$

(c) $\frac{2}{\pi}\int_0^{\pi} f(x)\sin n\pi\,dx$ (d) $\frac{2}{\pi}\int_0^{\pi} f(x)\sin nx\,dx$

Ans: (d)

49. The Euler's integral formula for the constants a_n for the function $f(x)$ defined in the interval $(0, 2l)$ is $a_n =$

(a) $\frac{1}{l}\int_0^{2l} f(x)\cos\frac{n\pi x}{l}\,dx$

(b) $\frac{2}{l}\int_0^{2l} f(x)\cos\frac{n\pi x}{l}\,dx$

(c) $\frac{1}{l}\int_0^{l} f(x)\cos\frac{n\pi x}{l}\,dx$

(d) $\frac{1}{l}\int_{-l}^{l} f(x)\cos\frac{n\pi x}{l}\,dx$ **Ans:** (a)

50. If $f(x) = \sqrt{1-\cos x}$ is expanded as a Fourier series in $(-\pi, \pi)$ then the value of $a_0 =$

(a) $\frac{\pi}{2}$ (b) $\frac{2}{\pi}$ (c) $\frac{2\sqrt{2}}{\pi}$ (d) $\frac{\pi\sqrt{2}}{4}$ **Ans:** (c)

51. The leading term in the Fourier series expansion of $f(x) = e^x$ in $(-\pi, \pi)$ is

(a) $\sinh\pi$ (b) $\frac{\sinh\pi}{\pi}$

(c) $\pi\sinh\pi$ (d) $\frac{\pi}{\sinh\pi}$ **Ans:** (b)

CHAPTER 9 FOURIER INTEGRAL TRANSFORMS

1. $\int_0^{\infty} e^{-ax}\sin bx\,dx =$

(a) $-\frac{b}{a^2+b^2}$ (b) $\frac{a}{a^2+b^2}$

(c) $\frac{b}{a^2+b^2}$ (d) $\frac{-a}{a^2+b^2}$ **Ans:** (c)

2. $\int_0^{\infty} e^{-ax}\cos bx\,dx =$

(a) $-\frac{a}{a^2+b^2}$ (b) $\frac{a}{a^2+b^2}$

(c) $\frac{b}{a^2+b^2}$ (d) $-\frac{b}{a^2+b^2}$ **Ans:** (b)

3. $\int_{-\infty}^{\infty} e^{-x^2}\,dx =$

(a) $\sqrt{\pi}$ (b) $\frac{\sqrt{\pi}}{2}$ (c) $\frac{\pi}{2}$ (d) $\sqrt{\frac{\pi}{2}}$

Ans: (a)

4. $\int_0^{\infty} \frac{e^{ax} - e^{-ax}}{e^{\pi x} - e^{-\pi x}} dx =$

(a) $\frac{1}{2}\tan\frac{a}{2}$ (b) $\frac{1}{2}\tanh\frac{a}{2}$

(c) $\tan\frac{a}{2}$ (d) $\frac{1}{2}\tan a$ **Ans:** (a)

5. $\int_0^{\infty} \frac{e^{ax} + e^{-ax}}{e^{\pi x} - e^{-\pi x}} dx =$

(a) $\frac{1}{2}\cos\frac{a}{2}$ (b) $\frac{1}{2}\tan\frac{a}{2}$

(c) $\frac{1}{2}\sec\frac{a}{2}$ (d) $\frac{1}{2}\cot\frac{a}{2}$ **Ans:** (c)

6. $\int_0^{\infty} \frac{e^{-ax}}{x}\sin bx\, dx = (a > 0, b > 0)$

(a) $\frac{1}{2}\cot^{-1}\frac{b}{a}$ (b) $\tan^{-1}\frac{b}{a}$

(c) $\frac{1}{2}\tan^{-1}\frac{b}{a}$ (d) $\cot^{-1}\frac{b}{a}$ **Ans:** (b)

7. $\int_0^{\infty} \frac{\sin ax}{x} dx, (a > 0) =$

(a) π (b) $\frac{\pi}{2}$ (c) $\frac{\pi}{4}$ (d) $\frac{\pi a}{2}$ **Ans:** (b)

8. The Fourier transform of

$$f(x) = \begin{cases} \frac{1}{2a}, & |x| \le a \\ 0, & |x| > a \end{cases} \text{ is}$$

(a) $\frac{\sin(\alpha a)}{\pi a}$ (b) $\frac{\sin(\alpha a)}{\pi a \alpha}$

(c) $\frac{\sin(\alpha a)}{2\pi a \alpha}$ (d) $\frac{\sin a}{2\pi a \alpha}$ **Ans:** (c)

9. The Fourier cosine transform of $f(x) = e^{-ax}$, $(x \ge 0, a \ge 0)$ is

(a) $\frac{a}{a^2 + \alpha^2}$ (b) $\frac{1}{a^2 + \alpha^2}$

(c) $\frac{a}{a^2 - \alpha^2}$ (d) $\frac{2a}{a^2 - \alpha^2}$ **Ans:** (a)

10. $\int_0^{\infty} \frac{\cos sx}{s^2 + 1} ds =$

(a) πe^{-x} (b) $\frac{\pi}{2}e^{-x}$ (c) $\frac{\pi}{2}e^{x}$ (d) $\frac{\pi}{4}e^{-x}$

Ans: (b)

CHAPTER 10 PARTIAL DIFFERENTIAL EQUATIONS

1. The partial differential equation obtained by eliminating a and b from $(x - a)^2 + (y - b)^2 = z^2 \cot^2\alpha$ is

(a) $p^2 - q^2 = \tan^2\alpha$ (b) $p^2 + q^2 = \tan^2\alpha$

(c) $p^2 + q^2 = \tan\alpha$ (d) $p^2 - q^2 = \cot^2\alpha$

Ans: (b)

2. The partial differential equation obtained by eliminating a and b from $x^2 + y^2 + (z - a)^2 = b^2$ is

(a) $py = qx$ (b) $py + qx = 0$
(c) $px = qy$ (d) $px + qy = 0$ **Ans:** (a)

3. The partial differential equation obtained by eliminating a and b from $z = ax + by + \frac{a-b}{a}$ is

(a) $px + qy + \frac{p-q}{p} = 0$ (b) $p = q$

(c) $z = px + qy + \frac{p-q}{p}$ (d) $p + q = z$

Ans: (c)

4. The partial differential equation obtained by eliminating the arbitrary function f from $z = f(x^2 - y^2)$ is

(a) $py + qx = 0$ (b) $py = qx$
(c) $px = qy$ (d) $px + qy = 0$ **Ans:** (a)

5. Solving the partial differential equation $p = xy$ we obtain

(a) $\frac{x^2}{2}y = z$ (b) $xy = z$

(c) $\frac{x^2}{2}+f(y)=z$ (d) $\frac{x^2}{2}y+f(y)=z$
Ans: (d)

6. The general solution of $px + qy = z$ is

(a) $F\left(\frac{x}{z},\frac{y}{z}\right)=0$ (b) $F(x, y) = 0$
(c) $F(x-z, y-z)=0$ (d) $F(x, y, z) = 0$
Ans: (a)

7. The general solution of $x(y-z)p + y(z-x)q = z(x-y)$ is

(a) $F(x+y+z, xyz) = 0$
(b) $F(xy+yz+zx, xyz) = 0$
(c) $F\left(\frac{1}{x}+\frac{1}{y}+\frac{1}{z}, xyz\right)$ (d) none of these
Ans: (a)

8. The general solution of $(y-z)p + (z-x)q = x-y$ is

(a) $F(x+y+z, (x+y+z)^2) = 0$
(b) $F(xy+yz+zx, xyz) = 0$
(c) $F(x+y+z, x^2+y^2+z^2) = 0$
(d) $F(x+y+z, x^2+y^2-z^2) = 0$ **Ans:** (c)

9. The complete integral of $pq = k$ is

(a) $z=ax+\frac{k}{a}y+c$ (b) $z=ax+\frac{k}{a}y$
(c) $z = ax + c$ (d) $z = kx + ay$ **Ans:** (a)

10. The complete integral of $z = px + qy + pq$ is

(a) $z = ax + by + ab$ (b) $z = ax + by$
(c) $z = bx + ay + c$ (d) $z = kx + ay$
Ans: (a)

11. The partial differential equation obtained by eliminating a and b from $z = ax + by$ is

(a) $px = qy$ (b) $py = qx$
(c) $z = px + qy$ (d) $px + qy = 0$ **Ans:** (c)

12. The partial differential equation obtained by eliminating a and b from $z = ax + by + ab$ is

(a) $z = px + qy$ (b) $z = px + qy + pq$
(c) $z = px - qy$ (d) $px = qy$ **Ans:** (b)

13. The partial differential equation obtained by eliminating a and b from $z = (x^2 + a^2)(y^2 + b^2)$ is

(a) $4xy = z(p+q)$ (b) $pq = 4xy$
(c) $pq = 4xyz$ (d) $pq = 2xyz$ **Ans:** (c)

14. The result of eliminating the constants h and k from $(x-h)^2 + (y-k)^2 + z = a^2$ is the partial differential equation

(a) $z^2(p^2+q^2) = a^2$ (b) $z^2(p^2+q^2+1) = a^2$
(c) $z^2p^2 + z^2q^2 + z = a^2$ (d) None of these
Ans: (b)

15. Eliminating a and b from $z = ae^{bt} \sin bx$ we obtain the partial differential equation

(a) $\frac{\partial^2 z}{\partial x^2}+\frac{\partial^2 z}{\partial t^2}=0$ (b) $\frac{\partial^2 z}{\partial x^2}=\frac{\partial^2 z}{\partial t^2}$
(c) $\frac{\partial^2 z}{\partial x^2}=\frac{\partial z}{\partial t}$ (d) $\frac{\partial z}{\partial x}=\frac{\partial z}{\partial t}$ **Ans:** (a)

16. Eliminating a and b from $z = ax + (1-a)y + b$ we obtain the partial differential equation

(a) $py = qx$ (b) $px = qy$
(c) $p = q$ (d) $p + q = 1$ **Ans:** (d)

17. The partial differential equation of all planes whose x and y intercepts are always equal is

(a) $p = q$ (b) $p + q = 0$
(c) $px = qy$ (d) $py = qx$ **Ans:** (a)

18. By eliminating the arbitrary function from $z=x^n f\left(\frac{xy}{z}\right)$ we obtain the partial differential equation

(a) $px + qy = z$ (b) $px + qy = zn$
(c) $px + qy = nz$ (d) $py + qx = nz$ **Ans:** (c)

19. By eliminating the arbitrary function f from $z = f\left(\frac{xy}{z}\right)$ we obtain the partial differential equation

(a) $py = qx$ (b) $px = qy$
(c) $px + qy = z$ (d) $py + qx = z$ **Ans:** (b)

20. The complex solution of $z = px + qy + \log pq$ is

(a) $z = ax + by + \log ab$ (b) $z = x + y + ab$
(c) $z = ax - by + e^{ab}$ (d) $z = ax + by$
Ans: (a)

21. The complete integral of $yp + xq + pq = 0$ is given by

(a) $2z = \frac{x^2}{a^2-1} - \frac{y^2}{a} + b$

(b) $z = \frac{x^2}{a^2-1} - \frac{y^2}{a} + b|$

(c) $z = \frac{x^2}{a^2-1} + \frac{y^2}{a} + b$

(d) $z = \frac{x^2}{1-a^2} - \frac{y^2}{a} + b$ **Ans:** (a)

22. The complete integral of $4z = p + q$ is

(a) $z = (x + ay + b)$ (b) $az = (x + ay + b)^2$
(c) $z = (x + ay + b)$ (d) $z = x + ay + b$
Ans: (b)

23. The general solution of $p + q = 1$ is

(a) $\phi(x, z) = 0$ (b) $\phi(x - y - z, x + y) = 0$
(c) $\phi(x - y, y - z) = 0$ (d) $\phi(x + y, y + z) = 0$
Ans: (c)

24. The general integral of $p + q = z$ is

(a) $F\left(x - y, \frac{e^y}{z}\right) = 0$ (b) $F\left(x + y, \frac{e^y}{z}\right) = 0$
(c) $F(x - y, ze^y) = 0$ (d) None of these
Ans: (a)

CHAPTER 11 Z-TRANSFORMS AND SOLUTION OF DIFFERENCE EQUATIONS

1. $Z\{\langle a^n \rangle\} =$

(a) $\frac{z}{z-1}$ (b) $\frac{z}{z+a}$ (c) $\frac{z}{a-z}$ (d) $\frac{z}{z-a}$
Ans: (d)

2. $Z\{\langle l^n \rangle\} =$

(a) $\frac{z}{z-1}$ (b) $\frac{z}{z+1}$ (c) $\frac{z}{1-z}$ (d) $\frac{z^2}{1-z}$
Ans: (a)

3. $Z\{\langle (-1)^n \rangle\} =$

(a) $\frac{z}{z-1}$ (b) $\frac{z}{z+1}$ (c) $\frac{z}{1-z}$ (d) $\frac{z^2}{1-z}$
Ans: (b)

4. $Z\left\{\left\langle \left(-\frac{1}{2}\right)^n \right\rangle\right\} =$

(a) $\frac{z}{z-1}$ (b) $\frac{2z}{2z+1}$ (c) $\frac{z}{2z+1}$ (d) $\frac{2z}{z+1}$
Ans: (b)

5. $Z\{\langle n \rangle\} =$

(a) $\frac{z^2}{(z-1)^2}$ (b) $\frac{z^2}{z-1}$ (c) $\frac{z}{(z-1)^2}$
(d) $\frac{z}{(z+1)^2}$ **Ans:** (c)

6. $Z\{\langle -1, 0, 1 \rangle\} =$

(a) $\frac{1}{z^2} - 1$ (b) $\frac{1}{z^2} + 1$ (c) $1 - \frac{1}{z^2}$ (d) $1 + \frac{1}{z}$
Ans: (a)

7. $Z\{\langle n^2 \rangle\} =$

(a) $\frac{z^2 - z}{(z-1)^3}$ (b) $\frac{z^2 + z}{(z-1)^3}$

(c) $\frac{z^2+z}{(z+1)^3}$ (d) $\frac{z^2+z}{(z-1)^2}$ **Ans:** (b)

8. $Z\left\{\left\langle \frac{1}{n+1}\right\rangle\right\} =$

(a) $ze^{1/z}+1$ (b) $ze^{1/z}-1$

(c) $z(e^{1/z}-1)$ (d) $z(e^{1/z}+1)$ **Ans:** (c)

9. $Z\{\langle \cos nt\rangle\} =$

(a) $\frac{z^2-\cos t}{z^2-2z\cos t+1}$ (b) $\frac{z(z-\cos t)}{z^2-2z\cos t+1}$

(c) $\frac{z(z+\cos t)}{z^2-2z\cos t+1}$ (d) $\frac{z(z-\cos t)}{z^2+2z\cos t+1}$

Ans: (b)

10. $Z\{\langle \sin nt\rangle\} =$

(a) $\frac{z\sin t}{z^2+2z\cos t+1}$ (b) $\frac{z\sin t}{z^2-2z\sin t+1}$

(c) $\frac{z\sin t}{z^2-2z\cos t+1}$ (d) $\frac{z\sin t}{z^2-z\cos t+1}$

Ans: (a)

11. $Z\left\{\left\langle \frac{1}{n!}\right\rangle\right\} =$

(a) $e^{1/z}$ (b) $e^{-1/z}$ (c) e^z (d) $\frac{1}{e^{1/z}}$ **Ans:** (a)

12. $Z\left\{\left\langle \frac{1}{n+1}\right\rangle\right\} =$

(a) $\log\left(\frac{z}{z-1}\right)$ (b) $z\log\left(\frac{z}{z-1}\right)$

(c) $z\log\left(\frac{z-1}{z}\right)$ (d) $z\log\left(\frac{z+1}{z}\right)$ **Ans:** (b)

CHAPTER 12 SPECIAL FUNCTIONS

A. Beta and Gamma Functions

1. If $\Gamma(x)=1$, then $x =$

(a) 1 (b) 0 (c) –1/2 (d) 1/2 **Ans:** (a)

2. $\Gamma(1/2) =$

(a) $\sqrt{\pi/2}$ (b) $\pi/2$

(c) $\sqrt{\pi}$ (d) $2/\sqrt{\pi}$ **Ans:** (c)

3. If $p=1/4$, then $\Gamma(p)\,\Gamma(1-p) =$

(a) $\sqrt{\pi}$ (b) $2\sqrt{\pi}$

(c) $\sqrt{(2\pi)}$ (d) $\sqrt{2}\pi$ **Ans:** (d)

4. $\int_0^{\pi/2} \sin\theta \cos^2\theta \, d\theta =$

(a) 2/3 (b) 1/3 (c) 3/2 (d) 3 **Ans:** (b)

5. $\int_0^{\pi/2} \sqrt{\cot\theta}\, d\theta =$

(a) $\pi/\sqrt{2}$ (b) $\pi/2$

(c) $\sqrt{\pi/2}$ (d) $\sqrt{(\pi)}/2$ **Ans:** (a)

6. If $\sqrt{\pi}\,\Gamma(2p) = a\Gamma(p)\,\Gamma(p+1/2)$, then $a =$

(a) 2^{2p-1} (b) 2^{2p+1}

(c) 2^{2p} (d) 2^{p-1} **Ans:** (a)

7. $B(m+1, n)/B(m, n) =$

(a) $n/(m+n)$ (b) $(m+n)/m$

(c) $m/(m+n)$ (d) $(m+n)/n$ **Ans:** (c)

8. $\Gamma(p)\,\Gamma(1-p) = \pi \Rightarrow p =$

(a) 1/4 (b) $\sqrt{2}$ (c) 2 (d) ½ **Ans:** (d)

9. $\int_0^{\infty} e^{-ax} x^{n-1}\, dx =$

(a) $\Gamma(n)/a^n$ (b) $\Gamma(n)/a$

(c) $\Gamma(n+1)/a^n$ (d) $\Gamma(n)/a^{n-1}$ **Ans:** (a)

10. $\int_0^{\infty} \exp\,(-\sqrt{x})\, dx =$

(a) 2 (b) 1 (c) 0 (d) $\sqrt{\pi/2}$ **Ans:** (a)

11. $p\int_0^{\infty} e^{-x^p}\, dx =$

(a) $\Gamma(p)$ (b) $\Gamma(1/p)$

(c) $\Gamma(p^2)$ (d) $\Gamma(1-p)$ **Ans:** (b)

12. $\int_0^{\pi/2} \sqrt{\sin\theta}\, d\theta \times \int_0^{\pi/2} \frac{d\theta}{\sqrt{\sin\theta}}$

(a) $3\pi/2$ (b) $\pi/2$ (c) π (d) 0 **Ans:** (c)

13. $\int_0^{\pi/2} \sqrt{\tan\theta} d\theta =$

(a) $\sqrt{\pi}$ (b) π

(c) $\pi/2$ (d) $\pi/\sqrt{2}$ **Ans:** (d)

14. $4\int_0^{\infty} \frac{x^2 dx}{1+x^4} =$

(a) 2π (b) $\pi\sqrt{2}$

(c) $\pi/2$ (d) $\pi/\sqrt{2}$ **Ans:** (b)

15. $\int_0^{\infty} e^{-x^2} dx =$

(a) $\sqrt{\pi}/2$ (b) $\sqrt{\pi}$

(c) $\pi/2]$ (d) π **Ans:** (a)

B. Legendre Functions

1. $\int_{-1}^{1} P_n^2(x) dx =$

(a) $2/(2n-1)$ (b) $2/(2n+1)$

(c) $1/(2n+1)$ (d) $1/(2n-1)$ **Ans:** (b)

2. The generating function for $P_n(x)$ is

(a) $(1-2xt-t^2)^{-1/2}$

(b) $(1-2xt+t^2)^{1/2}$

(c) $(1-2xt+t^2)^{-1/2}$

(d) $(1+2xt+t^2)^{-1/2}$ **Ans:** (c)

3. $1/(1+i) = a+bi \Rightarrow (a, b) =$

(a) $\left(\frac{1}{2}, -\frac{1}{2}\right)$ (b) $\left(\frac{1}{2}, \frac{1}{2}\right)$

(c) (, – 1) (d) $\left(-\frac{1}{2}, \frac{1}{2}\right)$ **Ans:** (a)

4. $P_0(x)$

(a) 1 (b) 3/2 (c) 0 (d) 2/3 **Ans:** (a)

5. $\frac{d}{dx}(P_1(-x)) =$

(a) 0 (b) 1 (c) –1 (d) 1/2 **Ans:** (c)

6. ByRodrigues' formula, $P_n(x) = CD^n\{(x^2-1)^n\}$ where the constant $C =$

(a) $1/[2^n(2n)!]$ (b) $1/n!$

(c) $1/2^n$ (d) $1/(2^n\, n!)$ **Ans:** (d)

7. (1 + x) when expressed in terms of Legendre polynomials

(a) $P_0(x) - P_1(x)$ (b) $P_0(x) + P_1(x)$

(c) $2P_0(x) - P_1(x)$ (d) $P_0(x) + 2P_1(x)$

Ans: (b)

8. $xP'_{n-1}(x) + n\, P_{n-1}(x) =$

(a) $P'_{n-1}(x)$ (b) $P_n(x)$

(c) $P'_n(x)$ (d) $P_{n+1}(x)$ **Ans:** (c)

9. The degree of the polynomial $P_4(x)$ is

(a) 3 (b) 2 (c) 4 (d) 1 **Ans:** (c)

10. $P_n(-x) = a^n P_n(x)$ where $a =$

(a) 2 (b) 0 (c) 1 d) –1 **Ans:** (d)

11. $(3x^2-1)$ when expressed in Legendre polynomials

(a) $P_2 + 2P_1$ (b) $3P_2 - P_1$

(c) $2P_2$ (d) $3P_2 - P_1$ **Ans:** (c)

12. $\int_{-1}^{1} P_0(x) dx$

(a) 1 (b) 2 (c) 0 (d) 3 **Ans:** (b)

13. $\int_{-1}^{1} P_m(x) P_n(x) dx\ (m \neq n) =$

(a) $2/(2n+1)$ (b) 0

(c) $2/(2m+n)$ (d) $2/(m+n)$ **Ans:** (b)

14. $P_0(x)$

(a) 1 (b) 3/2 (c) 0 (d) 2/3 **Ans:** (a)

15. $\frac{d}{dx}(P_1(-x)) =$

(a) 0 (b) 1 (c) –1 (d) 1/2 **Ans:** (c)

16. By Rodrigues' formula $P_n(x) = CD^n\{(x^2-1)^n\}$ where the constant $C =$

(a) $1/(2^n(2n)!)$ (b) $1/n!$

(c) $1/2^n$ (d) $1/(2^n\ n!)$ **Ans:** (d)

17. $\int_{-1}^{1} P_n^2(x)dx =$

(a) $2/(2n-1)$ (b) $2/(2n+1)$

(c) $1/(2n+1)$ (d) $1/(2n-1)$ **Ans:** (b)

18. The generating function of Legendre function $P_n(x)$ is

(a) $(1-2xt-t^2)^{-1/2}$ (b) $(1-2xt-t^2)^{1/2}$

(c) $(1-2xt+t^2)^{-1/2}$ (d) $(1-2xt+t^2)^{1/2}$

Ans: (c)

C. Bessel Functions

1. $\frac{d}{dx}(x^n J_n(x)) =$

(a) $x^{n-1}J_{n-1}(x)$ (b) $x^n J_{n-1}(x)$

(c) $x^{n+1}J_n(x)$ (d) $x^{n-1}J_n(x)$ **Ans:** (b)

2. $\frac{d}{dx}(x^{-n}J_n(x)) =$

(a) $-x^{-n}J_m(x)$ (b) $-x^{-n+1}J_n(x)$

(c) $-x^{-n}J_{n+1}(x)$ (d) $x^{-n}J_{n+1}(x)$ **Ans:** (c)

3. $J_{-1/2}(x) =$

(a) $\sqrt{\frac{2}{\pi x}}\cos x$ (b) $\sqrt{\frac{2}{\pi}}\cos x$

(c) $\sqrt{\frac{4}{\pi x}}\cos x$ (d) $\sqrt{\frac{1}{2\pi x}}\cos x$ **Ans:** (a)

4. $J_{n-1}(x) - J_{n+1}(x) =$

(a) $J_n(x)$ (b) $2J_n(x)$

(c) $2J_n'(x)$ (d) $J_n'(x)$ **Ans:** (c)

5. $\frac{d}{dx}(xJ_1(x)) =$

(a) $x^n J_{n+1}(x)$ (b) $x^{-n}J_{n-1}(x)$

(c) $x^n J_{n-1}(x)$ (d) $J_{n-1}(x)$ **Ans:** (d)

6. $J_n(-x) =$

(a) $(-1)^n J_n(x)$ (b) $J_n(x)$

(c) $(-1)^{n-1}J_n(x)$ (d) $(-1)^n J_{n-1}(x)$

Ans: (a)

7. $\int J_0(x)J_1(x)dx =$

(a) $[(z-a)^2 f(z)]_{z=a}$ (b) $2J_0^2(x)$

(c) $-2J_0^2(x)$ (d) $-\frac{1}{3}J_0^2(x)$ **Ans:** (a)

8. $J_{-\frac{1}{2}}(x)[J_{\frac{1}{2}}(x)]^{-1} =$

(a) cosec x (b) cot x

(c) cos x (d) sec x **Ans:** (b)

9. $J_1(0) =$

(a) 0 (b) 1 (c) –1 (d) 2 **Ans:** (a)

10. $f(z) = \frac{p(z)}{q(z)}$

(a) 1 (b) 0 (c) –1 (d) x **Ans:** (b)

11. $J_0(x)$ is a solution of

(a) $(xy')' + xy = 0$ (b) $(y'/x)' + y = 0$

(c) $(x^2y') + xy = 0$ (d) $x(y'/x)' + y = 0$

Ans: (a)

12. $\{f(z):a\}$ where $p(x) =$

(a) $\sqrt{(\pi x)}$ (b) $\sqrt{(\pi x/2)}$

(c) $\sqrt{(2/\pi x)}$ (d) $\sqrt{(2\pi x)}$ **Ans:** (c)

13. $[x^n J_n(x)]' =$

(a) $x^n J_{n+1}(x)$ (b) $x^{-n}J_{n-1}(x)$

(c) $x^n J_{n-1}(x)$ (d) $J_{n-1}(x)$ **Ans:** (c)

14. $J_{1/2}^2(x) + J_{-1/2}^2(x) =$

(a) $1/(\pi x)$ (b) $\pi x/2$

(c) $2\pi x$ (d) $2/(\pi x)$ **Ans:** (d)

15. $\int xJ_0(x)dx =$

(a) $xJ_1(x) - J_0(x)$ (b) $xJ_1(x)$

(c) $J_1(x)$ (d) $x^2 J_n(x)$ **Ans:** (b)

16. $\int \frac{1}{x^2} J_3(x) dx =$

(a) $\frac{1}{x^2} J_2(x)$ (b) $x^2 J_2(x)$

(c) $g^{(m)}(a)/(m-1)!$ (d) $-x^{-1} J_3(x)$

Ans: (a)

17. $g^{(m)(a)}/m!$

(a) $J_1(x)$ (b) $g^{(m-1)}(a)/(m-1)!$

(c) $(-1)^n J_n(x)$ (d) $xJ_1(x)$ **Ans:** (b)

18. $[xJ_1(x)]' =$

(a) $2J_2(x)$ (b) $(-1)J_1(x)$

(c) $xJ_0(x)$ (d) $J_1(x)$ **Ans:** (c)

19. $J_n(-x) =$

(a) $J_n(x)$ (b) $\frac{1}{2}$

(c) $2J_n(x)$ (d) $J_{-n}(x)$ **Ans:** (b)

20. $\frac{1}{n}\int_0^{\pi} \cos(n\theta - x\sin\theta) d\theta =$

(a) $J_n(x)$ (b) $J_{-n}(x)$

(c) $(-1)^n J_n(x)$ (d) $J_n(-x)$ **Ans:** (a)

CHAPTER 13 FUNCTIONS OF A COMPLEX VARIABLE

1. $f(z) = (z+i)/(z-i)$ is analytic

(a) for all z (b) for all $z \neq i$

(c) for all $z \neq -i$ (d) for no z **Ans:** (b)

2. $f(z) = \overline{z}$ is analytic

(a) at $z = 0$ only (b) everywhere

(c) nowhere (d) if $z \neq 0$ **Ans:** (c)

3. $f(z) = |z|^2$ is analytic

(a) at $z = 0$ only (b) everywhere

(c) nowhere (d) if $z \neq 0$ **Ans:** (a)

4. The harmonic conjugate of $u(x, y) = x^2 - y^2$ is $v =$

(a) $x^3 - y^3$ (b) $x^2 + y^2$

(c) $2xy$ (d) $x^2 y^2$ **Ans:** (c)

5. The analytic function whose real part is $u(x, y) = x^2 - y^2$ is $f(z) =$

(a) $|z|^2$ (b) z^3

(c) $\frac{1}{2\pi}$ (d) $\overline{z}^2$ **Ans:** (c)

6. The analytic function whose imaginary part is $\frac{y}{x^2 + y^2}$ is $f(z) =$

(a) $\frac{1}{z^2} + c$ (b) $\frac{1}{z} + c$

(c) $z^2 + c$ (d) $z^3 + c$ **Ans:** (b)

7. Cauchy--Riemann equation in polar coordinates

(a) $u_r = \frac{1}{r} v_\theta, \quad v_r = \frac{1}{r} u_\theta$

(b) $u_r = -\frac{1}{r} v_\theta, \quad v_r = \frac{1}{r} u_\theta$

(c) $u_r = \frac{1}{r} v_\theta, \quad v_r = -\frac{1}{r} u_\theta$

(d) $-\pi i$ **Ans:** (c)

8. Singularity is a point where $f(z)$ is not

(a) defined (b) having the limit

(c) continuous (d) differentiable

Ans: (d)

9. The singularities of cosec z are at $z =$

(a) $(n+1/2)\pi$ (b) $(2n-1)\pi$

(c) $2n\pi$ (d) $n\pi$ **Ans:** (d)

10. $f(z) = e^z$ has a singularity at

(a) the origin (b) $= \frac{e^{-\pi i}}{2(-\pi i)} \Rightarrow a = -\pi i$

(c) $z = n\pi i$ (d) no point **Ans:** (d)

11. If $f(z) = z/|z| (z \neq 0)$ and $f(0) = 1$, then f is

(a) continuous of $z = 0$ only

(b) continuous for all z

(c) discontinuous for all z

(d) discontinuous at $z = 0$ only **Ans:** (d)

12. If $u(x, y) = 2x - x^2 + my^2$ is harmonic in D, then $m =$

(a) 0 (b) 1 (c) 2 (d) 3 **Ans:** (b)

13. If $f(z) = z\bar{z}$, then $f'(z)$ exists

(a) for all z (b) nowhere

(c) at $z = 0$ only (d) at $z = 1$ only **Ans:** (c)

14. A function $u(x, y)$ having continuous second partial derivatives and satisfying $\nabla^2 u = 0$ is called a

(a) harmonious function

(b) harmonic function

(c) holomorphic function

(d) regular function **Ans:** (b)

15. The analytic function whose imaginary part is $v(x, y) = 2xy$ is

(a) $|z|^2$ (b) z (c) $1/z^2$ (d) z^2 **Ans:** (d)

16. The harmonic conjugate of $v(x, y) = 2xy$ is

(a) $x^2 - y^2$ (b) $x - y$

(c) $x^2 + y^2$ (d) $x^3 - y^3$ **Ans:** (a)

17. $1/(1+i) = a + bi \Rightarrow (a, b) =$

(a) $(1/2, -1/2)$ (b) $(-1, 1)$

(c) $(1, -1)$ (d) $(-1/2, 1/2)$ **Ans:** (a)

18. $(1+i)/(1-i) = a + bi \Rightarrow (a, b) =$

(a) (1, –1)) (b) (–1, 1)

(c) (0,1) (d) (1,0) **Ans:** (c)

19. $\text{Re}(e^z) =$

(a) $C_2 \cup C_3$ (b) e^{-x}

(c) $e^x \cos y$ (d) $e^{-x} \cos y$ **Ans:** (c)

20. The complex conjugate of $(1+i)^2$ is

(a) i (b) $2/i$ (c) $2i$ (d) $2 + i$ **Ans:** (b)

21. A function $u(x, y)$ having continuous second partial derivatives and satisfying $\nabla^2 u = 0$ is called a

(a) harmonious function

(b) harmonic function

(c) holomorphic function

(d) regular function **Ans:** (b)

22. The analytic function whose imaginary part is $v(x, y) = 2xy$ is

(a) $|z|^2$ (b) z (c) $1/z^2$ (d) z^2 **Ans:** (b)

23. The harmonic conjugate of $v(x, y) = 2xy$ is

(a) $x^2 - y^2$ (b) $x - y$

(c) $x^2 + y^2$ (d) $x^3 - y^3$ **Ans:** (a)

24. $\text{Re}(e^z) =$

(a) e^x (b) e^{-x}

(c) $-8\pi i$ (d) $e^{-x} \cos y$ **Ans:** (c)

25. If $f(z) = z/|z| (z \neq 0)$, then f is

(a) continuous at $z = 0$ only

(b) continuous for all z

(c) discontinuous for all z

(d) discontinuous at $z = 0$ only **Ans:** (d)

26. If $u(x, y) = 2x - x^2 + my^2$ is harmonic in D, then $m =$

(a) 0 (b) 1 (c) 2 (d) 3 **Ans:** (b)

27. If $f(z) = z\bar{z}$, then $f'(z)$ exists

(a) for all z (b) nowhere

(c) at $2\tan 1$ only (d) at $2\pi i \tan 1$ only

Ans: (c)

28. The complex conjugate of $(1+i)^2$ is

(a) i (b) $2/i$ (c) $2i$ (d) $2+i$ **Ans:** (b)

29. If $f(z) = z(2-z)$, then $f(1+i) =$

(a) 0 (b) i (c) $-i$ (d) 2 **Ans:** (b)

30. If $f(z) = |z|$ then $f(3-4i) =$

(a) 0 (b) 5 (c) -5 (d) 12 **Ans:** (b)

31. $e^{2n\pi i} =$

(a) 0 (b) -1 (c) 1 (d) i **Ans:** (b)

32. If $f(z) = u + iv$ is analytic then

(a) $u_x = v_y, u_y = v_x$ (b) $u_x = v_y, u_y = -v_x$

(c) $u_x = -v_y, u_y = v_x$ (d) $u_x = v_x, u_y = -v_y$

Ans: (b)

Chapter 15 Complex Integration

1. A line integral of any complex function depends

(a) only on the initial point of the path

(b) only on the terminal point of the path

(c) only on the end points of the path

(d) on the end points as well as the choice of the path **Ans:** (d)

2. The line integral of a complex function is independent of path if

(a) the function is analytic in a domain containing the path

(b) the domain is simply-connected

(c) the function is analytic in a simply-connected domain containing the path

(d) the function is continuous in a domain containing the path **Ans:** (c)

3. The integral of a complex function $f(z)$ vanishes over a path C if $f(z)$ is

(a) analytic and C is any curve in domain D

(b) any complex function and C is any curve in domain D

(c) non-analytic but C is any closed path

(d) analytic and C is a closed path in a simply-connected domain D **Ans:** (d)

4. A bounded domain D is one which lies

(a) between two parallel lines

(b) in some circle about the origin

(c) outside the unit circle

$\text{Res}_{z=2i}[ze^{\pi z}/(z^2-16)] =$

(d) between a pair of intersecting lines **Ans:** (b)

5. Among the following results, the one that does not *follow* from Cauchy's Integral Theorem is

(a) $\int_C e^z\, dz = 0$ (b) $\int_C \sin z\, dz = 0$

(c) $\int_C (z^3 + 3z^2)\, dz = 0$ (d) $\int_C \frac{dz}{z^2} = 0$

Ans: (d)

6. $\int_C \frac{dz}{z} = 2\pi i$ where C lies in $D = \left\{ z \,\middle|\, \frac{1}{2} < |z| < \frac{3}{2} \right\}$. Cauchy's Theorem is not applicable because

(a) $f(z) \notin H(D)$

(b) D is not simply-connected though $f(z) \in H(D)$

(c) $f(z) \in H(d)$ but $f(z)$ is not single-valued

(d) $f(z) \notin H$ (nbd of zero) **Ans:** (b)

7. The value of $\int_C \frac{\cos z dz}{(z-\pi i)^2}$ over a simple closed curve C enclosing πi is

(a) $-2i\pi \sinh \pi$ (b) $\pi \sinh \pi$

(c) $2\pi \sinh \pi$ (d) $2\pi i \sinh \pi$ **Ans:** (c)

8. If C is the circle $|z|=4$, then the value of the integral of $f(z)=\dfrac{\cos z}{z-\pi}$ over C (counter-clockwise) is

(a) 0 (b) $-2\pi i$ (c) $2\pi i$ (d) 1 **Ans:** (b)

9. If $\int_C (z-a)^m\, dz = 2\pi i$ over positively-oriented simple closed curve C about 'a', then $m =$

(a) -1 (b) 0 (c) $\neq -1$ (d) 1 **Ans:** (a)

10. If C is the st. line segment from 0 to $1+i$, then an upper bound (by ML-inequality) for the absolute value of the integral of $f(z)=z^2$ over C is

(a) 2 (b) 1 (c) $2\sqrt{2}$ (d) ∞ **Ans:** (c)

11. If C is a simple closed curve enclosing the origin, then $\int_C \dfrac{e^{az}}{z^2} = dz$

(a) πia (b) $2\pi ia$ (c) $2\pi a$ (d) $\dfrac{1}{2}\pi ia$

Ans: (b)

12. If C is a simple closed curve enclosing the origin, then $\int_C \dfrac{e^{iz}}{z^2}\, dz$

(a) 2π (b) π (c) $-2\pi i$ (d) $2\pi i$ **Ans:** (c)

13. $\int_C \bar{z} dz$, where C is a closed path, is

(a) 0 (b) πi (c) $2\pi i$ (d) 2π **Ans:** (c)

14. If C is any simple closed path $\int_C f(z)\, dz = 0$ is true for $f(z) =$

(a) $\sec z$ (b) $\bar{z}$ (c) e^z (d) $\dfrac{1}{z}$ **Ans:** (c)

15. $\int_C \sec z\, dz = 0$ is true if C is

(a) $|z|=2$ (b) $|z|=\pi$

(c) $\dfrac{x^2}{4}+\dfrac{y^2}{2}=1$ (d) $|z|=1$ **Ans:** (d)

16. $\int_C [e^z/(z-1)^2(z^2+4)]\, dz$, where $C: |z|=1.5$, is

(a) $\dfrac{6e\pi}{25}i$ (b) $\dfrac{6e\pi}{25}$ (c) $\dfrac{6\pi}{25}i$ (d) $\dfrac{6e}{25}i$ **Ans:** (a)

17. $\int_0^{1+i} z^2 dz =$

(a) $\dfrac{1}{3}+\dfrac{i}{3}$ (b) πi (c) $\dfrac{2}{3}-\dfrac{2i}{3}$ (d) $-\dfrac{2}{3}-\dfrac{2i}{3}$

Ans: (b)

18. $\int_{-\pi i}^{\pi i} \cos z\, dz =$

(a) $2i \sinh \pi$ (b) $2\sinh \pi$

(c) $2i \sin \pi$ (d) $\dfrac{2}{i}\sinh \pi$ **Ans:** (a)

19. $\int_{8+\pi i}^{8-3\pi i} e^{z/2}\, dz =$

(a) 1 (b) $2\pi i$ (c) $-i$ (d) 0 **Ans:** (d)

20. $\int_{-i}^{i} \dfrac{dz}{z} =$

(a) $\dfrac{\pi}{2}i$ (b) $2\pi i$ (c) πi (d) $4\pi i$ **Ans:** (c)

[Hint: $|z|=1$**]**

21. $\int_C [(z^2+4)/(z-3)]\, dz = 0$ is true if C is the circle

(a) $|z|=1$ (b) $|z|=2$

(c) $|z|=4$ (d) none of these **Ans:** (a) or (b)

22. If C is the straight line from $z=0$ to $z=i$, then $\int_C (y-x-3x^2 i)\, dz$

(a) $i/2$ (b) $1/2$ (c) i (d) $-i/2$ **Ans:** (a)

23. By ML-inequality, if C is the straight line from 0 to $1+i$ we have $\left|\int_C z^2 dz\right| \le$

(a) $2\sqrt{2}$ (b) 2 (c) $\sqrt{2}$ (d) 4 **Ans:** (a)

24. $\int_{-\pi i}^{\pi i} (1+\cos z) dz$

(a) $2\sinh \pi i + 2\pi i$ (b) $zi \sinh \pi i + \pi$

(c) $2\sinh \pi + \pi i$ (d) $2i(\sinh \pi + \pi)$

Ans: (d)

25. $\int_0^{\pi i} e^z\, dz =$

(a) 2 (b) -2 (c) $\sqrt{2}$ (d) π **Ans:** (b)

26. If C is the unit circle $C:|z-i|=1$ touching the z-axis at the origin, then $\int_C (z^2+1)/(z^2-1)dz =$

(a) 0 (b) $2\pi i$ (c) πi (d) $\pi i/2$ **Ans:** (a)

27. Among the following, which one is an entire function?

(a) e^z/z (b) $z\,e^z$

(c) $\sec z$ (d) $z\,\text{cosec}\,z$ **Ans:** (b)

28. The singularities of $(z^2+1)/(z^2-1)$ are at $z =$

(a) ± 1 (b) $\pm i$ (c) $1, i$ (d) $-1, -i$ **Ans:** (a)

29. If $f(a)=\int_C (3z^2+7z+1)/(z-a)$ where C is the circle $|z|=2$, then $f(3)=$

(a) πi (b) $2\pi i$ (c) -1 (d) 0 **Ans:** (d)

30. If $f(a)=\int_C (z^2+1)/(z-a)\,dz$ where C is the circle $2\pi i\alpha, 2\pi i$ then $f(1)=$

(a) $2\pi i$ (b) πi (c) $4\pi i$ (d) 0 **Ans:** (c)

31. If C is the circle $|z|=\pi$ traced in counter-clockwise direction $\int_C 1/(z-3i)dz =$

(a) $2\pi i$ (b) $2\pi i$ (c) πi (d) $4\pi i$ **Ans:** (b)

32. If C is the unit circle $|z|=1$, then $\int_C \sec z\,dz =$

(a) 0 (b) 1 (c) $2\pi i$ (d) $4\pi i$ **Ans:** (a)

33. If C is the unit circle $|z|=1$, then $\tan z$

(a) 0 (b) 1 (c) $2\pi i$ (d) $4\pi i$ **Ans:** (c)

34. If C is the unit circle $|z|=1$ described in the counter-clockwise direction, then $\int_C [z^3/(2z-i)]\,dz =$

(a) πi (b) $2\pi i$ (c) $\pi/4$ (d) $\pi/8$ **Ans:** (d)

35. If C is the circle $|z|=2$ described in the anti-clockwise direction, then $\int_C [e^z/(z-\pi i)^2]\,dz = 2\pi i a$ where $a =$

(a) 1 (b) -1 (c) 0 (d) 2 **Ans:** (b)

36. $\int_{-i}^{i} (1/z)\,dz =$

(a) π (b) $i\pi/2$ (c) $i\pi$ (d) 0 **Ans:** (c)

37. The order of the pole of $(e^z-1)/z^4$ is

(a) 3 (b) 1 (c) 2 (d) 4 **Ans:** (a)

38. The residue of $f(z)=z^2/(z-1)^2(z+2)$ at the simple pole $z=-2$ is

(a) 0 (b) 1 (c) 2/9 (d) 4/9 **Ans:** (d)

39. $\int_{8+\pi i}^{8-3\pi i} e^{z/2}dz =$

(a) $2\pi i$ (b) 0 (c) 2π (d) $4\pi i$ **Ans:** (b)

40. If C is the unit circle $|z|=1$, then $\int_C dz =$

(a) 2π (b) π (c) $-\pi$ (d) -2π **Ans:** (a)

41. $\int_{-i}^{i} \frac{dz}{z} =$

(a) $2\pi i$ (b) πi (c) $\frac{\pi i}{2}$ (d) $-\pi i$ **Ans:** (b)

42. $\oint_C e^z\,dz =$

(a) π (b) -1 (c) 0 (d) 1 **Ans:** (c)

43. Along $AB: y=0, 0\le x\le 1$, $\int_{AB} (x^2+y^2)(dx+idy) =$

(a) 1/3 (b) 1/2 (c) 2/3 (d) 1 **Ans:** (a)

44. The set $\{z\in C/1<|z|<2\}$ is a

(a) domain (b) closed region

(c) simply-connected (d) none of these

Ans: (a)

45. $\int_0^{\pi i} \cos z\,dz =$

(a) $\sin \pi i$ (b) $i\sinh \pi$

(c) $i\sin \pi$ (d) $\sin h\pi$ **Ans:** (a) or (b)

46. $\int_0^{1+i} z^2\,dz =$

(a) $-\frac{2}{3}+\frac{2}{3}i$ (b) $\frac{2}{3}+\frac{2}{3}i$

(c) $-\frac{2}{3}-\frac{2}{3}i$ (d) $\frac{2}{3}-\frac{2}{3}i$ **Ans:** (a)

47. $\int_{-i}^{i} (1/z)\, dz =$

(a) π (b) $i\,\pi/2$ (c) $i\pi$ (d) 0 **Ans:** (b)

48. The order of the pole of $(e^z - 1)/z^4$ is

(a) 3 (b) 1 (c) 2 (d) 4 **Ans:** (a)

49. The residue of $f(z) = z^2/(z-1)^2(z+2)$ at the simple pole $z = -2$ is

(a) 0 (b) 1 (c) 2/9 (d) 4/9 **Ans:** (d)

50. $\int_{8+\pi i}^{8-3\pi i} e^{z/2} dz =$

(a) $2\pi i$ (b) 0 (c) 2π (d) $4\pi i$ **Ans:** (b)

51. If C is the unit circle $|z| = 1$, then $\int_C dz =$

(a) 2π (b) π (c) $-\pi$ (d) -2π **Ans:** (b)

52. If C is the circle $|z| = \pi$ traced in counter clockwise direction $\int_C 1/(z-3i)dz =$

(a) 2π (b) $2\pi i$ (c) πi (d) $4\pi i$ **Ans:** (b)

53. If C is the unit circle $|z| = 1$, then $\int_C \sec z\, dz$

(a) 0 (b) 1 (c) $2\pi i$ (d) $4\pi i$ **Ans:** (a)

54. If C is the unit circle $|z| = 1$, then $\int_C \bar{z}\, dz$

(a) 0 (b) 1 (c) $2\pi i$ (d) $4\pi i$ **Ans:** (c)

55. If C is the unit circle $|z| = 1$ described in the counter-clockwise direction, then $\int_C \frac{z^3}{(2z-i)} dz =$

(a) πi (b) $2\pi i$ (c) $\pi/4$ (d) $\pi/8$ **Ans:** (d)

56. If C is the circle $|z| = 2$ described in the anti-clockwise direction, then $\int_C [e^z/(z-\pi i)^2]\, dz = 2\pi i a$ where $a =$

(a) 1 (b) −1 (c) 0 (d) 2 **Ans:** (b)

57. If C is the unit circle $C: |z - i| = 1$ touching the x-axis at the origin, then $\int_C (z^2+1)/(z^2-1)\, dz =$

(a) 0 (b) $2\pi i$ (c) πi (d) $\pi i/2$ **Ans:** (a)

58. Among the following, which one is an entire function?

(a) e^z/z (b) ze^z (c) $\sec z$ (d) z cosec z **Ans:** (b)

59. The singularities of $(z^2+1)/(z^2-1)$ are at $z =$

(a) ±1 (b) $\pm i$ (c) $1, i$ (d) $-1, -i$ **Ans:** (a)

60. If $f(a) = \int_C (3z^2 + 7z + 1)/(z-a)$ where C is the circle $|z| = 2$, then $f(3) =$

(a) πi (b) $2\pi i$ (c) −1 (d) 0 **Ans:** (d)

61. If $f(a) = \int_C (z^2+1)/(z-a)\, dz$ where C is the circle $|z| = 2$, then $f(1) =$

(a) $2\pi i$ (b) πi (c) $4\pi i$ (d) 0 **Ans:** (c)

62. $\int_C [(z^2+4)/(z-3)]\, dz = 0$ is true if C is the circle

(a) $|z| = 1$ (b) $|z| = 2$

(c) $|z| = 4$ (d) none of these **Ans:** (b)

63. If C is the straight line from $z = 0$ to $z = i$, then $\int_C (y - x - 3x^2 i)\, dz$

(a) $i/2$ (b) 1/2 (c) i (d) $-i/2$ **Ans:** (a)

64. By ML-inequality, if C is the straight line from 0 to $1 + i$ we have $\left|\int_C z^2 dz\right| \le$

(a) $2\sqrt{2}$ (b) 2 (c) $\sqrt{2}$ (d) 4 **Ans:** (a)

65. $\int_{-\pi i}^{\pi i} \cos z dz$

(a) $2\sinh \pi i$ (b) $zi \sinh \pi i$

(c) $2\sinh \pi$ (d) $2i \sinh \pi$ **Ans:** (d)

66. $\int_0^{\pi i} e^z dz =$

(a) 2 (b) −2 (c) $\sqrt{2}$ (d) π **Ans:** (b)

Chapter 16 Complex Power Series

1. The power series $|z| = 1$ represents the function

(a) $e^{(z-i)}$ (b) $\log(z-i)$

(c) $\log\left(\frac{1}{i}+z\right)$ (d) $\frac{1}{z}$ **Ans:** (d)

2. The series $\frac{-(z-\pi/2)}{1!}+\frac{(z-\pi/2)^3}{3!}+\frac{(z-\pi/2)^5}{5!}+\cdots$ represents the function

(a) $\sin z$ (b) $\cos z$ (c) $\tan z$ (d) $\sinh z$ **Ans:** (b)

3. Taylor's series expansion for $f(z)=(z-1)/(z+1)$ about $z=0$ is

(a) $1+2z+2z^2+2z^3+\cdots$

(b) $-1+z-z^2+z^3-\cdots$

(c) $-1+2z-2z^2+2z^3-\cdots$

(d) $1-z+z^2-z^3+\cdots$ **Ans:** (b)

4. The region of convergence of the power series $\frac{1}{z}(z-1)-\frac{1}{z^2}(z-1)^2+\frac{1}{z^3}(z-1)^3-\cdots$ is

(a) $|z-1|<\frac{2}{3}$ (b) $|z-1|<\frac{1}{2}$

(c) $|z-1|<2$ (d) $|z-1|<1$ **Ans:** (c)

5. Laurent's series expansion of $f(z)=z^2e^{1/z}$ at $z=0$is

(a) $z^2+\frac{1}{2}+\frac{1}{3!z}+\frac{1}{4!z^2}+\cdots$

(b) $z+\frac{1}{2}+\frac{1}{3!z}+\frac{1}{4!z^2}+\cdots$

(c) $z^2+z+\frac{1}{2}+\frac{1}{3!z}+\frac{1}{4!z^2}+\cdots$

(d) $z^2-z+\frac{1}{2}-\frac{1}{3!z}+\frac{1}{4!z^2}-\cdots$ **Ans:** (c)

6. Laurent's series expansion of $f(z)=\frac{1}{z-1}-\frac{1}{z-2}$ in the region $1<|z|<2$ is

(a) $\sum_{n=0}^{\infty}\frac{1+z^n}{z^n}$ (b) $\sum_{n=0}^{\infty}\left(\frac{1}{z^{n+1}}+\frac{z^n}{2^{n+1}}\right)$

(c) $\sum_{n=0}^{\infty}\left(\frac{1}{z^n}+\frac{z^n}{2^n}\right)$ (d) $\sum_{n=0}^{\infty}\left(\frac{1}{z^{n+1}}+\frac{z^{n-1}}{2^{n+1}}\right)$

Ans: (b)

7. $\frac{1}{1-z}=\sum_{n=0}^{\infty}z^n$ (Taylor's) $=-\sum_{n=0}^{\infty}z^{-n-1}$ (Laurent's). Give the regions of convergence of the two series respectively:

(a) $|z|\le 1, |z|\ge 1$ (b) $|z|<1, |z|>1$

(c) $|z|\le 1, |z|>1$ (d) $|z|<1, |z|\ge 1$

Ans: (b)

8. Laurent series of $f(z)=\frac{\sinh 3z}{z^3}$ for $0<|z|<\infty$

(a) $\frac{9}{z^2}+\frac{9}{4}+\frac{81}{40}z^2+\cdots$

(b) $\frac{3}{z^2}+\frac{9}{2}+\frac{81}{40}z^2+\cdots$

(c) $\frac{1}{z^2}+\frac{1}{2}+\frac{9}{40}z^2+\cdots$

(d) $\frac{3}{z^2}-\frac{9}{2}+\frac{81}{40}z^2-\cdots$ **Ans:** (b)

Zeros and Singularities

1. $z=n\pi$ (n: integer) are the zeros of the function

(a) $\cos z$ (b) $\sinh z$ (c) $\sin z$ (d) $\cosh z$

Ans: (c)

2. $z=(2n+1)\frac{\pi}{2}$ (n: integer) are the zeros of the function

(a) $\cos z$ (b) $\sinh z$

(c) $\sin z$ (d) $\sinh z$ **Ans:** (a)

3. The singularities of $\coth z$ are the zeros of

(a) $\tan z$ (b) $\sinh z$

(c) $\coth z$ (d) $\cosh z$ **Ans:** (b)

4. The singularities of e^{-z} are the zeros of

(a) e^z (b) e^{-z} (c) $\log z$ (d) $\tan z$

Ans: (a)

5. The singularity of $e^z/z^2(1-z)^3$ at $z=0$ is a/an

(a) simple pole

(b) pole of order two

(c) essential singularity

(d) non-isolated singularity **Ans:** (b)

6. The function $f(z) = \dfrac{\sin^2 z}{z}$ has at $z = 0$ a/an

(a) essential singularity

(b) removable singularity

(c) pole of order two

(d) a point of continuity **Ans:** (b)

7. Among the following, the function which has simple pole at $z = 0$ is $f(z) =$

(a) $(z+1)/z(z+2)$

(b) $= \dfrac{48+40}{128} = \dfrac{11}{16} < 1$

(c) $\sin z/\cos z$

(d) $(z^2-1)/z^2(1+z^2)$ **Ans:** (a)

8. Among the following, the function which has a removable singularity at $z = 0$ is

(a) $e^{\frac{1}{z}}$ (b) $\sin\dfrac{1}{z}$

(c) $\left|\dfrac{g(z)}{f(z)}\right| = \dfrac{1|z|^3}{|z|^7} = \dfrac{80}{81} < m|z| = 3/2$

(d) $\tanh\dfrac{1}{z}$ **Ans:** (c)

9. Among the following, the function which has an essential singularity at $z = \infty$ is

(a) $\sin\dfrac{1}{z}$ (b) $\cos\dfrac{1}{z}$ (c) e^z (d) $\sinh\dfrac{1}{z}$

Ans: (c)

10. Among the following, the function which has a pole of order 3 at $z = \infty$ is

(a) $\sin z$ (b) $z + e^z$ (c) $z^2 + 2e^z$ (d) z^3 **Ans:** (d)

11. If the principal part of Laurent's expansion of $f(z)$ contains no term, then the singularity $z = a$ of $f(z)$ is called ____ singularity.

(a) essential (b) isolated

(c) removable (d) non-essential

Ans: (c)

12. The singularity of $f(z) = (\sin z - \cos z)^{-1}$ is at $z =$

(a) $\dfrac{\pi}{6}$ (b) $\dfrac{\pi}{3}$ (c) $\dfrac{\pi}{2}$ (d) $\dfrac{\pi}{4}$ **Ans:** (d)

13. The simple poles of $f(z) = (\tan z)/z$, which lie inside the circle $|z| = 2$, are

(a) $0, -\dfrac{\pi}{2}$ (b) $0, \dfrac{\pi}{2}$ (c) $0, \pm\dfrac{\pi}{2}$ (d) $\pm\dfrac{\pi}{2}$

Ans: (c)

14. The number of singularities of $f(z) = e^{-z}$ is

(a) one (b) two (c) infinity (d) zero **Ans:** (d)

15. The simple poles of the function $f(z) = (z+1)/z^3(z^2+1)$ are

(a) $0, -1$ (b) $\pm i$ (c) $-1, \pm i$ (d) ± 1 **Ans:** (b)

16. The number of simple poles of $f(z) = z^4/(1+z^4)$ is

(a) 1 (b) 2 (c) 3 (d) 4 **Ans:** (d)

17. The zeros of $\sin\dfrac{1}{z}$ are at $z =$

(a) $n\pi$ $(n \in z)$ (b) $|w| = c^2$

(c) $n\pi$ $(n = 1, 3, 5, \ldots)$ (d) $\pm n\pi$ $(n = 1, 3, 5, \ldots)$

Ans: (b)

18. $z = 0$ is a zero of

(a) $\sin\dfrac{1}{z}$ (b) $\cos\dfrac{1}{z}$ (c) $\sin z$ (d) $\cos z$

Ans: (c)

19. The function $f(z) = z^k \sin z$ has a zero of second order if $k =$

(a) 0 (b) 1 (c) 2 (d) -1 **Ans:** (b)

[**Hint:** for $k = 1$, $f(0) = 0$, $f'(0) = (\sin z + 2\cos z)_{z=0} = 0$, $f''(0) = (2\cos z - \sin z)_{z=0} \neq 0$]

20. The function $f(z) = z^4 \sin^2\dfrac{1}{z}$ has a pole of order at $z = 0$

(a) one (b) two (c) three (d) four

Ans: (b)

$$\left[\textbf{Hint: } z^4\left(\frac{2\sin^2\frac{1}{z}}{2}\right) = z^4\left(1 - \cos\frac{2}{z}\right)\right]$$

$$= z^4\left[1-\left(1-\frac{2^2}{z^2}+\frac{2^4}{z^4}-\cdots\right)\right]$$

$$= 4z^2-16+\frac{64}{z^2}-\cdots$$

21. The type of singularity that $f(z)=(1-\cos z)/z$ has at $z=0$ is a/an ____ singularity.

(a) essential (b) irremovable

(c) removable (d) non-isolated

Ans: (c)

22. The zeros of $\sin\frac{1}{z}$ are at $z=$

(a) $(n\pi)^{-1}$ $(n\in Z))$

(b) $(2n+1)\pi$ $(n\in Z)$

(c) $(n\pi)^{-1}$ $(n=1,2,3,\ldots)$

(d) $\pm(n\pi)^{-1}$ $(n=1,2,3,\ldots)$ **Ans:** (d)

23. The singularity of $f(z)=(z+2)\sin\frac{1}{z-3}$ is at $z=$

(a) –2 (b) 2 (c) 3 (d) 0 **Ans:** (a)

24. The type of singularity at $z=3$ for $f(z)=(z+2)\sin[(z-3)^{-1}]$ is ____ singularity.

(a) removable (b) essential

(c) pole of order m (d) non isolated

Ans: (b)

25. The type of singularity at $z=0$ for $f(z)=(z-\sin z)/z^2$ is ____ singularity.

(a) removable (b) essential

(c) pole of order m (d) non isolated

Ans: (b)

26. The circle inside which all the zeros of $\sin\frac{1}{z}$ lie is $|z|=$

(a) $\frac{1}{4\pi}$ (b) $\frac{1}{3\pi}$ (c) $\frac{1}{2\pi}$ (d) $\frac{1}{\pi}$

Ans: (d)

27. The circle inside which all the zeros of $\cos\frac{1}{z}$ lie is $|z|=$

(a) $\frac{2}{3\pi}$ (b) $\frac{2}{\pi}$ (c) $\frac{1}{2\pi}$ (d) $\frac{1}{3\pi}$

Ans: (b)

28. $z=0$ is a removable singularity for the function $f(z)=$

(a) $e^{\frac{1}{z}}$ (b) $\sin\frac{1}{z}$ (c) $\frac{\sin z}{z}$ (d) $\frac{\cos z}{z}$

Ans: (c)

29. $z=0$ is an essential singularity for the function $f(z)=$

(a) $e^{\frac{1}{z}}$ (b) $\sin\frac{1}{z}$ (c) $\frac{\sin z}{z}$ (d) $\frac{\cos z}{z}$

Ans: (a)

30. Among the following, the function which has infinite number of isolated singularities is

(a) $e^z/(z^2+1)$ (b) $(\sin z)^{-1}$ (c) $e^{\frac{1}{z}}$ (d) e^{-z}

Ans: (b)

CHAPTER 17 CALCULUS OF RESIDUES

1. $\text{Res}_{z=0}\frac{e^z}{z^2}=$

(a) 2 (b) 0 (c) 1 (d) 3 **Ans:** (c)

2. If $z=a$ is a simple pole of $f(z)$, then $\text{Res}\,[f(z):a]=$

(a) $[(z-a)^2 f(z)]_{z=a}$

(b) $\lim_{z\to a}\frac{d}{dx}[(z-a)f(z)]$

(c) $\lim_{z\to a}f(z)$

(d) $\lim_{z\to a}(z-a)f(z)$ **Ans:** (d)

3. If $z=a$ is a simple pole of $f(z)=\frac{p(z)}{q(z)}$ where $p(z)$ is analytic at a and $p(a)\neq 0$ and $q(a)\neq 0$, then $\text{Res}\,\{f(z):a\}=$

(a) $p'(a)/q(a)$ (b) $p(a)/q(a)$

(c) $p(a)/q'(a)$ (d) $p'(a)/q'(a)$ **Ans:** (c)

4. If $z=a$ is a pole of order $m>1$ of $f(z)=\dfrac{g(z)}{(z-a)^m}$ where $g(z)$ is analytic at a and $g(a)\neq 0$, then Res $\{f(z):a\}=$

(a) $g^{(m)}(a)/(m-1)!$ (b) $g^{(m)(a)}/m!$
(c) $g^{(m-1)}(a)/(m-1)!$ (d) $g^{(m-1)}(a)$
Ans: (c)

5. Res $\{(z+1)/(z^2-2z):0\}=$

(a) $-\dfrac{1}{2}$ (b) $\dfrac{1}{2}$ (c) -1 (d) 1 **Ans:** (a)

6. If $f(z)=(1+e^z)/(z\cos z+\sin z)$, then Res $\{f(z):0\}=$

(a) 1 (b) 0 (c) -1 (d) 2 **Ans:** (a)

7. The residue of $\cot z$ at $z=0$ is

(a) 0 (b) 1 (c) -1 (d) 2 **Ans:** (b)

8. If $f(z)=\dfrac{e^{2z}}{(z-1)^2}$ then, Res $\{f(z):1\}=$

(a) 2 (b) e^z (c) 1 (d) $2e^2$ **Ans:** (d)

[**Hint:** Res $\{f(z):1\}=\dfrac{g'(1)}{1!}=\left[\dfrac{2e^2z}{1!}\right]_{z=1}=2e^z$;

$g(z)=e^{2z}$]

9. If $f(z)=\dfrac{e^z}{z^2+\pi^2}$, then Res $\{f(z):\pi i\}=$

(a) $\dfrac{1}{2\pi}$ (b) $\dfrac{i}{\pi}$ (c) $\dfrac{i}{2\pi}$ (d) 1 **Ans:** (c)

10. If $\text{Res}_{z=a}\dfrac{e^z}{z^2-a^2}=-\dfrac{i}{2\pi}$, then $a=$

(a) $-\dfrac{\pi i}{2}$ (b) $\dfrac{\pi i}{2}$ (c) πi (d) $-\pi i$
Ans: (d)

$$\left[\textbf{Hint: } \frac{e^z}{z+a}\Big|_{z=a}=\frac{e^a}{2a}=-\frac{i}{2\pi}=-\frac{1}{2\pi(-i)}\right.$$
$$\left.=\frac{e^{-\pi i}}{2(-\pi i)}\Rightarrow a=-\pi i\right]$$

11. If $\int_C \dfrac{4-3z}{z^2-z}dz=0$, then C is a simple closed path such that $C=$

(a) $C_1:|z|=.5$ (b) $C_2:|z|=1.5$
(c) $C_3:|z|=2$ (d) $C_2\cup C_3$ **Ans:** (d)

12. If C is any simple closed path described in counter-clockwise direction such that 0 and 1 lie inside C, then $\int_C \dfrac{4-3z}{z^2-z}dz=$

(a) $2\pi i$ (b) $-2\pi i$ (c) $-6\pi i$ (d) $-8\pi i$ **Ans:** (c)

13. If C is any positively-oriented single closed path such that 0 is inside and 1 is outside, then $\int_C \dfrac{4-3z}{z^2-z}dz=$

(a) $2\pi i$ (b) $-2\pi i$ (c) $-6\pi i$ (d) $-8\pi i$ **Ans:** (d)

14. If C is any positively-oriented simple closed path such that 1 is inside and 0 is outside, then $\int_C \dfrac{4-3z}{z^2-z}dz=$

(a) $2\pi i$ (b) $-2\pi i$ (c) $-6\pi i$ (d) $-8\pi i$ ..**Ans:** (a)

15. The value of $I=\int_C \dfrac{\tan z}{z^2-1}dz$ where $C:|z|=3/2$ is

(a) $\tan 1$ (b) $2\tan 1$
(c) $2\pi i\tan 1$ (d) 0 **Ans:** (c)

$$\left[\textbf{Hint: } I=2\pi i\left(\text{Res}_{z=1}\frac{\tan z}{z^2-1}+\text{Res}_{z=-1}\frac{\tan z}{z^2-1}\right)\right.$$
$$=2\pi i\left(\frac{\tan z}{2z}\Big|_{z=1}+\frac{\tan z}{2z}\Big|_{z=-1}\right)$$
$$\left.=2\pi i\tan 1\right]$$

16. $\text{Res}_{z=0}(ze^{\pi/z})=$

(a) π^2 (b) $\pi^2/2$ (c) $3\pi^2/2$ (d) 0
Ans: (b)

$$\left[\textbf{Hint: } ze^{\pi/2}=z\left(1+\frac{\pi}{2}+\frac{\pi^2}{2!z^2}+\cdots\right)\right.$$
$$\left.=z+\pi+\frac{\pi^2}{2}\cdot\frac{1}{z}+\cdots\right]$$

17. $\text{Res}_{z=2i}[ze^{\pi z}/(z^2-16)]=$

(a) $\dfrac{1}{8}$ (b) $\dfrac{1}{16}$ (c) $-\dfrac{1}{16}$ (d) $\dfrac{1}{32}$
Ans: (c)

$$\left[\textbf{Hint: } (ze^{\pi z}/4z^3)_{z=2i}=-\frac{1}{16}\right]$$

18. The value of $\int_C \frac{ze^{\pi z}}{z^2-16}\,dz$ where C is a positively-oriented simple closed curve enclosing $z=\pm 2i$ only is

(a) $\pi i/4$ (b) $-\pi i/4$ (c) $-\pi i/8$ (d) $-\pi i/16$ **Ans:** (b)

19. The value of $\int_C e^z\,dz$ where C is any simple closed path is

(a) 0 (b) $2\pi i$ (c) $4\pi i$ (d) $6\pi i$ **Ans:** (a)

20. $\text{Res}_{z=-2}[z^2/(z-1)(z+2)^2]=$

(a) 1 (b) 0 (c) 8/9 (d) 4/9 **Ans:** (c)

$$\left[\textbf{Hint: } \lim_{z\to -2}\frac{d}{dz}\left(\frac{z^2}{z-1}\right)=\left.\frac{z^2-2z}{(z-1)^2}\right|_{z=-2}=8/9\right]$$

21. $\text{Res}_{z=2}[z^2/(z-1)(z-2)^2]=$

(a) 1 (b) 0 (c) 4 (d) 2 **Ans:** (b)

$$\left[\textbf{Hint: } \lim_{z\to 2}\frac{d}{dz}\left(\frac{z^2}{z-1}\right)=\left.\frac{z^2-2z}{(z-1)^2}\right|_{z=2}=0\right]$$

22. For evaluating an integral of the type $\int_0^{2\pi} f(\cos\theta,\sin\theta)d\theta$, the contour to be used is

(a) the semicircle $s: z=\text{Re}^{i\theta}(0\le\theta\le\pi)$ in the upper half-plane with the line segment $L[-R,R]$ along the real axis

(b) the rectangle with vertices at $\pm|a|\pm|b|i$

(c) the sector $z=\text{Re}^{i\theta}(0\le\theta\le\alpha)$

(d) the unit circle $|z|=1$ **Ans:** (d)

23. For evaluating an integral of the type $\int_{-\infty}^{\infty} f(x)dx$, the contour to be used is

(a) the semicircle $s: z=\text{Re}^{i\theta}(0\le\theta\le\pi)$ in the upper half-plane with the line segment $L[-R,R]$ along the real axis

(b) the rectangle with vertices at $\pm|a|\pm|b|i$

(c) the sector $s=z=\text{Re}^{i\theta}(0\le\theta\le\alpha)$

(d) the unit circle $|z|=1$ **Ans:** (a)

24. If C is the circle $|z|=\pi$ described in counter-clockwise direction, then $\int_C \frac{dz}{z-3i}=$

(a) πi (b) $2\pi i$ (c) 0 (d) -1 **Ans:** (b)

25. If C is the circle $|z|=\frac{1}{2}$, then $\int_C [(z^2-z+1)/(z-1)]dz=$

(a) πi (b) $2\pi i$ (c) 0 (d) -1 **Ans:** (c)

26. If C is the circle $|z|=1$, then $\int_C [(z^2-z+1)/(z-1)]dz=$

(a) πi (b) $2\pi i$ (c) 0 (d) -1 **Ans:** (b)

27. If C is the circle $|z|=3$, then $I=\int_C [(e^{2z})/(z-1)(z-2)]dz=$

(a) $\pi i(e^4-e^2)$ (b) $2\pi i(e^4-e^2)$

(c) 0 (d) -1 **Ans:** (b)

$$\left[\textbf{Hint: } I=\int_c \frac{e^{2z}}{z-2}dz-\int_c \frac{e^{2z}}{z-1}dz = 2\pi i e^4-2\pi i e^2 \text{ by C.I.F}\right]$$

28. Let $f(\alpha)=\int_C \frac{z^2+1}{z-\alpha}dz$. If $|z|=2$, $\alpha=1$, then $F(1)=$

(a) $4\pi i$ (b) $2\pi i$ (c) 0 (d) $6\pi i$ **Ans:** (a)

29. If $|z|=\frac{1}{2}$, $\alpha=1$, then $F(1)$ in Qn 28 is

(a) $4\pi i$ (b) $2\pi i$ (c) 0 (d) $6\pi i$ **Ans:** (c)

30. For any α lying inside C (counter-clockwise), we have $F(\alpha)=\int_C \frac{z^2+1}{z-\alpha}dz=$

(a) $-2\pi i(\alpha^2+1)$ (b) $2\pi i\alpha, 2\pi i$

(c) 0 (d) $4\pi i(\alpha^2+1)$ **Ans:** (b)

31. $F'(\alpha)$ and $F''(\alpha)$ in Qn 30 are

(a) $4\pi i\alpha, 4\pi i$ (b) $2\pi i\alpha, 2\pi i$

(c) 0 (d) $4\pi i(\alpha^2+1)$ **Ans:** (a)

Chapter 18 Argument Principle and Rouche's Theorem

1. A function C whose only singularities are poles is called ________ function

(a) isomorphic (b) meromorphic

(c) endomorphic (d) epimorphic **Ans:** (b)

2. Among the following, the function which is meromorphic is

(a) e^z (b) $\sin z$ (c) $\tan z$ (d) $\cos z$ **Ans:** (c)

3. Let $f, g \in H(D)$ where D contains a closed curve C. If _____ on C, then $f(z)$ and $f(z)+g(z)$ have the same number of zeros inside C.

(a) $|g(z)|<|f(z)|$ (b) $|g(z)|=|f(z)$

(c) $|g(z)|>|f(z)|$ (d) $|f(z)||g(z)|>1$

Ans: (a)

4. Let $f, g \in H(D)$ where D contains a closed curve C. If $|g(z)|<|f(z)|$ on C, then $f(z)$ and _____ have the same number of zeros inside C.

(a) $g(z)$ (b) $f(z)+g(z)$

(c) $f(z)-g(z)$ (d) $f(z)g(z)$ **Ans:** (b)

5. The name of the theorem stated in Qn. 4 is_______ theorem.

(a) Rouche's

(b) Argument

(c) Liouville

(d) Fundamental theorem of Algebra **Ans:** (a)

6. The number of roots of $z^6+16z+1=0$ that lie inside $|z|=2$ is

(a) 2 (b) 4 (c) 5 (d) 6 **Ans:** (d)

[**Hint:** $f(z)=z^6$ (has 6 zeros) $g(z)=16z+1$

$$\left|\frac{g(z)}{f(z)}\right|=\frac{16z+1}{|z|^6}=\frac{33}{64}<1 \text{ on } |z|=2;\; z^6+$$

$16z+1$ has the same no. of zeros as f, *i.e.*, 6.]

7. The number of zeros of z^6+z^5+1 inside C: $|z|=2$ is

(a) 4 (b) 3 (c) 6 (d) 0 **Ans:** (c)

[**Hint:** $f(z)=z^6$ (six zeros) $g(z)=z^5+1$

$$\left|\frac{g(z)}{f(z)}\right|=\frac{|z|^5+1}{|z|^6}=\frac{2^5+1}{2^6}=\frac{33}{64}<1;\; z^6+z^5+1$$

has the same no. of zeros as f, *i.e.*, 6.]

8. The number of zeros of z^6+z^5+1 inside $C:|z|=2$ is

(a) 4 (b) 3 (c) 6 (d) 0 **Ans:** (b)

[**Hint:** $f(z)=2z^5+1, g(z)=z^6$

$$\left|\frac{g(z)}{f(z)}\right|=\frac{z^6}{2|z|^5+1}=\frac{z^6}{2.2^5+1}=\frac{64}{65}<1:$$

z^6+5z^5+1 has the same no. of zeros as f, *i.e.*, 5]

9. The number of zeros of z^n+z+1 inside C: $|z|=1$ is

(a) 0 (b) 1 (c) 2 (d) n **Ans:** (d)

[**Hint:** $f(z)=z^n$ (it as n zeros); $g(z)=z+1$

$$\left|\frac{g(z)}{f(z)}\right|=\frac{|z|+1}{|z|^n}=\frac{3}{2^n}<1 \text{ if } n\geq z;\; z^n+z+1$$

has the same no. of zeros as $f(z)$, *i.e.*, n.]

10. The number of zeros of z^n+z+1 inside C: $|z|=1$ is

(a) 0 (b) 1 (c) 2 (d) n **Ans:** (b)

[**Hint:** $f(z)=z+1$ (are zero) $g(z)=z^n$

$$\left|\frac{g(z)}{f(z)}\right|=\frac{|z|^n}{|z|+1}=\frac{1}{2}<1;\; z^n+z+1 \text{ has the same}$$

no. of zeros as $f(z)$, *i.e.*, 1.]

11. The number of zeros of z^n+z+1 in the annulus $1\leq|z|\leq 2$ is

(a) 2 (b) n (c) 3 (d) $(n-1)$ **Ans:** (d)

12. The number of zeros of $3z^n-e$ that lie inside $|z|=1$ is

(a) 1 (b) 2 (c) n (d) $n-1$ **Ans:** (c)

13. All the roots of $\pi z^2-e^z+0.2$ lie inside the circle $|z|=$

(a) 0.1 (b) 0.25 (c) 0.5 (d) 1 **Ans:** (d)

[**Hint:** $\left|\frac{g(z)}{f(z)}\right|=\frac{1-ez+0.2}{\pi|z|^2}$

$$=\frac{e+0.2}{\pi}<\text{Im}\,|z|=1]$$

14. $z^7 - 5z^3 + 12z^2$ has _____ zeros in the annulus $1 \le |z| \le 2$.

(a) 7 (b) 6 (c) 5 (d) 0 **Ans:** (c)

$$\left[\textbf{Hint:} \left|\frac{f(z)}{f(z)}\right| = \frac{|12z^2 - 5z^3|}{|z|^7} = \frac{48+40}{128} = \frac{11}{16} < 1\right]$$

15. The number of roots of $z^7 - 5z^3 = 0$ that lie in the punctured disc $0 < |z| < 3/2$ is

(a) 7 (b) 4 (c) 3 (d) 0 **Ans:** (b)

$$\left[\textbf{Hint:} \left|\frac{g(z)}{f(z)}\right| = \frac{1|z|^3}{|z|^7} = \frac{80}{81} < m\,|z| = 3/2\right]$$

CHAPTER 19 CONFORMAL MAPPING

1. A mapping that preserves angles between oriented curves both in magnitude and in sense is called a/an _______ mapping.

(a) informal (b) isogonal
(c) conformal (d) formal **Ans:** (c)

2. The mapping defined by an analytic function $f(z)$ is conformal at all points z except at points where _______

(a) $f'(z) = 0$ (b) $f'(z) \neq 0$
(c) $f'(z) > 0$ (d) $f'(z) < 0$ **Ans:** (a)

3. Under the mapping $w = z^2$, a circle $|z| = c$ is transformed to a/an

(a) ellipse $a^2u^2 + b^2v^2 = 1$ (b) circle $|w| = c^2$
(c) vertical line (d) horizontal line
Ans: (b)

4. The mapping $w = z^2$ transforms a vertical line to a/an

(a) horizontal line (b) circle
(c) ellipse (d) parabola **Ans:** (d)

5. Under the mapping $w = f(z)$ if the angle $\theta = \pi/n$ is transformed to an angle π in the w-plane, then $f(z)$

(a) z^n (b) z^{2n} (c) $z^{1/n}$ (d) z/n **Ans:** (a)

6. Under the transformation $w = iz$, the st. line $y = x$ in the z-plane is rotated through an angle ______ in the w-plane.

(a) $\pi/4$ (b) 2π (c) $\pi/2$ (d) π **Ans:** (c)

7. The mapping $w = \bar{z}$ preserves angles in

(a) size but not in sense
(b) sense but not in size
(c) size as well as sense
(d) neither size nor sense **Ans:** (a)

8. The mapping $w = |z|^2$ is not conformal at

(a) any point (b) any point except $z = 0$
(c) the origin (d) any point along the real axis
Ans: (b)

9. e^z maps

(a) vertical lines into circles
(b) circles into vertical lines
(c) circles into circles
(d) straight lines into straight lines **Ans:** (a)

10. e^z maps

(a) circles into circles
(b) circles into horizontal lines
(c) horizontal lines into rays through O
(d) rays through O into circles **Ans:** (c)

11. The two points where the mapping $w = z + \frac{1}{z}$ is not conformal are $z =$

(a) 0, 1 (b) $\pm\infty$ (c) 0, 0 (d) ± 1 **Ans:** (d)

12. The images of the vertical lines under the mapping $w = \sin z$ are

(a) hyperbolas (b) ellipes
(c) circles (d) parabolas **Ans:** (a)

13. The images of the horizontal lines under the mapping $w = \sin z$ are

(a) hyperbolas (b) ellipes
(c) circles (d) parabolas **Ans:** (b)

14. The condition for the map $w=(az+b)/(cz+d)$ to be conformal for all z is that

(a) $\begin{vmatrix} a & 1 \\ 1 & d \end{vmatrix}=0$ (b) $\begin{vmatrix} a & b \\ c & 1 \end{vmatrix}=0$

(c) $\begin{vmatrix} a & b \\ c & d \end{vmatrix}\neq 0$ (d) $\begin{vmatrix} a & b \\ c & d \end{vmatrix}=0$ **Ans:** (c)

15. The transformation $w=(az+b)/(cz+d)$ maps the unit circle $|w|=1$ into a st. line in the z-plane if

(a) $|a|=|c|$ (b) $|a|=|d|$

(c) $|b|=|c|$ (d) $|b|=|d|$ **Ans:** (a)

16. The fixed points of the transformation $w=z^2$ are

(a) 0, 1 (b) 0, –1 (c) –1, 1 (d) $-i, i$ **Ans:** (a)

17. The invariant points of the mapping $w=\dfrac{z}{2-z}$ are

(a) 1, –1 (b) 0, –1 (c) 0, 1 (d) –1, –1 **Ans:** (c)

18. The fixed points of $w=\dfrac{z-1}{z+1}$ are

(a) ±1 (b) $\pm i$ (c) 0, –1 (d) 0, 1 **Ans:** (b)

19. The mapping $w=z+\dfrac{1}{z}$ transforms circles of constant radius into

(a) confocal ellipses (b) hyperbolas

(c) circles (d) parabolas **Ans:** (a)

20. Among the following, the one that is not a critical point of the mapping $w=\cos z$ is

(a) π (b) 2π (c) 3π (d) $\pi/2$ **Ans:** (d)

21. The linear fractional transformation that maps the points $\infty, 1, 0$, respectively, into $0, 1, \infty$ is $w=$

(a) z (b) z^{-1} (c) iz (d) $-iz$ **Ans:** (b)

22. The mobius transformation that maps the points $0, i, \infty$, respectively, into $0, 1, \infty$ is $w=$

(a) z^{-1} (b) $-z$ (c) $-iz$ (d) iz **Ans:** (c)

23. The images of the points $z_1=\infty, z_2=i, z_3=0$ under the transformation $w=-1/z$ are

(a) $0, i, \infty$ (b) $0, -i, \infty$

(c) $0, -i, -\infty$ (d) $0, -i, 1$ **Ans:** (c)

24. The fixed points of the transformation $w=\dfrac{z-1+i}{z+2}$ are

(a) $1, -1, -i$ (b) $i, 1, -i$

(c) $1, -1, i$ (d) $i, -1, -i$ **Ans:** (d)

25. Under the transformation $w=\dfrac{1}{z}$, the point $\left(\cos\dfrac{\pi}{3}, \sin\dfrac{\pi}{3}\right)$ is mapped into the point

(a) $\left(\cos\dfrac{\pi}{6}, \sin\dfrac{\pi}{6}\right)$ (b) $\left(\cos\dfrac{\pi}{3}, -\sin\dfrac{\pi}{3}\right)$

(c) $(\cos\pi, \sin\pi)$ (d) $\left(\cos\dfrac{\pi}{3}, \sin\dfrac{\pi}{3}\right)$ **Ans:** (b)

26. The image of the line $x=1$ under the transformation $w=\sin z$ is a/an

(a) circle (b) ellipse

(c) hyperbola (d) parabola **Ans:** (c)

27. The image of the line $y=c$ under the transformation $w=\sin z$ is a/an

(a) circle (b) ellipse

(c) hyperbola (d) parabola **Ans:** (b)

28. The mapping under which a vertical line $z=c$ transforms into a circle is $w=$

(a) $\sin z$ (b) $\cosh z$ (c) z^2 (d) e^z **Ans:** (d)

29. $z=0$ is a critical point of the transformation $w=$

(a) $\sin z$ (b) $z^n\ (n>1)$

(c) e^z (d) $z+\dfrac{1}{z}$ **Ans:** (b)

30. The mapping under which vertical lines are transformed into a circle is $w=$

(a) z^2 (b) $z+\dfrac{1}{z}$ (c) e^z (d) $\sin z$ **Ans:** (c)

31. Under the mapping $w = z^2$, a st. line through the origin is transformed into a/an

(a) circle (b) ellipse

(c) parabola (d) st. line through O

Ans: (d)

32. Under the mapping $w = z^2$, a circle of radius 3 is transformed into a circle with radius

(a) $\sqrt{3}$ (b) 3 (c) 9 (d) 6 **Ans:** (c)

33. Under the mapping, the circle of radius ________ is transformed into a circle with radius $\frac{1}{2}$.

(a) $\frac{1}{4}$ (b) $\frac{1}{z}$ (c) $\frac{1}{\sqrt{2}}$ (d) 1 **Ans:** (c)

34. The image of the circles $|w-1| = k$ under the mapping $w = z^2$ is given by

(a) $|z-1||z+1| = k$ (b) $|z-1| = k$

(c) $|z+1| = k$ (d) $z^2 = k$ **Ans:** (a)

35. The fixed points of the transformation $w = z/(z+1)$ are

(a) 0, −1 (b) 0, 1 (c) 1, −1 (d) 2, 1

Ans: (a)

36. The sector $0 \le \theta \le \frac{\pi}{3}$ is mapped onto the upper half-plane by $w = z^n$ if $n =$

(a) 3 (b) 6 (c) 3 (d) 4 **Ans:** (b)

37. The number of fixed points of the mapping $w = \frac{1}{z}$ is

(a) 1 (b) 2 (c) 3 (d) 4 **Ans:** (b)

38. The equation of the line whose image under the map $w = e^z$ is the circle with radius with radius 'e' is

(a) $x = 0$ (b) $y = 1$

(c) $y = 0$ (d) $x = 1$ **Ans:** (d)

39. The _____ mapping has every point of it as a fixed point.

(a) rotation (b) inverse

(c) identity (d) translation **Ans:** (c)

40. $w = \bar{z}$ has _____ fixed points.

(a) no (b) infinitely many

(c) finitely many (d) some **Ans:** (b)

41. The translation mapping $w = z + c\ (c \ne 0)$ has _____ fixed points.

(a) no (b) infinitely many

(c) finitely many (d) some **Ans:** (a)

42. The rotation mapping $w = cz$ has ______ fixed point(s).

(a) no (b) infinitely many

(c) finitely many (d) one **Ans:** (d)

[Hint: $z = cz =)(c-1)z = 0 \Rightarrow z = 0$]

Fill in the Blanks

Chapter 1 Eigenvalues and Eigenvectors

1. The sum of the characteristic roots of $A=\begin{bmatrix}0 & 1\\ 1 & 0\end{bmatrix}$ is ______. **Ans:** 0

2. If the sum and the product of the eigenvalues of $\begin{bmatrix}a+1 & -1\\ 1 & a-1\end{bmatrix}$ are s_1 and s_2, respectively, then $s_1^2 =$ ______. **Ans:** $4s_2$

3. By diagonalisation of $A=\begin{bmatrix}2 & -2 & -4\\ -1 & 3 & 4\\ 1 & -2 & -3\end{bmatrix}$ we obtain $D=\begin{bmatrix}0 & 0 & 0\\ 0 & 1 & 0\\ 0 & 0 & 1\end{bmatrix}$. From this we conclude that A is ______. **Ans:** Singular

4. A is an n-square matrix with tr. $A = 0$ then the coefficient of λ^{n-1} in its characteristic equation is ______. **Ans:** 0

5. If the characteristic equation of matrix A of order 2 is $ax^2 + bx + c = 0$ then the characteristic equation of A^{-1} is ______. **Ans:** $cx^2 + bx + a = 0$

6. If the eigenvalues of A are 3, −1, 7 then the eigenvalues of $(A + I)$ are ______. **Ans:** 4, 0, 8

7. The eigenvalues a, b, c of A are in A.P and those of A^2 are in A.P too. Then $b =$ ______. **Ans:** $\dfrac{a^2+c^2}{a+c}$

8. If the product of the eigenvalues of A is zero then A is a ______ matrix. **Ans:** Singular

9. If the product of the eigenvalues of A is positive then A is a ______ matrix. **Ans:** Nonsingular

10. If 2 and $\frac{1}{2}$ are the eigenvalues of a 2-square matrix A then those of A^{-1} are ______. **Ans:** 2, $\frac{1}{2}$

11. If the characteristic equation of a 2-square matrix A is $ax^2 + bx + c = 0$ then that of $3A$ is ______. **Ans:** $ax^2 + 3bx + 9c = 0$

12. If 2 is an eigenvalue of $\begin{bmatrix}3 & -9 & -12\\ 1 & a & 4\\ 0 & 0 & 1\end{bmatrix}$ then $a =$ ______. **Ans:** −7

    $\left[\textbf{Hint: } |A-2I| = 0 \;\Rightarrow\; (-1)\begin{vmatrix}3-2 & -9\\ 1 & a-2\end{vmatrix}; \;\Rightarrow\; a=-7.\right]$

13. The latent roots of $\begin{bmatrix}1 & 1 & -1\\ 0 & 2 & -1\\ 0 & 0 & -2\end{bmatrix}$ are ______. **Ans:** 1, 2, −2

14. A is similar to $D = \text{diag}(1, -1, 3)$ then the proper values of A^2 are ______. **Ans:** 1, 1, 9

15. If '0' is a latent root of $\begin{bmatrix} 1 & 2 & 2 \\ 0 & a & 1 \\ 0 & 1 & -1 \end{bmatrix}$ then $a =$ ______. **Ans:** –1

$$\left[\textbf{Hint}: 0 = \begin{vmatrix} a & 1 \\ 1 & -1 \end{vmatrix} \Rightarrow a = -1. \right]$$

16. The eigenvalues of $\begin{bmatrix} 2 & 1 \\ 1 & 5 \end{bmatrix}$ are the roots of the equation ______. **Ans:** $x^2 - 7x + 9 = 0$

$$\left[\textbf{Hint}: x^2 - (2+5)x + \begin{vmatrix} 2 & 1 \\ 1 & 5 \end{vmatrix} = 0. \right]$$

17. If X_1 and X_2 are the eigenvectors of $A = \begin{bmatrix} 0 & 2 \\ 2 & 0 \end{bmatrix}$ then X_1 and X_2 are ______. **Ans:** Orthogonal

18. The algebraic multiplicity of the eigenvalue $\lambda = 2$ of the matrix $\begin{bmatrix} 1 & 0 & 2 \\ -1 & 2 & 2 \\ -1 & 0 & 4 \end{bmatrix}$ is ______. **Ans:** 2

$$\left[\textbf{Hint}: \text{Expanding by } C_2, |A - \lambda I.| = (2-\lambda)\begin{vmatrix} 1-\lambda & 2 \\ -1 & 4-\lambda \end{vmatrix} = -(\lambda-2)^2(\lambda-3). \right]$$

CHAPTER 2 QUADRATIC FORMS

1. The rank of the quadratic form $2x^2 + 4xy + z^2$ is ______. **Ans:** 3

2. The column vectors X_1, X_2 and X_3 are pairwise orthogonal if, for all $i, j = 1, 2, 3$ ______. **Ans:** $X_i X_j^T = 0\ (i \neq j)$

3. If X_1, X_2 and X_3 are three linearly independent eigenvectors of a symmetric matrix A and $\|X_i\| = \alpha_i$ $(i = 1, 2, 3)$ then the modal matrix of A is = ______. **Ans:** $\left[\frac{X_1}{\alpha_1}, \frac{X_2}{\alpha_2}, \frac{X_3}{\alpha_3} \right]$

4. The method of reduction of a quadratic form $q = X^T AX$ to its canonical form $Y^T DY$ by applying a linear transformation $X = PY$ where $P = [X_1, X_2, X_3]$, X_1, X_2, X_3 being normalised eigenvectors of A is called ______. **Ans:** Orthogonalisation

5. The method of reduction of a quadratic form $q = X^T AX$ to its canonical form $Y^T DY$ by applying row and column operations simultaneously on $A = I_3AI_3$ to convert A into a diagonal matrix is called ______. **Ans:** Diagonalisation by congruent row/column operations

6. The method of reduction of the quadratic form $q = X^T AX$ to canonical form by grouping the terms and completing the squares is called ______. **Ans:** Lagrange's reduction

7. The normalised modal matrix P corresponding to the symmetric matrix $A = \begin{bmatrix} 1 & 2 \\ 2 & 1 \end{bmatrix}$ is ______.

Ans: $\begin{bmatrix} \frac{1}{\sqrt{2}} & \frac{1}{\sqrt{2}} \\ \frac{1}{\sqrt{2}} & -\frac{1}{\sqrt{2}} \end{bmatrix}$

8. The modal matrix of $A = \begin{bmatrix} 2 & 4 \\ 1 & 5 \end{bmatrix}$ is $P = \begin{bmatrix} -4 & 1 \\ 1 & 1 \end{bmatrix}$. Then the normalised modal matrix is ______.

Ans: $\begin{bmatrix} \frac{-4}{\sqrt{17}} & \frac{1}{\sqrt{2}} \\ \frac{1}{\sqrt{17}} & \frac{1}{\sqrt{2}} \end{bmatrix}$

9. The modal matrix of $A = \begin{bmatrix} 0 & 1 & 0 \\ 1 & 0 & 1 \\ 0 & 1 & 0 \end{bmatrix}$ is

$\begin{bmatrix} -1 & 1 & -1 \\ 0 & \sqrt{2} & \sqrt{2} \\ 1 & 1 & -1 \end{bmatrix}$. Then the normalised modal matrix $P =$ _______. **Ans:** $\begin{bmatrix} -\frac{1}{\sqrt{2}} & \frac{1}{2} & -\frac{1}{2} \\ 0 & \frac{1}{\sqrt{2}} & \frac{1}{\sqrt{2}} \\ \frac{1}{\sqrt{2}} & \frac{1}{2} & -\frac{1}{2} \end{bmatrix}$

Real quadratic form in canonical form (Questions 10–14)

10. $2y_1^2 + y_2^2 - 3y_2^2$. Its rank is _______. **Ans:** 3
    Its index is _______. **Ans:** 2
    Its signature is _______. **Ans:** 1
    Its nature is _______. **Ans:** Indefinite

11. $y_1^2 + 2y_2^2 + 3y_3^2$. Its rank is _______. **Ans:** 3
    Its index is _______. **Ans:** 3
    Its signature is _______. **Ans:** 3
    Its nature is _______. **Ans:** Positive definite

12. $y_1^2 - 2y_3^2$. Its rank is _______. **Ans:** 2
    Its index is _______. **Ans:** 1
    Its signature is _______. **Ans:** 0
    Its nature is _______. **Ans:** Indefinite

13. $3y_1^2 + y_3^2$. Its rank is _______. **Ans:** 3
    Its index is _______. **Ans:** 2
    Its signature is _______. **Ans:** 2
    Its nature is _______. **Ans:** Positive semidefinite

14. $-2y_1^2 - y_2^2 - y_3^2$. Its rank is _______. **Ans:** 3
    Its index is _______. **Ans:** 0
    Its signature is _______. **Ans:** –3
    Its nature is _______. **Ans:** Negative definite

15. Quadratic form $2x_1x_2 + 2x_1x_3 + 2x_2x_3$; Its rank is _______. **Ans:** 3
    Its index is _______. **Ans:** 1
    Its signature is _______. **Ans:** –1
    Its nature is _______. **Ans:** Indefinite

16. $x^2 + y^2 + z^2 + 2yz$. Its rank is _______. **Ans:** 2
    Its index is _______. **Ans:** 2
    Its signature is _______. **Ans:** 2
    Its nature is _______. **Ans:** Positive semidefinite

CHAPTER 3 SOLUTION OF ALGEBRAIC AND TRANSCENDENTAL EQUATIONS

1. In Newton-Raphson method we approximate the graph of f by suitable _______. **Ans:** Tangents

2. Newton's iterative formula for finding a root of $f(x) = 0$ is $x_{n+1} =$ _______. **Ans:** $x_n + \dfrac{f(x_n)}{f'(x_n)}$

3. Newton's iterative formula for finding the square root of a positive number N is $x_{n+1} =$ _______. **Ans:** $\frac{1}{2}(x_n - N/x_n)$

4. Newton's iterative formula for finding the reciprocal of a number $x_{n+1} =$ _______. **Ans:** $x_n (2 - Nx_n)$

5. Newton iterative formula for finding the pth root of a positive number N is $x_{n+1} =$ _______. **Ans:** $\frac{1}{p}\left[(p-1)x_n + N/(x_n^{p-1})\right]$

6. Newton's iterative formula for finding the cube root of a number N is $x_{n+1} =$ _______. **Ans:** $\frac{1}{3}\left[2x_n + \frac{N}{x_n^2}\right]$

7. The cube root of 24 by Newton's formula, taking $x_0 = 3$ is _______. **Ans:** 2.889

8. The square root of 35 by Newton's formula taking $x_0 = 6$ is _______. **Ans:** 5.916

9. The Regula Falsi method is related to _______ at a point of the curve. **Ans:** Chord

10. The Newton-Raphson method is related to _______ at a point of the curve. **Ans:** Tangent

CHAPTER 4 INTERPOLATION

1. $\Delta^3 (1-x)(1-2x)(1-3x) =$ ________.
Ans: –36

2. $\Delta^n (e^{ax+b}) =$ ________. **Ans:** $(e^a - 1)^n e^{ax+b}$

3. $\Delta(u_x v_x) =$ ________. **Ans:** $u_x (\Delta v_x) + v_{x+1} (\Delta u_x)$

4. $\Delta x^{(r)} =$ ________. **Ans:** $rh\, x^{(r-1)}$

5. $\Delta(u_x/v_x) =$ ________.
Ans: $[v_x (\Delta u_x) - u_x (\Delta v_x)]/v_x v_{x+1}$

6. $(1 + \Delta)(1 - \nabla) =$ ________. **Ans:** 1

7. If $x^{(r)} = x(x-h)(x-2h) \cdots (x - rh + h)$ is the factorial function then the values of $\Delta^2 x^{(r)}$ and $\Delta^{r+1} x^{(r)}$ are, respectively, ________, ________.
Ans: $r(r-1)\, h^2 x^{(r-2)}$ and 0

8. If $\Delta^n f(x) = 0$ for $n \geq 5$ then $f(x)$ is a polynomial function of degree ________. **Ans:** 4

9. Newton's forward interpolation formula is used for interpolating the values of a function f near the ________ of the tabulated values of f.
Ans: Beginning

10. Newton's backward interpolation formula is used for interpolating the values of a function f near the ______ of the tabulated values of f.
Ans: End

11. Stirling or Bessel central difference interpolation formula is used for interpolating the values of a function f near the ________ of the tabulated values of f.
Ans: Middle

12. Lagrange's interpolation formula is used for interpolating the values of a function f near ________ of the tabulated values of f.
Ans: Any point

13. The missing value of y in the following table

x	100	101	102	103
y	2.0000	2.0043	–	2.0128

Ans: 2.0086

[**Hint:** Here three values of y are given; the polynomial degree 2 so that $\Delta^3 y_0 = 0 \Rightarrow y_3 - 3y_2 + 3y_1 - y_0 = 0$. Substituting the given values and solving we get the missing $y_2 = 2.0086$.]

14. From the data

x	0	1	2	3
y	7	5	y_2	7

$y_2 =$ ________. **Ans:** 5

[**Hint:** Since 3 values are given, the polynomial is of degree 2 so that $\Delta^3 y_0 = 0$
$\Rightarrow y_3 - 3y_2 + 3y_1 - y_0 = 0 \Rightarrow 3y_2 = 7 + 3 \times 5 - 7 \Rightarrow y_2 = 5$.]

15. From the data

x	0	5	10	15	20	25
y	6	10	y_2	17	y_4	31

y_2 and y_4 are, respectively, ______ and ______.
Ans: 13.25, 22.50

16. Compute the missing values: $(y_2, y_4) =$ (________, ________).

x	45	50	55	60	65
y	3	y_1	2	y_3	–2.4

[**Hint:** $\Delta^3 y_0 = 0 \Rightarrow 3y_1 + y_3 - 9 = 0$, $\Delta_3 y_1 = 0 \Rightarrow y_1 + 3y_3 - 3.6 = 0$; solving $y_1 = 2.925$, $y_3 = 0.225$.]
Ans: 2.925, 0.225

17. By Lagrange's formula $y(1) =$ ________ from the data.

x	0	2	3
y	3	1	2

Ans: 1.3333

$$\left[\textbf{Hint:}\quad y(x) = \frac{(x-x_1)(x-x_2)}{(x_0-x_1)(x_0-x_2)} y_0 + \frac{(x-x_0)(x-x_2)}{(x_1-x_0)(x_1-x_2)} y_1 + \frac{(x-x_0)(x-x_1)}{(x_2-x_0)(x_2-x_1)} y_2 \,(x=1)\right.$$

$$= \frac{(1-2)(1-3)}{(0-2)(0-3)} \cdot 3 + \frac{(1-0)(1-3)}{(2-0)(2-3)} \cdot 1 + \frac{(1-0)(1-2)}{(3-0)(3-2)} \cdot 2$$

$$= \frac{1}{3}\cdot 3 + \frac{1}{2}\cdot\frac{(-2)}{(-1)}1 + \frac{-1}{3}\cdot 2.\Bigg]$$

18. By Lagrange's interpolation formula $y(2)$ = ________ from the data.

x	0	1	3	4
y	−12	0	6	12

$y(2)$ = ________ **Ans:** 4

$$\Bigg[\textbf{Hint: } y(x) = \frac{(x-x_1)(x-x_2)(x-x_3)}{(x_0-x_1)(x_0-x_2)(x_0-x_3)}y_0 + \cdots + \cdots + \cdots \Rightarrow$$

$$y(2) = \frac{(2-1)(2-3)(2-4)}{(0-1)(0-3)(0-4)}(-12) + \frac{(2-0)(2-3)(2-4)}{(1-0)(1-3)(1-4)}\cdot 0 + \frac{(2-0)(2-1)(2-4)}{(3-0)(3-1)(3-4)}6 + \frac{(2-0)(2-1)(2-3)}{(4-0)(4-1)(4-3)}12$$

$$= \frac{2}{3.4}12 + 0 + \frac{2.2}{3.2}6 - \frac{2}{4.3}12 = 4.\Bigg]$$

19. Computing x for $y = 1$ by Lagrange's inverse interpolation formula from the data we have x = ________.

x	−2	−1	1
y	−7	2	0

Ans: 0.1587

$$\Bigg[\textbf{Hint: } x = \frac{(y-y_1)(y-y_2)}{(y_0-y_1)(y_0-y_2)}x_0 + \cdots + \cdots;\ x(1) = \frac{(1-2)(1-0)}{9.7}(-2) + \frac{(1+7)(1-0)}{(2+7)(2-0)}(-1) + \frac{(1+7)(1-2)}{(0+7)(0-2)}$$

$$= \frac{1.1}{9.7}\cdot 2 - \frac{8}{9.2}\cdot 1 + \frac{8.1}{7.2}\cdot 1$$

$$= \frac{2}{63} - \frac{4}{9} + \frac{4}{7} = \frac{10}{63} = 0.1587.\Bigg]$$

20. Computing $x(15)$ by Lagrange's inverse interpolation using the data

x	5	6	9	11
y	12	13	14	16

we get $x(15)$ = ________. **Ans:** 11.5

$$\Bigg[\textbf{Hint: } x(y) = \frac{(y-y_1)(y-y_2)(y-y_3)}{(y_0-y_1)(y_0-y_2)(y-y_3)}x_0 + \cdots + \cdots +$$

$$x(15) = \frac{2\cdot 1(-1)}{(-1)\cdot(-2)\cdot(-4)}\times 5 + \frac{3\cdot 1\cdot(-1)}{1\cdot(-1)\cdot(-3)}\times 6 + \frac{3\cdot 2\cdot(-1)}{2\cdot 1\cdot(-2)}\times 9 + \frac{3\cdot 2\cdot 1}{4\cdot 3\cdot 2}11$$

$$= \frac{5}{4} - 6 + \frac{27}{2} + \frac{11}{4} = 11.5.\Bigg]$$

CHAPTER 5 CURVE-FITTING

1. Given a set of data points (x_i, y_i) $(i = 0, 1, 2, \dots n)$ in an interval $I = [x_0, x_n]$ the process of estimating $y = f(x)$ at a point outside I is called ________.
Ans: Extrapolation

2. Given a set of data points (x_i, y_i) $(i = 0, 1, 2, \dots n)$ in an interval $I = [x_0, x_n]$ the process of estimating $y = f(x)$ at a point $\bar{x} \in (x_i, x_{i+1})$ is called ________.
Ans: Interpolation

3. The method of least squares minimizes the ________ of the deviations (errors) of expected values from observed data.
Ans: Sum of the squares

4. The equations $\sum y_i = na + b\sum x_i$ and $\sum x_i y_i = a\sum x_i + b\sum x_i^2$ are the normal equation of the ________ fit for the data.
Ans: Linear or straight line

5. The equations $\sum y_i = na + b\sum x_i + c\sum x_i^2$; $\sum x_i y_i = a\sum x_i + b\sum x_i^2 - c\sum x_i^3$ and $\sum x_i^2 y_i = a\sum x_i^2 + b\sum x_i^3 + c\sum x_i^4$ are the normal equations of the ________ fit for the data.
Ans: Parabolic or quadratic

6. The equations $\sum y_i = nA + B\sum x_i$ and $\sum x_i y_i = A\sum x_i + B\sum x_i^2$ where $X = \log x$, $Y = \log y$, $A = \log a$ and $B = b$ are the normal equations of

the ________ fit for the data.

Ans: Geometric or power curve

7. The equations $\sum y_i = nA + B\sum x_i$ and $\sum x_i y_i = A\sum x_i + B\sum x_i^2$ where $Y = \log y$, $A = \log a$ and $B = \log b$ are the normal equations of the ________ fit for the data.

Ans: Exponential curve

8. The equations $\sum y_i^* = na + b\sum x_i$ and $\sum x_i y_i^* = a\sum x_i - b\sum x_i^2$ where $y^* = \frac{1}{y}$ are the normal equations of the ________ fit for the data.

Ans: Hyperbolic

9. The least squares st. line passes through the point $(\overline{x}, \overline{y}) =$ ________.

Ans: $\left(\sum x_i / N_1, \sum y_i / N\right)$

10. The best fitting curve for a set of given data is the one having the least or minimum of the ________ of expected values from the observed data. **Ans:** Sum of the squares of derivations

11. The least squares straight line for the following data

x	1	2	3	4
y	6	4	3	5

is $y = a + bx$ where (a, b) is ________.

Ans: 5.5, –0.4

[**Hint:** Normal equations $4a + 10b = 18$, $10a + 30b = 43$.]

12. The least squares straight line for the data

x	0	1	3	6	8
y	1	3	2	5	4

is $y = a + bx$ where $(a, b) =$ ________.

Ans: 0.376, 1.646

[**Hint:** Normal equations are $5a + 18b = 15$; $18a + 110b = 71$.]

Chapter 6 Numerical Differentiation and Integration

1. By using Newton's forward differences formula, if $x_0 = 1.2$, $p = 0$, $h = 0.2$, $\Delta y_0 = 0.416$, $\Delta^2 y_0 = 0.336$ and $\Delta^3 y_0 = 0.048$ then $y'(1.2)$ ________.

Ans: 1.32

$$\left[\textbf{Hint: } y'(x_0) = \frac{1}{h}\left[\Delta y_0 - \frac{1}{2}\Delta^2 y_0 + \frac{1}{3}\Delta^3 y_0\right]\right.$$

$$= \frac{1}{0.2}\left[0.416 - \frac{1}{2}(0.336)\right.$$

$$\left.\left. + \frac{1}{3}(0.048)\right] = 1.32.\right]$$

2. Using the data in Question 1, $y''(1.2) =$ ________.

Ans: 7.2

$$\left[\textbf{Hint: } y''(x_0) = \frac{1}{h^2}\left[\Delta^2 y_0 - \Delta^3 y_0\right]\right.$$

$$\left. = \frac{1}{(0.2)^2}[0.336 - 0.048] = 7.2.\right]$$

3. Using Stirling's central difference formula if $p = 0$, $h = 0.1$, $\Delta y_{-1} = 0.10117$, $\Delta y_0 = 0.10318$, $\Delta^3 y_{-2} = 0.00101$ and $\Delta^3 y_{-1} = 0.00104$ then $y'(0.2) =$ ________.

Ans: 1.02

$$\left[\textbf{Hint: } y'(x_0 + h) = \frac{1}{h}\left[\frac{\Delta y_{-1} + \Delta y_0}{2} - \frac{1}{6}\frac{\Delta^3 y_{-2} + \Delta^3 y_{-1}}{2}\right]\right.$$

$$= \frac{1}{0.1}\left[\frac{0.10117 + 0.10318}{2}\right.$$

$$\left.\left. - \frac{1}{12}(0.00101 + 0.00104) = 1.02\right]\right]$$

4. Using Stirling's central difference formula with the data of Question 3 above. $y''(0.2) =$ ________.

Ans: 0.201

$$\left[\textbf{Hint: } y''(x_0 + h) = \frac{1}{h^2}\left[\Delta^2 y_{-1} - \frac{1}{12}\Delta^4 y_{-2}\right]\right.$$

$$\left. = \frac{1}{(0.1)^2}\left[0.00201 - \frac{0.00003}{12}\right] = 0.201.\right]$$

5. If $x_0 = 1.0$, $h = 0.2$, $\Delta y_0 = 0.128$, $\Delta^2 y_0 = 0.288$ and $\Delta^3 y_0 = 0.048$ find x for which y is maximum,

using Newton's forward formula; $x =$ ________.
Ans: −1.2

[**Hint**: Newton's forward interpolation formula

$$y = y_0 + p\Delta y_0 + \frac{p(p-1)}{2}\Delta^2 y_0 + \frac{p(p-1)(p-2)}{6}\Delta^3 y_0 + \cdots$$]

For maximum or minimum of tabulated function

$$\frac{dy}{dp} = 0 \Rightarrow \Delta y_0 + \left(p - \frac{1}{2}\right)\Delta^2 y_0 + \left(\frac{p^2}{2} - \frac{p}{2} + \frac{1}{3}\right)\Delta^3 y_0 = 0$$

neglecting $\Delta^n y_0$ for $n \geq 4$.

$\Rightarrow ap^2 + bp + c = 0$ where $a = \frac{1}{2}\Delta^3 y_0$;

$b = \Delta^2 y_0 - \frac{1}{2}\Delta^3 y_0; c = \frac{1}{2}\Delta^2 y_0 + \frac{1}{3}\Delta^3 y_0$.

Here $a = 0.024, b = 0.288 - \frac{1}{2}(0.048) = 0.264$,

$$c = 0.128 - \frac{1}{2}(0.288) + \frac{1}{3}(0.048) = 0$$

$ap^2 + bp + c = 0 \quad \Rightarrow \quad p = 0$ or $p = -11$

$\Rightarrow \quad x = x_0 + px = 1.0 - 11(0.2) = -1.2$

6. In Question 5 above, $y_{max} =$ ________.
Ans: 3.87

[**Hint:** $y = 0 + (-11)(0.128) + \frac{132}{2}(0.288) - \frac{11 \cdot 12 \cdot 13}{6}(0.048) = 3.87.$]

7. Newton–Cote's quadrature formula is $I =$ ________.
Ans: $nh\left[y_0 + \frac{n}{2}\Delta y_0 + \frac{n(2n-3)}{12}\Delta^2 y_0 + \frac{n(n-2)^2}{24}\Delta^3 y_0 + \cdots\right]$

8. The composite trapezoidel rule is $\int_{x_0}^{x_n} y\,dx =$ ________.
Ans: $\frac{h}{2}[y_0 + 2(y_1 + y_2 + \cdots + y_{n-1}) + y_n]$

9. For applying the Simpson's $\frac{1}{3}$ rule the minimum number of y-values required is ________.
Ans: 3

10. For applying the Simpson's $\frac{3}{8}$ ths rule the minimum number of y-values required is ________.
Ans: 4

11. For applying the Weddle's rule the minimum number of y-values required is ________.
Ans: 7

12. The value of $\int_{0.1}^{0.7} dx/x$ (with $h = 0.1$) by Weddle's rule is ________.
Ans: 1.953

13. $\int_0^4 e^x dx$ by Simpson's rule, given that $e = 2.72$, $e^2 = 7.39$, $e^3 = 20.09$, $e^4 = 54.6$ ________.
Ans: 53.87

14. The value of $\int_0^1 dx/x$ by Simpson's $\frac{1}{3}$ rule with $h = \frac{1}{8}$ is ________.
Ans: 0.6931

15. The value of $\int_0^1 dx/(1+x^2)$ by Simpson's $\frac{3}{8}$ rule with $h = \frac{1}{6}$ is ________.
Ans: 0.6932

16. The value of $\int_0^1 dx/(1+x^2)$ by Weddle's rule is ________.
Ans: 0.7854

17. To apply Simpson's $\frac{1}{3}$ and $\frac{3}{8}$ rules both, the interval must be divided into minimum of ________ sub intervals.
Ans: 6

18. To apply Boole's rule the minimum number of ordinates required is _______. **Ans:** 5

CHAPTER 7 NUMERICAL SOLUTION OF ORDINARY DIFFERENTIAL EQUATIONS

1. The formula $y_{n+1} = y_n + \frac{h}{1!}y'_n + \frac{h^2}{2!}y''_n + \frac{h^3}{3!}y'''_n + \cdots (n = 0,1,3,\cdots)$ is ______ iteration formula for the solution of $y' = f(x, y), y(x_0) = y_0.$ **Ans:** Taylor's

2. The formula $y_{n+1} = y_0 + \int_{x_0}^{x} f(x, y_n)\, dx, n = 0, 1,$ 2, 3,... is ________ iteration formula for the solution of $y' = f(x, y), y(x_0) = y_0$. **Ans:** Picard's

3. The formula $y_{n+1} = y_n + hf(x_n, y_n), n = 0, 1, 2, 3, \ldots$ is ________ iteration formula for the solution of $y' = f(x, y), y(x_0) = y_0$. **Ans:** Euler's

4. The formula $y_{n+1}^{(k+1)} = y_n + \frac{h}{2}\left[f(x_n, y_n) + f(x_{n+1}, y_{n+1}^{(k)})\right] (k = 0, 1, 2,\ldots;\ \text{in} = 0, 1, 2\cdots)$ is ______ iteration formula for the solution of $y' = f(x, y), y(x_0) = y_0$. **Ans:** Euler's modified method

5. To solve $y' = xy + 1, y(0) = 1$ numerically by Taylor's series method, we compute the values of y'_0, y''_0, y'''_0 and y_0^{iv} as______. **Ans:** 1, 1, 2, 3

6. To solve $y' = x + y, y(1) = 0$ numerically by Taylor's series method we compute the values of y'_0, y''_0, y'''_0 and y_0^{iv} as ______. **Ans:** 1, 2, 2, 2

7. In the numerical solution of $y' = f(x, y), y(0) = y_0$ by Picard's method the formula for the first approximation y_1 is $y_1 =$______. **Ans:** $\left(y_0 + \int_{x_0}^{x} f(x, y_0)\, dx\right)$

8. In the numerical solution of $y' = xy + 1, y(0) = 1$ by Picard's method the first approximation is $y_1 =$______. **Ans:** $\left(1 + x + \frac{1}{2}x^2\right)$

9. In the numerical solution of $y' = xy + 1, y(0) = 1$ by Picard's method the first approximation for $y_{0.1} =$______. **Ans:** 1.105

10. In the numerical solution of $y' = x - y, y_0 = 1$ by Picard's method the first approximation $y_1 =$______. **Ans:** $\left(1 - x + \frac{1}{2}x^2\right)$

11. In the numerical solution of $y' = x - y, y_0 = 1$ by Picard's method the first approximation y_1 at $x = 0.2$ is ______. **Ans:** 0.820

12. Solving the equation $y' = x + y, y(0) = 1$ are obtains $y(0.1)$ as ______. **Ans:** 1.1

CHAPTER 8 FOURIER SERIES

1. If $f(x) = \begin{cases} 0 & [-\pi, 0] \\ x & [0, \pi] \end{cases}$ has Fourier series expansion $f(x) = \frac{\pi}{4} - \frac{2}{\pi}\sum_{n=1}^{\infty}\frac{\cos(2n-1)x}{(2n-1)^2} + \sum_{n=1}^{\infty}\frac{(-1)^{n-1}\sin nx}{n}$ then $1 + \frac{1}{3^2} + \frac{1}{5^2} + \frac{1}{7^2} + \cdots =$ _______. **Ans:** $\left(\frac{\pi^2}{8}\right)$

2. If $x^2 = \frac{\pi^2}{3} - 4\sum_{n=1}^{\infty}\frac{(-1)^n \cos nx}{n^2}$ in $(-\pi, \pi)$ then $1 - \frac{1}{2^2} + \frac{1}{3^2} - \frac{1}{4^2} + \frac{1}{5^2} - \frac{1}{6^2} + \cdots =$ _______. **Ans:** $\left(\frac{\pi^2}{12}\right)$

3. If $1 - \frac{1}{2^2} + \frac{1}{3^2} - \frac{1}{4^2} + \cdots = \frac{\pi^2}{12}$ and $1 + \frac{1}{2^2} + \frac{1}{3^2} + \frac{1}{4^2} + \cdots = \frac{\pi^2}{6}$ then $\frac{1}{1^2} + \frac{1}{3^2} + \frac{1}{5^2} + \frac{1}{7^2} + \cdots =$ _______. **Ans:** $\frac{\pi^2}{8}$

4. If the half-range sine series in $(0, \pi)$ for $f(x) = x(\pi - x) = \frac{8}{\pi}\sum_{n=1}^{\infty}\frac{\sin(2n-1)x}{(2n-1)^3}$ then $1 - \frac{1}{3^3} + \frac{1}{5^3} - \frac{1}{7^3} + \cdots =$ ______. **Ans:** $\left(\frac{\pi^3}{32}\right)$

5. If the half-range cosine series in $(0, \pi)$ for $f(x) = \sin x = \frac{2}{\pi} - \frac{4}{\pi}\sum_{n=1}^{\infty}\frac{\cos 2nx}{4n^2 - 1}$ then $\frac{1}{1.3} + \frac{1}{3.5} - \frac{1}{5.7} + \frac{1}{7.9} \cdots =$ ______. **Ans:** $\left(\frac{1}{2}\right)$

6. If the half-range sine series of e^x in $(0, \pi)$ is $e^x = \frac{2}{\pi}\left[\frac{1+e^x}{1^2+1}\sin x + \frac{2(1-e^x)}{2^2+1}\sin 2x + \frac{3(1+e^x)}{3^2+1}\sin 3x + \cdots\right]$ then $\frac{1}{1^2+1} - \frac{3}{3^2+1} + \frac{5}{5^2+1} - \cdots =$ ______. **Ans:** $\left(\frac{\pi}{4}\text{sech}\frac{\pi}{2}\right)$

7. If the half-range cosine series in $(0, \pi)$ for $x\sin x = 1 - \frac{1}{2}\cos x - \frac{2}{1.3} + \cos 2x + \frac{2}{2.4}\cos 3x - \frac{2}{3.5}\cos 4x + \cdots$ then $\frac{1}{1.3} - \frac{1}{3.5} + \frac{1}{5.7} - \frac{1}{7.9} + \cdots =$ ______. **Ans:** $\left(\frac{\pi}{4} - \frac{1}{2}\right)$

8. If $\frac{1}{1.3} - \frac{1}{3.5} + \frac{1}{5.7} - \frac{1}{7.9} + \cdots = \frac{\pi}{4} - \frac{1}{2}$ and $\frac{1}{1.3} + \frac{1}{3.5} + \frac{1}{5.7} + \frac{1}{7.9} + \cdots = \frac{1}{2}$ then $\frac{1}{1.3} + \frac{1}{5.7} + \frac{1}{9.11} + \cdots =$ ______. **Ans:** $\left(\frac{\pi}{8}\right)$

9. If $e^{-ax} = \frac{2\sinh a\pi}{\pi}\left[\left(\frac{1}{2a} - \frac{a\cos x}{1^2+a^2} + \frac{a\cos 2x}{2^2+a^2} - \frac{a\cos 3x}{3^2+a^2} + \cdots\right) - \left(\frac{\sin x}{1^2+a^2} - \frac{2\sin 2x}{2^2+a^2} + \frac{3\sin 3x}{3^2+a^2} + \cdots\right)\right]$ in $(-\pi, \pi)$ then $\frac{1}{2^2+1} - \frac{1}{3^2+1} + \frac{1}{4^2+1} \cdots =$ ______. **Ans:** $\left(\frac{\pi}{2\sinh \pi}\right)$

10. If the Fourier series expansion of $f(x) = x^2$ in $(-l, l)$ is $x^2 = \frac{l^2}{3} - \frac{4l^2}{\pi^2}\sum_{n=1}^{\infty}\frac{(-1)^{n-1}}{n^2}\cos\frac{n\pi x}{l}$ then $\frac{1}{1^2} - \frac{1}{2^2} + \frac{1}{3^2} - \frac{1}{4^2} + \cdots =$ ______. **Ans:** $\left(\frac{\pi^2}{12}\right)$

11. If the Fourier series expansion of $\cos ax$ in $(-\pi, \pi)$ is $\cos ax = \frac{2a\sin a\pi}{\pi}\frac{1}{2a^2} + \sum_{n=1}^{\infty}\frac{(-1)^{n-1}\cos nx}{n^2 - a^2}$, $(a \notin z)$ then $\frac{1}{2a^2} + \sum_{n=1}^{\infty}\frac{(-1)^{n-1}}{n^2 - a^2} =$ ______. **Ans:** $\left(\frac{\pi}{2a\sin a\pi}\right)$

12. If n is an integer and 'a' is not an integer then $\frac{\sin(n+a)\pi}{\sin a\pi} =$ ______. **Ans:** $(-1)^n$

13. If n is an integer and 'a' is not an integer then $\frac{\sin(n-a)\pi}{\sin a\pi} =$ ______. **Ans:** $(-1)^{n-1}$

14. If the Fourier series for e^x $(0, 2\pi)$ is given by $e^x = \frac{e^{2\pi} - 1}{\pi}\left[\frac{1}{2} + \sum_{n=1}^{\infty}\frac{\cos nx}{n^2+1} - \sum_{n=1}^{\infty}\frac{n\sin nx}{1+n^2}\right]$ then $\frac{1}{2} + \sum_{n=1}^{\infty}\frac{(-1)^n}{1+n^2} =$ ______. **Ans:** $\left(\frac{\pi}{2}\text{cosech}\,\pi\right)$

[**Hint:** Put $x = \pi$.]

15. If the Fourier series for e^x $(0, 2\pi)$ is given by $e^x = \frac{e^{2\pi} - 1}{\pi}\left[\frac{1}{2} + \sum_{n=1}^{\infty}\frac{\cos nx}{n^2+1} - \sum_{n=1}^{\infty}\frac{n\sin nx}{n^2+1}\right]$ then $\frac{1}{2} - \sum_{n=1}^{\infty}\frac{(-1)^{n-1}}{4n^2+1} =$ ______. **Ans:** $\left(\frac{\pi e^{-\pi/2}}{2\sinh \pi}\right)$

$\left[\textbf{Hint: } \text{Put } x = \frac{\pi}{2}, n \text{ even.}\right]$

16. If the Fourier series of $f(x)=\begin{cases}0, & [-\pi, 0]\\ \sin x & [0, \pi]\end{cases}$

is $f(x)=\frac{1}{\pi}-\frac{2}{\pi}\sum_{n=1}^{\infty}\frac{\cos 2nx}{4n^2-1}+\frac{1}{2}\sin x$ then

$\frac{1}{1.3}+\frac{1}{5.7}-\frac{1}{9.11}+\cdots=$ ______. **Ans:** $\left(\frac{\pi}{8}\right)$

[**Hint:** Put $x = 0$ and $\frac{p}{2}$ and obtain

respectively $\frac{1}{1.3}+\frac{1}{3.5}+\frac{1}{5.7}+\cdots=\frac{1}{2}$

(i) and $\frac{1}{1.3}-\frac{1}{3.5}+\frac{1}{5.7}\cdots=\frac{\pi}{4}-\frac{1}{2}$

(ii) Then $\frac{1}{1.3}+\frac{1}{5.7}+\frac{1}{9.11}+\cdots=\frac{\text{(i)}+\text{(ii)}}{2}=\frac{\pi}{8}$.]

17. From Question 16, $\frac{1}{3.5}+\frac{1}{7.9}+\frac{1}{11.13}+\cdots=$

______. **Ans:** $\left(\frac{1}{2}-\frac{\pi}{8}\right)$

[**Hint:** In Question 16, $\frac{1}{3.5}+\frac{1}{7.9}+\frac{1}{11.13}+\cdots$

$=\frac{\text{(ii)}-\text{(i)}}{2}=\frac{1}{2}-\frac{\pi}{8}$.]

18. If n is an even number then $\cos\frac{n\pi}{2}=$ ______

Ans: $(-1)^{n/2}$

19. $\frac{\cos (n+a)\pi}{\cos a\pi}, (n\in Z, a\notin Z), =$ ______.

Ans: $(-1)^n$

20. $\sin n\pi=$ ______. **Ans:** 0

21. The function $f(x)\begin{cases}\frac{1}{2}(x-x^2), & 0<x<l\\ -\frac{1}{2}(x+x^2), & -l<x<0\end{cases}$ is

______ function in $(-l, l)$. **Ans:** Even

22. If $f(x) = |x|$ in $(-\pi, \pi)$ then the Fourier coefficient $a_0 =$ ______. **Ans:** $\left(\frac{\pi}{2}\right)$

23. If $f(x) = 2x$ is represented by Fourier series in $(-\pi, \pi)$ then the value of $a_0 =$ ______. **Ans:** 0

24. The Euler's integral formula for determining the constants b_n for a function $f(x)$ defined in $(-\pi, \pi)$

is $b_n =$ ______. **Ans:** $\frac{2}{\pi}\int_0^{\pi} f(x)\sin nx\, dx$

25. The Euler's integral formula for the constants a_n for the function $f(x)$ defined in the interval $(0, 2l)$

is $a_n =$ ______. **Ans:** $\frac{1}{l}\int_0^{2l} f(x)\cos\frac{n\pi x}{l}\, dx$

26. If $f(x) = |\sin x|$ is expanded as a Fourier series in $(-\pi, \pi)$ then the value of $a_0 =$ ______.

Ans: $\left(\frac{4}{\pi}\right)$

27. The leading term in the Fourier series expansion of $f(x) = e^x$ in $(-\pi, \pi)$ is ______. **Ans:** $\left(\frac{\sinh \pi}{\pi}\right)$

CHAPTER 9 FOURIER INTEGRAL TRANSFORMS

1. An integral transform of a function $f(x)$ is defined by $I\{f(x)\}=\int_a^b f(x)\, k(s, x)\, dx=\bar{f}(s)$. Here $k(s, x)$ is called the ______ of the transform. **Ans:** Kernel

2. In the case of Fourier transform $k(s, x) =$ ______. **Ans:** e^{isx}

3. In the case of Fourier transform the limits are $(a, b) =$ (______, ______). **Ans:** $-\infty, \infty$

4. The Fourier transform of a function $f(x)$ is $\bar{F}(s) =$ ______. **Ans:** $\int_{-\infty}^{\infty} e^{isx} f(x)dx$

5. The Fourier sine transform of a function $f(x)$ is ______. **Ans:** $\int_0^{\infty} f(x)\sin sx\, dx$

6. The Fourier cosine transform of a function $f(x)$ is $\bar{F}_c(s) =$ ______. **Ans:** $\int_0^{\infty} f(x)\cos sx\, dx$

7. The Fourier inverse transform of $\overline{F}(s)$ is $f(x) =$ ______. **Ans:** $\int_{-\infty}^{\infty} e^{-isx} \overline{F}_c(s)\, ds$

8. The inverse Fourier sine transform of $\overline{F}_s(s)$ is $f(x) =$ ______. **Ans:** $\int_0^{\infty} \overline{F}_s(s) \sin sx\, ds$

9. The inverse Fourier cosine transform of $\overline{F}_c(s)$ is $f(x) =$ ______. **Ans:** $\int_0^{\infty} \overline{F}_c(s) \cos sx\, ds$

10. If the complex Fourier transform of $f(x)$ is $\overline{F}(s)$ then that of $f(ax)$ is $F\{f(ax)\} =$ ______. **Ans:** $\left(\frac{1}{a}\overline{F}\left(\frac{s}{a}\right)\right)$

11. The complex Fourier transform of $f(x - a)$ is $F\{f(x-a)\} =$ ______. **Ans:** $e^{isa}\overline{F}(s)$

Chapter 10 Partial Differential Equations

1. The partial differential equation obtained by eliminating a and b from $(x - a)^2 + (y - b)^2 = z^2 \cot^2 \alpha$ is ______. **Ans:** $p^2 + q^2 = \tan^2 \alpha$

2. The partial differential equation obtained by eliminating a and b from $x^2 + y^2 + (z - a)^2 = b^2$ is ______. **Ans:** $py = qx$

3. The partial differential equation obtained by eliminating a and b from $z = ax + by + \frac{a-b}{a}$ is ______. **Ans:** $\left(z = px + qy + \frac{p-q}{p}\right)$

4. The partial differential equation obtained by eliminating the arbitrary function f from $z = f(x^2 - y^2)$ is ______. **Ans:** $py + qx = 0$

5. Solving the partial differential equation $p = xy$ we obtain ______. **Ans:** $\left(z = \frac{1}{2}x^2 y + f(y)\right)$

6. The general solution of $px + qy = z$ is ______. **Ans:** $F\left(\frac{x}{z}, \frac{y}{z}\right) = 0$

7. The general solution of $x(y - z)p + y(z - x)q = z(x - y)$ is ______. **Ans:** $F(x + y + z, xyz) = 0$

8. The general solution of $(y - z)p + (z - x)q = x - y$ is ______. **Ans:** $F(x + y + z, x^2 + y^2 + z^2) = 0$

9. The complete integral of $pq = k$ is ______. **Ans:** $\left(z = ax + \frac{k}{a}y + c\right)$

10. The complete integral of $z = px + qy + pq$ is ______. **Ans:** $z = ax + by + ab$

11. The partial differential equation obtained by eliminating a and b from $z = ax + by$ is ______. **Ans:** $z = px + qy$

12. The partial differential equation obtained by eliminating a and b from $z = ax + by + ab$ is ______. **Ans:** $z = px + qy + pq$

13. The partial differential equation obtained by eliminating a and b from $z = (x^2 + a^2)(y^2 + b^2)$ is ______. **Ans:** $pq = 4xyz$

14. The result of eliminating the constants h and k from $(x - h)^2 + (y - k)^2 + z = a^2$ is the partial differential equation ______. **Ans:** $z(p^2 + q^2 + 1) = a^2$

15. Eliminating a and b from $z = ae^{bt} \sin bx$ we obtain the partial differential equation ______. **Ans:** $\left(\frac{\partial^2 z}{\partial x^2} + \frac{\partial^2 z}{\partial t^2} = 0\right)$

16. Eliminating a and b from $z = ax + (1 - a)y + b$ we obtain the partial differential equation ______. **Ans:** $p + q = 1$

17. The partial differential equation of all planes whose x and y intercepts are always equal is ______. **Ans:** $p = q$

18. By eliminating the arbitrary function from $z = x^n f\left(\frac{y}{z}\right)$ we obtain the partial differential equation _______. **Ans:** $px + qy = nz$

19. By eliminating the arbitrary function f from $z = f\left(\frac{xy}{z}\right)$ we obtain the partial differential equation _______. **Ans:** $px = qy$

20. The complete solution of $z = px + qy + \log pq$ is _______. **Ans:** $z = ax + by + \log ab$

21. The complete integral of $yp + xq + pq = 0$ is given by _______. **Ans:** $(2z = (a^2 - 1)^{-1}x^2 - a^{-1}y^2 + b)$

22. The complete integral of $4z = pq$ is _______. **Ans:** $\left(\frac{(x+ay+b)^2}{a}\right) = \frac{1}{z}$

23. The general solution of $p + q = 1$ is _______. **Ans:** $\phi(x - y, y - z) = 0$

24. The general integral of $p + q = z$ is _______. **Ans:** $F(x - y, z^{-1}e^y) = 0$

CHAPTER 11 Z-TRANSFORMS AND SOLUTION OF DIFFERENCE EQUATIONS

1. $Z\{\langle \cos nt \rangle\} =$ _____. **Ans:** $\left(\frac{z(z - \cos t)}{z^2 - 2z\cos t + 1}\right)$

2. $Z\{\langle \sin nt \rangle\} =$ _____. **Ans:** $\left(\frac{z \sin t}{z^2 - 2z\cos t + 1}\right)$

3. $Z\{\langle {}^nC_k \mid 0 \le k \le n \rangle\} =$ _____. **Ans:** $\left(\left(1 + \frac{1}{z}\right)^n\right)$

4. $Z\{\langle 4.2^n + 5(-1)^n \rangle\} =$ _____. **Ans:** $\left(4\left(\frac{z}{z-2}\right) + 5\left(\frac{z}{z+1}\right)\right)$

5. $Z\{\langle n - 1 \rangle\} =$ _____. **Ans:** $\left(\frac{2z - z^2}{(z-1)^2}\right)$

6. $Z\{\langle \delta(n) \rangle\} =$ _____. **Ans:** 1

7. $Z^{-1}\left\{\frac{z^2}{(z-2)(z-3)}\right\} =$ _____. **Ans:** $\left(\left\langle 3^n \sum_{m=0}^{n} \left(\frac{2}{3}\right)^m \right\rangle\right)$

8. The complementary function of the difference equation $u_{n+2} - 7u_{n+1} - 8u_n = 2^n n^2$ is _____. **Ans:** $A\langle 8^n \rangle + B\langle (-1)^n \rangle$

9. The particular integral of the difference equation $u_{n+2} - 4u_{n+1} + 3u_n = 5^n$ is _____. **Ans:** $\left(\frac{1}{8}\langle 5^n \rangle\right)$

10. Initial Value Theorem: If $Z\{< u_n >\} = \bar{u}(z)$ then $u_0 =$ _____. **Ans:** $\lim_{z\to\infty} \bar{u}(z)$

CHAPTER 12 SPECIAL FUNCTIONS

A. Beta and Gamma Functions

1. $\int_0^{-\infty} e^{-x^2} dx =$ _______. **Ans:** $\sqrt{\pi}/2$

2. $\Gamma(p) = \int_0^{\infty} e^{-x^2} dx \Rightarrow p =$ _______. **Ans:** 3/2

3. $\int_0^{\infty} e^{-\sqrt{x}} dx =$ _______. **Ans:** 2

4. $2\int_0^{\pi/2} \sin^p\theta \cos^q\theta \, d\theta = B$ _______. **Ans:** $\left(\frac{p+1}{2}, \frac{q+1}{2}\right)$

5. $\int_0^{\pi/2} \tan^p \theta \, d\theta =$ _______. **Ans:** $\frac{\pi}{2}\sec\frac{p\pi}{2}$

6. $\int_0^\infty \frac{x^{p-1}}{1+x}dx =$ ______. **Ans:** $\pi \operatorname{cosec} p\pi$

7. $B(1/3, 2/3) =$ ______. **Ans:** $2\pi/\sqrt{3}$

8. $B(1/2, 1/2) =$ ______. **Ans:** π

9. $\Gamma\left(-\frac{1}{2}\right) = a\sqrt{\pi} \Rightarrow a =$ ______. **Ans:** –2

10. $\theta = \Gamma(1/2) \Rightarrow \cos\frac{\theta^2}{3} =$ ______. **Ans:** 1/2

11. $B(1, 1/2) =$ ______. **Ans:** 2

12. $\Gamma(3/2)\Gamma(1/2) =$ ______. **Ans:** $\pi/2$

13. $B(p+1, q) + B(p, q+1) = B$ ______. **Ans:** (p, q)

14. $B(p+1, q)/B(p, q) =$ ______. **Ans:** $p/(p+q)$

15. $\Gamma(1/2) =$ ______. **Ans:** $\sqrt{\pi}$

16. $\Gamma(3/2) =$ ______. **Ans:** $\frac{1}{2}\sqrt{\pi}$

17. $\Gamma(-1/2) =$ ______. **Ans:** $-2\sqrt{\pi}$

18. $\Gamma(x) = x \Rightarrow x =$ ______. **Ans:** 1

19. $\Gamma(-3/4) =$ ______ $\Gamma(1/4)$. **Ans:** –4/3

20. $\Gamma(1/4)\Gamma(3/4) =$ ______. **Ans:** $\sqrt{2}\pi$

21. $\Gamma(1/3)\Gamma(2/3) =$ ______. **Ans:** $2\pi/\sqrt{3}$

22. $\int_0^{\pi/2} \sin\theta \cos^2\theta d\theta =$ ______. **Ans:** 1/3

23. $\int_0^{\pi/2} \sqrt{\cot\theta}\, d\theta =$ ______. **Ans:** $\pi/\sqrt{2}$

24. $\sqrt{\pi}\,\Gamma(2p)/\Gamma(p)\,\Gamma(p+1/2) =$ ______. **Ans:** 2^{2p-1}

25. $\Gamma(p)\Gamma(1-p) = \pi \Rightarrow p =$ ______. **Ans:** 1/2

26. $\int_0^\infty e^{-ax}x^{n-1}dx =$ ______. **Ans:** $\Gamma(n)/a^n$

27. $\int_0^\infty \exp\left(-\sqrt{x}\right)dx =$ ______. **Ans:** 2

28. $\int_0^\infty e^{-x^p}dx =$ ______. **Ans:** $\Gamma(1/p+1)$

29. If $\tan\alpha = 1\,(0 < \alpha < \pi/2)$ and $\Gamma^2(x) = \alpha$ then $x =$ ______. **Ans:** 3/2

30. $\int_{-\infty}^\infty e^{-x^2}dx =$ ______. **Ans:** $\sqrt{\pi}$

31. $\int_0^\infty \frac{dx}{\sqrt{x}(1+x)} =$ ______. **Ans:** π

32. $(m+n)\ B\,(m+1, n) =$ ______. **Ans:** $mB(m, n)$

33. $\int_0^{\pi/2} \sqrt{\sin\theta}d\theta \int_0^{\pi/2} \frac{d\theta}{\sqrt{\sin\theta}} =$ ______. **Ans:** π

34. $\int_0^{\pi/2} \sqrt{\tan\theta}d\theta \int_0^{\pi/2} \sqrt{\cot\theta}d\theta =$ ______. **Ans:** $\frac{1}{2}\pi^2$

35. $2\int_0^1 \frac{dx}{\sqrt{x}(1+x)} = B$ ______. **Ans:** (1/2, 1/2)

36. $2^3 \cdot 3^2 \int_0^\infty \frac{xdx}{(2x+3)^5} =$ ______. **Ans:** $B(2, 3)$

37. $\int_0^1 \frac{(x-x^2)dx}{(x+1)^4} =$ ______. **Ans:** $\frac{1}{4}B(2,2)$

38. $\int_1^2 (x-1)^4(2-x)^5 dx =$ ______. **Ans:** $B(5, 6)$

B. Legendre Functions

1. $[P_1(-x)]^1 =$ ______. **Ans:** –1

2. $\int_{-1}^1 P_n^2(x)dx =$ ______. **Ans:** $2/(2n+1)$

3. Rodringue's Formula is $P_n(x) = CD^n[(x^2-1)^n]$ with $C =$ ______. **Ans:** $1/(2^n n!)$

4. The generating function for $P_n(x)$ is ______. **Ans:** $(1-2xt+t^2)^{-1/2}$

5. The sum of the series $1+t+t^2+t^3+\cdots$ in terms of Legendre function is ______. **Ans:** $P_n(1)$

6. $P_0(x) =$ ______. **Ans:** 1

7. $\int_{-1}^{1} P_m(x)P_n(x)dx =$ _______ $(m \neq n)$. **Ans:** 0

8. $P_1^1(x) =$ _______. **Ans:** $P_0(x)$

9. $P_n(1) =$ _______. **Ans:** 1

10. $P_n(-x) =$ _______ $P_n(x)$. **Ans:** $(-1)^n$

11. $P_n(-1) =$ _______. **Ans:** $(-1)^n$

12. $\int_{-1}^{1} P_3^2(x)dx =$ _______. **Ans:** 2/7

13. $\int_{-1}^{1} P_5(x)P_8(x)dx =$ _______. **Ans:** 0

14. If $x^2 + x = aP_0 + bP_1 + cP_2$, then $(a, b, c) =$ _______. **Ans:** $\left(\frac{1}{3}, 1, \frac{2}{3}\right)$

15. $P_{2n}(0) =$ _______. **Ans:** $(-1)^n(2n)!/2^{2n}(n!)^2$

16. $\int_{-1}^{1} P_m(x)P_n(x)dx =$ _______. **Ans:** $\frac{2}{2n+1}\delta_{mn}$

17. Legendre's differential equation is _______.
Ans: $[(1-x^2)y']' + x(x+1)y = 0$

18. $(n+1)P_{n+1}(x) + nP_{n-1}(x) =$ _______.
Ans: $(2n+1)xP_n(x)$

19. $P'_{n+1}(x) - P'_{n-1}(x) =$ _______.
Ans: $(2n+1)P_n(x)$

20. $\int_{-1}^{1} (1-x^2)P_n'^2(x)dx =$ _______.
Ans: $2n(n+1)/(2n+1)$

C. Bessel Functions

1. Physical and Engineering Problems involving vibrations or heat conduction in cylindrical regions give rise to _______ equation.
Ans: Bessel's differential

2. Bessel's differential equation of order n is _______
Ans: $\left(x^2\frac{d^2y}{dx^2} + x\frac{dy}{dx}\right) + (x^2 - x^2)y = 0$

3. Solutions of Bessel's differential equation are called _______ of order n.
Ans: Bessel's functions

4. The series expansion of $J_n(x) =$ _______.
Ans: $\sum_{r=0}^{n} \frac{(-1)^r}{r!\gamma(n+r+1)}\left(\frac{x}{r}\right)^{n+2r}$

5. $J_n(x)$ is called _______.
Ans: Bessel function of the first kind of order n

6. When n is an integer a second linearly independent solution of Bessel's differential equation is _______. **Ans:** $Y_n(x) = J_n(x)\int \frac{dx}{xJ_n^2(x)}$

7. _______ is called the Bessel function of the second kind of order n or Neumann function.
Ans: $y_m(x)$

8. $J_{1/2}(x) =$ _______. **Ans:** $\sqrt{\frac{2}{\pi x}}\sin x$

9. $J_{-1/2}(x) =$ _______. **Ans:** $\sqrt{\frac{2}{\pi x}}\cos x$

10. $\frac{2\pi}{n}J_n(x) - J_{n-1}(x) =$ _______. **Ans:** $J_{n+1}(x)$

11. The generating function for $J_n(x)$ is _______.
Ans: $e^{\frac{x}{2}(t-\frac{1}{t})}$

12. $\frac{1}{\pi}\int_0^{\pi} \cos(n\theta - x\sin\theta)d\theta =$ _______.
Ans: $J_n(x)$

13. $\frac{1}{\pi}\int_0^{\pi} \cos(x\sin\theta)\cos 2m\theta d\theta =$ _______.
Ans: $J_{2m}(x)$

14. $J_0^2 + 2(J_1^2 + J_2^2 + \cdots) =$ _______. **Ans:** 1

Chapter 13 Functions of a Complex Variable

1. $f(z) = \bar{z}$ is _______.
Ans: not analytic for any z

2. The Cauchy–Riemann equation is only _______ but not _______ for analytic of $f(z)$
Ans: necessary, sufficient

3. $f(z) = |z|^2$ is _______.
Ans: continuous for all z but derivable at $z = 0$ only

4. $f(z) = \dfrac{1+z}{1-z}$ is _______.
Ans: analytic for all finite $z \neq 1$

5. The area of the triangle formed by the points $1+i, i-1, 2i$ (in sq. units) is _______. **Ans:** 1

6. The triangle formed by joining the points $\left(\pm 1 + \sqrt{3}i\right)$ and the origin is _______.
Ans: equilateral

7. The discontinuity of $f(z) = \dfrac{z^2 + 3iz - 2}{z+i}$ at $z = -i$ may be removed by defining $f(-i) =$ _______.
Ans: i

8. An analytic function of constant absolute value is _______. **Ans:** constant

9. A function $u(x, y)$ having continuous second partial derivatives and satisfying $\nabla^2 u = 0$ is called a _______ function. **Ans:** harmonic

10. The analytic function whose imaginary part is $v(x, y) = 2xy$ is _______. **Ans:** z^2

11. The harmonic conjugate of $v(x, y) = 2xy$ is _______. **Ans:** $x^2 - y^2$

12. If $f(z) = z/|z|$ $(z \neq 0)$ and $f(0) = 0$, then f is _______. **Ans:** discontinuous at $z = 0$ only

13. If $u(x, y) = 2x - x^2 + my^2$ is harmonic in D, then $m =$ _______. **Ans:** 1

14. If $f(z) = z\bar{z}$, then $f'(z)$ exists _______.
Ans: $z = 0$ only

15. If $f(z) = u + iv$ is analytic, then _______.
Ans: $U_x = V_y,\ U_y = -V_x$

16. $f(z) = (1+z)/(1-z)$ is differentiable at _______. **Ans:** all z except $z = 1$

17. $f(z) = \text{Re}(z)$ is differentiable at _______.
Ans: no z

18. The curves $u(x, y) = e^x \cos y = c'$ are orthogonal to the curves $v(x, y) = c_2$ where $v(x, y) =$ _______. **Ans:** $e^x \sin y$

Chapter 14 Elementary Functions

1. $\sin(x+iy) =$ _______.
Ans: $(\sin x \cosh y + i \cos x \sinh y)$

2. $\cos(x+iy) =$ _______.
Ans: $(\cos x \cosh y - i \sin x \sinh y)$

3. $\text{Re}[\tan(x+iy)] =$ _______.
Ans: $\sin 2x/(\cos 2x + \cosh y)$

4. $|\cos(x+iy)|^2 =$ _______. **Ans:** $(\cos^2 x + \sin^2 y)$

5. Sin z is periodic with period _______. **Ans:** 2

6. Tan z is periodic with period _______. **Ans:** π

7. e^z is periodic with period _______. **Ans:** $2\pi i$

8. $\text{Re}\, e^{z^2} =$ _______. **Ans:** $e^{x^2-y^2} \cos 2xy$

9. $\text{Im}\, e^{z^2} =$ _______. **Ans:** $e^{x^2-y^2} \sin 2xy$

10. The solutions of $\cos z = 0$ are $z =$ _______.
Ans: $(2n+1)\dfrac{\pi}{2}$

11. When expressed in terms of natural logarithm, $\sin^{-1} z =$ _______. **Ans:** $[-i \ \ln (iz + \sqrt{1-z^2})]$

12. When expressed in terms of natural logarithm, $\cos^{-1} z =$ _______. **Ans:** $[\ln (z + \sqrt{z^2 - 1})]$

13. All the roots of $\cos z = 2$ are given by $z =$ _______. **Ans:** $[2n\pi \pm i \log(2+\sqrt{3})]$

14. All the roots of $\sinh z = i$ are given by $z =$ _______. **Ans:** $\left[\dfrac{\pi i}{2} \pm 2n\pi i\right]$

15. When expressed in terms of logarithm, $\tan^{-1}z =$ ______. **Ans:** $\frac{1}{2i}\ln\left(\frac{1+iz}{1-iz}\right)$

16. When expressed in terms of logarithm, $\sec^{-1}z =$ ______. **Ans:** $\frac{1}{i}\ln\left(\frac{1+\sqrt{1-z^2}}{z}\right)$

Chapter 15 Complex Integration

1. $\int_{z=0}^{1+i}(y-x-3ix^2)dz =$ ______. **Ans:** $1-i$

2. $\int_C \bar{z}\,dz$ where C is the upper half of the unit circle taken in the clockwise direction, is ______. **Ans:** $-\pi i$

3. If C is the circle $|z-a|=r$, then $\int_C \frac{dz}{z-a} =$ ______. **Ans:** $2\pi i$

4. If C is the circle $|z-a|=r$ and $n \neq -1$, then $\int_C (z-a)^n\,dz =$ ______. **Ans:** 0

5. If C is the straight path $y=x$ from $z=0$ to $1+i$, then $\int_C (x^2+iy)dz =$ ______. **Ans:** $\frac{1}{6}(-1+5i)$

6. $\int_0^1 (t-it)^2\,dt =$ ______. **Ans:** $\frac{1}{2}-\frac{1}{3}i$

7. $\int_0^1 (t-i)^{-1}\,dt =$ ______. **Ans:** $\frac{1}{2}\log 2+\frac{1}{4}\pi i$

8. $\int_0^1 t\,e^{-t^2}\,dt =$ ______. **Ans:** $\frac{1}{2}(1-e^{-1})$

9. If C is the arc of the circle $|z|=2$ from $\theta=0$ to $\theta=\pi/3$, then $\int_C z^2 dz =$ ______. **Ans:** $-\frac{16}{3}$

10. If C is the left half of the unit circle in the clockwise direction, then $\int_C |z|\,dz =$ ______. **Ans:** $2i$

11. By the ML-inequality, an upper bound for $\left|\int_C e^{2z}dz\right|$ where $C: y=\frac{\pi}{6}x$, $x=0$ to $x=1$ is ______. **Ans:** $e^{\sqrt{3}}$

12. By the Cauchy's Integral Theorem, $\int_C \frac{dz}{e^z-e^3} =$ ______ $(C:|z|=1)$. **Ans:** 0

13. By the Cauchy's Integral Theorem, $\int_C \sec z\,dz =$ ______ $(C:|z|=1)$. **Ans:** 0

14. By the Extended Cauchy Integral Theorem for multiply-connected domains, $\int_C \frac{2z-1}{z^2-z}dz; (C:|z|=2) =$ ______. **Ans:** $4\pi i$

15. By the Cauchy's Integral Theorem, $\int_c \frac{dz}{z^2+4}(c:|z-i|=2) =$ ______ $(C:|z|=1)$. **Ans:** $\frac{\pi}{2}$

16. If C is the circle $|z|=2$, $\int_C \frac{e^{2z}}{(z+1)^4}dz =$ ______. **Ans:** $\frac{8\pi i}{2e^2}$

Chapter 16 Complex Power Series

1. For $|z|>1$, $(1+z)^{-1} =$ ______. **Ans:** $\sum_{n=0}^{\infty}(-1)^n z^{-n-1}$

2. Taylor's series about $z=0$ of $f(z)=\frac{z-1}{z+1}$ is ______. **Ans:** $-1+2\sum_{n=1}^{\infty}(-1)^{n-1}z^n$

3. The region of convergence of the series in Qn. 2 is ______. **Ans:** $R:|z|<1$

4. Laurent's series expansion about $z=1$ of $f(z)=\frac{z}{(z-1)(z-3)}$ is ______. **Ans:** $-\frac{1}{2(z-1)}-\frac{3}{4}\sum_{n=0}^{\infty}\left(\frac{z-1}{2}\right)^n$

5. The region of convergence of the series in Qn. 4 is ______. **Ans:** $R:0<|z-1|<2$

6. Taylor's series for $f(z)=\frac{1}{(z-1)(z-2)}$ about $z=0$ is ______. **Ans:** $\frac{1}{2}+\frac{3}{4}z+\frac{7}{8}z^2+\frac{15}{6}z^3+\cdots$

7. The region of convergence of the series in Qn. 6 is ______. **Ans:** $R:0<|z|<1$

8. Laurent's series expansion for $|z|>1$ of $f(z)=\frac{z+1}{z-1}$ is ______. **Ans:** $1+2\sum_{n=1}^{\infty}\frac{1}{z^n}$

9. Laurent's series expansion for $|z-1|>1$ of $f(z)=\frac{z-1}{z}$ is ______.
Ans: $\sum_{n=0}^{\infty}(-1)^n(z-1)^{-n}$

10. Laurent's series expansion for $|z|>2$ of $f(z)=\frac{1}{z^2-z-2}$ is ______.
Ans: $\frac{1}{3z}\sum_{n=0}^{\infty}\frac{2^n-1}{z^n}$

11. If $C:|z|=3/2$, then $\int_C\frac{\pi z^2dz}{(z-1)(z-2)}=$ ______.
Ans: $2\pi i$

12. If $C:|z|=5$, then $\int_C\frac{ze^zdz}{z^2+9}=$ ______.
Ans: $2\pi i\cos 3$

13. $\int_0^{2\pi}\frac{d\theta}{a+b\cos\theta}=$ ______. **Ans:** $2\pi/\sqrt{a^2-b^2}$

14. $\int_0^{\infty}\frac{dx}{(x^2+1)(x^2+4)}=$ ______. **Ans:** $\pi/12$

15. $\int_0^{\infty}\frac{x^2dx}{(x^2+1)(x^2+4)}=$ ______. **Ans:** $\pi/6$

CHAPTER 17 CALCULUS OF RESIDUES

1. The poles of $f(z)=\frac{z}{z^2+1}$ are $z=$ ______.
Ans: $\pm i$

2. The Residues of $f(z)=\frac{z}{z^2+1}$ at $z=i$ is ______.
Ans: $\frac{1}{2}$

3. The Residues of $f(z)=\frac{z}{z^2+1}$ at $z=-i$ is ______.
Ans: $\frac{1}{2}$

4. If C is the circle $|z|=2$, then by the Residue Theorem $\int_C\frac{zdz}{z^2+1}=$ ______. **Ans:** $2\pi i$

5. The pole of $f(z)=\frac{ze^z}{(z-1)^3}$ is $z=$ ______.
Ans: 1

6. The order of the pole in Qn. 5 is ______.
Ans: 3

7. The Residue at $z=1$ of $f(z)=\frac{ze^z}{(z-1)^3}$ is ______.
Ans: $\frac{3e}{2}$

8. By the Residue Theorem, $\int_c\frac{ze^z}{(z-1)^3}dz$ where $C:|z|=2$ is ______. **Ans:** $3\pi ie$

9. The poles of $f(z)=\frac{\cos\pi z^2}{(z-1)(z-2)}$ are $z=$ ______. **Ans:** 1, 2

10. The Residue of $f(z)$ at $z=1$ in Qn. 9 is ______.
Ans: 1

CHAPTER 18 ARGUMENT PRINCIPLE AND ROUCHE'S THEOREM

1. A complex function $f(z)$ which is differentiable for all z in a domain D is called a/an ______ in D.
Ans: analytic or holomorphic or regular

2. A point $z=a$ where $f(z)$ is not differentiable is called a/an ______ of $f(z)$. **Ans:** singularity

3. A point $z=a$ at which $f(z)$ is not analytic but there exists a nbd of 'a' such that at each point of it except $z=a$, $f(z)$ is analytic is called a/an ______ of 'a'. **Ans:** isolated singularity

4. If $z=a$ is an isolated singularity for $f(z)$ such that $(z-a)^m f(z)=\phi(D)\in H(D)$ for some positive integer m and $\phi(a)\neq 0$ is called a ______.
Ans: pole of order m

5. If $m=1$ in Qn. 4, then it is called a ______.
Ans: simple pole

6. A complex function in a domain D whose singularities are only poles is called ______ function.
Ans: meromorphic

7. The change in the number of radius in the argument of $f(z)$ as the point z makes a cycle in the positive direction is given by ______.
Ans: $\frac{1}{2\pi}\delta_c[\arg f]$

8. By Argument Principle, if $f(z)$ is meromorphic in a domain D and C is any simple closed curve in D, then $\frac{1}{2\pi i}\int_C \frac{f'(z)dz}{f(z)} dz =$ ______. **Ans:** $[N-P]$

9. Rouche's theorem states that if $f, g \in H(D)$ and C is a closed curve in D and $|g(z)|<|f(z)|$ on C then ______ and ______ have the same number of zeros in C. **Ans:** $f(z), f(z)+g(z)$

10. Let D be a bounded domain $f(z) \in H(D)$ and nonconstant in D. Then max $|f(z)|$ occurs on the boundary of D. This is the statement of the ______. Theorem. **Ans:** Maximum modulus

Chapter 19 Conformal Mapping

1. A mapping that preserves the angles between oriented curves in magnitude and sense is called a ______. **Ans:** conformal mapping

2. The points which are mapped onto themselves under a conformal mapping are called ______ points. **Ans:** fixed invariant

3. The fixed points of $w = \frac{z}{2-z}$ are $z =$ ______. **Ans:** [0, 1]

4. The fixed points of $w = \frac{z-1}{z+1}$ are $z =$ ______. **Ans:** $\pm i$

5. The fixed points of $w = \frac{3(z-3)}{z}$ are $z =$ ______. **Ans:** [3, 3]

6. The fixed points of $w = \frac{(2z-5)}{z+4}$ are $z =$ ______. **Ans:** $-1 \pm 2i$

7. The critical points of the mapping $w = z^2$ is $z =$ ______. **Ans:** 0

8. The critical points of the mapping $w = z + \frac{1}{z}$ are $z =$ ______. **Ans:** ± 1

9. The invariant points of the transformation $w = \frac{1+z}{1-z}$ are $z =$ ______. **Ans:** $\pm i$

10. The invariant points of the transformation $w = \frac{3iz+13}{z-3i}$ are $z =$ ______. **Ans:** $3i \pm 2$

11. The transformation $w = \sin z$ maps st. lines $x = c$ into the family of ______. **Ans:** confocal hyperbolas

12. For the transformation $w = f(z)$, the point where $f(z) = 0$ is called a ______ point. **Ans:** critical

13. The transformation $w = \cosh z$ maps horizontal lines $y = c$ into the family of ______. **Ans:** confocal hyperbolas

14. The transformation $w = \cosh z$ is not conformal at points where $z =$ ______. **Ans:** $n\pi i$

15. The image of $|z-1|=1$ under the mapping $w = \frac{1}{z}$ is $u =$ ______. **Ans:** $\frac{1}{2}$

16. Under the bilinear transformation maps circles into ______ or ______. **Ans:** circles, st. lines

17. Under the transformation $w = z^2$, the image of the line $z = 4$ is a ______ whose equation is ______. **Ans:** parabola, $v^2 = -64(u-16)$

18. The Linear Fractional Transformation that maps $-1, i, 1$ into $0, 1, \infty$ is $w =$ ______. **Ans:** $(z+1)/(z-1)$

Match the Following

Chapter 1 Eigenvalues and Eigenvectors

1. λ_i are eigenvalues of A and $|A| \neq 0$
2. One of the eigenvalues of A is zero
3. λ_i are eigenvalues of A and $\lambda_i \neq 0$ for all i
4. λ_i are eigenvalues of orthogonal matrix A
5. A, B, P are n-square matrices $|P| \neq 0$ and $B = P^{-1} AP$
6. $D = [d_{ij}]$ is an n-square matrix with $d_{ij} = 0$ for $i \neq j$
7. Characteristic polynomials of A and A^2 are same
8. Characteristic equation of A is $\lambda^m = 0$
9. A is orthogonal

a. A is singular
b. $1/\lambda_i$ are also eigenvalues of A
c. A and B are similar
d. $|A|/\lambda_i$ are eigenvalues of Adj A
e. A is nonsingular
f. Characteristic polynomials of A and B are same
g. d_{ii} are eigenvalues of D
h. Characteristic equation of A is a reciprocal equation
i. A is idempotent
j. A is nilpotent with index m

Ans: 1. d; 2. a; 3. e; 4. b; 5. c; 6. g; 7. i; 8. j; 9. h

Chapter 2 Quadratic Forms

A (Quadratic Form)	B (Matrix of the Quadratic Form)
1. $x^2 - 2xy + 2y^2$	a. $\begin{bmatrix} 1 & 0 & 0 \\ 0 & 3 & -1 \\ 0 & -1 & 3 \end{bmatrix}$
2. $x^2 + 3y^2 + 3z^2 - 2yz$	b. $\begin{bmatrix} 0 & 1 & 1 \\ 1 & 0 & 1 \\ 1 & 1 & 0 \end{bmatrix}$
3. $2xy + 2yz + 2zx$	c. $\begin{bmatrix} 1 & -1 \\ -1 & 2 \end{bmatrix}$
4. $x^2 - 18xy + 5y^2$	d. $\begin{bmatrix} 1 & 2 & 3 \\ 2 & 0 & 3 \\ 3 & 3 & 1 \end{bmatrix}$

5. $x^2 + z^2 + 4xy + 6xz + 6yz$

e. $\begin{bmatrix} 0 & 2 & 2 \\ 2 & 0 & 2 \\ 2 & 2 & 0 \end{bmatrix}$

6. $x^2 + 2y^2 + 3z^2 + 7xy + xz$

f. $\begin{bmatrix} 1 & -9 \\ -9 & 5 \end{bmatrix}$

g. $\begin{bmatrix} 1 & 7/2 & 1/2 \\ 7/2 & 2 & 0 \\ 1/2 & 0 & 3 \end{bmatrix}$

Ans: 1. c; 2. a; 3. b; 4. f; 5. d; 6. g

CHAPTER 3 SOLUTION OF ALGEBRAIC AND TRANSCENDENTAL EQUATIONS

A (Equation)	B (Solution)
1. Algebraic equation $f(x) = \sum_{r=0}^{n} a_r x^r \,(a_r \in R, a_n \neq 0)$	a. Exactly three roots
Transcendental equation	
2. $\sin x = 2$ has	b. No root
3. $2\sin x = 1$ has	c. $\{n$ real/complex roots (distinct or repeated)
4. $2\sin x = x$ has	d. $\begin{cases}\text{One root lies in } [a,b] \text{ or} \\ \text{An odd number of roots of } f(x) \text{ lie in } [a,b]\end{cases}$
5. An equation $f(x) = 0$ is such that $f(a)f(b) < 0$	e. Infinitely many roots
6. An equation $f(x) = 0$ is such that $f(a)f(b) > 0$	f. $\begin{cases}\text{No root lies in } [a,b] \text{ or} \\ \text{an even number of roots of } f(x) \text{ lie in } [a,b]\end{cases}$
	g. Finitely many roots

Ans: 1. c; 2. b; 3. e; 4. a; 5. d; 6. f

CHAPTER 4 INTERPOLATION

1. $\Delta\nabla$	a. $E - 1$
2. $\frac{\Delta}{\nabla} - \frac{\nabla}{\Delta}$	b. Δ
3. ∇E	c. $1 - E^{-1}$
4. Δ	d. $\Delta + \nabla$
5. ∇	e. $\Delta - \nabla$
	f. $1 + E^{-1}$

Ans: 1. e; 2. d; 3. b; 4. a; 5. c

CHAPTER 6 NUMERICAL DIFFERENTIATION AND INTEGRATION

	$\int_a^b f(x)\,dx =$
1. Trapezoidal rule	a. $\frac{3h}{10}[(y_0+5y_1+y_2+6y_3+y_4+5y_5)+\cdots+\cdots]$
2. Simpson's $\frac{1}{3}$ rule	b. $\frac{3h}{8}[(y_0+y_n)+3(y_1+y_2+\cdots+y_{n-1}) +2(y_3+y_6+\cdots+y_{n-3})]$
3. Simpson's $\frac{3}{8}$ rule	c. $\frac{2h}{5}[7y_0+32y_1+12y_2+32y_3+7y_4+\cdots+\cdots]$
4. Boole's rule	d. $\frac{h}{2}[y_0+y_n+2(y_1+y_2+\cdots+y_{n-1})]$
5. Weddle's rule	e. $\frac{h}{4}[y_0+y_n+3(y_1+y_2+\cdots+y_{n-1})]$
	f. $\frac{h}{3}[(y_0+y_n)+4(y_1+y_3+\cdots+y_{n-1}) +2(y_2+y_4+\cdots+y_{n-2})]$

Ans: 1. d; 2. f; 3. b; 4. c; 5. a

CHAPTER 7 NUMERICAL SOLUTION OF ORDINARY DIFFERENTIAL EQUATIONS

A (Method of Solution)	B (Merits/Demerits)
1. Taylor's	a. Average of slope at (x_0, y_0) and $(x_1, y_1^{(0)})$ is used instead of slope at (x_0, y_0)
2. Picard's	b. Sequence of iterations may deviate considerably from exact value
3. Euler's	c. Evaluation of integrals may become tedious
4. Euler's modified	d. Computation of higher derivatives may become difficult
5. Runge–Kutta	e. Four previous values of y are required, which can be computed by these formulae only
6. Milne's or Adams–Bashforth–Moulton	f. Self-starting, easy to programme but $f(x, y)$ is to be evaluated for several values of x and y at each step
	g. Four previous values of y required, which are to be computed by other methods

Ans: 1. d; 2. c; 3. b; 4. a; 5. f; 6. g

CHAPTER 8 FOURIER SERIES

A (Numerical Series)	B (Sum)
1. $1-\frac{1}{3}+\frac{1}{5}-\frac{1}{7}+\cdots\left(\sum_{n=1}^{\infty}\frac{(-1)^{n-1}}{2n-1}\right)$	a. $\pi^2/8$
2. $\frac{1}{1.3}+\frac{1}{3.5}+\frac{1}{5.7}+\cdots\left(\sum\frac{1}{(2n-1)(2n+1)}\right)$	b. $\frac{1}{2}-\frac{\pi}{8}$
3. $\frac{1}{1.3}-\frac{1}{3.5}+\frac{1}{5.7}-\frac{1}{7.9}+\cdots\left(\sum\frac{(-1)^{n-1}}{(2n-1)(2n+1)}\right)$	c. $\frac{1}{2}$
4. $\frac{1}{1.3}+\frac{1}{5.7}+\frac{1}{9.11}+\cdots\left(\sum\frac{1}{(4n-3)(4n-1)}\right)$	d. $\pi/4$
5. $\frac{1}{3.5}+\frac{1}{7.9}+\frac{1}{11.13}+\cdots\left(\sum\frac{1}{(4n-1)(4n+1)}\right)$	e. $\frac{\pi-2}{4}$
6. $\frac{1}{1^2}+\frac{1}{2^2}+\frac{1}{3^2}+\frac{1}{4^2}+\cdots\left(\sum\frac{1}{n^2}\right)$	f. $\frac{\pi}{8}$
7. $\frac{1}{1^2}+\frac{1}{3^2}+\frac{1}{5^2}+\frac{1}{7^2}+\cdots\left(\sum\frac{1}{(2n-1)^2}\right)$	g. $\pi^2/6$
8. $\frac{1}{1^2}-\frac{1}{2^2}+\frac{1}{3^2}-\frac{1}{4^2}+\cdots\left(\sum\frac{(-1)^{n-1}}{n^2}\right)$	h. $\pi^2/9$
	i. $\pi^2/12$

Ans: 1. d; 2. c; 3. e; 4. f; 5. b; 6. g; 7. a; 8. i

CHAPTER 9 FOURIER INTEGRAL TRANSFORMS

1. $\int_0^{\infty}\frac{\sin x}{x}dx$	a. $\frac{\pi}{2}e^{-a}$
2. $\int_0^{\infty}\frac{x\cos x-\sin x}{x^3}\cos\frac{x}{2}dx$	b. $-\pi/4$
3. $\int_0^{\infty}\frac{x\cos x-\sin x}{x^3}dx$	c. $-3\pi/16$
4. $\int_0^{\infty}\left(\frac{\sin x}{2x}\right)^2 dx$	d. $\frac{\pi}{2}e^{-ax}$

5. $\int_0^\infty \frac{s}{a^2+s^2}\sin s x\,ds$

6. $\int_0^\infty \frac{x\sin ax}{1+x^2}dx$

e. $\pi/8$

f. $\pi/2$

g. πe^{-a}

Ans: 1. f; 2. c; 3. b; 4. e; 5. d; 6. a

CHAPTER 10 PARTIAL DIFFERENTIAL EQUATIONS

A (Relation)

1. $z = ax + by + ab$
2. $z = ax^2 + by^2$
3. $z = (x + a)(y + b)$
4. $z = x^2 f(y/x)$
5. $z = f(x^2 - y^2)$
6. $z = xy + f(x^2 + y^2)$

B (Partial Differential Equation)

a. $2z = px + qy$

b. $z = pq$

c. $z = px + qy + pq$

d. $px + qy = 0$

e. $3z = px + qy$

f. $py - qx = y^2 - x^2$

g. $px^2 + qy^2 = 1$

Ans: 1. c; 2. a; 3. b; 4. e; 5. d; 6. f

A (Partial Differential Equation)

1. $px + qy = z$
2. $p\tan x + q\tan y = \tan z$
3. $(mz - ny)p + (nx - lz)q = ly - mx$
4. $x(y - z)p + y(z - x)q = z(x - y)$
5. $(y - z)p + (z - x)q = x - y$
6. $p + q = 1$

B (General Solution)

a. $lx + my + nz = f(x^2 + y^2 + z^2)$

b. $x + y + z = f(x^2 + y^2 + z^2)$

c. $\frac{x}{y} = f\left(\frac{y}{z}\right)$

d. $f\left(\frac{\sin x}{\sin y}, \frac{\sin z}{\sin y}\right) = 0$

e. $x + y + z = f(xyz)$

f. $x + y = f(y + z)$

g. $x - y = f(y - z)$

Ans: 1. c; 2. d; 3. a; 4. e; 5. b; 6. g

CHAPTER 11 Z-TRANSFORMS AND SOLUTION OF DIFFERENCE EQUATIONS

A (Sequence $\langle u_n\rangle$)

1. $\langle a^n\rangle$
2. $\langle n^p\rangle$
3. $\langle n\rangle$
4. $\left\langle \frac{1}{n+1}\right\rangle$

B (Z-transform)

a. $z/(z-1)^3$

b. $e^{1/z}$

c. $-z\frac{d}{dz}(z\langle n^{p-1}\rangle)$

d. $z/(z-a)$

5. $\left\langle \frac{1}{n!} \right\rangle$

6. $\langle \cos n\,\theta \rangle$

7. $\langle \sin n\,\theta \rangle$

e. $-z\log\left(\frac{z-1}{z}\right)$

f. $\frac{z}{(z-1)^2}$

g. $\frac{z\sin\theta}{z^2-2z\cos\theta+1}$

h. $\frac{z(z-\cos\theta)}{z^2-2z\cos\theta+1}$

Ans: 1. d; 2. c; 3. f; 4. e; 5. b; 6. h; 7. g

CHAPTER 12 SPECIAL FUNCTIONS

A. BETA AND GAMMA FUNCTIONS

A	B
1. $B(p,q)$	a. $\Gamma(p+1)$
2. $p\,\Gamma(p)$	b. $\pi\sin\pi/4$
3. $\Gamma(p)\,\Gamma(1-p)$	c. $2^{2p-1}\Gamma(p)\ \Gamma(p+1/2)$
4. $\int_0^{\pi/2}\sqrt{\cot\theta}\,d\theta$	d. π cosec $p\pi$
5. $\sqrt{\pi}\,\Gamma(2p)$	e. $\Gamma(p)\Gamma(q)/\Gamma(p+q)$
	f. $\pi\sin p\pi$

Ans: 1. c; 2. a; 3. d; 4. b; 5. c

B. LEGENDRE FUNCTIONS

A (Function)	B (Value)
1. $P_{2n+1}(0)$	a. $(-1)^n(2n)!/2^{2n}(n!)^2$
2. $P_{2n}(0)$	b. $(-1)^n$
3. $P_n(-x)/P_n(x)$	c. $\frac{2}{2n+1}\delta_m n$
4. $P_n^1(1)$	d. $\frac{2n}{2n+1}$
5. $\int_{-1}^{1}P_m(x)P_n(x)dx$	e. $1+2+3+\cdots+n$
	f. 0

Ans: 1. f; 2. a; 3. b; 4. e; 5. c

C. Bessel Functions

A

1. $(J_n^2 + J_{n+1}^2)'$
2. $(xJ_nJ_{n+1})'$
3. $\int_0^1 xJ_n(dx)J_n(\beta x)dx$
4. $J_0^2 + 2(J_1^2 + J_2^2 + \cdots)$
5. $J_{1/2}^2 + J_{-1/2}^2$

B

a. $x(J_n^2 - J_{n+1}^2)$

b. $\frac{1}{2}J_{n+1}^2(\alpha)\delta_{\alpha\beta}$

c. $\frac{2}{x}[nJ_n^2 - (n+1)J_{n+1}^2]$

d. $2/\pi x$

e. $(2/pix)^2$

f. 1

Ans: 1. c; 2. a; 3. b; 4. f; 5. d

Chapter 13 Functions of a Complex Variable

A

1. $\nabla^2 u = 0$
2. $f(z) = u(x, y) + iv(x, y) \in H(D)$
3. $u(x,y) = 2\log(x^2 + y^2) + k$
4. $u(r,\theta) = r^2 \cos 2\theta = c$
5. $u(x,y) = e^{x^2-y^2} \cos 2xy$

B

a. $f(z) = e^{-z^2} + c$

b. $f(z) = e^{z^2} + c$

c. $u_x = v_y;\quad u_y = -v_x$

d. u is a harmonic function

e. $v(r,\theta) = r^2 \sin 2\theta = k$

f. $v(x, y) = 4\tan^{-1}\frac{y}{x} + c$

Ans: 1. d; 2. c; 3. f; 4. e; 5. b

Chapter 14 Elementary Functions

A (Function/Equation)

1. $\sin z = \cosh 4$
2. $\log(1+i)$
3. $(1+i)^i$
4. i^i

B (General Value)

a. $e^{i\log\sqrt{2} - \left(\frac{\pi}{4} + 2n\pi\right)}$

b. $e^{2n + \frac{1}{2}\pi}$

c. $\log\sqrt{2} + i(2n\pi + \pi/4)$

d. $z = n\pi + (-1)^n\left(\frac{\pi}{2} - 4i\right)$

5. $\sin^{-1}z$

e. $-\ln\left(iz+\sqrt{1-z^2}\right)$

f. $-i\ln\left(iz+\sqrt{1-z^2}\right)$

Ans: 1. d; 2. c; 3. a; 4. b; 5. f

CHAPTER 15 COMPLEX INTEGRATION

A (Integral)	B (Value)
1. $\int_C \frac{dz}{z^2(z-3)}$ $(C:\|z\|=2)$	a. 0
2. $\int_{8+\pi i}^{8-3\pi i} e^{z/2}dz$	b. $-\pi$
3. $\int_C \frac{z^3}{2z-i}$ $(C:\|z\|=1)$	c. $2\pi i/9$
4. $\int_C \frac{dz}{z^2+1}$ $(C:\|z+i\|=1)$	d. $\pi/8$
5. $\int_{-i}^{i} \frac{dz}{z}$	e. $2\pi i$
	f. πi

Ans: 1. c; 2. a; 3. d; 4. b; 5. f

CHAPTER 16 COMPLEX POWER SERIES

A (Function)	B (Type of Singularity)
1. $f(z)=\frac{\sin z}{z^2+1}$	a. Removable singularity
2. $f(z)=\frac{1}{\sin z}$	b. Simple pole
3. $f(z)=\frac{1-\cos z}{z}$	c. Infinite number of isolated singularities
4. $f(z)=e^z$	d. Pole of higher order
5. $f(z)=\frac{z^3}{z-1}$	e. Essential singularities at $z=\infty$
	f. Finite number of isolated singularities

Ans: 1. f; 2. c; 3. a; 4. e; 5. b

CHAPTER 17 CALCULUS OF RESIDUES

A (Function/Point)	B (Residue)
1. $\text{Res}_{z=-2}\frac{z^2}{(z-1)^2(z+2)}$	a. $-\frac{1}{2}$

2. $\text{Res}_{z=0} \dfrac{1-e^{2z}}{z^4}$

3. $\text{Res}_{z=-1} \dfrac{z^2-2z}{(z+1)^2(z^2+1)}$

4. $\text{Res}_{z=\pi/2} \tan z$

5. $\text{Res}_{z=i} \dfrac{1}{(z^2+1)(z^2-4)}$

b. – 4/3

c. –1

d. 4/9

e. $(10i)^{-1}$

f. $\dfrac{i}{10}$

Ans: 1. d; 2. b; 3. a; 4. c; 5. f

CHAPTER 18 ARGUMENT PRINCIPLE AND ROUCHE'S THEOREM

A (Equation)

1. $z^5+15z+1=0$
2. $e^z = az^4 \ (a > e)$
3. $z^4+z+1=0$
4. $z^4+4(1+i)+1=0$
5. $z^4+z^3+1=0$

B (Location of Roots)

a. 1 root in each quadrant

b. 1 root in the first quadrant

c. 4 roots in $\dfrac{3}{2} < |z| < 2$

d. 4 roots lie in the first quadrant

e. 4 roots in $|z| \le 1$

f. 1 complex root in the first quadrant

Ans: 1. e; 2. c; 3. b; 4. a; 5. f

CHAPTER 19 CONFORMAL MAPPING

A (Mapping of Points)

1. $(-1, i, 1) \to (0, i, \infty)$
2. $(\infty, i, 0) \to (-1, -i, 1)$
3. $(-1, 0, 1) \to (0, i, 3i)$
4. $(\infty, i, 0) \to (0, i, \infty)$
5. $(1, i, -1) \to (i, 0, -i)$

B (Bilinear Transformation)

a. $(1+iz)/(1-iz)$

b. $w = 3i(z+1)/(3-z)$

c. $w = -1/z$

d. $w = (1-z)/1(1+z)$

e. $w = -(z+1)/(z-1)$

f. $w = (-6z+2i)/(iz-3)$

Ans: 1. e; 2. d; 3. b; 4. c; 5. a

A (Transformation)	B (Fixed Points)
1. $w=(z-2i)^{-1}$	a. i
2. $w=(6z-9)/z$	b. ± 1
3. $w=(z-1+i)/(z+2)$	c. (3, 3)
4. $w=\dfrac{1}{z}$	d. 1, 2
5. $w=\dfrac{1}{3}(z^2+2)$	e. 2, 2
	f. i

Ans: 1. f; 2. c; 3. a; 4. b; 5. d

True or Flase Statements

Chapter 1 Eigenvalues and Eigenvectors

1. A matrix and its inverse have the same eigenvalues if A is orthogonal. **Ans:** T

2. Two linearly independent eigenvectors may correspond to an eigenvalue λ of a matrix. **Ans:** T

3. Cayley–Hamilton Theorem is applicable to every matrix. **Ans:** F

4. A square matrix with repeated eigenvalues is not diagonalizable. **Ans:** F

5. Powers of a square matrix can be found using diagonalization. **Ans:** T

6. Every real square matrix can be uniquely expressed as the sum of a symmetric and a skew-symmetric matrix. **Ans:** T

7. If A and B are symmetric then AB is symmetric. **Ans:** F

Chapter 2 Quadratic Forms

8. If A is a symmetric matrix with distinct eigenvalues then its eigenvectors are orthogonal. **Ans:** T

9. A square matrix with two repeated eigenvalues is not diagonalizable. **Ans:** F

10. A quadratic form is semidefinite if some of its eigenvalues are positive and the others negative. **Ans:** F

11. A quadratic form is positive definite if the nonzero coefficients in its canonical form are positive. **Ans:** F

Chapter 3 Solution of Algebraic and Transcendental Equations

12. We may solve $f(x) = x^3 - 3x + 1 = 0$ for a root near $x = 3$ by writing if as $x = \phi(x) = \dfrac{-1}{x^2 - 3}$. **Ans:** T

13. Newton-Raphson method for solving an equation is $x_1 = x_0 + \dfrac{f(x_0)}{f'(x_0)}$. **Ans:** F

14. Regula Falsi method has faster convergence than Newton's method. **Ans:** F

15. Bisection method can be used for finding a root of a positive number. **Ans:** F

Chapter 4 Interpolation

16. Interpolation is an estimation of desired quantity at a specified value $\bar{x}$, of the argument in an interval $[a, b]$. **Ans:** T

17. Newton's forward interpolation formula can be used when the data points are not evenly distributed. **Ans:** F

18. Lagrange's formula can be used for inverse interpolation. **Ans:** T

19. Lagrange's formula can be used even though the data points are not evenly distributed. **Ans:** T

CHAPTER 5 CURVE-FITTING

20. Curve-fitting methods have wider applicability than those of interpolation. **Ans:** T

21. To fit a curve of the form $y = ae^{bx}$ transformed equation of the form $\log y = a + b \log x$ is to be used. **Ans:** F

22. To fit a curve of the form $y = a_1 e^{b_1 x} + a_2 e^{b_2 x}$ the equation is transformed by application of logarithms. **Ans:** F

CHAPTER 6 NUMERICAL DIFFERENTIATION AND INTEGRATION

23. Numerical differentiation is used to know the rate of growth of a quantity $f(x)$ whose values a points $x_i (i = 1, 2, \ldots, n)$ are given. **Ans:** T

24. Numerical differentiation is used to know the maxima or minima of a tabulated quantity $y = f(x)$. **Ans:** T

25. Simpson's $\frac{3}{8}$ rule can be applied when n is even. **Ans:** F

26. For application of Weddle's rule the minimum number of values of y required is 6. **Ans:** F

27. Weddle's rule gives more accurate result than any other integration formula. **Ans:** T

CHAPTER 7 NUMERICAL SOLUTION OF ORDINARY DIFFERENTIAL EQUATIONS

28. Taylor series method to solve a differential equation is a single-step method which works well as long as successive derivatives of y can be easily calculated. **Ans:** T

29. Runge–Kutta methods for solution of differential equations are multi step methods. **Ans:** F

30. In Milne's method of solution of a differential equation predictor and corrector formulas are used. **Ans:** T

31. In Adam's method one required values of y at 4 previous stations are needed to compute a predictor value. **Ans:** T

CHAPTER 8 FOURIER SERIES

32. Fourier series expansion is possible for a function which is discontinuous. **Ans:** T

33. The function $f(x) = \begin{cases} 1 + \frac{2x}{\pi} & \text{in}[-\pi, 0] \\ 1 - \frac{2x}{\pi} & \text{in}[0, \pi] \end{cases}$ is an odd function in $[-\pi, \pi]$. **Ans:** F

34. The function $f(x) = \begin{cases} -k & \text{in}[-\pi, 0] \\ k & \text{in}[0, \pi] \end{cases}$ is an even function. **Ans:** F

35. Half-range Fourier series expansions are obtained by setting $a_n = 0$ or $b_n = 0$. **Ans:** F

36. $f(x) = 1$ has a Fourier sine series expansion in $[0, \pi]$. **Ans:** T

37. A function $f(x)$ defined in an interval $[0, l]$ has only sine or cosine series expansion. **Ans:** F

CHAPTER 9 FOURIER INTEGRAL TRANSFORMS

38. $F_s\{e^{-ax}\} = \dfrac{s}{a^2 + s^2}$. **Ans:** T

39. $\int_0^\infty \frac{s}{a^2+s^2}\sin sx\, ds = \frac{\pi}{2}e^{-ax}$. **Ans:** T

40. $\int_0^\infty \frac{\sin x}{x}dx = \pi$. **Ans:** F

41. If a function $f(x)$ and its transform $F(s)$ are same then it is called self-reciprocal. **Ans:** T

CHAPTER 10 PARTIAL DIFFERENTIAL EQUATIONS

42. The partial differential equation whose solution is $Z = ax + by + a/b$ is $Z = px + qy + p/q$. **Ans:** T

43. The partial differential equation of all unit spheres with their centres at the variable points (h, k, o) is $z^2(p^2 + q^2 + 1) = 1$. **Ans:** T

44. The general solution of $p + q = 0$ is $\phi(x - y, z) = 0$. **Ans:** T

45. $p^2 = a^2$ is a linear partial differential equation. **Ans:** F

46. The complete solution of $pq = 1$ is $z = ax + \frac{1}{a}y + c$. **Ans:** T

47. For solving the equation $Pp + Qq - R = 0$ the Lagrange's auxiliary equations are $\frac{dx}{P} = \frac{dy}{Q} = \frac{dz}{-R}$. **Ans:** F

CHAPTER 11 Z-TRANSFORMS AND SOLUTION OF DIFFERENCE EQUATIONS

48. Z-transform is to a function of natural number n as a Laplace transform is to a piecewise continuous function of t. **Ans:** T

49. $Z(-1)^n = \frac{z}{z+1}$. **Ans:** T

50. $Z\langle n^2\rangle = \frac{z(z-1)}{(z+1)^3}$. **Ans:** F

51. $Z^{-1}\left\{\frac{z^2}{(z-1)(z-3)}\right\} = \frac{1}{3}(3^{n+1}-1)$. **Ans:** F

52. The solution of $y_{n+1} - 2y_n = 0;\ y_0 = 1$ is $y_n = \langle 2^n\rangle$. **Ans:** T

CHAPTER 12 SPECIAL FUNCTIONS

A. Beta and Gamma Functions

1. $\int_0^{\pi/2} \sin^n x dx = \frac{\sqrt{\pi}}{2}\Gamma\left(\frac{n+1}{2}\right)/\Gamma\left(\frac{n+2}{2}\right)$ **Ans:** T

2. $\int_0^1 \left(\log\frac{1}{y}\right)^{-1/2} dy = \sqrt{\pi}/2$ **Ans:** F

3. $B(p,q) = 2\int_0^{\pi/2} \sin^{2p-1}\theta\cos^{2q-1}\theta d\theta$ **Ans:** T

4. $\int_0^1 \frac{x^{p-1}+x^{q-1}}{(1+x)^{p+q-2}}dx = B(p-1, q-1)$ **Ans:** F

5. $\Gamma(p)\Gamma(1+p) = \pi \operatorname{cosec} p\pi$ **Ans:** F

6. $B(p, q+1) = \frac{q}{p+q}B(p,q)$ **Ans:** T

B. Legendre Functions

1. $x^3 = \frac{1}{5}[2P_3(x) + 3P_1(x)]$ **Ans:** T

2. Rodrigues' Formula is $P_n(x) = \frac{1}{2^n n!}D^n(x^2-1)^n$ **Ans:** T

3. $\int_{-1}^{1} xP_n(x)P_{n-1}(x)dx = n/(2n-1)(2n+1)$ **Ans:** F

4. $\int_0^1 P_n(x)dx = 0$ if n is odd **Ans:** F

5. $P_n(1) = 1$ **Ans:** T

6. $P_n(-x) = (-1)^n P_n(x)$ **Ans:** T

C. Bessel Functions

1. The differential equation satisfied by $J_0(x)$ is $xy'' + y' + xy = 0$ **Ans:** T

2. $\frac{n}{x}J_n(x)+J_n'(x)=J_n(x)$ **Ans:** F

3. $J_{-n}(x)=(-1)^n J_n(x)$ **Ans:** T

4. $J_{1/2}^2+J_{-1/2}^2=\frac{2}{\pi x}$ **Ans:** T

5. $J_0'(x)=J_1(x)$ **Ans:** F

6. $\cos x=J_0-2J_2+2J_4$ **Ans:** T

CHAPTER 13 FUNCTIONS OF A COMPLEX VARIABLE

1. $f(z)=\bar{z}/z$ is not continuous at $z=0$ **Ans:** T

2. If a complex function is derivable at a point then it is continuous there. **Ans:** T

3. If a function $f(z)=u+iv$ satisfies Cauchy–Riemann equations, then $f(z)$ is analytic **Ans:** F

4. Laplace equation in polar coordinates $\frac{\partial^2 u}{\partial r^2}+\frac{1}{r}\frac{\partial u}{\partial r}+\frac{1}{r^2}\frac{\partial^2 u}{\partial \theta^2}=0$ **Ans:** T

5. $f(x+iy)=2xy+i(x^2-y^2)$ is analytic **Ans:** F

6. $f(x+iy)=\frac{1}{2}\log(x^2+y^2)+i\tan^{-1}\frac{x}{y}$ is analytic **Ans:** F

CHAPTER 14 ELEMENTARY FUNCTIONS

1. sin $(iz)=i$ sinh z **Ans:** T

2. $(\coth z)'=\operatorname{cosech}^2 z$ **Ans:** F

3. Principal value of i^i is $e^{\pi/2}$ **Ans:** F

4. A solution of $e^z=1+2i$ is $z=\frac{1}{2}\log 5+i\tan^{-1}2$ **Ans:** T

5. The principal value of $\log(1+i)+\log(1-i)$ is $\log 2$ **Ans:** T

6. $\tan^{-1}z=\frac{1}{2}\ln\left(\frac{i+z}{i-z}\right)$ **Ans:** F

CHAPTER 15 COMPLEX INTEGRATION

1. If C is the line segment from $z = 0$ to $z = i$, $\int_C (y-x-3ix^2)dz=i$ **Ans:** F

2. $\int_0^{\pi i} z\cos z\,dz=-\pi\sinh\pi+\cosh\pi-1$ **Ans:** T

3. If C is the unit circle $|z|=1$ then $\int_C \frac{(z^2+1)dz}{z(2z+1)}=-2\pi i$ **Ans:** F

4. $\int_0^{1-i} z^2 dz=\frac{1}{3}(1-i)^3$ **Ans:** T

5. $\int_C f(z)dz$ is independent of path c of integration **Ans:** F

6. If C is $|z|=3$ then $\int_C \frac{z^3+z^2+2z-1}{(z-1)^3}dz=8\pi i$ **Ans:** T

CHAPTER 16 COMPLEX POWER SERIES

1. A series of the form $\sum_{n=0}^{\infty} a_n(z-a)^n$ is called a power series **Ans:** T

2. If a series $\sum u_n(x)$ converges for every z in $C:|z-a|=R$, then C is called the circle of convergence **Ans:** T

3. Taylor's series expansion of a function $f(z)$ consists of both positive and negative powers of $(z-a)$ **Ans:** F

4. If $f(z)=\frac{1}{z^2-3z+2}$ is expanded in the annular region $1<|z|<2$ the series we obtain is Taylor's series **Ans:** F

5. Taylor's series is a particular case of Laurent's series **Ans:** T

6. If $f(z)$ has a Taylor's series expansion valid in a region D, then $f(z)\in H(D)$ **Ans:** T

CHAPTER 17 CALCULUS OF RESIDUES

1. By Cauchy's Residue Theorem, real integrals can be evaluated. **Ans:** T

2. If $f(z)=\sum_{n=0}^{\infty} a_n(z-a)^n$, then b_1 is called the residue of $f(z)$ at $z = a$ **Ans:** T

3. Res $\{f(z):z=a\}$ where $f(z)=\dfrac{p(z)}{q(z)}$ is $\dfrac{p'(a)}{q(a)}$ **Ans:** F

4. $\text{Res}_{z=0}\dfrac{\sin z}{z}=1$ **Ans:** F

5. $\text{Res}_{z=1}\dfrac{z^2}{1-z^4}=-\dfrac{1}{4}$ **Ans:** T

6. If C is $|z|=2$, then $\int_C \dfrac{z^3 dz}{(z-1)^2(z-3)}=-3.5\pi i$ **Ans:** T

Chapter 18 Argument Principle and Rouche's Theorem

1. Liouville's Theorem states that if $f(z)\in H(D)$ and $|f(z)|$ is bounded in D then $f(z)$ is constant in D **Ans:** T

2. Every polynomial of degree n has n roots **Ans:** T

3. If $f(z)\in H(D)$ then min $|f(z)|$ or max $|f(z)|$ can occur at any point of D **Ans:** F

4. $z^5+15z+1=0$ has all its 5 roots in the unit circle $|z|=1$ **Ans:** F

5. $e^z = z^n$ has n roots inside the unit circle **Ans:** F

6. One root of $z^4+z^3+1=0$ lies in the first quadrant **Ans:** T

Chapter 19 Conformal Mapping

1. The fixed points of the mapping $w=\dfrac{1}{z-2i}$ is $z=i$ **Ans:** T

2. The mapping $w=z^{1/n}$ maps sectors into half planes **Ans:** T

3. The image of the circle $|z+1|=1$ under the mapping $w=u+iv=\dfrac{1}{z}$ is the circle $|z|=\dfrac{1}{2}$ **Ans:** F

4. A bilinear transformation preserves the cross-ratio of four points **Ans:** T

5. The critical points of $w = \cos z$ are at $z = n\pi$ **Ans:** F

6. If the mapping $w = f(z)$ is conformal in D, then $f(z)\in H(D)$ **Ans:** T

Index